ELSEVIER'S ENCYCLOPÆDIA

of

ORGANIC CHEMISTRY

In this volume the literature has been consulted up to
and including 1946, the literature concerning the structure and configuration
of compounds up to 1958

ISBN 978-3-662-23761-8   ISBN 978-3-662-25863-7 (eBook)
DOI 10.1007/978-3-662-25863-7

© Springer-Verlag Berlin Heidelberg 1959
Ursprünglich erschienen bei Springer-Verlag Berlin • Göttingen • Heidelberg 1959
Softcover reprint of the hardcover 1st edition 1959

ELSEVIER'S ENCYCLOPÆDIA

of

ORGANIC CHEMISTRY

Series III

CARBOISOCYCLIC CONDENSED COMPOUNDS

Volume 14 — Supplement

Pages 2215 S – 2990 S

STEROIDS

OXO-COMPOUNDS

Under the auspices of Beilstein-Institut für Literatur der Organischen Chemis

Edited by F. RADT

SPRINGER-VERLAG BERLIN HEIDELBERG GMBH 1959

PREFACE

When, a few years ago, the publication of Elsevier's Encyclopædia of Organic Chemistry was suspended, it seemed desirable to complete – on account of its monographic character – the steroid section in accordance with the original conception. One consideration in favour thereof, among others, was that the steroid section of Beilstein's Handbuch will not be available for some years to come, so that also to those using Beilstein the present publication is a valuable help in bridging the gap. In view of the undiminished activity in this special field – equally important to chemists, biochemists, and the medical world – the hope that the completion of this section will meet the needs of a wide circle of readers appears to be justified.

Altogether three more volumes are planned for the Supplement, the first of which, now presented, covers the steroid Oxo-compounds, the literature having been consulted up to the end of 1946 in conformity with the previous volumes of the steroid Supplement. Where, however, important additions and corrections are concerned, literature up to the most recent date has been taken into account. In treating the subject matter the same principles have been followed as in the previous volumes of this section. Editorship remains in the competent hands of Dr. F. RADT, who has contributed a considerable proportion of the matter treated in this volume. He is assisted by Dr. A. GEORG and Dr. DORA STERN.

Frankfurt a. M., May, 1959   F. RICHTER

CONTENTS

Pages

TABLE OF ABBREVIATIONS EMPLOYED
IN THE TEXT

$[\alpha]$	specific rotation ($[\alpha]_D^{20}$, for 20^0 and sodium light)	ca.	about
		cal.	calorie(s)
Å	Ångström unit(s)	calcd.	calculated
abs.	absolute	c.c.	cubic centimetre(s)
absorpt.	absorption	cf.	compare
Ac	acetyl (AcCl, acetyl chloride; Ac$_2$O, acetic anhydride)	cm.	centimetre(s)
		coeff.	coefficient
		compd.	compound
addn.	addition	compn.	composition
alc.	alcohol, alcoholic	conc.	concentrated
alk.	alkaline	concn.	concentration
alkali	caustic alkali	cond.	conductivity
Am	amyl	condens.	condensation
amp.	ampere(s)	const.	constant
amt.	amount	contg.	containing
anh.	anhydrous	cor.	corrected
approx.	approximate, approximately	corresp.	corresponding
		crit.	critical
aq.	aqueous	cryst.	crystalline, crystals
assoc.	associate(s)	crystd.	crystallized
assocd.	associated	crystn.	crystallization
assocn.	association	D	Debye unit (10^{-18} e.s.u. $\times$ cm.)
asym.	asymmetrical		
at.	atom(s), atomic	d	density (d$_4^{13}$, specific gravity at 13^0 referred to water at 4^0)
atm.	atmosphere(s), atmospheric		
av.	average		
b.	(followed by a figure denoting temperature) boils at, boiling at	dec.	decomposed, decomposition
		deriv.	derivative
		det.	determine
bacteriol.	bacteriological	detd.	determined
biol.	biological	detn.	determination
b.p.	boiling point	dextroroty.	dextrorotatory
Bu	n-butyl	diazotd.	diazotized
Bz	benzoyl (BzCl, benzoyl chloride)	diazotn.	diazotization
		dil.	dilute

dild.	diluted	M.	molar
diln.	dilution	M	(with subscript) molecular refraction
dispn.	dispersion		
dissoc.	dissociate(s)	m.	(followed by a figure denoting temperature) melts at, melting at; also metre(s)
dissocd.	dissociated		
dissocn.	dissociation		
distd.	distilled		
distg.	distilling	ma.	milliampere(s)
distn.	distillation	max.	maximum, maxima
elec.	electric, electrical	Me	methyl (MeOH, methanol)
e.m.f.	electromotive force	meth.	methyl alcoholic
equil.	equilibrium	mg.	milligram(s)
equiv.	equivalent	min.	minimum; also minute(s)
Et	ethyl (Et_2O, ethyl ether)	mixt.	mixture
evap.	evaporate	ml.	millilitre(s)
evapd.	evaporated	mm.	millimetre(s)
evapg.	evaporating	$m\mu$	10^{-6} mm. (10 Å)
evapn.	evaporation	mmt.	moment
evoln.	evolution	mol.	molecule, molecular, mole
examn.	examination	m.p.	melting point
expt.	experiment	mutarotn.	mutarotation
exptl.	experimental	mv.	millivolt(s)
extrd.	extracted	n	index of refraction (n_D^{20}, for 20^0 and sodium light); also number of molecules in the crystallographic elementary cell
extrg.	extracting		
extrn.	extraction		
fluoresc.	fluorescence		
fmn.	formation		
f.p.	freezing point	N	normal (as applied to concentration)
fwd.	followed		
g.	gram(s)	neg.	negative
γ	microgram(s)	No.	number
H_2SO_4	conc. sulphuric acid	obtd.	obtained
hr.	hour	occ.	occurrence
hydrol.	hydrolysis	opt. act.	optically active
hydrold.	hydrolysed	org.	organic
inorg.	inorganic	oxid.	oxidized
insol.	insoluble	oxidn.	oxidation
isold.	isolated	p.	page
isoln.	isolation	Pat.	Patent; see Table of Periodicals under country of issue
^{0}K	^{0}Kelvin		
kg.	kilogram(s)		
kg.-cal.	kilogram calorie(s)	pathol.	pathological
kv.	kilovolt(s)	petrol.ether	petroleum ether
kw.	kilowatt(s)	Ph	phenyl
l.	litre(s)	pharmacol.	pharmacological
lævoroty.	lævorotatory	phys.	physical
lb.	pound(s)	physiol.	physiological

pos.	positive	sapond.	saponified
powd.	powdered	sapong.	saponifying
pp.	pages	satd.	saturated
ppt	precipitate	satg.	saturating
pptd.	precipitated	satn.	saturation
pptg.	precipitating	sec.	second(s), secondary
pptn.	precipitation	sepd.	separated
Pr	propyl	sepg.	separating
prepd.	prepared	sepn.	separation
prepg.	preparing	sol.	soluble
prepn.	preparation	soln.	solution
prim.	primary	soly.	solubility
priv. comm.	private communication	sp.	specific
prod.	product	sp. gr.	specific gravity
pyr.	pyridine	spectr.	spectrum
qual.	qualitative	sq. in.	square inch
quant.	quantitative	subl.	sublimes
rac.	racemic	subln.	sublimation
recrystd.	recrystallized	subst.	substance
recrystn.	recrystallization	sym.	symmetrical
red.	reduced	synth.	synthesis
redn.	reduction	temp.	temperature
ref.	references	tert.	tertiary
resoln.	resolution	unsatd.	unsaturated
resp.	respectively	v.	volt(s)
rn.	reaction	vac.	vacuum, vacuo
rotn.	rotation	vol.	volume
roty.	rotatory	w.	watt(s)
sapon.	saponification	wt.	weight

REPRODUCTION OF STRUCTURAL
FORMULÆ

In the structural formulæ a single line always means a single bond; e.g., the formula   represents

For the expression of stereochemical configuration by the use of dotted and wavy lines, see the Introduction, p. 1357 s.

TABLE OF PERIODICALS

ABBREVIATED TITLE	FULL TITLE
Am. Perfumer Essent. Oil Rev.	The American Perfumer and Essential Oil Review
Am. Rev. Tuberc.	The American Review of Tuberculosis
Anal. Chem.	Analytical Chemistry
Anales asoc. quím. Argentina	Anales de la asociación química Argentina
Anales fís. quím.	Anales de física y química (Madrid)
Anales soc. cient. argentina	Anales de la sociedad cientifica argentina
Anales soc. españ. fís. quím.	Anales de la sociedad española de física y química
Analyst	The Analyst
Anat. Record	The Anatomical Record
Anesthesiology	Anesthesiology
Angew. Botan.	Angewandte Botanik
Angew. Ch.	Angewandte Chemie; before 1932 Zeitschrift für angewandte Chemie; between 1942 and 1944 incl. Die Chemie
Ann.	Annalen der Chemie, Justus Liebigs; before 1840 Annalen der Pharmacie; from 1840–1874 (Vol. 172) Annalen der Chemie und Pharmacie
Ann. Acad. Sci. Fennicae	Annales Academiae Scientiarum Fennicae
Ann. acad. sci. tech. Varsovie	Annales de l'académie des sciences techniques à Varsovie
Ann. Applied Biol.	The Annals of Applied Biology
Ann. Botany	Annals of Botany
Ann. chim.	Annales de chimie; before 1914 Annales de chimie et de physique
Ann. chim. anal.	Annales chimie analytique
Ann. chim. anal. chim. appl.	Annales de chimie analytique et de chimie appliquée et Revue de chimie analytique réunies
Ann. chim. applicata	Annali di chimica applicata
Ann. combustibles liquides	Annales de l'office national des combustibles liquides
Ann. faculté sci. Marseille	Annales de la faculté des sciences de Marseille
Ann. fals.	Les annales des falsifications et des fraudes
Ann. inst. anal. phys.-chim. (U.S.S.R.)	Annales de l'institut d'analyse physico-chimique (U.S.S.R.)
Ann. inst. Pasteur	Annales de l'institut Pasteur
Ann. Internal Med.	Annals of Internal Medicine
Ann. jardin bot. Buitenzorg	Annales du Jardin botanique de Buitenzorg
Ann. méd. légale	Annales de médecine légale de criminologie, police scientifique, médecine sociale, et toxicologie
Ann. méd. (Paris)	Annales de médecine (Paris)
Ann. mines & carburants	Annales des mines & carburants
Ann. N. Y. Acad. Sci.	Annals of the New York Academy of Sciences
Ann. Philosophy	Annals of Philosophy
Ann. phys.	Annales de physique
Ann. Physik	Annalen der Physik; before 1900 Annalen der Physik und Chemie (Poggendorff; Wiedemann)
Ann. Rept. Agr. and Hort. Research Sta., Long Ashton, Bristol	The Annual Report of the Agricultural and Horticultural Research Station (The National Fruit and Cider Institute), Long Ashton, Bristol
Ann. Rept., East Malling Research Sta., Kent	Annual Report, East Malling Research Station, near Maidstone, Kent
Ann. Rev. Biochem.	Annual Review of Biochemistry
Ann. sci. École normale	Annales scientifiques de l'École Normale Supérieure
Ann. sci. univ. Jassy	Annales scientifiques de l'université de Jassy
Ann. soc. sci. Bruxelles	Annales de la société scientifique de Bruxelles
Ann. Surgery	Annals of Surgery
Ann. Trop. Med. Parasitol.	Annals of Tropical Medicine and Parasitology
Ann. univ. Lyon	Annales de l'université de Lyon
Ann. zymol.	Annales de zymologie
Annual Reports	Annual Reports on the Progress of Chemistry, London

Abbreviated Title	Full Title
Apoth. Ztg.	Apotheker Zeitung
Arb. biol. Reichsanstalt Land- u. Forstw.	Arbeiten aus der biologischen Reichsanstalt für Land- und Forstwirtschaft, Berlin-Dahlem
Arb. kais. Gesundh.	Arbeiten aus dem kaiserlichen Gesundheitsamte
Arb. Pharm. Inst. Univ. Berlin	Arbeiten aus dem Pharmazeutischen Institut der Universität Berlin
Arb. physiol. u. angew. Entomol.	Arbeiten über physiologische und angewandte Entomologie aus Berlin-Dahlem
Arb. Reichsgesundh.	Arbeiten aus dem Reichsgesundheitsamte
Arb. Staatsinst. exptl. Therap. u. Georg Speyer-Hause Frankfurt a. M.	Arbeiten aus dem Staatsinstitut für experimentelle Therapie und dem Georg Speyer-Hause zu Frankfurt a. M.
Arbeitsschutz	Arbeitsschutz, Unfallverhütung, Gewerbehygiene. Sonderausgabe des Reichsarbeitsblattes.
Arch. Anat. Physiol. u. wiss. Med.	Archiv für Anatomie, Physiologie und wissenschaftliche Medicin
Arch. belges dermatol. et syphiligr.	Archives belges de dermatologie et de syphiligraphie
Arch. Biochem.	Archives of Biochemistry
Arch. Biol. Hung.	Archiva Biologica Hungarica
Arch. Chem. i Farm.	Archiwum Chemji i Farmacji (Warsaw)
Arch. Dermatol. Syphilol.	Archives of Dermatology and Syphilology
Arch. Dermatol. u. Syphilis	Archiv für Dermatologie und Syphilis
Arch. exptl. Path. Pharmakol.	Archiv für experimentelle Pathologie und Pharmakologie
Arch. exptl. Zellforsch.	Archiv für experimentelle Zellforschung
Arch. farmacol. sper.	Archivio di Farmacologia sperimentale e Scienze affini
Arch. farm. y bioquim. Tucumán	Archivos de farmacia y bioquímica de Tucumán
Arch. fisiol.	Archivio di fisiologia
Arch. ges. Physiol.	Archiv für die gesamte Physiologie des Menschen und der Tiere (Pflügers)
Arch. Hyg. Bakt.	Archiv für Hygiene und Bakteriologie
Arch. Internal Med.	The Archives of Internal Medicine
Arch. intern. pharmacodynamie	Archives internationales de pharmacodynamie et de thérapie
Arch. intern. physiol.	Archives internationales de physiologie
Arch. ital. biol.	Archives italiennes de biologie
Arch. Math. Naturvidenskab	Archiv for Mathematik og Naturvidenskab
Arch. néerland. physiol.	Archives néerlandaises de physiologie de l'homme et des animaux
Arch. Ophthalmol. (Chicago)	Archives of Ophthalmology (Chicago)
Arch. Path.	Archives of Pathology
Arch. Pharm.	Archiv der Pharmazie; before 1832 Archiv des Apotheker-Vereins im nördlichen Teutschland
Arch. phys. biol.	Archives de physique biologique et de chimie-physique des corps organisés; before 1930 Archives de physique biologique
Arch. Schiffs- u. Tropen-Hyg.	Archiv für Schiffs- und Tropen-Hygiene, Pathologie und Therapie exotischer Krankheiten
Arch. sci.	Archives des sciences, Genève; before 1948 Archives des sciences physiques et naturelles, Genève
Arch. sci. biol. (Italy)	Archivio di scienze biologiche (Italy)
Arch. sci. biol. (U.S.S.R.)	Archives des sciences biologiques (U.S.S.R.)
Arch. sci. phys. nat.	Archives des sciences physiques et naturelles, Genève
Arch. Surg.	Archives of Surgery
Arch. wiss. u. prakt. Tierheilk.	Archiv für wissenschaftliche und praktische Tierheilkunde
Arhiv Hem. i Farm.	Arhiv za Hemiju i Farmaciju
Arkiv Kemi Mineral. Geol.	Arkiv för Kemi, Mineralogi och Geologi

ABBREVIATED TITLE	FULL TITLE
Arkiv Zool.	Arkiv för Zoologi, utgivet av K. Svenska Vetenskaps-akademien
Asphalt u. Teer, Strassenbautech.	Asphalt und Teer, Straßenbautechnik
Astrophys. J.	The Astrophysical Journal
Atti accad. Italia, Rend.	Atti della reale accademia d'Italia. Rendiconti della classe di scienze fisiche, matematiche e naturali
Atti accad. Lincei	Atti della reale accademia dei Lincei. Rendiconti della classe di scienze fisiche, matematiche e naturali; since 1921 Atti della reale accademia nazionale dei Lincei; since 1946 Atti della accademia nazionale dei Lincei
Atti accad. Lincei, Mem.	Atti della reale accademia dei Lincei. Memorie della classe di scienze fisiche, matematiche e naturali
Atti accad. Lincei, Transunti	Atti della reale accademia dei Lincei. Transunti
Atti accad. Torino	Atti della reale accademia delle scienze di Torino
Atti congr. intern. chim.	Atti del congresso internazionale di chimica
Atti congr. naz. chim. pura applicata	Atti del congresso nazionale di chimica pura ed applicata
Atti ist. veneto sci. lettere ed arti	Atti del (reale) istituto veneto di scienze, lettere ed arti
Atti soc. nat. e mat. Modena	Atti della società dei naturalisti e matematici di Modena
Australian Chem. Inst., J. & Proc.	The Australian Chemical Institute Journal & Proceedings
Australian J. Exptl. Biol. Med. Sci.	The Australian Journal of Experimental Biology and Medical Science
Australian Pat.	Australian Patent
Austrian Pat.	Austrian Patent
Automobiltech. Z.	Automobiltechnische Zeitschrift
Auto-Tech.	Auto-Technik
Beitr. ch. Physiol. Pathol.	Beiträge zur chemischen Physiologie und Pathologie
Beitr. pathol. Anat. u. allgem. Pathol.	Beiträge zur pathologischen Anatomie und zur allgemeinen Pathologie
Belg. Pat.	Belgian Patent
Ber.	Berichte der deutschen chemischen Gesellschaft; since 1947 Chemische Berichte
Ber. Afdeel. Handelsmuseum Ver. Kolon. Inst.	Berichten van de Afdeeling Handelsmuseum van de Koninklijke Vereeniging Koloniaal Instituut
Ber. deut. botan. Ges.	Berichte der deutschen botanischen Gesellschaft
Ber. deut. pharm. Ges.	Berichte der deutschen pharmazeutischen Gesellschaft
Ber. Ges. Kohlentech.	Berichte der Gesellschaft für Kohlentechnik
Ber. ges. Physiol. exptl. Pharmakol.	Berichte über die gesamte Physiologie und experimentelle Pharmakologie
Ber. K. Sächs. Ges. Wiss. Math.-phys. Kl.	Berichte über die Verhandlungen der Königl. Sächsischen Gesellschaft der Wissenschaften zu Leipzig, Mathematisch-physische Klasse; since 1919 Berichte über die Verhandlungen der Sächsischen Akademie der Wissenschaften zu Leipzig, Mathematisch-physische Klasse
Ber. Schimmel & Co.	Bericht der Schimmel & Co. Aktien-Gesellschaft Miltitz Bz. Leipzig über ätherische Öle, Riechstoffe usw.
Ber. ungar. pharm. Ges.	Berichte der ungarischen pharmazeutischen Gesellschaft
Berlin. klin. Wochschr.	Berliner klinische Wochenschrift
Berlin. tierärztl. Wochschr.	Berliner tierärztliche Wochenschrift
Berlin. u. Münch. tierärztl. Wochschr.	Berliner und Münchener tierärztliche Wochenschrift
Berzelius-Jahresber.	Jahresberichte über die Fortschritte der physischen Wissenschaften 1822-'41, Jahresberichte über die Fortschritte der Chemie 1842-'51 von Jacob Berzelius
Bihang K. Svenska Vetenskaps-Akad. Handl.	Bihang till Kongl. Svenska Vetenskaps-Akademiens Handlingar
Biochem. J.	The Biochemical Journal

ABBREVIATED TITLE	FULL TITLE
Bull. Agr. Ch. Soc. Japan	Bulletin of the Agricultural Chemical Society of Japan
Bull. assoc. franç. étude cancer	Bulletin de l'association française pour l'étude du cancer
Bull. biol. méd. exptl. U.R.S.S.	Bulletin de biologie et de médecine expérimentale de l'U.R.S.S.
Bull. Ch. Soc. Japan	Bulletin of the Chemical Society of Japan
Bull. classe sci., Acad. roy. Belg.	Bulletin de la classe des sciences, Académie royale de Belgique (Mededeelingen van de Klasse der Wetenschappen, Koninklijke Belgische Academie)
Bull. Imp. Inst.	Bulletin of the Imperial Institute
Bull. inst. océanog.	Bulletin de l'institut océanographique
Bull. Inst. Phys. Ch. Research	Bulletin of the Institute of Physical and Chemical Research (Tokyo)
Bull. inst. pin	Bulletin de l'institut du pin
Bull. inst. recherches biol. Perm	Bulletin de l'institut des recherches biologiques de Perm
Bull. intern. acad. polon.	Bulletin international de l'académie polonaise des sciences et des lettres, Classe des sciences mathématiques et naturelles; before 1919 Bulletin international de l'académie des sciences de Cracovie
Bull. Jardin bot. Buitenzorg	Bulletin du Jardin botanique de Buitenzorg
Bull. Johns Hopkins Hosp.	Bulletin of the Johns Hopkins Hospital
Bull. mat. grasses	Bulletin des matières grasses de l'institut colonial de Marseille
Bull. sci. acad. roy. Belg.	Bulletin de la classe des sciences, académie royale de Belgique
Bull. sci. école polytech. Timisoara	Bulletin scientifique de l'école polytechnique de Timisoara
Bull. sci. ind. Roure-Bertrand fils	see *Recherches Roure-Bertrand fils*
Bull. sci. pharmacol.	Bulletin des sciences pharmacologiques
Bull. sect. sci. acad. roumaine	Bulletin de la section scientifique de l'académie roumaine
Bull. soc. chim.	Bulletin de la société chimique de France
Bull. soc. chim. Belg.	Bulletin de la société chimique de Belgique et Recueil des travaux chimiques belges
Bull. soc. chim. biol.	Bulletin de la société de chimie biologique
Bull. soc. franç. minéral.	Bulletin de la société française de minéralogie
Bull. soc. ind. Mulhouse	Bulletin de la société industrielle de Mulhouse
Bull. soc. minéral. France	Bulletin de la société minéralogique de France
Bull. soc. naturalistes Moscou	Bulletin de la société des naturalistes de Moscou
Bull. soc. pharm. Bordeaux	Bulletin des travaux de la société de pharmacie de Bordeaux
Bull. soc. roumaine phys.	Bulletin de la société roumaine de physique
Bull. soc. roy. sci. Liège	Bulletin de la société royale des sciences de Liège
Bull. soc. sci. Bretagne	Bulletin de la société scientifique de Bretagne. Sciences mathématiques, physiques, et naturelles
Bull. soc. sci. Bucarest	Bulletin de la société des sciences de Bucarest-Roumanie
Bull. soc. vaudoise sci. natur.	Bulletin de la société vaudoise des sciences naturelles
Bull. Torrey Botan. Club	Bulletin of the Torrey Botanical Club
Bull. univ. Asie centrale	Bulletin de l'université de l'Asie centrale
Bur. Standards J. Research	Bureau of Standards Journal of Research
Bur. Standards, Technol. Papers	Bureau of Standards, Technological Papers
C. A.	Chemical Abstracts
Cancer Research	Cancer Research
Can. J. Research	Canadian Journal of Research
Can. Med. Assoc. J.	The Canadian Medical Association Journal
Can. Pat.	Canadian Patent
Caoutchouc & gutta-percha	Le caoutchouc & la gutta-percha
Časopis Českoslov. Lékárnictva	Časopis Československého Lékárnictva

ABBREVIATED TITLE	FULL TITLE
Cellulosechemie	Cellulosechemie
Cereal Chem.	Cereal Chemistry
Ch. Eng.	Chemical Engineering with Chemical & Metallurgical Engineering
Ch. Industrie	Die Chemische Industrie
Ch. Met. Eng.	Chemical and Metallurgical Engineering
Ch. News	The Chemical News and Journal of Industrial Science
Ch. Reviews	Chemical Reviews
Ch. Trade J.	The Chemical Trade Journal
Ch. Umschau Fette, Öle	Chemische Umschau auf dem Gebiete der Fette, Öle Wachse und Harze
Ch. Weekbl.	Chemisch Weekblad
Ch. Zbl.	Chemisches Zentralblatt
Ch. Ztg.	Chemiker-Zeitung
Chem. Age (London)	The Chemical Age (London)
Chem. App.	Chemische Apparatur
Chem. Ind. (Germany)	Die Chemische Industrie
Chem. Industries	Chemical Industries
Chem. Listy	Chemické Listy pro vědu a průmysl
Chem. Zelle u. Gewebe	Chemie der Zelle und Gewebe
Chemist and Druggist	The Chemist and Druggist
Chemistry & Industry	Chemistry & Industry
Chimia (Switz.)	Chimia (Switzerland)
Chimica e industria (Milan)	La Chimica e l'industria (Milan)
Chimie & industrie	Chimie & industrie
Chinese J. Physiol.	The Chinese Journal of Physiology
Ciencia (Mex.)	Ciencia (Mexico)
CIOS Report	Report prepared by Combined Intelligence Objectives Sub-committee
Clin. Sci.	Clinical Science Incorporating Heart
Coke Smokeless-Fuel Age	Coke and Smokeless-Fuel Age Incorporating Coal Carbonisation
Cold Spring Harbor Symp. Quant. Biol.	Cold Spring Harbor Symposia on Quantitative Biology
Col. Ind.	Colour Index (F. M. Rowe), 1st Ed. (1924) with Supplement (1928)
Collection Czechoslov.	Collection of Czechoslovak Chemical Communications
Collegium	Collegium
Colloid Symposium Annual	Colloid Symposium Annual; formerly Colloid Symposium Monograph
Colloid Symposium Monograph	Colloid Symposium Monograph
Color Trade J.	Color Trade Journal; since 1923 Color Trade Journal and Textile Chemist
Communs. Phys. Lab. Univ. Leiden	Communications from the Physical Laboratory of the University of Leiden
Compt. rend.	Comptes rendus hebdomadaires des séances de l'académie des sciences
Compt. rend. acad. sci. U.R.S.S.	Comptes rendus (Doklady) de l'académie des sciences de l'U.R.S.S.
Compt. rend. congr. soc. savantes Paris	Comptes rendus du congrès des sociétés savantes de Paris et des départements
Compt. rend. soc. biol.	Comptes rendus hebdomadaires des séances de la société de biologie et de ses filiales et associées
Compt. rend. soc. phys. hist. nat. Genève	Compte rendu des séances de la société de physique et d'histoire naturelle de Genève
Compt. rend. trav. labor. Carlsberg	Comptes-rendus des travaux du laboratoire Carlsberg

ABBREVIATED TITLE	FULL TITLE
Congr. chim. ind., Compt. rend. ....ème Congr.	Congrès de chimie industrielle, Compte-rendu du ....ème Congrès
Congr. intern. quím. pura y apl.	Congreso internacional de química pura y aplicada
Contrib. Boyce Thompson Inst.	Contributions from Boyce Thompson Institute
Crell's Ch. Ann.	Chemische Annalen für die Freunde der Naturlehre, Arzneygelahrtheit, Haushaltungskunst und Manufacturen: von D. Lorenz Crell
Current Science	Current Science
Dan. Pat.	Danish Patent
Dansk Tids. Farm.	Dansk Tidsskrift for Farmaci
Dermatologica	Dermatologica
Dermatol. Wochschr.	Dermatologische Wochenschrift
Deut. Apoth. Ztg.	Deutsche Apotheker Zeitung
Deut. Arch. klin. Med.	Deutsches Archiv für klinische Medizin
Deut. Färber-Ztg.	Deutsche Färber-Zeitung vereinigt mit der Leipziger Färber- und Zeugdrucker-Zeitung
Deut. med. Wochschr.	Deutsche medizinische Wochenschrift
Deut. Parfüm. Ztg.	Deutsche Parfümerie-Zeitung
Deut. Z. Chir.	Deutsche Zeitschrift für Chirurgie
Deut. Zuckerind.	Die Deutsche Zuckerindustrie
Dinglers Polytech. J.	Dinglers Polytechnisches Journal
Drug Cosmetic Ind.	Drug and Cosmetic Industry
Dutch Pat.	Dutch Patent
Econ. Botany	Economic Botany
Elektroch. Z.	Elektrochemische Zeitschrift
Endocrinology	Endocrinology
Endokrinologie	Endokrinologie
Eng. Mining J.	Engineering and Mining Journal
Enzymologia	Enzymologia
Ergeb. Physiol. exptl. Pharmakol.	Ergebnisse der Physiologie und experimentelle Pharmakologie
Ergeb. Vitamin- u. Hormonforsch.	Ergebnisse der Vitamin- und Hormonforschung
Experientia	Experientia
Färber-Ztg.	Färber-Zeitung. Zeitschrift für Färberei, Zeugdruck und den gesamten Farbenverbrauch
Farbe u. Lack	Farbe und Lack
Farben-Chem.	Der Farben-Chemiker
Farben-Ztg.	Farben-Zeitung
Farbstofftab.	Farbstofftabellen (G. Schultz), 7th. Ed., Vol. I (1931) with Supplement I (1934) and Supplement II (1939)
Farm. sci. e tec. (Pavia)	Il Farmaco scienza e tecnica (Pavia)
Federation Proc.	Federation Proceedings
Fermentforschung	Fermentforschung
Fettchem. Umschau	Fettchemische Umschau
Fette u. Seifen	Fette und Seifen
FIAT Final Report	Report prepared by Field Information Agency, Technical (United States Group Control Council for Germany)
Flora (Ger.)	Flora (Germany)
Food Research	Food Research
Foreign Petroleum Technol.	Foreign Petroleum Technology
Forschungsdienst	Der Forschungsdienst
Fortschr. Chem. Physik u. physik. Chem.	Fortschritte der Chemie, Physik und physikalischen Chemie

ABBREVIATED TITLE	FULL TITLE
Fortschr. Geb. Röntgenstrahlen	Fortschritte auf dem Gebiete der Röntgenstrahlen
Frankfurt. Z. Path.	Frankfurter Zeitschrift für Pathologie
Fr. Pat.	French Patent
Fuel	Fuel in Science and Practice
Gann	Gann. The Japanese Journal of Cancer Research
Gartenbauwiss.	Die Gartenbauwissenschaft
Gas	Gas (Engineering, Utilization, Load-Building, Management)
Gas, Het	Het Gas
Gas Age-Record	Gas Age-Record
Gas- u. Wasserfach	Das Gas- und Wasserfach
Gas World	The Gas World
Gazz.	Gazzetta chimica italiana
Ger. Pat.	German Patent
Ges. Abhandl. Kenntnis Kohle	Gesammelte Abhandlungen zur Kenntnis der Kohle
Giorn. chim. ind. applicata	Giornale di chimica industriale ed applicata
Glückauf	Glückauf berg- und hüttenmännische Zeitschrift
Graefe's Arch. Ophthalmol.	Graefe's Archiv für Ophthalmologie
Growth	Growth
Gummi-Ztg.	Gummi-Zeitung
Gummi-Ztg. u. Kautschuk	Gummi-Zeitung und Kautschuk
Harvey Lectures	The Harvey Lectures
Helv.	Helvetica Chimica Acta
Helv. Phys. Acta	Helvetica Physica Acta
Helv. Physiol. et Pharmacol. Acta	Helvetica Physiologica et Pharmacologica Acta
Hereditas	Hereditas (Genetiskt Arkiv)
Het Gas	Het Gas
Hung. Pat.	Hungarian Patent
Ind. chim. belge	Industrie chimique belge (Formerly Bulletin de la fédération des industries chimiques de Belgique)
Ind. Eng. Ch.	Industrial and Engineering Chemistry; before 1923 Journal of Industrial and Engineering Chemistry (*Ind. Ed.* = Industrial Edition; *Anal. Ed.* = Analytical Edition)
Indian Forest Records	The Indian Forest Records
Indian J. Med. Research	The Indian Journal of Medical Research
Indian J. Phys.	Indian Journal of Physics
Indian J. Physiol. and Allied Sci.	The Indian Journal of Physiology and Allied Sciences
Indian Pat.	Indian Patent
Industria Chimica	Industria Chimica
Intern. Sugar J.	The International Sugar Journal
Intern. Z. physik.-chem. Biol.	Internationale Zeitschrift für physikalisch-chemische Biologie
Iowa State Coll. J. Sci.	Iowa State College Journal of Science
Ital. Pat.	Italian Patent
Jahrb. prakt. Pharm.	Jahrbuch für praktische Pharmazie und verwandte Fächer
Jahrb. wiss. Botan.	Jahrbücher für wissenschaftliche Botanik
Jahresber. Ch.	Jahresbericht über die Fortschritte der Chemie und verwandter Teile anderer Wissenschaften
Jahresber. ch.-techn. Reichsanstalt	Jahresbericht der chemisch-technischen Reichsanstalt
Jahresber. Pharm.	Jahresbericht der Pharmazie
Japan. J. Med. Sci.	Japanese Journal of Medical Sciences
Japan. Pat.	Japanese Patent

ABBREVIATED TITLE	FULL TITLE
J. Agr. Ch. Soc. Japan	Journal of the Agricultural Chemical Society of Japan
J. Agr. Res.	Journal of Agricultural Research
J. Agr. Sci.	The Journal of Agricultural Science
J. Allergy	The Journal of Allergy
J. Am. Ceramic Soc.	Journal of the American Ceramic Society
J. Am. Ch. Soc.	The Journal of the American Chemical Society
J. Am. Leather Chemists' Assoc.	The Journal of the American Leather Chemists' Association
J. Am. Med. Assoc.	The Journal of the American Medical Association
J. Am. Pharm. Assoc.	Journal of the American Pharmaceutical Association (*Sci. Ed.* = Scientific Edition)
J. Annamalai Univ.	Journal of the Annamalai University
J. Appl. Phys.	Journal of Applied Physics
J. Applied Ch. U.S.S.R.	Journal of Applied Chemistry U.S.S.R.
J. Assoc. Offic. Agr. Chemists	Journal of the Association of Official Agricultural Chemists
J. Bact.	Journal of Bacteriology
J. Biochem.	The Journal of Biochemistry (Japan)
J. Biol. Ch.	The Journal of Biological Chemistry (Proc. = Proceedings of the American Society of Biological Chemists)
J. Cellular Comp. Physiol.	Journal of Cellular and Comparative Physiology
J. Ch. Education	Journal of Chemical Education
J. Ch. Physics	The Journal of Chemical Physics
J. Ch. Soc.	Journal of the Chemical Society (London)
J. Ch. Soc. Japan	Journal of the Chemical Society of Japan
J. chim. phys.	Journal de chimie physique et revue générale des colloïdes; since 1939 Journal de chimie physique et de physico-chimie biologique
J. Chinese Ch. Soc.	Journal of the Chinese Chemical Society
J. Clin. Endocrinol.	The Journal of Clinical Endocrinology
J. Clin. Investigation	The Journal of Clinical Investigation
J. Colloid Sci.	Journal of Colloid Science
J. Council Sci. Ind. Research	Journal of the Council for Scientific and Industrial Research
J. Econ. Entomol.	Journal of Economic Entomology
J. Endocrinol.	The Journal of Endocrinology
J. Exptl. Biol.	The Journal of Experimental Biology
J. Exptl. Med.	The Journal of Experimental Medicine
J. Exptl. Zoöl.	The Journal of Experimental Zoölogy
J. Franklin Inst.	Journal of the Franklin Institute
J. Gen. Ch. U.S.S.R.	Journal of General Chemistry U.S.S.R.
J. Gen. Physiol.	The Journal of General Physiology
J. Ind. Hyg. Toxicol.	The Journal of Industrial Hygiene and Toxicology; before 1936 The Journal of Industrial Hygiene
J. Indian Ch. Soc.	Journal of the Indian Chemical Society; vols. 2–4 The Quarterly Journal of ...
J. Indian Inst. Sci.	Journal of the Indian Institute of Science
—	Journal of Industrial and Engineering Chemistry, see Industrial and Engineering Chemistry
J. Inst. Petroleum	Journal of the Institute of Petroleum
J. Inst. Petroleum Tech.	Journal of the Institution of Petroleum Technologists
J. Invest. Dermatol.	The Journal of Investigative Dermatology
J. Lab. Clin. Med.	The Journal of Laboratory and Clinical Medicine
J. Landwirtsch.	Journal für Landwirtschaft
J. Mt. Sinai Hosp., N.Y.	Journal of the Mount Sinai Hospital, New York
J. Natl. Cancer Inst.	Journal of the National Cancer Institute
J. Nutrition	The Journal of Nutrition
J. Oil & Colour Chemists' Assoc.	Journal of the Oil & Colour Chemists' Association
J. Optical Soc. Am.	Journal of the Optical Society of America

Abbreviated Title	Full Title
J. Org. Ch.	The Journal of Organic Chemistry
J. Path. Bact.	The Journal of Pathology and Bacteriology
J. pharm. Belg.	Journal de pharmacie de Belgique
J. pharm. Belg. Rev. sci. pharm.	Journal de pharmacie de Belgique, Revue des sciences pharmaceutiques
J. pharm. chim.	Journal de pharmacie et de chimie
J. Pharm. Soc. Japan	Journal of the Pharmaceutical Society of Japan (*Trans.* = Transactions)
J. Pharmacol.	The Journal of Pharmacology and Experimental Therapeutics
J. Phys. Ch.	The Journal of Physical Chemistry
J. Phys. Ch. U.S.S.R.	Journal of Physical Chemistry U.S.S.R.
J. phys. radium	Journal de physique et le radium
J. Physiol.	The Journal of Physiology (London)
J. physiol. path. gén.	Journal de physiologie et de pathologie générale
J. Physiol. U.S.S.R.	The Journal of Physiology of U.S.S.R.
J. pr. Ch.	Journal für praktische Chemie
J. Proc. Roy. Soc. N.S. Wales	Journal and Proceedings of the Royal Society of New South Wales
J. Research Natl. Bur. Standards	Journal of Research of the National Bureau of Standards. Formerly Bureau of Standards Journal of Research
J. Rheol.	Journal of Rheology
J. Roy. Hort. Soc.	Journal of the Royal Horticultural Society
J. Roy. Soc. W. Australia	Journal of the Royal Society of Western Australia
J. Roy. Tech. Coll. Glasgow	Journal of the Royal Technical College (Glasgow)
J. Russ. Phys. Ch. Soc.	Journal of the Russian Physical-Chemical Society
J. Sci. Ind. Research (India)	Journal of Scientific and Industrial Research (India)
J. Soc. Ch. Ind.	Journal of the Society of Chemical Industry (*Trans.* = Transactions; *Rev.* = Review)
J. Soc. Ch. Ind. Japan	Journal of the Society of Chemical Industry of Japan
J. Soc. Dyers Colourists	The Journal of the Society of Dyers and Colourists
J. Textile Inst.	The Journal of the Textile Institute (*Trans.* = Transactions, *Proc.* = Proceedings)
J. Univ. Bombay	Journal of the University of Bombay (*Physical Sciences*)
J. Urol.	The Journal of Urology
J. usines gaz	Journal des usines à gaz
J. Wash. Acad. Sci.	Journal of the Washington Academy of Sciences
Kautschuk	Kautschuk
Kgl. Danske Vidensk. Selsk.	Det Kongelige Danske Videnskabernes Selskab (*Biol. Medd.* = Biologiske Meddelelser; *Mat.-fys. Medd.* = Matematisk-fysiske Meddelelser)
Kgl. Fysiograf. Sällskap. Lund, Förh.	Kungliga Fysiografiska Sällskapets i Lund, Förhandlingar
Kgl. Norske Vidensk. Selsk.	Kongelige Norske Videnskabers Selskabs (*Forh.* = Forhandlinger; *Skrifter* = Skrifter)
Kgl. Vetenskaps-Acad. Handl.	Kongl. Vetenskaps-Academiens Handlingar (Stockholm)
Kinotechnik	Die Kinotechnik
Klepzig's Textil-Z.	Klepzig's Textil-Zeitschrift
Klin. Wochschr.	Klinische Wochenschrift; before 1922 Berliner klinische Wochenschrift
Kolloid-Beihefte	Kolloid-Beihefte; before 1931 Kolloidchemische Beihefte
Kolloid-Z.	Kolloid-Zeitschrift
Krankheitsforsch.	Krankheitsforschung
Kunststoffe	Kunststoffe
Kunststoffe ver. Kunststoff-Tech. u. -Anwend.	Kunststoffe vereinigt mit Kunststoff-Technik und -Anwendung

ABBREVIATED TITLE	FULL TITLE
Lack- u. Farben-Z.	Lack- und Farben-Zeitschrift
Lait	Le Lait
Lancet	The Lancet
Landw. Vers.-Sta.	Die landwirtschaftlichen Versuchs-Stationen
Ledertech. Rundschau	Ledertechnische Rundschau
Leeuwenhoek-Vereeniging	Leeuwenhoek-Vereeniging
Leipzig. Monatschr. Textil-Ind.	Leipziger Monatschrift für Textil-Industrie
London Hosp. Gaz.	The London Hospital Gazette
Lunds Univ. Årsskr.	Lunds Universitets Årsskrift (Kungl. Fysiografiska Sällskapets i Lund, Handlingar)
Medd. Vetenskapsakad. Nobelinst.	Meddelanden fran K. Vetenskapsakademiens Nobelinstitut
Mededel. Vlaam. Chem. Ver.	Mededelingen van de Vlaamse Chemische Vereniging; before 1948 Mededeelingen van de Vlaamsche Chemische Vereeniging
Medicina (Buenos Aires)	Medicina (Buenos Aires)
Medicine	Medicine
Medizin und Chemie	Medizin und Chemie
Med. Klinik	Medizinische Klinik
Melliand Textilber.	Melliand Textilberichte
Mém. acad. roy. sci.	Mémoires de mathématique et de physique, tirés des registres de l'académie royale des sciences
Mem. accad. Italia	Memorie della reale accademia d'Italia. Classe di scienze fisiche, matematiche e naturali
Mem. accad. scienze Bologna	Memorie dell' accademia delle scienze dell' istituto di Bologna
Mem. Coll. Agr. Kyoto Imp. Univ.	Memoirs of the College of Agriculture, Kyoto Imperial University
Mem. Coll. Sci. Kyoto	Memoirs of the College of Science, Kyoto Imperial University
Mem. inst. Butantan (São Paulo)	Memórias do Instituto Butantan (São Paulo)
Mém. poudres	Mémorial des poudres; Mémorial des poudres et salpetres
Mem. Proc. Manchester Lit. Phil. Soc.	Memoirs and Proceedings of the Manchester Literary & Philosophical Society
Merck's Jahresber.	E. Merck's Jahresbericht
Metall u. Erz	Metall und Erz
Mfg. Chemist	The Manufacturing Chemist
Mikrochemie	Mikrochemie; since 1938 (Vol. 25) Mikrochemie vereinigt mit Mikrochimica Acta
Mikrochim. Acta	Mikrochimica Acta
Mitt. Forstwirtsch. Forstwiss.	Mitteilungen aus Forstwirtschaft und Forstwissenschaft
Mitt. Gebiete Lebensm. u. Hyg.	Mitteilungen aus dem Gebiete der Lebensmitteluntersuchung und Hygiene
Mitt. med. Ges. Okayama	Mitteilungen der medizinischen Gesellschaft zu Okayama
Mitt. technol. Gewerbe-Mus. Wien	Mittheilungen des Kaiserl. Königl. Technologischen Gewerbe-Museums in Wien
Monatsber. königl. preuss. Akad. Wiss.	Monatsberichte der königlichen preussischen Akademie der Wissenschaften zu Berlin
Monatschr. Textil-Ind.	Monatschrift für Textil-Industrie
Monatsh.	Monatshefte für Chemie und verwandte Teile anderer Wissenschaften
Mon. sci.	Moniteur scientifique du Docteur Quesneville
Münch. med. Wochschr.	Münchener medizinische Wochenschrift
Nachr. Akad. Wiss. Göttingen, math.-phys. Klasse	Nachrichten von der Akademie der Wissenschaften zu Göttingen, mathematisch-physikalische Klasse

ABBREVIATED TITLE	FULL TITLE
Nachr. Ges. Wiss. Göttingen, math.-phys. Klasse	Nachrichten von der Gesellschaft der Wissenschaften zu Göttingen, mathematisch-physikalische Klasse
Nachrbl. deut. Pflanzenschutzdienst	Nachrichtenblatt für den deutschen Pflanzenschutzdienst
Nature	Nature (London)
Nature, La	La Nature
Naturwiss.	Die Naturwissenschaften
Naturw. Museumsh.	Naturwissenschaftliche Museumshefte
Natuurw. Tijdschr.	Natuurwetenschappelijk Tijdschrift
Ned. Tijdschr. Geneesk.	Nederlandsch Tijdschrift voor Geneeskunde
Neues Repert. Pharm.	Neues Repertorium für Pharmazie
New Orleans Med. Surg. J.	New Orleans Medical and Surgical Journal
New Zealand J. Sci. Tech.	The New Zealand Journal of Science and Technology
News Ed.	News Edition (American Chemical Society)
Nieuw Tijdschr. Pharm. Nederland	Nieuw Tijdschrift voor de Pharmacie in Nederland
Nitrocellulose	Nitrocellulose
Norw. Pat.	Norwegian Patent
Nova Acta Regiae Soc. Sci. Upsaliensis	Nova Acta Regiae Societatis Scientiarum Upsaliensis
Nuovo cimento	Nuovo cimento
Nutrition Abstr. & Revs.	Nutrition Abstracts & Reviews
Öffentl. Gesundheitspflege	Öffentliche Gesundheitspflege
Öfversigt Finska Vetenskaps-Soc. Förhandl.	Öfversigt af Finska Vetenskaps-Societetens Förhandlingar
Öfversigt K. Vetenskaps-Akad. Förhandl.	Öfversigt af Kongl. Vetenskaps-Akademiens Förhandlingar (Stockholm)
Oel u. Kohle	Oel und Kohle (Oel und Kohle vereinigt mit Erdoel und Teer; Oel und Kohle vereinigt mit Petroleum
Oesterr. botan. Ztschr.	Oesterreichische botanische Zeitschrift
Oesterr. Chem. Ztg.	Oesterreichische Chemiker-Zeitung
Oil Colour Trades J.	Oil and Colour Trades Journal
Oil Gas J.	The Oil and Gas Journal
Oil & Soap	Oil & Soap
Onderstepoort J. Vet. Sci.	Onderstepoort Journal of Veterinary Science and Animal Industry
Org. Chem. Ind. (U.S.S.R.)	The Organic Chemical Industry (U.S.S.R.)
Org. Synth.	Organic Syntheses (*Coll. Vol.* = Collective Volume)
Paint Oil Chem. Rev.	Paint, Oil, and Chemical Review
Papers Michigan Acad. Sci.	Papers of the Michigan Academy of Science, Arts, and Letters
Parfumerie moderne	La Parfumerie moderne
Parfums France	Les parfums de France
Perfumery Essential Oil Record	The Perfumery and Essential Oil Record
Petroleum (London)	Petroleum (London)
Petroleum Z.	Petroleum, Zeitschrift für die gesamten Interessen der Mineralöl-Industrie und des Mineralöl-Handels
Pharm. Acta Helv.	Pharmaceutica Acta Helvetiae
Pharm. Arch.	Pharmaceutical Archives
Pharm. J.	The Pharmaceutical Journal; before 1895 The Pharmaceutical Journal and Transactions
Pharm. Monatsh.	Pharmazeutische Monatshefte
Pharm. Post	Pharmazeutische Post
Pharm. Rev.	Pharmaceutical Review; before 1896 Pharmaceutische Rundschau
Pharm. Weekbl.	Pharmaceutisch Weekblad
Pharm. Z. Russland	Pharmaceutische Zeitschrift für Russland

ABBREVIATED TITLE	FULL TITLE
Pharm. Zentralh.	Pharmazeutische Zentralhalle
Pharm. Ztg.	Pharmazeutische Zeitung
Pharmacia (Estonia)	Pharmacia (Estonia)
Pharmazie	Die Pharmazie
Philippine J. Sci.	The Philippine Journal of Science
Philos. Mag.	The Philosophical Magazine
Philos. Trans. Roy. Soc. London	Philosophical Transactions of the Royal Society London
Phot. Ind.	Die Photographische Industrie
Phot. Korr.	Photographische Korrespondenz
Phys. Rev.	The Physical Review
Physica	Physica (Nederlandsch Tijdschrift voor Natuurkunde); since 1933 Physica (Series IV A of Archives néerlandaises des sciences exactes et naturelles)
Physics	Physics
Physik. Ber.	Physikalische Berichte
Physik. Z.	Physikalische Zeitschrift
Physik. Z. Sowjetunion	Physikalische Zeitschrift der Sowjetunion
Physiol. Rev.	Physiological Reviews
Physiol. Zoöl.	Physiological Zoölogy
Planta	Planta
Poggendorff's Ann.	see *Ann. Physik*
Praktika Akad. Athenon	Praktika tes Akademias Athenon
Presse méd.	La Presse médicale
Proc. Am. Acad. Arts Sci.	Proceedings of the American Academy of Arts and Sciences
Proc. Am. Pharm. Assoc.	Proceedings of the American Pharmaceutical Association
Proc. Am. Soc. Hort. Sci.	Proceedings of the American Society for Horticultural Science
Proc. Cambridge Philos. Soc.	Proceedings of the Cambridge Philosophical Society
Proc. Ch. Soc.	Proceedings of the Chemical Society (London)
Proc. Imp. Acad. Japan	Proceedings of the Imperial Academy Japan
Proc. Indian Acad. Sci.	Proceedings of the Indian Academy of Sciences
Proc. Indian Assoc. Cultivation Sci.	Proceedings of the Indian Association for the Cultivation of Science
Proc. Indiana Acad. Sci.	Proceedings of the Indiana Academy of Science
Proc. Intern. Conf. Bituminous Coal	Proceedings of the International Conference on Bituminous Coal
Proc. Iowa Acad. Sci.	Proceedings of the Iowa Academy of Science
Proc. Japan Acad.	Proceedings of the Japan Academy
Proc. Kon. Akad. Wet. Amsterdam	Proceedings of the Koninklijke Akademie van Wetenschappen te Amsterdam; since 1938 Proceedings of the Koninklijke Nederlandsche Akademie van Wetenschappen
Proc. Linnean Soc. N. S. Wales	The Proceedings of the Linnean Society of New South Wales
Proc. Nat. Acad. Sci.	Proceedings of the National Academy of Sciences U.S.A.
Proc. Oklahoma Acad. Sci.	Proceedings of the Oklahoma Academy of Science
Proc. Phys.-Math. Soc. Japan	Proceedings of the Physico-Mathematical Society of Japan
Proc. Phys. Soc. London	The Proceedings of the Physical Society London
Proc. Roy. Irish Acad.	Proceedings of the Royal Irish Academy
Proc. Roy. Soc. Edinburgh	Proceedings of the Royal Society of Edinburgh
Proc. Roy. Soc. London	Proceedings of the Royal Society London
Proc. Roy. Soc. Queensland	Proceedings of the Royal Society of Queensland
Proc. Rubber Technol. Conf.	Proceedings of the Rubber Technology Conference
Proc. Soc. Exptl. Biol. Med.	Proceedings of the Society for Experimental Biology and Medicine
Proc. Staff Meetings Mayo Clinic	Proceedings of the Staff Meetings of the Mayo Clinic

ABBREVIATED TITLE	FULL TITLE
Proc. Trans. Roy. Soc. Canada	Proceedings and Transactions of the Royal Society of Canada
Protoplasma	Protoplasma
Przemysl Chemiczny	Przemysl Chemiczny
Pub. fac. sci. univ. Masaryk	Publications de la faculté des sciences de l'université Masaryk
Pub. sci. tech. ministère air	Publications scientifiques et techniques du ministère de l'air (France)
Quart. Bull. Northwestern Univ. Med. School	Quarterly Bulletin of Northwestern University Medical School
Quart. J. Exptl. Physiol.	Quarterly Journal of Experimental Physiology
Quart. J. Med.	The Quarterly Journal of Medicine
Quart. J. Microscop. Sci.	The Quarterly Journal of Microscopical Science
Quart. J. Pharm. Pharmacol.	Quarterly Journal of Pharmacy and Pharmacology
Quart. J. Sci.	Quarterly Journal of Science, Literature, and the Arts
Radiology	Radiology
Rasāyanam	Rasāyanam
Recherches Roure-Bertrand fils	Recherches (Continuation of Bulletin scientifique et industriel de la maison Roure-Bertrand fils)
Rec. trav. chim.	Recueil des travaux chimiques des Pays-Bas
Reichsamt Wirtschaftsausbau, Chem. Ber.	Reichsamt für Wirtschaftsausbau, Chemische Berichte
Reichsamt Wirtschaftsausbau, 1. Tagung der Arbeitsgemeinschaft "Teerverwertung"	Reichsamt für Wirtschaftsausbau, 1. Tagung der Arbeitsgemeinschaft "Teerverwertung"
Rend. accad. sci. fis. mat. Napoli	Rendiconti dell' accademia di scienze fisiche, matematiche e naturali della società reale di Napoli
Rend. ist. lombardo sci.	Rendiconti reale istituto lombardo di scienze e lettere
Rend. sem. fac. sci. reale univ. Cagliari	Rendiconti del seminario della facoltà di scienze della reale università di Cagliari
Rend. soc. chim. ital.	Rendiconti della società chimica italiana; before 1909 Rendiconti della società chimica di Roma
Repert. Pharm.	Repertorium für die Pharmazie
Rept. Australian New Zealand Assoc. Advancement Sci.	Report of the Australian and New Zealand Association for the Advancement of Science
Rev. argentina dermato-sifilol.	Revista argentina de dermato-sifilología
Rev. can. biol.	Revue canadienne de biologie
Rev. españ. fisiol.	Revista española de fisiología
Rev. gén. caoutchouc	Revue générale du caoutchouc
Rev. gén. colloïdes	Revue générale des colloïdes
Rev. gén. mat. color.	Revue générale des matières colorantes, (du blanchiment), de la teinture, de l'impression et des apprêts
Rev. gén. sci.	Revue générale des sciences pures et appliquées
Rev. gén. teint. impression blanchiment apprêt	Revue générale de teinture, impression, blanchiment, apprêt et de chimie textile et tinctoriale
Rev. mét.	Revue de métallurgie
Rev. prod. chim.	La Revue des produits chimiques et l'actualité scientifique
Rev. sci.	La Revue scientifique
Ricerca sci.	La Ricerca scientifica; from 1945–1947 Ricerca scientifica e ricostruzione; before 1944 La Ricerca scientifica ed il progresso tecnico (nell'economia nazionale)
Riechstoffind.	Die Riechstoffindustrie
Riv. ital. essenze e profumi	Rivista italiana delle essenze e profumi
Roczniki Chemji	Roczniki Chemji; since 1936 Roczniki Chemii

ABBREVIATED TITLE	FULL TITLE
Rubber Age (N.Y.)	The Rubber Age (New York)
Russ. Pat.	Russian Patent
S. African J. Med. Sci.	The South African Journal of Medical Sciences
Samml. Vergiftungsfällen	Sammlung von Vergiftungsfällen (Fühner-Wieland's)
Sborník České Akad. Zemědělské	Sborník České Akademie Zemědělské
Schweiz. Apoth. Ztg.	Schweizerische Apotheker Zeitung
Schweiz. med. Wochschr.	Schweizerische medizinische Wochenschrift
Schweiz. Ver. Gas- u. Wasserfach., Monats-Bull.	Schweizerischer Verein von Gas- und Wasserfachmännern, Monats-Bulletin
Schweiz. Wochschr. Ch. Pharm.	Schweizerische Wochenschrift für Chemie und Pharmazie
Schweiz. Z. Biochem.	Schweizerische Zeitschrift für Biochemie
Sci. Papers Inst. Phys. Ch. Research	Scientific Papers of the Institute of Physical and Chemical Research, Tokyo
Sci. Proc. Roy. Dublin Soc.	The Scientific Proceedings of the Royal Dublin Society
Sci. Repts. Moscow State Univ.	Scientific Reports of the Moscow State University (Uchenye Zapiski Moskovskogo Gosudarstvennogo Universiteta)
Science	Science
Science and Culture	Science and Culture
Science et ind.	Science et industrie
Science Repts. Nat. Tsing Hua Univ. Ser. A	The Science Reports of National Tsing Hua University, Series A. Mathematical, Physical, and Engineering Sciences
Science Repts. Nat. Univ. Peking	The Science Reports of the National University of Peking
Science Repts. Tôhoku Imp. Univ.	The Science Reports of the Tôhoku Imperial University
Scientia Pharm.	Scientia Pharmaceutica
Seifensieder-Ztg.	Seifensieder-Zeitung
Silliman's Am. J.	The American Journal of Science and Arts. Conducted by B. Silliman
Sitzber. Abhandl. naturforsch. Ges. Rostock	Sitzungsberichte und Abhandlungen der naturforschenden Gesellschaft zu Rostock
Sitzber. Akad. Wiss. Wien	Sitzungsberichte der Akademie der Wissenschaften in Wien, mathematisch-naturwissenschaftliche Klasse
Sitzber. Ges. Beförder. ges. Naturw. Marburg	Sitzungsberichte der Gesellschaft zur Beförderung der gesamten Naturwissenschaften zu Marburg
Sitzber. Heidelberger Akad. Wiss.	Sitzungsberichte der Heidelberger Akademie der Wissenschaften (Math.-naturw. Kl. = Mathematisch-naturwissenschaftliche Klasse)
Sitzber. königl. böhm. Ges. Wiss.	Sitzungsberichte der königlichen böhmischen Gesellschaft der Wissenschaften
Sitzber. königl. preuss. Akad. Wiss.	Sitzungsberichte der königlichen preussischen Akademie der Wissenschaften
Sitzber. math.-phys. Klasse bayer. Akad. Wiss. München	Sitzungsberichte der mathematisch-physikalischen Klasse der bayerischen Akademie der Wissenschaften zu München
Sitzber. Naturforsch. Ges. Univ. Tartu	Sitzungsberichte der Naturforscher-Gesellschaft bei der Universität Tartu
Sitzber. Vereins Beförder. Gewerbfleisses	Sitzungsberichte des Vereins zur Beförderung des Gewerbfleisses in Preussen
Skand. Arch. Physiol.	Skandinavisches Archiv für Physiologie
Soap	Soap; since 1938 Soap and Sanitary Chemicals
Soap, Perfumery & Cosmetics	Soap, Perfumery & Cosmetics
Southern Med. J.	Southern Medical Journal
Sperimentale	Lo Sperimentale
Stain Technol.	Stain Technology
Strahlentherapie	Strahlentherapie
Süddeut. Apoth. Ztg.	Süddeutsche Apotheker Zeitung

Abbreviated Title	Full Title
U.S. Pat.	United States Patent
U.S. Pub. Health Repts.	United States Public Health Reports
Verhandl. deut. physik. Ges.	Verhandlungen der deutschen physikalischen Gesellschaft
Verhandl. Ges. deut. Naturf. Ärzte	Verhandlungen der Gesellschaft deutscher Naturforscher und Ärzte
Verhandl. naturforsch. Ges. Basel	Verhandlungen der naturforschenden Gesellschaft in Basel
Verhandl. physic. med. Ges. Würzburg	Verhandlungen der physicalisch-medicinischen Gesellschaft in Würzburg
Verhandl. Ver. Schweiz. Physiol.	Verhandlungen des Vereins der Schweizer Physiologen
Versl. Kon. Akad. Wet. Amsterdam	Koninklijke Akademie van Wetenschappen te Amsterdam. Verslag van de Gewone Vergadering der Wis- en Natuurkundige Afdeeling. Since 1925 Verslag van de Gewone Vergadering der Afdeeling Natuurkunde
Vierteljahrsschr. naturforsch. Ges. Zürich	Vierteljahrsschrift der naturforschenden Gesellschaft in Zürich
Virchow's Arch. path. Anat.	Virchow's Archiv für pathologische Anatomie und Physiologie und für klinische Medizin
Vitamine und Hormone	Vitamine und Hormone
Vitamins and Hormones	Vitamins and Hormones
Wien. klin. Wochschr.	Wiener klinische Wochenschrift
Wien. med. Wochschr.	Wiener medizinische Wochenschrift
Wilhelm Roux' Arch. Entwicklungs-mech. Organ.	Wilhelm Roux' Archiv für Entwicklungsmechanik der Organismen
Wiss. Ber. Moskau. Staatsuniv.	Wissenschaftliche Berichte der Moskauer Staatsuniversität
Wiss. Veröffentl. Siemens-Konzern	Wissenschaftliche Veröffentlichungen aus dem Siemens-Konzern
Wiss. Veröffentl. Siemens-Werken	Wissenschaftliche Veröffentlichungen aus den Siemens-Werken
Wochschr. Brau.	Wochenschrift für Brauerei
Yale J. Biol. Med.	Yale Journal of Biology and Medicine
Yug. Pat.	Yugoslavian Patent
Z. allgem. oesterr. Apoth.-Ver.	Zeitschrift des allgemeinen oesterreichischen Apotheker-Vereines
Z. anal. Ch.	Zeitschrift für analytische Chemie
Z. ang. Ch.	Zeitschrift für angewandte Chemie, see Angewandte Chemie
Z. anorg. Ch.	Zeitschrift für anorganische und allgemeine Chemie
Z. Biol.	Zeitschrift für Biologie
Z. deut. Öl- u. Fett-Ind.	Zeitschrift der deutschen Öl- und Fett-Industrie
Z. Elektroch.	Zeitschrift für Elektrochemie und angewandte physikalische Chemie
Zentr. allgem. Path. u. path. Anat.	Zentralblatt für allgemeine Pathologie und pathologische Anatomie
Zentr. Gynäkol.	Zentralblatt für Gynäkologie
Zentr. inn. Med.	Zentralblatt für innere Medizin
Z. exptl. Path. Therapie	Zeitschrift für experimentelle Pathologie und Therapie
Z. Farbenind.	Zeitschrift für Farbenindustrie; before 1905 Zeitschrift für Farben- und Textilchemie, and Zeitschrift für Farben- und Textilindustrie
Z. f. Ch.	Zeitschrift für Chemie
Z. ges. exptl. Med.	Zeitschrift für die gesamte experimentelle Medizin
Z. ges. Schiess- u. Sprengstoffw.	Zeitschrift für das gesamte Schiess- und Sprengstoffwesen
Z. ges. Textil-Ind.	Zeitschrift für die gesamte Textil-Industrie

Abbreviated Title	Full Title
Z. Hyg. Infektionskrankh.	Zeitschrift für Hygiene und Infektionskrankheiten
Z. Immunitätsf.	Zeitschrift für Immunitätsforschung und experimentelle Therapie
Z. klin. Med.	Zeitschrift für klinische Medizin
Z. Krebsforsch.	Zeitschrift für Krebsforschung
Z. Krist.	Zeitschrift für Kristallographie, Kristallgeometrie, Kristallphysik, Kristallchemie
Z. mikroskop.-anat. Forsch.	Zeitschrift für mikroskopisch-anatomische Forschung
Z. Naturforsch.	Zeitschrift für Naturforschung
Z. Naturwissenschaften	Zeitschrift für Naturwissenschaften
Z. Pflanzenkrankh. Pflanzenschutz	Zeitschrift für Pflanzenkrankheiten und Pflanzenschutz
Z. Physik	Zeitschrift für Physik
Z. physik. Ch.	Zeitschrift für physikalische Chemie
Z. physik. chem. Unterricht	Zeitschrift für den physikalischen und chemischen Unterricht
Z. physiol. Ch.	Hoppe-Seyler's Zeitschrift für physiologische Chemie
Z. Spiritusind.	Zeitschrift für Spiritusindustrie
Ztbl. Bakt. Parasitenk.	Zentralblatt für Bakteriologie, Parasitenkunde und Infektionskrankheiten
Z. Untersuch. Lebensm.	Zeitschrift für Untersuchung der Lebensmittel; before 1926 Zeitschrift für Untersuchung der Nahrungs- und Genussmittel
Z. Ver. deut. Zucker-Ind.	Zeitschrift des Vereins der deutschen Zucker-Industrie; since 1935 Zeitschrift der Wirtschaftsgruppe Zuckerindustrie (*Allgem. Tl.* = Allgemeiner Teil; *Tech. Tl.* = Technischer Teil)
Z. Vitaminforsch.	Zeitschrift für Vitaminforschung
Z. wiss. Mikroskopie	Zeitschrift für wissenschaftliche Mikroskopie und mikroskopische Technik
Z. wiss. Phot.	Zeitschrift für wissenschaftliche Photographie, Photophysik und Photochemie
Z. Zellforsch. u. mikroskop. Anat.	Zeitschrift für Zellforschung und mikroskopische Anatomie
Z. Zuckerind. Böhmen-Mähren	Zeitschrift für die Zuckerindustrie der Böhmen-Mähren

Oxosteroids

I. CO IN SIDE CHAIN ONLY

A. Mono-oxo-Compounds

1. COMPOUNDS WITHOUT ANOTHER FUNCTIONAL GROUP

$\Delta^{3(?).5.16}$-**Pregnatrien-20-one** $C_{21}H_{28}O$. Needles (methanol), m. 142–3° cor. after softening at ca. 138° cor.; $[\alpha]_D^{13}$ —106° (acetone). — **Fmn.** Is formed in small amount, along with the acetate of $\Delta^{5.16}$-pregnadien-3β-ol-20-one (p. 2230 s), from the 3-acetate 17-benzoate of 17-iso-Δ^5-pregnene-3β.17β-diol-20-one (p. 2337 s) on melting and distillation at 250–280° and 10 mm. pressure. — **Rns.** Gives mainly allopregnan-20-one (p. 2217 s) on hydrogenation in glacial acetic acid in the presence of platinum oxide, followed by oxidation with CrO_3 in glacial acetic acid at room temperature (1943 Shoppee).

$\Delta^{3.11}$-**Pregnadien-20-one** $C_{21}H_{30}O$ (I). Colourless rhombohedra (methanol or pentane), m. 125–7° cor. Very readily soluble in ether. Gives a strong yellow coloration with tetranitromethane. — **Fmn.** Along with the anthraquinone-2-carboxylates of Δ^{11}-pregnen-3α-ol-20-one (p. 2239 s) and, probably, Δ^3-pregnen-12α*-ol-20-one (p. 2251 s), from the di-(anthraquinone-2-carboxylate) of pregnane-3α.12α*-diol-20-one (p. 2329 s) on thermal cleavage at 290–320° and 0.02 mm. pressure (small yield); analogously, along with the acetate of Δ^{11}-pregnen-3α-ol-20-one, from the 3-acetate 12-(anthraquinone-2-carboxylate) of the same diolone (small yield) (1944 v. Euw).

$\Delta^{4.16}$-**Pregnadien-20-one** (II) and $\Delta^{5.16}$-**Pregnadien-20-one** (III), $C_{21}H_{30}O$. Properties and mode of formation have not been described. — **Rns.** May be converted into Δ^4-androsten-17-one and Δ^5-androsten-17-one, resp. (pp. 2402 s and 2403 s, resp.) by treatment of their oximes with p-toluenesulphonyl chloride in cold pyridine, followed by hydrolysis with dil. H_2SO_4 at room temperature (1941 a PARKE, DAVIS & Co.).

* For configuration, see 1947 Sarett.

E.O.C. XIV s. 156

Δ^2-**Allopregnen-20-one** $C_{21}H_{32}O$. — **Fmn.** From 2.3-dibromocholestane (not described) on oxidation with CrO_3 in glacial acetic acid, followed by debromination with zinc dust in 95 % acetic acid at room temperature. — **Rns.** On reduction, followed by treatment with propionic anhydride, it reacts like Δ^4-pregnen-20-one (see below) (1940a, c CIBA).

Δ^3-**Allopregnen-20-one** $C_{21}H_{32}O$. — **Fmn.** From 3.4-dibromocholestane (not described) on oxidation with CrO_3 in glacial acetic acid, followed by debromination with zinc dust in 95 % acetic acid at room temperature. — **Rns.** Gives allopregnane-3.4-diol-20-one (p. 2322 s) on oxidation with 1.1 mols. OsO_4 in ether at room temperature, followed by heating with Na_2SO_3 in aqueous alcohol. On reduction, followed by treatment with propionic anhydride, it reacts like Δ^4-pregnen-20-one (see below) (1940a, c CIBA).

Δ^4-**Pregnen-20-one,** *Desoxoprogesterone* $C_{21}H_{32}O$ (p. 113). — **Fmn.** From Δ^4-pregnen-20α-ol (p. 1487 s) on bromination in glacial acetic acid, followed by oxidation with CrO_3 and then by debromination with zinc dust in acetic acid (1940 PARKE, DAVIS & Co.). From Δ^4-cholestene (p. 1423 s) or other analogous unsaturated hydrocarbons on oxidation with CrO_3 in glacial acetic acid after temporary protection of the double bond by bromine (1939, 1940a CIBA). From 5-chlorocholestane (p. 1460 s) in carbon tetrachloride-acetic acid on heating with $Ca(MnO_4)_2$ and aqueous H_2SO_4 at 50°, followed by refluxing of the resulting 5-chloroallopregnan-20-one (p. 2222 s) with sodium acetate solution (1940a CIBA; 1940 PARKE, DAVIS & Co.).

Rns. On oxidation with CrO_3 in strong acetic acid it gives Δ^4-pregnene-3.20-dione (progesterone, p. 2782 s) (1940b, c CIBA; 1940 PARKE, DAVIS & Co.), Δ^4-pregnene-6.20-dione (p. 2801 s), and Δ^4-pregnene-3.6.20-trione (p. 2965 s) (1940c CIBA). On hydrogenation in alcohol in the presence of Rupe nickel or Raney nickel, followed by treatment with propionic anhydride and pyridine, it yields Δ^4-pregnen-20β(?)-ol propionate (p. 1487 s) (1940b, c CIBA).

Δ^5-**Pregnen-20-one** $C_{21}H_{32}O^*$. — **Fmn.** From Δ^5-cholestene (p. 1425 s) or similar Δ^5-unsaturated hydrocarbons on chlorination or bromination, followed by oxidation with CrO_3 and dehalogenation with zinc dust in acetic acid (1940 PARKE, DAVIS & Co.; cf. 1940a CIBA). — **Rns.** On catalytic hydrogenation it reacts similarly to the preceding compound (1940c CIBA).

---

* After the closing date of this volume, a crystalline compound, m. 111–2°, which is probably Δ^5-pregnen-20-one has been described by Karrer (1951).

References, p. 2220 s

Δ^{16}-Pregnen-20-one $C_{21}H_{32}O$. Crystals (dil. acetone), m. 129–131°. — **Fmn.**
From pseudodesoxysarsasapogenin (or dihydropseudo-
desoxysarsasapogenin) on refluxing with acetic anhydride, oxidation of the resulting acetate with CrO_3
in acetic acid at 28°, and refluxing of the oxidation
product (which did not crystallize) with 2 % alcoholic
KOH; yield, 24 %. — **Rns.** Gives pregnan-20-one
(below) on hydrogenation in alcohol-ether in the presence of palladium-
barium sulphate (1941 Marker).

Δ^{16}-Allopregnen-20-one $C_{21}H_{32}O$. Crystals (methanol or acetone), m. 156–8°
(1942 b Marker). — **Fmn.** From 17α-bromoallopregnan-
20-one (p. 2222 s) on refluxing with pyridine; yield,
93 % (1942 b Marker). From desoxychlorogenin (des-
oxytigogenin) on heating with acetic anhydride at
200°, followed by refluxing with methanolic KOH,
oxidation of the resulting (not isolated) pseudodesoxy-
chlorogenin with CrO_3 in acetic acid at 30°, and refluxing with aqueous
methanolic K_2CO_3 (1942 a Marker).

Rns. Gives allopregnan-20-one (below) on hydrogenation in the presence
of palladium-barium sulphate in methanol-ether or ethanol-dioxane (1942 a, b
Marker). May be converted into androstan-17-one (p. 2403 s) by treatment
of its oxime (not described) with p-toluenesulphonyl chloride in cold pyridine,
followed by hydrolysis with dil. H_2SO_4 (1941 a PARKE, DAVIS & Co.).

Pregnan-20-one $C_{21}H_{34}O$ (p. 113). Needles (dil. alcohol), m. 116° (1939
Marker). — **Fmn.** From pregnan-20-ol (p. 1487 s) on
oxidation with excess CrO_3 in 95 % acetic acid on the
steam-bath (1 hr.); excellent yield (1939 Marker).
From Δ^{16}-pregnen-20-one (above) on hydrogenation
in alcohol-ether in the presence of palladium-barium
sulphate (1941 Marker). From pregnane-3.20-dione
(p. 2796 s) on refluxing with zinc and aqueous alcoholic HCl; yield, 52 %
(1939 Marker). — **Rns.** Gives pregnan-20β-ol (p. 1488 s; crude yield, ca. 80 %)
and a mixture of hydrocarbons on shaking with hydrogen, platinum oxide,
and acetic acid under 2.7 atm. pressure, followed by heating with aqueous
alcoholic NaOH (1939 Marker).

Semicarbazone $C_{22}H_{37}ON_3$. Crystals (methanol), m. 237° dec. (1939 Marker).
2.4-Dinitrophenylhydrazone $C_{27}H_{38}O_4N_4$, m. 240° dec. (1939 Marker).

Allopregnan-20-one $C_{21}H_{34}O$. For the configuration at C-17 see 1950 Shoppee. — Crystals (methanol), m. 129° (1939 Marker),
132° (1942 b Marker), 134° (1939 Mamoli), 136–9° cor.
(1945 Meystre); $[\alpha]_D^{26} + 102°$ (chloroform) (1945 Meystre;
see also 1950 Shoppee).

Fmn. From allopregnan-20α-ol (p. 1488 s) on oxidation with CrO_3 in acetic acid; yield, 40 % (1939
Marker). From $\Delta^{3(?).5.16}$-pregnatrien-20-one (p. 2215 s) on hydrogenation in

156*

glacial acetic acid in the presence of platinum oxide, followed by oxidation with CrO_3 in glacial acetic acid (1943 Shoppee). From $\varDelta^{16}$-allopregnen-20-one (p. 2217 s) on hydrogenation in the presence of palladium-barium sulphate in methanol-ether or in ethanol-dioxane (yield in the latter case, 90 %) (1942 a, b Marker). From 17α-bromoallopregnan-20-one (p. 2222 s) on heating in glacial acetic acid on the steam-bath with powdered zinc or iron filings (yields, 55 % and 60 %, resp.) or on shaking with hydrogen, palladium-barium sulphate, and pyridine in methanol at room temperature under 2 atm. pressure (yield, 82 %) (1942 b Marker). From 17α.21-dibromoallopregnan-20-one (p. 2223 s) on heating in glacial acetic acid on the steam-bath with zinc dust (yield, 51 %) or powdered iron (yield, 41 %) or on heating with potassium formate in formic acid in a sealed tube at 130⁰ (yield, 84 %) (1942 b Marker). From allopregnane-3.20-dione (p. 2797 s) on refluxing with zinc in aqueous alcoholic HCl; yield, 55 % (1939 Marker).

From 20-diphenylmethylene-allopregnane (p. 1444 s) on oxidation with ozone in chloroform; yield, 54 % (1939 Mamoli). From 24.24-diphenyl-$\varDelta^{23}$-allocholene (p. 1445 s) on refluxing with N-bromosuccinimide in carbon tetrachloride while irradiating, boiling (3 hrs.) of the filtrated solution which contains 24.24-diphenyl-22-bromo-$\varDelta^{23}$-allocholene, and oxidation of the resulting crude 24.24-diphenyl-$\varDelta^{20\ 22).23}$-allocholadiene (p. 1445 s) with CrO_3 in chloroform-acetic acid; yield, 44 % (1945 Meystre).

Rns. On keeping with persulphuric acid in acetic acid at 25⁰ for 7 days it gives the acetates of allopregnan-21-ol-20-one (p. 2225 s; main product) and androstan-17β-ol (p. 1512 s) (1940 b Marker); refluxing with persulphuric acid in acetic acid yields ætioallocholanic acid (p. 166) and a mixture of acetylated epimeric androstan-17-ols (1940 a Marker). Gives allopregnan-20β-ol (p. 1488 s) on shaking with hydrogen, platinum oxide, and acetic acid under 3 atm. pressure, followed by heating with alcoholic NaOH (1939 Marker). On treatment with 1 mol. bromine in acetic acid in the presence of conc. HBr it gives 17α-bromoallopregnan-20-one (p. 2222 s); with 2 mols. bromine, 17α.21-dibromoallopregnan-20-one (p. 2223 s) is formed (1942 b Marker; 1941 b PARKE, DAVIS & Co.).

Semicarbazone $C_{22}H_{37}ON_3$, m. 260⁰ dec. (1939 Marker; 1945 Meystre).

2.4-Dinitrophenylhydrazone $C_{27}H_{38}O_4N_4$. Crystals (benzine b. 120–130⁰), needles (from the melt), m. 220–3⁰ (1939 Marker; 1943 Shoppee).

Norcholanyl methyl ketone, Homocholan-24-one $C_{25}H_{42}O$ (p. 114: *Norcholyl methyl ketone*). Needles (acetone and methanol), m. 114–5⁰ cor.; $[\alpha]_D^{27} + 22.5⁰$ (chloroform) (1946 Hollander); $[\alpha]_D^{15} + 38⁰$ (1944 Okasaki). —**Fmn.** From norcholanyl chloromethyl ketone (p. 2221 s) on refluxing with granulated zinc in glacial acetic acid (1946 Hollander). For formation from cholanic acid, see also 1944 Okasaki. — **Rns.** Gives 23-bromonorcholanyl methyl ketone (p. 2221 s) on treatment with bromine in acetic acid in the presence of HBr (1946 Hollander).

Semicarbazone $C_{26}H_{45}ON_3$ (p. 114). Needles (alcohol), crystals (glacial acetic acid), m. 223° (1944 Okasaki).

Norcholanyl ethyl ketone, "*Norcholyl ethyl ketone*" $C_{26}H_{44}O$ (p. 114). Prisms (acetone), m. 94°; $[\alpha]_D^{16} + 24°$ (chloroform). Very readily soluble in ether, ethyl acetate, glacial acetic acid, chloroform, and alcohol. — **Fmn.** From cholanic acid amide on heating with ethylmagnesium bromide in anhydrous ether-benzene for 5 hrs. at 70–80°. — **Rns.** On Clemmensen reduction it gives "**bisnorsterocholane**" $C_{26}H_{46}$, prisms (alcohol - ether, or glacial acetic acid-ether), m. 74°, $[\alpha]_D^{15} + 27°$ (chloroform); very readily soluble in ether and chloroform, very sparingly in ethyl acetate, alcohol, and glacial acetic acid (1944 Okasaki).

Semicarbazone $C_{27}H_{47}ON_3$ (p. 114). Scales (abs. alcohol and benzene), m. 201°; almost insoluble in ether, ethyl acetate, acetone, and chloroform (1944 Okasaki).

27-Nor-Δ^4-cholesten-25-one $C_{26}H_{42}O$. — **Fmn.** Along with 21-nor-Δ^4-cholesten-20-one (below) and other compounds, from Δ^4-cholestene (p. 1423 s) on bromination, followed by oxidation of the resulting dibromide (p. 1472 s) with CrO_3 in glacial acetic acid and then by debromination with zinc dust in 95 % acetic acid (1939 CIBA; cf. 1940a CIBA; 1940 PARKE, DAVIS & Co.). From 5-chloro-27-nor-cholestan-25-one (p. 2222 s) on treatment with potassium acetate (1940a CIBA).

21-Nor-Δ^4-cholesten-20-one $C_{26}H_{42}O$. — **Fmn.** For its formation from Δ^4-cholestene, see the preceding compound. From 5-chloro-21-nor-cholestan-20-one (p. 2222 s) on treatment with potassium acetate (1940a CIBA).

2-Formyl-Δ^2-cholestene $C_{28}H_{46}O$. Needles (alcohol), m. 130–2° cor.; $[\alpha]_D + 75°$ (chloroform). Absorption max. in alcohol at 300 (1.85) and 235 (4.1) $m\mu$ (log ε). — **Fmn.** From cholestan-3-ol-2-glyoxylic acid lactone on distillation in vacuo. — **Rns.** On hydrogenation in alcohol in the presence of a platinum catalyst at 21° it gives 2-hydroxymethyl-cholestane m. 124–6° cor. (p. 1490 s) and, probably, a stereoisomer (1941 Plattner).

Oxime $C_{28}H_{47}ON$. Crystals (alcohol), m. 163–4° cor.; $[\alpha]_D + 52°$ (chloroform). Absorption max. in alcohol at 233 $m\mu$ (log ε 4.3). — **Rns.** On heating with sodium acetate and acetic anhydride at 75° for 1 hr. it gives its *acetate* $C_{30}H_{49}O_2N$ [crystals (alcohol), m. 122–3° cor.] whereas on refluxing with sodium acetate and a large excess of acetic anhydride for 2 hrs., the nitrile of Δ^2-cholestene-2-carboxylic acid is formed (1941 Plattner).

Norcholanyl phenyl ketone $C_{30}H_{44}O$. Crystals (diisopropyl ether), m. 137.5° to 138.5°; $[\alpha]_D^{25} + 27.5°$ (dioxane). — **Fmn.** From cholanic acid chloride (p. 170) on refluxing with diphenylcadmium in benzene, followed by decomposition of the resulting complex with dil. HCl. — *Oxime* $C_{30}H_{45}ON$, m. 133.5–135°; $[\alpha]_D^{25} + 15°$ (dioxane) (1945 Hoehn).

1939 CIBA*, *Swiss. Pat.* 235910 (issued 1945); *C.A.* **1949** 7056.
     Mamoli, *Gazz.* **69** 240, 242, 244.
     Marker, Lawson, *J. Am. Ch. Soc.* **61** 852.
1940 a CIBA* (Miescher, Wettstein), *Swed. Pat.* 103744 (issued 1942); *Ch. Ztbl.* **1942** II 2294; *Belgian Pat.* 440403 (issued 1941); *Ch. Ztbl.* **1942** I 2037; *Fr. Pat.* 886410 (issued 1943); *Ch. Ztbl.* **1944** I 448; CIBA PHARMACEUTICAL PRODUCTS (Miescher, Wettstein), *U.S. Pat.* 2319012 (issued 1943); *C.A.* **1943** 6096.
     b CIBA*, *Swiss Pat.* 235279 (issued 1945); *C.A.* **1949** 7649.
     c CIBA*, *Fr. Pat.* 886415 (issued 1943); *Ch. Ztbl.* **1944** I 450; (Miescher, Wettstein), *Swed. Pat.* 105136 (1941, issued 1942); *Ch. Ztbl.* **1944** II 143; CIBA PHARMACEUTICAL PRODUCTS (Miescher, Wettstein), *U.S. Pat.* 2323277 (issued 1943); *C.A.* **1944** 222, 223.
     a Marker, Rohrmann, Wittle, Crooks, Jones, *J. Am. Ch. Soc.* **62** 650.
     b Marker, *J. Am. Ch. Soc.* **62** 2543, 2545.
     PARKE, DAVIS & Co. (Marker, Wittle), *U.S. Pats.* 2397424-5-6 (issued 1946); *C.A.* **1946** 3570 to 3572.
1941 Marker, Turner, Wagner, Ulshafer, Crooks, Wittle, *J. Am. Ch. Soc.* **63** 779, 781.
     a PARKE, DAVIS & Co. (Tendick, Lawson), *U.S. Pat.* 2335616 (issued 1943); *C.A.* **1944** 3095; *Brit. Pat.* 563889 (issued 1944); *C.A.* **1946** 2942.
     b PARKE, DAVIS & Co. (Marker, Crooks), *U.S. Pat.* 2359773 (issued 1944); *C.A.* **1945** 4198; *Ch. Ztbl.* **1945** II 1233.
     Plattner, Jampolsky, *Helv.* **24** 1459, 1462 et seq.
1942 a Marker, Turner, Wittbecker, *J. Am. Ch. Soc.* **64** 809, 811.
     b Marker, Crooks, Wagner, Shabica, Jones, Wittbecker, *J. Am. Ch. Soc.* **64** 822.
1943 Shoppee, Prins, *Helv.* **26** 1004, 1006, 1009–1013.
1944 v. Euw, Lardon, Reichstein, *Helv.* **27** 821, 825, 836, 837.
     Okasaki, *J. Biochem.* **36** 77.
1945 Hoehn, Moffett, *J. Am. Ch. Soc.* **67** 740, 742.
     Meystre, Miescher, *Helv.* **28** 1497, 1505, 1506; CIBA PHARMACEUTICAL PRODUCTS (Miescher, Frey, Meystre, Wettstein), *U.S. Pat.* 2461910 (1946, issued 1949); *C.A.* **1949** 3474; *Ch. Ztbl.* **1949** E 1870, 1871.
1946 Hollander, Gallagher, *J. Biol. Ch.* **162** 549, 551.
1947 Sarett, *J. Am. Ch. Soc.* **69** 2899, 2901.
1950 Shoppee, Lewis, Elks, *Chemistry & Industry* **1950** 454.
1951 Karrer, Asmis, Sareen, Schwyzer, *Helv.* **34** 1022, 1027.

* Société pour l'industrie chimique à Bâle; Gesellschaft für chemische Industrie in Basel.

2. HALOGENO-MONOOXOSTEROIDS (CO IN SIDE CHAIN)

a. Halogen in side chain only

21-Chloropregnan-20-one $C_{21}H_{33}OCl$. Prisms (alcohol), m. 103–5⁰ cor.; $[\alpha]_D^{25}$ + 125⁰ (chloroform). — **Fmn.** From 21-diazopregnan-20-one (p. 2224 s) on passage of dry HCl into its solution in ether at 0⁰. — **Rns.** On refluxing with sodium benzoate in alcohol, the benzoate of pregnan-21-ol-20-one (p. 2225 s) is obtained (1941 Linville; 1942 Fried).

23-Bromonorcholanyl methyl ketone, 23-Bromohomocholan-24-one $C_{25}H_{41}OBr$.
Needles (methanol), m. 94.5–95.5⁰cor.; $[\alpha]_D^{25} + 87^0$ (chloroform). — **Fmn.** From norcholanyl methyl ketone (p. 2218 s) on treatment with bromine in acetic acid in the presence of aqueous HBr. — **Rns.** On refluxing with collidine, followed by oxidation of the resulting oily α.β-unsaturated ketone with CrO_3 in acetic acid, it gives small amounts of two isomeric bisnor-cholanic acids (m. 171–5⁰ cor. and m. 196–201⁰ cor.) (1946 Hollander).

Norcholanyl chloromethyl ketone, 25-Chlorohomocholan-24-one $C_{25}H_{41}OCl$.
Needles (methanol and acetone), m. 109 to 110⁰ cor. (1942 Knowles; 1946 Hollander); $[\alpha]_D$ in chloroform + 24⁰ (1942 Knowles), + 22⁰ (1946 Hollander). — **Fmn.** From norcholanyl diazomethyl ketone (p. 2224 s) on treatment with dry HCl in ether at 0⁰ (1942 Knowles; 1946 Hollander). — **Rns.** Gives norcholanyl methyl ketone (p. 2218 s) on refluxing with zinc in glacial acetic acid (1946 Hollander). On refluxing with potassium acetate in acetic acid, norcholanyl acetoxymethyl ketone (p. 2226 s) is obtained (1942 Knowles).

1941 Linville, Fried, Elderfield, *Science* [N. S.] **94** 284.
1942 Fried, Linville, Elderfield, *J. Org. Ch.* **7** 362, 368.
 Knowles, Fried, Elderfield, *J. Org. Ch.* **7** 383, 385.
1946 Hollander, Gallagher, *J. Biol. Ch.* **162** 549, 550, 552, 553.

b. Halogen in the ring system

3-Chloro-$\Delta^{5,16}$-pregnadien-20-one $C_{21}H_{29}OCl$ (I). Properties and mode of formation have not been described. — **Rns.** May be converted into 3-chloro-

Δ^5-androsten-17-one (p. 2475 s) by treatment of its oxime with p-toluenesulphonyl chloride in cold pyridine, followed by hydrolysis with dil. H_2SO_4 at room temperature (1941a PARKE, DAVIS & Co.).

3β-Chloro-Δ^5-pregnen-20-one $C_{21}H_{31}OCl$ (II) (p. 114). For the configuration at C-3, see 1946 Shoppee. — **Fmn.** From Δ^5-pregnen-3β-ol-20-one (p. 2233 s) on treatment with PCl_5 (1940 PARKE, DAVIS & Co.). — **Rns.** Gives Δ^5-pregnen-20α-ol (p. 1487 s) on reduction with sodium and amyl alcohol (1940 PARKE, DAVIS & Co.).

3β-Chloro-allopregnan-20-one $C_{21}H_{33}OCl$ (III). For this compound (not isolated), see under the reactions of 3β-chloroternorallocholanyl-diphenyl-carbinol, p. 1491 s.

5-Chloro-allopregnan-20-one $C_{21}H_{33}OCl$. Colourless crystals (1939 CIBA). — **Fmn.** Along with other compounds, from 5-chloro-cholestane (p. 1460 s) on oxidation with CrO_3 in acetic acid at 20⁰ (1939, 1940 CIBA) or with $Ca(MnO_4)_2$, aqueous H_2SO_4, and acetic acid in carbon tetrachloride at 50⁰ (1940 PARKE, DAVIS & Co.). — **Rns.** Gives Δ^4-pregnen-20-one (p. 2216 s) on refluxing with alkali acetate (1940 CIBA; 1940 PARKE, DAVIS & Co.).

17α-Bromo-allopregnan-20-one $C_{21}H_{33}OBr$. Crystals (acetone), m. 127–9⁰. — **Fmn.** From allopregnan-20-one (p. 2217 s) on treatment with 1 mol. bromine in acetic acid in the presence of conc. HBr; yield, ca. 80% (1942 Marker; 1941b PARKE, DAVIS & Co.). — **Rns.** Reverts to allopregnan-20-one on heating on the steam-bath with powdered zinc or iron filings in glacial acetic acid or on shaking with hydrogen, palladium-barium sulphate, and pyridine in methanol (1942 Marker). Gives 17α.21-dibromo-allopregnan-20-one (p. 2223 s) on treatment with 1 mol. bromine in acetic acid in the presence of 45% HBr at 35⁰ (1942 Marker; 1941b PARKE, DAVIS & Co.). On refluxing with pyridine, Δ^{16}-allopregnen-20-one (p. 2217 s) is formed (1942 Marker).

5-Chloro-27-nor-cholestan-25-one $C_{26}H_{43}OCl$ [I, R = $CH(CH_3)\cdot CH_2\cdot CH_2\cdot CH_2\cdot CO\cdot CH_3$]. — **Fmn.** Along with 5-chloro-21-nor-cholestan-20-one (below) and other compounds, from 5-chloro-cholestane (p. 1460 s) on oxidation with CrO_3 in acetic acid at 20⁰. — **Rns.** Gives 27-nor-Δ^4-cholesten-25-one (p. 2219 s) on treatment with potassium acetate (1940 CIBA).

5-Chloro-21-nor-cholestan-20-one $C_{26}H_{43}OCl$ [I, R = $CO\cdot CH_2\cdot CH_2\cdot CH_2\cdot CH(CH_3)_2$]. — **Fmn.** See that of the preceding compound. — **Rns.** Gives 21-nor-Δ^4-cholesten-20-one (p. 2219 s) on treatment with potassium acetate (1940 CIBA).

17α.21-Dibromo-allopregnan-20-one $C_{21}H_{32}OBr_2$. Crystals (acetone), m. 128⁰

to 130⁰. — **Fmn.** From allopregnan-20-one (page 2217 s) on treatment with 2 mols. bromine in acetic acid in the presence of conc. HBr, first at room temperature, then at 40⁰; similarly from 17α-bromo-allopregnan-20-one (p. 2222 s) on using 1 mol. bromine at 35⁰ (1942 Marker; 1941 b Parke, Davis & Co.). — **Rns.** Reverts to allopregnan-20-one on heating with zinc dust or powdered iron in glacial acetic acid on the steam-bath or with potassium formate in formic acid in a sealed tube at 130⁰ (1942 Marker). Gives $\Delta^{17(20)}$-allopregnen-21-oic acid on heating with methanolic KOH (1942 Marker; 1941 b Parke, Davis & Co.).

1939 Ciba*, *Swiss Pat.* 235 912 (issued 1945); *C.A.* **1949** 7056.
1940 Ciba* (Miescher, Wettstein), *Swed. Pat.* 103 744 (issued 1942); *Ch. Ztbl.* **1942** II 2294; *Fr. Pat.* 886 410 (issued 1943); *Ch. Ztbl.* **1944** I 448; Ciba Pharmaceutical Products (Miescher, Wettstein), *U.S. Pat.* 2 319 012 (issued 1943); *C.A.* **1943** 6096.
      Parke, Davis & Co. (Marker, Wittle), *U.S. Pats.* 2 397 424-5-6 (issued 1946); *C.A.* **1946** 3570, 3572.
1941 a Parke, Davis & Co. (Tendick, Lawson), *U.S. Pat.* 2 335 616 (issued 1943); *C.A.* **1944** 3095; *Brit. Pat.* 563 889 (issued 1944); *C.A.* **1946** 2942.
      b Parke, Davis & Co. (Marker, Crooks), *U.S. Pats.* 2 369 065 (issued 1945), 2 359 773 (issued 1944); *C.A.* **1945** 4197, 4198; *Ch. Ztbl.* **1945** II 1232, 1233.
1942 Marker, Crooks, Wagner, Shabica, Jones, Wittbecker, *J. Am. Ch. Soc.* **64** 822.
1946 Shoppee, *J. Ch. Soc.* **1946** 1147, 1149.

* Société pour l'industrie chimique à Bâle; Gesellschaft für chemische Industrie in Basel.

3. DIAZO-MONOOXOSTEROIDS (CO IN SIDE CHAIN)

21-Diazo-Δ^2-allopregnen-20-one $C_{21}H_{30}ON_2$. Crystals (ether-methanol), m. 140⁰ cor. dec.; $[\alpha]_D + 178^0$ (chloroform). — **Fmn.** From Δ^2-ætioallocholenic acid chloride and diazomethane in ether-benzene. — **Rns.** On heating with glacial acetic acid on the water-bath it gives Δ^2-allopregnen-21-ol-20-one acetate (p. 2225 s) (1945 Plattner).

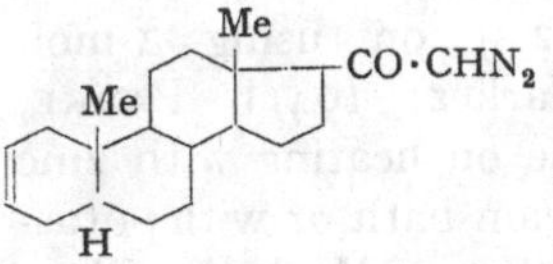

21-Diazopregnan-20-one $C_{21}H_{32}ON_2$. The crude substance forms a yellow crystalline solid (not analysed), m. 80–106⁰. — **Fmn.** From ætiocholanic acid chloride and diazomethane in ether-benzene at — 14⁰ (2 hrs.) and then at room temperature (12 hrs.). — **Rns.** Gives 21-chloro-pregnan-20-one (p. 2221 s) with dry HCl in ice-cold abs. ether. On warming with glacial acetic acid on the steam-bath, nitrogen is evolved and a brown oil is formed (1941 Linville; 1942 Fried).

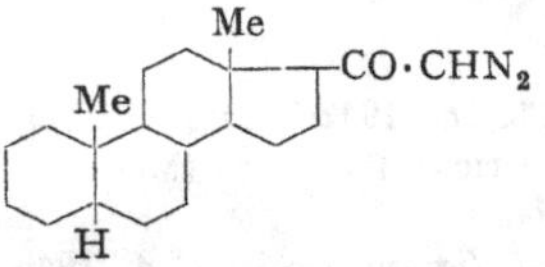

21-Diazo-allopregnan-20-one $C_{21}H_{32}ON_2$. Crystals (methanol), m. 120–121⁰ cor. dec.; $[\alpha]_D + 151^0$ (chloroform). — **Fmn.** From ætioallocholanic acid chloride and diazomethane in ether-benzene in the cold. — **Rns.** On heating with glacial acetic acid on the water-bath it gives allopregnan-21-ol-20-one acetate (p. 2225 s) (1943 Plattner).

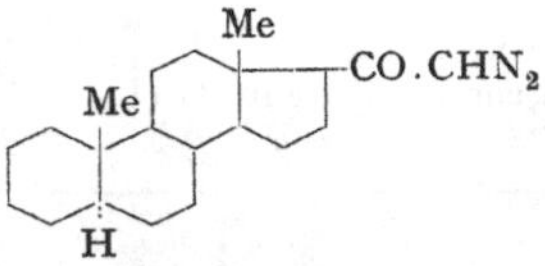

25-Diazo-homocholan-24-one, Norcholanyl diazomethyl ketone $C_{25}H_{40}ON_2$. Amorphous (not analysed). — **Fmn.** From cholanic acid chloride and diazomethane in ether-benzene at —5⁰. — **Rns.** Gives norcholanyl chloromethyl ketone (p. 2221 s) with dry HCl in ether at 0⁰ (1942 Knowles; 1946 Hollander).

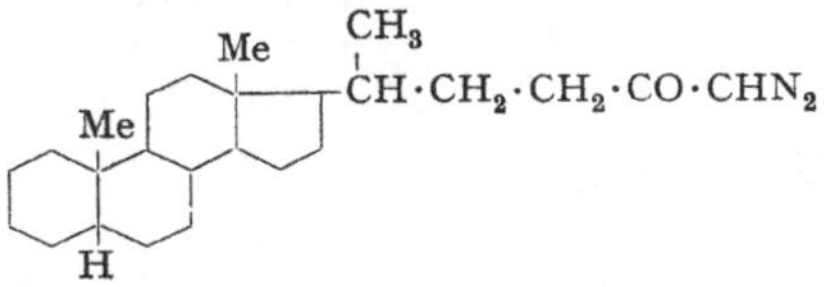

1941 Linville, Fried, Elderfield, *Science* [N. S.] **94** 284.
1942 Fried, Linville, Elderfield, *J. Org. Ch.* **7** 362, 368.
　　　 Knowles, Fried, Elderfield, *J. Org. Ch.* **7** 383, 385.
1943 Plattner, Ruzicka, Fürst, *Helv.* **26** 2274, 2277.
1945 Plattner, Fürst, *Helv.* **28** 173, 176.
1946 Hollander, Gallagher, *J. Biol. Ch.* **162** 549, 550.

4. HYDROXY-MONOOXOSTEROIDS (CO IN SIDE CHAIN)

a. OH in side chain only

Δ^2-Allopregnen-21-ol-20-one $C_{21}H_{32}O_2$.

Acetate $C_{23}H_{34}O_3$. Crystals (benzene-hexane), m. 180° cor. $[\alpha]_D$ +139° (chloroform). — **Fmn.** From 21-diazo-Δ^2-allopregnen-20-one (p. 2224 s) on heating with glacial acetic acid on the water-bath. — **Rns.** Gives 21-hydroxy-$\Delta^{2,2C(22)}$-norallocholadienic acid lactone (partial formula I) on boiling with ethyl bromoacetate and zinc in benzene, followed by boiling with acetic anhydride (1945 Plattner).

Pregnan-21-ol-20-one $C_{21}H_{34}O_2$.

Benzoate $C_{28}H_{38}O_3$. Crystals (abs. alcohol), m. 158–159° cor.; $[\alpha]_D^{25}$ +113° (chloroform). — **Fmn.** From 21-chloropregnan-20-one (p. 2221 s) on refluxing with sodium benzoate in alcohol. — **Rns.** Gives 21-hydroxy-$\Delta^{20(22)}$-norcholenic acid lactone (partial formula I, above) on refluxing with ethyl bromoacetate and zinc in benzene, followed by heating with glacial acetic acid containing hydrogen bromide at 135–140° (1941 Linville; 1942 Fried).

Allopregnan-21-ol-20-one $C_{21}H_{34}O_2$. Crystals (methanol), m. 115–7°. — **Fmn.**

The acetate is obtained from allopregnan-20-one (p. 2217 s) on keeping with persulphuric acid in acetic acid at 25° for 7 days (yield, 30–35 %); it is hydrolysed by boiling with $KHCO_3$ in methanol (1940 Marker). The acetate is also formed from 21-diazo-allopregnan-20-one (p. 2224 s) on heating with glacial acetic acid on the water-bath (1943 Plattner). — **Rns.** Gives ætioallocholanic acid (p. 166) on oxidation with CrO_3 in acetic acid (1940 Marker). The acetate may be converted into 21-hydroxy-$\Delta^{20(22)}$-norallocholenic acid lactone (partial formula I, see above) by boiling with ethyl bromoacetate and zinc in benzene-dioxane, followed by refluxing with acetic anhydride (1943 Plattner).

Acetate $C_{23}H_{36}O_3$. Crystals (acetone-methanol), m. 197–200° (1940 Marker), 200° cor. (1943 Plattner); $[\alpha]_D$ +102° (chloroform) (1943 Plattner). — **Fmn.** From the above hydroxy-ketone with acetic anhydride and pyridine at room temperature (1940 Marker). See also under formation of the hydroxy-ketone, above. — **Rns.** See above. — *Semicarbazone* $C_{24}H_{39}O_3N_3$, crystals (dil. alcohol), m. 242–4° dec. (1940 Marker).

References, p. 2226 s

Norcholanyl hydroxymethyl ketone, Homocholan-25-ol-24-one $C_{25}H_{42}O_2$.

$$Me \quad CH_3$$
$$Me \quad \overset{|}{CH} \cdot CH_2 \cdot CH_2 \cdot CO \cdot CH_2 \cdot OH$$

Acetate $C_{27}H_{44}O_3$. Crystals (alcohol) m. 81.5–82.5° cor.; $[\alpha]_D^{27} + 22°$ (chloroform). — **Fmn.** From norcholanyl chloromethyl ketone (p. 2221 s) on refluxing with potassium acetate in acetic acid; yield, 81 %. — **Rns.**
Gives β-norcholanyl-$\Delta^{\alpha,\beta}$-butenolide (cf. partial formula I, p. 2225 s) on refluxing with ethyl bromoacetate and zinc in benzene, followed by heating with acetic acid containing hydrobromic acid at 135–140° or by refluxing with acetic anhydride and then by sublimation at 190–210°/0.5 (1942 Knowles).

Bisnorcholanyl α-hydroxyisopropyl ketone, 24-Methylhomocholan-24-ol-23-one

$$Me \quad CH_3$$
$$Me \quad \overset{|}{CH} \cdot CH_2 \cdot CO \cdot C(CH_3)_2 \cdot OH$$

$C_{26}H_{44}O_2$. For this compound [tablets (alcohol), m. 96–97°] and its *acetate* $C_{28}H_{46}O_3$ [leaflets (alcohol), m. 147°], see p. 170, lines 14–15 from bottom.

1940 Marker, *J. Am. Ch. Soc.* **62** 2543, 2545, 2546.
1941 Linville, Fried, Elderfield, *Science* [N. S.] **94** 284.
1942 Fried, Linville, Elderfield, *J. Org. Ch.* **7** 362, 368, 369.
     Knowles, Fried, Elderfield, *J. Org. Ch.* **7** 383, 385, 386.
1943 Plattner, Ruzicka, Fürst, *Helv.* **26** 2274, 2277, 2278.
1945 Plattner, Fürst, *Helv.* **28** 173, 176.

---

b. Hydroxy-monooxosteroids Containing CO in Side Chain and at least One OH in the Ring System

1. MONOHYDROXY-COMPOUNDS WITHOUT OTHER FUNCTIONAL GROUPS

α. Compounds with two aliphatic side chains

19-Nor-14-iso-17-iso-$\Delta^{5(10?)}$-pregnen-3β-ol-20-one * $C_{20}H_{30}O_2$.

$$Me$$
$$CO \cdot CH_3$$
$$HO$$

For configurations and for the position of the double bond, see 21-diazo-19-nor*-14-iso-17-iso-$\Delta^{5(10?)}$-pregnen-3β-ol-20-one (p. 2292 s) prepared from the same acid chloride. — Has been obtained crystalline after the closing date for this volume: needles (aqueous methanol and ether · petroleum ether), m. 90°; $[\alpha]_D^{23} + 71°$ (chloroform) (1957 Barber). — **Fmn.** The acetate is formed from 3β-acetoxy-19-nor-14-iso-17-iso-$\Delta^{5(10?)}$-ætiocholenic acid chloride and methylzinc iodide in toluene-benzene; it is saponified by refluxing with K_2CO_3 in aqueous methanol (1944 Ehrenstein). — **Rns.** On refluxing with aluminium tert.-butoxide and acetone in benzene it gives 19-nor*-14-iso-17-iso-progesterone (p. 2777 s) (1944 Ehrenstein).

Acetate $C_{22}H_{32}O_3$. Slightly yellow, viscous resin; $[\alpha]_D + 81°$ (acetone) (1944 Ehrenstein). — **Fmn.** See above.

* The original designation 10-nor was later changed to 19-nor (see 1951 Ehrenstein).

3-(3β-Hydroxy-Δ⁵·⁷·⁹-œstratrien-17β-yl)-butan-2-one, Methyl [1-(3β-hydroxy-Δ⁵·⁷·⁹-œstratrien-17β-yl)-ethyl] ketone

$C_{22}H_{30}O_2$. Plates (aqueous acetone), m. 177 to 181° cor.; $[\alpha]_D^{19}$ —22° (chloroform). — **Fmn.** From 2-methyl-3-(3β-acetoxy-Δ⁵·⁷·⁹-œstratrien-17β-yl)-Δ¹-butene (p. 1515 s) on keeping with OsO₄ in ether at room temperature, followed by refluxing with aqueous alcoholic Na₂SO₃, oxidation of the reaction product (p. 1515 s) with HIO₄ in ether-methanol, and saponification. — **Rns.** Gives 1.1.2-trimethyl-2-(3β-hydroxy-Δ⁵·⁷·⁹-œstratrien-17β-yl)-ethanol (p. 1917 s) on refluxing with methylmagnesium iodide in ether-toluene (1943 Jacobsen).

Acetate $C_{24}H_{32}O_3$. Leaflets (aqueous methanol), m. 148–152° cor. — **Fmn.** From the above hydroxy-ketone on warming with acetic anhydride and pyridine (1943 Jacobsen).

1943 Jacobsen, *J. Am. Ch. Soc.* **65** 1789, 1792.
1944 Ehrenstein, *J. Org. Ch.* **9** 435, 452–454.
1951 Ehrenstein, Barber, Gordon, *J. Org. Ch.* **16** 349, 355 footnote 7.
1952 Ehrenstein, *Chimia (Switz.)* **6** 287, 288.
1954 Barber, Ehrenstein, *J. Org. Ch.* **19** 365, 366.
1957 Barber, Ehrenstein, *Ann.* **603** 89, 96, 104.

β. Compounds C₂₀ with three aliphatic side chains

17β-Formyl-Δ⁵-androsten-3β-ol, 3β-Hydroxy-Δ⁵-ætiocholenaldehyde C₂₀H₃₀O₂

Crystals, m. 148–153° cor. (after sublimation at 130 to 140°/₀.₀₁); $[\alpha]_D^{21}$ —14.5° (chloroform). Readily soluble in the common organic solvents except low-boiling petroleum ether. Shows an intense aldehyde reaction with ammoniacal silver solution. Gives a strong raspberry-red colour with 1.4-dihydroxynaphthalene (1940 Miescher).

Fmn. From the acetate of Δ⁵-androsten-3β-ol-17-one (dehydroepiandrosterone, p. 2609 s) on boiling with methoxymethyl-magnesium chloride in ether, followed by treatment with sulphuric acid and heating with anhydrous formic acid at 110–5° (1936 SCHERING-KAHLBAUM A.-G.); for probable formation in the reaction of the same acetate with ethyl dichloroacetate and amalgamated magnesium in ether, followed by treatment with alkali and decarboxylation, see 1939 Miescher; 1938d CIBA. As acetate, from 17β(?)-vinyl-Δ⁵-androsten-3β-yl acetate (p. 1519 s) on oxidation with ozone in chloroform and some pyridine at 0° after temporary protection of the nuclear double bond by bromine (1938 SCHERING A.-G.). From the mixture of C-20-epimeric Δ⁵-pregnene-3β.20.21-triols (p. 1953 s) on oxidation with periodic acid in aqueous dioxane at room temperature in an atmosphere of carbon dioxide; crude yield, 92 % (1940 Miescher). From the chloride of 3β-acetoxy-Δ⁵-ætiocholenic acid (p. 173) on reduction with palladium-barium sulphate, followed by saponification (1937 CIBA).

Rns. Gives 3-keto-Δ^4-ætiocholenaldehyde (p. 2777 s) on oxidation with aluminium tert.-butoxide and acetone (1939 CIBA). On treatment of the hydroxy-aldehyde or its acetate with methylmagnesium iodide in ether it gives a Δ^5-pregnene-3β.20-diol (p. 1922 s) (1936 SCHERING-KAHLBAUM A.-G.; 1938b CIBA). Gives Δ^5-pregnen-3β-ol-20-one (p. 2233 s) on treatment of its triphenylmethyl ether (not described) with diazomethane in ether, followed by heating with alcoholic HCl; the free hydroxy-aldehyde reacts with diazo-ethane and diazopropane in ether to give 21-methyl- and 21-ethyl-Δ^5-pregnen-3β-ol-20-ones (p. 2258 s and p. 2259 s, resp.); Δ^5-pregnen-3β-ol-20-one is also obtained on treatment of the hydroxy-aldehyde with ethyl diazomalonate in ether, followed by refluxing with 5 % methanolic KOH and then with 2 N H_2SO_4 (1938a, c CIBA).

Dimethyl acetal $C_{22}H_{36}O_3$. Crystals (ether), m. 185–9° cor. Gives a strong wine-red colour with 1.4-dihydroxynaphthalene. — **Fmn.** From the above hydroxy-aldehyde on keeping in 5 % methanolic HCl. From the mixture of 20-epimeric Δ^5-pregnene-3β.20.21-triols (p. 1953 s) on keeping with periodic acid in aqueous methanol in an atmosphere of CO_2 (1940 Miescher).

Semicarbazone $C_{21}H_{33}O_2N_3$. Crystals (aqueous methanol), dec. 226–8° cor. (1940 Miescher).

2.4-Dinitrophenylhydrazone $C_{26}H_{34}O_5N_4$. Yellow crystals (aqueous methanol), dec. 207–9° cor. (1940 Miescher).

Acetate, 3β-Acetoxy-Δ^5-ætiocholenaldehyde $C_{22}H_{32}O_3$. Needles (hexane), m. 169–171° cor.; $[\alpha]_D^{23}$ —13.5° (chloroform). — **Fmn.** From the above hydroxy-aldehyde on keeping with acetic anhydride and pyridine at room temperature (1940 Miescher).

17β-Formylandrostan-3α-ol, 3α-Hydroxyætioallocholanaldehyde $C_{20}H_{32}O_2$. —

Fmn. For its probable formation from esters or ethers of androsterone (p. 2629 s) in the reaction with alkyl dihalogenoacetates and amalgamated magnesium in ether, followed by treatment with alkali and decarboxylation, see 1938d CIBA; cf. 1939 Miescher.

1936 SCHERING-KAHLBAUM A.-G., *Brit. Pat.* 493128 (issued 1938); *Ch. Ztbl.* **1939** I 2829.
1937 CIBA*, *Swiss Pat.* 207497 (issued 1940); *C.A.* **1941** 3038; *Fr. Pat.* 841058 (1938, issued 1939); *C.A.* **1940** 1685; *Ch.Ztbl.***1940** I 916; CIBA PHARMACEUTICAL PRODUCTS (Reichstein), *U.S. Pat.* 2337271 (issued 1943); *C.A.* **1944** 3424.
1938 a CIBA*, *Fr. Pat.* 840513 (issued 1939); *C.A.* **1939** 8627; *Ch. Ztbl* **1939** II 1125.
    b CIBA*, *Fr. Pat.* 840783 (issued 1939); *C.A.* **1939** 8627; *Ch. Ztbl.* **1940** I 915; *Swiss Pats.* 211652, 215288 (issued 1941); *C.A.* **1942** 3637 and **1948** 3538, resp.; CIBA PHARMACEUTICAL PRODUCTS (Miescher, Wettstein), *U.S. Pat.* 2350792 (issued 1944); *C.A.* **1944** 5048.
    c CIBA* (Miescher, Wettstein), *Ger. Pat.* 739083 (issued 1943); *C.A.* **1945** 392; *Ch. Ztbl.* **1944** I 110.
    d CIBA* (Oppenauer, Miescher, Kägi), *Swed. Pat.* 96375 (issued 1939); *Ch. Ztbl.* **1940** I 1231. SCHERING A.-G., *Dutch Pat.* 50747 (issued 1941); *Ch. Ztbl.* **1941** II 2708.
1939 CIBA*, *Fr. Pat.* 851017 (issued 1940); *C.A.* **1942** 2090; *Ch. Ztbl.* **1940** II 932.
    Miescher, Kägi, *Helv.* **22** 184, 189.
1940 Miescher, Hunziker, Wettstein, *Helv.* **23** 1367.

    * Société pour l'industrie chimique à Bâle; Gesellschaft für chemische Industrie in Basel.

γ. Monohydroxy-monooxosteroids (CO in side chain, OH in the ring system) C_{21} with three aliphatic side chains

Summary of Pregnan-(Pregnen-etc.)-3-ol-20-ones

For values other than those chosen for the summary, see the compounds themselves

Config. at		Double bond	Config. of OH	M. p. °C.	$[\alpha]_D$*	Acetate		Page
C-5	C-17					m. p. °C.	$[\alpha]_D$*	
—	β	$\Delta^{3.5}$	—	—	—	138	-42^0 (chl.)	2229 s
—	—	$\Delta^{5.16}$	β	216	$\begin{cases} -27.5^0 \\ -38^0 \text{ (chl.)} \end{cases}$	176	-33^0	2230 s
—	β	Δ^5	α	152	$+54.5^0$	147	$+57^0$	2232 s
—	β	Δ^5	β	187	$+28^0$	149	$+20^0$	2233 s
norm.	β	Δ^{11}	α	126	—	137	—	2239 s
norm.	—	Δ^{16}	α	196	—	99	—	2239 s
norm.	—	Δ^{16}	β	190	—	146	—	2240 s
allo	—	Δ^{16}	α	222	$+54^0$ (chl.)	159	$+57^0$ (chl.)	2241 s
allo	—	Δ^{16}	β	207	$+50^0$	164	$+34^0$ to $+42^0$ (chl.)	2242 s
norm.	β	saturated	α	151	$+114$	$\begin{cases} 103 \\ 114 \end{cases}$	$\begin{matrix} +124^0 \\ +83.5^0 \text{ (chl.)} \end{matrix}$	2244 s
norm.	β	saturated	β	149	$+102$	121	$+86.5^0$	2245 s
allo	β	saturated	α	176	$\begin{cases} +91^0 \text{ to } 97^0 \\ +96^0 \text{ (chl.)} \end{cases}$	140	$+95^0$ to 112^0	2246 s
allo	β	saturated	β	194	$\begin{cases} +91^0 \\ +93^0 \text{ (chl.)} \end{cases}$	152	$\begin{cases} +80^0 \\ +74^0 \text{ (chl.)} \end{cases}$	2247 s
—	α	Δ^5	β	173	-141^0	171	-126^0	2249 s
norm.	α	saturated	α	144	-41^0 (di.)	159	-28^0 (MeOH)	2250 s
allo	α	saturated	β	139	-78^0	122	-75^0	2250 s

* In alcohol, unless otherwise stated; chl. = chloroform, di. = dioxane.

$\Delta^{3.5}$-Pregnadien-3-ol-20-one, 3-Enol form of progesterone $C_{21}H_{30}O_2$ (p. 151). —

Fmn. Its ester or ethers may be obtained from the corresponding derivatives of the 3-enol form of 3-keto-Δ^4-ætiocholenaldehyde (p. 2777 s) on treatment with diazomethane in ether (1938b, g CIBA). The acetate (below) is obtained from progesterone (p. 2782 s) on heating with acetic anhydride-acetyl chloride (2:3) under nitrogen at 100–150° (bath temperature) for 5 hrs. (yield, 47 %); the propionate (p. 151) is similarly obtained (120°, 1 hr.), the butyrate (p.151) by heating with sodium butyrate in butyric anhydride at 195° for 2 hrs. (yield, 43 %) (1937 Westphal), the benzoate (below) by benzoylation of progesterone, e.g., by refluxing with benzoyl chloride in benzine for 40 hrs. (1936 CIBA), the benzyl ether (below) by heating of progesterone with benzyl alcohol in the presence of p-toluenesulphonic acid in benzene (1940c SCHERING A.-G.). The acetate is also formed when the 3-enol acetates of the 21-halogenoprogesterones (p. 2277 s) are reduced with zinc in glacial acetic acid (1937e, h CIBA).

Rns. Gives Δ^4-pregnen-20α(?)-ol-3-one (p. 2510 s) on heating of the benzoate with aluminium isobutoxide in isobutyl alcohol, followed by saponification (1938f CIBA). On treatment of the acetate with alcoholic alkali it gives progesterone (p. 2782 s) (1937e, h CIBA).

Acetate $C_{23}H_{32}O_3$ (p. 151). Absorption max. in chloroform 239 mμ (ε 17 400) (1939 Dannenberg; see also 1937 Bugyi). — **Fmn.** and **Rns.** See above.

For the *propionate* and the *butyrate*, see p. 151.

Benzoate $C_{28}H_{34}O_3$. Crystals (dil. ethanol or methanol), m. 190–2^0 dec. Gives a yellow-brown coloration with tetranitromethane (1936 CIBA). — **Fmn.** and **Rns.** See above.

Benzyl ether, 3-Benzyloxy-$\Delta^{3,5}$-pregnadien-20-one $C_{28}H_{36}O_2$, m. 192^0; $[\alpha]_D^{20}$ —50^0 (dioxane) (1940c SCHERING A.-G.). — **Fmn.** From progesterone (p. 2782 s) on heating with benzyl alcohol in benzene, in the presence of p-toluenesulphonic acid (1940c SCHERING A.-G.).

$\Delta^{5,16}$-Pregnadien-3β-ol-20-one, *16-Dehydropregnenolone* $C_{21}H_{30}O_2$. Crystals

(methanol), m. 211–3^0 cor. (1939 Goldberg); need-les (ethyl acetate), m. 216^0 (1939a Butenandt); was once obtained as leaflets, m. 178^0 (1938c Butenandt; cf. 1939a Butenandt); $[\alpha]_D$ —27.5^0 (alcohol) (1946 Velluz), —38^0 (chloroform) (1945 SCHERING A.-G.). Absorption max. in ether at 237 mμ (1938c, 1939a Butenandt), 234 mμ (ε 9300) (1939 Dannenberg); in alcohol(?) at 240 mμ (log ε 3.95) (1939 Goldberg).

Fmn. As acetate, from the acetate of 17-ethynyl-$\Delta^{5,16}$-androstadien-3β-ol (p. 1517 s) on keeping with glacial acetic acid, acetic anhydride, HgO, and BF_3-ether at room temperature (yield, ca. 15%) or on heating with aniline and $HgCl_2$ in benzene and water at 60^0 (yield, 8%) (1943 Shoppee; see also 1938c SCHERING A.-G.). From 17α-ethynyl-Δ^5-androstene-3β.17β-diol (page 2030 s) on refluxing with mercuric acetamide (see 1943 Goldberg) in abs. alcohol (crude yield, 65%)(1939 Goldberg; cf. 1943 Shoppee). As acetate, from the diacetate of Δ^5-pregnene-3β.17α-diol-20-one (p. 2333 s) on heating with fuller's earth at 150–160^0 under 0.1 mm. pressure (1939b CIBA), from the 3-acetate of 17-iso-Δ^5-pregnene-3β.17β-diol-20-one (p. 2337 s) on heating with $POCl_3$ in pyridine at 100^0 (yield, ca. 20%), or from the 3-acetate 17-ben-zoate of the same diolone on fusion and distillation at 250–280^0 and 10 mm. (yield, ca. 20%)(1943 Shoppee).

From 3β-acetoxy-17-cyano-$\Delta^{5,16}$-androstadiene on boiling with methyl-magnesium bromide in ether (yield, 75%) (1939a Butenandt; see also 1945 SCHERING A.-G.); on using methylmagnesium iodide (1938c Butenandt), smaller yields are obtained (1939a Butenandt). As acetate, from 3β-acetoxy-17β-benzoyloxy-17-iso-Δ^5-ætiocholenic acid on heating with thionyl chloride in benzene on the water-bath, followed by treatment with methylzinc iodide in toluene and sublimation of the reaction product at 180^0 in a high vac. (1938c CIBA).

References, pp. 2252–2257 s

From diosgenin acetate (p. 279) on heating with acetic anhydride at 200°
for 10 hrs., to give pseudodiosgenin diacetate, keeping the latter with CrO_3
in acetic acid at 25–28°, and hydrolysing the oxidation product with boiling
alcoholic KOH (yield, 38%) (1940k Marker; see also 1940d, 1947 Marker).
From the oxidation product $C_{31}H_{46}O_7$ (p. 2332 s) of pseudodiosgenin diacetate
on hydrolysis with boiling alcoholic KOH, K_2CO_3, or HCl (quantitative yield)
(1941a Marker). From trillin (diosgenin glucoside) tetraacetate on heating
with acetic anhydride at 200° to give pseudotrillin acetate, keeping the latter
with CrO_3 in acetic acid at 25°, and hydrolysing the oxidation product with
boiling alcoholic HCl (good yield) (1940j Marker).

Rns.* Boiling with aluminium isopropoxide and cyclohexanone in toluene
affords $\Delta^{4.16}$-pregnadiene-3.20-dione (p. 2780 s) (1939a Butenandt; 1939c
SCHERING A.-G.). Oxidation of the acetate with excess aqueous KOI in
dioxane, first at room temperature and then at 80°, followed by alkaline
hydrolysis on the steam-bath, gives 3β-hydroxy-$\Delta^{5.16}$-ætiocholadienic acid
(1942g Marker). Gives Δ^5-pregnen-3β-ol-20-one (p. 2233 s) on hydrogenation
in the presence of palladium-$BaSO_4$ in ether at 0.1 atm. pressure (1940j
Marker) or in the presence of Raney nickel in aqueous alcoholic NaOH
(1939a Butenandt) or in methanol (1945 SCHERING A.-G.); the acetate reacts
similarly on hydrogenation in alcohol in the presence of Raney nickel or
on reduction with zinc in glacial acetic acid (1938c, 1939b CIBA). Reduction
with aluminium isopropoxide in boiling isopropyl alcohol, followed by refluxing
with alcoholic KOH, gives $\Delta^{5.16}$-pregnadiene-3β.20β-diol (p. 1918 s) (1941c
Marker). On reduction with sodium and boiling abs. alcohol, Δ^5-pregnene-
3β.20α-diol (p. 1920 s) is formed (1940d, 1941b Marker). Yields water-
insoluble hydrazones (see p. 2232 s) with Girard's reagents T and P (1946
Velluz).

Keeping with diazomethane in ether at room temperature gives the **pyrazoline compound** $C_{22}H_{32}O_2N_2$ (I, R = H) [crystals (acetone), m. 178° cor.
dec.] whose thermal decomposition at 175° and 0.04 mm. pressure produces
16-methyl-$\Delta^{5.16}$-pregnadien-3β-ol-20-one (p. 2267 s) (1944 Wettstein); the
acetate reacts in the same manner with diazomethane yielding the **pyrazoline
compound** $C_{24}H_{34}O_3N_2$ (I, R = CO·CH₃) [colourless crystals (acetone),
m. 168–9° cor. dec.] (1944 Wettstein) whose thermal decomposition at 180°

and 0.02 mm. pressure gives, according to Wettstein (1944) and Sandoval
(1951), 16-methyl-$\Delta^{5.16}$-pregnadien-3β-ol-20-one acetate as the main product,
together with a small amount of **16.17-methylene-Δ^5-pregnen-3β-ol-20-one
acetate** $C_{24}H_{34}O_3$ (II) [crystals (acetone), m. 202.5–205° cor. (1944 Wettstein);
crystals (ethyl acetate), m. 194–6° cor. (Kofler block); $[\alpha]_D^{20} + 33°$ (chloro-

---

* See also reactions of the acetate oxime, p. 2232 s.

E.O.C. XIV s. 157

form) (1951 Sandoval); strong terminal absorption near 220 mμ (alcohol) (1951 Sandoval; see also 1944 Wettstein); infra-red bands at 1736 cm^{-1} and 1685 cm^{-1} (carbon disulphide) (1951 Sandoval)].

On heating with methyl-magnesium iodide (using the acetate) in toluene on the steam-bath it yields 16α-methyl-Δ^5-pregnen-3β-ol-20-one (p. 2268 s) and $\Delta^{5.16}$-bisnorcholadiene-3β.20-diol (p. 1937 s); with isopropylmagnesium bromide and tert.-butyl magnesium chloride, 16α-isopropyl- and 16α-tert.-butyl-Δ^5-pregnen-3β-ol-20-one, resp., are obtained (1942e Marker).

$\Delta^{5.16}$-*Pregnadien-3β-ol-20-one oxime* C$_{21}$H$_{31}$O$_2$N. Needles (acetic acid), m. 215–220° dec. (1938c, 1939a Butenandt).

$\Delta^{5.16}$-*Pregnadien-3β-ol-20-one hydrazone* C$_{21}$H$_{32}$ON$_2$. — *Hydrazone formed with Girard's reagent T* (the hydrazide of trimethylcarboxymethylammonium chloride), C$_{26}$H$_{42}$O$_2$N$_3$Cl = C$_{21}$H$_{30}$O : N · NH · CO · CH$_2$ · N (Cl) (CH$_3$)$_3$. Crystals (methanol). Is not stable. Insoluble in water. **Fmn.** From the above hydr- oxy-ketone on boiling with Girard's reagent T in alcohol-acetic acid (91 % yield). **Rns.** Difficultly hydrolysable in the cold, easily hydrolysed by aqueous H$_2$SO$_4$ (1946 Velluz).

Hydrazone formed with Girard's reagent P (the hydrazide of N-carboxy-methylpyridinium chloride) C$_{28}$H$_{38}$O$_2$N$_3$Cl = C$_{21}$H$_{30}$O : N · NH · CO · CH$_2$ · N(Cl)C$_5$H$_5$ (no analysis). Properties, formation and reactions are similar to those of the preceding compound (1946 Velluz).

$\Delta^{5.16}$-*Pregnadien-3β-ol-20-one acetate* C$_{23}$H$_{32}$O$_3$. Needles (aqueous acetone), m. 176°; [α]$_D^{20}$ —33° (alcohol) (1939a Butenandt); [α]$_D^{12}$ —29° (acetone) (1943 Shoppee). Absorption max. in ether at 234 mμ (ε 9600) (1939 Dannenberg). — **Fmn.** From the above hydroxy-ketone on keeping with acetic anhydride and pyridine at ca. 18° (1939a Butenandt). See also under formation of the hydroxyketone. — **Rns.** See under those of the hydroxyketone.

$\Delta^{5.16}$-*Pregnadien-3β-ol-20-one acetate oxime* C$_{23}$H$_{33}$O$_3$N. Crystals (alcohol), m. 210–215° dec. — **Fmn.** From the above acetate and hydroxylamine hydro-chloride in pyridine or, in the presence of sodium acetate, in aqueous alcohol.— **Rns.** Treatment with an aromatic sulphonyl chloride in pyridine, followed by hydrolysis with dil. H$_2$SO$_4$, gives the acetate of Δ^5-androsten-3β-ol-17-one (p. 2609 s) (1941a PARKE, DAVIS & Co.).

Δ^5-**Pregnen-3α-ol-20-one** C$_{21}$H$_{32}$O$_2$. Needles (dil. alcohol), m. 148–152° after sintering at 144°; [α]$_D^{20}$ +54.5° (alcohol). Not precipitated by digitonin. — **Fmn.** Along with Δ^5-pregnen-3β-ol-20-one (below), from Δ^5-preg-nene-3.20-dione (p. 2793 s) on partial hydro-genation in alcohol in the presence of Raney nickel; crude yield, 28 % (1939c Butenandt).

Acetate $C_{23}H_{34}O_3$. Crystals (dil. alcohol), m. 147°; $[\alpha]_D^{20} + 57°$ (alcohol?). — **Fmn.** From the above hydroxy-ketone on keeping with acetic anhydride and pyridine at room temperature (1939c Butenandt).

Δ^5-Pregnen-3β-ol-20-one, *"Pregnenolone"* $C_{21}H_{32}O_2$ (I, R = H) (p. 114). Prisms (CHCl$_3$-petroleum ether), m. 186–7° cor. (1947 Prelog), 193° (1937c, h, 1938b, c CIBA); $[\alpha]_D^{20} + 28°$ (alcohol) (1937a Butenandt). — **Occ.** Occurs, along with allopregnan-3α-ol-20-one (p. 2246 s), allopregnan-3β-ol-20-one (p. 2247 s), testalolone (see p. 2381 s), and other compounds, in swine testes (1943 Ruzicka; 1947 Prelog).

(I)

FORMATION OF Δ^5-PREGNEN-3β-OL-20-ONE

Formation from other C_{21} steroids. From the 3-acetate of 3β-hydroxy-Δ^5-ternorcholenylamine (p. 1907 s) on treatment with HOCl in ether at 10° in the presence of some anhydrous Na$_2$SO$_4$, followed by refluxing of the resulting N-chloro-derivative with sodium ethoxide in alcohol and hydrolysis of the resulting imide with aqueous H$_2$SO$_4$ (1937b I. G. FARBENIND.; 1939 Ehrhart; 1942 Ruschig; see also 1946 I. G. FARBENIND.) [yield, 70% (1942 Ruschig)]. From $\Delta^{5.20}$-pregnadiene-3β.20-diol-20-acetate (p. 1918 s) on saponification with 0.1 N methanolic KOH, together with 17-iso-Δ^5-pregnen-3β-ol-20-one (p. 115) (1939c CIBA). From a Δ^5-pregnene-3β.17.20-triol (p. 2117 s) on heating with 30% H$_2$SO$_4$ in propyl alcohol, with H$_2$SO$_4$ in dioxane, or with KHSO$_4$ (1938a CIBA). From $\Delta^{5.16}$-pregnadien-3β-ol-20-one (p. 2230 s) on hydrogenation in the presence of palladium-BaSO$_4$ in ether at 0.1 atm. pressure (1940j Marker) or in the presence of Raney nickel in aqueous alcoholic NaOH (crude yield, 87%) (1939a Butenandt) or in methanol (yield, 81–86%) (1945 SCHERING A.-G.). As acetate, from $\Delta^{5.16}$-pregnadien-3β-ol-20-one acetate on hydrogenation in alcohol in the presence of Raney nickel or on reduction with zinc in glacial acetic acid (1938c, 1939b CIBA). As acetate, from i-pregnen-6-ol-20-one methyl ether (p. 2251 s) on refluxing with zinc acetate, glacial acetic acid, and acetic anhydride (crude yield, 70%) (1946 Riegel). From 21-chloro-Δ^5-pregnen-3β-ol-20-one (p. 2277 s) (or its acetate) on treatment with zinc or alkali iodide in glacial acetic acid or in dioxane + conc. HCl. From 21.21-dibromo-Δ^5-pregnen-3β-ol-20-one (p. 2283 s) on treatment with a zinc-copper alloy in glacial acetic acid containing HBr; the acetate is similarly formed (1937c, h CIBA). From 5-chloro-allopregnan-3β-ol-20-one acetate (p. 2284 s) on refluxing with methanolic K$_2$CO$_3$ (yield, 43%) or, less pure, on boiling with dimethylaniline, followed by saponification with methanolic K$_2$CO$_3$ (1946b Meystre). As acetate, from 5α.6β.17.21-tetrabromo-allopregnan-3β-ol-20-one acetate (p. 2290 s) on heating with iron and acetic acid on the steam-bath (1942d Marker). From the acetate of 21-diazo-Δ^5-pregnen-3β-ol-20-one (p. 2293 s) on treatment with NaI and conc. HCl in alcohol, first in the cold, then at boiling temperature (1937h CIBA). Together with Δ^5-pregnen-3α-ol-20-one, from Δ^5-pregnene-3.20-dione

157*

(p. 2793 s) on partial hydrogenation in the presence of Raney nickel in alcohol (yield, 53%) (1939c Butenandt).

Formation from C_{19}-steroids. Together with a small amount of an isomeric ketone*, from Δ^5-androsten-3β-ol-17-one (p. 2609 s) on refluxing with ethyl α-chloropropionate in ether in the presence of sodium ethoxide, followed by boiling with alcoholic NaOH (1937, 1939 Yarnall). Along with "neopregnenolone" (17a-methyl-D-homo-Δ^5-androsten-3β-ol-17-one, see 1940 Ruzicka), from the acetate of 5-dehydroepiandrosterone (Δ^5-androsten-3β-ol-17-one, p. 2609 s) on condensation with ethyl $\alpha.\alpha$-dichloropropionate in ether in the presence of amalgamated magnesium, followed by boiling with methanolic NaOH and decarboxylation at ca. 200° (1939a Miescher).

Formation from C_{20}-steroids. From 3β-hydroxy-Δ^5-ætiocholenaldehyde (p. 2227 s) on treatment with diethyl diazomalonate in ether, followed by refluxing with 5% methanolic KOH for 2 hrs. and then with dil. aqueous methanolic H_2SO_4 for 2 hrs. (yield, 35%); from the triphenylmethyl ether of the same hydroxy-aldehyde on treatment with diazomethane in ether, followed by heating with alcoholic HCl (yield, 40%) (1938b, g CIBA). As acetate, from 3β-acetoxy-Δ^5-ætiocholenic acid nitrile by the action of methylmagnesium iodide, followed by treatment with ice-cold dil. H_2SO_4 (1937c, 1938a I. G. FARBENIND.); is obtained in the same manner, from 3β-acetoxy-Δ^5-ætiocholenic acid diethylamide and, similarly, from 3β-acetoxy-Δ^5-ætiocholenic acid chloride and a methylcadmium compound (1938a I. G. FARBENIND.). From methyl 3β-hydroxy-Δ^5-ætiocholenate (p. 173) on keeping with ethyl acetate and sodium at room temperature, followed by saponification with dil. NaOH and decarboxylation by vac. distillation (1938 N. V. ORGANON). From the chloride of 3β-acetoxy-Δ^5-ætiocholenic acid on boiling with diethyl sodiomalonate in benzene, followed by saponification with boiling methanolic KOH and decarboxylation in boiling aqueous methanolic H_2SO_4 (1940b Wettstein; cf. 1937i CIBA).

Formation from C_{22}-steroids (directly or via intermediates). From the two 20-epimeric 20-methyl-Δ^5-pregnene-3β.20.21-triols (pp. 1955 s, 1956 s) on keeping with HIO_4 in methanol (1941 Hegner). From 3β-acetoxy-Δ^5-bisnorcholenic acid (p. 174) on heating with 1 mol. lead tetraacetate in glacial acetic acid, followed by saponification with 2 N methanolic KOH (refluxing for 3 hrs.) and oxidation with excess lead tetraacetate in glacial acetic acid at room temperature (1939a, 1940d SCHERING A.-G.). From 22.22-diphenyl-$\Delta^{5.20(22)}$-bisnorcholadien-3β-ol acetate (p. 1845 s) on treatment with bromine in carbon tetrachloride at —15°, ozonolysis at 0°, debromination with zinc dust in glacial acetic acid at 10–30° and finally at 50–55°, boiling with semicarbazide acetate in methanol, and cleavage of the acetate semicarbazone with aqueous H_2SO_4 in methanol-ether at 70° (yield: 5 kg. $\rightarrow$ 2.3 kg. acetate semicarbazone) (p. 114; 1940a SCHERING A.-G.); the nuclear double bond of 22.22-diphenyl-$\Delta^{5.20(22)}$-bisnorcholadien-3β-yl acetate may also be temporarily protected by

---

* This ketone (not characterized) was assumed by Yarnall (1937, 1939) to be 17-iso-Δ^5-pregnen-3β-ol-20-one (p. 2249 s) [which is formed when the present pregnenolone is heated with alcoholic alkali]; (see the following formation) (Editor).

the addition of hydrogen halide which is split off after the ozonization during
the saponification of the resulting acetate (1934c SCHERING A.-G.). As methyl
ether, from 20-diphenylmethylene-6β-methoxy-i-pregnene (p. 1846 s) on ozonolysis in carbon tetrachloride at room temperature, followed by refluxing
with methanol containing conc. H_2SO_4 (yield of crude methyl ether, 17 %)
(1946 Riegel).

    Formation from C_{24} and higher steroids. From 24.24-diphenyl-$\Delta^{5\cdot 20(22)\cdot 23}$-
cholatrien-3β-ol acetate (p. 1847 s) on oxidation with CrO_3 in chloroform-
acetic acid at 0–3⁰, followed by saponification with methanolic K_2CO_3 (yield,
44 %) (1946b MEYSTRE; cf. 1946 CIBA PHARMACEUTICAL PRODUCTS). The
formation from cholesteryl acetate dibromide with CrO_3 in acetic acid (at
28–30⁰) (p. 115) was confirmed by Ruzicka (1937); this oxidation is advantageously carried out in the presence of free H_2SO_4 at room temperature
(1938b SCHERING A.-G.). The separation from Δ^5-androsten-3β-ol-17-one and
other oxidation products of cholesterol can be accomplished by acetylation
of the mixture and precipitation of the digitonide of Δ^5-pregnen-3β-ol-20-one
acetate (1938 SCHERING CORP.); for separation from the other oxidation
products of cholesterol, see also 1938e CIBA. Is also formed from stigmasteryl
acetate (p. 88) on oxidation with CrO_3 after temporary protection of the
nuclear double bond by bromine (1934b SCHERING A.-G.). Regarding the
formation of the acetate on oxidation of pseudodiosgenin diacetate with CrO_3
in acetic acid, followed by treatment of the solution with Zn and $ZnBr_2$
claimed by 1945 PARKE, DAVIS & Co.; cf. 1947 Marker.

REACTIONS OF Δ^5-PREGNEN-3β-OL-20-ONE
(see also the reactions of the derivatives, p. 2238 s)

    Oxidation. Δ^5-Pregnen-3β-ol-20-one is converted into progesterone (p. 2782 s)
by heating with platinum black at 250—300⁰ in an atmosphere of CO_2 (1940j
Marker), with copper powder (or palladium, platinum, or silver) in a vac.
at 200⁰ or with a zinc-copper couple in the presence of cylohexanone or
quinoline (1936/37 CIBA), or with aluminium isopropoxide and cyclohexanone
or acetone in toluene or xylene, resp. (1937 SCHERING-KAHLBAUM A.-G.;
1940a, 1945 SCHERING A.-G.), or by the action of dehydrogenating bacteria
(Corynebacterium mediolanum, Flavobacterium dehydrogenans Arn., Micrococcus dehydrogenans) isolated from yeast (1938, 1939 Mamoli; 1940 Ercoli;
1942 Arnaudi; 1942 Molina). The oxidation to progesterone with CrO_3 via
the dibromide (see also "Bromination", p. 2237 s) may be carried out in acetic
acid-benzene, followed by debromination by shaking with zinc powder and
NaI in benzene-acetone (1941b I. G. FARBENIND.) or by treatment under CO_2
with aqueous $CrCl_2$ in acetone at 26⁰ (1945 Julian); in the oxidation to
progesterone with CrO_3 or $KMnO_4$ the double bond may also be protected
by the addition of dry HBr or HCl in ether-alcohol, and after the oxidation
the hydrogen halide may be removed by heating with potassium acetate or
pyridine (1935a SCHERING-KAHLBAUM A.-G.); unchanged Δ^5-pregnen-3β-ol-
20-one can be separated from the produced progesterone by means of its
betaine ester (1939 LABORATOIRES FRANÇAIS DE CHIMIOTHÉRAPIE). Pregnenolone imide (as obtained, e. g., from 3β-acetoxy-Δ^5-ternorcholenylamine)

may be converted into progesterone by refluxing with aluminium tert.-butoxide and acetone in benzene, followed by shaking with 0.5 N H_2SO_4 (1938b I. G. FARBENIND.).

On shaking with CrO_3 in glacial acetic acid at 27⁰ it gives allopregnan-5α-ol-3.6.20-trione (p. 2976 s) (1939, 1948 Ehrenstein); on keeping with CrO_3 in acetic acid at 20⁰, followed by refluxing with zinc dust, allopregnane-3.6.20-trione (p. 2966 s) is obtained (1940i Marker). For the action of aluminium isopropoxide + ketones giving progesterone, see above; on heating with aluminium tert.-butoxide in toluene, a Δ^5-pregnene-3β.20-diol (p. 1922 s) and Δ^4-pregnen-20-ol-3-one (p. 2510 s) are also formed (1937 N. V. ORGANON); on boiling with aluminium tert.-butoxide and benzoquinone in toluene it gives 6-dehydroprogesterone (p. 2778 s) (1940a Wettstein; 1941 b CIBA). On heating of the acetate with 3 mols. of lead tetraacetate in glacial acetic acid in the presence of some acetic anhydride at 68–70⁰ it gives the diacetate of Δ^5-pregnene-3β.21-diol-20-one (p. 2301 s) (1939 Ehrhart; 1939 Reichstein) and a small amount of the triacetate (m. 182–185⁰ cor.) of Δ^5-pregnene-3β.17α.21-triol-20-one (p.2358 s) (1939 Reichstein); on refluxing with 1 mol. of lead tetraacetate in glacial acetic acid for 2 hrs., SCHERING A.-G. (1940d) obtained a crystalline compound (methanol), m. 156–158⁰, assumed to contain a 17-acetoxy-group. On keeping with OsO_4 in dioxane at room temperature allopregnane-3β.5α.6α-triol-20-one (p. 2364 s) is formed; the acetate does not react with OsO_4 (1939, 1948 Ehrenstein). Heating of the acetate with H_2O_2 in acetic acid, followed by saponification with boiling methanolicKOH,affords allopregnane-3β.5α.6β-triol-20-one (p. 2365 s) (1939, 1948 Ehrenstein). On keeping with perbenzoic acid in chloroform it gives mainly 5α.6α-epoxyallopregnan-3β-ol-20-one (p. 2364 s) (1941, 1948 Ehrenstein). On oxidation of the acetate with $KMnO_4$ in acetic acid at 50⁰ it gives the 3-acetates of 5α.6α-epoxyallopregnan-3β-ol-20-one (p. 2364 s), allopregnane-3β.5α.6α-triol-20-one (p. 2364 s) and allopregnane-3β.5-diol-6.20-dione (p. 2866 s) (1940, 1948 Ehrenstein). On oxidation of the acetate with excess aqueous KOI in methanol at room temperature it gives (3β-hydroxy-Δ^5-ætiocholen-17β-yl)-glyoxylic acid (1941 Goldschmidt), whereas on oxidation in dioxane, first at room temperature and then at 80⁰, followed by alkaline hydrolysis on the steam-bath, 3β-hydroxy-Δ^5-ætiocholenic acid (p. 173) is obtained (1942g Marker). Oxidation of the acetate with persulphuric acid in acetic acid at 25⁰ (after temporary protection of the double bond with bromine), followed by hydrolysis with boiling alcoholic KOH, affords Δ^5-androstene-3β.17β-diol (p. 1999 s) (1940f Marker).

Pregnenolone may be converted into Δ^5-pregnen-3β-ol-20-on-21-al (p. 2378 s) by heating with SeO_2 in alcohol in a sealed tube at 120–130⁰ (1936 SCHERING A.-G.), by treatment in the cold with ethyl nitrate + sodium ethoxide in alcohol-benzene, with amyl nitrite + sodium ethoxide in alcohol, or with ethyl nitrite + potassium in ether, or by refluxing with p-nitroso-dimethylaniline + 30% NaOH in alcohol, in all cases followed by acid hydrolysis of the resulting condensation products (1937d, g CIBA), or by condensation of the acetate with aldehydes in the presence of alkali, followed by oxidation with ozone,

CrO$_3$, lead or manganese tetraacetate after temporary protection of the nuclear double bond by halogen or hydrogen halide (1937b SCHERING A.-G.).

Reduction. On hydrogenation in acetic acid in the presence of platinum oxide it gives allopregnane-3β.20β-diol (p. 1934 s) (1946b Pearlman); on hydrogenation of the acetate under the same conditions, followed by oxidation with CrO$_3$ in glacial acetic acid at room temperature, allopregnan-3β-ol-20-one acetate (p. 2249 s) and a small amount of pregnan-3β-ol-20-one acetate (p. 2246 s) are obtained (1946 Plattner). For alleged conversion into Δ^4-pregnene-3β.20-diol 3-acetate on hydrogenation of the acetate in alcohol in the presence of Raney nickel (1937 SCHERING CORP.), see the footnote on p. 1919 s. For reduction, see also under "Biological reactions", p. 2238 s.

Bromination, action of PCl$_5$. Gives the dibromide, 5α.6β-dibromo-allopregnan-3β-ol-20-one (p. 2289 s), with 1 mol. bromine, e. g., in glacial acetic acid (1934 Butenandt; 1935b SCHERING-KAHLBAUM A.-G.), chloroform (1934 Fernholz; 1945 Julian), or methylene chloride (1941b I. G. FARBENIND.). On treatment with excess bromine and some HBr in chloroform-glacial acetic acid, followed by boiling of the resulting oily 5.6.21-tribromopregnan-3β-ol-20-one with potassium acetate in alcohol-benzene, it gives the 21-acetate of Δ^5-pregnene-3β.21-diol-20-one (p. 2301 s) (1940 SCHERING CORP.). On treatment of the free hydroxyketone or its acetate (propionate) with 3 mols. of bromine in acetic acid (propionic acid) in the presence of hydrobromic acid it gives the acetate (propionate) of 5α.6β.17.21-tetrabromo-allopregnan-3β-ol-20-one (p. 2290 s) (1942d Marker; 1941b PARKE, DAVIS & Co.; 1950 Julian; see also 1940 SCHERING CORP.). Gives 3β-chloro-Δ^5-pregnen-20-one (p. 2222 s) on treatment with PCl$_5$ (1940 PARKE, DAVIS & Co.).

Action of organic compounds. The acetate reacts with methylmagnesium iodide in ether to give 20-methyl-Δ^5-pregnene-3β.20-diol (p. 1937 s) (1942d Marker; cf. 1934a SCHERING A.-G.). For conversion into 20-methyl-Δ^5-pregnene-3β.20-diol-21-al by reaction of the acetate with sodium acetylide etc., see the aldehyde (p. 2314 s). Gives 21-benzylidene-Δ^5-pregnen-3β-ol-20-one (p. 2271 s) on keeping with benzaldehyde and sodium ethoxide in abs. alcohol at room temperature (1937b SCHERING A.-G.; cf. 1942f Marker) or on refluxing with benzaldehyde and aqueous methanolic KHCO$_3$ (1945 Velluz); the acetate reacts similarly with benzaldehyde and some piperidine in abs. alcohol on boiling (1937b SCHERING A.-G.); for condensation with aldehydes, followed by oxidation, see under "Oxidation", p. 2236 s. Gives 3β.20-dihydroxy-Δ^5-norcholenic acid on refluxing of the acetate with ethyl bromoacetate and iodine-activated zinc in benzene, followed by hydrolysis with boiling aqueous methanolic K$_2$CO$_3$ (1942 Ruzicka). Forms no addition product with oxalic acid when heated with the latter in ethyl acetate (1941 Miescher). On heating with diethyl oxalate and sodium ethoxide in alcohol it gives ethyl Δ^5-pregnen-3β-ol-20-one-21-glyoxylate (1941a, 1943, 1946 I. G. FARBENIND.; 1948 Ruschig). On heating with pyridine and iodine on the steam-bath it gives N-(3β-hydroxy-20-keto-Δ^5-pregnen-21-yl)-pyridinium iodide (p. 2279 s) (1944 King).

Biological reactions. For the action of dehydrogenating bacteria, see under "Oxidation", p. 2235 s. After oral administration of Δ^5-pregnen-3β-ol-20-one to men and rabbits, pregnane-3α.20α-diol (p. 1923 s) has been found in the urine (men and rabbits) and in the bile (men) (1946a Pearlman).

Colour reactions. Colours are produced with the acetate on underlayering with conc. H_2SO_4: of the alcoholic solution (1939 Woker; 1939 Scherrer), of a mixture with benzaldehyde in alcohol (1939 Scherrer), and of a mixture with furfural in alcohol (1939 Woker). For the absorption spectrum in the Zimmermann test with m-dinitrobenzene and KOH, see 1940 McCullagh; 1943 Hansen.

Physiological properties. See pp. 1367 s, 1368 s, and the literature cited there.

DERIVATIVES OF Δ^5-PREGNEN-3β-OL-20-ONE

Δ^5-Pregnen-3β-ol-20-one acetate, 3β-Acetoxy-Δ^5-pregnen-20-one $C_{23}H_{34}O_3$ (I, p. 2233 s; R = CO·CH$_3$) (p. 115). Crystals (ethyl acetate-hexane), m. 149° cor. (1939b CIBA); $[\alpha]_D$ + 20° (alcohol) (1937a Butenandt). Is precipitated by digitonin (1938 SCHERING CORP.). — **Fmn.** and **Rns.** See under the free hydroxyketone, above. Its *oxime* (dec. ca. 197°) gives 17-amino-Δ^5-androsten-3β-ol (p. 1910 s) on refluxing with thionyl chloride in benzene, followed by hydrolysis with alcoholic HCl and then with aqueous NaOH (1937a I. G. FARBENIND.; 1939 Ehrhart). — *Semicarbazone* $C_{24}H_{37}O_3N_3$ (p. 115); on treatment with NaNO$_2$ in glacial acetic acid it gives the above acetate in quantitative yield (1946 Goldschmidt).

Δ^5-Pregnen-3β-ol-20-one propionate, 3β-Propionyloxy-Δ^5-pregnen-20-one $C_{24}H_{36}O_3$ (I, p. 2233 s; R = CO·C$_2$H$_5$). Crystals (methanol), m. 119—120°. — **Fmn.** From the free hydroxyketone on refluxing with propionic anhydride (1942d Marker).

Δ^5-Pregnen-3β-ol-20-one benzoate, 3β-Benzoyloxy-Δ^5-pregnen-20-one $C_{28}H_{36}O_3$ (I, p. 2233 s; R = CO·C$_6$H$_5$). Crystals (acetone-methanol), m. 191—2° (1946 Masson), 192—3° (1950 Giral; 1950 Mancera), 197° (1948 Barton); $[\alpha]_D$ in CHCl$_3$ + 40° (1948 Barton), + 47.4° (1950 Mancera), + 56° (1950 Giral).

Δ^5-Pregnen-3β-ol-20-one p-toluenesulphonate, 3β-(p-Toluenesulphonyl-oxy)-Δ^5-pregnen-20-one $C_{28}H_{38}O_4S$ (I, p. 2233 s; R = p-SO$_2$·C$_6$H$_4$·CH$_3$). Crystals (acetone-water), m. 139–140°; $[\alpha]_D^{20}$ + 9° (chloroform). — **Rns.** Gives the methyl ether of Δ^5-pregnen-3β-ol-20-one (below) on boiling with methanol; on heating with methanol and potassium acetate, the methyl ether of i-pregnen-6β-ol-20-one (p. 2251 s) is obtained (1937b Butenandt).

Δ^5-Pregnen-3β-ol-20-one methyl ether, 3β-Methoxy-Δ^5-pregnen-20-one $C_{22}H_{34}O_2$ (I, p. 2233 s; R = CH$_3$). Crystals (aqueous methanol), m. 123–4°; $[\alpha]_D^{18}$ + 18° (chloroform). — **Fmn.** From the above p-toluenesulphonate on boiling with methanol (1937b Butenandt). For formation by rearrangement of i-pregnen-6β-ol-20-one methyl ether, see this ether (p. 2251 s).

Δ^{11}-**Pregnen-3α-ol-20-one** $C_{21}H_{32}O_2$ (no analysis). Colourless needles (acetone-ether) m. 125–6⁰ cor. (1944 v. Euw). — **Fmn.** As acetate, together with some $\Delta^{3 \cdot 11}$-pregnadien-20-one (p. 2215 s), from pregnane-3α.12α-diol-20-one 3-acetate 12-(anthraquinone-β-carboxylate) (p. 2331 s) on thermal cleavage in a high vac. at 290⁰ (yield, ca. 30%); the acetate is saponified by keeping with methanolic KOH at 15⁰. As anthraquinone-β-carboxylate, together with Δ^3-pregnen-12α-ol-20-one anthraquinone-β-carboxylate (p. 2251 s) and $\Delta^{3 \cdot 11}$-pregnadien-20-one, from pregnane-3α.12α-diol-20-one (di-anthraquinone-β-carboxylate) on thermal cleavage at 290–320⁰ and 0.02 mm. (yield, ca. 8%) (1944 v. Euw). Together with other compounds, from methyl 11.12-dibromo-3α-acetoxycholanate on oxidation with CrO_3 in strong acetic acid at 70⁰, followed by debromination with zinc dust in acetic acid and saponification with boiling methanolic KOH (1943 Reichstein).

Rns. Oxidation with CrO_3 in glacial acetic acid yields Δ^{11}-pregnene-3.20-dione (p. 2793 s). On treatment of the acetate with N-bromoacetamide and sodium acetate in aqueous acetone-acetic acid at 16⁰ it gives 12α-bromo-pregnane-3α.11β-diol-20-one 3-acetate (p. 2351 s) and, probably, 9.11-dibromo-pregnane-3α.12-diol-20-one 3-acetate* (not isolated; convertible by oxidation, followed by heating with zinc dust and sodium acetate in glacial acetic acid, into a compound which is probably $\Delta^{9(11)}$-pregnen-3α-ol-12.20-dione acetate, p. 2849 s) (1944 v. Euw).

Acetate $C_{23}H_{34}O_3$. Colourless needles (methanol or low-boiling petroleum ether), m. 136–7⁰ cor. (1944 v. Euw). — **Fmn.** and **Rns.** See above.

Anthraquinone-β-carboxylate $C_{36}H_{38}O_5$. Yellowish needles (chloroform-acetone), m. 240–2⁰ cor. — **Fmn.** From the above hydroxy-ketone on treatment with anthraquinone-β-carboxylic acid chloride in pyridine-benzene (1944 v. Euw). See also under the hydroxy-ketone.

Δ^{16}-**Pregnen-3α-ol-20-one** $C_{21}H_{32}O_2$. Crystals (ether), m. 194–6⁰. — **Fmn.** From epi-sarsasapogenin acetate (p. 284) on heating with acetic anhydride at 200⁰ for 10 hrs., oxidation of the resulting crude pseudo-epi-sarsasapogenin diacetate with CrO_3 in acetic acid at 28⁰, and hydrolysis with boiling alcoholic KOH (yield, 52%). From dihydropseudo-epi-sarsasapogenin diacetate on oxidation with CrO_3 in acetic acid at 28⁰, and hydrolysis with boiling alcoholic KOH (yield, 61%). — **Rns.** Oxidation with CrO_3 in acetic acid at room temperature affords Δ^{16}-pregnene-3.20-dione (p. 2794 s). On hydrogenation in the presence of palladium-BaSO$_4$ in ether-alcohol it gives pregnan-3α-ol-20-one (p. 2244 s). On reduction with sodium and alcohol it gives pregnane-3α.20α-diol (p. 1923 s), on hydrogenation in the presence of a platinum oxide catalyst in acetic acid under 3 atm. pressure it yields pregnene-3α.20β-diol (p. 1929 s) (1940k Marker). The propionate oxime (not

---

* For similar reactions, see 1943 Reich; 1943 Hegner.

described) may be converted into the propionate of ætiocholan-3α-ol-17-one
(p. 2637 s) by treatment with p-toluenesulphonyl chloride in cold pyridine,
followed by hydrolysis with dil. H_2SO_4 at room temperature (1941a PARKE,
DAVIS & Co.).

Acetate $C_{23}H_{34}O_3$. Needles (dil. methanol and dil. acetone), m. 96–99°. —
Fmn. From the above hydroxy-ketone on refluxing with acetic anhydride
(1940k Marker).

Δ^{16}-**Pregnen-3β-ol-20-one** $C_{21}H_{32}O_2$. Crystals (ether-pentane), m. 188–190°
(1940k Marker); needles with 1 mol. EtOH
(alcohol), m. 207–9° (1940a Marker). — **Fmn.**
From 17-bromopregnan-3β-ol-20-one (p. 2285 s)
on refluxing with pyridine (yield, 75%); the
acetate is formed in like manner from 17-bromo-
pregnan-3β-ol-20-one acetate (1942a Marker).
As acetate, from 21-bromopregnan-3β-ol-20-one acetate (p. 2281 s) on refluxing with potassium acetate in glacial acetic acid, together with pregnane-
3β.21-diol-20-one diacetate (p. 2310 s) (1942b Marker). As acetate, from
16.17-dibromopregnan-3β-ol-20-one acetate (p. 2289 s) on shaking in dioxane
and pyridine with palladium-$BaSO_4$ catalyst and hydrogen at room temperature under 3 atm. pressure (yield, 72%), on refluxing with pyridine (yield,
14%), with NaI in methanol (yield, 82%), or with potassium acetate in acetic
acid (1942i Marker; 1941c PARKE, DAVIS & Co.). From sarsasapogenin
acetate (p. 282) on heating with acetic anhydride at 200° for 10 hrs., oxidation
of the resulting crude pseudosarsasapogenin diacetate with CrO_3 in acetic
acid at 28°, and hydrolysis with boiling alcoholic KOH; yield, 48% (1940k
Marker). As acetate, from pseudosarsasapogenin diacetate on oxidation with
CrO_3 in acetic acid at room temperature (1940a Marker). From pseudosarsasapogenin on oxidation with ozone in chloroform; the acetate is obtained
analogously by ozonolysis of pseudosarsasapogenin diacetate (1940e Marker).
From pseudosarsasapogenin diacetate on heating with H_2O_2 in acetic acid
at 70°, followed by hydrolysis with boiling methanolic KOH (1942c Marker).
From dihydropseudosarsasapogenin diacetate on oxidation with CrO_3 in
acetic acid at 28°, followed by hydrolysis with boiling alcoholic KOH; yield,
47% (1940a, k Marker).

Rns. Heating of the acetate with H_2O_2 in acetic acid at 70° affords 16.17-
epoxy-pregnan-3β-ol-20-one acetate (p. 2241 s) (1942c Marker; cf. 1947 Plattner). On oxidation with CrO_3 in acetic acid at room temperature it gives Δ^{16}-preg-
nene-3.20-dione (p. 2794 s); oxidation of the acetate with CrO_3 in acetic acid
at room temperature, followed by alkaline hydrolysis, gives 3β-hydroxy-
ætiobilianic acid (I) (1940a Marker). On oxidation of the acetate with excess
aqueous KOI in dioxane, first at room tempera-
ture and then at 80°, followed by alkaline hydrolysis on the steam-bath, it gives 3β-hydroxy-
Δ^{16}-ætiocholenic acid (1942g Marker). On
hydrogenation in the presence of palladium it

gives pregnan-3β-ol-20-one (p. 2245 s) (1942 a Marker); on hydrogenation of the acetate in alcohol in the presence of Adams's catalyst, followed by oxidation with CrO_3 in acetic acid at room temperature, the acetate of pregnan-3β-ol-20-one is obtained (1940a Marker). Reduction of the acetate with sodium and boiling alcohol affords pregnane-3β.20α-diol (p. 1930 s) (1940a Marker). Treatment of the acetate with bromine in acetic acid gives the acetate of 16.17-dibromopregnan-3β-ol-20-one (p. 2289 s) (1942 i Marker; 1941 b, c PARKE, DAVIS & Co.). See also the acetate oxime, below.

Δ^{16}-*Pregnen-3β-ol-20-one semicarbazone* $C_{22}H_{35}O_2N_3$. Crystals (alcohol), m. 240^0 dec. (1940a Marker)*.

Δ^{16}-*Pregnen-3β-ol-20-one acetate, 3β-Acetoxy-Δ^{16}-pregnen-20-one* $C_{23}H_{34}O_3$. Colourless plates (aqueous methanol), m. 144–146^0. — **Fmn.** From the above hydroxy-ketone on boiling with acetic anhydride (1940a Marker). See also under formation of the hydroxy-ketone. — **Rns.** See under those of the free hydroxy-ketone. — *Oxime* $C_{23}H_{35}O_3N$, m. 197—205^0 dec. Treatment with p-toluenesulphonyl chloride in cold pyridine, followed by hydrolysis with dil. H_2SO_4 at room temperature, gives the acetate of ætiocholan-3β-ol-17-one (p. 2638 s) (1941 a PARKE, DAVIS & Co.). — *Semicarbazone* $C_{24}H_{37}O_3N_3$, crystals (aqueous alcohol), m. 250–2^0 (1940a Marker)**.

Δ^{16}-Pregnen-3β-ol-20-one 16.17-oxide, 16.17-Epoxypregnan-3β-ol-20-one

$C_{21}H_{32}O_3$. For the structure, cf. 1947 Plattner. — Crystals (aqueous methanol), m. 223—225^0 (1942c Marker). — **Fmn.** The acetate is obtained from the acetate of Δ^{16}-pregnen-3β-ol-20-one (above) on heating with 30% H_2O_2 in acetic acid at 70^0 for 5 hrs. (yield, 28%); it is hydrolysed by boiling with alcoholic NaOH (1942c Marker). — *Acetate* $C_{23}H_{34}O_4$. Crystals (methanol), m. 179—180^0. **Fmn.** See above. May also be obtained from the above oxide with hot acetic anhydride (1942c Marker).

Δ^{16}-Allopregnen-3α-ol-20-one

$C_{21}H_{32}O_2$. Crystals (dil. methanol), m. 219–222^0 (1940g Marker); crystals (dil. dioxane) m. 226^0; $[\alpha]_D^{24} + 54^0$ (chloroform); absorption max. in ether at 234 mμ ($\varepsilon = 9100$) (1939 d Butenandt). — **Fmn.** From 3α-acetoxy-17-cyano-Δ^{16}-androstene on boiling with methylmagnesium bromide in ether, followed by boiling with acetic acid; crude yield, 54% (1939 d Butenandt). From epi-tigogenin acetate on heating with acetic anhydride at 200^0 for 10 hrs., oxidation of the resulting crude pseudo-epi-tigogenin diacetate with CrO_3 in acetic acid at 28^0, and hydrolysis with boiling alcoholic KOH; yield, 56% (1940k Marker). From acetylated pseudo-epi-tigogenin on oxidation with CrO_3 in acetic acid at 25^0, followed by hydro-

---

* The m. p.'s 256^0 for the semicarbazone and 237^1 for the acetate semicarbazone (1940e Marker) are probably given in error.

** 1940e Marker gives m. 237^0 which is probably a mistake.

lysis with boiling alcoholic KOH (yield, 1.5 g. pseudo-epi-tigogenin → 410 mg.)
(1940 g Marker). From dihydropseudo-epi-tigogenin diacetate on oxidation
with CrO_3 in acetic acid at 28°, followed by hydrolysis with boiling alcoholic
KOH; yield, 56% (1940 g, k Marker).

Rns. On keeping with CrO_3 in glacial acetic acid it gives Δ^{16}-allopregnene-
3.20-dione (p. 2795 s) (1939 d Butenandt; 1940 g Marker). Hydrogenation in
the presence of a palladium-BaSO$_4$ catalyst in alcohol-ether at room temperature affords allopregnan-3α-ol-20-one (p. 2246 s); hydrogenation of the acetate
in the presence of platinum oxide in alcohol, followed by oxidation with CrO_3
in acetic acid at room temperature, gives allopregnan-3α-ol-20-one acetate
(1940 g Marker). Its butyrate oxime (not described) may be converted into
the butyrate of androsterone (p. 2633 s) by treatment with p-toluenesulphonyl
chloride in cold pyridine, followed by hydrolysis with dil. H_2SO_4 at room
temperature (1941 a PARKE, DAVIS & Co.).

Acetate $C_{23}H_{34}O_3$. Leaflets (dil. methanol), m. 156–8° (1940 Marker), 159°;
$[\alpha]_D^{24} + 57°$ (chloroform) (1939 d Butenandt). — **Fmn.** From the above hydroxy-
ketone on boiling with acetic anhydride (1939 d Butenandt; 1940 g Marker). —
Rns. See above.

Δ^{16}-Allopregnen-3β-ol-20-one $C_{21}H_{32}O_2$. Crystals (dil. methanol or ether),

m. 202–4° (1940 h Marker); leaflets (aqueous
alcohol), m. 205–7° cor.; $[\alpha]_D^{17} + 50°$ (alcohol);
precipitable from 90% alcohol with digitonin
(1945, 1948 Klyne).

Occ. In the urine of pregnant mares (isolation
as the potassium and p-toluidine salts of the
hydrogen sulphate, below) (1945, 1948 Klyne).

Fmn. From 17-bromoallopregnan-3β-ol-20-one (p. 2286 s) on refluxing with
pyridine (yield, 63%); the acetate is obtained in like manner from 17-bromo-
allopregnan-3β-ol-20-one acetate (yield, 73%) (1942 h Marker). The acetate is
obtained from the 3-acetate of allopregnane-3β.17α-diol-20-one (p. 2335 s) on
heating with fuller's earth in vacuo (1939 b CIBA). From tigogenin acetate on
heating with acetic anhydride at 200° for 10 hrs., oxidation of the resulting
crude pseudotigogenin diacetate with CrO_3 in acetic acid at 28°, and hydrolysis
of the oxidation product with boiling alcoholic KOH; yield, 49% (1940 k
Marker). From the isolated oxidation product $C_{31}H_{48}O_7$ (p. 2333 s) of pseudotigogenin diacetate on refluxing with alcoholic KOH, K_2CO_3, or HCl; quantitative yield (1941 a Marker). From pseudotigogenin diacetate on heating
with H_2O_2 in acetic acid at 70°, followed by hydrolysis with hot methanolic
KOH (1942 c Marker). From dihydropseudotigogenin diacetate on oxidation
with CrO_3 in acetic acid at 28°, followed by hydrolysis with boiling alcoholic
KOH; yield, 60% (1940 h, k Marker).

Rns. On heating of the acetate with H_2O_2 in acetic acid at 70° it gives
16.17-epoxyallopregnan-3β-ol-20-one acetate (p. 2243 s) (1942 c Marker; 1947
Plattner). Oxidation with CrO_3 in acetic acid affords Δ^{16}-allopregnene-

3.20-dione (p. 2795 s) (1940h Marker). On reduction with aluminium iso-propoxide in boiling isopropyl alcohol, followed by refluxing with methanolic KOH it gives Δ^{16}-allopregnene-3β.20β-diol (p. 1922 s) (1941 c Marker). Hydro-genation in the presence of palladium-BaSO$_4$ in alcohol affords allopregnan-3β-ol-20-one (p. 2247 s) (1940h Marker), heating of the acetate with zinc dust in acetic acid on the steam-bath the acetate of this compound (1942h Marker). Its benzoate oxime (not described) may be converted into the benzoate of epiandrosterone (p. 2633 s) by treatment with p-toluenesulphonyl chloride in cold pyridine, followed by hydrolysis with dil. H$_2$SO$_4$ at room temperature (1941 a PARKE, DAVIS & Co.). Forms no semicarbazone (1945, 1948 Klyne; cf. 1951 Reich).

Hydrogen sulphate C$_{21}$H$_{32}$O$_5$S. — *Potassium salt* C$_{21}$H$_{31}$O$_5$SK. Needles with 1 H$_2$O (water and alcohol), browning at 234^0 cor., sintering at 241^0 cor., m. 244–5^0 cor. dec. Does not lose its water of crystallization at 80^0 in vac. over P$_2$O$_5$. Solubilities (mg./ml.): in water at b. p. 10–11, at 0^0 3; in alcohol at b. p. 9, at 0^0 2.5; in butanol at b. p. 10, at 0^0 7; in chloroform at b. p. and at 0^0 <0.1. Is dimorphous, sometimes prisms (water), m. 216–7^0 cor. dec. (not analysed) being obtained. — *p-Toluidine salt* C$_{28}$H$_{41}$O$_5$NS. White needles (water), softening at 186^0 cor., m. 195–7^0 cor.; $[\alpha]_D^{19}$ + 31.7^0 (chloroform). Solubilities (mg./ml.): in water at b. p. 3.5, at 15–20^0 0.5; in chloroform at 15–20^0 >30 (1945, 1948 Klyne).

Acetate C$_{23}$H$_{34}$O$_3$. Crystals (methanol), m. 162–4^0 (1940h Marker); $[\alpha]_D^{20}$ + 42^0 (chloroform) (1947 Plattner); $[\alpha]_D^{22}$ + 34^0 (chloroform) (1945, 1948 Klyne); absorption max. (solvent not specified, probably alcohol) at 242 (4.2) and 318 (2.1) mμ (log ε) (1947 Plattner; see also 1945, 1948 Klyne). — **Fmn.** From the above hydroxy-ketone on refluxing with acetic anhydride (1940h Marker). — **Rns.** See under those of the hydroxy-ketone; see also under formation of the hydroxy-ketone.

Δ^{16}-**Allopregnen-3β-ol-20-one 16.17-oxide, 16.17-Epoxyallopregnan-3β-ol-20-one** C$_{21}$H$_{32}$O$_3$. For the structure, see 1947 Plattner. — Crystals (methanol), m. 181–2^0 (1942c Marker). — **Fmn.** Its acetate is obtained from the acetate of Δ^{16}-allopregnen-3β-ol-20-one (above) on heating with 30% H$_2$O$_2$ in acetic acid at 70^0 for 5 hrs. (yield, 46%) (1942c Marker; cf. 1947 Plattner) or on treatment with perbenzoic acid in chloroform at room temperature (crude yield, 56%) (1947 Plattner); it is hydrolysed by boiling with alcoholic NaOH (1942c Marker).

Acetate C$_{23}$H$_{34}$O$_4$. Crystals (methanol), m. 185–6^0 (1942c Marker); crystals (acetone-alcohol), m. 186–7^0 cor. vac.; $[\alpha]_D^{20}$ + 51.6^0 (chloroform); absorption max. (solvent not stated) at 299 mμ (log ε 1.6) (1947 Plattner). — **Fmn.** See above. May also be obtained from the above oxide with hot acetic anhydride (1942c Marker).

Pregnan-3α-ol-20-one $C_{21}H_{34}O_2$ (p. 115). Crystals (acetone), m. 150–1⁰ cor. (1944 Moffett); needles (hexane or ether-petroleum ether), m. 151–4⁰ cor. (1946a Meystre); exists in two polymorphic forms, m. 134⁰ and m. 148⁰ (1938b Marker); $[\alpha]_D^{19}+114^0$ (abs. alcohol) (1938b Butenandt); $[\alpha]_D^{32}+112^0$ (methanol) (1944 Moffett); $[\alpha]_D^{22}+109.5^0$ (chloroform) (1946a Meystre); $[\alpha]_{5461}^{25}+129^0$ (alcohol) (1940 Hoehn).

Has no effect in the capon comb test (1938b Butenandt).

Occ. In the urine of pregnant sows (1939e Marker). For occurrence (as glucuronide) in the urine of non-pregnant women, see 1946 Marrian; cf. 1946, 1947 Sutherland.

Fmn. From 17-iso-pregnan-3α-ol-20-one (p. 2250 s) on partial isomerization by refluxing with methanolic NaOH or methanolic HCl (the equilibrium mixture contains about 71 % of the normal compound) (1944 Moffett). From Δ^{16}-pregnen-3α-ol-20-one (p. 2239 s) on hydrogenation in the presence of palladium-BaSO₄ catalyst in alcohol-ether (1940k Marker). From 3α-acetoxy-ternorcholanylamine (p. 1908 s) on converting with HOCl into its N-chloroderivative, eliminating HCl by alkali treatment, and hydrolysing the resulting imine with aqueous H_2SO_4 (1937b I.G. FARBENIND.). From 20-diphenyl-methylene-pregnan-3α-ol acetate (p. 1845 s) on ozonolysis, followed by hydrolysis (1934a SCHERING-KAHLBAUM A.-G.; 1940 Hoehn). From (crude) 24.24-diphenyl-$\Delta^{20(22)\cdot23}$-choladien-3α-ol acetate (p. 1849 s) (prepared from 24.24-diphenyl-Δ^{23}-cholen-3α-ol acetate) on oxidation with CrO_3 in chloroform-acetic acid at 20⁰, followed by saponification with boiling aqueous methanolic K_2CO_3; yield based on 24.24-diphenyl-Δ^{23}-cholen-3α-ol acetate, 53 % (1946a Meystre).

Rns.* On refluxing with methanolic NaOH or methanolic HCl partial isomerization to 17-iso-pregnan-3α-ol-20-one (p. 2250 s) takes place (1944 Moffett). Keeping of the acetate with persulphuric acid in acetic acid at 25⁰ for 8 days, followed by hydrolysis with boiling methanolic KOH, gives 5-iso-androstane-3α.17β-diol (p. 2017 s) and a small amount of 3α-hydroxyætiocholanic acid (p. 173) (1940f Marker). Gives pregnane-3α-20α-and-3α.20β-diols (pp. 1923 s and 1929 s, resp.) on reduction, the 20α-compound preponderating on using sodium and boiling alcohol (1938b Butenandt; 1946a Meystre), the 20β-compound preponderating on hydrogenation with a nickel catalyst and sodium ethoxide in alcohol or, even more, with a PtO₂ catalyst in glacial acetic acid (1946a Meystre). Refluxing with amalgamated zinc, conc. HCl, and acetic acid affords pregnan-3α-ol (p. 1521 s) (mostly as acetate) (1938b Marker); pregnan-3α-ol is also obtained on heating of the semicarbazone with sodium ethoxide in alcohol at 180⁰ (1939a Marker). On treatment with benzaldehyde and sodium ethoxide in abs. alcohol at 25⁰ it gives 21-benzylidene-pregnan-3α-ol-20-one (p. 2272 s) (1939c Marker). Heating with methylmagnesium iodide in ether affords 20-methylpregnane-3α-20-diol (p. 1938 s) (1938b Butenandt).

---

* See also "pregnan-3-ol-20-one" with unknown configuration at C-3 (p. 2246 s).

Oxime $C_{21}H_{35}O_2N$ (from alcohol), m. 224-6⁰ (1938b Butenandt).

2.4-*Dinitrophenylhydrazone* $C_{27}H_{38}O_5N_4$. Crystals (alcohol), m. 229⁰ (1938b Marker).

Acetate, **3α-*Acetoxy-pregnan-20-one*** $C_{23}H_{36}O_3$ (p. 115), m. 99⁰ (1937b Marker; 1938b Butenandt), 102-3⁰ cor. (1946a Meystre); 112-4⁰ (1940k Marker); $[\alpha]_D^{20} + 124⁰$ (abs. alcohol) (1938b Butenandt); $[\alpha]_D^{22} + 83.5⁰$ (chloroform) (1946a Meystre). — **Rns.** (See also above). Saponification by boiling with methanolic NaOH gives pregnan-3α-ol-20-one (above) and 17-iso-pregnan-3α-ol-20-one (p. 2250 s) (1944 Moffett).

β-D-*Glucuronide*, β-D-*Glucosiduronic acid* $C_{27}H_{42}O_8$. — *Sodium salt* $C_{27}H_{41}O_8Na$ (precipitated from the solution in 50% acetone with dry acetone). Not obtained quite free from inorganic material; m. 257-260⁰ cor. dec. *Isolation.* From human pregnancy urine, together with sodium pregnane-3α.20α-diol 3β-D-glucosiduronate (p. 1928 s, q. v.), from which it can be separated by means of Girard's reagent T (1946, 1947 Sutherland).

Pregnan-3β-ol-20-one $C_{21}H_{34}O_2$ (p. 115). $[\alpha]_D^{20} + 102⁰$ (abs. alcohol) (1938b Butenandt; but cf. 1946 Barton). — Has no effect in the capon comb test (1938b Butenandt).

     Fmn. From pregnane-3β.20β-diol (p. 1931 s) on partial acetylation by boiling with glacial acetic acid and acetic anhydride, followed by oxidation with CrO_3 in acetic acid and saponification with alcoholic KOH (1938 PARKE, DAVIS & Co.). As acetate, in small amounts, together with much allopregnan-3β-ol-20-one acetate, from the acetate of Δ^5-pregnen-3β-ol-20-one (p. 2233 s) on hydrogenation in the presence of platinum oxide in glacial acetic acid, followed by oxidation with CrO_3 in glacial acetic acid at room temperature (1946 Plattner). From Δ^{16}-pregnen-3β-ol-20-one (p. 2240 s) on hydrogenation in the presence of palladium (1942a Marker). As acetate, from Δ^{16}-pregnen-3β-ol-20-one acetate on hydrogenation in the presence of a platinum catalyst in alcohol, followed by oxidation with CrO_3 in acetic acid at room temperature (1940a Marker). As acetate, from that of 17-bromopregnan-3β-ol-20-one (p. 2285 s) on heating with zinc dust or powdered iron and glacial acetic acid on the steam-bath (yields, 80 and 73%, resp.) or on shaking with hydrogen, palladium-$BaSO_4$ catalyst, pyridine, methanol, and ether under 4 atm. pressure at room temperature (yield, 61%); the free hydroxy-ketone is formed in an analogous manner from 17-bromo-pregnan-3β-ol-20-one (1942a Marker). From 21-bromo-Δ^{16}-pregnen-3β-ol-20-one acetate (p. 2280 s) on heating with zinc dust and acetic acid on the steam-bath or on hydrogenation in the presence of palladium in methanol containing pyridine, followed in both cases by hydrolysis with boiling methanolic $KHCO_3$; yields, 44 and 72%, resp. As acetate, from 17.21-dibromo-pregnan-3β-ol-20-one acetate (p. 2288 s) on heating with zinc dust or powdered iron and glacial acetic acid on the steam-bath (yields, ca. 85%); the free hydroxy-ketone is formed in an analogous manner from 17.21-dibromo-pregnan-3β-ol-20-one (1942b Marker).

References, pp. 2252–2257 s

Rns.* Oxidation of the acetate with excess aqueous KOI in dioxane, first at room temperature and then at 80⁰, followed by alkaline hydrolysis on the steam-bath, gives 3β-hydroxyætiocholanic acid (1942g Marker). Reduction with sodium and boiling isopropyl alcohol affords a diol $C_{21}H_{36}O_2$, m. 189 — 190.5⁰, which may be pregnane-3β.20α-diol (p. 1930 s) or a C-17 epimer (1938b Butenandt; cf. 1946b Pearlman). Monobromination in ether or in glacial acetic acid containing some hydrogen bromide gives a mixture of 17-bromopregnan-3β-ol-20-one (p. 2285 s) and, probably, 21-bromopregnan-3β-ol-20-one (p. 2281 s); the acetate reacts similarly (1938 Masch; cf. 1942a Marker); dibromination of the pregnanolone (or its acetate) in glacial acetic acid containing some hydrogen bromide affords 17.21-dibromopregnan-3β-ol-20-one (or its acetate) (1942b Marker; 1941b PARKE, DAVIS & Co.; cf. 1938 Masch). On treatment with benzaldehyde and sodium ethoxide in abs. alcohol at room temperature it gives 21-benzylidenepregnan-3β-ol-20-one (p. 2272 s) (1939c Marker). On heating with methylmagnesium iodide in ether, 20-methylpregnane-3β.20-diol (p. 1938 s) is obtained (1938b Butenandt).

Digitonide (precipitated in 75 % alcohol), m. 199–208⁰ (1938b Butenandt).

Oxime $C_{21}H_{35}O_2N$ (from alcohol), m. 179⁰ (1938b Butenandt).

Acetate, 3β-Acetoxypregnan-20-one $C_{23}H_{36}O_3$ (p. 115), m. 116.5⁰ (1938b Butenandt); $[\alpha]_D^{20} + 86.5⁰$ (alcohol) (1938b Butenandt; but cf. 1946 Barton).

Pregnan-3-ol-20-one(s) $C_{21}H_{34}O_2$ **for which the configuration at C-3 cannot be derived from the available data.** — **Rns.** Heating with copper powder at 200⁰ affords pregnane-3.20-dione (p. 2796 s) (1936/37 CIBA). Gives pregnan-3-ol-20-on-21-al (p. 2272 s) on refluxing with SeO_2 in isoamyl alcohol or on warming of the acetate with sodium nitrite in strong acetic acid, followed by hydrolysis (1936 SCHERING A.-G.).

Allopregnan-3α-ol-20-one $C_{21}H_{34}O_2$ (p. 115). Is readily purified by crystallization from carbon tetrachloride (1939b Marker); $[\alpha]_D + 87.7⁰$ (abs. alcohol) (1938 Fleischer; but cf. 1946 Barton); $[\alpha]_D^{25} + 96.6⁰$ (alcohol) (1948 Lieberman); $[\alpha]_D^{16} + 96⁰$ (chloroform) (1947 Prelog). Shows no androgenic activity (1938a Butenandt).

Occ. In swine testes (1947 Prelog).

Fmn. From 17-iso-allopregnane-3α.17β.20-triol (? p. 2122 s) on heating with 30 % H_2SO_4 in propyl alcohol, with H_2SO_4 in dioxane, or with $KHSO_4$ (1938a CIBA). From Δ^{16}-allopregnen-3α-ol-20-one (p. 2241 s) on hydrogenation in the presence of palladium-$BaSO_4$ in alcohol-ether at room temperature (yield, 70 %); as acetate, from Δ^{16}-allopregnen-3α-ol-20-one acetate on hydrogenation in the presence of platinum oxide in alcohol, followed by oxidation with CrO_3 in acetic acid at room temperature (1940g Marker). From 21-diazo-allopregnan-3α-ol-20-one (p. 2295 s) on hydrogenation in alcohol in the presence

* See also "pregnan-3-ol-20-one" with unknown configuration at C-3 (above).

of palladium (1937 h CIBA). From allopregnane-3.20-dione (p. 2797 s) on partial hydrogenation in the presence of platinum and hydrobromic acid in acetic acid, followed by hydrolysis with boiling methanolic KOH (together with some allopregnan-3β-ol-20-one) (crude yield, ca. 44%) (1938 Fleischer; see also 1939b Marker), or in the presence of Raney nickel in abs. alcohol (together with much allopregnan-3β-ol-20-one) (yield, 5—10%) (1938a Butenandt); the isomers are separated by precipitating allopregnan-3β-ol-20-one with digitonin (1938 Fleischer; 1938a Butenandt).

From 3β-chloroternorallocholanyl-diphenyl-carbinol (p. 1491 s) on boiling with glacial acetic acid-acetic anhydride, ozonization of the resulting unsaturated hydrocarbon in chloroform at 0⁰, boiling the ozonolysis product with potassium acetate in valeric acid, and hydrolysis with boiling alcoholic KOH (1937a MARKER; see also 1946 Shoppee). The acetate is obtained from the methyl ester of 3α-acetoxybisnorallocholanic acid (p. 174) on heating with a phenylmagnesium bromide solution, followed by distillation in a high vac. and ozonization of the resulting unsaturated compound in chloroform (1936 I. G. FARBENIND.; 1936 WINTHROP CHEMICAL Co.). From epi-cholestanyl acetate (p. 1703 s) on oxidation with CrO_3, followed by saponification (1937a CIBA).

Rns*. Oxidation of the acetate with persulphuric acid in glacial acetic acid at 25⁰, followed by hydrolysis with alcoholic KOH, gives androstane-3α.17β-diol (p. 2011 s) (1940g Marker). Gives 3α-hydroxyætioallocholanic acid (p. 173) on keeping with benzaldehyde and sodium ethoxide in abs. alcohol, followed by acetylation with boiling acetic anhydride, subsequent oxidation with CrO_3 in acetic acid at 100⁰, and hydrolysis with boiling 5% KOH (1942f Marker). On grignardization followed by dehydration, acetylation, and ozonization, it gives the acetate of androsterone (p. 2629 s) (1934b SCHERING-KAHLBAUM A.-G.; 1936 WINTHROP CHEMICAL Co.; 1936 I. G. FARBENIND.). For absorption spectrum in the Zimmermann colour test with m-dinitrobenzene and KOH, see 1938 Callow.

Acetate, 3α-Acetoxyallopregnan-20-one $C_{23}H_{36}O_3$ (p. 115). $[\alpha]_D^{22} + 94.5^0$ (abs. alcohol) (1938 Fleischer; but cf. 1946 Barton). — **Fmn.** and **Rns.** See above.

Allopregnan-3β-ol-20-one $C_{21}H_{34}O_2$ (p. 115). $[\alpha]_D^{22} + 93^0$ (chloroform) (1946a Meystre). — **Occ.** In human labour and delivery urine (1942 Pearlman). In the urine of pregnant mares (1938a Marker; 1939 Heard; 1941 Oppenauer) and pregnant sows (1939e Marker). In swine testes (1943 Ruzicka; 1947 Prelog). In ox adrenals (1938a, b Beall; 1941 v. Euw).

Fmn. From allopregnane-3β.17α.20β-triol (substance J, p. 2119 s) on refluxing with aqueous methanolic H_2SO_4 (yield, 10%) (1938 Steiger); from the same(?) triol on heating with propyl alcohol and H_2SO_4 (1938a CIBA) or with zinc dust at 150–170⁰ under 10⁻⁴ mm pressure (1940b SCHERING A.-G.). As acetate, together with a small amount of pregnan-3β-ol-20-one

---

* See also "allopregnan-3-ol-20-one" with unknown configuration at C-3 (p. 2249 s).

E.O.C. XIV s. 158

acetate, from the acetate of Δ^5-pregnen-3β-ol-20-one (p. 2233 s) on hydrogenation in the presence of platinum oxide in glacial acetic acid, followed by oxidation with CrO_3 in glacial acetic acid at room temperature; yield, 85 % (1946 Plattner). From Δ^{16}-allopregnen-3β-ol-20-one (p. 2242 s) on hydrogenation in the presence of palladium-$BaSO_4$ in alcohol (1940h Marker); as acetate, from Δ^{16}-allopregnen-3β-ol-20-one acetate on heating with zinc dust in acetic acid on the steam-bath (yield, 87 %) (1942h Marker). From 21-chloro-allopregnan-3β-ol-20-one (p. 2282 s) on treatment with zinc or alkali iodide in glacial acetic acid (1937f CIBA). From 17-bromoallopregnan-3β-ol-20-one (p. 2286 s) on heating with powdered iron in glacial acetic acid on the steam-bath (yield, 75 %) or on shaking with hydrogen, palladium-$BaSO_4$, pyridine, and dioxane under 2.7 atm. pressure (yield, 65 %); the acetate is also obtained by the latter method from 17-bromoallopregnan-3β-ol-20-one acetate (yield, 60 %) (1942h Marker). As acetate, from 17.21-dibromoallopregnan-3β-ol-20-one acetate (p. 2288 s) on heating with zinc dust and glacial acetic acid on the water-bath (1946 Plattner). From Δ^1-allopregnene-3.20-dione (p. 2781 s) on hydrogenation in the presence of a partially inactivated platinum catalyst in alcohol at 3 atm. pressure (1939d Marker). From allopregnane-3.20-dione (p. 2797 s) on partial hydrogenation in the presence of platinum in acetic acid (yield: 2 g. → 1.4 g. digitonide) (1939b Marker), or in the presence of platinum and hydrobromic acid in acetic acid, followed by hydrolysis with boiling methanolic KOH (together with much allopregnan-3α-ol-20-one) (yield: 6.75 g. → 7.8 g. digitonide → 1.5 g. crude compound) (1938 Fleischer; see also 1939b Marker), or in the presence of Raney nickel in abs. alcohol (together with 5–10 % allopregnan-3α-ol-20-one) (yield, 73 %) (1938a Butenandt).

From 20-methyl-Δ^{20}-allopregnen-3β-yl acetate (p. 1525 s) on ozonization in chloroform at —10°, followed by boiling with methanolic KOH (1944 Koechlin). From (crude 24.24-diphenyl-$\Delta^{20(22).23}$-allocholadien-3β-yl acetate (p. 1849 s) (prepared from 24.24-diphenyl-Δ^{23}-allocholen-3β-yl acetate) on oxidation with CrO_3 in chloroform-acetic acid at 20°; was isolated partly as acetate and partly as free hydroxy-ketone (total yield based on 24.24-diphenyl-Δ^{23}-allocholen-3β-yl acetate, 60.6 %) (1946a Meystre). From pseudotigogenin acetate on oxidation with CrO_3 in acetic acid at 28°, followed by hydrogenation under 3 atm. pressure in acetic acid in the presence of Adams' catalyst, subsequent oxidation with CrO_3 in acetic acid, and hydrolysis with boiling alcoholic KOH (1940c Marker). Regarding the formation, in small amount, from a 17α-methyl-17-amino-D-homoandrostane-3β.17aβ-diol by the action of HNO_2, see 1946 Prins; 1952 Klyne; cf. also 1953 Cremlyn.

Rns.* Oxidation with CrO_3 in acetic acid at room temperature gives allopregnane-3.20-dione (p. 2797 s) and ca. 15 % of 20-ketoallopregnane-2‖3-dioic acid (see p. 2799 s) (1943 Plattner). On heating of the acetate with lead tetraacetate in glacial acetic acid in the presence of some acetic anhydride at 68—70° it gives allopregnane-3β.21-diol-20-one diacetate (p. 2312 s) and a small amount of allopregnane-3β.17α.21-triol-20-one triacetate (p. 2359 s) (1939 Reichstein). Keeping of the acetate with persulphuric acid in glacial

---

* See also "allopregnan-3-ol-20-one" with unknown configuration at C-3 (p. 2249 s).

acetic acid at 25⁰ for 10 days yields allopregnane-3β.21-diol-20-one diacetate
(p. 2312 s) and, after hydrolysis with alcoholic KOH, 3β-hydroxyætioallo-
cholanic acid (p. 173) and androstane-3β.17β-diol (p. 2014 s) (1940 h Marker;
see also 1940 b Marker; 1944 Koechlin); the last-mentioned compound is also
formed in small amount on boiling of the acetate with perbenzoic acid in
chloroform, followed by saponification with boiling methanolic KOH (1942
Burckhardt).

Reduction with sodium and boiling alcohol gives allopregnane-3β.20α-diol
(p. 1933 s) (1946 a Meystre). On treatment of the hydroxy-ketone (or its
acetate) with 1 mol. of bromine in acetic acid at room temperature it gives
17-bromoallopregnan-3β-ol-20-one (or its acetate) (p. 2286 s) (1942 h Marker;
1941 b PARKE, DAVIS & Co.); reacts similarly with 1 mol. of chlorine in acetic
acid-chloroform (1941 b PARKE, DAVIS & Co.); on treatment of the acetate
with 2 mols. of bromine in glacial acetic acid in the presence of HBr, 17.21-di-
bromoallopregnan-3β-ol-20-one acetate (p. 2288 s) is obtained (1942 h Marker;
1944 Koechlin; 1946 Plattner; see also 1948 Plattner).

The hydroxy-ketone (1939 c Marker) or its acetate (1944 Koechlin) react
with benzaldehyde and sodium ethoxide in abs. alcohol at 25⁰ to give 21-ben-
zylideneallopregnan-3β-ol-20-one (p. 2273 s); by treatment of the crude condensation product with acetic anhydride and pyridine at room temperature,
two (probably cis-trans stereoisomeric) 21-benzylideneallopregnan-3β-ol-
20-one acetates could be isolated (1944 Koechlin).

Gives no colour with alcoholic m-dinitrobenzene and KOH (Zimmermann
test) (1939 b Marker); for absorption spectrum in this test, see 1942 Pearlman.

Allopregnan-3β-ol-20-one acetate, 3β-Acetoxyallopregnan-20-one $C_{23}H_{36}O_3$
(p. 116). Leaflets (methanol), m. 150–2⁰ cor. (1946 a Meystre); $[\alpha]_D^{22}+80^0$
(abs. alcohol) (1938 Fleischer), +74⁰ (chloroform)* (1946 a Meystre). — **Fmn.**
and **Rns.** See above.

Allopregnan-3β-ol-20-one benzoate, 3β-Benzoyloxy-allopregnan-20-one
$C_{28}H_{38}O_3$. Crystals (acetone), m. 197–8⁰ (1948 Barton), 202⁰ cor. (1941
Oppenauer); $[\alpha]_D + 76^0$ (chloroform) (1948 Barton). — **Fmn.** From the
above hydroxy-ketone on treatment with benzoyl chloride and pyridine
(1948 Barton).

Allopregnan-3-ol-20-one $C_{21}H_{34}O_2$ for which the configuration at C-3 cannot be derived from the available data.

— **Rns.** Heating with SeO_2 in alcohol
on the water-bath gives allopregnan-3-ol-20-on-21-al (p. 2381 s) (1936
SCHERING A.-G.).

17-iso-Δ⁵-Pregnen-3β-ol-20-one $C_{21}H_{32}O_2$ (p. 115).

— **Fmn.** Together with
Δ⁵-pregnen-3β-ol-20-one (p. 2233 s), from Δ⁵·²⁰·
pregnadiene-3β.20-diol 20-acetate (p. 1918 s) on
saponification with 0.1 N methanolic KOH
(1939 c CIBA). As acetate, from Δ⁵-pregnene-
3β.17α.20β-triol 3.20-diacetate (p. 2116 s) on

---

* The value +92.8⁰ (*Helv.* **29** 470) is incorrect (1956 Plattner).

158*

heating with zinc dust at 120° and 0.01 mm. pressure (yield, 64%) (1939b Butenandt; cf. 1949 Fieser; cf. also 1939b SCHERING A.-G.; 1940a CIBA). — **Rns.** Oxidation with CrO_3 in glacial acetic acid, after temporary protection of the double bond with bromine, gives progesterone (p. 2782 s) (1937a SCHERING A.-G.; 1939b Butenandt). Gives 17-iso-progesterone (p. 2800 s) on oxidation with aluminium isopropoxide and cyclohexanone in boiling toluene (1939b Butenandt) or on treatment with the Corynebacterium mediolanum (isolated from yeast) (1942 Butenandt).

Acetate, 3β-Acetoxy-17-iso-Δ⁵-pregnen-20-one $C_{23}H_{34}O_3$ (p. 115). $[\alpha]_D^{20}$—126° (alcohol). — **Fmn.** From the above hydroxy-ketone on heating with acetic anhydride (1937a Butenandt). See also above.

17-iso-Pregnan-3α-ol-20-one $C_{21}H_{34}O_2$.

Crystals (ether), m. 142.5–144° cor.; $[\alpha]_D^{32}$—41° (dioxane). Is not precipitated by digitonin. — **Fmn.** From pregnan-3α-ol-20-one (p. 2244 s) on partial isomerization by refluxing with methanolic NaOH or methanolic HCl (the equilibrium mixture contains about 29% of the 17-iso compound). — **Rns.** On refluxing with methanolic NaOH or methanolic HCl partial isomerization to pregnan-3α-ol-20-one takes place (1944 Moffett). — *Oxime* $C_{21}H_{35}O_2N$. White crystals (methanol), m. 192–7° cor. dec. (1944 Moffett).

Acetate, 3α-Acetoxy-17-iso-pregnan-20-one $C_{23}H_{36}O_3$. Needles (methanol), m. 157–9° cor.; $[\alpha]_D^{34}$—28° (methanol). — **Fmn.** From the above hydroxy-ketone on refluxing with acetic anhydride and acetic acid (1944 Moffett).

17-iso-Allopregnan-3β-ol-20-one $C_{21}H_{34}O_2$ (p. 116).

The compound described on p. 116 was not pure but contaminated with allopregnan-3β-ol-20-one; the pure 17-iso compound forms prisms (ether-pentane), m. 139° (Kofler block), shows $[\alpha]_D^{22}$—77.7° (alcohol), and is not precipitated by digitonin (1948, 1949 Shoppee). — **Fmn.** As acetate from the 3.20-di-acetate of allopregnane-3β.17α.20α-triol (p. 2118 s) on refluxing with zinc dust in toluene for some days; yield, 46% allowing for recovered material (1948, 1949 Shoppee).

Acetate $C_{23}H_{36}O_3$ (cf. p. 116). Prisms (methanol), m. 119–122° (Kofler block); $[\alpha]_D^{21}$—75° (alcohol) (1948, 1949 Shoppee).

Allopregnan-5-ol-20-one $C_{21}H_{34}O_2$.

— **Fmn.** Along with other compounds, from cholestan-5-ol (p. 1723 s) on oxidation with CrO_3. — **Rns.** Gives a mixture of the two epimeric allopregnane-5.20-diols (p. 1937 s) and other compounds on reduction, e. g. by hydrogenation in alcohol in the presence of a nickel catalyst until 1 mol. hydrogen has been absorbed (1940b, 1941a CIBA).

i-Pregnen-6β-ol-20-one $C_{21}H_{32}O_2$. For configuration at C-3, C-5 and C-6, see i-cholesterol (p. 1723 s).

Methyl ether, 6β-Methoxy-i-pregnen-20-one $C_{22}H_{34}O_2$. Crystals (ether), m. 124–5⁰; [α]$_D^{20}$ + 132⁰ (chloroform) (1937b Butenandt). — **Fmn.** From $Δ^5$-pregnen-3β-ol-20-one p-toluenesulphonate (p. 2238 s) on heating with methanol and potassium acetate on the water-bath (1937b Butenandt). From 20-diphenylmethylene-6β-methoxy-i-pregnene (p. 1846 s) on ozonolysis in carbon tetrachloride at room temperature; was not isolated in this procedure but rearranged to $Δ^5$-pregnen-3β-ol-20-one methyl ether by refluxing with methanol containing conc. H_2SO_4 (1946 Riegel). — **Rns.** Refluxing with zinc acetate, glacial acetic acid, and acetic anhydride gives $Δ^5$-pregnen-3β-ol-20-one acetate (1946 Riegel). On keeping with conc. HCl in glacial acetic acid at room temperature it gives 3β-chloro-$Δ^5$-pregnen-20-one (p. 2222 s) (1937b Butenandt).

$Δ^3$-Pregnen-12α-ol-20-one $C_{21}H_{32}O_2$. For configuration at C-12, see pregnane-3α.12α-diol-20-one (p. 2329 s).

Anthraquinone - β - carboxylate $C_{36}H_{38}O_5$. Yellow prisms (acetone-ether), m. 190–2⁰ cor., assuming a dark violet colour in daylight, becoming again yellow on heating. — **Fmn.** Together with $Δ^{11}$-pregnen-3α-ol-20-one anthraquinone-β-carboxylate (p. 2239 s) and $Δ^{3.11}$-pregnadien-20-one (p. 2215 s), from pregnane-3α.12α-diol-20-one di-(anthraquinone-β-carboxylate) (p. 2331 s) by thermal cleavage at 290–320⁰ and 0.02 mm. pressure (1944 v. Euw).

$Δ^{5.17(20)}$-Pregnadien -3β- ol -21- al, 3β - Hydroxy -$Δ^{5.17(20)}$- pregnadien -21- al $C_{21}H_{30}O_2$. Needles (ether), m. ca. 175⁰ cor. (1941 Fuchs); crystals (ether-hexane), m. 180–1⁰ cor.; [α]$_D^{19}$ —65⁰ (chloroform); absorption max. in alcohol at 244 mμ (log ε 4.42) (1950 Heusser).

Fmn. As acetate, from the N-(p-dimethyl-aminophenyl)-isoxime of its acetate (p. 2252 s) on decomposition with aqueous HCl in ether, followed by keeping with acetic anhydride and pyridine at room temperature (1940 Reich). As acetate, from 17α-allyl-$Δ^5$-androstene-3β.17β-diol 3-acetate (p. 2041 s) on ozonization in glacial acetic acid after temporary protection of the nuclear double bond by bromine. As acetate, from the same starting compound via the C-21 stereoisomeric 21-hydroxymethyl-17-iso-$Δ^5$-pregnene-3β.17β.21-triols (page 2147 s) and 17-iso-$Δ^5$-pregnene-3β.17β-diol-21-al 3-acetate (p. 2345 s) on boiling the latter with glacial acetic acid under nitrogen for 4 hrs. (1939b Miescher; 1939a CIBA PHARMACEUTICAL PRODUCTS). From the two C-21 stereoisomeric ω-homo-$Δ^{5.17(20)}$-pregnadiene-3β.21.22-triols (p. 1954 s) on keeping with HIO_4 in aqueous dioxane at room temperature (1941 Fuchs).

Rns. Acetalization of the acetate with ethyl orthoformate and $NH_4 \cdot NO_3$ in methanol, followed by alkaline saponification of the ester group and oxidation with a ketone in the presence of a metal alkoxide, yields the acetal of $\Delta^{4.17(20)}$-pregnadien-3-on-21-al (p. 2802 s)(1939d, e CIBA). Treatment of the acetate with diazomethane in methylene chloride gives the acetate of 21-methyl-$\Delta^{5.17(20)}$-pregnadien-3β-ol-21-one (p. 2258 s) (1938g CIBA).

Acetate, 3β-Acetoxy-$\Delta^{5.17(20)}$-pregnadien-21-al $C_{23}H_{32}O_3$. Needles (dil. acetone), m. 185–7° cor. (1939b Miescher); $[\alpha]_D^{19}$ —60° (chloroform); absorption max. in alcohol at 244 mμ (log ε 4.44) (1950 Heusser). Reduces ammoniacal silver solution. Gives a red colour on heating with 1.4-dihydroxynaphthalene in glacial acetic acid + conc. HCl (1939b Miescher). — **Fmn.** From the above hydroxy-ketone on keeping with acetic anhydride and pyridine at room temperature (1941 Fuchs). See also above. — **Rns.** See above. — *Semicarbazone* $C_{24}H_{35}O_3N_3$, crystals (methanol), m. 245–6° cor. (1939b Miescher).

3β-Acetoxy-$\Delta^{5.17(20)}$-pregnadien-21-al N-(p-dimethylaminophenyl)-isoxime $C_{31}H_{42}O_3N_2 = $ AcO·$C_{19}H_{27}$:CH·CH:N(O)·C_6H_4·N(CH$_3$)$_2$. Dark yellow needles (alcohol-water), m. ca. 170°. — **Fmn.** From N-(3β-acetoxy-$\Delta^{5.17(20)}$-pregnadien-21-yl)-pyridinium bromide (p. 1869 s) and p-nitrosodimethylaniline in aqueous alcoholic NaOH at room temperature, together with crystals, m. 133–5°, which are probably deacetylated isoxime. — **Rns.** Treatment with aqueous HCl in ether, followed by acetylation gives the preceding acetate (1940 Reich).

$\Delta^{17(20)}$-Pregnen-3-ol-21-al, 3-Hydroxy-$\Delta^{17(20)}$-pregnen-21-al $C_{21}H_{32}O_2$ (no formation). Configuration at C-3 not stated. — **Rns.** For transformation of the acetate to a pregnane-3.17.20-triol-21-al (p. 2362 s) by means of OsO$_4$, see 1939e CIBA.

Allopregnan-3β-ol-21-al $C_{21}H_{34}O_2$.

Acetate $C_{23}H_{36}O_3$. Colourless crystals. — **Fmn.** From 3β-acetoxyallopregnan-21-oic acid chloride on hydrogenation with Pd-BaSO$_4$ in xylene at 150°. — **Rns.** Treatment with bromine and some HBr in glacial acetic acid and then with potassium acetate in abs. alcohol gives the diacetate of allopregnane-3β.20-diol-21-al (p. 2313 s) (1939a, b CIBA PHARMACEUTICAL PRODUCTS).

1934 Butenandt, Westphal, *Ber.* **67** 2085, 2087.
Fernholz, *Ber.* **67** 2027, 2029.
a SCHERING A.-G. (Butenandt), *Ger. Pat.* 695774 (issued 1940); *C. A.* **1941** 5649; *Ch. Ztbl.* **1940** II 2924; cf. also SCHERING-KAHLBAUM A.-G., *Swiss Pat.* 194942 (1935, issued 1938); *Ch. Ztbl.* **1938** II 1995, 1996.
b SCHERING A.-G. (Schoeller, Serini, Hildebrandt, Kathol), *Ger. Pat.* 698910 (issued 1940); *C. A.* **1941** 6603; *Ch. Ztbl.* **1941** I 1444; SCHERING-KAHLBAUM A.-G., *Brit. Pat.* 449379 (issued 1936); *C. A.* **1936** 8533; *Ch. Ztbl.* **1937** I 3370; *Fr. Pat.* 806468 (issued 1936); *C. A.* **1937** 4677.
c SCHERING A.-G. (Butenandt), *Ger. Pat.* 699309 (issued 1940); *C. A.* **1941** 7121; *Ch. Ztbl.* **1941** I 1196.

a SCHERING-KAHLBAUM A.-G., *Dutch Pat.* 39469 (issued 1936); *Ch. Ztbl.* **1937** I 3520; SCHERING A.-G. (Butenandt, Hildebrandt), *U.S. Pat.* 2156275 (issued 1939); *C. A.* **1939** 5869; *Ger. Pat.* 681868 (issued 1939); *C. A.* **1942** 2379.

b SCHERING-KAHLBAUM A.-G., *Austrian Pat.* 160242 (issued 1941); *Ch. Ztbl.* **1941** II 2708.

1935 a SCHERING-KAHLBAUM A.-G., *Brit. Pat.* 464397 (issued 1937); *C. A.* **1937** 6254; *Ch. Ztbl.* **1938** I 375; *Australian Pat.* 3962/1936 (issued 1937); *Ch. Ztbl.* **1937** I 4991, 4992.

b SCHERING-KAHLBAUM A.-G., *Swiss Pat.* 188987 (issued 1937); *Ch. Ztbl.* **1937** II 3041.

1936 CIBA*, *Swiss Pat.* 214331 (issued 1941); *C. A.* **1942** 5960; *Brit. Pat.* 477400 (issued 1938); *C. A.* **1938** 3914; *Ch. Ztbl.* **1938** II 119, 120.

I. G. FARBENIND., *Brit. Pat.* 478583 (issued 1938); *C. A.* **1938** 5161; *Ch. Ztbl.* **1938** II 1086, 1087.

SCHERING A.-G. (Butenandt), *Ger. Pat.* 709617 (issued 1941); *C. A.* **1943** 3454; *Ch. Ztbl.* **1942** I 642.

SCHERING-KAHLBAUM A.-G., *Brit. Pat.* 493128 (issued 1938); *Ch. Ztbl.* **1939** I 2829.

WINTHROP CHEMICAL Co. (Bockmühl, Ehrhart, Ruschig), *U.S. Pat.* 2188330 (issued 1940); *C. A.* **1940** 3884; *Ch. Ztbl.* **1940** II 1053.

1936/37 CIBA*, *Brit. Pat.* 476749 (issued 1938); *C. A.* **1938** 3770; *Ch. Ztbl.* **1938** II 120; *Fr. Pat.* 823139 (1937, issued 1938); *C. A.* **1938** 5585; *Ch. Ztbl.* **1938** II 120.

1937 Bugyi, *C. A.* **1938** 2191; *Ch. Ztbl.* **1938** I 1144.

a Butenandt, Fleischer, *Ber.* **70** 96, 98, 100, 101.

b Butenandt, Grosse, *Ber.* **70** 1446, 1448, 1449.

a CIBA*, *Swiss Pat.* 199449 (issued 1938); *C. A.* **1939** 3393; *Ch. Ztbl.* **1939** I 3930.

b CIBA*, *Swiss Pat.* 207497 (issued 1940); *C. A.* **1941** 3038; *Fr. Pat.* 841058 (1938, issued 1939); *C. A.* **1940** 1685; *Ch. Ztbl.* **1940** I 916; CIBA PHARMACEUTICAL PRODUCTS (Reichstein), *U.S. Pat.* 2337271 (issued 1943); *C. A.* **1944** 3424.

c CIBA*. *Swiss Pat.* 212191 (issued 1941); *C. A.* **1942** 3638; *Ch. Ztbl.* **1942** I 231; *Swiss Pat.* 215139 (issued 1941); *C. A.* **1948** 3144; *Swiss Pat.* 222121 (1937, issued 1942); *C. A.* **1949** 365; *Ch. Ztbl.* **1943** I 2113; *Swiss Pat.* 222689 (1937, issued 1942); *C. A.* **1949** 824; *Ch. Ztbl.* **1943** I 2113; *Brit. Pat.* 512954 (issued 1939); *C. A.* **1941** 1800.

d CIBA*, *Swiss Pat.* 212193 (issued 1941); *C. A.* **1942** 3636; *Swiss Pat.* 220924 (1937, issued 1942); *C. A.* **1949** 2377; *Ch. Ztbl.* **1943** I 1190; *Swiss Pat.* 225102 (issued 1942); *C. A.* **1949** 1916; *Fr. Pat.* 845802 (1938, issued 1939); *C. A.* **1941** 1186; *Ch. Ztbl.* **1940** I 2827; (Miescher, Wettstein, Fischer), *U.S. Pat.* 2188914 (issued 1940); *C. A.* **1940** 3882.

e CIBA*, *Swiss Pat.* 222123 (issued 1942); *C. A.* **1949** 366; *Ch. Ztbl.* **1943** I 2113.

f CIBA*, *Swiss Pat.* 222690 (issued 1942); *C. A.* **1949** 824; *Ch. Ztbl.* **1943** I 2113.

g CIBA*, *Swiss Pat.* 223205 (issued 1942); *C. A.* **1949** 1916; *Ch. Ztbl.* **1943** I 2321.

h CIBA*, *Brit. Pat.* 512954 (issued 1939); *C. A.* **1941** 1800; *Fr. Pat.* 835814 (1938, issued 1939); *C. A.* **1939** 5008; *Ch. Ztbl.* **1939** I 2827, 2828.

i CIBA*, *Swiss Pat.* 215285 (issued 1941); *C. A.* **1948** 3538; *Ch. Ztbl.* **1942** I 2296.

a I. G. FARBENIND., *Brit. Pat.* 501421 (issued 1939); *C. A.* **1939** 6340; *Ch. Ztbl.* **1939** II 170; *Fr. Pat.* 847487 (1938, issued 1939); *C. A.* **1941** 5654; *Ch. Ztbl.* **1940** I 2828.

b I. G. FARBENIND. (Bockmühl, Ehrhart, Ruschig, Aumüller), *Ger. Pat.* 693351 (issued 1940); *C. A.* **1941** 4921; *Brit. Pat.* 502666 (issued 1939); *C. A.* **1939** 7316; *Ch. Ztbl.* **1939** II 1126; *Fr. Pat.* 842026 (1938, issued 1939); *C. A.* **1940** 4743; *Ch. Ztbl.* **1940** I 429; *Swiss Pat.* 211731 (1938, issued 1941).

c I. G. FARBENIND., *Fr. Pat.* 820537 (issued 1937); *C. A.* **1938** 2956; *Ch. Ztbl.* **1938** I 2404.

a Marker, Kamm, Jones, Wittle, Oakwood, Crooks, *J. Am. Ch. Soc.* **59** 768; Marker, Kamm, McGinty, Jones, Wittle, Oakwood, Crooks, *J. Am. Ch. Soc.* **59** 1367.

b Marker, Kamm, *J. Am. Ch. Soc.* **59** 1373, 1374.

N. V. ORGANON, *Fr. Pat.* 823618 (issued 1938); *C. A.* **1938** 5854; *Ch. Ztbl.* **1938** II 355; *Brit. Pat.* 488987 (1937, issued 1938); *C. A.* **1939** 324; *Ch. Ztbl.* **1939** II 474; *Dutch Pat.* 50942 (1938, issued 1941); *Ch. Ztbl.* **1942** I 513; *Swiss Pat.* 223018 (1937, issued 1942); *Ch. Ztbl.* **1943** I 1798.

Ruzicka, Fischer, *Helv.* **20** 1291, 1296.

a SCHERING A.-G., *Brit. Pat.* 508305 (issued 1939); *C. A.* **1940** 59; *Ch. Ztbl.* **1940** I 1232.

---

* Société pour l'industrie chimique à Bâle; Gesellschaft für chemische Industrie in Basel.

b SCHERING A.-G. (Logemann), *Ger. Pat.* 737023 (issued 1943); *C. A.* **1944** 3781; *Fr. Pat.* 843082 (1938, issued 1939); *C. A.* **1940** 6412; *Ch. Ztbl.* **1940** I 759; SCHERING CORP. (Logemann), *U.S. Pat.* 2321690 (1938, issued 1943); *C. A.* **1943** 6824; *Ch. Ztbl.* **1945** II 687.

SCHERING CORP. (Schwenk, Whitman, Fleischer), *U.S. Pat.* 2212104 (issued 1940); *C. A.* **1941** 468; *Ch. Ztbl.* **1943** I 653.

SCHERING-KAHLBAUM A.-G., *Fr. Pat.* 822551 (issued 1938); *C. A.* **1938** 4174; *Ch. Ztbl.* **1938** II 120; SCHERING CORP. (Serini, Köster, Strassberger), *U.S. Pat.* 2379832 (issued 1945); *C. A.* **1945** 5053.

Westphal, *Ber.* **70** 2128, 2134, 2135.

Yarnall, Wallis, *J. Am. Ch. Soc.* **59** 951.

1938 a Beall, Reichstein, *Nature* **142** 479.

b Beall, *Biochem. J.* **32** 1957.

a Butenandt, Heusner, *Z. physiol. Ch.* **256** 236, 239 et seq.

b Butenandt, Müller, *Ber.* **71** 191, 194 et seq.

c Butenandt, Schmidt-Thomé, *Naturwiss.* **26** 253.

Callow, Callow, Emmens, *Biochem. J.* **32** 1312, 1323.

a CIBA*, *Fr. Pat.* 838916 (issued 1939); *C. A.* **1939** 6875; *Ch. Ztbl.* **1939** II 169.

b CIBA*, *Fr. Pat.* 840513 (issued 1939); *C. A.* **1939** 8627; *Ch. Ztbl.* **1939** II 1125.

c CIBA*, *Fr. Pat.* 840515 (issued 1939); *C. A.* **1939** 8627; *Ch. Ztbl.* **1939** II 1721.

d CIBA*, *Fr. Pat.* 840783 (issued 1939); *C. A.* **1939** 8627; *Ch. Ztbl.* **1940** I 915; *Swiss Pat.* 211652 (issued 1941); *C. A.* **1942** 3637; *Swiss Pat.* 215288 (issued 1941); *C. A.* **1948** 3538; CIBA PHARMACEUTICAL PRODUCTS (Miescher, Wettstein), *U.S. Pat.* 2350792 (issued 1944); *C. A.* **1944** 5048.

e CIBA* (Miescher, Fischer), *U.S. Pat.* 2172590 (issued 1939); *C. A.* **1940** 450; *Ch. Ztbl.* **1940** I 428; *Fr. Pat.* 841242 (issued 1939); *C. A.* **1940** 4240; *Ch. Ztbl.* **1940** I 916.

f CIBA* (Miescher, Fischer), *Ger. Pat.* 712857 (issued 1941); *C. A.* **1943** 4534; *Ch. Ztbl.* **1942** I 1282.

g CIBA* (Miescher, Wettstein), *Ger. Pat.* 739083 (issued 1943); *C. A.* **1945** 392; *Ch. Ztbl.* **1944** I 110.

h CIBA* (Oppenauer, Miescher, Kägi), *Swed. Pat.* 96375 (issued 1939); *Ch. Ztbl.* **1940** I 1231.

Fleischer, Whitman, Schwenk, *J. Am. Ch. Soc.* **60** 79.

a I.G. FARBENIND., *Brit. Pat.* 491798; *C. A.* **1939** 1106.

b I.G. FARBENIND., *Swiss Pat.* 225070 (issued 1943); *Ch. Ztbl.* **1944** II 675.

Mamoli, *Ber.* **71** 2701.

a Marker, Lawson, Wittle, Crooks, *J. Am. Ch. Soc.* **60** 1559.

b Marker, Lawson, *J. Am. Ch. Soc.* **60** 2438.

Masch, *Über Bromierungen in der Pregnanreihe*, Thesis, Danzig, pp. 12, 17, 18, 21, 23, 28.

N. V. ORGANON, *Fr. Pat.* 840994 (issued 1939); *C. A.* **1940** 1821; *Ch. Ztbl.* **1940** I 249; ROCHE-ORGANON (Reichstein), *U.S. Pat.* 2296572 (issued 1942); *C. A.* **1943** 1230.

PARKE, DAVIS & CO. (Marker), *Brit. Pat.* 516845 (issued 1940); *Ch. Ztbl.* **1940** II 2648.

a SCHERING A.-G., *Dutch Pat.* 50747 (issued 1941); *Ch. Ztbl.* **1941** II 2708.

b SCHERING A.-G., *Fr. Pat.* 834941 (issued 1938); *C. A.* **1939** 4602; *Ch. Ztbl.* **1939** I 2827.

c SCHERING A.-G., *Fr. Pat.* 845473 (issued 1939); *C. A.* **1941** 1185; *Ch. Ztbl.* **1943** I 653; SCHERING CORP. (Inhoffen, Logemann, Dannenbaum), *U.S. Pat.* 2280236 (issued 1942); *C. A.* **1942** 5322.

SCHERING CORP. (Whitman, Schwenk), *U.S. Pat.* 2221826 (issued 1940); *C. A.* **1941** 1412; *Ch. Ztbl.* **1941** I 3114; *Brit. Pat.* 544186 (issued 1942); *C. A.* **1942** 6315.

Steiger, Reichstein, *Helv.* **21** 546, 555.

1939 a Butenandt, Schmidt-Thomé, *Ber.* **72** 182.

b Butenandt, Schmidt-Thomé, Paul, *Ber.* **72** 1112, 1114, 1117, 1118.

c Butenandt, Heusner, *Ber.* **72** 1119.

d Butenandt, Mamoli, Heusner, *Ber.* **72** 1614.

a CIBA*, *Fr. Pat.* 851017 (issued 1940); *C. A.* **1942** 2090; *Ch. Ztbl.* **1940** II 932.

b CIBA*, *Fr. Pat.* 860278 (issued 1941); *Ch. Ztbl.* **1941** I 3258.

* Société pour l'industrie chimique à Bâle; Gesellschaft für chemische Industrie in Basel.

c CIBA*, *Brit. Pat.* 532262 (issued 1941); *C. A.* **1942** 874; *Fr. Pat.* 863162 (1940, issued 1941); *Ch. Ztbl.* **1941** II 3218; CIBA PHARMACEUTICAL PRODUCTS (Ruzicka), *U. S. Pat.* 2315817 (1940, issued 1943); *C. A.* **1943** 5558.

d CIBA*, *Fr. Pat.* 863257 (issued 1941); *Ch. Ztbl.* **1942** I 81.

e CIBA* (Miescher, Wettstein), *Swed. Pat.* 101201 (issued 1941); *Ch. Ztbl.* **1941** II 3101; *Norwegian Pat.* 60152 (issued 1942); *Ch. Ztbl.* **1943** I 2421.

a CIBA PHARMACEUTICAL PRODUCTS (Miescher, Fischer, Scholz, Wettstein), *U.S. Pat.* 2275790 (issued 1942); *C. A.* **1942** 4289; CIBA*, *Fr. Pat.* 857122 (issued 1940); *Ch. Ztbl.* **1941** I 1324.

b CIBA PHARMACEUTICAL PRODUCTS (Miescher, Fischer, Scholz, Wettstein), *U.S. Pat.* 2276543 (issued 1942); *C. A.* **1942** 4674; CIBA* (Miescher, Fischer, Scholz, Wettstein), *Swed. Pat.* 101148 (issued 1941); *Ch. Ztbl.* **1941** II 1651.

Dannenberg, *Abhandl. Preuss. Akad. Wiss. Math.-naturwiss. Kl.* **1939** No. 21, pp. 11, 16, 51, 56.

Ehrenstein, *J. Org. Ch.* **4** 506, 515, 516, 518.

Ehrhart, Ruschig, Aumüller, *Angew. Ch.* **52** 363, 364, 366.

Goldberg, Aeschbacher, *Helv.* **22** 1185; CIBA*, *Fr. Pat.* 876213 (1940, issued 1942), *Ch. Ztbl.* **1943** I 2320.

Heard, McKay, *J. Biol. Ch.* **131** 371, 378.

LABORATOIRES FRANÇAIS DE CHIMIOTHÉRAPIE (Sandulesco, Girard), *Dutch Pat.* 53405 (issued 1942); *Ch. Ztbl.* **1943** I 2220.

Mamoli, *Ber.* **72** 1863.

a Marker, Lawson, *J. Am. Ch. Soc.* **61** 586.

b Marker, Lawson, *J. Am. Ch. Soc.* **61** 588.

c Marker, Wittle, *J. Am. Ch. Soc.* **61** 1329.

d Marker, Wittle, Plambeck, *J. Am. Ch. Soc.* **61** 1333, 1335.

e Marker, Rohrmann, *J. Am. Ch. Soc.* **61** 3476.

a Miescher, Kägi, *Helv.* **22** 184, 189, 190, 193.

b Miescher, Wettstein, Scholz, *Helv.* **22** 894, 899, 906, 907.

Reichstein, Montigel, *Helv.* **22** 1212, 1216, 1218.

a SCHERING A.-G., *Swiss Pat.* 226686 (issued 1943); *Ch. Ztbl.* **1944** II 563.

b SCHERING A.-G. (Logemann, Hildebrand), *Ger. Pat.* 737420 (issued 1943); *C. A.* **1945** 5051; *Ch. Ztbl.* **1944** I 366.

c SCHERING A.-G., *Fr. Pat.* 871942 (issued 1942); *Ch. Ztbl.* **1942** II 2504.

Scherrer, *Helv.* **22** 1329, 1338.

Woker, Antener, *Helv.* **22** 1309, 1319.

Yarnall, Wallis, *J. Org. Ch.* **4** 270, 281, 282.

1940 a CIBA* (Miescher, Heer), *Ger. Pat.* 737570 (issued 1943); *C. A.* **1946** 179; *Fr. Pat.* 876214 (issued 1942); *Ch. Ztbl.* **1943** I 1496; *Dutch Pat.* 55050 (issued 1943); *Ch. Ztbl.* **1944** I 367; CIBA PHARMACEUTICAL PRODUCTS (Miescher, Heer), *U.S. Pat.* 2372841 (issued 1945); *C. A.* **1945** 4195; *Ch. Ztbl.* **1945** II 1773.

b CIBA*, *Fr. Pat.* 886415 (issued 1943); *Ch. Ztbl.* **1944** I 450.

Ehrenstein, Decker, *J. Org. Ch.* **5** 544, 557.

Ercoli, *C. A.* **1944** 1540; *Ch. Ztbl.* **1941** II 1028.

Hoehn, Mason, *J. Am. Ch. Soc.* **62** 569.

McCullagh, Schneider, Emery, *Endocrinology* **27** 71, 75, 77.

a Marker, Rohrmann, *J. Am. Ch. Soc.* **62** 521, 523 et seq.

b Marker, Rohrmann, Wittle, Crooks, Jones, *J. Am. Ch. Soc.* **62** 650.

c Marker, Rohrmann, *J. Am. Ch. Soc.* **62** 898, 899.

d Marker, Tsukamoto, Turner, *J. Am. Ch. Soc.* **62** 2525, 2531.

e Marker, Jones, Krueger, *J. Am. Ch. Soc.* **62** 2532, 2535, 2536.

f Marker, *J. Am. Ch. Soc.* **62** 2543, 2546.

g Marker, *J. Am. Ch. Soc.* **62** 2621, 2623, 2624.

h Marker, Turner, *J. Am. Ch. Soc.* **62** 3003, 3005.

i Marker, Jones, Turner, Rohrmann, *J. Am. Ch. Soc.* **62** 3006, 3009.

j Marker, Krueger, *J. Am. Ch. Soc.* **62** 3349.

k Marker, *J. Am. Ch. Soc.* **62** 3350.

Miescher, Hunziger, Wettstein, *Helv.* **23** 1367.

---

* Société pour l'industrie chimique à Bâle; Gesellschaft für chemische Industrie in Basel.

PARKE, DAVIS & Co. (Marker, Wittle), *U.S. Pat.* 2 397 424, -425, -426 (issued 1946); *C. A.*
  1946 3570, 3572.
Reich, *Helv.* **23** 219–221; cf. N. V. ORGANON, *Fr. Pat.* 864 315 (issued 1941); *Ch. Ztbl.* **1942** I 642.
Ruzicka, Meldahl, *Helv.* **23** 364.
a SCHERING A.-G., *BIOS Final Report* No. 449, pp. 195, 300–303; *FIAT Final Report* No. 996,
  pp. 28, 112–115.
b SCHERING A.-G., *Fr. Pat.* 856 641 (issued 1940); *Ch. Ztbl.* **1941** I 407.
c SCHERING A.-G., *Swiss Pat.* 229 776 (issued 1944); *Ch. Ztbl.* **1944** II 1301; *Fr. Pat.* 875 070
  (1941, issued 1942); *Ch. Ztbl.* **1943** I 1190; SCHERING CORP. (Köster), *U.S. Pat.* 2 363 338
  (issued 1944); *C. A.* **1945** 3127; *Ch. Ztbl.* **1945** II 1638.
d SCHERING A.-G., *Fr. Pat.* 868 396 (issued 1941); *Ch. Ztbl.* **1942** I 2560.
SCHERING CORP. (Inhoffen), *U.S. Pat.* 2 409 043 (issued 1946); *C. A.* **1947** 1398; *Ch. Ztbl.*
  1947 620.
a Wettstein, *Helv.* **23** 388, 395. — b Wettstein, *Helv.* **23** 1371, 1373.
1941 a CIBA* (Miescher, Wettstein), *Swed. Pat.* 105 137 (issued 1942); *Ch. Ztbl.* **1944** II 143; CIBA
  PHARMACEUTICAL PRODUCTS (Miescher, Wettstein), *U.S. Pat.* 2 323 276 (issued 1943); *C. A.*
  1944 222.
b CIBA*, *Fr. Pat.* 877 821 (issued 1943); *Dutch Pat.* 54 147 (issued 1943); *Ch. Ztbl.* **1943** II 147.
Ehrenstein, Stevens, *J. Org. Ch.* **6** 908, 912.
v. Euw, Reichstein, *Helv.* **24** 879, 880, 886.
Fuchs, Reichstein, *Helv.* **24** 804, 814, 816.
Goldschmidt, Middelbeek, Boasson, *Rec. trav. chim.* **60** 209.
Hegner, Reichstein, *Helv.* **24** 828, 835.
a I. G. FARBENIND., *Fr. Pat.* 869 653 (issued 1942); *Ch. Ztbl.* **1942** II 1154.
b I. G. FARBENIND., *Swiss Pat.* 227 974 (issued 1943); *Ch. Ztbl.* **1945** I 196; *Fr. Pat.* 886 984
  (1942, issued 1943); *Ch. Ztbl.* **1944** II 1091.
a Marker, Turner, Wagner, Ulshafer, Crooks, Wittle, *J. Am. Ch. Soc.* **63** 774–776.
b Marker, Crooks, Wittbecker, *J. Am. Ch. Soc.* **63** 777.
c Marker, Turner, Wagner, Ulshafer, Crooks, Wittle, *J. Am. Ch. Soc.* **63** 779, 781.
Miescher, Kägi, *Helv.* **24** 986.
Oppenauer, *Z. physiol. Ch.* **270** 97, 100, 102.
a PARKE, DAVIS & Co. (Tendick, Lawson), *U.S. Pat.* 2 335 616 (issued 1943); *C. A.* **1944** 3095;
  Brit. Pat. 563 889 (issued 1944); *C. A.* **1946** 2942.
b PARKE, DAVIS & Co. (Marker, Crooks), *U.S. Pat.* 2 359 773 (issued 1944); *C. A.* **1945** 4198;
  Ch. Ztbl. **1945** II 1233.
c PARKE, DAVIS & Co. (Marker, Crooks), *U.S. Pat.* 2 343 311 (issued 1944); *C. A.* **1944** 3094.
1942 Arnaudi, *C. A.* **1946** 3152; *Ztbl. Bakt. Parasitenk.* II. Abt. **105** 352.
Burckhardt, Reichstein, *Helv.* **25** 1434, 1443.
Butenandt, *Naturwiss.* **30** 4, 7.
a Marker, Crooks, Wagner, *J. Am. Ch. Soc.* **64** 210.
b Marker, Crooks, Wagner, *J. Am. Ch. Soc.* **64** 213.
c Marker, Jones, Wittbecker, *J. Am. Ch. Soc.* **64** 468.
d Marker, Crooks, Jones, Shabica, *J. Am. Ch. Soc.* **64** 1276, 1278, 1279.
e Marker, Crooks, *J. Am. Ch. Soc.* **64** 1280.
f Marker, Wittle, Jones, Crooks, *J. Am. Ch. Soc.* **64** 1282.
g Marker, Wagner, *J. Am. Ch. Soc.* **64** 1842.
h Marker, Crooks, Wagner, Wittbecker, *J. Am. Ch. Soc.* **64** 2089.
i Marker, Wagner, Wittbecker, *J. Am. Ch. Soc.* **64** 2093–2095.
Molina, *C. A.* **1946** 3152.
Pearlman, Pincus, Werthessen, *J. Biol. Ch.* **142** 649.
Ruschig, *Medizin und Chemie* **4** 327, 339; *C. A.* **1944** 4954; *Ch. Ztbl.* **1943** I 2689.
Ruzicka, Plattner, Pataki, *Helv.* **25** 425, 429.
1943 Goldberg, Aeschbacher, Hardegger, *Helv.* **26** 680, 682 footnote 5.
Hansen, Cantarow, Rakoff, Paschkis, *Endocrinology* **33** 282, 286.
Hegner, Reichstein, *Helv.* **26** 721, 723.
I. G. FARBENIND., *Fr. Pat.* 891 441 (issued 1944); *Ch. Ztbl.* **1946** I 1025.

* Société pour l'industrie chimique à Bâle; Gesellschaft für chemische Industrie in Basel.

Plattner, Fürst, *Helv.* **26** 2266, 2268, 2270.
Reich, Reichstein, *Helv.* **26** 562, 564.
Reichstein, *U.S. Pat.* 2387706 (issued 1945); *C. A.* **1946** 994, 995; *Brit. Pat.* 567393 (issued 1945); *C. A.* **1947** 3264.
Ruzicka, Prelog, *Helv.* **26** 975, 979, 989.
Shoppee, Prins, *Helv.* **26** 1004, 1007, 1009, 1011, 1013, 1014.
1944 v. Euw, Lardon, Reichstein, *Helv.* **27** 821, 836–838.
King, *J. Am. Ch. Soc.* **66** 1612.
Koechlin, Reichstein, *Helv.* **27** 549, 553, 557, 560, 562, 564.
Moffett, Hoehn, *J. Am. Ch. Soc.* **66** 2098.
Wettstein, *Helv.* **27** 1803, 1811.
1945 Julian, Cole, Magnani, Meyer, *J. Am. Ch. Soc.* **67** 1728.
Klyne, Marrian, *Biochem. J.* **39** xlv.
Parke, Davis & Co. (Crooks, Jones), *U.S. Pat.* 2383472; *C. A.* **1946** 178; *Ch. Ztbl.* **1948** I 1142.
Schering A.-G., *BIOS Final Report* No. 449, pp. 304, 308–317; *FIAT Final Report* No. 996, pp. 60, 64–71.
Velluz, Petit, *Bull. soc. chim.* [5] **12** 949.
1946 Barton, *J. Ch. Soc.* **1946** 1116, 1117.
Ciba Pharmaceutical Products (Miescher, Frey, Meystre, Wettstein), *U.S. Pat.* 2461910 (issued 1949); *C. A.* **1949** 3474; *Ch. Ztbl.* **1949** E 1870, 1871.
Goldschmidt, Veer, *Rec. trav. chim.* **65** 796.
I. G. Farbenind., *FIAT Final Report* No. 996, pp. 10–12, 23–26, 123–126.
Marrian, Gough, *Nature* **157** 438; *Biochem. J.* **40** 376.
Masson, *Am. J. Med. Sci.* **212** 1; *C. A.* **1946** 6144.
a Meystre, Miescher, *Helv.* **29** 33, 39 footnote 3, 42, 43, 46 et seq.
b Meystre, Frey, Neher, Wettstein, Miescher, *Helv.* **29** 627, 632, 633.
a Pearlman, Pincus, *Federation Proc.* **5** 79.
b Pearlman, *J. Biol. Ch.* **166** 473, 475.
Plattner, Heusser, Angliker, *Helv.* **29** 468, 470, 471.
Prins, Shoppee, *J. Ch. Soc.* **1946** 494, 497.
Riegel, Meyer, *J. Am. Ch. Soc.* **68** 1097, 1099.
Shoppee, *J. Ch. Soc.* **1946** 1138, 1143.
Sutherland, Marrian, *Biochem. J.* **40** lxi.
Velluz, Rousseau, *Bull. soc. chim.* **1946** 498.
1947 Marker, Lopez, *J. Am. Ch. Soc.* **69** 2380, 2382, 2383.
Plattner, Ruzicka, Heusser, Angliker, *Helv.* **30** 385–387, 389, 390.
Prelog, Tagmann, Lieberman, Ruzicka, *Helv.* **30** 1080, 1084, 1086, 1089.
Sutherland, Marrian, *Biochem. J.* **41** 193.
1948 Barton, Cox, *J. Ch. Soc.* **1948** 783, 790, 791.
Ehrenstein, *J. Org. Ch.* **13** 214, 219, 220.
Klyne, Schachter, Marrian, *Biochem. J.* **43** 231.
Lieberman, Dobriner, Hill, Fieser, Rhoads, *J. Biol. Ch.* **172** 263, 282.
Plattner, Heusser, Boyce, *Helv.* **31** 603, 606.
Ruschig, *Angew. Ch.* **60** A 247.
Shoppee, *Experientia* **4** 418, 420.
1949 Fieser, Huang-Minlon, *J. Am. Ch. Soc.* **71** 1840.
Shoppee, *J. Ch. Soc.* **1949** 1671, 1677 et seq.
1950 Giral, *J. Am. Ch. Soc.* **72** 1913, 1914.
Heusser, Eichenberger, Plattner, *Helv.* **33** 370, 374, 375.
Julian, Karpel, *J. Am. Ch. Soc.* **72** 362, 365.
Mancera, *J. Am. Ch. Soc.* **72** 5752, 5753.
1951 Reich, Collins, *J. Am. Ch. Soc.* **73** 1374.
Sandoval, Rosenkranz, Djerassi, *J. Am. Ch. Soc.* **73** 2383.
1952 Klyne, Shoppee, *Chemistry & Industry* **1952** 470, 471.
1953 Cremlyn, Garmaise, Shoppee, *J. Ch. Soc.* **1953** 1847.
1956 Plattner, *private communication.*

δ. Monohydroxy-monooxo-steroids (CO in side chain, OH in the ring system) C_{22} and higher with three aliphatic side chains

21 - Methyl-Δ^5- pregnen - 3β - ol - 20 - one,   3β-Hydroxy-17β-propionyl-Δ^5-androstene

$C_{22}H_{34}O_2$. Needles (dil. acetone), m. 170 to 171° cor. (1940 Wettstein; 1937b, 1938a Ciba). — **Fmn.** From 3β-hydroxy-Δ^5-ætiocholenaldehyde (p. 2227 s) by the action of diazoethane in ether (1938a Ciba). As acetate, on keeping the chloride of 3β-acetoxy-Δ^5-ætiocholenic acid (p. 173) with diethylzinc in benzene (1937b Ciba; 1941 Wettstein). From the aforementioned chloride on refluxing with the sodium or the magnesium derivative of diethyl methylmalonate in xylene or benzene, followed in both cases by saponification with boiling methanolic KOH and decarboxylation with boiling aqueous H_2SO_4; in the second case a large amount of methyl 3β-hydroxy-Δ^5-ætiocholenate is also produced (1940 Wettstein). — **Rns.** Boiling with aluminium isopropoxide and cyclohexanone in toluene gives 21-methylprogesterone (p. 2803 s) (1940 Wettstein).

Acetate $C_{24}H_{36}O_3$. Needles (hexane), m. 175.5–176.5° cor. (1940 Wettstein). — **Fmn.** See above.

21-Methyl-$\Delta^{5.17(20)}$-pregnadien-3β-ol-21-one,   17-Acetonylidene-Δ^5-androsten-3β-ol

$C_{22}H_{32}O_2$. For trans-configuration (CO·CH$_3$:18-CH$_3$), see 1957 Romo. — Crystals (methanol), m. 168–9° cor.; $[\alpha]_D$ —65° (dioxane) (1941 Plattner). — **Fmn.** The acetate is obtained from that of $\Delta^{5.17(20)}$-pregnadien-3β-ol-21-al (p. 2251 s) by the action of diazomethane in methylene chloride; yield, 50% (1938a Ciba). The acetate is formed on keeping 3β-acetoxy-$\Delta^{5.17(20)}$-pregnadien-21-oic acid chloride and methylzinc iodide in ethyl acetate-toluene-benzene; it is saponified by boiling methanolic K_2CO_3 (1941 Plattner). — **Rns.** Boiling with aluminium tert.-butoxide and acetone in benzene gives 17-acetonylidene-Δ^4-androsten-3-one (p. 2804 s). Hydrogenation of the acetate in the presence of Raney nickel in alcohol yields the acetate of 17β-acetonyl-Δ^5-androsten-3β-ol (below) (1941 Plattner).

Acetate $C_{24}H_{34}O_3$. Crystals (methanol), m. 189–190° cor.; $[\alpha]_D$ —63° (dioxane); absorption max. in alcohol (from the graph) at 240 (4.30) and 310 (2.1) mμ (log ε) (1941 Plattner). — **Fmn.** and **Rns.** See above.

21 - Methyl - Δ^5 - pregnen - 3β - ol - 21 - one,   17β - Acetonyl - Δ^5 - androsten - 3β - ol

$C_{22}H_{34}O_2$. For the configuration at C-17, see 1944 Plattner. — Crystals (methanol), m. 177–8° cor.; $[\alpha]_D$ —48° (dioxane) (1941 Plattner). — **Fmn.** The acetate is formed on hydrogenation of the acetate of 17-acetonylidene-Δ^5-androsten-3β-ol (above) in the presence of Raney nickel in alcohol;

it is saponified by boiling methanolic K_2CO_3 (1941 Plattner). — **Rns.** Boiling with aluminium tert.-butoxide and acetone in benzene gives 17β-acetonyl-Δ^4-androsten-3-one (p. 2804 s) (1941 Plattner).

Acetate $C_{24}H_{36}O_3$. Crystals (methanol), m. 156–7° cor.; $[\alpha]_D$ —49° (dioxane) (1941 Plattner). — **Fmn.** See above.

Bisnorcholan-3-ol-22-al, 3-Hydroxybisnorcholanaldehyde $C_{22}H_{36}O_2$, m. 128°. —

Fmn. From 21-amino-3-hydroxybisnorcholane 3-acetate (p. 1909 s) on treatment with HOCl in ether, heating of the reaction product with sodium ethoxide in alcohol, and hydrolysis of the resulting imine with aqueous H_2SO_4 (1937 I. G. FARBENIND.).

21-Ethyl-Δ^5-pregnen-3β-ol-20-one, 17β-Butyryl-Δ^5-androsten-3β-ol $C_{23}H_{36}O_2$.

Crystals (dil. acetone), m. 125–7° cor. (1940 Wettstein; 1938a CIBA). — **Fmn.** From 3β-hydroxy-Δ^5-ætiocholenaldehyde (p.2227 s) by the action of diazopropane in ether (1938a CIBA). As acetate, on keeping the chloride of 3β-acetoxy-Δ^5-ætiocholenic acid (p. 173) with propylzinc iodide in toluene-benzene at room temperature; from the same chloride and the sodium or magnesium derivative of diethyl ethylmalonate in the same manner as 21-methyl-Δ^5-pregnen-3β-ol-20-one (p. 2258 s) (1940 Wettstein). — **Rns.** Boiling with aluminium isopropoxide and cyclohexanone in toluene gives 21-ethylprogesterone (p. 2804 s) (1940 Wettstein).

Acetate $C_{25}H_{38}O_3$. Crystals (hexane), m. 114–5° cor. (1940 Wettstein). — **Fmn.** See above.

Δ^5-Norcholen-3β-ol-22-one, 3β-Hydroxy-Δ^5-ternorcholenyl methyl ketone, 3β-Hydroxy-Δ^5-pregnen-20α-yl methyl ketone* $C_{23}H_{36}O_2$.

Colourless plates (acetone), m. 176–7° (1945 Cole), 179–181° cor. (1941 Wettstein); $[\alpha]_D^{25}$ —70° (chloroform) (1945 Cole; but cf. 1946 Barton); hydrated plates (moist ether), m. 143–6° (1945 Cole). Is precipitated in alcohol by digitonin (1945 Cole).

Fmn. The acetate is formed from 3β-acetoxy-Δ^5-bisnorcholenic acid chloride with methylzinc iodide in toluene-benzene at room temperature (1941 Wettstein), with dimethylzinc in benzene (1937c CIBA), or with dimethylcadmium or dimethylzinc in ether-benzene, first at room temperature, followed by refluxing (yield with 2.5 mols. of dimethylcadmium → 95 % of acetate, with 2.5 mols. dimethylzinc → 75 % of acetate) (1945 Cole; cf. also 1941b GLIDDEN Co.); it is saponified by refluxing with aqueous methanolic alkali carbonate (1941 Wettstein; 1945 Cole).

---

* For the configuration of this compound and its stereoisomer at C-20, see 1949 Wieland.

Rns. Gives 3-keto-$\mathit{\Delta}^4$-ternorcholenyl methyl ketone (p. 2804 s) on boiling in toluene with cyclohexanone and aluminium isopropoxide (1941 Wett-stein) or aluminium tert.-butoxide (1945 Cole). On oxidation of the acetate with CrO_3 in acetic acid at 25^0, after temporary protection of the double bond by bromine, a small amount (12%) of 3β-acetoxy-$\mathit{\Delta}^5$-bisnorcholenic acid (p. 174) is formed (1945 Cole). Treatment of the acetate with methyl-magnesium bromide yields (3β-hydroxy-$\mathit{\Delta}^5$-ternorcholenyl)-dimethyl-carbinol (p. 1941 s) (1945 Julian; cf. also 1941 b GLIDDEN Co.).

Acetate, 3β-Acetoxy-$\mathit{\Delta}^5$-norcholen-22-one, 3β-Acetoxy-$\mathit{\Delta}^5$-ternorcholenyl methyl ketone $C_{25}H_{38}O_3$. Crystals (hexane), m. 177–8^0 cor. (1941 Wettstein); prisms (acetone), m. 175–6^0 (1945 Cole); $[\alpha]_D^{29}$ —80^0 (chloroform) (1945 Cole; but cf. 1946 Barton). — **Fmn.** From 3β-acetoxy-20-iso-$\mathit{\Delta}^5$-ternorcholenyl methyl ketone (below) on partial isomerization by refluxing with sodium methoxide in benzene (1945 Cole). See also above. — **Rns.** (See also above). Saponification with boiling aqueous methanolic Na_2CO_3 gives the corresponding hydroxy-ketone, whereas on hydrolysis with boiling aqueous methanolic KOH partial isomerization to the 20-iso-hydroxy-ketone occurs. On refluxing with sodium methoxide in benzene partial isomerization to the 20-iso-acetoxy-ketone takes place (1945 Cole). — *Semicarbazone* $C_{26}H_{41}O_3N_3$, needles, m. 236–7^0 dec. (1945 Cole).

20-iso-$\mathit{\Delta}^5$-Norcholen-3β-ol-22-one, **3β-Hydroxy-20-iso-$\mathit{\Delta}^5$-ternorcholenyl methyl ketone**, **3β-Hydroxy-$\mathit{\Delta}^5$-pregnen-20β-yl methyl ketone*** $C_{23}H_{36}O_2$. Colourless

plates (acetone), m. 188–9^0; $[\alpha]_D^{25}$ —51^0 (chloroform). Is precipitated in alcohol by digitonin (1945 Cole). — **Fmn.** From 3β-acetoxy-$\mathit{\Delta}^5$-ternorcholenyl methyl ketone (above) on refluxing with aqueous methanolic KOH, followed by slow crystallization from the warm alkaline solution (1945 Cole). See also the methyl ether (below). — **Rns.** Boiling with aluminium tert.-butoxide and cyclohexanone in toluene gives 3-keto-20-iso-$\mathit{\Delta}^4$-ternorcholenyl methyl ketone (p. 2805 s) (1945 Cole). Heating of the acetate with methylmagnesium bromide in toluene on the steam-bath yields (3β-hydroxy-20-iso-$\mathit{\Delta}^5$-ternorcholenyl)-dimethyl-carbinol (p. 1941 s) (1945 Julian).

Acetate, 3β-Acetoxy-20-iso-$\mathit{\Delta}^5$-norcholen-22-one, 3β-Acetoxy-20-iso-$\mathit{\Delta}^5$-ternor-cholenyl methyl ketone $C_{25}H_{38}O_3$. White flakes (glacial acetic acid), m. 205–6^0; $[\alpha]_D^{29}$ —50^0 (chloroform). — **Fmn.** From the above hydroxy-ketone on heating with acetic anhydride. From the stereoisomeric 3β-acetoxy-$\mathit{\Delta}^5$-ternorcholenyl methyl ketone (above) on partial isomerization by refluxing with sodium methoxide in benzene. — **Rns.** (See also above). On refluxing with sodium methoxide in benzene, partial isomerization to the stereoisomeric acetoxy-ketone takes place (1945 Cole).

Methyl ether, 3β-Methoxy-20-iso-$\mathit{\Delta}^5$-norcholen-22-one, 3β-Methoxy-20-iso-$\mathit{\Delta}^5$-ternorcholenyl methyl ketone $C_{24}H_{38}O_2$. White plates (hexane), m. 152–3^0;

---

* See the footnote on p. 2259 s.

$[\alpha]_D^{29}$ —52⁰ (chloroform). — **Fmn.** From the above hydroxy-ketone on treat-ment with p-toluenesulphonyl chloride and pyridine, followed by refluxing of the resulting p-toluenesulphonate (which was not isolated) with anhydrous methanol. From 2-(6-methoxy-20-iso-i-ternorcholenyl)-propene (p. 1528 s) on ozonization in carbon tetrachloride at 0⁰, followed by refluxing with dry methanol containing some conc. H_2SO_4. — **Semicarbazone** $C_{25}H_{41}O_2N_3$, crystalline powder (methanol), m. 208–210⁰ dec. (1945 Julian).

Norallocholan-3β-ol-22-one, 3β-Hydroxyternorallocholanyl methyl ketone, 3β-Hydroxyallopregnan-20α-yl methyl ketone*

$C_{23}H_{38}O_2$. Crystals (acetone), m. 164–5⁰; $[\alpha]_D^{26}$ —15.7⁰ (chloroform). — **Fmn.** The ace-tate is formed from the chloride of 3β-acet-oxybisnorallocholanic acid (p. 175) with di-methylcadmium in ether. The acetate is also obtained by ozonolysis of 2-(3β-acetoxy-ternorallocholanyl)-propene (p. 1527 s) in glacial acetic acid; it is saponified by refluxing with aqueous methanolic Na_2CO_3. — **Rns.** Heating of the acetate with methylmagnesium bromide in toluene on the steam-bath gives (3β-hydroxyternorallocholanyl)-dimethyl-carbinol (p. 1941 s) (1945 Julian).

Acetate, 3β-Acetoxyternorallocholanyl methyl ketone $C_{25}H_{40}O_3$. Crystals (meth-anol), m. 167–8⁰; $[\alpha]_D^{26}$ —23.6⁰ (chloroform). — **Fmn.** See above. — **Rns.** (See also above). Refluxing with aqueous methanolic Na_2CO_3 gives the above hydroxy-ketone, whereas refluxing with aqueous methanolic KOH yields 3β-hydroxy-20-iso-ternorallocholanyl methyl ketone (below) (1945 Julian).

20-iso-Norallocholan-3β-ol-22-one, 3β-Hydroxy-20-iso-ternorallocholanyl methyl ketone, 3β-Hydroxyallopregnan-20β-yl methyl ketone*

$C_{23}H_{38}O_2$. Crystals (methanol), m. 185–6⁰ (1945 Julian); $[\alpha]_D^{26}$ +12⁰ (chloroform) (1945 Julian; but cf. 1946 Barton). — **Fmn.** From 3β-acetoxyternor-allocholanyl methyl ketone (above) on reflux-ing with aqueous methanolic KOH, followed by slow crystallization from the warm alkaline solution. As acetate by ozonolysis of 2-(3β-acetoxy-20-iso-ternor-allocholanyl)-propene (p. 1527 s) in glacial acetic acid (1945 Julian). — **Rns.** Heating of the acetate with methylmagnesium bromide in toluene on the steam-bath gives (3β-hydroxy-20-iso-ternorallocholanyl)-dimethyl-carbinol (p. 1941 s) (1945 Julian).

Acetate, 20-iso-ternorallocholanyl methyl ketone $C_{25}H_{40}O_3$. Crystals (methanol), m. 180–1⁰ (1945 Julian); $[\alpha]_D^{26}$ +11.6⁰ (chloroform) (1945 Julian; but cf. 1946 Barton). — **Fmn.** From the above hydroxy-ketone on refluxing with acetic anhydride (1945 Julian). See also above.

---

* See the footnote on p. 2259 s.

Δ^5-Cholen-3β-ol-22-one, 3β-Hydroxy-Δ^5-ternorcholenyl ethyl ketone, 3β-Hydr-

oxy-Δ^5-pregnen-20α-yl ethyl ketone* $C_{24}H_{38}O_2$. Needles (acetone), m. 175–6°; $[\alpha]_D^{29}$ —66° (chloroform). — **Fmn.** The acetate is formed from 3β-acetoxy-Δ^5-bisnorcholenic acid chloride and diethylcadmium in ether-benzene; it is saponified by refluxing with aqueous methanolic Na_2CO_3. — **Rns.** Refluxing with aluminium tert.-butoxide and cyclohexanone in toluene gives 3-keto-Δ^4-ternorcholenyl ethyl ketone (p. 2805 s) (1945 Cole).

Acetate, 3β-Acetoxy-Δ^5-ternorcholenyl ethyl ketone $C_{26}H_{40}O_3$. Prisms (acetone), m. 171–2°. $[\alpha]_D^{29}$ —67° (chloroform). — **Fmn.** From the above hydroxyketone on boiling with acetic anhydride. See also above. — **Rns.** Saponification with boiling aqueous methanolic Na_2CO_3 gives the above hydroxy-ketone, whereas on hydrolysis with boiling aqueous methanolic KOH partial isomerization to 3β-hydroxy-20-iso-Δ^5-ternorcholenyl ethyl ketone (below) occurs. — *Semicarbazone* $C_{27}H_{43}O_3N_3$, m. 253–6°. **Fmn.** From the above acetoxyketone or (on long boiling) from the 20-iso-acetoxy-ketone with alcoholic semicarbazide (1945 Cole).

20-iso-Δ^5-Cholen-3β-ol-22-one, 3β-Hydroxy-20-iso-Δ^5-ternorcholenyl ethyl

ketone, 3β-Hydroxy-Δ^5-pregnen-20β-yl ethyl ketone* $C_{24}H_{38}O_2$. Plates (methanol), m. 158° to 160°. — **Fmn.** From its acetate on hydrolysis with Na_2CO_3 in wet methanol (1945 Cole).

Acetate, 3β-Acetoxy-20-iso-Δ^5-ternorcholenyl ethyl ketone $C_{26}H_{40}O_3$. Colourless plates (alcohol), m. 176–7°; $[\alpha]_D^{25}$ —53° (chloroform). — **Fmn.** From the above hydroxy-ketone on acetylation. From the stereoisomeric 3β-acetoxy-Δ^5-ternorcholenyl ethyl ketone (above) on refluxing with aqueous methanolic KOH, reacetylation of the resulting mixture of the two 20-epimeric hydroxy-ketones by heating with acetic anhydride at 100°, and subsequent chromatographic separation on Al_2O_3 (benzene eluant). — **Rns.** On boiling with alcoholic semicarbazide for 12 hrs., a 50% yield of 3β-acetoxy-Δ^5-ternorcholenyl ethyl ketone semicarbazone (above) is obtained (1945 Cole).

Allocholan-3α-ol-24-al, 3α-Hydroxyallocholanaldehyde $C_{24}H_{40}O_2$.

Acetate $C_{26}H_{42}O_3$. — **Fmn.** Is formed in small amount from epicholestanyl acetate (p. 1703 s) on oxidation with CrO_3 in acetic acid at 90°. — *Semicarbazone* $C_{27}H_{45}O_3N_3$, needles (methanol), m. 224.5–225.5° cor. (1937 a Ruzicka).

* See the footnote on p. 2259 s.

Δ^5-Bisnorcholesten-3β-ol-24-one, 3β-Hydroxy-Δ^5-bisnorcholesten-24-one

$C_{25}H_{40}O_2$, m. 116–129^0. — **Fmn.** From the acetate of 25-chloro-Δ^5-bisnorcholesten-3β-ol-24-one (p. 2283 s) on dehalogenation with zinc and glacial acetic acid, followed by hydrolysis. — **Rns.** Oxidation with aluminium phenoxide and acetone in benzene gives Δ^4-bisnorcholestene-3.24-dione (p. 2805 s). — *Oxime* $C_{25}H_{41}O_2N$. Druses (benzene), m. 173–9^0 dec. — *Semicarbazone* $C_{26}H_{43}O_2N_3$, dec. 194^0 (1938 Kuwada).

Acetate, *3β-Acetoxy-Δ^5-bisnorcholesten-24-one* $C_{27}H_{42}O_3$, m. 145–151^0. — **Fmn.** From the above hydroxy-ketone with acetic anhydride. — *Oxime* $C_{27}H_{43}O_3N$, m. 152–5^0. — *Semicarbazone* $C_{28}H_{45}O_3N_3$, dec. 211^0 (1938 Kuwada).

27-Nor-Δ^5-cholesten-3β-ol-24-one, 3β-Hydroxy-27-nor-Δ^5-cholesten-24-one

$C_{26}H_{42}O_2$. Sinters at 90^0, m. 114^0. — **Fmn.** From sugar-cane sitosteryl acetate dibromide on oxidation with CrO_3, followed by debromination. From 3β-acetoxy-Δ^5-cholenamide by the action of ethylmagnesium iodide. — **Rns.** Clemmensen reduction gives 27-nor-Δ^5-cholesten-3β-ol (p. 1528 s) (1940 Mitui). — *Acetate* $C_{28}H_{44}O_3$, m. 167–8^0 (1940 Mitui).

27-Nor-Δ^5-cholesten-3β-ol-25-one, 3β-Hydroxy-27-nor-Δ^5-cholesten-25-one

$C_{26}H_{42}O_2$. Leaflets (methanol) (1937 b Ruzicka), with 1 H_2O (1938 Hattori); sinters at ca. 114^0 cor., m. 126–7^0 cor. (1937 b Ruzicka), 127–9^0 cor. (1938 Hattori; 1938 b Ciba). The solvent-free compound, obtained by drying in a high vac. at 80^0, 100^0, and 110^0, is hygroscopic. With the Liebermann-Burchardt reagent it shows the same colour shift from violet through blue to green as cholesterol (1937 b Ruzicka).

Is inactive in the capon and rat tests and in the Corner-Clauberg test (1937 b Ruzicka).

Fmn. From cholesteryl acetate dibromide on oxidation with CrO_3 in acetic acid at 28–30^0, debromination, transformation into the semicarbazone, cleavage of the latter by heating with aqueous alcoholic H_2SO_4, and saponification with boiling aqueous methanolic K_2CO_3 (1937 b Ruzicka; 1938 Hattori); for separation from the other oxidation products of cholesterol, see also 1937, 1938 b Ciba. From the acetate of 26-chloro-27-nor-Δ^5-cholesten-3β-ol-25-one (p. 2283 s) on treatment with zinc dust and glacial acetic acid, followed by hydrolysis of the resulting acetate with methanolic KOH (1938 Hattori).

Rns. Oxidation with CrO_3 (after temporary protection of the double bond by bromine) gives 27-nor-Δ^4-cholestene-3.25-dione (p. 2805 s) (1937b Ruzicka), which is also formed by the action of dehydrogenating bacteria (Micrococcus dehydrogenans, Flavobacterium dehydrogenans Arn., Corynebacterium mediolanum) (1940 Ercoli; 1942 Arnaudi; 1942 Molina). Hydrogenation with platinum oxide in glacial acetic acid, followed by oxidation with CrO_3 in acetic acid, yields 27-nor-cholestane-3.25-dione (p. 2806 s) (1937b Ruzicka). Clemmensen reduction gives 27-nor-Δ^5-cholesten-3β-ol (p. 1528 s) (1940 Mitui). The hydroxy-ketone as well as its acetate form dibromides (dec. 130–1° and dec. 125–6°, resp., p. 2290 s) (1938 Hattori). On refluxing with benzaldehyde and aqueous methanolic $KHCO_3$, 26-benzylidene-27-nor-Δ^5-cholesten-3β-ol-25-one (p. 2276 s) is obtained (1945 Velluz). Gives no adduct with oxalic acid (1941 Miescher).

2.4-Dinitrophenylhydrazone $C_{32}H_{46}O_5N_4$, m. 159–160° (1938 Hattori).

Acetate, 3β-Acetoxy-27-nor-Δ^5-cholesten-25-one $C_{28}H_{44}O_3$. Leaflets (methanol or ethanol), m. 141.5–142° cor. (1937b Ruzicka; see also 1938 Hattori). — **Fmn.** From the above hydroxy-ketone on keeping with acetic anhydride and pyridine (1937b Ruzicka). See also above. — **Rns.** See above. — *Oxime* $C_{28}H_{45}O_3N$, m. 185° cor. (1938 Hattori). — *Semicarbazone* $C_{29}H_{47}O_3N_3$, needles (abs. alcohol), m. 237–8° cor. dec. (1937b Ruzicka ; 1938 Hattori); solubility in abs. alcohol ca. 1:1000 (1937b Ruzicka).

Benzoate, 3β-Benzoyloxy-27-nor-Δ^5-cholesten-25-one $C_{33}H_{46}O_3$. Leaflets (alcohol), m. 144–5° cor. — **Fmn.** From the above hydroxy-ketone on keeping with benzoyl chloride and pyridine (1937b Ruzicka).

27-Norcholestan-3α-ol-25-one, 3α-Hydroxy-27-norcholestan-25-one $C_{26}H_{44}O_2$.

For the structure, see 1937a Ruzicka. — Colourless needles (methanol or hexane), m. 181° to 182.5° cor. (1937a Ruzicka), 175–7° (1936 Zelinskiĭ; 1937 Ushakov). — Shows very little action in the capon comb test and is inactive in the Allen-Doisy and Corner-Clauberg tests (1937a Ruzicka).

Fmn. From epi-cholestanyl acetate on oxidation with CrO_3 in acetic acid, advantageously at 25–30°, transformation into the semicarbazone, cleavage of the latter by refluxing with aqueous alcoholic H_2SO_4, and saponification with boiling methanolic KOH (1937a Ruzicka; see also 1937a, 1938b CIBA; 1936 Zelinskiĭ; 1937 Ushakov).

Rns. Does not form iodoform with an alkaline solution of iodine in methanol (1937 Ushakov; cf. 1937a Ruzicka). Oxidation with CrO_3 in acetic acid at room temperature gives 27-norcholestane-3.25-dione (p. 2806 s). Gives 3α-hydroxyallocholanic acid (p. 177) on keeping of the acetate with benzaldehyde and hydrochloric acid in glacial acetic acid, followed by oxidation of the resulting benzylidene compound (which was not isolated) with CrO_3

in acetic acid, and saponification with boiling alcoholic KOH. On boiling the acetate with methylmagnesium iodide in ether, followed by refluxing with alcoholic NaOH, cholestane-3α.25-diol (p. 1943 s) is formed (1937a Ruzicka).

Semicarbazone $C_{27}H_{47}O_2N_3$. Crystals (alcohol), m. 221–3° cor. (1937a Ruzicka; see also 1936 Zelinskiĭ; 1937 Ushakov).

Acetate, 3α-Acetoxy-27-norcholestan-25-one $C_{28}H_{46}O_3$. Crystals (methanol), m. 111° cor. (1937a Ruzicka), 135–6° cor. (1937 Ushakov). — **Fmn.** From the above hydroxy-ketone on heating with acetic anhydride at 90° (1937a Ruzicka) or on boiling with acetic anhydride in benzene for 9 hrs. (1937 Ushakov).

21-Nor-Δ⁵-cholesten-3β-ol-20-one, 3β-Hydroxy-21-nor-Δ⁵-cholesten-20-one, 21-Isoamyl-Δ⁵-pregnen-3β-ol-20-one $C_{26}H_{42}O_2$.

Me

Me —CO·CH₂·CH₂·CH₂·CH(CH₃)₂

HO

Crystals (acetone), m. 136–8° cor. — **Fmn.** From the chloride of 3β-acetoxy-Δ⁵-ætiocholenic acid (p. 173) on boiling with diethyl sodio-isoamylmalonate in benzene, followed by saponification and decarboxylation (1940 Wettstein).

Acetate $C_{28}H_{44}O_3$. Needles (hexane), m. 142–3° cor. — **Fmn.** On keeping the above hydroxy-ketone with acetic anhydride and pyridine at room temperature (1940 Wettstein).

Δ⁵-Cholesten-3β-ol-22-one, 3β-Hydroxy-Δ⁵-ternorcholenyl isoamyl ketone, 3β-Hydroxy-Δ⁵-pregnen-20α-yl isoamyl ketone, 22-Ketocholesterol $C_{27}H_{44}O_2$.

Me   CH₃

Me —CH·CO·CH₂·CH₂·CH(CH₃)₂

HO

For configuration at C-20, see 1945 Cole; 1949 Wieland. — White needles (acetone), m. 142–3° (1945 Cole); [α]$_D^{30}$ —55° (chloroform) (1945 Cole; but cf. 1946 Barton); hydrated plates (not analysed) (aqueous acetone or aqueous alcohol), m. 123–5° (1945 Cole). Is precipitated in alcohol by digitonin (1945 Cole). — **Fmn.** The acetate is formed from 3β-acetoxy-Δ⁵-bisnorcholenic acid chloride and diisoamylcadmium in ether-benzene; it is saponified by refluxing with aqueous methanolic Na₂CO₃ (1945 Cole). — **Rns.** Heating with aluminium tert.-butoxide and cyclohexanone in toluene at 100° gives 3-keto-Δ⁴-ternorcholenyl isoamyl ketone (p. 2806 s). The acetate fails to yield an oxime or a semicarbazone in alcoholic solution; heating with hydroxylamine hydrochloride in pyridine gave an amorphous oxime which could not be purified (1945 Cole).

Acetate, 3β-Acetoxy-Δ⁵-ternorcholenyl isoamyl ketone $C_{29}H_{46}O_3$. Needles (alcohol), m. 152° (1945 Cole); [α]$_D^{29}$ —63° (chloroform) (1945 Cole; but cf. 1946 Barton). Very soluble in acetone, ether, benzene, and chloroform, only moder-

159*

ately soluble in acetic acid and in alcohols. With H_2SO_4 it gives a bright red coloration, with H_2SO_4-acetic anhydride a violet coloration (1945 Cole). — **Fmn.** and **Rns.** See above.

Δ^5-Cholesten-3β-ol-24-one, 3β-Hydroxy-Δ^5-norcholenyl isopropyl ketone, 24 - Ketocholesterol $C_{27}H_{44}O_2$.

Crystals (aqueous methanol), m. 137–138.5° cor.; $[\alpha]_D^{26}$ —37° (chloroform). — **Fmn.** The acetate is formed from 3β-acetoxy-Δ^5-cholenic acid chloride and diisopropylcadmium in ether-benzene; it is saponified with methanolic KOH. — **Rns.** On refluxing its acetate with aluminium isopropoxide in isopropyl alcohol it yields 24-hydroxycholesterol (p. 1943 s). Heating with hydrazine hydrate and sodium ethoxide in alcohol at 200° gives cholesterol. On refluxing the p-toluenesulphonate with potassium acetate in methanol, 24-keto-i-cholesteryl methyl ether (below) is obtained (1944 Riegel).

Semicarbazone $C_{28}H_{47}O_2N_3$. Crystals (dil. alcohol), m. 166–8°cor. (1944 Riegel).

Acetate $C_{29}H_{46}O_3$. Crystals (alcohol), m. 127.5–128° cor., becoming turbid at 129–130° cor., and finally m. 131° cor.; $[\alpha]_D^{26}$ —41° (chloroform). — **Fmn.** and **Rns.** See above. —*Oxime* $C_{29}H_{47}O_3N$, crystals (methanol); softens at 155° cor., m. 156–158.5° cor. (1944 Riegel).

p-Toluenesulphonate $C_{34}H_{50}O_4S$. Crystals (petroleum ether), softens at 115° cor., m. 119–120° cor. dec.; $[\alpha]_D^{26}$ —35° (chloroform). — **Fmn.** From the above hydroxy-ketone on treatment with p-toluenesulphonyl chloride and pyridine (1944 Riegel). — **Rns.** See above.

i-Cholesten-6β-ol-24-one, 6β-Hydroxy-i-norcholenyl isopropyl ketone, 24-Keto-i-cholesterol $C_{27}H_{44}O_2$.

For the configuration at C-6, see 24-hydroxy-i-cholesterol (p. 1943 s).

Methyl ether $C_{28}H_{46}O_2$. Crystals (acetone), m. 90.5–91.5° cor.; $[\alpha]_D^{26}$ +52° (chloroform). — **Fmn.** From 24-ketocholesteryl p-toluenesulphonate (above) on refluxing with potassium acetate in methanol. — **Rns.** Gives 24-hydroxy-i-cholesteryl methyl ether (p. 1943 s) on refluxing with aluminium isopropoxide in isopropyl alcohol (1944 Riegel).

1936 Zelinskiĭ, Ushakov, *Bull. acad. sci. U.R.S.S. Sér. chim.* **1936** 879, 885, 886; *C. A.* **1937** 5372; *Ch. Ztbl.* **1937** II 2534.
1937 a CIBA*, *Swiss Pat.* 199449 (issued 1938); *C. A.* **1939** 3393; *Ch. Ztbl.* **1939** I 3930.
   b CIBA*, *Swiss Pat.* 214536 (issued 1941); *C. A.* **1943** 1231; *Ch. Ztbl.* **1942** I 2296.
   c CIBA*, *Swiss Pat.* 214537 (issued 1941); *C. A.* **1942** 4978; *Ch. Ztbl.* **1942** I 2296.

* Société pour l'industrie chimique à Bâle; Gesellschaft für chemische Industrie in Basel.

I. G. FARBENIND. (Bockmühl, Ehrhart, Ruschig, Aumüller), *Ger. Pat.* 693351 (issued 1940); *C. A.* **1941** 4921; *Brit. Pat.* 502666 (issued 1939); *C. A.* **1939** 7316; *Ch. Ztbl.* **1939** II 1126; *Fr. Pat.* 842026 (1938, issued 1939); *C. A.* **1940** 4743; *Ch. Ztbl.* **1940** I 429.

a Ruzicka, Oberlin, Wirz, Meyer, *Helv.* **20** 1283, 1286 et seq.

b Ruzicka, Fischer, *Helv.* **20** 1291–1296.

Ushakov (Ouchakov), Epifanskiĭ, Chinayeva, *J. Gen. Ch. U.S.S.R.* **7** 1825; *Bull. soc. chim.* [5] **4** 1390; *C. A.* **1938** 583; *Ch. Ztbl.* **1938** I 3926.

1938 a CIBA* (Miescher, Wettstein), *Ger. Pat.* 739083 (issued 1943); *C. A.* **1945** 392; *Ch. Ztbl.* **1944** I 110.

b CIBA* (Miescher, Fischer), *U.S. Pat.* 2172590 (issued 1939); *C. A.* **1940** 450; *Ch. Ztbl.* **1940** I 428; *Fr. Pat.* 841242 (issued 1939); *C. A.* **1940** 4240; *Ch. Ztbl.* **1940** I 916.

Hattori, *C. A.* **1938** 7473; *Ch. Ztbl.* **1939** I 1182.

Kuwada, Yoshiki; *C. A.* **1938** 8432; *Ch. Ztbl.* **1939** I 1372.

1940 Ercoli, *C. A.* **1944** 1540; *Ch. Ztbl.* **1941** II 1028.

Mitui, *C. A.* **1941** 6263; *Ch. Ztbl.* **1942** I 878.

Wettstein, *Helv.* **23** 1371, 1374 et seq.

1941 GLIDDEN Co. (Julian, Cole), *U.S. Pat.* 2304101 (issued 1942); *C. A.* **1943** 2889.

Miescher, Kägi, *Helv.* **24** 986.

Plattner, Schreck, *Helv.* **24** 472.

Wettstein, *Helv.* **24** 311, 314, 315.

1942 Arnaudi, *C. A.* **1946** 3152; *Ztbl. Bakt. Parasitenk.* II. Abt. **105** 352; *Biological Abstracts* **21** 889 (1947).

Molina, *C. A.* **1946** 3152.

1944 Plattner, Bucher, Hardegger, *Helv.* **27** 1177, 1179.

Riegel, Kaye, *J. Am. Ch. Soc.* **66** 723.

1945 Cole, Julian, *J. Am. Ch. Soc.* **67** 1369, 1373, 1374.

Julian, Cole, Meyer, Herness, *J. Am. Ch. Soc.* **67** 1375, 1378, 1380, 1381.

Velluz, Petit, *Bull. soc. chim.* [5] **12** 949.

1946 Barton, *J. Ch. Soc.* **1946** 1116, 1117.

1949 Wieland, Miescher, *Helv.* **32** 1922.

1957 Romo, Romo de Vivar, *J. Am. Ch. Soc.* **79** 1118, 1120.

ε. Monohydroxy - monooxo - steroids (CO in side chain, OH in the ring system) with four aliphatic side chains

16 - Methyl - $\varDelta^{5,16}$ - pregnadien - 3β - ol - 20 - one

$C_{22}H_{32}O_2$. Leaflets or needles (acetone), m. 197–8° cor.; $[\alpha]_D^{21}$ —78° (alcohol) (1944 Wettstein); $[\alpha]_D^{20}$ —105° (dioxane); absorption max. (solvent not stated) at 250 mμ (log ε 4.26) (1952 Romo). With digitonin in hot alcohol a sparingly soluble addition product is quickly formed (1944 Wettstein). — **Fmn.** From $\varDelta^{5,16}$-pregnadien-3β-ol-20-one (p. 2230 s) on treatment with diazomethane in ether at room temperature, followed by thermal decomposition of the resulting pyrazoline compound (I, R = H; p. 2231 s) at 175° and 0.04 mm. pressure (1944 Wettstein); the acetate is similarly obtained (see also p. 2231 s) (1944 Wettstein; see also 1951 Sandoval) and is saponified by boiling with aqueous methanolic K_2CO_3 (1944 Wettstein). From 16-methyldiosgenin by side-chain degradation (1952 Romo).

* Société pour l'industrie chimique à Bâle; Gesellschaft für chemische Industrie in Basel.

Rns. The acetate immediately reduces dil. alcoholic permanganate solution. Gives 16-methyl-$\Delta^{4\cdot16}$-pregnadiene-3.20-dione (p. 2807 s) on refluxing with aluminium isopropoxide and cyclohexanone in toluene (1944 Wettstein). On partial hydrogenation in the presence of Raney nickel it gives 16β-methyl-Δ^5-pregnen-3β-ol-20-one (below) (1952 Romo); the acetate reacts similarly in alcohol in the presence of Rupe nickel at room temperature giving the corresponding acetate (1944 Wettstein; cf. 1952 Romo) and, at the same time, a compound containing 71.3% C and 9.0% H [needles (acetone), dec. 180° cor.] (1944 Wettstein).

Acetate $C_{24}H_{34}O_3$. Needles (acetone), m. 177–8° cor. (1944 Wettstein); [α]$_D$ in alcohol —83° (1944 Wettstein), —76° (1951 Sandoval), in chloroform —106° (1951 Sandoval), —95° (dioxane) (1952 Romo). Absorption max. in alcohol at 315* mμ (log ε 2.35*) (1944 Wettstein) and 250 mμ (log ε 4.06) (1944 Wettstein; 1951 Sandoval; cf. also 1952 Romo). Gives a strong yellow coloration with tetranitromethane (1944 Wettstein). — **Fmn.** From the above hydroxy-ketone on keeping with acetic anhydride and pyridine at room temperature (1944 Wettstein). See also above. — **Rns.** See above. — *Semicarbazone* $C_{25}H_{37}O_3N_3$, crystals (acetone), m. 276–8° cor. dec.; absorption max. in alcohol at 265* mμ (log ε 4.23*) (1944 Wettstein).

16α-Methyl-Δ^5-pregnen-3β-ol-20-one $C_{22}H_{34}O_2$. For α-configuration at C-16, see 1952 Romo. — Crystals containing solvent of crystallization (acetone), m. 191–192° (1942b Marker). — **Fmn.** From the acetate of $\Delta^{5\cdot16}$-pregnadien-3β-ol-20-one (p. 2230 s) on heating with methylmagnesium iodide in toluene on the steambath (1942b Marker; cf. 1952 Romo). — **Rns.** Gives 16α-methylprogesterone (m. 133–5°; p. 2807 s) on refluxing with aluminium tert.-butoxide and acetone in toluene (1942b Marker).

Semicarbazone $C_{23}H_{37}O_2N_3$, m. 245° dec. (1942b Marker).

Acetate $C_{24}H_{36}O_3$. Crystals (methanol), m. 177.5–178.5°. — **Fmn.** From the above hydroxy-ketone with acetic anhydride + pyridine (1942b Marker).

16β-Methyl-Δ^5-pregnen-3β-ol-20-one $C_{22}H_{34}O_2$. For configuration at C-16, see 1952 Romo. — Crystals (acetone), m. 205° to 207° cor.; [α]$_D^{23}$ —14° (alcohol) (1944 Wettstein); [α]$_D^{20}$ —13° (dioxane) (1952 Romo). With digitonin in hot alcohol a rather readily soluble addition product is slowly formed (1944 Wettstein). — **Fmn.** From 16-methyl-$\Delta^{5\cdot16}$-pregnadien-3β-ol-20-one (p. 2267 s) on partial hydrogenation in the presence of Raney nickel (1952 Romo); the acetate is similarly obtained from the corresponding acetate in alcohol in the presence of Rupe nickel at room temperature (1944 Wettstein). — **Rns.** Gives 16β-methylprogesterone (m. 210–211° cor.; p. 2807 s) on refluxing with

* From the graph.

References, p. 2270 s

aluminium isopropoxide and cyclohexanone in toluene (1944 Wettstein; 1952 Romo). On refluxing of its acetate with $KHCO_3$ in aqueous methanol it gives the above hydroxy-ketone (1944 Wettstein), whereas on hydrolysis with aqueous alcoholic K_2CO_3 it rearranges to 16β-methyl-17-iso-Δ^5-pregnen-3β-ol-20-one (below) (1944 Wettstein; cf. 1952 Romo); this rearrangement occurs also on acid hydrolysis of the acetate (1952 Romo).

Acetate $C_{24}H_{36}O_3$. Needles (hexane), m. 155–155.5° cor. (1944 Wettstein), crystals (methanol), m. 150–153° (1952 Romo); $[\alpha]_D^{20}$ —20° (alcohol) (1944 Wettstein); $[\alpha]_D^{20}$ —24° (dioxane) (1952 Romo). Gives a strong yellow coloration with tetranitromethane (1944 Wettstein). — **Fmn.** From the above hydroxy-ketone on treatment with acetic anhydride and pyridine at room temperature (1944 Wettstein). See also above. — **Rns.** See above. — *Oxime* $C_{24}H_{37}O_3N$, m. 205°; $[\alpha]_D^{20}$ —12° (dioxane) (1952 Romo).

16β-Methyl-17-iso-Δ^5-pregnen-3β-ol-20-one $C_{22}H_{34}O_2$. Crystals (methanol),

m. 226–228° (1952 Romo), 230–232° cor. (1944 Wettstein); $[\alpha]_D^{20}$ —117° (dioxane) (1952 Romo), $[\alpha]_D^{23}$ —113° (alcohol) (1944 Wettstein). — **Fmn.** From the acetate of the preceding normal pregnenolone on hydrolysis with aqueous alcoholic K_2CO_3 (1944 Wettstein; cf. 1952 Romo) or with acids (1952 Romo). — **Rns.** Gives 16β-methyl-17-iso-progesterone (m. 164–6°; p. 2808 s) on Oppenauer oxidation (1952 Romo).

Acetate $C_{24}H_{36}O_3$, m. 178–180°; $[\alpha]_D^{20}$ —115° (dioxane) (1952 Romo).

17α-Methylpregnan-3β-ol-20-one $C_{22}H_{36}O_2$. Regarding the configuration at

C-17, see 1949 Plattner; 1950a, b Heusser; 1952 Günthard. — Needles (acetone), m. 169–171° and 184–187° (polymorphic forms) (1942a Marker). — **Fmn.** From 17α-methyl-20-methylenepregnan-3β-ol (p. 1531 s) on oxidation with ozone in chloroform; the acetate is obtained from the acetate of the same compound on oxidation with CrO_3 in acetic acid on the steam-bath or from 17α-methyl-3β-acetoxyætiocholanic acid chloride on treatment with dimethylzinc in tetralin; the acetate is saponified by methanolic KOH (1942a Marker). — **Rns.** Gives 17α-methylpregnane-3.20-dione (p. 2808 s) on oxidation with CrO_3 in acetic acid at room temperature (1942a Marker).

16α-Isopropyl-Δ^5-pregnen-3β-ol-20-one $C_{24}H_{38}O_2$. For configuration at C-16,

see 1952 Romo. — Crystals (acetone), m. 157° to 158° (1942b Marker). — **Fmn.** From $\Delta^{5.16}$-pregnadien-3β-ol-20-one (p. 2230 s) on heating with isopropylmagnesium bromide in toluene on the steam-bath (1942b Marker). — **Rns.** Gives 16α-isopropylprogesterone (p. 2808 s) on refluxing with aluminium tert.-butoxide and acetone in toluene. On reduction with sodium and alcohol it gives 16α-isopropyl-Δ^5-pregnene-3β.20-diol (p. 1946 s). Does not form a semicarbazone (1942 b Marker).

Acetate $C_{26}H_{40}O_3$. Crystals, m. 131–2⁰. — **Fmn.** From the above hydroxy-ketone on treatment with acetic anhydride and pyridine (1942 b Marker).

16α-tert.-Butyl-Δ^5-pregnen-3β-ol-20-one $C_{25}H_{40}O_2$.

For configuration at C-16, see 1952 Romo. — Crystals (acetone), m. 189⁰ to 192⁰ (1942 b Marker). — **Fmn.** From $\Delta^{5.16}$-pregnadien-3β-ol-20-one (p. 2230 s) on heating with tert.-butylmagnesium chloride in toluene on the steam-bath (1942 b Marker). — **Rns.** Gives 16α-tert.-butyl-progesterone (p. 2808 s) on refluxing with aluminium tert.-butoxide and acetone in toluene. On reduction with sodium and alcohol it gives 16α-tert.-butyl-Δ^5-pregnene-3β.20-diol (p. 1946 s). Does not form a semicarbazone (1942 b Marker).

Acetate $C_{27}H_{42}O_3$, m. 156–8⁰ (from methanol). — **Fmn.** From the above hydroxy-ketone on treatment with acetic anhydride and pyridine (1942 b Marker).

1942 a Marker, Wagner, *J. Am. Ch. Soc.* **64** 1273, 1275.
   b Marker, Crooks, *J. Am. Ch. Soc.* **64** 1280.
1944 Wettstein, *Helv.* **27** 1803, 1807 et seq.
1949 Plattner, Heusser, Herzig, *Helv.* **32** 270.
1950 a Heusser, Engel, Herzig, Plattner, *Helv.* **33** 2229.
   b Heusser, Engel, Plattner, *Helv.* **33** 2237.
1951 Sandoval, Rosenkranz, Djerassi, *J. Am. Ch. Soc.* **73** 2383.
1952 Günthard, Beriger, Engel, Heusser, *Helv.* **35** 2437.
   Romo, Lepe, Romero, *C. A.* **1954** 9399.

3-Acetylcholestan-3-ol $C_{29}H_{50}O_2$, m. 173⁰.

— **Fmn.** Along with the two C-3 epimeric 3-methylcholestan-3-ols (p. 1776 s), from cholestan-3-one cyanohydrin on treatment with methylmagnesium iodide; isolated via the *semicarbazone* $C_{30}H_{53}O_2N_3$, dec. 250⁰ (1938 Kuwada).

1938 Kuwada, Miyasaka, *C. A.* **1938** 7474, 7475; *Ch. Ztbl.* **1939** I 1372.

ζ. Monohydroxy - monooxo - steroids (CO in side chain, OH in the ring system) containing cyclic radicals in side chain

3α-Hydroxyandrostan-17β*-yl phenyl ketone, 17β*-Benzoylandrostan-3α-ol

$C_{26}H_{36}O_2$. — **Fmn.** From the benzoate of androsterone (p. 2629 s) on treatment with ethyl α.α-dibromo-phenylacetate in the presence of zinc in ether, followed by heating of the reaction product with alcoholic NaOH and decarboxylation by heating in quinoline at 180–200° (1938 CIBA).

3β-Hydroxyandrostan-17β*-yl phenyl ketone, 17β*-Benzoylandrostan-3β-ol

$C_{26}H_{36}O_2$. — **Fmn.** Similar to that of the preceding compound, starting with the benzoate of androstan-3β-ol-17-one (epiandrosterone, page 2633 s) (1938 CIBA).

21-Benzylidene-Δ⁵-pregnen-3β-ol-20-one

$C_{28}H_{36}O_2$. Crystals (aqueous acetone), m. 130–131° (1942 Marker); $[\alpha]_D$ —18.5° (chloroform) (1945 Velluz). — **Fmn.** From Δ⁵-pregnen-3β-ol-20-one (p. 2233 s) and benzaldehyde in the presence of sodium ethoxide in abs. alcohol on keeping for a day (isolated as acetate; yield, 90%) (1937 SCHERING A.-G.) or similarly from the acetate of the pregnenolone (1942 Marker); from the same pregnenolone and benzaldehyde on refluxing with aqueous methanolic $KHCO_3$ (1945 Velluz). The acetate is obtained from the pregnenolone acetate and benzaldehyde on boiling with some piperidine in abs. alcohol for 5 hrs. with exclusion of moisture (1937 SCHERING A.-G.).

Rns. Gives 21-benzylidene-Δ⁴-pregnene-3.20-dione (21-benzylidenepro-gesterone, p. 2809 s) on refluxing with aluminium tert.-butoxide and acetone in toluene (1942 Marker). After temporary protection of the nuclear double bond by bromination the acetate gives the acetate of Δ⁵-pregnen-3β-ol-20-on-21-al (p. 2378 s) on ozonization in chloroform at —10° (1937 SCHERING A.-G.), and 3β-acetoxy-Δ⁵-ætiocholenic acid on oxidation with CrO_3 in acetic acid at 50° (1942 Marker). On hydrogenation in the presence of Raney nickel it gives 21-benzyl-Δ⁵-pregnen-3β-ol-20-one (below) (1945 Velluz); the acetate reacts similarly on hydrogenation in the presence of palladium-barium sulphate in dioxane (1942 Marker). Absorbs 1 mol. of bromine in chloroform. Does not react with Girard's reagent T (1945 Velluz).

Semicarbazone $C_{29}H_{39}O_2N_3$; m. 184–185° (1945 Velluz).

Acetate $C_{30}H_{38}O_3$. Crystals (ether, m. 184–6° (1937 SCHERING A.-G.); crystals (aqueous acetone), m. 180–2° (1942 Marker); $[\alpha]_D$ —5° (chloroform)

---

* For β-configuration at C-17, assigned by the editor, see the analogous formation of Δ⁵-pregnen-3β-ol-20-one from Δ⁵-androsten-3β-ol-17-one acetate with ethyl α.α-dichloropropionate (p. 2234 s).

(1945 Velluz). — **Fmn.** From the above hydroxyketone on refluxing with acetic anhydride for 1 hr. (1937 SCHERING A.-G.) or on treatment with acetic anhydride in pyridine (1942 Marker). See also above. — **Rns.** See under those of the hydroxyketone (p. 2271 s).

21-Benzyl-*Δ⁵*-pregnen-3*β*-ol-20-one $C_{28}H_{38}O_2$. Crystals (methanol), m. 135⁰ to 136⁰ (1942 Marker); $[\alpha]_D + 6.5^0$ (chloroform) (1945 Velluz). — **Fmn.** From 21-benzylidene-*Δ⁵*-pregnen-3*β*-ol-20-one (p. 2271 s) on hydrogenation in the presence of Raney nickel (1945 Velluz); the acetate is obtained from the acetate of the same compound on hydrogenation in the presence of palladium-barium sulphate in dioxane at room temperature under 3 atm. pressure, and is hydrolysed by refluxing with methanolic $KHCO_3$ (1942 Marker). — **Rns.** On refluxing with aluminium tert.-butoxide and acetone in toluene it gives 21-benzyl-*Δ⁴*-pregnene-3.20-dione (21-benzylprogesterone, p. 2809 s) (1942 Marker). Absorbs 1 mol. of bromine in chloroform (1945 Velluz).

Acetate $C_{30}H_{40}O_3$. Crystals (methanol), m. 127–8⁰ (1945 Velluz), 128–9⁰ and 143–5⁰ (two forms) (1942 Marker). — **Fmn.** See above.

21-Benzylidenepregnan-3*α*-ol-20-one $C_{28}H_{38}O_2$. Crystals (alcohol), m. 230–232⁰ (1939 Marker); $[\alpha]_{5461}^{25} + 181^0$ (alcohol) (1940 Hoehn). Sparingly soluble in acetone and ether (1939 Marker). — **Fmn.** From pregnan-3*α*-ol-20-one (page 2244 s) on keeping with benzaldehyde and sodium ethoxide in abs. alcohol at 25–30⁰ (1939 Marker). — **Rns.** Oxidation with CrO_3 in acetic acid at 25⁰ gives 21-benzylidenepregnane-3.20-dione (p. 2809 s). Oxidation of the acetate with CrO_3 in acetic acid at 50–70⁰ yields 3*α*-acetoxyætiocholanic acid (p. 173) (1939 Marker).

Acetate $C_{30}H_{40}O_3$. Plates (acetone), m. 152⁰. — **Fmn.** On refluxing the above hydroxy-ketone with acetic anhydride (1939 Marker). — **Rns.** See above.

21-Benzylidenepregnan-3*β*-ol-20-one $C_{28}H_{38}O_2$. Crystals (acetone), m. 179⁰ (1939 Marker). — **Fmn.** From pregnan-3*β*-ol-20-one (p. 2245 s) on keeping with benzaldehyde and sodium ethoxide in abs. alcohol (1939 Marker; 1942 Fried). — **Rns.** Oxidation with CrO_3 in acetic acid at 25⁰ gives 21-benzylidenepregnane-3.20-dione (p. 2809 s) (1939 Marker). Oxidation of the acetate with CrO_3 in acetic acid at 60–90⁰ yields 3*β*-acetoxy-ætiocholanic acid (1942 Marker; 1942 Fried).

Acetate $C_{30}H_{40}O_3$. Crystals (acetone), m. 175⁰. — **Fmn.** On refluxing of the above hydroxy-ketone with acetic anhydride (1939 Marker). — **Rns.** See above.

21-Benzylideneallopregnan-3β-ol-20-one $C_{28}H_{38}O_2$. Crystals (methanol), m.

185–7⁰ (1939 Marker). — **Fmn.** From allopregnan-3β-ol-20-one (p. 2247 s) on keeping with benzaldehyde and sodium ethoxide in abs. alcohol (1939 Marker); on treating the crude condensation product with acetic anhydride and pyridine at room temperature, the acetate was obtained in two (probably cis-trans stereoisomeric) forms (see below) which were separated by chromatography on Al_2O_3 (1944 Koechlin). — **Rns.** Oxidation of the high-melting acetate with CrO_3 in acetic acid at 100⁰ gives 3β-acetoxyætioallocholanic acid (p. 174) (1939 Marker) which is also obtained from both acetates by ozonization (1944 Koechlin). Heating of the acetates with PCl_5 in abs. benzene at 50⁰ and ozonization of the reaction product in chloroform, followed by alkaline saponification, gives androstan-3β-ol-17-one (epiandrosterone, p. 2633 s) (1944 Koechlin).

High-melting acetate $C_{30}H_{40}O_3$. Crystals (methanol), m. 207–9⁰ (1939 Marker); leaflets (benzene-ether), m. 211–4⁰; $[\alpha]_D^{16} + 75.5⁰$ (dioxane) (1944 Koechlin). — **Fmn.** (See also above.) On treatment of the above hydroxy-ketone with acetic anhydride (1939 Marker). — **Rns.** See above.

Low-melting acetate $C_{30}H_{40}O_3$. Prisms or leaflets (ether-petroleum ether), m. 150–2⁰; $[\alpha]_D^{16} + 56.5⁰$ (dioxane) (1944 Koechlin). — **Fmn.** and **Rns.** See under those of the free hydroxyketone, above.

3β-Hydroxy-Δ⁵-ternorcholenyl phenyl ketone, 3β-Hydroxy-Δ⁵-pregnen-20α-yl

phenyl ketone* $C_{28}H_{38}O_2$. Prisms (benzene), m. 192–3⁰; $[\alpha]_D^{29} —8⁰$ (chloroform). — **Fmn.** The acetate is obtained from 3β-acetoxy-Δ⁵-bisnorcholenic acid chloride with diphenylcadmium in ether-benzene at room temperature (yield, 93%) or with phenylzinc chloride under similar conditions (yield, 86%). The acetate is also formed from its dibromide (p. 2290 s) on treatment with zinc dust in ether-acetic acid at 40⁰; it is saponified by refluxing with aqueous methanolic Na_2CO_3 in benzene (1945 Cole).

Rns. On refluxing with methanolic KOH in benzene, partial isomerization to 3β-hydroxy-20-iso-Δ⁵-ternorcholenyl phenyl ketone (below) takes place. Refluxing with aluminium tert.-butoxide and cyclohexanone in toluene gives 3-keto-Δ⁴-ternorcholenyl phenyl ketone (p. 2809 s). Reduction (of crude hydroxy-ketone, m. 172–6⁰) with sodium and propanol yields a mixture of

---

* See the footnote on p. 2259 s.

diols from which (3β-hydroxy-Δ^5-ternorcholenyl)-phenyl-carbinol (p. 1947 s) could be isolated via the dibenzoate. With bromine in chloroform the acetate affords 5α.6β-dibromo-3β-acetoxyternorallocholanyl phenyl ketone (p. 2290 s). The acetate fails to form an oxime or a semicarbazone. Heating of the acetate with phenylmagnesium bromide in toluene at 95^0, followed by refluxing with acetic anhydride and acetic acid, gives 1-methyl-2.2-diphenyl-1-(3β-acetoxy-Δ^5-ætiocholen-17β-yl)-ethylene (p. 1845 s) (1945 Cole).

Acetate, 3β-Acetoxy-Δ^5-ternorcholenyl phenyl ketone $C_{30}H_{40}O_3$. Colourless bars (acetone), m. 221^0; [α]$_D^{28}$ —14^0 (chloroform). — **Fmn.** From the above hydroxy-ketone on heating with acetic anhydride and acetic acid at 110^0. See also above. — **Rns.** (See also above). Saponification by refluxing with aqueous methanolic Na_2CO_3 in benzene gives the above hydroxy-ketone, whereas on hydrolysis with KOH partial isomerization to the 20-iso-hydroxy-ketone (below) occurs (1945 Cole).

3β-Hydroxy-20-iso-Δ^5-ternorcholenyl phenyl ketone, 3β-Hydroxy-Δ^5-pregnen-20β-yl phenyl ketone* $C_{28}H_{38}O_2$.

Needles (acetone), m. 200–2^0 (1945 Cole); [α]$_D^{26}$ —47^0 (chloroform) (1945 Cole; but cf. 1946 Barton). — **Fmn.** From the stereoisomeric 3β-hydroxy-Δ^5-ternorcholenyl phenyl ketone (or its acetate) (above) on partial isomerization by refluxing with methanolic KOH in benzene (1945 Cole).

Acetate, 3β-Acetoxy-20-iso-Δ^5-ternorcholenyl phenyl ketone $C_{30}H_{40}O_3$. White needles, m. 212–3^0 (1945 Cole); [α]$_D^{28}$ —43^0 (chloroform) (1945 Cole; but cf. 1946 Barton). — **Fmn.** From the above 20-iso-hydroxy-ketone on heating with acetic anhydride on the steam-bath (1945 Cole). — **Rns.** Mild hydrolysis with Na_2CO_3 gives the corresponding 20-iso-hydroxy-ketone (above), whereas on hydrolysis with KOH or on prolonged hydrolysis with Na_2CO_3 partial isomerization to the stereoisomeric hydroxy-ketone (p. 2273 s) takes place (1945 Cole).

3α-Hydroxyternorcholanyl phenyl ketone, 3α-Hydroxypregnan-20-yl phenyl ketone $C_{28}H_{40}O_2$.

Crystals with $^1/_2$ $CH_3 \cdot$OH (methanol), m. 150–5^0; [α]$_D^{25}$ + 52.5^0 (dioxane). — **Fmn.** From the chloride of 3α-formoxy-bisnorcholanic acid on refluxing with diphenylcadmium in ether-benzene, followed by hydrolysis with boiling methanolic NaOH (1945 Hoehn).

Acetate, 3α-Acetoxyternorcholanyl phenyl ketone $C_{30}H_{42}O_3$. Crystals, m. 174.5^0 to 180^0; [α]$_D^{25}$ + 57.5^0 (dioxane). — **Fmn.** From the above hydroxy-ketone on refluxing with acetic acid and acetic anhydride (1945 Hoehn).

---

* See the footnote on p. 2259 s.

References, p. 2276 s

3α-Hydroxybisnorcholanyl phenyl ketone $C_{29}H_{42}O_2$.

Crystals (methanol), m. 121.5–123°; $[\alpha]_D^{25} + 27.5°$ (dioxane). — **Fmn.** From 3α-formoxynorcholanic acid chloride on refluxing with diphenyl-cadmium in ether-benzene, followed by hydrolysis with boiling methanolic NaOH (1945 Hoehn).

Acetate, 3α-Acetoxybisnorcholanyl phenyl ketone $C_{31}H_{44}O_3$. Crystals (methanol), m. 95–99°; $[\alpha]_D^{25} + 45°$ (dioxane). — **Fmn.** From the above hydroxyketone on refluxing with acetic acid and acetic anhydride (1945 Hoehn).

3β-Hydroxy-Δ^5-norcholenyl phenyl ketone $C_{30}H_{42}O_2$.

Crystals (ethanol), m. 129–134°; $[\alpha]_D^{25} - 30°$ (dioxane). — **Fmn.** From 3β-formoxy-Δ^5-cholenic acid chloride on refluxing with diphenylcadmium in ether-benzene, followed by hydrolysis with boiling methanolic NaOH (1945 Hoehn).

Formate, 3β-Formoxy-Δ^5-norcholenyl phenyl ketone $C_{31}H_{42}O_3$. Crystals (formic acid), m. 150–3°; $[\alpha]_D^{25} - 32.5°$ (dioxane). — **Fmn.** From the above hydroxy-ketone on heating with formic acid (1945 Hoehn).

Acetate, 3β-Acetoxy-Δ^5-norcholenyl phenyl ketone $C_{32}H_{44}O_3$. Crystals, m. 157–8°; $[\alpha]_D^{25} - 30°$ (dioxane). — **Fmn.** From the above hydroxy-ketone on refluxing with acetic acid and acetic anhydride (1945 Hoehn).

3α-Hydroxynorcholanyl phenyl ketone $C_{30}H_{44}O_2$.

Crystals (ether-petroleum ether), m. 146–7°; $[\alpha]_D^{25} + 40°$ (dioxane). — **Fmn.** From 3α-formoxycholanic acid chloride on refluxing with diphenylcadmium in ether-benzene, followed by hydrolysis with boiling methanolic NaOH (1945 Hoehn).

Acetate, 3α-Acetoxynorcholanyl phenyl ketone $C_{32}H_{46}O_3$. Crystals, m. 171–2°; $[\alpha]_D^{25} + 60°$ (dioxane). — **Fmn.** From the above hydroxy-ketone on refluxing with acetic acid and acetic anhydride (1945 Hoehn).

3β-Hydroxy-Δ^5-ternorcholenyl mesityl ketone, 3β-Hydroxy-Δ^5-pregnen-20α-yl 2.4.6-trimethylphenyl ketone $C_{31}H_{44}O_2$.

For configuration at C-20, see 1945 Cole; 1949 Wieland. — White needles (acetone), m. 211–2° (1945 Cole); $[\alpha]_D^{28} - 62°$ (chloroform) (1945 Cole; but cf. 1946 Barton). — **Fmn.** The acetate is formed from 3β-acetoxy-Δ^5-bisnorcholenic acid chloride with mesitylcadmium chloride in ether-toluene, first at room temperature, then at 110° (crude yield, 46%) or by the action of mesitylmagnesium bromide, followed by acetylation (yield, 40%); it is saponified by refluxing with aqueous methanolic Na$_2$CO$_3$ (1945 Cole). — **Rns.** Refluxing of the acetate with ethylmagnesium bromide in ether, followed by keeping with benzoyl chloride, yields merely the benzoate (see below) of the above hydroxy-ketone (1945 Cole).

Acetate, 3β-Acetoxy-Δ⁵-ternorcholenyl mesityl ketone C$_{33}$H$_{46}$O$_3$. White needles (acetone), m. 169–170° (1945 Cole); [α]$_D^{29}$ —54° (chloroform) (1945 Cole; but cf. 1946 Barton). When melted, then cooled, the melt displays a transient blue fluorescence while crystallizing (1945 Cole). — **Fmn.** From the above hydroxy-ketone on warming with acetic anhydride (1945 Cole). See also above. — **Rns.** See above.

Benzoate, 3β-Benzoyloxy-Δ⁵-ternorcholenyl mesityl ketone C$_{38}$H$_{48}$O$_3$. Crystals (alcohol), m. 154–6° (1945 Cole). — **Fmn.** See above under the reactions of the free hydroxy-ketone.

26-Benzylidene-27-nor-Δ⁵-cholesten-3β-ol-25-one C$_{33}$H$_{46}$O$_2$. Solid which melts

first at 120°, then at 154–5°; [α]$_D$ —39° (chloroform). — **Fmn.** From 27-nor-Δ⁵-cholesten-3β-ol-25-one (p. 2263 s) on refluxing with benzaldehyde and aqueous methanolic KHCO$_3$. — **Rns.** The benzylidene double bond and the CO group are reduced on hydrogenation in the presence of Raney nickel. —*Semicarbazone* C$_{34}$H$_{49}$O$_2$N$_3$, m. 248–9°. — *Acetate* C$_{35}$H$_{48}$O$_3$, m. 128–9° (1945 Velluz).

1937 Schering A.-G. (Logemann), *Ger. Pat.* 737023 (issued 1943); *C. A.* **1944** 3781; *Ch. Ztbl.* **1944** I 1457; *Fr. Pat.* 843082 (1938, issued 1939); *C. A.* **1940** 6412; *Ch. Ztbl.* **1940** I 759; Schering Corp. (Logemann), *U.S. Pat.* 2321690 (1938, issued 1943); *C. A.* **1943** 6824; *Ch. Ztbl.* **1945** II 687.
1938 Ciba (Soc. pour l'ind. chim. à Bâle; Ges. f. chem. Ind. Basel) (Oppenauer, Miescher, Kägi), *Swed. Pat.* 96375 (issued 1939); *Ch. Ztbl.* **1940** I 1231; *Ital. Pat.* 362454; *Ch. Ztbl.* **1939** II 3151.
1939 Marker, Wittle, *J. Am. Ch. Soc.* **61** 1329.
1940 Hoehn, Mason, *J. Am. Ch. Soc.* **62** 569.
1942 Fried, Linville, Elderfield, *J. Org. Ch.* **7** 362, 370.
      Marker, Wittle, Jones, Crooks, *J. Am. Ch. Soc.* **64** 1282.
1944 Koechlin, Reichstein, *Helv.* **27** 549, 557, 564 et seq.
1945 Cole, Julian, *J. Am. Ch. Soc.* **67** 1369, 1372, 1375.
      Hoehn, Moffett, *J. Am. Ch. Soc.* **67** 740, 742.
      Velluz, Petit, *Bull. soc. chim.* [5] **12** 949.
1946 Barton, *J. Ch. Soc.* **1946** 1116, 1117.
1949 Wieland, Miescher, *Helv.* **32** 1922.

2. Halogeno-monohydroxy-monooxo-steroids
(CO in Side Chain, OH in the Ring System)

α. Halogen in side chain only

Summary of 21-Halogeno-3-hydroxy-20-keto-compounds with Pregnane or Allopregnane Skeleton

Halogen	Configuration		Position of double bond	M. p. °C.	Acetate m. p. °C.	Page
	of OH	at C-5				
Cl	β	—	Δ^5	165–7	157–8	2277 s
Br	β	—	Δ^5	159	—	2278 s
I	β	—	Δ^5	141	129–131	2279 s
Br	β	norm.	Δ^{16}	155–7	151–4	2280 s
Cl	α	norm.	saturated	95–100	—	2280 s
Cl	β	norm.	saturated	184–5	115–6	2280 s
Br	β	norm.	saturated	127–8	{ 145–7 / 139–141 }	2281 s
Br	β	allo	Δ^{16}	—	146–8	2281 s
Cl	β	allo	saturated	157–9	—	2282 s
Br	β	allo	saturated	145.5	—	2282 s

21-Chloro-$\Delta^{3.5}$-pregnadien-3-ol-20-one $C_{21}H_{29}O_2Cl$ (I, R = Cl). **Enolic form of 21-chloroprogesterone** (p. 2810 s); for position of the double bonds in enol derivatives of progesterone, see 1937 Westphal.

Acetate $C_{23}H_{31}O_3Cl$. — **Fmn.** From 3-acetoxy-$\Delta^{3.5}$-ætiocholadienic acid chloride by reaction with diazomethane, followed by treatment with gaseous HCl. — **Rns.** On heating with zinc dust and glacial acetic acid it gives $\Delta^{3.5}$-pregnadien-3-ol-20-one acetate (p. 2229 s) which on saponification yields progesterone (page 2782 s) (1937 f, 1938 Ciba).

21-Bromo-$\Delta^{3.5}$-pregnadien-3-ol-20-one $C_{21}H_{29}O_2Br$ (I, R = Br) and **21-Iodo-$\Delta^{3.5}$-pregnadien-3-ol-20-one** $C_{21}H_{29}O_2I$ (I, R = I), **enolic forms of 21-bromo-progesterone and 21-iodoprogesterone**, resp. Their *acetates*, $C_{23}H_{31}O_3Br$ and $C_{23}H_{31}O_3I$, are formed similarly to the preceding compound and give the same pregnadienolone acetate on reduction (1937 f, 1938 Ciba).

21-Chloro-Δ^5-pregnen-3β-ol-20-one $C_{21}H_{31}O_2Cl$. Needles (alcohol), crystals (ether), m. 165–7° cor. (1939 Reich; cf. 1937 Steiger; 1941 Roche-Organon, Inc.). Is precipitated by digitonin in 80% alcohol. Reduces ammoniacal silver salt solutions in the cold (1937 Steiger; 1941 Roche-Organon, Inc.). — **Fmn.** From 21-diazo-Δ^5-pregnen-3β-ol-20-one (p. 2293 s) on treatment with gaseous HCl in abs. ether at room temperature (yield, more than 76%)

(1939 Reich; cf. 1938a N. V. ORGANON; 1941 ROCHE-ORGANON, INC.); the acetate is similarly obtained from the acetate of the diazoketone and is also obtained in small yield in the preparation of this acetate from 3β-acet-oxy-Δ^5-ætiocholenic acid chloride with diazomethane in ether; it is hydrolysed by refluxing with conc. HCl in alcohol (1937 Steiger; cf. 1938a N. V. ORGANON; 1941 ROCHE-ORGANON, INC.). The acetate is obtained from Δ^5-pregnene-3β.21-diol-20-one 3-acetate 21-p-toluenesulphonate (p. 2307 s) on refluxing with tetramethylammonium chloride in acetone for 2 hrs. (1940c Reichstein).

Rns. On oxidation with CrO$_3$ in glacial acetic acid at room temperature it gives mainly 21-chloro-Δ^4-pregnene-3.6.20-trione (p. 2971 s) (1937 Steiger); similar oxidation, but after protection of the double bond by bromine, affords progesterone (p. 2782 s) when the dehalogenation is carried out with zinc dust or potassium iodide in glacial acetic acid (1937 Steiger; cf. 1937c, 1938 CIBA); when the oxidation product is heated with fused sodium acetate in glacial acetic acid on the water-bath for several hours and then debrominated with zinc dust in glacial acetic acid, it gives the acetate of Δ^4-pregnen-21-ol-3.20-dione (desoxycorticosterone, p. 2820 s) (1938b N. V. ORGANON; 1941 ROCHE-ORGANON, INC.). On oxidation with aluminium tert.-butoxide and acetone in abs. benzene at room temperature (20 days) it gives 21-chloro-progesterone (p. 2810 s) (1940a Reichstein), also obtained by use of the same reagents with refluxing (24 hrs.) together with a small amount of an **acid** C$_{20}$H$_{28}$O$_3$(?), possibly impure **3-keto-17-iso-Δ^4-ætiocholenic acid** [crystals (dioxane-ether), m. 297–9° cor.] which with diazomethane gives the *methyl ester* C$_{21}$H$_{30}$O$_3$(?), crystals (acetone-pentane, ether-pentane, or methanol), m. 170–174° cor., absorption max. in alcohol at 244 mμ (log ε 4.16) (1939 Reich).

Gives Δ^5-pregnen-3β-ol-20-one (p. 2233 s) on treatment with zinc or alkali iodide in glacial acetic acid or with zinc and conc. HCl in dioxane; the acetate reacts similarly (1937d, g, 1938 CIBA). On heating with lead acetate in dioxane or with potassium acetate in alcohol or glacial acetic acid it gives the 21-acetate of Δ^5-pregnene-3β.21-diol-20-one (p. 2301 s) (1938a N. V. ORGANON). Heating with abs. pyridine on the steam-bath affords N-(3β-hydr-oxy-20-keto-Δ^5-pregnen-21-yl)-pyridinium chloride (p. 2279 s) (1939 Reich).

21-Chloro-Δ^5-pregnen-3β-ol-20-one acetate C$_{23}$H$_{33}$O$_3$Cl. Needles (abs. alcohol), m. 157–8° cor. (1937 Steiger). — **Fmn.** and **Rns.** See under those of the free chloropregnenolone (above).

21-Bromo-Δ^5-pregnen-3β-ol-20-one C$_{21}$H$_{31}$O$_2$Br. Crystals (acetone-ether), m. 159–159.5° cor. (1939 Reich). — **Fmn.** From

21-diazo-Δ^5-pregnen-3β-ol-20-one (p. 2293 s) on treatment with gaseous HBr in abs. ether at room temperature; yield, 58% (1939 Reich). From the 21-p-toluenesulphonate of Δ^5-preg-nene-3β.21-diol-20-one (p. 2307 s) on refluxing with NaBr in methanol or acetone (1940c Reichstein). — **Rns.** On refluxing with aluminium tert.-butoxide and acetone in benzene it gives 21-bromoprogesterone (p. 2810 s) and a small amount of an acid (cf. the analogous reaction of the chloro-compound, p. 2278 s) (1939 Reich). Condensation with diethyl sodiomalonate in boiling benzene, followed by refluxing with methanolic KOH, affords (3β-hydroxy-20-keto-Δ^5-pregnen-21-yl)-malonic acid (1946a Plattner). Gives N-(3β-hydroxy-20-keto-Δ^5-pregnen-21-yl)-pyridinium bromide (below) on heating with abs. pyridine (1939 Reich; cf. 1940 N. V. Organon).

21-Iodo-Δ^5-pregnen-3β-ol-20-one $C_{21}H_{31}O_2I$. Crystals (aqueous methanol), dec. 141° (1943 I. G. Farbenind.). Very unstable, easily losing iodine, e. g. on contact with acetic acid (1940c Reichstein). — **Fmn.** Its acetate is obtained from the 3-acetate 21-p-toluenesulphonate of Δ^5-pregnene-3β.21-diol-20-one (p. 2307 s) on treatment with NaI in acetone at room temperature, followed by brief boiling; yield, 82% (1940c Reichstein). From the disodium derivative of Δ^5-pregnen-3β-ol-20-one-21-glyoxylic acid (obtained from Δ^5-pregnen-3β-ol-20-one with diethyl oxalate + sodium ethoxide, followed by hydrolysis with NaOH) on treatment with iodine in aqueous methanolic KOH at room temperature (1943 I. G. Farbenind.).

Rns. Gives Δ^5-pregnen-3β-ol-20-one (p. 2233 s) on reduction with HI (1937c Ciba). On treatment with potassium acetate in acetone it gives the 21-acetate of Δ^5-pregnene-3β.21-diol-20-one (p. 2305 s) (1943 I. G. Farbenind.).

Acetate $C_{23}H_{33}O_3I$. Colourless crystals (ether-pentane), m. 129–131°; becomes yellowish on keeping (1940c Reichstein). — **Fmn.** See above.

N-(3β-Hydroxy-20-keto-Δ^5-pregnen-21-yl)-pyridinium hydroxide $C_{26}H_{37}O_3N$ (II, R = OH).

Chloride $C_{26}H_{36}O_2NCl$ (II, R = Cl). Crystals (aqueous alcohol), m. 289–290° cor. dec.; almost insoluble in ether and benzene, sparingly soluble in water, readily in hot alcohol. — **Fmn.** From 21-chloro-Δ^5-pregnen-3β-ol-20-one (p. 2277 s) on heating with abs. pyridine on the water-bath for 1 hr.; yield, 82%. — **Rns.** With p-nitroso-dimethylaniline and 1 mol. aqueous NaOH in alcohol with cooling in a freezing mixture it gives the 21-N-(p-dimethylaminophenyl)-isoxime of Δ^5-pregnen-3β-ol-20-one-21-al (p. 2378 s) (1939 Reich).

Bromide $C_{26}H_{36}O_2NBr$ (II, R = Br). Colourless crystalline powder, m. ca. 300° cor. dec.; sparingly soluble in water, more readily in alcohol, readily in ether. — **Fmn.** and **Rns.** Similar to those of the above chloride (1939 Reich; 1940 N. V. Organon).

Iodide $C_{26}H_{36}O_2NI$ (II, R = I). Crystals (methanol), m. 248–250°. — **Fmn.** From Δ^5-pregnen-3β-ol-20-one (p. 2233 s) on heating with iodine in pyridine on the steam-bath; yield, 50%. — **Rns.** On heating with NaOH in water or aqueous alcohol it gives 3β-hydroxy-Δ^5-ætiocholenic acid (1944 King).

E.O.C. XIV s. 160

Perchlorate $C_{26}H_{36}O_6NCl$ (II, R = ClO_4). Crystals with $^1/_2$ H_2O (methanol), m. 255–260⁰ dec. — **Fmn.** From the above iodide and perchloric acid in hot water. — **Rns.** Reacts with sodium hydroxide as does the above iodide (1944 King).

21-Bromo-Δ^{16}-pregnen-3β-ol-20-one $C_{21}H_{31}O_2Br$. Crystals, m. 155–7⁰. — **Fmn.** From 17.21-dibromopregnan-3β-ol-20-one (p. 2288 s) on refluxing with excess potassium acetate in glacial acetic acid for 90 min.; the acetate is similarly obtained from the corresp. acetate; yields, 48%. — **Rns.** On hydrogenation in dioxane in the presence of a palladium-barium sulphate catalyst at room temperature under 3 atm. pressure it gives 21-bromopregnan-3β-ol-20-one (p. 2281 s); the acetate reacts similarly. The acetate is converted into the acetate of pregnan-3β-ol-20-one (p. 2245 s) by heating with zinc dust and acetic acid on the steam-bath for 30 min. or by hydrogenation in the presence of palladium in methanol containing pyridine (1942 Marker).

Acetate $C_{23}H_{33}O_3Br$. Plates (methanol), m. 151–4⁰ (1942 Marker). — **Fmn.** and **Rns.** See above.

21-Chloropregnan-3α-ol-20-one $C_{21}H_{33}O_2Cl$ (not analysed). Crystals, m. 95⁰ to 100⁰ (1940b Reichstein). — **Fmn.** From 21-diazopregnan-3α-ol-20-one (p. 2294 s) on treatment with dry hydrogen chloride in abs. ether; yield, 39% (1940b Reichstein; cf. 1942 Marker). — **Rns.** Gives 21-chloropregnane-3.20-dione (p. 2811 s) on oxidation with CrO_3 in glacial acetic acid at room temperature (1940b Reichstein).

21-Chloropregnan-3β-ol-20-one $C_{21}H_{33}O_2Cl$ (III, R=Cl). Crystals (ethyl acetate), m. 184–5⁰ cor.; $[\alpha]_D$ + 116.5⁰ (chloroform) (1943 Ruzicka). — **Fmn.** From 21-diazopregnan-3β-ol-20-one (p. 2295 s) on treatment with dry HCl in abs. ether* (1943 Ruzicka); the acetate is similarly obtained from the corresp. acetate (1942 Fried). — **Rns.** See the pyridinium chloride, below.

Acetate $C_{23}H_{35}O_3Cl$. Prisms (methanol), m. 115–6⁰ cor. (1942 Fried). — **Fmn.** See above.

N-(3β-Hydroxy-20-ketopregnan-21-yl)-pyridinium chloride $C_{26}H_{38}O_2NCl$ [III, R = $\overset{+}{N}(C_5H_5)\overset{-}{Cl}$]. Leaflets (alcohol), m. 284⁰ cor. dec. — **Fmn.** From the above chloropregnanolone on heating with dry pyridine; yield, 88%. —

---

* For this reaction, cf. also 1942 Marker.

References, pp. 2283 s, 2284 s

Rns. On treatment with p-nitrosodimethylaniline and NaOH in alcohol it gives pregnan-3β-ol-20-on-21-al 21-N-(p-dimethylaminophenyl)-isoxime,which is converted into the aldehyde (p. 2380 s) by shaking with 2 N HCl in ether (1943 Ruzicka).

21-Bromopregnan-3β-ol-20-one $C_{21}H_{33}O_2Br$ (III, R = Br). Crystals, m. 127° to 128° (1942 Marker). — **Fmn.** (See also below). From 21-bromo-Δ^{16}-pregnen-3β-ol-20-one (p. 2280 s) on hydrogenation in dioxane in the presence of palladium-barium sulphate at room temperature under 3 atm. pressure (yield, 80%); the acetate is similarly obtained from the corresponding acetate (yield, 90%) (1942 Marker). The acetate is obtained from that of 21-diazo-pregnan-3β-ol-20-one (p. 2295 s) in benzene on treatment with a solution of hydrogen bromide in ether (1942 Fried). — **Rns.** (See also below). On refluxing with fused potassium acetate in glacial acetic acid for 3 hrs. it gives pregnane-3β.21-diol-20-one 21-acetate (p. 2310 s) and some Δ^{16}-pregnen-3β-ol-20-one (p. 2240 s); the acetate reacts similarly (1942 Marker).

Acetate $C_{23}H_{35}O_3Br$. Needles (methanol), m. 145–7° (1942 Marker); prisms (methanol), m. 139–141° cor.; $[\alpha]_D^{29} + 100° \pm 5°$ (chloroform) (1942 Fried). — **Fmn.** and **Rns.** See above.

21-Bromopregnan-3β-ol-20-one (see above) is probably contained in the mixture, mainly consisting of 17-bromopregnan-3β-ol-20-one (p. 2285 s), obtained from pregnan-3β-ol-20-one (p. 2245 s) with 1 mol. of bromine in ether on exposure to sunlight; a similar mixture of the acetates is obtained from pregnan-3β-ol-20-one acetate with 1 mol. of bromine and a little HBr in glacial acetic acid. On boiling of these mixtures with hydroxylamine acetate in alcohol, the present bromopregnanolone (or its acetate) is destroyed and pure 17-bromopregnan-3β-ol-20-one (or its acetate) remains. On oxidation of the mixture of bromopregnanolones with CrO_3 in glacial acetic acid it gives 17-bromopregnane-3.20-dione (p. 2814 s) and, probably, 21-bromopregnane-3.20-dione (p. 2811 s). On boiling with pyridine for 10 min., the mixture of bromopregnanolones gives 15–20% yield of a compound $C_{26}H_{38}O_2NBr$ [yellow blocks (methanol-ethyl acetate), m. 277°; only soluble in methanol and ethanol], probably **N-(3β-hydroxy-20-ketopregnan-21-yl)-pyridinium bromide** [III, p. 2280 s; R = $\overset{+}{N}(C_5H_5)\overset{-}{Br}$] (1938 Masch).

21-Bromo-Δ^{16}-allopregnen-3β-ol-20-one $C_{21}H_{31}O_2Br$. — **Fmn.** Its acetate is

obtained from the acetate of 17.21-dibromoallo-pregnan-3β-ol-20-one (p. 2288 s) on refluxing with fused potassium acetate in glacial acetic acid for 2 hrs.; crude yield, 34%. — **Rns.** The acetate gives the diacetate of Δ^{16}-allopregnene-3β.21-diol-20-one (p. 2309 s) on refluxing with fused potassium acetate and acetic anhydride in glacial acetic acid (1946b Plattner).

160*

Acetate $C_{23}H_{33}O_3Br$. Crystals (aqueous methanol), m. 146–8⁰ cor. dec.; sublimes partly at 90⁰ in a high vacuum with incipient decomposition; $[\alpha]_D^{18} + 21^0$ (chloroform) (1946b Plattner).

21-Chloroallopregnan-3β-ol-20-one $C_{21}H_{33}O_2Cl$ (IV, R = Cl) (not analysed).

Crystals (benzene-ether), m. 157–9⁰ cor. (1940b Reichstein). — **Fmn.** From 21-diazoallopregnan-3β-ol-20-one (p. 2295 s) on treatment with a solution of dry HCl in abs. ether; yield, 66% (1940b Reichstein). — **Rns.** Gives 21-chloro-allopregnane-3.20-dione (p. 2811 s) on oxidation with CrO_3 in glacial acetic acid at room temperature (1940b Reichstein). Gives allopregnan-3β-ol-20-one (p. 2247 s) on reduction with zinc dust or alkali iodide in glacial acetic acid; the acetate (not described in detail) reacts similarly (1937h CIBA). See also the pyridinium chloride, below.

N-(3β-Hydroxy-20-ketoallopregnan-21-yl)-pyridinium chloride $C_{26}H_{38}O_2NCl$

[IV, $R = \overset{+}{N}(C_5H_5)\overset{-}{Cl}$]. Leaflets (methanol-ether) with 1 H_2O (after drying in high vac. at 80–90⁰ for 6 hrs.), m. 273–4⁰ cor. dec. — **Fmn.** From the preceding compound on heating with dry pyridine; yield, 94%. — **Rns.** On treatment with p-nitrosodimethylaniline and NaOH in alcohol it gives allopregnan-3β-ol-20-on-21-al 21-N-(p-dimethylaminophenyl)-isoxime, which is converted into the aldehyde (p. 2380 s) by shaking with 2 N HCl in ether (1943 Ruzicka).

21-Bromoallopregnan-3β-ol-20-one $C_{21}H_{33}O_2Br$ (IV, R = Br). Rodlets

[acetone-ether], m. 144–145.5⁰ cor. — **Fmn.** From 21-diazoallopregnan-3β-ol-20-one (p. 2295 s) on treatment with dry HBr in abs. ether; yield, 62%. — **Rns.** Gives 21-bromoallopregnane-3.20-dione (p. 2811 s) on oxidation with CrO_3 in glacial acetic acid at room temperature (1940b Reichstein).

22-Chloro-ω-homo-Δ^5-pregnen-3β-ol-21-one, 21-Chloromethyl-Δ^5-pregnen-3β-ol-21-one $C_{22}H_{33}O_2Cl$. Crystals (metha-

nol), m. 128⁰ cor. — **Fmn.** Its acetate is obtained from the acetate of 21-diazomethyl-Δ^5-pregnen-3β-ol-21-one (p. 2296 s) on treatment with dry HCl in ether at 20⁰, and also as a by-product in the preparation of this diazo-compound from 3β-acetoxy-Δ^5-pregnen-21-oic acid chloride and diazomethane. — **Rns.** The acetate yields the diacetate of 21-hydroxymethyl-Δ^5-pregnen-3β-ol-21-one (p. 2314 s) on heating with anhydrous potassium acetate in glacial acetic acid (1945 Plattner).

Acetate $C_{24}H_{35}O_3Cl$. Crystals (ethyl acetate), m. 184–184.5⁰ cor.; $[\alpha]_D$ —43⁰ (chloroform) (1945 Plattner). — **Fmn.** and **Rns.** See above.

25-Chloro-Δ^5-homocholen-3β-ol-24-one,   25-Chloro-Δ^5-bisnorcholesten-3β-ol-24-one $C_{25}H_{39}O_2Cl$. — Fmn.

Its acetate is obtained from the acetate of 25-diazo-Δ^5-bisnorcholesten-3β-ol-24-one (p. 2297 s) on treatment with hydrogen chloride in abs. alcohol or chloroform. —

CH₃ — Me — CH·CH₂·CH₂·CO·CH₂Cl — Me — HO

Rns. The acetate gives the acetate of Δ^5-bisnorcholesten-3β-ol-24-one (p. 2263 s) on reduction with zinc and glacial acetic acid on the water-bath (1938 Kuwada).

Acetate $C_{27}H_{41}O_3Cl$. Needles (abs. alcohol), m. 187⁰ (1938 Kuwada). — **Fmn.** and **Rns.** See above.

26-Chloro-27-nor-Δ^5-cholesten-3β-ol-25-one $C_{26}H_{41}O_2Cl$. — Fmn.

Its acetate is obtained from the acetate of 26-diazo-27-nor-Δ^5-cholesten-3β-ol-25-one (prepared from 3β-acetoxy-Δ^5-homocholenic acid chloride and diazomethane) on treatment with dry HCl in ether. — **Rns.** The acetate gives

CH₃ — Me — CH·[CH₂]₃·CO·CH₂Cl — Me — HO

the acetate of 27-nor-Δ^5-cholesten-3β-ol-25-one (p. 2263 s) on reduction with zinc dust and glacial acetic acid (1938 Hattori).

Acetate $C_{28}H_{43}O_3Cl$. Plates, m. 182–4⁰ (1938 Hattori). — **Fmn.** and **Rns.** See above.

21.21-Dibromo-Δ^5-pregnen-3β-ol-20-one $C_{21}H_{30}O_2Br_2$.

Unstable (1937 b, 1938 CIBA). — **Fmn.** From 21-diazo-Δ^5-pregnen-3β-ol-20-one (p. 2293 s) on reaction with bromine (1937 a, b, 1938 CIBA); the acetate is similarly obtained from the corresp. acetate (1937 e CIBA). — **Rns.** Gives Δ^5-pregnen-3β-ol-20-one

Me — Me — CO·CHBr₂ — HO

(p. 2233 s) on reduction with zinc-copper alloy in glacial acetic acid containing HBr (1937 b, 1938 CIBA); the acetate behaves similarly (1937 e CIBA). On hydrogenation in a neutral medium in the presence of nickel or platinum it gives the two 20-epimeric Δ^5-pregnene-3β.20-diols (pp. 1920 s, 1922 s) (1937 b, 1938 CIBA). On boiling with $CaCO_3$ in aqueous propanol or with Na_2CO_3 in aqueous ethanol it yields Δ^5-pregnen-3β-ol-20-on-21-al (p. 2378 s) (1937 a CIBA).

1937 a CIBA*, *Swiss Pat.* 211427 (issued 1940); *C. A.* **1942** 3636.
   b CIBA*, *Swiss Pat.* 212191 (issued 1941); *C. A.* **1942** 3638; *Ch. Ztbl.* **1942** I 231.
   c CIBA*, *Swiss Pat.* 215138 (issued 1941); *C. A.* **1948** 3145.
   d CIBA*, *Swiss Pat.* 215139 (issued 1941); *C. A.* **1948** 3144.
   e CIBA*, *Swiss Pat.* 222121 (issued 1942); *C. A.* **1949** 365; *Ch. Ztbl.* **1943** I 2113.
   f CIBA*, *Swiss Pat.* 222123 (issued 1942); *C. A.* **1949** 366; *Ch. Ztbl.* **1943** I 2113.
   g CIBA*, *Swiss Pat.* 222689 (issued 1942); *C. A.* **1949** 824; *Ch. Ztbl.* **1943** I 2113.
   h CIBA*, *Swiss Pat.* 222690 (issued 1942); *C. A.* **1949** 824; *Ch. Ztbl.* **1943** I 2113.

* Société pour l'industrie chimique à Bâle; Gesellschaft für chemische Industrie in Basel.

Steiger, Reichstein, *Helv.* **20** 1164, 1172, 1174, 1175.
Westphal, *Ber.* **70** 2128, 2130.
1938 CIBA*, *Fr. Pat.* 835814 (issued 1939); *C. A.* **1939** 5008; *Ch. Ztbl.* **1939** I 2827, 2828; *Brit. Pat.*
512954 (issued 1939); *C. A.* **1941** 1800; *Fr. Pat.* 840417 (issued 1939); *Ch. Ztbl.* **1939** II
1125.
Hattori, *C. A.* **1938** 7473; *Ch. Ztbl.* **1939** I 1182.
Kuwada, Yosiki, *J. Pharm. Soc. Japan* **58** 187; *C. A.* **1938** 8432; *Ch. Ztbl.* **1939** I 1372.
Masch, *Über Bromierungen in der Pregnanreihe*, Thesis, Danzig, pp. 14, 24, 27.
a N. V. ORGANON, *Fr. Pat.* 835589 (issued 1938); *Brit. Pat.* 500767 (issued 1939); *C. A.* **1939**
5872; *Ch. Ztbl.* **1939** I 2829; ROCHE-ORGANON, INC. (Reichstein), *U.S. Pat.* 2312480
(issued 1943); *C. A.* **1943** 4862.
b N. V. ORGANON, *Fr. Pat.* 835669 (issued 1938); *C. A.* **1939** 8626; *Ch. Ztbl.* **1939** I 4089;
Indian Pat. 25064 (issued 1938); *Ch. Ztbl.* **1939** I 2458; *Brit. Pat.* 502289 (issued 1939);
C. A. **1939** 6346; ROCHE-ORGANON, INC. (Reichstein), *U.S. Pat.* 2312481 (issued 1943);
C. A. **1943** 5201; *Ch. Ztbl.* **1945** II 687, 688.
1939 Reich, Reichstein, *Helv.* **22** 1124, 1128 et seq., 1134, 1136.
1940 N. V. ORGANON, *Fr. Pat.* 864315 (issued 1941); *Ch. Ztbl.* **1942** I 642.
a Reichstein, v. Euw, *Helv.* **23** 136.
b Reichstein, Fuchs, *Helv.* **23** 658, 666, 668, 669.
c Reichstein, Schindler, *Helv.* **23** 669, 673, 674; ROCHE-ORGANON, INC. (Reichstein), *U.S. Pat.*
2357224 (1941, issued 1944); *C. A.* **1945** 391; *Ch. Ztbl.* **1945** II 1232.
1941 ROCHE-ORGANON, INC. (Reichstein), *U.S. Pat.* 2312483 (issued 1943); *C. A.* **1943** 4862;
Ch. Ztbl. **1945** II 687.
1942 Fried, Linville, Elderfield, *J. Org. Ch.* **7** 362, 371.
Marker, Crooks, Wagner, *J. Am. Ch. Soc.* **64** 213; PARKE, DAVIS & Co. (Marker, Crooks),
U.S. Pat. 2359272 (issued 1944); *C. A.* **1945** 4198; *Ch. Ztbl.* **1945** II 1232.
1943 I. G. FARBENIND., *Fr. Pat.* 891441 (issued 1944); *Ch. Ztbl.* **1946** I 1025; FIAT *Final Report*
No. 996, pp. 10–12; cf. also Ruschig, *Angew. Ch.* **60** A 247 (1948).
Ruzicka, Prelog, Wieland, *Helv.* **26** 2050.
1944 King, *J. Am. Ch. Soc.* **66** 1612.
1945 Plattner, Hardegger, Bucher, *Helv.* **28** 167, 170, 171.
1946 a Plattner, Heusser, Oeschger, *Helv.* **29** 253, 255.
b Plattner, Heusser, Angliker, *Helv.* **29** 468, 472.

β. Halogeno-monohydroxy-monooxo-steroids
(CO in side chain, OH and halogen in the ring system)

5-Chloro-allopregnan-3β-ol-20-one** $C_{21}H_{33}O_2Cl$. Configuration at C-5 ascribed
by analogy with 5-chlorocholestan-3β-ol (p.1877s)
(Editor). — Prisms (ethyl acetate), m. 164–174°
dec. (1947 Meystre; cf. 1946 CIBA PHARMA-
CEUTICAL PRODUCTS). — **Fmn.** From the acetate
of 24.24-diphenyl-5-chloro-$\Delta^{20(22)\cdot 23}$-allocholadien-
3β-ol** (p.1884s) by oxidation in chloroform with
CrO_3 and 80% acetic acid, initially at 5–10° then rising to 17–20°; the acetate
so obtained (yield, 75%) (1946 Meystre) is hydrolysed with slightly aqueous
methanolic HCl on the water-bath (1947 Meystre; cf. 1946 CIBA PHARMA-
CEUTICAL PRODUCTS).

* Société pour l'industrie chimique à Bâle; Gesellschaft für chemische Industrie in Basel.
** Described in the originals as derivatives of pregnane and choladiene, resp.; for allo-configuration, see footnote** on p. 1884 s.

Rns. Oxidation with CrO_3 in 90% acetic acid at 20^0 yields 5-chloroallo-pregnane-3.20-dione* (p. 2813 s) (1947 Meystre; cf. 1946 CIBA PHARMA-CEUTICAL PRODUCTS). On refluxing the acetate with K_2CO_3 in aqueous methanol, Δ^5-pregnen-3β-ol-20-one (p. 2233 s) is obtained; boiling of the acetate with dimethylaniline gives the acetate of Δ^5-pregnen-3β-ol-20-one (1946 Meystre).

Acetate $C_{23}H_{35}O_3Cl$. Crystals (acetone + diisopropyl ether), m. 190–196^0 cor.; $[\alpha]_D^{25} + 6.8^0 \pm 4^0$ (chloroform) (1946 Meystre). — **Fmn.** and **Rns.** See those of the parent ketonol, above.

17-Bromo-Δ^5-pregnen-3β-ol-20-one $C_{21}H_{31}O_2Br$.

Acetate $C_{23}H_{33}O_3Br$. The compound m. 148^0 to 150^0 cor., $[\alpha]_D^{21} - 179^0$ (dioxane), obtained by Ruzicka (1939) from a supposed "17-hydroxy-3β-acetoxy-Δ^5-pregnen-20-one" by reaction with PBr_3, was later shown to be 17a-methyl-17a-bromo-D-homo-Δ^5-androsten-3β-ol-17-one acetate (1940 Ruzicka). Genuine 17-bromo-Δ^5-pregnen-3β-ol-20-one acetate [m. 147–8^0 cor., $[\alpha]_D^{24} - 105^0$ (chloroform)] was obtained (1950 Heusser; cf. 1950 Julian) after the closing date for this volume.

17-Bromopregnan-3β-ol-20-one $C_{21}H_{33}O_2Br$. Crude**: white crystals (ether),

m. 169–171^0 (1942a Marker; 1941b PARKE, DAVIS & Co.; cf. 1938 Masch); pure (?)**: needles (dil. acetone or methanol), m. 192^0; $[\alpha]_D^{22} - 31^0$ (chloroform) (1938 Masch; cf. 1943 Shoppee).

Fmn. From pregnan-3β-ol-20-one (p. 2245 s) in glacial acetic acid by dropwise addition of bromine (1 mol.) in glacial acetic acid, containing small amounts of HBr, at 30^0 (yield, 80%) (1942a Marker; 1941b PARKE, DAVIS & Co.), or by treatment with bromine (1 mol.) in ether at room temperature (1938 Masch); the acetate is obtained similarly from the corresponding acetate (1938 Masch; 1942a Marker; 1941b PARKE, DAVIS & Co.) and is hydrolysed with 1 N methanolic HCl at room temperature (1938 Masch).

Rns. Gives 17-bromopregnane-3.20-dione (p. 2814 s) on oxidation with CrO_3 in glacial acetic acid (1938 Masch; cf. also 1942a Marker). Is reduced to pregnan-3β-ol-20-one (p. 2245 s) by heating with zinc or iron in glacial acetic acid on the steam-bath or by hydrogenation (at room temperature and 3 atm. pressure) in methanol containing pyridine in the presence of palladium-$BaSO_4$; the acetate reacts similarly; attempted hydrogenation of the acetate in the presence of palladium but in the absence of pyridine was without effect (1942a Marker). Treatment of the acetate with 1 mol. of bromine

---

* See footnote** on p. 2284 s.

** The substance m. ca. 170^0 (correct analysis) contains ca. 15–20% of an isomer, probably the 21-bromo-isomer (1938 Masch; cf. 1943 Shoppee); the substance, m. 192^0, free from that isomer, showed a too high C, H and a too low Br content (1938 Masch), and so probably contained some debrominated compound as impurity (Editor).

in carbon tetrachloride at room temperature gives the acetate of 17.21-di-
bromopregnan-3β-ol-20-one (p. 2288 s) (1938 Masch). Treatment of the acetate
with 5 % methanolic KOH at room temperature (20 hrs.) yields a compound
$C_{21}H_{34}O_3$ [needles (aqueous acetone or methanol), m. 197^0; unchanged by
HIO_4 in aqueous alcohol; oxidized by CrO_3 in glacial acetic acid at 5–10^0
to give, in low yield, a neutral product m. 114^0] considered te be *pregnane-*
3β.17-diol-20-one (1938 Masch), but for which the structure 17aβ-methyl-
D-homo-ætiocholane-3β.17aα-diol-17-one was later thought to be more
probable (1944 Shoppee). On refluxing of the acetate with aqueous methanolic $KHCO_3$ for 5 hrs. it gives 17-methyl-3β-hydroxy-ætiocholanic acid methyl
ester (1942 c Marker). Refluxing with dry pyridine yields $\varDelta^{16}$-pregnen-3β-ol-
20-one (p. 2240 s); the acetate behaves similarly (1942 a Marker); for similar
treatment of the crude bromopregnanolone, containing probably the 21-bro-
mo-isomer (1938 Masch), see the latter (p. 2281 s).

Acetate $C_{23}H_{35}O_3Br$. Needles (alcohol or aqueous acetone), m. 167^0 (1938
Masch); the crystals (methanol), m. 152–4^0, described by Marker (1942 a)
(cf. also 1941 b PARKE, DAVIS & Co.) are a mixture of the present acetate
with that of (probably) 21-bromopregnan-3β-ol-20-one (1938 Masch; cf. 1943
Shoppee); $[\alpha]_D^{21}$ —40^0 (chloroform). — **Fmn.** From the free bromo-ketonol
(above) on boiling with acetic anhydride. See also under formation of the
above bromo-ketonol and under the 21-bromo-isomer (p. 2281 s). — **Rns.**
See above.

17-Chloro-allopregnan-3β-ol-20-one $C_{21}H_{33}O_2Cl$ (I, below; R = Cl). — *Acetate*
$C_{23}H_{35}O_3Cl$. Crystals (dil. acetone). — **Fmn.** From the acetate of allopregnan-
3β-ol-20-one (p. 2247 s) in glacial acetic acid on addition of a 1 % solution
of chlorine (1 mol.) in chloroform at 30^0 (1941 b PARKE, DAVIS & Co.).

17-Bromo-allopregnan-3β-ol-20-one $C_{21}H_{33}O_2Br$ (I, R = Br). Crystals (ether-

pentane), m. 93–96^0 (1942 f Marker; 1941 b
PARKE, DAVIS & Co.). — **Fmn.** From allo-
pregnan-3β-ol-20-one (p. 2247 s) on treatment
with 1 mol. of bromine in glacial acetic acid at
room temperature (yield, 42 %); the acetate is
obtained similarly from the corresponding acetate
(1942 f Marker; 1941 b PARKE, DAVIS & Co.; cf. 1944 Koechlin).

Rns. Oxidation with CrO_3 in glacial acetic acid at room temperature yields
impure17-bromo-allopregnane-3.20-dione which gives $\varDelta^{16}$-allopregnene-3.20-di-
one (p. 2795 s) on refluxing with dry pyridine or with anhydrous potassium
acetate in glacial acetic acid, and allopregnane-3.20-dione (p. 2797 s) on
heating on the steam-bath with iron dust and acetic acid. Hydrogenation
under 2$^2/_3$ atm. pressure, in dioxane containing some pyridine, using palla-
dium-$BaSO_4$ catalyst, gives allopregnan-3β-ol-20-one (p. 2247 s); the acetate
is similarly hydrogenated in methanol-dioxane. Reduction to allopregnan-
3β-ol-20-one also occurs on heating with iron and glacial acetic acid on the
steam-bath (1942 f Marker). Treatment of the bromo-hydroxyketone (1941 b

PARKE, DAVIS & Co.) or its acetate (1942f Marker) with bromine (1 mol.) in glacial acetic acid at 40° yields the acetate of 17.21-dibromo-allopregnan-3β-ol-20-one (p. 2288 s). On refluxing with pyridine Δ^{16}-allopregnen-3β-ol-20-one (p. 2242 s) is formed; the acetate undergoes similar dehydrobromination (1942f Marker).

The acetate m. 200–2°, obtained by acetylation of the crude compound m. 170–8°, which was isolated from the neutral fraction after refluxing a crude mixture of 17-bromo- and 17.21-dibromo-allopregnan-3β-ol-20-one acetates with 5% methanolic KOH, was considered to be the methyl ester $C_{24}H_{38}O_4$ of 17-methyl-3β-acetoxy-ætioallocholanic acid arising by rearrangement of the 17-bromo-compound (1944 Koechlin), but more recent investigation (1948 Plattner) has shown these deductions to be erroneous; for the revised formula and structure of the methyl ester, see reactions of the acetate of 17.21-dibromo-allopregnan-3β-ol-20-one (p. 2288 s), from which in fact it arises.

Acetate $C_{23}H_{35}O_3Br$. Crystals (methanol), m. 155° (1942f Marker; 1941b PARKE, DAVIS & Co.). — **Fmn.** and **Rns.** See those of the parent ketonol, above.

17.21-Dibromo-Δ^5-pregnen-3β-ol-20-one $C_{21}H_{30}O_2Br_2$ (II, R = Br).

Acetate $C_{23}H_{32}O_3Br_2$. The mixture obtained by Marker (1942e) (cf. 1941b PARKE, DAVIS & Co.) (see below) must have contained the dibromo-compound as chief constituent, together with some 17-bromo-21-iodo-Δ^5-pregnen-3β-ol-20-one acetate, described below; the two compounds give the same reaction with KOH (1950 Julian). The product m. 148–9°, assigned this structure by SCHERING CORP. (1940), was in fact the impure 17-bromo-21-iodo-analogue (below) (1950 Julian). — No constants given. — **Fmn.** From the acetate of 5α.6β.17.21-tetrabromo-allopregnan-3β-ol-20-one (p. 2290 s) by reaction with the calculated amount of NaI (for re-establishing the 5:6 double bond) in boiling alcohol (1 hr.) (1942e Marker; cf. 1941b PARKE, DAVIS & Co.). — **Rns.** Heating with slightly aqueous methanolic KOH gives 3β-hydroxy-$\Delta^{5.17(20)}$-pregnadien-21-oic acid (1942e Marker; 1941b PARKE, DAVIS & Co.; cf. 1950 Julian).

17-Bromo-21-iodo-Δ^5-pregnen-3β-ol-20-one $C_{21}H_{30}O_2BrI$ (II, R = I).

Acetate $C_{23}H_{32}O_3BrI$. Structure ultimately assigned to the impure compound (1950 Julian) described by SCHERING CORP. (1940) as 17.21-dibromo-Δ^5-pregnen-3β-ol-20-one acetate, m. 148–9° (above). — Crystals (ether-methanol), m. 158° dec.; $[\alpha]_D^{28}$ —54° (chloroform) (1950 Julian). —**Fmn.** From the acetate of 5α.6β.17.21-tetrabromo-allopregnan-3β-ol-20-one (p. 2290 s) by refluxing with NaI (ca. 3.6 mols.) in benzene-alcohol (2 hrs.) (1940 SCHERING CORP.); a purer product is obtained by effecting the reaction at room temperature

(24 hrs.) with a greater excess of NaI (1950 Julian). — **Rns.** Refluxing with potassium benzoate in butanol-toluene gives the 3-acetate 21-benzoate of $\Delta^{5.16}$-pregnadiene-3β.21-diol-20-one (p. 2301 s) (1940 SCHERING CORP.).

17.21-Dibromopregnan-3β-ol-20-one $C_{21}H_{32}O_2Br_2$. White needles (ether), m. 190–2° (depressed by admixture with the acetate, below) (1942 b Marker; 1941 b PARKE, DAVIS & Co.). — **Fmn.** From pregnan-3β-ol-20-one (p. 2245 s) in glacial acetic acid at 40° by treatment with a 1 M solution of bromine (2 mols.) in glacial acetic acid (containing small amounts of HBr); the acetate is obtained similarly (yield, 62%) (1942 b Marker; 1941 b PARKE, DAVIS & Co.; cf. 1938 Masch). The acetate is also obtained from that of 17-bromopregnan-3β-ol-20-one (p. 2285 s) on bromination (1 mol. bromine) in carbon tetrachloride at room temperature (1938 Masch).

Rns. Reduction with iron and glacial acetic acid on the steam-bath regenerates pregnan-3β-ol-20-one; the acetate can be reduced similarly, or by use of zinc instead of iron (1942 b Marker). Boiling with excess methanolic KOH yields 3β-hydroxy-$\Delta^{17(20)}$-pregnen-21-oic acid (1942 d Marker; 1941 b PARKE, DAVIS & Co.). On refluxing with fused potassium acetate in glacial acetic acid, 21-bromo-Δ^{16}-pregnen-3β-ol-20-one (p. 2280 s) is formed; the acetate behaves similarly (1942 b Marker).

Acetate $C_{23}H_{34}O_3Br_2$. Crystals (acetone), m. 190–1° (1942 b Marker; 1941 b PARKE, DAVIS & Co.). — **Fmn.** and **Rns.** See those of the parent ketonol, above.

17.21-Dibromo-allopregnan-3β-ol-20-one $C_{21}H_{32}O_2Br_2$.

Acetate $C_{23}H_{34}O_3Br_2$. Crystals (methanol), m. 174–5° cor.; $[\alpha]_D^{21}$ —14 to —15° (chloroform) (1946 Plattner; cf. 1942 f Marker; 1941 b PARKE, DAVIS & Co.). — **Fmn.** From the acetate of allopregnan-3β-ol-20-one (p. 2247 s) by treatment with bromine (2.2 mols.) in glacial acetic acid containing some conc. HBr at room temperature; crude yield, 73% (1946 Plattner; cf. 1942 f Marker; 1941 b PARKE, DAVIS & Co.; 1944 Koechlin). From the acetate of 17-bromo-allopregnan-3β-ol-20-one (p. 2286 s) by treatment with a 1 M solution of bromine (1 mol.) in glacial acetic acid at 40° (yield, 57%) (1942 f Marker); is similarly obtained from the free bromohydroxyketone (1941 b PARKE, DAVIS & Co.).

Rns. By reduction with zinc dust and glacial acetic acid on the steam-bath allopregnan-3β-ol-20-one acetate is regenerated (1946 Plattner). Boiling with methanolic KOH (2 hrs.) gives 3β-hydroxy-$\Delta^{17(20)}$-allopregnen-21-oic acid (1942 f Marker; 1941 b PARKE, DAVIS & Co.); by treatment with the same reagent (1 hr.), followed by esterification of the acidic fraction with diazomethane at 0° and acetylation with acetic anhydride and pyridine at room temperature, the acetate methyl ester of that acid was isolated (1944 Koechlin); the neutral fraction isolated by Koechlin (1944) was subsequently shown to be, after acetylation, the same acetate methyl ester (1948 Plattner). On boiling with fused potassium acetate and glacial acetic acid, the acetate of 21-bromo-Δ^{16}-allopregnen-3β-ol-20-one (p. 2281 s) is formed; if acetic anhydride is added to the reaction mixture, the diacetate of Δ^{16}-allopregnene-3β.21-diol-20-one (p. 2309 s) is obtained (1946 Plattner).

5α.6β-Dibromo-allopregnan-3β-ol-20-one, Δ^5-Pregnen-3β-ol-20-one dibromide

$C_{21}H_{32}O_2Br_2$. Configuration at C-5 and C-6 is ascribed by analogy with that of (ordinary) cholesterol dibromide (p. 1889 s), similarly prepared (Editor). — White amorphous resin (1945 Julian; cf. 1934 Fernholz).

Fmn. From Δ^5-pregnen-3β-ol-20-one (p. 2233 s) by reaction with bromine (1 mol., avoiding excess) at room temperature or in the cold in chloroform (1934 Fernholz; 1945 Julian), glacial acetic acid (1934b, 1936 Butenandt; 1937 SCHERING CORP.), or methylene chloride (1941 I. G. FARBENIND.). From 20-diphenylmethylene-Δ^5-pregnen-3β-ol (p. 1844 s) by bromination (1 mol.) in chloroform, followed by ozonization at 0° in the same solvent (1935 SCHERING CORP.; 1937 Butenandt; cf. 1934a Butenandt).

Rns. Oxidation with CrO_3 in glacial acetic acid at 20° yields 5α.6β-dibromo-allopregnane-3.20-dione* (p. 2814 s) (1937 SCHERING CORP.; 1945 Julian; cf. reactions of Δ^5-pregnen-3-ol-20-one, p. 115; p. 2235 s). Debromination with zinc powder in glacial acetic acid yields Δ^5-pregnen-3β-ol-20-one (1937 Butenandt; cf. 1934a Butenandt). Gives Δ^4-pregnene-3.20-dione (progesterone, p. 2782 s) on boiling with collidine (1942 SCHERING A.-G.).

16.17-Dibromopregnan-3β-ol-20-one $C_{21}H_{32}O_2Br_2$.

Acetate $C_{23}H_{34}O_3Br_2$. Crystals (methanol), m. 137–140° (1942g Marker). — **Fmn.** From the acetate of Δ^{16}-pregnen-3β-ol-20-one (p. 2240 s) on treatment with 1 mol. of bromine in glacial acetic acid at room temperature (yield, 66%) (1942g Marker); the bromination may also be effected in carbon tetrachloride or nitrobenzene (1941a, b PARKE, DAVIS & Co.).

Rns. Reverts to Δ^{16}-pregnen-3β-ol-20-one acetate on shaking with hydrogen and palladium-barium sulphate in dioxane-pyridine at room temperature under 3 atm. pressure, on refluxing with pyridine for 3 hrs., on refluxing with fused potassium acetate in glacial acetic acid for 1 hr., or on refluxing with sodium iodide in methanol for 1 hr. (1942g Marker; 1941a PARKE, DAVIS & Co.). On refluxing with methanolic KOH for 1 hr. it gives 3β-hydroxy-$\Delta^{17(20)}$-pregnen-21-oic acid and a small amount of its methyl ester (1942g Marker; cf. also 1941a, b PARKE, DAVIS & Co.).

---

* Indicated as 5.6-dibromopregnane-3.20-dione in the original.

5α.6β-Dibromo-norcholestan-3β-ol-25-one, Δ⁵-Norcholesten-3β-ol-25-one
dibromide $C_{26}H_{42}O_2Br_2$. Configuration
at C-5 and C-6 by analogy with that
of cholesterol dibromide (p. 1889 s). —
Crystals, m. 130–1⁰ dec. — **Fmn.**
From 27-nor-Δ⁵-cholesten-3β-ol-25-one
(p. 2263 s) on bromination; the acetate
is similarly formed (1938 Hattori).

Acetate $C_{28}H_{44}O_3Br_2$, m. 125–6⁰ dec. (1938 Hattori).

5α.6β-Dibromo-3β-hydroxy-ternorallocholanyl phenyl ketone, 3β-Hydroxy-
Δ⁵-ternorcholenyl phenyl ketone dibromide
$C_{28}H_{38}O_2Br_2$. Configuration at C-3 deduced
from that of the precursor 3β-acetoxy-
Δ⁵-ternorcholenyl phenyl ketone (p. 2273 s);
that at·C-5 and C-6 by analogy with cholesterol dibromide (p. 1889 s) similarly prepared (Editor).

Acetate $C_{30}H_{40}O_3Br_2$. Plates (acetone), m. 170–1⁰. — **Fmn.** From the
acetate of 3β-hydroxy-Δ⁵-ternorcholenyl phenyl ketone (p. 2273 s) with
bromine (1 mol.) in chloroform. — **Rns.** By heating with zinc in ether-
glacial acetic acid at 40⁰, the precursor is regenerated (1945 Cole).

5α.6β.17.21-Tetrabromo-allopregnan-3β-ol-20-one $C_{21}H_{30}O_2Br_4$. The configuration at C-5 and C-6 is ascribed by analogy
with that of cholesterol dibromide (p. 1889 s)
similarly prepared (Editor). — Only esters are
known. — **Fmn.** The acetate is obtained from
Δ⁵-pregnen-3β-ol-20-one (p. 2233 s) or its acetate by treatment with a 1 M solution of
bromine (3 mols.) in glacial acetic acid, containing some HBr, at 40⁰ (1942 e
Marker; 1941 b PARKE, DAVIS & Co.; cf. 1950 Julian; 1940 SCHERING CORP.);
yield from the acetate, 64%; the propionate is obtained similarly using
Δ⁵-pregnen-3β-ol-20-one or its propionate in chloroform-propionic acid, containing some HBr (1942 e Marker; 1941 b PARKE, DAVIS & Co.).

Rns. Reduction of the acetate with iron powder and acetic acid on the
steam-bath regenerates Δ⁵-pregnen-3β-ol-20-one acetate. On boiling with
sodium iodide (ca. 2 mols.) in alcohol (1 hr.) the acetate gives the impure
acetate of 17.21-dibromo-Δ⁵-pregnen-3β-ol-20-one, which see (p. 2287 s) for
alkaline rearrangement to 3β-hydroxy-Δ⁵·¹⁷⁽²⁰⁾-pregnadien-21-oic acid (1942 e
Marker; 1941 b PARKE, DAVIS & Co.; 1950 Julian); using sodium iodide
(ca. 3.6 mols.) in boiling benzene-alcohol (2 hrs.) the product isolated is the
impure acetate of 17-bromo-21-iodo-Δ⁵-pregnen-3β-ol-20-one (p. 2287 s) (1940
SCHERING CORP.; cf. 1950 Julian).

Acetate $C_{23}H_{32}O_3Br_4$. Crystals (glacial acetic acid), m. 176–7⁰ dec.; sparingly
soluble in ether (1950 Julian; cf. 1942 e Marker; 1941 b PARKE, DAVIS & Co.;
1940 SCHERING CORP.). — **Fmn.** and **Rns.** See above.

References, p. 2291 s

Propionate $C_{24}H_{34}O_3Br_4$. Crystals (ether), m. 175° dec.; sparingly soluble in ether (1942e Marker; 1941b PARKE, DAVIS & Co.). — **Fmn.** See above.

1934 a Butenandt, Westphal, Cobler, *Ber.* **67** 1611, 1616.
    b Butenandt, Westphal, *Ber.* **67** 2085, 2087.
    Fernholz, *Ber.* **67** 2027, 2069.
1935 SCHERING CORP. (Butenandt), *U.S. Pat.* 2313732 (issued 1943); *C. A.* **1943** 5201; *Ch. Ztbl.*
    1944 II 1346; SCHERING-KAHLBAUM A.-G., *Fr. Pat.* 793505 (issued 1936); *C. A.* **1936** 4628;
    Ch. Ztbl. **1936** II 1209.
1936 Butenandt, Westphal, *Ber.* **69** 443, 447.
1937 Butenandt, Fleischer, *Ber.* **70** 96, 100.
    SCHERING CORP. (Inhoffen, Butenandt, Schwenk), *U.S. Pat.* 2340388 (issued 1944); *C. A.*
    1944 4386.
1938 Hattori, *C. A.* **1938** 7473; *Ch. Ztbl.* **1939** I 1182.
    Masch, *Über Bromierungen in der Pregnanreihe*, Thesis Techn. Hochschule Danzig, pp. 12,
    17, 21, et seq.
1939 Butenandt, Mamoli, Heusner, *Ber.* **72** 1614.
    Ruzicka, Meldahl, *Helv.* **22** 421.
1940 Ruzicka, Meldahl, *Helv.* **23** 364; cf. *Helv.* **23** 513.
    SCHERING CORP. (Inhoffen), *U.S. Pat.* 2409043 (issued 1946); *C. A.* **1947** 1398; *Ch. Ztbl.*
    1947 620.
1941 I. G. FARBENIND., *Swiss Pat.* 227974 (issued 1943); *Ch. Ztbl.* **1945** I 196; *Fr. Pat.* 886984
    (1942, issued 1943); *Ch. Ztbl.* **1944** II 1091.
    a PARKE, DAVIS & Co. (Marker, Crooks), *U.S. Pat.* 2343311 (issued 1944); *C. A.* **1944** 3094.
    b PARKE, DAVIS & Co. (Marker, Crooks), *U.S. Pats.* 2359773 (issued 1944), 2369065 (issued
    1945); *C. A.* **1945** 4197, 4198; *Ch. Ztbl.* **1945** II 1232, 1233; **1950** I P 8 (erratum).
1942 a Marker, Crooks, Wagner, *J. Am. Ch. Soc.* **64** 210.
    b Marker, Crooks, Wagner, *J. Am. Ch. Soc.* **64** 213.
    c Marker, Wagner, *J. Am. Ch. Soc.* **64** 216.
    d Marker, Crooks, Wagner, *J. Am. Ch. Soc.* **64** 817.
    e Marker, Crooks, Jones, Shabica, *J. Am. Ch. Soc.* **64** 1276.
    f Marker, Crooks, Wagner, Wittbecker, *J. Am. Ch. Soc.* **64** 2089.
    g Marker, Wagner, Wittbecker, *J. Am. Ch. Soc.* **64** 2093.
    SCHERING A.-G., *Fr. Pat.* 884085 (issued 1943); *Ch. Ztbl.* **1944** I 450.
1943 Shoppee, Prins, *Helv.* **26** 1004, 1008 (footnote 2).
1944 Koechlin, Reichstein, *Helv.* **27** 549, 562.
    Shoppee, *Helv.* **27** 8, 9.
1945 Cole, Julian, *J. Am. Ch. Soc.* **67** 1369, 1372.
    Julian, Cole, Magnani, Meyer, *J. Am. Ch. Soc.* **67** 1728.
1946 CIBA PHARMACEUTICAL PRODUCTS (Miescher, Frey, Meystre, Wettstein), *U.S. Pats.* 2461910,
    2461911 (issued 1949); *C. A.* **1949** 3474, 3475; *Ch. Ztbl.* **1949** E 1870, 1871.
    Meystre, Frey, Neher, Wettstein, Miescher, *Helv.* **29** 627, 632, 633.
    Plattner, Heusser, Angliker, *Helv.* **29** 468.
1947 Meystre, Wettstein, Miescher, *Helv.* **30** 1022, 1027.
1948 Plattner, Heusser, Boyce, *Helv.* **31** 603, 605, 611.
1950 Heusser, Engel, Herzig, Plattner, *Helv.* **33** 2229, 2234.
    Julian, Karpel, *J. Am. Ch. Soc.* **72** 362, 365.

3. Amino-monohydroxy-monooxo-steroids (CO in Side Chain, OH in Ring System)

21-Amino-Δ^5-pregnen-3β-ol-20-one $C_{21}H_{33}O_2N$. — **Fmn.** From Δ^5-pregnen-3β-ol-20-one (p. 2233 s) on treatment with

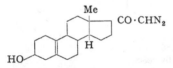

ethyl nitrite and potassium in ether or with amyl nitrite and sodium ethoxide in alcohol, followed by reduction of the resulting 21-iso-nitroso-Δ^5-pregnen-3β-ol-20-one (21-oxime of Δ^5-pregnen-3β-ol-20-on-21-al with sodium in boiling alcohol, with addition of acetic acid for maintenance of slight acidity. — **Rns.** Gives Δ^5-pregnen-3β-ol-20-on-21-al (p. 2378 s) on oxidation, e. g., with sodium nitrite in glacial acetic acid (1937, 1938 Ciba).

1937 Ciba*, *Swiss Pat.* 212193 (issued 1941); *C. A.* **1942** 3636; *Swiss Pat.* 223205 (issued 1942); *C. A.* **1949** 1916; *Ch. Ztbl.* **1943** I 2321; *Fr. Pat.* 845802 (1938, issued 1939); *C. A.* **1941** 1186; *Ch. Ztbl.* **1940** I 2827.
1938 Ciba* (Miescher, Wettstein, Fischer), *U.S. Pat.* 2188914 (issued 1940); *C. A.* **1940** 3882.

---

4. Diazo-monohydroxy-monooxo-steroids (CO in Side Chain, OH in Ring System)

21-Diazo-14-iso-17-iso-19-nor **-$\Delta^{5(10?)}$- pregnen-3β-ol-20-one $C_{20}H_{28}O_2N_2$. For configuration at C-3 (of the parent strophan-thidin), see 1947 Plattner; 1947 Speiser; for configuration at C–14 and C–17, see 1946, 1951, 1952 Ehrenstein; cf. also, e.g., 1936 Elderfield; 1948 Buzas; 1948 Plattner. The position (5:10) of the double bond has been assigned arbitrarily, position 5:6 also being possible (1954 Barber); position 4:5, also taken into account by Ehrenstein (1944), is most unlikely (1954 Barber).

Has been obtained crystalline after the closing date for this volume: pale yellow platelets, becoming turbid at ca. 110°, m. 151–2° dec. (1957 Barber). — **Fmn.** From 14-iso-œstrane-3β.5β-diol-10β.17α-dicarboxylic acid (14-iso-17-iso-21-nor-pregnane-3β.5β-diol-19.20-dioic acid) [obtained from strophan-thidin via several intermediates (1944 Ehrenstein; cf. also 1951, 1952 Ehrenstein; 1957 Barber)] on boiling with acetic anhydride, followed by distillation of the resulting glassy mass consisting mainly of the diacetate (1944, 1952 Ehrenstein; 1951 Barber) in a high vacuum to give 3β-acetoxy-14-iso-17-iso-19-nor **-$\Delta^{5(10?)}$-ætiocholenic acid (1944, 1952 Ehrenstein); this acid is converted by thionyl chloride into its chloride, treated with diazomethane in dry ether, and then saponified with slightly aqueous methanolic KOH at room temperature (1944 Ehrenstein; 1957 Barber). — **Rns.** Gives 21-diazo-14-iso-17-iso-19-nor **-progesterone (p. 2816 s) on refluxing with dry acetone and aluminium tert.-butoxide in dry benzene (1944 Ehrenstein).

---

* Gesellschaft für chemische Industrie in Basel; Société pour l'industrie chimique à Bâle.
** For explanation of the change from designation 10-nor-, as used by 1944 Ehrenstein, see 1951 Ehrenstein; cf. also p. 1355 s.

Acetate $C_{22}H_{30}O_3N_2$. Amber-coloured, viscous mass, not obtained pure (1944 Ehrenstein). — **Fmn.** See above.

21-Diazo-$\Delta^{3\cdot5}$-pregnadien-3-ol-20-one $C_{21}H_{28}O_2N_2$. **Enolic form of 21-diazo-progesterone** (p. 2816 s).

Acetate $C_{23}H_{30}O_3N_2$. No constants given. — **Fmn.** From 3-acetoxy-$\Delta^{3\cdot5}$-ætiocholadienic acid chloride by reaction with diazomethane (1937 e CIBA). — **Rns.** Treatment with HCl, HBr, or HI gives the corresponding 21-halogeno-progesterone 3-enol acetate (p. 2277 s) (1937 e CIBA).

21-Diazo-Δ^5-pregnen-3β-ol-20-one $C_{21}H_{30}O_2N_2$ (p. 117). Crystals, m. 148–9°; $[\alpha]_D + 66°$ (methylene chloride?) (1942/44 SCHERING A.-G.; cf. 1939 SCHERING A.-G.). — **Fmn.** From 3β-acetoxy-Δ^5-ætiocholenic acid (p. 173) by treatment with SOCl$_2$ for conversion into its crude chloride which is treated with diazomethane in dry ether at —10°, then at room temperature (yield, 57%) (1937 Steiger), ether being replaceable by methylene chloride (1942/44 SCHERING A.-G.); the acetoxy-diazoketone so obtained is then saponified in methanol at room temperature, either with KOH (1937 Steiger; 1942/44 SCHERING A.-G.) or with K$_2$CO$_3$ (yield, quantitative) (1938 N. V. ORGANON).

Rns. Oxidation with acetone and aluminium tert.-butoxide in abs. benzene, at room temperature (20 days) or on refluxing (14 hrs.), gives 21-diazo-progesterone (p. 2816 s) (1940a Reichstein). Reduction with alcoholic H$_2$S in the presence of 2 N ammonium hydroxide solution yields the corresponding hydrazone, which by acid hydrolysis gives Δ^5-pregnen-3β-ol-20-on-21-al (p. 2378 s); H$_2$S may be replaced by hydrogen catalysed by colloidal platinum, or by other reducing agents (1937 c, 1938 b CIBA). Treatment with bromine (1 mol.; conditions not stated) gives the unstable 21.21-dibromo-derivative (p. 2283 s) convertible by hydrolysis into Δ^5-pregnen-3β-ol-20-on-21-al (p. 2378 s) (1937 a, b CIBA); the acetate behaves similarly (1937 d CIBA). Reaction with 2 mols. of bromine (conditions not stated) gives a tetrabromo-compound (not isolated), which by CrO$_3$ oxidation in strong acetic acid, followed by debromination with zinc or NaI in the same solvent, yields progesterone (p. 2782 s) (1938a CIBA). With HCl (1937 Steiger; 1939 Reich) or HBr (1939 Reich), in abs. ether at room temperature, 21-chloro- or 21-bromo-Δ^5-pregnen-3β-ol-20-one (pp. 2277 s, 2278 s) is obtained. With HI (from NaI and HCl) in alcohol, first at room temperature and then on boiling, the corresponding 21-iodo-compound first formed is immediately reduced to Δ^5-pregnen-3β-ol-20-one (p. 2233 s) (1937f CIBA), also obtained by similar treatment of the acetate (1938a CIBA). Hydrolysis with aqueous H$_2$SO$_4$ yields Δ^5-pregnene-3β.21-diol-20-one (p. 2301 s); aqueous phosphoric or toluenesulphonic acid may also be used (1938 N. V. ORGANON; cf. 1937 Steiger). On heating with pure glacial acetic acid at 95°, the 21-acetate of Δ^5-pregnene-

$3\beta.21$-diol-20-one (p. 2301 s) is obtained (1937 Steiger); with bromoacetic acid in benzene the corresponding 21-bromoacetate results; the acetate behaves similarly (1945 b Plattner). Heating the acetate with ethyl bromoacetate and activated zinc turnings in benzene, followed by treatment with dil. alcoholic HCl, yields $3\beta.21$-dihydroxy-$\Delta^{5.20(22)}$-norcholadienic acid lactone (1942 CIBA PHARMACEUTICAL PRODUCTS). Acidolysis to give a 21-monoester of Δ^5-pregnene-$3\beta.21$-diol-20-one can be effected with benzoic acid by heating in dioxane at 120° (1937 Steiger) or with p-toluenesulphonic acid in abs. benzene at room temperature, then at 45–50°; the 3-acetate behaves similarly (1940 c Reichstein; 1941 ROCHE-ORGANON, INC.); other acids may be used (1939 SCHERING A.-G.).

Acetate $C_{23}H_{32}O_3N_2$. Pale yellow leaflets (ether-pentane), m. 150–3° cor. dec. (1937 Steiger); $[\alpha]_D^{20} + 66°$ (methylene chloride?) (1942/44 SCHERING A.-G.). Not precipitated by alcoholic digitonin (1937 Steiger). — **Fmn.** and **Rns.** See those of the parent diazolone (above).

21-Diazo-Δ^{11}-pregnen-3α-ol-20-one $C_{21}H_{30}O_2N_2$. See formation of Δ^{11}-pregnene-3.21-diol-20-one 21-acetate (p. 2308 s).

21-Diazopregnan-3α-ol-20-one $C_{21}H_{32}O_2N_2$. Crystals (ether), m. 174–8° dec. (crude). — **Fmn.** From the acetate of ætiolithocholic acid (p. 173) by treatment with thionyl chloride at 0°, then at 20° (1940 b Reichstein), or in dry benzene at room temperature (1942 a Ruzicka), for conversion into the crude acid chloride which is allowed to react with diazomethane in benzene-ether at —10° to —15°, rising to room temperature (1940 b Reichstein; 1942 a Ruzicka); the crude acetate so obtained is saponified with KOH in slightly aqueous methanol at 20° (1940 b Reichstein). — **Rns.** Treatment with dry HCl in abs. ether gives 21-chloropregnan-3α-ol-20-one (p. 2280 s). Heating with glacial acetic acid at 95° gives the 21-acetate of pregnane-3α.21-diol-20-one (p. 2309 s) (1940 b Reichstein); the acetate behaves similarly (1942 a Ruzicka); similar acidolysis occurs with bromoacetic acid (1942 a CIBA; 1943 CIBA PHARMACEUTICAL PRODUCTS). Heating the acetate with ethyl bromoacetate and activated zinc turnings in benzene, followed by treatment with dil. alcoholic HCl, yields 3α.21-dihydroxy-$\Delta^{20(22)}$-norcholenic acid lactone* (1942 CIBA PHARMACEUTICAL PRODUCTS).

Acetate $C_{23}H_{34}O_3N_2$. Yellow crystals (from the benzene-ether reaction mixture, crude), m. 76–84° dec. (1940 b Reichstein; cf. 1942 a Ruzicka); crystals (ether), m. 140–2° cor. dec. (1942 a Ruzicka). Shows a broad absorption zone (solvent not stated) between 240 and 290 mμ (log ε ca. 4.0) and a low absorption max. at ca. 380 mμ (log ε 1.5); $[\alpha]_D + 189°$ (chloroform) (1942 a Ruzicka). — **Fmn.** and **Rns.** See those of the parent diazolone (above).

---

* This lactone is erroneously designated as 14-desoxy-digitoxigenin in the patent cited; it is in fact the C-3, C-14-epimer of this compound (Editor; cf. 1949 Fieser).

21-Diazopregnan-3β-ol-20-one $C_{21}H_{32}O_2N_2$. Crystals (ether?), m. 128–132° dec. (crude). — **Fmn.** From the chloride of 3β-acetoxy-ætiocholanic acid in benzene on treatment with diazomethane in abs. ether, first at —15°, then at room temperature, followed by saponification of the crude *acetate* (yellow syrup) obtained, using slightly aqueous methanolic KOH at 20° (1940b Reichstein; 1942 Fried). — **Rns.** Treatment with dry HCl in abs. ether gives 21-chloropregnan-3β-ol-20-one (p. 2280 s) (1943 Ruzicka); the acetate (reacting at 0°) behaves similarly or, by using HBr in benzene-ether, yields the acetate of 21-bromopregnan-3β-ol-20-one (p. 2281 s) (1942 Fried). Acetolysis of the diazolone, or its acetate, with glacial acetic acid at 95°, gives the 21-acetate, or the diacetate, of pregnane-3β.21-diol-20-one (p. 2310 s) (1940b Reichstein; 1942 Fried).

21-Diazoallopregnan-3α-ol-20-one, *21-Diazo-3-epi-hydroxy-allopregnan-20-one* $C_{21}H_{32}O_2N_2$. — **Fmn.** The free compound has not been described nor its mode of formation indicated. The acetate has been obtained from the chloride of 3α-acetoxy-alloætiocholanic acid (cf. p. 173) in benzene by treatment with diazomethane in ether at —10° (1943 Plattner). — **Rns.** Hydrogenation in alcohol in the presence of palladium gives allopregnan-3α-ol-20-one (p. 2246 s) (1938a CIBA). Acetolysis of the acetate, using glacial acetic acid on the steam-bath, gives the diacetate of allopregnane-3α.21-diol-20-one (p. 2311 s) (1943 Plattner); reaction of the diazolone with bromoacetic acid is similar (1942c CIBA).

Acetate $C_{23}H_{34}O_3N_2$. Crystals (ethyl acetate-hexane), dec. 156–8° cor.; $[\alpha]_D$ + 140° (average) (chloroform). Shows the ultra-violet absorption spectrum typical for α-diazoketones (1943 Plattner; cf. 1942a Ruzicka).

21-Diazoallopregnan-3β-ol-20-one $C_{21}H_{32}O_2N_2$. Crystals (ether), m. 170–2° cor. dec. (1939 Reichstein). — **Fmn.** From the chloride of 3β-acetoxy-ætioallocholanic acid (cf. p. 173), by treatment with diazomethane in dry ether, first at —10° then at room temperature, followed by saponification of the acetate so formed, using slightly aqueous methanolic KOH at room temperature (1939 Reichstein). — **Rns.** Treatment with dry HCl or HBr in absolute ether gives 21-chloro- or 21-bromo-allopregnan-3β-ol-20-one (p. 2282 s) (1940b Reichstein; cf. 1943 Ruzicka). Acetolysis of the diazolone, or its acetate, using glacial acetic acid at 95–100° yields the 21-acetate, or diacetate, of allopregnane-3β.21-diol-20-one (p. 2311 s) (1939 Reichstein); the diazolone reacts similarly with bromoacetic acid (1942b CIBA).

E.O.C. XIV s. 161

References, pp. 2298 s, 2299 s

Acetate $C_{23}H_{34}O_3N_2$. Crystals (slightly aqueous methanol), m. 134–134.5° cor. dec. (1939 Reichstein); crystals (methanol), m. 131–2° cor.; $[\alpha]_D$ (average) +133° (chloroform) (1943 Plattner); readily soluble in ether (1939 Reichstein). — **Fmn.** and **Rns.** See those of the parent diazolone (above).

21-Diazomethyl-Δ^5-pregnen-3β-ol-21-one $C_{22}H_{32}O_2N_2$.

Acetate $C_{24}H_{34}O_3N_2$. Crystals (benzene-hexane), m. 149.5° cor. dec., $[\alpha]_D$ —49° (chloroform). — **Fmn.** From the chloride of 3β-acetoxy-Δ^5-pregnen-21-oic acid by reaction with diazomethane in abs. benzene-ether, first in the cold then at room temperature; crude yield, quantitative. — **Rns.** Treatment with 5 % ethereal HCl at 20° gives 21-chloromethyl-3β-acetoxy-Δ^5-pregnen-21-one (p. 2282 s). Heating with glacial acetic acid on the steam-bath yields the diacetate of ω-homo-Δ^5-pregnene-3β.22-diol-21-one (p. 2314 s) (1945 a Plattner).

21-Diazomethyl-allopregnan-3β-ol-21-one $C_{22}H_{34}O_2N_2$.

Acetate $C_{24}H_{36}O_3N_2$. Crystals (benzene-hexane), m. 162° cor. dec., $[\alpha]_D$ —8° (chloroform). — **Fmn.** From 3β-acetoxy-allopregnan-21-oic acid, which by treatment with thionyl chloride in benzene in the cold yields a mixture of the acid chloride and the acid anhydride; by reaction of this mixture with diazomethane in abs. benzene-ether, first in the cold then at room temperature, the diazolone acetate is obtained together with ca. 50% of the methyl ester of the starting acid, separable by chromatography on Al_2O_3. — **Rns.** Heating on the steam-bath with glacial acetic acid gives the diacetate of ω-homo-allopregnane-3β.22-diol-21-one (p. 2314 s) (1945 a Plattner).

23-Diazo-Δ^5-norcholen-3β-ol-22-one, 3β-Hydroxy-Δ^5-ternorcholenyl diazomethyl ketone $C_{23}H_{34}O_2N_2$.

Yellow powder (crude product from methylene chloride-diisopropyl ether-hexane), no m. p. indicated, no analysis given (1941 Wettstein). — **Fmn.** From the acid chloride of 3β-acetoxy-Δ^5-bisnorcholenic acid (p. 174) by reaction with diazomethane in methylene chloride, first at 0° then at room temperature; the acetate so obtained is saponified with methanolic KOH at room temperature (1941 Wettstein; cf. 1938 I. G. FARBENIND.). — **Rns.** Heating with glacial acetic acid and fused potassium acetate at 98° gives the 23-acetate of Δ^5-norcholene-3β.23-diol-22-one (p. 2314 s) (1941 Wettstein); similarly the acetate on boiling with glacial acetic acid (1938 I. G. FARBENIND.) or by treatment with bromoacetic acid (1943 CIBA PHARMACEUTICAL PRODUCTS) gives the diacetate or 23-bromoacetate 3-acetate, resp., of that diolone.

References, pp. 2298 s, 2299 s

Acetate $C_{25}H_{38}O_3N_2$. Yellowish needles (diisopropyl ether), not quite pure, m. ca. 260–5° cor.* (unsharp) (1941 Wettstein). — **Fmn.** and **Rns.** See those of the parent diazolone (above).

24-Diazocholan-3α-ol-23-one, 3α-Hydroxy-bisnorcholanyl diazomethyl ketone

$C_{24}H_{38}O_2N_2$.

Acetate $C_{26}H_{40}O_3N_2$. Crystals (benzene-ether), no m. p. given. — **Fmn.** From the chloride of 3α-acetoxy-norcholanic acid (p. 175) in benzene by treatment with ethereal diazomethane. — **Rns.** Boiling with glacial acetic acid yields the diacetate of cholane-3α.24-diol-23-one (p. 2315 s) (1938 I. G. FARBENIND.).

25-Diazo-Δ⁵-bisnorcholesten-3β-ol-24-one, 25-Diazo-Δ⁵-homocholen-3β-ol-24-one, 3β-Hydroxy-Δ⁵-norcholenyl diazomethyl ketone $C_{25}H_{38}O_2N_2$.

Acetate $C_{27}H_{40}O_3N_2$. Pale yellow leaflets (1938 Hattori), m. 155° to 155.5° cor. dec.; $[\alpha]_D$ —52° (chloroform) (1942 b Ruzicka). — **Fmn.** From the acetate of 3β-hydroxy-Δ⁵-cholenic acid (p. 175) by reaction with thionyl chloride for conversion into the acid chloride, which is treated with dry ethereal diazomethane in the cold; yield, 72% (1938 Kuwada; cf. 1938 Hattori). — **Rns.** Treatment with Ag₂O in abs. alcohol gives the ethyl ester of 3β-acetoxy-Δ⁵-homocholenic acid; reaction with NH₃ in alcohol, in presence of AgNO₃, yields the corresponding amide (1938 Hattori). With HCl in abs. alcohol or chloroform, the acetate of 25-chloro-Δ⁵-homocholen-3β-ol-24-one (p. 2283 s) is obtained (1938 Kuwada). Acidolysis by means of anhydrous glacial acetic acid at 95° yields the diacetate of Δ⁵-homocholene 3β.25-diol-24-one (p. 2316 s) (1942 b Ruzicka); similarly the corresponding 25-bromoacetate 3-acetate or the 25-(α-bromopropionate) 3-acetate is obtained by use of bromoacetic acid in benzene (1945 b Plattner) or α-bromopropionic acid (1943 CIBA PHARMACEUTICAL PRODUCTS), resp. Heating with ethyl bromoacetate and zinc in benzene, followed by treatment with dil. alcoholic HCl, heating with acetic anhydride, and hydrolysis, yields β-(3β-hydroxy-Δ⁵-norcholenyl)-Δ^{α.β}-butenolide (1942 CIBA PHARMACEUTICAL PRODUCTS; cf. 1942 b Ruzicka).

25-Diazo-Δ⁵-norcholesten-3β-ol-24-one, 3β-Hydroxy-Δ⁵-norcholenyl α-diazoethyl ketone $C_{26}H_{40}O_2N_2$.

Acetate $C_{28}H_{42}O_3N_2$. No constants given. — **Fmn.** From the chloride of 3β-acetoxy-Δ⁵-cholenic acid (cf. p. 175) by

---

* The m. p. 153–4° indicated in *U.S. Pat.* 2202619 (see 1938 I. G. FARBENIND.) is questioned by Wettstein (1941).

161*

reaction with diazoethane (1942, 1943 CIBA PHARMACEUTICAL PRODUCTS). —
Rns. Boiling with bromoacetic acid yields the 25-bromoacetate 3-acetate of
Δ^5-norcholestene-3β.25-diol-24-one (p. 2316 s) (1943 CIBA PHARMACEUTICAL
PRODUCTS). Condensation with α-bromopropionic ester or nitrile in the presence of zinc gives, after the usual working up, β-(3β-hydroxy-Δ^5-norcholenyl)-
α.β-dimethyl-$\Delta^{\alpha.\beta}$-butenolide (1942 CIBA PHARMACEUTICAL PRODUCTS).

26-Diazo-Δ^5-norcholesten-3β-ol-25-one, 3β-Hydroxy-Δ^5-cholenyl diazomethyl ketone $C_{26}H_{40}O_2N_2$.

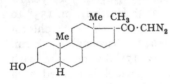

Acetate $C_{28}H_{42}O_3N_2$. No constants
given — **Fmn.** From the chloride of
3β*-acetoxy-Δ^5-homocholenic acid on
treatment with diazomethane. —
Rns. Treatment with dry HCl in
ether gives the acetate of 26-chloro-27-nor-Δ^5-cholesten-3β-ol-25-one (page
2283 s) (1938 Hattori).

17α-Methyl-21-diazopregnan-3β-ol-20-one $C_{22}H_{34}O_2N_2$.

Acetate $C_{24}H_{36}O_3N_2$. No constants given; obtained only as crude intermediate. —**Fmn.** From
the chloride of 17α-methyl-3β-acetoxy-ætiochol-
anic acid by treatment with ethereal diazomethane in the cold. — **Rns.** Treatment with
gaseous HCl in moist ether at room temperature
yields 17α-methylpregnane-3β.21-diol-20-one (p. 2316 s) (1942 Marker).

1936 Elderfield, *J. Biol. Ch.* **113** 631.
1937 a CIBA**, *Swiss Pat.* 211 427 (issued 1940); *C. A.* **1942** 3636.
   b CIBA**, *Swiss Pat.* 212 191 (issued 1941); *C. A.* **1942** 3638; *Ch. Ztbl.* **1942** I 231.
   c CIBA**, *Swiss Pat.* 213 049 (issued 1941); *C. A.* **1942** 4674; *Ch. Ztbl.* **1942** I 385.
   d CIBA**, *Swiss Pat.* 222 121 (issued 1942); *C. A.* **1949** 365; *Ch. Ztbl.* **1943** I 2113.
   e CIBA**, *Swiss Pat.* 222 123 (issued 1942); *C. A.* **1949** 366; *Ch. Ztbl.* **1943** I 2113; *Fr. Pat.*
     840 417 (1938, issued 1939); *Ch. Ztbl.* **1939** II 1125.
   f CIBA**, *Swiss Pat.* 226 816 (issued 1943); *C. A.* **1949** 2376.
   Steiger, Reichstein, *Helv.* **20** 1164, 1168, 1172 et seq., 1176; N. V. ORGANON, *Swiss Pat.* 204 844
     (issued 1939); ROCHE-ORGANON, INC. (Reichstein), *U.S. Pat.* 2 312 483 (1941, issued 1943);
     C. A. **1943** 4862; *Ch. Ztbl.* **1945** II 687.
1938 a —, *Fr. Pat.* 835 814 (issued 1939); *C. A.* **1939** 5008; *Ch. Ztbl.* **1939** I 2827; *Brit. Pat.*
     512 954 (issued 1939); *C. A.* **1941** 1800.
   b CIBA**, *Fr. Pat.* 840 514 (issued 1939); *C. A.* **1939** 8627; *Ch. Ztbl.* **1939** II 1125.
   Hattori, *C. A.* **1938** 7473; *Ch. Ztbl.* **1939** I 1182.
   I. G. FARBENIND., *Fr. Pat.* 847 129 (issued 1939); *C. A.* **1941** 5653; *Ch. Ztbl.* **1940** I 2827;
     WINTHROP CHEMICAL CO. (Bockmühl, Ehrhart, Ruschig, Aumüller), *U.S. Pat.* 2 202 619
     (issued 1940); *C. A.* **1940** 6772.
   Kuwada, Yosiki, *J. Pharm. Soc. Japan* **58** 187 (German abstract); *C. A.* **1938** 8432; *Ch. Ztbl.*
     1939 I 1372.
   N. V. ORGANON, *Fr. Pat.* 835 589 (issued 1938); *Brit. Pat.* 500 767 (issued 1939); *C. A.* **1939**
     5872; *Ch. Ztbl.* **1939** I 2829; ROCHE-ORGANON, INC. (Reichstein), *U.S. Pat.* 2 312 480 (issued
     1943); *C. A.* **1943** 4862.

  * 3β-Configuration ascribed on the basis of that of its ultimate precursor, 3β-hydroxy-Δ^5-cholenic
acid (p. 175) (Editor).
  ** Gesellschaft für chemische Industrie in Basel; Société pour l'industrie chimique à Bâle.

1939 Reich, Reichstein, *Helv.* **22** 1124, 1128.
Reichstein, v. Euw, *Helv.* **22** 1209.
SCHERING A.-G. (Serini, Eysenbach), *Swed. Pat.* 100374 (issued 1940); *Ch. Ztbl.* **1941** I 3259; *Fr. Pat.* 856659 (1939, issued 1940); *Ch. Ztbl.* **1941** I 1196.
1940 a Reichstein, v. Euw, *Helv.* **23** 136.
    b Reichstein, Fuchs, *Helv.* **23** 658, 664, 666 et seq.
    c Reichstein, Schindler, *Helv.* **23** 669.
1941 ROCHE-ORGANON, INC. (Reichstein), *U.S. Pat.* 2357224 (issued 1944); *C. A.* **1945** 391; *Ch. Ztbl.* **1945** II 1232.
Wettstein, *Helv.* **24** 311, 313 (footnote 1), 315.
1942 a CIBA*, *Swiss Pat.* 242988 (issued 1946); *C. A.* **1949** 7977c.
    b CIBA*, *Swiss Pat.* 242989 (issued 1946); *C. A.* **1949** 7977e.
    c CIBA*, *Swiss Pat.* 242990 (issued 1946); *C. A.* **1949** 7977f.
CIBA PHARMACEUTICAL PRODUCTS (Ruzicka), *U.S. Pat.* 2361967 (issued 1944); *C. A.* **1945** 2626.
Fried, Linville, Elderfeld, *J. Org. Ch.* **7** 362, 371.
Marker, Wagner, *J. Am. Ch. Soc.* **64** 1273.
    a Ruzicka, Plattner, Balla, *Helv.* **25** 65, 76.
    b Ruzicka, Plattner, Steusser, *Helv.* **25** 435.
1942/44 SCHERING A.-G., *BIOS Final Report* No. 449, pp. 195, 238 et seq.; *FIAT Final Report* No. 996 p. 91 et seq.
1943 CIBA PHARMACEUTICAL PRODUCTS (Ruzicka), *U.S. Pat.* 2386749 (issued 1945); *C. A.* **1946** 1181; CIBA*, *Brit. Pat.* 579169 (issued 1946); *C. A.* **1947** 2450.
Plattner, Ruzicka, Fürst, *Helv.* **26** 2274.
Ruzicka, Prelog, Wieland, *Helv.* **26** 2050, 2052, 2055.
1944 Ehrenstein, *J. Org. Ch.* **9** 435, 439 et seq., 447 et seq.
1945 a Plattner, Hardegger, Bucher, *Helv.* **28** 167.
    b Plattner, Heusser, *Helv.* **28** 1044.
1946 Ehrenstein, Johnson, *J. Org. Ch.* **11** 823, 829 footnote 8.
1947 Plattner, Segre, Ernst, *Helv.* **30** 1432, 1435.
Speiser, Reichstein, *Helv.* **30** 2143, 2146, 2148.
1948 Buzas, Reichstein, *Helv.* **31** 84.
Ehrenstein, *Ch. Reviews* **42** 457, 475 footnote 5a.
Plattner, Heusser, Segre, *Helv.* **31** 249, 250, 251.
1949 L. F. Fieser, M. Fieser, *Natural Products related to Phenanthrene*, 3rd Ed., New York, pp. 516 et seq., 535, 645, 649.
1951 Barber, Ehrenstein, *J. Org. Ch.* **16** 1622.
Ehrenstein, Barber, Gordon, *J. Org. Ch.* **16** 349, 355 et seq.
1952 Ehrenstein, *Chimia (Switz.)* **6** 287.
1954 Barber, Ehrenstein, *J. Org. Ch.* **19** 365, 366.
1957 Barber, Ehrenstein, *Ann.* **603** 89, 96, 105.

* Gesellschaft für chemische Industrie in Basel; Société pour l'industrie chimique à Bâle.

5. Dihydroxy-monooxo-steroids Containing CO and One OH in Side Chain, the other OH in the Ring System

Summary of Pregnane-(Pregnene-, etc.)-3.21-diol-20-ones

For values other than those chosen for the summary, see the compounds themselves

Config. at		Double bond(s)	M. p. °C.	$[\alpha]_D$*	21-Acetate		Diacetate		Page
C-5	C-3				m. p. °C.	$[\alpha]_D$*	m. p. °C.	$[\alpha]_D$*	
—	—	$\Delta^{3.5}$	—	—	—	—	112	—	2300 s
—	β?	$\Delta^{5.8}$	—	—	—	—	—	—	2300 s
—	β?	$\Delta^{5.9(11)}$	—	—	—	—	—	—	2300 s
—	β	$\Delta^{5.16}$	—	—	—	—	—	—	2301 s
—	β	Δ^5	160	—	185	$+33^0$	167	$+31^0$	2301 s
allo†	β	$\Delta^{9(11)(?)}$	—	—	—	—	147	$+33^0$(ac.)	2308 s
norm.	α	Δ^{11}	—	—	—	—	—	—	2308 s
allo	β	Δ^{14}	159	—	—	—	—	—	2309 s
allo	β	Δ^{16}	—	—	—	—	127	$+36^0$	2309 s
norm.	α	saturated	—	—	181	$+109^0$	88	$+106^0$	2309 s
norm.	β	saturated	—	—	123	—	$\begin{cases} 112 \\ 146 \end{cases}$	$+91^0$	2310 s
allo	α	saturated	—	—	—	—	165	$+92^0$	2311 s
allo	β	saturated	171	$+84^0$(alc.)	204	—	154	$\begin{cases} +82^0 \\ +77^0 \text{(ac.)} \end{cases}$	2311 s

* In chloroform, unless otherwise stated; ac. = acetone.
† 17-iso?

$\Delta^{3.5}$-**Pregnadiene-3.21-diol-20-one** $C_{21}H_{30}O_3$. *3-Enolic form of desoxycortico-sterone* (p. 2820 s). Double bond positions

assigned by analogy with $\Delta^{3.5}$-cholestadien-3-ol (p. 1542 s) and $\Delta^{3.5}$-androstadiene-3.17β-diol (p. 1992 s), similarly obtained (Editor). — Only diesters are known; they give a brown coloration with tetranitromethane (1936a, b Ciba).

Fmn. The diesters are obtained from esters of desoxycorticosterone (p. 2820 s) by boiling with the appropriate acid anhydride and fused potassium acylate, or heating with the corresponding acid chloride and pyridine in an inert solvent (1936a, b Ciba).

Diacetate $C_{25}H_{34}O_5$. Crystals (hexane), m. 111–2^0 (1936a Ciba).

3-Propionate 21-acetate $C_{26}H_{36}O_5$. Colourless needles (hexane) (1936a Ciba).

3-Acetate 21-propionate $C_{26}H_{36}O_5$. Colourless crystals (hexane) (1936b Ciba).

Dipropionate $C_{27}H_{38}O_5$. Colourless needles (hexane) (1936b Ciba).

$\Delta^{5.8}$- **and** $\Delta^{5.9(11)}$-**Pregnadiene-3**(β?)**.21-diol-20-ones** $C_{21}H_{30}O_3$ (I and II, resp.; p. 2301 s). — *21-Acetates* $C_{23}H_{32}O_4$. No constants given. **Fmn.** From the 21-acetate of Δ^5-pregnene-3(β?).11β.21-triol-20-one (p. 2353 s) by heating under reflux with a mixture (4:1) of glacial acetic acid and conc. HCl (1942a Reichstein).

and

$\varDelta^{5.16}$-**Pregnadiene-3β.21-diol-20-one** $C_{21}H_{30}O_3$.

3-Acetate 21-benzoate $C_{30}H_{36}O_5$. Crystals. — **Fmn.** From the acetate of "17.21-di-bromo-$\varDelta^5$-pregnen-3β-ol-20-one" (which in fact was 17-bromo-21-iodo-$\varDelta^5$-pregnen-3β-ol-20-one acetate, see p. 2287 s) by refluxing in toluene with potassium benzoate and butanol. — **Rns.** Partial hydrogenation, using a nickel or palladium catalyst, yields the 3-acetate 21-benzoate of $\varDelta^5$-pregnene-3β.21-diol-20-one (p. 2307 s) (1940 SCHERING CORP.).

$\varDelta^5$-**Pregnene-3β.21-diol-20-one**, *21-Hydroxypregnenolone* $C_{21}H_{32}O_3$ (p. 116).

Colourless leaflets (acetone-ether), which may contain some solvent, m. ca. 155–160°; often a second m. p. can be observed (1938a N. V. ORGANON; 1941 ROCHE-ORGANON, INC.; cf. 1937 Steiger; 1938c N. V. ORGANON). Can be sublimed in high vac. (1937 Steiger). Sparingly soluble in ether (1940b Reichstein). — Is precipitated by digitonin (1938a N. V. ORGANON). Strongly reducing towards methanolic alkaline silver diammine at room temperature (1937 Steiger; 1938a N. V. ORGANON). For colorimetric estimation by reduction of phosphomolybdic acid (Folin-Wu reagent), in acetic acid medium, to molybdenum blue, see 1946 Heard.

FORMATION OF $\varDelta^5$-PREGNENE-3β.21-DIOL-20-ONE

From the 3-acetate by refluxing with conc. HCl in 50% aqueous methanol (1938a N. V. ORGANON; cf. 1937 Steiger), or similarly from the diacetate (1937 Steiger); from the 21-p-toluenesulphonate by saponification with K_2CO_3 in slightly aqueous methanol at room temperature (1940b Reichstein).

From 17-iso-$\varDelta^5$-pregnene-3β.17β.20b.21-tetrol (p. 2140 s) on dehydration with 90% formic acid at 120°, followed by saponification of the ester so formed with aqueous methanolic K_2CO_3 and distillation at 200° in a high vacuum (1937 SCHERING A.-G.); for alternative dehydration of the tetrol*, see 1938a CIBA.

From $\varDelta^5$-pregnen-3β-ol-20-one (p. 2233 s) by reaction with diethyl oxalate in presence of sodium ethoxide to give, after saponification and acidification, $\varDelta^5$-pregnen-3β-ol-20-one-21-glyoxylic acid, which is then treated with lead tetraacetate in alcohol-benzene, in two stages, and the final product of oxidation hydrolysed with aqueous alcoholic H_2SO_4 under reflux (1941 I. G.

* Presumed to be the same as that used in the 1937 SCHERING A.-G. patent (Editor).

FARBENIND.). The 21-acetate is obtained from Δ^5-pregnen-3β-ol-20-one by reaction at room temperature with bromine in chloroform-glacial acetic acid (containing a few drops of HBr), followed by boiling of the resulting oily 5.6.21-tribromopregnan-3β-ol-20-one with potassium acetate in benzene-alcohol (1940 SCHERING CORP.). The diacetate is obtained from the acetate of Δ^5-pregnen-3β-ol-20-one by heating with lead tetraacetate in glacial acetic acid (containing small amounts of acetic anhydride) at 68–70°, followed by chromatography on Al$_2$O$_3$ (yield, 28% of starting material not recovered) (1939b Reichstein; cf. 1939 Ehrhart). The 21-acetate is obtained from 21-chloro-Δ^5-pregnen-3β-ol-20-one (p. 2277 s) by heating with potassium acetate in alcohol or in glacial acetic acid, or with lead acetate in dioxane (1938a N. V. ORGANON; cf. 1938b N. V. ORGANON), or from the 21-iodo-analogue (p. 2279 s) with potassium acetate in acetone on shaking for 3 days at room temperature or refluxing for 6 hrs. (1943 I. G. FARBENIND.). From 21-diazo-Δ^5-pregnen-3β-ol-20-one (p. 2293 s) by hydrolysis with aqueous sulphuric acid (1938a N. V. ORGANON); for formation of the two monoacetates from the same diazo-compound or its acetate, see p. 116; the diacetate (p. 2306 s) is obtained when the acetate of the diazo-compound is heated with glacial acetic acid (1937 Steiger); see also the formation of other esters (pp. 2305 s–2307 s).

From 3β-hydroxy-Δ^5-ætiocholenic acid (p. 173) by condensation of its methyl ester with methoxyacetic ester, using sodium or other alkali metal or alkaline earth metal, followed by hydrolysis of the ester product with dil. alkali or boiling aqueous alcoholic HCl and finally by decarboxylation (1938c N. V. ORGANON). From the acetate of the same starting acid by condensation of the acid chloride with diethyl sodio-chloromalonate in benzene, with subsequent alkaline saponification of the condensation product, followed by acidification and decarboxylation by heating, to give a residue which may be purified by chromatography or via a derivative; an alkoxy-malonic ester metal derivative may also be used in this reaction instead of the chloro-malonic ester (1938b CIBA).

PREPARATION OF Δ^5-PREGNENE-3β.21-DIOL-20-ONE

The 21-acetate is obtained as follows: 3β-Acetoxy-Δ^5-ætiocholenic acid (400 g.) is treated with thionyl chloride (800 c.c.) at —15°, then at +15° and finally at 30°, and after evaporation in vacuo of excess thionyl chloride; the crude acid chloride, dissolved in methylene chloride (1^1/$_2$ l.), and a conc. aqueous solution (2 l.) of KOH·K$_2$CO$_3$ (1:3; d 1.46) are added to a solution of nitrosomethylurea (600 g.) in methylene chloride (60 l.) at —10° over a period of 2 hrs. with stirring; after agitation for a further 2 hrs., the temperature is allowed to rise slowly to +5°. After separation, the methylene chloride solution is treated with Al$_2$O$_3$ (for destruction of excess diazomethane and elimination of resins), concentrated (below 50°) to ca. 2 l., and the resulting solution of 21-diazo-Δ^5-pregnen-3β-ol-20-one acetate is saponified directly by reaction with methanolic KOH (4 l., 50 g.) at ca. 15° for 18 hrs. The crude, solvent-containing product (600–800 g.), obtained by methylene chloride extraction after dilution of the alkaline solution with water, concentration at 50^0, and final evaporation in vac. below 30^0, is heated in glacial acetic acid (2 l.) containing anhydrous potassium acetate (20 g.) at ca. 75^0 for 15–20 min. rising to 93^0 (ca. 30 min.) for conversion into 21-acetoxy-Δ^5-pregnen-3β-ol-20-one, which crystallizes from the reaction mixture and is purified by trituration with warm water and recrystallization from acetone. The over-all yield of pure substance, including further isolation from mother liquors, is 75–76 % (1942/44 Schering A.-G.).

Reactions of Δ^5-Pregnene-3β.21-diol-20-one

Oxidation. Perbenzoic acid oxidation of the 21-acetate in dry chloroform in the cold and then at room temperature yields a mixture of the two 5.6-epoxides ("α" and "β") in which the "α"-oxide (the 21-acetate of $5\alpha.6\alpha$-epoxy-allopregnane-3β.21-diol-20-one, p. 2372 s) largely preponderates (1941 Ehrenstein; cf. 1944b Plattner); a similar oxidation of the diacetate at -10^0 yields a mixture of the diacetate analogues also consisting chiefly of the "α"-oxide isomer (1946a Ruzicka). The 21-esters are converted into the corresponding esters of desoxycorticosterone (p. 2820 s) by treatment with 1 mol. bromine in chloroform or glacial acetic acid, followed by oxidation of the resulting dibromide with CrO_3 in strong acetic acid and then by debromination using zinc dust + sodium acetate on the water-bath (preferably followed by heating with glacial acetic acid) (1937 Steiger; 1937a N. V. Organon; 1938, 1941 Roche-Organon, Inc.), or sodium iodide (1938b N. V. Organon; 1938, 1941 Roche-Organon, Inc.), or aqueous chromous chloride solution (containing zinc chloride and hydrochloric acid) in acetone under CO_2 at room temperature (1945 Julian); the triphenylmethyl ether (p. 2307 s) gives free desoxycorticosterone when brominated and oxidized as above and debrominated by heating with zinc dust + sodium acetate in alcohol on the water-bath, followed by heating with aqueous alcoholic HCl (1938b N. V. Organon; 1938, 1941 Roche-Organon, Inc.).

Oxidation of Δ^5-pregnene-3β.21-diol-20-one by heating with aluminium isopropoxide and acetone in benzene yields Δ^4-pregnene-3.20-dion-21-al (p. 2870 s) (1938 Schering A.-G.). Oxidation of a 21-ester to the corresponding desoxycorticosterone ester is effected by means of cyclohexanone and aluminium tert.-butoxide in boiling toluene, using the 21-acetate or 21-palmitate (1939b Schering A.-G.), with cyclohexanone and aluminium isopropoxide in boiling toluene, using the 21-acetate (1942/44 Schering A.-G.), the 21-propionate, 21-butyrate (1937b N. V. Organon), 21-palmitate (1937c N. V. Organon), or the 21-benzoate (1939b Schering A.-G.; cf. 1940 Schering Corp.), or with acetone and aluminium tert.-butoxide in boiling benzene using the 21-acetate (1937b N. V. Organon). Oxidation of the 21-acetate with p-benzoquinone and aluminium tert.-butoxide in boiling toluene yields 6-dehydrodesoxycorticosterone acetate (p. 2818 s) together with a not identified by-product (m. 175^0 dec.) (1940 Wettstein; 1941 Ciba). The 21-acetate can be oxidized biochemically (with simultaneous hydrolysis) to desoxycorticosterone, using Corynebacterium mediolanum in sterile yeast water (phosphate buffer) at $36–37^0$ in an atmosphere of oxygen (1939 Mamoli).

Reduction. On hydrogenation of Δ^5-pregnene-3β.21-diol-20-one 21-acetate*
in alcohol-glacial acetic acid in the presence of platinum oxide at room
temperature under atmospheric pressure, then in methanol in the presence
of Raney nickel at 80° under 70 atm. pressure, followed by boiling with
methanolic KOH, it gives allopregnane-3β.20.21-triol (probably a mixture of
the C-20 epimers; p. 1954 s) (1938a Steiger). Catalytic hydrogenation of the
diacetate in glacial acetic acid at room temperature, using platinum oxide
catalyst, gives a mixture of 20-epimeric allopregnane-3β.20.21-triol 3.21-di-
acetates (not isolated), converted by CrO_3 oxidation in glacial acetic acid
at room temperature into the 3.21-diacetate of allopregnane-3β.21-diol-20-one
(p. 2311 s) (1946 Plattner). Reduction of the diolone (1939 N. V. ORGANON)
or of its 21-acetate (1938b Steiger; cf. 1940 Miescher; 1943 Ehrenstein;
1944 Fieser) by means of isopropyl alcohol and aluminium isopropoxide gives
a mixture of 20-epimeric Δ^5-pregnene-3β.20.21-triols (p. 1953 s).

Reactions with organic compounds. Reaction of the 21-acetate with methylmagnesium bromide in abs. ether-benzene, and then benzene under reflux,
yields a mixture of C-20 epimeric 20-methyl-Δ^5-pregnene-3β.20.21-triols
(pp. 1955 s, 1956 s) together with by-products which, after acetylation, could
be separated chromatographically and gave two *crystalline products*, m. 184–8°
and 194–6°, resp. (both strongly reducing towards methanolic alkaline silver
diammine solution in the cold) (1941 Hegner). Reaction of the diacetate with
zinc and bromoacetic ester in boiling dry dioxane or benzene yields, after
chromatography, 3β.21-dihydroxy-$\Delta^{5.20(22)}$-norcholadienic acid lactone together with its 3-acetate (1941 b Ruzicka); if the crude reaction product is
refluxed with acetic anhydride, the 3-acetate is practically the sole product
(1942 c Ruzicka), but if reacetylation is effected with acetic anhydride and
pyridine, the same 3-acetate together with the 23→21-lactone of 20.21-di-
hydroxy-3β-acetoxy-Δ^5-norcholenic acid can be isolated (1941 a Ruzicka).
Condensation with zinc and α-bromopropionic ester under similar conditions,
followed by dehydration of the product with boiling acetic anhydride, yields
22-methyl-21-hydroxy-3β-acetoxy-$\Delta^{5.20(22)}$-norcholadienic acid lactone (1944
Ruzicka). No 20.21-monoacetone methyl-lactolide could be isolated after
prolonged shaking with methanol-acetone in presence of anhydrous $CuSO_4$
(1937 Steiger).

PHYSIOLOGICAL PROPERTIES OF Δ^5-PREGNENE-3β.21-DIOL-20-ONE

Hormonal activity. The 21-acetate shows a *corticoid* activity as great
or greater than $^1/_{10}$ of that shown by desoxycorticosterone acetate, the
activity ratio varies in the different tests applied: Waterman dog test (1939
Waterman; 1943 Cleghorn), a modification of this on rats (1941 Selye), work
performance of adrenalectomized rats (1939 Waterman; 1940 Ingle), survival
of adrenalectomized guinea pigs (1946 Bruzzone) or rats (1942 Segaloff);
in contrast to desoxycorticosterone acetate, the compound caused no adrenal

---

* This starting material is given on p .163 of the original paper; the statement on p. 169 of the
same paper, that the reaction was carried out with desoxycorticosterone acetate, is very probably
erroneous (1957 Reichstein).

cortical atrophy on intact female rats (1941 Selye). For effect of the 21-hydroxy-(acetoxy-) group in relation to hormonal activity, see 1942b Selye. The 21-acetate has a weak *folliculoid* (œstrogenic) effect (1942a Selye; 1942 Masson) and weak *luteoid* effect ($^1/_{16}$ that of progesterone) (1942 Mixner; cf. 1942b Selye; 1943 Lipschütz) but is without *testoid* (androgenic) effect (1942a Selye).

Other physiological activities. The 21-acetate exerts a strong anæsthetic effect (rats) (1942a Selye; cf. 1945 Wycis; 1945 Spiegel) and is also anti-convulsant (rats) (1945 Wycis; 1945 Spiegel). The 21-sodium hydrogen phosphate of 3-acetoxy-Δ^5-pregnen-21-ol-20-one exerts a swelling effect (sphering) on the red blood corpuscles of mammals (1941 Netsky).

For biological oxidation, see p. 2303 s.

DERIVATIVES OF Δ^5-PREGNENE-3β.21-DIOL-20-ONE

Δ^5-*Pregnene-3β.21-diol-20-one 3-acetate* $C_{23}H_{34}O_4$ (p. 116). Crystals (ether) (1937 Steiger; 1941 ROCHE-ORGANON, INC.), m. 150–155°; very hygroscopic (1938a N. V. ORGANON; 1941 ROCHE-ORGANON, INC.); sublimes in high vacuum. Strongly reducing towards alkaline silver diammine; gives no precipitate with digitonin (1937 Steiger; 1941 ROCHE-ORGANON, INC.). — **Rns.** Treatment with triphenylmethyl chloride and pyridine in benzene at room temperature gives the 21-triphenylmethyl ether 3-acetate (p. 2308 s) (1938a N. V. ORGANON; 1941 ROCHE-ORGANON, INC.). Hydrolysis to the parent diolone occurs with boiling methanolic HCl (1937 Steiger; 1941 ROCHE-ORGANON, INC.). — For physiological activity of the *3-acetate 21-sodium hydrogen phosphate* $C_{23}H_{34}O_7PNa$ (no properties or formation given), see that of the parent diolone, above.

Δ^5-*Pregnene-3β.21-diol-20-one 21-acetate* $C_{23}H_{34}O_4$ (p. 116). Colourless needles (acetone) (1937 Steiger), needles (glacial acetic acid) containing solvent of crystallization, become opaque at 80° or on standing in air (1938a N. V. ORGANON; 1941 ROCHE-ORGANON, INC.; 1937 Steiger), m. 185°; [α]$_D$ + 34° (acetone) (1948 Pesez; cf. 1942a Selye), [α]$_D$ + 33° (chloroform) (1942/44 SCHERING A.-G.); can be sublimed in high vacuum (1937 Steiger). Is precipitated by digitonin (1938a N. V. ORGANON; 1941 ROCHE-ORGANON, INC.), but not by nicotinic acid hydrazide in methanol or ethanol (1945 Velluz). Is reducing towards alkaline silver diammine (1938a N. V. ORGANON; 1941 ROCHE-ORGANON, INC.). — **Fmn.** For formation from Δ^5-pregnen-3β-ol-20-one, or from its 21-chloro- or 21-iodo-derivatives, see that of the parent diolone (p. 2301 s); for preparation from 3β-acetoxy-Δ^5-ætiocholenic acid, see p. 2302 s. — **Rns.** For oxidation and reduction, see reactions of the parent diolone, p. 2303 s. Attempted saponification with alcoholic KOH, in presence of air or oxygen, causes breakdown to give sodium formate and, after acidification, 3β-hydroxy-Δ^5-ætiocholenic acid (p. 173) (1947 Velluz). Acetylation, using acetic anhydride and pyridine at room temperature, yields the diacetate (see below) (1941a Ruzicka). For other reactions, see those of the parent diolone (p. 2304 s) For physiological properties, see p. 2304 s.

Δ^5-Pregnene-3β.21-diol-20-one diacetate $C_{25}H_{36}O_5$. Six-sided leaflets (ether-pentane or methanol), crystals (acetone-methanol), m. 167⁰ cor. (1939b Reichstein; cf. 1941 Hegner; 1941a Ruzicka; 1948 Ernst); $[\alpha]_D^{22} + 30.5^0$ (chloroform) (1948 Ernst; cf. 1946a Ruzicka). Its methanolic solution reduces alkaline silver diammine rapidly at room temperature (1939b Reichstein). — **Fmn.** From the parent diolone (1940b Reichstein) or from the 21-acetate (above) (1941a Ruzicka) by treatment with acetic anhydride and pyridine at room temperature. As a by-product by transesterification in the acetylation of the 21-bromoacetate (see below); yield, 24 % (1945b Plattner). For formation from the acetate of Δ^5-pregnen-3β-ol-20-one, see formation of the parent diolone (p. 2302 s). From the acetate of 21-diazo-Δ^5-pregnen-3β-ol-20-one (p. 2293 s) on warming with glacial acetic acid (1937 Steiger). — **Rns.** Gives the parent diolone by acid hydrolysis (1937 Steiger). For oxidation, hydrogenation, and Reformatsky condensations, see reactions of the parent diolone, pp. 2303 s, 2304 s.

Δ^5-Pregnene-3β.21-diol-20-one 21-propionate $C_{24}H_{36}O_4$, *21-butyrate* $C_{25}H_{38}O_4$ (1937a, b N. V. ORGANON) and *21-palmitate* $C_{37}H_{62}O_4$ (1937c N. V. ORGANON; 1939b SCHERING A.-G.). No constants given. — **Fmn.** From 21-diazo-Δ^5-pregnen-3β-ol-20-one (p. 2293 s) by acidolysis, no details given (1939b SCHERING A.-G.). — **Rns.** For oxidation to the corresponding esters of Δ^4-pregnen-21-ol-3.20-dione (desoxycorticosterone, p. 2820 s), see oxidation of the parent diolone (p. 2303 s).

Δ^5-Pregnene-3β.21-diol-20-one 21-bromoacetate $C_{23}H_{33}O_4Br$. Leaflets (methanol) containing solvent of crystallization, m. 112–3⁰ cor.; the *monohydrate* is obtained after heating over P_2O_5 at 85⁰ in a high vacuum; $[\alpha]_D^{24} + 25^{0*}$ (chloroform). — **Fmn.** From 21-diazo-Δ^5-pregnen-3β-ol-20-one (p. 2293 s) by reaction with bromoacetic acid in abs. benzene, first at room temperature, then under reflux; yield, 48 %. — **Rns.** Heating with acetic anhydride on the water-bath yields the 3-acetate (below) together with Δ^5-pregnene-3β.21-diol-20-one diacetate (see above) (1945b Plattner).

Δ^5-Pregnene-3β.21-diol-20-one 3-acetate 21-bromoacetate $C_{25}H_{35}O_5Br$. Needles (methanol), m. 137–8⁰ cor.; $[\alpha]_D^{24} + 26^{0*}$ (1945b Plattner), $[\alpha]_D + 22.5^0$ (1942a CIBA) (both in chloroform). — **Fmn.** From the 3-acetate of 21-diazo-Δ^5-pregnen-3β-ol-20-one (p. 2293 s) by reaction with bromoacetic acid in dry benzene (1945b Plattner; cf. 1942a CIBA). From the 21-bromoacetate as described above; yield, 70 % (1945b Plattner). — **Rns.** Treatment with zinc turnings in boiling benzene, in the presence of some bromoacetic ester, gives 20.21-dihydroxy-3β-acetoxy-Δ^5-norcholenic acid 23→21-lactone, which on heating in vacuo at 240⁰ or prolonged boiling with acetic anhydride yields 21-hydroxy-3β-acetoxy-$\Delta^{5.20(22)}$-norcholadienic acid lactone; both these lactones were obtained, after a similar condensation, by chromatography of the first product of condensation; the norcholadienic acid lactone was the sole product isolated

---

* The molecular rotation difference between the 3-acetate 21-bromoacetate and the parent 21-bromoacetate is + 16, whereas a value of ca. 30 would be anticipated, which suggests that one (or both) of the rotations observed is incorrect (1946 Barton).

from a similar reaction in boiling dioxane, when the crude product of condensation was allowed to react with acetic anhydride and pyridine at 60°, followed
by chromatography on Al_2O_3 (1942 a CIBA; cf. 1945 b Plattner).

Δ⁵-Pregnene-3β.21-diol-20-one 21-benzoate $C_{28}H_{36}O_4$. Spherical aggregates
(methanol or acetone-methanol) containing solvent of crystallization, m. ca.
140° (unsharp); resolidifying to lustrous granules, m. 171–3° cor. (1937
Steiger; 1938 a N. V. ORGANON; 1941 ROCHE-ORGANON, INC.). — **Fmn.** From
21-diazo-*Δ⁵*-pregnen-3β-ol-20-one by heating with benzoic acid in dioxane at
120° (1937 Steiger; 1941 ROCHE-ORGANON, INC.). From the 3-acetate 21-ben-
zoate (see below). — **Rns.** For oxidation, see reactions of the parent diolone
(p. 2303 s).

Δ⁵-Pregnene-3β.21-diol-20-one 3-acetate 21-benzoate $C_{30}H_{38}O_5$. No constants
given. — **Fmn.** From the 3-acetate 21-benzoate of *Δ⁵·¹⁶*-pregnadiene-3β.21-
diol-20-one (p. 2301 s) by partial hydrogenation, using a nickel or palladium
catalyst. — **Rns.** Partial saponification yields the 21-benzoate, above (1940
SCHERING CORP.).

Δ⁵-Pregnene-3β.21-diol-20-one 21-p-toluenesulphonate $C_{28}H_{38}O_5S$. Colourless
crystals (ether-pentane), m. 123–4°. — **Fmn.** From 21-diazo-*Δ⁵*-pregnen-
3β-ol-20-one (p. 2293 s) by reaction with p-toluenesulphonic acid in abs. benzene
at room temperature, then warming to 40–50°, followed by chromatography
on Al_2O_3; yield, 44%. — **Rns.** Saponification to the free diolone occurs on
standing with aqueous methanolic K_2CO_3 at room temperature; under similar
conditions there is no appreciable reaction with $KHCO_3$. Treatment with
NaI in acetone, first at room temperature and then short heating, yields
21-iodo-*Δ⁵*-pregnen-3β-ol-20-one (p. 2279 s); similarly refluxing with NaBr in
acetone suspension or in methanolic solution yields the 21-bromo-analogue
(p. 2278 s); the 3-acetate (see below) behaves similarly; the latter also gives
the acetate of 21-chloro-*Δ⁵*-pregnen-3β-ol-20-one (p. 2277 s) by refluxing with
an acetone suspension of tetramethylammonium chloride (1940 b Reichstein).

Δ⁵-Pregnene-3β.21-diol-20-one 3-acetate 21-p-toluenesulphonate $C_{30}H_{40}O_6S$.
Colourless crystals (ether-pentane), m. 120–1°. — **Fmn.** From 21-diazo-
Δ⁵-pregnen-3β-ol-20-one acetate as for the 21-p-toluenesulphonate (above);
yield, 53%. — **Rns.** Treatment with pyridine on the water-bath ($^1/_2$ hr.) or

$$Me \quad Me \quad -CO \cdot CH_2 \cdot \overset{+}{N}(C_5H_5) \cdot \overset{-}{O}SO_2 \cdot C_6H_4 \cdot CH_3 (p)$$

HO (I)

at room temperature (several hrs.) yields **N-(3β-acetoxy-20-keto-*Δ⁵*-preg-
nen-21-yl)-pyridinium p-toluenesulphonate** $C_{35}H_{45}O_6NS$ (I) (no analysis given)
[white crystalline powder (benzene), m. 228° cor. dec.; gives a yellow coloration with aqueous NaOH] (1940 b Reichstein).

Δ⁵-Pregnene-3β.21-diol-20-one 21-triphenylmethyl ether $C_{40}H_{46}O_3$. Almost
colourless resin, slightly contaminated with triphenylcarbinol; only very
weakly reducing towards ammoniacal silver solution in the cold. — **Fmn.**

From Δ^5-pregnene-3β.21-diol-20-one 3-acetate (p. 2305 s) by reaction with triphenylmethyl chloride in pyridine-benzene first at 0°, then standing at room temperature; the *3-acetate* so obtained, as a thick syrup, is saponified with methanolic KOH at room temperature (1938a N. V. ORGANON; 1941 ROCHE-ORGANON, INC.). — **Rns.** For oxidation, see that of the parent diolone (p. 2303 s).

Δ^5-Pregnene-3β.21-diol-20-one di-D-glucoside* $C_{33}H_{52}O_{13}$. No constants given. — **Fmn.** From the parent diolone by reaction with acetobromoglucose (2 mols.) and silver oxide in ether at room temperature, followed by saponification with alcoholic NaOH (1939a CIBA).

(17-iso?)-$\Delta^{9(11)(?)}$-Allopregnene-3β.21-diol-20-one $C_{21}H_{32}O_3$. Position of the double bond is doubtful; the 11:12 position advocated by Shoppee (1940) should probably be revised to 9:11 (1943, 1946 Shoppee), a further shift due to the influence of strong acid not being excluded (1940 Shoppee). The configuration at C-17 is uncertain; partial or total isomerization may occur under the strong acid conditions of formation (1940 Shoppee).

Diacetate $C_{25}H_{36}O_5$. Colourless needles (methanol), not quite pure; m. 146° to 147° cor., with previous softening at 142°; $[\alpha]_D^{17} + 33° \pm 6°$ (acetone) for a less pure specimen. Strongly reducing towards alkaline silver diammine solution at room temperature. Gives a yellow coloration with tetranitromethane in chloroform. — **Fmn.** From 17-iso-allopregnane-3β.11β.21-triol-20-one 3.21-diacetate ("iso-R-diacetate", p. 2355 s) by heating under reflux with glacial acetic acid and conc. HCl (9:1 by vol.), evaporation to dryness in vacuo, and acetylation of the residue with acetic anhydride and pyridine at room temperature; yield, 31% (1940 Shoppee).

Δ^{11}-Pregnene-3α.21-diol-20-one $C_{21}H_{32}O_3$.

21-Acetate $C_{23}H_{34}O_4$. No constants given. — **Fmn.** From 3α-acetoxy-Δ^{11}-ætiocholenic acid by treatment with thionyl chloride to give the acid chloride, which by reaction with diazomethane in ether, followed by saponification with NaOH in methanol, gives **21-diazo-Δ^{11}-pregnen-3α-ol-20-one** $C_{21}H_{30}O_2N_2$ (not described in detail) which is then heated with glacial acetic acid (1942c Reichstein). — **Rns.** Oxidation to the acetate of Δ^{11}-pregnen-21-ol-3.20-dione (p. 2830 s) occurs on boiling with aluminium phenoxide and acetone in benzene, or on standing with CrO_3 in glacial acetic acid. May be directly converted into pregnan-21-ol-3.11.20-trione acetate (p. 2973 s) on treatment with 3 mols. of N-bromoacetamide, followed by oxidation and debromination (1942c Reichstein). When the above-mentioned

---

* Referred to in the patent (abstract) as being derived from 3.21-dihydroxy-Δ^4-pregnen-20-one, but no compound of that structure has been described (Editor).

diazo-hydroxyketone is oxidized with acetone and aluminium phenoxide in abs. benzene and the resulting **21-diazo-Δ^{11}-pregnene-3.20-dione** $C_{21}H_{28}O_2N_2$ (not isolated pure) is heated with glacial acetic acid at 100°, the acetate of Δ^{11}-pregnen-21-ol-3.20-dione (p. 2830 s) is obtained (1942 N. V. ORGANON; cf. also 1943 Reichstein).

Δ^{14}-Allopregnene-3β.21-diol-20-one $C_{21}H_{32}O_3$.

In view of the now accepted structure (1947 Ruzicka) of the precursor "α"-anhydrouzarigenin (p. 232), the position of the double bond originally ascribed at 5:6 (1942a Ruzicka) must be revised to 14:15 (Editor). — M. p. 159°. — **Fmn.** From "α"-anhydrouzarigenin acetate by bromine addition in dil. acetic acid, followed by ozonization, debromination with zinc in alcohol, and hydrolysis with alcoholic $NaHCO_3$ (1942a Ruzicka).

Δ^{16}-Allopregnene-3β.21-diol-20-one $C_{21}H_{32}O_3$.

Diacetate $C_{25}H_{36}O_5$. Plates (methanol), needles (aqueous methanol), m. 126–7° cor.; $[\alpha]_D^{20} + 36°$ (chloroform) (1946 Plattner); absorption max. (solvent not stated) at 243 (4.05) and 310 (1.8) mμ (log ε) (1946 Plattner; cf. 1946b Ruzicka). — **Fmn.** From the acetate of 21-bromo-Δ^{16}-allopregnen-3β-ol-20-one (p. 2281 s) by refluxing with potassium acetate in acetic anhydride-glacial acetic acid; yield, 55%. From the acetate of 17.21-dibromo-allopregnan-3β-ol-20-one (p. 2288 s) by refluxing with potassium acetate in glacial acetic acid containing some acetic anhydride; yield, 20%. From the diacetate of 17-bromo-allopregnane-3β.21-diol-20-one (p. 2317 s) on refluxing with pyridine; crude yield, 95% (1946 Plattner).

Rns. Hydrogenation in 95% alcohol using palladium-$CaCO_3$ catalyst, discontinued when 1 mol. of hydrogen has been absorbed, gives the diacetate of allopregnane-3β.21-diol-20-one (p. 2311 s) (1946 Plattner). Reaction with bromoacetic ester and zinc foil in benzene-dioxane yields β-(3β-acetoxy-Δ^{16}-ætiocholen-17-yl)-$\Delta^{\alpha.\beta}$-butenolide (21-hydroxy-3β-acetoxy-$\Delta^{16.20(22)}$-nor-allocholadienic acid lactone) (1946b Ruzicka).

Pregnane-3α.21-diol-20-one $C_{21}H_{34}O_3$. — Occ.

Is a possible intermediate in the biological reduction of desoxycorticosterone (p. 2820 s) to pregnane-3α.20α-diol (p. 1923 s) (1944 Horwitt),

21-Acetate $C_{23}H_{36}O_4$. Needles (ether-pentane), colourless rodlets (acetone), m. 179.5° to 181°; sublimes at 190–200° (bath temperature) and 0.01 mm. pressure; $[\alpha]_D^{18} + 109°$, $[\alpha]_{5461}^{18} + 136°$ (both in chloroform). — **Fmn.** From 21-diazopregnan-3α-ol-20-one (p. 2294 s) on heating with abs. glacial acetic acid at 95°; yield, 29% pure + 41% crude. — **Rns.**

Oxidation with CrO_3 in glacial acetic acid at 20° gives the acetate of pregnan-21-ol-3.20-dione (p. 2831 s) (1940a Reichstein).

Diacetate $C_{25}H_{38}O_5$. Colourless crystals (ether-hexane) probably containing solvent of crystallization, m. 60–70°; recrystallized from the melt, m. 86–88°; $[\alpha]_D + 106°$ (chloroform). — **Fmn.** From the crude acetate of 21-diazopregnan-3α-ol-20-one by heating on the water-bath with abs. glacial acetic acid; yield, 55%. — **Rns.** Refluxing in benzene-dioxane with bromoacetic ester and zinc, in the presence of iodine, followed by acetylation of the product with warm acetic anhydride and pyridine, gives 21-hydroxy-3α-acetoxy-$\Delta^{20(22)}$-norcholenic acid lactone together with 20.21-dihydroxy-3α-acetoxy-norcholanic acid lactone (1942b Ruzicka).

21-Bromoacetate $C_{23}H_{35}O_4Br$. No constants given. — **Fmn.** From 21-diazo-pregnan-3α-ol-20-one (p. 2294 s) by treatment with bromoacetic acid. — **Rns.** Treatment with zinc foil in benzene, containing small amounts of bromo-acetic ester, followed by acetylation of the product in warm acetic anhydride and pyridine, gives 21-hydroxy-3α-acetoxy-$\Delta^{20(22)}$-norcholenic acid lactone together with 20.21-dihydroxy-3α-acetoxy-norcholanic acid γ-lactone (1942b CIBA).

Pregnane-3β.21-diol-20-one $C_{21}H_{34}O_3$.

Me

Me ———CO·CH$_2$·OH

HO———H

21-Acetate $C_{23}H_{36}O_4$. Crystals with $^1/_2 H_2O$ (ether-pentane), m. 119–123°; occasionally a second form, m. 136–8°, is obtained (1940a Reichstein); m. 121–3° (1942a Marker). — **Fmn.** From crude 21-diazopregnan-3β-ol-20-one (p. 2295 s) by heating with glacial acetic acid at 95°; crude yield, 97% (1940a Reichstein). Together with Δ^{16}-pregnen-3β-ol-20-one (p. 2240 s) (isolated from the mother liquors) from 21-bromopregnan-3β-ol-20-one (p. 2281 s) by refluxing with fused potassium acetate in glacial acetic acid; yield, 37% (1942a Marker). — **Rns.** Oxidation with CrO_3 in glacial acetic acid at 20° gives the acetate of pregnan-21-ol-3.20-dione (p. 2831 s) (1940a Reichstein).

Diacetate $C_{25}H_{38}O_5$. Prisms (methanol) containing $^1/_2$ MeOH, soften at 50–60° and melt with vapour evolution; m. 111–2° cor. after drying at 80° under 15 mm. pressure (1942 Fried); m. 145–6° (1942a Marker); $[\alpha]_D^{27} + 91 \pm 4°$ (chloroform) (1942 Fried). — **Fmn.** Together with the acetate of Δ^{16}-pregnen-3β-ol-20-one (p. 2240 s), from the acetate of 21-bromopregnan-3β-ol-20-one (p. 2281 s) by refluxing with fused potassium acetate in glacial acetic acid; yield, 28% (1942a Marker). From 3β-acetoxy-ætiocholanic acid by treatment with thionyl chloride to give the acid chloride, which is treated with diazo-methane in benzene-ether; the oily diazo-compound so obtained is then heated on the steam-bath with glacial acetic acid; crude yield, 52% after chromatography on Al_2O_3 (1942 Fried). — **Rns.** Heating with zinc and bromoacetic ester in boiling benzene, followed by further refluxing after addition of alcohol, gives 14-desoxy-14-iso-thevetigenin together with 20.21-dihydroxy-3β-acetoxy-norcholanic acid 23→21-lactone (1942 Fried).

Allopregnane-3α.21-diol-20-one $C_{21}H_{34}O_3$.

Diacetate $C_{25}H_{38}O_5$. Colourless needles (benzene-hexane), m. 165° cor., $[\alpha]_D + 92°$ (chloroform). — **Fmn.** From the acetate of 21-diazoallopregnan-3α-ol-20-one (p. 2295 s) by heating with glacial acetic acid on the water-bath. — **Rns.** Boiling in dry benzene-dioxane with zinc and bromoacetic ester, in the presence of iodine, followed by acetylation of the product with acetic anhydride and pyridine, gives the γ-lactone of 20.21-dihydroxy-3α-acetoxy-norallocholanic acid together with that of 21-hydroxy-3α-acetoxy-$Δ^{20(22)}$-norallocholenic acid lactone (1942 d CIBA).

Allopregnane-3β.21-diol-20-one $C_{21}H_{34}O_3$. Solid, m. 150–2°*. — **Fmn.** From

17-iso-allopregnane-3β.17.20.21-tetrol (page 2144 s) by sublimation with zinc dust at 150–170° in a high vacuum; the diacetate is similarly obtained from the 3.20.21-tri-acetate of the same tetrol (1939 a SCHERING A.-G.). From the aforesaid tetrol on dehy-dration by heating with 90 % formic acid at 120° for 2 hrs., refluxing of the resulting product with aqueous methanolic K_2CO_3, and distillation at 200° in a high vacuum (yield, 0.8 %) (1937 SCHERING A.-G.). The diacetate is obtained from $Δ^{17(20)}$-allopregnene-3β.21-diol (p. 1936 s) by treatment with monoperphthalic acid in chloroform in the cold, followed by heating of the resulting epoxide with HCl in acetone in an atmosphere of nitrogen, acetylation with acetic anhydride in pyridine, and chromatography on Al_2O_3 (1939 c SCHERING A.-G.). The diacetate is obtained from the acetate of allopregnan-3β-ol-20-one (p. 2247 s) on heating with lead tetraacetate and some acetic anhydride in glacial acetic acid at 68–70° (yield, 53 %) (1939 b Reichstein) or on treatment with Caro's acid in glacial acetic acid at 25° for 10 days, along with the acetate of "dihydroisoandrosterone" (androstane-3β.17β-diol, p. 2014 s), from which it is separated by means of Girard's reagent (1940, Marker; cf. 1944 Koechlin). The 21-acetate is obtained from crude 21-diazo-allopregnan-3β-ol-20-one (p. 2295 s) by heating with glacial acetic acid at 95–100° (yield, 58 %); the diacetate is similarly obtained from the diazolone acetate (1939 a Reichstein). The diacetate is obtained from that of $Δ^5$-preg-nene-3β.21-diol-20-one (p. 2301 s) by catalytic hydrogenation, using platinum oxide in glacial acetic acid at 70°, followed by reoxidation of the product, partially triol, with CrO_3 in glacial acetic acid at room temperature (yield, 75 %); the diacetate is also obtained from that of $Δ^{16}$-allopregnene-3β.21-diol-20-one (p. 2309 s) by catalytic hydrogenation, using palladium-calcium carbonate in 95 % alcohol at room temperature, discontinued after absorption of 1 mol. of hydrogen (yield, 86 %) (1946 Plattner).

---

* In view of the constants (m. 170–1°, $[\alpha]_D^{23} + 84°$ in abs. alcohol) indicated for this compound (1952 Schneider) after the closing date for this volume, it is probable that this m. p., quoted in *Swiss Pat.* 229604, is that of the diacetate (cf. the parallel *Fr. Pat.* 856641) (Editor).

E.O.C. XIV s. 162

Rns. Oxidation of the 21-acetate with CrO_3 in glacial acetic acid at room temperature gives the acetate of allopregnan-21-ol-3.20-dione (p. 2832 s) (1939a Reichstein). On bromination of the diacetate with 1 mol. of bromine in carbon tetrachloride-glacial acetic acid at 60° it gives the 17-bromo-derivative (p. 2317 s) (1946 Plattner). The crude diacetate, as obtained from allopregnan-3β-ol-20-one acetate with Caro's acid in glacial acetic acid (see above), gives 3β-hydroxy-ætioallocholanic acid (p. 173) on refluxing with 5 % alcoholic KOH (1940 Marker). Gives β-(3β-acetoxy-ætioallocholan-17-yl)·$\Delta^{\alpha,\beta}$-butenolide (21-hydroxy-3β-acetoxy-$\Delta^{20(22)}$-norallocholenic acid lactone) on condensation of the diacetate with bromoacetic ester and zinc in benzene, followed by dehydration of the product by distillation with Al_2O_3 at 180° under 0.01 mm. pressure (1938c CIBA) or by boiling with acetic anhydride (1943 Plattner; 1946b Ruzicka).

21-Acetate $C_{23}H_{36}O_4$. Needles (glacial acetic acid or methanol), leaflets (acetone-pentane), m. 202–204° cor.; can be sublimed in a high vacuum (1939a Reichstein). — **Fmn.** and **Rns.** See above.

Diacetate $C_{25}H_{38}O_5$. Leaflets (ether-pentane), crystals (benzene-hexane), needles (methanol, or acetone-methanol), m. 152–153.5° cor. (1939a, b Reichstein; 1942 v. Euw; 1943, 1946 Plattner); the needles (from methanol) become opaque at 90–100° (1939a Reichstein); $[\alpha]_D +81.5°$ (average, in chloroform) (1943, 1946 Plattner), $[\alpha]_D^{17} +77°$, $[\alpha]_{5461}^{17} +93°$ (both in acetone), $[\alpha]_D^{19} +78°$, $[\alpha]_{5461}^{19} +94°$ (both in dioxane) (1942 v. Euw). — **Fmn.** From the above diolone on treatment with acetic anhydride and pyridine at room temperature (1937 SCHERING A.-G.); similarly from the 21-acetate (1939a Reichstein). See also formation of the parent diolone, above. — **Rns.** Is not further oxidized by lead tetraacetate in glacial acetic acid containing some acetic anhydride, at 70° (1939b Reichstein). See also reactions of the parent diolone, above.

21-Bromoacetate $C_{23}H_{35}O_4Br$. No constants given. — **Fmn.** From 21-diazo-allopregnan-3β-ol-20-one (p. 2295 s) by treatment with bromoacetic acid. — **Rns.** Treatment with zinc foil in benzene gives 21-hydroxy-3β-acetoxy-$\Delta^{20(22)}$-norallocholenic acid lactone together with 20.21-dihydroxy-3β-acetoxy-norallocholanic acid γ-lactone (1942c CIBA).

Δ^5-Pregnene-3β.20 (β ?)-diol-21-al $C_{21}H_{32}O_3$. Configuration at C-3 based on that of its ultimate precursor, 3β-hydroxy-Δ^5-cholenic acid (p. 173); 20β-configuration seems probable in view of the rotational difference between the diol and its 20-monoacetate, as compared with that for known 20β-hydroxy-pregnanes (Editor).

Dimethyl acetal $C_{23}H_{38}O_4$. Needles (methanol), m. 135–6° cor., $[\alpha]_D^{16} -48°$ (methanol); only general absorption starting from ca. 260 mμ, appreciable only below 234 mμ. The acetal and its esters give a red coloration on heating with 1.4-dihydroxynaphthalene and a little conc. HCl in glacial acetic acid on the water-bath (Raudnitz-Puluj reaction). — **Fmn.** From the dimethyl-acetal of Δ^5-pregnen-3β-ol-20-on-21-al (p. 2379 s) by reduction with aluminium isopropoxide and isopropyl alcohol with slow distillation; separation from the 20(α?)-epimer (not isolated) also formed is effected by repeated crystallization or, better, by conversion to the diacetate using glacial acetic acid and pyridine first at room temperature, then at 70° (yield, 68%), followed by boiling with methanolic KOH. — **Rns.** Gives the dimethyl acetal of Δ^4-pregnen-20(β?)-ol-3-on-21-al (p. 2379 s) on oxidation with acetone and aluminium tert.-butoxide in boiling dry benzene or with cyclohexanone and aluminium isopropoxide in boiling toluene; the 20-acetate reacts similarly (1941 Schindler).

3-Acetate dimethyl acetal $C_{25}H_{40}O_5$. Crystals (ether-petroleum ether), m. 122.5–123° cor., $[\alpha]_D^{20}$ —21.4° ± 3°(?*) (acetone). — **Fmn.** From the above dimethyl acetal by acetylation with acetic anhydride and pyridine at 20°, followed by chromatography on Al_2O_3 for separation from some diacetate (1941 Schindler).

20-Acetate dimethyl acetal $C_{25}H_{40}O_5$. Colourless needles (methanol), m. 151° to 152° cor., $[\alpha]_D^{17}$ —17° ± 2° (methanol) (1941 Schindler), $[\alpha]_D$ —26.0° ± 4° (acetone) (1956 Reichstein, *private communication*). — **Fmn.** From the diacetate (see below) by saponification with K_2CO_3 in slightly aqueous methanol at 20°; yield, 68% (1941 Schindler).

Diacetate dimethyl acetal $C_{27}H_{42}O_6$. Flat needles (ether-pentane) (1941 Schindler), plates (acetone) (1956 Reichstein, *private communication*), m. 185° to 186° cor.; $[\alpha]_D^{17}$ —21.5° ± 2° (1941 Schindler), —25.8° ± 2° (1956 Reichstein, *private communication*) (both in acetone). — **Fmn.** From the diolal dimethyl acetal or its 3-acetate (above) on acetylation with heating (1941 Schindler). See also formation of the diolal dimethyl acetal, above. — **Rns.** Gives the above 20-acetate dimethyl acetal on hydrolysis with K_2CO_3 in slightly aqueous methanol at 20° (1941 Schindler).

Allopregnane-3β.20-diol-21-al** $C_{21}H_{34}O_3$. Configuration at C-3 and C-5 based on that of the precursor, allopregnan-3β-ol-21-al (p. 2252 s). — No constants given. — **Fmn.** From the acetate of allopregnan-3β-ol-21-al (p. 2252 s) by reaction with bromine in glacial acetic acid containing some HBr at room temperature, followed by treatment of the product with potassium acetate in boiling abs. alcohol; after chromatography the resulting *diacetate* $C_{25}H_{38}O_5$ [crystals (hexane or dil. acetone)] is saponified with aqueous alcoholic K_2CO_3 (1939 b CIBA).

---

* A rotation slightly more laevorotatory than that of the free dimethyl acetal would be expected (Editor; cf. 1946 Barton, and the rotations quoted for the 21-acetate and the diacetate, below); owing to lack of material, the determination of this rotation could not be repeated (1956 Reichstein, *private communication*).

** Referred to as pregnane-3.20-diol-21-al in the patents cited.

162*

References, pp. 2317–2319 s

ω-**Homo-** Δ^5-**pregnene-3**β.**22-diol-21-one,**   **21-Hydroxymethyl-**Δ^5-**pregnen-**
3β-**ol-21-one** $C_{22}H_{34}O_3$.

—CH$_2$·CO·CH$_2$·OH   *22-Acetate* $C_{24}H_{36}O_4$. M. p. 138–139°
(1944 Ciba).

Diacetate $C_{26}H_{38}O_5$. Crystals (ethyl acetate-petroleum ether), m. 143°, [α]$_D$ —48°
(chloroform). — **Fmn.** From the acetate
of 21-chloromethyl-Δ^5-pregnen-3β-ol-21-one (p. 2282 s) by heating with anhydrous potassium acetate in glacial acetic acid at 100°. From the crude acetate of 21-diazomethyl-Δ^5-pregnen-3β-ol-21-one (p. 2296 s) by heating on the water-bath with glacial acetic acid, followed by chromatography on Al_2O_3; yield, 59%. — **Rns.** Boiling with bromoacetic ester and zinc foil in dry benzene containing some pyridine, followed by dehydration with boiling acetic anhydride, gives β-(3β-acetoxy-21-nor-Δ^5-pregnen-20-yl)-$\Delta^{\alpha,\beta}$-butenolide (1945 a Plattner).

ω-**Homo-allopregnane-3**β.**22-diol-21-one,**   **21-Hydroxymethyl-allopregnan-**
3β-**ol-21-one** $C_{22}H_{36}O_3$.

—CH$_2$·CO·CH$_2$·OH   *Diacetate* $C_{26}H_{40}O_5$. Crystals (benzene-petroleum ether), m. 129° cor.; [α]$_D$ +4°
(chloroform). — **Fmn.** From the acetate
of 21-diazomethyl-allopregnan-3β-ol-21-one (p. 2296 s) by heating on the water-bath with glacial acetic acid. — **Rns.** Boiling with bromoacetic ester and zinc foil in dry benzene, followed by reacetylation with acetic anhydride and pyridine at room temperature, gives β-(3β-acetoxy-21-nor-allopregnan-20-yl)-β-acetoxy-γ-butyrolactone (1945 a Plattner).

20-Methyl-Δ^5-**pregnene-3**β.**20-diol-21-al** $C_{22}H_{34}O_3$. Crystals (ether); strongly reducing towards alkaline silver diammine; gives a red colour on heating with 1.4-dihydroxynaphthalene in glacial acetic acid + conc. HCl. — **Fmn.** From the acetate of Δ^5-pregnen-3β-ol-20-one (p. 2233 s) by reaction with sodium acetylide, followed by partial hydrogenation of the triple bond using palladium-$CaCO_3$ catalyst, then by oxidation with OsO_4 to give **20-methyl-21-hydroxymethyl-**Δ^5-**pregnene-3.20.21-triol** $C_{23}H_{38}O_4$ (no constants given); this tetrol is then further oxidized using lead tetraacetate in benzene at 40° in an atmosphere of nitrogen, or by means of 1 mol. HIO_4 (1942 b Reichstein).

Δ^5-**Norcholene-3**β.**23-diol-22-one,**   **3**β-**Hydroxy-**Δ^5-**ternorcholenyl hydroxy-**
methylketone $C_{23}H_{36}O_3$. Crystals (acetone),
m. 227.5° (1938 I. G. Farbenind.). —
Fmn. From the acetate of 23-diazo-Δ^5-norcholen-3β ol-22-one (p. 2296 s) by heating to boiling with glacial acetic acid,

followed by saponification of the resulting diacetate with boiling 5 % methanolic alkali (1938 I. G. FARBENIND.); the 23-acetate is obtained from the
same diazo-compound by first saponifying with methanolic KOH at 17°,
then treatment of the crude hydroxy-diazoketone with anhydrous potassium
acetate in glacial acetic acid at 98°, followed by chromatography on Al_2O_3
(1941 Wettstein). — **Rns.** Oxidation of the 23-acetate with cyclohexanone
and aluminium isopropoxide in boiling dry toluene gives the acetate of
Δ^4-norcholen-23-ol-3.22-dione (p. 2834 s) (1941 Wettstein).

23-Acetate $C_{25}H_{38}O_4$. Crystals (methanol), m. 152–3° cor. (1941 Wettstein);
crystals (methanol-ethyl acetate), m. 278° (1938 I. G. FARBENIND.; but cf.
1941 Wettstein); strongly reducing towards ammoniacal silver nitrate (1941
Wettstein).—**Fmn.** From the diacetate by partial saponification with 3 % methanolic alkali at room temperature (1938 I. G. FARBENIND.). See also formation
of the parent diolone; above. — **Rns.** See those of the parent diolone, above.

Diacetate $C_{27}H_{40}O_5$. Crystals (methanol), m. 159° (1938 I. G. FARBENIND.);
crystals (hexane), dimorphous; m. ca. 164–5° and 171–2° cor.; strongly
reducing towards ammoniacal silver solution (1941 Wettstein). — **Fmn.** From
the 23-acetate (above) by treatment with acetic anhydride and pyridine at
room temperature (1941 Wettstein). See also formation of the parent diolone,
above.

3-Acetate 23-bromoacetate $C_{27}H_{39}O_5Br$. No constants given. — **Fmn.** From
the acetate of 23-diazo-Δ^5-norcholen-3β-ol-22-one (p. 2296 s) by reaction with
bromoacetic acid. — **Rns.** Refluxing with zinc-copper in benzene containing
small amounts of bromoacetic ester, followed by dehydration with boiling
acetic anhydride, gives β-(3β-acetoxy-Δ^5-pregnen-20-yl)-$\Delta^{\alpha.\beta}$-butenolide
(1943 CIBA).

Cholane-3α.24-diol-23-one, 3α-Hydroxy-bisnorcholanyl hydroxymethyl ketone

$C_{24}H_{40}O_3$.
Diacetate $C_{28}H_{44}O_5$. Flat prisms
(aqueous acetic acid), crystals (acetone or methanol), m. 95°. — **Fmn.**
From the acetate of norlithocholic
acid (p. 175) on treatment with thio-
nyl chloride and conversion of the resulting acid chloride with diazomethane
into the acetate of 24-diazocholan-3α-ol-23-one (p. 2297 s), which is then
boiled with glacial acetic acid; yield, 37 % (1938 I. G. FARBENIND.).

Δ^5-Homocholene-3β.25-diol-24-one, Δ^5-Bisnorcholestene-3β.25-diol-24-one

$C_{25}H_{40}O_3$.
Diacetate $C_{29}H_{44}O_5$. Colourless
needles (alcohol), m. 125.5–126°
cor.; $[\alpha]_D$ —45°(chloroform)(1942d
Ruzicka); for comparison of molec-

ular rotation with those of compounds with similar structure, see 1944a Platt-ner. — **Fmn.** From the acetate of 25-diazo-Δ^5-homocholen-3β-ol-24-one (page 2297 s) by heating with pure glacial acetic acid at 95°; yield, 78% (1942d Ruzicka). — **Rns.** Treatment with bromoacetic ester and iodine-activated zinc foil in boiling abs. benzene-dioxane, followed by dehydration with boiling acetic anhydride, gives β-(3β-acetoxy-Δ^5-norcholenyl)-$\Delta^{\alpha.\beta}$-butenolide (1942d Ruzicka); similar condensation with α-bromopropionic ester, followed by dehydration, yields β-(3β-acetoxy-Δ^5-norcholenyl)-α-methyl-$\Delta^{\alpha.\beta}$-butenolide (1944 Ruzicka).

3-Acetate 25-bromoacetate $C_{29}H_{43}O_5Br$. Needles (ethanol), m. 128–130°, $[\alpha]_D^{ca.23}$ —36° (chloroform) (1945b Plattner). — **Fmn.** From the acetate of 25-diazo-Δ^5-homocholen-3β-ol-24-one (p. 2297 s) by reaction with bromo-acetic acid in benzene at room temperature; yield, 83% (1945 Plattner). — **Rns.** Boiling with zinc foil in benzene containing small amounts of bromo-acetic ester, gives β-(3β-hydroxy-Δ^5-norcholenyl)-$\beta.\gamma$-dihydroxybutyric acid γ-lactone, together with its 3-monoacetate (1943 CIBA); a similar condensation, using 0.75 equiv. of the bromoacetic ester and followed by dehydration with acetic anhydride, gives β-(3β-acetoxy-Δ^5-norcholenyl)-$\Delta^{\alpha.\beta}$-butenolide (1945 Plattner).

3-Acetate 25-α-bromopropionate $C_{30}H_{45}O_5Br$. No constants given. — **Rns.** Boiling with zinc foil in benzene containing small amounts of α-bromo-propionic ester, followed by dehydration with boiling acetic anhydride, gives β-(3β-acetoxy-Δ^5-norcholenyl)-α-methyl-$\Delta^{\alpha.\beta}$-butenolide (1943 CIBA).

Δ^5-Norcholestene-3β.25-diol-24-one $C_{26}H_{42}O_3$.

3-Acetate 25-bromoacetate $C_{30}H_{45}O_5Br$. No constants given. — **Fmn.** From 3β-acetoxy-Δ^5-cholenic acid chloride by reaction with diazoethane, followed by boiling of the resulting diazo-compound with bromoacetic acid. — **Rns.** Self-condensation by means of zinc in boiling benzene, followed by dehydration, gives β-(3β-acetoxy-Δ^5-norcholenyl)-γ-methyl-$\Delta^{\alpha.\beta}$-butenolide (1943 CIBA).

17α-Methylpregnane-3β.21-diol-20-one $C_{22}H_{36}O_3$.

White crystals (methanol), m. 140–142°. — **Fmn.** From 17α-methyl-3β-acetoxy-ætiocholanic acid methyl ester, which is successively hydrolysed, reacetylat-ed, and then treated with $SOCl_2$ at 0–5° to give the crude acid chloride acetate, which is subsequently converted into the acetate of 17α-methyl-21-diazopregnan-3β-ol-20-one by reaction with diazo-methane in dry cold ether; the crude diazolone ester is treated with moist ethereal HCl at room temperature; yield, 33% (1942b Marker).

17-Bromoallopregnane-3β.21-diol-20-one $C_{21}H_{33}O_3Br$. Configuration at C-17 uncertain in view of mode of formation (Editor).

Diacetate $C_{25}H_{37}O_5Br$. Needles (acetone-methanol), m. 144–144.5° cor. dec.; $[\alpha]_D^{17}$ —28.5° (chloroform). — **Fmn.** From the diacetate of allopregnane-3β.21-diol-20-one (p. 2311 s) by short warming at 60° with 1 mol. bromine in carbon tetrachloride-glacial acetic acid; yield, 60 %. — **Rns.** Boiling with pyridine gives the diacetate of Δ^{16}-allopregnene-3β.21-diol-20-one (p. 2309 s) (1946 Plattner).

1936 a Ciba*, *Swiss Pats.* 215558, –559 (issued 1941); *C. A.* **1948** 3784; *Ch. Ztbl.* **1942** I 2296.
  b Ciba*, *Swiss Pats.* 216104, –105 (issued 1941); *C. A.* **1948** 5057; *Ch. Ztbl.* **1942** II 197, 1037**.
1937 a N. V. Organon, *Swiss Pats.* 216127, –128, –129 (issued 1941); *Ch. Ztbl.* **1942** II 567; *Swiss Pat.* 226174 (issued 1943); *Ch. Ztbl.* **1944** I 449.
  b N. V. Organon, *Swiss Pats.* 217990, 219346, 219347 (issued 1942); *Ch. Ztbl.* **1943** I 1389.
  c N. V. Organon, *Swiss Pat.* 225509 (issued 1943); *Ch. Ztbl.* **1944** I 36.
  Schering A.-G. (Serini, Logemann), *Ger. Pat.* 736848 (issued 1943); *C. A.* **1944** 3094; *Ch. Ztbl.* **1944** I 300.
  Steiger, Reichstein, *Helv.* **20** 1164, 1167–8, 1173, 1176.
1938 a Ciba*, *Fr. Pat.* 838916 (issued 1939); *C. A.* **1939** 6875; *Ch. Ztbl.* **1939** II 169.
  b Ciba*, *Swiss Pat.* 215560 (issued 1941); *C. A.* **1948** 3784; *Ch. Ztbl.* **1942** I 2296.
  c Ciba*, *Swiss Pat.* 242987 (issued 1946); *C. A.* **1949** 7977; *Fr. Pat.* 860252 (1939, issued 1941); *Ch. Ztbl.* **1941** I 3409; (Ruzicka, Reichstein), *Swed. Pat.* 99785 (1939, issued 1940); *Ch. Ztbl.* **1941** I 1844.
  I. G. Farbenind., *Fr. Pat.* 847129 (issued 1939); *C. A.* **1941** 5653; *Ch. Ztbl.* **1940** I 2827; Winthrop Chemical Co. (Bockmühl, Ehrhart, Ruschig, Aumüller), *U.S. Pat.* 2202619 (issued 1940); *C. A.* **1940** 6772.
  a N. V. Organon, *Indian Pat.* 25060 (issued 1938); *Ch. Ztbl.* **1939** I 1805; *Fr. Pat.* 835589 (issued 1938), *Brit. Pat.* 500767 (issued 1939); *C. A.* **1939** 5872; *Ch. Ztbl.* **1939** I 2829; Roche-Organon, Inc. (Reichstein), *U.S. Pat.* 2312480 (issued 1943); *C. A.* **1943** 4862.
  b N. V. Organon, *Fr. Pat.* 835669 (issued 1938); *C. A.* **1939** 8626; *Ch. Ztbl.* **1939** I 4089; *Brit. Pat.* 502289 (issued 1939); *C. A.* **1939** 6346.
  c N. V. Organon, *Fr. Pat.* 840994 (issued 1939); *C. A.* **1940** 1821; *Ch. Ztbl.* **1940** I 249; Roche-Organon, Inc. (Reichstein), *U.S. Pat.* 2296572 (issued 1942); *C. A.* **1943** 1230.
  Roche-Organon, Inc. (Reichstein), *U.S. Pat.* 2312481 (issued 1943); *C.A.* **1943** 5201; *Ch. Ztbl.* **1945** II 687, 688.
  Schering A.-G., *Fr. Pat.* 857832 (issued 1940); *Ch. Ztbl.* **1941** I 1324.
  a Steiger, Reichstein, *Helv.* **21** 161, 163, 169.
  b Steiger, Reichstein, *Helv.* **21** 171, 176.
1939 a Ciba*, *Fr. Pat.* 856329 (issued 1940); *Ch. Ztbl.* **1941** I 928.
  b Ciba* (Miescher, Fischer, Scholz, Wettstein), *Swed. Pat.* 101148 (issued 1941); *Ch. Ztbl.* **1941** II 1651; Ciba Pharmaceutical Products (Miescher, Fischer, Scholz, Wettstein), *U.S. Pats.* 2275790, 2276543 (issued 1942); *C. A.* **1942** 4289, 4674.
  Ehrhart, Ruschig, Aumüller, *Angew. Ch.* **52** 363, 365.
  Mamoli, *Ber.* **72** 1863.
  N. V. Organon, *Fr. Pat.* 848798 (issued 1939); *C. A.* **1941** 6068; *Ch. Ztbl.* **1940** I 2828.
  a Reichstein, v. Euw, *Helv.* **22** 1209.
  b Reichstein, Montigel, *Helv.* **22** 1212, 1214 (footnote 4), 1216 et seq.
  a Schering A.-G., *Fr. Pat.* 856641 (issued 1940); *Ch. Ztbl.* **1941** I 407; *Swiss Pat.* 229604 (issued 1944); *Ch. Ztbl.* **1944** II 1203.

* Gesellschaft für chemische Industrie in Basel; Société pour l'industrie chimique à Bâle.
** In *Ch. Ztbl.* **1942** II 1037 (referring to *Swiss Pat.* 216104) the starting material is erroneously given as corticosterone 21-propionate instead of desoxycorticosterone 21-propionate.

b Schering A.-G., *Fr. Pat.* 856659 (issued 1940); *Ch. Ztbl.* **1941** I 1196; (Serini, Eysenbach), *Swed. Pat.* 100374 (issued 1940); *Ch. Ztbl.* **1941** I 3259.

c Schering A.-G. (Logemann), *Swed. Pat.* 102639 (issued 1941); *Ch. Ztbl.* **1942** II 74; *Fr. Pat.* 868336 (1940, issued 1941); *Ch. Ztbl.* **1942** I 2038.

Waterman, Danby, Gaarenstroom, Spanhoff, Uyldert, *C. A.* **1940** 1062; *Ch. Ztbl.* **1939** II 2346.

1940 Ingle, *Proc. Soc. Exptl. Biol. Med.* **44** 450.

Marker, Turner, *J. Am. Ch. Soc.* **62** 3003, 3005.

Miescher, Hunzicker, Wettstein, *Helv.* **23** 1367, 1368.

a Reichstein, Fuchs, *Helv.* **23** 658, 664, 667.

b Reichstein, Schindler, *Helv.* **23** 669; see also Roche-Organon (Reichstein), *U.S. Pat.* 2357224 (1941, issued 1944); *C. A.* **1945** 391; *Ch. Ztbl.* **1945** II 1232.

Schering Corp. (Inhoffen), *U.S. Pat.* 2409043 (issued 1946); *C. A.* **1947** 1398; *Ch. Ztbl.* **1947** 620.

Shoppee, *Helv.* **23** 740, 743; cf. Shoppee, Reichstein, *Helv.* **23** 729, 737.

Wettstein, *Helv.* **23** 388, 396.

1941 Ciba*, *Fr. Pat.* 877821 (issued 1943); *Dutch Pat.* 54147 (issued 1943); *Ch. Ztbl.* **1943** II 147, 148.

Ehrenstein, *J. Org. Ch.* **6** 626, 642.

Hegner, Reichstein, *Helv.* **24** 828, 833.

I. G. Farbenind., *Fr. Pat.* 869653 (issued 1942); *Ch. Ztbl.* **1942** II 1154.

Netsky, Jacobs, *Biol. Bull.* **81** 295.

Roche-Organon, Inc. (Reichstein), *U.S. Pat.* 2312483 (issued 1943); *C. A.* **1943** 4862; *Ch. Ztbl.* **1945** II 687.

a Ruzicka, Reichstein, Fürst, *Helv.* **24** 76.

b Ruzicka, Plattner, Fürst, *Helv.* **24** 716, 722.

Schindler, Frey, Reichstein, *Helv.* **24** 360, 363, 365 et seq.

Selye, *Science* [N. S.] **94** 94.

Wettstein, *Helv.* **24** 311, 315 et seq.

1942 a Ciba*, *Swiss Pat.* 238516 (issued 1945); *C. A.* **1949** 5810; Ciba Pharmaceutical Products (Ruzicka), *U.S. Pat.* 2386749 (1943, issued 1945); *C. A.* **1946** 1181.

b Ciba*, *Swiss Pat.* 242988 (issued 1946); *C. A.* **1946** 7977c; Ciba Pharmaceutical Products, *U.S. Pat.* 2386749 (issued 1945); *C. A.* **1946** 1181.

c Ciba*, *Swiss Pat.* 242989 (issued 1946); *C. A.* **1949** 7977e.

d Ciba*, *Swiss Pat.* 242990 (issued 1946); *C. A.* **1949** 7977f.

v. Euw, Reichstein, *Helv.* **25** 988, 1021.

Fried, Linville, Elderfield, *J. Org. Ch.* **7** 362, 371.

a Marker, Crooks, Wagner, *J. Am. Ch. Soc.* **64** 213, 215.

b Marker, Wagner, *J. Am. Ch. Soc.* **64** 1273.

Masson, Borduas, Selye, *C. A.* **1942** 2317.

Mixner, Turner, *Endocrinology* **30** 706, 709.

N. V. Organon, *Swiss Pat.* 254992 (issued 1949); *Ch. Ztbl.* **1950** I 2388.

a Reichstein, *Fr. Pat.* 52035 (issued 1943); *Ch. Ztbl.* **1945** I 195.

b Reichstein, *U.S. Pat.* 2389325 (issued 1945); *C. A.* **1946** 1973.

c Reichstein, *Fr. Pat.* 887641 (issued 1943); *Ch. Ztbl.* **1944** II 878; *U.S. Pat.* 2403683 (1943, issued 1946); *C. A.* **1946** 6216, 6222; *Ch. Ztbl.* **1947** 232.

a Ruzicka, *Fr. Pat.* 882377 (issued 1943); *Ch. Ztbl.* **1944** I 567.

b Ruzicka, Plattner, Balla, *Helv.* **25** 65, 76, 77.

c Ruzicka, Plattner, Fürst, *Helv.* **25** 79, 81.

d Ruzicka, Plattner, Heusser, *Helv.* **25** 435.

Segaloff, Nelson, *Endocrinology* **31** 592.

a Selye, *Endocrinology* **30** 437, 443.

b Selye, Masson, *Science* [N. S.] **96** 358.

1942/44 Schering A.-G., *BIOS Final Report* No. 449, pp. 195, 238, 239, 244 et seq., 250; *FIAT Final Report* No. 996, pp. 91, 92, 96 et seq., 100.

* Gesellschaft für chemische Industrie in Basel; Société pour l'industrie chimique à Bâle.

1943 CIBA* (Ruzicka), *Swed. Pat.* 109956 (issued 1944); *Ch. Ztbl.* **1945** II 74; *Brit. Pat.* 579169 (issued 1946); *C. A.* **1947** 2450; CIBA PHARMACEUTICAL PRODUCTS (Ruzicka), *U.S. Pat.* 2386749 (issued 1945); *C. A.* **1946** 1181.

Cleghorn, *Endocrinology* **32** 165.

Ehrenstein, *J. Org. Ch.* **8** 83, 89.

I. G. FARBENIND., *Fr. Pat.* 891441 (issued 1944); *Ch. Ztbl.* **1946** I 1025; *FIAT Final Report* No. **996**, p. 12; cf. also Ruschig, *Angew. Ch.* **60** A 247 (1948).

Lardon, Reichstein, *Helv.* **26** 607, 618.

Lipschütz, Bruzzone, Fuenzalida, *Proc. Soc. Exptl. Biol. Med.* **54** 303.

Plattner, Ruzicka, Fürst, *Helv.* **26** 2274.

Reichstein, *U.S. Pat.* 2401775 (issued 1946); *C. A.* **1946** 5884; *Ch. Ztbl.* **1947** 74; *U.S. Pat.* 2404768 (1945, issued 1946); *C. A.* **1946** 6222; *Ch. Ztbl.* **1947** 233.

Shoppee, Reichstein, *Helv.* **26** 1316, 1319 footnote 1.

1944 CIBA*, *Brit. Pat.* 560632 (issued 1944); *C. A.* **1946** 2943.

Fieser, Fields, Liebermann, *J. Biol. Ch.* **156** 191, 199.

Horwitt, Dorfman, Shipley, Fish, *J. Biol. Ch.* **155** 213, 217.

Koechlin, Reichstein, *Helv.* **27** 549, 553.

a Plattner, Heusser, *Helv.* **27** 748, 750.

b Plattner, Lang, *Helv.* **27** 1872, 1873.

Ruzicka, Plattner, Heusser, *Helv.* **27** 1173.

1945 Julian, Cole, Magnani, Meyer, *J. Am. Ch. Soc.* **67** 1728, 1729; GLIDDEN Co. (Julian, Cole, Magnani, Conde), *U.S. Pat.* 2374683 (1944, issued 1945); *C. A.* **1946** 1636; *Ch. Ztbl.* **1946** I 372.

a Plattner, Hardegger, Bucher, *Helv.* **28** 167.

b Plattner, Heusser, *Helv.* **28** 1044, 1045, 1048, 1049.

Spiegel, Wycis, *C. A.* **1946** 952.

Velluz, Petit, *Bull. soc. chim.* [5] **12** 951.

Wycis, Spiegel, *C. A.* **1945** 3066.

1946 Barton, *J. Ch. Soc.* **1946** 1116, 1117.

Bruzzone, Borel, Schwarz, *Endocrinology* **39** 194, 197 et seq.

Heard, Sobel, *J. Biol. Ch.* **165** 687.

Plattner, Heusser, Angliker, *Helv.* **29** 468.

a Ruzicka, Plattner, Heusser, Ernst, *Helv.* **29** 248, 250.

b Ruzicka, Plattner, Heusser, *Helv.* **29** 473.

Shoppee, *J. Ch. Soc.* **1946** 1134.

1947 Ruzicka, Plattner, Fürst, Heusser, *Helv.* **30** 694.

Velluz, Petit, Pesez, Berret, *Bull. soc. chim.* [5] **14** 123.

1948 Ernst, *Beitrag zur Stereochemie des Strophanthidins und Periplogenins*, Thesis E. T. H. Zürich, p. 33.

Pesez, Herbain, *Bull. soc. chim.* [5] **15** 104, 105.

1952 Schneider, *J. Biol. Ch.* **199** 235.

1956 Reichstein, *private communication*.

* Gesellschaft für chemische Industrie in Basel; Société pour l'industrie chimique à Bâle.

6. DIHYDROXY-MONOOXO-STEROIDS (CO IN SIDE CHAIN, OH IN THE RING SYSTEM)

17β-Formylandrostane-3β.11β-diol $C_{20}H_{32}O_3$ (not analysed). The crude aldehyde m. 159–178⁰. — **Fmn.** From allopregnane-
3β.11β.20.21-tetrol (p. 2138 s) on oxidation with
periodic acid in aqueous methanol at 20⁰. —**Rns.** On
boiling with methylmagnesium bromide in ether-
toluene it gives allopregnane-3β.11β.20-triol (probably a mixture of C-20 stereoisomers, p. 2109 s) (1938
Steiger).

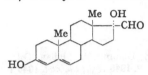

**17β-Formyl-Δ³·⁵-androstadiene-3.17α-diol, 3-Enol of 17β-formyl-Δ⁴-androsten-
17α-ol-3-one** $C_{20}H_{28}O_3$. — **Rns.** The diacetate gives
progesterone (p. 2782 s) on treatment with dimethyl
zinc in ether, followed by saponification with alcoholic sodium hydroxide solution and dehydration
of the resulting Δ⁴-pregnene-17α.20β-diol-3-one
(p. 2721 s) by heating with hydrochloric acid or sulphuric acid in propanol
or dioxane (1938 CIBA).

17α-Formyl-Δ⁵-androstene-3β.17β-diol, *3β.17-Dihydroxy-17-iso-Δ⁵-ætio-
cholenaldehyde* $C_{20}H_{30}O_3$. — **Fmn.** Its diacetate is
obtained from the diacetate of 17α-vinyl-Δ⁵-andro-
stene-3β.17β-diol (p. 2032 s) on bromination with
1 mol. bromine in carbon tetrachloride + a little
pyridine, followed by ozonization and then by debro-
mination with zinc dust in glacial acetic acid, first at 10–20⁰, then at 95–100⁰;
similarly from 17α-styryl-Δ⁵-androstene-3β.17β-diol (obtained from 5-dehydro-
epiandrosterone, styryl bromide, and lithium in ether) (1938a SCHERING
A.-G.). — **Rns.** Gives 17α-formyl-Δ⁴-androsten-17β-ol-3-one on Oppenauer
oxidation (1938a SCHERING A.-G.). Gives Δ⁵-pregnene-3β.17.20-triol (see
p. 2117 s) on treatment with methylmagnesium iodide (1938b SCHERING
A.-G.).

17α-Formylandrostane-3β.17β-diol $C_{20}H_{32}O_3$. — **Fmn.** As a by-product from
17α-vinylandrostane-3β.17β-diol (p. 2037 s) on oxi-
dation with OsO_4, followed by boiling with aqueous
alcoholic Na_2SO_3 (1937 SCHERING-KAHLBAUM A.-G.);
its diacetate is obtained from the diacetate of the
same diol on ozonization in ethyl acetate, followed
by shaking with hydrogen and palladium-calcium
carbonate (1939 Ruzicka) or by treatment with zinc in glacial acetic acid
(1940 Prins). — **Rns.** Gives mainly 17-iso-allopregnane-3β.17β.20b-triol (see
p. 2122 s) on treatment of the diacetate with methylmagnesium bromide in
ether, followed by boiling with methanolic KOH (1940 Prins).

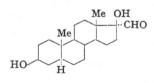

17β-Formylandrostane-3β.17α-diol $C_{20}H_{32}O_3$. Needles with $^1/_2\,H_2O$ (ether), m. 150–3° cor.; $[\alpha]_D^{17}$ —16.6° ± 5° (alcohol) (1941 a Prins); anhydrous prisms (benzene), m. 187–190° (1941 b Prins). Gives a strong red coloration on heating with 1.4-dihydroxynaphthalene in acetic-hydrochloric acid. Reduces alkaline silver diammine solution at 40–50°, but scarcely at room temperature (1941 a, b Prins). — **Fmn.** From allopregnane-3β.17α.20β.21-tetrol (p. 2141 s) on oxidation with periodic acid in aqueous dioxane (1941 a, b Prins). — **Rns.** With methylmagnesium bromide in ether-benzene it gives a mixture of the C-20 epimeric allopregnane-3β.17α.20-triols (pp. 2118 s, 2119 s) (1941 a Prins). Gives allopregnane-3β.17α-diol-20-one (p. 2335 s) on treatment with diazomethane in ether (1941 b Prins).

1937 SCHERING-KAHLBAUM A.-G., *Brit. Pat.* 488814 (issued 1938); *C. A.* **1939** 172; *Ch. Ztbl.* **1939** I 3591.

1938 CIBA (Soc. pour l'ind. chim. à Bâle; Ges. f. chem. Ind. Basel), *Fr. Pat.* 840783 (issued 1939); *C. A.* **1939** 8627; *Ch. Ztbl.* **1940** I 915, 916.

  a SCHERING A.-G., *Fr. Pat.* 49586 (issued 1939); *C. A.* **1942** 2689; *Ch. Ztbl.* **1940** I 428; SCHERING CORP. (Logemann, Dannenbaum), *U.S. Pat.* 2344992 (issued 1944); *C. A.* **1944** 3783.

  b SCHERING A.-G., *Fr. Pat.* 839568 (issued 1939); *C. A.* **1939** 7819; *Ch. Ztbl.* **1939** II 1126. Steiger, Reichstein, *Helv.* **21** 161, 166, 167.

1939 Ruzicka, Hofmann, *Helv.* **22** 150, 153.

1940 Prins, Reichstein, *Helv.* **23** 1490, 1495.

1941 a Prins, Reichstein, *Helv.* **24** 396.

  b Prins, Reichstein, *Helv.* **24** 945, 948.

Summary of Pregnane-(Pregnene-,etc.)-diol-20-ones

For values other than those chosen for the summary, see the compounds themselves.

C-5	C-17 (CO·CH₃)	Double bond	Position of OH	M. p. °C.	$[\alpha]_D$*	3-Acetate m. p. °C.	3-Acetate $[\alpha]_D$*	Diacetate m. p. °C.	Diacetate $[\alpha]_D$*	Page
allo	β	satd.	$3\xi.4\xi$	—	—	—	—	—	—	2322 s
norm.	β	satd.	$3\beta.5\beta$	—	—	—	—	—	—	2323 s
allo	—	Δ^{16}	$3\beta.6\alpha$	145→205	$+69^0$	—	—	115	$+64^0$	2323 s
allo	—	Δ^{16}	$3\beta.6\beta$	216	—	—	—	{ 235 / 165	-18^0	2324 s
norm.	β	satd.	$3\alpha.6\alpha$	194	$+66^0$ (alc.)	—	—	130	$+53^0$ (di.)	2324 s
allo	β	satd.	$3\alpha.6\alpha$	196	$+105^0$ (alc.)	—	—	156	$+92^0$ (alc.)	2325 s
allo	β	satd.	$3\beta.6\alpha$	210	$+95^0$	—	—	103	{ $+88^0$ / $+74.5^0$	2326 s
allo	β	satd.	$3\beta.6\beta$	—	—	—	—	177	$+18^0$	2326 s
norm.	β	satd.	$3\alpha.7\alpha$	172	—	—	—	—	—	2327 s
norm.	β	satd.	$3\alpha.11\alpha$	184	$+96^0$	—	—	148	$+74^0$	2327 s
norm.	β	satd.	$3\alpha.11\beta$	225	—	184	$+148^0$ (ac.)	—	—	2327 s
norm.	β	satd.	$3\beta.11\beta$	260	—	164	$+115^0$ (ac.)	—	—	2328 s
norm.	β	satd.	$3\alpha.12\alpha$	170	$+165^0$ (alc.)+	95—110	—	122	$+190^0$ (alc.)+	2329 s
norm.	β	satd.	$3\alpha.12\beta$	192	{ $+8^0$ / $+55^0$ (ac.)	—	—	—	—	2331 s
norm.	α	satd.	$3\alpha.12\beta$	233	-46^0	—	—	—	—	2331 s
allo	α	satd.	$3\beta.14\xi$	—	—	166	—	—	—	2332 s
—	β	Δ^5	$3\beta.16\beta$	—	—	—	—	—	—	2332 s
allo	β	satd.	$3\beta.16\beta$	—	—	—	—	—	—	2333 s
—	β	Δ^5	$3\beta.17\alpha$	273	-37^0 (di.)	235	-41^0 (di.)	—	—	2333 s
norm.	β	satd.	$3\alpha.17\alpha$	219	$+64^0$ (alc.)	202	—	—	—	2335 s
allo	β	satd.	$3\beta.17\alpha$	266	$+31^0$ (alc.)	192	$+16^0$ (ac.)	—	—	2335 s
—	α	Δ^5	$3\beta.17\beta$	176	-65^0	188	-61^0	195	-56^0 (di.)	2337 s
norm.	α	satd.	$3\beta.17\beta$	—	—	154	—	171	—	2342 s
allo	α	satd.	$3\beta.17\beta$	210	-23^0 (di.)	183	-24^0 (di.)	229	{ -4^0 (di.) / $+2.5^0$ (ac.)	2342 s

* In chloroform, unless otherwise stated; ac. = acetone; di. = dioxane. + $[\alpha]_{5461}$.

Allopregnane-3.4-diol-20-one $C_{21}H_{34}O_3$. Crystals (acetone or dil. methanol).
— **Fmn.** From Δ^3-allopregnen-20-one (see page 2216 s) on treatment with 1.1 mols. of OsO_4 in ether at room temperature for 5 days, followed by heating with Na_2SO_3 in aqueous alcohol (1940 CIBA; 1941 CIBA PHARMACEUTICAL PRODUCTS).

1940 CIBA (Soc. pour l'ind. chim. à Bâle; Ges. f. chem. Ind. in Basel), *Fr. Pat.* 886 415 (issued 1943); *Ch. Ztbl.* **1944** I 450.
1941 CIBA PHARMACEUTICAL PRODUCTS (Miescher, Wettstein), *U.S. Pat.* 2 323 277 (issued 1943); *C. A.* **1944** 222, 223.

Pregnane-3β.5β-diol-20-one $C_{21}H_{34}O_3$. Not characterized. — **Fmn.** From periplogenin (p. 234; cf. 1941 Paist for α.β-position of the double bond in the lactone ring) on dehydration, followed by partial hydrogenation, acetylation, ozonolysis, fission, and hydrolysis. From periplogenin acetate on ozonolysis, followed by decarboxylation, hydrolysis, treatment of the resulting 14-iso-pregnane-3β.5β.14β-triol-20-one with 5% methanolic HCl at room temperature, and catalytic reduction. — **Rns.** Gives Δ^4-pregnene-3.20-dione (progesterone, p. 2782 s) on oxidation with CrO_3 in glacial acetic acid, followed by dehydration with a 5% solution of hydrogen chloride in glacial acetic acid (1937 CIBA).

1937 CIBA (Ges. f. chem. Ind. in Basel; Soc. pour l'ind. chim. à Bâle), *Brit. Pat.* 482321 (issued 1938); *Ch. Ztbl.* **1938** II 3272.
1941 Paist, Blout, Uhle, Elderfield, *J. Org. Ch.* **6** 273.

Δ^{16}-**Allopregnene-3β.6α-diol-20-one** $C_{21}H_{32}O_3$. Crystals (hexane-acetone), m. 143–5°; on cooling the melt and remelting it melts at 202–5° (1951 Mancera; cf. 1954 Salamon); $[\alpha]_D^{20} +69°$ (chloroform) (1951 Mancera). Absorption max. in alcohol at 240 mμ (log ε 4.09) (1951 Mancera), 239 mμ (log ε 3.93) (1954 Salamon); infra-red max. in carbon disulphide at 1670 cm.⁻¹ (1954 Salamon).

Fmn. From chlorogenin diacetate on heating with acetic anhydride in a sealed tube at 200° for 10 hrs., followed by oxidation of the resulting "pseudochlorogenin" triacetate (partial formula I) with CrO_3 in strong acetic acid at room temperature and hydrolysis by boiling with methanolic potassium hydroxide or aqueous methanolic potassium carbonate (1951 Mancera; cf. 1954 Salamon; 1940b Marker); is similarly obtained from "dihydro-pseudochlorogenin diacetate" (1940b Marker) which is probably identical with the above triacetate (1954 Salamon).

Rns. Gives allopregnane-3β.6α-diol-20-one (p. 2326 s) on shaking with hydrogen and palladium-barium sulphate in ethyl acetate at room temperature (1951 Mancera); the diacetate may be similarly reduced to the diacetate of the saturated ketone or by shaking with hydrogen and platinum oxide in acetic acid, followed by oxidation with CrO_3 in strong acetic acid at 28° (1954 Salamon; cf. 1940b Marker).

Diacetate $C_{25}H_{36}O_5$. Prisms (ether-pentane, followed by sublimation in a high vacuum at 150°), m. 109–115°; $[\alpha]_D^{25} +63.7°$ (chloroform). Absorption max. in alcohol at 239 mμ (log ε 3.98); infra-red max. in carbon disulphide at 1670 cm.⁻¹. — **Fmn.** From the above compound on treatment with glacial acetic acid in pyridine at 26° for 16 hrs. (1954 Salamon).

Δ^{16}-Allopregnene-3β.6β-diol-20-one $C_{21}H_{32}O_3$. Crystals (methanol or acetone), m. 214–6⁰ (1942 Marker). — **Fmn.** The di-acetate is obtained from "β"-chlorogenin. di-acetate on heating with acetic anhydride in an autoclave at 185–190⁰ for 8 hrs., followed by oxidation of the resulting "pseudo-β-chlorogenin" triacetate (partial formula I) with CrO_3 in strong acetic acid at room temperature and hydrolysis by refluxing with 2 % alcoholic KOH for 15 min. or with $KHCO_3$ in aqueous methanol for 35 min.; over-all yield, 63 % (1954 Romo; cf. 1942 Marker); on hydrolysis by refluxing with 2 % alcoholic KOH for 15 min., the free diolone is obtained (1942 Marker).

Rns. Gives Δ^{16}-allopregnene-3.6.20-trione (p. 2965 s) on oxidation with CrO_3 in strong acetic acid at room temperature (1942 Marker). On hydro-genation in glacial acetic acid in the presence of platinum oxide under 3 atm. pressure, allopregnane-3β.6β.20β-triol (p. 2018 s) is formed (1942 Marker); on hydrogenation of the diacetate in ethyl acetate in the presence of palladium-charcoal at room temperature and atmospheric pressure it gives the diacetate of allopregnane-3β.6β-diol-20-one (p. 2326 s) (1954 Romo). The diacetate oxime (not described) may be converted into the diacetate of androstane-3β.6β-diol-17-one (p. 2747 s) by treatment with p-toluenesulphonyl chloride in cold pyridine, followed by hydrolysis with dil. H_2SO_4 at room temperature (1941 PARKE, DAVIS & Co.).

Diacetate $C_{25}H_{36}O_5$. The diacetate obtained from the above diolone by boiling with acetic anhydride forms crystals (methanol), m. 233–5⁰ (1942 Marker); the diacetate obtained from "β"-chlorogenin diacetate, as described above, forms crystals (ether-pentane), m. 164–5⁰, and shows $[\alpha]_D^{20}$ —18⁰ (chloro-form), absorption max. in 95 % alcohol at 238 mμ (log ε 4.03) and infra-red max. in chloroform at 1718 and 1660 cm.$^{-1}$ (1954 Romo).

Pregnane-3α.6α-diol-20-one $C_{21}H_{34}O_3$. For the configuration at C-6, see 1947 Moffett. — Crystals with 1 C_6H_6 (benzene), m. 102–6⁰ (1946 Moffett; cf. 1948, 1950 Lieber-man); after drying in a high vacuum at 80⁰ to 100⁰ or on crystallization from ethyl acetate it melts at 187–190⁰ (1946 Moffett); needles (acetone, or acetone-diisopropyl ether), m. 192–4⁰ (1948, 1950 Lieberman); needles (ethyl acetate), m. 198⁰ (1939 Kimura*); $[\alpha]_D^{26}$ +70⁰ (dioxane), +62⁰ (methanol) (1946 Moffett); $[\alpha]_D^{24}$ +66⁰ (alcohol) (1948, 1950 Lieberman); $[\alpha]_D$ +6.5⁰ (alcohol) (1939 Kimura*).

Occ. In human pregnancy urine (1948, 1950 Lieberman).

Fmn. Its diacetate is obtained from the methyl ester of bisnorhyodesoxy-cholic acid (3α.6α-dihydroxy-bisnorcholanic acid) by the action of ethyl-magnesium bromide, acetylation of the resulting carbinol (22-ethylcholane-3α.6α.22-triol, p. 2132 s), and oxidation with CrO_3 in glacial acetic acid (1939

---

* The structure of the compounds described by Kimura (1939) and Marker (1940a) seems doubtful (1946 Moffett).

Kimura*); from the same ester on refluxing with phenylmagnesium bromide in benzene, acetylation of the resulting carbinol (not described) with acetic anhydride in pyridine at 100°, dehydration of the resulting diacetate by refluxing with glacial acetic acid for 6 hrs., and ozonization of the resulting ethylene compound (not described) in chloroform at 0° (1940a Marker*) or in glacial acetic acid + a little acetic anhydride at 10–15° (crude yield, 25%) (1946 Moffett). The diacetate is also obtained from 2.2-diphenyl-1-(3α.6α-di-acetoxy-bisnorcholanyl)-ethylene (p. 2092 s) by refluxing for 15 min. with N-bromosuccinimide in carbon tetrachloride with illumination with electric light, followed by refluxing with dimethylaniline for 10 min., reacetylation by refluxing with glacial acetic acid-acetic anhydride for 2 hrs., and oxidation of the crude 1-methyl-4.4-diphenyl-1-(3α.6α-diacetoxy-ætiocholanyl)-$\Delta^{1.3}$-butadiene with CrO_3 in chloroform-acetic acid at 0–23° (1946 Moffett). The diacetate is hydrolysed by warming with 2 N methanolic NaOH for 15 min. (1946 Moffett).

Rns. Gives pregnane-3.6.20-trione (p. 2966 s) on oxidation with CrO_3 in strong acetic acid at 30° (1946 Moffett). May be converted into progesterone (p. 2782 s) by treatment of the diacetate with ca. 1 mol. KOH in methanol at 20° (48 hrs.), neutralization with dil. H_2SO_4, oxidation with CrO_3 in strong acetic acid at room temperature (1 hr.), refluxing of the resulting pregnan-6α-ol-3.20-dione acetate with 2% methanolic KOH for 75 min., and dehydration by heating with fused $KHSO_4$ under 4 mm. pressure at 130° for 1.5 hrs. and then at 180° for 4 hrs. (1940a Marker*). For conversion into ætiocholane-3α.6α-diol-17-one, see p. 2747 s.

2.4-Dinitrophenylhydrazone $C_{27}H_{38}O_6N_4$ (not analysed). Crystals (methanol), m. 234–8° (1948, 1950 Lieberman).

Diacetate $C_{25}H_{38}O_5$. Crystals (methanol), m. 131–3° (1946 Moffett); crystals (acetone-ether), m. 128–9° (1948, 1950 Lieberman); crystals (ether-pentane), m. 100° (1940a Marker*); $[\alpha]_D^{24} +53° \pm 3°$ (dioxane) (1946 Moffett). — **Fmn.** and **Rns.** See above.

Allopregnane-3α.6α-diol-20-one $C_{21}H_{34}O_3$. For the configuration at C-6, see

1950 Lieberman. — Needles (benzene or acetone), m. 195–6°; $[\alpha]_D^{26} +105°$ (alcohol) (1948 Lieberman). — **Occ.** In human pregnancy urine (1948 Lieberman). — **Rns.** Gives allopregnane-3.6.20-trione (p. 2966 s) on oxidation with CrO_3. On treatment with sodium hypoiodite it gives iodoform and an acid m. 278–282°. Is unchanged by treatment with periodic acid in methanol at room temperature and by heating in aqueous HCl-acetic acid (1948 Lieberman).

Oxime $C_{21}H_{35}O_3N$. Crystals (methanol), m. 288–291° (1948 Lieberman).

2.4-Dinitrophenylhydrazone $C_{27}H_{38}O_6N_4$. Crystals (aqueous methanol), m. 235° (1948 Lieberman).

* See the footnote on p. 2324 s.

Diacetate $C_{25}H_{38}O_5$. Needles (ligroin), m. 155–6⁰; $[\alpha]_D^{24}$ +92⁰ (alcohol). —
Fmn. From the above diolone on refluxing with acetic anhydride in pyridine
for 1 hr. (1948 Lieberman).

Allopregnane-3β.6α-diol-20-one $C_{21}H_{34}O_3$. *"Compound B 7"* (1948 Lieberman;
see 1954 Salamon). — Crystals (ether-acetone),
m. 208–210⁰ (1940b Marker; 1954 Salamon);
needles (aqueous methanol), m. 209–210⁰ (1948
Lieberman); crystals (acetone), m. 202–3⁰ (1951
Mancera); $[\alpha]_D^{25}$ +99⁰ (alcohol) (1948 Lieberman);
$[\alpha]_D^{20}$ +95⁰ (chloroform) (1951 Mancera). — **Occ.**
In human pregnancy urine (1948 Lieberman). —**Fmn.** From Δ^{16}-allopregnene-
3β.6α-diol-20-one (p. 2323 s) on shaking with hydrogen and palladium-barium
sulphate in ethyl acetate at room temperature (1951 Mancera); its diacetate is
similarly obtained from the corresp. diacetate or by shaking of the latter diacetate with hydrogen and platinum oxide in acetic acid, followed by oxidation
with CrO_3 in strong acetic acid at 28⁰ (1954 Salamon; cf. 1940b Marker); the
diacetate is hydrolysed by refluxing with alcoholic KOH (1940b Marker).

Rns. Gives allopregnane-3.6.20-trione (p. 2966 s) on oxidation with CrO_3
in strong acetic acid (1940b Marker; 1954 Salamon; 1948 Lieberman).

Oxime $C_{21}H_{35}O_3N$. Crystals (dil. alcohol), m. 275⁰ dec. (1948 Lieberman).

Diacetate $C_{25}H_{38}O_5$. Crystals (dil. methanol), m. 102.5–103.5⁰ (1948 Lieberman); crystals (hexane), m. 101–2⁰ (1951 Mancera); prisms (ether-pentane),
m. 101–4⁰ (1954 Salamon); $[\alpha]_D$ in chloroform +88.4⁰ (1951 Mancera),
+74.5⁰ (1954 Salamon). — **Fmn.** From the above diolone on treatment with
acetic anhydride in pyridine at room temperature (1951 Mancera; 1954
Salamon). See also above.

Allopregnane-3β.6β-diol-20-one $C_{21}H_{34}O_3$. For the *diacetate* $C_{25}H_{38}O_5$ [m. 175⁰
to 177⁰; $[\alpha]_D^{20}$ +18⁰ (chloroform)] obtained from the diacetate of Δ^{16}-allo-
pregnene-3β.6β-diol-20-one (p. 2324 s) on hydrogenation in ethyl acetate in
the presence of palladium-charcoal (after the closing date for this volume), see
1954 Romo.

1939 Kimura, Sugiyama, *J. Biochem.* **29** 409, 418.
1940 a Marker, Krueger, *J. Am. Ch. Soc.* **62** 79; PARKE, DAVIS & Co. (Marker), *U.S. Pats.* 2 337 563,
    2 337 564 (issued 1943); *C. A.* **1944** 3423; PARKE, DAVIS & Co. (Marker, Lawson), *U.S. Pat.*
    2 366 204 (1941, issued 1945); *C. A.* **1945** 1649; *Ch. Ztbl.* **1945** II 1385, 1386.
    b Marker, Jones, Turner, Rohrmann, *J. Am. Ch. Soc.* **62** 3006, 3008.
1941 PARKE, DAVIS & Co. (Tendick, Lawson), *U.S. Pat.* 2 335 616 (issued 1943); *C. A.* **1944** 3095.
1942 Marker, Turner, Wittbecker, *J. Am. Ch. Soc.* **64** 809, 812.
1946 Moffett, Stafford, Linsk, Hoehn, *J. Am. Ch. Soc.* **68** 1857, 1859.
1947 Moffett, Hoehn, *J. Am. Ch. Soc.* **69** 1995.
1948 Lieberman, Dobriner, Hill, Fieser, Rhoads, *J. Biol. Ch.* **172** 263, 287, 288, 292.
1950 Lieberman, Fukushima, Dobriner, *J. Biol. Ch.* **182** 299, 301, 311.
1951 Mancera, Rosenkranz, Djerassi, *J. Org. Ch.* **16** 192, 194, 195.
1954 Romo, Rosenkranz, Sondheimer, *J. Am. Ch. Soc.* **76** 5169.
    Salamon, Dobriner, *J. Biol. Ch.* **207** 323.

Pregnane-3α.7α-diol-20-one $C_{21}H_{34}O_3$. Prisms with $^1/_2$ H_2O (alcohol), m. 170^0

to 172^0. Readily soluble in methanol, ethanol, and ethyl acetate, rather sparingly in ether and glacial acetic acid. Gives a cherry-red Liebermann colour reaction. — **Fmn.** Its diacetate is obtained from the methyl ester of bisnorchenodesoxycholic acid (3α.7α-dihydroxy-bisnorchol-anic acid) on treatment with ethylmagnesium bromide in ether-benzene at 50^0, followed by acetylation of the resulting 22-ethylcholane-3α.7α.22-triol (p. 2132 s) with acetic anhydride in dry pyridine on the water-bath for 24 hrs. and oxidation of the resulting syrupy product with CrO_3 in glacial acetic acid at 80–90^0 for 1.5 hrs. (small yield); the diacetate is isolated as semicarbazone (below) which is hydrolysed by boiling with aqueous alcoholic HCl for $^1/_2$ hr. and then with 10% alcoholic KOH for 2 hrs. (1938 Ishihara).

Diacetate semicarbazone $C_{26}H_{41}O_5N_3$, needles (alcohol), m. $271-2^0$; gives a weak, brown-red Liebermann colour reaction (1938 Ishihara).

1938 Ishihara, *J. Biochem.* **27** 265, 276.

Pregnane-3α.11α-diol-20-one $C_{21}H_{34}O_3$. Platelets (ethyl acetate-petroleum

ether), m. $182.5-184^0$ cor.; $[\alpha]_D^{23}$ $+96^0$ (chloroform). — **Fmn.** The diacetate is obtained from the methyl ester of 3α.11α-dihydroxy-bisnor-cholanic acid on refluxing with phenylmagnesium bromide in ether-benzene, dehydration of the resulting carbinol by continuous distillation of a benzene solution with iodine, acetylation of the resulting 1-methyl-2.2-diphenyl-1-(3α.11α-dihydroxyætiocholanyl)-ethylene (p. 2088 s), ozonization of the diacetate in anhydrous methanol-ethyl acetate (1:1) at -30^0, followed by shaking with hydrogen and palladium-$CaCO_3$ in the same solvent mixture (yield from the last-mentioned diacetate, 76%); the diacetate is hydrolysed with aqueous alcoholic NaOH at room temperature. — **Rns.** Gives pregnane-3.11.20-trione (p. 2967 s) on oxidation with CrO_3 in strong acetic acid at 4^0. With benzaldehyde and sodium ethoxide in abs. alcohol at -15^0 to -3^0 it gives the 21-benzylidene derivative (p. 2348 s) (1946 Long).

Diacetate $C_{25}H_{38}O_5$. Prisms (ethyl acetate), m. $147-148.5^0$ cor.; $[\alpha]_D^{29}$ $+74^0$ (chloroform) (1946 Long). — **Fmn.** See above.

Pregnane-3α.11β-diol-20-one * $C_{21}H_{34}O_3$ (not analysed). Crystals (acetone-

ether), m. $222-5^0$ cor. (1944b v. Euw). — **Fmn.** The 3-acetate is obtained from the 3-acetate of pregnane-3α.11β.20β-triol (p. 2108 s) on heating with aluminium phenoxide and acetone in a partially evacuated tube on the water-bath

* See the footnote on p. 2328 s.

for 40 hrs. (yield, 50% allowing for starting material recovered). The
3-acetate is also obtained from the methyl ester of 3α.11β-dihydroxy-bisnor-
cholanic acid * on refluxing with phenylmagnesium bromide in ether-benzene,
followed by acetylation of the resulting carbinol, dehydration by refluxing
with glacial acetic acid, and ozonization of the resulting 1-methyl-2.2 - di-
phenyl-1-(3α-acetoxy-11β-hydroxyætiocholanyl)-ethylene (p. 2089 s) in ethyl
acetate at —80°, using ca. 1 mol. O_3, immediately followed by fission of the
ozonide with zinc dust and glacial acetic acid (yield from the ethylene compound, 70%); the acetate is then hydrolysed by treatment with methanolic
KOH at 20° for 16 hrs. (1944b v. Euw). — **Rns.** May be converted into the
21-acetate of pregnane-3α.11β.21-triol-20-one (p. 2354 s) in the same way as
the 3β-isomer (below) (1944c v. Euw). The 3-acetate gives the acetate of
pregnan-3α-ol-11.20-dione (p. 2847 s) on oxidation with CrO_3 in glacial acetic
acid at 20° (1944b v. Euw).

3-Acetate $C_{23}H_{36}O_4$. Granula (ether-petroleum ether), m. 182–4° cor.; $[\alpha]_D^{13}$
$+147.5°$ (acetone) (1944b v. Euw). — **Fmn.** and **Rns.** See above.

Pregnane-3β.11β-diol-20-one * $C_{21}H_{34}O_3$ (not analysed). Granula probably
containing water of crystallization, m. 255° to
260° cor. (1944a v. Euw). — **Fmn.** The 3-ace-
tate is obtained similarly to that of the above
3-epimer, starting from the corresp. 3-epimers;
yields, 46% and 63%, resp. (1944a v. Euw). —
Rns. On heating with lead tetraacetate in 98%
to 98.5% acetic acid in a sealed tube at 55° for 24 hrs. it gives the 21-acetate
of pregnane-3β.11β.21-triol-20-one (p. 2354 s), which on heating with aluminium phenoxide and acetone in benzene in a sealed tube at 100° yields the
21-acetate of pregnane-11β.21-diol-3.20-dione (p. 2856 s); when the diolone is
treated with lead tetraacetate in glacial acetic acid containing traces of
acetic anhydride, some pregnane-3β.21-diol-11.20-dione 21-acetate (p. 2861 s)
is formed (1944c v. Euw). The 3-acetate gives the acetate of pregnan-3β-ol-
11.20-dione (p. 2848 s) on oxidation with CrO_3 in glacial acetic acid at 20° for
8 hrs. (1944a v. Euw).

3-Acetate $C_{23}H_{36}O_4$. Needles (ether-petroleum ether), m. 163–4° cor.; $[\alpha]_D^{12}$
$+115°$ (acetone) (1944a v. Euw). — **Fmn.** and **Rns.** See above.

1944 a v. Euw, Lardon, Reichstein, *Helv.* **27** 821, 826, 827, 833.
     b v. Euw, Lardon, Reichstein, *Helv.* **27** 821, 828, 839.
     c v. Euw, Lardon, Reichstein, *Helv.* **27** 1287, 1290, 1292, 1293; Reichstein, *Fr. Pat.* 898140
        (1943, issued 1945); *Ch. Ztbl.* **1946** I 1747, 1748; *Brit. Pat.* 594878 (issued 1947); *C. A.*
        1948 2404; *U.S. Pat.* 2440874 (1943, issued 1948); *C. A.* **1948** 5622; *Ch. Ztbl.* **1948** E 456.
1946 Long, Marshall, Gallagher, *J. Biol. Ch.* **165** 197, 203.

---

* For reversal of configuration at C-11 and C-17 as compared with the original, see pp. 1362 s,
1363 s.

Pregnane-3α.12α-diol-20-one* $C_{21}H_{34}O_3$. Crystals (50% alcohol), m. 165–6⁰ (1938 Hoehn), 166–8⁰ cor. (1940 Reichstein), 167⁰ to 170⁰ cor. (1944 Meystre), 176–8⁰ cor. (1943 Reich); crystals (ether, and ethyl acetate-ligroin), m. 178–180⁰ (Turner, see 1946 Ettlinger); $[\alpha]_{5461}^{25}$ +165⁰ ± 5⁰ (alcohol) (1938 Hoehn). — **Fmn.** From 3α.12α - diacetoxy - ternorcholanylamine (p. 2104 s) on treatment with HOCl, followed by treatment with alcoholic sodium ethoxide and then with dil. H_2SO_4 (1939 Ehrhart; 1941 I. G. FARBEN-IND.). The diacetate is obtained from the diacetate of 22.22-diphenyl-$\varDelta^{20(22)}$-bisnorcholene-3α.12α-diol (p. 2089 s) on ozonization in chloroform at 0⁰ (1938 Hoehn; 1943 Hegner), from the diacetate of 24.24-diphenyl-$\varDelta^{20(22).23}$-chola-diene-3α.12α-diol (p. 2093 s) on oxidation with CrO_3 in strong acetic acid (1944 Meystre) or on ozonization in methanol-ethyl acetate, followed by decomposition of the ozonide with zinc dust in acetic acid and then by oxidation with CrO_3 in acetic acid at room temperature (1946 Ettlinger). In small amount from the diacetate methyl ester of desoxycholic acid (3α.12α-dihydroxycholanic acid) on oxidation with CrO_3 in strong acetic acid at 70–85⁰ (1943 Reich).

Rns. (See also the reactions of the esters, below). Gives pregnane-3.12.20-trione on oxidation with CrO_3 in glacial acetic acid at 20⁰ (1940 Reichstein); the compound m. 189–191⁰, obtained in this reaction and assumed to be ætiocholane-3.12.17-trione (1938 Hoehn), was probably impure pregnane-3.12.20-trione (1940 Reichstein). On oxidation of the 12-acetate with CrO_3 in glacial acetic acid at 20⁰ it gives the acetate of pregnan-12α-ol-3.20-dione (p. 2841 s) (1938 I. G. FARBENIND.; 1938 WINTHROP CHEMICAL Co.; 1939 Ehrhart; 1940 Reichstein; 1941 Shoppee); on similar oxidation of the 3-acetate, the acetate of pregnan-3α-ol-12.20-dione (p. 2849 s) is obtained (1944 Katz). On treatment of the diacetate with lead tetraacetate in glacial acetic acid-acetic anhydride at 68–72⁰ it yields the triacetate of pregnane-3α.12α.21-triol-20-one (p. 2356 s) (1944 Ruzicka). Gives ætiocholane-3α.12α-diol-17-one (p. 2749 s) on treatment with ethyl nitrite and sodium ethoxide in abs. alcohol at room temperature, followed by refluxing with aqueous alcoholic HCl (1946 Ettlinger). For degradation to ætiodesoxycholic acid via the 21-benzylidene derivative, see the latter (p. 2348 s).

On treatment of the diacetate with 3 mols. of bromine in glacial acetic acid not above 38⁰, followed by boiling of the resulting product** with 5% methanolic KOH for 1 hr., it gives 17-methyl-3α.12α-dihydroxyætiocholanic acid methyl ester (isolated as the diacetate) and, as the main product, an acid fraction [containing 3α.12α-dihydroxy-$\varDelta^{17(20)}$-pregnen-21-oic acid (cf. also 1956 Engel)] which yields a small amount of the diacetate of ætiocholane-3α.12α-diol-17-one (p. 2749 s) on methylation with diazomethane, followed by

---

* For configuration at C-12 and C-17, see pp. 1362 s, 1363 s.

** According to Koechlin (1944) this bromination product is a mixture of 17-bromo- and 17.21-di-bromo-3α.12α-diacetoxypregnan-20-ones; according to Julian (1950, 1956) it is for the most part 17.21.21-tribromo-3α.12α-diacetoxypregnan-20-one.

163*

acetylation with acetic anhydride in pyridine on the water-bath, ozonization in chloroform, heating with zinc dust in glacial acetic acid on the water-bath, boiling with 3 % methanolic KOH, and reacetylation with acetic anhydride in pyridine on the water-bath (1944 Koechlin).

Gives the 21-benzylidene derivative (p. 2348 s) on treatment with benz-aldehyde and sodium ethoxide in abs. alcohol at 20° (1944 Koechlin; cf. 1937 SCHERING A.-G.; 1938 Hoehn; 1940 Reichstein). On treatment of the diacetate with ethyl bromoacetate and zinc in benzene, followed by refluxing with methanolic KOH, it gives 3α.12α.20-trihydroxynorcholanic acid (1944 Ruzicka).

3-Acetate $C_{23}H_{36}O_4$. Needles (benzene, ether, or acetone), m. 95–110° cor. — **Fmn.** From the above diolone on refluxing with 1 mol. acetic anhydride in abs. benzene for some hrs.; yield, 44 % (1944 v. Euw). — **Rns.** On treatment with β-anthraquinonecarboxylic acid chloride in abs. benzene it gives the 3-acetate 12-(β-anthraquinonecarboxylate) (below), a compound containing 76.7 % C and 6.6 % H [yellowish needles (acetone-ether), m. 225–6° cor.], a compound m. 282–3° [crystals (dioxane-acetone)], and other compounds (1944 v. Euw).

12-Acetate $C_{23}H_{36}O_4$. Needles (methanol), m. 208–210° cor. (1940 Reich-stein; cf. 1944 Ruzicka), 212–4° cor. (1944 Meystre); $[\alpha]_D^{17}$ ca. +150° (acetone) (1940 Reichstein; 1944 Meystre), +158° (chloroform) (1944 Ruzicka); $[\alpha]_{5461}^{17}$ ca. +193° (acetone) (1940 Reichstein). — **Fmn.** From the diacetate (below) on treatment with 1 mol. K_2CO_3 in aqueous methanol at 20° for 22 hrs. (1940 Reichstein; cf. also 1944 Meystre) or with 1 mol. NaOH in aqueous methanol at room temperature (1938 I. G. FARBENIND.; 1938 WIN-THROP CHEMICAL Co.). — **Rns.** Is hydrolysed by refluxing with methanolic KOH for 2 hrs. (1940 Reichstein).

Diacetate $C_{25}H_{38}O_5$. Crystals (methanol), m. 121–122.5° (1938 Hoehn), 114–5° cor. (1944 Ruzicka), 116–8° cor. (1944 Meystre); crystals (petroleum ether), m. 140–1° (1946 Ettlinger); $[\alpha]_{5461}^{25}$ +190° (alcohol) (1938 Hoehn). — **Fmn.** From the diolone on heating with acetic anhydride and pyridine on the water-bath for 3 hrs. (1943 Reich; cf. 1938 WINTHROP CHEMICAL Co.). From the 12-acetate (above) with acetic anhydride and pyridine (1944 Meystre; 1944 Ruzicka). See also under the formation of the diolone (p. 2329 s). — **Rns.** For partial hydrolysis, see the 12-acetate (above).

12-Benzoate $C_{28}H_{38}O_4$. Granula (acetone-ether), m. 160–1° cor. — **Fmn.** From the dibenzoate (below) on refluxing with 1 mol. of methanolic KOH for 5 min., followed by keeping at 12° for 16 hrs. — **Rns.** Gives a syrupy acetate on acetylation (1944 v. Euw).

Dibenzoate $C_{35}H_{42}O_5$. Prisms (acetone-ether), m. 183–4° cor. — **Fmn.** From the free diolone on treatment with benzoyl chloride and pyridine in abs. benzene at 20° for 20 hrs. (1944 v. Euw).

12-(β-Anthraquinonecarboxylate) $C_{36}H_{40}O_6$ (no analysis). Yellow granula (acetone-ether), m. 230–1° cor.; rapidly becomes green on exposure to the

air. — **Fmn.** From the di-(β-anthraquinonecarboxylate) (below) on heating
on the water-bath with potassium glycinate in dioxane-abs. alcohol for 13 hrs.
or with potassium phenoxide in phenol for 3.5 hrs. (1944 v. Euw). — **Rns.**
Gives the corresp. ester of pregnan-12α-ol-3.20-dione (p. 2841 s) on oxidation
with CrO_3 in glacial acetic acid at 18° (1944 v. Euw). For acetylation, see the
following compound.

3-Acetate 12-(β-anthraquinonecarboxylate) $C_{38}H_{42}O_7$. Yellow needles (benzene-
ether, and acetone-ether), m. 174–5° cor. — **Fmn.** From the preceding ester
on treatment with acetic anhydride and pyridine at 20° for 16 hrs. See also
the reactions of the 3-acetate (p. 2330 s) (1944 v. Euw). — **Rns.** On heating
in a high vacuum at 290° it gives the acetate of Δ^{11}-pregnen-3α-ol-20-one
(p. 2239 s) and a very small amount of $\Delta^{3.11}$-pregnadien-20-one (p. 2215 s)
(1944 v. Euw).

Di-(β-anthraquinonecarboxylate) $C_{51}H_{46}O_9$. Leaflets (dioxane-acetone),
m. 283–4° cor. Sparingly soluble in most solvents. — **Fmn.** From the free
diolone on heating with β-anthraquinonecarboxylic acid chloride and pyridine
in benzene on the water-bath for 2 hrs. — **Rns.** On heating at 290–320°
(bath temperature) under 0.02 mm. pressure it gives small amounts of
$\Delta^{3.11}$-pregnadien-20-one (p. 2215 s; main product), Δ^{11}-pregnen-3α-ol-20-one
β-anthraquinonecarboxylate, and a *compound* $C_{36}H_{38}O_5$ [yellow prisms or
granula (acetone-ether), m. 190–2° cor., becoming dark violet in daylight and
yellow again on heating], probably the β-anthraquinonecarboxylate of Δ^3-preg-
nen-12α-ol-20-one (1944 v. Euw). For partial hydrolysis, see the 12-(β-anthra-
quinonecarboxylate), above.

Pregnane-3α.12β-diol-20-one* $C_{21}H_{34}O_3$. Plates (ether-petroleum ether),
m. 191–2° cor.; $[\alpha]_D^{20}$ + 55° (acetone), + 8° (chloro-
form). — **Fmn.** From the diacetate of 1-methyl-
2.2-diphenyl -1-(3α.12β-dihydroxyætiocholanyl)-
ethylene (p. 2090 s) on ozonization in chloroform
at —10°, followed by boiling in aqueous methanol
with K_2CO_3 and then with KOH (yield, 14%),
along with an even smaller amount of 17-iso-pregnane-3α.12β-diol-20-one
(below). — **Rns.** On oxidation with CrO_3 in glacial acetic acid it gives preg-
nane-3.12.20-trione (p. 2969 s) (1945 Sorkin).

17-iso-Pregnane-3α.12β-diol-20-one* $C_{21}H_{34}O_3$. Platelets (chloroform-ether),
m. 231–3° cor.; $[\alpha]_D^{15}$ —46° (chloroform). — **Fmn.**
See under the formation of the above 17-epimer. —
Rns. Gives 17-iso-pregnane-3.12.20-trione (page
2969 s) on oxidation with CrO_3 in glacial acetic
acid (1945 Sorkin).

1937 SCHERING A.-G. (Logemann), *Ger. Pat.* 737023 (issued 1943); *C. A.* **1944** 3781; *Fr. Pat.*
843082 (1938, issued 1939); *C. A.* **1940** 6412; *Ch. Ztbl.* **1940** I 759.

* For reversal of configuration at C-12 and C-17 as compared with the original, see pp. 1362 s,
1363 s.

1938 Hoehn, Mason, *J. Am. Ch. Soc.* **60** 1493, 1495, 1496.
   I.G. FARBENIND., *Fr. Pa·.* 834072 (issued 1938); *C. A.* **1939** 3394; *Ch. Ztbl.* **1939** I 3931.
   WINTHROP CHEMICAL Co. (Bockmühl, Ehrhart, Ruschig, Aumüller), *U.S. Pat.* 2142170
      (issued 1939); *C. A.* **1939** 3078; *Ch. Ztbl.* **1939** II 170.
1939 Ehrhart, Ruschig, Aumüller, *Angew. Ch.* **52** 363, 364.
1940 Reichstein, Arx, *Helv.* **23** 747, 749–751.
1941 I.G. FARBENIND., *Swiss Pat.* 226006 (issued 1943); *Ch. Ztbl.* **1944** II 563.
   Shoppee, Reichstein, *Helv.* **24** 351, 357.
1943 Hegner, Reichstein, *Helv.* **26** 715, 719.
   Reich, Reichstein, *Helv.* **26** 2102, 2105.
1944 v. Euw, Lardon, Reichstein, *Helv.* **27** 821, 834–837.
   Katz, Reichstein, *Pharm. Acta Helv.* **19** 231, 261.
   Koechlin, Reichstein, *Helv.* **27** 549, 563, 566.
   Meystre, Frey, Wettstein, Miescher, *Helv.* **27** 1815, 1822, 1823; CIBA PHARMACEUTICAL
      PRODUCTS (Miescher, Frey, Meystre, Wettstein), *U.S. Pats.* 2461563 (1943, issued 1949),
      2461910 (1946, issued 1949); *C. A.* **1949** 3474; *Ch. Ztbl.* **1949** E 1870, 1871.
   Ruzicka, Plattner, Pataki, *Helv.* **27** 988, 991, 992.
1945 Sorkin, Reichstein, *Helv.* **28** 875, 890, 891.
1946 Ettlinger, Fieser, *J. Biol. Ch.* **164** 451.
1950 Julian, Karpel, *J. Am. Ch. Soc.* **72** 362, 364.
1956 Engel, Jennings, Just, *J. Am. Ch. Soc.* **78** 6153.
   Julian, Cochrane, Magnani, Karpel, *J. Am. Ch. Soc.* **78** 3153, 3154.

17-iso-Allopregnane-3β.14-diol-20-one $C_{21}H_{34}O_3$.

3-Acetate $C_{23}H_{36}O_4$, m. 165–6°. — **Fmn.**
From the acetate of 14.15-epoxy-Δ^{16}-allopreg-
nen-3β-ol-20-one on shaking with hydrogen and
a platinum oxide catalyst in alcohol (1946 CIBA).

1946 CIBA (Ges. f. chem. Ind. in Basel, Soc. pour l'industrie chim. à Bâle), *Austrian Pat.* 162906
   (issued 1949); *Ch. Ztbl.* **1949** E 1699.

Δ^5-Pregnene-3β.16β-diol-20-one $C_{21}H_{32}O_3$ (I, R = R'= H). For the con-

figuration at C-16, see 1949 Hirschmann.

3-Acetate 16-(δ-acetoxyisocaproate), *Diosone
diacetate** $C_{31}H_{46}O_7$ [I, R = CO·CH$_3$, R'= CO·
CH$_2$·CH$_2$·CH(CH$_3$)·CH$_2$·O·CO·CH$_3$]. Crystals
(methanol), m. 84–86°. — **Fmn.** From pseudo-
diosgenin diacetate on oxidation with CrO$_3$ in strong acetic acid at 15–28°
for 45 min.; yield, 39%. — **Rns.** On reduction with sodium in dry isopropyl
alcohol it gives Δ^5-pregnene-3β.20α-diol (p. 1920 s). On refluxing with aluminium isopropoxide in dry isopropyl alcohol for 7 hrs., followed by refluxing
with 2% methanolic KOH for 30 min., Δ^5-pregnene-3β.16β.20β-triol (p. 2110 s)

---

* For this name, see 1947 Marker.

is obtained. Gives the corresponding diester of allopregnane-3β.16β-diol-20-one (below) on hydrogenation in the presence of platinum oxide in ether under 2 atm. pressure, and of allopregnane-3β.16β.20β-triol (p. 2114 s) on similar hydrogenation in glacial acetic acid under 3 atm. pressure. On refluxing with alcoholic K_2CO_3, KOH, or HCl it gives $\Delta^{5.16}$-pregnadien-3β-ol-20-one (p. 2230 s) (1941 Marker).

Allopregnane-3β.16β-diol-20-one $C_{21}H_{34}O_3$ (II, R = R' = H). For the configuration at C-16, see 1949 Hirschmann.

*3-Acetate 16-(δ-acetoxyisocaproate), Tigone diacetate** $C_{31}H_{48}O_7$ [II, R = CO·CH$_3$, R' = CO· CH$_2$·CH$_2$·CH(CH$_3$)·CH$_2$·O·CO·CH$_3$]. Crystals (methanol), m. 102–104°. — **Fmn.** From dihydropseudotigogenin diacetate on oxidation with CrO_3 in strong acetic acid at 30° for 100 min. (yield, 37 %); is similarly obtained from pseudotigogenin diacetate. From the corresp. diester of Δ^5-pregnene-3β.16β-diol-20-one (above) on hydrogenation in ether in the presence of platinum oxide under 2 atm. pressure for 20 min. — **Rns.** Gives 3β-hydroxy-ætioallobilianic acid on oxidation with CrO_3 in strong acetic acid at 25° for 20 hrs., followed by alkaline hydrolysis. On reduction with sodium in dry isopropyl alcohol it gives allopregnane-3β.20α-diol (p. 1933 s). Gives allopregnane-3β.16β.20β-triol (p. 2114 s) on refluxing with aluminium isopropoxide in dry isopropyl alcohol for 7 hrs. or on hydrogenation in glacial acetic acid in the presence of platinum oxide under 2 atm. pressure for 2 hrs. at room temperature and then for 90 min. at 70°, in both cases followed by hydrolysis with methanolic or alcoholic KOH. Is converted into Δ^{16}-allopregnen-3β-ol-20-one (p. 2242 s) by refluxing with alcoholic K_2CO_3, KOH, or HCl (1941 Marker).

1941 Marker, Turner, Wagner, Ulshafer, Crooks, Wittle, *J. Am. Ch. Soc.* **63** 774; PARKE, DAVIS & Co. (Marker, Crooks, Wittle), *U.S. Pat.* 2352851 (issued 1944); *C. A.* **1945** 1022; PARKE, DAVIS & Co. (Marker), *U.S. Pats.* 2352648 (issued 1944), 2409293 (issued 1946); *C. A.* **1945** 1022; **1947** 1396; *Ch. Ztbl.* **1947** 1393.
1947 Marker, Wagner, Ulshafer, Wittbecker, Goldsmith, Ruof, *J. Am. Ch. Soc.* **69** 2167, 2172.
1949 Hirschmann, Hirschmann, Daus, *J. Biol. Ch.* **178** 751.

Δ^5-Pregnene-3β.17α-diol-20-one $C_{21}H_{32}O_3$. For the configuration at C-17, see

p. 1363 s. — Short needles (alcohol), m. 271–3° cor.; sometimes it melts only partly at this temperature, changing to long needles m. ca. 287° cor. (1941 Hegner; 1941 Fuchs; cf. also 1947 Hirschmann); $[\alpha]_D^{15}$ —37° ±3° (dioxane) (1941 Hegner; 1941 Fuchs). For the absorption curve of the pigment formed in the Zimmermann colour reaction (with alkaline m-dinitrobenzene), see 1947 Hirschmann. Does not reduce alkaline silver diammine solution at room temperature (1941 Fuchs).

* For this name, see 1947 Marker.

Occ. In the urine of a boy with adrenocortical carcinoma (1947 Hirschmann).

Fmn. From the 3.21-diacetate of 20-methyl-Δ^5-pregnene-3β.20a.21-triol (p. 1955 s) on refluxing with POCl$_3$ in abs. pyridine for $^1/_2$ hr., followed by treatment of the resulting mixture with OsO$_4$ in abs. ether at room temperature for 56 hrs., refluxing with Na$_2$SO$_3$ in aqueous alcohol for 5 hrs., and oxidation of the resulting mixture (m. 200–220°) with HIO$_4$ in aqueous methanol at room temperature for 15 hrs.; similarly from the 3.21-diacetate of 20-methyl-Δ^5-pregnene-3β.20b.21-triol (p. 1956 s) (1941 Hegner). Its 3-acetate is obtained from the 3-acetate 17-triphenylmethyl ether (not described in detail) of Δ^5-pregnene-3β.17α.20β-triol (see p. 2115 s) on oxidation, followed by acid hydrolysis (1939c CIBA). From a mixture of 20-methyl-Δ^5-pregnene-3β.17α. 20.21-tetrols (p. 2147 s) on oxidation with HIO$_4$ in aqueous dioxane at room temperature for 16 hrs. (1941 Fuchs).

Rns. (See also the scheme on p. 2340 s). On oxidation of its dibromide with CrO$_3$ in glacial acetic acid at room temperature, followed by debromination with zinc dust and potassium acetate on the water-bath and then by heating with glacial acetic acid, it gives small amounts of Δ^4-androstene-3.17-dione (p. 2880 s), 17-hydroxyprogesterone (p. 2843 s), and other compounds (1941 Hegner); similar oxidation of the 3-acetate, with temporary protection of the double bond by bromine, yields the acetate of 5-dehydro-epiandrosterone (p. 2609 s) (1947 Hirschmann). On heating with aluminium tert.-butoxide and acetone in abs. benzene in a sealed tube at 100° for 20 hrs. it gives a small amount of 17aβ-methyl-D-homo-Δ^4-androsten-17aα-ol-3.17-dione (V, p. 2340 s) (1941 Hegner). On hydrogenation of the 3-acetate in glacial acetic acid in the presence of platinum oxide, followed by acetylation with glacial acetic acid in pyridine at 65°, it gives the 3.20-diacetates of allo-pregnane-3β.17.20α- and -3β.17.20β-triols (pp. 2118 s, 2119 s; ratio 1:20) (1941 Hegner).

On heating the diacetate with fuller's earth at 150–160° under 0.1 mm. pressure it gives the acetate of $\Delta^{5.16}$-pregnadien-3β-ol-20-one (p. 2230 s) (1939c CIBA). The 3-acetate is rearranged to the 3-acetate of 17aβ-methyl-D-homo-Δ^5-androstene-3β.17aα-diol-17-one (III, p. 2340 s) by the action of Al$_2$O$_3$ in moist benzene (1947 Hirschmann).

3-Acetate C$_{23}$H$_{34}$O$_4$. Needles (benzene-pentane, or from ether on rapid crystallization), leaflets or flat needles (from ether on slow crystallization), both m. 234–5° cor.; sometimes a change occurs at this temperature and then the crystals are completely molten only at ca. 270° (1941 Hegner; 1941 Fuchs; cf. also 1947 Hirschmann); [α]$_D^{15}$ —41° (dioxane) (1941 Hegner; cf. 1941 Fuchs). Absorption max. in 95% alcohol at 294 mμ (1947 Hirschmann). — **Fmn.** From the above diolone on treatment with acetic anhydride in abs. pyridine at room temperature (1941 Hegner; 1941 Fuchs). — **Rns.** Reverts to the diolone on treatment with K$_2$CO$_3$ in ca. 97% methanol at room temperature for 48 hrs. (1941 Hegner).

Pregnane-3α.17α-diol-20-one $C_{21}H_{34}O_3$. Crystals (toluene or acetone), m. 212°
to 213° cor. (1943 Strickler; 1947 Mason), 219° to
219.5° cor. (1945, 1948 Lieberman); $[\alpha]_D$ in alcohol + 52° (1947 Mason), +64.5° ± 4° (1948
Lieberman). Very soluble in acetone and methanol, less soluble in toluene (1943 Strickler).
Gives a weak Zimmermann colour reaction (with
alkaline m-dinitrobenzene) (1943 Strickler; 1945 Lieberman). — **Occ.** As
glucuronide (below) in the urine of a young female hermaphrodite (1943
Strickler; 1947 Mason). In the urine of women with adrenal cortical hyperplasia and adrenal cortical tumour, in the urine of a cryptorchid male (1945,
1948 Lieberman), and in the urine of a eunuchoid male given testosterone
(p. 2580 s) by injection (1945 Lieberman). — **Rns.** Gives ætiocholane-3.17-dione
(p. 2895 s) with CrO_3 in strong acetic acid at room temperature (1945 Lieberman); on similar oxidation of the 3-acetate, the acetate of ætiocholan-3α-ol-
17-one (p. 2637 s) is formed (1945 Lieberman; 1947 Mason).

Oxime $C_{21}H_{35}O_3N$. Needles (aqueous methanol), dec. 247–250° cor. (1945
Lieberman), m. 223–5° (1943 Strickler).

3-Acetate $C_{23}H_{36}O_4$. Rosettes (acetone-ligroin), m. 201–2° cor. (1945 Lieberman); crystals (methanol), m. 196–8° cor. (1947 Mason). — **Fmn.** From the
above diolone and glacial acetic acid in pyridine on refluxing for 1 hr. (1945
Lieberman) or on heating at 90° for 30 min. (1947 Mason). — **Rns.** See above.

Glucuronide $C_{27}H_{42}O_9$. *Sodium salt* $C_{27}H_{41}O_9Na$ (not obtained pure), crystals
(95 % alcohol), m. 266–8° dec. (1947 Mason; cf. 1943 Strickler). In the
Zimmermann test it gives a colour similar to that of the free diolone (1947
Mason). Gives the free diolone on hydrolysis with aqueous HCl or with rat
liver enzyme (1947 Mason).

Pregnane-3β.17α-diol-20-one $C_{21}H_{34}O_3$. For a compound, originally considered
to possess this structure (1938 Masch), for which,
however, the structure 17aβ-methyl-D-homo-
ætiocholane-3β.17aα-diol-17-one is more probable
(1944 Shoppee), see under the reactions of 17-bro-
mopregnan-3β-ol-20-one (p. 2286 s).

Allopregnane-3β.17α-diol-20-one, *Reichstein's Substance L* $C_{21}H_{34}O_3$. For
configuration at C-17, see p. 1363 s. For
identity with *Compound G* (1936 Wintersteiner),
see 1938a, c Reichstein. — Rhombohedral
crystals (abs. alcohol), m. 264–6° cor., partially
changing to needles m. ca. 272° cor. (1938 c
Reichstein; cf. 1941 b v. Euw); platelets (alcohol),
m. 264° cor. dec. after softening at 254° (1936 Wintersteiner); $[\alpha]_D^{21}$ + 30.6°
± 3° (abs. alcohol) (1938 c Reichstein), $[\alpha]_D^{26}$ + 38° (95 % alcohol) (1936 Winter-

steiner), $[\alpha]_D^{14} +31.5^0$ (dioxane) (1941 b v. Euw). Soluble in 30 % methanol and in ether, sparingly soluble in boiling benzene (1938 c Reichstein) and in hot acetone, insoluble in water (1936 Wintersteiner).

Occ. In adrenal cortex extract (1936 Wintersteiner; 1938 b, c Reichstein; 1941 b v. Euw); it is purified by treatment with acetic anhydride in abs. pyridine at room temperature for 16 hrs.*, followed by refluxing of the result- ing 3-acetate with methanolic KOH for 15 min. (1938 c Reichstein).

Fmn. From 20-methylallopregnane-3β.20a.21-triol 3.21-diacetate (p. 1956 s) on refluxing with $POCl_3$ in abs. pyridine for $^1/_2$ hr., oxidation of the resulting mixture with OsO_4 in abs. ether at room temperature for 24 hrs., refluxing with Na_2SO_3 in aqueous alcohol for 5 hrs., and oxidation of the resulting mixture, containing the C-20 epimeric 20-methylallopregnane-3β.17.20.21- tetrols, with HIO_4 in aqueous methanol at room temperature for 15 hrs. (crude yield, 38 %) (1941 Hegner); the afore-mentioned mixture of tetrols may also be obtained in another way (see p. 2147 s) and may be oxidized with HIO_4 in aqueous dioxane at room temperature for 5 hrs. (1941 a v. Euw). From 17β-formylandrostane-3β.17α-diol (p. 2321 s) on treating with diazo- methane in ether at room temperature for 2 days (1941 Prins).

Rns. Does not reduce alkaline silver diammine solution. On oxidation with CrO_3 in glacial acetic acid at room temperature it gives a **compound** $C_{21}H_{32}O_3$ [leaflets (abs. alcohol), m. 270–2^0 cor.], assumed to be allopregnan- 17-ol-3.20-dione (q. v., p. 2845 s) and a small amount of androstane-3.17-dione (p. 2892 s) (1938 c Reichstein). Is not attacked by periodic acid (1941 Reich). On hydrogenation of the 3-acetate in methanol in the presence of Raney nickel at 90^0 under 135–170 atm. pressure it gives the 3-acetates of allo- pregnane-3β.17α.20α- and 20β-triols (pp. 2118 s, 2119 s; ratio 1:3.4) (1938 c Reichstein). On heating the 3-acetate with fuller's earth in vacuo it gives the acetate of Δ^{16}-allopregnen-3β-ol-20-one (p. 2242 s) (1939 c CIBA). Reacts incompletely with Girard's reagent at room temperature (1938 c Reichstein). For rearrangement to D-homo-steroids, investigated after the closing date of this volume, see, e. g., 1955 Fukushima; 1957 Turner.

Semicarbazone $C_{22}H_{37}O_3N_3$. Plates (alcohol), m. 263–5^0 cor. dec. (1936 Wintersteiner).

3-Acetate $C_{23}H_{36}O_4$. Needles or leaflets (depending on the rate of crystalli- zation from ether or ether-pentane), m. 191–2^0 cor.; $[\alpha]_D$ +15^0 to +16^0 (acetone).— **Fmn.** From the above diolone on treatment with acetic anhydride in abs. pyridine at room temperature (1938 c Reichstein; 1941 a v. Euw; 1941 Hegner; 1941 Prins). — **Rns.** See above.

---

* On chromatography of the resulting mixture of acetates on Al_2O_3, a **compound** $C_{22}H_{38}O_5$ was obtained as a by-product [needles (ether), m. 182–182.5^0 cor. after sublimation at 150^0 in high vac.; $[\alpha]_D^{18}$ +19^0 (acetone); *semicarbazone* $C_{24}H_{41}O_5N_3$, crystals (abs. alcohol), m. 255–9^0 cor.], probably the diacetate of a dihydroxyketone $C_{21}H_{31}O_3$; on refluxing with methanolic KOH it gives prisms (abs. alcohol), m. 200–211^0 cor., which on oxidation with CrO_3 in glacial acetic acid at room temperature yield crystals (ether-pentane), m. 179–185^0 cor. (1938 c Reichstein).

17-iso-Δ^5-Pregnene-3β.17β-diol-20-one $C_{21}H_{32}O_3$. For the configuration at C-17, see p. 1363 s. — Crystals (ethyl acetate), m. 162.5–163° cor. (1945 Hardegger); crystals (aqueous methanol), m. 174–6° (1941 Stavely); crystals (benzene), m. 176–9° cor. (1943b Shoppee); leaflets (ethyl acetate, on cooling slowly), needles (ethyl acetate, on cooling rapidly or on addition of hexane), both m. 190–1° cor.; on keeping the needles for 2 years they melt at 262–276° cor. and after recrystallization from ethyl acetate at 186–190° cor. (1943 Goldberg; cf., however, 1945 Hardegger); [α]$_D$ in chloroform —65.5° (1941 Stavely), —60° (1943b Shoppee), in dioxane —83° to —88° (1943 Goldberg; cf. also 1945 Hardegger). Absorption max. in 95% alcohol at 295 mμ (1947 Hirschmann). — From the absorption curve of the pigment formed in the Zimmermann reaction (with alkaline m-dinitrobenzene) it is assumed that the 3-acetate rearranges to the 3-acetate of 17aβ-methyl-D-homo-Δ^5-androstene-3β.17aα-diol-17-one (III, p. 2340 s) prior to condensation with the dinitrobenzene (1947 Hirschmann).

Fmn. (See also the scheme on p. 2340 s). From 17α-ethynyl-Δ^5-androstene-3β.17β-diol (p. 2030 s) on stirring with HgCl$_2$, aniline, benzene, and water for 20 hrs. at 60°; the diolone anil primarily formed is at once partly hydrolysed to the diolone and partly rearranged to 17a-methyl-17a-anilino-D-homo-Δ^5-androsten-3β-ol-17-one (XI, p. 2340 s); yield of diolone, 40% (1941 Stavely; cf. 1940 Stavely; 1939 Goldberg; 1943b Shoppee); the diacetate is obtained in 70% yield from the diol diacetate by the same method (1943b Shoppee). The diacetate is formed from the afore-mentioned diol, its 3-acetate, or its diacetate on treatment with HgO and BF$_3$-ether in glacial acetic acid + some acetic anhydride at room temperature (1938, 1939a Ruzicka*; 1939a CIBA; 1939 SCHERING CORP.); the corresp. diesters may be similarly obtained from the 3-acetate 17-benzoate of the diol (1938 Ruzicka; 1939a CIBA) and from the 3-acetate 17-stearate (1945 Hardegger); in this reaction (starting from the diacetate) BF$_3$ may be replaced by SnCl$_4$, SiCl$_4$ or, especially, FeCl$_3$ (1940b Ruzicka*). The diolone is also obtained from the afore-mentioned diol by boiling with mercury p-toluenesulphonamide in 96% alcohol for 72 hrs., and similarly its diacetate from the diol diacetate (1943 Goldberg; but cf. 1945 Hardegger). The 17-acetate is obtained from the same diol on shaking with 2 mols. of mercuric acetate in abs. alcohol or, better, in ethyl acetate for 24 hrs., followed by treatment with H$_2$S (1939b Ruzicka*; 1939b CIBA); the 17-propionate is similarly formed using mercuric propionate in ethyl propionate (1939b CIBA). The diacetate is obtained from 3β.17β-diacetoxy-17-iso-Δ^5-ætiocholenic acid chloride by the action of dimethylzinc in benzene, followed by hydrolysis with dil. HCl (1937 CIBA).

Rns. *Oxidation* (see also the scheme on p. 2340 s). On refluxing with aluminium isopropoxide and cyclohexanone in toluene it gives 17aα-methyl-

---

* The acyl derivatives of the diolones here described have been formulated by Ruzicka (1939a,b, 1940a, b) as derivatives of D-homo-steroids; for their true structure, see 1943a, b Shoppee. The D-homo-steroid formulæ of the diolones obtained by hydrolysis of the afore-mentioned acyl derivatives have been revised by Ruzicka (1940a).

D-homo-Δ^4-androsten-17aβ-ol-3.17-dione (VI, p. 2340 s) (1941 Stavely; cf. 1943b Shoppee; cf. also 1943 Goldberg*; 1938 Ciba). On refluxing of the 17-acetate (as obtained from the diacetate with K_2CO_3 in aqueous methanol at room temperature) with aluminium tert.-butoxide and acetone in benzene it gives 17β-acetoxy-17-iso-progesterone (p. 2844 s) which on hydrolysis with dil. KOH yields 17aβ-methyl-D-homo-Δ^4-androsten-17aα-ol-3.17-dione (V, p. 2340 s) (1938 Ruzicka; 1939 Ciba Pharmaceutical Products; cf. 1943a Shoppee). On bromination of the 3-acetate in glacial acetic acid, followed by oxidation with CrO_3 at 45° and then by debromination with zinc dust on the steam-bath, the acetate of 5-dehydroepiandrosterone (p. 2609 s) is obtained (1940 Stavely).

Reduction. The diacetate is reduced to the diacetate of 17-iso-allopregnane-3β.17β-diol-20-one (p. 2342 s) on hydrogenation in glacial acetic acid in the presence of reduced platinum oxide at room temperature under atmospheric pressure (1938, 1939a Ruzicka) or in methanol in the presence of Raney nickel at 105° under 137 atm. pressure (1945 Hardegger). On hydrogenation of the diolone in the presence of platinum oxide in abs. alcohol it gives 17-iso-allopregnane-3β.17β.20a-triol (p. 2121 s) (1941 Stavely), whereas in glacial acetic acid the 20-isomer (p. 2122 s) is formed (1943b Shoppee). On hydrogenation of the diacetate in dibutyl ether in the presence of copper chromite at 200° under 210 atm. pressure it gives a **compound** $C_{25}H_{38}O_4$ [leaflets (methanol), m. 130.5–131.5° cor.; $[\alpha]_D$ —82° (dioxane); yellow coloration with tetranitromethane] which on hydrogenation in glacial acetic acid, using platinum, gives the saturated **compound** $C_{25}H_{40}O_4$ [leaflets (ether-methanol), m. 134–6° cor.; $[\alpha]_D$ —36° (dioxane)] (1945 Hardegger).

Rearrangement (see also the scheme on p. 2340 s). The diolone undergoes rearrangement to 17aα-methyl-D-homo-Δ^5-androstene-3β.17aβ-diol-17-one (IV, p. 2340 s) on chromatography on Al_2O_3 in benzene (1941 Stavely); the rearrangement occurs to an extent of 50% on brief contact with Al_2O_3 and using dry benzene, and is quantitative on longer contact and using moist benzene; the 3-acetate reacts similarly giving the corresp. ester; when the diolone is treated with HgO and BF_3-ether in glacial acetic acid + some acetic anhydride at room temperature, the diacetate of the same D-homo-steroid is obtained (1943b Shoppee). The isopregnenediolone undergoes rearrangement to 17aβ-methyl-D-homo-Δ^5-androstene-3β.17aα-diol-17-one (III, p. 2340 s) when the following esters are hydrolysed by refluxing with methanolic KOH: the 3-acetate (1940 Stavely; cf. also 1941 Stavely), the 17-acetate (1939b Ruzicka**; 1939b Ciba), the diacetate (1938 Ruzicka; 1939 Schering Corp.; 1943b Shoppee), the 3-acetate 17-stearate (1945 Hardegger), and the 3-acetate 17-benzoate (1938 Ruzicka). On attempted replacement of the 17-OH group by Br by treatment of the 3-acetate with ca. $^1/_3$ mol.

---

* The product m. 176–8° obtained by Goldberg on refluxing the diolone with aluminium tert.-butoxide and acetone in benzene is according to this author probably impure 17a α-methyl-D-homo-Δ^5-androstene-3β.17aβ-diol-17-one; the molecular formula $C_{21}H_{30}O_3$, however, the absorption band at 240 mμ, and the mode of formation make it more probable that it is identical with the afore-mentioned hydroxydiketone (Editor).

** See the footnote on p. 2337 s.

PBr$_3$ and $^1/_4$ mol. pyridine in benzene at room temperature, finally at 40°, the 3-acetates of the two afore-mentioned D-homosteroids are formed (1943c Shoppee).

Dehydration and similar reactions. The diacetate is almost unchanged by distillation with or without zinc at 210–240° bath temperature and 10 mm. pressure, by heating with zinc in toluene at 110° or in xylene at 140°, or with pyridine-formamide (1:1) in a sealed tube at ca. 180° for 15 hrs.; in the last case a small amount of the 17-acetate is obtained (1943c Shoppee). On heating the 3-acetate with POCl$_3$ in pyridine at 100° for 3 hrs. it gives the acetate of $\Delta^{5.16}$-pregnadien-3β-ol-20-one (p. 2230 s) which is also obtained, along with a small amount of $\Delta^{3(7).5.16}$-pregnatrien-20-one (p. 2215 s), on distillation of the 3-acetate 17-benzoate at 250–280° under 10 mm. pressure (1943c Shoppee). The 3-acetate 17-stearate remains unchanged on brief boiling under atmospheric pressure; on continued boiling it decomposes, but no $\Delta^{5.16}$-pregnadien-3β-ol-20-one acetate could be isolated (1945 Hardegger).

DERIVATIVES OF 17-ISO-Δ^5-PREGNENE-3β.17β-DIOL-20-ONE

Oxime C$_{21}$H$_{33}$O$_3$N. Needles (methanol), m. 255–260° cor. (1943 Goldberg).

3-Acetate C$_{23}$H$_{34}$O$_4$. Needles (benzene-petroleum ether), m. 196–8° (1940 Stavely); plates (chloroform-pentane), crystals (ethyl acetate), m. 187–8° cor. (1943b Shoppee; 1943 Goldberg); [α]$_D$ in chloroform —61° (1940 Stavely; 1943b Shoppee), in dioxane —80.5° (1943 Goldberg). Very sparingly soluble in pyridine (1943b Shoppee). — **Fmn.** From the diolone on treatment with acetic anhydride in pyridine at room temperature (1940 Stavely; 1943b Shoppee; 1943 Goldberg). — **Rns.** See under the reactions of the diolone. — *Oxime* C$_{23}$H$_{35}$O$_4$N, crystals (benzene-petroleum ether), m. 254–6° (1940 Stavely); crystals (methanol), m. 235–240° cor. dec. (1943 Goldberg).

17-Acetate C$_{23}$H$_{34}$O$_4$. Crystals (acetone), m. 221–2° cor.; [α]$_D$ —53°(dioxane) (1939b Ruzicka*; 1939b CIBA; cf. also 1943c Shoppee). — **Fmn.** See under formation and reactions of the diolone.

Anil C$_{27}$H$_{37}$O$_2$N. Is the product primarily formed, but not isolated, in the reaction of 17α-ethynyl-Δ^5-androstene-3β.17β-diol with aniline + HgCl$_2$; it rearranges to 17a-methyl-17a-anilino-D-homo-Δ^5-androsten-3β-ol-17-one (XI, p. 2340 s) and, in the presence of water, is hydrolysed to give the free isopregnenediolone (1941 Stavely; 1943a, b Shoppee). Its *3.17-diacetate* C$_{31}$H$_{41}$O$_4$N [crystals (aqueous alcohol), m. 207–9° cor.] is obtained from the diacetate of the afore-mentioned diol by heating with aniline and HgCl$_2$ at 70–80° for 24 hrs. (1939 Goldberg); when prepared in the presence of water it is at once hydrolysed to give the diacetate of the isopregnenediolone (1943b Shoppee).

Diacetate C$_{25}$H$_{36}$O$_5$. Leaflets (ether-pentane), platelets (acetone-hexane), crystals (ether, methanol, or ethyl acetate), m. 193–5° cor.; [α]$_D$ —52° to —56° (dioxane) (1938, 1939a, b Ruzicka*; 1943b Shoppee; 1943 Goldberg).

---

* See the footnote on p. 2337 s.

Rearrangement of Δ⁵-Pregnene-3β.17α-diol-20-one (I) and 17-iso-Δ⁵-Pregnene-3β.17β-diol-20-one (II) to D-Homo-steroids, and Related Reactions

a means R = R' = H   c means R = H, R' = CO·CH₃
b means R = CO·CH₃, R' = H   d means R = R' = CO·CH₃

(I), p. 2333 s

Ia dibromide with CrO₃, followed by debromination

Ib on Al₂O₃ (IIIb ←)

Ia on Oppenauer oxidation

(small yield)

(III)

IIIa on Oppenauer oxidation

(V)

(IIIa ←)

(IIIb ←)

Me OH CO·CH₃

KOH

(VII), p. 2843 s

18%

41%

VIIIa or b with KOH

Me OR' CO·CH₃

IVa on Oppenauer oxidation

(IV)

IIa with BF₃, AcOH-Ac₂O (IVd ←)

(VI)

(→ IVb)

IIa, b on Al₂O₃ (IVa, b, resp. ←)

IIb with PBr₃-pyridine

IIb, c, or d with alkali

(II), p. 2337 s

IIc on Oppenauer oxidation

VIIIa on Al₂O₃

(VIII), p. 2844 s

IXa with HgSO₄ + water (→ IIIa)

IXa, b, or d with HgO-BF₃ in AcOH-Ac₂O (→ IId)

IXa with Hg(OAc)₂ (→ IIc)

IXa or d with Hg p-toluene-sulphonamide (→ IIa or d)

IIa or d with water (IIa or d, resp. ←)

Xa or d with water

(IX), p. 2030 s

IXa or d with aniline + HgCl₂ (→ Xa or d, resp.)

(X)

Xa by rearrangement

(XI)

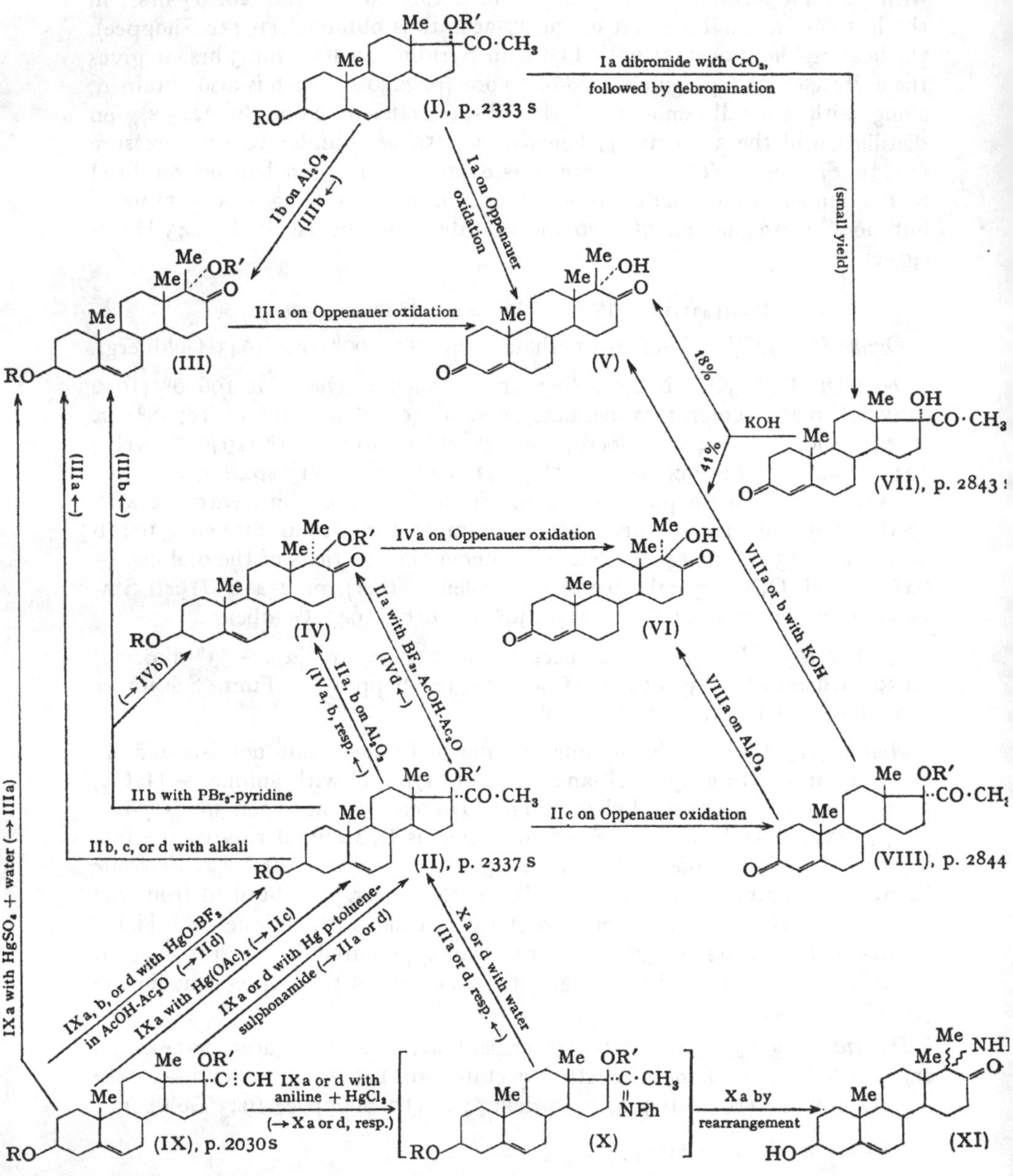

Rearrangement of 17-iso-Pregnane-3β.17β-diol-20-one (V) and 17-iso-Allopregnane-3β.17β-diol-20-one (XI) to D-Homo-steroids, and Related Reactions

a means R = R′ = H

b means R = CO·CH₃, R′= H

c means R = H, R′= CO·CH₃

d means R = R′ = CO·CH₃

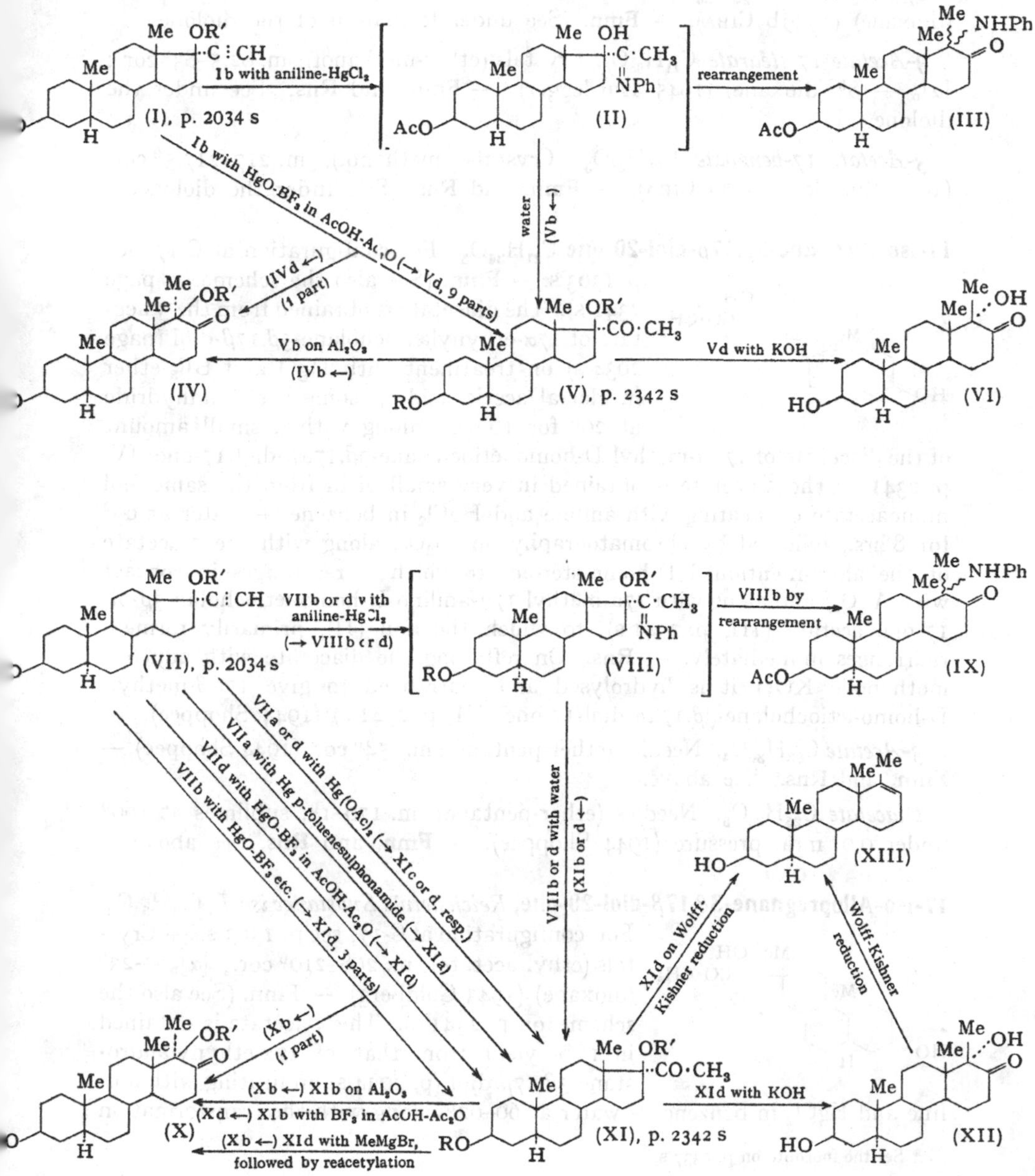

Absorption max. (solvent not stated) at ca. 280 mμ (log ε 1.78) (1938 Ruzicka). — **Fmn.** From the diolone on heating with acetic anhydride in pyridine at 120° for 2 hrs. (1943b Shoppee) or at 105° for 48 hrs. (1943 Goldberg). From the 17-acetate (above) on treatment with acetic anhydride in pyridine at room temperature (1939b Ruzicka*). See also under formation of the diolone. — **Rns.** See under the reactions of the diolone.

17-Propionate $C_{24}H_{36}O_4$. Crystals (ether-hexane), m. 147°; $[\alpha]_D$ —50° (dioxane) (1939b CIBA). — **Fmn.** See under formation of the diolone.

3-Acetate 17-stearate $C_{41}H_{68}O_5$. Crystals (ether-methanol), m. 82.5–83° cor.; $[\alpha]_D$ —18° (dioxane) (1945 Hardegger). — **Fmn.** and **Rns.** See under the diolone.

3-Acetate 17-benzoate $C_{30}H_{38}O_5$. Crystals (methanol), m. 217–217.5° cor. (1938 Ruzicka; 1939a CIBA). — **Fmn.** and **Rns.** See under the diolone.

17-iso-Pregnane-3β.17β-diol-20-one $C_{21}H_{34}O_3$.

For configuration at C-17, see p. 1363 s. — **Fmn.** (See also the scheme on page 2341 s). The diacetate is obtained from the 3-acetate of 17α-ethynylætiocholane-3β.17β-diol (page 2034 s) on treatment with HgO and BF₃-ether in glacial acetic acid + some acetic anhydride at 20° for 16 hrs., along with a small amount of the diacetate of 17aα-methyl-D-homo-ætiocholane-3β.17aβ-diol-17-one (IV, p. 2341 s); the 3-acetate is obtained in very small yield from the same diol monoacetate on heating with aniline and HgCl₂ in benzene + water at 60° for 8 hrs., followed by chromatography on Al₂O₃, along with the 3-acetate of the afore-mentioned D-homo-steroid (to which it rearranges in contact with Al₂O₃) and along with 17a-methyl-17a-anilino-D-homo-ætiocholan-3β-ol-17-one acetate (III, p. 2341 s), to which the anil (II), primarily formed, rearranges immediately. — **Rns.** On refluxing the diacetate with aqueous methanolic KOH it is hydrolysed and rearranged to give 17aβ-methyl-D-homo-ætiocholane-3β.17aα-diol-17-one (VI, p. 2341 s) (1944 Shoppee).

3-Acetate $C_{23}H_{36}O_4$. Needles (ether-pentane), m. 154° cor. (1944 Shoppee).— **Fmn.** and **Rns.** See above.

Diacetate $C_{25}H_{38}O_5$. Needles (ether-pentane), m. 170–1°; sublimes at 160° under 0.01 mm. pressure (1944 Shoppee). — **Fmn.** and **Rns.** See above.

17-iso-Allopregnane-3β.17β-diol-20-one, *Reichstein's Substance iso-L*, $C_{21}H_{34}O_3$.

For configuration at C-17, see p. 1363 s. — Crystals (ethyl acetate), m. 208–210° cor.; $[\alpha]_D$ —23° (dioxane) (1943 Goldberg). — **Fmn.** (See also the scheme on p. 2341 s). The 3-acetate is obtained in 17% yield from that of 17α-ethynylandrostane-3β.17β-diol (p. 2034 s) on heating with aniline and HgCl₂ in benzene + water at 60–62°, along with the transformation

. * See the footnote on p. 2337 s.

product, 17a-methyl-17a-anilino-D-homo-androstan-3β-ol-17-one 3-acetate
(IX, p. 2341 s), of the anil (VIII, p. 2341 s) formed primarily; the diacetate
is similarly obtained, but in 78% yield, from the diacetate of the same diol
(1943a Shoppee). The diacetate is also obtained from the 3-acetate or the
diacetate of the afore-mentioned diol on treatment with HgO and BF₃-ether
in glacial acetic acid + some acetic anhydride at room temperature for
16 hrs., in the first case along with a smaller amount of the diacetate of
17aα-methyl-D-homo-androstane-3β.17aβ-diol-17-one (X, p. 2341 s) (1939a
Ruzicka*; 1939a CIBA; cf. 1943a Shoppee). The free diolone is obtained in
good yield from the diol (see above) on heating with mercury p-toluene-
sulphonamide in 96% alcohol on the water-bath for 60 hrs. (1943 Goldberg).
The 17-acetate or the diacetate is obtained from the same diol or its diacetate,
resp., on shaking with mercuric acetate in ethyl acetate at room temperature
for 24 hrs., followed by treatment with hydrogen sulphide (1939b Ruzicka*;
1939b CIBA; cf. 1943a Shoppee). The diacetate may also be obtained from
the diacetate of 17-iso-Δ⁵-pregnene-3β.17β-diol-20-one (p. 2337 s) on hydrogenation in glacial acetic acid in the presence of reduced platinum oxide at
room temperature (1938 Ruzicka).

Rns. (See also the scheme on p. 2341 s). On oxidation of the 3-acetate
with CrO₃ in glacial acetic acid at room temperature it gives the acetate of
epiandrosterone (p. 2633 s) (1943a Shoppee); the diacetate is unattacked
under the same conditions (1939a Ruzicka*). On hydrogenation of the
3-acetate in alcohol in the presence of reduced platinum oxide, followed
by acetylation with acetic anhydride in pyridine at room temperature, it
gives the 3.20-diacetate of 17-iso-allopregnane-3β.17β.20b-triol (p. 2122 s)
(1943a Shoppee); the diacetate is unattacked by heating with hydrogen in
the presence of Raney nickel in methanol at 100–120° or in the presence
of platinum oxide in glacial acetic acid at 100° (1943a Shoppee; 1939a
Ruzicka*). On heating the diacetate with hydrazine hydrate and sodium
ethoxide in alcohol in a sealed tube at 180° for 10 hrs. it gives 17a-methyl-
D-homo-Δ¹⁷-androsten-3β-ol (1943a Shoppee).

The 3-acetate rearranges to the 3-acetate of 17aα-methyl-D-homo-andro-
stane-3β.17aβ-diol-17-one (X, p. 2341 s) in contact with Al₂O₃ (1943a Shoppee)
and gives the diacetate of the same D-homo-steroid on treatment with BF₃-
ether in glacial acetic acid + some acetic anhydride at room temperature
(1943b Shoppee). The diacetate is hydrolysed and rearranged to give 17aβ-
methyl-D-homo-androstane-3β.17aα-diol-17-one (XII, p. 2341 s). On warming
with aqueous methanolic K₂CO₃ for 3 hrs. (1939b Ruzicka*) or on boiling
with methanolic KOH for 30 min. (1943a Shoppee; cf. 1939a Ruzicka*;
1939a, b CIBA). On refluxing of the diacetate with methylmagnesium bromide
in ether-dioxane, followed by reacetylation with acetic anhydride in pyridine
at room temperature, it gives 17aα-methyl-D-homo-androstane-3β.17aβ-diol-
17-one 3-acetate (X, p. 2341 s) and small amounts of 20-methyl-17-iso-allo-
pregnane-3β.17β.20-triol 3-acetate (p. 2128 s) and of a compound m. ca. 265°
(1943d Shoppee).

* See the footnote on p. 2337 s.

E.O.C. XIV s. 164

3-Acetate $C_{23}H_{36}O_4$. Flat needles (acetone-hexane), m. 181–3⁰ cor. and 192–4⁰ cor. (1943a Shoppee); crystals (acetone-petroleum ether), m. 173–5⁰ cor. (1953 Turner); $[\alpha]_D^{23}$ —24⁰ ± 3⁰, $[\alpha]_{5461}^{23}$ —29⁰ ± 3⁰ (dioxane) (1943a Shoppee). — **Fmn.** and **Rns.** See under those of the above diolone.

17-Acetate $C_{23}H_{36}O_4$. Crystals (ether-hexane), m. 202–4⁰ cor.; $[\alpha]_D$ 0⁰ ± 2⁰ (dioxane) (1939b Ruzicka*). — **Fmn.** See under the diolone.

Diacetate $C_{25}H_{38}O_5$. Leaflets (ether-pentane or chloroform-pentane), crystals (methanol), m. 227–9⁰ cor. (1939a Ruzicka*; 1943a Shoppee); $[\alpha]_D$ +2.5⁰ ± 2⁰ (acetone) (1943a Shoppee; cf. 1939a Ruzicka*), —4⁰ (dioxane) (1938 Ruzicka). — **Fmn.** See under the free diolone. May also be obtained from the 3-acetate on heating with acetic anhydride in pyridine at 100⁰ for 2 hrs.; small yield (1943a Shoppee). — **Rns.** See under those of the free diolone.

1936 Wintersteiner, Pfiffner, *J. Biol. Ch.* **116** 291, 298, 302.
1937 Ciba**, *Swiss Pat.* 211651 (issued 1941); *C.A.* **1942** 3637; *Fr. Pat.* 840515 (issued 1939); *C.A.* **1939** 8627; *Ch. Ztbl.* **1939** II 1721.
1938 Ciba**, *Swiss Pat.* 228644 (issued 1943); *C.A.* **1949** 3476; *Ch. Ztbl.* **1944** II 1300; *Brit. Pat.* 536621 (1939, issued 1941); *C.A.* **1942** 1739; *Ch. Ztbl.* **1943** II 147; *Fr. Pat.* 861318 (1939, issued 1941); *Ch. Ztbl.* **1941** I 3259.
Masch, *Über Bromierungen in der Pregnanreihe*, Thesis, Danzig, pp. 15, 25.
a Reichstein in Ruzicka, Stepp, *Ergebnisse der Vitamin- und Hormonforschung*, Vol. **1**, Leipzig, pp. 364, 365.
b Reichstein, v. Euw, *Helv.* **21** 1197, 1200, 1207.
c Reichstein, Gätzi, *Helv.* **21** 1497.
Ruzicka, Meldahl, *Helv.* **21** 1760, 1762, 1764, 1765, 1767, 1769.
1939 a Ciba**, *Fr. Pat.* 858127 (issued 1940); *Ch. Ztbl.* **1941** I 3258.
b Ciba**, *Brit. Pat.* 532262 (issued 1941); *C.A.* **1942** 874; *Fr. Pat.* 863162 (1940, issued 1941); *Ch. Ztbl.* **1941** II 3218; Ciba Pharmaceutical Products (Ruzicka), *U.S. Pat.* 2315817 (1940, issued 1943); *C.A.* **1943** 5558.
c Ciba**, *Fr. Pat.* 860278 (issued 1941); *Ch. Ztbl.* **1941** I 3258.
Ciba Pharmaceutical Products (Ruzicka), *U.S. Pat.* 2365292 (issued 1944); *C.A.* **1945** 4199; *Ch. Ztbl.* **1945** II 1233.
Goldberg, Aeschbacher, *Helv.* **22** 1188.
a Ruzicka, Gätzi, Reichstein, *Helv.* **22** 626, 631–634.
b Ruzicka, Goldberg, Hunziker, *Helv.* **22** 707, 711–714.
Schering Corp. (Inhoffen, Logemann, Dannenbaum), *U.S. Pat.* 2280236 (issued 1942); *C.A.* **1942** 5322.
1940 a Ruzicka, Meldahl, *Helv.* **23** 364. — b Ruzicka, Meldahl, *Helv.* **23** 513–515.
Stavely, *J. Am. Ch. Soc.* **62** 489; Squibb & Sons (Stavely), *U.S. Pat.* 2411172 (issued 1946); *C.A.* **1947** 1396; *Ch. Ztbl.* **1947** 232.
1941 a v. Euw, Reichstein, *Helv.* **24** 418.
b v. Euw, Reichstein, *Helv.* **24** 879, 883.
Fuchs, Reichstein, *Helv.* **24** 804, 821, 822.
Hegner, Reichstein, *Helv.* **24** 828, 836 et seq.
Prins, Reichstein, *Helv.* **24** 945, 948, 949.
Reich, Reichstein, *Arch. intern. pharmacodynamie* **65** 415, 417.
Stavely, *J. Am. Ch. Soc.* **63** 3127.
1943 Goldberg, Aeschbacher, Hardegger, *Helv.* **26** 680, 682–684, 686.
a Shoppee, Prins, *Helv.* **26** 185, 194, 195, 197–200.
b Shoppee, Prins, *Helv.* **26** 201, 212, 214–219.

* See the footnote on p. 2337 s.
** Société pour l'industrie chimique à Bâle; Gesellschaft für chemische Industrie in Basel.

c Shoppee, Prins, *Helv.* **26** 1004, 1009, 1013, 1015, 1016.
d Shoppee, Prins, *Helv.* **26** 2089, 2092.
  Strickler, Shaffer, Wilson, Strickler, *J. Biol. Ch.* **148** 251.
1944 Shoppee, *Helv.* **27** 8, 9, 14, 17, 18.
1945 Hardegger, Scholz, *Helv.* **28** 1355.
  Lieberman, Dobriner, *J. Biol. Ch.* **161** 269.
1947 Hirschmann, Hirschmann, *J. Biol. Ch.* **167** 7, 9, 11, 12, 19 et seq.
  Mason, Strickler, *J. Biol. Ch.* **171** 543.
1948 Lieberman, Dobriner, Hill, Fieser, Rhoads, *J. Biol. Ch.* **172** 263, 269, 287.
1953 Turner, *J. Am. Ch. Soc.* **75** 3484, 3487.
1955 Fukushima, Dobriner, Heffler, Kritchevsky, Herling, Roberts, *J. Am. Ch. Soc.* **77** 6585.
1957 Turner, Perelman, Park, *J. Am. Ch. Soc.* **79** 1108, 1112.

17-iso-Δ^5-Pregnene-3β.17β-diol-21-al $C_{21}H_{32}O_3$. For configuration at C-17, see p. 1363 s. — The crude substance (not analysed) m. ca. 152–7° cor. (1939 Miescher). — **Fmn.** From a mixture of the 21-epimeric 21-hydroxymethyl-17-iso-Δ^5-pregnene-3β.17β. 21-triols (see p. 2147 s) on oxidation with $KIO_4 + H_2SO_4$ in aqueous methanol under nitrogen at 30° (3 hrs.) (1939 Miescher). — **Rns.** On oxidation with SeO_2, followed by reduction, it gives a mixture of 17-iso-Δ^5-pregnene-3β.4β.17β.20.21-pentol and 17-iso-Δ^4-pregnene-3β.17β.20.21-pentol (see p. 2201 s) (1937 CIBA). On refluxing the 3-acetate (below) with glacial acetic acid under nitrogen for 4 hrs. it gives the acetate of $\Delta^{5.17(20)}$-pregnadien-3β-ol-21-al (p. 2251 s) (1939 Miescher).

3-Acetate $C_{23}H_{34}O_4$. The crude substance (not analysed) m. ca. 121–5° cor.— **Fmn.** From the preceding compound on treatment with acetic anhydride in abs. pyridine at room temperature for 12 hrs. (1939 Miescher).

1937 CIBA*, *Swiss Pat.* 212337 (issued 1941); *C. A.* **1942** 3637; *Ch. Ztbl.* **1942** I 643; *Brit. Pat.* 497394 (issued 1939); *C. A.* **1939** 3812; *Ch. Ztbl.* **1939** I 2828; CIBA PHARMACEUTICAL PRODUCTS (Miescher, Wettstein), *U.S. Pat.* 2229813 (1938, issued 1941); *C. A.* **1941** 3040; *Ch. Ztbl.* **1941** II 2467, 2468.
1939 Miescher, Wettstein, Scholz, *Helv.* **22** 894, 907; cf. CIBA PHARMACEUTICAL PRODUCTS (Miescher, Fischer, Scholz, Wettstein), *U.S. Pat.* 2275790 (issued 1942); *C. A.* **1942** 4289; CIBA*, *Fr. Pat.* 857122 (issued 1940); *Ch. Ztbl.* **1941** I 1324.

Cholane-3α.6α-diol-23-one, *Bisnorhyodesoxycholyl methyl ketone* $C_{24}H_{40}O_3$. Needles with 1 H_2O (ethyl acetate), m. 233°; $[\alpha]_D^{28}$ —3.2° (alcohol). Readily soluble in methanol, ether, and acetone, rather sparingly in ethanol and ethyl acetate. Gives a brownish red Liebermann reaction. — **Fmn.** Its diacetate is obtained, along with bisnorhyodesoxycholic acid diacetate, from 23-methyl-cholane-3α.6α.23-triol 3.6-diacetate (p. 2130 s) on oxidation with CrO_3 in

---

* Société pour l'industrie chimique à Bâle; Gesellschaft für chemische Industrie in Basel.
164*

glacial acetic acid on the water-bath; it is hydrolysed by heating with 10% alcoholic KOH on the water-bath. — **Rns.** The diacetate is practically not attacked by CrO_3 in glacial acetic acid (1939 Kimura).

Diacetate $C_{28}H_{44}O_5$. Needles (alcohol), m. 178⁰ (1939 Kimura). — **Fmn.** See above.

Cholane-3α.7α-diol-23-one, *Bisnorchenodesoxycholyl methyl ketone* $C_{24}H_{40}O_3$.

Needles with 1 H_2O (dil. alcohol); sinters at 85⁰, m. 160–1⁰; $[\alpha]_D^{22} +3.5⁰$ (alcohol). Readily soluble in methanol, ethanol, acetone, ether, and ethyl acetate, rather sparingly in dil. acetic acid. Gives a violet-red Liebermann reaction. — **Fmn.**

Its diacetate is obtained, along with bisnorchenodesoxycholic acid diacetate, from 23-methylcholane-3α.7α.23-triol 3.7-diacetate (p. 2131 s) on oxidation with CrO_3 in glacial acetic acid on the water-bath (yield, 8%); it is hydrolysed by heating with 10% alcoholic KOH on the water-bath. — **Rns.** The diacetate is oxidized to a small extent by CrO_3 to give bisnorchenodesoxycholic acid diacetate (1938 Ishihara).

Diacetate $C_{28}H_{44}O_5$. Needles (alcohol), m. 189–190⁰ (1938 Ishihara). — **Fmn.** and **Rns.** See above.

Cholane-3α.12α-diol-23-one, *Bisnordesoxycholyl methyl ketone* $C_{24}H_{40}O_3$

(wrongly formulated in the original paper as the next lower homologue). — **Fmn.** Its diacetate is obtained, along with bisnordesoxycholic acid diacetate, from the 3.12-diacetate of 23-methylcholane-3α.12α.23-triol (p. 2131 s) on oxidation with CrO_3 in glacial acetic acid on the water-bath; very small yield (1939 Kazuno).

Diacetate $C_{28}H_{44}O_5$ (not analysed). Prisms (methanol), m. 148–150⁰ (1939 Kazuno). — **Fmn.** See above.

Homocholane-3α.6α-diol-24-one, *Norhyodesoxycholyl methyl ketone* $C_{25}H_{42}O_3$.

Needles with $^1/_2$ H_2O (ethyl acetate), m. 183⁰; $[\alpha]_D^{15} +20⁰$ (alcohol). Readily soluble in methanol, ethanol, and ether, rather sparingly in ethyl acetate. Gives a violet-red Liebermann reaction. — **Fmn.** Its diacetate is obtained, along with norhyodesoxycholic acid diacetate, from 24.24-dimethylcholane-3α.6α.24-triol 3.6-diacetate (p. 2131 s) on oxidation with CrO_3 in glacial acetic acid on the water-bath (yield, 2%); it is hydrolysed by heating with 10% alcoholic KOH on the water-bath for 2 hrs. — **Rns.** The diacetate is partly oxidized to norhyodesoxycholic acid diacetate by heating with CrO_3 in glacial acetic acid on the water-bath (1939 Kimura).

Diacetate $C_{29}H_{46}O_5$. Needles with 1 H_2O (alcohol), m. 163° (1939 Kimura). — **Fmn.** and **Rns.** See above.

Homocholane-3α.7α-diol-24-one *Norchenodesoxycholyl methyl ketone* $C_{25}H_{42}O_3$.

Needles (alcohol), m. 175–6°; $[\alpha]_D^{22}$ +9° (alcohol). Readily soluble in methanol and ethanol, rather sparingly in glacial acetic acid, ether, and ethyl acetate. Gives a violet-red Liebermann reaction. — **Fmn.**

Its diacetate is obtained, along with norchenodesoxycholic acid diacetate, from the 3.7-diacetate of 24.24-dimethylcholane-3α.7α.24-triol (p. 2131 s) on oxidation with CrO_3 in glacial acetic acid on the water-bath (very small yield); it is hydrolysed by heating with 10% alcoholic KOH on the water-bath. — **Rns.** The diacetate is oxidized to a small extent by CrO_3 in glacial acetic acid on the water-bath to give norchenodesoxycholic acid diacetate (1938 Ishihara).

Diacetate $C_{29}H_{46}O_5$. Needles (abs. alcohol), m. 132–3° (1938 Ishihara). — **Fmn.** and **Rns.** See above.

Homocholane-3α.12α-diol-24-one, *Nordesoxycholyl methyl ketone* $C_{25}H_{42}O_3$. —

Fmn. Its diacetate is obtained, along with nordesoxycholic acid diacetate, from the 3.12-diacetate of 24.24-dimethylcholane-3α.12α.24-triol (p. 2132 s) on oxidation with CrO_3 in glacial acetic acid on the water-bath; very small yield (1939 Kazuno).

Diacetate $C_{29}H_{46}O_5$. Prisms (methanol), m. 141° (1939 Kazuno). — **Fmn.** See above.

1938 Ishihara, *J. Biochem.* **27** 265, 274–276.
1939 Kazuno, Shimizu, *J. Biochem.* **29** 421, 432, 433.
   Kimura, Sugiyama, *J. Biochem.* **29** 409, 416–418.

17β-Benzoyl-ætiocholane-3α.12α-diol, 3α.12α-Dihydroxyætiocholanyl phenyl ketone $C_{26}H_{36}O_3$. For configuration at C-12 and C-17, see pp. 1362 s, 1363 s.

Diacetate $C_{30}H_{40}O_5$. A substance [crystals (ether-petroleum ether), m. 152–3°], which possibly possesses this structure, is obtained in very small yields from the methyl ester of ætiodes-

oxycholic acid (3α.12α-dihydroxyætiocholanic acid) on refluxing with phenylmagnesium bromide in benzene, followed by refluxing with aqueous methanolic KOH, acetylation with acetic anhydride in pyridine on the water-bath,
and chromatography of the resulting mixture containing mainly 3α.12α-diacet-
oxyætiocholanyl-diphenyl-carbinol, and also from this carbinol on refluxing
with glacial acetic acid (1944 Koechlin).

21-Benzylidenepregnane-3α.11α-diol-20-one $C_{28}H_{38}O_3$. Needles (benzene or
acetone), m. 219–220⁰ cor.; $[α]_D^{24} + 107⁰$
(abs. alcohol), $+ 86⁰$ (chloroform). —
Fmn. From pregnane-3α.11α-diol-20-one
(p. 2327 s) and benzaldehyde in abs.
alcohol in the presence of sodium ethoxide
at —15⁰ to —3⁰; yield, 83 %. — **Rns.** May
be converted into 3α.11α-dihydroxyætiocholanic acid by ozonization of its
diacetate in anhydrous methanol-ethyl acetate (1:1) at —45⁰, followed by
shaking of the solution with hydrogen and palladium-calcium carbonate,
oxidation of the resulting product with periodic acid in alcohol, and
hydrolysis with 1 N NaOH on the steam-bath (1946 Long).

Diacetate $C_{32}H_{42}O_5$. Needles (ethyl acetate-petroleum ether), m. 162–163⁰
cor.; $[α]_D^{24} + 55⁰$ (abs. alcohol), $+71⁰$ (chloroform). Absorption max. in 95 %
alcohol at 294 mμ (ε 24000). — **Fmn.** From the above diolone on heating
with acetic anhydride and pyridine (1946 Long). — **Rns.** See above.

21-Benzylidenepregnane-3α.12α-diol-20-one $C_{28}H_{38}O_3$. For configuration at
C-12 and C-17, see pp. 1362 s, 1363 s. —
The crude product (not analysed) forms
crystals (alcohol-ether), m. 213–5⁰ (1944
Koechlin; cf. 1940 Reichstein). — **Fmn.**
From pregnane-3α.12α-diol-20-one (page
2329 s) with benzaldehyde and sodium
ethoxide in abs. alcohol at 20⁰ (1940 Reichstein; 1944 Koechlin; cf. 1937
Schering A.-G.); is similarly obtained from the diacetate of this diolone
(1938 Hoehn). — **Rns.** May be converted into ætiodesoxycholic acid (3α.12α-
dihydroxyætiocholanic acid) by ozonization of its diacetate in chloroform at
—10⁰, followed by oxidation with HIO_4 in alcohol and hydrolysis with
2 N NaOH at 100⁰ (1938 Hoehn; cf. 1940 Reichstein). On treatment of the
diacetate with PCl_5 in benzene at 50⁰, followed by ozonization and reacetylation of the resulting neutral fraction, a crystalline substance, m. 175–184⁰,
is obtained which is not identical with ætiocholane-3α.12α-diol-17-one diacetate
(1944 Koechlin).

Diacetate $C_{32}H_{42}O_5$. Slightly yellowish needles (from methanol at —10⁰),
m. 119–121⁰; $[α]_D^{16} + 201⁰$ (dioxane). — **Fmn.** From the above diolone on
heating with acetic anhydride in pyridine on the water-bath for 3 hrs. (1944
Koechlin).

3α.12α-Dihydroxyternorcholanyl phenyl ketone $C_{28}H_{40}O_3$.

Not obtained crystalline. — **Fmn.** From the diformate of bis-nordesoxycholic acid chloride on treatment with diphenylcadmium in boiling benzene, followed by decomposition with dil. HCl and hydrolysis of the reaction product by boiling with 5% methanolic NaOH (1945 Hoehn).

Diformate $C_{30}H_{40}O_5$. Crystals (formic acid), m. 242–5°; $[\alpha]_D^{25}$ +125° (dioxane). — **Fmn.** From the above crude diolone on heating with formic acid (d 1.20) on the water-bath (1945 Hoehn).

Diacetate $C_{32}H_{44}O_5$. Crystals (methanol), m. 193–195.5°; $[\alpha]_D^{25}$ +122.5° (dioxane). — **Fmn.** From the diolone on refluxing with glacial acetic acid-acetic anhydride (2:3) (1945 Hoehn).

3α.12α-Dihydroxybisnorcholanyl phenyl ketone $C_{29}H_{42}O_3$.

Crystals with $\frac{1}{2}$ $CH_3 \cdot OH$ (methanol), m. 105–115°; $[\alpha]_D^{25}$ +59° (dioxane). — **Fmn.** As above, starting from the diformate of nordesoxycholic acid chloride (1945 Hoehn).

Oxime $C_{29}H_{43}O_3N$. Crystals, m. 280° to 285° dec.; insoluble in the common organic solvents (1945 Hoehn).

Diacetate $C_{33}H_{46}O_5$. Crystals (alcohol), m. 141–2°; $[\alpha]_D^{25}$ +90° (dioxane). **Fmn.** From the diolone on refluxing with glacial acetic acid-acetic anhydride (2:3) (1945 Hoehn).

3α.11α-Dihydroxynorcholanyl phenyl ketone $C_{30}H_{44}O_3$.

Needles (alcohol), m. 196–197.5° cor.; $[\alpha]_D^{23}$ +16° (chloroform). — **Fmn.** From methyl 3α.11α-dihydroxycholanate on refluxing with phenylmagnesium bromide in ether-benzene, followed by refluxing of the resulting crude carbinol with glacial acetic acid, then with acetic anhydride in glacial acetic acid, addition of perchloric acid in the cold, oxidation with CrO_3 in strong acetic acid, saponification, and separation of the ketonic fraction from the non-ketonic by means of the Girard reagent T; small yield (1946 Long).

Diacetate $C_{34}H_{48}O_5$. Plates (methanol or ethanol), m. 133–134° cor.; $[\alpha]_D^{25}$ +9° (chloroform). Absorption max. in 95% ethanol at 242.5 (12500) and 280 (980) m$\mu(\varepsilon)$. — **Fmn.** From the above diolone by the action of acetic anhydride and perchloric acid (1946 Long).

3α.12α-Dihydroxynorcholanyl phenyl ketone, *Desoxycholophenone* $C_{30}H_{44}O_3$.

Crystals (methanol), m. 203–5° (1945 Hoehn); needles (acetone), m. 203–4° (1945 Julian); $[\alpha]_D^{25} + 47.5°$ (dioxane) (1945 Hoehn). — **Fmn.** Similar to that of the lower homologues, starting from the diformate of desoxycholic acid chloride; yield, ca. 80% (1945 Hoehn; 1945 Julian). The diacetate is regenerated from its 23-bromo-derivative (stable form; p. 2351 s) by treatment in acetone with a 1 N solution of $CrCl_2$ under CO_2 at room temperature; yield, 90% (1945 Julian). — **Rns.** On oxidation with CrO_3 in strong acetic acid below room temperature it gives 3.12-diketonor-cholanyl phenyl ketone (p. 2970 s); on oxidation of the diacetate with CrO_3 in acetic acid + some H_2SO_4 at 50°, followed by refluxing with aqueous NaOH, a small amount of nordesoxycholic acid is formed. On heating with hydrazine hydrate and methanolic sodium methoxide in a bomb at 180–190° for 3 hrs. it gives 24-phenylcholane-3α.12α-diol (only obtained amorphous) which on oxidation with CrO_3 in strong acetic acid at room temperature yields 24-phenylcholane-3.12-dione (p. 2914 s) (1945 Hoehn). On treatment of the diacetate with 1 mol. bromine in glacial acetic acid containing a little hydrogen bromide it yields the 23-bromo-derivative (p. 2351 s) (1945 Julian).

Oxime $C_{30}H_{45}O_3N$. Crystals (benzene), m. 194.5–198°; $[\alpha]_D^{25} + 30°$ (dioxane) (1945 Hoehn).

Diformate $C_{32}H_{44}O_5$. Crystals (methanol), m. 122–5°; $[\alpha]_D^{25} + 87.5°$ (dioxane). — **Fmn.** From the above diolone on heating with formic acid (d 1.20) on the water-bath (1945 Hoehn).

Diacetate $C_{34}H_{48}O_5$. Crystals (acetone), prisms (glacial acetic acid), m. 136° to 137° (1945 Hoehn; 1945 Julian); $[\alpha]_D^{25} + 92.5°$ (dioxane) (1945 Hoehn). — **Fmn.** From the diolone on refluxing with glacial acetic acid-acetic anhydride (2:3) (1945 Hoehn).

1937 Schering A.-G. (Logemann), *Ger. Pat.* 737023 (issued 1943); *C. A.* **1944** 3781; *Fr. Pat.* 843082 (1938, issued 1939); *C. A.* **1940** 6412; *Ch. Ztbl.* **1940** I 759; Schering Corp. (Logemann), *U.S. Pat.* 2321690 (1938, issued 1943); *C. A.* **1943** 6824; *Ch. Ztbl.* **1945** II 687.
1938 Hoehn, Mason, *J. Am. Ch. Soc.* **60** 1493, 1496.
1940 Reichstein, v. Arx, *Helv.* **23** 747, 751.
1944 Koechlin, Reichstein, *Helv.* **27** 549, 558, 559, 566.
1945 Hoehn, Moffet, *J. Am. Ch. Soc.* **67** 740.
Julian, Cole, Magnani, Meyer, *J. Am. Ch. Soc.* **67** 1728.
1946 Long, Marshall, Gallagher, *J.Biol.Ch.* **165** 197, 200, 203, 207 208.

7. HALOGENO-DIHYDROXY-MONOOXO-STEROIDS CONTAINING CO IN SIDE CHAIN AND OH IN THE RING SYSTEM

a. Halogen in Side Chain

23-Bromo-3α.12α-dihydroxy-norcholanyl phenyl ketone $C_{30}H_{43}O_3Br$. For con-

figuration at C-3 and C-12, see that of its ultimate precursor, desoxy-cholic acid (Introduction to steroids, pp. 1361 s, 1362 s).

Diacetate $C_{34}H_{47}O_5Br$. Exists in two forms, differing probably by the configuration at C-23. (a) *Labile form:* white crystals (acetone), m. 106° to 108°, $[\alpha]_D^{29} +91°$ (chloroform). (b) *Stable form:* white plates (alcohol), m. 165–175°, $[\alpha]_D^{28} +105°$ (chloroform). — **Fmn.** From 3α.12α-diacetoxy-nor-cholanyl phenyl ketone (p. 2350 s) by treatment with 1 mol. bromine in glacial acetic acid, in the presence of HBr; crystallization of the reaction mixture from acetone yields the labile form; the stable form is isolated from the mother liquor residue after boiling with alcohol. Conversion of the labile to the stable form occurs readily in boiling alcohol; in this change no re-arrangement of carbon-carbon linkages is involved. — **Rns.** Reduction of the stable form in acetone with a 1 N solution of $CrCl_2$ in an atmosphere of CO_2 at room temperature regenerates 3α.12α-diacetoxy-norcholanyl phenyl ketone.

1945 Julian, Cole, Magnani, Meyer, *J. Am. Ch. Soc.* **67** 1728, 1730.

---

b. Halogen in the Ring System

12α-Bromopregnane-3α.11β-diol-20-one $C_{21}H_{33}O_3Br$. For reversal of the 11α-and 12β*-configuration originally ascribed, see Introduction to steroids, p. 1362 s.

3-Acetate $C_{23}H_{35}O_4Br$. Colourless needles (ether or acetone-ether), m. 213–4° cor. — **Fmn.** From the acetate of Δ^{11}-pregnen-3α-ol-20-one (p. 2239 s) by treatment in acetone with an aqueous solution of N-bromoacetamide, sodium acetate, and some acetic acid, at 16°; yield of pure, crystalline substance, 29%. — **Rns.** Oxidation with CrO_3, in glacial acetic acid at room temperature, gives the acetate of 12α-bromo-pregnan-3α-ol-11.20-dione (p. 2851 s) (1944 v. Euw).

17-Bromopregnane-3α.12α-diol-20-one $C_{21}H_{33}O_3Br$, **17.21-Dibromopregnane-3α.12α-diol-20-one** $C_{21}H_{32}O_3Br_2$, and **17.21.21-Tribromopregnane-3α.12α-diol-20-one** $C_{21}H_{31}O_3Br_3$. — For their diacetates $C_{25}H_{37}O_5Br$, $C_{25}H_{36}O_5Br_2$, and $C_{25}H_{35}O_5Br_3$, resp., see the reactions of pregnane-3α.12α-diol-20-one, p. 2329 s.

1944 v. Euw, Lardon, Reichstein, *Helv.* **27** 821, 822, 825, 838.
1949 Fieser, Fieser, *Natural Products Related to Phenanthrene*, 3rd Ed., New York, p. 455.

---

* Addition of HOBr at the 11.12-double bond in trans-position has been assumed by 1944 v. Euw; cf. also 1949 Fieser.

8. Diazo-dihydroxy-monooxo-steroids containing CO in Side Chain and OH
in the Ring System

21-Diazopregnane-3α.12α-diol-20-one $C_{21}H_{32}O_3N_2$. For inversion of the
12β-configuration originally ascribed, see Introduction to steroids, p. 1362 s. — Amorphous,
crude product; no constants given. — **Fmn.**
From the diacetate of ætio-desoxycholic acid
by treatment with thionyl chloride for conversion to its acid chloride which, without
purification, is dissolved in benzene and treated with an abs. ethereal solution
of diazomethane first at —15°, then at room temperature; the diacetate so
obtained is saponified with aqueous methanolic KOH at room temperature;
crude yield, quantitative (1943 Fuchs). — **Rns.** Treatment of the 12-acetate
with aluminium phenoxide and acetone or cyclohexanone yields **21-diazo-
pregnan-12α-ol-3.20-dione acetate** $C_{23}H_{32}O_4N_2$ (not described in detail)
which on heating with glacial acetic acid gives the diacetate of pregnane-
12α.21-diol-3.20-dione (p. 2858 s) (1943 Reichstein). Heating with pure glacial
acetic acid at 100° yields the 21-acetate of pregnane-3α.12α.21-triol-20-one
(p. 2356 s); by similar treatment, the 12-acetate yields the corresponding
12.21-diacetate (1943 Fuchs).

12-Acetate $C_{23}H_{34}O_4N_2$. Brown, amorphous product, containing some free
diazo-diolone; no constants given. — **Fmn.** From the diacetate (below) by
partial saponification using K_2CO_3-$KHCO_3$ in aqueous methanol at room
temperature (1943 Fuchs). — **Rns.** See those of the parent diazo-diolone,
above.

Diacetate $C_{25}H_{36}O_5N_2$. Golden yellow oil; no constants given (1943 Fuchs).
— **Fmn.** See that of the parent diazo-diolone, above. — **Rns.** For partial and
complete saponification, see formation of the 12-acetate and the parent
diazo-diolone, resp. (above).

1943 Fuchs, Reichstein, *Helv.* **26** 511, 516, 518.
   Reichstein, *U.S. Pat.* 2401775 (issued 1946); *C. A.* **1946** 5884; *Ch. Ztbl.* **1947** 74; cf. *U. S.
   Pat.* 2404768 (issued 1946); *C. A.* **1946** 6222; *Ch. Ztbl.* **1947** 233.

9. TRIHYDROXY-MONOOXO-STEROIDS CONTAINING CO AND ONE OH-GROUP IN SIDE CHAIN AND TWO OH-GROUPS IN THE RING SYSTEM

Summary of Dihydroxy-pregnan (pregnen, etc.)-21-ol-20-ones

For values other than those chosen for the summary, see the compounds themselves

Configuration at C-5	C-17 (CO·CH₃)	Double bond(s)	Position of OH	M.p. °C.	[α]D*	Diacetate Position	Diacetate M.p. °	Diacetate [α]D*	Triacetate M.p. °	Triacetate [α]D*	Page
allo	β	satd.	3β.5α.21	—	—	3.21	173	+81⁰ (chl.)	—	—	2353 s
—	β	Δ⁵	3β.11β.21	—	—	—	—	—	—	—	2353 s
norm.	β	satd.	3α.11β.21	—	—	—	—	—	—	—	2354 s
norm.	β	satd.	3β.11β.21	—	—	—	—	—	—	—	2354 s
allo	β	satd.	3β.11β.21	204	—	3.21	174	+101⁰	—	—	2354 s
allo	α	satd.	3β.11β.21	—	—	3.21	133→148	−60⁰	—	—	2355 s
norm.	β	satd.	3α.12α.21	—	—	12.21	95→158	+151⁰	153	+160⁰	2356 s
—	α?	Δ³·⁵	3.17.21	—	—	—	—	—	—	—	2357 s
—	β	Δ⁵	3β.17α.21	242	−16⁰ (alc.)	3.21	175→200	−15⁰ (chl.)	185	—	2357 s
—	α	Δ⁵	3β.17β.21	—	—	—	—	—	—	—	2358 s
allo	β	satd.	3β.17α.21	239	+48⁰ (alc.)	3.21	211	+28⁰	192	—	2358 s
allo	α	satd.	3β.17β.21	—	—	3.21	161	−56⁰	180	−13⁰	2360 s

* In acetone, unless otherwise stated; alc. = alcohol; chl. = chloroform.

Allopregnane-3β.5.21-triol-20-one C₂₁H₃₄O₄.

3.21-Diacetate C₂₅H₃₈O₆. Crystals (methanol), m. 172–173.5⁰ cor.; [α]$_D^{21}$ + 81⁰ (average) (chloroform). — **Fmn.** From the crude 20-epimeric mixture (m. 145–7⁰) of allopregnane-3β.5.20.21-tetrol 3.21-diacetate (p. 2138 s) by oxidation with CrO₃ in 90% acetic acid at room temperature, followed by chromatography on Al₂O₃; yield, 76%. — **Rns.** Heating on the water-bath with bromoacetic ester and zinc turnings, in the presence of iodine, followed by acetylation with acetic anhydride and pyridine at room temperature, gives β-(3β-acetoxy-5-hydroxy-ætiocholan-17-yl)-Δ^{α.β}-butenolide (1946 Ruzicka).

Δ⁵-Pregnene-3(β?).11(β?).21-triol-20-one C₂₁H₃₂O₄.

For a Δ⁵-compound without indication of epi-configuration, 3β-configuration seems likely; 11β-configuration may be assumed as no true 11α-compounds had been prepared up to 1946 (Editor; cf. 1947 v. Euw).

21-Acetate C₂₃H₃₄O₅. Neither constants nor mode of formation given. — **Rns.** Refluxing with glacial acetic acid-conc. HCl (4:1) yields a mixture of the 21-acetates of Δ⁵·⁸- and Δ⁵·⁹(¹¹)-pregnadiene-3(β?).21-diol-20-ones (page 2300 s) (1942a Reichstein).

References, p. 2363 s

Pregnane-3α.11β.21-triol-20-one $C_{21}H_{34}O_4$. For inversion of the 11α-configuration originally assigned, see Introduction to steroids, p. 1362 s.

21-Acetate $C_{23}H_{36}O_5$. Obtained only as crude product, no constants given. — **Fmn.** From pregnane-3α.11β-diol-20-one (p. 2327 s) by oxidation with lead tetraacetate in 98% acetic acid at 55°. — **Rns.** Oxidation with acetone and aluminium phenoxide in benzene (sealed tube), on the boiling water-bath, gives the 21-acetate of pregnane-11β.21-diol-3.20-dione (p. 2856 s) (1944 v. Euw).

Pregnane-3β.11β.21-triol-20-one $C_{21}H_{34}O_4$. For inversion of the 11α-configuration originally assigned, see Introduction to steroids, p. 1362 s.

21-Acetate $C_{23}H_{36}O_5$. Granules and prisms (ether containing some petroleum ether), m. 199–200° cor.; $[\alpha]_D^{14} + 129°$, $[\alpha]_{5461}^{14} + 145°$ (both in acetone); sparingly soluble in ether. Strongly reducing towards methanolic alkaline silver diammine at room temperature (1944 v. Euw.) — **Fmn.** From pregnane-3β.11β-diol-20-one (p. 2328 s) by oxidation with lead tetraacetate in 98% acetic acid at 55°; yield, 30%, of not recovered starting material (1944 v. Euw; cf. 1943 Reichstein). — **Rns.** Oxidation with CrO_3 (probably in glacial acetic acid at room temperature) gives the acetate of pregnan-21-ol-3.11.20-trione (p. 2973 s) (1943 Reichstein; cf. 1944 v. Euw). Oxidation with aluminium phenoxide and acetone in benzene in a sealed tube at 100° gives the 21-acetate of pregnane-11β.21-diol-3.20-dione (p. 2856 s) (1944 v. Euw; cf. 1943 Reichstein).

Allopregnane-3β.11β.21-triol-20-one, *Reichstein's Substance R* $C_{21}H_{34}O_4$. For structure, see 1938c Reichstein, 1940a Shoppee. For configuration at C-3, C-11, and C-17, see 1947 v. Euw; cf. 1940a Shoppee. — Colourless needles (abs. alcohol), m. 202–4° cor. (1938c Reichstein).

Occ. and **Isolation.** Occurs in adrenal cortex and is isolated in the following manner: the alcoholic extract of the whole glands (cattle) is treated with permutit for the removal of adrenaline (1936a Reichstein), then evaporated to dryness, and the residue partitioned between pentane and 30% aqueous methanol; the aqueous layer, after concentration and acidification, is extracted with ether; after deacidification by shaking with a small amount of aqueous $KHCO_3$ solution and re-extraction with large amounts of distilled water, the ether solution is evaporated to give an "ether residue" (10.4 g. from 1000 kg. glands), which is separated by means of Girard's reagent T into a ketonic and "non-ketonic" fraction (1936b Reichstein); the former is further fractionated, dissolved in ether-acetone for elimination of compounds which crystallize from that medium, giving finally an amorphous mother-liquor residue (ca. 1.8 g.) which is treated with succinic anhydride-pyridine for separation of hydroxy-ketones as hydrogen succinates; the product regenerated from the latter is submitted to further fractional crystallization from acetone-ether, and the uncrystallizable residue is acetylated with acetic anhydride-pyridine before chromatography on Al_2O_3; from the benzene eluates crude R-diacetate is isolated together with dehydrocorticosterone acetate, from which it is separated by alternate crystallization from methanol and ether; the pure R-diacetate so obtained (yield, ca. 7.5 mg.) is saponified with $KHCO_3$ in boiling aqueous methanol (1938b Reichstein).

Fmn. An allopregnane-3.11.21-triol-20-one is obtained from allopregnane-3.11.17.20.21-pentol (p. 2202 s? *) on boiling with H_2SO_4 in aqueous propanol-dioxane, or by heating with $KHSO_4$ (1938 CIBA) *.

Rns. Oxidation with CrO_3 in glacial acetic acid at room temperature gives 3.11-diketo-ætioallocholanic acid; similar oxidation of the 3.21-diacetate gives the diacetate of allopregnane-3β.21-diol-11.20-dione (p. 2861 s) (1938c Reichstein). Oxidation with HIO_4 in aqueous dioxane at 22^0 gives 3β.11β-dihydr-oxy-ætioallocholanic acid (1947 v. Euw).

3.21-Diacetate $C_{25}H_{38}O_6$. Colourless needles or rodlets (methanol) (1938b, c Reichstein), m. 173–4^0 cor.; sublimes in high vacuum; $[\alpha]_D^{17} + 84^0$, $[\alpha]_{5461}^{17} + 103^0$ (both in dioxane), $[\alpha]_D^{18} + 92^0$, $[\alpha]_{5461}^{18} + 114^0$ (both in acetone) (1942 v. Euw), $[\alpha]_D^{20} + 101^0$ (acetone) (1952 Pataki). Strongly reducing towards alkaline silver diammine at room temperature; does not give a green fluorescence with conc. H_2SO_4 (1938b Reichstein). — **Fmn.** Obtained as intermediate in the isolation of the parent triolone (see above). — **Rns.** See above.

17-iso-Allopregnane-3β.11β.21-triol-20-one, Reichstein's Substance iso-R

$C_{21}H_{34}O_4$.

3.21-Diacetate $C_{25}H_{38}O_6$. Colourless prisms (ether-pentane), m. 133^0, resolidifies, and finally melts at 147–8^0 cor.; undergoes molecular distillation at 160–5^0 (bath temperature) and 0.01 mm. pressure; $[\alpha]_D^{18} -60^0$ (acetone). Strongly reducing towards methanolic alkaline silver diammine at room temperature. Does not give a fluorescence reaction with conc. H_2SO_4 or coloration with tetranitromethane (1940a Shoppee). — **Fmn.** From the 3.20.21-triacetate of allopregnane-3β.11β.17α.20β.21-pentol (p. 2202 s) by boiling with zinc dust in abs. toluene; after crystallization from ether, the remaining unchanged starting material is submitted to further similar treatment, and after four such operations, the united crude product is chromatographed on Al_2O_3; crude yield, 31 % (1940a Shoppee; 1940 CIBA PHARMACEUTICAL PRODUCTS). — **Rns.** Oxidation with CrO_3 in glacial acetic acid at room temperature gives the diacetate of 17-iso-allopregnane-3β.21-diol-11.20-

---

* The identity of the pentol starting material with Reichstein's Substance A is highly probable, as no configurational isomer appears in the literature up to their date; likewise the identity of the product with Substance R seems probable (Editor).

References, p. 2363 s

dione (p. 2862 s) (1940a Shoppee). Unchanged by boiling with glacial acetic acid or short boiling with 1% HCl in glacial acetic acid (1940a Shoppee); boiling with glacial acetic acid + conc. HCl (9:1 by vol.) yields 17-iso(?)-$\Delta^{9(11)}$(?)-allopregnene-3β.21-diol-20-one (p. 2308 s) (1940a, b Shoppee).

Pregnane-3α.12α.21-triol-20-one $C_{21}H_{34}O_4$. For inversion of the 12β-configuration originally ascribed, see Introduction to steroids, p. 1362 s. Only esters (acetates) have been obtained. — **Fmn.** The triacetate is obtained from the diacetate of pregnane-3α.12α-diol-20-one (p. 2329 s) by oxidation with lead tetraacetate in glacial acetic acid containing some acetic anhydride, at 68–72°, followed by chromatography on Al_2O_3; yield, 10% of the not recovered starting material (1944 Ruzicka). The 21-monoacetate and the 12.21-diacetate are obtained from the diacetate of ætiodesoxycholic acid by treatment with thionyl chloride at room temperature for conversion to the acid chloride, which reacts with diazomethane in ether-benzene to give the oily diacetate of 21-diazopregnane-3α.12α-diol-20-one (p. 2352 s); this crude diacetate is saponified with aqueous methanolic KOH at room temperature to give the diazo-diolone, or with aqueous methanolic K_2CO_3 (containing $KHCO_3$) at room temperature, to give the diazo-diolone 12-acetate, which, on heating with glacial acetic acid at 100–5°, followed by chromatography on Al_2O_3, give the triolone 21-acetate (over-all yield, 52%) or the 12.21-diacetate (over-all yield, 42%), resp. (1943 Fuchs).

Rns. Oxidation of the 21-acetate with CrO_3 (1 mol.) in glacial acetic acid at room temperature gives chiefly the 21-acetates of pregnane-3α.21-diol-12.20-dione (p. 2863 s; main product), of pregnane-12α.21-diol-3.20-dione (p. 2858 s; small amount), and of pregnan-21-ol-3.12.20-trione (p. 2974 s); the latter is the sole product isolated when 2 mols. of CrO_3 are used. Similar oxidation of the 12.21-diacetate yields the diacetate of pregnane-12α.21-diol-3.20-dione (p. 2858 s). Oxidation of the 21-acetate with acetone and aluminium phenoxide in boiling abs. benzene gives the 21-acetate of pregnane-12α.21-diol-3.20-dione (p. 2858 s) (1943 Fuchs). By heating the triacetate with bromoacetic ester and zinc turnings on the water-bath, condensation occurs to give 21-hydroxy-3α.12α-diacetoxy-$\Delta^{20(22)}$-norcholenic acid lactone (12-epi-14-desoxy-14-iso-digoxigenin 3.12-diacetate) (1944 Ruzicka).

For the rate of reduction of phosphomolybdic acid (Folin-Wu reagent) by the 21-acetate in glacial acetic acid at 100° to give molybdenum blue, and for colorimetric estimation, see 1946 Heard.

21-Acetate $C_{23}H_{36}O_5$. Needles (ether-methanol) *(methanolate?)*, m. 94–110°, resolidifying to melt finally at 144–148.5° cor.; *monohydrate:* needles (aqueous methanol, air-dried), dehydrated by heating at 70° in high vacuum, m. 149.5° to 150.5°; $[\alpha]_D^{18}$ +140° ± 4° (acetone). Readily soluble in ether, appreciably less so in presence of some methanol (1943 Fuchs). — **Fmn.** and **Rns.** See above.

12.21-Diacetate $C_{25}H_{38}O_6$. Cubes (benzene-ether), rodlets (ether-pentane 1:1), m. 72–95°, resolidifying to melt finally at 156–8° cor.; crystals (acetone-ether), m. 156–8°; the low-melting form slowly changes into the stable high-melting form on standing; $[\alpha]_D^{19}$ +151° (acetone) (1943 Fuchs). — **Fmn.** and **Rns.** See above.

Triacetate $C_{27}H_{40}O_7$. Colourless needles (ether-pentane), m. 114–5° cor. (1943 Fuchs); crystals (ether), m. 150.5–151° cor. (1944 Ruzicka); crystals (hexane), m. 153–153.5° cor. (1947 Meystre); $[\alpha]_D^{14.5}$ +157° (chloroform), $[\alpha]_D^{19}$ +153° (acetone) (1944 Ruzicka), $[\alpha]_D^{25}$ +160° ± 4° (acetone) (1947 Meystre). — **Fmn.** From the 21-acetate or the 12.21-diacetate by heating with acetic anhydride and pyridine at 90° (1943 Fuchs). See also formation of the triolone, above. — **Rns.** See above.

17-iso-$\Delta^{3.5}$-Pregnadiene-3.17β.21-triol-20-one, 3-Enolic form of 17β.21-dihydr-oxy-17-iso-progesterone $C_{21}H_{30}O_4$.

For configuration at C-17, see that of the ultimate precursor, 3β.17β-diacetoxy-17-iso-Δ^5-ætio-cholenic acid.

Triacetate $C_{27}H_{36}O_7$. Not characterized. — **Fmn.** From the diacetate of 17-iso-Δ^4-pregnene-17β.21-diol-3.20-dione (17.21-di-hydroxy-17-iso-progesterone, p. 2859 s) by the action of acetic anhydride. — **Rns.** Refluxing with aluminium isobutoxide in anhydrous isobutyl alcohol gives 17-iso-$\Delta^{3.5}$-pregnadiene-3.17.20.21-tetrol 3.17.21-triacetate $C_{27}H_{38}O_7$ (no constants given), which by saponification with 2 % alcoholic alkali hydroxide yields a 17-iso-Δ^4-pregnene-17β.20.21-triol-3-one (p. 2728 s); the reduction can also be effected by use of isopropyl alcohol or cyclohexanol with aluminium isopropoxide, by hydrogenation with a weakly active nickel catalyst, or biochemically by means of fermenting yeast (1937 a CIBA).

Δ^5-Pregnene-3β.17α.21-triol-20-one $C_{21}H_{32}O_4$.

For inversion of the 17β-configuration previously assigned to the OH-group, see Introduction to steroids, p. 1363 s. Obtained only as crude product in solution*. **Fmn.** From crude Δ^5-pregnene-3β.17α.20β-triol-21-al (p. 2361 s) by heating (6 hrs.) in abs. pyridine in a boiling toluene-bath; the solution of the triolone so formed is treated directly with acetic anhydride at room temperature for isolation as the 3.21-diacetate (yield, 34%). — **Rns.** Treatment of the diacetate with methylmagnesium bromide in abs. ether-toluene, with subsequent heating on the water-bath after evaporation of ether, yields a mixture of the 20-epimeric 20-methyl-Δ^5-pregnene-3β.17α.20.21-tetrols (p. 2147 s), together with some hydrolysed starting product (1941 Fuchs).

3.21-Diacetate $C_{25}H_{36}O_6$. Needles (ether), m. (partially) ca. 175° with slow transformation into lancets, m. 193–200° cor.; colourless rodlets or rhombo-hedra (benzene), which undergo change without melting at ca. 178°, and

* Pure Δ^5-pregnene-3β.17α.21-triol-20-one [crystals (ethyl acetate-methanol), m. 240–2° cor., $[\alpha]_D^{20}$ —16° (abs. alcohol)] (1954 Florey; cf. 1951 Heer) has been obtained after the closing date for this volume.

finally melt at 198–200° cor.; $[\alpha]_D^{16}$ —15.3° ± 3° (chloroform). Strongly reducing towards silver diammine solution at room temperature (1941 Fuchs). — **Fmn.** See that of the parent triolone, above.

Triacetate (?) $C_{27}H_{38}O_7$. Identity very likely though not definitely established. — Colourless needles (ether), m. 182–5° cor. (1939b Reichstein). — **Fmn.** Together with the diacetate of Δ^5-pregnene-3β.21-diol-20-one (p. 2301 s), from the acetate of Δ^5-pregnen-3β-ol-20-one (p. 2233 s) by oxidation with lead tetraacetate in pure glacial acetic acid at 68–70° and subsequent separation by chromatography on Al_2O_3 (1939b Reichstein; cf. 1940 SCHERING A.-G.).

17-iso-Δ^5-Pregnene-3β.17β.21-triol-20-one $C_{21}H_{32}O_4$. — Fmn.

The 17.21-diacetate (not characterized) is obtained from 3β.17β-diacetoxy-17-iso-Δ^5-ætiocholenic acid chloride by reaction with diazomethane, followed by saponification of the 3-acetoxy-group and then by heating with acetic acid. — **Rns.** Dehydrogenation of the 17.21-diacetate with cyclohexanone in the presence of aluminium isopropoxide affords the diacetate of 17-iso-Δ^4-pregnene-17β.21-diol-3.20-dione (17.21-dihydroxy-17-iso-progesterone, p. 2859 s). On reduction, e.g. by boiling with aluminium isopropoxide and cyclohexanol, it gives a 17-iso-Δ^5-pregnene-3β.17β.20.21-tetrol (1937a, b CIBA).

Allopregnane-3β.17α.21-triol-20-one, *Reichstein's Substance P* $C_{21}H_{34}O_4$.

For structure, see 1938a Reichstein; for configuration at C-3, see 1939a Reichstein; for configuration at C-17, see 1947 v. Euw; cf. 1939a Reichstein; 1942 v. Euw. — Needles (acetone-ether, or abs. alcohol), m. 230–9° cor. dec.; $[\alpha]_D^{20}$ +48° (abs. alcohol). Readily soluble in alcohol or acetone, sparingly in ether or water. Strongly reducing towards methanolic alkaline silver diammine in the cold. Readily precipitated by digitonin in hot 60% methanol (1938a Reichstein).

Occ. and **Isolation.** Allopregnane-3β.17.21-triol-20-one is one of the steroids found in the adrenal cortex and is isolated from the "ether residue" (10.4 g. obtained from 1000 kg. cattle adrenals), for which see isolation of Reichstein's Substance R (p. 2354 s). After treatment of this "ether residue" with Girard's reagent T, the triolone remains in the "non-ketonic" fraction which, after elimination of some phenolic products, is submitted to fractional crystallization from slightly aqueous acetone-toluene; Substance P* accumulates in the most soluble crystal fraction, from which it is isolated via its 3.21-diacetate* (yield, 97 mg.) (1938 Steiger); from the acetone-toluene mother liquors a further yield of ca. 22 mg. of diacetate is obtained after acetylation and chromatography on Al_2O_3. The 3.21-diacetate is saponified with $KHCO_3$ in boiling aqueous methanol (1938a Reichstein).

---

* The substance is designated as "new reducing substance" (or its acetate) in this original.

Fmn. From allopregnane-3β.17α.20β-triol-21-al (p. 2361 s) by heating a well dried sample with pyridine in a bath of boiling toluene; the product is isolated as 3.21-diacetate after treating the cooled pyridine solution with acetic anhydride (yield, 16%) (1941 a v. Euw; 1941 Reichstein).

Rns. Oxidation with CrO_3 in glacial acetic acid at room temperature gives androstane-3.17-dione (p. 2892 s). Oxidation with HIO_4 in 50% methanol containing some H_2SO_4, at room temperature, gives 3β.17-dihydroxy-ætioallo-cholanic acid. Hydrogenation of the 3.21-diacetate, using Raney nickel catalyst at 95° and 200 atm. pressure, gives a mixture of the 3.21-diacetates of allopregnane-3β.17α.20β.21-tetrol (p. 2141 s) and its 20α-epimer (p. 2143 s) (1938a Reichstein). By heating the 3.21-diacetate with methylmagnesium bromide in ether-toluene on the water-bath, a mixture of C-20 epimeric 20-methylallopregnane-3β.17.20.21-tetrols (p. 2147 s) is obtained, together with some hydrolysed starting material (1941b v. Euw). For attempted transformation into allopregnane-3β.17-diol-20-one via the 21-p-toluenesulphonate and the 21-iodide, see 1940 Prins.

For the rate of reduction of phosphomolybdic acid (Folin-Wu reagent) by the 3.21-diacetate in glacial acetic acid at 100° to give molybdenum blue, and for colorimetric estimation, see 1946 Heard.

3.21-Diacetate $C_{25}H_{38}O_6$. Leaflets appearing like needles (benzene) (1938 Steiger*), needle aggregates (acetone-ether) (1941a v. Euw), colourless needles (chloroform) (1942 v. Euw), m. 210–1° cor. (1938a Reichstein; cf. 1938 Steiger); sublimes at 195–200° under 0.003 mm. pressure (1938 Steiger); $[\alpha]_D^{17} + 38^0$, $[\alpha]_{5461}^{17} + 54^0$ (1940 Prins), $[\alpha]_D^{12} + 46^0$ (1941a v. Euw) (all in chloroform)**; $[\alpha]_D^{18} + 44.5^0$, $[\alpha]_{5461}^{18} + 57^0$ (both in dioxane) (1942 v. Euw); $[\alpha]_D^{16} + 28^0$ (acetone) (1942 Prins). Very sparingly soluble in methanol (1938a Reichstein). Strongly reducing towards methanolic alkaline silver diammine at room temperature; unchanged on standing with CrO_3 in glacial acetic acid at room temperature (1938 Steiger). — **Fmn.** and **Rns.** See above.

Triacetate(?) $C_{27}H_{40}O_7$, for which the identity is very probable but not established. Needles (ether-pentane, or ether), not quite pure, m. 190–2° cor.; sparingly soluble in methanol; strongly reducing towards alkaline silver diammine in methanol at room temperature. — **Fmn.** Together with the diacetate of allopregnane-3β.21-diol-20-one (p. 2311 s), from the acetate of allopregnan-3β-ol-20-one (p. 2247 s) by oxidation with lead tetraacetate in glacial acetic acid containing some acetic anhydride, at 68–70°, with subsequent separation by chromatography on Al_2O_3; yield, ca. 2%. — **Rns.** Partial saponification was effecte dby aqueous methanolic $KHCO_3$ (44 hrs.) to give a product (not characterized) which was oxidized with HIO_4 in aqueous methanol containing H_2SO_4; complete saponification of this oxidation product, using boiling methanolic KOH, yielded 3β.17-dihydroxy-ætioallocholanic acid (1939b Reichstein).

---

* See the footnote on p. 2358 s.

** Values for $[\alpha]_D$ in chloroform ranging from $+31^0$ to $+48^0$ are quoted (1950 Rosenkranz; 1950 Wagner; 1951 Kritchevsky) after the closing date for this volume; average of all quoted values: $+42^0$ (Editor).

17-iso-Allopregnane-3β.17β.21-triol-20-one, *Reichstein's Substance 17-iso-P*

$C_{21}H_{34}O_4$. — Obtained only admixed with its 3-acetate (below). — **Fmn.** From the diacetate of 17-iso-Δ^{20}-allopregnene-3β.17β-diol (p. 2037 s) by oxidation with OsO_4 in abs. ether, followed by decomposition of the osmic esters formed using $KClO_3$ and dil. H_2SO_4, all at room temperature; for separation of the (not isolated) *3.17-diacetate* so formed from the main product, the 3.17-diacetate of 17-iso-allopregnane-3β.17β.20b.21-tetrol (p. 2144 s), the mixture is acetylated with acetic anhydride and pyridine in benzene at room temperature before chromatography on Al_2O_3, whereby the 3.17.21-triacetate (yield, 8%) is isolated from the benzene eluates; in one case appreciable transesterification (17 → 21) occurred, resulting in the isolation of the 3.21-diacetate (yield, 17%); a similar oxidation of the 3-acetate of 17-iso-Δ^{20}-allopregnene-3β.17β-diol, followed by acetylation of the reaction products and chromatography, also led to the isolation of the 3.21-diacetate (yield, 6%). The triacetate is obtained from the corresponding ester of 17-iso-allopregnane-3β.17β.20b.21-tetrol by oxidation with CrO_3 in glacial acetic acid at room temperature. Attempted isolation of the free triolone from its 3.21-diacetate by prolonged saponification with $KHCO_3$ in aqueous methanol at 20° yielded a mixture of the triolone with its 3-acetate, from which only the latter could be crystallized (1942 Prins). — **Rns.** Oxidation of the crude triolone (using mother liquors of the 3-acetate) with HIO_4 in aqueous methanol at room temperature yielded 3β.17-dihydroxy-17-iso-ætioallocholanic acid. Catalytic hydrogenation of the triacetate, using Raney nickel in methanol at 100° and 150–155 atm. pressure, followed by saponification with boiling methanolic KOH and reacetylation with acetic anhydride and pyridine at room temperature, yields the 3.20.21-triacetate of 17-iso-allopregnane-3β.17β.20b.21-tetrol (1942 Prins).

3-Acetate $C_{23}H_{36}O_5$. Colourless prisms (methanol-ether), m. 195–7°, $[\alpha]_D^{19}$ —43° (acetone). — **Fmn.** From the 3.21-diacetate (below) by saponification with aqueous methanolic $KHCO_3$ (3 days), followed by crystallization from ether-methanol for separation from admixed triolone; yield, 27% (1942 Prins).

17-Acetate(?) $C_{23}H_{36}O_5$. Identity seems highly probable (Editor). — Granules (ether), m. 216–7°, perhaps not quite pure (1942 Prins). — **Fmn.** See reactions of the triacetate (below).

3.21-Diacetate $C_{25}H_{38}O_6$. Needle rosettes (ether-pentane) (1942 Prins), crystals (ether), m. 159–161° cor. (1942 v. Euw); can be sublimed in high vacuo; $[\alpha]_D^{19}$ —56°, $[\alpha]_{5461}^{19}$ —67° (both in acetone) (1942 Prins), $[\alpha]_D^{19}$ —68°, $[\alpha]_{5461}^{19}$ —90° (both in dioxane) (1942 v. Euw). Strongly reducing towards methanolic alkaline silver diammine at room temperature (1942 Prins). — **Fmn.** See that of the parent triolone, above. — **Rns.** See formation of the 3-acetate, above.

Triacetate $C_{27}H_{40}O_7$. Colourless leaflets or needles (ether-pentane), m. 178° to 180° with previous sintering at 173°; sublimes at 165–175° (bath temperature) under 0.01 mm. pressure; $[\alpha]_D^{15}$ —13° ±5°, $[\alpha]_{5461}^{15}$ —19° ± 5° (both in acetone). Solutions in ether-pentane readily give gels which crystallize only on long standing or heating. Reduces methanolic alkaline silver diammine at room temperature. — **Fmn.** See that of the parent triolone, above. — **Rns.** For catalytic hydrogenation, see reactions of the parent triolone, above. Saponification with aqueous methanolic $KHCO_3$ for 24 hrs. yields a not-separated mixture (m. 195–205°) of the 17-acetate (see above) with the *3.17-diacetate*, from which the triacetate can be regenerated by acetylation; by prolonging the saponification for 3 days, some 17-acetate(?) can be isolated (1942 Prins), cf. saponification of the 3.21-diacetate, above; oxidation of the crude saponification product with HIO_4 in aqueous methanol at room temperature, followed by complete saponification of the oxidation product with boiling methanolic KOH, yields 3β.17-dihydroxy-17-iso-ætioallocholanic acid (1942 Prins).

Δ⁵ - Pregnene - 3β.17α.20β - triol - 21 - al* $C_{21}H_{32}O_4$.

Crystalline precipitate (aqueous methanol) (1941 Fuchs), needles (solvent not indicated), m. 235° (1942 b Reichstein). — **Fmn.** From ω-homo-Δ⁵-pregnene-3β.17α.20β.21 b.22-pentol (p. 2149 s) by oxidation with HIO_4 (1 mol.) in ca. 90 % aqueous dioxane at room temperature; this oxidation may also be effected with lead tetraacetate, tetrapropionate, or tetrabenzoate (1942 b Reichstein); the 3.20-diacetate is obtained by similar HIO_4 oxidation of the pentol 3.20-diacetate (yield, 54 %) or of its 21a-epimer (p. 2148 s) (yield, 34 %) and is saponified by aqueous methanolic $KHCO_3$ at room temperature (1941 Fuchs). — **Rns.** Heating with pyridine in a boiling toluene-bath, followed by addition of acetic anhydride to the cooled solution for acetylation of the product at room temperature, yields the 3.21-diacetate of Δ⁵-pregnene-3β.17α.21-triol-20-one (p. 2357 s) (1941 Fuchs).

3.20-Diacetate $C_{25}H_{36}O_6$. Colourless granules or rodlets (acetone-ether), m. 164–5° cor. with slight dec.; $[\alpha]_D^{15}$ —25° (average) (dioxane); reducing towards methanolic alkaline silver diammine at room temperature; gives a red coloration with 1.4-dihydroxynaphthalene in glacial acetic acid containing HCl (1941 Fuchs). — **Fmn.** See that of the triolal, above.

Allopregnane-3β.17α.20β-triol-21-al* $C_{21}H_{34}O_4$.

Crystalline powder (methanol-ether), m. 201–7° cor. dec. (1941 a v. Euw; cf. 1942 b Reichstein). — **Fmn.** From ω-homo-allo-pregnane - 3β.17α.20β.21 b.22 - pentol (p. 2150 s) by oxidation with HIO_4 (1 mol.) in ca. 90 % aqueous dioxane at room temperature (1942 b Reichstein); the 3.20-diacetate is obtained by similar oxidation of the pentol 3.20-diacetate (yield, 53 %) or that of the

---

* For configuration at C-17, the reverse of that assigned by the investigators, and that at C-20 see Introduction to steroids, p. 1363 s and 1364 s.

165*

pentol 21a-epimer (p. 2150 s) (yield, 45%) and is saponified with aqueous methanolic $KHCO_3$ at room temperature (1941a v. Euw). — **Rns.** Heating with pyridine in a boiling toluene-bath, followed by addition of acetic anhydride to the cooled solution for acetylation at room temperature, yields the 3.21-diacetate of allopregnane-3β.17α.21-triol-20-one (p. 2358 s) (1941a v. Euw; 1941 Reichstein).

3.20-Diacetate $C_{25}H_{38}O_6$. Colourless spherical aggregates (ether-acetone), m. 181–2° cor. dec.; $[\alpha]_D^{16}$ +36° (dioxane); reducing towards methanolic ammoniacal silver diammine; gives a red coloration with 1.4-dihydroxynaphthalene in glacial acetic acid containing HCl (1941a v. Euw). — **Fmn.** See that of the parent triolal, above.

(Allo?)-Pregnane-3.17.20-triol-21-al for which the configuration at C-3, C-5, C-17 and C-20 has not been ascertained, *17.20-Dihydroxypregnan-3-ol-21-al* $C_{21}H_{34}O_4$. 17α.20β-Configuration seems likely in view of the mode of formation; identity with the allopregnanetriolal described above is possible (Editor). — No constants given. — **Fmn.** From the acetate of $\Delta^{17(20)}$-pregnen-3-ol-21-al by treatment with orthoformic ester in methanol containing some NH_4NO_3, for conversion to the dimethyl acetal, which is then oxidized with OsO_4 at room temperature; the osmic ester formed is decomposed with hot alcoholic $NaHSO_3$ and the resulting acetal then hydrolysed with alcoholic HCl (1939 Ciba).

24.24 - Diethylcholane - 3α.12α.24 - triol - 23 - one $C_{28}H_{48}O_4$. — A compound $C_{34}H_{54}O_7$ [needles (aqueous alcohol), m. 180°] said to be the *tri-acetate* of this triolone has been obtained from 24.24-diethylcholane-3α.12α.24-triol triacetate (p. 2133 s) on oxidation with CrO_3 in glacial acetic acid on the water-bath, along with nordesoxycholic acid diacetate and a compound $C_{23}H_{34}O_5$, m. 205–6°, said to be ætiocholane-3α.12α-diol-17-one diacetate which, however, melts at ca. 162° (see p. 2750 s) (1939 Kazuno).

6-Methylallopregnane-3β.5α.21-triol-20-one $C_{22}H_{36}O_4$. The alternative 3β.6.21-triol-20-one structure, is not excluded (1943 Ehrenstein). For assumption of the allopregnane structure, see the starting compound, 5α.6α-epoxy-allopregnane-3β.20.21-triol (p. 1953 s).

3.21-Diacetate $C_{26}H_{40}O_6$. Obtained only as a crude resin; strongly reducing towards methanolic alkaline silver diammine. — **Fmn.** From the 3.21-diacetate of 6-methylallopregnane-3β.5α.20.21-tetrol (p. 2147 s) by oxidation with CrO_3 in ca. 93% acetic acid at 26° (1943 Ehrenstein).

1936 a Reichstein, *Helv.* **19** 29, 33.
b Reichstein, *Helv.* **19** 1107, 1120, 1122.

1937 a CIBA*, *Swiss Pat.* 207496 (issued 1940); *C. A.* **1942** 3635; *Brit. Pat.* 517288 (1938, issued 1940); *Ch. Ztbl.* **1940** II 1328; CIBA PHARMACEUTICAL PRODUCTS (Miescher, Wettstein), *U.S. Pat.* 2239012 (1938, issued 1941); *C. A.* **1941** 4921.
b CIBA*, *Swiss Pat.* 229124 (issued 1944); *C. A.* **1949** 3476; *Ch. Ztbl.* **1944** II 1202.

1938 CIBA*, *Fr. Pat.* 838916 (issued 1939); *C. A.* **1939** 6875; *Ch. Ztbl.* **1939** II 169.
a Reichstein, Gätzi, *Helv.* **21** 1185, 1190 et seq.
b Reichstein, v. Euw, *Helv.* **21** 1197, 1200 et seq., 1207.
c Reichstein, *Helv.* **21** 1490.
Steiger, Reichstein, *Helv.* **21** 546, 550 et seq., 560.

1939 CIBA* (Miescher, Wettstein), *Swed. Pat.* 101201 (issued 1941); *Ch. Ztbl.* **1941** II 3101; *Norw. Pat.* 60152 (issued 1942); *Ch. Ztbl.* **1943** I 2421.
Kazuno, Shimizu, *J. Biochem.* **29** 421, 432.
a Reichstein, Meystre, *Helv.* **22** 728.
b Reichstein, Montigel, *Helv.* **22** 1212, 1214, 1216 et seq.

1940 CIBA PHARMACEUTICAL PRODUCTS (Miescher, Heer), *U.S. Pat.* 2372841 (issued 1945); *C. A.* **1945** 4195; *Ch. Ztbl.* **1945** II 1773; CIBA* (Miescher, Heer), *Ger. Pat.* 737570 (issued 1943); *C. A.* **1946** 179; *Fr. Pat.* 876214 (issued 1942); *Ch. Ztbl.* **1943** I 1496; *Dutch Pat.* 55050 (issued 1943); *Ch. Ztbl.* **1944** I 367.
Prins, Reichstein, *Helv.* **23** 1490, 1500.
SCHERING A.-G., *Fr. Pat.* 868396 (issued 1941); *Ch. Ztbl.* **1942** I 2560.
a Shoppee, Reichstein, *Helv.* **23** 729, 735, 737.
b Shoppee, *Helv.* **23** 740, 743.

1941 a v. Euw, Reichstein, *Helv.* **24** 401, 411, 412, 415.
b v. Euw, Reichstein, *Helv.* **24** 418.
Fuchs, Reichstein, *Helv.* **24** 804, 819 et seq., 826.
Reichstein, *Fr. Pat.* 888825 (issued 1943); *Ch. Ztbl.* **1944** II 1203.

1942 v. Euw, Reichstein, *Helv.* **25** 988, 995, 1020, 1021.
Prins, Reichstein, *Helv.* **25** 300, 303, 305, 306 et seq., 312, 313 et seq.
a Reichstein, *Fr. Pat.* 52035 (issued 1943); *Ch. Ztbl.* **1945** I 195.
b Reichstein, *Fr. Pat.* 888228 (issued 1943); *Ch. Ztbl.* **1945** I 68; *U.S. Pat.* 2389325 (issued 1945); *C. A.* **1946** 1973.

1943 Ehrenstein, *J. Org. Ch.* **8** 83, 85, 93.
Fuchs, Reichstein, *Helv.* **26** 511, 517 et seq.; Reichstein, *U.S. Pat.* 2401775 (issued 1946); *C. A.* **1946** 5884; *Ch. Ztbl.* **1947** 74; *U.S. Pat.* 2404768 (1945, issued 1946); *C. A.* **1946** 6222; *Ch. Ztbl.* **1947** 233.
Reichstein, *Fr. Pat.* 898140 (issued 1945); *Ch. Ztbl.* **1946** I 1747; *Brit. Pat.* 594878 (issued 1947); *C. A.* **1948** 2404; *U.S. Pat.* 2440874 (issued 1948); *C. A.* **1948** 5622; *Ch. Ztbl.* **1948** E 456.

1944 v. Euw, Lardon, Reichstein, *Helv.* **27** 1287.
Ruzicka, Plattner, Pataki, *Helv.* **27** 988, 992, 993.

1946 Heard, Sobel, *J. Biol. Ch.* **165** 687.
Ruzicka, Plattner, Heusser, Ernst, *Helv.* **29** 248.

1947 v. Euw, Reichstein, *Helv.* **30** 205, 207 (footnote 4), 212, 213, 215.
Meystre, Wettstein, *Helv.* **30** 1037, 1043.

1950 Rosenkranz, Pataki, Kaufmann, Berlin, Djerassi, *J. Am. Ch. Soc.* **72** 4081, 4084.
Wagner, Moore, *J. Am. Ch. Soc.* **72** 5301, 5304.

1951 Heer, Miescher, *Helv.* **34** 359, 370.
Kritchevsky, Gallagher, *J. Am. Ch. Soc.* **73** 184, 188.

1952 Pataki, Rosenkranz, Djerassi, *J. Biol. Ch.* **195** 751.

1954 Florey, Ehrenstein, *J. Org. Ch.* **13** 1331, 1343.

---

* Gesellschaft für chemische Industrie in Basel; Société pour l'industrie chimique à Bâle.

10. Trihydroxy-monooxo-steroids containing CO in Side Chain,
OH in the Ring System

17β-Formylandrostane-3β.11β.17α-triol $C_{20}H_{32}O_4$. See the **aldehyde** (I) on
p. 2203 s.

Δ¹⁶-Allopregnene-2α.3β.12-triol-20-one $C_{21}H_{32}O_4$. For structure of the ultimate precursor, agavogenin, see 1947 Marker;
for its configuration at C-2 and C-3, see 1952
Wendler; cf. 1951 Pataki. — Crystals (methanol).
— **Fmn.** From pseudo-agavogenin tetraacetate
with CrO_3 in strong acetic acid at 20°, followed
by saponification (1944 Parke, Davis & Co.).

Allopregnane-3β.5α.6α-triol-20-one $C_{21}H_{34}O_4$. For revised nomenclature and
configuration at C-5 and C-6, see 1948 Ehren-
stein. — White needles (ethyl acetate), m. 231°
to 232.5° cor., with previous sintering at 229°
(1939 Ehrenstein; cf. 1943 Ruzicka); $[\alpha]_D^{26} +60°$
(methanol) (1939 Ehrenstein). — **Fmn.** From
Δ^5-pregnen-3β-ol-20-one (p. 2233 s) by oxidation
with OsO_4 in abs. dioxane at room temperature, followed by decomposition
of the resulting osmic ester with Na_2SO_3 in boiling aqueous alcohol (yield,
53%) (1939, 1940a Ehrenstein); a similar oxidation of an inseparable mixture
of Δ^5-pregnen-3β-ol-20-one with allopregnan-3β-ol-20-one (from swine testicles)
was carried out in methylene chloride-ether containing traces of pyridine,
and the osmic ester intermediate decomposed with mannitol in 0.1 N KOH
before chromatographic separation on Al_2O_3 from the unchanged allopregnan-
3β-ol-20-one (1943 Ruzicka). — **Rns.** Oxidation with CrO_3 in strong acetic
acid gives allopregnan-5α-ol-3.6.20-trione (p. 2976 s) (1939 Ehrenstein).

3-Acetate (?) $C_{23}H_{36}O_5$. Crystalline powder (ether), m. 226–8° cor.; sparingly soluble in ether. — **Fmn.** See that of 5α.6α-epoxy-allopregnan-3β-ol-
20-one, below. — **Rns.** Refluxing with acetic anhydride gave a product
m. 234–5°, which gave no m. p. depression with the 3.6-diacetate (below). —
Oxime $C_{23}H_{37}O_5N$, crystals (aqueous alcohol), m. 221–3° (1940b Ehrenstein).

3.6-Diacetate $C_{25}H_{38}O_6$. Scales (95% alcohol), m. 251.5–252° cor.; $[\alpha]_D^{17.5}$
$+57°$ (acetone). — **Fmn.** From the triolone by refluxing with acetic anhydride;
crude yield, 87%. — **Rns.** Saponification with KOH (ca. 1 mol.) in abs. alcohol
at room temp. gives a mixture of monoacetates (1940a Ehrenstein).

5α.6α-Epoxy-allopregnan-3β-ol-20-one, **Δ⁵-Pregnen-3β-ol-20-one α-oxide**
$C_{21}H_{32}O_3$. For revised nomenclature and con-
figuration at C-5 and C-6, see 1944 Plattner;
1948 Ehrenstein. — Crystals (acetone, acetone-
ether or -methanol), m. 180–4° cor., clear at 187°
(1941 Ehrenstein), m. 190–190.5° cor. (1950
Davis); $[\alpha]_D^{24} +1.0°$ (acetone) (1941 Ehrenstein).

Fmn. Together with some *β-oxide* (not isolated from the mother liquors) from Δ^5-pregnen-3β-ol-20-one (p. 2233 s) by oxidation with perbenzoic acid in chloroform in the cold; yield, 65% (1941 Ehrenstein). The acetate is obtained from Δ^5-pregnen-3β-ol-20-one acetate by oxidation with $KMnO_4$ in strong acetic acid at 50°, followed by chromatography on Al_2O_3 and isolation from the benzene-petroleum ether eluate; allopregnane-3β.5α.6α-triol-20-one 3-acetate(?) (see above) is obtained from a subsequent ether-methanol eluate (1940b Ehrenstein).

Rns. Oxidation with CrO_3 in strong acetic acid at room temperature, or with $KMnO_4$ in acetic acid at 50°, gives allopregnan-5α-ol-3.6.20-trione (p. 2976 s). Hydrolysis of a mother liquor product admixed with the β-oxide, using H_2SO_4 in aqueous acetone at room temperature, gives allopregnane-3β.5α.6β-triol-20-one (below); acetolysis by refluxing with glacial acetic acid yields chiefly the 6-acetate together with some 3.6-diacetate of that triolone (1941 Ehrenstein).

3-Acetate $C_{23}H_{34}O_4$. Prisms (ether-petroleum ether) (1940b Ehrenstein), needles (acetone-ether), m. 167–8° cor. (1941 Ehrenstein); readily soluble in ether (1940b Ehrenstein). — **Fmn.** From the parent epoxy-olone (above) by refluxing with acetic anhydride; yield, 91% (1941 Ehrenstein). See also formation of the epoxy-olone, above. — *Oxime* $C_{23}H_{35}O_4N$, crystals (aqueous alcohol), m. 219–221° cor. (1940b Ehrenstein).

Allopregnane-3β.5α.6β-triol-20-one $C_{21}H_{34}O_4$. For revised nomenclature and configuration, see 1944 Plattner; 1948 Ehrenstein. — Rectangular platelets (acetone), m. 256° to 258° cor. (1939 Ehrenstein), $[\alpha]_D^{20} + 59°$ (chloroform) (1951 Mancera). — **Fmn.** From the acetate of Δ^5-pregnen-3β-ol-20-one (p. 2233 s) by treatment with H_2O_2 in strong acetic acid on the water-bath, followed by saponification of the resinous *3-acetate* so formed, using boiling methanolic KOH; yield, 30% (1939, 1940a Ehrenstein). From a mixture of Δ^5-pregnen-3β-ol-20-one α- and β-oxides (mother liquors of the pure α-oxide, p. 2364 s) by hydrolysis with H_2SO_4 in aqueous acetone at room temperature (yield, 75%); by acetolysis of the α-oxide alone, on refluxing with glacial acetic acid, a mixture of the 6-acetate (yield, 61%) with the 3.6-diacetate (yield, 11%) is obtained, separable by fractionation from ether and chromatography on Al_2O_3 (1941 Ehrenstein).

Rns. Oxidation with CrO_3 in strong acetic acid at room temperature gives allopregnan-5α-ol-3.6.20-trione (p. 2976 s) (1939 Ehrenstein); similar oxidation of the 6-acetate yields the corresponding ester of allopregnane-5α.6β-diol-3.20-dione (p. 2865 s) (1940a, 1941 Ehrenstein). Heating with $KHSO_4$ gives Δ^4-pregnene-3.20-dione (p. 2782 s) (1937b CIBA)*.

3-Acetate $C_{23}H_{36}O_5$. See above under "Formation".

* The identity of the 3.5.6-trihydroxy-pregnan-20-one mentioned in the abstract of *Ch. Ztbl.* is assumed (Editor).

6-Acetate $C_{23}H_{36}O_5$. Crystals (methanol), white platelets (ether containing petroleum ether) (1940a Ehrenstein), crystals (acetone), m. 247–248.5° cor.; $[\alpha]_D^{23} + 8.0°$ to $+ 12°$ (acetone) (1941 Ehrenstein), $[\alpha]_D^{18} + 18° \pm 3°$ (chloroform) (1950 Keller); sparingly soluble in ether (1941 Ehrenstein). — **Fmn.** Together with small amounts of the parent triolone by partial saponification of the 3.6-diacetate (below) using abs. alcoholic KOH (1 mol.) at room temperature; separation by fractional crystallization from ether and from acetone, and by chromatography on Al_2O_3; yield, 42 % (1941 Ehrenstein; cf. 1940a Ehrenstein). See also formation of the parent triolone, above. — **Rns.** Refluxing with acetic anhydride gives the 3.6-diacetate (below) (1941 Ehrenstein); see also reactions of the parent triolone (above).

3.6-Diacetate $C_{25}H_{38}O_6$. Plates (methanol) (1940a Ehrenstein), crystals (ether), m. 217–9° cor. (1941 Ehrenstein); $[\alpha]_D$ in acetone —2.0° (1940a Ehrenstein), —11° (1950 Keller); in chloroform —12° ± 3° (1950 Keller), —7° (1951 Mancera); readily soluble in ether (1941 Ehrenstein). — **Fmn.** From the free triolone (1940a Ehrenstein) or from the 6-acetate (1941 Ehrenstein) by refluxing with acetic anhydride; see also formation of the parent triolone (above). — **Rns.** For partial saponification, see formation of the 6-acetate (above).

14-iso-Pregnane-3β.5β.14β-triol-20-one $C_{21}H_{34}O_4$. Not characterized. — **Fmn.**

From the acetate of periplogenin (p. 234; cf. 1941 Paist for α.β-position of the double bond in the lactone ring) on ozonolysis in glacial acetic acid or carbon tetrachloride, followed by decarboxylation and hydrolysis (1937a CIBA). — **Rns.** Gives pregnane-3β.5β-diol-20-one (p. 2323 s) on dehydration with 5 % methanolic HCl at room temperature, followed by catalytic reduction (1937a CIBA).

Pregnane-3α.7α.12α-triol-20-one, 3α.7α.12α-Trihydroxy-ætiocholan-17β-yl

methyl ketone $C_{21}H_{34}O_4$ (p. 117: *Pregnan-20-one-3.7.12-triol*)*. Configuration at C-3, C-7, and C-12 deduced from that of its ultimate precursor, cholic acid, for which see Introduction to steroids, pp. 1361 s, 1362 s. — Crystals with 1 H_2O (acetone-petroleum ether, moist acetone-ether, moist ether, or moist ethyl acetate), m. 120–6° dec. (1936 N. V. ORGANON; 1937 Morsman; 1940c Ehrenstein; 1945 Meystre); the anhydrous form has been obtained as an uncrystallizable oil by heating the hydrate at 100–110° and then shortly at 135°, under 0.05 mm. pressure (1937 Morsman). Rather sparingly soluble in ether and benzene, readily soluble in methanol, ethanol, glacial acetic acid, and ethyl acetate (1937 Morsman).

Fmn. Together with triacetyl-norcholic acid (p. 190), from the tetraacetate of 24.24-dimethylcholane-3α.7α.12α.24-tetrol (p. 2199 s) by oxidation with

---

* For another "pregnane-3α.7α.12α-triol-20-one", see p. 2368 S.

CrO$_3$ in strong acetic acid on the water-bath; the crude ketone triacetate is separated from the neutral products of reaction by means of phenylhydrazine-sulphonic acid and subsequently purified via the semicarbazone (1936 N. V. ORGANON). Together with ætiocholic acid, from (3α.7α.12α-triacetoxyternorcholanyl)-diphenyl-carbinol (p. 2200 s) by oxidation with CrO$_3$ and strong acetic acid on the water-bath; treatment of the neutral reaction products with Girard's reagent T gives the crude triolone triacetate (yield, 8%), which after purification is saponified (yield, 72%) with aqueous methanolic KOH. From the same starting material the triacetate has been obtained by dehydration, followed by ozonization (cf. p. 117), with an improved yield of 62% (1940c Ehrenstein). From the triacetate of 24.24-diphenyl-$\Delta^{20(22).23}$-choladiene-3α.7α.12α-triol (p. 2194 s) by oxidation with CrO$_3$ in 80% acetic acid + chloroform; saponification with K$_2$CO$_3$ in slightly aqueous methanol, on the water-bath, of the resinous triacetate so obtained gives the 7-acetate (see below; yield, 38%), which is further saponified with KOH in slightly aqueous alcohol, on the water-bath, to give the triolone (1945 Meystre); by treatment of the crude triacetate of the present triolone with Girard's reagent P, saponification of the resulting ketone fraction by refluxing with K$_2$CO$_3$ in 84% methanol for 2 hrs., and fractional crystallization from ethyl acetate · isopropyl ether, the more soluble 7.12-diacetate can also be isolated; yields, from crude triacetate: 11% of the 7-acetate, 8.5% of the 7.12-diacetate (1946 Meystre).

Rns. Oxidation with CrO$_3$ in strong acetic acid, at room temperature, gives pregnane-3.7.12.20-tetrone (p. 2985 s); similar oxidation of the 7-acetate yields the corresponding ester of pregnan-7α-ol-3.12.20-trione (p. 2976 s) (1940c Ehrenstein; 1951 Ruff). Oxidation of the 7.12-diacetate with CrO$_3$, in 90% acetic acid at 20°, gives the corresponding ester of pregnane-7α.12α-diol-3.20-dione (p. 2866 s) (1946 Meystre). Oppenauer oxidation of the 7-acetate using cyclohexanone and aluminium isopropoxide in boiling toluene, followed by acetylation with acetic anhydride and pyridine on the water-bath, gives the diacetate of pregnane-7α.12α-diol-3.20-dione (p. 2866 s) (1940c Ehrenstein).

Semicarbazone C$_{22}$H$_{37}$O$_4$N$_3$. Leaflets (abs. alcohol), m. 270–2° cor. dec. (1936 N. V. ORGANON).

7 (formerly 12)-Acetate C$_{23}$H$_{36}$O$_5$. For revision of structure, see 1947 Lardon; 1948 Ehrenstein; 1951 Ruff. — Crystals (chloroform-benzene-petroleum ether, or acetone-ether) (1940c Ehrenstein), transparent needles (isopropyl ether), m. 232–4° cor., cloudy at 140° (1945 Meystre), m. 236–8° cor. (1951 Ruff); [α]$_D^{23}$ +58° ± 4° (chloroform) (1945 Meystre), [α]$_D^{27}$ +82° (acetone)* (1940c Ehrenstein). — **Fmn.** From the triacetate (below) by saponification with abs. alcoholic KOH (ca. 2 mols.) at room temperature, followed by chromatography on Al$_2$O$_3$ (yield, ca. 78%) (1940c Ehrenstein; cf. 1945, 1946 Meystre); see also formation of the parent triolone (above).

---

* The sample probably contained some diacetate (Editor; cf. 1946 Meystre).

7.12-Diacetate $C_{25}H_{38}O_6$. Crystals (ethyl acetate-isopropyl ether), m. 190°
to 191° cor., $[\alpha]_D^{21}$ + 126° ± 4° (chloroform) (1946 Meystre). — **Fmn.** and
Rns. See those of the parent triolone, above.

Triacetate $C_{27}H_{40}O_7$. Leaflets (ether-pentane) containing water of crystallization, m. 134–5° cor. (1937 Morsman; 1945 Meystre); prism clusters (ether-
petroleum ether), m. 150–2° cor. (1940c Ehrenstein); transparent needles
(hexane), m. 156–161° cor. (1945 Meystre); distils at 180–190° (bath temperature) and 0.05 mm. pressure (1937 Morsman); $[\alpha]_D$ + 127° ± 4° (alcohol)
(1945 Meystre); readily soluble in the usual organic solvents except petroleum
ether (1937 Morsman). — **Fmn.** From the 7-acetate by acetylation with acetic
anhydride and pyridine on the water-bath (1945 Meystre); see also formation
of the parent triolone. — **Rns.** Complete saponification is effected by refluxing
with excess aqueous methanolic KOH (1937 Morsman); for alternative
procedure, also partial saponification, see formation of the parent diolone
(above). — *Semicarbazone* $C_{28}H_{43}O_7N_3$, crystals (abs. alcohol), m. ca. 180° cor.
(1936 N. V. ORGANON).

A **pregnane-3α.7α.12α-triol-20-one** $C_{21}H_{34}O_4$ [prisms with $^1/_2$ H_2O (acetone),
m. 210–211°; Liebermann reaction violet-red → cherry-red; positive Hammarsten reaction] is said to be formed, along with "3α.7α.12α-trihydroxy-
ætiocholanic acid" (m. 171°*),
from 3α.7α.12α.22-tetrahydr-
oxysterocholanic acid δ-lac-
tone (I) on treatment with
methylmagnesium iodide in
ether-benzene, followed by

acetylation of the resulting carbinol (II), oxidation of the triacetate
with CrO_3 in glacial acetic acid, and saponification with 5% KOH on the
water-bath (1942 Kanemitu). On oxidation with CrO_3 in glacial acetic acid,
this pregnanetriolone gives "pregnane-3.7.12.20-tetrone" m. 185–187° (1942
Kanemitu); according to Ehrenstein (1940c), however, this tetrone (p. 2985 s)
melts at 238–242°. — *Oxime* $C_{21}H_{35}O_4N$, scales (aqueous alcohol), dec. 249°
(1942 Kanemitu).

Pregnane- and **Allopregnane-3β.16α.17α-triol-20-ones** $C_{21}H_{34}O_4$. — For their
anhydro-derivatives, see Δ^{16}-pregnen - and Δ^{16}-allopregnen-3β-ol-20-one 16.17-
oxides (pp. 2241 s and 2243 s, resp.).

25-Methylcoprostane-3α.7α.12α-triol-24-one, 3α.7α.12α-Trihydroxy-24-keto-

isobufostane $C_{28}H_{48}O_4$. Crystals
(ethyl acetate), m. 145–150°, crystals with $^1/_2$ H_2O (aqueous alcohol), m. 161°; easily soluble in
ethyl acetate, rather sparingly
soluble in alcohol. Liebermann-
Burchard reaction: cherry-red; Hammarsten reaction: positive. — **Fmn.**

---

* See also 1942 Shimizu; according to Ehrenstein (1940c) this acid melts at 254–8°, according
to Lardon (1947) at 266–270° cor. (Kofler block).

From the triacetate of pentahydroxybufostane (p. 2204 s) by treatment with CrO$_3$ in glacial acetic acid on the water-bath. The ether insoluble neutral fraction is purified via the *triacetoxyketo-isobufostane semicarbazone* C$_{35}$H$_{57}$O$_7$N$_3$ (crystalline, no constants given), which is hydrolysed, first with hot alcoholic HCl to give the oily *triacetate* C$_{34}$H$_{54}$O$_7$, then with hot 10% alcoholic NaOH to the free triolone. — **Rns.** Oxidation with CrO$_3$ in glacial acetic acid at room temperature gives tetraketo-isobufostane (p. 2985 s) (1940 Kazuno).

24-Phenylcholane-3α.7α.12α-triol-24-one, 3α.7α.12α-Trihydroxynorcholanyl phenyl ketone, *"Cholophenone"*

C$_{30}$H$_{44}$O$_4$. Configuration at C-3, C-7, and C-12 deduced from that of the ultimate precursor, cholic acid, for which see Introduction to steroids, pp. 1361 s, 1362 s (Editor). — Crystals (methanol) containing ½ CH$_3$OH (1945 Hoehn) or ½ H$_2$O (1944 Jacobsen), m. 175–7⁰, [α]$_D$ +39⁰ (dioxane) (1945 Hoehn), [α]$_D^{20}$ +26⁰ (chloroform) (1944 Jacobsen). — **Fmn.** From the triformate of cholic acid by reaction with thionyl chloride at room temperature for conversion into its acid chloride, which is allowed to react with diphenylcadmium in boiling dry benzene; after decomposition with dil. HCl of the complex so formed, the crude ketone formate resulting is saponified with boiling methanolic NaOH (1945 Hoehn; cf. 1944 Jacobsen). — **Rns.** Bromination yields a mixture of the 23-epimeric 24-phenyl-23-bromocholane-3α.7α.12α-triol-24-ones, from which only the 23"β"-epimer (p. 2370 s) has been isolated (1944 Jacobsen).

Oxime C$_{30}$H$_{45}$O$_4$N, m. 214–7⁰ dec. (1944 Jacobsen).

2.4-Dinitrophenylhydrazone C$_{36}$H$_{48}$O$_7$N$_4$, m. 221–222.5⁰ (1944 Jacobsen).

Triacetate C$_{36}$H$_{50}$O$_7$. Crystals (methanol), m. 122–3⁰, [α]$_D^{25}$ +78.5⁰ (dioxane) (1945 Hoehn), [α]$_D^{20}$ +79⁰ (chloroform) (1944 Jacobsen).

1936 N. V. Organon, *Dutch Pat.* 42782 (issued 1938); *C. A.* **1938** 6809; *Ch. Ztbl.* **1938** II 1085.
1937 a Ciba*, *Brit. Pat.* 482321 (issued 1938); *Ch. Ztbl.* **1938** II 3272.
b Ciba*, *Brit. Pat.* 497394 (issued 1939); *C. A.* **1939** 3812; *Ch. Ztbl.* **1939** I 2828; Ciba Pharmaceutical Products (Miescher, Wettstein), *U.S. Pat.* 2229813 (1938, issued 1941); *C. A.* **1941** 3040; *Ch. Ztbl.* **1941** II 2467.
Morsman, Steiger, Reichstein, *Helv.* **20** 3, 14, 15.
1939 Ehrenstein, *J. Org. Ch.* **4** 506, 515 et seq.
1940 a Ehrenstein, Stevens, *J. Org. Ch.* **5** 318, 323, 324, 325, 327.
b Ehrenstein, Decker, *J. Org. Ch.* **5** 544, 557 et seq.
c Ehrenstein, Stevens, *J. Org. Ch.* **5** 660, 667 et seq.
Kazuno, *Z. physiol. Ch.* **266** 11, 27, 28.
1941 Ehrenstein, Stevens, *J. Org. Ch.* **6** 908, 911, 912 et seq., 918.
Paist, Blout, Uhle, Elderfield, *J. Org. Ch.* **6** 273.
1942 Kanemitu (Kanemitsu), *J. Biochem.* **35** 173, 179, 180.
Shimizu, Kazuno, *J. Biochem.* **35** 184.
1943 Ruzicka, Prelog, *Helv.* **26** 975, 989 et seq.

* Gesellschaft für chemische Industrie in Basel; Société pour l'industrie chimique à Bâle.

1944 Jacobsen, *J. Am. Ch. Soc.* **66** 662.
PARKE, DAVIS & Co. (Wagner), *U.S. Pat.* 2408830 (issued 1946); *C. A.* **1947** 1253; *Ch. Ztbl.* **1947** 1011.
Plattner, Lang, *Helv.* **27** 1872, 1873.
1945 Hoehn, Moffett, *J. Am. Ch. Soc.* **67** 740.
Meystre, Miescher, *Helv.* **28** 1497, 1503; CIBA PHARMACEUTICAL PRODUCTS (Miescher, Frey, Meystre, Wettstein), *U.S. Pat.* 2461910 (1946, issued 1949); *C. A.* **1949** 3474; *Ch. Ztbl.* **1949**E 1870, 1871.
1946 Meystre, Frey, Neher, Wettstein, Miescher, *Helv.* **29** 627, 630, 634.
1947 Lardon, *Helv.* **30** 597, 598 (footnote 5), 608.
Marker, Wagner, Ulshafer, Wittbecker, Goldsmith, Ruof, *J. Am. Ch. Soc.* **69** 2167, 2169, 2182 et seq.
1948 Ehrenstein, *J. Org. Ch.* **13** 214, 219, 220, 222.
1950 Davis, Petrow, *J. Ch. Soc.* **1950** 1185, 1188.
Keller, Weiss, *J. Ch. Soc.* **1950** 2709, 2713, 2714.
1951 Mancera, Rosenkranz, Djerassi, *J. Org. Ch.* **16** 192, 195.
Pataki, Rosenkranz, Djerassi, *J. Am. Ch. Soc.* **73** 5375.
Ruff, Reichstein, *Helv.* **34** 70, 71, 78.
1952 Wendler, Slates, Tishler, *J. Am. Ch. Soc.* **74** 4894.

---

11. HALOGENO- AND DIAZO-TRIHYDROXY-MONOOXO-STEROIDS CONTAINING CO IN SIDE CHAIN, OH IN THE RING SYSTEM

25 - Chloro - homocholane - 3α.7α.12α - triol - 24 - one, 3α.7α.12α - Trihydroxy-norcholanyl chloromethyl ketone

$C_{25}H_{41}O_4Cl$. Crystals (acetone-petroleum ether), m. 191.5–192.5° cor., $[\alpha]_D^{27}$ +39° (methanol). — **Fmn.** From 25-diazo-homocholane-3α.7α.12α-triol-24-one (page 2371 s) by treatment with dry HCl in ether at 0°; the *triformate* $C_{28}H_{41}O_7Cl$ (white amorphous product) and the *triacetate* $C_{31}H_{47}O_7Cl$ (oil) are similarly obtained from the corresponding diazo-triolone esters. — **Rns.** Refluxing the triacetate with sodium benzoate in aqueous alcohol yields the 25-benzoate 3.7.12-triacetate of homocholane-3α.7α.12α.25-tetrol-24-one (p. 2376 s) (1942 Knowles).

24 - Phenyl - 23 - bromocholane - 3α.7α.12α - triol - 24 - one, 23-Bromo-3α.7α.12α-trihydroxy-norcholanyl phenyl ketone, *23 - Bromo - "cholophenone"* $C_{30}H_{43}O_4Br$.

Triacetate $C_{36}H_{49}O_7Br$. *23"β"-Epimer* (configuration arbitrarily assigned): Crystals with 1 H_2O (solvent not indicated), m. 108.5–111.5°, which are dehydrated by drying in vacuo; $[\alpha]_D^{20}$ +95° (chloroform). — **Fmn.** From the triacetate of 24-phenyl-cholane-3α.7α.12α-triol-24-one (p. 2369 s) by bromination; the 23"β"-epimer crystallizes from the 23-epimeric mixture so obtained. — **Rns.** Acetolysis (of the 23"β"-epimer or of the mixed epimers) yields the tetraacetate(s) of 24-phenylcholane-3α.7α.12α.23-tetrol-20-one (p. 2377 s) (1944 Jacobsen).

25-Diazo-homocholane-3α.7α.12α-triol-24-one, 3α.7α.12α-Trihydroxy-norcho-lanyldiazomethyl ketone $C_{25}H_{40}O_4N_2$.

Configuration at C-3, C-7, and C-12, based on that of cholic acid, its ultimate precursor, for which see Introduction to steroids pp. 1361 s, 1362 s; this implies reversal of the 12β-configuration originally assigned by Ruzicka (1944) (Editor). — Crystals (moist chloroform) behaving as *hydrate*, impure, no constants given (1942 Knowles); yellow uncrystallizable oil (1944 Ruzicka). — **Fmn.** From triformylcholic acid (p. 194) by treatment with thionyl chloride in boiling abs. benzene for conversion into its acid chloride which is allowed to react with excess diazomethane in abs. benzene at —10°, and then left to stand at room temperature; the crystalline triformate so obtained (yield, 70%) is saponified with methanolic KOH at room temperature (1944 Ruzicka). A crude triformate is obtained by a similar reaction using diazomethane (1 equiv.) in ether at 0°; by using excess of the reagent (6 mols.), a methylated amorphous product results, from which the crude diazo-triolone hydrate is obtained by saponification with cold methanolic KOH; a crude *triacetate* is obtained by similar treatment of triacetylcholic acid chloride with diazomethane in ether-benzene (1942 Knowles).

Rns. Treatment with dry HCl in ether at 0° gives 25-chloro-homocholane-3α.7α.12α-triol-24-one (p. 2370 s); the crude triformate or triacetate yields the corresponding triformyl or triacetyl ester (1942 Knowles). Acetolysis of the crude diazo-triolone with glacial acetic acid at room temperature, followed by short heating at 95°, gives the (crude) 25-acetate of homocholane-3α.7α.12α.25-tetrol-24-one (p. 2376 s); acetolysis of the triformate with glacial acetic acid on the water-bath gives the corresponding 3.7.12-triformyl ester (1944 Ruzicka); similar treatment of the crude triacetate gives the analagous tetraacetate (1942 Knowles).

Triformate $C_{28}H_{40}O_7N_2$. Needles (methanol), m. 128–9° dec.; $[\alpha]_D^{21}$ +87° (chloroform) (1944 Ruzicka). — **Fmn.** and **Rns.** See those of the parent diazo-triolone (above).

1942 Knowles, Fried, Elderfield, *J. Org. Ch.* **7** 383, 387, 388.
1944 Jacobsen, *J. Am. Ch. Soc.* **66** 662.
    Ruzicka, Plattner, Heusser, *Helv.* **27** 186, 190, 191.

12. Tetrahydroxy-monooxo-steroids Containing CO and One OH-Group in Side Chain, the others in the Ring System

Allopregnane-3β.5α.6α.21-tetrol-20-one $C_{21}H_{34}O_5$.

5α.6α-Epoxy-allopregnane-3β.21-diol-20-one, Δ⁵-Pregnene-3β.21-diol-20-one
α-oxide $C_{21}H_{32}O_4$. For revised nomenclature and configuration at C-5 and C-6, see 1944b Plattner; 1948 Ehrenstein. — Only acetates have been obtained. — **Fmn.** The 21-acetate is obtained from the 21-acetate of Δ⁵-pregnene-3β.21-diol-20-one (p. 2301 s) by treatment with perbenzoic acid in chloroform, first in the cold, then at room temperature; separation from the (not isolated) β-oxide acetate simultaneously formed is effected by crystallization from acetone (minimum yield, 64%) or by chromatography (1941 Ehrenstein). Similarly a mixture of the α- (82%) and β- (18%) oxide diacetates is obtained from the diacetate of the same diolone using perbenzoic acid in chloroform at —10°; the α-isomer (yield, 48%) is largely separated by crystallization from ethyl acetate, and after chromatography of the mother liquor on Al_2O_3, the β-oxide is first eluted and finally a further quantity of the α-isomer can be isolated (1946 Ruzicka). — **Rns.** Catalytic hydrogenation of the diacetate, using platinum oxide in glacial acetic acid at room temperature, yields a mixture of the 3.21-diacetates of the 20-epimeric allopregnane-3β.5.20.21-tetrols (p. 2138 s) (1946 Ruzicka). Refluxing the 21-acetate with glacial acetic acid gives a mixture of the 6.21-diacetate (preponderating) and the 3.6.21-triacetate of allopregnane-3β.5α.6β.21-tetrol-20-one (below) (1941 Ehrenstein).

21-Acetate $C_{23}H_{34}O_5$. Needles (acetone), m. 195–7°, $[\alpha]_D^{26}$ +15.6° (acetone) (1941 Ehrenstein). — **Fmn.** and **Rns.** See above.

Diacetate $C_{25}H_{36}O_6$. Needles (ethyl acetate), m. 180–2° cor., $[\alpha]_D^{22}$ +22.6° (average) (chloroform) (1946 Ruzicka). — **Fmn.** and **Rns.** See above.

Allopregnane-3β.5α.6β.21-tetrol-20-one $C_{21}H_{34}O_5$. For revised nomenclature and configuration at C-5 and C-6, see 1944b Plattner; 1948 Ehrenstein. — Only acetates have been obtained. — **Fmn.** A mixture of the 6.21-diacetate and the 3.6.21-triacetate is obtained from the 21-acetate of 5α.6α-epoxy-allopregnane-3β.21-diol-20-one (above) by refluxing with glacial acetic acid; after 45 min. refluxing, followed by chromatographic separation on Al_2O_3, the crude yields are: 70% diacetate and 17% triacetate; after 105 min. refluxing, a 38% yield of crystalline triacetate and a 45% yield of non-crystalline crude diacetate results (1941 Ehrenstein). — **Rns.** Oxidation of the 6.21-diacetate with CrO_3 in ca. 94% acetic acid at room temperature gives the 6.21-diacetate of allopregnane-5α.6β.21-triol-3.20-dione (p. 2867 s) (1941 Ehrenstein).

6.21-Diacetate $C_{25}H_{38}O_7$. Irregular plates (partly rosettes) (ether containing petroleum ether), m. 118°, clear at 126°; $[\alpha]_D^{26} +16.7°*$ (acetone); sparingly soluble in ether. — **Fmn.** The crude product, obtained as described above, is treated with small amounts of ether for removal of the chief impurities (1941 Ehrenstein). — **Rns.** See above.

3.6.21-Triacetate $C_{27}H_{40}O_8$. Crystals (ether-petroleum ether), m. 176–177.5° cor.; $[\alpha]_D^{26} +3.5°$ (acetone) (1941 Ehrenstein), $[\alpha]_D^{19.5} -0.3°$ (chloroform) (1951 Herzig); readily soluble in ether (1941 Ehrenstein). — **Fmn.** As described above; also by acetylation of the 6.21-diacetate with acetic anhydride and pyridine at room temperature (crude yield, 97%) (1941 Ehrenstein).

Pregnane-3β.5β.6β.21-tetrol-20-one $C_{21}H_{34}O_5$.

5β.6β-Epoxy-pregnane-3β.21-diol-20-one, Δ^5-Pregnene-3β.21-diol-20-one β-oxide $C_{21}H_{32}O_4$.

Diacetate $C_{25}H_{36}O_6$. Crystals (alcohol), m. 160–160.5° cor., $[\alpha]_D^{22} +70.5°$ (chloroform). — **Fmn.** Together with the diacetate of the isomeric α-oxide (82%) (p. 2372 s), from the diacetate of Δ^5-pregnene-3β.21-diol-20-one by oxidation with perbenzoic acid in chloroform at −10°; after crystallization of most of the α-oxide diacetate from ethyl acetate, the mother liquors are chromatographed on Al_2O_3 from which the β-isomer is eluted first with benzene-petroleum ether (1946 Ruzicka).

Allopregnane-3α.11β.17α.21-tetrol-20-one, *Reichstein's Substance C* $C_{21}H_{34}O_5$

(p. 117). For identity of *Kendall's Compound C* (1936 Mason) and *Wintersteiner's Compound D* (1935 Wintersteiner) with Reichstein's Substance C, see 1936 Wintersteiner; 1937 Reichstein; 1938 Mason. — For structure, see 1938 Mason; 1940 Shoppee; 1942 v. Euw; for configuration at C-3, C-11, and C-17, see 1947 v. Euw; cf. 1942 v. Euw.

Colourless skew-pointed needles (alcohol), m. 273–6° cor. dec. (1942 v. Euw); $[\alpha]_{5461}^{25} +84° \pm 5°$ (1938 Mason), $[\alpha]_{5461}^{15} +90.2° \pm 4°$, $[\alpha]_D^{15} +73.1° \pm 4°$ (1942 v. Euw), all in alcohol; $[\alpha]_{5461}^{18} +75.7° \pm 5°$, $[\alpha]_D^{18} +59.2° \pm 5°$ (both in dioxane) (1942 v. Euw). Sparingly soluble in cold acetone (1938 Mason; 1942 v. Euw). The pure compound gives no precipitate with digitonin**. With conc. H_2SO_4 a red-brown solution is obtained which slowly changes to blue-green, but is not fluorescent (1942 v. Euw). Reducing towards Benedict's solution on heating (1935 Wintersteiner).

Isolation from adrenal glands. The alcoholic extract of the whole adrenal glands (cattle) is treated with permutit for removal of adrenaline (1936a

---

* Sample not quite pure, probably containing some triacetate.

** The precipitation previously reported (p. 117) was due to small amounts of admixed substance V (1942 v. Euw).

Reichstein), then evaporated to dryness and the residue partitioned between pentane and 30% aqueous methanol. The aqueous layer, after concentration and acidification, is extracted with ether, the ether solution (after deacidification with small amounts of KHCO₃ solution) is re-extracted with large amounts of distilled water. The aqueous solution is again concentrated, acidified to Congo red and re-extracted with ether, the ether solution (again deacidified) being evaporated to dryness. The residue of this last ether solution is separated by means of Girard's reagent into a "non-ketonic" fraction, a more reactive and a less reactive ketone fraction; yield of the last: 1.54 g. from 1000 kg. of glands (1936b Reichstein). By slow crystallization at 0° from aqueous acetone-ether (or -toluene), ca. 25% of the less reactive ketone material separates, consisting of substances C, D (p. 2869 s), and V (p. 2375 s) (1936b Reichstein; 1942 v. Euw). Substance C is separated from the more soluble D and V by boiling with small amounts of acetone; further small amounts can be isolated as diacetate from the amorphous mother-liquor product of the acetone-ether (or -toluene) crystallization, by acetylation before chromatographic separation (1942 v. Euw). For improvements in the extraction method (first ether extraction replaced by ethyl acetate), and in the separation of ketonic and non-ketonic fractions, see 1941 Reichstein. For an alternative extraction method, see 1935 Pfiffner; 1935 Wintersteiner; yield of crude product, 6 mg. from 100 kg. glands (1935 Wintersteiner).

Rns. Oxidation of the 3.21-diacetate with CrO_3 in glacial acetic acid at 15° gives the 3.21-diacetate of allopregnane-3α.17.21-triol-11.20-dione (page 2869 s) (1942 v. Euw). From the HIO_4 oxidation to 3α.11β.17-trihydroxy-ætioallocholanic acid (cf. p. 117), carried out at room temperature in 40% alcohol containing H_2SO_4, formaldehyde can also be isolated (1938 Mason); for this oxidation see also 1942 v. Euw. Oxidation with alkaline silver solution in aqueous alcohol (1936, 1938 Mason) gives an acid $C_{21}H_{34}O_6$, probably **3α.11β.17.20-tetrahydroxy-allopregnan-21-oic acid** (I) ["*Acid* 3" (1936 Mason), structure deduced from that of the parent tetrolone since established (Editor; cf. 1938 Mason); blades (ethyl acetate containing some alcohol), m. 240–2° dec. (sharp); determination of active hydrogen indicates 4 OH groups; oxidation with $K_2Cr_2O_7$ and H_2SO_4

gives a ketone (1936 Mason), later recognized to be impure androstane-3.11.17-trione (page 2980 s) (1938 Mason)].

3.21-Diacetate $C_{25}H_{38}O_7$. Rectangular leaflets (acetone-ether), m. 204–5° cor.; $[\alpha]_D^{15}$ +74°, $[\alpha]_{5461}^{15}$ +90.5° (both in dioxane) (1942 v. Euw). — **Fmn.** Obtained as intermediate in the isolation of the tetrolone (above); also from the pure tetrolone by treatment with acetic anhydride and pyridine at room temperature (1942 v. Euw; cf. 1937 Reichstein).

A **compound** $C_{21}H_{34}O_5$, considered to be probably identical with Reichstein's Substance C, has been isolated from adrenal glands (cattle); yield, 24 mg. from 880 kg. glands. — Fine needles (95% alcohol), m. 250–2° cor. dec.;

$[\alpha]_D^{25}$ +84.5° (abs. alcohol); very sparingly soluble in 95 % alcohol. — The compound showed an activity similar to that of dehydrocorticosterone (page 2972 s) in the survival and growth test on adrenalectomized rats (1939 Kuizenga).

Allopregnane-3β.11β.17.21-tetrol-20-one, *Reichstein's Substance V* $C_{21}H_{34}O_5$.

For reversal of the configuration originally ascribed at C-17, see 1947 v. Euw. — Hygroscopic needles (aqueous methanol), possibly not quite pure, containing water of crystallization, m. 220–5° cor., after becoming opaque, with loss of water, at ca. 100°; $[\alpha]_D^{13}$ +51°, $[\alpha]_{5461}^{13}$ +68° (both in dioxane). A sparingly soluble digitonide is rapidly precipitated from 73 % aqueous methanol at 18°. Strongly reducing towards methanolic alkaline silver diammine at room temperature. Gives readily a non-fluorescent red-brown solution with conc. H_2SO_4 (1942 v. Euw).

Isolation. From adrenal glands (cattle), together with its 3α-epimer, Reichstein's Substance C (which see, p. 2373 s). The crude crystalline mixture of Substances C, D, and V (p. 2374 s) is boiled with small amounts of acetone, in which C is largely insoluble, and the mother-liquor product is acetylated with acetic anhydride and pyridine at room temperature; after chromatography of the acetate mixture on Al_2O_3, V-diacetate is isolated from the intermediate ether-benzene eluates; saponification, accompanied by some decomposition to give acid products, is effected with aqueous methanolic K_2CO_3 at 15° (1942 v. Euw).

Fmn. For synthesis, see (after the closing date for this volume) 1955 Chamberlin.

Rns. Oxidation with excess CrO_3, in glacial acetic acid at room temperature gives androstane-3.11.17-trione (p. 2980 s); similar treatment of the 3.21-diacetate yields the corresponding ester of allopregnane-3β.17.21-triol-11.20-dione (p. 2869 s). Oxidation with periodic acid in aqueous dioxane at 15° gives 3β.11β.17-trihydroxy-ætioallocholanic acid. Catalytic hydrogenation of the diacetate, using platinum oxide in glacial acetic acid at room temperature, yields an inseparable mixture of the 3.21-diacetate of allopregnane-3β.11β.17α.20β.21-pentol (p. 2202 s) with its 20-epimer (1942 v. Euw).

3.21-Diacetate $C_{25}H_{38}O_7$. Lancets (ether), m. 231–6° cor. (1937 Reichstein); colourless, hard, prismatic needles (acetone-ether), m. 225–7° cor.; $[\alpha]_D^{18}$ +63°, $[\alpha]_{5461}^{17}$ +77° (dioxane) (1942 v. Euw); crystals (methanol), m. 229–231°, $[\alpha]_D^{20}$ +77° (acetone) (1952 Pataki). — **Fmn.** From the mother liquors of the acetate mixture resulting by acetylation of slightly impure Substance C (p. 2373 s), using acetic anhydride and pyridine at room temperature (1937 Reichstein; cf. 1942 v. Euw). See also isolation of the parent tetrolone, above.

A **compound** $C_{21}H_{34}O_5$, considered as being possibly an optical isomer of Reichstein's Substance C, has been isolated from adrenal glands (cattle); yield,

E.O.C. XIV s. 166

42 mg. from 880 kg. glands (1939 Kuizenga). In view of the mode of isolation (obtained from the acetone mother liquors of the isomeric compound m. 250⁰ to 252⁰; cf. p. 2374 s) and the constants recorded, identity with Reichstein's Substance V seems possible (Editor). — Heavy needles (acetone), m. 231–4⁰ cor. dec.; $[\alpha]_D^{25}$ +71.4⁰ (abs. alcohol). — The compound was inactive in the survival and growth test on adrenalectomized rats (1939 Kuizenga).

Homocholane-3α.7α.12α.25-tetrol-24-one, 3α.7α.12α-Trihydroxy-norcholanyl hydroxymethyl ketone $C_{25}H_{42}O_5$.

Configuration at C-3, C-7, and C-12 is based on that of cholic acid, its ultimate precursor, for which see Introduction to steroids, pp. 1361 s, 1362 s; this implies reversal of the 12β-configuration originally ascribed by Ruzicka (1944) (Editor). — Only esters have been prepared. For comparison of their molecular rotations with those of steroids having similar structure and configuration in the tetracyclic system, see 1944 a Plattner. — **Fmn.** The 25-acetate 3.7.12-triformate is obtained from the triformate of 25-diazohomocholane-3α.7α.12α-triol-24-one (p. 2371 s) by heating with glacial acetic acid on the water-bath; yield, 56% (1944 Ruzicka). Similarly the tetraacetate is obtained from the crude diazo-triolone by reaction with glacial acetic acid at room temperature, followed by slow heating to 95⁰, with subsequent acetylation of the resulting crude *25-acetate* with acetic anhydride and pyridine under reflux (yield, 31%) (1944 Ruzicka); the tetraacetate is also obtained by similar acetolysis of the crude triacetyl-diazo-tetrolone (1942 Knowles). The 25-benzoate 3.7.12-triacetate is obtained from the 7.12-diacetate of cholic acid (p. 191) by acetylation, followed by treatment with thionyl chloride to give the acid chloride triacetate which, by reaction with diazomethane and then with HCl, is converted into the crude triacetate of 25-chlorohomocholane-3α.7α.12α-triol-24-one (p. 2370 s); the chloro-ketone triacetate is refluxed with sodium benzoate in aqueous alcohol (7 hrs.) and the product crystallized from ligroin; a further yield is obtained by renewed refluxing with sodium benzoate and then reacetylation of the ligroin mother liquor product; over-all yield, 26.5% (1942 Knowles).

Rns. Refluxing the tetraacetate with bromoacetic ester and zinc turnings (activated by iodine) in anhydrous benzene-dioxane, followed by dehydration with boiling acetic anhydride, gives β-(3α.7α.12α-triacetoxy-norcholanyl)-$\Delta^{\alpha:\beta}$-butenolide; by submitting the 25-acetate 3.7.12-triformate to a similar condensation and then dehydration by formic acid at 55–60⁰, β-(3α.7α.12α-triformoxy-norcholanyl)-$\Delta^{\alpha:\beta}$-butenolide is obtained; if in the latter case the zinc complex first formed is heated with aqueous alcoholic HCl for 3 hrs., β-(3α-hydroxy-7α.12α-diformoxy-norcholanyl)-β-hydroxy-butanolide, or after 18 hrs. heating β-(3α.7α.12α-trihydroxy-norcholanyl)-β-hydroxy-butanolide, is obtained (1944 Ruzicka). For similar condensation of the 25-benzoate 3.7.12-triacetate, from which no crystalline product was isolated, see 1942 Knowles.

References, p. 2377 s

25-Acetate 3.7.12-triformate $C_{30}H_{44}O_9$. Colourless platelets (alcohol), crystals (ether), m. 118–9⁰ cor., $[\alpha]_D^{18}$ +77.5⁰ (chloroform) (1944 Ruzicka). — **Fmn.** and **Rns.** See above.

Tetraacetate $C_{33}H_{50}O_9$. Lancets (ethyl acetate-hexane), m. 132–132.5⁰ cor., $[\alpha]_D^{17}$ +77⁰ (chloroform) (1944 Ruzicka). — **Fmn.** and **Rns.** See above.

25-Benzoate 3.7.12-triacetate $C_{38}H_{52}O_9$. White crystals (alcohol), m. 178⁰ to 180.5⁰ cor.; $[\alpha]_D^{27}$ +63⁰ (chloroform) (1942 Knowles). — **Fmn.** and **Rns.** See above.

24-Phenylcholane-3α.7α.12α.23-tetrol-24-one, 3α.7α.12α.23-Tetrahydroxy-nor-cholanyl phenyl ketone, *23-Hydr-oxy-"cholophenone"* $C_{30}H_{44}O_5$.

Only a crude product has been obtained. — **Fmn.** A mixture of the 23-epimeric tetraacetates is obtained by acetolysis of the mixed 23-epimeric 24-phenyl-23-bromocholane-3α.7α.12α-triol-24-ones (page 2370 s); similar acetolysis of the 23"β"-bromo-compound yields the tetra-acetate of the 23"α"-form, described below. — **Rns.** Oxidation of the crude hydrolysis product of the tetraacetate, by means of $CuSO_4$ in aqueous pyridine, gives 24-phenylcholane-3α.7α.12α-triol-23.24-dione (p. 2381 s) (1944 Jacobsen).

24-Phenylcholane-3α.7α.12α.23"α"-tetrol-24-one tetraacetate $C_{38}H_{52}O_9$. The "α"-configuration at C-23 is arbitrarily assigned on the basis of its formation with Walden inversion from a "β"-bromo-compound (see above). The *hemi-hydrate* $C_{38}H_{52}O_9 + \frac{1}{2} H_2O$ is crystalline (solvent not indicated), m. 180–2⁰; $[\alpha]_D^{20}$ —10⁰ (chloroform) (1944 Jacobsen).

1935 Pfiffner, Wintersteiner, Vars, *J. Biol. Ch.* **111** 585, 587 et seq.
Wintersteiner, Pfiffner, *J. Biol. Ch.* **111** 599, 603.
1936 Mason, Myers, Kendall, *J. Biol. Ch.* **114** 613, 619, 620.
a Reichstein, *Helv.* **19** 29, 38, 47, 58.
b Reichstein, *Helv.* **19** 1107, 1109. 1110, 1120 et seq.
Wintersteiner, Pfiffner, *J. Biol. Ch.* **116** 291.
1937 Reichstein, *Helv* **20** 978, 983, 990.
1938 Mason, Hoehn, Kendall, *J. Biol. Ch.* **124** 459, 464, 467, 471.
1939 Kuizenga, Cartland, *Endocrinology* **24** 526, 529, 530.
1940 Shoppee, *Helv.* **23** 740.
1941 Ehrenstein, *J. Org. Ch.* **6** 626, 642 et seq.
Reichstein, v. Euw, *Helv.* **24** 247 E, 250 E et seq.
1942 v. Euw, Reichstein, *Helv.* **25** 988, 992, 994, 1002 et seq., 1010 et seq.
Knowles, Fried, Elderfield, *J. Org. Ch.* **7** 383, 387.
1944 Jacobsen, *J. Am. Ch. Soc.* **66** 662.
a Plattner, Heusser, *Helv.* **27** 748, 750.
b Plattner, Lang, *Helv.* **27** 1872, 1873.
Ruzicka, Plattner, Heusser, *Helv.* **27** 186, 191 et seq.
1946 Ruzicka, Plattner, Heusser, Ernst, *Helv.* **29** 248.
1947 v. Euw, Reichstein, *Helv.* **30** 205, 206 et seq., 210 et seq.
1948 Ehrenstein, *J. Org. Ch.* **13** 214, 221.
1951 Herzig, Ehrenstein, *J. Org. Ch.* **16** 1050, 1058.
1952 Pataki, Rosenkranz, Djerassi, *J. Biol. Ch.* **195** 751, 753.
1955 Chamberlin, Chemerda, *J. Am. Ch. Soc.* **77** 1221.

B. Dioxo-Compounds (both CO-Groups in Side Chain)

Δ^5-Pregnen-3β-ol-20-on-21-al, (3β-Hydroxy-Δ^5-androsten-17β-yl)-glyoxal, (3β-Hydroxy-Δ^5-ætiocholen-17β-yl)-glyoxal

—CO·CHO $C_{21}H_{30}O_3$. Pale brownish crystals with 1 H_2O (after drying in a high vacuum at room temperature for 10 min.), m. 135–6° cor.; sometimes a pale brownish crystalline powder (another hydrate or a polymer?), m. 170° cor., was also obtained (1939 Reich); colourless crystals (two forms), m. 140° and 172° (1937 a, c, d, f Ciba).

Fmn. * From Δ^5-pregnen-3β-ol-20-one (p. 2233 s) on treatment with amyl nitrite and sodium ethoxide in cold alcohol, followed by hydrolysis of the resulting isonitroso-compound with sodium nitrite in strong acetic acid (1937 a, 1938 a Ciba), on treatment with ethyl nitrite and potassium in ether, followed by reduction of the resulting isonitroso-compound with sodium in boiling alcohol to 21-amino-Δ^5-pregnen-3β-ol-20-one which is then oxidized with nitrous acid (1937 e, 1938 a Ciba). From the same hydroxyketone on treatment with ethyl nitrate and sodium ethoxide in alcohol-benzene with cooling, followed by refluxing with 5 % alcoholic H_2SO_4 (1937 f Ciba). From the same hydroxyketone on refluxing with p-nitrosodimethylaniline and 30 % NaOH in alcohol, followed by refluxing of the resulting 21-N-(p-dimethyl-aminophenyl)-isoxime (p. 2379 s) with 5 % alcoholic H_2SO_4 (1937 d, 1938 a Ciba). From the same hydroxyketone on condensation with aldehydes, e. g. benzaldehyde, followed by oxidation of the condensation products (see, e. g. p. 2271 s) with ozone, CrO_3, lead or manganese tetraacetate after temporary protection of the nuclear double bond with halogen or hydrogen halide (1937 Schering A.-G.).

From N-(3β-hydroxy-20-keto-Δ^5-pregnen-21-yl)-pyridinium chloride or bromide (p. 2279 s) on treatment with p-nitrosodimethylaniline and 1 N NaOH in alcohol in a freezing mixture, followed by shaking of the resulting 21-N-(p-dimethylaminophenyl)-isoxime (p. 2379 s) with 2 N HCl in ether; yield, ca. 43 % (1939 Reich). From 21-diazo-Δ^5-pregnen-3β-ol-20-one (p. 2293 s) on treatment with H_2S and ammonia in alcohol, followed by heating with sulphuric-acetic acid (1937 c, 1938 b Ciba).

Rns. Reduces alkaline silver diammine solution at room temperature (1939 Reich). Its dimethyl acetal (below) is oxidized to the corresponding acetal of Δ^4-pregnene-3.20-dion-21-al (p. 2870 s) on refluxing with aluminium tert.-butoxide and acetone in abs. benzene (1939 Reich); the diethyl mercaptal reacts similarly (1941 Schindler). The dimethyl acetal is reduced to the acetal of Δ^5-pregnene-3β.20-diol-21-al (m. 135–6° cor.; p. 2312 s) by boiling with aluminium isopropoxide in abs. isopropyl alcohol with removal (by continuous distillation) of the acetone formed (1941 Schindler). Is reduced to Δ^5-pregnene-3β.20.21-triol (p. 1953 s) on hydrogenation in alcohol in the

---

* See also Δ^5-pregnen-3-ol-20-on-21-al with unknown configuration at C-3 (p. 2379 s).

presence of a cobalt-nickel catalyst under a small over-pressure, until 2 mols. of hydrogen have been absorbed (1937b CIBA; 1938 CIBA PHARMACEUTICAL PRODUCTS).

On treatment with ethylmercaptan and gaseous dry HCl at 16⁰ it gives 3β-hydroxy-Δ⁵-ætiocholenic acid and the diethyl mercaptal (below) (1941 Schindler). On refluxing with o-phenylenediamine in abs. alcohol for 1 hr. it gives the corresponding **quinoxaline-compound** $C_{27}H_{34}ON_2$, prisms (ether-pentane) changing to needles at 200⁰, m. 229–231⁰, subliming at ca. 240⁰/₀.₀₂ (1939 Reich).

Dioxime $C_{21}H_{32}O_3N_2$. Slightly pink, crystalline powder (alcohol-benzene), changing to needles at ca. 225⁰, m. 285–290⁰. On heating under 0.01 mm. pressure it gives a sublimate at 210⁰ (bath temperature) which melts un-sharply on heating to 300⁰. With nickel acetate in aqueous alcohol it gives a light brown precipitate not melting on heating to 300⁰ (1939 Reich).

Dianil $C_{33}H_{40}ON_2$. Colourless crystals (ether-pentane), m. 85⁰ to ca. 90⁰. — **Fmn.** From the above pregnenolonal on heating with 2 mols. aniline in alcohol on the water-bath for 10 min. (1939 Reich).

21-(p-Dimethylaminoanil) N-oxide, 21-N-(p-Dimethylaminophenyl)-isoxime $C_{29}H_{40}O_3N_2$ [side chain, CO·CH:N(O)·C_6H_4·NMe_2]. Orange-yellow needles (aqueous alcohol), m. 133–4⁰, which contain ca. 1 H_2O after drying in a high vacuum at 85⁰ for 1 hr. (1939 Reich). — **Fmn.** and **Rns.** See under those of the above pregnenolonal.

Dimethyl acetal $C_{23}H_{36}O_4$. Colourless crystals (ether-pentane), m. 112–3⁰ cor.; $[\alpha]_D^{22} + 39^0$, $[\alpha]_{5461}^{22} + 52^0$ (methanol) (1939 Reich). Absorption max. in alcohol (from the graph) at 305 mμ (log ε 1.86) (1941 Schindler). — **Fmn.** From the above pregnenolonal and methanolic HCl on keeping at room temperature for 24 hrs. or on refluxing for 1 hr. (1939 Reich). — **Rns.** On treatment with hydrochloric acid in acetic acid it is slowly hydrolysed at room temperature, but decomposed on heating on the water-bath for 1 hr. (1939 Reich). For other reactions, see those of the above pregnenolonal.

Diethyl mercaptal $C_{25}H_{40}O_2S_2$. Colourless leaflets (ether-pentane), m. 124⁰ to 125⁰ cor. (1941 Schindler); $[\alpha]_D^{21} + 138^0$ (acetone) (1941 Schindler; but cf. 1946 Barton). — **Fmn.** From the above pregnenolonal on treatment with ethylmercaptan and dry gaseous HCl; yield, 27% (1941 Schindler). — **Rns.** For oxidation, see under that of the above pregnenolonal. Gives the acetate (below) on treatment with acetic anhydride in pyridine at 20⁰ for 16 hrs. (1941 Schindler).

Acetate diethyl mercaptal $C_{27}H_{42}O_3S_2$. Crystals (ether-pentane), m. 130–2⁰ cor. (1941 Schindler); $[\alpha]_D^{17} + 150^0$ (acetone) (1941 Schindler; but cf. 1946 Barton). — **Fmn.** See the preceding compound.

A **Δ⁵-pregnen-3-ol-20-on-21-al**, the C-3 configuration of which is not indi-cated, is obtained from a Δ⁵-pregnen-3-ol-20-one on heating with SeO_2 in alcohol in a sealed tube at 120–130⁰ (yield, ca. 20%). Its *dioxime* forms crystals (alcohol), m. 232⁰ (1936 SCHERING A.-G.).

Pregnan-3β-ol-20-on-21-al, (3β-Hydroxyætiocholan-17β-yl)-glyoxal $C_{21}H_{32}O_3$.

Yellowish crystals with 1 H_2O (after drying in a high vacuum at room temperature for 20 min.), sintering at 127°, m. 143° cor.; $[\alpha]_D$ +103° ± 3° (pyridine); after drying in a high vacuum at 90° for 15 hrs. it changes to crystals with $^1/_2$ H_2O, almost without change of m. p. and $[\alpha]_D$ value.

Fmn. From N-(3β-hydroxy-20-ketopregnan-21-yl)-pyridinium chloride (page 2280 s) on treatment with p-nitrosodimethylaniline and 1 N NaOH in alcohol in a freezing mixture, followed by hydrolysis of the resulting 21-N-(p-dimethyl-aminophenyl)-isoxime (not described in detail) with 2 N HCl in ether at room temperature; crude yield, 84%. — **Rns.** Gives 3-ketoætiocholanic acid on oxidation with CrO_3 in glacial acetic acid, and 3β-hydroxyætiocholanic acid on oxidation with periodic acid in aqueous alcoholic H_2SO_4 (1943 b Ruzicka).

Dioxime $C_{21}H_{34}O_3N_2$. Almost colourless leaflets (aqueous pyridine), m. 217° to 223° cor. dec. (1943 b Ruzicka).

Dimethyl acetal $\dot{C}_{23}H_{38}O_4$. Colourless crystals (ligroin), m. 126–9° cor.; $[\alpha]_D$ +132° ± 10° (chloroform). — **Fmn.** From the above pregnanolonal on refluxing with 1% methanolic HCl for 1 hr. (1943 b Ruzicka).

A **pregnan-3-ol-20-on-21-al**, the C-3 configuration of which is not indicated, is obtained from a pregnan-3-ol-20-one on refluxing with SeO_2 in isoamyl alcohol for $3^1/_2$ hrs.; its acetate is obtained from the acetate of the same hydroxyketone on warming with sodium nitrite in strong acetic acid. Its *dioxime* forms leaflets which sinter at ca. 227° and melt on heating to 240° with decomposition (1936 Schering A.-G.).

Allopregnan-3β-ol-20-on-21-al, (3β-Hydroxyandrostan-17β-yl)-glyoxal $C_{21}H_{32}O_3$. Yellowish crystals with 1 H_2O (after drying in a high vacuum at room temperature for 20 min.), sintering at 136°, m. 155° cor.; $[\alpha]_D$ +92.7° ± 3° (pyridine); after drying in a high vacuum at 90° for 24 hrs. it changes to crystals with $^1/_2$ H_2O with nearly unchanged m.p. and with $[\alpha]_D$ +87.5° ± 3° (pyridine). — **Fmn.** From N-(3β-hydroxy-20-keto-allopregnan-21-yl)-pyridinium chloride (p. 2282 s) on treatment with p-nitroso-dimethylaniline and 1 N NaOH in alcohol in a freezing mixture, followed by hydrolysis of the resulting 21-N-(p-dimethylaminophenyl)-isoxime (below) with 2 N HCl in ether at room temperature; crude over-all yield, 62%. — **Rns.** Gives 3-ketoætioallocholanic acid on oxidation with CrO_3 in glacial acetic acid, and 3β-hydroxyætioallocholanic acid on oxidation with periodic acid in aqueous alcoholic H_2SO_4 (1943 b Ruzicka).

Dioxime $C_{21}H_{34}O_3N_2$. Crystals (aqueous pyridine), m. 246–9° cor.dec. (1943 b Ruzicka).

References, p. 2382 s

21-(p-Dimethylaminoanil) -N-oxide, *21-N-(p-Dimethylaminophenyl)-isoxime*
$C_{29}H_{42}O_3N_2$ (not analysed) [side chain, $CO \cdot CH:N(O) \cdot C_6H_4 \cdot NMe_2$]. Orange
crystals (alcohol), m. ca. 119–120° cor. (1943b Ruzicka). — **Fmn.** and **Rns.**
See above.

Dimethyl acetal $C_{23}H_{38}O_4$. Colourless needles (ether-petroleum ether), m.
113–5° cor.; $[\alpha]_D$ +112° (chloroform). — **Fmn.** From the above allopreg-
nanolonal on refluxing with 1% methanolic HCl for 1 hr. (1943b Ruzicka).

An **allopregnan-3-ol-20-on-21-al**, the C-3 configuration of which is not indicated, is obtained in small yield from an allopregnan-3-ol-20-one on refluxing
with SeO_2 in alcohol for $2^1/_2$ hrs.; with phenylhydrazine it gives a condensation product [yellow crystals (alcohol), m. 257–260°] (1936 SCHERING A.-G.).

Testalolone $C_{21}H_{32}O_3$ (p. 117). Is not identical with pregnan- or allopregnan-
3β-ol-20-on-21-al (1943a, b Ruzicka) as was assumed by Hirano (1936) and
Marker (1938). — Crystals, m. 258–264° dec.; $[\alpha]_D^{24}$ —67.5° (pyridine) (1936
Hirano). Platelets (methanol), sintering at 257°, m. 268° cor.; $[\alpha]_D$ in pyridine
—48° after 30 min., —39° after 48 hrs. (1943a Ruzicka). Soluble in pyridine,
glacial acetic acid, and chloroform when hot, insoluble in hot alcohol, benzene,
ether, water, mineral acids, and alkali (1936 Hirano). Gives no coloration
with tetranitromethane. Is slowly reduced by ammoniacal silver solution
(1943a Ruzicka). — Is physiologically inactive (1936 Hirano).

Occ. In boar testes (1936 Hirano; 1943a Ruzicka). — **Rns.** On oxidation
with CrO_3 in glacial acetic acid it gives crystals (hexane-alcohol), m. 246°
to 249° dec. On heating testalolone above the m. p. (evolution of gas) it
resolidifies, begins to remelt at ca. 273° and is completely molten at 277°
giving a crystalline powder, m. 262–272° dec., and crystals (dil. alcohol),
sintering at 165°, m. 171–173.5°. On treatment of testalolone with semicarbazide it gives a **compound** $C_{22}H_{33}O_2N_3$, crystals (aqueous pyridine), dec.
237.5–238.5°; the benzoate reacts similarly to give a **compound** $C_{29}H_{37}O_3N_3$,
crystals, sintering at 216°, m. 226–235.5° dec. (1936 Hirano).

Digitonide $C_{21}H_{32}O_3 + C_{56}H_{92}O_{29} + 6 H_2O$. Crystals (dil. alcohol), m. ca.
250°, foaming at 264° (1936 Hirano).

Dioxime $C_{21}H_{34}O_3N_2$. Leaflets (alcohol), m. 238–9° cor. dec. (1943a Ruzicka;
see also 1936 Hirano). — **Rns.** On heating with acetic anhydride on the
water-bath it gives crystals (dil. alcohol), m. 153–4° dec. (1936 Hirano).

Benzoate $C_{28}H_{36}O_4$. Leaflets (alcohol), m. 218–224° dec. (1936 Hirano). —
Rns. See above.

24-Phenylcholane-3α.7α.12α-triol-23.24-dione, *"Bisnorcholyl benzoyl ketone"*

$C_{30}H_{42}O_5$. — **Fmn.** The triacetate is
obtained from the 3.7.12-triacetate
of 24-phenylcholane-3α.7α.12α.23α(?)-
tetrol-24-one (p. 2377 s) on oxidation
with copper sulphate in aqueous pyri-

References, p. 2382 s

dine. — **Rns.** On acetylation of the triacetate it gives the corresponding 23-enol acetate (not described in detail). The triacetate reacts with o-phenylenediamine to give the **quinoxaline derivative** $C_{42}H_{52}O_6N_2$ (no analysis given), m. 217–218.5° (1944 Jacobsen).

7.12-Diacetate $C_{34}H_{46}O_7$, m. 201–203.5°; $[\alpha]_D^{20}$ +80° (chloroform) (1944 Jacobsen).

Triacetate $C_{36}H_{48}O_8$, m. 166–9° (after drying, 161.5–166°); $[\alpha]_D^{20}$ +92° (chloroform) (1944 Jacobsen).

1936 Hirano, *C. A.* **1937** 3125; *Ch. Ztbl.* **1937** I 1451.

SCHERING A.-G. (Butenandt), *Ger. Pat.* 709617 (issued 1941); *C. A.* **1943** 3454; *Ch. Ztbl.* **1942** I 642.

1937 a CIBA*, *Swiss Pat.* 212193 (issued 1941); *C. A.* **1942** 3636.

b CIBA*, *Swiss Pat.* 212338 (issued 1941); *C. A.* **1942** 3636; *Brit. Pat.* 517288 (1938, issued 1940); *Ch. Ztbl.* **1940** II 1328.

c CIBA*, *Swiss Pat.* 213049 (issued 1941); *C. A.* **1942** 4674; *Ch. Ztbl.* **1942** I 385; *Fr. Pat.* 840514 (1938, issued 1939); *C. A.* **1939** 8627; *Ch. Ztbl.* **1939** II 1125, 1126.

d CIBA*, *Swiss Pat.* 220924 (issued 1942); *C. A.* **1949** 2377; *Ch. Ztbl.* **1943** I 1190.

e CIBA*, *Swiss Pat.* 223205 (issued 1942); *C. A.* **1949** 1916; *Ch. Ztbl.* **1943** I 2321.

f CIBA*, *Swiss Pat.* 225102 (issued 1943); *C. A.* **1949** 1916.

SCHERING A.-G. (Logemann), *Ger. Pat.* 737023 (issued 1943); *C. A.* **1944** 3781; *Fr. Pat.* 843082 (1938, issued 1939); *C. A.* **1940** 6412; *Ch. Ztbl.* **1940** I 759; SCHERING CORP. (Logemann), *U.S. Pat.* 2321690 (1938, issued 1943); *C. A.* **1943** 6824; *Ch. Ztbl.* **1945** II 687.

1938 a CIBA* (Miescher, Wettstein, Fischer), *U.S. Pat.* 2188914 (issued 1940); *C. A.* **1940** 3882; *Fr. Pat.* 845802 (issued 1939); *C. A.* **1941** 1186; *Ch. Ztbl.* **1940** I 2827.

b CIBA*, *Fr. Pat.* 840514 (issued 1939); *C. A.* **1939** 8627; *Ch. Ztbl.* **1939** II 1125, 1126.

CIBA PHARMACEUTICAL PRODUCTS (Miescher, Wettstein), *U.S. Pat.* 2239012 (issued 1941); *C. A.* **1941** 4921.

Marker, *J. Am. Ch. Soc.* **60** 1725, 1728.

1939 Reich, Reichstein, *Helv.* **22** 1124, 1129-1133; cf. also N. V. ORGANON, *Fr. Pat.* 864315 (1940, issued 1941); *Ch. Ztbl.* **1942** I 642.

1941 Schindler, Frey, Reichstein, *Helv.* **24** 360, 365, 373, 374.

1943 a Ruzicka, Prelog, *Helv.* **26** 975, 980, 992.

b Ruzicka, Prelog, Wieland, *Helv.* **26** 2050.

1944 Jacobsen, *J. Am. Ch. Soc.* **66** 662.

1946 Barton, *J. Ch. Soc.* **1946** 1116, 1117.

---

* Gesellschaft für chemische Industrie in Basel; Société pour l'industrie chimique à Bâle.

II. OXOSTEROIDS (CO IN THE RING SYSTEM)

A. Monooxo-Compounds

I. COMPOUNDS WITHOUT OTHER FUNCTIONAL GROUPS

1. Compounds Without Side Chain

4-Keto-1.2-cyclopentano-1.2.3.4-tetrahydrophenanthrene $C_{17}H_{16}O$ (I, below). Crystals (alcohol), m. 119–120⁰. — **Fmn.** From 3-(β-naphthyl)-cyclopentanone-2-acetic acid semicarbazone on Wolff-Kishner reduction, followed by conversion of the resulting crude 2-(β-naphthyl)-cyclopentane-1-acetic acid into the acid chloride with thionyl chloride and cyclization in carbon disulphide with $AlCl_3$ at 0⁰ (yield, 23%) or with $SnCl_4$, first at —10⁰, then at room temperature (yield, 52%) (1946a Butenandt).

Rns. (See also the reaction scheme, below). Its potassium enolate (obtained by boiling with potassium tert. butoxide in tert.-butyl alcohol) reacts with methyl iodide to give the lower-melting 3-methyl-4-keto-1.2-cyclopentano-1.2.3.4-tetrahydrophenanthrene (p. 2387 s). The ketone reacts with methylmagnesium iodide to give 4-methyl-4-hydroxy-1.2-cyclopentano-1.2.3.4-tetrahydrophenanthrene [not purified; partial formula (III), below], which affords 4-methyl-1.2-cyclopentenophenanthrene (p. 1387 s) on heating with

Scheme for conversion of 4-keto-1.2-cyclopentano-1.2.3.4-tetrahydrophenanthrene into methyl- and dimethyl-1.2-cyclopentenophenanthrenes
(according to 1946a Butenandt)

* The α- and β-configurations are those arbitrarily assigned by Butenandt.

References, p. 2386 s

platinum-charcoal at 310°. Reacts with dimethyl oxalate and sodium methoxide in boiling benzene to give methyl 4-keto-1.2-cyclopentano-1.2.3.4-tetra-
hydro-3-phenanthrylglyoxylate (partial formula II, p. 2383 s). Gives no semicarbazone under the usual conditions (1946a Butenandt).

3-Keto-1.2-cyclopentano-1.2.3.9.10.11-hexahydrophenanthrene $C_{17}H_{18}O$ (p. 118).

Crystals (alcohol), m. 171° (1941 Robinson). — **Fmn.** The
condensation of sodio-α-tetralone with 1-acetyl-Δ^1-cyclo-
pentene (cf. p. 118) is best carried out by refluxing in dry
ether for ca. 12 hrs.; yield, 33% (1941 Robinson). — **Rns.**
Gives 3-hydroxy-1.2-cyclopentano-1.2.3.9.10.11-hexahydro-
phenanthrene (p. 1494 s) on boiling with aluminium isopropoxide in isopropyl
alcohol with removal (by slow distillation) of the acetone formed (1941 Robinson). On treatment with methylmagnesium iodide in abs. ether, followed
by hydrolysis with cold dil. HCl, it gives a methylcarbinol which sponta-
neously dehydrates to give 3-methyl-1.2-cyclopentano-1.9.10.11-tetrahydro-
phenanthrene (p. 1387 s) (1946a Butenandt).

Semicarbazone $C_{18}H_{21}ON_3$, m. 235° dec. (not quite pure); very sparingly
soluble in alcohol (1946b Butenandt).

5′-Keto-1.2-cyclopentenophenanthrene $C_{17}H_{12}O$ (p. 118). Colourless needles

(alcohol-acetone) (1937 Bachmann), m. 188.6–189.4° cor. (1943
Riegel); sublimes at 200° and 0.3 mm. pressure (1937 Bachmann). — **Fmn.** From β-(2-phenanthryl)-propionyl chloride
on treatment with $AlCl_3$ in nitrobenzene, first at room temperature, then at 80°; yield, 92% (1943 Riegel). From 1.2-cyclopentenophen-
anthrene (page 1382 s) on oxidation with CrO_3 in acetic acid at room temperature; yield, ca. 50% (1938 Hoch; cf. also 1946b Butenandt).

Oxime $C_{17}H_{13}ON$. Needles (alcohol-dioxane), m. 235–6° cor. dec. (1943
Riegel).

4′-Keto-1.2-($\Delta^{1(5')}$-cyclopenteno)-1.2.3.4-tetrahydrophenanthrene $C_{17}H_{14}O$.

Cream-coloured plates (benzene or acetone), m. 185–185.5°;
distils at ca. 200° and 0.2 mm. pressure (1942 Wilds).
Absorption max. in abs. alcohol at 218.5 (4.32), 238 (4.120),
245 (4.125), [255 (4.27)], 266 (4.56), 276 (4.63), 316 (4.47),
and [360 (3.69)] mμ (log ε) (1947 Wilds). Gives an orange-yellow solution
with conc. H_2SO_4 (1942 Wilds). — **Fmn.** From (1-keto-1.2.3.4-tetrahydro-
2-phenanthryl)-acetone and from ethyl α-(1-keto-1.2.3.4-tetrahydro-2-phen-
anthryl)-acetoacetate on refluxing with 5% aqueous KOH in an atmosphere
of nitrogen for 6 hrs.; yields, 90% and 84%, resp. (1942 Wilds). — **Rns.**
Gives 1.2-cyclopentenophenanthrene on Clemmensen reduction, followed by
dehydrogenation with palladium-charcoal at 300–320°. On partial hydrogenation in dioxane in the presence of palladium-charcoal at room temperature, a mixture of two stereoisomeric forms of 4′-keto-1.2-cyclopentano-1.2.3.4-
tetrahydrophenanthrene (p. 2385 s) is obtained (1942 Wilds).

Oxime $C_{17}H_{15}ON$. Pale yellow needles (alcohol), m. 247–250° (1942 Wilds).

4'-Keto-1.2-cyclopentano-1.2.3.4-tetrahydrophenanthrene $C_{17}H_{18}O$. Two diastereoisomers exist: Isomer A, plates (methanol), m. 115° to 116°; Isomer B, needles (methanol), m. 146–147.5°. The mixed isomers give a pale yellow coloration with conc. H_2SO_4. — **Fmn.** A mixture of the two isomers is obtained in 91% yield from the preceding ketone on hydrogenation (1 mol. H_2) in dioxane in the presence of palladium-charcoal at room temperature and atmospheric pressure; the mixture can be separated by fractional crystallization from methanol, isomer B being the less soluble (1942 Wilds).

Oxime (mixture) $C_{17}H_{17}ON$, m. 155–160°; recrystallization from alcohol raises the m. p. to 163–168° (1942 Wilds).

Semicarbazone (mixture) $C_{18}H_{19}ON_3$, m. 240–243° (1942 Wilds).

3'-Keto-1.2-cyclopentenophenanthrene $C_{17}H_{12}O$ (p. 118). Colourless plates (acetone-methanol), m. 196–7°. — **Fmn.** From the dimethyl ester of phenanthrene-2-carboxylic-1-(β-propionic) acid on cyclization by warming with sodium methoxide in benzene, followed by hydrolysis and decarboxylation by refluxing with hydrochloric-acetic acids (1943 Bachmann).

3'-Keto-1.2-cyclopenteno-3.4-dihydrophenanthrene $C_{17}H_{14}O$ (p. 118). Colourless prisms (by sublimation at 170–200° and 0.01 mm. pressure), m. 214–216° (1943 Bachmann); pale yellow plates (chloroform-acetone), m. 222.5–223.3° cor. (1945 Johnson). Absorption max. in abs. alcohol at 219 (4.58), 241 (3.82), [250 (3.86), 260 (4.23)], 270 (4.58), 280 (4.67), 324 (4.15), 335 (4.19), and [360 (3.95)] mμ (log ε) (1947 Wilds).

Fmn. From methyl 7-(α-naphthyl)-4-ketoheptanoate on treatment with alcohol-free sodium ethoxide in ether, followed by decomposition with water, precipitation with dil. H_2SO_4, and boiling of the resulting syrupy material with P_2O_5 in moist benzene; yield, 14.5% (1939 Robinson). From the dimethyl ester of 3.4-dihydrophenanthrene-2-carboxylic-1-(β-propionic) acid on refluxing with sodium methoxide in benzene for 11 hrs., followed by refluxing of the resulting methyl 3'-keto-1.2-cyclopenteno-3.4-dihydrophenanthrene-4'-carboxylate (yield, 89%) with hydrochloric-acetic acid for 3 hrs. (yield, 98%) (1943 Bachmann). From β-carbethoxy-β-(3.4-dihydro-1-phenanthryl)-propionic acid on cyclization by refluxing with anhydrous $ZnCl_2$ in acetic anhydride-glacial acetic acid for 4.5 hrs., followed by addition of hydrochloric acid and continued refluxing for 1 hr.; yield, 50%. From β-(3.4-dihydro-1-phenanthryl)-propionic acid on cyclization by refluxing with anhydrous $ZnCl_2$ in acetic anhydride - glacial acetic acid (yield, 53%), or from the lactone of β-(1-hydroxy-1.2.3.4-tetrahydro-1-phenanthryl)-propionic acid on similar cyclization or using P_2O_5 in boiling benzene (yield, 32%) (1945 Johnson).

References, p. 2386 s

Rns. May be converted into 1.2-cyclopentenophenanthrene (p. 1382 s) by Clemmensen reduction, followed by dehydrogenation with palladium-charcoal at 300–320⁰ (1945 Johnson).

Oxime $C_{17}H_{15}ON$. Almost colourless needles (methanol), m. 253–4⁰ dec.; m. 254.5–255⁰ in an evacuated tube (1945 Johnson).

Semicarbazone $C_{18}H_{17}ON_3$. Needles, m. 317–8⁰ dec. (evacuated tube); very sparingly soluble in most solvents (1945 Johnson).

2.4-Dinitrophenylhydrazone $C_{23}H_{18}O_4N_4$ (no analysis given), crimson-coloured (1939 Robinson).

3'-Keto-1.2-cyclopentano-1.2.3.4-tetrahydrophenanthrene $C_{17}H_{16}O$.

dl-cis-Form, dl-cis-Norequilenan-17-one (1945 Birch). Prisms (methanol or aqueous ethanol), m. 111–2⁰ (1941 Koebner); colourless plates (by distillation at 145–175⁰ and 0.01 mm. pressure), m. 112–3⁰ (1943 Bachmann). — **Fmn.** From 4.3'-diketo-1.2-cyclopentano-1.2.3.4-tetrahydrophenanthrene (p. 2872 s) on hydrogenation in alcohol at room temperature in the presence of platinized charcoal and $PdCl_2$ (1941 Koebner). From methyl 3'-keto-1.2-cyclopentano-1.2.3.4-tetrahydro-phenanthrene-4'-carboxylate on refluxing with hydrochloric-acetic acids; yield, 92% (1943 Bachmann).

Rns. Gives 3'-methyl-3'-hydroxy-1.2-cyclopentano-1.2.3.4-tetrahydro-phenanthrene (p. 1500 s) on treatment with methylmagnesium iodide in ether (1941 Koebner). May be converted into the 2-methyl homologue (cis-isomer; dl-3-desoxyisoequilenin, p. 2391 s) by treatment with ethyl formate and sodium in benzene, followed by conversion of the resulting 4'-hydroxy-methylene-derivative (p. 2510 s) by heating with N-methylaniline in toluene into the corresponding 4'-(N-methyl-anilinomethylene)-compound (p. 2507 s), methylation of the latter compound by heating with methyl iodide and sodamide in benzene, and hydrolysis by refluxing first with 10% HCl and then with 5% NaOH (1945 Birch). Gives the 4'-piperonylidene derivative (p. 2515 s) on heating with piperonal and sodium ethoxide in alcohol (1941 Koebner).

2.4-Dinitrophenylhydrazone $C_{23}H_{20}O_4N_4$. Yellow needles (acetic acid), m. 255⁰ to 256⁰ (1941 Koebner).

1937 Bachmann, Kloetzel, *J. Am. Ch. Soc.* **59** 2207, 2211, 2212.
1938 Hoch, *Compt. rend.* **207** 921.
1939 Robinson, Thompson, *J. Ch. Soc.* **1939** 1739.
1941 Koebner, Robinson, Cardwell, *J. Ch. Soc.* **1941** 566, 573, 575.
  Robinson, Slater, *J. Ch. Soc.* **1941** 376, 379.
1942 Wilds, *J. Am. Ch. Soc.* **64** 1421, 1425; **66** 2135 (1944).
1943 Bachmann, Gregg, Pratt, *J. Am. Ch. Soc.* **65** 2314, 2316.
  Riegel, Gold, Kubico, *J. Am. Ch. Soc.* **65** 1772.
1945 Birch, Jaeger, Robinson, *J. Ch. Soc.* **1945** 582, 585.
  Johnson, Petersen, *J. Am. Ch. Soc.* **67** 1366.
1946 a Butenandt, Dannenberg, v. Dresler, *Z. Naturforsch.* **1** 151.
  b Butenandt, Dannenberg, v. Dresler, *Z. Naturforsch.* **1** 222.
1947 Wilds, Beck, Close, Djerassi, Johnson, Johnson, Shunk, *J. Am. Ch. Soc.* **69** 1985, 1993.

2. Monoketo-steroids (CO in the Ring System) with One Aliphatic Side Chain

3-Methyl-4-keto-1.2-cyclopentano-1.2.3.4-tetrahydrophenanthrene $C_{18}H_{18}O$.

Two diastereoisomers are known, differing by the configuration at the C-atom marked by an asterisk.

Lower-melting isomer. Crystals (alcohol), m. 85–6°. — **Fmn.** (See also the reaction scheme on p. 2383 s). From 4-keto-1.2-cyclopentano-1.2.3.4-tetrahydrophenanthrene on boiling with potassium tert.-butoxide in tert.-butyl alcohol, and treatment of the resulting potassium enolate with methyl iodide; yield, 46%. From the higher-melting isomer (below) on boiling with aqueous methanolic KOH. — **Rns.** (See also the scheme on p. 2383 s). Gives 3-methyl-1.2-cyclopentenophenanthrene (p. 1387 s) on Clemmensen reduction, followed by dehydrogenation with platinum charcoal at 300°. On treatment with methyllithium in ether it gives 3.4-dimethyl-4-hydroxy-1.2-cyclopentano-1.2.3.4-tetrahydrophenanthrene (p. 1501 s). Forms no semicarbazone under the usual conditions (1946a Butenandt).

Higher-melting isomer. Needles (methanol-acetone), m. 117–8°. — **Fmn,** From methyl 4-keto-1.2-cyclopentano-1.2.3.4-tetrahydrophenanthrene-3-carboxylate (obtained from 4-keto-1.2-cyclopentano-1.2.3.4-tetrahydrophenanthrene; see the scheme on p. 2383 s) on treatment with sodium methoxide, followed by boiling of the resulting sodium enolate with methyl iodide in benzene and then by acid hydrolysis; yield, 51%. — **Rns.** It is isomerized to the above lower-melting isomer by boiling with aqueous methanolic KOH (1946a Butenandt).

6-Methyl-3-keto-1.2-cyclopentano-1.2.3.9.10.11-hexahydrophenanthrene

$C_{18}H_{20}O$. Crystals (chloroform-benzene), m. 205–6°. Absorption max. in chloroform (from the graph) at 297 mμ (ε 18800). — **Fmn.** From 7-methyl-1-tetralone on condensation with 1-acetylcyclopentene by means of sodamide in ether in an atmosphere of hydrogen, followed by acidification with dil. H_2SO_4; yield, 35%. — **Rns.** Gives 6-methyl-1.2-cyclopentano-1.2.3.9.10.11-hexahydrophenanthrene (p. 1388 s) on Wolff-Kishner reduction of the semicarbazone (1946b Butenandt).

Semicarbazone $C_{19}H_{23}ON_3$, m. 245–8° dec.; very sparingly soluble in the usual solvents (1946b Butenandt).

3'-Methyl-5'-keto-1.2-cyclopentenophenanthrene $C_{18}H_{14}O$ (p. 118).

Oxime $C_{18}H_{15}ON$. Exists in two (syn- and anti-) forms: "α-Form", colourless plates (alcohol), m. 169–171° cor. dec. (bath preheated to 160°). — "β-Form", yellow needles (dil. alcohol), yellow prisms (methanol), m. 165–170° cor. dec. (bath preheated to 160°); more soluble in alcohol than the isomer (1943 Riegel).

1- Methyl- 5'- keto -1.2 - cyclopentano -1.2.3.9.10.11-hexahydrophenanthrene (I)

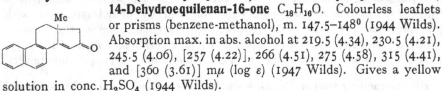

or   **2 - Methyl - 3' - keto - 1.2 - cyclopentano -
1.2.3.9.10.11- hexahydrophenanthrene** (II),
$C_{18}H_{20}O$ (not analysed). Oil, b. $160-180^0/_{0\cdot05}$.
Fmn. From 1-vinyl-3.4-dihydronaphthalene
and 1-methyl-Δ^1-cyclopenten-5-one on heat-
ing at 160^0 under nitrogen for 22 hrs. (1937, 1938 Bockemüller).

2.4-Dinitrophenylhydrazone $C_{24}H_{24}O_4N_4$, yellow crystals, m. 186–9^0 with
darkening (1937, 1938 Bockemüller).

**2 - Methyl - 4'-keto -1.2 - ($\Delta^{1'(5')}$- cyclopenteno) -1.2.3.4-tetrahydrophenanthrene,
14-Dehydroequilenan-16-one** $C_{18}H_{16}O$. Colourless leaflets
or prisms (benzene-methanol), m. 147.5–148^0 (1944 Wilds).
Absorption max. in abs. alcohol at 219.5 (4.34), 230.5 (4.21),
245.5 (4.06), [257 (4.22)], 266 (4.51), 275 (4.58), 315 (4.41),
and [360 (3.61)] mμ (log ε) (1947 Wilds). Gives a yellow
solution in conc. H_2SO_4 (1944 Wilds).

Fmn. From (2-methyl-1-keto-1.2.3.4-tetrahydro-2-phenanthryl)-acetone on
refluxing with 5% aqueous KOH; yield, 96% (1944 Wilds).

Rns. On hydrogenation in dioxane in the presence of palladium-charcoal
at room temperature and atmospheric pressure (18 hrs.), it gives mainly
trans-equilenan-16-one (below) (1944, 1946 Wilds); on hydrogenation in the
presence of a palladium-carbon catalyst and potassium hydroxide (or potassium tert.-butoxide) in abs. alcohol (or dry tert.-butyl alcohol, resp.) at room
temperature and atmospheric pressure, it gives cis-equilenan-16-one (p. 2389 s)
(1950 Wilds). With butyl nitrite and potassium tert.-butoxide in tert.-butyl
alcohol it gives 14-dehydroequilenan-16.17-dione 17-oxime (p. 2875 s). Condensation with dimethyl oxalate in benzene in the presence of sodium methoxide affords methyl 14-dehydroequilenan-16-one-17-glyoxylate (1944 Wilds).

Oxime $C_{18}H_{17}ON$. Crystals (alcohol), m. 197–205^0 cor. dec. (1944 Wilds).

**2-Methyl-4'-keto-1.2-cyclopentano-1.2.3.4-tetrahydrophenanthrene, Equilenan-
16-one** $C_{18}H_{18}O$. Exists in two racemic
diastereoisomeric forms, the trans-form
and the cis-form; for the configuration,
see 1946, 1950 Wilds.
      trans - Equilenan - 16 - one, *trans-
16-Equilenone* $C_{18}H_{18}O$ (IIIa). Leaflets (methanol), m. 168.5–169^0 cor. Gives
no coloration with conc. H_2SO_4 (1944 Wilds). — **Fmn.** From the preceding
14-dehydro-ketone on hydrogenation in dioxane in the presence of a palladium-charcoal catalyst at room temperature and atmospheric pressure
(18 hrs.); yield, ca. 60% (1944 Wilds). — **Rns.** It gives trans-dl-equilenan
(β-equilenan, p. 1386 s) on Clemmensen reduction, or on Wolff-Kishner
reduction of its semicarbazone (1946 Wilds).

Oxime $C_{18}H_{19}ON$, plates (benzene-methanol) m. 211–3° cor. (vac.) (1944 Wilds).

Semicarbazone $C_{19}H_{21}ON_3$, leaflets (tert.-butyl alcohol), m. 251.5–253° cor. dec. (1946 Wilds).

cis-Equilenan-16-one, *cis-16-Equilenone* $C_{18}H_{18}O$ (IIIb). Needles (methanol), m. 94.5–95° cor. — **Fmn.** From the 14-dehydro-ketone (p. 2388 s) on hydrogenation with a palladium-charcoal catalyst at room temperature and atmospheric pressure in the presence of potassium hydroxide in abs. alcohol (yield, 89%) or in the presence of potassium tert.-butoxide in dry tert.-butyl alcohol (yield, 78%). — **Rns.** On Clemmensen reduction it gives cis-dl-equilenan (α-equilenan, p. 1386 s) (1950 Wilds).

Oxime $C_{18}H_{19}ON$, needles (alcohol), m. 165–166.5° cor. (1950 Wilds).

2-Methyl-3'-keto-1.2-cyclopentano-1.2.3.4-tetrahydro-phenanthrene, Equilenan-17-one* $C_{18}H_{18}O$.

Exists in two racemic diastereoisomeric forms, the trans-form, **3-desoxyequilenin** (IVa) and the cis-form, **3-desoxyisoequilenin** (IVb); the trans-isomer was originally designated β-form, the cis-isomer α-form (1940, 1951 Bachmann). For d-3-desoxy-equilenin, see p. 2391 s; d- and l-3-desoxyisoequilenin have been described after the closing date of this volume by Bachmann (1950a, 1951).

dl-3-Desoxyequilenin, dl-trans-Equilenan-17-one* $C_{18}H_{18}O$ (IVa), formerly called β-*dl-17-equilenone* (1940, 1951 Bachmann). Colourless plates (acetone-alcohol), m. 188.5–189.5° (vac.) (1940 Bachmann). — Shows no œstrogenic activity in rats (1940 Bachmann). — **Fmn.** From the dimethyl ester of the β-form of 2-methyl-1.2.3.4-tetrahydrophenanthrene-2-carboxylic-1-(β-propionic) acid on heating with sodium methoxide in benzene, followed by refluxing of the resulting methyl dl-trans-equilenan-17-one-16-carboxylate ("β-dl-16-carbomethoxy-17-equilenone") with conc. HCl and strong acetic acid in an atmosphere of nitrogen for 1 hr. (1940 Bachmann). From dl-1.2.3.4-tetrahydro-3-desoxyequilenin (p. 2392 s) on heating with sulphur at 210° for 2 hrs.; yield, 49% (1944 Bachmann).

Rns. Clemmensen reduction, or Wolff-Kishner reduction of the semicarbazone, affords trans-dl-equilenan (β-dl-equilenan, p. 1386 s) (1946 Wilds). On heating with palladium-charcoal at 250° (bath temperature) in an atmosphere of nitrogen for 8 min., it is almost quantitatively epimerized to dl-3-desoxyisoequilenin (p. 2391 s); on heating at 350° (bath temperature) for 20 min., 2-methyl-1-ethylphenanthrene is formed (1950a Bachmann).

Addition compound with 1.3.5-trinitrobenzene (1:1), $C_{24}H_{21}O_7N_3$, yellow needles (alcohol), m. 153–4° cor. (1950a Bachmann).

---

* For the name equilenan and its numbering, see p. 1385 s; cf. p. 1348 s.

3-Desoxyequilenin and 3-Desoxy-1.2.3.4-tetrahydroequilenin and Their 14-Epimers

(Reactions marked with an asterisk have been carried out with the d-compounds, the others with the dl-compounds)

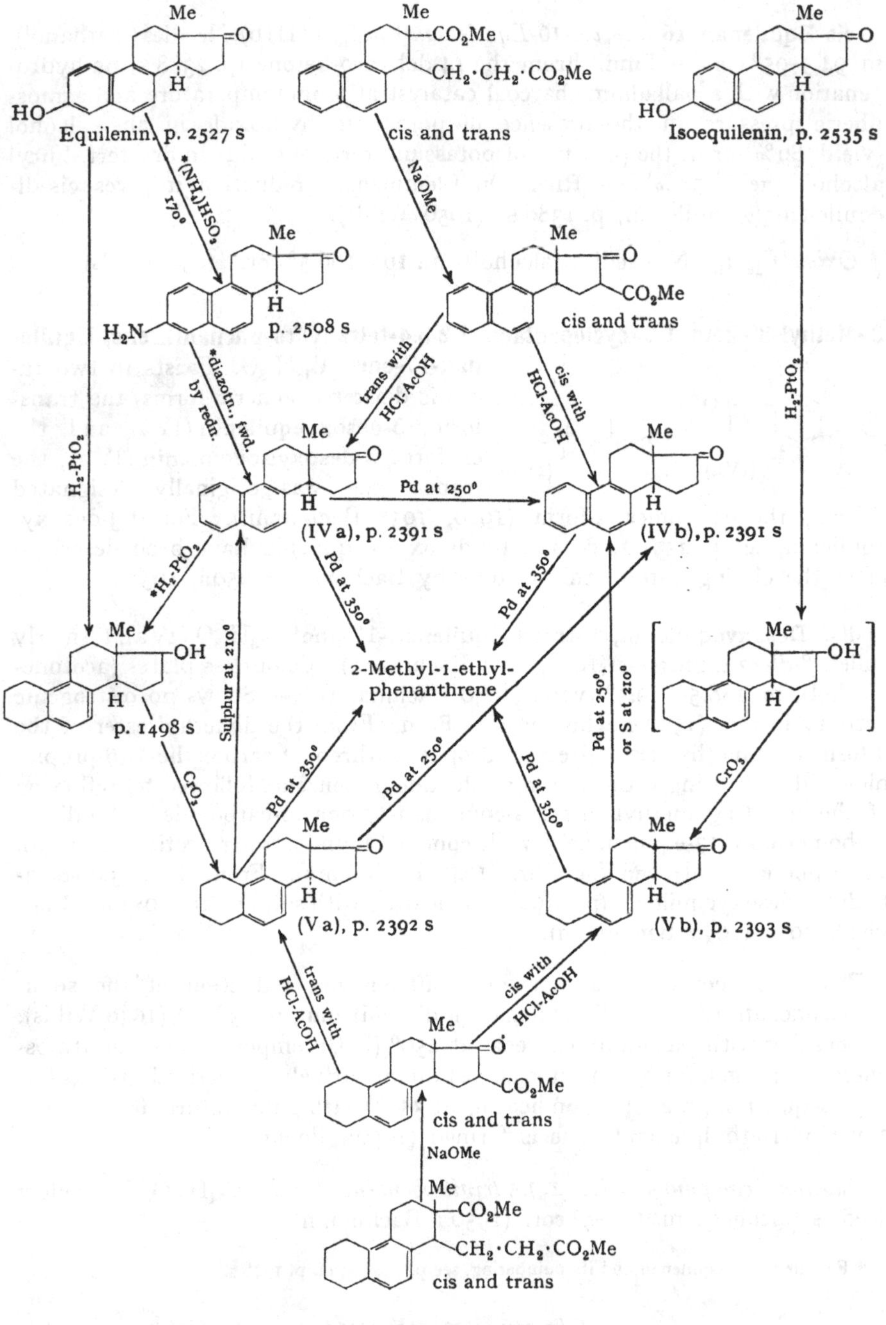

Equilenin, p. 2527 s

cis and trans

Isoequilenin, p. 2535 s

*(NH₄)HSO₃
170°

NaOMe

p. 2508 s

cis and trans

trans with
HCl·AcOH

cis with
HCl·AcOH

H₂·PtO₂

H₂N

*diazotn., fwd.
by redn.

(IV a), p. 2391 s

Pd at 250°

(IV b), p. 2391 s

H₂·PtO₂

*H₂·PtO₂

Pd at 350°

Pd at 350°

p. 1498 s

Sulphur at 210°

2-Methyl-1-ethyl-
phenanthrene

Pd at 250°,
or S at 210°

CrO₃

CrO₃

Pd at 350°

Pd at 250°

Pd at 350°

(V a), p. 2392 s

(V b), p. 2393 s

trans with
HCl·AcOH

cis with
HCl·AcOH

cis and trans

NaOMe

cis and trans

Oxime $C_{18}H_{19}ON$, colourless needles (methanol), m. 217–217.5° cor. with previous softening and partial sublimation at 213° (1950a Bachmann).

Semicarbazone $C_{19}H_{21}ON_3$, colourless needles (pyridine-alcohol), m. 256.5° to 257.5° cor. dec. (1946 Wilds).

d-3-Desoxyequilenin $C_{18}H_{18}O$ (IVa). Needles (ether-pentane), m. 156–8° (1939 Marker*); needles (methanol or by sublimation), m. 155–7° cor. (1945 Prelog); crystals (petroleum ether b. 60–75°), m. 160–1° cor. (1950b Bachmann); $[\alpha]_D^{22}$ +117° (chloroform) (1945 Prelog), $[\alpha]_D^{27}$ +115° (chloroform) (1950b Bachmann). Absorption max. (from the graph) in alcohol at 282.5 (3.7) and 322 (2.75) mμ (log ε) (1945 Prelog). — Exerts a weak œstrogenic action in the Allen-Doisy test (1945 Prelog).

Occ. In pregnant mares' urine; it is isolated in small amounts from the neutral by-products in the course of the œstrone extraction (1945 Prelog). — **Fmn.** From d-equilenin (p. 2527 s) on heating with ammonium disulphite in a sealed tube at 170° for 12 hrs., followed by diazotization of the resulting d-3-amino-3-desoxyequilenin and decomposition of the diazonium chloride with hypophosphorous acid (1950b Bachmann). From 3-desoxy-11-keto-equilenin (obtained by oxidation of the carbinol fraction from pregnant mares' urine; p. 2874 s) on refluxing with zinc and hydrochloric acid in 95% alcohol for 3.5 hrs. (1939 Marker*). — **Rns.** On hydrogenation in glacial acetic acid + some conc. HCl in the presence of platinum oxide at 20° and atmospheric pressure it gives 3-desoxy-hexahydroequilenin($\Delta^{5\cdot 7\cdot 9}$-œstratrien-17β-ol, p. 1498 s) (1945 Prelog).

Addition compound with 1.3.5-trinitrobenzene (1:1), $C_{24}H_{21}O_7N_3$, orange-yellow needles (alcohol), m. 155–6° cor. —*Oxime* $C_{18}H_{19}ON$, needles (petroleum ether b. 60–75°), m. 179.5–180.5° cor. — *2.4-Dinitrophenylhydrazone* $C_{24}H_{22}O_4N_4$, orange prisms (ethyl acetate-alcohol), m. 250–2° cor. dec. (1950b Bachmann).

dl-3-Desoxyisoequilenin, dl-cis-Equilenan-17-one** $C_{18}H_{18}O$ (IVb, p. 2390 s), formerly called α-dl-17-equilenone (1940, 1951 Bachmann). Colourless plates (methanol or ethanol), crystals (methanol-acetone); m. 101–2° cor. (1940, 1950a Bachmann; 1945 Birch). — Shows no œstrogenic activity in rats (1940 Bachmann). — **Fmn.** From the α-form of 2-methyl-1.2.3.4-tetrahydro-phenanthrene-2-carboxylic-1-(β-propionic) acid on pyrolysis of its lead salt or on heating with acetic anhydride, followed by sublimation under reduced pressure; in better yield from the dimethyl ester of the same acid on heating with sodium methoxide in benzene, followed by refluxing of the resulting methyl dl-cis-equilenan-17-one-16-carboxylate ("α-dl-16-carbomethoxy-17-equilenone") with conc. HCl and strong acetic acid in an atmosphere of nitrogen for 1 hr. (1940 Bachmann). From dl-3-desoxyequilenin (p. 2389 s), 1.2.3.4-tetrahydro-3-desoxyequilenin and 1.2.3.4-tetrahydro-3-desoxyisoequilenin (p. 2392 s) on heating with palladium-charcoal at 250° (bath temperature) for 8 min. (very good yields) (1950a Bachmann); from the last-mentioned

---

* Cf. 1950b Bachmann, footnote 4.
** For the name equilenan and its numbering, see p. 1385 s; cf. p. 1348 s.
E.O.C. XIV s. 167

References, pp. 2394 s, 2395:

compound also on dehydrogenation with sulphur by heating at 210° for 2 hrs. (yield, 41 %) (1944 Bachmann). For its formation from 3'-keto-1.2-cyclopentano-1.2.3.4-tetrahydro-phenanthrene by methylation with temporary protection of the CH_2-group adjacent to the CO-group (1945 Birch), see under the reactions of this ketone (p. 2386 s).

Rns. Clemmensen reduction affords cis-dl-equilenan (α-dl-equilenan, p. 1386 s) (1940 Bachmann; 1946 Wilds). Is resistant to heating with palladium-charcoal at 250° (bath temperature); at 350° (bath temperature; 20 min.) it gives 2-methyl-1-ethylphenanthrene (1950a Bachmann). Gives the 16-piperonylidene derivative (p. 2515 s) with piperonal (1945 Birch).

Addition compound with 1.3.5-trinitrobenzene (1:1), $C_{24}H_{21}O_7N_3$, yellow needles (alcohol), m. 133–4° cor. (1950a Bachmann). — *Picrate* (1:1) $C_{24}H_{21}O_8N_3$, yellow needles (alcohol), m. 109.5–110.5° (1940 Bachmann). — *Oxime* $C_{18}H_{19}ON$, prisms (methanol), m. 192–3° cor. (1950a Bachmann). — *Semicarbazone* $C_{19}H_{21}ON_3$, crystals (butanol), m. 264–266.5° cor. dec. (1946 Wilds).

2-Methyl-3'-keto-1.2-cyclopentano-1.2.3.9.10.11-hexahydrophenanthrene

$C_{18}H_{20}O$ (formula II on p. 2388 s). For a compound which may possess this structure, see p. 2388 s.

2-Methyl-3'-keto-1.2-cyclopentano-1.2.3.4.5.6-hexahydrophenanthrene,

$\Delta^{3.5.7.9}$-**Oestratetraen-17-one** $C_{18}H_{20}O$. Colourless needles (aqueous alcohol), m. 114–6°; can be distilled at 130–140° and 0.01 mm. pressure (1941 Heard). Absorption max. in abs. alcohol at 268 mμ (ε 4600) (1941 Heard; cf. 1938 Chakravorty). — **Fmn.** From $\Delta^{5.7.9}$-œstratrien-3β-ol-17-one (p. 2561 s) on dehydration by heating with $KHSO_4$ at 150–5°; low yield (1941 Heard). From epi-neoergosteryl acetate (p. 1534 s) on oxidation with CrO_3 in glacial acetic acid at 60–5°; low yield (1938 Chakravorty; cf. 1941 Heard).

Semicarbazone $C_{18}H_{23}ON_3$, m. 255° dec. (1938 Chakravorty).

2-Methyl-3'-keto-1.2-cyclopentano-1.2.3.4.5.6.7.8-octahydrophenanthrene,

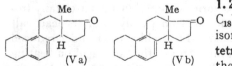

1.2.3.4-**Tetrahydroequilenan-17-one** $C_{18}H_{22}O$. Exists in two racemic diastereoisomeric forms, the trans-form, **1.2.3.4-tetrahydro-3-desoxyequilenin** (Va), and the cis-form, **1.2.3.4-tetrahydro-3-desoxyisoequilenin** (Vb); the trans-isomer was originally designated β-form, the cis-isomer α-form (1944, 1951 Bachmann). For an optically active 1.2.3.4-tetrahydro-3-desoxyequilenin, see p. 2392 s; d- and l-1.2.3.4-tetrahydro-3-desoxyisoequilenin have been described after the closing date of this volume by Bachmann (1950a, 1951).

dl-1.2.3.4-Tetrahydro-3-desoxyequilenin, dl-trans-1.2.3.4-Tetrahydroequilenan-17-one $C_{18}H_{22}O$ (Va), formerly called *β-dl-1.2.3.4-tetrahydro-17-equilenone* (1944, 1951 Bachmann). Crystals (methanol), m. 114–5° (vac.); distils at

110⁰ and 0.05 mm. pressure (1944 Bachmann). — **Fmn.** From the dimethyl ester of the β-form of 2-methyl-1.2.3.4.5.6.7.8-octahydrophenanthrene-2-carboxylic-1-(β-propionic) acid on heating with sodium methoxide in benzene, followed by refluxing of the resulting methyl dl-trans-1.2.3.4-tetrahydroequilenan - 17 - one - 16 - carboxylate ("β - 1.2.3.4 - tetrahydro - 16-carbomethoxy-17-equilenone") with conc. HCl and strong acetic acid in an atmosphere of nitrogen for 2 hrs. (good yield) (1944 Bachmann). From dl-equilenin (page 2534 s) on hydrogenation in the presence of Adams's PtO_2 catalyst in methanol containing some conc. HCl under 1 atm. pressure at room temperature (1 hr.), followed by oxidation of the resulting dl-2-methyl-3'-hydroxy-1.2-cyclopentano-1.2.3.4.5.6.7.8-octahydrophenanthrene (m. 126–8⁰; cf. p. 1498 s, where the opt. act. compound is described) with CrO_3 in strong acetic acid at room temperature for 2 hrs. (1951 Bachmann). — **Rns.** Dehydrogenation with sulphur at 210⁰ (2 hrs.) gives dl-3-desoxyequilenin (p. 2389 s) (1944 Bachmann), whereas on heating with palladium-charcoal at 250⁰ (bath temperature) for 8 min., dl-3-desoxyisoequilenin (p. 2391 s) is formed; on heating with palladium-charcoal at 350⁰ (bath temperature) for 20 min. it gives 2-methyl-1-ethylphenanthrene (1950a Bachmann).

Semicarbazone $C_{19}H_{25}ON_3$, prisms, m. 274–5⁰ (vac.) (1944 Bachmann).

Opt. act. 1.2.3.4-Tetrahydro-3-desoxyequilenin, opt. act. trans-1.2.3.4-Tetrahydroequilenan-17-one, $Δ^{5.7.9}$-Oestratrien-17-one $C_{18}H_{22}O$ (Va). Colourless needles (aqueous methanol), m. 107–9⁰ (1939 Marker; cf. 1950a Bachmann). — **Fmn.** From $Δ^{5.7.9}$-œstratrien-17β-ol (p. 1498 s) on oxidation with CrO_3 in acetic acid at 25⁰ (1939 Marker). — *Oxime* $C_{18}H_{23}ON$, needles (acetone), m. 203–5⁰ dec. (1939 Marker).

dl-1.2.3.4-Tetrahydro-3-desoxyisoequilenin, dl-cis-1.2.3.4-Tetrahydroequilenan-17-one $C_{18}H_{22}O$ (Vb, p. 2392 s), formerly called α-*dl-1.2.3.4-tetrahydro-17-equilenone* (1944, 1951 Bachmann). Prisms (methanol), m. 72–3⁰; distils at 100⁰ and 0.05 mm. pressure (1944 Bachmann). — **Fmn.** From the dimethyl ester of the α-form of 2-methyl-1.2.3.4.5.6.7.8-octahydrophenanthrene-2-carboxylic-1-(β-propionic) acid (1944 Bachmann) and from dl-isoequilenin (page 2536 s) (1951 Bachmann) similarly to the analogous formations of dl-1.2.3.4-tetrahydro-3-desoxyequilenin (see above). — **Rns.** Is dehydrogenated to dl-3-desoxyisoequilenin (p. 2391 s) by heating with sulphur at 210⁰ for 2 hrs. (1944 Bachmann) or with palladium-charcoal at 250⁰ (bath temperature) for 8 min. (1950a Bachmann); on heating with palladium-charcoal at 350⁰ (bath temperature) for 20 min., 2-methyl-1-ethylphenanthrene is obtained (1950a Bachmann).

Semicarbazone $C_{19}H_{25}ON_3$, prisms (alcohol), m. 243–4⁰ (1944 Bachmann).

For a ketone $C_{18}H_{20}O$ (m. 85⁰; *semicarbazone* $C_{19}H_{23}ON_3$, m. 209⁰) which may be **2-methyl-3'-keto-1.2-cyclopentano-2.3.4.9.10.12-hexahydrophenanthrene** (II, p. 1386 s) or **2-methyl-3'-keto-1.2-($Δ^{1(5')}$-cyclopenteno)-1.2.3.4.9.10.11.12-octahydrophenanthrene** (III, p. 1386 s), described by Nenitzescu (1942), see under "Nenitzescu's equilenan", p. 1386 s.

167*

3′-Ethyl-5′-keto-1.2-cyclopentenophenanthrene $C_{19}H_{16}O$. Prisms (acetone-methanol), m. 110–111.2° cor. — **Fmn.** From β-(2-phen-anthryl)-valeric acid chloride on cyclization by means of $AlCl_3$ in nitrobenzene, first at room temperature, then at 80°; yield, 78.5 %. — **Rns.** Gives 3′-ethyl-1.2-cyclo-pentenophenanthrene (p. 1392 s) on Clemmensen reduction (1943 Riegel).

Oxime $C_{19}H_{17}ON$. Exists in two (syn- and anti-) forms: "*α-Form*", colourless crystals (95 % alcohol), m. 172.5–174.5° cor. dec. — "*β-Form*", yellow crystals (dil. alcohol), m. 169–170.8° cor. dec. (1943 Riegel).

3′-Isopropyl-4-keto-1.2-cyclopentano-1.4.9.10.11.12-hexahydrophenanthrene
$C_{20}H_{24}O$. Glassy mass, b. 193–198°/$_{0.1}$. Absorption max. (in alcohol?) at 241 mμ (ε 8080). — **Fmn.** From 1-acetyl-3.4-dihydronaphthalene and the sodium deri-vative (or, better, the lithium derivative) of 2-iso-propylcyclopentanone in boiling ether, followed by heating of the resulting diketone [probably 1-acetyl-2-(3-isopropyl-2-keto-1-cyclopentyl)-tetralin] with alcoholic sodium ethoxide on the steam-bath for 1 hr.; over-all yield, using the lithium derivative, 60 %. — **Rns.** On heating with palladium-charcoal at 260° it gives a compound m. 162°, probably 3′-isopropyl-4-hydroxy-1.2-cyclo-pentenophenanthrene (p. 1500 s). On reduction with sodium in alcohol, followed by heating with selenium at 320°, it gives 3′-isopropyl-1.2-cyclo-pentenophenanthrene (p. 1392 s). It is converted into 4-methyl-3′-isopropyl-1.2-cyclopentano-1.9.10.11-tetrahydrophenanthrene (p. 1399 s) by the action of methylmagnesium iodide or bromide in the presence of cuprous bromide or chloride, resp., followed by heating of the non-ketonic fraction with a trace of iodine to 180° (1944 Birch).

2.4-Dinitrophenylhydrazone $C_{26}H_{28}O_4N_4$, probably a mixture of cis- and trans-isomers. Dark red, amorphous solid, m. ca. 85–7° (1944 Birch).

3′-Isopropyl-5′-keto-1.2-cyclopentenophenanthrene $C_{20}H_{18}O$. Needles (acetone), m. 143.6–144.4° cor. — **Fmn.** From β-(2-phenanthryl)-isocaproic acid chloride on cyclization by means of $AlCl_3$ in nitrobenzene, first at room temperature, then at 80°; yield, 75 %. — **Rns.** Gives 3′-isopropyl-1.2-cyclopentenophenanthrene (p. 1392 s) on Clemmensen reduction (1943 Riegel).

Oxime $C_{20}H_{19}ON$. Colourless needles (alcohol), m. 205–211° cor. dec. (1943 Riegel).

1937 Bockemüller, *Ger. Pat.* 711471 (issued 1941); *C. A.* **1943** 4076; *Ch. Ztbl.* **1942** I 540; Winthrop Chemical Co. (Bockemüller), *U.S. Pat.* 2179809 (issued 1939); *C. A.* **1940** 1823; *Ch. Ztbl.* **1940** I 3179.
1938 Bockemüller, *Angew. Ch.* **51** 188.
  Chakravorty, Wallis, *J. Am. Ch. Soc.* **60** 1379.
1939 Marker, Rohrmann, *J. Am. Ch. Soc.* **61** 3314.
1940 Bachmann, Wilds, *J. Am. Ch. Soc.* **62** 2084, 2087.

1941 Heard, Hoffmann, *J. Biol. Ch.* **138** 651.
1942 Nenitzescu, Ciorănescu, *Ber.* **75** 1765, 1770.
1943 Riegel, Gold, Kubico, *J. Am. Ch. Soc.* **65** 1772.
1944 Bachmann, Morin, *J. Am. Ch. Soc.* **66** 553.
    Birch, Robinson, *J. Ch. Soc.* **1944** 503, 505, 506.
    Wilds, Beck, *J. Am. Ch. Soc.* **66** 1688, 1692, 1694.
1945 Birch, Jaeger, Robinson, *J. Ch. Soc.* **1945** 582, 585.
    Prelog, Führer, *Helv.* **28** 583.
1946 a Butenandt, Dannenberg, v. Dresler, *Z. Naturforsch.* **1** 151.
    b Butenandt, Dannenberg, v. Dresler, *Z. Naturforsch.* **1** 227.
    Wilds, Beck, Johnson, *J. Am. Ch. Soc.* **68** 2161.
1947 Wilds, Beck, Close, Djerassi, Johnson, Johnson, Shunk, *J. Am. Ch. Soc.* **69** 1985, 1993.
1950 a Bachmann, Dreiding, *J. Am. Ch. Soc.* **72** 1323, 1327, 1328.
    b Bachmann, Dreiding, *J. Am. Ch. Soc.* **72** 1329.
    Wilds, Johnson, Sutton, *J. Am. Ch. Soc.* **72** 5524, 5526.
1951 Bachmann, Dreiding, Stephenson, *J. Am. Ch. Soc.* **73** 2765.

---

3. Monoketo-steroids (CO in the Ring System) C_{19} with Two Aliphatic Side Chains

2.13 - Dimethyl - 7 - keto - 1.2 - cyclopentano - 1.2.3.4.5.6.7.9.10.11.12.13 - dodeca-hydrophenanthrene, 10.13 - Dimethyl - Δ^4 - steren - 3 - one,

"*Desoxytestosterone*" $C_{19}H_{28}O$. Pale yellow, highly viscous oil, b. 150–155°/0.08 (bath temperature). — **Fmn.** From 2.5 - dimethyl - 6 - keto - 1.2 - cyclopentanoperhydronaphthalene (I) and 4-diethylamino-butan-2-one methiodide on condensation by means of sodamide in alcohol - ether under nitrogen, first at room temperature (several hrs.), then on refluxing for 1 hr.; the resulting product, containing the diketone

(II), is treated with dry sodium ethoxide in benzene at room temperature overnight and then at 60° for 1 hr. (1945/46, 1947 Mukharji).

2.13 - Dimethyl - 9 - keto - 1.2 - cyclopentano - 1.2.3.4.5.6.7.8.9.12.13.14 - dodeca-hydrophenanthrene, 10.13-Dimethyl-Δ^7-steren-6-one $C_{19}H_{28}O$.

Yellow oil with a characteristic odour; b. 153–157° below 1 mm. pressure. — **Fmn.** From equimolecular amounts of 8-methylhydrindan-4-one and 2-methyl-1-acetyl-Δ^1-cyclohexene on condensation by means of potassium isopropoxide in dry pyridine - ether, first in the cold (2 days), then on refluxing for 1 hr.; yield, 28% (1946 Bagchi).

Semicarbazone $C_{20}H_{31}ON_3$. Crystals (dil. alcohol), shrinking at 158°, m. 164–166° (1946 Bagchi).

1945/46 Mukharji, *Science and Culture* **11** 574.
1946 Bagchi, Banerjee, *J. Indian Ch. Soc.* **23** 397, 401.
1947 Mukharji, *J. Indian Ch. Soc.* **24** 91, 92, 97.

Summary of Androstan-(Androsten-, etc.) -ones

For values other than those chosen for the summary, see the compounds themselves

Position of CO	Configuration at C-5	Double bond(s)	M. p. °C.	$[\alpha]_D$*	Oxime m. p. °C.	Semi-carbazone m. p. °C.	Page
3	—	$\Delta^{4\cdot16}$	133	$+123°$	—	—	2396 s
3	norm.	Δ^{16}	141	$+38°$	165	—	2396 s
3	norm.	satd.	105	$+25°$	—	240	2397 s
3	iso	Δ^{16}	86	$+31°$	—	—	2398 s
3	iso	satd.	60	$+17°$	—	—	2398 s
6	—	$i^{(3\cdot5)}$	122	$+35°$ (alc.)	—	—	2400 s
11	norm.	$\Delta^{2 \text{ or } 3.16}$	74	—	—	—	2400 s
11	norm.	satd.	52	—	—	—	119
17	—	$\Delta^{3\cdot5}$	89	$-31°$(alc.)	166	292	2400 s
17	norm.	$\Delta^{2 \text{ or } 3}$	$\begin{cases}105\\114\end{cases}$	$+152°$ (alc.)	154	$\begin{cases}275\\295\end{cases}$	2401 s
17	—	Δ^4	80	—	—	—	2402 s
17	—	Δ^5	$\begin{cases}88\\107\end{cases}$	—	—	287	2403 s
17	norm.	satd.	122	$+89°$	—	288	2403 s
17	iso	satd.	105	—	—	265	119

* In chloroform, unless otherwise stated.

$\Delta^{4\cdot16}$**-Androstadien-3-one** $C_{19}H_{26}O$. Crystals (hexane), m. 131.5-133.5° cor.; $[\alpha]_D^{16} +123° \pm 3.5°$ (chloroform). Absorption max. (solvent not stated) at 240 mμ (log ε 4.25). Gives an intense Kägi-Miescher colour reaction (cf. p. 1740 s). Has a strong urine-like odour. — **Fmn.** From the benzoate of Δ^4-androsten-17α-ol-3·one (epitestosterone, p. 2579 s) on heating at 300° in an atmosphere of nitrogen (1945 Prelog).

Δ^{16}**-Androsten-3-one** $C_{19}H_{28}O$. Six-sided plates (pentane), m. 140-1° cor.; sublimes at 115° and 0.05 mm. pressure; $[\alpha]_D^{17} +38°$ (chloroform). Gives a blue colour reaction in the Kägi-Miescher test (cf. p. 1740 s). Develops an intense urine-like odour (1944 Prelog). — **Fmn.** From the hexahydrobenzoate of androstan-17α*-ol-3-one (p. 2596 s) on heating at 300° in an atmosphere of nitrogen; yield, 62% (1944 Prelog). From cholestan-3-one (p. 2436 s) on heating at 350° (1937 SCHERING-KAHLBAUM A.-G.; cf. 1937 SCHERING A.-G.)**.

Rns. On refluxing with aluminium isopropoxide in isopropyl alcohol it gives a mixture of the 3-epimeric Δ^{16}-androsten-3-ols (p. 1506 s) (1944 Prelog). Gives androstan-3-one (below) on hydrogenation in glacial acetic acid in the

* For reversion of the configuration given in the original paper, see p. 1363 s.

** The identity of the compound obtained by this method (no constants given) seems very doubtful; cf. footnote * to Δ^{16}-androsten-3α-ol, p. 1506 s.

presence of platinum oxide, followed by oxidation with CrO_3 (1945 Prelog). On heating with hydrazine hydrate and sodium methoxide in methanol in a sealed tube at 190° it gives Δ^{16}-androstene (p. 1396 s) (1944 Prelog).

Oxime $C_{19}H_{29}ON$. Crystals (methanol), m. 163–5° cor. — **Rns.** Gives 3-amino-Δ^{16}-androstene (p. 1480 s) on boiling with sodium in alcohol (1945 Prelog).

Androstan-3-one $C_{19}H_{30}O$ (Ia, p. 2399 s). Colourless needles (hexane, or by sublimation at 96° in a high vac.), m. 104.5–105.5° cor.; $[\alpha]_D^{20} +25.4° \pm 3°$ (chloroform). Has an urine-like odour (1945 Prelog). — **Fmn.** From andro-stan-3β-ol (p. 1507 s) on oxidation with CrO_3 in strong acetic acid at 4° (yield, 80%) (1946 Heard); on similar oxidation at 60°, Ruzicka (1945) obtained 35% yield of the ketone and 37% yield of androstane-2‖3-dioic acid (p. 1508 s). From Δ^{16}-androsten-3-one (above) on hydrogenation in glacial acetic acid in the presence of platinum oxide, followed by oxidation with CrO_3; yield, 93% (1945 Prelog). From the semicarbazone of Δ^5-androsten-3β-ol-17-one (dehydroepiandrosterone, p. 2609 s) on heating with sodium ethoxide in alcohol at 180° (for the resulting products, see under formation of Δ^5-androsten-3β-ol, p. 1504 s), followed by oxidation with CrO_3 at room temperature; small yield (1937 Burrows).

Rns.* Gives 3‖4-androstan-4-ol-3-oic acid lactone (below) on oxidation with perbenzoic acid in chloroform at −10° (1945 Prelog). By the action of methylmagnesium iodide in ether the two 3-epimeric 3-methylandrostan-3-ols (pp. 1515 s, 1516 s) are formed (1947 Ruzicka).

Semicarbazone $C_{20}H_{33}ON_3$. Crystals (dioxane), m. 238–240° (1937 Burrows).

3‖4-Androstan-4-ol-3-oic acid lactone $C_{19}H_{30}O_2$ (Va, p. 2399 s). Colourless leaflets (ether), m. 185.5–186° cor.; sublimes at 124–5° and 0.01 mm. pressure; $[\alpha]_D^{17} -38° \pm 3°$ (chloroform). Has a urine-like odour when hot (1945 Prelog). **Fmn.*** From androstan-3-one (above) on oxidation with perbenzoic acid in chloroform at −10°. From 3‖4-Δ^{16}-androsten-4-ol-3-oic acid lactone (IIa, below) on hydrogenation in alcohol in the presence of platinum oxide (1945 Prelog). — **Rns.*** The free hydroxy-acid, obtained by refluxing with metha-nolic KOH, gives androstane-3‖4-dioic acid (IVa, p. 2398 s) on oxidation with CrO_3 in strong acetic acid (1945 Ruzicka).

3‖4-Δ^{16}-Androsten-4-ol-3-oic acid lactone $C_{19}H_{28}O_2$ (IIa, p. 2399 s). Leaflets (ether-petroleum ether), m. 172.5–173° cor. (after sublimation at 145°/0.01); $[\alpha]_D^{17} -26.6° \pm 3°$ (chloroform). Gives a strong yellow coloration with tetra-nitromethane and a violet Kägi-Miescher reaction. Has a distinct urine-like odour, especially when hot. — **Fmn.*** From the 17-hexahydrobenzoate of 3‖4-androstane-4.17α-diol-3-oic acid lactone (p. 2597 s; formula IIIa on page 2399 s) on distillation in an atmosphere of nitrogen through a tube heated at 310°. — **Rns.*** On hydrogenation in alcohol in the presence of platinum oxide it gives 3‖4-androstan-4-ol-3-oic acid lactone (Va, above) (1945 Prelog).

_____

* See also the scheme on p. 2399 s.

Androstane-3‖4-dioic acid $C_{19}H_{30}O_4$ (IVa, p. 2399 s). Crystals (ether), m. 242⁰ to 244⁰ cor.; $[\alpha]_D^{21}$ —29⁰ (alcohol). — **Fmn.*** From 3‖4-androstan-4-ol-3-oic acid lactone (p. 2397 s) on refluxing with methanolic KOH, followed by oxidation with CrO_3 in strong acetic acid. — **Rns.*** Its anhydride, obtained by refluxing with acetic anhydride, gives A-nor-ætiocholan-3-one (p. 1373 s; formula VI on p. 2399 s) on distillation under 15 mm. pressure (1945 Ruzicka).

5-iso-Δ^{16}-Androsten-3-one, Δ^{16}-Aetiocholen-3-one $C_{19}H_{28}O$. Crystals (petroleum ether), m. 85–6⁰ cor.; sublimes at 80–90⁰ and 0.01 mm. pressure; $[\alpha]_D^{16}$ +31⁰ ±3⁰ (chloroform). Gives a yellow coloration with tetranitromethane and a violet colour in the Kägi-Miescher test (cf. p. 1740 s). Has a distinct urine-like odour, but much weaker than that of Δ^{16}-androsten-3-one (p. 2396 s). — **Fmn.** From the benzoate of ætiocholan-17α**-ol-3-one (p. 2599 s) on distillation in an atmosphere of nitrogen through a tube heated at 300⁰. — **Rns.** On boiling with aluminium isopropoxide in isopropyl alcohol it gives a mixture of the two 3-epimeric Δ^{16}-ætiocholen-3-ols (p. 1508 s). Gives ætiocholan-3-one (below) on partial hydrogenation in alcohol in the presence of palladium-$BaCO_3$ (1945 Prelog).

5-iso-Androstan-3-one, Aetiocholan-3-one $C_{19}H_{30}O$ (Ib, p. 2399 s). Crystals (aqueous acetone), m. 59–60⁰ cor.; distils at 70–5⁰ and 0.02 mm. pressure; $[\alpha]_D^{15}$ +17⁰ ±3⁰ (chloroform). Has a urine-like odour, much weaker than androstan-3-one (p. 2397 s) (1945 Prelog). — **Fmn.** From ætiocholan-3α-ol (p. 1508 s) on oxidation with CrO_3 in 90% acetic acid at 4⁰; yield, 59% (1946 Heard). From Δ^{16}-ætiocholen-3-one (above) on partial hydrogenation in alcohol in the presence of palladium-$BaCO_3$; yield, 37% (1945 Prelog). — **Rns.*** Gives 3‖4-ætiocholan-4-ol-3-oic acid lactone (below) on oxidation with perbenzoic acid in chloroform at —10⁰. Oxidation with CrO_3 in strong acetic acid at 60⁰ affords ætiocholane-3‖4-dioic acid (p. 2400 s) (1945 Ruzicka).

3‖4-Aetiocholan-4-ol-3-oic acid lactone $C_{19}H_{30}O_2$ (Vb, p. 2399 s). Crystals (petroleum ether), m. 142–3⁰ cor.; sublimes at 100–5⁰ and 0.01 mm. pressure; $[\alpha]_D^{20}$ +33.4⁰ ±2⁰ (chloroform). — **Fmn.*** From ætiocholan-3-one (above) on oxidation with perbenzoic acid in chloroform at —10⁰. From 3‖4-Δ^{16}-ætiocholen-4-ol-3-oic acid lactone (IIb, below) on hydrogenation in alcohol in the presence of platinum oxide. — **Rns.*** The free hydroxy-acid, obtained by refluxing with methanolic KOH, gives ætiocholane-3‖4-dioic acid (IVb, p. 2400 s) on oxidation with CrO_3 in acetic acid (1945 Ruzicka).

3‖4-Δ^{16}-Aetiocholen-4-ol-3-oic acid lactone $C_{19}H_{28}O_2$ (IIb, p. 2399 s), m. 127 to 128⁰ cor.; sublimes at 90–5⁰ under 0.05 mm. pressure; $[\alpha]_D^{22}$ +43.8⁰ ±2⁰ (chloroform). Gives a yellow coloration with tetranitromethane and a violet Kägi-Miescher reaction. — **Fmn.*** From the 17-benzoate of 3‖4-ætiocholane-4.17α-diol-3-oic acid lactone (p. 2600 s) on distillation in a stream of nitrogen

---

    * See also the scheme on p. 2399 s.
    ** For reversion of the configuration given in the original paper, see p. 1363 s.

Degradation of Androstan-3-one and Aetiocholan-3-one, and Related Reactions

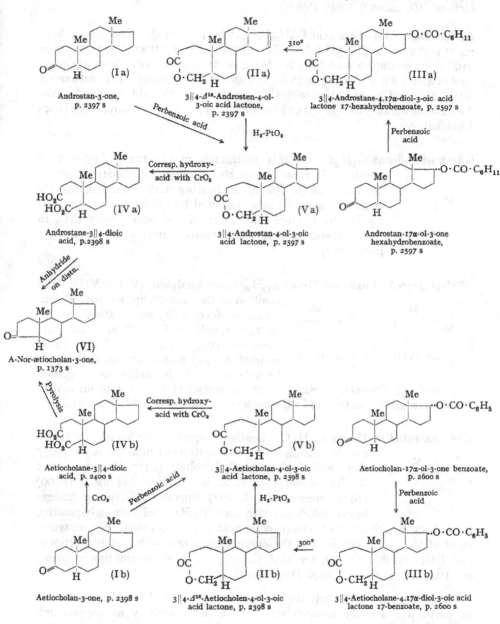

(I a) Androstan-3-one, p. 2397 s

(II a) 3‖4-Δ^{16}-Androsten-4-ol-3-oic acid lactone, p. 2397 s

310°

(III a) 3‖4-Androstane-4.17α-diol-3-oic acid lactone 17-hexahydrobenzoate, p. 2597 s

Perbenzoic acid

H_2-PtO_2

Perbenzoic acid

(IV a) Androstane-3‖4-dioic acid, p. 2398 s

Corresp. hydroxy-acid with CrO_3

(V a) 3‖4-Androstan-4-ol-3-oic acid lactone, p. 2397 s

(III a side) Androstan-17α-ol-3-one hexahydrobenzoate, p. 2597 s

Anhydride on distn.

(VI) A-Nor-ætiocholan-3-one, p. 1373 s

Pyrolysis

(IV b) Aetiocholane-3‖4-dioic acid, p. 2400 s

Corresp. hydroxy-acid with CrO_3

(V b) 3‖4-Aetiocholan-4-ol-3-oic acid lactone, p. 2398 s

Aetiocholan-17α-ol-3-one benzoate, p. 2600 s

Perbenzoic acid

CrO_3

Perbenzoic acid

H_2-PtO_2

(I b) Aetiocholan-3-one, p. 2398 s

(II b) 3‖4-Δ^{16}-Aetiocholen-4-ol-3-oic acid lactone, p. 2398 s

300°

(III b) 3‖4-Aetiocholane-4.17α-diol-3-oic acid lactone 17-benzoate, p. 2600 s

through a tube heated at 300°. — **Rns.*** On hydrogenation in alcohol in the presence of a platinum oxide catalyst it gives 3‖4-ætiocholan-4-ol-3-oic acid lactone (Vb, above) (1945 Prelog).

Aetiocholane-3‖4-dioic acid $C_{19}H_{30}O_4$ (IVb, p. 2399 s).   Crystals (ether), m. 253–5° cor.; $[\alpha]_D^{22} + 32°$ (alcohol). — **Fmn.*** From ætiocholan-3-one (page 2398 s) on oxidation with CrO_3 in strong acetic acid at 60°. From 3‖4-ætiocholan-4-ol-3-oic acid lactone (Vb, p. 2398 s) on refluxing with methanolic KOH, followed by oxidation with CrO_3 in strong acetic acid. — **Rns.*** Gives A-nor-ætiocholan-3-one (p. 1373 s; formula VI on p. 2399 s) on pyrolysis (1945 Ruzicka).

i-Androsten-6-one $C_{19}H_{28}O$.   Crystals (methanol), m. 122–122.5°; $[\alpha]_D^{20} + 34.5°$ (alcohol). — **Fmn.** From the p-toluenesulphonate of Δ^5-androsten-3β-ol (p. 1504 s) on heating with potassium acetate in 50% aqueous acetone, followed by oxidation of the resulting i-androsten-6-ol (not obtained crystalline) with CrO_3 in glacial acetic acid at room temperature; yield, 10% (1942 Butenandt).

$\Delta^{2.16}$-or $\Delta^{3.16}$-Androstadien-11-one $C_{19}H_{26}O$ (not analysed) (VII or VIII).   For position of the keto-group, see 1937 Reichstein. — Needles (by sublimation), m. 72° to 74°; distils at 80–120° and 0.05 mm. pressure (1937 Steiger). — **Fmn.** From androstane-3.17-diol-11-one (p. 149) on dehydration by Tschugaeff's xanthogenate method (1937 Steiger). — **Rns.** Gives androstan-11-one (p. 119) on hydrogenation in glacial acetic acid in the presence of platinum oxide (1937 Steiger).

$\Delta^{3.5}$-Androstadien-17-one $C_{19}H_{26}O$.   Leaflets (aqueous methanol), m. 88–9°; can be distilled at 100° and 0.001 mm. pressure (1937 Burrows); $[\alpha]_D^{20} - 30.4°$ (alcohol) (1937 Burrows; 1941 Wolfe).   Absorption max. in ether at 234 mμ (ε 18000) (1939 Dannenberg; cf. 1937 Burrows).   Gives an orange-brown coloration with conc. H_2SO_4, and a strongly positive Rosenheim reaction with trichloroacetic acid (1937 Burrows).   For measurement of the colour obtained in the Zimmermann test with m-dinitrobenzene and KOH in abs. alcohol, see 1938 Callow; for its determination in urine, see 1944 Zimmermann; 1946 Dingemanse.

Shows no œstrogenic activity towards mice in the Allen-Doisy test and no cortin-like activity towards rats, but a weak activity on capon-comb growth (1937 Burrows).   It has no anæsthetic effect when injected into rats (1942 Selye).

* See also the scheme on p. 2399 s.

Occ. The ketone has been isolated by acid hydrolysis (aqueous HCl) from the urine of a male patient with a tumour of the adrenal gland cortex; it may have been only a secondary product, however, formed from a precursor during the hydrolysis (see its formation from 5-dehydroepiandrosterone, below) (1937 Burrows). In the urine of a girl with a cortico-adrenal tumour (1941 Wolfe), in the urine (even non-hydrolysed) of patients with adrenal hyperplasia, and in small amounts (and not always) in the urine of healthy persons (1942 Dobriner; cf. also 1941 Pincus).

Fmn. From Δ^5-androsten-3β-ol-17-one (5-dehydroepiandrosterone, page 2609 s) on heating with anhydrous $CuSO_4$ at 200° for 15 min. (1937 Burrows), on boiling with P_2O_5 in benzene (1945 Ross), or on addition to urine, devoid of the present dienone, followed by acid hydrolysis (1942 Dobriner).

Rns. Decolorizes pyridine sulphate dibromide and boiling $KMnO_4$ solution (1937 Burrows). Gives $\Delta^{3.5}$-androstadien-17β-ol (p. 1509 s) on reduction with aluminium isopropoxide in isopropyl alcohol (1938 Butenandt). Hydrogenation in alcohol in the presence of a palladium black catalyst yields androstan-17-one (p. 2403 s); using a platinum oxide catalyst, nearly 3 mols. of hydrogen are absorbed (formation of androstan-17-ol?) (1937 Burrows). On refluxing with acetone in the presence of methanolic KOH it gives 16-isopropylidene-$\Delta^{3.5}$-androstadien-17-one (p. 2412 s) (1945 Ross).

Oxime $C_{19}H_{27}ON$. Colourless needles (methanol), m. 164–170° (mixture of the syn- and anti-isomers?) (1937 Burrows), m. 164–6° (1941 Wolfe).

Semicarbazone $C_{20}H_{29}ON_3$. Microscopic needles (dioxane), m. 291–2° on rapid heating, m. 275–280° on slow heating; sparingly soluble in alcohol (1937 Burrows).

Δ^2 and/or Δ^3-Androsten-17-one $C_{19}H_{28}O$ (I and/or II, resp.) (p. 128: *unsaturated ketone, m. 104°*). The position of the double bond is not definitely established; the products isolated might be varying mixtures of the two isomers; cf. the m. p.'s recorded below (1940 Hirschmann; 1942 Venning; 1946 Shoppee).

Rectangular plates (methanol), m. 105° cor. (1946 Shoppee), 107–9°, 111–4° (1940 Hirschmann), 105.5–107° (1942 Venning), 114–114.5° (1942 Dobriner); colourless leaflets (petroleum ether), m. 104.5–105.5° cor.; sublimes at 83° and 0.02 mm. pressure (1945 Prelog); $[\alpha]_D^{28} + 152°$ (alcohol) (1940 Hirschmann), $[\alpha]_D^{17} + 146° \pm 9°$ (alcohol) (1945 Prelog; cf. 1942 Dobriner). Has a urine-like odour, especially when hot (1945 Prelog). Gives a purple coloration in the Zimmermann test with m-dinitrobenzene and alkali (1940 Hirschmann); for the effect of alcohol and KOH concentration on this reaction, see 1945 Wilson.

Occ. The ketone has been isolated from the urine of normal men and women (1940 Engel; cf. 1940 Hirschmann; 1942 Dobriner; 1942 Pearlman), of ovariectomized women (1940 Hirschmann), of patients with malignant

References, pp. 2404 s, 2405 s

tumours (1942 Pearlman; but cf. 1942 Dobriner), and of persons with adrenal hyperplasia (1942 Dobriner); also (in larger amounts) from the urine of a woman subsequent to injection of the propionate of Δ^4-androsten-17β-ol-3-one (testosterone, p. 2580 s) (1945 Schiller) and from the urine of male chimpanzees to which testosterone propionate had been given orally (but it might have been formed by dehydration of androsterone during the extraction process) (1944 Fish).

Fmn. Along with the acetate of androstan-3β-ol-17-one (epiandrosterone, p. 2633 s), from 3α-chloroandrostan-17-one (p. 2476 s) on heating with potassium acetate and glacial acetic acid in a bomb tube at 180⁰ (1946 Shoppee); along with the acetate of androstan-3α-ol-17-one (androsterone, p. 2629 s), from 3β-chloroandrostan-17-one (p. 2476 s) by the same method (yield, 35 % to 80 %) (1934 Butenandt; cf. 1946 Shoppee); from the latter chloroketone on heating with quinoline (1937 Marker; cf. 1946 Shoppee) or on refluxing with sodium iodide and pyridine (yield, 55 %) (1940 Hirschmann). From androsterone (p. 2629 s) on distillation with boric anhydride at 300⁰ (yield, 48%) (1945 Prelog); together with androsterone, from sodium androsterone sulphate (p. 2632 s) by hydrolysis with aqueous HCl in carbon tetrachloride under reflux (1942 Venning).

Rns. Gives mainly 2-hydroxyandrosterone (p. 2745 s) on oxidation with H_2O_2 in acetic acid at 100⁰ (1939 Marker). On reduction with sodium in boiling propyl alcohol, Δ^2 or Δ^3-androsten-17β-ol (p. 1510 s) is formed (1937 Marker). Hydrogenation in alcohol in the presence of a palladium catalyst yields androstan-17-one (p. 2403 s); the reaction proceeds similarly in glacial acetic acid (1934 Butenandt; 1940 Hirschmann), but in this case a mixture of the 17-epimeric androstan-17-ols is also obtained (1934 Butenandt). Bromine addition occurs giving a bromo-derivative m. ca. 160⁰ (1934 Butenandt).

Oxime $C_{19}H_{29}ON$. Crystals (aqueous alcohol), m. 153–4⁰ (1942 Venning; 1942 Pearlman).

Semicarbazone $C_{20}H_{31}ON_3$, m. 275⁰ (1934 Butenandt); the melting point varies between 283⁰ and 295⁰ cor. for different samples (1940 Hirschmann). Sparingly soluble (1934 Butenandt).

Δ^4-Androsten-17-one $C_{19}H_{28}O$. Crystals (aqueous methanol), m. 78–80⁰ (1940 Marker). — **Fmn.** Along with various other compounds, from Δ^4-cholestene (coprostene, p. 1423 s) on protection of the double bond with bromine or HCl, followed by oxidation of the resulting dibromide (p. 1472 s) or hydrochloride (5-chlorocholestane, p. 1460 s), resp., with CrO_3 or other oxidants and then by heating with zinc dust in acetic acid or by refluxing with sodium acetate in acetic acid, resp. (1939a CIBA). From the oxime of $\Delta^{4.16}$-pregnadien-20-one (not described in detail) on treatment with p-toluenesulphonyl chloride in cold pyridine, followed by hydrolysis with dil. H_2SO_4 at room temperature (1941 PARKE, DAVIS & Co.). From Δ^4-androsten-17β-ol (desoxotestosterone; p. 1510 s) on oxidation with CrO_3 in acetic acid at room temperature after protective bromination, followed by debromination with zinc dust in acetic acid on the steam-bath (1940 Marker). From Δ^5-androsten-17-one (below) on partial isomerization by means of hydrogen chloride; the resulting mixture of isomers is difficult to separate (1940 Marker).

Rns. Gives Δ^4-androstene-3.17-dione (p. 2880 s) on oxidation with CrO_3 in glacial acetic acid (1940 Marker). Hydrogenation in alcohol with Rupe or Raney nickel catalyst gives Δ^4-androsten-17β-ol (p. 1510 s). Reaction with alkyl (or alkenyl) halides by Grignard's method yields the corresponding 17α-alkyl (or alkenyl)-Δ^4-androsten-17β-ols (1939b CIBA).

Δ^5-Androsten-17-one $C_{19}H_{28}O$. Crystals (aqueous alcohol), m. 105–7° (1940 Marker; 1940 PARKE, DAVIS & Co.), 88° (1937 Kuwada). — **Fmn.** From Δ^5-cholestene (p. 1425 s) on bromination, followed by oxidation of the resulting dibromide (p. 1473 s) with CrO_3 and strong acetic acid in carbon tetrachloride at 48–50° and then by debromination with zinc dust in glacial acetic acid at 95° (1940 Marker; 1940 PARKE, DAVIS & Co.). From the oxime of $\Delta^{5.16}$-pregnadien-20-one (not described in detail) on treatment with p-toluenesulphonyl chloride in cold pyridine, followed by hydrolysis with dil. H_2SO_4 at room temperature (1941 PARKE, DAVIS & Co.). From Δ^5-androsten-17β-ol (p. 1511 s) on oxidation after temporary protection of the double bond with bromine (1937 Kuwada). — **Rns.** Gives Δ^5-androsten-17β-ol (p. 1511 s) on reduction with sodium in boiling propyl alcohol. On treatment with hydrogen chloride it is partially isomerized to Δ^4-androsten-17-one (p. 2402 s) (1940 Marker; 1940 PARKE, DAVIS & Co.).

Semicarbazone $C_{20}H_{31}ON_3$. White powder, m. 285–7° (1940 Marker; 1940 PARKE, DAVIS & Co.).

An **androsten-17-one** (the position of the double bond is not indicated) gives an intense blue colour with a highly concentrated solution of $SbCl_3$ in glacial acetic acid - acetic anhydride (9:1) (1943 Pincus).

Androstan-17-one $C_{19}H_{30}O$ (p. 119). Leaflets (methanol), six-sided platelets (aqueous acetone), m. 121–2° (1940 Hirschmann); cube-like crystals (ethyl acetate-hexane), m. 119.5–120.5° cor. (1944 Ruzicka); $[\alpha]_D$ in chloroform +88° to +95° (1944 Ruzicka), +89° (1947 Djerassi). For measurement of its colour in the Zimmermann test with m-dinitrobenzene and KOH in abs. alcohol, see 1938 Callow.

Fmn. From androstan-17β-ol (p. 1512 s) on oxidation with CrO_3 in glacial acetic acid at room temperature (1936 Reichstein; 1944 Ruzicka). From $\Delta^{3.5}$-androstadien-17-one (p. 2400 s) on hydrogenation in alcohol in the presence of palladium (1937 Burrows). From Δ^2 or Δ^3-androsten-17-one (p. 2401 s) on hydrogenation in glacial acetic acid or alcohol in the presence of palladium (1934 Butenandt; 1940 Hirschmann). Along with androstan-17β-ol and the 3-acetate of androstane-3β.6β.17β-triol (p. 2164 s), from the acetate of $5\beta.6\beta$-epoxy-ætiocholan-3β-ol-17-one (see p. 2770 s) on partial hydrogenation in glacial acetic acid using Adams's PtO_2 catalyst for 2 hrs. (1944 Ruzicka). From the oxime of Δ^{16}-allopregnen-20-one (p. 2217 s) on treatment with p-toluenesulphonyl chloride in cold pyridine, followed by hydrolysis with dil. H_2SO_4 at room temperature (1941 PARKE, DAVIS & Co.). From $\Delta^{17(20)}$-allopregnen-21-oic acid on ozonolysis in ice-cold chloroform (1942 Marker).

Rns. On heating of its semicarbazone with sodium ethoxide in alcohol for 18 hrs. it gives androstane (p. 1396 s) and androstan-17β-ol (p. 1512 s) (1937 Burrows).

Semicarbazone $C_{20}H_{33}ON_3$ (p. 119). Microcrystalline powder (dioxane), m. 287–8° on rapid heating (1937 Burrows); crystals (chloroform), m. 284–5° dec. (1942 Marker).

1934 Butenandt, Dannenbaum, *Z. physiol. Ch.* **229** 192, 206.
1936 Reichstein, *Helv.* **19** 979, 983.
1937 Burrows, Cook, Roe, Warren, *Biochem. J.* **31** 950, 954 et seq.
     Kuwada, Miyasaka, Yosiki, *C. A.* **1938** 1275; *Ch. Ztbl.* **1938** II 1611.
     Marker, Kamm, Jones, Mixon, *J. Am. Ch. Soc.* **59** 1363.
     Reichstein, *Helv.* **20** 978.
     SCHERING A.-G., *Dutch Pat.* 47962 (issued 1940); *Ch. Ztbl.* **1940** II 374.
     SCHERING CORP. (Inhoffen, Butenandt, Schwenk), *U.S. Pat.* 2340388 (issued 1944); *C. A.*
       1944 4386; *Ch. Ztbl.* **1947** 76; SCHERING-KAHLBAUM A.-G., *Fr. Pat.* 835524 (issued 1938)
       and *Brit. Pat.* 500353 (issued 1939); *Ch. Ztbl.* **1939** I 5010.
     SCHERING-KAHLBAUM A.-G., *Brit. Pat.* 494773 (issued 1938); *C. A.* **1939** 2537; *Ch. Ztbl.*
       1939 I 4652.
     Steiger, Reichstein, *Helv.* **20** 817, 825, 826.
1938 Butenandt, Heusner, *Ber.* **71** 198, 204.
     Callow, Callow, Emmens, *Biochem. J.* **32** 1312, 1321, 1322.
1939 a CIBA*, *Swiss Pat.* 230497 (issued 1944); *C. A.* **1949** 4432; *Ch. Ztbl.* **1944** II 1203; *Fr. Pat.*
       886410 (1940, issued 1943); *Ch. Ztbl.* **1944** I 448; CIBA* (Miescher, Wettstein), *Swed. Pat.*
       103744 (1940, issued 1942); *Ch. Ztbl.* **1942** II 2294; *Brit. Pat.* 550478 (issued 1943); *C. A.*
       1944 1612; CIBA PHARMACEUTICAL PRODUCTS (Miescher, Wettstein), *U.S. Pat.* 2319012
       (issued 1943); *C. A.* **1943** 6096.
     b CIBA*, *Swiss Pat.* 241646 (issued 1946); *C. A.* **1949** 7978; CIBA* (Miescher, Wettstein),
       Swed. Pats. 105134, 105135 (1940, issued 1942); *Ch. Ztbl.* **1943** I 1695; *Brit. Pat.* 550684
       (issued 1943); *C. A.* **1944** 1611; CIBA PHARMACEUTICAL PRODUCTS (Miescher, Wettstein),
       U.S. Pat. 2374369 (issued 1945); *C. A.* **1945** 5412.
     Dannenberg, *Abhandl. Preuss. Akad. Wiss. Math.-naturwiss. Klasse* No. 21, pp. 8, 51.
     Marker, Plambeck, *J. Am. Ch. Soc.* **61** 1332.
1940 Engel, Thorn, Lewis, *Am. J. Physiol.* **129** 352.
     Hirschmann, *J. Biol. Ch.* **136** 483, 484, 492.
     Marker, Wittle, Tullar, *J. Am. Ch. Soc.* **62** 223.
     PARKE, DAVIS & Co. (Marker, Wittle), *U.S. Pats.* 2397424 (issued 1946), 2397425 (1943, issued
       1946), 2397426 (1944, issued 1946); *C. A.* **1946** 3570.
1941 PARKE, DAVIS & Co. (Tendick, Lawson), *U.S. Pat.* 2335616 (issued 1943); *C. A.* **1944** 3095;
       Brit. Pat. 563889 (issued 1944); *C. A.* **1946** 2942.
     Pincus, Pearlman, *Endocrinology* **29** 413, 423.
     Wolfe, Fieser, Friedgood, *J. Am. Ch. Soc.* **63** 582.
1942 Butenandt, Surányi, *Ber.* **75** 591, 597.
     Dobriner, Gordon, Rhoads, Lieberman, Fieser, *Science* [N. S.] **95** 534.

* Société pour l'industrie chimique à Bâle; Gesellschaft für chemische Industrie in Basel.

Marker, Crooks, Wagner, Shabica, Jones, Wittbecker, *J. Am. Ch. Soc.* **64** 822, 824; PARKE, DAVIS & Co. (Marker, Crooks), *U.S. Pat.* 2359773 (1941, issued 1944); *C. A.* **1945** 4198; *Ch. Ztbl.* **1945** II 1233.

Pearlman, *Endocrinology* **30** 270.

Selye, *Endocrinology* **30** 437, 442.

Venning, Hoffman, Browne, *J. Biol. Ch.* **146** 369, 374 et seq.

1943 Pincus, *Endocrinology* **32** 176.

1944 Fish, Dorfman, *Endocrinology* **35** 23.

Prelog, Ruzicka, Wieland, *Helv.* **27** 66.

Ruzicka, Muhr, *Helv.* **27** 503, 511, 512.

Zimmermann, *Vitamine und Hormone* **5** [1952] 10, 271.

1945 Prelog, Ruzicka, Meister, Wieland, *Helv.* **28** 618, 623 et seq.

Ross, *J. Ch. Soc.* **1945** 25.

Ruzicka, Prelog, Meister, *Helv.* **28** 1651.

Schiller, Dorfman, Miller, *Endocrinology* **36** 355.

Wilson, Nathanson, *Endocrinology* **37** 208, 211.

1946 Dingemanse, Huis in't Veld, *Acta Brevia Neerland.* **14** 34.

Heard, McKay, *J. Biol. Ch.* **165** 677, 682, 684.

Shoppee, *J. Ch. Soc.* **1946** 1138, 1143, 1147.

1947 Djerassi, *J. Org. Ch.* **12** 823, 824.

Ruzicka, *Helv.* **30** 867, 871.

---

4. Monoketo-steroids (CO in the Ring System) C_{20}–C_{26} With Three Aliphatic Side Chains

17-Methylene-Δ^4-androsten-3-one $C_{20}H_{28}O$. For confirmation of the structure tentatively assigned to this compound by Miescher (1939), see 1955 Sondheimer. — Crystals (hexane), m. 135–6° cor.; $[\alpha]_D^{21}$ +137° (alcohol) (1939 Miescher); crystals (acetone-hexane, or methanol), m. 134–5°; $[\alpha]_D^{20}$ +124° (alcohol), +131° (chloroform); absorption max. in 95% alcohol at 240 mμ (log ε 4.23) (1955 Sondheimer). — **Fmn.** From 17-methyltestosterone (p. 2645 s) or 17-methyl-epitestosterone (p. 2646 s) on sublimation with anhydrous $CuSO_4$ at 135–150° and 0.01 mm. pressure; yield, 89% (1939 Miescher). From 17-methylene-Δ^5-androsten-3β-ol on re-fluxing with aluminium isopropoxide and cyclohexanone in dry toluene for 1 hr.; yield, 79% (1955 Sondheimer). — **Rns.** Oxidation in abs. ether with OsO_4 at —10° gives 17β-hydroxymethyl-Δ^4-androsten-17α-ol-3-one (p. 2719 s) (1939 Miescher; 1955 Sondheimer).

Semicarbazone $C_{21}H_{31}ON_3$. Crystals (alcohol), m. 230–1° cor. dec. (1939 Miescher).

17-Methylene-androstan-3-one $C_{20}H_{30}O$. Crystals (petroleum ether), m. 127–8° cor., subliming at 95° and 0.01 mm. pressure, $[\alpha]_D^{21}$ +62° (chloroform) (1947 Ruzicka); crystals (methanol), m. 130–131.5°, $[\alpha]_D$ +41° (chloroform), infra-red absorption max. at 882, 1653, and 1706 cm.$^{-1}$ (1957 Sondheimer). The odour of the compound resembles

that of cedar oil (1947 Ruzicka). — **Fmn.** From 17-methylene-androstan-3β-ol
(p. 1516 s) on oxidation with the CrO_3-pyridine complex in pyridine at room
temperature (1957 Sondheimer). From 17α-methylandrostan-17β-ol-3-one
(p. 2647 s) on distillation with anhydrous $CuSO_4$ under 0.1 mm. pressure (yield,
35 %) (1947 Ruzicka); for the homogeneity of the product so obtained see,
however, 1957 Sondheimer. — **Rns.** Oxidation with OsO_4 in abs. ether +
pyridine at room temperature, yields 17-hydroxymethyl-androstan-17-ol-3-one
(mixture of the two 17-epimers?, p. 2720 s). Hydrogenation in alcohol in the
presence of a palladium-$BaCO_3$ catalyst yields 17β-methylandrostan-3-one
(see below). On boiling with aluminium isopropoxide in abs. isopropyl alcohol, a mixture of the 3-epimeric 17-methylene-androstan-3-ols (p. 1516 s) is
produced (1947 Ruzicka).

17β-Methylandrostan-3-one $C_{20}H_{32}O$. Crystals (petroleum ether), m. 130–2°

cor., subliming at 85° under 0.01 mm. pressure; $[\alpha]_D^{21}$
+ 31.6° ± 2° (chloroform). — **Fmn.** From 17-methyl-
ene-androstan-3-one (of doubtful homogeneity; see
above) on hydrogenation in alcohol in the presence
of a palladium-$BaCO_3$ catalyst (1947 Ruzicka).

Δ⁴·¹⁶.T²⁰- Pregnadienyn-3-one(?), 17-Ethynyl-Δ⁴·¹⁶-androstadien-3-one(?)

$C_{21}H_{26}O$. The structure ascribed to this compound by
Inhoffen (1938) is questioned by Hardegger (1945).—
Needles (methanol), m. 166° (1938 Inhoffen). — Has
no hormonal activity in the Allen-Corner test (1938
Inhoffen).

Fmn. From 17-ethynyltestosterone (p. 2648 s) on refluxing with 90 % formic acid; yield, 43 % (1938 Inhoffen). — **Rns.** Can be hydrated to Δ⁴·¹⁶-preg-
nadiene-3.20-dione (p. 2780 s) by heating with $HgSO_4$ in 70 % acetic acid
and some conc. H_2SO_4 or by refluxing with $HgO + BF_3$-ether in glacial acetic
acid and some acetic anhydride, followed by hydrolysis with boiling methanolic KOH (1938 Schering A.-G.).

Δ⁴·¹⁶·²⁰- Pregnatrien-3-one, 17-Vinyl-Δ⁴·¹⁶-androstadien-3-one $C_{21}H_{28}O$. Was

not obtained quite pure. — Needles (methanol),
m. 122–9° cor.; needles (aqueous methanol; from
the mother liquor), m. 124–135°. — **Fmn.** From
17-vinyltestosterone (p. 2650 s) on treatment with
PBr_3 and then with pyridine, followed by subli-
mation of the resulting pyridinium bromide compound (p. 2473 s) at 215–8°
under 0.02 mm. pressure (1940 Reich).

Δ⁴·¹⁷⁽²⁰⁾-Pregnadien-3-one, 17-Ethylidene-Δ⁴-androsten-3-one $C_{21}H_{30}O$ (p. 145)
(I, p. 2408 s). The ketone prepared by Ruzicka (1939) was later recognized
(1942a Ruzicka) to be identical with that of Butenandt (p. 145); for probable

trans-configuration of the methyl group at C-20 with respect to C-13, see $\Delta^{5 \cdot 17(20)}$-pregnadien-3β-ol (p. 1518 s). — Crystals (methanol), m. 142–3° cor.; [α]$_D$ +117.5° (chloroform) (1939 Ruzicka).

Fmn. From $\Delta^{5 \cdot 17(20)}$-pregnadien-3β-ol (p. 1518 s) on refluxing with aluminium tert.-butoxide and acetone in benzene; yield, 88% (1939 Ruzicka). From $\Delta^{4 \cdot 17(20)}$-pregnadien-3-one enol-benzyl ether ($\Delta^{3 \cdot 5 \cdot 17(20)}$-pregnatrien-3-yl benzyl ether, p. 1517 s) on hydrolysis with hydrochloric acid (1939 SCHERING A.-G.). From 17α-ethyl-Δ^4-androsten-17β-ol-3-one (17-ethyltestosterone, page 2651 s) on dehydration by sublimation in a high vacuum at 115–120° in the presence of anhydrous $CuSO_4$ (yield, 45%) (1938a Butenandt; for other methods of dehydration, see 1936 SCHERING-KAHLBAUM A.-G.).

Rns. Oxidation with OsO_4 in abs. ether at 0° gives Δ^4-pregnene-17α.20β-diol-3-one (p. 2721 s) (1938a Butenandt; 1939 SCHERING A.-G.; cf. 1942a Ruzicka). Oxidation with monoperphthalic acid in chloroform at 0°, then at room temperature (see also the scheme on p. 2408 s), gives a mixture of variable m. p., consisting of compounds all having the molecular formula $C_{21}H_{30}O_2$, from which three pure individual compounds, A, B, and C, may be isolated by fractional crystallization and chromatography; the B-isomer is obviously 17α.20β-epoxy-Δ^4-pregnen-3-one (II, below), the A-isomer very probably an epimeric 17.20-epoxy-Δ^4-pregnen-3-one (below), whereas the C-isomer (p. 2410 s) seems to be a doubly unsaturated keto-alcohol (1942a Ruzicka; cf. 1942b Ruzicka; 1943 Shoppee).

Semicarbazone $C_{22}H_{33}ON_3$ (p. 145). Crystals (methanol), m. 224–6° cor. dec. (1942a Ruzicka).

enol-Benzyl ether $C_{28}H_{36}O$. See $\Delta^{3 \cdot 5 \cdot 17(20)}$-pregnatrien-3-yl benzyl ether (page 1517 s).

"A-Isomer", probably epimeric with the "B-isomer" (below), $C_{21}H_{30}O_2$. Granular crystals (chloroform-ether), m. 174.5–175.5° cor.; [α]$_D$ +82° (chloroform). Absorption max. (solvent not stated) at 240 (4.2) and 315 (1.8) mμ (log ε). — **Fmn.** See under the B-isomer (below). — **Rns.** Is very slowly oxidized by monoperphthalic acid. Is not changed on standing at room temperature with 0.5 N methanolic KOH*. — *Semicarbazone* $C_{22}H_{33}O_2N_3$, leaflets (aqueous methanol), m. 227–8° cor. dec. (1942a Ruzicka).

"B-Isomer", 17α.20β-Epoxy-Δ^4-pregnen-3-one $C_{21}H_{30}O_2$ (II, p. 2408 s). The configurations at C-17 and C-20 have been assigned by the editor in view of the configuration of the parent compound (I, p. 2406 s) and of the stereochemical course of the reaction involved (see 1950 Shoppee). — Crystals (ethyl acetate), m. 188.5–190° cor.; [α]$_D$ +106° (chloroform); absorption max. (solvent not stated) at 240 (4.2) and 315 (1.8) mμ (log ε) (1942a Ruzicka).

Shows no progestational hormonal activity (1942a Ruzicka).

* In the table on p. 1299 of the original paper (1942a Ruzicka), the reactions of the isomers A and B with methanolic KOH and with monoperphthalic acid have been inverted (1955 Hardegger).

E.O.C. XIV s. 168

Reaction of $\Delta^{4,17(20)}$-Pregnadien-3-one (I) with Monoperphthalic Acid

Fmn.* As the main product along with the A-isomer (p. 2407 s) and the C-isomer (p. 2410 s), from $\Delta^{4,17(20)}$-pregnadien-3-one (I, p. 2406 s) on oxidation with monoperphthalic acid in chloroform at room temperature in the dark; the isomers are separated by fractional crystallization from ethyl acetate (the B-isomer being the less soluble) and by chromatography in benzene-petroleum ether on Al_2O_3 (1942a Ruzicka).

Rns.* Does not react with monoperphthalic acid. Isomerization to prog-esterone (p. 2782 s) or 17-iso-progesterone (p. 2800 s) could not be effected by heating in the presence of catalysts (1942a Ruzicka). Treatment with glacial acetic acid at room temperature gives a rearrangement product (III, p. 2409 s); with anhydrous $ZnCl_2$ in acetic anhydride at room temperature, the acetate of this product is formed in low yield, together with a **compound** $C_{23}H_{32}O_4$ [needles (aqueous acetone), m. 136° cor.] (1942b Ruzicka). No reaction occurs with acetic anhydride and pyridine at room temperature. No hydrolysis to the corresponding 17.20-diol could be effected with water-dioxane, with dil. acids, or with anhydrous $MgBr_2$. With methanolic KOH** at room temperature, the B-isomer is partly transformed into a waxy product (1942a Ruzicka).

* See also the scheme, above.
** See the footnote on p. 2407 s.

Semicarbazone $C_{22}H_{33}O_2N_3$, needles (aqueous methanol), m. 217–8⁰ cor. dec. (1942a Ruzicka).

17-Methyl-18-nor-$\Delta^{4.13}$-pregnadien-20-ol-3-one, $\Delta^{4.13}$-Retropregnadien-20-ol-3-one (?) $C_{21}H_{30}O_2$ (III, p. 2408 s). The structure of a similar compound is uncertain (1951 Lardon). — Needles with $1/2$ mol. of acetone (aqueous acetone), m. 125.5–126.5⁰ cor.; becomes acetone-free on drying in a current of air at 100⁰; $[\alpha]_D$ +30.5⁰ (chloroform). Shows an intense yellow coloration with tetranitromethane (1942b Ruzicka). — **Fmn.*** From the preceding compound on treatment with glacial acetic acid at room temperature (yield, 84%) (1942b Ruzicka); for the probable mechanism of this reaction, see 1943 Shoppee. — **Rns.*** With monoperphthalic acid in chloroform at room temperature it forms an epoxide (IV, below). Oxidation with OsO_4 in abs. ether at room temperature gives a Δ^4-retropregnene-13.14.20-triol-3-one (? V, p. 2410 s). Hydrogenation, first in alcohol in the presence of a palladium-$CaCO_3$ catalyst (1 mol. of hydrogen absorbed), then in glacial acetic acid in the presence of a PtO_2 catalyst to saturation (two more mols. of hydrogen absorbed), followed by oxidation of the saturated diol mixture with CrO_3 in glacial acetic acid at room temperature, gives 34% yield of a saturated **diketone** [**retropregnane-3.20-dione?** (VI, p. 2408 s; R = O)] $C_{21}H_{32}O_2$, leaflets (aqueous acetone), m. 80–80.5⁰ cor., $[\alpha]_D$ +4⁰ (chloroform) (1942b Ruzicka).

Semicarbazone $C_{22}H_{33}O_2N_3$, crystals (alcohol), m. 213–4⁰ cor. (1942b Ruzicka).

Acetate $C_{23}H_{32}O_3$. Needles with $1/2$ H_2O (ethyl acetate), m. 172⁰ cor.; $[\alpha]_D$ +59⁰ (chloroform). Gives a yellow coloration with tetranitromethane in chloroform. — **Fmn.*** From the above hydroxy-ketone on treatment with acetic anhydride in pyridine at room temperature; quantitative yield. From 17α.20β-epoxy-Δ^4-pregnen-3-one (II, p. 2407 s) on treatment with acetic anhydride and anhydrous $ZnCl_2$ at room temperature; low yield (1942b Ruzicka). — **Rns.*** Regenerates $\Delta^{4.13}$-retropregnadien-20-ol-3-one (III), when heated with methanolic KOH. Oxidation with monoperphthalic acid at room temperature in chloroform yields an epoxide, m. 220–1⁰ cor., probably the acetate of the epoxide (IV), below. No perceptible oxidation with OsO_4 occurs. Hydrogenation, first in alcohol in the presence of a palladium-$CaCO_3$ catalyst, then in glacial acetic acid in the presence of a PtO_2 catalyst, followed by oxidation with CrO_3 in glacial acetic acid, affords 50% yield of an **acetoxy-ketone** [**retropregnan-20-ol-3-one acetate?** (VI, p. 2408 s; CR = CH·OAc)] $C_{23}H_{36}O_3$ [prisms (pentane), m. 116–7⁰ cor.; $[\alpha]_D$ +14.3⁰ (chloroform)], which on heating with hydrazine hydrate and alcoholic sodium ethoxide at 200⁰ gives a "pregnanol" $C_{21}H_{36}O$ (retropregnan-20-ol?, p. 1488 s) (1942b Ruzicka).

13.14-Epoxy-Δ^4-retropregnen-20-ol-3-one, $\Delta^{4.13}$-Retropregnadien-20-ol-3-one 13.14-oxide (?) $C_{21}H_{30}O_3$ (IV, p. 2408 s). Needles (ethyl acetate-hexane), m. 162⁰ cor. — **Fmn.*** From $\Delta^{4.13}$-retropregnadien-20-ol-3-one (above) on treatment with monoperphthalic acid in chloroform at room temperature. —

---

* See also the scheme on p. 2408 s.

168*

Rns. Treatment with acetic anhydride in pyridine gives a crystalline compound, m. 148–9⁰ cor. (1942 b Ruzicka).

Acetate (?) $C_{23}H_{32}O_4$. Leaflets (ethyl acetate-chloroform), m. 220–1⁰ cor. — **Fmn.*** From the acetate of $\Delta^{4 \cdot 13}$-retropregnadien-20-ol-3-one (p. 2409 s) on treatment with monoperphthalic acid in chloroform at room temperature (1942 b Ruzicka).

Δ^4-Retropregnene-13.14.20-triol-3-one (?) $C_{21}H_{32}O_4$ (V, p. 2408 s). Quadratic crystals (aqueous methanol), m. 227–8⁰ cor., subliming at 220⁰ in a high vacuum. — **Fmn.*** From $\Delta^{4 \cdot 13}$-retropregnadien-20-ol-3-one (III, p. 2409 s) on treatment with OsO_4 in abs. ether at room temperature. — **Rns.** Oxidation with HIO_4 in aqueous methanol at 50⁰ yielded neither volatile nor crystalline products (1942 b Ruzicka).

Monoacetate $C_{23}H_{34}O_5$. Pale yellow resin, distilling at 170⁰ and 0.01 mm. pressure. — **Fmn.** From the above triolone with acetic anhydride in pyridine at room temperature (1942 b Ruzicka).

"C-Isomer" $C_{21}H_{30}O_2$. Doubly unsaturated keto-alcohol; structure of ring A as in the isomers A and B (p. 2407 s). — Needles (ether), m. 189–190⁰ cor.; $[\alpha]_D$ +111⁰ (chloroform); absorption max. (solvent not stated) at 240 (4.2) and 315 (1.8) mμ (log ε). Gives a yellow coloration with tetranitromethane. — **Fmn.** See under formation of the B-isomer (p. 2408 s); the C-isomer is perhaps not contained in the crude reaction mixture, but formed during the working up of the latter. — **Rns.** Consumes 1 atom of oxygen on treatment with monoperphthalic acid. Is not changed on standing at room temperature with methanolic KOH. — *Semicarbazone* $C_{22}H_{33}O_2N_3$, granular crystals (aqueous methanol), m. 207⁰ cor., beginning to decompose at 200⁰. — *Acetate* $C_{23}H_{32}O_3$, crystals (carbon tetrachloride-hexane), m. 152.5–153.5⁰ cor.; prepared with acetic anhydride + pyridine (1942 a Ruzicka).

$\Delta^{4 \cdot 20}$-Pregnadien-3-one (?), 17β-Vinyl-Δ^4-androsten-3-one (?) $C_{21}H_{30}O$. Neither
analysis nor constants are given; mode of formation not recorded. — **Rns.** Oxidation with OsO_4 in abs. ether at room temperature gives Δ^4-pregnene-20.21-diol-3-one (m. 166–7⁰, p. 2514 s) (1937 SCHERING A.-G.).

Δ^4-Pregnen-3-one $C_{21}H_{32}O$. Needles (aqueous alcohol), m. 90⁰; very soluble in the usual solvents. — **Fmn.** From 4β-bromopregnan-3-one (p. 2477 s) on refluxing with pyridine; yield, 27 % (1939 a Marker).

Semicarbazone $C_{22}H_{35}ON_3$. Crystals (dil. alcohol), m. 216⁰; fairly soluble in most organic solvents (1939 a Marker).

2.4-Dinitrophenylhydrazone $C_{27}H_{36}O_4N_4$. Dark red crystals (alcohol), m. 198⁰ (1939 a Marker).

* See also the scheme on p. 2408 s.

$\Delta^{17(20)}$-**Allopregnen-3-one, 17-Ethylidene-androstan-3-one** $C_{21}H_{32}O$. Solid, m. 112–9° cor.; sublimes at 85° in a high vacuum; $[\alpha]_D^{22}$ +55.5° (chloroform) (1947 Ruzicka). — **Fmn.** From allopregnan-20β-ol-3-one (p. 2511 s) on refluxing with $ZnCl_2$ in glacial acetic acid for 3 hrs. (not isolated in a pure state) (1937 Marker; 1938 PARKE, DAVIS & Co.). From 17-iso-allopregnan-17β-ol-3-one (p. 2652 s) on sublimation in the presence of anhydrous $CuSO_4$ at 140° under 0.01 mm. pressure; yield, 93% (1947 Ruzicka).

Rns. Gives androstane-3.17-dione (p. 2892 s) on ozonolysis in chloroform at 0°, followed by oxidation with CrO_3 in glacial acetic acid at room temperature (1937 Marker; 1938 PARKE, DAVIS & Co.). Treatment with OsO_4 and some pyridine in abs. ether, followed by shaking with mannitol and 1 N aqueous KOH, affords allopregnane-17α.20β-diol-3-one (p. 2721 s) (1947 Ruzicka). On hydrogenation in glacial acetic acid in the presence of palladium-barium sulphate it gives allopregnan-3-one (below), whereas in the presence of previously reduced platinum oxide allopregnan-3-ol, chiefly the β-epimer (page 1522 s), is obtained (1947 Ruzicka).

Pregnan-3-one $C_{21}H_{34}O$. Needles (aqueous alcohol), m. 115°. Gives a red coloration with m-dinitrobenzene and KOH in aqueous alcohol (1938 Marker). — **Fmn.** From pregnan-3α-ol (p. 1521 s) on oxidation in acetic acid with CrO_3 at room temperature; yield, 71% (1938 Marker). — **Rns.** Catalytic hydrogenation in acetic acid + some conc. HBr (PtO_2 catalyst, at room temperature and 2.3 atm.), followed by heating with alcoholic NaOH, affords pregnan-3β-ol (p. 1522 s) (1938 Marker). Treatment with bromine and a little HBr in glacial acetic acid at room temperature yields 4β-bromopregnan-3-one (p. 2477 s) (1939a Marker).

Semicarbazone $C_{22}H_{37}ON_3$. The crude product melts at 133° and is very soluble in the ordinary solvents except ether (1938 Marker).

2.4-Dinitrophenylhydrazone $C_{27}H_{38}O_4N_4$, crystals (alcohol), m. 163° (1938 Marker).

Allopregnan-3-one $C_{21}H_{34}O$. Crystals (alcohol), m. 116–7° cor. (1939 Ruzicka), subliming at 80° and 0.005 mm. pressure; $[\alpha]_D^{21}$ +44° (chloroform) (1947 Ruzicka). — **Fmn.** From allopregnan-3β-ol (p. 1522 s) on oxidation with CrO_3 in glacial acetic acid at room temperature; yield, 80% (1939 Ruzicka). From $\Delta^{17(20)}$-allopregnen-3-one (see above) on hydrogenation in the presence of a palladium-$BaSO_4$ catalyst in glacial acetic acid (1947 Ruzicka). — **Rns.** Reduction with aluminium isopropoxide in boiling abs. isopropyl alcohol gives a mixture of the two epimeric allopregnan-3-ols (p. 1522 s) (1947 Ruzicka). On heating the hydrazone with hydrazine hydrate and sodium ethoxide in alcohol at 200°, allopregnane (p. 1403 s) is obtained (1939 Ruzicka).

References, pp. 2414 s, 2415 s

Hydrazone $C_{21}H_{36}N_2$. Crystals (chloroform-methanol), m. ca. 226° cor. dec. (1939 Ruzicka).

Semicarbazone $C_{22}H_{37}ON_3$. Crystals (chloroform-methanol), m. ca. 230° cor. dec. (1939 Ruzicka).

9-iso-Pregnan-11-one $C_{21}H_{34}O$. This structure was ascribed, as the most likely, to **uran-11-one** (1939b Marker). For probable structure of urane and its derivatives, see 9-iso-pregnane (p. 1403 s).

ω-Homo-$\Delta^{4\cdot17(20)\cdot21}$-pregnatrien-3-one, 17-Allylidene-Δ^4-androsten-3-one $C_{22}H_{30}O$. Needles (abs. alcohol-ethyl acetate), m. 172–4° (1938b Butenandt); crystals (acetone-ether), m. 173–5° cor., subliming at 140° to 150° (bath temperature) under 0.02 mm. pressure (1940 v. Euw). Absorption max. in ether at 236 mμ (ε 42 100) (1938b Butenandt; 1939 Dannenberg). — Exerts no hormonal activity in the Allen-Doisy test (on the castrated mouse) and in Fussgänger's capon comb test (1938b Butenandt).

Fmn. From 17-allyltestosterone (p. 2653 s) on dehydration by heating with $POCl_3$ and pyridine; yield, 62 % (1938b Butenandt; cf. 1940 v. Euw). — **Rns.** On oxidation with OsO_4 in abs. ether at room temperature it gives ω-homo-Δ^4-pregnene-17α.20β.21.22-tetrol-3-one (p. 2732 s) (1938b Butenandt) and a **compound** $C_{22}H_{32}O_3$ (m. 142–3° cor.), probably a diolone (1940 v. Euw).

Semicarbazone $C_{23}H_{33}ON_3$. Rectangular blocks (alcohol-chloroform), becoming slightly brown at 250°, not molten at 365° (1938b Butenandt). Absorption max. in chloroform at 240 mμ (ε ca. 30000; from the graph) and 270 mμ (ε 31 300) (1938b Butenandt; 1939 Dannenberg). Very sparingly soluble in alcohol and chloroform (1938b Butenandt).

16-Isopropylidene-$\Delta^{3\cdot5}$-androstadien-17-one $C_{22}H_{30}O$. Needles (hot methanol or ethanol, on slow cooling): prisms (ethanol, on rapid cooling), m. 193–5° (on slow heating). Absorption max. (from the graph; solvent not stated) at 235 (4.48), 244 (4.41), and 338 (1.9) mμ (log ε) — **Fmn.** From $\Delta^{3\cdot5}$-androstadien-17-one (p. 2400 s) on condensation with acetone by refluxing in methanolic KOH; excellent yield. From 16-isopropylidene-Δ^5-androsten-3β-ol-17-one (p. 2656 s) on boiling with P_2O_5 in benzene; is also obtained, along with the last-mentioned compound, from the residues of crystallization of 5-dehydroepiandrosterone acetate (p. 2610 s; obtained by oxidation of cholesteryl acetate dibromide) on refluxing for 2 hrs. with KOH in commercial methanol (containing some acetone). — **Rns.** Reduction of its hydrazone (not described) on heating at 210° with hydrazine hydrate and sodium ethoxide in alcohol affords 16-isopropylidene-$\Delta^{3.5}$-androstadiene (p. 1404 s). Forms no semicarbazone under the usual conditions (1945 Ross).

2 - (3 - Keto - Δ^4 - ternorcholenyl) - propene $C_{24}H_{36}O$. White plates (alcohol), m. 170–1°; $[\alpha]_D^{28} +55°$ (chloroform). — **Fmn.** From 2-(3β-hydroxy-Δ^5-ternorcholenyl)-propene (p. 1526 s) on oxidation with cyclohexanone and aluminium tert.-butoxide in toluene at 100°; yield, 67%. — **Rns.** Ozonization in chloroform at 0° readily gives 3-keto-Δ^4-ternorcholenyl methyl ketone (p. 2804 s), together with some acidic product (1945 Julian).

2-(3-Keto-20-iso-Δ^4-ternorcholenyl)-propene $C_{24}H_{36}O$. Pale yellow prisms (methanol), m. 142–3°; $[\alpha]_D^{25} +144°$ (chloroform). — **Fmn.** From 2-(3β-hydroxy-20-iso-Δ^5-ternorcholenyl)-propene (p. 1526 s) on refluxing with cyclohexanone and aluminium tert.-butoxide in dry toluene; yield, 70%. — **Rns.** Reacts sluggishly with ozone to form an ozonide, more difficult to hydrolyse than the corresponding ozonide of the above 20 n-isomer (1945 Julian).

Oxime $C_{24}H_{37}ON$. Crystals (alcohol), m. 224–7° dec. (1945 Julian).

23.24-Bisnor-coprostan-3-one, 23-Methylcholan-3-one $C_{25}H_{42}O$. For nomenclature and numbering, see 23.24-bisnor-coprostan-3β-ol, p. 1528 s. — Not obtained crystalline; neither analysis nor constants are given. — **Fmn.** From epi-23.24-bisnor-coprostanol (p. 1528 s) on oxidation with CrO_3 in glacial acetic acid at room temperature. — **Rns.** Hydrogenation in glacial acetic acid and conc. HCl in the presence of a platinum oxide catalyst gives a mixture of the two epimeric bisnor-coprostan-3-ols (p. 1528 s) (1936 Reindel).

Semicarbazone $C_{26}H_{45}ON_3$, m. 158–161° (1936 Reindel).

24.24-Dimethyl-$\Delta^{4.20(22).23}$-cholatrien-3-one $C_{26}H_{38}O$. Obtained only as crude product (no constants given). — **Fmn.** From the 3-acetate of 24.24-dimethyl-Δ^5-cholene-3β.24-diol (p. 1942 s) which is dehydrated with P_2O_5 in benzene, yielding **24.24-dimethyl-$\Delta^{5.23}$-choladien-3β-yl acetate** $C_{28}H_{44}O_2$ (m. 124–6°); this compound is treated with dry HCl in cold abs. alcohol-ether and the *hydrochloride* $C_{28}H_{45}O_2Cl$ so obtained (needles,

no constants given) boiled 20 min. with N-bromosuccinimide in carbon tetrachloride, yielding the *22-bromo-compound* (not purified) which is refluxed 1 hr.
with collidine; the trienyl acetate so formed (not isolated) is saponified with
boiling methanolic KOH and the crude *trienol* so obtained (mixture of the
$\varDelta^4$- and $\varDelta^5$-isomers?-Editor) oxidized by Oppenauer's method, e.g. with
aluminium isopropoxide and cyclohexanone. From 24.24-dimethyl-$\varDelta^{4\cdot23}$-chola-
dien-3-one (below) by bromination with N-bromosuccinimide, followed by
dehydrobromination, as described above. — **Rns.** Oxidation with CrO_3 in
strong acetic acid at room temperature yields $\varDelta^4$-pregnene-3.20-dione (progesterone, p. 2782 s) (1943 Ciba Pharmaceutical Products).

24.24-Dimethyl-$\varDelta^{4\cdot23}$-choladien-3-one $C_{26}H_{40}O$. No constants given. — **Fmn.**

24.24 - Dimethyl - $\varDelta^{5\cdot23}$ - choladien - 3β - yl
acetate (p. 2413 s) is saponified and the
hydroxy-compound (not described) oxidized with aluminium isopropoxide and
cyclohexanone. — **Rns.** Bromination
with N-bromosuccinimide in carbon tetrachloride, followed by dehydro-
bromination, e.g. by refluxing with collidine, yields 24.24-dimethyl-$\varDelta^{4\cdot20(22)\cdot23}$-
cholatrien-3-one (p. 2413 s) (1943 Ciba Pharmaceutical Products).

24-Nor-coprostan-3-one, 24.24-Dimethylcholan-3-one $C_{26}H_{44}O$. For nomen-

clature and numbering, see 24-nor-
coprostan-3β-ol, p. 1529 s. — Not
obtained crystalline; neither analysis nor constants are given. —
Fmn. From 24-nor-coprostan-3α-ol
(p. 1530 s) on oxidation with CrO_3
in glacial acetic acid at room temperature. — **Rns.** Hydrogenation in glacial
acetic acid + HCl in the presence of a PtO_2 catalyst gives a mixture of the
two epimeric 24-nor-coprostan-3-ols (1936 Reindel).

Semicarbazone $C_{27}H_{47}ON_3$, m. 167–8° dec. (1936 Reindel).

1936 Reindel, Niederländer, *Ann.* **522** 218, 228, 238.
Schering-Kahlbaum A.-G., *Swiss Pat.* 198704 (issued 1938); *Ch. Ztbl.* **1939** I 2642.
1937 Marker, Kamm, Jones, Oakwood, *J. Am. Ch. Soc.* **59** 614.
Schering A.-G., *Swiss Pat.* 223208 (issued 1942); *Ch. Ztbl.* **1943** I 2113.
1938 a Butenandt, Schmidt-Thomé, Paul, *Ber.* **71** 1313.
b Butenandt, Peters, *Ber.* **71** 2688, 2693.
Inhoffen, Logemann, Hohlweg, Serini, *Ber.* **71** 1024, 1029, 1032.
Marker, Lawson, *J. Am. Ch. Soc.* **60** 2438.
Parke, Davis & Co. (Marker, Jones, Oakwood), *Brit. Pat.* 512940 (issued 1939); *C. A.* **1941**
1187; *Ch. Ztbl.* **1940** I 2826, 2827.
Schering A.-G., *Dutch Pat.* 52579 (issued 1942); *Ch. Ztbl.* **1942** II 2615; *Fr. Pat.* 845473
(issued 1939); *C. A.* **1941** 1185; *Ch.Ztbl.* **1943** I 653; Schering Corp. (Inhoffen, Logemann,
Dannenbaum), *U.S. Pat.* 2280236 (1939, issued 1942); *C. A.* **1942** 5322.
1939 Dannenberg, *Abhandl. Preuss. Akad. Wiss. Math.-naturwiss. Klasse* **1939** No. 21, pp. 47–49, 65.
a Marker, Lawson, *J. Am. Ch. Soc.* **61** 586.
b Marker, Rohrmann, *J. Am. Ch. Soc.* **61** 2719, 2721.
Miescher, Klarer, *Helv.* **22** 962, 968.

Ruzicka, Goldberg, Hardegger, *Helv.* **22** 1294, 1297–1299.
SCHERING A.-G., *Fr. Pat.* 868721 (issued 1942); *Ch. Ztbl.* **1942** II 74; *Swed. Pat.* 103355 (issued 1941); *Ch. Ztbl.* **1942** II 1603.
1940 v. Euw, Reichstein, *Helv.* **23** 1114, 1118, 1122.
Reich, *Helv.* **23** 219, 223.
1942 a Ruzicka, Goldberg, Hardegger, *Helv.* **25** 1297.
b Ruzicka, Goldberg, Hardegger, *Helv.* **25** 1680.
1943 CIBA PHARMACEUTICAL PRODUCTS (Miescher, Frey, Meystre, Wettstein), *U.S. Pat.* 2461563 (issued 1949); *C. A.* **1949** 3474; *Ch. Ztbl.* **1949** E 1870.
Shoppee, Prins, *Helv.* **26** 1004, 1008, 1009.
1945 Hardegger, Scholz, *Helv.* **28** 1355, 1356.
Julian, Cole, Meyer, Herness, *J. Am. Ch. Soc.* **67** 1375, 1378, 1379.
Ross, *J. Ch. Soc.* **1945** 25.
1947 Ruzicka, Meister, Prelog, *Helv.* **30** 867, 873, 874, 876, 877.
1950 Shoppee, *Nature* **166** 107.
1951 Lardon, Reichstein, *Helv.* **34** 756, 757 (footnote 3), 763.
1955 Hardegger, *private communication.*
Sondheimer, Mancera, Urquiza, Rosenkranz, *J. Am. Ch. Soc.* **77** 4145, 4148.
1957 Sondheimer, Mechoulam, *J. Am. Ch. Soc.* **79** 5029, 5032, 5033.

5. Monoketo-steroids (CO in the ring system) C_{27} with aliphatic side chains

Summary of the Ketones with Cholestane (or Coprostane) Skeleton *

Position of CO	Config. at C-5	Position of double bond(s)	M.p. °C.	$[\alpha]_D$ (°) in chloroform	$\lambda_{max.}$ (mμ) in alcohol	Oxime m. p. °C.	Semi-carbazone, m. p. °C.	Page
1	α	satd.	86	+112	—	—	—	2417 s
2	—	$\Delta^{3.5}$	122	−62	290[1])	—	—	2417 s
2	α	satd.	131	+50	—	200[2])	—	2417 s
3	—	$\Delta^{1.4.6}$	83	±0	224, 258, 300	—	—	2418 s
3	—	$\Delta^{1.4}$	112	+30	245; 236[3])	—	231	2418 s
3	—	$\Delta^{4.6}$	81	+34	285	178	230 dec.	2420 s
3	—	$\Delta^{4.7}$	88	+34	239; 230[3])	—	240 dec.	2421 s
3	α	$\Delta^{8.24}$	105	+70	—	—	230 dec.	2422 s
3	α	$\Delta^{8(14).24}$ or $\Delta^{14.24}$ }	163	—	—	239	—	2422 s
3	α	Δ^1	100	+58	230	—	—	2423 s
3	β	Δ^1	101	+65[4])	230[3])	—	207	2423 s
3	—	Δ^4	82 and 88	+92	240.5, 242	60; 153	235	2424 s
3	—	Δ^5	126	−8	—	188 dec.	—	2435 s
3	α	Δ^8	125	+71	—	—	243 dec.	2436 s
3	α	satd.	129	+42	—	199	238 dec.	2436 s
3	β	satd.	62	+36	—	71	192 dec.	2443 s
4	—	Δ^5	112	−32[5])	241	147	226	2450 s
4	α	satd.	99	+30	—	206	—	2451 s
6	—	$\Delta^{2.4}$	130	—	314	—	—	2452 s
6	α	Δ^2	105	+28	285–300[1])	186	—	2452 s
6	—	i[(3:5)]	97	+45	290[1])	144	—	2452 s
6	α	satd.	99	−7	—	197	—	2454 s
7	—	$\Delta^{3.5}$	114.5	−306	280, 325	177	207	2454 s
7	—	Δ^5	130	−138	235–237[1])	180	225 dec.	2455 s
7	α	Δ^8	87	+4	251[1])	—	—	2456 s
7	α	satd.	116	−47	292	135	—	2456 s

* For values other than those chosen for the summary, see the compounds themselves.
[1]) Solvent not indicated.   [2]) 1949 Fürst.   [3]) In ether.   [4]) Purity doubtful (see text).
[5]) In alcohol.

19-Nor-$\Delta^{5.7.9}$-ergostatrien-3-one $C_{27}H_{40}O$. Crystals (methanol), m. 174°. —
Fmn. From 22.23-dihydro-neoergosterol (p. 1534 s: 19-nor-$\Delta^{5.7.9}$-ergostatrien-3β(?)-ol) by refluxing with copper bronze at 300° and 1–2 mm. pressure,

followed by distillation, the ketone is obtained in small amounts together
with the main product, a compound m. 140°, probably *19-nor-$\Delta^{1.3.5(10).6.8}$-
ergostapentaen-3-ol* $C_{27}H_{38}O$ (I) (no analysis; yields a sparingly soluble sodium
salt) (1937 Krekeler).

1937 Krekeler, *Einige Umsetzungen am Cholestenon*, Thesis Univ. Göttingen, pp. 10, 19.

Cholestan-1-one $C_{27}H_{46}O$. The compound [m. 120–120.5° cor., $[\alpha]_D$ +41° (chloroform)] to which this structure has been assigned, obtained by CrO_3 oxidation of a supposed cholestan-1-ol (cf. p. 1537 s) (1944b Ruzicka), is now considered to be probably an addition compound (1:1) of cholestan-2-one with cholestan-4-one (p. 2451 s) (Editor; cf. 1953 Fieser)*.

$\varDelta^{3.5}$-**Cholestadien-2-one** $C_{27}H_{42}O$. Needles (aqueous methanol), m. 121.5–122.5° cor.; sublimes in high vac. at 180°; $[\alpha]_D$ —62° (chloroform); absorption max. (solvent not stated) 290 mμ (log ε 4.1). — **Fmn.** From the p-toluenesulphonate of $\varDelta^3$-cholesten-3-ol-2-one (p. 2659 s) by heating with sodium iodide in dry acetone at 160° for 17 hrs. (sealed tube), followed by boiling of the resulting neutral dark brown oil with zinc dust in alcohol; crude yield, 49%. — **Rns.** Hydrogenation in glacial acetic acid using platinum oxide catalyst, followed by reoxidation with CrO_3 in strong acetic acid, all at room temperature, affords cholestan-2-one (below) (1944a Ruzicka).

Cholestan-2-one $C_{27}H_{46}O$. Crystals (methanol), m. 130.5–131.5° cor.; sublimes in a high vacuum; $[\alpha]_D$ +49° (chloroform) (1944a Ruzicka); $[\alpha]_D$ +50.7° (chloroform) (1949 Fürst). — **Fmn.** From either cholestan-2α- or -2β-ol (pp. 1537 s, 1538 s) by CrO_3 oxidation in glacial acetic acid at room temperature. From $\varDelta^{3.5}$-cholestadien-2-one (above) by hydrogenation in glacial acetic acid using platinum oxide catalyst, followed by reoxidation with CrO_3 in strong acetic acid, all at room temperature. From the p-toluenesulphonate of $\varDelta^3$-cholesten-3-ol-2-one (page 2659 s) by hydrogenation in alcohol at ca. 70° for 20 hrs. using Raney-nickel catalyst, followed by reoxidation with CrO_3 in strong acetic acid at room temperature; yield, 83% (1944a Ruzicka). From the "acetate of cholestan-2-ol-3-one" (p. 2665 s) by reaction with hydrazine hydrate and then sodium amyl oxide in amyl alcohol at 180–200°; the resulting mixture, mainly consisting of cholestane (p. 1429 s), is separated by chromatography on Al_2O_3, and the fraction eluted by ether-acetone (containing cholestan-2-ol) is oxidized with CrO_3 in glacial acetic acid-chloroform at room temperature; over-all yield, 3.7% (1944b Ruzicka).

Rns. Oxidation with CrO_3 in 90% acetic acid at 60° gives cholestane-2‖3-dioic acid (p. 1713 s). Hydrogenation in glacial acetic acid at room temperature using platinum oxide catalyst gives cholestan-2β-ol (p. 1538 s); reduction by means of sodium and alcohol affords cholestan-2α-ol (p. 1537 s) (1944a

---

* Authentic cholestan-1-one, prepared after the closing date for this volume, shows m. p. 85.5° to 86.5° cor., $[\alpha]_D^{24}$ +112° (chloroform) (1954 Striebel; cf. 1954 Plattner).

References, p. 2418 s

Ruzicka). Wolff-Kishner reduction of the (not isolated) hydrazone yields cholestane (1944b Ruzicka).

Compound with cholestan-4-one (1:1), see p. 2451 s.

1944 a Ruzicka, Plattner, Furrer, *Helv.* **27** 524, 529.
     b Ruzicka, Plattner, Furrer, *Helv.* **27** 727, 732 et seq.
1949 Fürst, Plattner, *Helv.* **32** 275, 282.
1953 Fieser, Romero, *J. Am. Ch. Soc.* **75** 4716.
1954 Plattner, Fürst, Els, *Helv.* **37** 1399, 1406.
     Striebel, Tamm, *Helv.* **37** 1094, 1109, 1112.

$\Delta^{1\cdot4\cdot6}$-**Cholestatrien-3-one** $C_{27}H_{40}O$. For structure, see 1950 Romo. For iden-

tity with the main constituent of the impure product formulated by Barkow (1940) as $\Delta^{4\cdot6\cdot8}$-*cholestatrien-3-one*, see 1950 Djerassi. — Crystals (acetone) (1940 Barkow), crystals (petroleum ether at dry ice temperature), m. 82–83°; $[\alpha]_D^{20} \pm 0°$ (chloroform); absorption max. in alcohol: 224, 258, 300 mμ (log ε 4.03, 3.97, 4.11) (1950 Djerassi; cf. 1940 Barkow). — **Fmn.** From 2.6-dibromo-$\Delta^{1\cdot4\cdot6}$-cholestatrien-3-one (p. 2497 s) by hydrogenation in boiling amyl alcohol containing KOH (>2 equivalents) using palladium-BaSO$_4$ catalyst; yield, 51% (1940 Barkow). — **Rns.** Hydrogenation in glacial acetic acid at room temperature using a platinum oxide catalyst (discontinued after 4 mols. of hydrogen absorbed) gives cholestan-3β-ol (p. 1695 s). Refluxing with acetic anhydride and acetyl chloride yields a product considered to be the *enol-acetate* $C_{29}H_{42}O_2$ [crystals (acetone), m. 114°, after sublimation in a high vacuum, 115°; $[\alpha]_D^{20} + 105°$ (chloroform); absorption max. (solvent not indicated) 265 mμ* ($k = 1.0$ mm.$^{-1}$); various attempts to eliminate acetic acid from this product were unsuccessful] (1940 Barkow).

$\Delta^{1\cdot4}$-**Cholestadien-3-one** $C_{27}H_{42}O$. White leaflets (aqueous alcohol or aqueous

acetone) (1936b SCHERING A.-G.), colourless prisms (methanol), m. 111.5° to 112.5° (1938 Inhoffen; cf. 1946 a Wilds); distils at 170–180° and 0.0003 mm. pressure (1936b SCHERING A.-G.); $[\alpha]_D^{23}$ + 28° (1938 Inhoffen), $[\alpha]_D^{24.5} + 31°$ (1946a Wilds; cf. 1936b SCHERING A.-G.; 1939b Butenandt), both in chloroform. Absorption max. in ether 236 mμ (log ε 4.20) (1939 Dannenberg; cf. 1938, 1940 Inhoffen), in alcohol 245 mμ (log ε 4.15) (1946a Wilds). Sparingly soluble in methanol, more soluble in ethanol, readily soluble in other usual solvents (1938 Inhoffen).

Fmn. From 2α.4α-dibromocholestan-3-one (p. 2495 s) by refluxing with collidine for 80 min., followed by chromatographic separation from the addition compound with the $\Delta^{4\cdot6}$-isomer (p. 2421 s) (yield, 60–70%) (1946a Wilds; cf. 1939b Butenandt); the dehydrobromination can also be effected by heating with potassium naphthoxide in boiling dry xylene, followed by

---

* The absorption max. quoted is not compatible with the $\Delta^{1\cdot3\cdot5\cdot7}$-tetraenol structure expected for such an enol-acetate (Editor).

heating at 220–240° and 2 mm. pressure (1936b SCHERING A.-G.), by refluxing with 2.4-dimethylpyridine (1940 Inhoffen), or with dry pyridine, via the intermediate N-(3-keto-Δ^4-cholesten-2-yl)-pyridinium bromide (p. 2496 s), which is decomposed to give the dienone on heating at 220° and 0.0006 mm. pressure in presence of anhydrous sodium acetate (yield, 11%) (1938, 1940 Inhoffen). From the benzoate of Δ^4-cholesten-2-ol-3-one (p. 2664 s) or from that of Δ^1-cholesten-4-ol-3-one (p. 2668 s) or the corresponding isovalerate by heating under 2 mm. pressure and subsequent distillation at 0.05 mm. pressure, all at 220° (1936b SCHERING A.-G.).

Rns. Ozonization (incomplete) in glacial acetic acid at room temperature affords an oil from which 5-keto-3‖5-A-nor-Δ^1-cholesten-3-oic acid (I, see below) and 2.5-dimethyl-3'-[1.5-dimethylhexyl]-6-keto-1.2-cyclopentano-perhydronaphthalene (II, p. 2420 s) have been isolated (1938, 1939b Inhoffen). Hydrogenation using palladium black catalyst in ether at room temperature (ca. 2.4 mols. of hydrogen absorbed), gives coprostanone (page 2443 s); using a Rupe-nickel catalyst in alcohol (discontinued after 1.13 mols. of hydrogen absorbed), the product is Δ^1-coprosten-3-one (p. 2423 s) (1938 Inhoffen). Reduction by means of aluminium isopropoxide and boiling iso-propanol affords $\Delta^{1.4}$-cholestadien-3-ol (p. 1542 s) (1937b SCHERING A.-G.), also formed by the action of fermenting yeast in aqueous solution at room temperature (1937 SCHERING CORP.). On heating at 300–320° for 3 hrs. in an atmosphere of CO_2, some methane is eliminated and 4-methyl-19-nor-$\Delta^{1.3.5(10)}$-cholestatrien-1-ol (p. 1535 s) is formed (1940, 1941, 1951b Inhoffen); this compound is also obtained on heating with 9.10-dihydrophenanthrene at 382–388° (sealed tube) (1946b Wilds), on heating with HI (d 1.96) in acetic anhydride on the water-bath (1941 Inhoffen), or on treatment with conc. H_2SO_4 and acetic anhydride at room temperature (1941 Inhoffen; cf. 1938 Inhoffen; 1946a Wilds); from the reaction product by heating at 300–320° in CO_2 (above), a second phenolic product (formed by loss of methane, not obtained pure) is also isolated, by distillation at 190–200° under 0.0003 mm. pressure, which after catalytic hydrogenation, acetylation, and CrO_3 oxidation in glacial acetic acid, gives a product showing œstrogenic activity (1937c, d SCHERING A.-G.). Acetylation (conditions not stated) yields an enol-acetate, $\Delta^{1.3.5}$(?)-cholestatrien-3-yl acetate (p. 1539 s) (1937c SCHERING A.-G.).

Addition compound with $\Delta^{4.6}$-cholestadien-3-one, see p. 2421 s.

Semicarbazone $C_{28}H_{45}ON_3$. Needles (benzene-alcohol), m. 230–1° (1938 Inhoffen). Absorption max. in chloroform at 248 and 302 mμ (log ε 10900 and 18800, resp.) (1939 Dannenberg; cf. 1938 Inhoffen).

5-Keto-3‖5-A-nor-Δ^1-cholesten-3-oic acid (I) $C_{26}H_{42}O_3$. Lancets (ace-

(I)

tone), m. 207–207.5°; absorption max. in ether (0.02%) at ca. 230 mμ (k = ca. 0.45 mm.$^{-1}$); reducing towards alkaline $KMnO_4$ (1938 Inhoffen). — **Fmn.** From $\Delta^{1.4}$-cholestadien-3-one by ozonization (see above); separation from neutral products by extraction of the mixed product in ether with 5% KOH; crude yield, 21% (1938 Inhoffen). — **Rns.** Hydrogenation in ether at room temperature, using palladium black catalyst, gives 5-keto-3‖5-A-nor-cholestan-3-oic acid (p. 1710 s) (1939b Inhoffen).

2.5-Dimethyl-3'-[1.5-dimethylhexyl]-6-keto-1.2-cyclopentano-perhydro-naphthalene (II) $C_{23}H_{40}O$. Oil, not obtained pure. — *Semicarbazone* $C_{24}H_{43}ON_3$.

Solid (alcohol-chloroform), m. 224–5°dec., giving strong depression with the semicarbazone of $\Delta^{1.4}$-cholestadien-3-one (p. 2419 s). **Fmn.** The neutral product of ozonization of $\Delta^{1.4}$-cholestadien-3-one (see p. 2419 s) is treated with semicarbazide in alcohol (1939b Inhoffen).

$\Delta^{4.6}$-Cholestadien-3-one, 6-Dehydrocoprostenone $C_{27}H_{42}O$ (p. 119).

Crystals (acetone) (1940 Barkow), plates (90% alcohol) (1942 Wintersteiner), needles (methanol), m. 80.5–81.5° cor. (1946 a Wilds; cf. 1943 Hardegger); $[\alpha]_D^{25} + 33°$ (1942 Bergström; cf. 1946a Wilds), + 35° (1943 Hardegger), both in chloroform. Absorption max. in alcohol at 285 mμ (ε 26000) (1942 Wintersteiner; cf. 1942 Bergström; 1943 Hardegger; 1946a Wilds), in dioxane at 283 mμ (1939 Dannenberg).

Occ. and **Isolation.** From the acetone extract of human arteriosclerotic aorta (yield, 0.06% of the cholesterol-free unsaponifiable material) (1943 Hardegger) or of swine spleen (yield, 23 mg. from 15.7 kg. of acetone extract) (1943 Prelog); in each case the compound is separated by a process of exhaustive chromatography and its appearance as an artefact arising from cholesterol in the course of extraction is not excluded.

Fmn. From $\Delta^{4.6}$-cholestadien-3β-ol (p. 1544 s) by refluxing with acetone and aluminium tert.-butoxide in benzene; yield, isolated as dinitrophenylhydrazone, 96% (1940 Petrow). From cholesterol (p. 1568 s) by oxidation in boiling toluene with benzoquinone and aluminium isopropoxide (yield, 44%) (1944 Ushakov) or aluminium tert.-butoxide (yield, 36–41%) (1946 a Wilds). From the 7-ethyl ether of Δ^5-cholestene-3β.7α-diol (p. 2066 s) by oxidation with acetone and aluminium phenoxide in boiling dry benzene, followed by chromatographic separation from other products (crude yield, 76%) and by isolation as oxime or semicarbazone (1942 Bergström; cf. 1946 Henbest). From 2-bromo-$\Delta^{1.4.6}$-cholestatrien-3-one (p. 2479 s) by hydrogenation in methanol containing potassium acetate, using a platinum oxide catalyst at room temperature (discontinued after 2 mols. of hydrogen are absorbed); yield, 52% (1940 Barkow). From 6β-bromo-Δ^4-cholesten-3-one (p. 2484 s) by heating with potassium acetate in acetic anhydride at 200° (bomb), followed by separation into an acidic and a neutral fraction and distillation of the latter in a high vacuum; isolation as oxime (1937 Krekeler). In very small yield, together with the 7-epimeric 7-hydroxycholesterols (pp. 2066 s, 2068 s), from the acetate of 7-ketocholesterol (p. 2682 s) by reduction with aluminium isopropoxide and isopropanol with slow distillation of the solvent; after complete saponification of the products, the diols are crystallized and the mother liquor product submitted to further crystallization and chromatography; the mechanism of the reaction is discussed (1942 Wintersteiner). See also formation of the addition compound with $\Delta^{1.4}$-cholestadien-3-one, described below.

Rns. Reduction by means of aluminium isopropoxide and boiling isopropanol affords a mixture of the 3-epimeric $\Delta^{4.6}$-cholestadien-3-ols (pp. 1544 s, 1545 s) together with small amounts (not isolated) of a cholestatriene (1940 Petrow). By heating the semicarbazone with sodium ethoxide and alcohol at 200° (sealed tube) an inseparable mixture containing mainly $\Delta^{4.6}$-cholestadiene (p. 1417 s) is obtained (1941 Eck). Irradiation (quartz lamp) in hexane solution under CO_2 for 75–80 hrs. gives ca. 30% yield of lumicholestadienone [probably 4-($\Delta^{4.6}$-cholestadien-3-on-4-yl)-Δ^4-cholesten-3-one] (1944 Ushakov). Refluxing with acetic anhydride and potassium acetate yields an oily enol-acetate, $\Delta^{2.4.6(?)}$-cholestatrien-3-yl acetate (p. 1539 s) (1937c, d SCHERING A.-G.).

Addition compound of $\Delta^{4.6}$-cholestadien-3-one with $\Delta^{1.4}$-cholestadien-3-one (1:1) [$C_{27}H_{42}O$]$_2$. Needle clusters (aqueous methanol), m. 68.5–70° cor., $[\alpha]_D^{24} +33°$ (chloroform). — **Fmn.** From the components by crystallizing equal amounts from aqueous methanol. Together with $\Delta^{1.4}$-cholestadien-3-one (p. 2418 s), from 2α.4α-dibromocholestan-3-one (p. 2495 s) by refluxing with collidine for 80 min.; after crystallization of the main product from methanol, the mother liquor residue is chromatographed on Al_2O_3 and fractionally eluted, followed by further fractional crystallization; yield, 8% (1946a Wilds).

Oxime $C_{27}H_{43}ON$. Crystals (ethyl acetate), m. 183° (1937 Krekeler); needle rosettes (methanol), m. 176–7° (1943 Hardegger); crystals (hexane), m. 178° (1940 Barkow), crystals (chloroform-methanol), m. 173.5–175° cor. (1943 Prelog; cf. 1942 Bergström). Absorption max. in chloroform at 280 mμ (ε 18 100) (1939 Dannenberg, quoting 1937 Krekeler; cf. 1943 Hardegger).

Semicarbazone $C_{28}H_{45}ON_3$ (p. 119). Crystals (alcohol), m. 228–230° dec.; absorption max. in dioxane at 305 mμ (ε 45000) (1942 Bergström; 1942 Wintersteiner; cf. 1940 Barkow).

2.4-Dinitrophenylhydrazone $C_{33}H_{46}O_4N_4$. Dark red needles (chloroform-alcohol), m. 231–2° (1940 Petrow).

$\Delta^{4.7}$ - Cholestadien - 3 - one, 7 - Dehydrocoprostenone, *"Dehydrocholestenone"*

$C_{27}H_{42}O$. For evidence in favour of the $\Delta^{4.7}$-, rather than the $\Delta^{4.8(14)}$-structure also envisaged by Windaus (1939), see 1952b Antonucci. — Needles (methanol), m. 88°; $[\alpha]_D^{17} +34°$ (chloroform); absorption max. in ether (from the graph) at 230 mμ (ε 16000) (1939 Windaus), in abs. alcohol at 238–239 mμ (ε 16400) (1952b Antonucci). — **Fmn.** From

$\Delta^{5\cdot7}$-cholestadien-3β-ol (7-dehydrocholesterol, p. 1547 s) by refluxing with dry acetone and aluminium tert.-butoxide in dry benzene; crude yield, 90% (by spectrographic assay) (1939 Windaus). — **Rns.** Reduction by means of aluminium isopropoxide and isopropanol, with slow distillation, gives chiefly allo-dehydrocholesterol (p. 1546 s) and its 3α-epimer (p. 1546 s), together with regenerated 7-dehydrocholesterol and small amounts of its (not isolated) 3α-epimer (p. 1551 s) (1939 Windaus).

Semicarbazone $C_{28}H_{45}ON_3$. Needles (chloroform-methanol), m. 240° dec.; absorption max. in chloroform at 274 mμ (ε ca. 30000) (1939 Windaus).

$\Delta^{8\cdot24}$-Cholestadien-3-one, Zymostadienone $C_{27}H_{42}O$.

For position of double bonds, see that of the parent zymosterol (p. 1559 s). — White needles (methanol), m. 104–5°; $[\alpha]_D^{20}$ +75.5° (1941 Wieland), $[\alpha]_D^{20}$ +65° (1952 Bladon), both in chloroform. — **Fmn.** From zymosterol (p. 1559 s) by oxidation using aluminium isopropoxide with cyclohexanone in boiling toluene (crude yield, 60%) or with acetone in boiling benzene (yield, 20%), or by heating with CuO at 300° (yield, 30%) (1941 Wieland). From epi-zymosterol (p. 1561 s) by Oppenauer oxidation (1942 Wieland). — **Rns.** Reduction by means of aluminium isopropoxide and isopropanol under reflux regenerates zymosterol and epi-zymosterol (1942 Wieland).

Semicarbazone $C_{28}H_{45}ON_3$. Crystals (alcohol), m. 230° dec.; solubility in boiling alcohol: 1 g. in 700 c.c. (1941 Wieland).

$\Delta^{8(14)\cdot24}$ or $\Delta^{14\cdot24}$(?)-Cholestadien-3-one, "Zymostadienone" $C_{27}H_{42}O$.

Is not identical with $\Delta^{8\cdot24}$-cholestadien-3-one (above) (1941 Wieland). A $\Delta^{8(14)\cdot24}$- or

$\Delta^{14\cdot24}$-dienone structure (nuclear double bond shift under the acidic conditions of formation, cf. the comparable transformation of δ-cholestenol into α- and β-cholestenols, pp. 1690 s, 1692 s, 1694 s) seems probable (Editor). — Crystals (methanol), m. 162–4° cor. (1930 Reindel). — **Fmn.** From zymosterol (p. 1559 s) by CrO_3 oxidation in 97% acetic acid at 90°, together with unidentified acid products; yield, 3.5% (1930 Reindel).

Oxime $C_{27}H_{43}ON$, m. 238–240° cor. (1930 Reindel).

$Δ^1$-**Cholesten-3-one** $C_{27}H_{44}O$ (p. 119). This structure was originally ascribed (1935 Butenandt) to the compound later referred to as "hetero-$Δ^1$-cholesten-3-one" when genuine $Δ^1$-cholesten-3-one was first obtained (1939b Butenandt); the "hetero"-isomer has since been identified as $Δ^5$-cholesten-4-one (p. 2450 s). — Blades (alcohol), m. 98–100° cor.*; $[α]_D^{23}$ +57.5° (chloroform), +62.6° (alcohol) (1947 Djerassi)*. Absorption max. in alcohol (from the graph) at 230 mμ (log ε 4.0) (1940 Jacobsen); for variations according to solvent, see 1941 Woodward. *Hydrate:* Crystals (aqueous methanol or aqueous acetone), m. 107–8° (1940 Jacobsen). For method of estimation, which permits distinction from $Δ^4$-cholesten-3-one, depending on condensation with Girard's reagent T, followed by polarographic analysis of the reaction mixture, see 1940 Wolfe. — **Fmn.** From 2α-bromocholestan-3-one (p. 2480 s) by boiling with 2.4-dimethylpyridine, followed by distillation in a high vacuum at 160–170° (1940 Inhoffen); the yield, using collidine (cf. p. 119), is almost quantitative (1939b Butenandt); for detailed investigation of this reaction and purification via $Δ^1$-cholesten-3-one dibromide, see 1940 Jacobsen; cf. 1947 Djerassi. — **Rns.** Conversion into $Δ^1$-androstene-3.17-dione (p. 2879 s) can be effected by standard procedure for degradation of the side chain (1939b SCHERING A.-G.). Treatment with bromine in glacial acetic acid in the cold gives $Δ^1$-cholesten-3-one dibromide (p. 2494 s) (1940 Jacobsen). $Δ^1$-Cholesten-3-one is not attacked by fermenting yeast (1940a Butenandt).

$Δ^1$-**Coprosten-3-one** $C_{27}H_{44}O$. For structure, see 1951 Djerassi. — M. p. 101°; absorption max. (solvent not stated) at 230 mμ (log ε 4.01) (1951 Djerassi). Leaflets (alcohol), m. 81–3°** (clearing at 85°); $[α]_D^{22}$ +64.6° (chloroform)** (1938 Inhoffen); absorption max. in ether ≤230 mμ (k ≥0.7 mm.$^{-1}$) (1938 Inhoffen; cf. 1943 Evans). — **Fmn.** From $Δ^{1.4}$-cholestadien-3-one (p. 2418 s) by hydrogenation in alcohol using Rupe nickel catalyst (discontinued after ca. 1 mol. of hydrogen is absorbed); yield, ca. 10% (1938 Inhoffen).

Semicarbazone $C_{28}H_{47}ON_3$. Needles (benzene-alcohol), m. 207° (1938 Inhoffen). Absorption max. in chloroform at 272 mμ (ε 9500) (1939 Dannenberg; cf. 1938 Inhoffen). For comparison of the absorption band of the semicarbazone with that of the parent ketone, see 1943 Evans.

---

* These authors point out that samples prepared earlier were all contaminated by the $Δ^4$-isomer.
** The authors doubt complete homogeneity of their product.

References, pp. 2445–2450 s

Δ^4-Cholesten-3-one, Coprostenone, *Cholestenone* $C_{27}H_{44}O$ (p. 120). Monoclinic prisms (ethyl acetate) (1908 Jaeger), prismatic needles from methanol (1938 Galinovsky; 1941 Wieland) or acetone-methanol (1941 Spring; cf. 1939 Spielman). Dimorphic: (I) needles (alcohol), m. 88°, (II) needles (alcohol), m. 82°; the two forms, obtained by fractional benzene elution after adsorption on Al_2O_3, are interconvertible by seeding the alcoholic solution of the alternative form (1943 Barton). Monoclinic needles*, a 10.61, b 7.86, c 19.98 Å, β 135°; space group $P2_1$, n = 2; d 1.078 ±0.002; for optical crystallographic data, see original (1940 Bernal; cf. 1908 Jaeger). Optical rotation in chloroform $[\alpha]_D^{20}$ ca. +89° (1938, 1941 Galinovsky; 1941 Wieland), +92° (1943 Barton), in alcohol $[\alpha]_D$ +87° (1938a Ruzicka); for rotatory dispersion of a sample having $[\alpha]_D$ +82.4° in chloroform, see 1939 Décombe. Absorption max. in alcohol at 240.5, 242, and 312 mμ (ε 18000, 19950, and 100, resp.) (1942 Jones; 1943 Evans; cf. 1937 Hogness); in ether at 234 mμ (ε 16400) (1939 Dannenberg); the absorption measurements of Mohler (1937) (cf. p. 120) are in hexane; for displacement of absorption maxima with variation of solvents, and for relationship between absorption spectrum and structure, see 1941, 1942 Woodward; 1946 Heard.

Polarographic reduction potential, in aqueous NH_4Cl solution 1.30 v. (1938 Adkins). For investigation of surface pressure of monolayer films on water (limiting area of the very compressible monolayer film at room temperature and zero pressure, 59 sq. Å) and their interpenetration by dissolved aliphatic alcohols, acids, or phenols, see 1939 Adam; 1941/42 Pankhurst.

OCCURRENCE AND ISOLATION OF Δ^4-CHOLESTENONE

Has been isolated from the lipoid material of the pituitary gland anterior lobe (yield, 0.042 g. from 0.82 kg. acetone extract); its formation as a secondary product, arising from cholesterol in the course of the isolation process, is not excluded (1945 Prelog). For occurrence in swine testicles, and possibly in swine spleen, see 1943 Prelog. In the fæces of dogs and rats fed on brain; isolated from the ketonic fraction of the lipoid material via the o-tolylsemicarbazone; yield, ca. 1 % of the unsaponifiable lipoids (1943 Rosenheim).

FORMATION OF Δ^4-CHOLESTENONE

From cholesterylamine (p. 1481 s) by treatment with HOCl in ether at 0°, followed by boiling with sodium ethoxide and hydrolysis of the resulting imine with dil. H_2SO_4 (yield, 78%); similarly from cholesterylmethylamine (1937 I.G. FARBENIND.). Together with cholestanone (p. 2436 s) and coprostanone (p. 2443 s), from cholesterol (p. 1568 s) by dismutation on shaking with reduced platinum catalyst in 50% acetic acid at 127° (sealed tube);

---

* Probably of form II, as form I does not seem to have been observed before 1943 (Editor).

References, pp. 2445-2450 s

the ketones are separated as p-carboxyphenylhydrazones from cholestane and coprostane also formed, the saturated ketones removed by treatment of the mixed derivatives with formaldehyde, and the remaining cholestenone hydrazone finally decomposed with pyruvic acid (yield, 5 %); the mechanism of this reaction is investigated (1944 Anker). From cholesterol on conversion with bromine in benzene into the dibromide, followed by oxidation with $KMnO_4$ and aqueous H_2SO_4 at room temperature and then by debromination with zinc dust in glacial acetic acid at room temperature; yield, 50 % (1939 Spielman; cf. 1941 Salvioni). From cholesterol dibromide (p. 1889 s) by oxidation with CrO_3 in strong acetic acid at 45–50°, followed by debromination of the crude product with $FeCl_2$ in boiling 96 % alcohol (atmosphere of CO_2) containing acetic acid and potassium acetate; yield, 67 % (1941 c Bretschneider). From cholesterol dibromide by boiling with collidine (yield, 65 %) (1941 Galinovsky) or by treatment with $AgNO_3$ in pyridine at room temperature in the dark, followed by chromatography of the resulting ether-soluble resin (yield, 22–27 %) (1941 Spring). For biological oxidation of cholesterol to give cholestenone, see "Physiological properties" (p. 2431 s). From epicholesterol (p. 1670 s) by refluxing with acetone and aluminium tert.-butoxide in dry benzene (1940 Barnett). From $Δ^5$-cholestene-$3β.4β$-diol (p. 2051 s) by heating with alcoholic HCl for 10 min. on the water-bath, or with water for 8 hrs. at 200° (sealed tube); from $Δ^4$-cholestene-$3β.6β$-diol (p. 2060 s) by the same methods or by heating with anhydrous $CuSO_4$ at 180° under reduced pressure (1937 Rosenheim) or on warming with mineral acids (1938 Petrow).

Together with $Δ^1$-coprosten-3-one (p. 2423 s), from $Δ^{1.4}$-cholestadien-3-one (p. 2418 s) by hydrogenation using Rupe nickel catalyst in alcohol (discontinued after absorption of 1.13 mols. of hydrogen); separation by fractional crystallization from alcohol (1938 Inhoffen). From $Δ^5$-cholesten-3-one (page 2435 s) by isomerization on heating with small amounts of mineral acids (dil. H_2SO_4, HCl, HBr) in methanol, ethanol, or glacial acetic acid, or with small amounts of NaOH in ethanol (1937 c SCHERING-KAHLBAUM A.-G.); from the ethylene ketal of $Δ^5$-cholesten-3-one (p. 2435 s) on refluxing with alcohol containing dil. HCl (yield, almost quantitative) (1941 SQUIBB & SONS).

From $2α$-bromocholestan-3-one (p. 2480 s) by treatment with quinoline, aniline, dimethylaniline, aminoethanol, or other organic bases (1937 b CIBA), or via its pyridinium compound, N-(3-ketocholestan-2-yl)-pyridinium bromide (p. 2482 s), by distillation at 250–300° and 10 mm. pressure, ether extraction of the distillate, and redistillation of the extract at 250° in a high vacuum (yield, 33 % + 27 % crude) (1938 a Ruzicka; cf. 1937 b CIBA). From 5-bromo-cholestan-3-one * (p. 2483 s) by heating with potassium acetate in alcohol (1943 Urushibara). From $Δ^5$-cholestenone dibromide (p. 2500 s) by catalytic hydrogenation, using platinum black in ether at room temperature (1939 Décombe). From 2.6-dibromo-$Δ^{1.4.6}$-cholestatrien-3-one (p. 2497 s) by catalytic hydrogenation in boiling alcohol containing KOH (1940 Barkow). From $2β.5α.6β$-tribromocholestan-3-one (p. 2504 s) by heating with zinc dust in glacial acetic acid on the water-bath; low yield (1936 Butenandt). From

---

* Erroneously referred to as "cholestenone bromide" in the *C. A.* abstract.
169*

cholestan-5-ol-3-one (p. 2671 s) by refluxing with acetic anhydride (crude yield, 42%) or by boiling with HCl gas in chloroform (1944 Plattner; cf. 1943 Chuman). From the benzoate of cholestan-6α-ol-3-one (p. 2673 s) by slow distillation at 190⁰ in a high vacuum (1941 PARKE, DAVIS & Co.). In good yield, from cholestan-6β-ol-3-one (p. 2673 s) by heating with KHSO₄ at 125⁰ and subsequent distillation at 185⁰ and 4 mm. (1940a Marker).

From 3-keto-*Δ⁴*-cholesten-2-yl-glyoxylic acid by distillation in vacuo (1938b Ruzicka). From cholestenone pinacol [3β.3'β-dihydroxy-3.3'-bi-(*Δ⁴*-cholestenyl)] by oxidation with lead tetraacetate and glacial acetic acid in benzene on the water-bath (10 min.) and then at room temperature (24 hrs.) (yield, 76%), or, together with cholestanone, by thermal decomposition at 220–230⁰ and 0.01 mm. pressure, followed by fractional crystallization from aqueous alcohol (1938 Galinovsky).

See also biological formation, described in "Physiological properties" (p. 2431 s).

<h4 align="center">PREPARATION OF Δ⁴-CHOLESTENONE</h4>

For (large scale) preparation from cholesterol by bromination in carbon tetrachloride, followed by oxidation of the (still dissolved) dibromide with 85% H₂CrO₄ and glacial acetic acid at 20⁰, and finally debromination with zinc in methanol (crude yield, quantitative), see 1945 C. F. BOEHRINGER & SOEHNE; 1938a CHINOIN.

For further details concerning the preparation from cholesterol by Oppenauer oxidation using aluminium tert.-butoxide and acetone in boiling benzene (cf. p. 120) (yield 70–81%), see 1941 Oppenauer; a 90–95% yield is indicated by 1937 N. V. ORGANON; cf. 1942 Jones. The following alternative procedures for this mode of preparation have been described: using cyclohexanone and aluminium isopropoxide at 100⁰ (yield, as semicarbazone, 90% (1937a SCHERING-KAHLBAUM A.-G.) or in boiling toluene (1941 Wieland); using methyl ethyl ketone and aluminium tert.-butoxide in boiling benzene (yield, 84%) (1941 Adkins); for detailed investigation of this reaction, see 1938, 1941 Adkins.

<h4 align="center">REACTIONS OF Δ⁴-CHOLESTENONE</h4>

Oxidation. Oxidation to mixtures containing *Δ⁴*-pregnene-3.20-dione (progesterone, p. 2782 s) can be effected with molecular oxygen in an air-bath at 170⁰ using V₂O₅ catalyst (1941a Bretschneider), or by means of air on more prolonged heating at 120–130⁰ using V₂O₅ or KMnO₄ catalysts, whereby acidic products and traces of *Δ⁴*-androstene-3.17-dione (p. 2880 s) have also been isolated (1939 CHINOIN). For ozonization to give 5-keto-3‖5-A-nor-cholestan-3-oic acid (p. 1710 s), see also 1938 Bolt. Oxidation with K₂S₂O₈ and 96% H₂SO₄ in glacial acetic acid at room temperature yields the lactone of 5-hydroxy-3‖5-A-nor-cholestan-3-oic acid (p. 1709 s) (1941/42 Salamon). Treatment with H₂O₂ and some OsO₄ in ether at 20⁰ gives cholestane-4.5-diol-3-one (p. 2754 s) (1938 Butenandt); the 4.5-osmic ester of possibly the same diolone, isolated as complex with 2 mols. of pyridine or 1 mol. of α.α'-dipyridyl, is obtained on treatment of 1 mol. of Δ^4-cholestenone in ether with 1 mol. of OsO$_4$ and 4 mols. of pyridine or with 0.5 mol. of OsO$_4$ and 2 mols. of $\alpha.\alpha'$-dipyridyl (1942 Criegee). Oxidation by means of lead tetraacetate in glacial acetic acid-acetic anhydride at 70° affords 2α-acetoxy-Δ^4-cholesten-3-one (p. 2664 s) (1944 Seebeck; cf. 1939 Ehrhart). In addition to progesterone and Δ^4-androstene-3.17-dione previously isolated (cf. p. 120) from the CrO$_3$ oxidation of Δ^4-cholestenone, 27-nor-Δ^4-cholestene-3.25-dione (p. 2805 s) is also obtained (1938 Ciba); by a similar oxidation in strong acetic acid at 20° (24 hrs.) 3-keto-Δ^4-cholenic and 3-keto-Δ^4-ætiocholenic acids are also isolated (1937a Ciba); similar oxidation is carried out in heterogeneous medium (aqueous H$_2$CrO$_4$ and carbon tetrachloride) at 20° (1939 C. F. Boehringer & Soehne), or with CrO$_3$ in H$_2$SO$_4$-acetic acid mixtures (containing more than 50% H$_2$SO$_4$) at —2° (1938b Chinoin). Oxidation to mixed products containing progesterone is also effected via the intermediate cholestenone dibromide, using CrO$_3$ in 85% acetic acid at 20° (1945 C. F. Boehringer & Soehne) or KMnO$_4$ (1935 Tavastsherna; but cf. 1939 Spielman), followed by debromination. Δ^4-Cholestenone reacts with SeO$_2$ in glacial acetic acid at 100°, but not in boiling alcohol, to give a compound forming an insoluble enolic potassium salt and giving a purple colour with alcoholic FeCl$_3$, to which the structure **Δ^4-cholestene-2.3-dione** C$_{27}$H$_{42}$O$_2$ may be ascribed by analogy with the corresponding reaction of cholestanone (1938 Stiller).

Hydrogenation and reduction. Catalytic hydrogenation in 95% alcohol, using 10% Raney nickel, yields epi-coprostanol (p. 1706 s) together with other hydroxylated products, whereas by use of 5% Raney nickel, and cessation after 1 mol. of hydrogen has been absorbed, cholestanol (p. 1695 s) is obtained (1942 Ruzicka; for investigation of similar hydrogenation in the presence of NaOH, or in abs. alcohol containing sodium ethoxide, see 1942 Ruzicka. The **compound** C$_{54}$H$_{90}$O$_2$ (p. 120) obtained by reduction of Δ^4-cholestenone with sodium amalgam in alcoholic acetic acid (1906 Windaus) is shown to be cholestenone pinacol [3β.3'β-dihydroxy-3.3'-bi-(Δ^4-cholestenyl)] (1938 Galinovsky; cf. 1958 Bladon). Reduction of the ketazine with sodium ethoxide and hydrazine hydrate in abs. alcohol at 200° affords Δ^4-cholestene* (1939 Dutcher; cf. 1940 Kon); by the analogous reduction of the semicarbazone at 210° cholesterol (p. 1568 s), epi-coprosterol (p. 1706 s), β-cholestanol (page 1695 s), and small amounts of allocholesterols (p. 1564 s, 1565 s) are also obtained, together with a **compound** C$_{28}$H$_{43}$ON$_3$ considered to be identical with that similarly obtained from Δ^4-cholestene-3.6-dione (cf. p. 158) (1939 Dutcher). Δ^4-Cholestenone is not reduced by Bacillus putrificus (Bienstock) in yeast suspension (1939 Mamoli).

Action of light. Irradiation with mercury arc light of a Δ^4-cholestenone solution in hexane (1938 E. Bergmann; 1940b Butenandt) or in benzene (1939a Butenandt), or with sunlight in either of these solvents (1939a Butenandt), air being excluded, affords "lumicholestenone" [4-(Δ^5-cholesten-3-on-4-yl)-

---

* In view of the isolation of previously unknown Δ^3-cholestene by Wolff-Kishner reduction of Δ^4-cholestenone (1949 Lardelli), the identity or purity of this reported Δ^4-cholestene seems uncertain (Editor).

cholestan-3-one] (1939a Inhoffen; 1940b Butenandt; but see also 1952 Butenandt) together with cholestenone pinacol [3β.3'β-dihydroxy-3.3'-bi-($Δ^4$-cholestenyl)] (1940b Butenandt); if oxygen is not rigidly excluded, a photo-oxidation product [m. p. 157°, [α]$_D$ + 23° (chloroform); quinoxaline derivative, m. 228°] is also isolated, thought to be possibly impure cholestane-3.4-dione (p. 2904 s) (1938 E. Bergmann).

Reactions with halogens. The structures of bromination products of $Δ^4$-cholestenone (cf. p. 120) have been revised as follows: 2α.6α- and 2α.6β-dibromo-$Δ^4$-cholesten-3-ones (p. 2499 s) (formerly stereoisomeric 4.6-dibromo-$Δ^4$-cholesten-3-ones, m. 133° and 162–3° resp.), 2.2.6β-tribromo-$Δ^4$-cholesten-3-one (p. 2503 s) (formerly 4.6.6-tribromo-$Δ^4$-cholesten-3-one), 2α.6(?)-dibromo-$Δ^{4.6}$-cholestadien-3-one (p. 2498 s) (formerly 4.6-dibromo-$Δ^{4.6}$-cholestadien-3-one), 2.2.6-tribromo-$Δ^{4.6}$-cholestadien-3-one (p. 2503 s) (formerly 4.6.7-tribromo-$Δ^{4.6}$-cholestadien-3-one, m. 165–6°), and 2.6-dibromo-$Δ^{1.4.6}$-cholestatrien-3-one (p. 2497 s) (formerly 4.6-dibromo-$Δ^{4.6.8}$-cholestatrien-3-one, m. 203°). For conditions of bromination, see also 1936a SCHERING A.-G.*; for general review of the bromination of $Δ^4$-cholestenone, see 1942 Butenandt; but cf. 1950 Rosenkranz; 1950 Djerassi. Direct conversion to 6β-bromo-$Δ^4$-cholesten-3-one (p. 2484 s) is possible, using bromine (1 mol.) in glacial acetic acid containing 2% of acetic anhydride at room temperature (1938 C. F. BOEHRINGER & SOEHNE). For investigation of the reaction mechanism of bromination by means of iodine monobromide in carbon tetrachloride, and the catalytic effect of HBr or glacial acetic acid, see 1938, 1940b Ralls.

Reactions with acids. Heating $Δ^4$-cholestenone with conc. HNO_3 in glacial acetic acid on the water-bath yields "trinitrocholesterilene" (p. 1600 s) (1906 Windaus). Treatment with conc. H_2SO_4 and acetic anhydride in a cooling mixture, or addition of acetic anhydride to $Δ^4$-cholestenone in glacial acetic acid-H_2SO_4 affords $Δ^4$-cholesten-3-one-6-sulphonic acid (1937 Windaus; cf. 1939 Kuhr).

Reactions with organic compounds. Reaction with methylmagnesium iodide in ether at room temperature (24 hrs.) and then short refluxing, followed by treatment with ice-cold aqueous NH_4Cl, yields **3-methyl-$Δ^4$-cholesten-3-ol** $C_{28}H_{48}O$ [needles (acetone), m. 118°] which by refluxing with acetic anhydride affords a **hydrocarbon** $C_{28}H_{46}$ [crystals (acetone), m. 80°, absorption max. in ether 238 mμ (k = 1.95 mm.$^{-1}$), easily soluble in petroleum ether, insoluble in methanol; not dehydrogenated by sunlight irradiation in alcohol-benzene in presence of eosin; does not react with maleic anhydride in boiling xylene] thought to be **3-methyl-$Δ^{3.5}$-cholestadiene** (1937 Krekeler). Condensation with phenylmagnesium bromide in ether, followed by treatment with NH_4Cl and iced water, gives 3-phenyl-$Δ^4$-cholesten-3-ol (p. 1851 s) (1939 W. Bergmann); the same condensation product treated with dil. H_2SO_4 yields 3-phenyl-$Δ^{3.5}$-cholestadiene (p. 1445 s); by an analogous reaction with α-naphthylmagnesium bromide, 3-(α-naphthyl)-$Δ^{3.5}$-cholestadiene (p. 1448 s) is formed (1937 Urushibara). Reaction with β-phenylethylmagnesium chloride in an-

* The structures assigned in the patent are those formerly ascribed, as cited above.

hydrous ether under reflux, followed by dehydration of the resulting (not isolated) carbinol with $KHSO_4$ at 150–160⁰ and then distillation in a high vacuum at ca. 270⁰, yields 3-(β-phenylethyl)-cholestadiene (p. 1447 s) (1938 Dansi). Condensation with bromoacetic ester and zinc in boiling anhydrous benzene, followed by dehydration with $KHSO_4$ at 150–160⁰ and saponification with alcoholic KOH, affords ($Δ^{2.4(?)}$-cholestadien-3-yl)-acetic acid (1938 Dansi; cf. 1936 SCHERING CORP.); by similar condensation with phenylbromoacetic acid ($Δ^{2.4(?)}$-cholestadien-3-yl)-phenylacetic acid is obtained (1936 SCHERING CORP.).

Short boiling with phenylhydrazine gives the phenylhydrazone (p. 2434 s) (1904 Diels), but on heating the reactants in glacial acetic acid, until the phenylhydrazone which first separates has redissolved, indolo-3'.2': 2.3-($Δ^{2.4}$-cholestadiene) (p. 2432 s) is obtained (1937 Rossner). For reaction with carboxymethylhydrazine, see 1936 Anchel.

Heating with butyl alcohol or cholesterol in presence of $CaCl_2$, Na_2SO_4, p-toluenesulphonic acid, or other acid catalysts gives the butyl or cholesteryl enol-ethers, derivatives of $Δ^{3.5}$-cholestadien-3-ol (p. 1542 s) (1940 SCHERING CORP.); heating with ethyl orthoformate and formic acid in the presence of small amounts of conc. H_2SO_4 at 50⁰ gives the corresponding ethyl ether (1938 Schwenk); similarly esters of $Δ^{3.5}$-cholestadien-3-ol result by reaction of $Δ^4$-cholestenone with suitable acid derivatives: heating with acetyl chloride and acetic anhydride at 95–100⁰ (atmosphere of nitrogen) yields the enol-acetate (1937 Westphal; cf. 1936a CIBA); short refluxing with benzoyl chloride alone (1946 Ross) or with pyridine in benzine (b. p. 100⁰) (1936a CIBA), or heating with benzoic anhydride at 360⁰ (1937 Krekeler) gives the enol-benzoate; heating with benzoyl chloride at 100⁰ (bomb) affords the enol-chloride (3-chloro-$Δ^{3.5}$-cholestadiene, p. 1450 s) (1936a CIBA). Refluxing with ethylene glycol in benzene in presence of p-toluenesulphonic acid yields the ethylene ketal of $Δ^5$-cholesten-3-one (p. 2435 s) (1939a SCHERING A.-G.; 1941 SQUIBB & SONS); analogous condensations occur with propylene and trimethylene glycols (1941 SQUIBB & SONS). By treatment with ethyl mercaptan, in presence of Na_2SO_4 and $ZnCl_2$ at 3–5⁰, the enol-thioether ($Δ^{3.5}$-cholestadiene-3-thiol ethyl ether*) is formed (1946 Bernstein; cf. 1949 Rosenkranz); the same product results by reaction at room temperature with ethyl mercaptan and trithio-orthoformic ester or formic ester, with benzenesulphonic acid present in each case (1940a, b CHINOIN). Reaction with 1 mol. of benzyl mercaptan in ethyl formate in the presence of benzenesulphonic acid affords $Δ^{3.5}$-cholestadiene-3-thiol benzyl ether*; similar reaction with 2 mols. of benzyl mercaptan gives $Δ^4$-cholestenone dibenzyl thioketal (p. 2432 s) (1940b CHINOIN). For reaction with ethylene dithioglycol, see $Δ^4$-cholestenone ethylenedithio ketal, p. 2432 s.

Heating with acetamide at 300⁰ yields $Δ^4$-cholestenone acetylimine (page 2432 s) (1937 Krekeler). Condensation with oxalic ester and sodium ethoxide in warm abs. alcohol, followed by saponification with aqueous alcoholic KOH, affords $Δ^{2.4}$-cholestadien-3-ol-2-glyoxylic acid and/or its enol derivative

---

* To be described in a later part of this supplement.

("cholestenone oxalic acid") (1938b Ruzicka). There is no reaction with p-tolyl or o-anisyl isocyanide in presence of benzoic acid (1945 Baker).

Heating with piperidine at 210–220° yields a **compound** $C_{32}H_{51 \text{ or } 53}N$ [crystals (benzene-methanol), m. 159°] (1906 Windaus). For reaction with pyridine and $POCl_3$, see N-($\Delta^{3.5}$-cholestadien-3-yl)-pyridinium chloride (page 2431 s). On heating with isatin and KOH in aqueous alcohol, degradation occurred and no condensation product could be isolated (cf. reactions of cholestanone) (1944 Buu-Hoi).

For colour reactions, see below. See also biological reactions in "Physiological properties" (p. 2431 s).

ANALYTICAL PROPERTIES OF Δ^4-CHOLESTENONE

Colour reactions and their photometrical use. Treatment of a chloroform solution with conc. H_2SO_4 gives a deep red acid layer and a yellow chloroform layer; with H_2SO_4-acetic anhydride the colour sequence is yellow → red → purple → green-blue (1904 Diels). Solutions of Δ^4-cholestenone in pure conc. or dil. H_2SO_4 show absorption max. at 290 mμ or 500 mμ, resp. (1938 Graff); in pure 96% H_2SO_4 there is a wide absorption band at 280–310 mμ (max. at 296 mμ, log k_{max} 4.4, from the curve) which on dilution to 68% (aqueous or alcoholic) H_2SO_4 narrows, becoming slightly less intense, and is accompanied by new bands at 480 mμ (log k ca. 3.9) and ca. 600 mμ (1939 Bandow). Gives a yellow-brown colour with $SbCl_3$ (1941 Spring). Reduces moderately phosphomolybdic acid (Folin-Wu reagent) in acetic acid medium to molybdenum blue; this reaction can be used for colorimetric estimation (1946 Heard). Gives a violet colour in the Zimmermann reaction (with m-dinitrobenzene and KOH in alcohol) (1937 Kaziro); spectrophotometric studies, with increased accuracy, of this reaction reveal that Δ^4-cholestenone is considerably less strongly absorbing than androsterone in the region of 520 mμ; the use of this reagent for differentiating 3- and 17-ketosteroids is discussed (1938 Callow; 1940 McCullagh; 1944 Zimmermann). With p-nitrobenzenediazonium chloride a dark coloration appears after 30 min. in glacial acetic acid, after 5 min. in glacial acetic acid - dioxane (1938 Fieser). For behaviour towards numerous sterol colour reagents, for which reactions are usually weak or negative, see 1928 Wokes.

Separation and estimation. Separation of Δ^4-cholestenone from cholestanone can be effected by chromotography on silica-gel or Al_2O_3 (1941 b Bretschneider), separation from progesterone and androsterone by extraction of their benzene solution with HCl (1937 C. F. BOEHRINGER & SOEHNE); for attempted separation from cholestatrienone dibromide by a thermal diffusion process, see 1940 Korsching; for use of the p-carboxyphenylhydrazone derivative for separation, see 1936 Anchel. For assay by use of Girard's reagent T, see 1941 Hughes, and for its application for differentiation from Δ^1-cholesten-3-one, see 1940 Wolfe. Titration with thiocyanogen in glacial acetic acid-carbon tetrachloride indicates only 0.11 double bonds (1937 Stavely); for estimation of carbonyl group basicity, measured by adsorption exchange, see 1944 Bersin.

PHYSIOLOGICAL PROPERTIES OF Δ^4-CHOLESTENONE

Evidence, based on parallel sapogenin feeding experiments, is furnished that Δ^4-cholestenone is an intermediate in the metabolic conversion of cholesterol to coprosterol (1942a, b Marker; but cf. 1941 Fieser), and is supported by the isolation of coprostenone from the fæces of dogs fed on brain (1943 Rosenheim), also by the isolation of deuteriocoprosterol after feeding of deuteriocoprostenone (below) to human beings (1938 Anchel); the conversion of cholesterol into Δ^4-cholestenone is biologically reversible (1942a, b Marker). The oxidation of cholesterol to Δ^4-cholestenone is brought about by Proactinomyces erythropolis (and other Proactinomyces species) which also further oxidize the ketone (1944 Turfitt).

Δ^4-Cholestenone has neither anæsthetic effect (rats) nor folliculoid effect (rats) (1942 Selye), nor does it exert antifibromatogenic action in regard to abdominal fibroids induced by œstrogen administration (female castrated guinea pigs) (1943/44, 1944 Lipschütz; 1944 Iglesias). The products obtained by ultra-violet irradiation of Δ^4-cholestenone, or by treatment with conc. H_2SO_4 and acetic anhydride in glacial acetic acid at 85–90°, do not show antirachitic activity (1937 Eck). Injection of sesame oil containing Δ^4-cholestenone, or progesterone contaminated with Δ^4-cholestenone, results in oleoma development at the injection site (this effect is not obtained with pure progesterone) (1946 Bischoff). Administration of Δ^4-cholestenone or its enol-acetate, to adrenalectomized rats is not followed by thymus involution, nor is any toxicity observed (1937 Schacher).

MODIFIED FORMS AND REACTION PRODUCTS OF Δ^4-CHOLESTENONE

Deuterio-Δ^4-cholesten-3-one, Deuteriocoprostenone $C_{27}H_{40}D_4O$. Contains ca. 4 atoms of deuterium. — Crystals (abs. alcohol), m. 81° (1938 Anchel). — **Fmn.** From Δ^5-cholesten-3-one (p. 2435 s) by refluxing with NaOH in alcohol containing 80% D_2O (3:1); yield, 82% (1938 Anchel). A product containing only ca. 2 atoms of deuterium can be isolated from mixed products which result by treatment of cholesterol with deuterated acetic acid, D_2O, and a platinum catalyst at 127° (cf. the analagous formation of Δ^4-cholestenone, p. 2424 s); yield, 0.25% (1944 Anker). — **Rns.** By recrystallization from a solution of NaOH in 75% aqueous alcohol, 96% of the deuterium content is lost (1938 Anchel).

N-($\Delta^{3.5}$-Cholestadien-3-yl)-pyridinium chloride $C_{32}H_{48}NCl$. The $\Delta^{3.5}$-structure is favoured by the authors but the $\Delta^{4.6}$-structure is not excluded. — Crystals (pyridine), dec. 220–225°; decomposes on standing. Readily soluble in alcohol, acetic anhydride, hot pyridine, and chloroform, insoluble in benzene, ether, and petroleum ether. — **Fmn.** From Δ^4-cholestenone by treatment with pyridine and $POCl_3$ at room temperature

$$* C_8H_{17} = \overset{\displaystyle CH_3}{\underset{\displaystyle CH \cdot [CH_2]_3 \cdot CH(CH_3)_2}{|}}$$

(4 days) with intermediate heating on the water-bath (3 hrs.); yield, 56%. — **Rns.** On heating in vacuo at 100°, pyridine hydrochloride is eliminated and on further heating at 150–200° in a high vacuum $\Delta^{3·5·7(?)}$-cholestatriene (p. 1407 s) is obtained (1944 Müller).

Indolo-3′.2′:2.3-($\Delta^{2·4}$-cholestadiene) $C_{33}H_{47}N$. White needles (alcohol), m. 195°. — **Fmn.** See reactions of Δ^4-chole-stenone (p. 2429 s). — **Rns.** By heating strongly with selenium cyclization is thought to occur giving rise to mixtures from which the following compounds have been isolated by chromatographic separation: **compound** $C_{29}H_{37}N$ [formulated as (I)**, white rosettes (methanol), m. 170°; obtained by heating at 320° for 40 hrs.; yield 3.5%] and **compound** $C_{28}H_{27}N$ [formulated as (II)**; yellow crystals (benzene, containing 1% of methanol), m. 303°; obtained by heating at 320° (16 hrs.) and then at 340° (30 hrs.); yield, 4%]; by further heating at 360° for 9 hrs. a nitrogen-free compound 20-methyl-cholanthrene (p. 531 s) is isolated (1937 Rossner).

DERIVATIVES OF Δ^4-CHOLESTENONE

The so called *cholestenone ethylene ketal* (1939a SCHERING A.-G.) has been shown (1941 Fernholz; 1952a Antonucci) to be a derivative of Δ^5-cholestenone, which see p. 2435 s.

Δ^4-*Cholestenone di-benzylthio ketal* $C_{41}H_{58}S_2$. Crystals (benzene - abs. alcohol), m. 126–127°; [α]$_D$ +110° (benzene) (1940b CHINOIN). — **Fmn.** See reactions of Δ^4-cholestenone, p. 2429 s.

Δ^4-*Cholestenone ethylenedithio ketal* $C_{29}H_{48}S_2$. For confirmation of the 4:5 position of the double bond, see 1952a Antonucci. — Needles (acetone), m. 106–7°; [α]$_D^{27}$ +119° (chloroform). — **Fmn.** By reaction of the parent ketone with ethylene dithioglycol in presence of $ZnCl_2$ and Na_2SO_4. — **Rns.** Reduction with Raney nickel in alcohol gives Δ^4-cholestene (p. 1423 s) (1944, 1947 Hauptmann).

Δ^4-*Cholestenone acetylimine* $C_{29}H_{47}ON$. Crystals (acetone-chloroform), m. 238°; can be distilled in a high vacuum (1937 Krekeler). Absorption max.

---

* $C_8H_{17} = \overset{\displaystyle CH_3}{\underset{\displaystyle CH \cdot [CH_2]_3 \cdot CH(CH_3)_2}{|}}$

** In view of the analogous selenium dehydrogenation of cholesterol, leading to Diels's second hydrocarbon "$C_{25}H_{24}$" (p. 50; p. 1599 s) without loss of the side chain carbon atoms C-24 to C-27, the structures ascribed to these compounds seem questionable (Editor).

in chloroform at 270 mμ (ε 16000) (1939 Dannenberg; cf. 1937 Krekeler). Very sparingly soluble in all usual solvents except chloroform and benzene (1937 Krekeler). — **Fmn.** From Δ^4-cholestenone by heating with acetamide at 300°, or from its oxime with acetic anhydride under reflux (1937 Krekeler). — **Rns.** Refluxing with sulphuric acid in benzene-alcohol regenerates the parent ketone (1937 Krekeler).

Δ^4-*Cholestenone oxime* C$_{27}$H$_{45}$ON (p. 121). For syn and anti designation, based on refractive indices for the oximes in alcoholic solution and for their non-crystalline acetyl derivatives, see 1940a Ralls.

Anti-form of Δ^4-*cholestenone oxime.* Crystals (pentane) (1937 Mohler), (benzene, alcohol, or carbon tetrachloride) (1940a Ralls), (ethyl acetate) (1942 Jones), m. 152–3° (1937 Mohler; 1942 Jones); can be distilled in a high vacuum (1937 Mohler). Absorption max. in alcohol at 240 and 242 mμ (ε 23000 and 17600, resp.) (1943 Evans; cf. 1942 Jones), in glacial acetic acid at 238 mμ (log ε 4.4) (1938 Ralls). Water-surface films of the oxime show small compressibility (contrast to those of the free ketone); molecular area (from the graph) ca. 43 and ca. 38 sq. Å at 0 and 25 dynes/cm. pressure resp., and room temperature (1930 Adam). — **Fmn.** Treatment of Δ^4-cholestenone with hydroxylamine hydrochloride in cold methanol affords an *addition product* C$_{27}$H$_{47}$O$_2$N (m. 142–147°, turning yellow at 135°; strongly reducing towards Fehling's solution), which on boiling with hydrochloric acid yields the oxime; the latter is obtained directly from the initial reactants at slightly higher temperatures (1904 Diels). From the ketone by refluxing with hydroxylamine acetate in aqueous alcohol (1942 Jones). From the syn-form (below) by recrystallization from benzene or carbon tetrachloride (1940a Ralls). — **Rns.** Reduction with sodium in boiling abs. alcohol, followed by acetylation, gives the N-acetyl derivatives of 3β-amino-Δ^4-cholestene (p. 1482 s: "β"-acetyl-cholesterylamine), its 3α-isomer (p. 1482 s: "γ"-acetyl-cholesterylamine), 3β-amino-Δ^5-cholestene (p. 1481 s), and 3β-aminocholestane (p. 1483 s: 3-aminocholestane) (1911 Windaus; 1956 Shoppee). Treatment with bromine in carbon tetrachloride affords a mixed product m. 123–6° (containing 20.6% bromine), which loses hydrogen bromide much less readily than the brominated syn-form (below) when treated with alcoholic KOH at 82° (1940a Ralls). For investigation of bromination rate under various conditions, see 1940b Ralls. Refluxing with acetic anhydride yields Δ^4-cholestenone acetylimine (p. 2432 s) (1937 Krekeler). Acetylation with acetic anhydride in dioxane at 25° or at 85° affords glassy solid products with differing refractive indices; that obtained at 85° appeared to be identical with the acetylation product of the syn-isomer (below) (1940a Ralls). For conversion into the syn-form, see below.

Syn-form of Δ^4-*cholestenone oxime.* Crystals (glacial acetic acid), m. 60°; resolidifies at 90° and melts finally at 151.8°. Absorption max. in glacial acetic acid at 238 mμ (log ε 4.4) (1938 Ralls). — **Fmn.** From the anti-form (above) by crystallization from glacial acetic acid (1938 Ralls), propionic acid, or slightly warmed butyric acid (1940a Ralls). — **Rns.** Treatment with bromine in glacial acetic acid gives a mixed product, m. 92^0 (containing 20 % or more of bromine), which loses HBr readily on treatment with alcoholic KOH at 82^0, or even on standing in the dry state, to give a *monobromo-derivative* $C_{27}H_{44}ONBr$ (m. 87^0; attempts to acetylate this product were unsuccessful) (1940a Ralls). For rates of bromination under various conditions, see 1940b Ralls. Reversion to the anti-form occurs on recrystallization from benzene or carbon tetrachloride. Attempted acetylation with acetic anhydride in glacial acetic acid carried out at 25^0 or at 80–85^0 gave non-crystalline products with identical refractive index (cf. reaction of the anti-form) (1940a Ralls).

$Δ^4$-Cholestenone O-(carboxymethyl)-oxime $C_{29}H_{47}O_3N$. Crystals (ethyl acetate), m. 158–9^0 cor. dec. — **Fmn.** From cholestenone on refluxing with O-(carboxymethyl)-hydroxylamine hydrochloride and sodium acetate in 90 % alcohol; quantitative yield (1936 Anchel).

$Δ^4$-Cholestenone semicarbazone $C_{28}H_{47}ON_3$ (p. 121). Crystals (alcohol-ethyl acetate), m. 234–5^0 (1942 Jones). Absorption max. in chloroform at 270.5 mμ (ε 26000) (1942 Jones); for absorption in alcohol (as quoted on p. 121), see also 1943 Evans. — For reactions, see those of the parent ketone (p. 2427 s).

$Δ^4$-Cholestenone o-tolylsemicarbazone $C_{35}H_{53}ON_3$ (p. 121). Feathery needles (butyl alcohol) (1937 Rosenheim), colourless needle clusters (dioxane), m. 242^0 to 243^0 (1943 Rosenheim).

$Δ^4$-Cholestenone phenylhydrazone $C_{33}H_{50}N_2$. Yellowish needles (ethyl acetate); sinters at ca. 142^0 and finally melts at 152^0 (1904 Diels). — **Fmn.** From the components by short boiling alone (1904 Diels) or in glacial acetic acid (1937 Rossner). — **Rns.** For further condensation, see reactions of the parent ketone (p. 2429 s).

$Δ^4$-Cholestenone p-nitrophenylhydrazone $C_{33}H_{49}O_2N_3$. Orange-yellow prisms (acetone); sinters at ca. 160^0 with complete melting (unsharp) at ca. 195^0 (1904 Diels).

$Δ^4$-Cholestenone 2.4-dinitrophenylhydrazone $C_{33}H_{48}O_4N_4$ (p. 121). Red needles (benzene-alcohol), plates (benzene), m. 233^0 (1942 Jones), red hair-like crystals or rosettes (benzene-alcohol) which on standing change to transparent prisms, m. 233–4^0 (1937 Rosenheim; 1944 Turfitt). Absorption max. in chloroform at 256 (21500), 281 (16000), 292 (11500), and 393 (29500) mμ (ε) (1945 Braude; cf. 1942 Jones). — **Fmn.** From $Δ^5$-cholestene-3β.4β-diol (p. 2051 s) or from $Δ^4$-cholestene-3β.6β-diol (p. 2060 s) by warming on the water-bath with 2.4-dinitrophenylhydrazine in alcohol containing HCl or H_2SO_4 (1937 Rosenheim). From $Δ^4$-cholestenone by mixing with 2.4-dinitrophenylhydrazine hydrochloride, both in hot alcoholic solution (1942 Jones).

$Δ^4$-Cholestenone p-carboxyphenylhydrazone $C_{34}H_{50}O_2N_2$. Crystals (alcohol), m. > 200^0. — **Rns.** This derivative is not decomposed by formaldehyde in alcohol; on refluxing with pyruvic acid in 95 % alcohol, a 78 % yield of the ketone can be recovered (1936 Anchel). For use in $Δ^4$-cholestenone separation, see 1936 Anchel.

Δ⁴-Cholestenone ketazine $C_{54}H_{88}N_2$. Amorphous yellow powder, dec. $> 190^0$ (1939 Dutcher). — For reactions, see those of the parent ketone (p. 2427 s).

For *Δ⁴-cholestenone-enol esters* and *ethers*, see derivatives of $Δ^{3.5}$-cholestadien-3-ol (p. 1542 s).

Δ⁵-**Cholesten-3-one** $C_{27}H_{44}O$ (p. 121). Crystals (methanol), m. 126^0, $[\alpha]_D^{ca.\ 20}$—8^0

(average value) (chloroform) (1948 Barton). — **Fmn.** For evidence of its formation as a primary reaction product in the dismutation of cholesterol in 50% acetic acid at 127^0, using a platinum catalyst, see 1944 Anker. Debromination of *Δ⁵*-cholestenone dibromide (cf. p. 121) with zinc dust can also be effected in methanol or acetone, but this occurs more slowly than in alcohol (1937b SCHERING KAHLBAUM A.-G.). — **Rns.** Treatment with perbenzoic acid in chloroform at room temperature gives two products, separated by crystallization from alcohol, referred to as *Δ⁵*-cholesten-3-one "α"- and "β"-oxides (1937 Ruzicka); the "β"-**oxide** $C_{27}H_{44}O_2$ (p. 121) is in fact 5α.6α-epoxycholestan-3-one (p. 2755 s) (1949 Urushibara; 1956 Ellis); the "α"-**oxide** $C_{27}H_{44}O_2$ (p. 121), obtained in ca. 14% yield (1937 Ruzicka), is now considered to be very probably **4-hydroxy-3‖4-*Δ⁵*-cholesten-3-oic acid lactone** $C_{27}H_{44}O_2$ (I) (1954 Mori). For conversion into coprostenone and into the dibromide (cf. p. 121), see also 1937c SCHERING KAHLBAUM A.-G. For conversion into deuteriocoprostenone, see this compound, p. 2431 s.

Ethylene ketal $C_{29}H_{48}O_2$. Originally considered to be the *Δ⁴*-isomer; for the now accepted *Δ⁵*-structure, see 1941 Fernholz; 1952a Antonucci. — Crystals (alcohol-pyridine) (1939a SCHERING A.-G.), leaflets (benzene-alcohol) (1941 SQUIBB & SONS), crystals (ether-methanol), m. $134-5^0$; $[\alpha]_D^{30}$—31^0 (chloroform) (1952a Antonucci), $[\alpha]_D^{20}$—11^0 (dioxane) (1939a SCHERING A.-G.), $[\alpha]_D$—28^0 (solvent not indicated) (1941 SQUIBB & SONS). No absorption in the near ultra-violet (1939a SCHERING A.-G.; 1952a Antonucci). — **Fmn.** From *Δ⁴*-cholesten-3-one (p. 2424 s) by refluxing with ethylene glycol in benzene containing traces of p-toluenesulphonic acid, with subsequent slow distillation of the azeotropic mixture; yield, 45% (1941 SQUIBB & SONS; cf. 1939a SCHERING A.-G.). — **Rns.** Treatment with perbenzoic acid affords the α-oxide (5α.6α-epoxycholestan-3-one ethylene ketal, p. 2755 s), which by mild CrO_3 oxidation gives cholestan-5-ol-3.6-dione (p. 2953 s) (1941 Fernholz). Hydrogenation, using palladium black, yields the ethylene ketal of cholestanone (p. 2442 s) (1941 SQUIBB & SONS; cf. 1954 Dauben). Refluxing with alcohol containing dil. HCl regenerates *Δ⁴*-cholestenone (1941 SQUIBB & SONS).

Propylene ketal $C_{30}H_{50}O_2$ (mixed stereoisomers, m. $135-140^0$) and *trimethylene ketal* $C_{30}H_{50}O_2$ (m. 137^0) are obtained analogously to the ethylene ketal (above) by reaction of *Δ⁴*-cholestenone with propylene glycol and trimethylene glycol, resp. (1941 SQUIBB & SONS).

Δ⁸-Cholesten-3-one, Zymostenone $C_{27}H_{44}O$. Double bond position based on that of its precursor, Δ^8-cholesten-3β-ol, which see p. 1690 s. — Needles (methanol), m. 124–5°, $[\alpha]_D^{20}$ +71° (chloroform) (1941, 1942 Wieland). — **Fmn.** From Δ^8-cholesten-3β-ol by refluxing with cyclohexane and aluminium isopropoxide in toluene, followed by conversion into the semicarbazone and regeneration with warm 25% H_2SO_4 (crude yield, 54%, still containing ca. 15% of starting material) (1941 Wieland); similarly from Δ^8-cholesten-3α-ol (p. 1691 s) (1942 Wieland).

Rns. Refluxing with isopropyl alcohol and aluminium isopropoxide in toluene gives Δ^8-cholesten-3β-ol (1941 Wieland) together with Δ^8-cholesten-3α-ol (1942 Wieland).

Semicarbazone $C_{28}H_{47}ON_3$. White solid (alcohol), not purified; m. 243° dec. (1941 Wieland).

Cholestan-3-one, Cholestanone, *β-Cholestanone* $C_{27}H_{46}O$ (p. 122). For identity with *zymostanone*, see 1940 Heath-Brown. — Needles (methanol) (1940 Heath-Brown), crystals (methanol-ethanol) (1941 b Bretschneider), (aqueous alcohol) (1938 Galinovsky), (acetone-alcohol) (1944 Anker), m. 129° to 129.5° cor. (1946a Wilds; cf. 1944 Anker); $[\alpha]_D^{ca.\ 20}$ +42°* (chloroform) (1940 Heath-Brown; 1941 b Bretschneider; 1944 Anker; 1948 Barton), $[\alpha]_D$ +43° (alcohol) (1946 Velluz).

FORMATION OF CHOLESTANONE

Together with coprostanone (p. 2443 s), from cholesterol (p. 1568 s) by heating with freshly reduced nickel at 220° (separation by fractional crystallization) (1927 Windaus) or by shaking with reduced platinum catalyst in 50% acetic acid at 127° (sealed tube); the mixed saturated ketones (yield, 25%) are separated together with Δ^4-cholestenone (p. 2424 s) from other products in the form of their p-carboxyphenylhydrazones, then from the unsaturated ketone by selective decomposition of the saturated ketone hydrazones with formaldehyde; finally the saturated ketones are fractionated by crystallization and chromatography (1944 Anker). Together with cholestane (p. 1429 s), from cholestan-3β-ol (p. 1695 s) by similar dismutation in deuterated 50% acetic acid; yield, 2.4%, no appreciable deuterium content (1944 Anker). From cholestan-3β-ol by oxidation with $KMnO_4$ in 96% acetic acid at room temperature for 7 hrs.; using twice the amount of $KMnO_4$ at 55° for 3 hrs., approximatively equal amounts of the ketone and of cholestane-2‖3-dioic acid (p. 1713 s) are formed (1940b Marker). From Δ^1-cholesten-3-one

* Average value, excluding the discordant value +48.5° given by Butenandt (1939b) for a product the purity of which seems unwarranted (Editor).

(p. 2423 s) by catalytic hydrogenation, using palladium-CaCO$_3$ catalyst, in methanol at room temperature (1939b Butenandt). From 2-bromo-$\Delta^{1.4.6}$-cholestatrien-3-one (p. 2479 s) by hydrogenation (discontinued after absorption of 4 mols. of hydrogen) in methanol containing potassium acetate, using platinum oxide catalyst (1940 Barkow). From 2α-bromocholestan-3-one (p. 2480 s) by hydrogenation (until 1 mol. of hydrogen has been absorbed) in dibutyl ether at room temperature, using platinum black catalyst (1932 Jakubowicz); from the same bromoketone, as main product, on refluxing with dimethylaniline (1937 Schwenk) or, in very small yield, together with Δ^1-cholesten-3-one, by refluxing with collidine for 12 hrs. (1940 Jacobsen). Together with Δ^4-cholesten-3-one (p. 2424 s), by thermal decomposition of cholestenone-pinacol [3β.3′β-dihydroxy-3.3′-bi-(Δ^4-cholestenyl)] at 220–230° and 0.01 mm. pressure; the distillate (70–80 % yield) is fractionally crystallized from aqueous alcohol (1938 Galinovsky).

PREPARATION OF CHOLESTANONE

Cholestanol (50 g.) in benzene (500 c.c.) is added slowly, with cooling, to a solution of sodium dichromate (68 g.) in H$_2$O (300 c.c.) containing H$_2$SO$_4$ (90 c.c.) and glacial acetic acid (50 c.c.) and the mixture is then agitated for 6 hrs. at 25–30°. The benzene solution is separated and washed, the solvent removed by distillation, and the residue recrystallized from alcohol; yield, 83–84 % (1943 Bruce).

REACTIONS OF CHOLESTANONE

Oxidation. Oxidation with (NH$_4$)$_2$S$_2$O$_8$ in slightly aqueous acetic acid at 80° yields, besides unidentified products, two lactones C$_{27}$H$_{46}$O$_2$, m. 201–2° and 184–6° resp. (1918 Ellis); probably the former is the lactone of 2-hydroxy-2‖3-cholestan-3-oic acid (p. 1711 s) and the latter the lactone of 4-hydroxy-3‖4-cholestan-3-oic acid (p. 1716 s) (Editor); for persulphuric acid oxidation, see also 1940c Marker. On standing with perbenzoic acid in chloroform in the dark at room temperature, the lactone of 4-hydroxy-3‖4-cholestan-3-oic acid (p. 1716 s) is formed (1942 Burckhardt).

Oxidation with CrO$_3$ in strong acetic acid at 40–50°, or with KMnO$_4$ and H$_2$SO$_4$ in aqueous acetic acid at room temperature, yields, in addition to carboxylic acids (cf. p. 122) and steam-volatile products, diketone products containing chiefly androstane-3.17-dione (p. 2892 s) (1936 C. F. BOEHRINGER & SOEHNE). On boiling with SeO$_2$ in 90 % alcohol, cholestane-2.3-dione form A (= Δ^3-cholesten-3-ol-2-one, p. 2659 s) is obtained (1938 Stiller).

Hydrogenation and reduction. Hydrogenation to give chiefly epi-cholestanol (p. 1702 s) is also effected in alcohol-ether using a slightly platinized Raney nickel catalyst; the rate of hydrogenation is greatly increased when small amounts of NaOH are present (1937 Délépine). Clemmensen reduction yields cholestane (p. 1429 s) (1917 Windaus). Reduction of the semicarbazone with sodium ethoxide at 180° gives cholestanol (p. 1695 s) and epi-cholestanol (p. 1702 s); by reaction at 200° cholestane is also formed; by use of excess hydrazine hydrate with sodium ethoxide at 200⁰, only cholestane is isolated; on replacing the semicarbazone by either the hydrazone or ketazine, the same products are obtained for either of the reactions at 200⁰; other aspects of the Wolff-Kishner reduction are discussed (1939 Dutcher). Reduction to cholestanol is also brought about by brewer's yeast in slightly alcoholic aqueous suspension at 35⁰ (1938 SCHERING CORP.).

Action of heat, halogens, and acids. Heating at ca. 350⁰ causes degradation to phenanthrene and a compound considered to be Δ^{16}-androsten-3-one * (p. 2396 s), with elimination of iso-octane and iso-octene (1937 d SCHERING-KAHLBAUM A.-G.). Treatment with bromine in chloroform at room tempera-ture affords 2α-bromocholestan-3-one (p. 2480 s) (1932 Jakubowicz); by reaction in glacial acetic acid, with slight warming, the product isolated is 2α.4α-dibromocholestan-3-one (p. 2495 s) (1946a Wilds), shown to arise by rearrangement of the 2.2-dibromo-isomer (p. 2494 s) in presence of HBr (1947 Djerassi). Using iodine monobromide in carbon tetrachloride - glacial acetic acid, 2α-bromo-, 2.2-dibromo-, and 2α.4α-dibromocholestan-3-ones can all be isolated; the reaction is monomolecular, catalysed by HBr, but in-hibited by large concentrations of the latter; reaction in carbon tetrachloride alone proceeds similarly (1938 Ralls; cf. 1947 Djerassi). By reaction with conc. H_2SO_4 in presence of acetic anhydride, with ice-water cooling, cholestan-3-one-2-sulphonic acid is formed (1937 Windaus).

Reactions with organic compounds. Refluxing with methylmagnesium iodide in ether gives a mixture of epimeric 3-methylcholestan-3-ols (p. 1776 s) from which one pure form can be separated (1937 Farmer; cf. 1937 Bolt); similar reactions carried out with the Grignard reagent obtainable from isopropyl, cyclohexyl, phenyl, α-naphthyl, and β-naphthyl bromides and from tert.-butyl chloride afford 3-isopropylcholestan-3-ol (p. 1843 s), 3-cyclohexylchole-stan-3-ol (p. 1852 s), 3-phenylcholestan-3-ol (p. 1852 s), 3-(α-naphthyl)- and 3-(β-naphthyl)-cholestan-3-ols (p. 1852 s), and 3-tert.-butylcholestan-3-ol (p. 1843 s), resp. (1937 Bolt). Treatment with β-phenylethylmagnesium chloride in boiling anhydrous ether, followed by heating with KHSO₄ at 150–160⁰ and high vacuum distillation, yields 3-(β-phenylethyl)-cholestene (p. 1448 s) (1938 Dansi). Reaction with ω-bromostyrene and lithium in ether, followed by hydrolysis, yields a compound formulated as **3-styrylcholestan-3-ol** $C_{35}H_{54}O$ (neither analysis nor constants given), which by ozonization in glacial acetic acid, followed by oxidation with CrO₃, gives 3-hydroxy-cholestane-3-carboxylic acid (1937a SCHERING A.-G.). Condensation with bromoacetic ester and zinc in boiling benzene, followed by dehydration with KHSO₄ at 150–160⁰ and finally saponification with alcoholic KOH, gives Δ^3-cholesten-3-yl-acetic acid and/or 3-carboxymethylene-cholestane (1938 Dansi). Reaction with benzoyl chloride, alone at 160–170⁰, or in boiling benzine (b. 100–110⁰), or in presence of pyridine, or reaction with benzoic anhydride affords the enol-benzoate, Δ^2-cholesten-3-yl benzoate (p. 1562 s)

---

* Structure doubtful; cf. footnotes pp. 1506 s and 1507 s concerning analogous degradation of cholestanyl methyl ether and epi-cholestanol (Editor).

(1936b CIBA). Condensation with isoamyl formate in the presence of sodium (1938 Stiller; cf. 1938 CIBA PHARMACEUTICAL PRODUCTS) or in the presence of sodium ethoxide (1943 Goldberg), both in ether at room temperature, followed by acidification, gives 2 or 4-hydroxymethylene-cholestan-3-one (p. 2513 s). By reaction with KCN in alcohol containing conc. HCl at room temperature, 3-hydroxycholestane-3-carboxylic acid nitrile (cholestanone cyanohydrin) is formed (1938 Kuwada; cf. 1943 Goldberg). Treatment with phenyl iso-cyanide and benzoic acid in ether at room temperature affords the anilide of 3-benzoyloxy-cholestane-3-carboxylic acid; reacts similarly with p-tolyl and o-anisyl isocyanides (1945 Baker). For reactions with cysteine, phenyl-hydrazine, benzaldehyde, isatin, and methylisatin, see the reaction products described on pp. 2440 s–2442 s.

ANALYTICAL PROPERTIES OF CHOLESTANONE

Colour reactions. Cholestanone gives no coloration with the Liebermann-Burchard reagent (1938 Petrow). With H_2SO_4 and acetic anhydride in glacial acetic acid at 85–90°, a brown coloration is developed; no anti-rachitically active product could be isolated (1939 Eck). Gives a violet colour in the Zimmermann reaction (with m-dinitrobenzene and KOH in alcohol) (1937 Kaziro); a more accurate investigation of this reaction shows that the colour development is more rapid than for 17-keto steroids, but the colour attained is less intense and there is subsequent fading; the maximum (at 560 mμ) is reached in 5 min. (1938 Callow; cf. 1944 Zimmer-mann). With p-nitrobenzenediazonium chloride in acetic acid cholestanone gives a positive reaction, but only slowly; in glacial acetic acid-dioxane a dark red colour develops in 30 min. (1938 Fieser). For colour sequence observ-ed when an alcoholic solution of the ketone and furfural is underlayered with conc. H_2SO_4, see 1939 Woker.

Estimation and separation. For gravimetric assay after reaction with Girard's reagent T, see 1941 Hughes; the polarographic assay used for Δ^4-cholestenone is not applicable to cholestanone (1940 Wolfe). For separation by chromato-graphy from Δ^4-cholestenone, see 1941 b Bretschneider; from coprostanone, see 1944 Anker; for separation by means of derivatives, see 1936 Anchel. Cholestanone gives a sparingly soluble *digitonide* (no analysis) on short boiling with a 1 % alcoholic solution of digitonin; the precipitate crystallizes only slowly; solubility ca. 0.09 % in alcohol at 20° (1946 Velluz; cf. 1937 Kawasaki).

MODIFIED FORMS AND REACTION PRODUCTS OF CHOLESTANONE

Deuteriocholestanone, with ca. 3.5 atoms of D. Crystals, m. 128–128.5°. — **Fmn.** From cholesterol and a mixture of deuterioacetic acid (60 % content of D) and D_2O (99 % D) on heating at 127° with finely divided platinum; deuterated cholesterol, coprostenone, coprostanone, cholestane, and probably coprostane, are formed at the same time; for separation, see the analogous formation of ordinary cholestanone, p. 2436 s; yield 0.5 % (1944 Anker).

E.O.C. XIV s. 170

2 (?)-Benzylidene-cholestan-3-one $C_{34}H_{50}O$. Two isomeric forms are known, considered to be either position isomers (2- and 4-benzylidene-cholestan-3-ones) or stereoisomers (1943 Goldberg); in view of their mode of interconversion (1932 Jakubowicz), the geometric isomerism seems the more likely (Editor). — *Form I.* Needles (ethyl acetate), m. 145–6° cor. (1943 Goldberg; cf. 1932 Jakubowicz), $[\alpha]_{578}^{13}$ —120°, $[\alpha]_{546}^{13}$ —141°, $[\alpha]_{436}^{13}$ —257°, all in chloroform (1932 Jakubowicz). — *Form II.* Crystals (acetone-alcohol), m. 126° to 128° cor. (1943 Goldberg), $[\alpha]_{578}^{13}$ —105°, $[\alpha]_{546}^{13}$ —127°, both in chloroform (1932 Jakubowicz). — **Fmn.** Both forms are obtained together with 2 (?)-(α-hydroxybenzyl)-cholestan-3-one (see below) from cholestan-3-one by condensation with benzaldehyde in alcohol containing small amounts of 10% aqueous NaOH at room temperature; by fractional crystallization from ethyl acetate form I separates first (yield, 4.1%), then the hydroxybenzyl-compound and from the mother liquors very small yields of form II are obtained by chromatography on Al_2O_3 (1943 Goldberg). By a similar reaction in ca. 98% alcohol at room temperature, only the two benzylidene compounds were isolated, and form I was the chief product (yield, 49%), whereas by reaction at 69° form II was almost the sole product (1932 Jakubowicz). — **Rns.** Attempts to effect further condensation of form I with benzaldehyde in alcohol containing sodium ethoxide were unsuccessful, but transformation into form II occurred (1932 Jakubowicz).

2 (?)-(α-Hydroxybenzyl)-cholestan-3-one $C_{34}H_{52}O_2$. The structure ascribed seems probable by its relation to 2(?)-benzylidene-cholestan-3-one (see above) (Editor). — Crystals (ethyl acetate), m. 184–6° cor. — **Fmn.** Together with the 2(?)-benzylidene-cholestan-3-ones (see above); yield, 3.5% (1943 Goldberg).

4-Carboxythiazolidine-(2 spiro 3)-cholestane, "*Cholestanone thiazolidine*" $C_{30}H_{51}O_2NS$. No analysis, no constants given. — High melting solid, insoluble in water, sparingly soluble in organic solvents. Stable in aqueous NH_3 or saturated $NaHCO_3$ solution, slowly decomposed in solutions of Na_2CO_3 or NaOH. — **Fmn.** From cholestanone by condensation with cysteine. — **Rns.** Oxidation by mild reagents, such as iodine or $K_3Fe(CN)_6$, gives cholestanone together with cystine. — *Ethyl ester* $C_{32}H_{55}O_2NS$, lower melting than the free acid, readily soluble in non-polar solvents; obtained from cholestanone by condensation with ethyl cysteinate (1946 Lieberman).

$$* C_8H_{17} = \overset{CH_3}{\underset{|}{CH}} \cdot [CH_2]_3 \cdot CH(CH_3)_2.$$

Indolo-3′.2′:2.3-(Δ^2-cholestene) (I) or **Indolo-2′.3′:3.4-(Δ^3-cholestene)** (II) $C_{33}H_{49}N$ (p. 122: *carbazole derivative*, m. 180–1°). On the basis of surface film

area measurements Dorée (1935) prefers the unsymmetrical structure (II), but the symmetrical structure (I) is adopted by Rossner (1937), presumably on the basis of the known greater tendency for reaction at C-2 in cholestanone. — White plates (benzene-alcohol), m. 180–1°; very soluble in benzene or ethyl acetate, sparingly soluble in methanol, ethanol, or glacial acetic acid; molecular area of surface film 43 sq. Å (1935 Dorée). — **Fmn.** From cholestanone by heating with phenylhydrazine in glacial acetic acid on the water-bath for 20 min.; yield, 60–70% (1935 Dorée). — **Rns.** Treatment with conc. H_2SO_4 in presence of acetic anhydride at ca. 0° gives 43% yield of a **sulphonic acid** $C_{33}H_{49}O_3NS$ [crystals (acetone or abs. ether), m. 235°; its aqueous solutions are strongly fluorescent; sodium salt readily, potassium salt less readily, water-soluble; *methyl ester* $C_{34}H_{51}O_3NS$, needles (acetone or ethyl acetate), m. 190°]. On heating with selenium, progressive cyclization and dehydrogenation occurs and products, for which formulæ (III) and (IV) are suggested, have been isolated from the resulting mixtures by chromatography: **compound** $C_{29}H_{39}N$ (III)** [white rosettes (alcohol), m. 203°; formed by heating for

40 hrs. at 320°; yield 4%]; **compound** $C_{21}H_{17}N$ (IV: *20-methyl-3-aminochol-anthrene?*)** [pale yellow needles (benzene), m. 225° turning brown; can be sublimed in a high vacuum; is without carcinogenic properties; formed by heating for 16 hrs. at 320° and then for 30 hrs. at 340°; yield, 3%]; after still further heating at 360° (9 hrs.), 20-methylcholanthrene (p. 531 s) is obtained (1937 Rossner).

(4′-Carboxyquinolino)-3′.2′:2.3-cholestane $C_{35}H_{49}O_2N$ (V, R = CO_2H, R′ = H). Almost colourless microcrystalline powder, m. > 310°, above this temperature decomposition occurs with evolution of CO_2. Solutions in alkali foam readily and are strongly hæmolytic. —

* See the footnote on p. 2440 s.
** An alternative structure for compound (III) might be based on the unsymmetrical structure (II) for the indolocholestene. — See also footnote ** on p. 2432 s.

170*

Fmn. From cholestanone by heating with isatin and KOH in aqueous alcohol on the water-bath, followed by acidification; yield, almost quantitative. — **Rns.** Cautious heating until no more CO_2 is evolved, followed by rapid distillation, all in a high vacuum, affords **quinolino-3'.2':2.3-cholestane** $C_{34}H_{49}N$ (V, R = R'= H) [colourless leaflets (alcohol), m. 193°; dissolves in conc. H_2SO_4 without coloration; its alcoholic solution does not fluoresce; on strongly heating with selenium, only nitrogenous products could be isolated; *hydrochloride* $C_{34}H_{50}NCl$, yellowish leaflets (alcohol-benzene), m. 191°; *picrate* (no analysis), bright yellow needles (alcohol), m. 201–2° dec.] (1944 Buu-Hoi).

(6'-Methyl-4'-carboxyquinolino)-3'.2':2.3-cholestane $C_{36}H_{51}O_2N$ (V, R = CO_2H, R'= CH_3). — Pale yellow microcrystalline powder, m. > 310°; aqueous alkaline solutions foam readily and are strongly hæmolytic. — **Fmn.** From cholestanone and 5-methylisatin by heating with KOH in aqueous alcohol on the water-bath, followed by acidification; yield, almost quantitative. — **Rns.** Cautious heating until no more CO_2 is evolved, followed by rapid distillation, all in a high vacuum, gives **6'-methylquinolino-3'.2':2.3-cholestane** $C_{35}H_{51}N$ (V, R = H, R'= CH_3) [colourless leaflets (abs. alcohol), m. 176°; dissolves in conc. H_2SO_4 without coloration; *picrate* (no analysis given), bright yellow needles, dec. > 230°] (1944 Buu-Hoi).

DERIVATIVES OF CHOLESTANONE

Cholestanone diethyl ketal $C_{31}H_{56}O_2$. Crystals (alcohol-pyridine), m. 68° to 69.5°; $[\alpha]_D$ +26° (dioxane). — **Fmn.** From the ketone with orthoformic ester in benzene containing alcoholic HCl at 25°; yield, 77%. — **Rns.** Boiling with alcohol containing dil. HCl regenerates the ketone; boiling in xylene affords the enol-ether, Δ^2-cholesten-3-yl ethyl ether (p. 1562 s) (1938 Serini).

Cholestanone ethylene ketal $C_{29}H_{50}O_2$. Crystals (alcohol), m. 115° (1941 SQUIBB & SONS), m. 113°, $[\alpha]_D^{24}$ +21.6° (chloroform) (1954 Dauben). — **Fmn.** From Δ^5-cholesten-3-one ethylene ketal (p. 2435 s) by catalytic hydrogenation using palladium black (1941 SQUIBB & SONS) or palladium-$BaSO_4$ in alcohol at room temperature (1954 Dauben). From cholestanone by reaction with ethylene glycol in benzene containing traces of p-toluenesulphonic acid, with slow distillation; yield, 82% (1941 SQUIBB & SONS).

Cholestanone di-ethylthio ketal $C_{31}H_{56}S_2$. Crystals (dil. acetone, below 50°), m. 80–82° (1946 Bernstein); m. 67°, $[\alpha]_D$ +34° (benzene) (1940b CHINOIN). — **Fmn.** From cholestanone by reaction with ethyl mercaptan in presence of $ZnCl_2$ and anhydrous Na_2SO_4 at 3–5° (crude yield, 78%) (1946 Bernstein) or in presence of benzenesulphonic acid and, preferably, ethyl formate at room temperature (yields, 90%) (1940b CHINOIN; cf. also 1940a CHINOIN). — **Rns.** Heating with Raney nickel in dioxane on the steam-bath affords cholestane (1946 Bernstein).

Cholestanone oxime $C_{27}H_{47}ON$, m. 199° (1938 Ralls).

Cholestanone O-(carboxymethyl)-oxime $C_{29}H_{49}O_3N$. Crystals (ethyl acetate), m. 151–2° cor. dec. — **Fmn.** From cholestanone on refluxing with O-(carboxymethyl)-hydroxylamine hydrochloride and sodium acetate in 90% alcohol; quantitative yield. — **Rns.** The ketone can be regenerated in 95% (crude) yield by refluxing with 95% alcohol containing HCl; its use for separation of cholestanone is described (1936 Anchel).

Cholestanone hydrazone $C_{27}H_{48}N_2$. Pale yellow needles (aqueous alcohol), m. 248° dec. with previous softening at 230°. — **Fmn.** From the ketone by refluxing with hydrazine hydrate in abs. alcohol; followed by addition of water to slight turbulence and standing; quantitative yield (1939 Dutcher).

Cholestanone semicarbazone $C_{28}H_{49}ON_3$ (p. 122). Colourless amorphous powder, m. 238° dec., sinters at 227°; obtainable by the usual method in quantitative yield (1939 Dutcher).

Cholestanone o-tolylsemicarbazone $C_{35}H_{55}ON_3$. Colourless needles (benzene-alcohol), m. 228–9° dec. (1938 Stiller).

Cholestanone p-carboxyphenylhydrazone $C_{34}H_{52}O_2N_2$. Six-sided plates (alcohol), dec. $> 200°$. — **Rns.** The ketone is regenerated in 89% yield by refluxing with formaldehyde in 95% alcohol; the use of this derivative for separation of cholestanone, particularly from Δ^4-cholestenone, is described (1936 Anchel).

Cholestanone ketazine $C_{54}H_{92}N_2$. Amorphous white powder, dec. ca. 200°. — **Fmn.** As for the hydrazone (above) but water is replaced by acetic acid after the initial reaction in abs. alcohol (1939 Dutcher).

Cholestanone-enol esters and ethers. See p. 1562 s.

Coprostan-3-one, Coprostanone $C_{27}H_{46}O$ (p. 121). For relation between configuration and optical rotation, see 1941 Bernstein. — Colourless leaflets (abs. alcohol), m. 61–62° cor. (1946 Lederer; cf. 1934 Ruzicka; 1938 Inhoffen); $[\alpha]_D +36°$ (benzene) (1934 Ruzicka), $+36°$ (chloroform) (1944 Anker), $[\alpha]_D^{16}$ $+36°$ (alcohol) (1946 Lederer). Gives no colour reaction with the Liebermann-Burchard reagent (1938 Petrow). — For evidence of coprostanone as intermediate in the metabolic conversion of cholesterol into coprosterol, see 1946 Lederer; see also deuteriocoprostanone (p. 2445 s).

Occ. and Isolation. In ambergris; the ketone constituents, separated by Girard's reagent T, are submitted to fractional distillation at 0.3 mm. pressure, and the products distilling at 210–215° are further fractionated by chromatography (yield, ca. 4%); the coprostanone so obtained may be an artefact (1946 Lederer). For isolation from mixtures by use of carboxylated derivatives (described below), see 1936 Anchel.

Fmn. Together with cholestanone (p. 2436 s), from cholesterol (p. 1568 s) by treatment with freshly reduced nickel at 220°; separation by fractional crystallization (1927 Windaus). From cholesterol, together with other products, by shaking with active platinum in dil. acetic acid at 127°; separation via its addition compound(?) with cholestanone, which see p. 2444 s (1944

Anker). From $\Delta^{1\cdot4}$-cholestadien-3-one (p. 2418 s) by complete hydrogenation in ether, using a platinum black catalyst (1938 Inhoffen). From 2-bromo-$\Delta^{1\cdot4\cdot6}$-cholestatrien-3-one (p. 2479 s) by hydrogenation in hot alcohol containing KOH, using palladium black or palladium-BaSO$_4$ catalyst. Together with unchanged starting material, from 2.6-dibromo-$\Delta^{1\cdot4\cdot6}$-cholestatrien-3-one (p. 2497 s) by hydrogenation in cold alcohol (1940 Barkow).

Rns. Oxidation with (NH$_4$)$_2$S$_2$O$_8$ in glacial acetic acid on the water-bath, gives the lactone of 4-hydroxy-3‖4-coprostan-3-oic acid (p. 1716 s), together with that of 2-hydroxy-2‖3-coprostan-3-oic acid (p. 1711 s) and small amounts of non-investigated products (1913 Gardner; cf. 1942 Burckhardt); by oxidation with perbenzoic acid in chloroform at 17°, the lactone of 4-hydroxy-3‖4-coprostan-3-oic acid is obtained (1942 Burckhardt). Oxidation with SeO$_2$ in alcohol occurs much less readily than for cholestanone (p. 2436 s), giving a "diosphenol" (not formulated; forms an insoluble potassium enol salt; purple coloration with FeCl$_3$) (1938 Stiller).

Hydrogenation in alcohol, using platinum oxide catalyst, yields epi-coprostanol (p. 1706 s) together with coprostanol (p. 1704 s) and coprostane (p. 1431 s) (1946 Lederer). On heating the semicarbazone with sodium ethoxide in abs. alcohol at 180° (sealed tube), epicoprostanol is formed together with coprostanol and small amounts of a hydrocarbon product (1939 Dutcher).

The disubstituted product of bromination (cf. p. 121), for which Inhoffen (1949) still favoured the 4.4-structure, has been shown to be 2β.4β-dibromo-coprostan-3-one (p. 2497 s) (1951 Djerassi; 1951 b Inhoffen). Treatment with conc. H$_2$SO$_4$ and acetic anhydride, with cooling in ice-water, gives coprostan-3-one-2-sulphonic acid, isolated in the form of its methyl ester (m. 171–2°) (1937 Windaus), from the mother liquors of which the methyl ester (m. 104–5°) of the isomeric 4-sulphonic acid (main product) can be isolated (1938 Windaus). — For reactions with heavy water and with phenylhydrazine, see the reaction products described below.

Addition compound of coprostanone with cholestanone (?) *. Crystals (ether-methanol, alcohol, or acetone-alcohol), m. 78–79°, [α]$_D$ +41° (chloroform); cannot be separated into the components by recrystallization from the solvents mentioned. — **Fmn.** From cholesterol (p. 1568 s) by shaking with active platinum in dil. acetic acid at 127°; the ketone products are separated in the form of p-carboxyphenylhydrazones, from which cholestanone (p. 2436s) and coprostanone are regenerated by treatment with formaldehyde; on fractional crystallization from ethanol and ether-methanol, cholestanone separates first and the addition compound is isolated from the mother liquors; the mechanism of this reaction is discussed. — **Rns.** Separation into the components by fractional elution after adsorption on Al$_2$O$_3$ (1944 Anker).

Coprostanone O-(carboxymethyl)-oxime C$_{29}$H$_{49}$O$_3$N. Crystals (ethyl acetate), m. 150–1° cor. dec. — **Fmn.** From coprostanone on refluxing with O-(carboxymethyl)-hydroxylamine hydrochloride and sodium acetate in 90% alcohol;

---

* If this product is a genuine addition compound, the ratio coprostanone/cholestanone = 1:3 seems the most likely from the rotation values recorded (Editor).

quantitative yield. — **Rns.** Gives a 100% yield of ketone on hydrolysis with 95% alcohol containing HCl (1936 Anchel).

Coprostanone semicarbazone $C_{28}H_{49}ON_3$ (p. 121). Colourless, amorphous powder, m. 192° dec. (sinters at 178°) (1939 Dutcher).

Coprostanone p-carboxyphenylhydrazone $C_{34}H_{52}O_2N_2$. Crystals (alcohol), dec. >200°; obtainable in 87% yield; on refluxing with formaldehyde in 95% alcohol the ketone regenerated amounts to 91% (1936 Anchel).

Deuteriocoprostanone (cf. p. 121), m. 60–62°, containing 1.28 atom % deuterium, situated at C-2 or C-4, is obtained from coprostanone by refluxing with alcoholic D_2O containing NaOH; its conversion into deuteriocoprosterol in the human metabolism is discussed (1938 Anchel). — A product m. 58–60°, containing ca. 3.5 D atoms, is obtained from cholesterol, together with other products, by shaking with activated platinum in deuterated acetic acid at 127° (cf. the analogous formation of coprostanone, above, and reactions of cholesterol, p. 1589 s) (1944 Anker).

Page 121, line 3 from bottom: for "coprostenol" read "coprostanol".

Indolo - 2'.3' : 3.4 - (Δ^3 - coprostene) $C_{33}H_{49}N$ (p. 121: *carbazole derivative,* m. 192°). Structure assigned by analogy with that proposed for the analogous indolocholestene (page 2441 s), and in view of the greater tendency for reaction at C-4 in the case of coprostanone (Editor). White rectangular plates (benzene-alcohol), m. 192° to give a red liquid; almost insoluble in alcohol and in petroleum ether; readily absorbs bromine from chloroform (1908, 1909 Dorée). — **Fmn.** From coprostanone on warming with phenylhydrazine in glacial acetic acid at 40° (yield, ca. 50%) or on heating directly with phenylhydrazine (1908 Dorée). — *N-Nitroso-derivative* $C_{33}H_{48}ON_2$. Rods or plates (benzene-petroleum ether); shrinks and turns red at 148°, melts at 158° to a dark red liquid; very soluble in benzene, very sparingly soluble in acetone, alcohol, or petroleum ether, insoluble in ether. **Fmn.** From the above indolocoprostene by treatment with KNO_2 in ether-glacial acetic acid in the cold; yield almost quantitative (1909 Dorée).

1904 Diels, Abderhalden, *Ber.* **37** 3092, 3099 et seq.
1906 Windaus, *Ber.* **39** 518, 520, 521.
1908 Dorée, Gardner, *J. Ch. Soc.* **1908** 1625, 1629.
       Jaeger, *Z. Krist.* **44** 561, 567.
1909 Dorée, *J. Ch. Soc.* **95** 638, 653, 654.
1911 Windaus, Adamla, *Ber.* **44** 3051, 3056.
1913 Gardner, Godden, *Biochem. J.* **7** 588.
1917 Windaus, *Ber.* **50** 133, 137.
1918 Ellis, Gardner, *Biochem. J.* **12** 72, 77.
1927 Windaus, *Ann.* **453** 101, 110.
1928 Wokes, *Biochem. J.* **22** 830, 832.
1930 Adam, Rosenheim, *Proc. Roy. Soc. London* A **126** 25, 26, 27, 31.
       Reindel, Weichmann, *Ann.* **482** 120, 122, 129.

* See the footnote on p. 2440 s.

1932 Jakubowicz, *Contribution à l'étude de l'isomérie des cholestanols et de leurs dérivés halogénés*, Thesis Univ. Nancy, pp. 128, 131, 135 et seq.

1934 Ruzicka, Brüngger, Eichenberger, Meyer, *Helv.* **17** 1407, 1414.

1935 Butenandt, Wolff, *Ber.* **68** 2091, 2093.

Dorée, Petrow, *J. Ch. Soc.* **1935** 1391, 1392.

Tavastsherna, *Arch. sci. biol. (U.S.S.R.)* **40** 141, 144; *C. A.* **1937** 6670; *Ch. Ztbl.* **1937** II 3323.

1936 Anchel, Schoenheimer, *J. Biol. Ch.* **114** 539, 541, 542, 543, 544, 545.

C. F. BOEHRINGER & SOEHNE (Dirscherl, Hanusch), *Ger. Pat.* 695638 (issued 1940); *C. A.* **1941** 5654; *Ch. Ztbl.* **1940** II 2784.

Butenandt, Schramm, *Ber.* **69** 2289, 2296.

a CIBA*, *Swiss Pats.* 211649, 214330 (issued 1941); *C. A.* **1942** 3633, 5960; *Brit. Pat.* 477400 (issued 1938); *Indian Pat.* 24038 (1937, issued 1938); *C. A.* **1938** 3914; *Ch. Ztbl.* **1938** II 119; CIBA PHARMACEUTICAL PRODUCTS (Ruzicka, Fischer), *U.S. Pat.* 2248438 (issued 1941); *C. A.* **1941** 6742; *Ch. Ztbl.* **1944** I 111.

b CIBA*, *Swiss Pat.* 214328 (issued 1941); *C. A.* **1942** 5960; *Ch. Ztbl.* **1942** I 2296.

a SCHERING A.-G., *Ger. Pat.* 699248 (issued 1940); *C. A.* **1941** 6742; *Ch. Ztbl.* **1941** I 1325.

b SCHERING A.-G. (Inhoffen), *Ger. Pat.* 722943 (issued 1942); *C. A.* **1943** 5201; SCHERING CORP. (Inhoffen, Butenandt, Schwenk), *U.S. Pat.* 2340388 (1937, issued 1944); *C. A.* **1944** 4386.

SCHERING CORP. (Schwenk, Whitman), *U.S. Pat.* 2247822 (issued 1941); *C. A.* **1941** 6396; SCHERING-KAHLBAUM A.-G. (Schwenk, Whitman), *Brit. Pat.* 501196 (1937, issued 1939); *C. A.* **1939** 5868; *Ch. Ztbl.* **1939** II 686.

1937 C. F. BOEHRINGER & SOEHNE, *Fr. Pat.* 835527 (issued 1938); *C. A.* **1939** 4600; *Ch. Ztbl.* **1939** I 2827; *Dutch Pat.* 50888 (issued 1941); *C. A.* **1942** 5322; *Ch. Ztbl.* **1941** II 3293.

Bolt, Backer, *Rec. trav. chim.* **56** 1139.

a CIBA*, *Swiss Pat.* 214603 (issued 1941); *C. A.* **1942** 4977; *Ch. Ztbl.* **1942** I 2561; *Swiss Pat.* 221804 (issued 1942); *C. A.* **1949** 700; *Fr. Pat.* 830043 (issued 1938); *C. A.* **1939** 179; *Ch. Ztbl.* **1939** I 2035.

b CIBA*, *Swiss Pat.* 222124 (issued 1942); *C. A.* **1949** 366; (Ruzicka), *Swed. Pat.* 98368 (1938, issued 1940); *Ch. Ztbl.* **1940** II 2185; CIBA PHARMACEUTICAL PRODUCTS (Ruzicka), *U.S. Pat.* 2232636 (1938, issued 1941); *C. A.* **1941** 3266; *Ch. Ztbl.* **1941** II 3100, 3101.

Délépine, Horeau, *Bull. soc. chim.* [5] **4** 31, 43.

Eck, Thomas, Yoder, *J. Biol. Ch.* **117** 655, 658.

Farmer, Kon, *J. Ch. Soc.* **1937** 414, 418.

Hogness, Sidwell, Zscheile, *J. Biol. Ch.* **120** 239, 244, 253.

I. G. FARBENIND. (Bockmühl, Ehrhart, Ruschig, Aumüller), *Ger. Pat.* 693351 (issued 1940); *C. A.* **1941** 4921; *Brit. Pat.* 502666 (issued 1939); *C. A.* **1939** 7316; *Ch. Ztbl.* **1939** II 1126; *Fr. Pat.* 842026 (1938, issued 1939); *C. A.* **1940** 4743; *Ch. Ztbl.* **1940** I 429.

Kawasaki, *C. A.* **1938** 3414; *Ch. Ztbl.* **1938** II 2944.

Kaziro, Shimada, *Z. physiol. Ch.* **249** 220, 222.

Krekeler, *Einige Umsetzungen am Cholestenon*, Thesis Univ. Göttingen, pp. 6, 9, 12, 17 et seq., 20, 21.

Mohler, *Helv.* **20** 289.

N. V. ORGANON, *Fr. Pat.* 827623 (issued 1938); *C. A.* **1938** 8437; *Ch. Ztbl.* **1938** II 3119; Alien Property Custodian (Oppenauer), *U.S. Pat.* 2384335 (issued 1945); *C. A.* **1946** 178; *Ch. Ztbl.* **1948** I 1039**.

Rosenheim, Starling, *J. Ch. Soc.* **1937** 377, 382, 383.

Rossner, *Z. physiol. Ch.* **249** 267, 268, 269 et seq.

Ruzicka, Bosshard, *Helv.* **20** 244, 248.

Schacher, Browne, Selye, *Proc. Soc. Exptl. Biol. Med.* **36** 488.

a SCHERING A.-G., *Fr. Pat.* 831131 (issued 1938); *C. A.* **1939** 1762; *Ch. Ztbl.* **1939** I 3590.

b SCHERING A.-G., *Fr. Pat.* 835526 (issued 1938); *C. A.* **1939** 4599; *Ch. Ztbl.* **1939** I 4088.

c SCHERING A.-G., *Fr. Pat.* 838704 (issued 1939); *Brit. Pat.* 508576 (issued 1939); *C. A.* **1940** 777; *Ch. Ztbl.* **1939** II 1722; SCHERING CORP. (Inhoffen), *U.S. Pat.* 2280828 (issued 1942); *C. A.* **1942** 5618.

* Gesellschaft für chemische Industrie in Basel; Société pour l'industrie chimique à Bâle.

** By error, Pat. No. is quoted here as 2384535.

d Schering A.-G., *Austrian Pat. Application; FIAT Final Report* No. 996, pp. 130, 139, 152, 155, 157, 158 (issued 1947).

Schering Corp. (Schoeller, Serini, Inhoffen), *U. S. Pat.* 2264861 (issued 1941); *C. A.* **1942** 1613.

a Schering-Kahlbaum A.-G., *Fr. Pat.* 822551 (issued 1938); *C. A.* **1938** 4174; *Ch. Ztbl.* **1938** II 120; Schering Corp. (Serini, Köster, Strassberger), *U.S. Pat.* 2379832 (issued 1945); *C. A.* **1945** 5053.

b Schering-Kahlbaum A.-G., *Brit. Pat.* 486992 (issued 1938); *C. A.* **1938** 8707; *Ch. Ztbl.* **1939** I 1603; *Swiss Pat.* 199448 (issued 1938); *Ch. Ztbl.* **1939** I 3931.

c Schering-Kahlbaum A.-G., *Brit. Pat.* 492725 (issued 1938); *C. A.* **1939** 1884; *Ch. Ztbl.* **1939** I 2642; Schering Corp. (Butenandt), *U.S. Pat.* 2248954 (issued 1941); *C. A.* **1941** 6742; *Ch. Ztbl.* **1942** I 3285.

d Schering-Kahlbaum A.-G., *Brit. Pat.* 494773 (issued 1938); *C. A.* **1939** 2537; *Ch. Ztbl.* **1939** I 4652; Schering A.-G., *Dutch Pat.* 47962 (issued 1940); *Ch. Ztbl.* **1940** II 374.

Schwenk, Whitman, *J. Am. Ch. Soc.* **59** 949.

Stavely, Bergmann, *J. Org. Ch.* **2** 580.

Urushibara, Ando, Araki, Ozawa, *Bull. Ch. Soc. Japan* **12** 353.

Westphal, *Ber.* **70** 2128, 2134.

Windaus, Kuhr, *Ann.* **532** 52, 57, 65, 67.

1938 Adkins, Cox, *J. Am. Ch. Soc.* **60** 1151, 1153.

Anchel, Schoenheimer, *J. Biol. Ch.* **125** 23.

E. Bergmann, Hirshberg, *Nature* **142** 1037.

C. F. Boehringer & Soehne (Hatzig), *Ger. Pat.* 712256 (issued 1941); *C. A.* **1943** 4408; *Ch. Ztbl.* **1942** I 1162.

Bolt, *Rec. trav. chim.* **57** 905, 906.

Butenandt, Wolz, *Ber.* **71** 1483, 1485.

Callow, Callow, Emmens, *Biochem. J.* **32** 1312, 1322, 1324.

a Chinoin**, *Fr. Pat.* 840964 (issued 1939); *C. A.* **1940** 1821, 5601; *Ch. Ztbl.* **1940** I 427.

b Chinoin** (Bretschneider), *Brit. Pat.* 518266 (issued 1940); *C. A.* **1941** 7662; *Ch. Ztbl.* **1941** I 928; (Bretschneider, Salamon), *U.S. Pat.* 2246341 (issued 1941); *C. A.* **1941** 6068; *Fr. Pat.* 866306 (1939, issued 1941); *Dutch Pat.* 51984 (1939, issued 1942); *Ch. Ztbl.* **1942** II 317.

Ciba* (Miescher, Fischer), *U.S. Pat.* 2172590 (issued 1939); *C. A.* **1940** 450; *Ch. Ztbl.* **1940** I 428; *Fr. Pat.* 841242 (issued 1939); *Ch. Ztbl.* **1940** I 916.

Ciba Pharmaceutical Products (Ruzicka), *U.S. Pat.* 2281622; *C. A.* **1942** 5958.

Dansi, *Gazz.* **68** 273.

Fieser, Campbell, *J. Am. Ch. Soc.* **60** 159, 168, 169.

Galinovsky, Bretschneider, *Monatsh.* **72** 190; *Sitzber. Akad. Wiss. Wien IIb* **147** 266.

Graff, *Biochem. Z.* **298** 179, 195.

Inhoffen, Huang-Minlon, *Ber.* **71** 1720, 1722, 1723, 1727.

Kuwada, Miyasaka, *C. A.* **1938** 7474, 7475; *Ch. Ztbl.* **1939** I 1372.

Petrow, Rosenheim, Starling, *J. Ch. Soc.* **1938** 677, 679, 681.

Ralls, *J. Am. Ch. Soc.* **60** 1744.

a Ruzicka, Plattner, Aeschbacher, *Helv.* **21** 866, 869.

b Ruzicka, Plattner, *Helv.* **21** 1717, 1719.

Schering Corp. (Mamoli), *U.S. Pat.* 2186906 (issued 1940); *C. A.* **1940** 3436; *Ch. Ztbl.* **1940** I 3426.

Schwenk, Fleischer, Whitman, *J. Am. Ch. Soc.* **60** 1702.

Serini, Köster, *Ber.* **71** 1766, 1767.

Stiller, Rosenheim, *J. Ch. Soc.* **1938** 353, 355.

Windaus, Mielke, *Ann.* **536** 116, 126.

1939 Adam, Askew, Pankhurst, *Proc. Roy. Soc. London* A **170** 485, 493, 497.

Bandow, *Biochem. Z.* **301** 37, 44, 45.

W. Bergmann, Hirschmann, *J. Org. Ch.* **4** 40, 46.

---

* Gesellschaft für chemische Industrie in Basel; Société pour l'industrie chimique à Bâle.
** Chinoin gyógyszer és vegyészeti termékek gyára r. t. (Dr. Kereszty & Dr. Wolf).

C. F. Boehringer & Soehne, *Fr. Pat.* 846099 (issued 1939); *C. A.* **1941** 1185; *Ch. Ztbl.* **1940** I 2830.

a Butenandt, Wolff, *Ber.* **72** 1121.

b Butenandt, Mamoli, Dannenberg, Masch, Paland, *Ber.* **72** 1617, 1621, 1623, 1627.

Chinoin* (Bretschneider, Fári), *U.S. Pat.* 2283411 (issued 1942); *C. A.* **1942** 6170; *Brit. Pat.* 530559 (issued 1940/41); *C. A.* **1942** 98; *Ch. Ztbl.* **1942** I 1025.

Dannenberg, *Abhandl. Preuss. Akad. Wiss. Math.-naturwiss. Kl.* No. 21, pp. 10, 14, 27, 43, 45, 53, 54, 59, 64, 65.

Décombe, Rabinowitch, *Bull. soc. chim.* [5] **6** 1510, 1519.

Dutcher, Wintersteiner, *J. Am. Ch. Soc.* **61** 1992, 1994, 1996–1998.

Eck, Thomas, *J. Biol. Ch.* **128** 257, 260, 261.

Ehrhart, Ruschig, Aumüller, *Ber.* **72** 2035, 2036.

a Inhoffen, Huang-Minlon, *Naturwiss.* **27** 167.

b Inhoffen, Huang-Minlon, *Ber.* **72** 1686.

Kuhr, *Ber.* **72** 929.

Mamoli, Koch, Teschen, *Z. physiol. Ch.* **261** 287, 289.

a Schering A.-G. (Köster, Inhoffen), *Dutch Pat.* 52656 (issued 1942); *Ch. Ztbl.* **1942** II 2615; *Swed. Pat.* 109502 (issued 1944); *Ch. Ztbl.* **1944** II 247; Schering Corp. (Köster, Inhoffen), *U.S. Pat.* 2302636 (issued 1942); *C. A.* **1943** 2388.

b Schering A.-G. (Butenandt), *Ger. Pat.* 736846 (issued 1943); *C. A.* **1944** 3094; *Fr. Pat.* 867697 (1940, issued 1941); *Ch. Ztbl.* **1942** I 2561; *Dan. Pat.* 61322 (1940, issued 1943); *Ch. Ztbl.* **1944** II 49; *Swed. Pat.* 111450 (1940, issued 1944); *Ch. Ztbl.* **1947** 1875; Schering Corp. (Butenandt), *U.S. Pat.* 2441560 (1940, issued 1948); *C. A.* **1948** 6862; *Ch. Ztbl.* **1949** II 681.

Spielman, Meyer, *J. Am. Ch. Soc.* **61** 893, 895.

Windaus, Kaufmann, *Ann.* **542** 218.

Woker, Antener, *Helv.* **22** 1309, 1315.

1940 Barkow, *Über bromfreie ungesättigte Sterinketone*, Thesis Techn. Hochschule Danzig, pp. 4, 7, 8, 17.

Barnett, Heilbron, Jones, Verrill, *J. Ch. Soc.* **1940** 1390, 1392.

Bernal, Crowfoot, Fankuchen, *Philos. Trans. Roy. Soc. London* A **239** 135, 140, 161.

a Butenandt, Dannenberg, Lázló, *Ber.* **73** 818, 819.

b Butenandt, Poschmann, *Ber.* **73** 893, 896.

a Chinoin* (Földi), *Hung. Pat.* 132575 (issued 1944); *C. A.* **1949** 3980.

b Chinoin* (Földi), *Hung. Pat.* 135687 (issued 1949); *C. A.* **1950** 4047.

Heath-Brown, Heilbron, Jones, *J. Ch. Soc.* **1940** 1482, 1487.

Inhoffen, Zühlsdorff, Huang-Minlon, *Ber.* **73** 451, 455.

Jacobsen, *J. Am. Ch. Soc.* **62** 1620.

Kon, Soper, *J. Ch. Soc.* **1940** 1335.

Korsching, Wirtz, *Ber.* **73** 249, 266.

McCullagh, Schneider, Emery, *Endocrinology* **27** 71, 73.

a Marker, Krueger, *J. Am. Ch. Soc.* **62** 79.

b Marker, Rohrmann, *J. Am. Ch. Soc.* **62** 516.

c Marker, *J. Am. Ch. Soc.* **62** 2543, 2544.

Petrow, *J. Ch. Soc.* **1940** 66.

a Ralls, *J. Am. Ch. Soc.* **62** 2459.

b Ralls, *J. Am. Ch. Soc.* **62** 3485.

Schering Corp. (Köster), *U.S. Pat.* 2363338 (issued 1944); *C. A.* **1945** 3127; *Ch. Ztbl.* **1945** II 1638; Schering A.-G., *Fr. Pat.* 875070 (1941, issued 1942); *Ch. Ztbl.* **1943** I 1190.

Wolfe, Hirschberg, Fieser, *J. Biol. Ch.* **136** 653, 678, 680.

1941 Adkins, Franklin, *J. Am. Ch. Soc.* **63** 2381, 2383.

Bernstein, Kauzmann, Wallis, *J. Org. Ch.* **6** 319.

a Bretschneider, *Ber.* **74** 1360.

b Bretschneider, *Monatsh.* **74** 53, 55; *Sitzber. Akad. Wiss. Wien* IIb **150** 127, 129.

c Bretschneider, Ajtai, *Monatsh.* **74** 57; *Sitzber. Akad. Wiss. Wien* IIb **150** 131.

Eck, Hollingsworth, *J. Am. Ch. Soc.* **63** 107, 110.

---

* See footnote ** on p. 2447 s.

Fernholz, Stavely, *Abstracts of the 102nd meeting of the American Chemical Society, Atlantic City*, p. M 39.
Fieser, Wolfe, *J. Am. Ch. Soc.* **63** 1485.
Galinovsky, *Ber.* **74** 1048.
Hughes, *J. Biol. Ch.* **140** 21.
Inhoffen, Zühlsdorff, *Ber.* **74** 604, 611, 612.
Oppenauer, *Organic Syntheses* Vol. **21**, New York, p. 18; Coll. Vol. **3** (1955) 207.
PARKE, DAVIS & Co. (Marker, Lawson), *U.S. Pat.* 2366204 (issued 1945); *C. A.* **1945** 1649; *Ch. Ztbl.* **1945** II 1385.
Salvioni, *Biochim. e terap. sper.* **28** 1.
Spring, Swain, *J. Ch. Soc.* **1941** 320.
SQUIBB & SONS (Fernholz), *U.S. Pat.* 2378918 (issued 1945); *C. A.* **1945** 5051.
Wieland, Rath, Benend, *Ann.* **548** 19, 27, 31, 33.
Woodward, *J. Am. Ch. Soc.* **63** 1123.

1941/42 Pankhurst, *Proc. Roy. Soc. London* A **179** 393, 396.
Salamon, *Z. physiol. Ch.* **272** 61.

1942 Bergström, Wintersteiner, *J. Biol. Ch.* **143** 503, 506.
Burckhardt, Reichstein, *Helv.* **25** 1434, 1439, 1441.
Butenandt, *Naturwiss.* **30** 4.
Criegee, Marchand, Wannowius, *Ann.* **550** 99, 102, 123, 125.
Jones, Wilkinson, Kerlogue, *J. Ch. Soc.* **1942** 391.
a Marker, Wittbecker, Wagner, Turner, *J. Am. Ch. Soc.* **64** 818.
b Marker, Wagner, Ulshafer, *J. Am. Ch. Soc.* **64** 1653.
Ruzicka, Plattner, Balla, *Helv.* **25** 65, 67, 71.
Selye, *Endocrinology* **30** 437, 444.
Wieland, Benend, *Ber.* **75** 1708, 1713.
Wintersteiner, Ruigh, *J. Am. Ch. Soc.* **64** 2453, 2455, 2456.
Woodward, *J. Am. Ch. Soc.* **64** 76.

1943 Barton, Jones, *J. Ch. Soc.* **1943** 602.
Bruce, *Org. Synth. Coll. Vol.* **2** (New York), p. 139.
Chuman, *C. A.* **1947** 3806.
Evans, Gillam, *J. Ch. Soc.* **1943** 565, 567.
Goldberg, Kirchensteiner, *Helv.* **26** 288, 289, 294–296, 299.
Hardegger, Ruzicka, Tagmann, *Helv.* **26** 2205, 2207, 2217.
Prelog, Ruzicka, Stein, *Helv.* **26** 2222, 2225, 2233, 2237.
Rosenheim, Webster, *Biochem. J.* **37** 513.
Urushibara, Mori, *C. A.* **1947** 3807.

1943/44 Lipschütz, *C. A.* **1944** 4990.

1944 Anker, Bloch, *J. Am. Ch. Soc.* **66** 1752.
Bersin, Meyer, *Angew. Ch.* **57** 117, 118.
Buu-Hoi, Cagniant, *Ber.* **77** 118, 121.
Hauptmann, *C. A.* **1946** 569.
Iglesias, Lipschütz, *Proc. Soc. Exptl. Biol. Med.* **55** 41.
Lipschütz, *Nature* **153** 260, 261.
Müller, Langerbeck, Neuhoff, *Ber.* **77** 141, 146, 151.
Plattner, Petrzilka, Lang, *Helv.* **27** 513, 514, 519.
Seebeck, Reichstein, *Helv.* **27** 948.
Turfitt, *Biochem. J.* **38** 492.
Ushakov, Kosheleva, *J. Gen. Ch. U.S.S.R.* **14** 1138; *C. A.* **1946** 4071.
Zimmermann, *Vitamine und Hormone* **5** [1952] 1, 8, 11, 13.

1945 Baker, Schlesinger, *J. Am. Ch. Soc.* **67** 1499.
C. F. BOEHRINGER & SOEHNE, *BIOS Final Report* No. 766, p. 136, 140; *FIAT Final Report* No. 71, p. 66.
Braude, Jones, *J. Ch. Soc.* **1945** 498, 501.
Prelog, Beyerman, *Experientia* **1** 64.

1946 Bernstein, Dorfman, *J. Am. Ch. Soc.* **68** 1152.

Bischoff, Rupp, *C. A.* **1947** 7505.
Heard, Sobel, *J. Biol. Ch.* **165** 687, 689 et seq.
Henbest, Jones, *Nature* **158** 950.
Lederer, Marx, Mercier, Pérot, *Helv.* **29** 1354, 1359, 1363.
Lieberman, *Experientia* **2** 411.
Ross, *J. Ch. Soc.* **1946** 737.
Velluz, Petit, Pesez, *Bull. soc. chim.* [5] **13** 558.
a Wilds, Djerassi, *J. Am. Ch. Soc.* **68** 1712.
b Wilds, Djerassi, *J. Am. Ch. Soc.* **68** 2125, 2132.
1947 Djerassi, Scholz, *J. Am. Ch. Soc.* **69** 2404, 2409.
Hauptmann, *J. Am. Ch. Soc.* **69** 562, 564.
1948 Barton, Cox, *J. Ch. Soc.* **1948** 783, 792.
1949 Inhoffen, Stoeck, Nebel, *Ann.* **563** 135.
Lardelli, Jeger, *Helv.* **32** 1817, 1824, 1833.
Rosenkranz, Kaufmann, Romo, *J. Am. Ch. Soc.* **71** 3689, 3691 footnote 17.
Urushibara, Chuman, *Bull. Ch. Soc. Japan* **22** 273; *C. A.* **1952** 129.
1950 Djerassi, Rosenkranz, Romo, Kaufmann, Pataki, *J. Am. Ch. Soc.* **72** 4534, 4539.
Romo, Djerassi, Rosenkranz, *J. Org. Ch.* **15** 896, 898.
Rosenkranz, Djerassi, Kaufmann, Pataki, Romo, *Nature* **165** 814.
1951 Djerassi, Rosenkranz, *Experientia* **7** 93.
a Inhoffen, *Angew. Ch.* **63** 297.
b Inhoffen, Kölling, Koch, Nebel, *Ber.* **84** 361, 368.
1952 a Antonucci, Bernstein, Littell, Sax, Williams, *J. Org. Ch.* **17** 1341, 1345.
b Antonucci, Bernstein, Heller, Williams, *J. Org. Ch.* **17** 1446, 1448.
Bladon, Henbest, Wood, *J. Ch. Soc.* **1952** 2737, 2740.
Butenandt, Karlson-Poschmann, Failer, Schiedt, Biekert, *Ann.* **575** 123.
1954 Dauben, Löken, Ringold, *J. Am. Ch. Soc.* **76** 1359, 1360, 1363.
Mori, Mukawa, *Bull. Ch. Soc. Japan* **27** 479; *C. A.* **1955** 10341.
1956 Ellis, Petrow, *J. Ch. Soc.* **1956** 4417.
Shoppee, Evans, Richards, Summers, *J. Ch. Soc.* **1956** 1649, 1651, 1653, 1654.
1958 Bladon, Cornforth, Jaeger, *J. Ch. Soc.* **1958** 863, 864.

Δ^5-**Cholesten-4-one** $C_{27}H_{44}O$ (p. 119: *hetero-Δ^1-Cholesten-3-one*). For structure of this compound, originally designated (1935 Butenandt) as Δ^1-*cholesten-3-one*, see 1944 Butenandt; cf. 1939 Butenandt. — Platelets (alcohol or acetone), m. 111–2°; absorption max. in alcohol at 241 mμ (ε 7200) (1944 Butenandt), in chloroform at 240 mμ (ε 7200) (1939 Dannenberg; cf. 1935 Butenandt); sublimes at 100–110° and 0.001 mm. pressure (1935 Butenandt). — **Fmn.** For modified procedure (cf. p. 119) giving an improved yield (18%), see 1944 Butenandt.

Rns. Boiling with $KMnO_4$ in acetone (1944 Butenandt) or long standing with perhydrol and OsO_4 in ether at 20° (1938 Butenandt) gives cholestane-5.6-diol-4-one (p. 2756 s). Hydrogenation in alcohol at room temperature, using palladium-$CaCO_3$ catalyst (discontinued after absorption of 2 mols. of hydrogen) affords a mixture of the two epimeric cholestan-4-ols (pp. 1722 s, 1723 s). After protection of the ketone group by ketal formation with orthoformic ester, hydrogenation in alcohol using a Raney nickel catalyst, followed by hydrolysis by boiling with addition of conc. HCl, affords cholestan-4-one (p. 2451 s). Reduction by means of zinc amalgam and conc. HCl in boiling alcohol yields a crystalline hydrocarbon, m. 71–73° (apparently impure coprostene, p. 1423 s, Editor), which by hydrogenation in glacial acetic acid, using platinum oxide catalyst, gives cholestane (p. 1429 s) (1944 Butenandt). On heating at 320–330° in a stream of CO_2, some methane is split off, and aromatization of ring A is said to occur (1937 Schering A.-G.). Treatment with benzoyl chloride in pyridine or boiling with benzoyl chloride in benzine (b. 100°) affords the enol-benzoate, $\Delta^{3\cdot5(\text{or } 4\cdot6?)}$-cholestadien-4-yl benzoate (p. 1722 s) (1936 Ciba).

Semicarbazone $C_{28}H_{47}ON_3$. M. p. 224–6°; sparingly soluble in organic solvents (1944 Butenandt).

Cholestan-4-one $C_{27}H_{46}O$ (p. 122).

For revised structure of this compound, originally designated as *cholestan-6-one* (1920 Windaus), see 1932 Tschesche. — Needles (acetone) (1944 Butenandt); crystals (methanol), m. 99–99.5° cor. after sublimation in a high vacuum at 145°; $[\alpha]_D$ +29.5° (chloroform) (1944 Ruzicka). —

Fmn. From either of the two 4-epimeric cholestan-4-ols (pp. 1722 s, 1723 s) by CrO_3 oxidation (1935 Tschesche); from cholestan-4α-ol by CrO_3 oxidation in glacial acetic acid (1944 Ruzicka). From Δ^5-cholesten-4-one (p. 2450 s) by hydrogenation in alcohol at room temperature, using palladium-$CaCO_3$ catalyst (discontinued after absorption of 2 mols. of hydrogen), followed by re-oxidation of the crude epimeric cholestan-4-ol mixture with acetone and aluminium phenoxide in benzene under reflux (yield, ca. 20%); from the same ketone, after protective ketal formation with orthoformic ester, on hydrogenation in alcohol in the presence of Raney nickel, followed by hydrolysis by boiling after addition of conc. HCl (yield, 60%) (1944 Butenandt). From cholestan-3α-ol-4-one (p. 2676 s) by heating with HBr in glacial acetic acid, followed by reduction with zinc dust and glacial acetic acid, both reactions at 100°, and finally re-oxidation with CrO_3 in glacial acetic acid at 20° (1944 Ruzicka). See also formation of the addition compound, below. — **Rns.** See those of the addition compound, below.

Compound with cholestan-2-one (1:1). Originally designated as cholestan-1-one, the precursor of which, the so-called cholestan-1-ol (1944 Ruzicka), has been shown to be a mixture or addition compound of cholestan-2α- and -4α-ols (1953 b Fieser). The equimolecular proportion assigned is based on its optical rotation and on that of the mixed cholestan-2α- and -4α-ol precursor (Editor). — Crystals (methanol), m. 120–120.5° cor.; $[\alpha]_D$ +41° (chloroform); sublimes in high vacuo at 155°. — **Fmn.** From so called cholestan-1-ol (p. 1537 s) by oxidation with CrO_3 and strong acetic acid in chloroform at room temperature; yield, 98% (1944 Ruzicka; 1953 b Fieser). — **Rns.** Treatment with hydrazine hydrate and then heating with sodium amyl oxide in amyl alcohol at 190° gives cholestane (p. 1429 s) (1944 Ruzicka).

Oxime $C_{27}H_{47}ON$ (p. 122). Crystals (alcohol), m. 205–7° (1944 Butenandt).

$\Delta^{2,4}$-Cholestadien-6-one $C_{27}H_{42}O$. Needles (acetone or abs. alcohol), m. 129° to 130° cor.; sublimes in a high vacuum at 135° (1951 Reich). Absorption max. in alcohol at 314 mμ (log ε 3.882) (1951 Reich; cf. 1946 Ross). — **Fmn.** Cholestane-3β.5-diol-6-one (p. 2757 s) is transformed via its 3-p-toluenesulphonate into Δ^2-cholesten-5-ol-6-one, which is dehydrated by refluxing with alcoholic H_2SO_4 or by treatment with dry HCl in cooled chloroform (1951 Reich).

Δ^2-Cholesten-6-one $C_{27}H_{44}O$. Formerly described as Δ^4-*cholesten-6-one* (1939 Ladenburg); for revised structure, see 1946 Blunschy. — Crystals (ethanol or methanol), m. 104–105° (1939 Ladenburg; 1946 Blunschy); [α]$_D$ +28° (chloroform); absorption max. (solvent not stated) at 285–300 mμ (log ε 1.7) (1946 Blunschy). — **Fmn.** From 3β-bromocholestan-6-one (p. 2488 s) by refluxing with quinoline in an atmosphere of nitrogen (yield, 79%) (1939 Ladenburg; cf. 1946 Blunschy); from 3α-chlorocholestan-6-one (p. 2484 s) or its 3β-epimer (p. 2486 s) similarly (yields, 24% and 57%, resp.), or from the latter by heating with collidine in a sealed tube at 250° (yield, 57%) (1946 Blunschy). — **Rns.** Oxidation with $KMnO_4$ in hot acetone affords 6-keto-cholestane-2‖3-dioic acid (p. 1714 s), also obtained by ozonization in glacial acetic acid at 100°, followed by oxidation of the intermediate aldehyde with CrO_3 in 85% acetic acid at 20°. Treatment with hydrazine hydrate, followed by heating with sodium ethoxide in alcohol at 160° (sealed tube), gives Δ^2-cholestene (p. 1422 s) (1946 Blunschy).

Oxime $C_{27}H_{45}ON$. Crystals (alcohol) (1939 Ladenburg), m. 185–6° (1946 Blunschy).

i-Cholesten-6-one* $C_{27}H_{44}O$ (p. 129: *Cholesten-6-one*, m. 96°). For evidence of the cyclopropane ring structure and identity with Windaus's (1919) "*heterocholestenone*", see 1938 Ford; 1939 Ladenburg; 1948 Dodson. For identity with Vanghelovici's (1935 a) *isocholestan-6-one*, see 1948 Shoppee. For evidence of structure based on absorption spectra, see 1944 Klotz; 1946 Riegel; 1951 Josien; for configuration, see 1939 Ladenburg; 1948 Dodson. — Needle rosettes (methanol) (1919 Windaus), needles (acetone) (1938b Heilbron), crystals (aqueous alcohol, aqueous methanol, or aqueous acetone) (1938 Ford), m. 97° (1938b Heilbron;

---

* The name 3.5-cyclocholestan-6-one has recently been proposed for this compound; see Introduction to steroids, p. 1352 s.

cf. 1939 Ladenburg); sublimes in a high vacuum (1938b Heilbron); $[\alpha]_D^{18} + 41^0$ (1938b Heilbron), $[\alpha]_D^{22} + 45^0$ (1948 Dodson) both in chloroform *. Absorption max. (solvent not stated) at ca. 290 mμ (log ε 1.3) (1944 Klotz; cf. 1938b Heilbron). Gives no coloration with tetranitromethane and no reaction with perbenzoic acid (1938b Heilbron); is not attacked by H_2O_2 in glacial acetic acid nor by neutral or alkaline $KMnO_4$ (1939 Ladenburg).

Fmn. From *i*-cholesterol (p. 1723 s) by oxidation with CrO_3 in glacial acetic acid at room temperature (1938b Heilbron); isolated by conversion into the oxime (yield, 20%) (1937 Wallis), followed by hydrolysis with dil. H_2SO_4 in alcohol (1938 Ford; 1939 Ladenburg); is similarly prepared from the acetate (more slowly) or the methyl ether of *i*-cholesterol (1938b Heilbron). From 3β-chlorocholestan-6-one (p. 2486 s) by vacuum distillation (1921 Windaus), or by reduction with sodium amalgam in hot alcohol (1935a Vanghelovici; cf. 1948 Shoppee).

Rns. For oxidation to (α_1-)*i*-cholestene-6‖7-dioic acid, see this acid (below). For hydrogenation (cf. p. 129), see also 1938 Ford. Reduction with aluminium isopropoxide and boiling isopropanol gives epi-*i*-cholesterol (p. 1725 s) (1938b Heilbron). Treatment with a 3% solution of HCl in glacial acetic acid at room temperature affords 3β-chlorocholestan-6-one (p. 2486 s) (1938 Ford; cf. 1938b Heilbron); similar reaction with 34% aqueous HBr in strong acetic acid gives 3β-bromocholestan-6-one (p. 2488 s) (1939 Ladenburg). Refluxing with glacial acetic acid containing 5 N H_2SO_4 (1 in 4), followed by treatment with alcoholic NaOH, gives cholestan-3β-ol-6-one (p. 2677 s) (1939 Ladenburg). Treatment with ethyl formate, in presence of small amounts of formic acid and conc. H_2SO_4, gives the enol-ether, Δ^6-**i-cholesta-dien-6-yl ethyl ether**(?) $C_{29}H_{48}O$, m. 120^0 (1938 SCHERING A.-G.).

Oxime $C_{27}H_{45}ON$. Leaflets (aqueous alcohol), m. 143–4^0 (1937 Wallis; 1939 Ladenburg); colourless prisms (methanol), m. 123^0 (1938b Heilbron); needles (alcohol), m. 157^0 (1935a Vanghelovici; cf. 1948 Shoppee).

i-Cholestene-6‖7-dioic acid $C_{27}H_{44}O_4$. For confirmation of the structure and configuration of this compound, referred to by Ladenburg (1939) as "α_1-*i-chole-stane*-6‖7-*dioic acid*", see 1955 Gates. — Crystals (dil. alcohol), m. 232–3^0; $[\alpha]_D^{25}$ + 18^0 (acetone). Does not give a positive Liebermann reaction. — **Fmn.** From i-cholesten-6-one (above) by shaking with bromine and 10% aqueous KOH in pyridine overnight at room temperature. — **Rns.** Does not decolorize bromine, nor add on water or hydrogen halide. All attempts to effect rupture of the cyclopropane ring were without success (1939 Ladenburg).

For "α_2- and β (β_2)-i-cholestane-6‖7-dioic acids", see pp. 2487 s and 2486 s, resp.

---

* The value $[\alpha]_D^{25} + 65^0$ (chloroform) quoted by Ford (1938) is for a sample to which a wrong m. p. 110–111^0 (cf. 1939 Ladenburg) is assigned.

References, pp. 2456–2458 s

Cholestan-6-one $C_{27}H_{46}O$ (p. 123). Formerly designated *heterocholestanone*, or *cholestan-7-one* (1920 Windaus). — Needle clusters (methanol) (1938 Jackson), prisms (acetone) (1935 b Vanghelovici), leaflets (acetone), m. 98–99° cor. (1944 Plattner); $[\alpha]_D$ —7° (chloroform) (1958 Shoppee). Liebermann-Burchard reaction: violet coloration, with a band at 550–600 mμ (1938 Petrow); practically no colour reaction with m-dinitrobenzene and KOH in alcohol (1938 Callow).

Fmn. From 3β-bromo-6-nitro-Δ^5-cholestene (p. 1479 s) on boiling with zinc dust in glacial acetic acid (1935 b Vanghelovici). From 5α.6α-epoxycholestane (p. 2076 s) by hydrogenation in glacial acetic acid at room temperature using platinum oxide catalyst (discontinued when 1.1 mols. of hydrogen have been absorbed), followed by chromatographic separation of cholestane (p. 1429 s) from the mixed cholestan-5-ol (p. 1723 s) and -6α-ol (p. 1725 s) formed, before oxidation with CrO_3 in 90% acetic acid at room temperature, and finally chromatographic separation from cholestan-5-ol; yield, 28% (1944 Plattner). From the acetate of Δ^4-cholesten-3β-ol-6-one (p. 2677 s) by hydrogenation in ether, using palladium black catalyst (discontinued after absorption of 1 mol. of hydrogen) (1938 Jackson). From cholestane-6-sulphonic acid by oxidation of the lithium salt with alkaline $KMnO_4$ at 70° (1938 Windaus).

Rns. Oxidation with SeO_2 occurs readily in glacial acetic acid, but not in alcohol (1938 Stiller). Reduction to cholestan-6α-ol (p. 1725 s) is more effectively carried out (cf. p. 123) with sodium in amyl alcohol (1938 a Heilbron). Clemmensen reduction yields cholestane (p. 1429 s) (1919 Windaus).

Oxime $C_{27}H_{47}ON$ (p. 123). Crystals (alcohol) m. 197° (1935 b Vanghelovici; cf. 1938 Ford).

$\Delta^{3.5}$-**Cholestadien-7-one, 7-Ketocholesterylene** $C_{27}H_{42}O$ (p. 123). Plates (methanol) (1941 Bergström), white leaflets (acetone), m. 114.5° cor. (1943 Ruzicka; 1943 Hardegger; cf. 1937 Marker); sublimes in a high vacuum at 115–125° (1943 Hardegger); $[\alpha]_D$ —306° (chloroform) (1943 Ruzicka; cf. 1941 Bergström), —305° (benzene) (1945 Daniel). Absorption max. in alcohol at 280 and 325 mμ (log ε 4.4 and 3.1, resp.) (1940 Jackson; 1943 Hardegger; cf. 1941 Bergström; 1943 Ruzicka). Gives a yellow coloration with tetranitromethane (1943 Ruzicka); with p-nitrobenzenediazonium chloride in glacial acetic acid-dioxane, only a faint colour is developed after 2 hrs. (1938 Fieser). With acetic anhydride and conc. H_2SO_4 in glacial acetic acid at 85–90°, a brown solution is obtained; the reaction product has no anti-rachitic effect (towards rats) (1939 Eck).

Occ. and **Isolation.** Has been isolated from the unsaponifiable material of swine testicles (yield, 114 mg. from 4.76 kg. acetone extract) (1943 Ruzicka; cf. 1944 Prelog), swine spleen (yield, 468 mg. from 15.7 kg. acetone extract)

(1943 Prelog), sclerotic aortas (yield, 1.24 g. from 127 g. of unsaponifiable material) (1943 Hardegger), and wool fat (yield, ca. 84 mg. from 12.7 g. of unsaponifiable material) (1945 Daniel), but in each case its appearance as an artefact, possibly arising from 7-ketocholesterol (p. 2682 s) or its esters, is considered to be probable; it appears in the "non-ketonic" fraction after treatment of the cholesterol-freed product with Girard's reagent and is separated from that fraction by exhaustive chromatography.

Fmn. Together with 7-ketocholesterol (p. 2682 s) and other products, from cholesterol (p. 1568 s) by oxygen aeration of a colloidal aqueous solution (stabilized by sodium stearate) at 85° and pH 8.5; after treatment of the mixed products with Girard's reagent, it is isolated from the "non-ketonic" fraction by fractional elution after adsorption on Al_2O_3 (1941 Wintersteiner; 1941 Bergström). Together with 7-keto-epi-cholesterol (p. 2681 s), from 7-keto-cholesteryl chloride (p. 2488 s) by heating with potassium acetate in valeric acid at 180°; after saponification with alcoholic KOH, the oily product is treated with succinic anhydride in pyridine for separation of the 7-keto-epi-cholesterol in the form of its hydrogen succinate soluble in aqueous Na_2CO_3 (1937 Marker); for mechanism of this reaction, see 1946 Shoppee. From 7-ketocholesterol on treatment with dry HCl in chloroform (1938 Ogata). From the acetate of 7-ketocholesterol by refluxing with methanolic NaOH (1946 Barnett; cf. also 1938 Hattori), or by treatment with HBr in glacial acetic acid either under reflux or on standing at 37° for 15 hrs. (yield in each case quantitative) (1940 Jackson); the analogous elimination of acetic acid from the acetate of 7-keto-epi-cholesterol occurs very readily (1939 Windaus). Together with the acetate of Δ^5-cholesten-3β-ol-7-one (p. 2681 s), from the acetate of 6β-bromo-cholestan-3β-ol-7-one (p. 2709 s) by refluxing with anhydrous pyridine; the dienone is isolated from the mother liquors after crystallizing the ketonyl acetate from methanol; can be obtained from the 6-epimeric bromoketonyl acetate (p. 2709 s), for which reaction occurs more slowly (1938 Barr). From the acetate of 7-keto-cholesterol 5.6-dibromide (p. 2715 s) by refluxing with dimethylaniline (1940 Jackson).

Rns. There is no reaction with SeO_2 in alcohol nor in glacial acetic acid, both at 100° (1938 Stiller). Hydrogenation in ethyl acetate, using platinum oxide catalyst, gives cholestan-7β-ol (p. 1727 s) together with cholestan-7-one (p. 2456 s) (1938a Heilbron). Gives pseudocholesterol (p. 1726 s) on reduction with sodium in dry isopropyl alcohol (1938 Ogata).

Oxime $C_{27}H_{45}ON$. Crystals (methanol), m. 176–8° (1943 Hardegger; cf. 1938 Hattori; 1938 Ogata).

Semicarbazone $C_{28}H_{45}ON_3$ (p. 123). Almost colourless needles (methanol), m. 206.5–207.5° (1943 Hardegger).

Δ^5-Cholesten-7-one, 7-Ketocholestene $C_{27}H_{44}O$ (p. 123).

White needles (acetone) (1939 Marker), m. 130–130.5° cor., $[\alpha]_D^{28}$ —138° (chloroform) (1939 Tominaga). Absorption max. (solvent not stated) at 235–237 mμ (1937 Burawoy). Heating with conc. H_2SO_4 and acetic anhydride in glacial

E.O.C. XIV s. 171

acetic acid at 85–90° develops a brown coloration; the reaction product formed shows no antirachitic activity (rats) (1939 Eck).

Fmn. From Δ^5-cholesten-7β-ol (p. 1726 s) by treatment with bromine in ether-glacial acetic acid, followed by oxidation of the resulting crude dibromide with CrO_3 in benzene - 80% acetic acid at room temperature and debromination with zinc dust and glacial acetic acid on the steam-bath (1939 Marker). — **Rns.** Treatment with conc. H_2SO_4 in acetic anhydride at —10° gives Δ^5-cholesten-7-one-4-sulphonic acid (methyl ester, m. 180–1°) (1937 Windaus; cf. 1938 Windaus).

Oxime $C_{27}H_{45}ON$. Needles (alcohol), m. 179–180° cor. (1939 Tominaga).

Δ^8-Cholesten-7-one $C_{27}H_{44}O$. Crystals (methanol), m. 86.5–87.5°; $[\alpha]_D^{21} + 3.8°$ (chloroform); absorption max. (solvent not stated) at 251 mμ (1941 Eck). — **Fmn.** From a mixture of Δ^7- and Δ^8-cholestene (p. 1427 s) by CrO_3 oxidation in benzene-glacial acetic acid containing dil. H_2SO_4 at room temperature, followed by chromatographic separation from unchanged starting material and from a diketone product*; yield, 13% (1941 Eck). — **Rns.** Reduction with sodium in boiling amyl alcohol, followed by oxidation with CrO_3, gives cholestan-7-one (below) (1941 Eck). Boiling with isopropanol and aluminium isopropoxide affords Δ^8-cholesten-7-ol (p. 1727 s) (1942 Eck).

Cholestan-7-one $C_{27}H_{46}O$ (p. 123). Formerly designated as *cholestan-8-one* (1920 Windaus). — Plates (alcohol) (1938a Heilbron), crystals (methanol), m. 115–6°** (1941 Eck), crystals (acetone), m. 117° (1937 Marker); $[\alpha]_D^{24}$ —47° (chloroform). Absorption max. in alcohol at 292 mμ (ε 40) (1950 Wintersteiner). — **Fmn.** From $\Delta^{3.5}$-cholestadien-7-one (p. 2454 s) by hydrogenation in ethyl acetate, using platinum oxide catalyst; yield, 53% (1938a Heilbron). From Δ^8-cholesten-7-one (above) by reduction with sodium and boiling amyl alcohol, followed by CrO_3 oxidation of the crude product (1941 Eck). — **Rns.** Clemmensen reduction yields cholestane (p. 1429 s) (1920 Windaus). Reduction with sodium and amyl alcohol (1938a Heilbron) or with aluminium isopropoxide (and isopropanol?) (1941 Eck) gives cholestan-7β-ol (p. 1727 s).

Oxime $C_{27}H_{47}ON$. M. p. 134–5° (1941 Eck).

1919 Windaus, Dalmer, *Ber.* **52** 162, 164, 168.
1920 Windaus, *Ber.* **53** 488, 495.
1921 Windaus, v. Staden, *Ber.* **54** 1059, 1062.

* Fieser (1953a) suggests that the pure "diketone" m. 74–75° (p. 1427 s) obtained from this fraction may be a mixture of the epoxides of Δ^8- and $\Delta^{8(14)}$-cholesten-7-one.

** The product m. 108–9°, obtained by Windaus's method (cf. p. 123) seems to be admixed with an isomeride, as deduced from the curve for rate of oxime formation (1948 Décombe).

1932 Tschesche, *Ber.* **65** 1842.
1935 Butenandt, Wolff, *Ber.* **68** 2091.
Tschesche, Hagedorn, *Ber.* **68** 2247.
a Vanghelovici, Angelescu, *Bul. soc. chim. România* **17** A 177.
b Vanghelovici, Vasiliu, *Bul. soc. chim. România* **17** A 249, 257.
1936 CIBA (Soc. pour l'ind. chim. à Bâle; Ges. f. chem. Ind. Basel), *Brit. Pat.* 477400 (issued 1938);
C. A. **1938** 3914; *Ch. Ztbl.* **1938** II 119, 120; *Indian Pat.* 211038 (1937, issued 1938); *Ch. Ztbl.*
1938 II 119, 120.
1937 Burawoy, *J. Ch. Soc.* **1937** 409, 410.
Marker, Kamm, Fleming, Popkin, Wittle, *J. Am. Ch. Soc.* **59** 619.
SCHERING A.-G., *Austrian Pat. Application; FIAT Final Report* No. 996, pp. 130, 159 (issued
1947).
Wallis, Fernholz, Gephart, *J. Am. Ch. Soc.* **59** 137, 140.
Windaus, Kuhr, *Ann.* **532** 52, 64.
1938 Barr, Heilbron, Jones, Spring, *J. Ch. Soc.* **1938** 334, 336.
Butenandt, Wolz, *Ber.* **71** 1483.
Callow, Callow, Emmens, *Biochem. J.* **32** 1312, 1323.
Fieser, Campbell, *J. Am. Ch. Soc.* **60** 159, 169.
Ford, Chakravorty, Wallis, *J. Am. Ch. Soc.* **60** 413.
Hattori, *C. A.* **1938** 7473; *Ch. Ztbl.* **1939** I 1182.
a Heilbron, Shaw, Spring, *Rec. trav. chim.* **57** 529, 534.
b Heilbron, Hodges, Spring, *J. Ch. Soc.* **1938** 759.
Jackson, Jones, *J. Ch. Soc.* **1938** 1406, 1408.
Ogata, *C. A.* **1939** 640.
Petrow, Rosenheim, Starling, *J. Ch. Soc.* **1938** 677, 681.
SCHERING A.-G., *Brit. Pat.* 494484 (issued 1938); *C. A.* **1939** 2535; *Ch. Ztbl.* **1939** I 3932.
Stiller, Rosenheim, *J. Ch. Soc.* **1938** 353, 355.
Windaus, Mielke, *Ann.* **536** 116, 122.
1939 Butenandt, Mamoli, Dannenberg, Masch, Paland, *Ber.* **72** 1617.
Dannenberg, *Abhandl. Preuss. Akad. Wiss. Math.-naturwiss. Kl.* **1939** No. 21, pp. 14, 54.
Eck, Thomas, *J. Biol. Ch.* **128** 257, 261.
Ladenburg, Chakravorty, Wallis, *J. Am. Ch. Soc.* **61** 3483.
Marker, Rohrmann, *J. Am. Ch. Soc.* **61** 3022.
Tominaga, *Bull. Ch. Soc. Japan* **14** 486, 488.
Windaus, Naggatz, *Ann.* **542** 204.
1940 Jackson, Jones, *J. Ch. Soc.* **1940** 659, 663.
1941 Bergström, Wintersteiner, *J. Biol. Ch.* **141** 597, 599, 603 et seq.
Eck, Hollingsworth, *J. Am. Ch. Soc.* **63** 2986, 2987, 2989, 2990.
Wintersteiner, Bergström, *J. Biol. Ch.* **137** 785.
1942 Eck, Hollingsworth, *J. Am. Ch. Soc.* **64** 140, 142.
1943 Hardegger, Ruzicka, Tagmann, *Helv.* **26** 2205, 2208, 2215–2217.
Prelog, Ruzicka, Stein, *Helv.* **26** 2222, 2226, 2236.
Ruzicka, Prelog, *Helv.* **26** 975, 981, 988, 993.
1944 Butenandt, Ruhenstroth-Bauer, *Ber.* **77** 397.
Klotz, *J. Am. Ch. Soc.* **66** 88.
Plattner, Petrzilka, Lang, *Helv.* **27** 513, 523.
Prelog, Ruzicka, *Helv.* **27** 61, 64.
Ruzicka, Plattner, Furrer, *Helv.* **27** 727, 734, 736.
1945 Daniel, Lederer, Velluz, *Bull. soc. chim. biol.* **27** 218, 223.
1946 Barnett, Ryman, Smith, *J. Ch. Soc.* **1946** 526, 528.
Blunschy, Hardegger, Simon, *Helv.* **29** 199, 202.
Riegel, Hager, Zenitz, *J. Am. Ch. Soc.* **68** 2562.
Ross (quoting E.R.H. Jones, unpublished work), *J. Ch. Soc.* **1946** 737, 738.
Shoppee, *J. Ch. Soc.* **1946** 1147, 1149.
1948 Décombe, Jacquemain, Rabinovitch, *Bull. soc. chim.* [5] **15** 447.
Dodson, Riegel, *J. Org. Ch.* **13** 424, 433.
Shoppee, *J. Ch. Soc.* **1948** 1032.

1950 Wintersteiner, Moore, *J. Am. Ch. Soc.* **72** 1923, 1930.
1951 Josien, Fuson, Cary, *J. Am. Ch. Soc.* **73** 4445, 4448.
      Reich, Walker, Collins, *J. Org. Ch.* **16** 1753, 1754, 1757.
1953 a Fieser, Ourisson, *J. Am. Ch. Soc.* **75** 4404, 4408.
      b Fieser, Romero, *J. Am. Ch. Soc.* **75** 4716.
1955 Gates, Wallis, *J. Org. Ch.* **20** 610.
1958 Shoppee, Jenkins, Summers, *J. Ch. Soc.* **1958** 1657, 1658, 1661.

---

6. Monoketo-steroids (CO in the ring system) C_{28} with aliphatic side chains

Summary of the Ketones (CO in 3-position) with Ergostane Skeleton*

Configuration		Position of double bond(s)	M. p. °C.	$[\alpha]_D$ (°) in $CHCl_3$	$\lambda_{max.}$ (mμ) in alcohol	Oxime m. p. °C.	Semi-carbazone m. p. °C.	Page
C-5	C-10							
—	norm.	$\Delta^{4.7.9(11).22}$	141	+190	242	—	244 dec.	2458 s
—	norm.	$\Delta^{4.6.22}$	108	−30	280, 335	—	246 dec.	2459 s
—	norm.	$\Delta^{4.7.22}$	132	−11	239, 320	—	236	2459 s
—	iso	$\Delta^{4.7.22}$	140	+49	229	—	247 dec.	2460 s
α	norm.	$\Delta^{7.9(11).22}$	205	+45	235, 243, 252[1])	245 dec.	—	2460 s
β	norm.	$\Delta^{7.9(11).22}$	131	+53	245[2])	—	—	2461 s
α	norm.	$\Delta^{8.14.22}$	150	−58	248[3])	189	—	2461 s
β	norm.	$\Delta^{8.14.22}$	120	−49	248[3])	167	—	2461 s
α	norm.	$\Delta^{7.22}$	185	+2	—	—	254 dec.	2461 s
?	iso	$\Delta^{7.22(?)}$	176	+32	—	211 dec.	—	2461 s
α	norm.	$\Delta^{8.14}$	148	−1	—	213	—	2461 s
α	norm.	$\Delta^{8(14).22}$	} 125	−5 to +4	—	193 dec.	226	2462 s
α	norm.	$\Delta^{14.22}$						
α	norm.	Δ^{7}	159	+22				[4])
α	norm.	$\Delta^{8(14)}$	130	+30	—	—	—	2462 s
α	norm.	Δ^{14}	150	+40	—	215	—	2462 s
α	norm.	satd.	160	+33	—	216	—	2463 s
β	norm.	satd.	94	—	—	—	—	2463 s
α	iso	satd.	122	−17.5	—	166	—	2463 s
α	norm.[5])	satd.	157	+42	—	—	—	2463 s

* For values other than those chosen for the summary, see the compounds themselves.
[1]) Solvent not indicated.   [2]) In ether.   [3]) In ether(?).   [4]) 1948a Barton.
[5]) 24-iso; probably impure.

$\Delta^{4.7.9(11).22}$-**Ergostatetraen-3-one,** *Ergostatetraenone* $C_{28}H_{40}O$. Plates (acetone), m. 140–2°; $[\alpha]_D^{20}$ +190° (chloro-form). Absorption max. in alcohol at 242 mμ (log ε 4.50). — **Fmn.** From dehydroergosterol (p. 1730 s) by oxidation with acetone and aluminium tert.-butoxide in boiling benzene. — **Rns.** Boiling with aluminium isopropoxide in isopropanol gives an addition compound of dehydroergosterol with its 3-epimer from which only the former has been satisfactorily separated. Refluxing with acetic anhydride and pyridine gives the enol-acetate, $\Delta^{3.5.7.9(11).22}$-ergostapentaen-3-yl acetate (p. 1729 s) (1938 Heilbron).

Semicarbazone $C_{29}H_{43}ON_3$. Needles (methanol-chloroform), m. 244° dec. (1938 Heilbron).

$\Delta^{4\cdot6\cdot22}$-**Ergostatrien-3-one,** *Isoergosterone* $C_{28}H_{42}O$ (p. 124). Colourless needles (methanol), m. 108°; $[\alpha]_D^{20}$ —30° (chloroform). Absorption max. in alcohol at 280 mμ (log ε 4.52) and 335 mμ. — **Fmn.** From $\Delta^{4\cdot7\cdot22}$-ergostatrienone enol-acetate (p. 1730 s) by boiling with methanolic KOH; yield, 88%. — **Rns.** Refluxing with acetic anhydride and acetyl chloride gives the enol-acetate, $\Delta^{2\cdot4\cdot6\cdot22}$-ergostatetraen-3-yl acetate (p. 1730 s) (1938 Heilbron).

Semicarbazone $C_{29}H_{45}ON_3$ (p. 124). Needles (methanol-chloroform), m. 245–6° dec.; absorption max. in alcohol at 304 mμ (log ε 4.67) (1938 Heilbron).

$\Delta^{4\cdot7\cdot22}$-**Ergostatrien-3-one,** *"Ergosterone"* $C_{28}H_{42}O$ (p. 124). Crystals (acetone-light petroleum), m. 132° (1938 Heilbron); $[\alpha]_D^{20}$ —0.8° (chloroform) (1938 Heilbron); the values $[\alpha]_D$ —0.52° and —0.84° (chloroform) (1937 Wetter) (cf. p. 124) are in error for —5.2° and —8.4°, resp. (Heilbron, see 1939 Windaus); the value ca. —16° given by Oppenauer (1937a) is ascribed to the presence of ergosterol (1938 Windaus)*. Absorption max. in alcohol at 230 and 320 mμ (log ε 4.30 and 1.61, resp.) (1938 Heilbron)*. Gives an intense green fluorescence with trichloroacetic acid (1939 v. Christiani).

Fmn. The Oppenauer oxidation of ergosterol (p. 1735 s) (cf. p. 124) can be effected in gasoline (b. 100–125°); yield, 75% (1937b Oppenauer). From epi-alloergosterol (p. 1734 s) by Oppenauer oxidation (1939 Windaus). For its probable formation from ergosterol by oxidation with lead tetraacetate, see 1939 v. Christiani. The yield from $\Delta^{5\cdot7\cdot22}$-ergostatrienone maleic acid adduct (cf. p. 124) is 18% (1937 Wetter).

Rns. Reduction with aluminium isopropoxide and boiling isopropanol gives chiefly $\Delta^{4\cdot7\cdot22}$-ergostatrien-3β-ol (p. 1732 s) and its 3α-epimer (p. 1734 s), together with some ergosterol and very small amounts of epi-ergosterol (p. 1746 s) which could only be detected spectroscopically; the supposed iso-lation of epi-ergosterol (cf. 1937 Marker) by this reaction is shown to be erroneous (1938, 1939 Windaus; cf. 1938 Heilbron). Refluxing with acetic anhydride-pyridine gives the enol-acetate, $\Delta^{3\cdot5\cdot7\cdot22}$-ergostatetraen-3-yl acetate (p. 1730 s) (1938 Heilbron). After conversion to the enol-acetate aromatization can be effected by heating at 310–320° in a stream of CO_2 (1937 SCHERING A.-G.).

---

* More recently, $[\alpha]_D$ values of —10° to —12° (chloroform) and a main absorption max. (in alcohol) of 238–239 mμ (ε 15000–17000) have been indicated for the pure compound (1952 Antonucci; 1952 Fieser).

References, pp. 2463 s, 2464 s

Semicarbazone $C_{29}H_{45}ON_3$ (p. 124). Needles (chloroform-methanol), m. 236^0 (1937 Wetter). Absorption max. in chloroform at 270 mμ (ε 35 800) (1939 Dannenberg; 1937 Wetter); for comparison of the absorption band of the semicarbazone with that of the parent ketone*, see 1943 Evans.

10-iso-$\Delta^{4\cdot7\cdot22}$-Ergostatrien-3-one, Lumistatrienone $C_{28}H_{42}O$.

For confirmatory evidence of the configuration at C-10, see that of its precursor, lumisterol (p. 1749 s). — Needles (acetone-light petroleum), m. 139^0 to 140^0; $[\alpha]_D^{20}$ $+49^0$ (chloroform); absorption max. in alcohol at 229 mμ (log ε 4.23). — **Fmn.** From lumisterol by oxidation with acetone and aluminium tert.-butoxide in boiling benzene. — **Rns.** Reduction with aluminium isopropoxide and boiling isopropanol gives the addition compound of lumisterol with its 3-epimer (p. 1750 s). Heating under reflux with acetic anhydride-pyridine yields the enol-acetate, 10-iso-$\Delta^{3\cdot5\cdot7\cdot22}$-ergostatetraen-3-yl acetate (p. 1730 s) (1938 Heilbron).

Semicarbazone $C_{29}H_{45}ON_3$. Needles (alcohol-chloroform), m. 247^0 dec. (1938 Heilbron).

$\Delta^{5\cdot7\cdot22}$-Ergostatrien-3-one $C_{28}H_{42}O$ (p. 124).

Maleic acid adduct $C_{32}H_{46}O_5$. Crystals (ethyl acetate-acetone), m. 188^0. — **Fmn.** Is isolated from the $NaHCO_3$ wash liquors obtained on working up the anhydride (see below). — **Rns.** Distillation at 210^0 gives the anhydride (1937 Wetter).

Maleic anhydride adduct $C_{32}H_{44}O_4$ (p. 124). Needles (methanol or glacial acetic acid), m. $178–9^0$ (1936 Dimroth). — **Fmn.** From ergosterol maleic anhydride adduct (p. 73) by oxidation with CrO_3 in ca. 99% acetic acid at $15–20^0$ (yield, 65%) (1936 Dimroth). From the maleic acid adduct (above) by distillation at 210^0 (1937 Wetter).

$\Delta^{7\cdot9(11)\cdot22}$-Ergostatrien-3-one, *Ergostatrienone-D* $C_{28}H_{42}O$ (p. 124).

For structure assigned, see that of its reduction product, ergosterol-D (page 1752 s; cf. p. 124). — **Rns.** By effecting the acid rearrangement to ergostatrienone-B_1 (cf. p. 124) in an ice-salt freezing mixture, a second, strongly lævorotatory product is obtained, which by reduction with sodium and alcohol gives an alcohol (not described; forms an adduct with maleic anhydride) (1932 Dithmar).

---

* Erroneously referred to as $\Delta^{4\cdot7}$-ergostadienone-3 in the original paper.

$\Delta^{7\cdot9(11)\cdot22}$-**Coproergostatrien-3-one,** *u-Ergostatrienone* $C_{28}H_{42}O$ (p. 124). For structure assigned, see that of its reduction product u-ergostatrienol (p. 1752 s; cf. p. 124).

$\Delta^{8\cdot14\cdot22}$-**Ergostatrien-3-one,** *Ergostatrienone-B$_1$* $C_{28}H_{42}O$ (p. 124). For structure assigned, see that of its reduction product, ergosterol-B$_1$ (p. 1753 s; cf. p. 78). — Needles (ether-methanol), m. 149–150°; $[\alpha]_D^{19}$ —57.5° (chloroform); readily soluble in ether, acetone, and ethanol, less soluble in methanol; gives a positive Liebermann-Burchard reaction (1932 Dithmar).

$\Delta^{8\cdot14\cdot22}$-**Coproergostatrien-3-one,** *u-Ergostatrienone-B* $C_{28}H_{42}O$ (p. 124). For structure assigned, see that of its reduction product u-ergostatrienol-B (p. 1753 s; cf. p. 124). — Needles (alcohol), m. 120°; $[\alpha]_D^{16}$ —49.1° (chloroform); readily soluble in chloroform, ether, and acetone, less soluble in alcohol (1932 Dithmar).

$\Delta^{7\cdot22}$-**Ergostadien-3-one,** *Ergostadienone-I,* *"α"-Ergostadienone* $C_{28}H_{44}O$ (page 125: α-*Ergostadienone*). For revised structure, see 1948b Barton; cf. also that of its reduction product, "α"-dihydroergosterol (page 1758 s). — Crystals (ethyl acetate-methanol), m. 184.5°; $[\alpha]_D^{ca.\ 20}$ +2° (chloroform). — **Fmn.** From "α"-dihydroergosterol by Oppenauer oxidation (1948b Barton).

10-iso-$\Delta^{7\cdot22}$-**Ergostadien-3-one (?),** **Lumistadienone** $C_{28}H_{44}O$ (p. 125). The structure assigned is based on that ascribed to its precursor, dihydrolumisterol, which see p. 1760 s (Editor).

$\Delta^{8\cdot14}$-**Ergostadien-3-one,** *Dehydroergostenone* $C_{28}H_{44}O$ (p. 81). Plates (alcohol) (1932 Heilbron), crystals (ethyl acetate-methanol), m. 148°; $[\alpha]_D^{ca.\ 20}$ —1° (chloroform) (1948a Barton); readily soluble in hot alcohol (1932 Heilbron). — **Fmn.** From $\Delta^{8\cdot14}$-

ergostadien-3β-ol (p. 1761 s) by Oppenauer oxidation (1948a Barton). From "bromodehydroergostenone" (p. 125) by heating with zinc dust in glacial acetic acid (1932 Heilbron).

$\Delta^{8(14)\cdot22}$-**Ergostadien-3-one** (I) and $\Delta^{14\cdot22}$-**Ergostadien-3-one** (II), "β"-**Ergo-stadienone**, "β"-*Dihydroergosterone, Ergostadienone-III* $C_{28}H_{44}O$ (p. 125:β-*Ergo-*

$$\text{(I)} \qquad\qquad \text{(II)}$$

stadienone). In view of the constitution now assigned to its reduction product "β"-dihydroergosterol (p. 1762 s), this product must be either an addition compound or a mixture of the (not isolated) $\Delta^{8(14)\cdot22}$- and $\Delta^{14\cdot22}$-ergostadien-3-ones; conflicting $[\alpha]_D$ values recorded and behaviour on chromatography rather favour the second alternative (1954 Barton). — Needles (glacial acetic acid), m. 125° (1929 Heilbron); needles (alcohol), m. 114°, $[\alpha]_D$ —5° (chloroform) (1932 Dithmar); crystals (alcohol), m. 125–6°, $[\alpha]_D$ +4° (chloroform) (1949 Barton). Readily soluble in ether, acetone, and chloroform, less soluble in alcohol (1932 Dithmar). — **Fmn.** From "β"-dihydroergosterol (p. 1762 s) by oxidation with CrO_3 in slightly aqueous acetic acid at 70° (1929 Heilbron) or by Oppenauer oxidation (1949 Barton). — **Rns.** Catalytic hydrogenation using palladium black in ether, yields α-ergostenol (p. 1765 s) (1929 Heilbron). Reverts to "β"-dihydroergosterol on reduction with sodium and abs. alcohol (1932 Dithmar).

$\Delta^{8(14)}$-**Ergosten-3-one,** α-**Ergostenone** $C_{28}H_{46}O$ (p. 125). For confirmation of structure assigned, see that of its reduction product, α-ergostenol (p. 1765 s). — Needles (methanol), m. 132–3° (1941 Wieland); crystals (chloroform-methanol), m. 129° to 130° (1948a Barton); $[\alpha]_D$ +37° (average value) (1941 Wieland), $[\alpha]_D^{ca.20}$ +30° (1948a Barton), both in chloroform. — **Fmn.** From α-ergostenol by oxidation with copper powder at 25° and 4 mm. pressure (1937 Marker) or by Oppenauer oxidation (1948a Barton). — **Rns.** Boiling with dry isopropanol and aluminium isopropoxide gives a mixture of α-ergostenol with its 3-epimer (p. 1766 s) (1937 Marker).

Δ^{14}-**Ergosten-3-one,** β-**Ergostenone** $C_{28}H_{46}O$ (p. 125). Crystals (methanol chloroform), m. 149–150°; $[\alpha]_D^{ca.20}$ +40° (chloroform). —**Fmn.** From β-ergostenol (p. 1767 s) by Oppen-auer oxidation (1948a Barton).

Ergostan-3-one, Ergostanone $C_{28}H_{48}O$ (p. 125). For relation between structure and optical rotation, see 1941 Bernstein; 1945 Barton. —Crystals (ethyl acetate-methanol), m. 160°; $[\alpha]_D^{ca.\ 20} + 33°$ (chloroform) (1949 Barton).

Coproergostan-3-one, *u-Ergostanone* $C_{28}H_{48}O$ (p. 86). Structure assigned on the basis of that now ascribed to its precursor, u-ergostanol (page 1767; cf. p. 86). — Crystals (methanol), m. 94° (1937 Laucht).

10-iso-Ergostan-3-one $C_{28}H_{48}O$. See lumistanone, p. 126.

24-iso-Ergostan-3-one, Campestanone $C_{28}H_{48}O$. Structure based on the probable identity of ostreasterol-I with chalinasterol (see p. 1771 s); the compound described by Bergmann (1934b) as *ostreastanone* (cf. p. 127) is probably an impure form of campestanone (Editor). — M. p. 157°; $[\alpha]_D^{19} + 42°$ (chloroform) (1934b Bergmann). —**Fmn.** From ostreasterol (p. 1771 s) by conversion into the acetate on boiling with acetic anhydride (1934a Bergmann), followed by hydrogenation in glacial acetic acid at 70° using platinum black catalyst; the *ostreastanyl acetate* so obtained [m. 137°, $[\alpha]_D^{19} + 14.6°$ (chloroform)] is saponified to give *ostreastanol* [m. 141°, $[\alpha]_D^{19} + 23.7°$ (chloroform)] which is oxidized with CrO_3 in ca. 95 % acetic acid on the water-bath (1934b Bergmann).

1929 Heilbron, Johnstone, Spring, *J. Ch. Soc.* **1929** 2248, 2254.
1932 Dithmar, Achtermann, *Z. physiol. Ch.* **205** 55, 59 et seq.
    Heilbron, Simpson, *J. Ch. Soc.* **1932** 2400.
1934 a Bergmann, *J. Biol. Ch.* **104** 317.
    b Bergmann, *J. Biol. Ch.* **104** 553.
1936 Dimroth, Trautmann, *Ber.* **69** 669, 673.
1937 Laucht, *Z. physiol. Ch.* **246** 171, 175.
    Marker, Kamm, Laucius, Oakwood, *J. Am. Ch. Soc.* **59** 1840.
    a Oppenauer, *Rec. trav. chim.* **56** 137.
    b Oppenauer, *U.S. Pat.* 2384335 (issued 1945); *C. A.* **1946** 178; *Ch. Ztbl.* **1948** I 1039.
    Schering A.-G., *Austrian Pat. Application*; *Fiat Final Report* No. 996, pp. 130, 154 (issued 1947).
    Wetter, Dimroth, *Ber.* **70** 1665, 1667, 1669, 1670.
1938 Heilbron, Kennedy, Spring, Swain, *J. Ch. Soc.* **1938** 869.
    Windaus, Buchholz, *Ber.* **71** 576.
1939 v. Christiani, Anger, *Ber.* **72** 1124.
    Dannenberg, *Abhandl. Preuss. Akad. Wiss. Math.-naturwiss. Kl.* **1939** No. 21, pp. 1, 43, 64.
    Windaus, Buchholz, *Ber.* **72** 597.
1941 Bernstein, Kauzmann, Wallis, *J. Org. Ch.* **6** 319.
    Wieland, Rath, Hesse, *Ann.* **548** 34, 43, 45.

1943 Evans, Gillam, *J. Ch. Soc.* **1943** 565, 567.
1945 Barton, *J. Ch. Soc.* **1945** 813, 814.
1948 a Barton, Cox, *J. Ch. Soc.* **1948** 783, 793.
    b Barton, Cox, *J. Ch. Soc.* **1948** 1354, 1356.
1949 Barton, Cox, *J. Ch. Soc.* **1949** 1771, 1776, 1779.
1952 Antonucci, Bernstein, Heller, Williams, *J. Org. Ch.* **17** 1446, 1449.
    Fieser, Rosen, Fieser, *J. Am. Ch. Soc.* **74** 5397, 5402.
1954 Barton, *private communication.*

6α-Methyl-Δ^4-cholesten-3-one $C_{28}H_{46}O$. For configuration assigned at C-6, see 1952 Turner. — Crystals (benzine, methanol-benzine, or methanol-ethyl acetate), m. 126.5–127.5° cor. (1939 Ushakov); crystals (methanol), m. 127–128.5° (1952 Turner); $[\alpha]_D$ +60.5° (dioxane) (1952 Turner). Absorption max. in alcohol at 240 mμ (log ε 4.3) (1939 Ushakov), at 241 mμ (log ε 4.22) (1952 Turner). Liebermann-Burchard reaction: grass-green, with pink-green fluorescence, on standing. Weak green-yellow colour reaction with tetranitromethane (1939 Ushakov). — **Fmn.** From 6β-methylcholestan-5-ol-3-one (p. 2697 s) by treatment with gaseous HCl in ice-cold chloroform; yield, 87 % (1939 Ushakov).

1939 Ushakov, Madaeva, *J. Gen. Ch. U.S.S.R.* **9** 436, 441; *C. A.* **1939** 9309; *Ch. Ztbl.* **1939** II 4488.
1952 Turner, *J. Am. Ch. Soc.* **74** 5362.

7. Monoketo-steroids (CO in the ring system) C_{29} with aliphatic side chains

Summary of the Ketones with Sitostane Skeleton*

Position of CO	Configuration		Position double bond(s)	M. p. °C.	$[\alpha]_D$ (°) in $CHCl_3$	Oxime m. °C.	Semi-carbazone m. °C.	Page
	C-5	C-24						
3	—	norm.	$\Delta^{4.22}$	94; 125	+63	188	239	2465 s
3	—	iso	$\Delta^{4.22}$	112	+57[1])	—	—	2466 s
3	—	norm.	$\Delta^{4.24(28)}$	94	+76	167	238 dec.	2466 s
3	α	norm.	$\Delta^{7.22}$	181	+20	257 dec.	280 dec.	2467 s
3	—	norm.	Δ^4	88; 102	+86	176	250 dec.	2467 s
3	—	iso	Δ^4	79	+80[1])	—	—	2468 s
3	α	norm.	$\Delta^{8(14)}$	119	+17	186	246 dec.	2469 s
3	α	norm.	satd.	157	+41	219	259	2469 s
3	β	norm.	satd.	114	+34	—	—	2470 s
3	α	iso	satd.	162	ca. +40	210	—	2470 s
6	—	norm.	i(3:5)	77	—	173	—	2470 s
6	α	norm.	satd.	106	—	192	—	127
7	—	norm.	$\Delta^{3.5}$	107	—	—	—	2470 s

* For values other than those chosen for the summary, see the compounds themselves.
[1]) Solvent not indicated.

$\Delta^{4.22}$-Stigmastadien-3-one, *Stigmastadienone* $C_{29}H_{46}O$ (p. 126: *Stigmastenone*).

For evidence of the dienone structure, see 1939 Fernholz. — Crystals (acetone, ethyl acetate, or ether-methanol), m. 94° (1938a Marker); leaflets (alcohol), m. 125° (1939 Fernholz); crystals (acetone-methanol), m. 124.5–125° (1942 Jones); apparently *dimorphic*, m. 94° and m. 127° (1941 PARKE, DAVIS & Co.); $[\alpha]_D^{25}$ +63° (1939 Fernholz; cf. 1948 Barton), $[\alpha]_D^{20}$ +56° (1942 Jones) both in chloroform. Absorption max. in alcohol at 241 and 307.5 mμ (ε 17000 and 75, resp.) (1942 Jones).

Fmn. From stigmasterol (p. 1784 s) by oxidation using aluminium isopropoxide and acetone in boiling benzene (yield, 50%) (1937 SCHERING-KAHLBAUM A.-G.) or cyclohexanone in boiling toluene (1939 Fernholz), or by means of aluminium tert.-butoxide and acetone in anhydrous benzene under reflux (yield, 58%) (1942 Jones). From $\Delta^{5.22}$-stigmastadien-3β.4β-diol (p. 2081 s) by refluxing with alcohol containing conc. HCl (1938a Marker); this dehydration can also be effected with H_2SO_4 or H_3PO_4, or starting from the diol mono- or di-esters (1938 PARKE, DAVIS & Co.). From stigmastadienone 22.23-dibromide (p. 2473 s) by heating with zinc dust and glacial acetic acid on the steam-bath, and purification via the semicarbazone; yield, 31% (1939 Fernholz). From Δ^{22}-stigmasten-6α-ol-3-one (p. 2694 s) by slow sublimation in a high vacuum in presence of fused $ZnCl_2$ (1941 PARKE, DAVIS & Co.).

Rns. For oxidation, alone or admixed with cinchone (p. 2467 s), by means of H_2CrO_4, $KMnO_4$, peracids, or metallic oxides, to give ketonic products

References, pp. 2470 s, 2471 s

with male or progestational hormonal activity, see 1935 C. F. BOEHRINGER & SOEHNE. Boiling with sodium in dry amyl alcohol gives the addition compound of Δ^{22}-stigmasten-3β-ol with Δ^{22}-coprostigmasten-3α-ol (p. 1826 s) (1937a Marker; 1950 Barton), together with a product m. 72° (thought to be a hydrocarbon, no analysis given; absorbs bromine) (1937a Marker). Refluxing with aluminium isopropoxide in dry isopropanol affords the two 3-epimeric $\Delta^{4\cdot22}$-stigmastadien-3-ols (p. 1781 s) together with stigmasterylene (p. 1440 s) (1937b Marker). Heating with ethylene glycol in benzene containing small amounts of p-toluenesulphonic acid, with slow distillation of the mixture, yields $\Delta^{5\cdot22}$-**stigmastadien-3-one ethylene ketal*** $C_{31}H_{50}O_2$, m.131° (1941 SQUIBB & SONS).

Oxime $C_{29}H_{47}ON$. Needle clusters (petroleum ether b. 40–60°), m. 187–8°; absorption max. in alcohol at 240 mμ (ε 22000) (1942 Jones).

Semicarbazone $C_{30}H_{49}ON_3$. Crystals (alcohol-chloroform), m. 238–9°; absorption max. in chloroform at 271 mμ (ε 29000) (1942 Jones; cf. 1939 Fernholz).

2.4-Dinitrophenylhydrazone $C_{35}H_{50}O_4N_4$. Red needles (alcohol-benzene), m. 244–5° dec. (1942 Jones); absorption max. in chloroform at 393 (31000), 293 (17000), 280 (18500), 265 (23000), and 242 mμ (ε) (1945 Braude; cf. 1942 Jones).

24-iso-$\Delta^{4\cdot22}$-Stigmastadien-3-one, $\Delta^{4\cdot22}$-Poriferastadien-3-one, "*Poriferastenone*"

$C_{29}H_{46}O$. The structure ascribed is based on that of the parent poriferasterol (p. 1838 s) (Editor). — M. p. 111–112.5°; $[\alpha]_D^{25}$ +57° (solvent not indicated). — **Fmn.** From poriferasterol by refluxing with aluminium isopropoxide and cyclohexanone in toluene (1942 Lyon).

$\Delta^{4\cdot24(28)}$-Stigmastadien-3-one, Fucostadienone

$C_{29}H_{46}O$. The structure assigned is based on that of the parent fucosterol (p. 1793 s) (Editor). — Plates (acetone), m. 94–94.5°; $[\alpha]_D^{20}$ +76° (chloroform). — **Fmn.** From fucosterol by oxidation with dry acetone and aluminium tert.-butoxide in boiling dry benzene; yield, 50% (1942 Jones).

Oxime $C_{29}H_{47}ON$. Needles (after sublimation in high vacuo), m. 166–7°; absorption max. in alcohol at 240.5 mμ (ε 23000) (1942 Jones).

---

* The $\Delta^{5\cdot22}$-structure is assigned on the basis of that now accepted for cholestenone ethylene ketal (p. 2435 s), obtained by a similar reaction which involves shift of the double bond from Δ^4 to Δ^5 position.

Semicarbazone $C_{30}H_{49}ON_3$. Crystals (alcohol-chloroform), m. 238° dec.; absorption max. in chloroform at 271.5 mμ (ε 27000) (1942 Jones).

2.4-Dinitrophenylhydrazone $C_{35}H_{50}O_4N_4$. Flat needles (benzene), m. 237°; absorption max. in chloroform at 392 mμ (ε 32000) (1942 Jones).

$\Delta^{7.22}$-Stigmastadien-3-one, "α"-Spinastadienone, *Bessistadienone, "Bessisten-*

one" $C_{29}H_{46}O$ (p. 87: α-*Spinasta-dienone*). For structure, see that of the parent "α"-spinasterol, page 1794 s. — Plates (acetic acid, alcohol, or acetone)* (1937 Simpson), needles (chloroform-methanol), m. 180–1°(1940a Kuwada); [α]$_D^{17}$ + 19.5°(chloroform)* (1937 Simpson). Absorption max. in hexane at 240 and 280–290 mμ (ε 410 and 22, resp.)**(1940a Kuwada). — **Fmn.** From "α"-spinasterol (bessisterol) by oxidation with CrO_3 in strong acetic acid at room temperature (crude yield, 88%) (1937 Simpson) or by Oppenauer oxidation (aluminium phenoxide catalyst) (1940a Kuwada). — **Rns.** Catalytic hydrogenation, using platinum oxide in ethyl acetate, gives $\Delta^{8(14)}$-stigmasten-3β-ol (p. 1823 s). Reduction by the method of Meerwein-Ponndorf yields "α"-spinasterol (p. 1794 s) (1940a Kuwada).

Oxime $C_{29}H_{47}ON$. Plates (chloroform-methanol), m. 253–5° dec. (1937 Simpson), m. 257° dec. (1940a Kuwada).

Semicarbazone $C_{30}H_{49}ON_3$. M. p. 279.5° dec.; sparingly soluble (1940a Kuwada).

Δ^4-Sitosten-3-one, "β"-*Sitostenone, Cinchone, "α"-Fucostenone* $C_{29}H_{48}O$

(p. 126). For identity of cinchone (p. 126) with "β"-sitostenone, see that of cinchol with "β"-sitosterol, p. 1805 s. For evidence of Δ^4-double bond position, see 1938 Dirscherl. — Crystals (ethyl acetate or acetone), m. 83° (1938a Marker); crystals (alcohol), m. 83–84° (1942 Jones); crystals (methanol or ethanol), m. 88° (1943 Barton); [α]$_D^{20}$ +86° (chloroform) (1943 Barton). Absorption max. in alcohol at 241 mμ (ε 17000–20000) and 307 mμ (ε 75) (1942 Jones; 1943 Barton).

Fmn. The yield from cinchol (p. 1808 s) by CrO_3 oxidation of its dibromide (cf. p. 126) is 32% (1938 Dirscherl). Together with another ketonic product, by oxidation of wheat germ sitosterol (p. 1801 s), using anhydrous acetone and aluminium tert.-butoxide in boiling dry benzene, followed by chromato-

---

* The identity of this product with that of Kuwada (1940a), for which no question of double bond shift arises, seems highly probable by comparison of the molecular rotation difference [M]"α"-spinastadienone — [M]"α"-spinasterol with the difference [M] $\Delta^{7.22}$-ergostadienone — [M] $\Delta^{7.22}$-ergostadien-3β-ol (Editor).

** The absorption max. are probably due to an impurity; cf. footnote ** on p. 1795 s (Editor).

graphy on Al_2O_3; yield, 15% (1942 Jones). From Δ^5-sitostene-3β.4β-diol (p. 2082 s) or from its acetate by refluxing in alcohol containing conc. HCl or other mineral acids (1938a Marker; 1938 PARKE, DAVIS & Co.). From sitostan-6β-ol-3-one (p. 2694 s) on heating with p-toluenesulphonic acid, followed by slow sublimation in a high vacuum (1941 PARKE, DAVIS & Co.).

Rns. Oxidation with CrO_3 in glacial acetic acid at 50–60° gives mixed products from which one with male, another with progestational hormonal activity can be separated (1935 C. F. BOEHRINGER & SOEHNE). Reduction with boiling isopropanol and aluminium isopropoxide affords the two 3-epimeric Δ^4-sitosten-3-ols (p. 1800 s) (1937b Marker).

Oxime $C_{29}H_{49}ON$. M. p. 175.5°; absorption max. in alcohol at 241 mμ (ε 23500) (1943 Barton; cf. 1942 Jones).

Semicarbazone $C_{30}H_{51}ON_3$ (p. 126). M. p. 250° dec.; absorption max. in chloroform at 273 mμ (ε 30000) (1943 Barton).

2.4-Dinitrophenylhydrazone $C_{35}H_{52}O_4N_4$. Red powder (benzene-alcohol) (1942 Jones), m. 253° dec. (1943 Barton); absorption max. in chloroform at 259 (21500), 282 (16000), 292 (14000), and 395 (31500) mμ (ε) (1945 Braude; cf. 1942 Jones; 1943 Barton).

A **ketone** $C_{29}H_{48}O$, characterized by its 2.4-dinitrophenylhydrazone, has been obtained from plantation rubber (1941 Heilbron). In view of the isolation of "β"-sitosterol from this source (cf. p. 1809 s) and the m. p. of the dinitrophenylhydrazone (below), it seems likely that the ketone is an impure sample of "β"-sitostenone (Editor). — *Isolation.* After separation of crystalline material from the alcoholic extract of the unsaponifiable fraction, a gummy product resulted on evaporation of the mother liquors; this was fractionally distilled at 10^{-3} mm. pressure and the fraction distilling at 140–150° was submitted to reaction with 2.4-dinitrophenylhydrazine in alcohol giving, after chromatography, a *2.4-dinitrophenylhydrazone* $C_{35}H_{52}O_4N_4$ [scarlet prisms (benzene-methanol), m. 239–240°]; from the hydrazone mother liquors a *second dinitrophenylhydrazone* was isolated [golden leaflets (light petroleum), m. 121–3°] for which the analytical data could not be interpreted (1941 Heilbron).

24-iso-Δ^4-Stigmasten-3-one, Δ^4-Poriferasten-3-one, *Clionastenone, γ-Sitostenone* $C_{29}H_{48}O$.

For structure, see 1942 Bergmann. — M. p. 79°; [α]$_D$ +80° (solvent not indicated); ultraviolet absorption spectrum indicates an $\alpha.\beta$-unsaturated ketone (1942 Bergmann; 1942 Kind). — **Fmn.** From clionasterol (p. 1839 s) by Oppenauer oxidation, using aluminium isopropoxide catalyst (1942 Bergmann). From Δ^4-clionastene-3β.6β-diol (p. 2084 s) by refluxing with acids (1942 Kind).

2.4-Dinitrophenylhydrazone $C_{35}H_{52}O_4N_4$, m. 230° (1942 Bergmann; 1942 Kind).

$\Delta^{8(14)}$-Stigmasten-3-one, α-Stigmastenone, α-*Bessistenone* $C_{29}H_{48}O$. Plates, m. 116.5–120.5°; $[\alpha]_D^{22} + 17.4°$ (chloroform). Absorption max. at 280 mμ (ε ca. 22). — **Fmn.** From $\Delta^{8(14)}$-stigmasten-3β-ol (p. 1823 s) by Oppenauer oxidation. — **Rns.** Regenerates $\Delta^{8(14)}$-stigmasten-3β-ol on catalytic hydrogenation (1940a Kuwada).

Oxime $C_{29}H_{49}ON$, m. 186°. — **Semicarbazone** $C_{30}H_{51}ON_3$, m. 245.5° dec. (1940a Kuwada).

Stigmastan-3-one, "β"-Sitostanone, *Spinastanone, Fucostanone* $C_{29}H_{50}O$ (p. 127). For identity of spinastanone and fucostanone, see 1938 Larsen; for that with α-*typhastanone*, see 1937 Kuwada; for identity with *dihydrocinchone* (p. 127), see that of cinchol (p. 1805 s). For probable nonidentity with *ostreastanone*, see p. 2463 s. For relation between configuration and optical rotation, see 1941 Bernstein; 1947 Bergmann. — Crystals (methanol or ethanol), m. 156.5–157° (1938 Larsen; cf. 1943 Barton), m. 161° (1940b Kuwada); n_α 1.512, n_β 1.529 (1937 Kuwada); $[\alpha]_D^{20} + 40.5°$ (chloroform) (1938 Larsen; cf. 1940b Kuwada).

Fmn. From "β"-sitosterol (p. 1808 s) by heating with freshly reduced nickel at 220° (8 hrs.) (1928 Bonstedt). From wheat germ or "Tallöl" sitosterol (p. 1803 s) by oxidation with anhydrous acetone and aluminium tert.-butoxide in boiling dry benzene (18 hrs.) (1943 Barton; cf. 1942 Jones). The yield from spinastanol by oxidation with CrO_3 (cf. p. 127) in 96% acetic acid at room temperature is 50% (1938 Larsen).

Rns. Oxidation with CrO_3 in strong acetic acid at 70° affords stigmastane-2‖3-dioic acid (p. 1828 s) (1928 Bonstedt). Reduction to stigmastane (page 1442 s) occurs on boiling with amalgamated zinc in glacial acetic acid containing HCl (1938 Larsen).

Oxime $C_{29}H_{51}ON$. Crystals (methanol or acetone), m. 219°, $[\alpha]_D^{20} + 29°$ (chloroform) (1938 Larsen; 1940b Kuwada).

Semicarbazone $C_{30}H_{53}ON_3$, m. 259° (1937 Kuwada).

2.4-Dinitrophenylhydrazone $C_{35}H_{54}O_4N_4$. Red amorphous powder (ethyl acetate), m. 208–9°* (1942 Jones); yellow solid, m. 233° dec. (1943 Barton); absorption max.* in chloroform at 255 (18000), 281 (14000), 291 (10000), and 368 (25500) mμ (ε) (1945 Braude; cf. 1942 Jones).

* The authors state that the sitostanone starting material was impure.

24 b-Ethylcoprostan-3-one, Coprositostanone $C_{29}H_{50}O$ (p. 127: *24-Ethylcopro-*

stanone). Crystals (ethyl acetate-methanol), m. 113–4°; $[\alpha]_D^{ca.20} + 34°$ (chloroform) (1950 Barton). — **Rns.** Treatment with bromine in glacial acetic acid containing small amounts of HBr gives 24 b-ethyl-4β-bromo-coprostan-3-one (p. 2489 s) (1937 a Marker).

24-iso-Stigmastan-3-one, Poriferastanone, *"γ"-Sitostanone* $C_{29}H_{50}O$ (p. 127).

For identity with the "stigmasta-none" of Mazur (1941), cf. that of spongillasterol (p. 1839 s). For relation between configuration and molecular rotation, see 1941, 1942 Bernstein; 1947 Bergmann. — Crystals (aqueous alcohol), m. 155°, $[\alpha]_D + 39°$ (chloroform) (1941 Mazur); crystals (alcohol), m. 161.5° cor. (1941 Valentine; cf. 1942 Kind); $[\alpha]_D + 47°$ (1941 Valentine), $+43°$ (1942 Kind), $+38°$, $+41°$ (1947 Bergmann) all in chloroform. — **Fmn.** From poriferastanol (p. 1841 s) by oxidation with CrO_3 in ca. 95 % acetic acid on the steam-bath (1941 Valentine; 1942 Kind; 1941 Mazur).

Oxime $C_{29}H_{51}ON$. Crystals (aqueous alcohol), m. 210° (1941 Mazur).

i-Sitosten-6-one $C_{29}H_{48}O$. For identity of this compound, described by Vanghelovici (1935) as "sitostan-6-one" (cf. p. 127), see 1948 Shoppee.— Crystals (acetone), m. 77° (1935 Vanghelovici). — **Fmn.** From 3β-chloro-sitostan-6-one (p. 2490 s) by reduction with sodium amalgam in hot alcohol (1935 Vanghelovici).

Oxime $C_{29}H_{49}ON$. Needles (alcohol), m. 173° (1935 Vanghelovici).

$\Delta^{3.5}$-Sitostadien-7-one, *7-Ketositosterylene* $C_{29}H_{46}O$. Pale yellow plates (acetone-alcohol), m. 106–7°. — **Fmn.** From 3β-chloro-Δ^5-sitosten-7-one (p. 2490 s) by refluxing with alcoholic KOH (1940 Marker).

1928 Bonstedt, *Z. physiol. Ch.* **176** 269, 275.
1935 C. F. BOEHRINGER & SOEHNE, *Ger. Pat.* 697757 (issued 1940); *C. A.* **1941** 6739, 9719; *Ch. Ztbl.*
     1943 I 653; RARE CHEMICALS INC. (Dirscherl, Hanusch), *U.S. Pat.* 2152625 (issued 1939);
     C. A. **1939** 5132; *Ch. Ztbl.* **1940** I 94.
     Vanghelovici, Angelescu, *Bul. soc. chim. România* **17** A 177.

1937 Kuwada, Morimoto, *J. Pharm. Soc. Japan* **57** 62 (German abstract); cf. *C. A.* **1937** 4984; *Ch. Ztbl.* **1937** II 1825.

a Marker, Wittle, *J. Am. Ch. Soc.* **59** 2704, 2706, 2707.

b Marker, Oakwood, *J. Am. Ch. Soc.* **59** 2708.

SCHERING-KAHLBAUM A.-G., *Fr. Pat.* 822551 (issued 1938); *C. A.* **1938** 4174; *Ch. Ztbl.* **1938** II 120; SCHERING CORP. (Serini, Köster, Strassberger), *U.S. Pat.* 2379832 (issued 1945); *C. A.* **1945** 5053.

Simpson, *J. Ch. Soc.* **1937** 730, 732.

1938 Dirscherl, Kraus, *Z. physiol. Ch.* **253** 64.

Larsen, *J. Am. Ch. Soc.* **60** 2431.

a Marker, Kamm, Wittle, *J. Am. Ch. Soc.* **60** 1071.

b Marker, Rohrmann, *J. Am. Ch. Soc.* **60** 1073, 1074.

PARKE, DAVIS & Co. (Marker), *U.S. Pat.* 2227839 (issued 1941); *C. A.* **1941** 2532; *Ch. Ztbl.* **1941** II 81.

1939 Fernholz, Stavely, *J. Am. Ch. Soc.* **61** 2956.

1940 a Kuwada, Yosiki, *J. Pharm. Soc. Japan* **60** 25 (German abstract); cf. *C. A.* **1940** 5088; *Ch. Ztbl.* **1940** II 630.

b Kuwada, Yosiki, *C. A.* **1941** 461; *Ch. Ztbl.* **1941** I 1038.

Marker, Rohrmann, *J. Am. Ch. Soc.* **62** 516.

1941 Bernstein, Kauzmann, Wallis, *J. Org. Ch.* **6** 319.

Heilbron, Jones, Roberts, Wilkinson, *J. Ch. Soc.* **1941** 344, 347.

Mazur, *J. Am. Ch. Soc.* **63** 2442.

PARKE, DAVIS & Co. (Marker, Lawson), *U.S. Pat.* 2366204 (issued 1945); *C. A.* **1945** 1649; *Ch. Ztbl.* **1945** II 1385.

SQUIBB & SONS (Fernholz), *U.S. Pat.* 2378918 (issued 1945); *C. A.* **1945** 5051.

Valentine, Bergmann, *J. Org. Ch.* **6** 452.

1942 Bergmann, Kind, *J. Am. Ch. Soc.* **64** 473.

Bernstein, Wilson, Wallis, *J. Org. Ch.* **7** 103.

Jones, Wilkinson, Kerlogue, *J. Ch. Soc.* **1942** 391.

Kind, Bergmann, *J. Org. Ch.* **7** 341.

Lyon, Bergmann, *J. Org. Ch.* **7** 428.

1943 Barton, Jones, *J. Ch. Soc.* **1943** 599.

1945 Braude, Jones, *J. Ch. Soc.* **1945** 498, 499, 501.

1947 Bergmann, Low, *J. Org. Ch.* **12** 67, 70, 71.

1948 Barton, Cox, *J. Ch. Soc.* **1948** 783, 793.

Shoppee, *J. Ch. Soc.* **1948** 1043, 1044.

1950 Barton, Brooks, *J. Am. Ch. Soc.* **72** 1633, 1636.

8. Monoketo-steroids (CO in the ring system) containing cyclic radicals in side chain

2 (?)-Benzylidenecholestan-3-one $C_{34}H_{50}O$. See p. 2440 s.

22.22-Diphenyl-$\Delta^{4 \cdot 20(22)}$-bisnorcholadien-3-one, 20-Diphenylmethylene-Δ^4-pregnen-3-one, 1-Methyl-2.2-diphenyl-1-(3-keto-Δ^4-ætiocholenyl)-ethylene $C_{34}H_{40}O$. Prisms (acetone), m. ca. 229°. — **Fmn.** From 1-methyl-2.2-diphenyl-1-(3β-hydroxy-Δ^5-ætiocholenyl)-ethylene (p. 1844 s) on Oppenauer oxidation using aluminium tert.-butoxide and cyclohexanone, or on oxidation of its 5.6-dibromide (p. 1844 s) with CrO_3 in strong acetic acid, followed by debromination of the resulting dibromo-ketone (page 2502 s) with zinc dust and glacial acetic acid in ether. From (3-keto-Δ^4-ternorcholenyl)-diphenyl-carbinol (p. 2513 s) on refluxing with glacial acetic acid for 20 min. (1941 GLIDDEN Co.).

24.24-Diphenyl-$\Delta^{4 \cdot 20(22) \cdot 23}$-cholatrien-3-one $C_{36}H_{42}O$. Jelly (from alcohol), m. 106–110° cor.; $[\alpha]_D^{21}$ +140° (chloroform) (1946 Meystre). Gives a brown coloration with tetranitromethane, a yellow-brown coloration with trichloroacetic acid, and a bordeaux-red solution in the Liebermann-Burchard reaction (1946 Miescher). — **Fmn.** From 24.24-diphenyl-$\Delta^{5 \cdot 20(22) \cdot 23}$-cholatrien-3$\beta$-ol (p. 1847 s) on boiling with aluminium isopropoxide and cyclohexanone in toluene; yield, 95 % (1946 Meystre; 1946 CIBA PHARMACEUTICAL PRODUCTS). — **Rns.** Gives Δ^4-pregnene-3.20-dione (progesterone, p. 2782 s) on oxidation with CrO_3 and 80 % acetic acid in chloroform at 0–3° (1946 Meystre; 1946 CIBA PHARMACEUTICAL PRODUCTS; 1947 Miescher).

Semicarbazone $C_{37}H_{45}ON_3$. Crystals (alcohol), m. 168–170° (1946 Meystre).

1941 GLIDDEN Co. (Julian, Cole), *U.S. Pat.* 2394551 (issued 1946); *C. A.* **1946** 2593.
1946 CIBA PHARMACEUTICAL PRODUCTS (Miescher, Frey, Meystre, Wettstein), *U.S. Pats.* 2461910, 2461911 (issued 1949), 2461912 (1947, issued 1949); *C. A.* **1949** 3474, 3475; *Ch. Ztbl.*
    1949 E 1870, 1871.
    Meystre, Frey, Neher, Wettstein, Miescher, *Helv.* **29** 627, 632.
    Miescher, *Helv.* **29** 743, 746.
1947 Miescher, Schmidlin, *Helv.* **30** 1405, 1408.

II. HALOGENO-MONOKETO-STEROIDS (CO IN THE RING SYSTEM)

1. Halogen in Side Chain Only

21-Bromo-$\Delta^{4.17(20)}$-pregnadien-3-one $C_{21}H_{29}OBr$ (I, R=Br). Crystals (acetone), m. 126–7° cor. (1939 Ruzicka). Gives a strong, red-violet coloration on heating with 1.4-naphthoquinone and conc. HCl in glacial acetic acid at 60–70° (1939 Miescher). — **Fmn.** From 17-iso-$\Delta^{4.20}$-pregnadien-17β-ol-3-one (17-vinyltestosterone, p. 2650 s) on treatment with PBr$_3$ and a little pyridine in abs. chloroform, first at —20°, then at room temperature; quantitative crude yield (1939 Ruzicka; 1939b CIBA; 1939 SCHERING A.-G.). — **Rns.** Gives the acetate of $\Delta^{4.17(20)}$-pregnadien-21-ol-3-one (p. 2512 s) on refluxing with anhydrous potassium acetate in dry acetone (1939 Ruzicka; 1939b CIBA; 1939 SCHERING A.-G.). For the reaction with pyridine, see the following compound.

N-(3-Keto-$\Delta^{4.17(20)}$-pregnadien-21-yl)-pyridinium bromide $C_{26}H_{34}ONBr$ [I, R=N(Br)C$_5$H$_5$]. Colourless crystals (alcohol-benzene), m. 213–4° cor. Readily soluble in alcohol and chloroform, very sparingly in ether and benzene (1940 Reich). — **Fmn.** From the preceding compound on treatment with pyridine at room temperature; yield, 51% (1940 Reich; 1940 N. V. ORGANON). — **Rns.** On sublimation at 215–8° under 0.02 mm. pressure, $\Delta^{4.16.20}$-pregnatrien-3-one (p. 2406 s) is probably formed (1940 Reich). Gives the 21-N-(p-dimethylaminophenyl)-isoxime of $\Delta^{4.17(20)}$-pregnadien-3-on-21-al (p. 2803 s) on treatment with p-nitroso-dimethylaniline in aqueous alcoholic NaOH at room temperature (1940 Reich; 1940 N. V. ORGANON).

21-Bromo-Δ^4-pregnen-3-one $C_{21}H_{31}OBr$. — **Fmn.** From the 3-propionate 21-acetate of $\Delta^{5.17(20)}$-pregnadiene-3β.21-diol (p. 1935 s) on partial saponification, followed by hydrogenation and bromination to give 21-bromo-Δ^5-pregnen-3β-ol propionate (not described in detail) which is then saponified and oxidized. — **Rns.** Gives Δ^4-pregnen-3-on-21-al (p. 2803 s) on refluxing with hexamethylenetetramine in aqueous alcohol (1939a CIBA).

22.23-Dibromo-Δ^4-stigmasten-3-one, $\Delta^{4.22}$-Stigmastadien-3-one 22.23-dibromide $C_{29}H_{46}OBr_2$. Crystals (acetone), m. 182–4°; $[\alpha]_D^{22}$ +53° (chloroform). — **Fmn.** From stigmasterol 22.23-dibromide (p. 1871 s) on refluxing with aluminium tert.-butoxide and acetone in dry benzene; crude yield, 72%. — **Rns.** Gives $\Delta^{4.22}$-stigmastadien-3-one (p. 2465 s) on heating with zinc dust in glacial acetic acid on the steam-bath (1939 Fernholz).

172*

1939 a CIBA*, Fr. Pat. 857 122 (issued 1940); Ch. Ztbl. 1941 I 1324; CIBA PHARMACEUTICAL PRO-
      DUCTS (Miescher, Fischer, Scholz, Wettstein), U.S. Pat. 2275790, 2276543 (both issued
      1942); C. A. 1942 4289, 4674.
    b CIBA*, Fr. Pat. 860323 (issued 1941); Ch. Ztbl. 1941 I 3550.
      Fernholz, Stavely, J. Am. Ch. Soc. 61 2956.
      Miescher, Wettstein, Scholz, Helv. 22 894, 903, 904.
      Ruzicka, Müller, Helv. 22 416.
      SCHERING A.-G. (Logemann), Swed. Pat. 102639 (issued 1941); Ch. Ztbl. 1942 II 74; Fr. Pat.
      868336 (1940, issued 1941); Ch. Ztbl. 1942 I 2038.
1940 N. V. ORGANON, Fr. Pat. 864315 (issued 1941); Ch. Ztbl. 1942 I 642.
      Reich, Helv. 23 219, 222.

2. Monohalogeno-monoketo-steroids (CO and Halogen in the Ring System)

a. COMPOUNDS C_{19}–C_{21}

17"α"-Chloro-Δ^4-androsten-3-one $C_{19}H_{27}OCl$. Colourless leaflets, m. 148°.

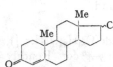

Very unstable in the crude state, much more stable when pure. Very soluble in most organic solvents (1939 Westphal). — Shows a high androgenic activity in the capon comb test (1939 Westphal). — **Fmn.** From 17"α"-chloro-Δ^5-androsten-3β-ol (p. 1873 s) on oxidation with CrO_3 in glacial acetic acid at room temperature, under temporary protection of the double bond by bromination; yield, 88% (1935 SCHERING-KAHLBAUM A.-G.). From Δ^4-androsten-17β-ol-3-one (testosterone, p. 2580 s) on treatment with PCl_5 and some dry $CaCO_3$ in dry chloroform at 0°, along with smaller amounts of testosterone dihydrogen phosphate (p. 2588 s) and 3.17-dichloro-$\Delta^{3.5}$-androstadiene (p. 1467 s) (1939 Westphal).

Rns. Readily loses HCl, more profound decomposition occurring at the same time (1939 Westphal). On heating with potassium acetate and glacial acetic acid in a sealed tube at 150–180° it gives the acetate of testosterone (p. 2580 s) (1935 SCHERING-KAHLBAUM A.-G.).

3-Chloro-$\Delta^{3.5}$-androstadien-17-one $C_{19}H_{25}OCl$. Needles (methanol), m. 143°

(1937 b Kuwada). Absorption max. in ether at 238 mμ (ε 23200) (1939 Dannenberg). — **Fmn.** From Δ^4-androstene-3.17-dione (p. 2880 s) by the action of acetyl chloride at room temperature (1937 b Kuwada). — **Rns.** Reduction with sodium in abs. alcohol affords $\Delta^{3.5}$-androstadien-17β-ol (p. 1509 s) (1937 b Kuwada).

3β-Chloro-Δ^5-androsten-17-one, 5-Dehydroepiandrosteryl chloride $C_{19}H_{27}OCl$

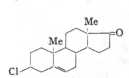

(p. 128: *Dehydroandrosteryl chloride*). For configuration at C-3 and discussion of the reaction mechanism involved in its formation from Δ^5-androsten-3β-ol-17-one (5-dehydroepiandrosterone, p. 2609 s) and in its reaction with benzoate ion, see 1946 b Shoppee; cf. 1937, 1938 Bergmann.

* Société pour l'industrie chimique à Bâle; Gesellschaft für chemische Industrie in Basel.

Colourless needles (methanol), m. 154.5–156° cor. (1941 Wolfe), 156–7° (1945 Mason); sublimes in a high vacuum at 100–110° (1941 Wolfe); $[\alpha]_D^{28} + 19°$ (alcohol) (1945 Mason), $[\alpha]_D^{25} + 13.5°$ (dioxane) (1950 Hershberg). For infra-red spectrum between 2 and 12.4 μ, see 1946 Furchgott. Readily soluble in ether, acetone, chloroform, dioxane, ethyl acetate, and hot alcohol, soluble in petroleum ether. With acetic anhydride and conc. H_2SO_4 it gives a yellow coloration, rapidly turning intensely pink (1934 Butenandt). For the Zimmermann colour test with m-dinitrobenzene and KOH in alcohol, see 1945 Wilson.

Is devoid of anæsthetic activity in rats (1942 Selye).

Occ. Chloroandrostenone has been isolated from the urine of women with cortico-adrenal tumours (1941 Wolfe; 1945 Mason) and from that of men with benign hypertrophy of the prostate (1942 Miller); the chloroketone isolated from urine (see also p. 128) is not considered to be a genuine product of metabolism, but formed from a precursor during the hydrolysis of the urine with hydrochloric acid (1934 Butenandt); 5-dehydroepiandrosterone (p. 2609 s) has been recognized as one such precursor (1935 Butenandt; 1941 Wolfe); a second precursor, suspected by Butenandt (1935), has been identified as i-androsten-6β-ol-17-one (page 2644 s) (1948 Dingemanse; 1948 Barton; cf. 1941 Rosenheim) which, however, according to Fieser (1949) may be formed from a conjugate of dehydroepiandrosterone during the isolation from urine.

Fmn. From 3β-chloro-Δ^5-androsten-17β-ol (p. 1874 s) on oxidation with CrO_3, after temporary protection of the double bond by bromination (1937 a Kuwada). For improvement of yield (56%) in the preparation from dehydroepiandrosterone (p. 2609 s) with thionyl chloride (cf. p. 128), by working in ether solution in the presence of $CaCO_3$, see 1939 Kuwada; is obtained as a by-product, along with 5-chloroandrostan-3β-ol-17-one (p. 2705 s), from dehydroepiandrosterone with dry HCl in chloroform at 25° (1950 Hershberg). From i-androsten-6β-ol-17-one (p. 2644 s) on treatment with dil. aqueous alcoholic HCl at room temperature (1948 Dingemanse; cf. 1948 Barton). From cholesteryl chloride hydrochloride (p. 1469 s) on oxidation with CrO_3 in strong acetic acid at 40°, followed by refluxing with alcoholic NaOH (1936 CIBA). For preparation from cholesteryl chloride dibromide (p. 1475 s) by CrO_3 oxidation (cf. p. 128), see also 1936 MERCK & Co.

A not characterized *3-chloro-Δ^5-androsten-17-one* was obtained from the oxime of a 3-chloro-$\Delta^{5.16}$-pregnadien-20-one (p. 2221 s) by treatment with p-toluenesulphonyl chloride in cold pyridine, followed by hydrolysis with dil. H_2SO_4 at room temperature (1941 PARKE, DAVIS & Co.).

Rns. Gives Δ^5-androsten-17β-ol (p. 1511 s) on reduction with sodium in abs. alcohol (1939 Kuwada). Hydrogenation in ether in the presence of a platinum catalyst affords 3β-chloro-androstan-17-ol (p. 1875 s) (1937 MERCK & Co.). On nitration with HNO_3 (d 1.5) in glacial acetic acid at room temperature it gives 3β-chloro-6-nitro-Δ^5-androsten-17-one (p. 2507 s) (1946 Blunschy). Gives organo-metallic compounds by the action of dispersed metals (sodium, potassium, lithium, magnesium) in ether, benzene or other hydrocarbons, or liquid ammonia, in an inert atmosphere, preferably with application of sound or ultrasonic waves; these compounds absorb CO_2 to form Δ^5-androsten-

References, pp. 2477 s, 2478 s

17-one-3-carboxylic acid (1938 SCHERING A.-G.). For conversion into the benzoate of 5-dehydroepiandrosterone (p. 2609 s) by heating with sodium benzoate, see also 1936 CIBA.

Semicarbazone $C_{20}H_{30}ON_3Cl$ (p. 128). Crystals, m. 278.5–280.5° cor. dec. (on rapid heating), 271–3° cor. (on slow heating). Sparingly soluble in boiling alcohol (1941 Wolfe).

3β-Bromo-Δ⁵-androsten-17-one $C_{19}H_{27}OBr$. For β-configuration at C-3, see the analogous formation of the above chloro-isomer. — Crystals, m. 174°. — **Fmn.** From i-androsten-6β-ol-17-one acetate (p. 2644 s) on treatment with 48% HBr in glacial acetic acid; quantitative yield. — **Rns.** Consumes 1 mol. of oxygen on titration with perbenzoic acid. Gives the acetate of Δ⁵-androsten-3β-ol-17-one (5-dehydroepiandrosterone, p. 2609 s) on refluxing with silver acetate in glacial acetic acid for 20 min. (1942 Butenandt).

3α-Chloroandrostan-17-one $C_{19}H_{29}OCl$ (p. 128: *3β-Chloroætioallocholan-17-one, 3β-Chloroandrostan-17-one*). For reversal of the 3β-configuration formerly ascribed, and for occurrence of Walden inversion in the formation from epiandrosterone with PCl₅ and on acetolysis, see 1946a Shoppee; cf. 1937 Bergmann.

Prisms (methanol), m. 128° cor.; [α]$_D^{22}$ +94° (chloroform) (1946a Shoppee). — **Fmn.** From androstan-3α-ol-17-one (androsterone, p. 2629 s) on treatment in dry ether with excess thionyl chloride at 5° in the presence of dry CaCO₃ (yield, 5%), along with a compound m. ca. 200° (main product), very probably androsterone sulphite. From androstan-3β-ol-17-one (epiandrosterone, p. 2633 s) on treatment in dry chloroform with 1 mol. PCl₅ at 0° in the presence of dry CaCO₃; yield, 67% (1946a Shoppee). — **Rns.** On heating with a 20% solution of fused potassium acetate in glacial acetic acid in a sealed tube at 180° it gives epiandrosterone acetate and Δ² ᵒʳ ³-androsten-17-one (p. 2401 s) (1946a Shoppee).

3β-Chloroandrostan-17-one $C_{19}H_{29}OCl$ (I, R = Cl) (p. 128: *3α-Chloroætioallocholan-17-one, 3α-Chloroandrostan-17-one*). For reversal of the 3α-configuration formerly ascribed, and for occurrence of Walden inversion in the formation from androsterone with PCl₅ and on acetolysis, see 1946a Shoppee; cf. 1937 Bergmann.

Needles (alcohol), m. 172–3° cor.; [α]$_D^{20}$ +92° (chloroform) (1946a Shoppee). Gives a pale yellow solution with conc. H₂SO₄ in acetic anhydride (1934 Butenandt). — **Fmn.** From 3β-chloroandrostan-17-ol (p. 1875 s) on oxidation with CrO₃ in glacial acetic acid (1937 MERCK & Co.). From androstan-3α-ol-17-one (androsterone, p. 2629 s) by the action of 1 mol. PCl₅ in chloroform at 0° in the presence of CaCO₃; yield, 70% (1940 Hirschmann; 1946a

Shoppee). From 3β-chlorocholestane (p. 1456 s) on oxidation with CrO_3 in strong acetic acid at 90°; yield, 13% (isolated as semicarbazone) (1935 PARKE, DAVIS & Co.).

Rns. Gives androsterone (p. 2629 s) on hydrolysis with aqueous NaOH (1935 PARKE, DAVIS & Co.). Is dehydrochlorinated to give $\Delta^{2 \, or \, 3}$-androsten-17-one (p. 2401 s) by refluxing with quinoline (1937 Marker) or with sodium iodide and pyridine (1940 Hirschmann). Gives organo-metallic compounds by the action of dispersed metals (sodium, potassium, lithium, magnesium) in ether, benzene or other hydrocarbons, or liquid ammonia, in an inert atmosphere, preferably with application of sound or ultrasonic waves; these compounds absorb CO_2 to give epimeric androstan-17-one-3-carboxylic acids (1938 SCHERING A.-G.).

3β - Bromoandrostan - 17 - one $C_{19}H_{29}OBr$ (I, p. 2476 s; R = Br) (p. 128: *3α-Bromoœtioallocholan-17-one*, *3α-Bromoandrostan-17-one*). For reversal of the 3α-configuration previously ascribed, see the parent "α"-cholestyl bromide, p. 1457 s.

5α-Chloroandrostan-17-one $C_{19}H_{29}OCl$. Colourless crystals; volatile with difficulty (1939 CIBA). — **Fmn.** Along with acids and other ketones, from 5-chlorocholestane (p. 1460 s) on oxidation with CrO_3 in strong acetic acid at room temperature; is separated from the other ketones with the aid of Girard's reagent P (1939 CIBA). — **Rns.** Gives 17α-methyl-Δ^4-androsten-17β-ol (p. 1517 s) on treatment with methylmagnesium iodide in ether, followed by boiling with 5% methanolic KOH (1942 CIBA PHARMACEUTICAL PRODUCTS).

4β-Bromopregnan-3-one $C_{21}H_{33}OBr$ (II, R = Br). For β-configuration at C-4, assigned by Editor, see that of 4β-bromocoprostan-3-one (p. 2483 s). —Needles (aqueous alcohol), m. 137°. — **Fmn.** From pregnan-3-one (p. 2411 s) on treatment with 1 mol. bromine and a little 48% HBr in glacial acetic acid with cooling; yield, 70%. — **Rns.** On refluxing with dry pyridine for 8 hrs. it gives Δ^4-pregnen-3-one (p. 2410 s) and the following pyridinium compound (1939 Marker).

N-(3-Ketopregnan-4β-yl)-pyridinium bromide $C_{26}H_{38}ONBr$ [II, R = N(Br)C_5H_5]*. Plates (dil. alcohol), m. 235°. Dissociates in aqueous solution and gives a yellow coloration with aqueous NaOH. — **Fmn.** From the preceding compound on refluxing with dry pyridine for 8 hrs.; yield, 33% (1939 Marker).

1934 Butenandt, Dannenbaum, *Z. physiol. Ch.* **229** 192, 194, 197, 202, 203.
1935 Butenandt, Dannenbaum, Hanisch, Kudszus, *Z. physiol. Ch.* **237** 57, 62, 63.

* For alternative formulation as a γ-substituted pyridine derivative, as suggested for similar pyridinium compounds, see 1940 Inhoffen.

PARKE, DAVIS & Co. (Marker), *U.S. Pat.* 2144726 (issued 1939); *C. A.* **1939** 2909; *Ch. Ztbl.*
1939 II 171.

SCHERING-KAHLBAUM A.-G., *Brit. Pat.* 464396 (issued 1937); *C. A.* **1937** 6418; *Ch. Ztbl.*
1937 II 3041; *Swiss Pat.* 202972 (1936, issued 1939); *Ch. Ztbl.* **1940** I 429.

1936 CIBA*, *Fr. Pat.* 804229 (issued 1936); *C. A.* **1937** 2613; *Ch. Ztbl.* **1937** I 3674; *Indian Pat.* 22679
(issued 1936); *Ch. Ztbl.* **1937** I 2819.

MERCK & Co. (Weijlard), *U.S. Pat.* 2176113 (issued 1939); *C. A.* **1940** 1038; *Ch. Ztbl.* **1945** I
825.

1937 Bergmann, *Helv.* **20** 590, 600.

a Kuwada, Miyasaka, *J. Pharm. Soc. Japan* **57** 232 (German abstract); *C. A.* **1938** 1275;
Ch. Ztbl. **1938** II 1610.

b Kuwada, Miyasaka, Yosiki, *J. Pharm. Soc. Japan* **57** 234 (German abstract); *C. A.* **1938**
1275; *Ch. Ztbl.* **1938** II 1611.

Marker, Kamm, Jones, Mixon, *J. Am. Ch. Soc.* **59** 1363.

MERCK & Co. (Weijlard), *U.S. Pat.* 2131082 (issued 1938); *C. A.* **1938** 9405; *Ch. Ztbl.* **1939** I
3771.

1938 Bergmann, *J. Am. Ch. Soc.* **60** 1997.

SCHERING A.-G., *Fr. Pat.* 844222 (issued 1939); *C. A.* **1940** 7545; *Ch. Ztbl.* **1940** I 1231.

1939 CIBA*, *Swiss Pat.* 235911 (issued 1945); *C. A.* **1949** 7056; *Fr. Pat.* 886410 (1940, issued
1943); *Ch. Ztbl.* **1944** I 448; (Miescher, Wettstein), *Swed. Pat.* 103744 (1940, issued 1942);
Ch. Ztbl. **1942** II 2294; CIBA PHARMACEUTICAL PRODUCTS (Miescher, Wettstein), *U.S. Pat.*
2319012 (1940, issued 1943); *C. A.* **1943** 6096.

Dannenberg, *Abhandl. Preuss. Akad. Wiss. Math.-naturwiss. Klasse* **1939** No. 21, pp. 10, 51.

Kuwada, Tutihasi, *C. A.* **1939** 8209; *Ch. Ztbl.* **1939** II 2433.

Marker, Lawson, *J. Am. Ch. Soc.* **61** 586.

Westphal, Wang, Hellmann, *Ber.* **72** 1233.

1940 CIBA PHARMACEUTICAL PRODUCTS (Miescher, Wettstein), *U.S. Pat.* 2311067 (issued 1943);
C. A. **1943** 4534.

Hirschmann, *J. Biol. Ch.* **136** 483, 485, 495.

Inhoffen, Zühlsdorff, Huang-Minlon, *Ber.* **73** 451, 452.

1941 PARKE, DAVIS & Co. (Tendick, Lawson), *U.S. Pat.* 2335616 (issued 1943); *C. A.* **1944** 3095.

Rosenheim, *Nature* **147** 776.

Wolfe, Fieser, Friedgood, *J. Am. Ch. Soc.* **63** 582.

1942 Butenandt, Surányi, *Ber.* **75** 591.

CIBA PHARMACEUTICAL PRODUCTS (Miescher, Wettstein), *U.S. Pat.* 2374370 (issued 1945);
C. A. **1945** 5412.

Miller, *C. A.* **1942** 6222.

Selye, *Endocrinology* **30** 437, 441.

1945 Mason, Kepler, *J. Biol. Ch.* **161** 235, 247.

Wilson, Nathanson, *Endocrinology* **37** 208, 211.

1946 Blunschy, Hardegger, Simon, *Helv.* **29** 199, 201.

Furchgott, Rosenkrantz, Shorr, *J. Biol. Ch.* **163** 375, 378.

a Shoppee, *J. Ch. Soc.* **1946** 1138, 1142 et seq., 1146, 1147.

b Shoppee, *J. Ch. Soc.* **1946** 1147, 1148.

1948 Barton, Klyne, *Nature* **162** 493.

Dingemanse, Huis in 't Veld, Hartogh-Katz, *Nature* **161** 848; **162** 492.

1949 Fieser, Fieser, *Natural Products Related to Phenanthrene*, 3rd Ed., New York, p. 494.

1950 Hershberg, Rubin, Schwenk, *J. Org. Ch.* **15** 292, 295.

---

* Gesellschaft für chemische Industrie in Basel; Société pour l'industrie chimique à Bâle.

b. MONOHALOGENO-MONOKETO-STEROIDS C_{27}–C_{29} (CO AND HALOGEN IN THE RING SYSTEM)

Summary of Monohalogeno-monoketones with Cholestane (or Coprostane) Skeleton

(including some data from papers published after the closing date for this volume)

Position of CO	Position of halogen	Config. at C-5	Position of double bond(s)	M.p. °C.	$[\alpha]_D(°)$ in chloroform	Absorpt. max. (mμ) in alcohol	Page
3	2 Br	—	$\Delta^{1.4.6}$	148–9 dec.	−16	224, 271, 308	2479 s
3	2 Br	α	Δ^1	92	+37	256	1)
3	2 Br	β	Δ^1	133.5–135.5	+53	256²)	3)
3	2α Br	—	Δ^4	{ 117–9, 132–4	{ +81, +94	243	2480 s
3	2α Br	α	satd.	170	+41	282	2480 s
3	4 Br	—	Δ^4	113–4	+110	261–2	2482 s
3	4β Br	β	satd.	110–1	+41	—	2483 s
3	5α Cl	α	satd.	135	—	—	2483 s
3	5α Br	α	satd.	110–4 dec.	—	—	2483 s
3	6α Cl	—	Δ^4	125–6	+59	239	4)
3	6β Cl	—	Δ^4	129–130 dec.	+14	241	2484 s
3	6α Br	—	Δ^4	113	+53	238	5)
3	6β Br	—	Δ^4	132	+6	244⁶)	2484 s
6	3α Cl	α	satd.	182	+8	—	2484 s
6	3β Cl	α	satd.	130	−0.6	—	2486 s
6	3α Br	α	satd.	173	+10	—	7)
6	3β Br	α	satd.	123	+3	—	2488 s
7	3β Cl	—	Δ^5	122; 145	—	270²),8)	2488 s
7	3β Cl	α	satd.	139	—	—	2489 s
7	6 Br	—	$\Delta^{3.5}$	117	—	279, 344	2489 s

1) 1947 Djerassi. — 2) Solvent not indicated. — 3) 1951 Djerassi. — 4) 1950a Barton. — 5) 1950b Barton. — 6) In ether. — 7) 1952b Shoppee. — 8) See footnote * on p. 2488 s.

2-Bromo-$\Delta^{1.4.6}$-cholestatrien-3-one $C_{27}H_{39}OBr$.

For structure, see 1955 Fieser; cf. 1950 Djerassi; the (impure) compound originally obtained by Barkow (1940) was then designated *4-bromo-$\Delta^{4.6.8}$-cholestatrien-3-one.* — Colourless needles (acetone), m. 160° (1940 Barkow); crystals (petroleum ether), m. 148–9° dec.; $[\alpha]_D$ −16° (chloroform); absorption max. in alcohol at 224, 271, and 308 mμ (ε 14780, 12000, and 9750, resp.) (1955 Fieser; cf. 1939 Dannenberg; 1940 Barkow).

Fmn. From 2.6-dibromo-$\Delta^{1.4.6}$-cholestatrien-3-one (p. 2497 s) in good yield, by hydrogenation in boiling amyl alcohol containing KOH (1 equivalent), using a palladium-BaSO$_4$ catalyst. From 2.2.6β-tribromo-Δ^4-cholesten-3-one (p. 2503 s) by heating (in bomb) with glacial acetic acid and potassium acetate at 230° for 7 hrs.; yield, 13.5 % (1940 Barkow).

Rns. Perhydrogenation, using platinum oxide catalyst in glacial acetic acid at room temperature, gives coprostane (p. 1431 s); partial hydrogenation, using the same catalyst in methanol containing potassium acetate, yields cholestanone (p. 2436 s) or, by interruption after uptake of 2 mols. of hydrogen, $\Delta^{4\cdot6}$-cholestadien-3-one (p. 2420 s); hydrogenation in hot alcohol containing KOH and using a palladium catalyst affords coprostanone (p. 2443 s) (1940 Barkow). Heating at 150° with o-phenylenediamine gives 13% yield of the **quinoxaline-compound** $C_{33}H_{44}N_2$ [crystals (ethyl acetate), m. 144°] (1940 Barkow) to which the revised structure (I) may be ascribed on the basis of that now accepted for the parent bromoketone (Editor).

2α-Bromo-Δ^4-cholesten-3-one, 2α-Bromocoprostenone $C_{27}H_{43}OBr$. For configuration at C-2, see 1955 Fieser; 1956 Ellis. — Colourless needles (ethanol), m. 117–9°, $[\alpha]_D^{29} + 81°$ (chloroform); absorption max. in alcohol at 243 mμ (log ε 4.15) (1949 Djerassi); needles (methanol), m. 132–4°, $[\alpha]_D^{19} + 94°$ (chloroform), absorption max. in isopropanol at 245 mμ (log ε 4.13) (1956 Ellis). — **Fmn.** From 2α.4α-dibromo-cholestan-3-one (p. 2495 s) on very short (40 sec.) refluxing with collidine; yield, 34% (1949 Djerassi). — **Rns.** Further refluxing with collidine yields $\Delta^{1\cdot4}$-cholestadien-3-one (1949 Djerassi).

The **α.β - unsaturated monobromoketone** $C_{27}H_{43}OBr$, needles (alcohol), m. 123°, obtained on dehydrochlorination of 2-bromo-5-chlorocholestan-3-one (p. 2497 s) (formerly: 5-chloro-4-bromocholestanone, p. 129) by refluxing with potassium acetate in benzene-alcohol for 30 min. (1937 SCHERING-KAHLBAUM A.-G.; cf. 1936 Butenandt), should now be formulated as 2-bromo-Δ^4-cholesten-3-one and may be identical with the 2α-epimer (above) (Editor). — **Rns.** Treatment with potassium acetate in butanol yields cholestane-3.4-dione (p. 2904 s) and cholestane - 3.6 - dione (p. 2907 s) (1937 SCHERING - KAHL-BAUM A.-G.).

2α-Bromocholestan-3-one $C_{27}H_{45}OBr$ (p. 128). For confirmation of structure, see 1938 Ruzicka; for identity with *2-bromozymostanone*, see 1940 Heath-Brown. For revised configuration at C-2, see 1952 Jones; 1953 a, b Corey; 1953 c Fieser; the opposite configuration had previously been assigned (1947 Djerassi; 1953 a Fieser).

Needles (alcohol-chloroform, or aqueous alcohol), m. 167.5–168° cor. (1947 Djerassi; cf. 1940 Heath-Brown); crystals (acetone-methanol or -ethanol), m. 171.5° (1938 Ralls; cf. 1932 Jakubowicz; 1953 b Corey). Rotation in chloroform: $[\alpha]_D + 41°$ (1947 Djerassi), $[\alpha]_{578}^{12} + 47°$, $[\alpha]_{546}^{12} + 55°$, $[\alpha]_{436}^{12} + 108°$ (1932 Jakubowicz). Absorption max. in alcohol at 282 mμ (log ε 1.59) (1954 Cookson).

Fmn. From cholestanone (p. 2436 s) by treatment with bromine at room temperature in chloroform (1932 Jakubowicz) or in glacial acetic acid containing small amounts of HBr (1940 Heath-Brown). From the same ketone by treatment with excess iodine monobromide in glacial acetic acid - carbon tetrachloride (yield, 60%); the mechanism of reaction has been investigated (1938 Ralls).

Rns. Oxidation does not occur with persulphuric acid in glacial acetic acid (cf. reactions of cholestanone, p. 2437 s) (1940b Marker). Hydrogenation in dibutyl ether, using a platinum black catalyst, gives cholestan-3-one (interruption after 1 mol. of hydrogen absorbed) or cholestan-3α-ol (p. 1702 s) (after 2 mols. absorbed) (1932 Jakubowicz). On boiling with aluminium isopropoxide in isopropanol, bromine is completely removed with formation of a gummy product (1938, 1940 Stevens). Reaction with bromine in glacial acetic acid (cf. p. 128) yields primarily 2.2-dibromocholestan-3-one (p. 2494 s) which subsequently isomerizes to the 2.4-isomer (p. 2495 s), the isomerization being favoured by HBr or heat, retarded by H_2O, sodium acetate, or H_2O_2; bromination by means of N-bromosuccinimide in carbon tetrachloride (with exposure to strong light) also yields the 2.2-dibromo-compound (1947 Djerassi). For the catalytic effect of 48% HBr, on bromination by means of iodine monobromide to give 2.2- and 2.4-dibromocholestan-3-ones, see 1938 Ralls. Treatment with conc. H_2SO_4 and acetic anhydride at 20⁰ gives the enol acetate, 2-bromo-$\Delta^{2(or\ 3)}$-cholestenyl acetate (p. 1876 s) (1937 Windaus). By treatment with 0.01 N KOH in abs. alcohol at room temperature, KBr is instantly eliminated, affording an oily product (1932 Jakubowicz). Reaction with a 6% solution of $AgNO_3$ in pyridine at room temperature yielded on one occasion a **compound** $C_{54}H_{90}O_2$(?) [prismatic needles (pyridine), m. 206–7⁰, no absorption in the near ultra-violet, easily soluble in chloroform, sparingly in other usual solvents] for which the structure *bi-(3-ketocholestan-2-yl)* is suggested; attempted repetition of this reaction was unsuccessful (1938 Wolff). For the velocity of reaction with 0.025 N CaI_2 in butanol at 100⁰ and with 0.01 N potassium formate in abs. alcohol at 69⁰, see 1932 Jakubowicz. Boiling with anhydrous sodium acetate in glacial acetic acid gives "2-acetoxy-cholestan-3-one" (1938 Ruzicka); later investigation of this reaction, carried out in an atmosphere of nitrogen, showed that this product is an addition compound of 2α- and 4α-acetoxy-cholestan-3-ones (p. 2665 s) (1953b Fieser). The reaction with potassium acetate in glacial acetic acid at 200–210⁰ (p. 128) has been shown to give Δ^5-cholesten-4-one (p. 2450 s) (1944 Butenandt); for this reaction a revised mechanism is proposed (1953b Fieser). Refluxing with potassium benzoate in toluene-butanol gives the benzoate of cholestan-2α-ol-3-one (m. 198⁰) (p. 2666 s) together with an isomer (m. 145–6⁰)* (1938 Inhoffen); the mixed reaction product obtained by treatment with potassium benzoate in toluene-butanol (or by using potassium acetate) undergoes aromatization by heating at 320–330⁰ in an atmosphere of CO_2, to give a phenolic substance which, after chromic acid oxidation, yields a product with positive Allen-Doisy reaction (1937a, b SCHERING A.-G.). Refluxing with 2.4-dimethyl-

---

* Cf. the analogous reaction with sodium acetate (above) giving the acetates of cholestan-2α- and 4α-ol-3-ones.

pyridine gives Δ^1-cholesten-3-one (p. 2423 s) (1940 Inhoffen; cf. 1939 SCHERING A.-G.); for reaction with 2.6-dimethylpyridine or with pyridine, see reaction products, below. The reaction with boiling collidine (cf. p. 128) is shown to be complex: incomplete after 6 hrs., and after 12 hrs. a mixture of cholestan-3-one with impure Δ^1-cholesten-3-one results (1940 Jacobsen). For conversion into Δ^4-cholesten-3-one (p. 2424 s) by use of quinoline, aniline, dimethylaniline, aminoethanol, or other organic bases, see 1937 CIBA.

N-(3-Ketocholestan-2-yl)-pyridinium bromide $C_{32}H_{50}ONBr$ (II,$R_1 = C_5H_5$, $R_2 = Br$). For alternative possible structures, see 1940 Inhoffen. — Needles (alcohol) (1937 CIBA); white crystals (alcohol or glacial acetic acid), m. 310° cor. dec.; insoluble in H_2O and in petroleum ether, sparingly soluble in alcohol or glacial acetic acid, readily in chloroform. Turns yellow in contact with NaOH; immediately gives a heavy precipitate with alcoholic $AgNO_3$ (1938 Ruzicka). — **Fmn.** From the bromo-ketone on brief refluxing with pyridine; yield, 66 % (1938 Ruzicka; cf. 1935 Butenandt; 1937 CIBA*). — **Rns.** Distillation at 250–300° under 10 mm. pressure gives Δ^4-cholesten-3-one together with smaller amounts of higher melting unidentified products (1938 Ruzicka; cf. 1937 CIBA*). Treatment with NaOH and p-nitrosodimethylaniline in chloroform-alcohol gives the 2-(p-dimethylamino-phenyl)-isoxime of cholestane-2.3-dione (page 2902 s) (1944 Ruzicka).

N-(3-Ketocholestan-2-yl)-2.6-dimethylpyridinium bromide $C_{34}H_{54}ONBr$ (II, $R_1 = C_7H_9$, $R_2 = Br$). Colourless leaflets (chloroform-acetone), m. 299–300°. — **Fmn.** From the parent bromoketone by refluxing with 2.6-dimethylpyridine. — **Rns.** Undergoes thermal decomposition with elimination of dimethylpyridinium bromide (1940 Inhoffen).

"4-Bromo-$\Delta^{4.6.8}$-cholestatrien-3-one" $C_{27}H_{39}OBr$. See 2-bromo-$\Delta^{1.4.6}$-cholestatrien-3-one, p. 2479 s.

4-Bromo-Δ^4-cholesten-3-one, 4-Bromocoprostenone $C_{27}H_{43}OBr$. The product m. 147° (below) to which this constitution was first assigned (1940 Barkow; cf. 1939 Dannenberg) was impure and its constitution cannot yet be assigned (1950 Djerassi)**; this is confirmed by the isolation (performed after the closing date for this volume), using a different method, of a pure compound

---

   * The analogous *N-(3-ketocholestan-2-yl)-pyridinium chloride* $C_{32}H_{50}ONCl$ (II, $R_1 = C_5H_5$, $R_2 = Cl$), obtained similarly from **2-chlorocholestan-3-one** $C_{27}H_{45}OCl$ (for which no information is available), is said to undergo similar thermal decomposition (1937 CIBA).

   ** In the light of recent structural formulations for its precursors, and in view of the absorption max. quoted, this product [analysis gave 19.7 % Br, requires 17.3 % (1940 Barkow)] might well be impure *2-bromo-$\Delta^{1.4}$-cholestadien-3-one* (Editor).

with this structure (1956 Kirk). — Needles (alcohol), m. 113–4°; $[\alpha]_D^2$ +110° (chloroform), absorption max. in alcohol at 261–2 mμ (log ε 4.08) (1956 Kirk).

Barkow's "4-Bromocholestenone $C_{27}H_{43}OBr$" *. Crystals (glacial acetic acid), m. 147° (1940 Barkow); absorption max. in ether at ca. 250 mμ (ε 15200) (1939 Dannenberg; cf. 1940 Barkow). — **Fmn.** From 2.6-dibromo-$\Delta^{1.4.6}$-cholestatrien-3-one (p. 2497 s) by hydrogenation (8 hrs.) in boiling amyl alcohol in absence of alkali, using a palladium-$BaSO_4$ catalyst; yield not stated, but 75 % of starting material was recovered. From 2.2.6β-tribromo-Δ^4-cholesten-3-one (p. 2503 s) by heating in glacial acetic acid saturated with potassium acetate, at 230° (bomb) (1940 Barkow). — **Rns.** Boiling with zinc in glacial acetic acid or heating with potassium acetate and glacial acetic acid in a bomb does not effect debromination (1940 Barkow).

4β-Bromocoprostan-3-one $C_{27}H_{45}OBr$ (p. 129). For configuration at C-4, see 1952 Jones; 1953a Corey. — Rotation in chloroform, $[\alpha]_D^{23}$ +40.5° (1948 Djerassi). — **Rns.** Is not attacked by persulphuric acid in glacial acetic acid (1940b Marker). Reaction with bromine (1 mol.) affords 2β.4β-dibromocoprostan-3-one (p. 2497 s) (1938 Wolff; cf. 1936 Butenandt). Boiling with fused potassium acetate in glacial acetic acid gives the acetate (m. 149°) of coprostan-4(?)-ol-3-one ** (p. 2670 s) (1939b Marker).

5-Chlorocholestan-3-one $C_{27}H_{45}OCl$ (III, R = Cl) (p. 129). Configuration at C-5 assigned on the basis of that of its precursor, 5-chlorocholestan-3β-ol (page 1877 s) (Editor). — The m.p.'s quoted on p. 129 refer to crystals (m. 102°) containing, and crystals (m. 135°) without, solvent of crystallization (1936 Butenandt). — **Rns.** The bromination product (cf. p. 129) is 2-bromo-5-chlorocholestan-3-one (p. 2497 s) (1955 Fieser).

5-Bromocholestan-3-one *** $C_{27}H_{45}OBr$ (III, R = Br). M. p. 110–4° dec. — **Fmn.** From cholesterol hydrobromide (p. 1879 s) by oxidation with CrO_3 in strong acetic acid. — **Rns.** Heating with potassium acetate in alcohol affords Δ^4-cholesten-3-one (p. 2424 s) (1943 Urushibara).

---

* See footnote** on p. 2482 s.

** Position of OH questionable (1954 Georg).

*** Erroneously designated as cholestenone bromide (instead of cholestenone hydrobromide) in *C. A.* (Editor).

References, pp. 2491 s, 2492 s

6β-Chloro-Δ⁴-cholesten-3-one, 6β-Chlorocoprostenone $C_{27}H_{43}OCl$ (IV, R = Cl).

For configuration at C-6, see 1950 Rivett; 1950a Barton. — Needles (ethyl acetate-methanol), m. 129–130⁰ dec.; $[\alpha]_D^{ca.20} + 14^0$ (chloroform); absorption max. in alcohol at 241 mμ (ε 15 100) (1950a Barton; cf. 1950 Rivett). — **Fmn.** From 5α.6β-dichlorocholestan-3-one (p. 2500 s) by treatment with potassium acetate in boiling aqueous benzene (1936 CIBA; cf. 1950a Barton; 1950 Rivett). — **Rns.** On refluxing with anhydrous potassium acetate in glacial acetic acid it yields the acetate of Δ⁴-cholesten-2α-ol-3-one (p. 2663 s) (1950 Rivett; cf. 1953b Fieser).

6β-Bromo-Δ⁴-cholesten-3-one, 6β-Bromocoprostenone $C_{27}H_{43}OBr$ (IV, R = Br) (p. 129). For configuration at C-6, see 1950b Barton. — Crystals (strong acetic acid), m. 122–3⁰ (1938b C. F. BOEHRINGER & SÖHNE); crystals (petroleum ether), m. 132⁰; $[\alpha]_D^{ca.25} + 6^0$ (chloroform) (1950b Barton; cf. 1946 Reich); absorption max. in ether at 244 mμ (ε 13700) (1939 Dannenberg; cf. 1936 Inhoffen). — **Fmn.** From the acetate of Δ³·⁵-cholestadien-3-ol (p. 1542 s) on standing with N-bromoacetamide in aqueous butanol (crude yield, almost quantitative) or in aqueous acetone (1946 Reich). From coprostenone (page 2424 s) by treatment in glacial acetic acid at room temperature with bromine (ca. 1.15 mols.) in the presence of acetic anhydride (1938b C. F. BOEHRINGER & SÖHNE). From Δ⁵-cholestenone dibromide (p. 2500 s) by refluxing with pyridine in benzene (1936 CIBA) or by treatment with potassium acetate in benzene-acetic anhydride on the water-bath (1937 Krekeler).

Rns. Oxidation with CrO_3 in glacial acetic acid - carbon tetrachloride at room temperature, followed by debromination using zinc in glacial acetic acid (alternatively with KI or by catalytic hydrogenation), gives Δ⁴-pregnene-3.20-dione (progesterone, p. 2782 s) together with smaller amounts of Δ⁴-androstene-3.17-dione (p. 2880 s) (1938a C. F. BOEHRINGER & SÖHNE). Hydrolysis with methanolic KOH at room temperature yields Δ⁴-cholesten-6β-ol-3-one (p. 2672 s) (1937 Dane). Refluxing with anhydrous potassium acetate in glacial acetic acid gives the acetate of Δ⁴-cholesten-2α-ol-3-one (p. 2663 s) (1950 Rivett; cf. 1953b Fieser). Heating with potassium acetate in acetic anhydride at 200⁰ (bomb) affords acidic and neutral products, and by distillation of the latter in high vacuo, 6-dehydrocoprostenone (p. 2420 s) is obtained (1937 Krekeler).

3α-Chlorocholestan-6-one $C_{27}H_{45}OCl$ (p. 129: β-Chlorocholestan-6-one). For

reversal of configuration at C-3, see 1948 Dodson; 1948a Shoppee; cf. 1939 Ladenburg. — M.p. 181.5–182.5⁰ cor.; $[\alpha]_D^{24} + 7.7^0$ (chloroform) (1948 Dodson). Liebermann-Burchard reaction: violet coloration with absorption band at 550–600 mμ (1938 Petrow). Decomposed on long standing in conc. H_2SO_4, the solution becoming fluorescent (1904 Windaus).

References, pp. 2491 s, 2492 s

Rns. Treatment with bromine in chloroform yields a **compound** $C_{27}H_{44}OClBr$ [plates or prisms (acetone-methanol), m. 116–7°, almost insoluble in methanol or ethanol, sparingly soluble in glacial acetic acid, readily in benzene or acetone] (1904 Windaus), for which the structure **3α-chloro-5(or 7?)-bromo-cholestan-6-one** seems highly probable (Editor; cf. 1937 Heilbron). Boiling with quinoline, in an atmosphere of nitrogen, yields Δ^2-cholesten-6-one (page 2452 s), but the yield is lower than that from the 3β-epimer (p. 2486 s) (1946 Blunschy).

3α-Chlorocholestane-6‖7-dioic acid $C_{27}H_{45}O_4Cl$ [p. 129: *β-chloro-dicarboxylic acid* (III); cf. p. 42]. For revised configuration at C-3, see 1939 Ladenburg; 1948a Shoppee. — Needle rosettes (glacial acetic acid), m. 243°; insoluble in water or petroleum ether, sparingly soluble in ether, chloroform, acetone, glacial acetic acid, or alcohol, even on heating; very sparingly soluble in benzene, the solution in which shows a blue fluorescence (1904 Windaus). — **Fmn.** From 3α-chlorocholestan-6-one (p. 2484 s) by heating with glacial acetic acid and fuming HNO_3 at 60–70°; yield, ca. 60 % (1904 Windaus). — **Rns.** Reduction by boiling with zinc and HCl in glacial acetic acid gives 74 % yield of a **hydroxy-acid** $C_{27}H_{48}O_3$ (V) or (VI) [white needles (acetone or dil. acetic acid), m. 212°, readily soluble in ether, less soluble in acetone, alcohol, or glacial acetic acid, almost insoluble in petroleum ether; does not decolorize $KMnO_4$ solution] together with its *lactone* $C_{27}H_{46}O_2$ [colourless needles (petroleum ether or acetic anhydride), m. 118°, readily soluble in ether, petroleum ether, glacial acetic acid, or acetic anhydride; can also be obtained from the parent hydroxy-acid (above) by vacuum distillation at 300–320° or on boiling with acetic anhydride; reconverted into the hydroxy-acid by heating with aqueous methanolic KOH on the water-bath, followed by acidification] (1921 Windaus). For conversion into 3β-hydroxycholestane-6‖7-dioic acid, see below. Refluxing with acetic anhydride yields a compound, m. 187° (1904 Windaus) or 191° (1921 Windaus), considered to be the *anhydride* $C_{27}H_{43}O_3Cl$, but recently shown (1954 Gut) to be 3α-chloro-B-nor-coprostan-6-one $C_{26}H_{43}OCl$ (VII).

Page 42: For "β · $C_{27}H_{45}O_4Cl$; *anhydride* $C_{27}H_{43}O_3Cl$, m. 142°" *read* "β-$C_{27}H_{45}O_4Cl$, m. 243°".

3β-Hydroxycholestane-6‖7-dioic acid $C_{27}H_{46}O_5$ (p. 42). For confirmation of configuration at C-3, previously assigned arbitrarily, see 1948a Shoppee; cf. 1935 Lettré. — *Monohydrate:* Plates (aqueous acetone), m. 239–240° on rapid heating (1904 Windaus); on slow heating dehydration occurs at 130–140° to give, after

recrystallization, the *anhydrous form:* needles (acetone-pentane), m. 241–2⁰ cor.
dec.* (1948a Shoppee; cf. 1904 Windaus). Readily soluble in acetone, methanol, and ethanol, sparingly in benzene or chloroform, insoluble in petroleum
ether (1904 Windaus). — **Fmn.** From 3α-chlorocholestane-6‖7-dioic acid
(p. 2485 s) by heating on the water-bath (1 hr. only) with 10% aqueous
KOH (1904 Windaus); this reaction involves Walden inversion (1948a
Shoppee); the acid $C_{27}H_{44}O_4 \cdot H_2O$ [m. 230–1⁰, $[\alpha]_D^{24} + 55^0$ (acetone)], obtained
from the same chloro-dioic acid by heating with sodium ethoxide and alcohol
for 1 hr. in a sealed tube at 120⁰ and considered to be "*β-i-cholestene-6‖7-dioic
acid*" (a stereoisomer of i-cholestene-6‖7-dioic acid, p. 2453 s) (1939 Ladenburg), has been shown to be the present hydroxy-dioic acid in an imperfectly
dried state (1955 Gates). — **Rns.** For oxidation to 3-ketocholestane-6‖7-dioic
acid and for conversion into the 6 → 3-lactone, see below.

A crystalline *magnesium salt* (readily soluble in alcohol, insoluble in water),
a crystalline *barium salt*, and a crystalline *ethyl ester* have been prepared
(1904 Windaus), for which no constants are available.

6 → 3 Lactone $C_{27}H_{44}O_4$ (p. 42). For confirmation of the γ-lactone structure
assigned by Lettré (1935), see 1948a Shoppee. Needles (ether-petroleum
ether) (1904 Windaus)**; needles (alcohol), m. 211–3⁰ (1935 Lettré); needles
(ether-pentane, or methanol), m. 214–5⁰ cor. (1948a Shoppee). Titrates for
a monobasic acid with 0.01 N NaOH (1935 Lettré). Fairly soluble in ether
and benzene, sparingly soluble in petroleum ether (1904 Windaus).

3-Ketocholestane-6‖7-dioic acid $C_{27}H_{44}O_5$ (p. 42). Needles (alcohol or glacial

acetic acid), m. ca. 255⁰ on rapid heating (1904, 1921
Windaus); readily soluble in methanol, ethanol, acetone, glacial acetic acid, sparingly soluble in chloroform or benzene, insoluble in petroleum ether (1904
Windaus). — **Fmn.** From 3α- and 3β-hydroxychole-
stane-6‖7-dioic acids (p. 2488 s, p. 2485 s) by oxidation with ca. 1 mol. of
CrO_3 in 75% acetic acid on the water-bath or in 83% acetic acid at 70⁰,
resp. (1904, 1921 Windaus). — **Rns.** Oxidation with excess CrO_3 in 85%
acetic acid at 75–80⁰ affords **cholestane-2‖3.6‖7-tetraoic acid** (p. 42: tetracarb-
oxylic acid $C_{27}H_{44}O_8$, m. 174⁰ dec.). Reduction with zinc amalgam in boiling
glacial acetic acid containing conc. HCl yields **cholestane-6‖7-dioic acid**
$C_{27}H_{46}O_4$, m. 273⁰ (p. 64: Dicarboxylic acid XII) (1921 Windaus). — *Oxime*
$C_{27}H_{45}O_5N$ (p. 42). Needle aggregates (alcohol), m. 213–4⁰ dec. (1904 Windaus).

3β-Chlorocholestan-6-one $C_{27}H_{45}OCl$ (p. 129: *α-Chlorocholestan-6-one*). For

reversal of configuration at C-3, see
1948 Dodson; 1948a Shoppee; cf. 1939
Ladenburg. — White needles (methanol), m. 131⁰ (1938 Heilbron); crystals
(alcohol), m. 129.5–130.5⁰ cor.; $[\alpha]_D^{24}$
—0.6⁰±0.6⁰ (chloroform) (1948 Dod-

---

* The "anhydride $C_{27}H_{44}O_4$", m. 241⁰, referred to on p. 42, is the anhydrous acid.
** Windaus erroneously considered this lactone to be the anhydride (1948a Shoppee).

References, pp. 2491 s, 2492 s

son; cf. 1938 Ford). Gives a violet coloration in the Liebermann-Burchard reaction with absorption band at 550–600 mμ (1938 Petrow). Forms a very compressible surface film on water, for which the molecular area is > 85 sq. Å at zero pressure (1935 Adam).

Fmn. From i-cholesten-6-one (p. 2452 s) by treatment with glacial acetic acid containing 3 % of HCl at room temperature; yield, 91 % (1938 Ford; cf. 1938 Heilbron).

Rns. Clemmensen reduction in boiling glacial acetic acid yields 3β-chloro-cholestane (p. 1456 s) (1935 Vanghelovici). Reduction by means of sodium amalgam in warm alcohol gives i-cholesten-6-one (p. 2452 s) (1935 Vanghe-lovici; cf. 1948a Shoppee). In contrast to the 3α-epimer, elimination of HCl can be effected by distillation in vacuo, regenerating i-cholesten-6-one (1921 Windaus); elimination of HCl to give Δ^2-cholesten-6-one occurs by refluxing with quinoline in an atmosphere of nitrogen (1946 Blunschy), though less satisfactorily than for the corresponding 3β-bromo-compound (p. 2488 s) (1939 Ladenburg), or with collidine at 250° (sealed tube) (1946 Blunschy). The product of reaction with boiling alcoholic KOH (1938 Ford; cf. p. 129) is i-cholesten-6-one (1948a Shoppee).

Oxime $C_{27}H_{46}ONCl$ (p. 129). Crystals (alcohol), m. 175° (1938 Ford).

3β-Chlorocholestane-6∥7-dioic acid $C_{27}H_{45}O_4Cl$ [p. 129: *α-chlorodicarboxylic acid* (III); cf. p. 42]. For configuration at C-3, see 1948a Shoppee. — Needles (glacial acetic acid), m. 263–4°; readily soluble in hot alcohol or glacial acetic acid, sparingly soluble in cold acetone (1921 Windaus). — **Fmn.** From 3β-chlorocholestan-6-one (above) by treatment with fuming HNO_3 in glacial acetic acid at 60–70° (1921 Windaus). — **Rns.** Not attacked by zinc and HCl in glacial acetic acid (cf. 3α-epimer, p. 2485 s) (1921 Windaus). There is no attack by 0.5 N KOH at 50°; for reaction with 25 % KOH (1921 Windaus), see 3α-hydroxychole-stane-6∥7-dioic acid (p. 2488 s). Heating with sodium ethoxide and alcohol for 8 hrs. in a sealed tube at 150° gives a **compound** $C_{27}H_{44}O_4$ [crystals (aqueous alcohol or acetic acid), m. 265°; $[\alpha]_D^{25}$ +46° (dioxane)] considered to be a con-figurational isomer of (α₁-)i-cholestene-6∥7-dioic acid (p. 2453 s) and designated "*α₂-i-cholestene-6∥7-dioic acid*" (1939 Ladenburg) which, however, on reinvesti-gation has been shown, by spectral analysis, to be an aromatic acid of unknown structure (1955 Gates); attempts to repeat the preparation of this acid failed, but resulted in the isolation of Δ^4-*cholestene-6∥7-dioic acid* $C_{27}H_{44}O_4$, crystals (aqueous methanol), m. 226–7° dec., $[\alpha]_D$ +74° (acetone), absorption max. <225 mμ (1955 Gates). Refluxing with acetic anhydride gives a compound $C_{27}H_{43}O_3Cl$ [plates (acetic anhydride), m. 124°; readily soluble in ben-zene, ether, and petroleum ether], considered to be the *anhydride* (p. 42) (1921 Windaus); cf., however, the analogous reaction of the 3α-epimer (p. 2485 s).

E.O.C. XIV s. 173

3α-Hydroxycholestane-6‖7-dioic acid $C_{27}H_{46}O_5$ (p.42). For confirmation of the 3α-configuration previously assigned arbitrarily, see 1948a Shoppee; cf. 1935 Lettré. — Leaflets (aqueous acetone) (1921 Windaus); prisms (dioxane) changing at 200–205° into needles, m. 218° to 221° cor. (1948a Shoppee); readily soluble in ether and in acetone (1921 Windaus). — **Fmn.** From 3β-chlorocholestane-6‖7-dioic acid (p. 2487 s) by heating for 8 hrs. with 25% KOH on the water-bath, followed by acidification (1921 Windaus); this reaction involves Walden inversion (1948a Shoppee). — **Rns.** Oxidation with CrO_3 in ca. 75% acetic acid on the water-bath for 75 min. affords 3-ketocholestane-6‖7-dioic acid (p. 2486 s) (1921 Windaus). In contrast to the 3β-epimer, this acid gives no lactone by refluxing with acetic anhydride, but probably forms an anhydride (1948a Shoppee).

3β-Bromocholestan-6-one $C_{27}H_{45}OBr$. For reversal of the 3α-configuration originally assigned (1939 Ladenburg), see 1948a Shoppee. — Crystals (dil. acetic acid and alcohol), m. 123° (1939 Ladenburg); needles (acetic acid), m. 123° (Kofler block), $[\alpha]_D +3° \pm 1°$ (chloroform) (1952a Shoppee). —**Fmn.** From i-cholesten-6-one (p. 2452 s) by reaction with 34% aqueous HBr in glacial acetic acid at room temperature; quantitative yield (1939 Ladenburg). — **Rns.** Refluxing with quinoline in an atmosphere of nitrogen gives Δ^2-cholesten-6-one (p. 2452 s) in good yield (1946 Blunschy; cf. 1939 Ladenburg).

3β-Chloro-Δ^5-cholesten-7-one, 7-Ketocholesteryl chloride $C_{27}H_{43}OCl$ (p. 130). For confirmatory evidence of structure, see 1938 Bergmann. Configuration at C-3 is based on that of its precursor, cholesteryl chloride (p. 1450s) (Editor). — Crystals (alcohol), m. 121–2° (1939 Eck); absorption max. (solvent not indicated) 270 mμ* (1938 Bergmann). — **Fmn.** From cholesteryl chloride by oxidation with CrO_3 in ca. 97% acetic acid at 55° (yield, 24%) (1937a Marker; see also 1928 Minovici**; cf. 1933 Windaus).

Rns. The products of hydrogenation (cf. p. 130) in glacial acetic acid at 3 atm. pressure, using a platinum catalyst, are ca. 80% of uncrystallizable 3β-chlorocholestan-7-ol (p. 1883 s) (presumably a mixture of C-7 epimers, Editor) together with 3β-chlorocholestane (p. 1456 s) (1937a Marker); using Adams's catalyst in ether at 3 atm. pressure and 25°, 3β-chlorocholestan-7-one

---

* Bergmann (1938) cites this absorption max. as evidence for αβ-unsaturated ketone structure. As all comparable αβ-unsaturated steroid ketones show absorption max. near 240 mμ (cf. 1949 Fieser) and in view of other cases cited by Bergmann (1938), it seems very likely that 270 mμ is a misprint for 240 mμ (Editor).

** Formula and structure assigned by these authors were incorrect (Editor).

(below) is formed (1939a Marker). Refluxing with potassium acetate in valeric acid at 180° gives the acetate of Δ^5-cholesten-3α-ol-7-one (p. 2681 s) together with some $\Delta^{3.5}$-cholestadien-7-one (p. 2454 s) (1937a Marker); for reaction mechanism, see 1938 Bergmann; 1946 Shoppee. Treatment with acetic anhydride and conc. H_2SO_4 in glacial acetic acid at 85–90° gives a brown solution from which the product isolated is without antirachitic effect (1939 Eck).

p-Nitrophenylhydrazone $C_{33}H_{48}O_2N_3Cl$. Needles (glacial acetic acid), m. 200° (1928 Minovici).

3β-Chlorocholestan-7-one, 7-*Ketocholestyl chloride* $C_{27}H_{45}OCl$. Configuration

at C-3 derived from that of its precursor, 3β-chloro-Δ^5-cholesten-7-one (page 2488 s) (Editor). — White prisms (acetone) (1939a Marker); crystals (alcohol), m. 139° (1937a Marker). — **Fmn.** From crude 3β-chlorocholestan-7-ol (p. 1883 s) (probably C-7 epimeric mixture, Editor) by oxidation with CrO_3 in glacial acetic acid at 50–60° (2 hrs.), followed by standing overnight at room temperature (1937a Marker). From 3β-chloro-Δ^5-cholesten-7-one (above) by hydrogenation in ether at 25° and 3 atm. pressure using Adams's catalyst (1939a Marker).

Oxime $C_{27}H_{46}ONCl$. Crystals (aqueous alcohol), m. 152.4° (1939a Marker).

6 - Bromo - $\Delta^{3.5}$ - cholestadien - 7 - one $C_{27}H_{41}OBr$. The alternative 8-bromo-

$\Delta^{3.5}$-cholestadien-7-one structure is not excluded, but is considered to be improbable. — Pale yellow needles (aqueous acetone), m. 117°; absorption max. (in alcohol?) 279, 344 mμ (log ε 4.28, 2.2); slow decomposition occurs on standing. — **Fmn.** From 3.4.6-tribromo-Δ^5-cholesten-7-one (p. 2504 s) by boiling (1 min.) with KI in acetone; yield, 67%. — **Rns.** Unchanged by boiling with zinc dust in methanol or acetic acid, with pyridine, or with dimethylaniline; attempted reconversion into the initial tribromide by bromine addition was unsuccessful (1940 Jacobsen).

24 b-Ethyl-4β-bromocoprostan-3-one, 4β-Bromo-coprositostanone $C_{29}H_{49}OBr$.

Configuration at C-4 assigned by analogy with 4β-bromocoprostan-3-one (p. 2483 s) similarly obtained, that at C-24 by its relationship to Δ^4-sitosten-3-one (p. 2467 s) (Editor). — Crystals (glacial acetic acid or alcohol), m. 149°. — **Fmn.** From 24b-ethylcoprostan-3-one (p. 2470 s) by

173*

treatment with 1 mol. of bromine in glacial acetic acid containing small amounts of HBr. — **Rns.** Refluxing with pyridine gives Δ^4-sitosten-3-one (p. 2467 s) (1937 b Marker).

3β- Chlorositostan - 6 - one, 3β - Chlorostigmastan - 6 - one $C_{29}H_{49}OCl$.

For configuration at C-3, see 1948 b Shoppee. — Prismatic needles (alcohol), m. 112° (1935 Vanghelovici). — **Fmn.** From 6-nitrositosteryl chloride (p. 1479 s) by heating with zinc and glacial acetic acid on the water-bath (1935 Vanghelovici). — **Rns.** Oxidation with fuming HNO_3 in glacial acetic acid at 65° gives **3β-chlorositostane-6‖7-dioic acid** $C_{29}H_{49}O_4Cl$ (I) [needles (glacial acetic acid), m. 277°; gives an anhydride (not described) on heating with acetic anhydride in vacuo]. Undergoes Clemmensen reduction to give 3β-chlorositostane (p. 1466 s) (1935 Vanghelovici). Reduction with sodium amalgam in alcohol gives i-sitosten-6-one (p. 2470 s) (1935 Vanghelovici; cf. 1948 b Shoppee).

Oxime $C_{29}H_{50}ONCl$. Needles (alcohol), m. 180° (1935 Vanghelovici).

Semicarbazone $C_{30}H_{52}ON_3Cl$. Crystals (alcohol), m. 207° (1935 Vanghelovici).

(p?)-Nitrophenylhydrazone $C_{35}H_{54}O_2N_3Cl$. Yellow prisms (glacial acetic acid), m. 188° (1935 Vanghelovici).

3β- Chloro -Δ^5-sitosten - 7 - one, 3β - Chloro -Δ^5-stigmasten - 7 - one, 7 - Ketositosteryl chloride $C_{29}H_{47}OCl$.

Configuration at C-3 based on that of the precursor, sitosteryl chloride (p. 1464 s) (Editor). — White plates (acetone), m. 155–6°. — **Fmn.** From sitosteryl chloride by CrO_3 oxidation in ca. 97% acetic acid at 55°. — **Rns.** Hydrogenation in ether at room temperature and 3 atm. pressure, using Adams's catalyst, affords 3β-chlorostigmastan-7-one (below). Reduction by means of aluminium isopropoxide and boiling isopropyl alcohol gives 7-hydroxysitosteryl chloride (p. 1884 s). Boiling with alcoholic KOH yields 7-ketositosterylene (p. 2470 s) (1940 a Marker).

3β- Chlorositostan -7- one, 3β- Chlorostigmastan -7- one, *7-Ketositostyl chloride*

$C_{29}H_{49}OCl$. Configuration at C-3 assigned on the basis of that of the precursor, 3β-chloro-Δ^5-stigmasten-7-one (above) (Editor). — White plates (acetone-alcohol),

m. 128–9⁰. — **Fmn.** From 3β-chloro-Δ^5-stigmasten-7-one by hydrogenation in ether at room temperature and 3 atm. pressure, using Adams's catalyst. — **Rns.** Resists oxidation by CrO_3 at 60⁰. Hydrogenation in glacial acetic acid at room temperature and 3 atm. pressure, using Adams's catalyst, gives 3β-chlorostigmastane (p. 1466 s) (1940a Marker).

1904 Windaus, Stein, *Ber.* **37** 3699, 3702 et seq.
1921 Windaus, v. Staden, *Ber.* **54** 1059, 1062 et seq.
1928 Minovici, Vanghelovici, *C. A.* **1929** 2721; *Ch. Ztbl.* **1929** I 2193.
1932 Jakubowicz, *Contribution à l'étude de l'isomérie des cholestanols et de leurs dérivés halogénés*, Thesis Univ. Nancy, p. 128 et seq.
1933 Windaus, Deppe, *Ber.* **66** 1563, 1565.
1935 Adam, Askew, Danielli, *Biochem. J.* **29** 1786, 1798, 1799.
  Butenandt, Wolff, *Ber.* **68** 2091, 2092.
  Lettré, *Ber.* **68** 766.
  Vanghelovici, Angelescu, *Bul. soc. chim. România* **17** A 177; *C. A.* **1936** 1064; *Ch. Ztbl.* **1936** I 356.
1936 Butenandt, Schramm, Wolff, Kudszus, *Ber.* **69** 2779, 2780.
  CIBA* (Ruzicka), *U.S. Pat.* 2085474 (issued 1937); *C. A.* **1937** 5950; *Ch. Ztbl.* **1938** I 374.
  Inhoffen, *Ber.* **69** 2141, 2142.
1937 CIBA*, *Swiss Pat.* 222124 (issued 1942); *C. A.* **1949** 366; (Ruzicka), *Swed. Pat.* 98368 (1938, issued 1940); *Ch. Ztbl.* **1940** II 2185; CIBA PHARMACEUTICAL PRODUCTS (Ruzicka), *U.S. Pat.* 2232636 (1938, issued 1941); *C. A.* **1941** 3266; *Ch. Ztbl.* **1941** II 3100, 3101.
  Dane, Wang, Schulte, *Z. physiol. Ch.* **245** 80, 88.
  Heilbron, Jones, Spring, *J. Ch. Soc.* **1937** 801, 803.
  Krekeler, *Einige Umsetzungen am Cholestenon*, Thesis Univ. Göttingen, p. 20.
  a Marker, Kamm, Fleming, Popkin, Wittle, *J. Am. Ch. Soc.* **59** 619.
  b Marker, Wittle, *J. Am. Ch. Soc.* **59** 2704, 2707.
  a SCHERING A.-G., *Austrian Pat. Application; FIAT Final Report* **996**, pp. 130, 157 et seq. (issued 1947).
  b SCHERING A.-G., *Fr. Pat.* 838704 (issued 1939); *Brit. Pat.* 508576 (issued 1939); *C. A.* **1940** 777, 778; *Ch. Ztbl.* **1939** II 1722, 1723; SCHERING CORP. (Inhoffen), *U.S. Pat.* 2280828 (issued 1942); *C. A.* **1942** 5618.
  SCHERING-KAHLBAUM A.-G., *Fr. Pat.* 835524 (issued 1938); *Brit. Pat.* 500353 (issued 1939); *C. A.* **1939** 5872; *Ch. Ztbl.* **1939** I 5010; SCHERING CORP. (Inhoffen, Butenandt, Schwenk), *U. S. Pat.* 2340388 (issued 1944); *C. A.* **1944** 4386.
  Windaus, Kuhr, *Ann.* **532** 52.
1938 Bergmann, *J. Am. Ch. Soc.* **60** 1997.
  a C. F. BOEHRINGER & SÖHNE, *Fr. Pat.* 844850 (issued 1939); *C. A.* **1940** 7932; *Ch. Ztbl.* **1940** I 1391.
  b C. F. BOEHRINGER & SÖHNE (Hatzig), *Ger. Pat.* 712256 (issued 1941); *C. A.* **1943** 4408; *Ch. Ztbl.* **1942** I 1162.
  Ford, Chakravorty, Wallis, *J. Am. Ch. Soc.* **60** 413.
  Heilbron, Hodges, Spring, *J. Ch. Soc.* **1938** 759.
  Inhoffen, Huang-Minlon, *Ber.* **71** 1720, 1724, 1729.
  Petrow, Rosenheim, Starling, *J. Ch. Soc.* **1938** 677, 681.
  Ralls, *J. Am. Ch. Soc.* **60** 1744.
  Ruzicka, Plattner, Aeschbacher, *Helv.* **21** 866.
  Stevens, *J. Am. Ch. Soc.* **60** 3089.
  Wolff, *Über die Bromierung von Cholestanon und Koprostanon*, Thesis Techn. Hochschule Danzig, pp. 8, 10, 17.
1939 Dannenberg, *Abhandl. Preuss. Akad. Wiss. Math.-naturwiss. Kl.* **1939** No. 21, pp. 12, 35, 53, 62.
  Eck, Thomas, *J. Biol. Ch.* **128** 257, 258, 261.
  Ladenburg, Chakravorty, Wallis, *J. Am. Ch. Soc.* **61** 3483.
  a Marker, Rohrmann, *J. Am. Ch. Soc.* **61** 3022.

* Gesellschaft für chemische Industrie in Basel; Société pour l'industrie chimique à Bâle.

b Marker, Wittle, Plambeck, Rohrmann, Krueger, Ulshafer, *J. Am. Ch. Soc.* **61** 3317.

Schering A.-G. (Butenandt), *Ger. Pat.* 736846 (issued 1943); *C. A.* **1944** 3094; *Ch. Ztbl.* **1944** I 1457; *Fr. Pat.* 867697 (1940, issued 1941); *Ch. Ztbl.* **1942** I 2561.

1940 Barkow, *Über bromfreie ungesättigte Sterinketone*, Thesis Techn. Hochschule Danzig, pp. 4, 10, 14, 18.

Heath-Brown, Heilbron, Jones, *J. Ch. Soc.* **1940** 1482, 1488.

Inhoffen, Zühlsdorff, Huang-Minlon, *Ber.* **73** 451, 454.

Jackson, Jones, *J. Ch. Soc.* **1940** 659, 663.

Jacobsen, *J. Am. Ch. Soc.* **62** 1620.

a Marker, Rohrmann, *J. Am. Ch. Soc.* **62** 516.

b Marker, *J. Am. Ch. Soc.* **62** 2543, 2544.

Stevens, Allenby, Du Bois, *J. Am. Ch. Soc.* **62** 1424, 1428.

1943 Urushibara, Mori, *C. A.* **1947** 3807.

1944 Butenandt, Ruhenstroth-Bauer, *Ber.* **77** 397.

Ruzicka, Plattner, Furrer, *Helv.* **27** 524, 527.

1946 Blunschy, Hardegger, Simon, *Helv.* **29** 199, 202.

Reich, Lardon, *Helv.* **29** 671, 674, 675.

Shoppee, *J. Ch. Soc.* **1946** 1147, 1149.

1947 Djerassi, Scholz, *J. Am. Ch. Soc.* **69** 2404, 2405, 2409; cf. *Experientia* **3** 107.

1948 Djerassi, Scholz, *J. Am. Ch. Soc.* **70** 417.

Dodson, Riegel, *J. Org. Ch.* **13** 424, 433.

a Shoppee, *J. Ch. Soc.* **1948** 1032–4, 1039 et seq.

b Shoppee, *J. Ch. Soc.* **1948** 1043, 1044.

1949 Djerassi, *J. Am. Ch. Soc.* **71** 1003, 1008.

L. F. Fieser, M. Fieser, *Natural Products related to Phenanthrene*, 3rd Ed., New York, p. 190.

1950 a Barton, Miller, *J. Am. Ch. Soc.* **72** 370, 373.

b Barton, Miller, *J. Am. Ch. Soc.* **72** 1066, 1070.

Djerassi, Rosenkranz, Romo, Kaufmann, Pataki, *J. Am. Ch. Soc.* **72** 4534, 4536 (footnote 16).

Rivett, Wallis, *J. Org. Ch.* **15** 35, 39, 40.

1951 Djerassi, Rosenkranz, *Experientia* **7** 93.

Inhoffen, Kölling, Koch, Nebel, *Ber.* **84** 361.

1952 Jones, Ramsay, Herling, Dobriner, *J. Am. Ch. Soc.* **74** 2828, 2829.

a Shoppee, Summers, *J. Ch. Soc.* **1952** 1786, 1789.

b Shoppee, Summers, *J. Ch. Soc.* **1952** 1790. 1793.

1953 a Corey, *Experientia* **9** 329.

b Corey, *J. Am. Ch. Soc.* **75** 4832.

a Fieser, Dominguez, *J. Am. Ch. Soc.* **75** 1704.

b Fieser, Romero, *J. Am. Ch. Soc.* **75** 4716.

c Fieser, Huang, *J. Am. Ch. Soc.* **75** 4837.

1954 Cookson, *J. Ch. Soc.* **1954** 282, 283.

Georg, *Arch. sci.* **7** 114.

Gut, *J. Am. Ch. Soc.* **76** 2261.

1955 M. Fieser, Romero, L. F. Fieser, *J. Am. Ch. Soc.* **77** 3305, 3307.

Gates, Wallis, *J. Org. Ch.* **20** 610.

1956 Ellis, Petrow, *J. Ch. Soc.* **1956** 1179, 1182.

Kirk, Patel, Petrow, *J. Ch. Soc.* **1956** 627.

3. Polyhalogeno-monoketo-steroids (CO and Halogen in the Ring System)

5.6-Dibromo-androstan (or -ætiocholan)-17-one, Δ^5-Androsten-17-one dibromide $C_{19}H_{28}OBr_2$.

Only obtained as a mixture with the other dibromides mentioned below. — **Fmn.** From Δ^5-cholestene dibromide (p. 1473 s) on oxidation with CrO_3 and strong acetic acid in carbon tetrachloride at 48–50°, along with the dibromides of Δ^5-pregnen-20-one and Δ^5-cholenic acid. — **Rns.** Gives Δ^5-androsten-17-one (p. 2403 s) on heating with zinc dust in glacial acetic acid at 95° (1940 Marker; 1940 PARKE, DAVIS & Co.).

1940 Marker, Wittle, Tullar, *J. Am. Ch. Soc.* **62** 223, 224.
    PARKE, DAVIS & Co. (Marker, Wittle), *U.S. Pats.* 2397424 (issued 1946), 2397425 (1943, issued 1946), 2397426 (1944, issued 1946); *C. A.* **1946** 3570, 3571.

Summary of Di- and Tri-halogeno-monoketones with Cholestane (or Coprostane) Skeleton

(including some data from papers published after the closing date for this volume)

Position of CO	Position of halogen	Config. at C-5	Position double bond(s)	M. p. °C.	$[\alpha]_D(°)$ in chloroform	Absorpt. max. (mμ) in alcohol	Page
3	1 Br · 2 Br	α	satd.	dec. 85	—	—	2494 s
3	2 Br · 2 Br	α	satd.	147	+104 to +111	294	2494 s
3	2 Br · 4 Br	—	$\Delta^{1.4.6}$	189	+23	228, 280, 326	[1])
3	2α Br · 4α Br	α	satd.	194	+3	—	2495 s
3	2β Br · 4β Br	β	satd.	138	+10	—	2497 s
3	2 Br · 5α Cl	α	satd.	122	—	—	2497 s
3	2 Br · 6 Br	—	$\Delta^{1.4.6}$	203	−38	<230, 267, 297[2])	2497 s
3	2 Br · 6 Br	—	$\Delta^{1.4}$	181 dec.	−55	253	[3])
3	2α Br · 6(?) Br	—	$\Delta^{4.6}$	153; 185	—	296[4])	2498 s
3	2α Br · 6α Br	—	Δ^4	133	—	248[2])	2499 s
3	2α Br · 6β Br	—	Δ^4	170	+50	248	2499 s
3	2β Br · 6β Br	—	Δ^4	124	−41	257[5])	[6])
3	5α Cl · 6β Cl	α	satd.	117	−27	—	2500 s
3	5α Br · 6β Br	α	satd.	80	−53	—	2500 s
7	4 Br · 6 Br	—	$\Delta^{3.5}$	190	—	303[7])	2501 s
3	2 Br · 2 Br · 4 Br	—	Δ^4	183	−112	277	2502 s
3	2 Br · 2 Br · 6 Br	—	$\Delta^{4.6}$	130; 166	−22	310[4])	2503 s
3	2 Br · 2 Br · 6β Br	—	Δ^4	186	−6	253[2])	2503 s
3	2 Br · 4 Br · 6 Br	—	$\Delta^{1.4}$	209 dec.	−101	274[8])	[1])
3	2α Br · 5α Br · 6β Br	α	satd.	140	−45	—	2504 s
3	2β Br · 5α Br · 6β Br	α	satd.	106	+9	—	2504 s
7	3 Br · 4 Br · 6 Br	—	Δ^5	dec. ca. 143	—	267[7])	2504 s

[1]) 1952 Inhoffen. — [2]) In ether. — [3]) 1955 Fieser. — [4]) In chloroform. — [5]) In isopropyl alcohol. [6]) 1956 Ellis. — [7]) Solvent not indicated. — [8]) In methanol.

1.2-Dibromocholestan-3-one, Δ^1**-Cholestenone dibromide** $C_{27}H_{44}OBr_2$. Only an impure sample has been obtained, dec. 85°. Liberates iodine from warm alcoholic NaI. — **Fmn.** From 2α-bromocholestan-3-one (p. 2480 s) by boiling with collidine (12 hrs.) to give a halogen-free mixed product (74% yield) which is fractionally crystallized from methanol; the intermediate fractions are treated with bromine in cold glacial acetic acid (1940 Jacobsen).

Rns. Debromination with sodium iodide or zinc dust in alcohol, or by means of potassium iodide in 80% acetone, gives Δ^1-cholesten-3-one (p. 2423 s) (1940 Jacobsen).

2.2-Dibromocholestan-3-one $C_{27}H_{44}OBr_2$. For structure, see 1947 a Djerassi; cf. 1943 Inhoffen. This structure had previously been assigned (1936 Ruzicka) to 2α.4α-dibromocholestan-3-one (page 2495 s). — White spherules (alcohol), m. 139–140° (1909 Dorée; cf. 1949 Djerassi); crystals (methanol-ethanol), m. 147° (1938 Ralls); $[\alpha]_D^{22} + 104°$ to $+ 111°$ (chloroform) (1947 a, b Djerassi). Absorption max. (inflexion) in alcohol at 294 mμ (log ε 2.13) (1954 Cookson). Very soluble in petroleum, sparingly soluble in alcohol (1909 Dorée).

Fmn. Together with the 2α.4α-isomer (p. 2495 s), from cholestan-3-one (p. 2436 s) by treatment with bromine in chloroform (1909 Dorée); the 2.2-dibromo-compound is the primary product of dibromination of cholestan-3-one in glacial acetic acid, but isomerization to 2α.4α-dibromocholestan-3-one readily occurs, favoured by heating or presence of HBr, retarded by water, H_2O_2, or sodium acetate (1947 a Djerassi). A mixture with 2α-bromocholestan-3-one (p. 2480 s) (chief product) and the 2α.4α-isomer is obtained by treatment of cholestan-3-one with excess iodine monobromide and glacial acetic acid in carbon tetrachloride, the products being fractionated from methanol-acetone (yield, 1%); similar treatment of 2α-bromocholestan-3-one yields a mixture of the 2.2- and 2α.4α-dibromo-compounds, the latter in relatively greater preponderance (1938 Ralls). From 2α-bromocholestan-3-one (p. 2480 s) on reaction with either bromine in glacial acetic acid, sodium acetate being added before working up, or with N-bromosuccinimide in carbon tetrachloride with exposure to strong light (1947 a Djerassi).

Rns. Standing with fused potassium acetate in benzene-alcohol at room temperature affords a *halogen-free compound* [needles (acetone), m. 119°, not further investigated] (1938 Ralls). Refluxing with potassium benzoate in toluene-butanol yields the benzoate of Δ^1-cholesten-4-ol-3-one (p. 2668 s) (1936b SCHERING A.-G.; cf. 1937b SCHERING-KAHLBAUM A.-G.). For isomerization to the 2α.4α-isomer, see formation, above.

2α.4α-Dibromocholestan-3-one $C_{27}H_{44}OBr_2$ (p. 130). For confirmation of the structure assigned (1936b Butenandt) to this compound, to which the 2.2-dibromo-structure had also been ascribed (1936 Ruzicka), see 1937 Inhoffen; 1938 Ralls. For configuration at C-2 and C-4, based on infra-red absorption max., see 1952 Jones.

White needles (light petroleum) (1909 Dorée; cf. 1943 Inhoffen), crystals (methanol-acetone) (1938 Ralls), needles (chloroform-methanol) (1938 Wolff), prismatic needles (petroleum ether), m. 194–194.5° cor. dec.; $[\alpha]_D^{24} + 3°$ (chloroform) (1946 Wilds). Ultra-violet absorption curve in alcohol almost linear from 265 to 300 mμ; log ε 1.76 at 290 mμ (1954 Cookson). Sparingly soluble in light petroleum, almost insoluble in alcohol (1909 Dorée).

Fmn. From cholestan-3-one (p. 2436 s) by treatment with slightly more than 2 mols. of bromine in glacial acetic acid with initial gentle warming and then standing (yield, 92%) (1946 Wilds; cf. 1947a Djerassi). Together with 2α-bromocholestan-3-one (p. 2480 s) and the 2.2-dibromo-isomer (page 2494 s), from cholestan-3-one by reaction with bromine in chloroform (1909 Dorée; cf. 1943 Inhoffen); a similar mixed product, separable by fractional crystallization from methanol-acetone, results by treatment of cholestan-3-one with iodine monobromide and glacial acetic acid in CCl_4 (yield, 2%); from 2α-bromocholestan-3-one by this mode of bromination the 2α.4α-isomer is obtained admixed with smaller amounts of the 2.2-isomer (1938 Ralls).

Rns. Oxidation with CrO_3 and H_2SO_4 in glacial acetic acid-ethylene chloride at 18° gives 2α.4α-dibromoandrostane-3.17-dione (p. 2924 s) (1942 SCHERING A.-G.). Refluxing with dry isopropanol and aluminium isopropoxide in benzene affords 2α.4α-dibromocholestan-3-ol (p. 1887 s: 2.4-dibromocholestan-3-ol)(1938 Inhoffen). Treatment with bromine in glacial acetic acid gives 2.2.4-tribromo-Δ^4-cholesten-3-one (p. 2502 s) (1937b SCHERING-KAHLBAUM A.-G.; cf. 1952 Inhoffen). Reaction with alkali hydroxide solutions (varying concentrations), in the cold or in the hot, gave uncrystallizable oils still containing bromine. Treatment with a 6% solution of $AgNO_3$ in pyridine, at room temperature for 48 hrs., yields a **compound** $C_{27}H_{43}O_2Br$ [prismatic needles (chloroform-alcohol), m. 151–2°, gives a violet-blue coloration with $FeCl_3$; not debrominated on boiling with zinc dust in alcohol; gives $\Delta^{4.6}$-cholestadien-4-ol-3-one (p. 2666 s) with potassium acetate in glacial acetic acid at 180°] to which the structure **2-bromo-Δ^4-cholesten-4-ol-3-one** (I) has been ascribed (1938 Wolff). Heating with potassium acetate in butanol yields, besides cholestane-3.4-dione (cf. p. 130), a *compound* m. 160–170° (not further investigated) (1938 Wolff; but cf. 1938 Ralls). The products of reaction with potassium benzoate in boiling toluene-butanol, m. 137–8° and 177° (cf. p. 130), are the benzoates of Δ^4-cholesten-2-ol-3-one (p. 2664 s) and Δ^1-cholesten-4-ol-3-one (p. 2668 s), resp. (1936b SCHERING A.-G.; 1937b SCHERING-KAHLBAUM A.-G.; cf. 1937 Inhoffen); when the dibromoketone is refluxed

with potassium benzoate in isovaleric acid, the isovalerate of Δ^1-chole-
sten - 4 - ol - 3 - one (p. 2668 s) is obtained (1936b SCHERING A.-G.; 1937b
SCHERING - KAHLBAUM A.-G.). Refluxing with dimethylaniline (5 hrs.)
gives a **compound** $C_{34}H_{49}N$ [white needles (ethyl acetate), m. 230–2⁰;
structure (II) is considered the most
probable of the alternatives postulated;
deep wine - red coloration with p - nitro -
diazobenzene in alcohol-glacial acetic acid]
(1937 Schwenk). The product of reaction
by heating with o-phenylenediamine at 140–150⁰ is the quinoxaline derivative
of cholestane-2.3-dione (p. 2903 s) (1937 Inhoffen), identical with that obtained by Ruzicka (1936) on refluxing the reactants in alcohol for 8 hrs.;
by allowing the same quantities to react in boiling alcohol for 5 hrs. and
with subsequent working-up at lower temperature, the product isolated
is *2α.4α-dibromo-3-(o-aminophenylimino)-cholestane* $C_{33}H_{50}N_2Br_2$ (III) [crys-
tals (alcohol), m. 184⁰] (1938 Ralls). By
heating with dry pyridine at 135⁰ (bath
temperature), $\Delta^{1.4}$ - cholestadien - 3 - one
(p. 2418 s) is formed together with N-(3-ke-
to-Δ^4-cholesten-2-yl)-pyridinium bromide
(see below) (1938 Inhoffen; cf. 1939 Inhoffen); dehydrobromination can also
be effected by use of piperidine, α-picoline, α.γ-lutidine, substituted anilines,
or potassium naphthoxide (1936b SCHERING A.-G.; 1937b SCHERING-KAHL-
BAUM A.-G.). Refluxing with collidine also gives $\Delta^{1.4}$-cholestadien-3-one, in
better yield (1939 Butenandt), together with smaller amounts of the $\Delta^{4.6}$-isomer
(p. 2420 s) (1946 Wilds), but with very short duration of reaction (40 sec.),
2α-bromo-Δ^4-cholesten-3-one (p. 2480 s) is formed (1949 Djerassi); if the crude
product of prolonged reaction with boiling collidine is left to stand with acetic
anhydride and conc. H_2SO_4, the product hydrolysed with methanolic alkali
and then etherified with dimethyl sulphate and NaOH in warm alcohol,
the methyl ether of 4-methyl-19-nor-$\Delta^{1.3.5(10)}$-cholestatrien-1-ol (p. 1535 s) is
obtained (1946 Wilds; cf. 1951a Inhoffen). For review concerning dehydrobromination of 2.4-dibromocholestanone, see 1947 Inhoffen.

p. 130, line 16 from bottom: for "dimethylaniline in pyr." *read* "dimethylaniline or pyridine".

N-(3-Keto-Δ^4-cholesten-2-yl)-pyridinium bromide $C_{32}H_{48}ONBr$. For possible
alternative structure, see 1940 Inhoffen. — Crys-
tals (aqueous alcohol), dec. 322⁰ (1937a SCHE-
RING A.-G.; cf. 1937b SCHERING-KAHLBAUM
A.-G.); sparingly soluble in water or ether (1938
Inhoffen). — **Fmn.** See reactions of 2α.4α-di-
bromocholestan-3-one, above. — **Rns.** Reduction by means of isopropanol
and aluminium isopropoxide yields **N-(3-hydroxy-Δ^4-cholesten-2-yl)-pyridi-
nium bromide** $C_{32}H_{50}ONBr$ (not described) (1937b SCHERING A.-G.). Heating
at 180–200⁰ under 6 mm. pressure, followed by distillation in a high vacuum
at 200–220⁰, effects decomposition to give pyridinium bromide and $\Delta^{1.4}$-chole-

stadien-3-one (p. 2418) (1937b SCHERING-KAHLBAUM A.-G.). Undergoes aromatization on distillation at 310–330° in high vacuum (1937a SCHERING A.-G.).

2β.4β - Dibromocoprostan - 3 - one $C_{27}H_{44}OBr_2$ (p. 130: 4.4?-Dibromocopro-

stanone). Further support for the 4.4-dibromo-structure (1949 Inhoffen) was subsequently rejected in favour of that now assigned (1951 Djerassi; 1951b Inhoffen), largely on the basis of infra-red absorption, from which also the configuration at C-2 and C-4 is deduced (1952 Jones). — Prisms (acetone containing some alcohol) (1909 Dorée; cf. 1936b Butenandt), prismatic needles (glacial acetic acid), crystals (chloroform-methanol) (1938 Wolff), crystals (acetone-alcohol), m. 137–8° (1951b Inhoffen; cf. 1951 Djerassi; 1936b Butenandt; 1938 Wolff); $[\alpha]_D^{20} + 10°$ (chloroform?) (1951 Djerassi).

Fmn. From coprostan-3-one (p. 2443 s) by bromination in chloroform (1909 Dorée). — **Rns.** Refluxing with anhydrous potassium acetate in butanol (5 hrs.) gives Δ^4-cholesten-4-ol-3-one (p. 2669 s) together with a by-product, m. 160–170°, also isolated by a similar treatment of 2α.4α-dibromocholestan-3-one (p. 2495 s); using ethanol instead of butanol, a yellow oily product, still containing bromine, resulted (1938 Wolff).

2 - Bromo - 5 - chloro - cholestan - 3 - one $C_{27}H_{44}OClBr$ (p. 129: 5-chloro-4-bromo-

cholestanone). For revised structure, see 1955 Fieser. — Crystals (aqueous acetone), m. ca. 122° (1937b SCHERING-KAHLBAUM A.-G.). — **Fmn.** From 5-chlorocholestan-3-one (p. 2483 s) by reaction with 1 mol. of bromine in glacial acetic acid at room temperature, followed by precipitation with water after decolorization of the solution (5 min.) (1937b SCHERING-KAHLBAUM A.-G.). — **Rns.** Refluxing with potassium acetate in benzene-alcohol for 30 min. yields an unsaturated bromoketone, m. 123° (1937b SCHERING-KAHLBAUM A.-G.; cf. 1936b Butenandt), probably identical with 2α-bromo-Δ^4-cholesten-3-one (p. 2480 s) (Editor); if the reaction with potassium acetate is carried out in glacial acetic acid at 160° for 6 hrs., Δ^4-cholesten-4-ol-3-one (p. 2669 s) is obtained, whereas at 220° for 5 hrs. cholestane-3.6-dione (page 2907 s) is formed (1937b SCHERING-KAHLBAUM A.-G.; cf. 1936b Butenandt).

2.6-Dibromo-$\Delta^{1.4.6}$-cholestatrien-3-one $C_{27}H_{38}OBr_2$ (p. 120: dibromocholesta-

trienone m. 203°). For revision of the 4.6 - dibromo - $\Delta^{4 \cdot 6 \cdot 8(9 \text{ or } 14)}$ - cholestatrien-3-one structure originally assigned (1937 Butenandt), see 1955 Fieser; cf. 1952 Inhoffen. — Needles (chloroform-glacial acetic acid) (1937 Dane; cf.

1937 Butenandt), leaflets (chloroform-methanol) (1940 Barkow), crystals (aqueous acetone) (1937 Butenandt), m. 203° (1937 Butenandt; 1937 Dane; 1940 Barkow); $[\alpha]_D^{20}$ —38° (chloroform) (1937 Butenandt). Absorption max. in ether at <230, 267, 297 mμ (ε >12000, 12200, 9300) (1939 Dannenberg; cf. 1937 Butenandt). — For attempted separation from Δ^4-cholesten-3-one (p. 2424 s) by thermal diffusion process in either benzene or carbon tetrachloride solution, see 1940 Korsching.

Fmn. From Δ^4-cholesten-3-one (p. 2424 s) by slow addition of bromine (4 mols.) in ether-glacial acetic acid, first leaving at room temperature overnight, finally heating under reflux (2 hrs.) (yield, 80%); this procedure is preferable to the isolation as by-product in the Butenandt (1937) preparation of 2.2.6-tribromo-$\Delta^{4.6}$-cholestadien-3-one (p. 2503 s) by use of bromine (6 mols.) in chloroform-glacial acetic acid (1940 Barkow; cf. 1936a SCHERING A.-G.). From 2.2.6-tribromo-$\Delta^{4.6}$-cholestadien-3-one by treatment with HBr-glacial acetic acid, first at room temperature, then heating on the water-bath; yield, 72% (1937 Butenandt).

Rns. Hydrogenation in boiling alcohol, in presence of KOH, gives Δ^4-cholesten-3-one (p. 2424 s); similar procedure in the cold yields some coprostan-3-one (p. 2443 s) together with unchanged material. On hydrogenation in boiling amyl alcohol, using palladium-BaSO$_4$ catalyst, the products are 2-bromo-$\Delta^{1.4.6}$-cholestatrien-3-one (p. 2479 s), $\Delta^{1.4.6}$-cholestatrien-3-one (page 2418 s), and so-called "4-bromocholestenone", m. 147° (p. 2483 s), when carried out in presence of 1 equiv., 2 equivs., or absence of KOH, resp. (1940 Barkow). Resistant towards SeO$_2$, OsO$_4$, or mercuric acetate oxidation, towards acids or further attack by bromine, towards zinc dust in glacial acetic acid, copper powder at 280°, and towards diazomethane or maleic anhydride in xylene; is not attacked by quinoline at 185° or by diethylaniline at 215° (1937 Butenandt; 1940 Barkow); with pyridine at 100° extremely slow elimination of HBr occurs, somewhat accelerated in presence of AgNO$_3$ (1937 Dane).

Oxime C$_{27}$H$_{39}$ONBr$_2$. White powder (aqueous methanol), m. 118°; obtained by the usual method in only 47% yield, together with unchanged material (1937 Butenandt).

2α.6(?) - Dibromo - $\Delta^{4.6}$ - cholestadien - 3 - one C$_{27}$H$_{40}$OBr$_2$. Originally describ-

ed as *2.4 - dibromo - Δ^4 - cholesten - 3 - one* (1936 Ruzicka), then formulated as *4.6 - dibromo - $\Delta^{4.6}$ - cholestadien - 3 - one* (1936b Butenandt; see p. 130) and later as *2.2 - dibromo - $\Delta^{4.6}$ - cholestadien - 3 - one* (1955 Fieser). The last-mentioned structure, however, is questionable; Fieser (1957) agrees with Editor's opinion that a 2α.6-dibromo-structure is more likely, especially in view of the absence of colour reaction with tetranitromethane and failure to form a quinoxaline derivative.

Colourless needles (abs. alcohol-benzine), m. 184–5° cor. (1936 Ruzicka; cf. 1936a SCHERING A.-G.); an allotropic form, colourless needles (chloroform-methanol), m. 150–3°, has sometimes been observed, which changes to the higher-melting form on dry rubbing (1936b Butenandt; 1937b SCHERING-KAHLBAUM A.-G.). Absorption max. in chloroform at 296 mμ (ε 19400) (1939 Dannenberg). Does not give a yellow coloration with tetranitromethane (1936 Ruzicka).

Fmn. From Δ^4-cholesten-3-one (p. 2424 s) by treatment with bromine (ca. 4 mols.) in glacial acetic acid-chloroform at room temperature (yield, 8%) (1937 Butenandt); if the reaction is carried out in ether-glacial acetic acid and the first two molecules of bromine added slowly, the yield can be raised to 39% (1936a SCHERING A.-G.). From 2.2.6β-tribromo-Δ^4-cholesten-3-one (p. 2503 s) by treatment with HBr in glacial acetic acid-ethyl acetate; yield, 17%. From 2.2.5α.6β-tetrabromocholestan-3-one (p. 2505 s) by reaction at room temperature either with HBr in glacial acetic acid-ether (yield, 39%) or with potassium acetate in alcohol-benzine (yield, 65%) (1937 Butenandt).

Rns. Treatment with bromine in glacial acetic acid, containing small amounts of conc. HBr, yields 2.2.6-tribromo-$\Delta^{4.6}$-cholestadien-3-one (p. 2503 s) (1937 Butenandt). Remains unchanged on refluxing in abs. alcohol containing benzene, with either potassium acetate or o-phenylenediamine (1936 Ruzicka), or on heating with acetic anhydride (1937 Butenandt). Gives an oily product with hydroxylamine (1936 Ruzicka).

Semicarbazone $C_{28}H_{43}ON_3Br_2$ (no analysis), m. 210–220°; the product was imperfectly crystalline and not easy to purify (1936 Ruzicka).

2α.6α-Dibromo-Δ^4-cholesten-3-one $C_{27}H_{42}OBr_2$ (p. 131: *4.6-Dibromo-Δ^4-chole-*

sten-3-one, isomer m. 133°). For revision of structure, and configuration now assigned, see 1950 Djerassi; 1955 Fieser. — Absorption max. in ether at 248 mμ (ε 13400) (1939 Dannenberg). More soluble in glacial acetic acid or in glacial acetic acid-alcohol than is the 6β-epimer (below) (1936a SCHERING A.-G.; cf. 1936b Inhoffen). — **Fmn.** Crystallizes from the mother-liquors of the 6β-epimer when prepared by bromination of Δ^4-cholesten-3-one (p. 2424 s) in ether-glacial acetic acid at room temperature (cf. p. 131) (1936a SCHERING A.-G.; 1937b SCHERING-KAHLBAUM A.-G.). — **Rns.** See the 6β-epimer, below.

2α.6β-Dibromo-Δ^4-cholesten-3-one $C_{27}H_{42}OBr_2$ (p. 131: *4.6-Dibromo-Δ^4-chole-*

sten-3-one, isomer m. 162–3°). For revision of structure and the configuration now assigned, see 1955 Fieser; cf. 1950 Djerassi. — Colourless needles (aqueous acetone), m. 169–170°; [α]$_D$ +50° (chloroform); absorption max. in alcohol at 248 mμ (ε 14100) (1955 Fieser; cf. 1939 Dannenberg; 1950 Djerassi). Sparingly soluble in glacial acetic acid, insoluble in methanol (1936a SCHERING A.-G.).

References, pp. 2505 s, 2506 s

Fmn. The yield from coprostenone (p. 2424 s) by bromination (ca. 2 mols. of bromine) in ether-glacial acetic acid at room temperature (cf. p. 131) is 21% (1936a SCHERING A.-G.; 1937b SCHERING-KAHLBAUM A.-G.; cf. 1936b Inhoffen). From 6β-bromo-Δ^4-cholesten-3-one (p. 2484 s) by treatment with bromine (1 mol.) in ether-glacial acetic acid (containing some HBr) at room temperature; yield, ca. 56% (1936b Inhoffen). — **Rns.** The tribromo-compound formed by treating the 6α- or the 6β-epimer (cf. p. 131) with bromine in ether-glacial acetic acid at room temperature, in the presence of HBr (1937 Butenandt), has been identified as 2.2.6β-tribromo-Δ^4-cholesten-3-one (p. 2503 s) (1955 Fieser). Both epimers are resistant to bromination in the presence of sodium acetate and to boiling with potassium acetate in glacial acetic acid. The 6β-epimer is also stable towards hot HBr and to AgNO₃ in pyridine at room temperature (the behaviour of the 6α-epimer towards these reagents was not investigated) (1937 Butenandt).

4.4(?)-Dibromocoprostanone $C_{27}H_{44}OBr_2$ (p. 130). See 2β.4β-dibromocoprostan-3-one, p. 2497 s.

4.6-Dibromo-$\Delta^{4.6}$-cholestadien-3-one $C_{27}H_{40}OBr_2$ (p. 130). See 2α.6-dibromo-$\Delta^{4.6}$-cholestadien-3-one, p. 2498 s.

5α.6β-Dichlorocholestan-3-one, Δ^5-Cholesten-3-one dichloride $C_{27}H_{44}OCl_2$. For structure, see 1950 Rivett; for reversal of the 5β-configuration there assigned, see 1950a Barton. — Crystals (ethanol-methanol), m. 110–111° (Maquenne block); softens at 108° if progressively heated (1939 Décombe); crystals (ethyl acetate-methanol), m. 116–7° dec., softening at 114° (1950a Barton). Rotation in chloroform: $[\alpha]_{578}$ —30°, $[\alpha]_{546}$ —35°, $[\alpha]_{436}$ —60° (1939 Décombe), $[\alpha]_D^{ca.\ 20}$ —27° (1950a Barton; cf. 1950 Rivett). — **Fmn.** From 5α.6β-dichlorocholestan-3β-ol (p. 1888 s) by CrO₃ oxidation in slightly dil. acetic acid at 55°; yield, 78% (1939 Décombe). — **Rns.** Hydrogenation in aqueous ether affords 5α.6β-dichlorocholestan-3β-ol, 6β-chlorocholestan-3β-ol (p. 1881 s), or cholestan-3β-ol (p. 1695 s), together with unidentified oily products, when the reaction is discontinued after absorption of 1, 2, or 3 mols. of hydrogen resp. (1939 Décombe). Boiling with aqueous potassium acetate in benzene yields 6β-chloro-Δ^4-cholesten-3-one (p. 2484 s) (1936 CIBA; cf. 1950a Barton; 1950 Rivett).

5α.6β-Dibromocholestan-3-one, Δ^5-Cholestenone dibromide $C_{27}H_{44}OBr_2$ (p. 131). For configuration at C-5 and C-6, see 1950b Barton. — Crystals (benzene-methanol), m. 88° dec. (1937 Krekeler); crystals (acetone), m. 80°, which lose HBr if recrystallized from alcohol or glacial acetic acid (1939 Décombe).

Rotation in chloroform: $[\alpha]_{578}$ —56°, $[\alpha]_{546}$ —60°, $[\alpha]_{436}$ —123° (1939 Décombe), $[\alpha]_D^{ca.\ 25}$ —53° (1950b Barton).

Fmn. For formation from cholesterol dibromide ($5\alpha.6\beta$-dibromocholestan-3β-ol, p. 1889 s) (cf. p. 131), see also 1936 Ruzicka; 1937 Krekeler; 1939 Décombe; 1941 Bretschneider; 1943 SQUIBB & SONS. From Δ^5-cholesten-3-one (p. 2435 s) by treatment with bromine (1 mol.) in glacial acetic acid at room temperature (1937a SCHERING-KAHLBAUM A.-G.).

Rns. Oxidation by means of CrO_3 or $KMnO_4$ in glacial acetic acid suitably diluted (e. g. with carbon tetrachloride), followed by dehalogenation, using zinc dust, KI, or by hydrogenation, is applied for conversion into Δ^4-androstene-3.17-dione (p. 2880 s) and Δ^4-pregnene-3.20-dione (progesterone, p. 2782 s) (1938, 1939 C. F. BOEHRINGER & SÖHNE); the latter product is shown (by biological assays) to be present in the resinous material which results by ozonization in ethyl chloride at 0° or in chloroform at 25°, in the presence of either $CaCO_3$ or piperidine, followed by treatment with zinc dust and glacial acetic acid (1943 SQUIBB & SONS). Hydrogenation in aqueous ether, using a platinum black catalyst, discontinued after absorption of 1 mol. of hydrogen, gives Δ^4-cholesten-3-one (p. 2424 s); the same product results when operating in presence of Na_2CO_3 (1939 Décombe). The reaction at room temperature with 1 mol. of bromine (cf. p. 131) gives the less stable $2\beta.5\alpha.6\beta$-tribromocholestan-3-one when effected in glacial acetic acid (containing small amounts of HBr), followed by precipitation with water after 5–10 min. (1936a Butenandt); the more stable 2α-epimeric tribromo-compound results by reaction in ether-acetic acid for 20 min., followed by evaporation of the ether in vacuo at 30° (1936a Inhoffen), but by bromination in this same solvent, a mixture of the two epimers can also be obtained (1937 Krekeler). Reaction with 2 mols. of bromine in glacial acetic acid (containing small amounts of HBr) at room temperature for 18 hrs. affords $2.2.5\alpha.6\beta$-tetrabromocholestan-3-one (1936a Butenandt). For debromination to Δ^4-cholesten-3-one using zinc dust, see also 1937a SCHERING-KAHLBAUM A.-G., by means of $FeCl_2$ in alcohol, see 1941 Bretschneider. For mechanism of the conversion into cholestane-3.6-dione (p. 2907 s) by boiling in alcohol (cf. p. 131), see 1941 Ando; prolonged refluxing in benzene with aqueous potassium acetate (bath temperature 160°) also yields cholestane-3.6-dione (1936 Ruzicka).

4.6-Dibromo-$\Delta^{3.5}$-cholestadien-7-one $C_{27}H_{40}OBr_2$.

Hairy needles (dil. acetic acid), m. 189–190°. Absorption max. (solvent not stated) at 303 mμ (log ε 4.05). — **Fmn.** From 3.4.6-tribromo-Δ^5-cholesten-7-one (p. 2504 s) by heating with fused potassium acetate and glacial acetic acid on the steam-bath (crude yield, 95 %), by reaction with $AgNO_3$ in pyridine at 15° (yield, 58 %), or by refluxing with dimethylaniline (lower yield) (1940 Jackson).

5α.6β-Dibromostigmastan-3-one, "β"-Sitostenone dibromide, 5.6-Dibromo-

cinchone, α-Fucostenone dibromide $C_{29}H_{48}OBr_2$. Configuration at C-5 and C-6 deduced from that of its precursor "β"-sitosterol dibromide (p. 1899 s). — Solid (no constants given) (1936 Coffey). — **Fmn.** From α-dihydrofucosterol (p. 1808 s) by treatment with bromine in glacial acetic acid, followed by CrO_3 oxidation in 98% acetic acid, all at room temperature (1936 Coffey). From "β"-sitosterol (= cinchol) dibromide (p. 1899 s) in benzene by oxidation at room temperature with $KMnO_4$ in aqueous H_2SO_4 (1915 Heiduschka) or by means of CrO_3 in strong acetic acid (1938 Dirscherl). — **Rns.** Heating with zinc dust and glacial acetic acid on the steam-bath gives Δ^4-sitosten-3-one (p. 2467 s) (1936 Coffey; cf. 1915 Heiduschka); for debromination using NaI, see 1938 Dirscherl.

20-Diphenylmethylene-5.6-dibromo-pregnan (or allopregnan)-3-one, 22.22-Di-

phenyl-5.6-dibromo-$\Delta^{20(22)}$-bisnorcholen (or bisnorallocholen)-3-one, 1-Methyl-2.2-diphenyl-1-[5.6-dibromo-3-keto-androstan-17β-yl (or ætiocholanyl)]-ethylene $C_{34}H_{40}OBr_2$. Colourless needles (methanol), m. ca. 173° dec. — **Fmn.** From the 5.6-dibromide of 1-methyl-2.2-diphenyl-1-(3β-hydroxy-Δ^5-ætiocholenyl)-ethylene (p. 1844 s) on oxidation with CrO_3 in strong acetic acid at 30°. — **Rns.** On warming with zinc dust and glacial acetic acid in ether it gives 1-methyl-2.2-diphenyl-1-(3-keto-Δ^4-ætiocholenyl)-ethylene (p. 2472 s) (1941 GLIDDEN Co.).

2.2.4-Tribromo-Δ^4-cholesten-3-one $C_{27}H_{41}OBr_3$ (p. 130: tribromocoprostanone).

For revised structure of this compound, originally formulated by Ruzicka (1936) as 2.4.4-tribromocoprostanone, later as 2.4.6-tribromo-Δ^4-cholesten-3-one (1948 Inhoffen), see 1952 Inhoffen. — Crystals (acetone) (1936 Ruzicka), refracting leaflets (chloroform-alcohol), m. 182.5–183°; $[\alpha]_D^{20}$ —112° (chloroform); absorption max. in alcohol at 277 mμ (ε 13150) (1952 Inhoffen). — **Fmn.** From 2β.4β-dibromocoprostan-3-one (p. 2497 s) by reaction with bromine (1 mol.) in carbon tetrachloride at 12° for 2 days, with illumination from a 1000 watt lamp; yield, 35% (1936 Ruzicka). From 2α.4α-dibromocholestan-3-one (p. 2495 s) by bromination in glacial acetic acid containing some HBr, at 90° (1937 b SCHERING-KAHLBAUM A.-G.; cf. 1952 Inhoffen). — **Rns.** Refluxing with pyridine yields a pyridinium compound, white powder, insoluble in ether, soluble in chloroform (1937 b SCHERING-KAHLBAUM A.-G.).

2.2.6-Tribromo-$\Delta^{4\cdot6}$-cholestadien-3-one $C_{27}H_{39}OBr_3$ (p. 120: *tribromocholesta-*
dienone). For revision of structure of this
compound, formulated by Butenandt
(1937) as the *4.6.7-tribromo-isomer*,
see 1955 Fieser. — Dimorphic: crystals
(chloroform-glacial acetic acid), m. 165°
to 166°, or (from the same solvent)
m. 130° (1937 Butenandt). Absorption max. in chloroform at 310 mμ (ε 18900)
(1939 Dannenberg). — **Fmn.** The yield from Δ^4-cholesten-3-one (p. 2424 s) by
bromination (cf. p. 120) varies between 5 and 16%. From 2α.6(?)-dibromo-
$\Delta^{4\cdot6}$-cholestadien-3-one (p. 2498 s) by treatment with bromine and glacial
acetic acid, containing small amounts of conc. HBr, in ether; yield, 60–70%
(1937 Butenandt). — **Rns.** With conc. HBr in glacial acetic acid-chloroform
at room temperature 2.6-dibromo-$\Delta^{1\cdot4\cdot6}$-cholestatrien-3-one (p. 2497 s) is obtained (1937 Butenandt).

2.2.6β-Tribromo-Δ^4-cholesten-3-one $C_{27}H_{41}OBr_3$ (p. 132: *4.6.6-Tribromo-*
Δ^4-cholesten-3-one). For revised struc-
ture and for configuration at C-6,
based on ultra-violet absorption, see
1955 Fieser. — Crystals (acetone)
m. 185–6° (1955 Fieser); the m. p. may
be as high as 194° by rapid heating
(1937 Butenandt); [α]$_D$ —6° (chloroform) (1955 Fieser). Absorption max.
in ether at 253 mμ (ε 13900) (1939 Dannenberg). Easily soluble in ether,
acetone, and benzene, sparingly soluble in glacial acetic acid, insoluble in
alcohol (1936a Butenandt). — **Fmn.** The yield by bromination of coprostenone in ether-glacial acetic acid (cf. p. 132) at room temperature is ca.
34% (1937 Butenandt).

Rns. Oxidation by means of KMnO$_4$ or CrO$_3$ in strong acetic acid diluted
by carbon tetrachloride, followed by debromination (zinc dust, KI, catalytic
hydrogenation) may be applied for obtaining Δ^4-pregnene-3.20-dione (progesterone, p. 2782 s) (1938, 1939 C. F. BOEHRINGER & SÖHNE). The compound
obtained by debromination with sodium iodide in boiling alcohol (p. 132)
is 2-bromo-Δ^4-cholesten-6-ol-3-one ethyl ether (see p. 2706 s); reacts similarly
in methanol to give the corresponding methyl ether (1937 Butenandt). Is
not altered by short heating in alcohol with potassium acetate or zinc dust
(1937 Butenandt); heating in a bomb at 230° (7 hrs.) in glacial acetic acid
saturated with potassium acetate gives so-called "4-bromocholestenone"
m. 147° (p. 2483 s) (1940 Barkow). The dibromocholestadienone obtained by
treatment of the present tribromo-ketone with HBr in glacial acetic acid-ethyl
acetate at room temperature (1937 Butenandt; cf. p. 132) is 2α.6(?)-dibromo-
$\Delta^{4\cdot6}$-cholestadien-3-one (see p. 2498 s).

2α.5α.6β-Tribromocholestan-3-one $C_{27}H_{43}OBr_3$ (p. 131: *4.5.6-Tribromochole-stan-3-one, isomer m. 137-8⁰*). Con-figuration at C-5 and C-6 derived from that of the precursor, 5α.6β-di-bromo-cholestan-3-one (p. 2500 s). The higher-melting epimer is the more stable, shown by its formation by epi-merization (see below), and so must have equatorial configuration (1953, 1954 Corey); for revision of the 4β.5α.6β-structure and configuration assigned by Corey (1953, 1954), see 1955 Fieser; 1956 Ellis.

Crystals (glacial acetic acid), m. 139.5–140⁰ (1954 Corey); $[\alpha]_D^{21}$ —44.5⁰ (chloroform) (1956 Ellis). — **Fmn.** From 5α.6β-dibromocholestan-3-one (cf. p. 131) by treatment with bromine (1 mol.) in ether-glacial acetic acid at 20⁰ for 20 min., followed by removal of ether in vacuo at 30⁰ (yield, 71%) (1936a Inhoffen); by similar procedure a product m. 117⁰ (probably a mixture of the 2α- and 2β-epimers, Editor) is obtained in 83% yield (1937 Krekeler). — **Rns.** Oxidation with $KMnO_4$ or CrO_3 in strong acetic acid, followed by debromination (zinc, KI, catalytic hydrogenation) is applied for its* con-version to Δ^4-pregnene-3.20-dione (progesterone, p. 2782 s) and Δ^4-androstene-3.17-dione (p. 2880 s) (1938, 1939 BOEHRINGER & SÖHNE).

2β.5α.6β-Tribromocholestan-3-one $C_{27}H_{43}OBr_3$ (p. 131: *4.5.6-Tribromochole-stan-3-one, isomer m. 106⁰*). For revis-ed structure and configuration, see those of the 2α-epimer, above. — Crys-tals (aqueous acetone), m. 106–106.3⁰ (1954 Corey); $[\alpha]_D^{21}$ +9⁰ (chloroform) (1956 Ellis). — **Fmn.** From 5α.6β-di-bromocholestan-3-one (p. 2500 s) (cf. p. 131) by reaction with 1 mol. of bromine in glacial acetic acid (containing small amounts of HBr) at room temperature, followed by precipitation by addition of water after 5–10 min., yield, 76% (1936a Butenandt). — **Rns.** Yields the 2α-epimer on standing at room temperature in ether saturated with HCl (1954 Corey).

3.4.6-Tribromo-Δ^5-cholesten-7-one $C_{27}H_{41}OBr_3$. The structure assigned is con-sidered to be the most probable. — Needles (chloroform-glacial acetic acid), dec. ca. 143⁰, but this temperature varies considerably with the rate of heating. Absorption max. (solvent not indicated) at 267 mμ (log ε 3.98). — **Fmn.** From the acetate of 7-ketocholesterol (p. 2682 s) by treatment with bromine and HBr in glacial acetic acid at 60⁰; yield, 35%. — **Rns.** Boiling with KI in acetone gives 6-bromo-$\Delta^{3.5}$-cholestadien-7-one (p. 2489 s). Treat-ment with $AgNO_3$ in pyridine at room temperature, with fused potassium

---

* Not indicated which epimer was used.

acetate in glacial acetic acid on the steam-bath, or (in lower yield) boiling
with dimethylaniline affords 4.6-dibromo-$\Delta^{3.5}$-cholestadien-7-one (p. 2501 s)
(1940 Jackson).

2.2.5α.6β-Tetrabromocholestan-3-one $C_{27}H_{42}OBr_4$ (p. 132: *4.4.5.6-Tetrabromo-cholestan-3-one*). For revision of structure and for the configuration assigned, see 1955 Fieser. — **Fmn.** From cholestenone dibromide (p. 2500 s) by reaction with 2 mols. of bromine in glacial acetic acid (containing some HBr) at
room temperature (18 hrs.); yield, 75% (1936a Butenandt). — **Rns.** The
dibromocholestadien-3-one obtained by the action of potassium acetate in
alcohol or by the action of HBr in ether - glacial acetic acid (see p. 132) is
2α.6(?)-dibromo-$\Delta^{4.6}$-cholestadien-3-one (see p. 2498 s).

1909 Dorée, *J. Ch. Soc.* **95** 638, 648, 649.
1915 Heiduschka, Gloth, *Arch. Pharm.* **253** 415, 421.
1936 a Butenandt, Schramm, *Ber.* **69** 2289, 2295, 2297.
   b Butenandt, Schramm, Wolff, Kudszus, *Ber.* **69** 2779.
   Ciba* (Ruzicka), *U.S. Pat.* 2085474 (issued 1937); *C. A.* **1937** 5950; *Ch. Ztbl.* **1938** I 374.
   Coffey, Heilbron, Spring, *J. Ch. Soc.* **1936** 738, 740.
   a Inhoffen, *Ber.* **69** 1134, 1137.
   b Inhoffen, *Ber.* **69** 2141, 2147.
   Ruzicka, Bosshard, Fischer, Wirz, *Helv.* **19** 1147 et seq., 1152, 1153.
   a Schering A.-G., *Ger. Pat.* 699248 (issued 1940); *C. A.* **1941** 6742; *Ch. Ztbl.* **1941** I 1325.
   b Schering A.-G. (Inhoffen), *Ger. Pat.* 722943 (issued 1942); *C. A.* **1943** 5201.
1937 Butenandt, Schramm, Kudszus, *Ann.* **531** 176, 179, 180, 182, 184, 192 et seq.
   Dane, Wang, Schulte, *Z. physiol. Ch.* **245** 80, 81, 85.
   Inhoffen, *Ber.* **70** 1695.
   Krekeler, *Einige Umsetzungen am Cholestenon,* Thesis Univ. Göttingen, pp. 10 et seq., 19 et seq.
   a Schering A.-G., *Austrian Pat. Application; FIAT Final Report* No. **996**, pp. 130, 139, 160, 166 (issued 1947).
   b Schering A.-G., *Fr. Pat.* 835526 (issued 1938); *C. A.* **1939** 4599; *Ch. Ztbl.* **1939** I 4088.
   a Schering-Kahlbaum A.-G., *Brit. Pat.* 486992 (issued 1938); *C. A.* **1938** 8707; *Ch. Ztbl.*
   1939 I 1602; *Brit. Pat.* 492725 (issued 1938); *C. A.* **1939** 1884; *Ch. Ztbl.* **1939** I 2642;
   Schering Corp. (Butenandt), *U.S. Pat.* 2248954 (issued 1941); *C. A.* **1941** 6742; *Ch. Ztbl.*
   1942 I 3285.
   b Schering-Kahlbaum A.-G., *Fr. Pat.* 835524 (issued 1938); *Brit. Pat.* 500353 (issued 1939);
   C. A. **1939** 5872; *Ch. Ztbl.* **1939** I 5010; Schering A.-G., *Dan. Pat.* 61375 (issued 1943);
   Ch. Ztbl. **1944** II 49; Schering Corp. (Inhoffen, Butenandt, Schwenk), *U.S. Pat.* 2340388
   (issued 1944); *C. A.* **1944** 4386.
   Schwenk, Whitman, *J. Am. Ch. Soc.* **59** 949.
1938 C. F. Boehringer & Söhne, *Fr. Pat.* 844850 (issued 1939); *C. A.* **1940** 7932; *Ch. Ztbl.* **1940** I
   1391.
   Dirscherl, Kraus, *Z. physiol. Ch.* **253** 64, 68.
   Inhoffen, Huang-Minlon, *Ber.* **71** 1720, 1724, 1726, 1728.
   Ralls, *J. Am. Ch. Soc.* **60** 1744, 1750, 1752.
   Wolff, *Über die Bromierung von Cholestanon und Koprostanon,* Thesis Techn. Hochschule
   Danzig, pp. 5, 8, 10, 14 et seq.
1939 C. F. Boehringer & Söhne, *Fr. Pat.* 860492 (issued 1941); *Ch. Ztbl.* **1941** I 3408.
   Butenandt, Mamoli, Dannenberg, Masch, Paland, *Ber.* **72** 1617, 1623.

* Gesellschaft für chemische Industrie in Basel; Société pour l'industrie chimique à Bâle.

Dannenberg, *Abhandl. Preuss. Akad. Wiss. Math.-naturwiss. Kl.* **1939** No. 21, pp. 27, 29, 35, 59, 62.
Décombe, Rabinowitch, *Bull. soc. chim.* [5] **6** 1510, 1516.
Inhoffen, Huang-Minlon, *Ber.* **72** 1686.
1940 Barkow, *Über bromfreie ungesättigte Sterinketone*, Thesis Techn. Hochschule Danzig, pp. 4, 14.
Inhoffen, Zühlsdorff, Huang-Minlon, *Ber.* **73** 451, 452.
Jackson, Jones, *J. Ch. Soc.* **1940** 659, 661, 663.
Jacobsen, *J. Am. Ch. Soc.* **62** 1620.
Korsching, Wirtz, *Ber.* **73** 249, 266.
1941 Ando, *C. A.* **1950** 4016.
Bretschneider, Ajtai, *Monatsh.* **74** 57; *Sitzber. Akad. Wiss. Wien II b* **150** 131.
GLIDDEN Co. (Julian, Cole), *U.S. Pat.* 2394551 (issued 1946); *C. A.* **1946** 2593.
1942 SCHERING A.-G., *Fr. Pat.* 878993 (issued 1943); *Ch. Ztbl.* **1943** II 250.
1943 Inhoffen, Zühlsdorff, *Ber.* **76** 233, 234 (footnote).
SQUIBB & SONS (Ruigh), *U.S. Pat.* 2413000 (issued 1946); *C. A.* **1947** 1396; *Ch. Ztbl.* **1947** 1102.
1946 Wilds, Djerassi, *J. Am. Ch. Soc.* **68** 1712, 1714.
1947 a Djerassi, Scholz, *J. Am. Ch. Soc.* **69** 2404, 2405, 2408; cf. *Experientia* **3** 107.
b Djerassi, *J. Org. Ch.* **12** 823, 827.
Inhoffen, *Angew. Ch.* A **59** 207.
1948 Inhoffen, Stoeck, *Experientia* **4** 426.
1949 Djerassi, *J. Am. Ch. Soc.* **71** 1003, 1008.
Inhoffen, Stoeck (with Nebel), *Ann.* **563** 127, 135.
1950 a Barton, Miller, *J. Am. Ch. Soc.* **72** 370, 373.
b Barton, Miller, *J. Am. Ch. Soc.* **72** 1066, 1069, 1670.
Djerassi, Rosenkranz, Romo, Kaufmann, Pataki, *J. Am. Ch. Soc.* **72** 4534.
Rivett, Wallis, *J. Org. Ch.* **15** 35, 38, 39.
1951 Djerassi, Rosenkranz, *Experientia* **7** 93.
a Inhoffen, *Angew. Ch.* **63** 297.
b Inhoffen, Kölling, Koch, Nebel, *Ber.* **84** 361, 368.
1952 Inhoffen, Becker, *Ber.* **85** 181.
Jones, Ramsay, Herling, Dobriner, *J. Am. Ch. Soc.* **74** 2828, 2829.
1953 Corey, *Experientia* **9** 329.
1954 Cookson, *J. Ch. Soc.* **1954** 282, 283.
Corey, *J. Am. Ch. Soc.* **76** 175, 179.
1955 M. Fieser, Romero, L. F. Fieser, *J. Am. Ch. Soc.* **77** 3305, 3307.
1956 Ellis, Petrow, *J. Ch. Soc.* **1956** 1179, 1181.
1957 Fieser, Fieser, *private communication.*

III. NITRO- AND AMINO-MONOKETOSTEROIDS
(CO IN THE RING SYSTEM)

3β-Chloro-6-nitro-Δ^5-androsten-17-one $C_{19}H_{26}O_3NCl$. For β-configuration at C-3, see the parent 3β-chloro-Δ^5-androsten-17-one (page 2474 s). — Crystals (methanol), m. 189–190°. — **Fmn.** From 3β chloro-Δ^5-androsten-17-one (p. 2474 s) on shaking with conc. HNO_3 (d 1.5) in glacial acetic acid; yield, 56%. — **Rns.** Gives 3β-chloroandrostane-6.17-dione (p. 2919 s) on heating with zinc dust in glacial acetic acid at 100°.

1946 Blunschy, Hardegger, Simon, *Helv.* **29** 199, 201, 202.

4'-(N-Methyl-anilinomethylene)-3'-keto-1.2-cyclopentano-1.2.3.4-tetrahydrophenanthrene $C_{25}H_{23}ON$ (I, R = H).

dl-cis-Form, 16-(N-Methyl-anilinomethylene)-dl-cis-norequilenan-17-one. Pale yellow prisms (toluene-light petroleum), m. 164°. — **Fmn.** From 4'-formyl-3'-keto-1.2-cyclopentano-1.2.3.4 tetrahydrophenanthrene (16-formyl-dl-cis-norequilenan-17-one, p. 2510 s) on heating with methylaniline in toluene on the steam-bath; yield 82% (1945 Birch). — **Rns.** For methylation, see the following compound.

2-Methyl-4'-(N-methyl-anilinomethylene)-3'-keto-1.2-cyclopentano-1.2.3.4-tetrahydrophenanthrene $C_{26}H_{25}ON$ (I, R = CH_3).

dl-cis-Form, 16-(N-Methyl-anilinomethylene)-dl-cis-equilenan-17-one, 16-(N-Methyl-anilinomethylene)-dl-3-desoxyisoequilenin. Pale yellow prisms (methanol), m. 149–150°. — **Fmn.** From the preceding compound (I, R = H) on heating in benzene with sodamide in a stream of hydrogen on the steam-bath, followed by addition of methyl iodide and further heating. — **Rns.** Gives dl-cis-2-methyl-3'-keto-1.2-cyclopentano-1.2.3.4-tetrahydrophenanthrene (dl-cis-equilenan-17-one, p. 2391 s) on refluxing with 10% HCl, followed by refluxing of the resulting product with 5% NaOH (1945 Birch).

17-Aminomethyl-Δ^4-androsten-3-one $C_{20}H_{31}ON$. Mode of formation not indicated. — **Rns.** Gives 17a-hydroxy-D-homo-Δ^4-androsten-3-one on treatment with sodium nitrite in very dil. acetic acid (1940 CIBA).

17-Aminomethyl-androstan-3-one $C_{20}H_{33}ON$. Mode of formation not indicated. **Rns.** Gives 17a-hydroxy-D-homo-androstan-3-one on treatment with sodium nitrite in very dil. acetic acid (1940 CIBA).

20-Amino-Δ^4-pregnen-3-one, 3-Keto-Δ^4-ternorcholenylamine $C_{21}H_{33}ON$. Crystals (ether) (1936 I. G. FARBENIND.). — **Fmn.** From 3β - hydroxy - Δ^5 - ternorcholenylamine (p. 1907 s) on refluxing with aluminium tert.-butoxide and acetone in benzene (yield, 98%) (1938 I. G. FARBENIND.) or from the "acetate" of the same amine on oxidation with CrO_3 in strong acetic acid at room temperature after temporary protection of the double bond by bromination (1935, 1936 I. G. FARBENIND.). — **Rns.** Gives Δ^4-pregnene-3.20-dione (progesterone, p. 2782 s) on treatment with HOCl solution in ether, followed by conversion of the resulting N-chloro-derivative with sodium methoxide in methanol into the ketimine, which is then hydrolysed with dil. H_2SO_4 (1936 I. G. FARBENIND.; 1939 Ehrhart; 1942 Ruschig); the dehydrohalogenation of the N-chloro-derivative may also be carried out with pyridine or $AgNO_3$ (1935 I. G. FARBENIND.). On reduction, using aluminium tert.-butoxide, it gives *3-hydroxy-Δ^4-ternorcholenylamine** $C_{21}H_{35}ON$, the "acetate" of which melts at 247–254° (1938 I. G. FARBENIND.).

Sulphate, colourless flakes, dec. ca. 311° (1938 I. G. FARBENIND.).

3-Amino-d-trans-equilenan-17-one, 3-Amino-d-desoxyequilenin $C_{18}H_{19}ON$. Prisms (after sublimation at 160–180°/$_{0.02}$), crystals (benzene or methanol), m. 219–220° cor. dec. (bath not preheated), 226–227° cor. (evacuated tube; bath preheated to 210°). — **Fmn.** From d-equilenin (p. 2527 s) on heating with aqueous ammonium disulphite solution in a sealed tube at ca. 170° for ca. 12 hrs.; yield, 71%. — **Rns.** Reverts to d-equilenin on treatment with sodium nitrite in hot 50% H_2SO_4. Gives d-3-desoxyequilenin (p. 2391 s) on diazotization with sodium nitrite in dil. HCl, followed by reduction of the resulting diazonium chloride with hypophosphorous acid, first at 2° (20 hrs.), then on the steam-bath (5 min.). — *Hydrochloride*, grey powder; insoluble in conc. HCl, slightly soluble in dil. HCl (1950 Bachmann).

N-Acetyl derivative $C_{20}H_{21}O_2N$. Prisms with 1 MeOH (methanol), m. 250° to 252° cor. vac. (bath not preheated), 264.5–265.5° cor. vac. (bath preheated to 200°). — **Fmn.** From the preceding compound on treatment with acetic anhydride and a little water at room temperature for 10 min. or on heating with acetic anhydride in benzene (1950 Bachmann).

17-Amino-Δ^4-androsten-3-one, 3-Keto-Δ^4-ætiocholenylamine $C_{19}H_{29}ON$. — **Fmn.** From the hydrochloride of 17-amino-Δ^5 androsten-3β-ol (p. 1910 s) on oxidation with CrO_3 in strong acetic acid in the cold after temporary protection of the double bond by bromination, followed by debromination with zinc and acetic acid in boiling benzene.

---

* Δ^4-Structure ascribed by Editor in view of the mode of formation; the compound may be a mixture of the C-3 epimers (cf. the analogous formation of coprostenol and epi-coprostenol, p. 37).

From the same amino-alcohol on refluxing with aluminium tert.-butoxide and acetone in benzene for 10–15 hrs. (1937a I. G. FARBENIND.). — **Rns.** Gives Δ^4-androstene-3.17-dione (p. 2880 s) on treatment with HOCl solution in ether in the presence of Na_2SO_4 at —5° to 0°, followed by conversion of the resulting N-chloro-derivative with sodium ethoxide in abs. alcohol on the steam-bath into the ketimine, which is then hydrolysed with dil. H_2SO_4 (1937b I. G. FARBENIND.; cf. 1939 Ehrhart). On treatment with sodium nitrite and acetic acid in dil. alcohol it gives Δ^4-androsten-17β-ol-3-one (testosterone, p. 2580 s) (1937a I. G. FARBENIND.).

17-Amino-ætiocholan-3-one, 3-Keto-ætiocholanylamine $C_{19}H_{31}ON$. — Fmn.

From 17-amino-ætiocholan-3-ol (p. 1911 s) on oxidation with CrO_3 in glacial acetic acid at room temperature (1937c I. G. FARBENIND.). — **Rns.** On bromination in glacial acetic acid in the presence of HBr it gives **4-bromo-3-keto-ætiocholanylamine** $C_{19}H_{30}ONBr$, which on treatment with $NaNO_2$ and acetic acid in alcohol is converted into **4-bromo-ætiocholan-17β-ol-3-one** $C_{19}H_{29}O_2Br$; the latter gives Δ^4-androsten-17β-ol-3-one (testosterone, p. 2580 s) on dehydrobromination with silver acetate and glacial acetic acid (1937a, c I. G. FARBENIND.).

1935 I. G. FARBENIND., *Fr. Pat.* 797960 (issued 1936); *C. A.* **1936** 7123; *Ch. Ztbl.* **1937** I 663; *Brit. Pat.* 465960 (issued 1937); *C. A.* **1937** 7444.
1936 I.G. FARBENIND., *Austrian Pat.* 160752 (issued 1942); *Ch. Ztbl.* **1942** II 1037.
1937 a I.G. FARBENIND., *Fr. Pat.* 819975 (issued 1937); *C. A.* **1938** 2956; *Ch. Ztbl.* **1938** I 2403; *Swiss Pat.* 207799 (issued 1940).
   b I.G. FARBENIND. (Bockmühl, Ehrhart, Ruschig, Aumüller), *Ger. Pat.* 693351 (issued 1940); *C. A.* **1941** 4921; *Brit. Pat.* 502666 (issued 1939); *C. A.* **1939** 7316; *Ch. Ztbl.* **1939** II 1126; *Fr. Pat.* 842026 (1938, issued 1939); *C. A.* **1940** 4743; *Ch. Ztbl.* **1940** I 429; *Swiss Pat.* 216340 (1938, issued 1941); *Ch. Ztbl.* **1942** II 197.
   c I.G. FARBENIND., *Swiss Pat.* 218515 (issued 1942); *Ch. Ztbl.* **1943** I 1389.
1938 I.G. FARBENIND., *Brit. Pat.* 508804 (issued 1939); *C. A.* **1940** 2864; *Ch. Ztbl.* **1940** I 1391; *Swiss Pat.* 215140 (issued 1941); *Ch. Ztbl.* **1942** I 1781.
1939 Ehrhart, Ruschig, Aumüller, *Angew. Ch.* **52** 363.
1940 CIBA (Soc. pour l'ind. chim. à Bâle; Ges. f. chem. Ind. Basel), *Fr. Pat.* 886414 (issued 1943); *Ch. Ztbl.* **1944** I 773, 774; *Dutch Pat.* 55220 (1941, issued 1943); *Ch. Ztbl.* **1944** I 301.
1942 Ruschig, *Medizin und Chemie* Vol. 4, pp. 327, 340.
1945 Birch, Jaeger, Robinson, *J. Ch. Soc.* **1945** 582, 585.
1950 Bachmann, Dreiding, *J. Am. Ch. Soc.* **72** 1329.

IV. HYDROXY-MONOKETO-STEROIDS (CO IN THE RING SYSTEM)

1. OH in Side Chain Only

4'-Hydroxymethylene-3'-keto-1.2-cyclopentano-1.2.3.4-tetrahydrophenan-
threne (I) and/or **4'-Formyl-3'-keto-1.2-cyclopentano-1.2.3.4-tetrahydrophenanthrene**
(II), $C_{18}H_{16}O_2$.

dl-cis-Form, 16-Hydroxymethylene-dl-cis-norequilenan-17-one (I) and/or
16-Formyl-dl-cis-norequilenan-17-one (II). Cream-coloured plates (ethyl
acetate-light petroleum b. 40–60⁰), m. 134⁰. Gives a purple-brown coloration
with alcoholic $FeCl_3$. — **Fmn.** From dl-cis-norequilenan-17-one (p. 2386 s)
on treatment with ethyl formate and sodium in benzene. — **Rns.** Gives
16-(N-methyl-anilinomethylene)-dl-cis-norequilenan-17-one (p. 2507 s) on
heating with methylaniline in toluene (1945 Birch).

17β-Hydroxymethyl-Δ⁴-androsten-3-one $C_{20}H_{30}O_2$. Rods (dil. acetone), m. 158⁰
to 159⁰ cor. (1939c Miescher). — Has no andro-
genic effect in the capon comb and rat tests (1940
Miescher). — **Fmn.** From 17β-hydroxymethyl-
Δ⁵-androsten-3β-ol (p. 1917 s) on refluxing with
aluminium isopropoxide and cyclohexanone in
toluene (1939c Miescher). — **Rns.** Gives 3-keto-Δ⁴-ætiocholenic acid (p. 199)
on oxidation with CrO_3 in acetic acid at 5⁰ (1939c Miescher). On heating with
oxalic acid in vacuo it gives D-homo-Δ⁴·¹⁷-androstadien-3-one(?) (1941 CIBA).

Acetate $C_{22}H_{32}O_3$. Crystals (hexane), m. 114–5⁰ cor. — **Fmn.** From the
above hydroxy-ketone with acetic anhydride and pyridine at room temperature. — *Semicarbazone* $C_{23}H_{35}O_3N_3$, crystals (methanol), m. 214–5⁰ cor. (1939c
Miescher).

17-Methyl-18-nor-Δ⁴·¹³-pregnadien-20-ol-3-one, Δ⁴·¹³-Retropregnadien-20-ol-
3-one $C_{21}H_{30}O_2$. For a compound which may
possess this structure (1942 Ruzicka), see
p. 2409 s; for the acetate of the corresponding
saturated compound (1942 Ruzicka), see the
acetoxy-ketone $C_{23}H_{36}O_3$, p. 2409 s.

Δ⁴-Pregnen-20α-ol-3-one $C_{21}H_{32}O_2$ (p. 132). Crystals (ether), m. 158–160⁰
(1941a Marker); crystals (aqueous alcohol), m. 162⁰
(1938 CIBA). — **Fmn.** From the 20-benzoate of
Δ⁴-pregnene-3.20α-diol (? p. 1919 s) on oxidation
with CrO_3 in glacial acetic acid at 15–20⁰ (after
temporary protection of the double bond with
bromine), followed by hydrolysis with methanolic KOH (1937 SCHERING
CORP.; cf. 1951 Hirschmann). From Δ⁵-pregnene-3β.4β.20α-triol (p. 2106 s)

on boiling with aqueous alcoholic HCl; yield, 62 % (1941 a Marker). From Δ^5-pregnen-3β-ol-20-one (p. 2233 s) on refluxing with aluminium tert.-butoxide in toluene (1937 N. V. Organon)*. From progesterone 3-enol-benzoate (page 2229 s) on refluxing with aluminium isobutoxide in isobutyl alcohol for 5 hrs., followed by refluxing with K_2CO_3 in aqueous methanol; yield, 50 % (1938 Ciba)*. From 17β-formyl-Δ^4-androsten-3-one (p. 2777 s) on treatment with dimethylzinc (1937 Ciba).

Rns. Gives progesterone (p. 2782 s) on oxidation with CrO_3 in acetic acid at room temperature (1941 a Marker; see also 1935 Schering-Kahlbaum A.-G.).

Pregnan-20α-ol-3-one $C_{21}H_{34}O_2$ (p. 132). — **Fmn.** From pregnane-3.20-dione 3-enol derivatives on refluxing with aluminium iso-butoxide in isobutyl alcohol, followed by refluxing with K_2CO_3 in aqueous methanol (1938 Ciba)*. — **Rns.** Gives pregnane-3.20-dione (p. 2796 s) on heating with copper powder in a vac. at 200° (1936/37 Ciba). On hydrogenation of the acetate in acetic-hydrobromic acid in the presence of platinum oxide it gives the 20-acetate of pregnane-3β.20α-diol (p. 1930 s) (1937 b Marker). On heating of the acetate semicarbazone with sodium ethoxide in abs. alcohol at 180°, pregnane-3α.20α-diol (p. 1923 s) is formed. On refluxing the acetate with zinc + conc. HCl in alcohol, followed by heating with alcoholic NaOH, it gives pregnan-20α-ol (p. 1487 s). The acetate gives the 4β-halogeno-derivatives (pp. 2513 s, 2514 s) on chlorination or bromination (1938 Parke, Davis & Co.).

Allopregnan-20α-ol-3-one $C_{21}H_{34}O_2$ (p. 133). The pure hydroxy-ketone has been described after the closing date for this volume; it melts at 178.5–181° cor., its acetate at 159–160.5° cor. (the solidified melt remelts at 162° to 163.5° cor.) (1951 Hirschmann). — **Fmn.** An allopregnan-20-ol-3-one (no constants given) may be obtained from allopregnane-3.20-dione 3-enol derivatives on refluxing with aluminium isobutoxide in isobutyl alcohol, followed by refluxing with K_2CO_3 in aqueous methanol (1938 Ciba). — **Rns.** On refluxing of the preparation m. 128° (see p. 133) with zinc strips and aqueous alcoholic HCl it gives allo-pregnan-20α-ol (p. 1488 s) and a *diol*** $C_{21}H_{36}O_2$, crystals (methanol), m. 229°, which gives a *diacetate* $C_{25}H_{40}O_4$, needles (dil. alcohol), m. 162° (1939 Marker).

Allopregnan-20β-ol-3-one $C_{21}H_{34}O_2$ (p. 133). — **Rns.** Gives allopregnan-20β-ol (p. 1488 s) on refluxing of the acetate with zinc strips and aqueous alcoholic HCl, followed by hydrolysis with alcoholic NaOH (1939 Marker). See also under formation of $\Delta^{17(20)}$-allopregnen-3-one (p. 2411 s).

---

* Marker (1941 b) obtained 20β-hydroxy-compounds on Meerwein-Ponndorff reduction of several 20-keto-compounds.

** This diol may be allopregnane-3β.20α-diol (p. 1933 s), m. 220 2° cor.; diacetate, m. 165° (Editor).

Δ**4·16·20-Pregnatrien-21-ol-3-one** $C_{21}H_{28}O_2$. **Enolic form of a pregnadien-3-on-21-al.**

Me

Me————CH:CH·OH   *Acetate* $C_{23}H_{30}O_3$. Leaflets (methanol), m. 192–4⁰ cor.; sublimes at 175⁰ in a high vac.

O=

On heating with 1.4-dihydroxy-naphthalene in glacial acetic acid + conc. HCl it gives a red coloration, on heating with 1.4-naphthoquinone in glacial acetic acid + conc. HCl a weakly red-violet coloration. — **Fmn.** From Δ4·17(20)-pregnadien-3-on-21-al (p. 2802 s) or from 17-iso-Δ4-pregnen-17β-ol-3-on-21-al (p. 2846 s) on boiling with acetic anhydride in glacial acetic acid under nitrogen; in the second case there was isolated in addition a *stereoisomeride (?)* (stereoisomeric at the 20.21-double bond?) $C_{23}H_{30}O_3$(?) [needles (methanol), m. 262–4⁰ cor. dec.], which was not obtained analytically pure (1939b Miescher).

Δ**4·17(20)-Pregnadien-21-ol-3-one** $C_{21}H_{30}O_2$. Needles (petroleum ether or aqueous methanol), crystals (hexane), m. 138–9⁰ (1939a

Me

Me————=CH·CH$_2$·OH   Miescher; 1939a Ruzicka); [α]$_D$ +116.5⁰ (alcohol) (1939a Ruzicka). With conc. H_2SO_4 in

O=

glacial acetic acid it gives a green fluorescence gradually becoming orange. On heating with 1.4-naphthoquinone in glacial acetic acid + conc. HCl it gives a red-violet colour; the acetate shows the same colour (1939b Miescher). — Is inactive in the corpus luteum hormone test (1939 Logemann).

Fmn. From 17-vinyltestosterone (p. 2650 s) on heating with acetic anhydride at 100⁰ and then with trichloroacetic acid in glacial acetic acid at 60⁰, followed by saponification with boiling methanolic K_2CO_3 (1939a Miescher; 1939b CIBA). The acetate is formed from 21-bromo-Δ4·17(20)-pregnadien-3-one (p. 2473 s) on boiling with potassium acetate in acetone (crude yield, 90%); it is saponified by boiling with methanolic KOH (1939a Ruzicka; 1939b CIBA; 1939 SCHERING A.-G.).

Rns. On shaking the benzene solution with $K_2Cr_2O_7$ and aqueous H_2SO_4 under cooling or on treatment with CrO_3 in glacial acetic acid, Δ4·17(20)-pregnadien-3-on-21-al (p. 2802 s) is formed (1939b Miescher; 1939a CIBA). Keeping of the acetate with monoperphthalic acid in ether at room temperature gives a *17.20-oxide*, **21-acetoxy-17.20-epoxy-Δ4-pregnen-3-one** $C_{23}H_{32}O_4$ [crystals (methanol), m. 125⁰ cor.; [α]$_D$ +99⁰ (dioxane)] (1939b Ruzicka; see also 1939 SCHERING A.-G.). Keeping of the acetate with OsO_4 in ether at room temperature, followed by heating of the reaction product with Na_2SO_3 in aqueous alcohol, yields Δ4-pregnene-17α.20β.21-triol-3-one (p. 2725 s) (1939b Ruzicka; 1939 Logemann).

Acetate, 21-Acetoxy-Δ4·17(20)-pregnadien-3-one $C_{23}H_{32}O_3$. Crystals (hexane), m. 107⁰ (1939a Miescher; 1939a Ruzicka). — **Fmn.** See above; also by acetylation of the above hydroxy-ketone with acetic anhydride and pyridine at room temperature (1939a Miescher). — **Rns.** See above.

$\Delta^{4.20}$-Pregnadien-21-ol-3-one, Enolic form of Δ^4-pregnen-3-on-21-al $C_{21}H_{30}O_2$.

Acetate $C_{23}H_{32}O_3$. — **Fmn.** From Δ^4-pregnen-3-on-21-al (p. 2803 s) on boiling with 20 parts of acetic anhydride and 1 part of anhydrous sodium acetate for 6 hrs. — **Rns.** On treatment with perbenzoic acid in ether, followed by hydrolysis with hot alcoholic HCl and then with aqueous alcoholic K_2CO_3, it gives Δ^4-pregnen-20-ol-3-on-21-al (p. 2833 s) (1939a CIBA).

2 or 4-Hydroxymethylene-cholestan-3-one (Ia or Ib) and/or **2 or 4-Formyl-cholestan-3-one** (IIa or IIb), $C_{28}H_{46}O_2$. The 2-position is more likely (see 1949 Fieser). — Microcrystalline powder (chloroform-methanol), m. 182–184°, clearing at 195° (1938 Stiller), m. 176–8° cor. (1943 Goldberg), 186° (1938 CIBA PHARMACEUTICAL PRODUCTS). Sparingly soluble in alcohol, ethyl acetate,

and glacial acetic acid, moderately soluble in dioxane, heptane, and benzene, readily in chloroform. Gives an intense purple coloration with alcoholic $FeCl_3$ (1938 Stiller). — **Fmn.** From cholestan-3-one (p. 2436 s) on treatment in ether with isoamyl formate and sodium (1938 Stiller; 1938 CIBA PHARMACEUTICAL PRODUCTS) or sodium ethoxide (1943 Goldberg).

For **2 or 4-(α-hydroxybenzyl)-cholestan-3-one** $C_{34}H_{52}O_2$ (1943 Goldberg), see p. 2440 s.

22.22-Diphenyl-Δ^4-bisnorcholen-22-ol-3-one, (3-Keto-Δ^4-ternorcholenyl)-di-phenyl-carbinol $C_{34}H_{42}O_2$. Not characterized. — **Fmn.** From (3β-hydroxy-Δ^5-ternor-cholenyl)-diphenyl-carbinol (p. 1947 s) on refluxing with aluminium tert.-butoxide and cyclohexanone in toluene for 20 min. —

Rns. Gives 1-methyl-2.2-diphenyl-1-(3-keto-Δ^4-ætiocholenyl)-ethylene (p. 2472 s) on refluxing with glacial acetic acid for 20 min. (1941 GLIDDEN Co.).

4β-Chloropregnan-20α-ol-3-one* $C_{21}H_{33}O_2Cl$ (III, R = Cl). — **Fmn.** Its acetate is obtained from the acetate of pregnan-20α-ol-3-one (p. 2511 s) by chlorination. — **Rns.** Gives the diacetate of pregnane-4(?).20α-diol-3-one (p. 2720 s) on refluxing of the acetate with potassium acetate in glacial acetic acid (1938 PARKE, DAVIS & Co.).

---

* For β-configuration at C-4, assigned by Editor, see that of 4β-bromocoprostan-3-one (p. 2483 s).

4β-Bromopregnan-20α-ol-3-one* $C_{21}H_{33}O_2Br$ (III, R = Br) (p. 133). — **Rns.**
On refluxing of its acetate with potassium acetate in glacial acetic acid it
gives the diacetate of pregnane-4(?).20α-diol-3-one (p. 2720 s) (1937 a Marker).

Δ^4-Pregnene-20.21-diol-3-one $C_{21}H_{32}O_3$. Was obtained in two forms, "α" and
"β", differing in the configuration of OH

Me

Me ——CH(OH)·CH$_2$·OH

O=

at C-20. — **Fmn.** The acetonides of the
two forms are obtained from a mixture of
stereoisomeric Δ^5-pregnene-3β.20.21-triol
20.21-acetonides (p. 1953 s) on refluxing
with aluminium tert.-butoxide in acetone-benzene, followed by chromatographic purification and separation on an Al_2O_3 column, the acetonide of the
"α"-form being the main product; the acetonides are then heated with
aqueous alcoholic acetic acid (1938 Steiger). The two forms are also produced
on keeping $\Delta^{4.20}$-pregnadien-3-one (p. 2410 s) with OsO_4 in ether at room
temperature, followed by refluxing with aqueous alcoholic Na_2SO_3 (1948
Julian; see also 1937 SCHERING A.-G.).

Δ^4-Pregnene-20"α".21-diol-3-one $C_{21}H_{32}O_3$. Needles (acetone) (1948 Julian)
m. 166–7° cor.; $[\alpha]_D^{20}$ +92.6° (abs. alcohol) (1938 Steiger); $[\alpha]_D^{27}$ +98° (chloroform) (1948 Julian). Easily soluble in alcohol and acetone, sparingly in ether,
nearly insoluble in petroleum ether. Is inactive in the Everse-de Fremery
test in the rat (1938 Steiger). — **Fmn.** See above. — **Rns.** On keeping with
HIO_4 in aqueous dioxane under CO_2 it gives 17β-formyl-Δ^4-androsten-3-one
(p. 2777 s) (1940 Miescher). — *Acetonide, 20.21-Isopropylidene-Δ^4-pregnene-
20"α".21-diol-3-one* $C_{24}H_{36}O_3$. Colourless leaflets (pentane), m. 126° cor.; $[\alpha]_D^{20}$
+91.5° (acetone); absorption max. in hexane at 232 mμ (log ε 4.32; from
the graph); very easily soluble in all organic solvents except petroleum ether,
nearly insoluble in water (1938 Steiger). **Fmn.** See above.

Δ^4-Pregnene-20"β".21-diol-3-one $C_{21}H_{32}O_3$. Colourless prisms (acetone),
m. 192.5–194.5°; $[\alpha]_D^{35}$ +97° (chloroform) (1948 Julian; cf. 1938 Steiger). —
Fmn. See above. — *Acetonide, 20.21-Isopropylidene-Δ^4-pregnene-20"β".21-diol-
3-one* $C_{24}H_{36}O_3$. Colourless leaflets (pentane), m. 132° cor.; $[\alpha]_D^{20}$ +70.5° (acetone); absorption max. in alcohol at 241 mμ (log ε 4.14; from the graph);
shows the same solubilities as the "α"-isomeride (1938 Steiger). **Fmn.** See
above.

6-Methyl-Δ^4-pregnene-20.21-diol-3-one $C_{22}H_{34}O_3$. *Diacetate, 6-Methyl-20.21-
diacetoxy-Δ^4-pregnen-3-one* $C_{26}H_{38}O_5$. Crys-

Me

Me ——CH(OH)·CH$_2$·OH

O=

Me

talline warts (ether + petroleum ether),
m. 165–170°; absorption max. in abs. alcohol at 245.5 mμ (molecular extinction
coefficient 13.83 × 10^{-3}). Does not reduce
complex silver ammonium salts. — **Fmn.**
By passing a stream of dry HCl through an ice-cold solution of 6-methylpreg-

* See the footnote on p. 2513 s.

nane-5.20.21-triol-3-one   20.21-diacetate   (p. 2730 s)   in   chloroform   (1943
Ehrenstein).

16-Piperonylidene-dl-cis-norequilenan-17-one $C_{25}H_{20}O_3$ (I, R = H). Pale yellow prisms (benzene-light petroleum b. 40^0
to 60^0), m. 173–4^0. — **Fmn.** On heating
dl-cis-norequilenan-17-one (p. 2386 s) with
piperonal and alcoholic sodium ethoxide. —
Rns. Refluxing with methyl iodide and potassium tert.-butoxide in tert.-butyl alcohol
gives 16-piperonylidene-dl-cis-equilenan-17-one (below) (1941 Koebner).

16-Piperonylidene-dl-cis-equilenan-17-one $C_{26}H_{22}O_3$ (I, R = CH$_3$). Yellow flat
prisms or plates [ethyl acetate + light petroleum (b. 40–60^0) or + alcohol],
m. 158–9^0 (1941 Koebner; 1945 Birch). — **Fmn.** From 16-piperonylidene-
dl-cis-norequilenan-17-one (above) on refluxing with methyl iodide and potassium tert.-butoxide in tert.-butyl alcohol (1941 Koebner). From dl-cis-
equilenan-17-one (p. 2391 s) and piperonal (1945 Birch).

1935 Schering-Kahlbaum A.-G., *Fr. Pat.* 806467 (issued 1936); *C. A.* **1937** 4682; *Ch. Ztbl.* **1937** I
4265.
1936/37 Ciba*, *Brit. Pat.* 476749 (issued 1938); *C. A.* **1938** 3770; *Ch. Ztbl.* **1938** II 120; *Fr. Pat.*
823139 (1937, issued 1938); *C. A.* **1938** 5585; *Ch. Ztbl.* **1938** II 120.
1937 Ciba*, *Swiss Pat.* 215287 (issued 1941); *C. A.* **1948** 3538; *Ch. Ztbl.* **1942** I 2430.
  a Marker, Kamm, Jones, *J. Am. Ch. Soc.* **59** 1595, 1596.
  b Marker, Kamm, Wittle, Oakwood, Lawson, Lautius, *J. Am. Ch. Soc.* **59** 2291, 2295.
  N. V. Organon, *Fr. Pat.* 823618 (issued 1938); *C. A.* **1938** 5854; *Ch. Ztbl.* **1938** II 355; *Brit.
  Pat.* 488987 (issued 1938); *C. A.* **1939** 324; *Ch. Ztbl.* **1939** II 474; *Dutch Pat.* 50942 (1938,
  issued 1941); *Ch. Ztbl.* **1942** I 513; *Swiss Pat.* 223018 (issued 1942); *Ch. Ztbl.* **1943** I 1798.
  Schering A.-G., *Swiss Pat.* 223208 (issued 1942); *Ch. Ztbl.* **1943** I 2113.
  Schering Corp. (Schwenk, Whitman, Fleischer), *U.S. Pat.* 2212104 (issued 1940); *C. A.*
  1941 468; *Ch. Ztbl.* **1943** I 653.
1938 Ciba* (Miescher, Fischer), *Ger. Pat.* 712857 (issued 1941); *C. A.* **1943** 4534; *Ch. Ztbl.* **1942** I 1282.
  Ciba Pharmaceutical Products (Ruzicka), *U.S.Pat.* 2281622 (issued 1942); *C.A.* **1942** 5958.
  Parke, Davis & Co. (Marker), *Brit. Pat.* 516846 (issued 1940); *Ch. Ztbl.* **1940** II 2648.
  Steiger, Reichstein, *Helv.* **21** 171, 175, 177 et seq.; Roche-Organon, Inc. (Reichstein),
  U.S. Pat. 2312482 (1939, issued 1943); *C. A.* **1943** 4863; *Ch. Ztbl.* **1945** II 687; *U.S. Pat.*
  2312483 (1941, issued 1943); *C. A.* **1943** 4862; *Ch. Ztbl.* **1945** II 687.
  Stiller, Rosenheim, *J. Ch. Soc.* **1938** 353, 354, 357.
1939 a Ciba*, *Fr. Pat.* 857122 (issued 1940); *Ch. Ztbl.* **1941** I 1324; Ciba Pharmaceutical Products (Miescher, Fischer, Scholz, Wettstein), *U.S. Pats.* 2275790, 2276543 (both issued
  1942); *C. A.* **1942** 4289, 4674.
  b Ciba*, *Fr. Pat.* 860223 (issued 1941); *Ch. Ztbl.* **1941** I 3550.
  Logemann, *Naturwiss.* **27** 196.
  Marker, Lawson, *J. Am. Ch. Soc.* **61** 852.
  a Miescher, Scholz, *Helv.* **22** 120, 124, 125.
  b Miescher, Wettstein, Scholz, *Helv.* **22** 894, 898 (footnote 1), 899, 900, 904, 905.
  c Miescher, Wettstein, *Helv.* **22** 1262, 1266 et seq.
  a Ruzicka, Müller, *Helv.* **22** 416, 420.
  b Ruzicka, Müller, *Helv.* **22** 755, 757.
  Schering A.-G. (Logemann), *Swed. Pat.* 102639 (issued 1941); *Ch. Ztbl.* **1942** II 74; *Fr. Pat.*
  868336 (1940, issued 1941); *Ch. Ztbl.* **1942** I 2038.

* Société pour l'industrie chimique à Bâle; Gesellschaft für chemische Industrie in Basel.

1940 Miescher, Hunziger, Wettstein, *Helv.* **23** 400, 402.
1941 CIBA*, *Dutch Pat.* 55 220 (issued 1943); *Ch. Ztbl.* **1944** I 301.
   GLIDDEN Co. (Julian, Cole), *U.S. Pat.* 2 394 551 (issued 1946); *C. A.* **1946** 2593.
   Koebner, Robinson, *J. Ch. Soc.* **1941** 566, 573.
   a Marker, Crooks, Wittbecker, *J. Am. Ch. Soc.* **63** 777.
   b Marker, Turner, Wagner, Ulshafer, Crooks, Wittle, *J. Am. Ch. Soc.* **63** 779.
1941/42 Duyvené de Wit, *Biochem. Z.* **310** 83, 90.
1942 Ruzicka, Goldberg, Hardegger, *Helv.* **25** 1680, 1683 et seq.
1943 Ehrenstein, *J. Org. Ch.* **8** 83, 88, 93, 94.
   Goldberg, Kirchensteiner, *Helv.* **26** 288, 299, 300.
1945 Birch, Jaeger, Robinson, *J. Ch. Soc.* **1945** 582, 585.
1948 Julian, Meyer, Printy, *J. Am. Ch. Soc.* **70** 887, 888, 891.
1949 Fieser, Fieser, *Natural Products Related to Phenanthrene*, 3rd Ed., New York, pp. 261–266.
1951 Hirschmann, Dans, Hirschmann, *J. Biol. Ch.* **192** 115, 116, 118, 119, 124, 126, 128.

2. Monohydroxy-monoketo-hydro-1.2-cyclopentenophenanthrenes Without a Side Chain

7 - Methoxy - 3 - keto - 1.2 - cyclopentano - 1.2.3.9.10.11 - hexahydrophenanthrene.

Form-A $C_{18}H_{20}O_2$ (p. 133). — **Rns.** Heating with aluminium isopropoxide and isopropyl alcohol gives 3-hydroxy-7-methoxy-1.2-cyclopentano-1.2.3.9.10.11-hexahydrophenanthrene (p. 1958 s). Reduction of the oxime (which is not described) with sodium and butyl alcohol, finally on the oil-bath, yields two isomeric 3-amino-7-methoxy-1.2-cyclopentano-1.2.3.4.9.10.11.12-octahydrophenanthrenes (p. 1910 s) (1941 Robinson).

7 - Methoxy - 3 - keto - 1.2 - cyclopentano - 1.2.3.4.9.10.11.12 - octahydrophenanthrenes-α and-β $C_{18}H_{22}O_2$ (p. 134). Regarding the methylation, cf. 1943 Johnson.

4-Hydroxy-3′-keto-1.2-cyclopentenophenanthrene $C_{17}H_{12}O_2$. Pale yellow needles (isoamyl alcohol); darkens from about 270°, softens at 300°, m. 310–315° to a brown tar. Very sparingly soluble in most organic solvents, except pyridine (1938 Robinson).

Fmn. As acetate, from 3-(2-naphthyl)-Δ^2-cyclopentenone-2-acetic acid on refluxing with acetic anhydride (yield, 90%); the acetate is hydrolysed by hot alcoholic NaOH (1938 Robinson). As acetate, from 4.3′-diketo-1.2-cyclopentano-1.2.3.4-tetrahydrophenanthrene (page 2872 s) on boiling with nitrobenzene and aqueous alcoholic NaOH for 3 min., followed by shaking with acetic anhydride (1941 Koebner).

Rns. On oxidation of the acetate with CrO_3 in boiling acetic acid it gives a small amount of an acid $C_{17}H_{10}O_7$ (perhaps mixed with the lower homologous acid $C_{16}H_8O_7$) which is probably **4-hydroxyphenanthrenequinone-2-carboxylic-**

---

* Société pour l'industrie chimique à Bâle; Gesellschaft für chemische Industrie in Basel.

1-acetic acid [orange prisms (ethyl acetate); does not melt at 350°; the solution in aqueous Na_2CO_3 is weak orange-yellow, in alcoholic KOH dull bluish green; forms a phenazine derivative with o-phenylenediamine in acetic acid]. Clemmensen reduction of the acetate in the presence of anisole affords a phenol which on distillation with zinc dust in a stream of hydrogen gives a mixture of hydrocarbons, m. about 160°. Couples with aromatic diazonium salts to give scarlet-red azo-compounds. Boiling with anisaldehyde and alcoholic KOH gives the 4′-anisylidene derivative (p. 2723 s). Condensation of the methyl ether with 2-hydroxy-3-methoxybenzaldehyde in ethyl acetate saturated with HCl yields "*8.4′-dimethoxyphenanthracyclopentadienochromylium chloride*" $C_{26}H_{19}O_3Cl$ [I; no analysis; bronze needles (alcoholic HCl); *FeCl₃ salt* $C_{26}H_{19}O_3Cl_4Fe$, needles, chocolate brown in mass (formic + acetic acids), shrinks at 245–8°, then gradually carbonizes; the red solution in H_2SO_4 is not fluorescent] (1938 Robinson).

The methyl ether condenses with ethyl bromoacetate, on refluxing with zinc wool in benzene - toluene, to give the ethyl ester of 4-methoxy-(1.2-cyclopenteno-phenanthrylidene)-3′-acetic acid and/or 4-methoxy-[1.2-($Δ^{1'·3'}$-cyclopentadieno)-phenanthrene]-3′-acetic acid (1941 Robinson).

Oxime $C_{17}H_{13}O_2N$. Colourless needles (ethyl acetate), m. 271° dec. (1938 Robinson).

Acetate $C_{19}H_{14}O_3$. Colourless needles (acetic acid), m. 207°. The yellow solution in H_2SO_4 exhibits a weak green fluorescence that becomes much stronger on gentle heating (1938 Robinson). — **Fmn.** and **Rns.** See above.

Methyl ether $C_{18}H_{14}O_2$. Colourless needles (alcohol), m. 179°. The solutions in neutral solvents exhibit a violet fluorescence. — **Fmn.** By treating an alcoholic suspension of the acetate (above) alternately with 40% KOH and dimethyl sulphate (1938 Robinson). — **Rns.** See above.

7-Hydroxy-3′-keto-1.2-cyclopenteno-3.4-dihydrophenanthrene, *Dehydronorequilenin* $C_{17}H_{14}O_2$.

Yellowish needles (glacial acetic acid), m. 319° dec. (bath preheated to 315°). Sparingly soluble in ether, moderately in alcohol to give a solution with a strong green fluorescence. Gives no colour with alcoholic $FeCl_3$. — **Fmn.** From the methyl ether (below) on heating with glacial acetic acid-HBr at 110° (1939 Chuang).

Methyl ether $C_{18}H_{16}O_2$. Pale yellow crystals (alcohol), m. 210–1°. Difficultly soluble in ether and cold alcohol, more easily soluble in acetone and benzene. The alcoholic solution exhibits a green fluorescence. — **Fmn.** From 2-[β-(6-methoxy-1-naphthyl)-ethyl]-cyclopentane-1.3-dione on boiling with P_2O_5 in benzene. — **Rns.** Decolorizes $KMnO_4$ in acetic acid. Gives a red precipitate with bromine in carbon tetrachloride. Refluxing with amalgamated zinc, conc. HCl, some toluene, and alcohol, followed by heating of the resulting oil with selenium at 300–320°, yields 7-methoxy-1.2-cyclopentenophenanthrene

(p. 1494 s). For hydrolysis, see above. On boiling with semicarbazide hydrochloride and potassium acetate in dil. alcohol, a difficultly soluble *semicarbazone* (dec. ca. 310⁰) is formed (1939 Chuang).

7 - Hydroxy - 3' - keto - 1.2 - cyclopentano - 1.2.3.4 - tetrahydro - phenanthrene

$C_{17}H_{16}O_2$ (I).

dl-cis Form, dl-18-Nor-isoequilenin, originally named
x-norequilenin (1938 Koebner). — For the cis-fusion
of rings C and D, see 1945 Birch; see also the references
quoted for the trans-fusion of these rings in equilenin (p. 2527 s). — Colourless
plates (methanol), m. 245–7⁰ vac. (1943 Bachmann). Very sparingly soluble
in most solvents (1941 Koebner).

Its acetate is œstrogenic in doses of 10 mg. (1941 Koebner).

Fmn. Its methyl ether is obtained from 7-methoxy-4.3'-diketo-1.2-cyclo-
pentano-1.2.3.4-tetrahydrophenanthrene (p. 2939 s) on hydrogenation in the
presence of a platinum-palladium-charcoal catalyst in alcohol at room temperature (1938 Koebner); it is demethylated by heating with HI (d. 1.75)
and acetic acid at 130–140⁰ for 15 min. (1941 Koebner). From 7-methoxy-
3'-keto-4'-carbomethoxy-1.2-cyclopentano-1.2.3.4-tetrahydrophenanthrene on
refluxing with acetic acid and hydrochloric acid in an atmosphere of nitrogen
(see synthesis below) (1943 Bachmann).

Synthesis (1943 Bachmann)*, see p. 2519 s.

Rns. The methyl ether gives 16-piperonylidene-dl-18-nor-isoequilenin methyl
ether (p. 2730 s) on heating with piperonal and aqueous alcoholic NaOH;
on similar condensation with benzaldehyde, but followed by exposure to the
air, 4'-benzylidene-7-methoxy-3'-keto-1.2-cyclopentenophenanthrene (page
2572 s) is obtained (1938, 1941 Koebner). On treatment of the methyl ether
with ethyl formate and powdered sodium in benzene under nitrogen at room
temperature it gives 16-hydroxymethylene-dl-18-nor-isoequilenin methyl ether
(p. 2719 s) (1945 Birch; see also 1947 Johnson).

Acetate $C_{19}H_{18}O_3$. Yellow needles (aqueous alcohol), m. 135–6⁰. Its yellow
solution in cold conc. H_2SO_4 develops a green fluorescence on heating. —
Fmn. From the above phenol on heating with acetic anhydride and pyridine
on the steam-bath. — **Rns.** Forms an orange-red *picrate* (1941 Koebner).

Methyl ether $C_{18}H_{18}O_2$. Prisms (methanol), m. 116–7⁰ (1938 Koebner). Its
yellow solution in cold conc. H_2SO_4 develops a green fluorescence on heating
(1939 Robinson; 1941 Koebner). — **Fmn.** See above under formation of the
phenol; may also be obtained from this compound by the action of dimethyl
sulphate and NaOH (1943 Bachmann). — **Rns.** See above under those of
the phenol. — *2.4-Dinitrophenylhydrazone* $C_{24}H_{22}O_5N_4$, orange-yellow plates
(acetic acid), m. 246–7⁰ dec. (1938 Koebner).

A **compound** $C_{18}H_{18}O_2$, yellow prisms m. 230⁰, claimed to be a *7-methoxy-
3'-keto-1.2-cyclopentano-1.2.3.4-tetrahydrophenanthrene*, was obtained from an

---

* For Robinson's synthesis of 18-nor-isoequilenin, see Robinson's synthesis of isoequilenin,
p. 2537 s.

Synthesis of dl-18-Nor-isoequilenin (I)
(1943 Bachmann)

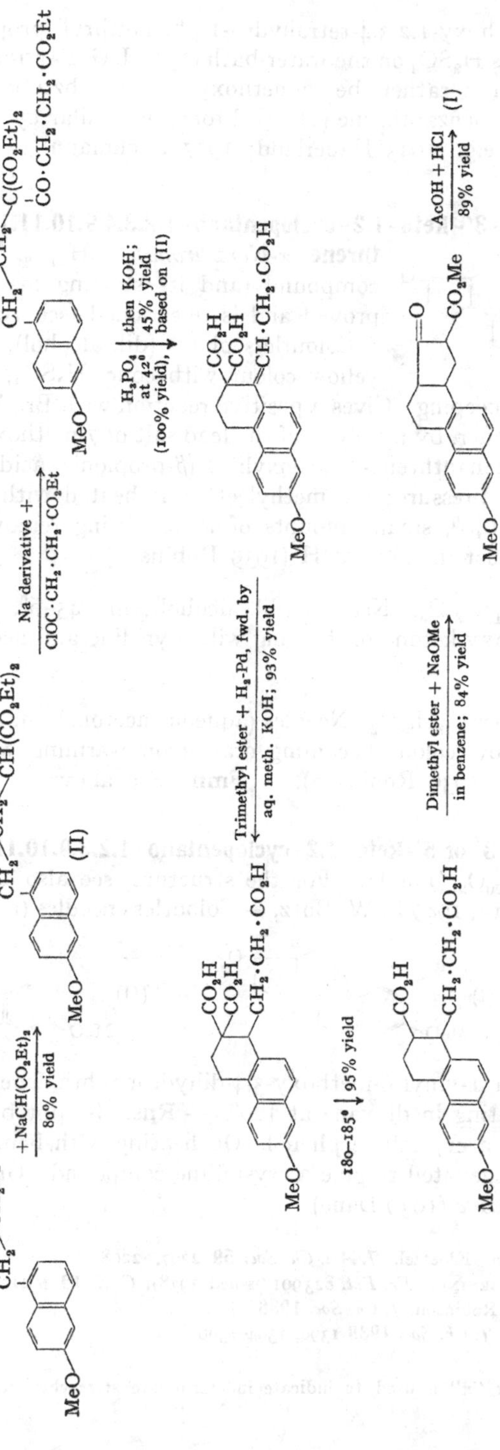

CH₂·CH₂·Br
+ NaCH(CO₂Et)₂
80% yield

CH₂·CH₂·CH(CO₂Et)₂ (II)

Na derivative +
ClOC·CH₂·CH₂·CO₂Et

CH₂·CH₂·C(CO₂Et)₂
·CO·CH₂·CH₂·CO₂Et

H₃PO₄ at 42°
(100% yield), then KOH;
45% yield based on (II)

CO₂H
CO₂H
CH·CH₂·CO₂H

Trimethyl ester + H₂-Pd, fwd. by
aq. meth. KOH; 93% yield

CO₂H
CO₂H
CH₂·CH₂·CO₂H

180–185° ; 95% yield

CO₂H
CH₂·CH₂·CO₂H

Dimethyl ester + NaOMe
in benzene; 84% yield

O
CO₂Me

AcOH + HCl
89% yield

(I)

oily β-(7-methoxy-1.2.3.4-tetrahydro-1-phenanthryl)-propionic acid on heating with aqueous H_2SO_4 on the water-bath (1937 I. G. FARBEN-IND.); it may rather be 6-methoxy-1.2.3.bz 1.bz 2.bz 3- (III) hexahydro-4-benzanthrone (III) (Editor; for similar cyclizations see, e.g., 1943 Haberland; 1937 Bachmann).

7 - Hydroxy - 3' - keto - 1.2 - cyclopentano - 1.2.3.4.9.10.11.12 - octahydrophenan-threne, *x-Norœstrone* * $C_{17}H_{20}O_2$. The structure of this compound (and its starting material) has not been proved and is questioned (see 1947 Stork). Colourless plates (dil. alcohol), m. 222°. Exhibits a yellow colour with conc. H_2SO_4, unaltered on gently warming or keeping. Gives a positive reaction with Brady's reagent. — **Fmn.** As methyl ether, by pyrolysis of the lead salt of 7-methoxy-1.2.3.4.9.10.11.12-octahydrophenanthrene-2-carboxylic-1-(β-propionic) acid (?) over a free flame at 0.25 mm. pressure; the methyl ether is heated with acetic acid and HI (d. 1.7) at 140°, small amounts of iodine being removed by boiling with some zinc dust in dil. NaOH (1939 Robinson).

Acetate $C_{19}H_{22}O_3$. Needles (dil. alcohol), m. 145–6°. — **Fmn.** From the above hydroxy-ketone on heating with pyridine and acetic anhydride (1939 Robinson).

Methyl ether $C_{18}H_{22}O_2$. Needles (aqueous acetone), m. 142–3°. Conc. H_2SO_4 gives a yellow colour becoming orange on warming without developing a fluorescence (1939 Robinson). — **Fmn.** See above.

7 - Methoxy - 3' or 5' - keto - 1.2 - cyclopentano - 1.2.3.9.10.11 - hexahydrophenan-threne $C_{18}H_{20}O_2$ (I or II). For the structure, see also 1939 Robinson; 1942 E. W. J. Butz; 1949 L. W. Butz. — Colourless needles (methanol), m. 141°. —

(I) MeO   (II) MeO

Fmn. From 1-vinyl-6-methoxy-3.4-dihydronaphthalene and Δ^2-cyclopente-none on heating in dioxane at 120°. — **Rns.** Is split by HBr-glacial acetic acid to give a crystalline phenol. On heating with benzoquinone in anisole it is dehydrogenated to give a crystalline compound. Gives a yellow *dinitro-phenylhydrazone* (1939 Dane).

1937 Bachmann, Kloetzel, *J. Am. Ch. Soc.* **59** 2207, 2208.
    I. G. FARBENIND., *Fr. Pat.* 823901 (issued 1938); *C. A.* **1938** 4176; *Ch. Ztbl.* **1938** II 556.
1938 Koebner, Robinson, *J. Ch. Soc.* **1938** 1994.
    Robinson, *J. Ch. Soc.* **1938** 1390, 1394–1396.

  * The prefix "x" is used to indicate indeterminate stereochemical configuration (see 1939 Robinson).

1939 Chuang, Ma, Tien, Huang, *Ber.* **72** 949, 951, 952.
Dane, Eder, *Ann.* **539** 207, 211, 212.
Robinson, Rydon, *J. Ch. Soc.* **1939** 1394, 1397, 1398, 1402, 1403.
1941 Koebner, Robinson, *J. Ch. Soc.* **1941** 566, 572–574.
Robinson, Slater, *J. Ch. Soc.* **1941** 376, 379, 380, 383.
1942 E. W. J. Butz, L. W. Butz, *J. Org. Ch.* **7** 199, 214.
1943 Bachmann, Gregg, Pratt, *J. Am. Ch. Soc.* **65** 2314, 2316, 2317.
Haberland, *Ber.* **76** 621.
Johnson, *J. Am. Ch. Soc.* **65** 1317.
1945 Birch, Jaeger, Robinson, *J. Ch. Soc.* **1945** 582, 585.
1947 Johnson, Posvic, *J. Am. Ch. Soc.* **69** 1361, 1366.
Stork, *J. Am. Ch. Soc.* **69** 576, 578.
1949 L. W. Butz, Rytina, *Organic Reactions*, Vol. V, New York, pp. 136, 139, 189, 190.

---

3. Monohydroxy-monoketo-hydro-1.2-cyclopentenophenanthrenes with One Side Chain

a. METHYL COMPOUNDS WITH CH_3 IN OTHER POSITIONS THAN AT C-13 OF THE STEROID RING SYSTEM

7-Methyl-4-hydroxy-3′-keto-1.2-cyclopentenophenanthrene $C_{18}H_{14}O_2$. Sandy powder, m. 290° dec.; sparingly soluble. — **Fmn.** From its acetate (below) on boiling with aqueous alcoholic NaOH or on attempted reduction by Clemmensen's method (1939 Kon).

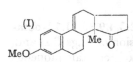

Acetate $C_{20}H_{16}O_3$. Needles (ethyl acetate), m. 224° dec. — **Fmn.** From 3-(6-methyl-2-naphthyl)-Δ^2-cyclopentenone-2-acetic acid on boiling with acetic anhydride. — **Rns.** Hydrogenation with Adams's PtO_2 catalyst in acetic acid at 80–5° until about 6 mols. hydrogen are absorbed, followed by dehydrogenation of the oily reduction product with palladium-charcoal at 320°, gives 7-methyl-1.2-cyclopentenophenanthrene (p. 1388 s) (1939 Kon).

Methyl ether $C_{19}H_{16}O_2$. Needles (benzene-petroleum ether), m. 190–1°. — **Fmn.** From the acetate (above) on boiling with aqueous alcoholic NaOH, followed by treatment with dimethyl sulphate. — **Rns.** Treatment with methylmagnesium iodide in ether yields 7.3′-dimethyl-4-methoxy-1.2-($\Delta^{1'.3'}$-cyclopentadieno)-phenanthrene (p. 1501 s; see also p. 2213 s) (1939 Kon).

1 - Methyl - 7 - methoxy - 5′ - keto-1.2-cyclopentano-1.2.3.9.10.11-hexahydrophenanthrene (I) or **2-Methyl-7-methoxy-3′-keto-1.2-cyclopentano - 1.2.3.9.10.11 - hexahydrophenanthrene** (X, p. 2541 s; R = CH_3), $C_{19}H_{22}O_2$. Yellow, very viscous oil, b. 178–212° in a high vacuum. — **Fmn.** From 1-vinyl-6-methoxy-3.4-dihydronaphthalene and excess 2-methyl-Δ^2-cyclopentenone on heating in the presence of pyrogallol under CO_2 in a closed tube at 200–205° for 24 hrs.; yield, 63%. — *2.4-Dinitrophenyl-hydrazone* $C_{25}H_{26}O_5N_4$. Yellow needles and orange prisms (benzine b. 70–80°), m. 141–144° (1937 Bockemüller; cf. also 1938 Bockemüller).

175*

cis *-1-Methyl-7-hydroxy-5'-keto-1.2-($\Delta^{3'}$-cyclopenteno)-1.2.3.4.9.10.11.12-octahydrophenanthrene

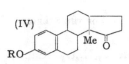

$C_{18}H_{20}O_2$ (II). Originally formulated as a stereoisomeric form of 15-dehydrooestrone (III) (1939 Dane); for structure (II), see 1942 E. W. J. Butz; 1949 L. W. Butz; cf. also 1938 Dane. — Crystals (alcohol), m. 244–5° (browning) (1939 Dane). — **Fmn.** From 1-methyl-7-methoxy-4'-hydroxy-5'-keto-1.2-cyclopentano-1.2.3.4.9.10.11.12-octahydrophenanthrene (p. 2737 s) on refluxing with glacial acetic acid and 48% HBr for $^1/_2$ hr.; crude yield, 70% (1939 Dane). — **Rns.** For reduction, see the following compound. Gives a red *2.4-dinitrophenylhydrazone* (1939 Dane).

cis *-1-Methyl-7-hydroxy-5'-keto-1.2-cyclopentano-1.2.3.4.9.10.11.12-octahydrophenanthrene

$C_{18}H_{22}O_2$. Originally formulated as a stereoisomeric form of oestrone (cf. the formula on p. 2543 s); for structure (IV, R = H), see 1942 E. W. J. Butz; 1949 L. W. Butz. — Prisms (alcohol), m. 210° (1939 Dane). — **Fmn.** From the preceding compound on shaking with hydrogen in methanol in the presence of palladium-charcoal (1939 Dane). — **Rns.** Gives a yellow *2.4-dinitrophenylhydrazone* (1939 Dane).

A **compound** $C_{18}H_{22}O_2$, m. ca. 140°, to which structure (IV, R = H) or that of a stereoisomeric form of oestrone (cf. the formula on p. 2543 s) was ascribed and which was said to possess the same hormonal activity as oestrone (1942 Breitner), is probably a mixture (1948 Heer; 1949 Miescher; 1948 Bachmann); it has been prepared from 1-vinyl-6-methoxy-3.4-dihydronaphthalene and citraconic anhydride (cf. also the formation of the following methyl ether), followed by reduction of the resulting 1 and/or 2-methyl-7-methoxyhexahydrophenanthrene-1.2-dicarboxylic anhydride (first with hydrogen over palladium in dioxane, then with sodium in boiling abs. alcohol), hydrolysis of the resulting phthalide to 1 and/or 2-methyl-2 and/or 1-hydroxymethyl-7-methoxy-octahydro-1 and/or 2-phenanthroic acid, successive treatment of the methyl ester with PBr_5 and ethyl sodiomalonate, hydrolysis, decarboxylation, cyclization of the resulting 1 and/or 2-methyl-7-methoxy-octahydrophenanthrene-1 and/or 2-carboxylic-2 and/or 1-(β-propionic) acid, and demethylation of the oily methyl ether (cf. the following ether) by heating with HBr-glacial acetic acid (1942 Breitner).

1-Methyl-7-methoxy-5'-keto-1.2-cyclopentano-1.2.3.4.9.10.11.12-octahydrophenanthrene

$C_{19}H_{24}O_2$ (IV, R = CH_3; with cis-fusion of rings C/D and, presumably, trans-fusion of rings B/C). Prisms (methanol), crystals (aqueous methanol), m. 100–101°. — **Fmn.** From 1-methyl-7-methoxy-1.2.3.4.9.10.11.12-octahydrophenanthrene-1-carboxylic-2-(β-propionic) acid (m. 215°; V) on heating with lead acetate in methanol, followed by distillation at 290–310°

---

* "cis" refers to the fusion of the rings C and D.

under 0.01 mm. pressure; the diacid (V) has been obtained by the following synthesis (1948 Bachmann; cf. 1951 Bachmann):

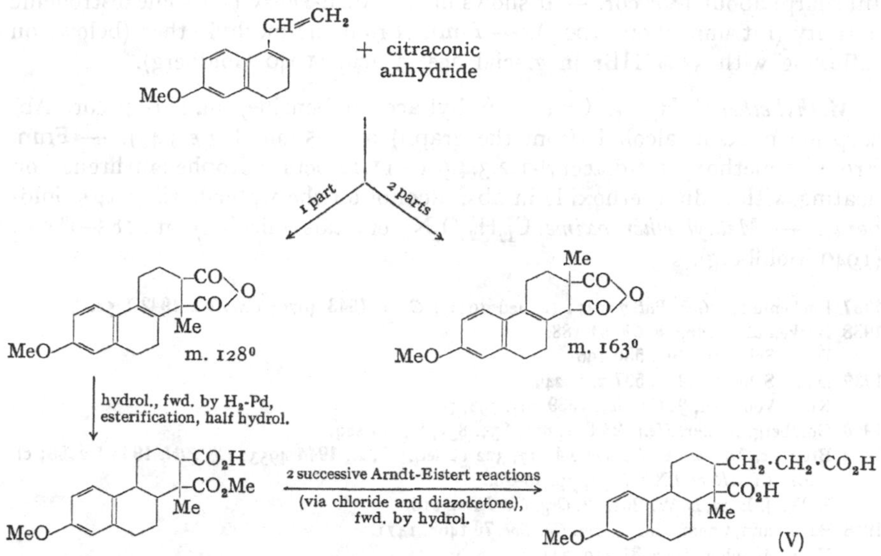

3′-Methyl-7-methoxy-5′-keto-1.2-(Δ3′-cyclopenteno)-1.2.3.4-tetrahydrophenanthrene (VI) or **5′-Methyl-7-methoxy-3′-keto-1.2-(Δ4′-cyclopenteno)-1.2.3.4-tetrahydrophenanthrene** *(15-Methyl-15-dehydro-x*-norequilenin methyl ether)* (VII), $C_{19}H_{18}O_2$. Crystals (chloroform-petroleum ether), m. 116–7⁰ cor. Absorption max. in alcohol (from the graph) at 340 (3.3), 282.5 (4.4), and 250 (4.3) mμ (log ε). — **Fmn.** From the 7-methoxy-1.2-diacetyl-1.2.3.4.9.10- or 1.2.3.9.10.11-hexahydrophenanthrene, m. 174–5⁰ cor. (obtained as main product, together with an isomer, m. 107–8⁰ cor., from 1-vinyl-6-methoxy-3.4-dihydronaphthalene and 1.2-diacetylethylene) on heating with sodium methoxide in methanol and some benzene on the water-bath with access of air (1940 Goldberg).

(VI)
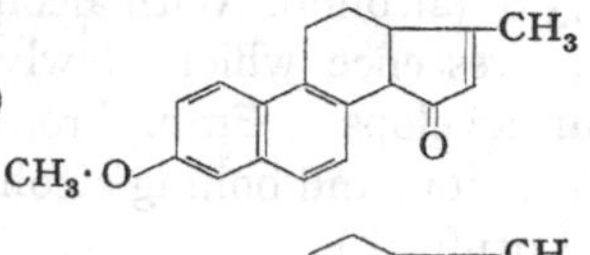

(VII)
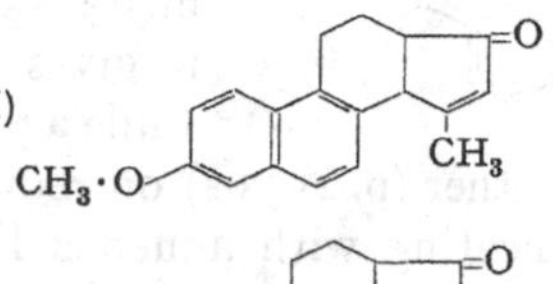

(VIII)
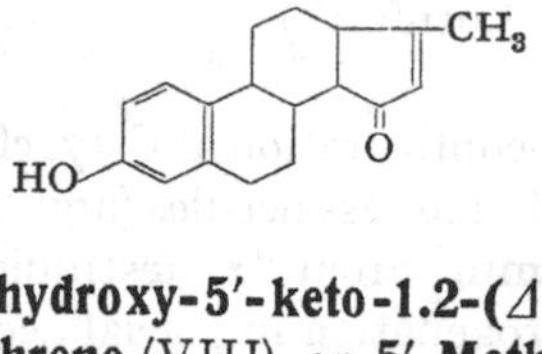

(IX)
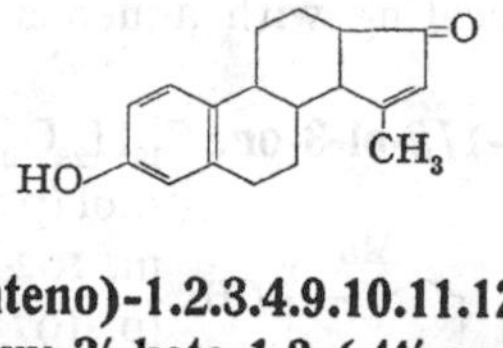

3′-Methyl-7-hydroxy-5′-keto-1.2-(Δ3′-cyclopenteno)-1.2.3.4.9.10.11.12-octahydrophenanthrene (VIII) or **5′-Methyl-7-hydroxy-3′-keto-1.2-(Δ4′-cyclopenteno)-1.2.3.4.9.10.11.12-octahydrophenanthrene** *(15-Methyl-15-dehydro-x*-norœstrone)* (IX), $C_{18}H_{20}O_2$. The free hydroxy-ketone is perhaps a mixture

* The prefix "x" is used by the authors to indicate indeterminate stereochemical correlation.

References, p. 2524 s

of isomers formed by partial shift of the isolated double bond or by partial change of configuration at C-13 or C-14. — Crystals (alcohol-water), m. (unsharp) about 180⁰ cor. — It shows in the Allen-Doisy test some œstrogenic activity (rat unit about 100 γ). — **Fmn.** From the methyl ether (below) on refluxing with 33% HBr in glacial acetic acid (1940 Goldberg).

Methyl ether $C_{19}H_{22}O_2$. Crystals (ethyl acetate-benzine), m. 181–3⁰ cor. Absorption max. in alcohol (from the graph) at 285 mμ (log ε 3.45). — **Fmn.** From 7-methoxy-1.2-diacetyl-1.2.3.4.9.10.11.12-octahydrophenanthrene on heating with sodium ethoxide in abs. alcohol on the water-bath (1940 Goldberg). — *Methyl ether oxime* $C_{19}H_{23}O_2N$, crystals (alcohol), m. 185–6⁰ cor. (1940 Goldberg).

1937 Bockemüller, *Ger. Pat.* 711471 (issued 1941); *C. A.* **1943** 4076; *Ch. Ztbl.* **1942** I 540.
1938 Bockemüller, *Angew. Ch.* **51** 188.
     Dane, Schmitt, *Ann.* **536** 196.
1939 Dane, Schmitt, *Ann.* **537** 246, 249.
     Kon, Woolman, *J. Ch. Soc.* **1939** 794, 798, 799.
1940 Goldberg, Müller, *Helv.* **23** 831, 833, 834, 835, 838 et seq.
1942 Breitner, *Medizin und Chemie* **4** 317, 322 et seq.; *C. A.* **1944** 4953; *Ch. Ztbl.* **1943** I 2688; cf. also *Cios Report* XXV–54, pp. 20–26.
     E. W. J. Butz, L. W. Butz, *J. Org. Ch.* **7** 199, 216.
1948 Bachmann, Chemerda, *J. Am. Ch. Soc.* **70** 1468, 1471.
     Heer, Miescher, *Helv.* **31** 219, 224.
1949 L. W. Butz, Rytina, *Organic Reactions*, Vol. V, New York, pp. 136, 139, 189, 190.
     Miescher, *Experientia* **5** 1, 9, 10.
1951 Bachmann, Controulis, *J. Am. Ch. Soc.* **73** 2636–2638.

b. **METHYL-MONOHYDROXY-MONOKETO-HYDRO-1.2-CYCLOPENTENOPHENANTHRENES WITH THE METHYL GROUP ATTACHED TO C-13 OF THE STEROID RING SYSTEM**

α. **CO in 3-, 12-, or 16-position**

$Δ^{5.7.9}$-**Oestratrien-17β-ol-3-one** $C_{18}H_{22}O_2$. The 17β-configuration is assigned

by analogy with various other 17-hydroxy-steroids, similarly obtained (Editor). — Platelets (benzene), m. 152–3⁰; $[α]_D^{22}$ + 33.6⁰ (alcohol). With alcoholic alkali it gives a green fluorescence which slowly vanishes while a purple colour develops. — **Fmn.** From equilenin methyl ether (p. 2533 s) on reduction with sodium and boiling alcohol, followed by heating with aqueous HCl (1942 Cornforth).

Oestran-17β-ol-3-one $C_{18}H_{28}O_2$. For the β-configuration at C-17, cf. "α"-œstradiol (p. 1966 s). — Colourless needles (aqueous acetone),

m. 102–104⁰. — **Fmn.** From "α"-œstradiol 17-acetate (p. 1973 s) on hydrogenation in glacial acetic acid-abs. alcohol in the presence of Adams's catalyst at room temperature under 0.7 atm. pressure, followed by oxidation with CrO_3 in acetic acid at room temperature and hydrolysis with alcoholic KOH (1940 Marker).

**2 - Methyl - 7 - methoxy - 3 - keto - 1.2 - cyclopentano - 1.2.3.4.9.10.11.12 - octahydro-
phenanthrene** $C_{19}H_{24}O_2$ (p. 134). Proof of the position
of the methyl group is lacking (1943 Johnson).

Stereoisomeride-β (p. 134). — The m. p. of the *dinitrophenylhydrazone* $C_{25}H_{28}O_5N_4$ is 171–2°, the m. p. of
the *semicarbazone* $C_{20}H_{27}O_2N_3$ is 226–7° (1937 Peak).

**2 - Methyl - 3' - hydroxy - 4' - keto - 1.2 - ($\Delta^{1(5')}$ - cyclopenteno) - 1.2.3.4 - tetrahydro-
phenanthrene** $C_{18}H_{16}O_2$. For the structure, see 1947
Wilds. — Light yellow leaflets (benzene), m. 207–208.5°
cor. (1944 Wilds). The absorption spectrum in abs.
alcohol shows max. at 219 (4.32), 239 (4.080), 246.5
(4.085), 268 (4.49), 277 (4.56), 317.5 (4.44) mμ (log ε)
and inflexions at 260 (4.25) and 360 (3.68) mμ (log ε) (1947 Wilds). The
crystals give a bright red colour with conc. H_2SO_4 and dissolve to an orange
solution (1944 Wilds).

Fmn. From 2-methyl-3'.4'-diketo-1.2-($\Delta^{1(5')}$-cyclopenteno)-1.2.3.4-tetra-
hydrophenanthrene (14-dehydroequilenan-16.17-dione, p. 2874 s) on hydrogenation in dioxane in the presence of palladium-charcoal at room temperature
under atmospheric pressure (1944 Wilds). — **Rns.** Oxidation with periodic
acid in aqueous dioxane at room temperature yields 2-methyl-2-formyl-1.2.3.4-
tetrahydro-1-phenanthrylidene-acetic acid (in its cyclized hydroxy-lactone
form) (1944, 1947 Wilds). For reduction at the dropping mercury electrode,
see 1944 Wilds.

1937 Peak, Robinson, *J. Ch. Soc.* 1937 1581, 1586.
1940 Marker, Rohrmann, *J. Am. Ch. Soc.* 62 73, 75.
1942 Cornforth, Cornforth, Robinson, *J. Ch. Soc.* 1942 689, 691.
1943 Johnson, *J. Am. Ch. Soc.* 65 1317.
1944 Wilds, Beck, *J. Am. Ch. Soc.* 66 1688, 1690, 1693.
1947 Wilds, Beck, Close, Djerassi, Johnson, Johnson, Shunk, *J. Am. Ch. Soc.* 69 1985, 1987,
      1988, 1993.

β. CO in 17-position of the steroid ring system

**2 - Methyl - 7 - methoxy - 3' - keto - 1.2 - ($\Delta^{1(5')}$ - cyclopenteno) - 1.2 - dihydrophenan-
threne** $C_{19}H_{16}O_2$ (I). A compound $C_{19}H_{16}O_2$ [yellow rods and needles (methanol
or ethanol), m. 170°], obtained from β-(2-methyl-7-methoxy-3.4-dihydro-
1-phenanthryl)-propionic acid and β-(2-methyl-7-methoxy-1.2.3.4-tetrahydro-
1-phenanthrylidene)-propionic acid, had been considered to possess structure
(I) or to be bz 1-methyl-6-methoxy-2.3-dihydro-4-benzanthrone (II); a

compound $C_{19}H_{18}O_2$ (needles, rhombohedra, and rods, m. 205–7°), obtained from the same acids, had been considered to be 2-methyl-7-methoxy-3'-keto-1.2 - ($\Delta^{1(5')}$ - cyclopenteno) - 1.2.3.4 - tetrahydrophenanthrene (IV, below) or bz 1-methyl-6-methoxy- 2.3.bz 2.bz 3 - tetrahydro-4-benzanthrone (III, above) (1939a, b, 1943 Haberland); for the structure of these compounds and their starting materials, see also 1947b, 1950 Johnson; 1951c Bachmann.

2-Methyl-7-methoxy-3'-keto-1.2-($\Delta^{1(5')}$-cyclopenteno)-1.2.3.4-tetrahydrophenanthrene, 14-Dehydroequilenin methyl ether $C_{19}H_{18}O_2$

(IV) (cf. also compound $C_{19}H_{18}O_2$, above).

dl-Form*. Colourless plates (acetone-methanol), m. 161.5–162.5° cor. (1945, 1947b Johnson); sublimes at 140° and 0.5–1 mm. pressure (1947b Johnson). Ultra-violet absorption max. in alcohol (1950 Johnson):

λ (mμ)	253	261.5	293.5	304.5	333	349.5
log ε	4.67	4.66	4.20	4.17	3.34	3.24

For its infra-red absorption spectrum, see 1948, 1952 Jones.

Fmn. (For *synthesis*, see p. 2530 s.) From 15-dehydroisoequilenin methyl ether (below) on partial rearrangement by refluxing with pyridine and conc. HCl; yield 44% (1947b Johnson). From 15-carboxy-15-dehydroisoequilenin methyl ether** on refluxing with pyridine and conc. HCl, whereby partly decarboxylation may be preceded by a shift of the double bond and partly 15-dehydroisoequilenin methyl ether, formed at the same time, is rearranged (1947b, 1951 Johnson); from the same acid on heating at 200° under 23 mm. pressure (1945, 1947b Johnson), along with a smaller amount of the 15-dehydro-isomer (below) (1947b Johnson).

Rns. On hydrogenation in ethyl acetate in the presence of palladium-charcoal at room temperature it gives 63% of dl-equilenin methyl ether (p. 2534 s) and 32% of dl-isoequilenin methyl ether (p. 2538 s) (1945, 1947b Johnson); the latter is the main product on catalytic hydrogenation in alcohol in the presence of piperidine or KOH (1947b Johnson). Is partly isomerized to the 15-dehydro-compound (below) by refluxing with pyridine and conc. HCl (1947b Johnson).

2-Methyl-7-methoxy-3'-keto-1.2-($\Delta^{4'}$-cyclopenteno)-1.2.3.4-tetrahydrophenanthrene $C_{19}H_{18}O_2$.

dl-cis-Form, dl-15-Dehydroisoequilenin methyl ether. Colourless blades (diisopropyl ether), m. 104.5–106° cor. (1947b Johnson). For its infra-red absorption spectrum, see 1948 Jones. — **Fmn.** (For *synthesis*, see p. 2530 s.) From the above 14-dehydro-isomer on partial rearrangement by

---

* A lævorotatory 14-dehydroequilenin methyl ether, corresponding in configuration to d-equilenin and also the above dl-form have been prepared by McNiven (1954) after the closing date of this volume.

** Originally formulated tentatively as the 14-dehydro-compound (1947b Johnson); for the true structure, see 1951 Johnson.

refluxing with pyridine and conc. HCl. From 15-carboxy-15-dehydroiso-equilenin methyl ether§ on heating at 200° under 23 mm. pressure, along with much of the 14-dehydro-isomer (cf. also the formation of the latter) (1947b Johnson). — **Rns.** On hydrogenation in ethyl acetate in the presence of palladium-charcoal it gives 96 % of dl-isoequilenin methyl ether (p. 2538 s). Is partly isomerized to the above 14-dehydro-compound by refluxing with pyridine and conc. HCl (1947b Johnson).

2-Methyl-7-hydroxy-3′-keto-1.2-cyclopentano-1.2.3.4-tetrahydrophenanthrenes $C_{18}H_{18}O_2$.

d-trans ("β")-Equilenan-3-ol-17-one, d-trans ("β")-3-Hydroxy-17-equi-lenone, d-Equilenin $C_{18}H_{18}O_2$ (Ia) (p. 134). For the trans-fusion of rings C and D in equilenin (d, l, and dl), see 1945 Birch; 1948 Shoppee; 1948 Klyne; 1948d Heer; 1949, 1950a, b, c, 1951a Bachmann; 1950 Stork. For the correspondence of the configuration both at C-13 and C-14 of equilenin and œstrone, see 1950a Bachmann; 1950 Rosen-kranz; 1950 Kaufmann.

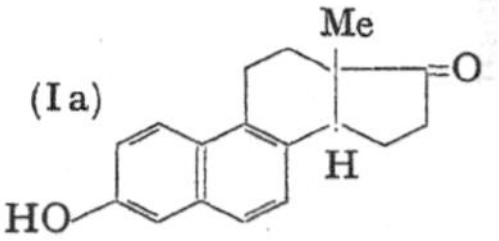

Colourless needles (alcohol), m. 258–9° cor. vac.; sublimes at 170–180° under 0.01 mm. pressure. Probably orthorhombic; refractive indices (D line) n_α 1.509, n_β 1.718, n_γ 1.800 (1932 Girard; 1940 Bachmann). $[\alpha]_D^{16} + 87°$ (dioxane) (1932 Girard).

Page 134 lines 19–18 from bottom: delete "$[\alpha]_D + 87°$ (alc.) (1933 Sandulesco)".

Ultra-violet absorption max. in alcohol:

     mμ 231†   270   282   292†   325   340 ⎱ 1935 Dirscherl; † 1936 Wintersteiner;
     log ε   4.78   3.94*   4.03*   3.58   3.68*   3.66*⎰ * from the graph

     mμ 230   270   280   292   328   340 ⎱ 1950 Kaufmann
     log ε* 4.67   3.70   3.76   3.52   3.27   3.30⎰ * from the graph

and in ether (1940 Bachmann):

     mμ 270   282   294   332   345
     log ε   3.85   3.87   3.74   3.58   3.68

Dirscherl (1935) found the same max. in ethereal solution as in alcoholic (see above).

For its infra-red absorption spectrum, see 1946 Furchgott.

Solubilities expressed in g. of d-equilenin per 100 g. solvent (1941 Doisy):

	Methanol	Ethanol 95%	Butanol	Dibutyl ether	Benzene	Toluene	Acetone	Chloroform	Dioxane
10°	0.665	0.753	0.500	0.066	0.093	0.058	1.70	0.18	4.69
30°	0.982	1.12	0.943	0.105	0.130	0.101	4.09	0.388	6.35

Solubility in petroleum ether (b. 89–92°) at 30°: 9.8 γ per c.c. (1941 Doisy). Partition ratio between 70 % alcohol and benzene, 1:6 (1938 Westerfeld).

§ See footnote** on p. 2526 s.

References, pp. 2541–2543 s

Bachmann's (1939, 1940) Synthesis of Equilenin and Isoequilenin*

(Percentages refer to yields)

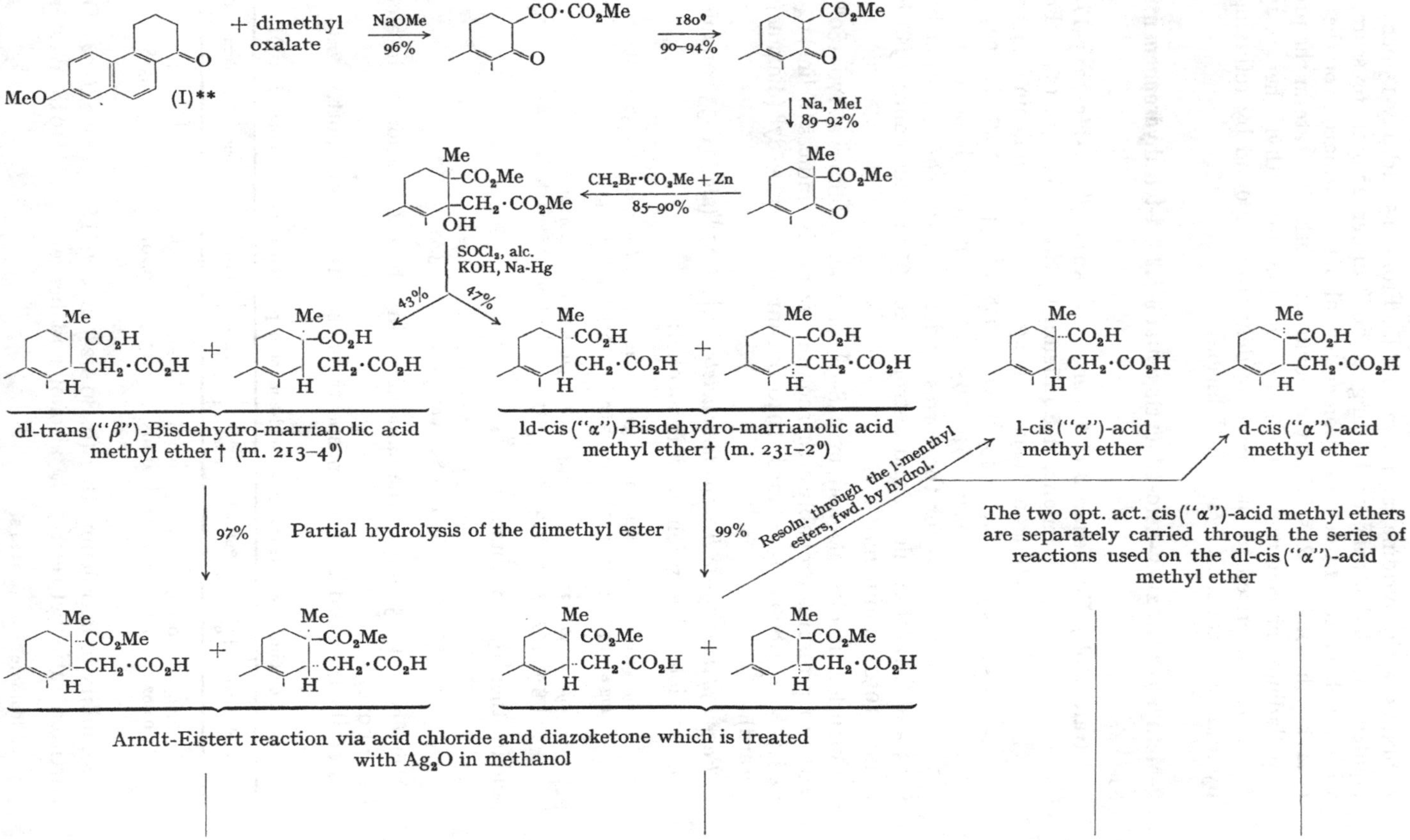
MeO
(I)**
+ dimethyl oxalate
NaOMe
96%
CO·CO₂Me
180°
90–94%
CO₂Me
Na, MeI
89–92%
Me
CO₂Me
CH₂Br·CO₂Me + Zn
85–90%
CH₂·CO₂Me
OH
SOCl₂, alc. KOH, Na-Hg
43%
47%
CO₂H
CH₂·CO₂H
H
dl-trans ("β")-Bisdehydro-marrianolic acid methyl ether† (m. 213–4°)
ld-cis ("α")-Bisdehydro-marrianolic acid methyl ether† (m. 231–2°)
l-cis ("α")-acid methyl ether
d-cis ("α")-acid methyl ether
97%
Partial hydrolysis of the dimethyl ester
99%
Resoln. through the l-menthyl esters, fwd. by hydrol.
The two opt. act. cis ("α")-acid methyl ethers are separately carried through the series of reactions used on the dl-cis ("α")-acid methyl ether
Arndt-Eistert reaction via acid chloride and diazoketone which is treated with Ag₂O in methanol

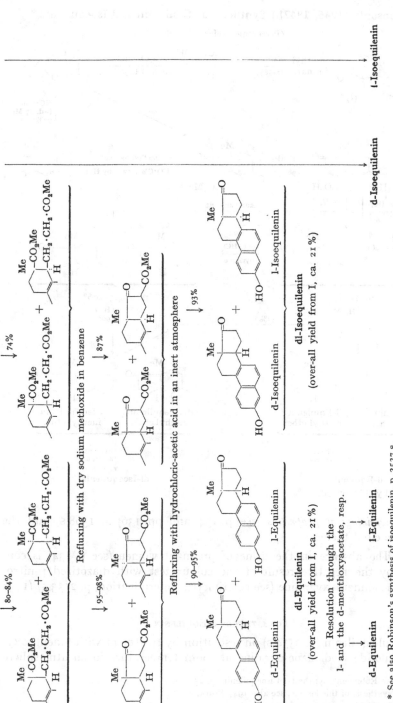

Refluxing with dry sodium methoxide in benzene

Refluxing with hydrochloric-acetic acid in an inert atmosphere

dl-Equilenin
(over-all yield from I, ca. 21 %)

Resolution through the
l- and the d-menthoxyacetate, resp.

d-Equilenin   l-Equilenin

dl-Isoequilenin
(over-all yield from I, ca. 21 %)

* See also Robinson's synthesis of isoequilenin, p. 2537 s.
** For synthesis of this ketone, see also 1947 Stork.
† For the name, see 1946 Heer; cis and trans refer to the relative orientation of H and Me at C-1 and C-2 of the phenanthrene ring system (see 1951 a Bachmann).

Johnson's (1945, 1947b) Synthesis of Equilenin and Isoequilenin *

(Percentages refer to yields)

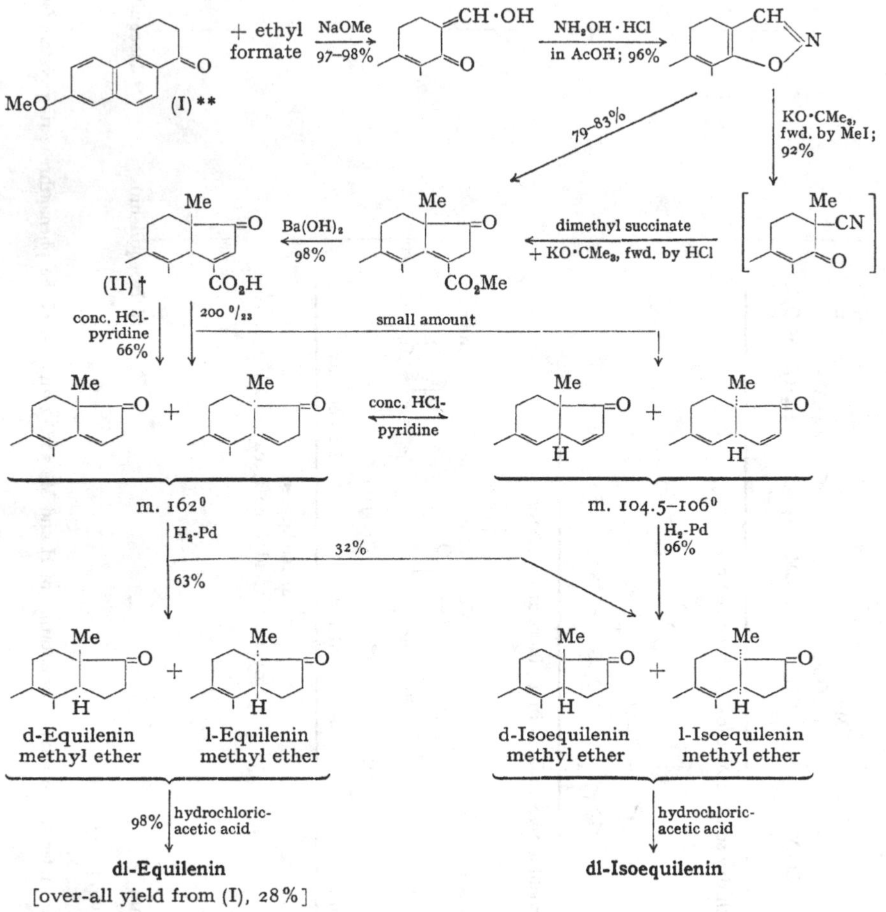

m. 162°   m. 104.5–106°

d-Equilenin   l-Equilenin   d-Isoequilenin   l-Isoequilenin
methyl ether   methyl ether   methyl ether   methyl ether

dl-Equilenin   **dl-Isoequilenin**
[over-all yield from (I), 28%]

For *physiological properties*, see p. 134 and pp. 1367 s, 1368 s, and the literature cited there.

Occ. The absence in the urine of pregnant women (see p. 134) is confirmed by the negative result of the colour test with diazotized p-nitrobenzeneazodimethoxyaniline (see colorimetric determination, p. 2533 s) (1938 Marx).

Formation of d-Equilenin

From dl-equilenin (p. 2534 s) on resolution by treatment with l-menthoxyacetyl chloride in dioxane-pyridine at room temperature in an atmosphere

---

* See also Robinson's synthesis of isoequilenin, p. 2537 s.
** For synthesis of this ketone, see also 1947 Stork.
† For the structure, see 1951 Johnson.

of nitrogen, isolation of d-equilenin l-menthoxyacetate (see p. 2533 s) from the mixture of diastereoisomeric esters by crystallization from petroleum ether-acetone and finally from acetone alone, and hydrolysis by refluxing with acetic acid and aqueous HCl in an atmosphere of carbon dioxide (1939, 1940 Bachmann; cf. 1947 b Johnson). As benzoate, from the 3-benzoate of "β"-dihydroequilenin (trans-equilenan-3.17α-diol, p. 1960 s) on oxidation with CrO_3 in acetic acid at 20° (1936 Wintersteiner; 1938a Hirschmann). The compound obtained from œstrane-3.17-diols-A and -B (p. 1984 s) on heating with platinum black in a stream of nitrogen at 215–225°, which was considered to be equilenin (1938a, c Marker), may have been impure d-iso-equilenin (p. 2535 s) (1950a Bachmann). The acetate is obtained in almost quantitative crude yield from that of 6-dehydroœstrone (p. 2538 s) on re-fluxing with SeO_2 in glacial acetic acid in a current of nitrogen for 10–15 min. (1950 Kaufmann).

REACTIONS OF d-EQUILENIN

Oxidation. On oxidation of the acetate with CrO_3 in acetic acid at 25° it gives 11-ketoequilenin acetate (p. 2940 s) in low yield* (1939 Marker). Treatment of the methyl or benzyl ether with KOI in methanol at room temperature followed by boiling with aqueous methanolic KOH, gives the corresp. ethers of d-†trans("β")-bisdehydro-marrianolic acid (I) (1944 Miescher; 1945 Heer; cf. 1946 Heer). Fusion with KOH at 275° (see also 1939, 1947 Hohlweg) yields d-†trans("β")-bisdehydro-doisynolic acid (II), l-†cis("α")-bisdehydro-doisynolic acid [(III), œstrogenically very active, formed by inversion at C-1 of the phenanthrene ring system (C-14 of equilenin) (see 1947 Heer)], and another acid (m. 230–2° cor.) about which see alkali fusion of "α"-dihydro-equilenin (trans-equilenan-3.17β-diol; p. 1961 s) (1944 Miescher; 1945 Heer; cf. 1946, 1948b, c Heer).

Reduction. For reduction to the two epimeric dihydroequilenins (p. 1960 s) by refluxing with aluminium isopropoxide in isopropyl alcohol, see also 1938b Marker. On hydrogenation with Rupe nickel in aqueous alkaline solution until 1 mol. hydrogen has been taken up, "α"-dihydroequilenin is formed (1945 Heer). Reduction with magnesium in 5 % NH_3 solution + 5 % NH_4Cl yields "α"(?)-dihydroequilenin** (1937 SCHERING-KAHLBAUM A.-G.). On hydro-genation with Adams' catalyst in ether-abs. alcohol at 3 atm. and 25°, "α"-di-hydroequilenin is obtained (1939 Marker). Hydrogenation, using platinum

---

* The main oxidation product is a 14-hydroxyequilenin 3-acetate (1954 McNiven).

† trans and cis refer to the relative orientation of H and CH_3 at C-1 and C-2 of the phenanthrene ring system.

** It is not mentioned in the patent abstracts whether they deal with "α"-or "β"-dihydroequilenin, but "α"-dihydroequilenin seems more likely.

oxide catalyst and some HCl in alcohol, gives "α"-dihydroequilenin, $\Delta^{5.7.9}$-oestratriene-3β.17β-diol (p. 1983 s), and $\Delta^{5.7.9}$-oestratrien-17β-ol (p. 1498 s) (1938 Ruzicka; see also 1939 Marker). Reduction with sodium in boiling ethyl alcohol or amyl alcohol yields $\Delta^{5.7.9}$-oestratriene-3α.17β-diol (p. 1983 s) (1938 Ruzicka; 1938 David) and $\Delta^{5.7.9}$-oestratriene-3β.17β-diol (p. 1983 s) (1938 David; 1938d Marker; cf. 1939 Marker; 1941 Heard), and a smaller phenolic fraction which on conversion into the benzoate, followed by oxidation with CrO_3 and hydrolysis, gives oestrone (p. 2543 s) (1938b Marker). The highly oestrogenic oil formed on reduction of equilenin with sodium in alcohol (1934/35 David) owes its high potency probably to the presence of some oestradiol (1938b Marker; 1938 David). On reduction of the methyl ether with sodium and boiling alcohol, followed by heating with aqueous HCl, $\Delta^{5.7.9}$-oestratrien-17-ol-3-one (p. 2527 s) is produced (1942 Cornforth). The Wolff-Kishner reduction of the methyl ether semicarbazone with sodium ethoxide in alcohol at 180° is accompanied by complete demethylation giving a gum which on methylation with methyl p-toluenesulphonate and aqueous KOH affords desoxoequilenin methyl ether (p. 1496 s) (1934 Cohen). The action of fermenting yeast gives "α"(?)-dihydroequilenin* (1937 Schering A.-G.).

Other reactions. On heating d-equilenin with aqueous ammonium bisulphite in a sealed tube at ca. 170° for 12 hrs. it gives d-3-amino-3-desoxyequilenin (p. 2508 s) which on diazotization with $NaNO_2$ in cold dil. HCl, followed by treatment with hypophosphorous acid, yields d-3-desoxyequilenin (page 2391 s) (1950b Bachmann). When the sodium salt (obtained with sodium ethoxide in abs. alcohol) is heated with CO_2 at 180°, it gives equilenin-2 or 4-carboxylic acid ("equileninsalicylic acid") (1940 Niederl). On reaction of the acetate, dissolved in benzene ether, with potassium acetylide in liquid ammonia, followed by saponification with methanolic KOH, 17α-ethynyl-dihydroequilenin (p. 1989 s) is obtained (1938 Inhoffen). For conversion of the methyl ether via its reaction product with methylmagnesium iodide into 3'.3'-dimethyl-7-methoxy-1.2-cyclopentenophenanthrene, see the reactions of 17α-methyl-17β-hydroxy-3-methoxy-$\Delta^{1.3.5(10).6.8}$-oestropentaene (p. 1987 s). Condensation of the methyl ether with piperonal in alcohol containing a little KOH gives 16-piperonylidene-equilenin methyl ether (p. 2731 s) (1941 Koebner).

Colour reactions. d-Equilenin gives a green fluorescent solution when warmed with conc. H_2SO_4 (1939 Robinson). On underlayering the alcoholic solution with conc. H_2SO_4, a pale orange fluorescence develops at the interface; in the presence of furfural in the alcoholic solution, a violet to raspberry colour appears on standing (1939 Woker). When to 1 c.c. of a suspension, prepared from 1 mg. equilenin dissolved in 3 drops of alcohol and diluted with water to 10 c.c., are added 1–2 drops of 0.1 % alcoholic 1-nitroso-2-naphthol and 5 drops of HNO_3 (d. 1.4) and the test tube is heated in boiling water, a blue-violet colour appears changing to green and finally yellow (1937 Voss).

---

* See footnote ** on p. 2531 s.

For *colorimetric determination* based on the blue colour of the azo-dye produced by coupling with diazotized p-nitrobenzeneazodimethoxyaniline, see 1938 Marx.

Compound of d-equilenin with trinitrobenzene (1:1) $C_{24}H_{21}O_8N_3$. Yellow needles (abs. alcohol), m. 205–7° vac. (1940 Bachmann).

d-Equilenin 2.4-dinitrophenylhydrazone $C_{24}H_{22}O_5N_4$. Orange-coloured crystals (alcohol-water), m. 268–270° dec. (1945 Veitch). — For its separation from the 2.4-dinitrophenylhydrazone of œstrone, see p. 2546 s.

d-Equilenin acetate $C_{20}H_{20}O_3$ (p. 135); $[\alpha]_D^{20} + 72°$ (chloroform) (1950 Rosenkranz; 1950 Kaufmann). For its ultra-violet absorption curve, see 1950 Kaufmann.

d-Equilenin l-menthoxyacetate $C_{30}H_{38}O_4$. Colourless plates (acetone), m. 174° to 174.5° vac. (1940 Bachmann); 177.2–177.8° cor. vac. (1947b Johnson); $[\alpha]_D^{30} + 18°$ (benzene) (1940 Bachmann). — **Fmn.** From d-equilenin and from dl-equilenin on treatment with l-menthoxyacetyl chloride (for details see the formation of d-equilenin by resolution of dl-equilenin, p. 2530 s) (1939, 1940 Bachmann).

d-Equilenin methyl ether $C_{19}H_{20}O_2$ (p. 135). $[\alpha]_D^{20} + 66°$ (alcohol) (1945 Heer). — **Rns.** See under the reactions of d-equilenin, pp. 2531 s, 2532 s. — *Semicarbazone* $C_{20}H_{23}O_2N_3$ (no analysis), amorphous; m. 273–5° dec.; very sparingly soluble in the usual media (1934 Cohen).

d-Equilenin carboxymethyl ether, d-Equilenin-O-acetic acid $C_{20}H_{20}O_4$. White needles (acetone-ether), m. 233–6°. — **Fmn.** Together with some of its ethyl ester (below), by refluxing equilenin with ethyl chloroacetate and sodium ethoxide in alcohol, finally with addition of KOH. — **Rns.** Hydrogenation with Adams' catalyst and some HCl in alcohol at 2 atm. and room temperature gives $\Delta^{5.7.9}$-œstratrien-17β-ol (p. 1498 s). — *Methyl ester* $C_{21}H_{22}O_4$. White plates (acetone-methanol), m. 180–2°. **Fmn.** From the acid with diazomethane. — *Ethyl ester* $C_{22}H_{24}O_4$. White crystals (aqueous acetone), m. 141.5° to 143°. **Fmn.** See the acid. **Rns.** Is hydrolysed to the acid with an excess of alcoholic KOH (1939 Marker).

d-Equilenin benzyl ether $C_{25}H_{24}O_2$. Crystals (alcohol), m. 178° cor.; $[\alpha]_D^{20}$ +68°, +71° (alcohol). Very sparingly soluble in alcohol. — **Fmn.** From the potassium salt of equilenin (obtained by evaporation of a solution of equilenin in methanolic KOH in vac.) on boiling with benzyl chloride in alcohol (yield, 83%), together with some neutral compound, m. 135° cor., of unknown composition (1945 Heer). — **Rns.** See p. 2531 s.

1-trans ("β")-Equilenan-3-ol-17-one, 1-trans ("β")-3-Hydroxy-17-equilenone, 1-Equilenin $C_{18}H_{18}O_2$ (I b). Colourless needles (acetone-methanol), m. 258–9° cor. vac.; $[\alpha]_D^{30}$ —85° (dioxane) (1940 Bachmann). — For *physiological properties*, see pp. 1367 s, 1368 s, and the literature cited there.

Fmn. (For *synthesis*, see p. 2528 s.) From dl-equilenin on resolution by treatment with d-menthoxyacetyl chloride, isolation of l-equilenin d-menthoxyacetate (below) from the mixture of diastereoisomeric esters in the same way
as for d-equilenin (see p. 2530 s), and hydrolysis (1940 Bachmann).

l - Equilenin   d - menthoxyacetate $C_{30}H_{38}O_4$. Colourless plates (acetone),
m. 174.5–175° vac.; $[\alpha]_D^{30}$ —16° (benzene) (1940 Bachmann). —**Fmn.** See above.

**dl-trans ("β")-Equilenan-3-ol-17-one, dl-trans ("β")-3-Hydroxy-17-equi-
lenone, dl-Equilenin** $C_{18}H_{18}O_2$ [I a (p. 2527 s) + I b (p. 2533 s)]. Crystals
(acetone-alcohol); melts usually at 276–8° vac., sometimes at 265°; in the
latter case the solid formed on cooling remelts at 276–8° vac. In one run
colourless leaflets, m. 287–8° vac., were obtained. Crystals (benzene) containing solvent of crystallization, m. about 180° with loss of solvent. Gives
a deep red liquid when melted in air (1939, 1940 Bachmann).

Fmn. (For *syntheses*, see p. 2528 s and p. 2530 s.) The methyl ether is
formed in 63% yield, together with 32% of dl-isoequilenin methyl ether, by
hydrogenation of dl-14-dehydroequilenin methyl ether (p. 2526 s) over palladium-charcoal in ethyl acetate at room temperature; it is demethylated by
refluxing with acetic acid and hydrochloric acid in an atmosphere of nitrogen
(98% yield) (1945, 1947b Johnson). From 16-carbomethoxy-dl-equilenin
methyl ether on refluxing with acetic acid and hydrochloric acid in an atmosphere of nitrogen or carbon dioxide for 10 hrs., whereby also the ether
linkage is completely cleaved (on heating for a shorter time some dl-equilenin
methyl ether is also obtained) (yield, 90–95%) (1939, 1940 Bachmann).

Rns. On hydrogenation in the presence of Adams's catalyst in methanol
containing some conc. HCl at room temperature it gives the racemic form
of $\Delta^{5.7\ 9}$-œstratrien-17β-ol (for the opt. act. form, see p. 1498 s) which on
oxidation with CrO_3 in strong acetic acid yields dl-1.2.3.4-tetrahydro-3-des-
oxyequilenin (p. 2392 s) (1951 b Bachmann). For resolution into d- and
l-equilenin, see these compounds. With methylmagnesium iodide the methyl
ether gives 17(trans)-methyl-17-hydroxy-3-methoxy-trans-dl-equilenan (page
1987 s) (1939, 1940 Bachmann).

dl-Equilenin acetate $C_{20}H_{20}O_3$. Colourless needles (methanol), m. 153–4° vac.,
solidifying, and remelting at 159.5–160° vac. — **Fmn.** From dl-equilenin on
treatment with acetyl chloride and pyridine in glacial acetic acid at room
temperature (1940 Bachmann).

dl-Equilenin benzoate $C_{25}H_{22}O_3$. Colourless plates (methanol-ethyl acetate),
m. 248.5–249.5° vac. — **Fmn.** From dl-equilenin by means of benzoyl chloride
and pyridine (1940 Bachmann).

dl-Equilenin dl-menthoxyacetate $C_{30}H_{38}O_4$, m. 151–2° vac. — **Fmn.** From
equal parts of d-equilenin-l-menthoxyacetate (p. 2533 s) and l-equilenin
d-menthoxyacetate (above) (1940 Bachmann).

dl-Equilenin methyl ether $C_{19}H_{20}O_2$. Colourless plates (acetone-methanol),
m. 185–186.5° vac. Sublimes at 180° and 0.01 mm. — **Fmn.** From dl-equi-

lenin on treatment with dimethyl sulphate in alkaline solution (1940 Bachmann). See also under formation of dl-equilenin (p. 2534 s). — **Rns.** See
under those of dl-equilenin.

d-cis ("α")-Equilenan-3-ol-17-one, d-cis ("α")-3-Hydroxy-17-equilenone,
d-Isoequilenin, "*14-epi-Equilenin*" $C_{18}H_{18}O_2$ (IIa). For
the cis-fusion of rings C and D in isoequilenin (d, l, and
dl), see the references quoted for the trans-fusion of
these rings in equilenin (p. 2527 s).

Colourless needles (dil. alcohol), m. ca. 263° cor. (vac.)
with previous softening (1938b Hirschmann; cf. 1940 Bachmann); colourless
leaflets or plates (strong acetic acid), m. 265–6° cor. (vac.); colourless plates
(acetone + alcohol), m. 272–3° uncor. (vac.) (1940 Bachmann); $[\alpha]_D^{30}$ + 160°
(alcohol) (1938b Hirschmann; see also 1940 Bachmann); $[\alpha]_D^{29}$ + 147° (dioxane)
(1940 Bachmann). Shows the same absorption spectrum as equilenin (1938b
Hirschmann); cf., however, Banes (1950) who gives the absorption max.
(in alcohol?) 267, 277.5, 289, 325, 338 mμ.

For *physiological properties*, see pp. 1367 s, 1368 s, and the literature cited
there.

Fmn. (For *synthesis*, see p. 2528 s). From isoequilin A (p. 2540 s) on
heating with palladium black in abs. alcohol at 80° (1938b Hirschmann;
cf. 1953 Banes). From œstrone (p. 2543 s) on heating with palladium black
at 260° (yield, 25 %) (1941 Butenandt). From 16-carbomethoxy-d-isoequilenin
methyl ether on refluxing with acetic acid and hydrochloric acid in an atmosphere of nitrogen for 10 hrs. (1940 Bachmann). Purification via the
picrate (light orange crystals) (1938b Hirschmann; 1941 Butenandt). — The
compound obtained from œstrane-3.17-diol-A (p. 1984 s) and from œstrane-
3.17β-diol-B (p. 1984 s) on heating with platinum black in a stream of nitrogen
at 215–225°, which was considered to be equilenin (1938a, c Marker), may
have been impure d-isoequilenin (1950a Bachmann).

d-Isoequilenin acetate $C_{20}H_{20}O_3$. Colourless prisms (methanol), m. 149°
to 149.5° (cor. vac.?), solidifying, and remelting at 127–8° (cor. vac.?);
$[\alpha]_D^{22}$ + 129° (abs. alcohol). — **Fmn.** From d-isoequilenin on treatment with
acetyl chloride and pyridine in glacial acetic acid or with acetic anhydride
and pyridine (1940 Bachmann).

d-Isoequilenin methyl ether $C_{19}H_{20}O_2$. Colourless plates (methanol), m. 118.5°
to 119.5° (1940 Bachmann); colourless needles (sublimation in high vac.),
m. 119–119.5° (1941 Butenandt); $[\alpha]_D^{23}$ + 176° (alcohol) (1947 Heer). — **Fmn.**
From d-isoequilenin on treatment with dimethyl sulphate in alkaline solution
(1940 Bachmann; 1941 Butenandt).

1-cis ("α")-Equilenan-3-ol-17-one, 1-cis ("α")-3-Hydroxy-17-equilenone,
1-Isoequilenin $C_{18}H_{18}O_2$ (IIb). Colourless plates (acetone-
alcohol), m. 257–8° vac.; when first obtained by acidification of the alkaline solution it melts at 272–3° vac.;
$[\alpha]_D^{28}$ — 162° (abs. alcohol), — 147° (dioxane) (1940 Bachmann).

E.O.C. XIV s. 176

References, pp. 2541–2543 s

For *physiological properties*, see pp. 1367 s, 1368 s, and the literature cited there.

Fmn. (For *synthesis*, see p. 2528 s.) From lumiœstrone (p. 2559 s) on heating with palladium black at 260° (yield, 40%) (1941 Butenandt). On cyclization of the d-dimethyl ester of cis ("α")-2-methyl-7-methoxy-1.2.3.4-tetrahydrophenanthrene-2-carboxylic-1-(β-propionic) acid (by refluxing with sodium methoxide in benzene) to 16-carbomethoxy-l-isoequilenin methyl ether (not isolated), followed by refluxing with acetic acid and hydrochloric acid in an atmosphere of nitrogen (1940 Bachmann).

l-Isoequilenin methyl ether $C_{19}H_{20}O_2$, m. 119°. — **Fmn.** From l-isoequilenin on heating with dimethyl sulphate and aqueous KOH (1941 Butenandt).

dl-cis("α")-Equilenan-3-ol-17-one, dl-cis ("α")-3-Hydroxy-17-equilenone, dl-Isoequilenin $C_{18}H_{18}O_2$ [II a (p. 2535 s) + II b (p. 2535 s)]. Crystals (strong acetic acid), m. 223–4° vac.; plates (acetone-alcohol), m. 206–206.5°, solidifying, and remelting at 221–2°. Sublimes under 0.01 mm. pressure (1940 Bachmann).

Fmn. (For *syntheses*, see p. 2528 s, p. 2530 s, and p. 2537 s.) From equal parts of d- and l-isoequilenin on crystallization from acetic acid (1940 Bachmann). As methyl ether in 32% yield, together with 63% of dl-equilenin methyl ether, by hydrogenation of dl-14-dehydroequilenin methyl ether (page 2526 s) over palladium-charcoal in ethyl acetate at room temperature (1945, 1947 b Johnson); higher yields are obtained when this hydrogenation is carried out in the presence of piperidine or KOH (57% yield) (1947 b Johnson). As methyl ether in 96% yield by hydrogenation of dl-15-dehydroisoequilenin methyl ether (p. 2526 s) over palladium-charcoal in ethyl acetate (1947 b Johnson). As methyl ether from dl-18-nor-isoequilenin methyl ether (p. 2518 s) on keeping with ethyl formate and sodium or sodium methoxide in benzene under nitrogen, whereby the 16-hydroxymethylene derivative (p. 2719 s) results in 95–97% yield (1945 Birch; 1947 a Johnson), refluxing of the latter with isopropyl iodide and K_2CO_3 in acetone to give an oily 16-isopropyloxy-methylene derivative, methylation at C-13 by heating with potassium amide and methyl iodide in ether, hydrolysis of the crude 16-(isopropyloxy-methylene)-dl-isoequilenin methyl ether by keeping with $FeCl_3 \cdot 6\ H_2O$ in ether and methanol at room temperature under nitrogen, and removal of the 16-hydroxymethylene group by boiling with aqueous alkali under nitrogen (yield, 15%) (1947 a Johnson). Is formed (as methyl ether) similarly from 16-(methyl-anilino-methylene)-dl-isoequilenin methyl ether (p. 2718 s) on refluxing with aqueous alcoholic H_2SO_4, followed by boiling of the reaction product with aqueous alcoholic KOH under nitrogen (yield, 55%) (1945 Birch). From 16-carbomethoxy-dl-isoequilenin methyl ether on refluxing with acetic acid and hydrochloric acid in an atmosphere of nitrogen for 10 hrs. (yield, 93%), together with some (4%) dl-isoequilenin methyl ether (1940 Bachmann). The methyl ether is demethylated by refluxing with acetic acid and hydrobromic or hydrochloric acid under nitrogen (1945 Birch; 1947 b Johnson).

Rns. On catalytic hydrogenation, followed by oxidation with CrO_3, it reacts similarly to dl-equilenin (see p. 2534 s) giving dl-1.2.3.4-tetrahydro-

Robinson's Synthesis of Isoequilenin*

(1938 Robinson; 1938 Koebner; 1945 Birch)

MeO—[naphthalene]—CO·CH₃ + [furan]—CHO →(NaOMe)→ MeO—[naphthalene]—CO·CH=CH—[furanone] →(alc. HCl)→

HO₂C—CH₂—CH₂——CO / CO—CH₂—CH₂ [naphthalene] MeO →(aq. KOH)→ HO₂C—CH₂—[cyclopentenone]—[naphthalene] MeO →(H₂, Pd-SrCO₃)→

HO₂C—CH₂—[cyclopentanone]—[naphthalene] MeO →(P₂O₅·H₃PO₄ (d 1.75), 120°)→ [phenanthrene diketone] MeO →(H₂, Pt-Pd-C)→

[tetracyclic ketone] MeO
dl-18-Nor-isoequilenin
methyl ether**
→(Na + HCO₂Et)→ [=O † / =CH·OH] MeO →(PhNHMe)→

MeO—[tetracyclic]—=O / =CH·N<Me / Ph →(NaNH₂, MeI)→

MeO—[Me tetracyclic]—=O / =CH·N<Me / Ph H + [Me / =O / =CH·N<Me / Ph] H

→(H₂SO₄; KOH)→ MeO—[Me tetracyclic]—=O H + [Me / =O] H →(AcOH + HBr under nitrogen)→ **dl-Isoequilenin**

3-desoxyisoequilenin (p. 2393 s) (1951 b Bachmann). With methylmagnesium iodide the methyl ether gives 17-methyl-17-hydroxy-3-methoxy-cis-dl-equilenan (p. 1987 s) (1940 Bachmann).

Compound of dl-isoequilenin with trinitrobenzene (1:1) $C_{24}H_{21}O_8N_3$. Yellow needles (abs. alcohol), m. 186–7° vac. (1940 Bachmann).

* See also Bachmann's and Johnson's Syntheses of Isoequilenin, p. 2528 s and p. 2530 s.
** See also Bachmann's Synthesis of 18-Nor-isoequilenin, p. 2519 s.
† See also 1947a Johnson, who proceeded via the corresp. isopropyloxymethylene derivative.
176*

dl-Isoequilenin acetate $C_{20}H_{20}O_3$. Colourless prisms (acetone-methanol), m. 159–160° vac. (1940 Bachmann).

dl-Isoequilenin methyl ether $C_{19}H_{20}O_2$. Colourless prisms (acetone-methanol), m. 127–127.5° vac., solidifying, and remelting at 130–130.5° vac. Sublimes at 160–180° and 0.01 mm. pressure (1940 Bachmann). — **Fmn.** and **Rns.** See pp. 2536 s, 2537 s.

6-Dehydrocœstrone, "$\varDelta^6$-Isoequilin" $C_{18}H_{20}O_2$. Plates (abs. alcohol), m. 265° to 266° cor. (1940 Pearlman); prisms (methanol), m. 261–3° cor. (Kofler block) (1950 Kaufmann); $[\alpha]_D$ in dioxane —120° (1950 Pearlman), —127° (1950 Kaufmann). Absorption max. in alcohol at 220 (4.49), 262 (3.95), and 304 (3.44) mμ (log ε) (1940 Pearlman; 1950 Kaufmann). — Exhibits about one-third the œstrogenic potency of œstrone (1940 Pearlman).

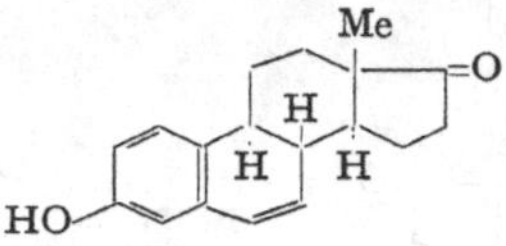

Fmn. From 7-chlorocœstrone benzoate (p. 2702 s) on heating in a sealed tube on a steam-bath with pyridine containing NaI, followed by boiling with methanolic KOH (1940 Pearlman). From $\varDelta^{1.4.6}$-androstatriene-3.17-dione (p. 2877 s) on passing slowly its solution in mineral oil through a glass tube filled with helices and heated to 600°; yield, 40 % (1950 Kaufmann).

Rns. On refluxing of the acetate with SeO_2 in glacial acetic acid in a current of nitrogen it gives the acetate of equilenin (p. 2527 s) (1950 Kaufmann). Hydrogenation in abs. alcohol or ethyl acetate in the presence of palladium affords œstrone (p. 2543 s) (1940 Pearlman; 1950 Kaufmann). On refluxing of the acetate with lithium aluminium hydride in ether it gives 6-dehydro-œstra-3.17β-diol (p. 1961 s) (1950 Kaufmann); the same (?) diol is obtained on reduction of the hydroxyketone with aluminium isopropoxide in isopropyl alcohol (1940 SCHERING CORP.).

Acetate $C_{20}H_{22}O_3$. Crystals (benzene-pentane 1:10), m. 140–141° cor. (1940 Pearlman); crystals (methanol), m. 140–140.5° cor.; $[\alpha]_D^{20}$ —114° (dioxane) (1950 Kaufmann). — **Fmn.** From the above 6-dehydrocœstrone with acetic anhydride in pyridine at room temperature (1940 Pearlman) or on heating for 2 hrs. on the steam-bath (1950 Kaufmann). — **Rns.** See above.

Benzoate $C_{25}H_{24}O_3$. Needles (abs. alcohol, or alcohol-chloroform), m. 202° cor. (1940 Pearlman), 200.5–202.5° cor. (Kofler block) (1950 Kaufmann). — **Fmn.** From 6-dehydrocœstrone (above) on treatment with benzoyl chloride and aqueous NaOH (1940 Pearlman; 1950 Kaufmann).

Regarding a *compound described as isoequilin* $C_{18}H_{20}O_2$ *of the above structure* [needles (alcohol-water), m. 248–250° with red coloration, decomposing at 252°; $[\alpha]_D^{19}$ +170° (dioxane); absorption max. 265, 275, and 334 mμ; nearly insoluble in ether; 60 γ active in the Allen-Doisy test on rats], which was obtained from a brominated androstane-3.17-dione by elimination of HBr with pyridine, followed by heating at 300° (1937, 1940 Inhoffen; 1937 SCHERING CORP.), cf. 1940 Pearlman; 1950 Hershberg; 1950 Kaufmann.

7-Dehydrooestrone, Equilin $C_{18}H_{20}O_2$ (p. 135). For the trans-fusion of rings C and D in equilin, see the references quoted for the trans-fusion of these rings in equilenin (p. 2527 s) to which equilin is dehydrogenated by heating with palladium at 80–90° under nitrogen (p. 135; see also 1950 a Bachmann). — For its absorption spectrum (max. in alcohol at 280 mμ, ε 2000), see also 1939 Dannenberg; for its infra-red absorption spectrum, see 1946 Furchgott. Partition ratio between 70 % alcohol and benzene, 1:2.5 (1938 Westerfeld).

For *physiological properties*, see also pp. 1367 s, 1368 s and the literature cited there.

Occ. For the *isolation* from the urine of pregnant mares, see also 1936 Beall.

For a *synthesis* of equilin, performed after the closing date for this volume, see 1958 Zderic.

Rns. Keeping of the acetate with OsO_4 in ether, followed by boiling with Na_2SO_3 in aqueous alcohol gives equilin-7.8-glycol (p. 2767 s) (1938 Serini; 1939 Pearlman). On alkali fusion at 275° it forms the same (or a similar) oestrogenically active mixture of carboxylic acids as equilenin (see the latter, p. 2531 s) (1939, 1947 Hohlweg; cf. 1944 Miescher; 1945 Heer). On treatment with palladium and hydrogen in alcohol, hydrogenation and dehydrogenation occur simultaneously (1935 Dirscherl; see also 1938 Serini). Reduction with sodium in abs. alcohol gives "α"-dihydroequilin (p. 1962 s) (1934/35 David) which is also formed on reduction with magnesium in 5 % ammonia solution containing NH_4Cl (1937 SCHERING-KAHLBAUM A.-G.) or by the action of fermenting yeast (1937 SCHERING A.-G.). Refluxing with conc. HCl and glacial acetic acid in an atmosphere of CO_2 gives "isoequilin A" (p. 2540 s) (1938b Hirschmann; cf. 1950, 1953 Banes). When the sodium salt (obtained with sodium ethoxide in abs. alcohol) is heated with CO_2 at 180°, it gives equilin-2 or 4-carboxylic acid *("equilinsalicylic acid")* (1940 Niederl). Equilin reacts in dioxane-benzene solution with potassium acetylide in liquid ammonia to give 17α-ethynyldihydroequilin (p. 1989 s) (1938 Inhoffen).

Colour reactions. Shows a yellowish green fluorescence in ultra-violet light when treated with conc. H_2SO_4 (1937 Kocsis; 1937 Bugyi). Colours appear at the interface on underlayering with conc. H_2SO_4: (1) the alcoholic solution (1939 Woker; 1939 Scherrer); (2) a mixture with benzaldehyde in alcohol (1939 Scherrer); (3) a mixture with furfural in alcohol (1939 Woker). For the Zimmermann test with m-dinitrobenzene and KOH, see 1944 Zimmermann. When 1–2 drops of 0.1 % alcoholic 1-nitroso-2-naphthol and 5 drops of HNO_3 (d 1.4) are added to 1 c.c. of a suspension, prepared from 1 mg. equilin dissolved in 3 drops of alcohol and diluted with water to 10 c.c., and the test tube is heated in boiling water, a bluish red colour appears changing to blue-violet, green, and finally yellow (1937 Voss).

Equilin oxime $C_{18}H_{21}O_2N$ (p. 135). For reduction with sodium and alcohol to give 17-amino-$\Delta^{1.3.5(10)}$-oestratrien-3-ol (p. 1910 s), see 1943 Lettré.

Equilin acetate $C_{20}H_{22}O_3$ (not analysed), m. 125.5–126.5° cor. (1939 Pearlman). — **Rns.** See p. 2539 s.

For a compound which may be an **isomer of equilin** (1938b Marker), see the *phenolic ketone* $C_{18}H_{20}O_2$ (m. 222–5°) under the reactions ot "*β*"-dihydroequilenin (p. 1960 s).

cis-2-Methyl-7-hydroxy-3'-keto-1.2-cyclopentano-1.2.3.4.9.10-hexahydrophen-anthrene, 8-Dehydro-14-iso-œstrone, "14-epi-Δ⁸-Equi-

lin" $C_{18}H_{20}O_2$*. The name **isoequilin A** was given by Hirschmann (1938b) to an 8-dehydro-14-iso-œstrone which was not quite pure, by Heer (1948d) to 8-dehydro-14-iso-œstrone which was purified via its methyl ether, and by Banes (1950, 1953) to the mixture of ketones obtained by acid isomerization of equilin (see "**Fmn.**", below).

Platelets (dil. methanol), m. 233–5° cor.; $[\alpha]_D$ +234° (in alcohol?); absorption max. of the methyl ether in alcohol (from the graph) at ca. 267 mμ (log ε ca. 4.33) (1948d Heer). Plates (alcohol), m. 231–2°; $[\alpha]_D$ + 232° (in alcohol?); absorption max. in alcohol (from the graph) at ca. 280 mμ (log ε ca. 4.20) (1953 Banes). The "isoequilin A" of Hirschmann (1938b) formed plates (alcohol), m. 231° cor., $[\alpha]_D^{25}$ +222° (alcohol), absorption max. in alcohol (from the graph) at ca. 272 mμ (log ε ca. 4.23) (cf. also 1939 Dannenberg), exhibited a green fluorescence in conc. H_2SO_4, and gave a brown-orange pigment on coupling with an acid solution of p-nitrodiazobenzene.

The pure compound has only a weak œstrogenic activity (1948d Heer).

Fmn. From equilin (p. 2539 s) on refluxing with conc. HCl and glacial acetic acid in an atmosphere of CO_2 for 80 min. (1938b Hirschmann), along with 8-dehydroœstrone [m. 228–230°; $[\alpha]_D$ +86° (alcohol)] and 9(11)-dehydroœstrone [X (p. 2541 s), R = H; m. 254–5°; $[\alpha]_D$ +311° (alcohol)] (1950, 1953 Banes); the last-mentioned two compounds are partly isomerized to 8-dehydro-14-iso-œstrone by heating with conc. HCl-glacial acetic acid (1953 Banes). From its methyl ether (p. 2541 s) on heating with pyridine hydrochloride at 180° for 5 hrs. (1948d Heer).

Rns. On treatment of its acetate with OsO_4 in ether, followed by refluxing with Na_2SO_3 in alcohol, it yields 8-hydroxy-9(11)-dehydro-14-iso-œstrone (p. 2740 s) (1938b Hirschmann). Gives d-isoequilenin (p. 2535 s) on refluxing with palladium black in abs. alcohol (1938b Hirschmann) or in benzene (1953 Banes).

Semicarbazone $C_{19}H_{23}O_2N_3$. Crystals with $^1/_2$ H_2O (alcohol), m. 230° cor. dec. (1938b Hirschmann).

Acetate $C_{20}H_{22}O_3$. Prisms (dil. alcohol), m. 95° cor. after softening at 83° (1938b Hirschmann), m. 93–4° (1953 Banes). Absorption max. in hexane at 268 mμ (1938b Hirschmann). For its infra-red absorption spectrum in carbon disulphide, see 1953 Banes. — **Fmn.** From the above hydroxyketone on

---

* After the closing date for this volume, there have been prepared the enantiomorph, 8-dehydro-13-iso-œstrone, by Hess (1953), the racemic form by Heer (1948d), and 8-dehydroœstrone (cf. also above under "Fmn.") by Banes (1953).

treatment with acetic anhydride and pyridine at room temperature (1938b Hirschmann; 1953 Banes). — **Rns.** See above.

Methyl ether $C_{19}H_{22}O_2$. Rhombohedra (methanol), m. 124–6° cor.; $[\alpha]_D^{21}$ + 201° (solvent not stated); for its ultra-violet absorption spectrum, see under the free hydroxyketone. — **Fmn.** From the hydroxyketone on treatment with dimethyl sulphate in aqueous NaOH. — **Rns.** Gives the free hydroxyketone on heating with pyridine hydrochloride at 180° (1948d Heer).

2 - Methyl -7- hydroxy -3'- keto -1.2- cyclopentano-1.2.3.9.10.11-hexahydrophenanthrene $C_{18}H_{20}O_2$ (X, R = H).

trans-Form, 9(11)-Dehydroœstrone $C_{18}H_{20}O_2$ (X, R = H). This compound has been described after the closing date for this volume by Banes (1950, 1953); see also under formation of 8-dehydro-14-iso-œstrone (page 2540 s). Its *methyl ether* $C_{19}H_{22}O_2$ (X, R = CH₃) melts at 143–4°. For an oily **compound** $C_{19}H_{22}O_2$, for which the same structure or that of a stereoisomer has been considered, see p. 2521 s.

15-Dehydroœstrone $C_{18}H_{20}O_2$ (III, p. 2522 s). For a compound which was originally formulated as a stereoisomeric 15-dehydroœstrone, see p. 2522 s.

1932 Girard, Sandulesco, Fridenson, Rutgers, *Compt. rend.* **195** 981.
1934 Cohen, Cook, Hewett, Girard, *J. Ch. Soc.* **1934** 653, 655.
1934/35 David, *Acta Brevia Neerland.* **4** 63.
1935 Dirscherl, Hanusch, *Z. physiol. Ch.* **233** 13, 19, 20, 25.
1936 Beall, Edson, *Biochem. J.* **30** 577, 580.
      Wintersteiner, Schwenk, Hirschmann, Whitman, *J. Am. Ch. Soc.* **58** 2652.
1937 Bugyi, *C. A.* **1938** 2191; *Ch. Ztbl.* **1938** I 1144.
      Inhoffen, *Naturwiss.* **25** 125; cf. SCHERING A.-G., *FIAT Final Report* No. 996, pp. 130, 150, 151.
      Kocsis, Bugyi, *C. A.* **1938** 2162; *Ch. Ztbl.* **1938** I 3353.
      SCHERING A.-G. (Schoeller), *Ger. Pat.* 722098 (issued 1942); *C. A.* **1943** 5200; *Fr. Pat.* 49185 (1938); *C. A.* **1939** 5132; *Ch. Ztbl.* **1939** I 4089; SCHERING CORP. (Schoeller), *U.S. Pat.* 2184167 (1938, issued 1939); *C. A.* **1940** 2538.
      SCHERING CORP. (Inhoffen), *U.S. Pat.* 2280828 (issued 1942); *C. A.* **1942** 5618; SCHERING A.-G., *Brit. Pat.* 508576 (issued 1939); *Fr. Pat.* 838704 (issued 1939); *C. A.* **1940** 777; *Ch. Ztbl.* **1939** II 1722.
      SCHERING-KAHLBAUM A.-G., *Fr. Pat.* 48262 (issued 1937); *C. A.* **1938** 4728; *Ch. Ztbl.* **1938** I 4501; *Brit. Pat.* 472612 (issued 1937); *C. A.* **1938** 1409.
      Voss, *Z. physiol. Ch.* **250** 218.
1938 David, *Acta Brevia Neerland.* **8** 211, 213, 214.
      a Hirschmann, Wintersteiner, *J. Biol. Ch.* **122** 303, 316.
      b Hirschmann, Wintersteiner, *J. Biol. Ch.* **126** 737, 738, 744 et seq.
      Inhoffen, Logemann, Hohlweg, Serini, *Ber.* **71** 1024, 1030.
      Koebner, Robinson, *J. Ch. Soc.* **1938** 1994, 1996, 1997.
      a Marker, Rohrmann, Wittle, Lawson, *J. Am. Ch. Soc.* **60** 1512.
      b Marker, *J. Am. Ch. Soc.* **60** 1897 et seq.
      c Marker, Rohrmann, Lawson, Wittle, *J. Am. Ch. Soc.* **60** 1901, 1903.
      d Marker, Rohrmann, Wittle, Tendick, *J. Am. Ch. Soc.* **60** 2440.
      Marx, Sobotka, *J. Biol. Ch.* **124** 693.
      Robinson, *J. Ch. Soc.* **1938** 1390, 1394 et seq.
      Ruzicka, Müller, Mörgeli, *Helv.* **21** 1394, 1398 et seq.
      Serini, Logemann, *Ber.* **71** 186, 188, 189.
      Westerfeld, Thayer, MacCorquodale, Doisy, *J. Biol. Ch.* **126** 181, 190.

1939 Bachmann, Cole, Wilds, *J. Am. Ch. Soc.* **61** 974.

Dannenberg, *Abhandl. Preuss. Akad. Wiss. Math.-naturwiss. Kl.* **1939** No. 21, pp. 38–41, 63, 64.

a Haberland, *Ber.* **72** 1215, 1217.

b Haberland, Heinrich, *Ber.* **72** 1222, 1224.

Hohlweg, Inhoffen (to SCHERING A.-G.), *Ger. Pat.* 719572 (issued 1942); *C. A.* **1943** 1836; *Ch. Ztbl.* **1942** II 566; Hohlweg, Inhoffen (to SCHERING CORP.), *U.S. Pat.* 2294616 (issued 1942); *C. A.* **1943** 1015.

Marker, Rohrmann, *J. Am. Ch. Soc.* **61** 3314, 3316, 3317.

Pearlman, Wintersteiner, *J. Biol. Ch.* **130** 35, 39.

Robinson, Rydon, *J. Ch. Soc.* **1939** 1394, 1397, 1398, 1402, 1403.

Scherrer, *Helv.* **22** 1329, 1340.

Woker, Antener, *Helv.* **22** 47, 57, 58.

1940 Bachmann, Cole, Wilds, *J. Am. Ch. Soc.* **62** 824, 827–830, 835 et seq.

Inhoffen, *Angew. Ch.* **53** 471, 474.

Niederl, *U.S. Pat.* 2322311 (issued 1943); *C. A.* **1944** 221.

Pearlman, Wintersteiner, *J. Biol. Ch.* **132** 605 et seq.

SCHERING CORP. (Schwenk), *U.S. Pat.* 2327376 (issued 1943); *C. A.* **1944** 753.

1941 Butenandt, Wolff, Karlson, *Ber.* **74** 1308, 1311, 1312.

Doisy, Huffman, Thayer, Doisy, *J. Biol. Ch.* **138** 283.

Heard, Hoffman, *J. Biol. Ch.* **138** 651, 656.

Koebner, Robinson, *J. Ch. Soc.* **1941** 566, 572 et seq.

1942 Cornforth, Cornforth, Robinson, *J. Ch. Soc.* **1942** 689, 691.

1943 Haberland, *Ber.* **76** 621, 622.

Lettré, *Z. physiol. Ch.* **278** 206.

1944 Miescher, *Helv.* **27** 1727, 1729, 1730, 1732.

Zimmermann, *Vitamine und Hormone* **5** [1952] 1, 124, 152, 158.

1945 Birch, Jaeger, Robinson, *J. Ch. Soc.* **1945** 582, 583, 585, 586.

Heer, Billeter, Miescher, *Helv.* **28** 991, 993, 997, 999, 1000, 1001, 1003.

Johnson, Petersen, Gutsche, *J. Am. Ch. Soc.* **67** 2274.

Veitch, Milone, *J. Biol. Ch.* **157** 417.

1946 Furchgott, Rosenkrantz, Shorr, *J. Biol. Ch.* **164** 621, 623.

Heer, Miescher, *Helv.* **29** 1895 et seq.

1947 Heer, Miescher, *Helv.* **30** 550, 553–555.

Hohlweg, Inhoffen, *Deut. med. Wochschr.* **72** 86.

a Johnson, Posvic, *J. Am. Ch. Soc.* **69** 1361, 1366.

b Johnson, Petersen, Gutsche, *J. Am. Ch. Soc.* **69** 2942, 2948, 2951, 2952, 2954.

Stork, *J. Am. Ch. Soc.* **69** 2936.

1948 a Heer, Miescher, *Helv.* **31** 219, 224.

b Heer, Miescher, *Helv.* **31** 229, 232.

c Heer, Miescher, *Helv.* **31** 405, 407.

d Heer, Miescher, *Helv.* **31** 1289, 1291, 1292, 1295.

Jones, Williams, Whalen, Dobriner, *J. Am. Ch. Soc.* **70** 2024, 2027, 2029.

Klyne, *Nature* **161** 434.

Shoppee, *Nature* **161** 207.

1949 Bachmann, Ramirez, *J. Am. Ch. Soc.* **71** 2273.

1950 a Bachmann, Dreiding, *J. Am. Ch. Soc.* **72** 1323–1325.

b Bachmann, Dreiding, *J. Am. Ch. Soc.* **72** 1329.

c Bachmann, Ramirez, *J. Am. Ch. Soc.* **72** 2527.

Banes, Carol, Haenni, *J. Biol. Ch.* **187** 557, 561, 567 footnote 7.

Hershberg, Rubin, Schwenk, *J. Org. Ch.* **15** 292, 293, 297.

Johnson, Stromberg, *J. Am. Ch. Soc.* **72** 505, 506, 510.

Kaufmann, Pataki, Rosenkranz, Romo, Djerassi, *J. Am. Ch. Soc.* **72** 4531 et seq.

Pearlman, Wintersteiner, *Nature* **165** 815.

Rosenkranz, Djerassi, Kaufmann, Pataki, Romo, *Nature* **165** 814.

Stork, Singh, *Nature* **165** 816.

1951 a Bachmann, Controulis, *J. Am. Ch. Soc.* **73** 2636–2638.

b Bachmann, Dreiding, Stephenson, *J. Am. Ch. Soc.* **73** 2765.

c Bachmann, Holmen, *J. Am. Ch. Soc.* **73** 3660, 3663, 3664.
     Johnson, Gutsche, Hirschmann, Stromberg, *J. Am. Ch. Soc.* **73** 322, 324.
1952 Jones, Ramsay, Keir, Dobriner, *J. Am. Ch. Soc.* **74** 80, 84.
1953 Banes, Carol, *J. Biol. Ch.* **204** 509.
     Hess, Banes, *J. Biol. Ch.* **200** 629, 631, 635.
1954 McNiven, *J. Am. Ch. Soc.* **76** 1725, 1727.
1958 Zderic, Bowers, Carpio, Djerassi, *J. Am. Ch. Soc.* **80** 2596.

---

2 - Methyl-7- hydroxy -3′- keto -1.2- cyclopentano -1.2.3.4.9.10.11.12- octahydrophenanthrenes, $\Delta^{1 \cdot 3 \cdot 5(10)}$-Oestratrien-3-ol-17-ones, Oestrones $C_{18}H_{22}O_2$. For a summary of the eight possible racemates possessing this structure, see 1952a, 1958 Johnson; most of them have been described after the closing date for this volume. In this volume there are described: natural œstrone (pp. 2543 s to 2558 s), 8-iso-œstrone (p. 2559 s), 13-iso-œstrone (lumiœstrone, p. 2559 s), and dl-œstrone-a of Bachmann (p. 2560 s). For compounds for which the œstrone structure has been considered, see p. 2522 s.

Natural Oestrone, Oestrone, *Follicular hormone, Theelin* $C_{18}H_{22}O_2$ (p. 135). For trade names of commercial preparations of œstrone, its hydrogen sulphate, and its benzoate, see 1949 Burrows; 1949 Selye. — For configuration of the ring system, see Introduction to Steroids, p. 1360 s; see also 1948 Shoppee; 1948 Klyne; 1949 Heer; 1950 Kaufmann; 1951 Bachmann; 1958 Johnson.

PHYSICAL PROPERTIES OF OESTRONE

For melting points of samples of œstrone of varying degrees of purity, see 1941 Bergel; cf. 1948b Velluz; 1950 Hershberg. — *Monoclinic metastable form* (p. 135): space group $P2_1$ (1936 Bernal); n_α 1.520, n_β 1.642, n_γ 1.690 (1930 Slawson; 1934 Kofler). *Orthorhombic metastable form* (p. 135): β 90°; space group $P2_1\,2_1\,2_1$; n = 4 (1936 Bernal); n_α 1.594, n_β 1.628, n_γ 1.647 (1934 Kofler); M_D 76.3; d (calculated from X-ray data) 1.24 (1934 Neuhaus). *Orthorhombic stable form* (p. 135): β 90°; space group $P2_1\,2_1\,2_1$ (1936 Bernal); n_α 1.512, n_β 1.619, n_γ 1.692 (1934 Kofler).

Optical rotation at 20–25° (1948b Velluz; 1950 Hershberg):

Solvent	$[\alpha]_{5461}$	$[\alpha]_{5791}$	$[\alpha]_D$	$[\alpha]_{6430}$
dioxane	+198°	+170°	+163°	+126°
alcohol.	+200°	+173°	+165°	

For a summary of older rotation values, see 1948b Velluz.

Absorption max. in alcohol at 224±1 (6100), 230±1 (5050), 281±0.5 (2045), and 287±1 (1940) mμ (ε) (1937 Hogness); see also 1936 Sureau; 1936 Callow; 1937 John; 1939 Butenandt; 1939 Dannenberg; 1939 Elvidge;

1941a Heard; 1942 Reynolds; 1948 Friedgood; 1948 Wilds; 1950 Anner. Absorption max. in ether (from the graph) at 280 mμ (1937 Windaus). Absorption max. in alcohol + some KOH (pH 9.5) at 300 mμ (1939 Chevallier), in 0.01 N alcoholic NaOH at 300 mμ (1939 Dannenberg), in 0.1 N aqueous NaOH at 239 and 293 mμ (1946 Carol). The shift in alkaline solution towards longer waves (see also p. 136) is due to salt formation by the hydroxyl group, the absorption of the methyl ether being unaffected (1936 Callow). The solution in conc. H_2SO_4 shows strong absorption at 303 and 450 mμ which changes to 500 mμ on dilution (1939 Bandow). For absorption spectrum of crystalline œstrone in the infra-red (λ = 2 to 12.4 μ), see 1946 Furchgott.

Solubilities expressed in g. of œsterone per 100 g. of solvent (1941 Doisy):

	MeOH	EtOH (95%)	Butanol	Dibutyl ether	Benzene	Toluene	Acetone	CHCl₃	Dioxane
10⁰	0.33	0.35	0.33	0.041	0.056	0.052	1.21	0.255	2.11
30⁰	0.63	0.666	0.60	0.071	0.13	0.099	1.78	0.53	2.84

Solubility in petroleum ether (b. 89–92⁰) at 30⁰: 8.9 γ per c.c. (1941 Doisy). Partition ratio between 70% alcohol and benzene = 1:3 (1938a Westerfeld). For distribution coefficients of œstrone between immiscible solvent pairs (various organic solvents on the one hand, and water or aqueous solutions of acids, alkalis, or salts on the other hand), see 1941 Bachman; 1942 Mather. Oestrone is soluble in aqueous solutions of sodium dehydrocholate, apparently with formation of a dialysable complex (1944 Cantarow). For solutions in aqueous urethan or ethylene glycol or a mixture of both, see 1936a RICHTER GEDEON; see also 1939 Mossini. For solutions in esters (acetate, benzoate) of benzyl alcohol, see 1937 I. G. FARBENIND. For solutions in aqueous veratryl alcohol, see 1938 HOFFMANN-LA ROCHE & Co. A.-G.

OCCURRENCE AND ISOLATION OF OESTRONE

Occ. According to 1939 Butenandt (see also 1949 Cohen), œstrone in human pregnancy urine is conjugated with H_2SO_4 to œstrone hydrogen sulphate (p. 2553 s); according to 1952 Oneson, however, a major portion is conjugated with glucuronic acid. There are indications of the presence of œstrone in the urine of non-pregnant human females (1945 Pincus; 1946b Stimmel). It has been isolated from the urine of a woman with an adrenal tumour (1945 Mason); from human male urine (6 mg. from 17000 l.) (1938 Dingemanse; see also 1945 Pincus; 1946b Stimmel); and from human placenta (12 mg. from 422 kg.) (1938b Westerfeld). For the isolation of the hydrogen sulphate (as alkali salt) from the urine of pregnant mares, see 1938 Schachter, see also 1939 Butenandt; from that of stallions, see 1945 Jensen. For the œstrone content of the urine of pregnant and non-pregnant mares, see 1938 Lapiner. For the occurrence in the urine of mature stallions and, in small amounts, in that of male foals and geldings, and for the œstrone content of sexual glands and adrenals of horses, see 1936 Häussler. For the occurrence in horse testes (0.360 mg. per kg.), see 1940a Beall; in swine testes, see 1943 Ruzicka; in sow ovaries (0.010 mg. per kg. of whole ovaries), see 1938a Westerfeld; in ox adrenals (0.008 mg. per kg.), see 1939, 1940a, b Beall; in the urine of bulls (9 mg. from 380 l.), see 1939a Marker; and of steers (in approximately the same amount), see 1939b Marker. After the administration of "α"-œstradiol (p. 1966 s), œstrone can be isolated from the urine of male or female guinea-pigs (1940, 1941 Fish), of the rabbit (together with "β"-œstradiol, p. 1964 s) (1941c Heard; 1942 Fish), of monkeys (1945 Dorfman; cf. 1937/38 Westerfeld), and of the human male (1941b Heard; see also 1943a Schiller). It is formed (together with œstriol, p. 2152 s) by perfusion of rat liver with "α"-œstradiol (1943b Schiller).

For occurrence of œstrone in oat, see 1936 Häussler; for occurrence of œstrogenic substances in oat, see also 1930 Walker; 1933 Butenandt.

Isolation. For the isolation from the urine of pregnant mares, see also 1936 Beall; 1936, 1937 van Stolk. For the isolation from stallion urine, see also 1936 HOFFMANN-LA ROCHE & Co. A.-G.; 1939 Einhorn; 1946 Dr. G. HENNING.

FORMATION OF ŒSTRONE *

The methyl ether is obtained from 2-methyl-7-methoxy-1.2.3.4.9.10.11.12-octahydrophenanthrene-2-carboxylic-1-(β-propionic) acid on distillation with acetic anhydride (1936 Bardhan) or on heating with $PbCO_3$ (1938 Litvan); it is demethylated by boiling with hydriodic and acetic acids (1938 Litvan). Regarding the formation of œstrone (p. 136/7) by degradation of ergosterol (p. 1735 s) via tetradehydro-neoergosterol and its phenolic tetrahydride (19-nor-$\Delta^{1.3.5(10).22}$-ergostatetraen 3-ol, p. 1533 s), see also 1938a Marker. Oestrone is formed in small amount from "β"-dihydroequilenin (p. 1960 s) or from equilenin (p. 2527 s) on reduction with sodium in boiling amyl alcohol, conversion of the resulting phenolic fraction into the benzoate, oxidation with CrO_3 in acetic acid, and hydrolysis with alcoholic KOH (yield from equilenin, 2.7%) (1938a Marker). From "α"-œstradiol (p. 1966 s) by the action of Flavobacterium dehydrogenans Arn. (Micrococcus dehydrogenans) isolated from yeast (1941 Ercoli; 1942 Arnaudi), or of pseudodiphtheria bacilli (1944/45 Zimmermann). From 6-dehydroœstrone (p. 2538 s) on hydrogenation in the presence of palladium black in abs. alcohol (yield, 80%) (1940 Pearlman) or in ethyl acetate (yield, 90%) (1950 Kaufmann). Together with "α"-œstradiol, by heating $\Delta^{1.4}$-androstadien-17β-ol-3-one (p. 2576 s) in an evacuated sealed tube at 325° (1941, 1947 Inhoffen; 1941 SCHERING A.-G.). For formation from 5-dehydroepiandrosterone (p. 2609 s) via $\Delta^{1.4}$-androstadien-3β-ol-17-one, see 1945 SCHERING A.-G.; but cf. 1950 Hershberg. From $\Delta^{1.4}$-androstadiene-3.17-dione (p. 2877 s) on heating in an evacuated tube at 325° (1937e SCHERING A.-G.).**

Separation. Oestrone can be quantitatively separated from non-ketonic œstrogens by transformation into its carboxymethyl-oxime (p. 2553 s) which is soluble in aqueous $NaHCO_3$ (1940 Huffman). For separation of œstrone,

---

* After the closing date for this volume, the first total synthesis of œstrone has been accomplished by Anner, Miescher (1948a, b); for other syntheses, see 1950, 1951b, 1952a, 1957a Johnson.
** See also, after the closing date for this volume, 1950 Hershberg.

"α"-œstradiol, and œstriol (p. 2152 s) by adsorption of the œstrogen mixture on a column of activated Al_2O_3 and fractional elution with methanol-benzene mixtures, see 1944, 1946a Stimmel. For chromatographic separation, see also 1943 Heintzberger. Oestrone can be separated from equilin (p. 2539 s) by treatment of the mixture in alcoholic solution with trinitrobenzene with which œstrone forms a sparingly soluble addition compound (p. 2553 s) while equilin remains in the mother liquor (1936a LES LABORATOIRES FRANÇAIS DE CHIMIOTHÉRAPIE). The 2.4-dinitrophenylhydrazones of œstrone (p. 2553 s) and of equilenin (p. 2533 s) are separated by passing the benzene solution of the mixture through a column of Al_2O_3; elution with 1% alcohol in petroleum ether removes the œstrone 2.4-dinitrophenylhydrazone only; the equilenin 2.4-dinitrophenylhydrazone may then be eluted by increasing the concentration of alcohol or by washing with acetone (1945a Veitch).

Purification. Via the addition compound with quinoline (p. 137) which is decomposed by shaking with aqueous HCl in ether (1934 Butenandt). Via its azobenzene-p-carboxylate (p. 2555 s) from which it is regenerated by hydrolysis with 2% alcoholic KOH (1941 Bergel).

REACTIONS OF OESTRONE

Contrary to the findings of Butenandt (1930) (see p. 137), it was shown that œstrone is stable in alcoholic solution (1935 Rowlands ; 1939 Pedersen-Bjergaard).

Oxidation and dehydrogenation. Oxidation with alkaline H_2O_2 at room temperature, followed by acidification with HCl, gives œstrololactone (page 2551 s) (1942 Westerfeld); oxidation of the acetate with H_2O_2 in acetic acid at 35° (1947a Jacobsen) or better with peracetic acid in glacial acetic acid in the presence of some p-toluenesulphonic acid at 10° (1947b Jacobsen) yields analogously œstrololactone acetate. Treatment of the benzyl ether with KOI in methyl alcohol at room temperature, followed by boiling with aqueous methanolic (I) KOH, gives the benzyl ether of d-*trans-marrianolic acid (I) (1944 Miescher; 1945a Heer; see also 1946 Heer). Heating of œstrone with palladium black at 260° yields d-isoequilenin (p. 2535 s) (1941 Butenandt).

Reduction. For the hydrogenation to "α"-œstradiol (p. 1966 s) in the presence of Adams's platinum oxide catalyst in abs. alcohol (p. 137), see also 1940 Marker. The esters yield "α"-œstradiol 3-esters on hydrogenation in the presence of platinum oxide catalyst in ethyl acetate (1937a, b, 1938b Miescher; 1937a CIBA) or in ether-alcohol (1939c Marker), in the presence of a nickel catalyst (1932 SCHERING-KAHLBAUM A.-G.; 1937d SCHERING A.-G.), or a Raney catalyst (nickel or cobalt) (1936b SCHERING-KAHLBAUM A.-G.; 1938 Mamoli). Hydrogenation of the crude acetate in alcohol in the presence of a nickel catalyst at 180° and 100 atm. pressure gives a mixture of the

---

* trans refers to the relative orientation of H and CH_3 at C-1 and C-2 of the phenanthrene ring system.

3-acetates of the epimeric œstradiols (1932, 1933c SCHERING-KAHLBAUM A.-G.). "α"-Œstradiol is obtained on reduction of œstrone with aluminium in 5% aqueous alkali (1935 SCHERING CORP.), with amalgamated aluminium foil in wet ethyl acetate, or with zinc dust in glacial acetic acid (1933a SCHERING-KAHLBAUM A.-G.), and seems also to be formed by heating material containing œstrone with $Na_2S_2O_4$ on the water-bath (1936b RICHTER GEDEON). Reduction of œstrone with aluminium isopropoxide in boiling isopropyl alcohol yields "α"-œstradiol and "β"-œstradiol (p. 1964 s) (1938c Marker); by the analogous reduction of the benzoate ,"α"- and "β"-œstradiol benzoates are formed (1938a Marker). On treatment of the benzoate with isopropyl-magnesium iodide, followed by saponification, "α"-œstradiol is said to result (1934b SCHERING-KAHLBAUM A.-G.); this diol is formed by the action of fermenting yeast on œstrone in aqueous solution + some dioxane at 37° (1939 Wettstein) or œstrone esters in aqueous solution + some alcohol at room temperature (1938 Mamoli) (see also 1937b CIBA; 1937c SCHERING A.-G.); it is also said to be produced by the enzymic reduction of œstrone benzoate by succinic dehydrogenase in the presence of sodium succinate (1940 Miwa; cf., however, 1954 Repke). Oestrone is not reduced by enzyme extracts of hog ovaries, beef suprarenal glands, or bull testes (1938b Marker). By hydrogenation of œstrone in acid solution (using platinum oxide catalyst in alcohol containing HCl) (cf. p. 137), œstrane-3.17β-diol-B (p. 1984 s) (1938b Marker; see also 1938c Marker) and œstran-17β-ol (p. 1498 s) (1940 Marker) have been obtained; the same compounds (or mixtures of stereoisomers) are formed, along with œstran-3-ol-17-ones (p. 2563 s), on hydrogenation in the presence of platinum oxide in glacial acetic acid, 80% or 60% acetic acid at room temperature, the last-mentioned compound as the main product on hydrogenation in 80% acetic acid at 40° (1936 Dirscherl); the mono-hydroxy-œstranes, obtained on hydrogenation in acid solution and originally assumed to be 3-hydroxy-œstranes (see p. 137 lines 9–10), are actually 17-hydroxy-œstranes (1940 Marker). Hydrogenation to a mixture of stereo-isomeric œstrane-3.17-diols (see p. 1984 s) is accomplished by using a nickel catalyst in hot alcohol (1931 SCHERING A.-G.), in cyclohexanol at 170° and 60–70 atm. pressure, or (with a nickel-copper catalyst) in 10% KOH at 160° and 100 atm. pressure (1933 SCHERING A.-G.). Hydrogenation of œstrone esters in 96% alcohol at 100° and 100 atm. pressure, using a nickel-chromium catalyst, affords a mixture of 3-esters of stereoisomeric œstrane-3.17-diols (1934a SCHERING-KAHLBAUM A.-G.).

Action of light. On exposure to ultra-violet radiation in abs. alcohol, the m. p. is lowered, the specific rotation diminished, the absorption spectrum in the ultra-violet and the fluorescence produced by H_2SO_4-glacial acetic acid in chloroform (see below) are altered, and the intensities of other colour reactions are reduced (1938 Laporta), while the œstrogenic activity remains unchanged (1936, 1938 Laporta; cf., however, 1941 Ito). In aqueous alcoholic NaOH, ultra-violet light was found to destroy the œstrogenic activity (1942/43, 1945 Figge; see also 1948 Herlant). Irradiation with ultra-violet light of wave length 313 mμ in dioxane gives lumiœstrone (p. 2559 s) (1941, 1942

Butenandt) and a **carboxylic acid** $C_{18}H_{24}O_3$ (m. 212–4° cor.) (1946 Heer); for calculation of the conversion energy in the transition of œstrone to lumi-œstrone, see 1944 Koyenuma. Irradiation with visible light in aqueous alcoholic solution does not decrease the biological strength of œstrone (1940 Sakanoue). The action of X-rays in alkaline or in aqueous acetic acid solution yields small amounts of œstrololactone (1951 Keller).

Halogenation, nitrosation, and sulphonation. Treatment of the acetate with bromine in chloroform-glacial acetic acid in the presence of some HBr gives the acetate of 16-bromo-œstrone (p. 2702 s)*; the benzoate reacts similarly with bromine in chloroform (1937b SCHERING A.-G.). — Reaction of the benzoate with potassium tert.-butoxide and isoamyl nitrite in tert.-butyl alcohol at room temperature under nitrogen, followed by saponification, yields 16-oximino-œstrone (p. 2941 s) (1944 Huffman; see also 1942 Huffman); the methyl ether on similar treatment with potassium tert.-butoxide and isoamyl nitrite gives 16-oximino-œstrone methyl ether* (1938 Litvan; see also 1946 Butenandt; 1947 Huffman); the benzyl ether reacts similarly (1948 Huffman). — Treatment with chlorosulphonic acid and pyridine in chloroform at room temperature, followed by addition of methanolic NaOH, yields the sodium salt of œstrone hydrogen sulphate (p. 2553 s); on boiling with chlorosulphonic acid in chloroform-carbon tetrachloride and some ether, œstrone-2 or 4-sulphonic acid is formed (1939 Butenandt).

Other reactions with inorganic compounds. The acid ($C_{17}H_{22}O_3$) formed on fusion with KOH (p. 137; see also 1947 Hohlweg), is actually d-**trans-doisynolic acid $C_{18}H_{24}O_3$ (I) (1944 Miescher; 1945a Heer; see also 1946, 1948 Heer). By the action of CO_2 at 200° and 50–60 atm. pressure, the sodium salt yields a carboxylated compound *("œstronesalicylic acid")* (1936 C. F. BOEHRINGER & SÖHNE; 1940 Niederl) which gives a benzoate of high hormone activity (1936 C. F. BOEHRINGER & SÖHNE).

Reactions with organic compounds. Reacts in dioxane-ether solution with potassium acetylide in liquid NH_3 to give 17α-ethynylœstra-3.17β-diol (page 1990 s) (1938 Inhoffen); reacts in the same way in dioxane-benzene solution (1937c CIBA). Treatment of œstrone (or its acetate) dissolved in abs. ether with acetylene in the presence of sodamide gives also 17α-ethynylœstra-3.17β-diol (or its acetate) (1935 Claude; 1936 SCHERING A.-G.; 1936a SCHE-RING-KAHLBAUM A.-G.), which is also formed by the action of ethynyl-magnesium bromide on œstrone in ether (1936c SCHERING-KAHLBAUM A.-G.). Reaction of œstrone (1934c SCHERING-KAHLBAUM A.-G.) or its benzoate (1934b SCHERING-KAHLBAUM A.-G.) with methylmagnesium iodide yields 17α-methylœstra 3.17β-diol (p. 1988 s); the methyl ether affords 17α-methyl-œstra-3.17β-diol 3-methyl ether (1935 Cohen). The action of ethylmagnesium iodide on the benzoate gives similarly 17α-ethylœstra-3.17β-diol (p. 1990 s) (1934b SCHERING-KAHLBAUM A.-G.), that of styrylmagnesium bromide on

---

* This is an intermediate in the conversion of œstrone into œstriol (see p. 2155 s).
** trans refers to the relative orientation of H and CH_3 at C-1 and C-2 of the phenanthrene ring system.

the acetate 17α-styryloestra-3.17β-diol (p. 2045 s) (1937a SCHERING A.-G.), whereas "α"-oestradiol is obtained in the reaction of isopropylmagnesium iodide with the benzoate, followed by saponification (1934b SCHERING-KAHL-BAUM A.-G.). Diazotized p-nitroaniline couples with oestrone in glacial acetic acid to give a red-brown amorphous substance, m. 145–155° dec.; with oestrone methyl ether, a greenish brown product is formed (1941 King); for application of the coupling reaction with diazotized amines in the colorimetric determination of oestrone, see p. 2550 s. Condensation of the methyl ether with m-nitrobenzaldehyde in alcohol-dioxane and some methanolic KOH at room temperature gives 16-(m-nitrobenzylidene)-oestrone methyl ether (p. 2717 s) (1943 Wettstein). Treatment with amyl formate and sodium in abs. ether yields 16-hydroxymethylene-(or formyl-)oestrone (p. 2719 s) (1939 HOFF-MANN-LA ROCHE & Co. A.-G.). On shaking the acetate with potassium cyanide in glacial acetic acid and alcohol, two 17-epimeric oestrone cyanohydrin 3-acetates are formed (1941a, b Goldberg). Heating with isatin and aqueous alcoholic KOH on the water-bath gives the **acid** $C_{26}H_{25}O_3N$ (I) [pale yellow microcrystalline powder; decomposes on heating with evolution of CO_2; easily soluble in aqueous alkali to strongly foaming solutions] (1944 Buu-Hoi).

Biological reactions. For the inactivation of oestrone by enzymes, see 1940 Westerfeld; 1942 Graubard; 1942, 1945 Zondek; for the inactivation by microorganisms, see 1943 Zondek. For microbiological reduction, see p. 2547 s.

The excretion of "α"-oestradiol (p. 1966 s) and oestriol (p. 2152 s) is strongly increased after injection of oestrone into men (1942 Pincus; 1942, 1943 Pearlman) and non-pregnant women (1942 Pincus). The excretion of oestrogens in the urine is increased after injection of oestrone into male and female dogs (1941 Dingemanse; 1942 Longwell); to a smaller extent the (non-ketonic) oestrogens are also excreted in the bile (1942 Longwell). For the metabolism of oestrone, see also 1949 Fieser.

PHYSIOLOGICAL PROPERTIES OF OESTRONE

For physiological properties of oestrone, see p. 136 and pp. 1367 s, 1368 s and the literature cited there.

ANALYTICAL PROPERTIES OF OESTRONE *
(see also "Separation", p. 2545 s)

Colour reactions and detection. For summaries, see 1934 Häussler; 1937 v. Novák; 1938 Zimmermann; 1945 Cheymol. For detection based on the absorption spectra in conc. and dil. H_2SO_4, see 1939 Bandow. In aqueous or alcoholic solution with conc. H_2SO_4 a yellowish green fluorescence is shown in ultra-violet light (1937 Kocsis; 1937b Bugyi). A chloroform solution shaken with an equal vol. of a H_2SO_4-glacial acetic acid mixture (1:2 by vol.)

---

* They are mostly the same as those for oestradiol and oestriol.

yields a yellow lower (H_2SO_4) phase with strong green-yellow fluorescence (max. ca. 574 mμ) (œstriol and œstradiol give the same reaction) (1936b Bierry; see also 1936a Bierry). For detection by fluorescence, see also 1942 Konstantinova-Schlezinger. For colour reactions with conc. H_2SO_4, see also 1939a, b Woker; 1939 Scherrer. Oestrone in chloroform solution with an equal vol. of a solution of titanium sulphate in conc. H_2SO_4 (0.5 to 1 g. in 100 c.c.) gives a greenish fluorescence and a wine-red colour in the H_2SO_4 layer only (absorption max. at about 500 mμ); the red colour is stable for several hrs. and finally changes to brown (distinction from œstradiol for which the red colour quickly changes to yellow) (1946a, b Devis). On heating with phthalic anhydride and anhydrous $SnCl_4$ at 116–120^0, a dark red to reddish brown mass is formed which dissolves in chloroform, to give a brilliant deep pink solution (absorption band 538.6 mμ) with greenish yellow fluorescence (max. 557 mμ) (œstradiol and œstriol show the same reaction) (1941 Kleiner; cf. 1947 Wülfert). With m-dinitrobenzene and KOH in aqueous alcoholic solution a violet colour appears (1935 Zimmermann; cf. also 1944 Zimmermann). If 1–2 drops of a 0.1 % alcoholic solution of 1-nitroso-2-naphthol and 5 drops of HNO_3 (d 1.4) are added to 1 c.c. of a suspension, prepared from 1 mg. œstrone dissolved in 3 drops of alcohol and diluted with water to 10 c.c., and the test tube is heated in boiling water, a deep red colour is produced which fades slowly; this colour reaction differs from those given by equilin (p. 2539 s) and equilenin (p. 2527 s) under the same conditions (1937 Voss). For colour reactions with benzaldehyde and conc. H_2SO_4, see 1939 Scherrer; with furfural and conc. H_2SO_4, see 1939a, b Woker. A chloroform solution heated with a solution of $ZnCl_2$ in glacial acetic acid and benzoyl chloride to 70^0 and then allowed to cool, gives rise to a coloured compound (absorption max. at 500 to 504 mμ) (œstradiol gives a similar but less intense colour while the reaction is negative with œstriol) (1936 Pincus).

Determination. For a summary, see 1938 Zimmermann. For modifications of the determination by the Kober colour test (p. 137), replacing phenolsulphonic acid by β-naphtholsulphonic acid, see 1938 Kober; by guaiacolsulphonic acid (whereby œstriol practically does not interfere with the estimation of œstrone and œstradiol), see 1943 Szego; for other modifications and improvements, see 1936 Pincus; 1937 Venning (study of the various factors affecting the stability and intensity of the colour produced in the Kober test); 1937a Lapiner; 1937 Wu; 1939, 1941 Bachman; 1941, 1942 Mather; 1942 Winkler; 1943 Jayle; 1946b Stimmel; 1946 Mayer. For determination based on the colour produced with benzoyl chloride (see above), see 1936 Pincus. For determination based on the use of the dye formed by coupling with diazotized sulphanilic acid, see 1936 Schmulovitz; 1937 Pincus; of the dye produced with diazotized dianisidine, see 1940 Talbot; 1944 Reifenstein. For determination based on the colour (specific for 17-ketosteroids) appearing with m-dinitrobenzene and KOH (see above), see 1935, 1936/37 Zimmermann; 1938 Callow; see also 1940 McCullagh; 1943 Hansen. For determination based on the colour of the 2.4-dinitrophenylhydrazone (p. 2553 s), see 1945b Veitch. Oestrone can be determined gravimetrically by precipitating the aqueous solution of the compound obtained with Girard's reagent T with HgI_2 (in aqueous KI) with which it forms a highly insoluble complex (1941 Hughes). The aqueous solution of the Girard derivative can also be used for the polarographic determination (1940 Wolfe; 1946 Björnson). For spectrophotometric determination of œstrone in commercial sex hormone preparations, see 1937 a, b Bugyi; 1939 Elvidge; 1946 Carol (here also a gravimetric method is described); in biological fluids (urine), see 1936 Sureau; 1939 Chevallier.

OXIDATION PRODUCTS OF OESTRONE *

Oestrolic acid, 13‖17-$\Delta^{1\cdot3\cdot5(10)}$-Oestratriene-3.13-diol-17-oic acid, β-(2-Methyl-2.7-dihydroxy-1.2.3.4.9.10.11.12-octahydro-1-phenanthryl)-propionic acid $C_{18}H_{24}O_4$ (I) and **Oestrololactone** $C_{18}H_{22}O_3$ (II). For recent literature con-

firming the structure as resulting from oxidative cleavage of ring D of œstrone between C-13 and C-17, already favoured by 1942 Westerfeld and 1947 a Jacobsen, see 1952 Picha; 1953 Fried; 1955 Wendler; 1956 Murray.

Oestrolic acid $C_{18}H_{24}O_4$ (I). Plates (acetone + water), m. 225° cor. dec. (aluminium block); if the melting point is determined in an open capillary, the acid lactonizes without sintering and m. ca. 330° (1947 a Jacobsen). — **Fmn.** From œstrololactone acetate (p. 2552 s) by heating with aqueous NaOH, cooling to 15°, and slowly acidifying with HCl to about pH 5 (1947 a Jacobsen). — *Sodium salt* $C_{18}H_{23}O_4Na + 2 H_2O$. Needles (alcohol), m. 225° cor. (aluminium block) (1947 a Jacobsen), 285° dec. (1946 Jacobsen). **Fmn.** From œstrololactone acetate by heating with aqueous methanolic NaOH, cooling, and saturating with CO_2 (1947 a Jacobsen).

Methyl ester $C_{19}H_{26}O_4$. Glistening needles (aqueous methanol) becoming opaque on standing in a dry atmosphere and then m. 95–97° cor.**; contains 1 mol. $CH_3 \cdot OH$ when dried at 65° for 1 hr.; this is lost on drying at 60–65° for 8 hrs. — **Fmn.** From the acid (above) with diazomethane in ether. — **Rns.** Refluxing with aqueous methanolic $KHCO_3$ gives œstrololactone. Heating with benzoyl chloride in pyridine yields œstrololactone benzoate (p. 2552 s) and a fraction which after alkaline hydrolysis gives crystals, m. 167–171° cor. (1947 a Jacobsen).

Lactone, Oestrololactone $C_{18}H_{22}O_3$ (II, above). For physical and chemical identity of the lactones obtained by the various methods, see 1949 Jacques; 1951 Keller. According to 1946, 1947 a Jacobsen, his lactone and that of 1942 Westerfeld cannot be identical because of their widely differing physio-

---

* For oxidation of œstrone to marrianolic acid, see p. 2546 s.
** According to 1946 Jacobsen, the methyl ester m. 95–97° contains $^1/_2$ mol. H_2O.

logical properties; it is suggested (1946 Jacobsen) that they may be stereoisomers. — Crystals (dil. pyridine), m. 335–340° (1942 Westerfeld); prisms
(cyclohexanone or "methyl cellosolve"), m. 339° cor. (aluminium block) (1947a
Jacobsen), 349° (gold block) (1949 Jacques); m. 309° dec. (Mather, see 1944
Smith); prisms (alcohol), m. 310–2° dec. (Kofler block ±2°); $[\alpha]_D^{21}$ +40°
(pyridine) (1951 Keller). Absorption max. in dioxane at 284 mμ (ε 2200)
(1942 Westerfeld). Very sparingly soluble in most organic solvents (1942
Westerfeld; 1947a Jacobsen).

Fmn. In small yield, by the action of X-rays on œstrone in alkaline or
aqueous acetic acid solution (1951 Keller). By treatment of œstrone in dil.
NaOH with 10% aqueous H_2O_2 at room temperature for 3 days, followed
by acidification with HCl (one-half of the œstrone is recovered; yield from
the oxidized œstrone, ca. 45%); the lactone is purified via its acetate (1942
Westerfeld). The acetate is formed from œstrone acetate on standing in
glacial acetic acid with 30% aqueous H_2O_2 in the dark at 35° for 60 hrs.
(yield of crude acetate, 57–63%) (1947a Jacobsen). The acetate is obtained
in better yield (80–85% of crude acetate) from œstrone acetate on standing
with peracetic acid in glacial acetic acid in the presence of some p-toluenesulphonic acid in the dark at 10° for 100 hrs. (1947b Jacobsen). The acetate
is saponified by heating with methanolic NaOH and the resulting œstrolic
acid relactonized by acidification with HCl and warming to 60–65° (1947a
Jacobsen; see also 1942 Westerfeld).

Rns. Dissolves in dil. NaOH (see also 1941b Heard) on heating and shaking
with formation of œstrolic acid (1942 Westerfeld; 1947a Jacobsen). On
refluxing with abs. methanol and some HCl or H_2SO_4, the lactone ring is
opened under esterification of the carboxyl group (the methyl ester has not
been obtained pure in this manner). Dissolves in conc. H_2SO_4 to a yellow
solution with a green fluorescence. Kober test (p. 137) and Zimmermann
test (with m-dinitrobenzene + KOH) are negative. With diazotized p-nitroaniline an orange-yellow colour is obtained (1942 Westerfeld).

Oestrololactone acetate $C_{20}H_{24}O_4$. Needles (alcohol), m. 143.5–145° (1942
Westerfeld); diamond-shaped crystals (methanol), m. 149–150.5° cor.; $[\alpha]_D^{24}$
+42° (chloroform) (1947a Jacobsen); ultra-violet absorption max. in 95%
alcohol at 269 mμ (ε 750), and 276 mμ (ε 720) (1942 Westerfeld). For infra-red
absorption spectrum, see 1950 Jones. — **Fmn.** See formation of œstrolo-
lactone, above. Also from œstrololactone by acetylation with acetic anhydride and pyridine (1942 Westerfeld; 1947a Jacobsen).

Oestrololactone propionate $C_{21}H_{26}O_4$. Needles (aqueous methanol), m. 146°
to 148.5° cor. — **Fmn.** From œstrololactone on heating with propionic anhydride and pyridine (1947a Jacobsen).

Oestrololactone benzoate $C_{25}H_{26}O_4$. Needles (alcohol or acetone), m. 241°
to 244° cor. — **Fmn.** From œstrololactone with benzoyl chloride and pyridine
(1947a Jacobsen).

Oestrolic acid* 7-methyl ether $C_{19}H_{26}O_4$. Tablets (aqueous acetone), m. 135°
to 136° cor. dec. — **Fmn.** From œstrololactone methyl ether (below) by

---

* The number 7 refers to the numbering of phenanthrene.

treatment with aqueous methanolic NaOH, removal of methanol, and careful acidification as in the preparation of œstrolic acid (p. 2551 s) (1947a Jacobsen).

Methyl ester of the 7-methyl ether $C_{20}H_{28}O_4$ (after drying at 65⁰). Glistening leaves (aqueous methanol), m. 64.5–65.5⁰ cor. becoming opaque after drying at 65⁰ and then m. 77.5–78.5⁰ cor. — **Fmn.** From the preceding compound with ethereal diazomethane (1947a Jacobsen).

Lactone of the 7-methyl ether, Oestrololactone methyl ether $C_{19}H_{24}O_3$. Needles (aqueous methanol), m. 172.5–174⁰ cor. (1947a Jacobsen), 166–8⁰ (1942 Westerfeld). — **Fmn.** From œstrololactone with dimethyl sulphate and alkali, followed by acidification (1942 Westerfeld; 1947a Jacobsen) or by refluxing with methyl iodide and Ag_2O (1947a Jacobsen).

ADDITION COMPOUNDS OF OESTRONE

Compound with trinitrobenzene (2:1) $C_{42}H_{47}O_{10}N_3$. Orange-yellow crystals (alcohol saturated with trinitrobenzene); sparingly soluble in alcohol. — **Rns.** Reduction with sodium and propyl alcohol gives œstradiol (ca. 80% "α"- and 20% "β"-œstradiol). — It is used in the separation of œstrone from equilin (see p. 2546 s) (1936a LES LABORATOIRES FRANÇAIS DE CHIMIOTHÉRAPIE).

Compound with quinoline (1:1) $C_{27}H_{29}O_2N$ (p. 137). For its use in the purification of œstrone, see p. 2546 s.

DERIVATIVES OF OESTRONE

Oestrone oxime $C_{18}H_{23}O_2N$ (p. 137). **Rns.** Reduction with sodium and alcohol gives 17-amino-$\Delta^{1.3.5(10)}$-œstratrien-3-ol (p. 1910 s) (1932 SCHERING A.-G.).

Oestrone O-(carboxymethyl)-oxime $C_{20}H_{25}O_4N$. White needles (aqueous alcohol) with 0.5 $C_2H_5 \cdot OH$, m. 188⁰; soluble in aqueous $NaHCO_3$ (1940 Huffman). — **Fmn.** From œstrone on refluxing with O-(carboxymethyl)-hydroxylamine hydrochloride, potassium acetate, and aldehyde-free propyl alcohol (1940 Huffman). — **Rns.** Refluxing with aqueous alcoholic HCl at 85–90⁰ regenerates œstrone (1940 Huffman). For the inactivation by enzymes, see 1942 Graubard. — It is used in the separation of œstrone from non-ketonic œstrogens (see p. 2545 s) (1940 Huffman).

Oestrone 2.4-dinitrophenylhydrazone $C_{24}H_{26}O_5N_4$. Yellow crystals (dioxane-alcohol), darkening at 268⁰, m. 278–280⁰ dec. (1941 King). Absorption max. in 0.1 N alcoholic KOH at 440 mμ (1945b Veitch). — It is used for the colorimetric determination of œstrone (1945b Veitch). For its use in the chromatographic separation of œstrone from equilenin, see p. 2546 s.

Oestrone hydrogen sulphate $C_{18}H_{22}O_5S$. Only its salts are known. For trade names of commercial preparations, see 1949 Selye. — **Occ.** It has been isolated as potassium salt (see below) from the urine of pregnant mares (1938 Schachter; see also 1939 Butenandt) and as sodium salt (see below) from the urine of stallions (1945 Jensen). For the occurrence in human pregnancy urine, see p. 2544 s. — **Fmn.** and **Rns.** See the sodium salt, below.

177*

Sodium salt $C_{18}H_{21}O_5SNa + H_2O$. Colourless needles (methanol + ether), m. 228–230° cor. with decomposition to œstrone and $NaHSO_4$; $[\alpha]_D^{20} + 110°$ (solvent not indicated) (1939 Butenandt). Absorption max. in methanol at 272 mμ (ε 850) (1939 Dannenberg). Soluble in water, methanol, ethanol, butanol, acetone, and pyridine, less soluble in ethyl acetate and chloroform, insoluble in ether and petroleum ether (1939 Butenandt). It is œstrogenically less active than œstrone (1939 Butenandt; cf. 1949 Grant). — **Fmn.** From œstrone on treatment with chlorosulphonic acid and pyridine in chloroform at room temperature, followed by neutralization with methanolic NaOH (1939 Butenandt). — **Rns.** Decomposes slowly on standing in the air (1939 Butenandt). Hydrogenation with PtO_2 catalyst in 0.2 M phosphate buffer at room temperature gives the sodium salt of "α"-œstradiol 3-(hydrogen sulphate) (p. 1971 s) (1944 Grant). Is rapidly hydrolysed to œstrone by heating with dil. HCl and by the enzyme phenolsulphatase (takadiastase) (1939 Butenandt).

Potassium salt $C_{18}H_{21}O_5SK$ (slightly impure). White crystals (butanol), m. 233–242° (1938 Schachter). — *Barium salt* (no analysis given). Amorphous; easily soluble in butanol (1939 Butenandt). — *Pyridinium salt* $C_{23}H_{27}O_5NS$. Needles (methanol), m. 173–5°; $[\alpha]_D^{23} + 84°$ (solvent not indicated). It is œstrogenically more active than the sodium salt (1939 Butenandt). — *Quinidinium salt* $C_{38}H_{46}O_7N_2S + 3 H_2O$. Colourless needles (dil. acetone), m. 167–170° (1939 Butenandt). Shows the same œstrogenic activity as the sodium salt. Is stable in the air (1939 Butenandt). — *Quininium salt* (acetone-ether) (no analysis given). Amorphous solid, m. 168–170° (1939 Butenandt).

Oestrone hydrogen sulphate semicarbazone $C_{19}H_{25}O_5N_3S$. Only its salts are known. — *Sodium salt* (no analysis given), dec. 258–260°; moderately soluble in water. — *Barium salt* (no analysis given). Colourless prisms (methanol), dec. above 270° (1939 Butenandt).

Esters of œstrone with organic acids

For the *œstrogenic potencies* and the duration of the effect of various esters (see also p. 136), see 1937b Lapiner; 1938a Miescher; 1938 Fischer (but cf. 1938b Miescher); 1938 Bergel; 1943 Parkes. — **Rns.** See reactions of œstrone, p. 2546 s et seq.

Oestrone acetate $C_{20}H_{24}O_3$ (p. 137), $[\alpha]_D + 128°$ (chloroform) (1948a Velluz), $+ 138°$ (chloroform) (1950 Velluz); $[\alpha]_D^{20} + 148°$ (dioxane) (1950 Kaufmann). Absorption max. in alcohol at 276, 270, 262 mμ (1948b Velluz; see also 1937 John; 1942 Westerfeld); in ether at 260 mμ (ε 1040) (1939 Dannenberg).

Oestrone trimethylacetate $C_{23}H_{30}O_3$. White needles (acetone-methanol), m. 164–6°. — **Fmn.** From œstrone and trimethylacetyl chloride in pyridine. — **Rns.** Hydrogenation in the presence of Adams's catalyst in ether-alcohol at room temperature gives "α"-œstradiol 3-trimethylacetate (p. 1973 s) (1939c Marker).

Oestrone tert.-butylacetate $C_{24}H_{32}O_3$. White plates (methanol), m. 148–150⁰. — **Fmn.** From œstrone and tert.-butylacetyl chloride in pyridine or in.10 % aqueous KOH. — **Rns.** Hydrogenation in the presence of Adams's catalyst in ether-alcohol gives "α"-œstradiol 3-tert.-butylacetate (p. 1973 s) (1939c Marker).

Oestrone œnanthate, Oestrone heptanoate $C_{25}H_{34}O_3$. Crystals (dil. alcohol), m. 54–54.5⁰. — **Fmn.** From œstrone and œnanthyl chloride in pyridine (1936 CIBA).

Oestrone caprylate, Oestrone octanoate $C_{26}H_{36}O_3$. Crystals (hexane and aqueous alcohol), m. 70–71⁰. — **Fmn.** From œstrone and caprylyl chloride in pyridine (1938a Miescher). — **Rns.** Hydrogenation in the presence of platinum oxide in ethyl acetate gives "α"-œstradiol 3-caprylate (p. 1974 s) (1938b Miescher).

Oestrone laurate $C_{30}H_{44}O_3$. Crystals (methanol and ethanol), m. 69.5–70⁰. — **Fmn.** From œstrone and lauroyl chloride in pyridine (1938a Miescher).

Oestrone benzoate, "Benztrone" (see 1949 Burrows) $C_{25}H_{26}O_3$ (p. 137). Crystals (chloroform-petroleum ether), m. 218⁰ (1937b Lapiner). Optical rotation in dioxane at 25⁰: $[\alpha]_{546} + 133$⁰ ; $[\alpha]_{589} + 112$⁰; $[\alpha]_{643} + 80$⁰ (1950 Hershberg). For solutions in benzyl acetate, see 1937 I. G. FARBENIND. — **Rns.** See those of œstrone, p. 2546 s. — *Di-ethylthio ketal* $C_{29}H_{36}O_2S_2$, m. ca. 85⁰, $[\alpha]_D + 38$⁰ (benzene); is formed from the benzoate and ethyl mercaptan in the presence of benzenesulphonic acid at room temperature (1940 CHINOIN).

Oestrone 3.5 - dinitrobenzoate $C_{25}H_{24}O_7N_2$. Crystals (benzene-alcohol), m. 193–4⁰. — **Fmn.** From œstrone and 3.5-dinitrobenzoyl chloride in pyridine (1940a, b Beall).

Oestrone p-benzeneazobenzoate, Oestrone azobenzene-p-carboxylate $C_{31}H_{30}O_3N_2$. Red rhombs (ethyl acetate), m. 226.5–227.5⁰ cor. — **Fmn.** By heating œstrone with p-benzeneazobenzoyl chloride in pyridine at 50⁰. — It is used for the purification of œstrone (see p. 2546 s) (1941 Bergel).

Oestrone α-naphthoate $C_{29}H_{28}O_3$ (p. 137), m. 210⁰ (1938b Westerfeld).

Oestrone β-naphthoate $C_{29}H_{28}O_3$. Crystals (dioxane) m. 262–4⁰. — **Fmn.** From œstrone and β-naphthoyl chloride in pyridine (1938 Bergel).

Ethers of Oestrone

Oestrone methyl ether $C_{19}H_{24}O_2$ (p. 138). $[\alpha]_D + 160$⁰ (dioxane) (1942 Goldberg). Absorption max. in 90 % alcohol (from the graph) at 279 mμ (log ε/c 0.79); the absorption spectrum is not changed by the addition of alkali (in contrast to that of free œstrone, see p. 2544 s) (1936 Callow; cf. 1939 Dannenberg). It gives no fluorescent solution in conc. H_2SO_4 (1939 Robinson). For its œstrogenic activity (p. 138), see also 1938 Zondek; 1941 Urushibara; 1945b Heer. — **Fmn.** From œstrone with dimethyl sulphate and aqueous alkali on shaking (1931 Thayer; 1932 Marrian) or boiling (1932 Butenandt), or with methyl p-toluenesulphonate and aqueous KOH on heating (1935 Cohen). For other formations, see those of œstrone, p. 136 and p. 2545 s. — **Rns.** On boiling with sodium in benzene, addition of dimethyl sulphate, and further boiling, "α"-œstradiol dimethyl ether (p. 1975 s) is obtained (1941 Urushibara). For other reactions, see those of œstrone, pp. 2548 s, 2549 s.

Oestrone ethyl ether $C_{20}H_{26}O_2$. M. p. 126°; $[\alpha]_D$ +150° (dioxane). — Its œstrogenic activity is a little less than that of free œstrone. — **Fmn.** By treatment of œstrone with ethyl p-toluenesulphonate and KOH (1945 Courrier).

Oestrone allyl ether $C_{21}H_{26}O_2$ (p. 138). — **Fmn.** From œstrone on heating with allyl bromide and sodium ethoxide in abs. alcohol at 80–5°. — **Rns.** Boiling in diethylaniline under nitrogen gives 2 or 4-allylœstrone (p. 2574 s) (1937 b Miescher).

Oestrone β-diethylamino-ethyl ether $C_{24}H_{35}O_2N$. Leaflets (dil. alcohol), m. 76° to 77°. — It shows no œstrogenic activity. — **Fmn.** By heating œstrone with β-diethylamino-ethyl chloride and aqueous KOH. — *Hydrochloride* (no analysis given). Crystals (abs. alcohol-ethyl acetate), m. 190–1°; soluble in water (1938 Bergel).

Oestrone methoxymethyl ether $C_{20}H_{26}O_3$. Needles (alcohol), m. 100°. — **Fmn.** By heating œstrone with sodium ethoxide in dioxane, cooling, and adding a benzene solution of chlorodimethyl ether. — **Rns.** Reduction with sodium in hot alcohol gives "α"-œstradiol 3-methoxymethyl ether (see p. 1972 s) (1936 b LES LABORATOIRES FRANÇAIS DE CHIMIOTHÉRAPIE).

Oestrone carboxymethyl ether, Oestrone-O-acetic acid $C_{20}H_{24}O_4$. Needles or white plates (aqueous acetone), m. 209° (1938 Ercoli), 209–211° (1939 d Marker). — For œstrogenic activity, see 1938 Ercoli. — **Fmn.** By refluxing œstrone with chloroacetic acid in 5 % aqueous KOH for 2 hrs. (yield, 74 % allowing for recovered œstrone) (1938 Ercoli) or with ethyl chloroacetate and an excess of alcoholic sodium ethoxide, finally under addition of KOH (yield, 73 % allowing for recovered œstrone) (1939 d Marker); lower yields are obtained by heating œstrone with ethyl diazoacetate at 130–140°, followed by saponification with methanolic KOH (1938 Ercoli). — *Oxime* $C_{20}H_{25}O_4N$. White crystals (alcohol), m. 230–2° dec. (1939 d Marker). — *Methyl ester* $C_{21}H_{26}O_4$. White scales (aqueous acetone), m. 126°* (1938 Ercoli), 130–2° (1939 d Marker). **Fmn.** From the acid with diazomethane in ether (1938 Ercoli) or ether-methanol (1939 d Marker).

Oestrone α-carboxyethyl ether, Oestrone-O-α-propionic acid $C_{21}H_{26}O_4$. White crystals (aqueous acetone), m. 195–8°. — **Fmn.** From œstrone and α-chloropropionic acid in a similar way as œstrone carboxymethyl ether (above), but using a much greater excess of sodium ethoxide. — *Methyl ester* $C_{22}H_{28}O_4$. White crystals (aqueous acetone), m. 137–9°. **Fmn.** From the acid with diazomethane in methanol-ether (1939 d Marker).

---

* On p. 143 of the original paper 126° is given and on p. 145, probably erroneously, 226° as the m. p.

Oestrone benzyl ether $C_{25}H_{28}O_2$, m. 135–6°. — **Fmn.** From œstrone and benzyl chloride in the presence of sodium ethoxide. — **Rns.** Reduction with aluminium isopropoxide in isopropyl alcohol, with sodium in alcohol, with $Na_2S_2O_4$ or with another hyposulphite in the presence of iron oxide gives "α"-œstradiol 3-benzyl ether (p. 1976 s) (1938 CHINOIN). For oxidation, see p. 2546 s; for nitrosation, see p. 2548 s.

Oestrone p-nitrophenyl ether $C_{24}H_{25}O_4N$. Light-yellow plates (acetone-methanol-water), m. 192–4°. — **Fmn.** On heating the potassium salt of œstrone with p-fluoronitrobenzene and a copper catalyst at 200–210°. — **Rns.** Reduction with a solution of $SnCl_2$ in glacial acetic acid gives œstrone p-amino-phenyl ether (1941 King).

Oestrone p-nitrobenzyl ether $C_{25}H_{27}O_4N$. Pale green-yellow needles (acetone-alcohol-water), m. 176.5–178.5°. — **Fmn.** On heating œstrone with p-nitro-benzyl bromide and sodium ethoxide in alcohol. — **Rns.** Reduction to the corresponding amine could not be effected. — *Semicarbazone* $C_{26}H_{30}O_4N_4$ (no analysis given). Colourless crystals (dioxane-ethylene chloride), m. 273–5° (1941 King).

Oestrone p-aminophenyl ether $C_{24}H_{27}O_2N$. Colourless plates (dil. methanol), m. 166.5–168.5°. — Possesses a moderately high œstrogenic activity. — **Fmn.** By reduction of œstrone p-nitrophenyl ether with a solution of $SnCl_2$ in glacial acetic acid at 20° and decomposition of the resulting $SnCl_2$ salt with NaOH in aqueous methanol. — **Rns.** Diazotization with $NaNO_2$ and dil. HCl at 0° gives a pale green-yellow solution which couples with β-naphthol (bright scarlet precipitate), l-tyrosine (orange precipitate), and casein (orange-yellow product exhibiting about one-tenth of the physiological activity of œstrone p-aminophenyl ether). Refluxing with phosgene in benzene-toluene yields œstrone p-isocyanatophenyl ether (p. 2558 s) (1941 King).

SnCl₂ salt. Almost colourless needles, m. ca. 295° dec. — *Picrate (I:I)* $C_{30}H_{30}O_9N_4$. Lemon-yellow crystals (benzene), darkening at 120°, m. ca. 160° dec. It is easily dissociated. — *2.4-Dinitrophenylhydrazone* $C_{30}H_{31}O_5N_5$. Orange-yellow crystals (pyridine-alcohol-water), m. 238–240° dec. — *Semi-carbazone* $C_{25}H_{30}O_2N_4$. Colourless needles (pyridine-dioxane), m. ca. 295° (1941 King).

Oestrone p-acetaminophenyl ether $C_{26}H_{29}O_3N$. Colourless needles (benzene-ligroin), softening at 170–2°, resolidifying, and m. 202–4°; hydrated colourless needles (aqueous methanol), softening at 100° and again at 170°, resolidifying, and m. 201–2°; dehydrated over P_2O_5. — **Fmn.** From œstrone p-aminophenyl ether on boiling with acetic acid and acetic anhydride or with the latter alone (1941 King).

Oestrone p-carbomethoxyaminophenyl ether $C_{26}H_{29}O_4N$. Colourless needles (ligroin), m. 210–2°. Very soluble in all organic solvents except petroleum ether. — **Fmn.** From œstrone p-isocyanatophenyl ether (below) on boiling with abs. methanol (1941 King).

Oestrone p-carbethoxyaminophenyl ether $C_{27}H_{31}O_4N$. Colourless crystals (ligroin), m. 163–5⁰. Very soluble in all organic solvents except petroleum ether. — **Fmn.** Similar to that of the preceding compound (1941 King).

Oestrone p-isocyanatophenyl ether $C_{25}H_{25}O_3N$ (no analysis given). Not obtained pure. Colourless plates (carbon tetrachloride-petroleum ether), m. 138⁰ to 143⁰. Extremely soluble in carbon tetrachloride and benzene. — **Fmn.** From œstrone p-aminophenyl ether on refluxing with phosgene in benzene-toluene. — **Rns.** Boiling with methanol or ethanol yields the corresponding œstrone p-carbalkoxyaminophenyl ethers (1941 King).

Oestrone D-glucoside $C_{24}H_{32}O_7$. Crystals (abs. alcohol), m. 223⁰. Insoluble in ether (1933 C. F. BOEHRINGER & SÖHNE). — For the œstrogenic effect, see 1940 Hagedorn; 1942 Bennekou. — **Fmn.** From œstrone by heating with pentaacetylglucose and anhydrous $ZnCl_2$ at 130⁰ or by treatment with acetochloroglucose and aqueous alcoholic potassium hydroxide, followed in both cases by shaking with aqueous barium hydroxide (1933 C. F. BOEHRINGER & SÖHNE).

Oestrone tetraacetyl-D-glucoside $C_{32}H_{40}O_{11}$. Needles (alcohol), m. 214⁰; $[\alpha]_D^{17}$ + 64.6⁰ (chloroform); sparingly soluble in ethanol and methanol, more easily soluble in chloroform, benzene, and ether (1940 Hagedorn). — For the œstrogenic effect, see 1937 Kongsted; 1940 Hagedorn; 1942 Bennekou. — **Fmn.** From œstrone and acetobromoglucose on allowing to stand with aqueous potassium hydroxide in acetone at room temperature (10 % yield) or on heating with silver carbonate in quinoline at 60⁰ (63 % yield) (1940 Hagedorn; see also 1937 Kongsted). — **Rns.** Gives "α"-œstradiol 3-D-glucoside (p. 1976 s) on hydrogenation in the presence of platinum oxide in abs. alcohol (1940 Hagedorn), on reduction with sodium and boiling propyl alcohol (1941 LøVENS), or on heating with aluminium isopropoxide in isopropyl alcohol (1938 SCHERING A.-G.).

Oestrone D-galactoside $C_{24}H_{32}O_7$. Yellowish powder (chloroform). — **Fmn.** By heating œstrone with pentaacetylgalactose and anhydrous $ZnCl_2$ at 120⁰, followed by shaking with aqueous barium hydroxide (1933 C. F. BOEHRINGER & SÖHNE).

Oestrone lactoside $C_{30}H_{42}O_{12}$. White powder. — **Fmn.** From œstrone on shaking with acetobromolactose, Ag_2O, and quinoline at room temperature or on treatment with acetylated lactose and $ZnCl_2$, followed by saponification with alcoholic NaOH (1939 CIBA).

For an *oestrone 6-(β-gentiobiosido)-D-glucoside,* see 1939 CIBA.

Oestrone β-D-glucuronide, Oestrone β-D-glucosiduronic acid $C_{24}H_{30}O_8$. For its possible occurrence in human pregnancy urine, see p. 2544 s.

Methyl (oestrone 2.3.4-tri-O-acetyl-β-D-glucosid)uronate $C_{31}H_{38}O_{11}$. Crystals (abs. alcohol-chloroform), m. 225.5–228⁰; $[\alpha]_D^{25}$ + 57⁰ (chloroform). — Produces œstrus. — **Fmn.** By shaking œstrone with methyl acetobromo-D-glucuronate and Ag_2CO_3 in benzene (1939 Schapiro).

8-iso-$\Delta^{1\cdot3\cdot5(10)}$-Oestratrien-3-ol-17-one, 8-iso-Oestrone $C_{18}H_{22}O_2$. Crystals (methanol), m. 247°; $[\alpha]_D^{20}$ +94° (dioxane). — 8-iso-Oestrone and its benzoate both have about one-third of the œstrogenic activity of natural œstrone and its benzoate, resp., in the Allen-Doisy test (rats). **Fmn.** The benzoate is obtained from the 3-benzoate of 8-iso-œstra-3.17β-diol (p. 1982 s) on oxidation with CrO_3 in glacial acetic acid; it is saponified by refluxing with methanolic KOH (1938 Serini).

Semicarbazone $C_{19}H_{25}O_2N_3$. Crystals (methanol), m. 270° (1938 Serini).

Benzoate $C_{25}H_{26}O_3$. Crystals (ether), m. 196°; $[\alpha]_D^{20}$ +61° (dioxane) (1938 Serini). — Oestrogenic activity, and formation, see above.

13-iso-$\Delta^{1\cdot3\cdot5(10)}$-Oestratrien-3-ol-17-one, 13-iso-Oestrone, Lumiœstrone $C_{18}H_{22}O_2$. For confirmation of the 13-iso-configuration, see 1951a, 1952a Johnson. — Prisms (alcohol), m. 268° to 269°; $[\alpha]_D^{21}$ —43° (dioxane); sparingly soluble in ether, chloroform, carbon tetrachloride, and benzene, soluble in boiling dioxane, alcohol, and acetone (1941 Butenandt). — It is inactive in the Allen-Doisy test (1941 Butenandt). — **Fmn.*** From œstrone dissolved in dioxane on irradiation with ultra-violet light of wave length 313 mμ in a one-quantum process (1941, 1942 Butenandt). For calculation of the conversion energy, see 1944 Koyenuma. It is separated from œstrone by means of Girard's reagent T with which it does not form a (water-soluble) compound (1942 Butenandt). — **Rns.** Treatment of the methyl ether with KOI in methyl alcohol at room temperature, followed by esterification with diazomethane in ether, gives the dimethyl ester methyl ether of d-**cis-lumimarrianolic acid (I) (1946 Heer). Heating with palladium black at 260° yields l-isoequilenin (p. 2535 s). Reduction of the methyl ether with sodium and boiling propanol affords the 3-methyl ether of lumiœstra-3.17β(?)-diol (p. 1982 s). On heating the semicarbazone with a solution of sodium in alcohol at 190–200°, desoxolumiœstrone (p. 1497 s) is formed (1941 Butenandt). Lumiœstrone is not reconverted to œstrone on irradiation of its solution in dioxane with ultra-violet light of wave length 313 mμ (1944 Butenandt).

Oxime $C_{18}H_{23}O_2N$. Needles (alcohol + water), m. 200–202° (1941 Butenandt).

Semicarbazone $C_{19}H_{25}O_2N_3$. Needles (alcohol + water), m. 273° (Kofler block) (1941 Butenandt). — **Rns.** See those of lumiœstrone, above.

---

* After the closing date for this volume, the total synthesis of dl-lumiœstrone has been accomplished by Johnson (1950, 1951a, 1952a).

** Cis refers to the relative orientation of H and CH_3 at C-1 and C-2 of the phenanthrene ring system; d refers to the optical rotation.

Acetate $C_{20}H_{24}O_3$. Crystals (alcohol), m. 89–90°. — **Fmn.** From the hydroxy-ketone on standing with pyridine-acetic anhydride (1941 Butenandt).

Benzoate $C_{25}H_{26}O_3$, m. 177–183° (crude, was not purified; no analysis given). — **Fmn.** From the hydroxy-ketone with benzoyl chloride in pyridine (1944 Butenandt).

Methyl ether $C_{19}H_{24}O_2$. Needles (alcohol), m. 129–130°; $[\alpha]_D^{23}$ —28° (chloroform). — **Fmn.** From the hydroxyketone with dimethyl sulphate and aqueous KOH (1941 Butenandt). — **Rns.** See those of lumiœstrone, p. 2559 s.

dl-Oestrone-a of Bachmann, $C_{18}H_{22}O_2$. On the basis of the close agreement of the m. p.'s of this stereoisomeric œstrone (and its benzoate) with those of an œstrone-a (and its benzoate) (1949 Anner, Miescher), which later was shown to be dl-14-iso-œstrone (1950 Anner, Miescher; 1957 a, b Johnson), Anner and Miescher (1949, 1950) assumed identity of these compounds; in view of the discrepancy in m. p.'s of the methyl ethers, however, Johnson (1957 a) believes that œstrone-a of Bachmann is another racemic stereoisomer of natural œstrone, very probably **dl-8-iso-13-iso-œstrone** (1958 Johnson).

Colourless leaflets (methanol or aqueous alcohol), m. 214–214.5°. Sublimes at 200° and 0.05 mm. Absorption max. in 95 % alcohol (from the graph) at 281.5 (3.32), inflexion at 286 (3.29) mμ (log ε). Gives the Kober test with phenol and sulphuric acid as does œstrone (1942 b Bachmann).

Exhibits only very little œstrogenic activity (1942 b Bachmann).

Fmn. Methyl 2-methyl-7-methoxy-2-carbomethoxy-1.2.3.4.9.10-hexahydro-1-phenanthrylideneacetate is hydrogenated in the presence of a palladium-charcoal catalyst in abs. alcohol at room temperature and atmospheric pressure and the resulting mixture of stereoisomeric methyl 2-methyl-7-methoxy-2-carbomethoxy-1.2.3.4.9.10.11.12-octahydro-1-phenanthrylacetates is hydrolysed to 2-methyl-7-methoxy-2-carbomethoxy-1.2.3.4.9.10.11.12-octahydro-1-phenanthrylacetic acid by refluxing with aqueous methanolic NaOH under nitrogen. By the procedures that are employed in Bachmann's synthesis of equilenin and isoequilenin (see p. 2528 s), without separation of the stereoisomers, this acid is converted via the chloride and diazoketone to methyl β-(2-methyl-7-methoxy-2-carbomethoxy-1.2.3.4.9.10.11.12-octahydro-1-phenanthryl)-propionate which is cyclized to a mixture of stereoisomeric 16-carbomethoxyœstrone methyl ethers by heating with sodium methoxide under nitrogen. The carbomethoxy group is removed by boiling with acetic acid and HCl under nitrogen, and demethylation is effected by refluxing with acetic acid and HBr under nitrogen. From the solution of the resulting mixture of stereoisomeric œstrones in methanol œstrone-a crystallizes (1942 b Bachmann).

Rns. On treatment of the methyl ether with methylmagnesium iodide, followed by dehydration of the resulting (not purified) methylcarbinol (see p. 1988 s) with $KHSO_4$, and dehydrogenation by heating with palladium-charcoal at 300°, it yields 3'.3'-dimethyl-7-methoxy-1.2-cyclopentenophenanthrene (p. 1502 s) (1942 b Bachmann).

Benzoate $C_{25}H_{26}O_3$. Colourless needles (acetone-alcohol), m. 175–6°. — **Fmn.** By treatment of the hydroxy-ketone with benzoyl chloride and pyridine in benzene (1942 b Bachmann).

Methyl ether $C_{19}H_{24}O_2$ (no analysis given by Bachmann). Exists in dimorphic forms: crystals (petroleum ether), m. 81.5–82°; colourless leaflets (dil. methanol), m. 101.5–102.5°. — **Fmn.** By the action of dimethyl sulphate on the sodium salt (not described) of the hydroxy-ketone in water (1942 b Bachmann). — **Rns.** See those of œstrone-a, p. 2560 s.

2 - Methyl -7- hydroxy -3′- keto-1.2-cyclopentano-1.2.3.4.5.6.7.8-octahydrophenanthrene, $\varDelta^{5\cdot7\cdot9}$ - Oestratrien-3β-ol-17-one $C_{18}H_{22}O_2$.

Needles and monoclinic prisms (aqueous alcohol), m. 138–139.5°; $[\alpha]_D^{27} +59°$ (alcohol); absorption max. in abs. alcohol at 269.5 (345) and 278 (240) mμ (ε) (1940, 1941 a Heard). — It is not precipitated by digitonin (1941 a Heard). — **Occ.** In the urine of pregnant mares; it is isolated via the acetate and benzoate (1940, 1941 a Heard). — **Rns.** Hydrogenation in the presence of Adams's platinum catalyst in alcohol gives $\varDelta^{5\cdot7\cdot9}$-œstratriene-3β.17β-diol (p. 1983 s). It is saturated to bromine (1940, 1941 a Heard). Heating with $KHSO_4$ at 150–155° in an atmosphere of nitrogen yields $\varDelta^{3\cdot5\cdot7\cdot9}$-œstratetraen-17-one (p. 2392 s) (1941 a Heard).

Colour reactions. With tetranitromethane a deep yellow colour appears. With HNO_3 in the cold an orange-yellow colour, not altered by ammonia, is produced (1940, 1941 a Heard). It shows the Liebermann-Burchard reaction and the reverse Salkowski reaction (see p. 71) (1941 a Heard). The red-violet colour with the Zimmermann reagent develops at the same rate as with œstrone (1940, 1941 a Heard).

Oxime $C_{18}H_{23}O_2N$. Needles (aqueous alcohol), m. 195–7° dec. (1940, 1941 a Heard).

Acetate $C_{20}H_{24}O_3$. Rectangular plates (aqueous acetone), m. 158°. — **Fmn.** From the hydroxy-ketone on treatment with acetic anhydride and pyridine at room temperature (1940, 1941 a Heard).

Benzoate $C_{25}H_{26}O_3$ (no analysis given). Crystals (benzene-ligroin), softening at 190°, m. 196–8°; sparingly soluble in alcohol. — **Fmn.** From the hydroxyketone on treatment with benzoyl chloride in pyridine at room temperature (1941 a Heard).

Folliculosterone $C_{18}H_{22}O_2$. The $\varDelta^{5\cdot7\cdot9}$-œstratrien-3-ol-17-one structure (see above) has also been claimed by Remesov (1936, 1937 a, b) for folliculosterone; however, this is not compatible with its absorption spectrum (cf. 1941 a Heard) and its high œstrogenic activity (cf. 1938 Ruzicka); see also (on p. 2562 s) the discrepancy between the melting point of one of the intermediary compounds in its formation and that found by other investigators for this substance.

Crystals (alcohol), m. 248–248.5°; mol. refraction 76.9; $[\alpha]_D^{18} + 162°$ (chloroform). Absorption max. in chloroform at 282 mμ (high intensity); the absorption spectrum remains unchanged in alkaline solution. Soluble in chloroform, ether, and acetone, less soluble in alcohol. It is not precipitated by digitonin. The Liebermann-Burchard and the Salkowski test (p. 44) are negative. It is as active œstrogenically as œstrone (1937a, b Remesov).

Fmn. By degradation of ergosterol (p. 1735 s) via the acetate of neoergosterol dibromide (p. 1870 s); the latter is ozonized in glacial acetic acid at room temperature, the ozonide decomposed, the methyl ester (m. 144–5°) * of the resulting α-(3β-acetoxy-$\Delta^{5·7·9}$-œstratrien-17β-yl)-propionic acid (m. 210° to 212°) treated with phenylmagnesium bromide, the crude diphenylcarbinol (see p. 1946 s) so obtained acetylated and oxidized with CrO_3 in glacial acetic acid; the crude carboxylic acid so formed is converted into the methyl ester, the ester treated again with phenylmagnesium bromide, the resulting resinous diphenylcarbinol subjected to thermal decomposition, followed by acetylation and ozonization in chloroform, the reaction product treated with zinc and glacial acetic acid, then with aqueous NaOH, and finally sublimed at 210° and 0.001 mm. pressure (1936, 1937b Remesov).

Oxime $C_{18}H_{23}O_2N$ (no analysis given). Crystals (alcohol), m. 236° (1937a Remesov).

Semicarbazone $C_{19}H_{25}O_2N_3$ (no analysis given). Crystals, m. 260–2° (1937a Remesov).

Acetate $C_{20}H_{24}O_3$. Needles (alcohol), m. 112–6°. — **Fmn.** From the hydroxyketone with acetic anhydride and pyridine in the cold (1937a, b Remesov).

Benzoate $C_{25}H_{26}O_3$. Needles (abs. alcohol), m. 202°. — **Fmn.** ·From the hydroxy-ketone with benzoyl chloride and pyridine (1937a, b Remesov).

2-Methyl-7-hydroxy-3'-keto-1.2-cyclopentanoperhydrophenanthrene, Oestran-3-ol-17-one

$C_{18}H_{28}O_2$. Crystalline œstran-3-ol-17-ones (no constants reported), exerting a male sex hormone effect in the capon comb test, may be obtained as follows: From œstrane-3.17-diol (octahydrofollicular hormone, see p. 1984 s) by oxidation with CrO_3 in H_2SO_4 at 60°, followed by hydrogenation of the resulting (not purified) diketone $C_{18}H_{26}O_2$, using a platinum oxide catalyst in alcohol or glacial acetic acid. From œstrane-3.17-diol by oxidation with aqueous $KMnO_4$. From œstrane-3.17-diol 3-benzoate by oxidation with Beckmann's mixture at 60°, followed by saponification of the resulting crystalline benzoate. From œstrane-3.16.17-triol ("hexahydrofollicular hormone hydrate", p. 108) by boiling with 5% H_2SO_4, by allowing to stand with 90% H_2SO_4, or by fusion with phthalic anhydride at 130–140°, followed by saponification (1933b SCHERING-KAHLBAUM A.-G.). — **Rns.** Reaction with ethynylmagnesium bromide in ether gives (not characterized) *17-ethynylœstrane-3.17-diol* $C_{20}H_{30}O_2$ (1936 SCHERING CORP.).

---

* According to 1943 Jacobsen, this methyl ester melts at 159.5–161.5° cor.; this m. p. was confirmed by 1952 Mosettig.

Oestran-3-ol-17-ones $C_{18}H_{28}O_2$ have also been obtained from œstrone (page 2543 s) on hydrogenation in acetic acid in the presence of platinum oxide; the sirupy mixture obtained on hydrogenation in glacial acetic acid at room temperature yields a *semicarbazone* $C_{19}H_{31}O_2N_3$, crystals (petroleum ether), m. 255⁰ after sintering at 253⁰, $[\alpha]_D^{22} +77^0$ (chloroform-alcohol, 3:1); the sirupy mixture obtained on hydrogenation in 80% acetic acid at 40⁰ gives a *semicarbazone* $C_{19}H_{31}O_2N_3$, crystals (benzene-petroleum ether), m. 254–5⁰ dec., $[\alpha]_D^{20} +46^0$ (abs. alcohol), more readily soluble in alcohol than the preceding semicarbazone, and a *2.4-dinitrophenylhydrazone* $C_{24}H_{32}O_5N_4$, crystals (glacial acetic acid), m. ca. 100–105⁰, very readily soluble (1936 Dirscherl).

2-Methyl-9-hydroxy-3'-keto-1.2-cyclopentano-1.2.3.4-tetrahydrophenanthrene, Equilenan-6-ol-17-one, 6-Hydroxy-17-equilenone $C_{18}H_{18}O_2$.

Exists in two geometric forms, "α" and "β", each being a racemate (1940 Bachmann). The "α"-form has probably the trans-, the "β"-form the cis-configuration (1951 Hirschmann). — Both forms are inactive when injected in amounts up to 500 γ into ovariectomized rats (1940 Bachmann). — **Fmn.*** "α"- and "β"-6-Hydroxy-17-equilenones are formed from the corresponding "α"- and "β"-6-methoxy-16-carbomethoxy-17-equilenones on refluxing with acetic and hydrochloric acids in an atmosphere of nitrogen for 10 hrs. for the "α"-form and for 2 hrs. for the "β"-form (yields, 74% "α"-form, 87% "β"-form); by refluxing for ¹/₂ hr. only, the methyl ethers (see below) are obtained besides the hydroxy-ketones (1940 Bachmann).

"α"-6-Hydroxy-17-equilenone $C_{18}H_{18}O_2$. Colourless prisms (dil. alcohol), m. 240–2⁰ vac. when placed in a bath at 220⁰. — *Methyl ether, "α"-6-Methoxy-17-equilenone* $C_{19}H_{20}O_2$. Colourless prisms (methanol), m. 147.5–149⁰ vac. (1940 Bachmann). — **Fmn.** See above.

"β"-6-Hydroxy-17-equilenone $C_{18}H_{18}O_2$. Colourless needles (dil. alcohol), m. 171.5⁰ to 172.5⁰ vac.; crystals with solvent of crystallization (benzene-petroleum ether), m. 101–2⁰ dec. — *Methyl ether, "β"-6-Methoxy-17-equilenone* $C_{19}H_{20}O_2$. Colourless needles (methanol), m. 112–3⁰ vac. (1940 Bachmann). — **Fmn.** See above.

2-Methyl-9-hydroxy-3'-keto-1.2-cyclopentano-1.2.3.4.5.6.7.8-octahydrophenanthrene, 6-Hydroxy-1.2.3.4-tetrahydro-17-equilenone

$C_{18}H_{22}O_2$. Exists in two geometric forms, "α" and "β", each being a racemate. — **Fmn.*** The methyl ethers (see below) of "α"- and "β"-6-hydroxy-1.2.3.4-tetrahydro-17-equilenones are obtained from the corresponding "α"- and "β"-6-methoxy-16-carbomethoxy-1.2.3.4-tetrahydro-17-equilenones on refluxing with acetic and hydrochloric acids in an atmosphere of nitrogen for 1 hr. (1942a Bachmann).

---

* See Bachmann's analogous synthesis of equilenin and isoequilenin, p. 2528 s.

References, pp. 2564–2570 s

"α"-6-Hydroxy-1.2.3.4-tetrahydro-17-equilenone $C_{18}H_{22}O_2$. Colourless prisms (dil. alcohol), m. 150–150.5⁰. — Fails to induce the œstrus response in ovariectomized rats in doses as high as 1000 γ. — **Fmn.** From the methyl ether (below) on refluxing with acetic and hydrobromic acids in an atmosphere of nitrogen for 3 hrs. — *Methyl ether, "α"-6-Methoxy-1.2.3.4-tetrahydro-17-equilenone* $C_{19}H_{24}O_2$. Colourless needles (methanol), m. 118–118.5⁰ vac. (1942a Bachmann). **Fmn.** See p. 2563 s.

"β"-6-Hydroxy-1.2.3.4-tetrahydro-17-equilenone $C_{18}H_{22}O_2$. Has not yet been obtained in crystalline form. — *Methyl ether, "β"-6-Methoxy-1.2.3.4-tetrahydro-17-equilenone* $C_{19}H_{24}O_2$. Colourless prisms (methanol), m. 108–9⁰ (1942a Bachmann). **Fmn.** See p. 2563 s.

1930 Butenandt, *Z. physiol. Ch.* **191** 140, 143.
Slawson, *J. Biol. Ch.* **87** 373.
Walker, Janney, *Endocrinology* **14** 389.
1931 SCHERING A.-G., *Ger. Pat.* 672958 (issued 1939); *C. A.* **1939** 6531; *Ch. Ztbl.* **1939** I 4812.
Thayer, Levin, Doisy, *J. Biol. Ch.* **91** 791, 796.
1932 Butenandt, Störmer, Westphal, *Z. physiol. Ch.* **208** 149, 167.
Marrian, Haslewood, *Biochem. J.* **26** 25, 29.
SCHERING A.-G., *Ger. Pat.* 711378 (issued 1941); *C. A.* **1943** 4208; *Ch. Ztbl.* **1942** I 3281;
SCHERING-KAHLBAUM A.-G., *Brit. Pat.* 428133 (1933, issued 1935); *C. A.* **1935** 6708;
Ch. Ztbl. **1935** II 2548.
SCHERING-KAHLBAUM A.-G. (Schwenk, Hildebrandt), *Ger. Pat.* 644390 (issued 1937); *C. A.*
1937 6418.
1933 C. F. BOEHRINGER & SÖHNE (Johannessohn, Rabald), *Ger. Pat.* 650089 (issued 1937); *C. A.*
1938 1053; *Ch. Ztbl.* **1937** II 3198; *U.S. Pat.* 2088792 (1934, issued 1937); *C. A.* **1937**
6826; *Ch. Ztbl.* **1937** II 3198.
Butenandt, Jacobi, *Z. physiol. Ch.* **218** 104, 109.
SCHERING A.-G., *Ger. Pat.* 734621 (issued 1943); *C. A.* **1944** 1328; SCHERING-KAHLBAUM
A.-G. (Butenandt, Schoeller, Hildebrandt, Schwenk), *Brit. Pat.* 427588 (1934, issued
1935); *C. A.* **1935** 6707; *Ch. Ztbl.* **1935** II 2548; *U.S. Pat.* 2086139 (Schoeller, Schwenk,
Hildebrandt) (1934, issued 1937); *C. A.* **1937** 5953.
a SCHERING-KAHLBAUM A.-G., *Australian Pat.* 15422 (issued 1934); *Ch. Ztbl.* **1935** I 752;
Brit. Pat. 428132 (issued 1935); *C. A.* **1935** 6707; *Ch. Ztbl.* **1935** II 4474.
b SCHERING-KAHLBAUM A.-G., *Brit. Pat.* 421681 (issued 1935); *C. A.* **1935** 3785; *Ch. Ztbl.*
1935 II 2547; *U.S. Pat.* 2075868 (Schoeller, Hildebrandt, Schwenk) (1933, issued 1937);
C. A. **1937** 3640.
c SCHERING-KAHLBAUM A.-G., *Swiss Pat.* 186463 (issued 1936); *Ch. Ztbl.* **1937** I 4263.
1934 Butenandt, Westphal, *Z. physiol. Ch.* **223** 147, 159.
Häussler, *Helv.* **17** 531, 534.
Kofler, Hauschild, *Z. physiol. Ch.* **224** 150; *Mikrochemie* **15** 55.
Neuhaus, *Ber.* **67** 1627; *Z. Krist.* **89** 505, 510.
a SCHERING-KAHLBAUM A.-G., *Austrian Pat.* 146908 (issued 1936); *Ch. Ztbl.* **1937** I 1479.
b SCHERING-KAHLBAUM A.-G., *Brit. Pat.* 446501 (issued 1936); *C. A.* **1936** 6759; *Ch. Ztbl.*
1936 II 1026, 1027.
c SCHERING-KAHLBAUM A.-G., *Norw. Pat.* 55506 (issued 1935); *Ch. Ztbl.* **1935** II 3798.
1935 Claude, *Fr. Pat.* 809819 (issued 1937); *C. A.* **1937** 8544; *Ch. Ztbl.* **1937** II 814.
Cohen, Cook, Hewett, *J. Ch. Soc.* **1935** 445, 449.
Rowlands, Callow, *Biochem. J.* **29** 837.
SCHERING CORP. (Schwenk, Whitman), *U.S. Pat.* 2072830 (issued 1937); *C. A.* **1937** 3213;
Ch. Ztbl. **1937** I 4535; SCHERING-KAHLBAUM A.-G., *Brit. Pat.* 467107 (issued 1937); *C. A.*
1937 8125.
Zimmermann, *Z. physiol. Ch.* **233** 257, 262.

1936 Bardhan, *J. Ch. Soc.* **1936** 1848, 1851.

Beall, Edson, *Biochem. J.* **30** 577.

Bernal, Crowfoot, *Z. Krist.* **93** 464.

a Bierry, Gouron, *Compt. rend.* **202** 686.

b Bierry, Gouron, *Compt. rend. soc. biol.* **122** 147.

C. F. BOEHRINGER & SÖHNE (Johannessohn, Hatzig, Rabald), *Ger. Pat.* 710966 (issued 1941); *C. A.* **1943** 3771; *Ch. Ztbl.* **1942** I 513.

Callow, *Biochem. J.* **30** 906.

CIBA*, *Swiss Pat.* 206114 (issued 1939); *C. A.* **1941** 3039; *Dutch Pat.* 46893 (1937, issued 1939); *Ch. Ztbl.* **1940** I 94; *Ger. Pat.* 695738 (Miescher, Scholz) (1937, issued 1940); *C. A.* **1941** 5652.

Dirscherl, *Z. physiol. Ch.* **239** 53, 65, 66.

Häussler, *Festschrift E. C. Barell*, Basel, p. 327; *C. A.* **1937** 2662; *Ch. Ztbl.* **1931** I 2393.

HOFFMANN-LA ROCHE & Co. A.-G., *Ger. Pat.* 702110 (issued 1941); *C. A.* **1941** 8217; *Swiss Pat.* 200841 (1937, issued 1939); *Ch. Ztbl.* **1939** II 1127.

Laporta, Vacca, *Atti accad. Lincei* [6] **23** 212.

a LES LABORATOIRES FRANÇAIS DE CHIMIOTHÉRAPIE (Girard, Sandulesco), *Ger. Pat.* 676115 (issued 1939); *C. A.* **1939** 6531; *Ch. Ztbl.* **1939** II 4032.

b LES LABORATOIRES FRANÇAIS DE CHIMIOTHÉRAPIE (Girard, Sandulesco), *Ger. Pat.* 702574 (issued 1941); *C. A.* **1941** 8217; *Ch. Ztbl.* **1941** I 2972.

Pincus, Wheder, Young, Zahl, *J. Biol. Ch.* **116** 253, 261 et seq.

Remesov, *Rec. trav. chim.* **55** 797.

a RICHTER GEDEON VEGYÉSZETI GYÁR R. T., *Hung. Pat.* 117519 (issued 1938); *C. A.* **1938** 3914; *Ch. Ztbl.* **1938** II 557; *Brit. Pat.* 487267 (1937, issued 1938); *C. A.* **1939** 325; *Ch. Ztbl.* **1938** II 2459.

b RICHTER GEDEON VEGYÉSZETI GYÁR R. T., *Hung. Pat.* 117667 (issued 1938); *C. A.* **1938** 3914; *Ch. Ztbl.* **1939** I 2035.

SCHERING A.-G., *Dan. Pat.* 58384 (issued 1941); *Ch. Ztbl.* **1941** I 3408.

SCHERING CORP. (Serini, Strassberger), *U.S. Pat.* 2243887 (issued 1941); *C. A.* **1941** 5654.

a SCHERING-KAHLBAUM A.-G., *Brit. Pat.* 468123 (issued 1937); *C. A.* **1937** 8836.

b SCHERING-KAHLBAUM A.-G., *Fr. Pat.* 811567 (issued 1937); *C. A.* **1938** 1284; *Ch. Ztbl.* **1937** II 3919.

c SCHERING-KAHLBAUM A.-G., *Swiss Pat.* 201866 (issued 1939); *Ch. Ztbl.* **1939** II 1337.

Schmulovitz, Wylie, *J. Biol. Ch.* **116** 415.

van Stolk, Pénau, de Leuchêre, *C. A.* **1937** 5831; *Ch. Ztbl.* **1937** I 925.

Sureau, Grandadam, *Compt. rend.* **203** 440.

1936/37 Zimmermann, *Z. physiol. Ch.* **245** 47, 52.

1937 a Bugyi, *C. A.* **1937** 6817; *Ch. Ztbl.* **1937** II 1837.

b Bugyi, *C. A.* **1938** 2191; *Ch. Ztbl.* **1938** I 1144.

a CIBA*, *Fr. Pat.* 826162 (issued 1938); *C. A.* **1939** 323; *Ch. Ztbl.* **1938** II 1448; *U.S. Pat.* 2156599 (Miescher, Scholz) (1937, issued 1939); *C. A.* **1939** 5998.

b CIBA*, *Swiss Pat.* 199043 (issued 1938); *C. A.* **1939** 3535; *Ch. Ztbl.* **1939** I 3591; *Swiss Pat.* 206109 (1937, issued 1939); *C. A.* **1941** 3039; *Ch. Ztbl.* **1940** I 3959.

c CIBA*, *Swiss Pat.* 202847 (issued 1939); *C. A.* **1939** 9554; *Ch. Ztbl.* **1940** I 249.

Hogness, Sidwell, Zscheile, *J. Biol. Ch.* **120** 239, 243, 249, 252.

I. G. FARBENIND., *Ger. Pat.* 696594 (Friedrich) (issued 1940); *C. A.* **1941** 6067; *Brit. Pat.* 490942 (1937, issued 1938); *C. A.* **1939** 816; *Ch. Ztbl.* **1939** I 3591; *Dutch Pat.* 47430 (1938, issued 1939); *Ch. Ztbl.* **1940** I 2984.

John, *Z. physiol. Ch.* **250** 11, 18.

Kocsis, Bugyi, *Microchim. Acta* **2** 291.

Kongsted, *Norw. Pat.* 59280 (issued 1938); *Ch. Ztbl.* **1938** II 355; *U.S. Pat.* 2275969 (1937, issued 1942); *C. A.* **1942** 4673.

a Lapiner, Leontovich, Kosheverova, *C. A.* **1939** 8228; *Ch. Ztbl.* **1938** I 920.

b Lapiner, Altman, Leontovich, *C. A.* **1941** 5988; *Ch. Ztbl.* **1939** II 4504.

a Miescher, Scholz, *Helv.* **20** 263, 265, 269.

b Miescher, Scholz, *Helv.* **20** 1237, 1240, 1243, 1244.

* Société pour l'industrie chimique à Bâle; Gesellschaft für chemische Industrie in Basel.

v. Novák, *C. A.* **1937** 6270; *Ch. Ztbl.* **1937** II 1381.

Pincus, Zahl, *J. Gen. Physiol.* **20** 879, 881.

a Remesov, *Rec. trav. chim.* **56** 1093.

b Remesov, *C. A.* **1937** 8542; *Ch. Ztbl.* **1937** II 3323.

a SCHERING A.-G., *Fr. Pat.* 831 131 (issued 1938); *C. A.* **1939** 1762; *Ch. Ztbl.* **1939** I 3590.

b SCHERING A.-G. (Inhoffen), *Ger. Pat.* 720015 (issued 1942); *C. A.* **1943** 2520; *Ch. Ztbl.* **1942** II 1037.

c SCHERING A.-G. (Schoeller), *Ger. Pat.* 722098 (issued 1942); *C. A.* **1943** 5200; *Ch. Ztbl.* **1942** II 2970; *Fr. Pat.* 49185 (1938, issued 1938); *C. A.* **1939** 5132; *Ch. Ztbl.* **1939** I 4089; *Fr. Pat.* 51438 (1941, issued 1942); *Ch. Ztbl.* **1942** II 2719; *Brit. Pat.* 511433 (1938, issued 1939); *C. A.* **1940** 6022; SCHERING CORP. (Schoeller), *U.S. Pat.* 2184167 (1938, issued 1939); *C. A.* **1940** 2538.

d SCHERING A.-G. (Hohlweg), *Ger. Pat.* 730909 (issued 1943); *C. A.* **1944** 621; *Ch. Ztbl.* **1943** I 2113.

e SCHERING A.-G., *Swiss Pat.* 230019 (issued 1944); *Ch. Ztbl.* **1944** II 1204.

van Stolk, de Leuchêre, *Compt. rend.* **205** 395.

Venning, Evelyn, Harkness, Browne, *J. Biol. Ch.* **120** 225.

Voss, *Z. physiol. Ch.* **250** 218.

Windaus, Deppe, *Ber.* **70** 76, 77.

Wu, Chou, *Chinese J. Physiol.* **11** 413.

1937/38 Westerfeld, Doisy, *Ann. Internal Med.* **11** 267, 270; *C. A.* **1937** 7975; *Ch. Ztbl.* **1938** I 348.

1938 Bergel, Todd, *Biochem. J.* **32** 2145.

Callow, Callow, Emmens, *Biochem. J.* **32** 1312, 1322.

CHINOIN GYÓGYSZER ÉS VEGYÉSZETI TERMÉKEK GYÁRA R. T., *Brit. Pat.* 503568 (issued 1939); *C. A.* **1939** 7049; *Ch. Ztbl.* **1940** I 430; *Brit. Pat.* 503569 (1938, issued 1939); *C. A.* **1939** 7049; *Ch. Ztbl.* **1940** I 430; *Brit. Pat.* 506252 (1938, issued 1939); *C. A.* **1939** 9554; *Ch. Ztbl.* **1940** I 430; *Fr. Pat.* 842722 (1938, issued 1939); *C. A.* **1940** 6022; *Ch. Ztbl.* **1940** I 430; *U.S. Pat.* 2208915 (Weisz) (1938, issued 1940); *C. A.* **1941** 137.

Dingemanse, Laqueur, Mühlbock, *Nature* **141** 927.

Ercoli, Mamoli, *Gazz.* **68** 142.

Fischer, *Arch. intern. pharmacodynamie* **58** 332.

HOFFMANN-LA ROCHE & Co. A.-G., *Ger. Pat.* 678115 (issued 1939); *C. A.* **1939** 7962; *Ch. Ztbl.* **1939** II 2814.

Inhoffen, Logemann, Hohlweg, Serini, *Ber.* **71** 1024, 1029.

Kober, *Biochem. J.* **32** 357.

Lapiner, Kosheverova, *C. A.* **1941** 1077.

Laporta, Lafrotta, *Arch. sci. biol.* (*Italy*) **24** 452.

Litvan, Robinson, *J. Ch. Soc.* **1938** 1997, 2000, 2001.

Mamoli, *Ber.* **71** 2696.

a Marker, *J. Am. Ch. Soc.* **60** 1897, 1900.

b Marker, Rohrmann, Lawson, Wittle, *J. Am. Ch. Soc.* **60** 1901, 1903.

c Marker, Rohrmann, *J. Am. Ch. Soc.* **60** 2927.

a Miescher, Scholz, Tschopp, *Biochem. J.* **32** 141, 142.

b Miescher, Scholz, Tschopp, *Biochem. J.* **32** 1273, 1274, 1279 footnote.

Ruzicka, Müller, Mörgeli, *Helv.* **21** 1394, 1398.

Schachter, Marrian, *J. Biol. Ch.* **126** 663, 667.

SCHERING A.-G. (Schoeller, Gehrke, Inhoffen), *Ger. Pat.* 725151 (issued 1942); *C. A.* **1943** 5830; *Fr. Pat.* 856348 1939, issued 1940); *Belg. Pat.* 434173 (1939); *Ch. Ztbl.* **1941** I 669; *Swed. Pat.* 103866 (Gehrke, Schoeller, Inhoffen) (1939, issued 1942); *Ch. Ztbl.* **1942** II 1720.

Serini, Logemann, *Ber.* **71** 186, 189 et seq.

a Westerfeld, Thayer, MacCorquodale, Doisy, *J. Biol. Ch.* **126** 181, 183, 190.

b Westerfeld, MacCorquodale, Thayer, Doisy, *J. Biol. Ch.* **126** 195, 199.

Zimmermann, *Klin. Wochschr.* **17** 1103.

Zondek, Bergmann, *Biochem. J.* **32** 641, 644.

1939 Bachman, *J. Biol. Ch.* **131** 455.

Bandow, *Biochem. Z.* **301** 37, 44, 54.

Beall, *Nature* **144** 76.

Butenandt, Hofstetter, *Z. physiol. Ch.* **259** 222, 226 (footnote 9), 228, 230, 232, 234.

Chevallier, Manuel, *Compt. rend. soc. biol.* **132** 521.

CIBA*, *Fr. Pat.* 856328 (issued 1940); *Ch. Ztbl.* **1941** I 927.

Dannenberg, *Abhandl. Preuss. Akad. Wiss. Math.-naturwiss. Kl.* **1939** No. 21, pp. 40, 63.

Einhorn, Kuorozova, *C. A.* **1940** 5914; *Ch. Ztbl.* **1941** I 2685.

Elvidge, *Quart. J. Pharm. Pharmacol.* **12** 347, 354, 581.

HOFFMANN-LA ROCHE & Co. A.-G., *Swiss Pat.* 211292 (issued 1940); *C. A.* **1942** 3635; *Ch. Ztbl.*
  1941 I 2972; *Brit. Pat.* 538478 (issued 1941); *C. A.* **1942** 3635; *Swed. Pat.* 100175 (Meyer)
  (1940, issued 1940); *Ch. Ztbl.* **1941** I 1572; HOFFMANN-LA ROCHE INC. (Meyer), *U. S. Pat.*
  2257701 (1940, issued 1941); *C. A.* **1942** 628.

a Marker, *J. Am. Ch. Soc.* **61** 944, 946.

b Marker, *J. Am. Ch. Soc.* **61** 1287.

c Marker, Rohrmann, *J. Am. Ch. Soc.* **61** 1922.

d Marker, Rohrmann, *J. Am. Ch. Soc.* **61** 2974.

Mossini, *C. A.* **1940** 2135.

Pedersen-Bjergaard, *Comparative Studies Concerning the Strength of Oestrogenic Substances,*
  Copenhagen and London, pp. 47, 66, 179.

Robinson, Rydon, *J. Ch. Soc.* **1939** 1394, 1402.

Schapiro, *Biochem. J.* **33** 385, 388.

Scherrer, *Helv.* **22** 1329, 1339, 1340.

Wettstein, *Helv.* **22** 250.

a Woker, Antener, *Helv.* **22** 47, 56.

b Woker, Antener, *Helv.* **22** 1309, 1321, 1325.

1940 Bachmann, Holmes, *J. Am. Ch. Soc.* **62** 2750, 2751, 2754.

a Beall, *Biochem. J.* **34** 1293, 1296, 1297.

b Beall, *J. Endocrinol.* **2** 81, 87.

CHINOIN GYÓGYSZER ÉS VEGYÉSZETI TERMÉKEK GYÁRA R. T. (Földi), *Hung. Pat.* 135687
  (issued 1940); *C. A.* **1950** 4047, 4048.

Fish, Dorfman, *Science* [N.S.] **91** 388.

Hagedorn, Johannessohn, Rabald, Voss, *Z. physiol. Ch.* **264** 23, 26, 27, 29.

Heard, Hoffman, *J. Biol. Ch.* **135** 801.

Huffman, MacCorquodale, Thayer, Doisy, Smith, Smith, *J. Biol. Ch.* **134** 591, 594, 595.

McCullagh, Schneider, Emery, *Endocrinology* **27** 71, 72, 77.

Marker, Rohrmann, *J. Am. Ch. Soc.* **62** 73, 74, 75.

Miwa, Ito, Hayazu, *Klin. Wochschr.* **19** 954.

Niederl, *U.S. Pat.* 2322311 (issued 1943); *C. A.* **1944** 221.

Pearlman, Wintersteiner, *J. Biol. Ch.* **132** 605, 611.

Sakanoue, *C. A.* **1942** 805.

Talbot, Wolfe, MacLachlan, Karush, Butler, *J. Biol. Ch.* **134** 319.

Westerfeld, *Biochem. J.* **34** 51, 55.

Wolfe, Hershberg, Fieser, *J. Biol. Ch.* **136** 653, 673.

1941 Bachman, Pettit, *J. Biol. Ch.* **138** 689, 690, 697.

Bergel, Cohen, *J. Ch. Soc.* **1941** 795, 796.

Butenandt, Wolff, Karlson, *Ber.* **74** 1308, 1310, 1311.

Dingemanse, Tyslowitz, *Endocrinology* **28** 450.

Doisy Jr., Huffman, Thayer, Doisy, *J. Biol. Ch.* **138** 283.

Ercoli, *C. A.* **1946** 7338.

Fish, Dorfman, *J. Biol. Ch.* **140** 83.

a Goldberg, Studer, *Helv.* **24** 478, 480.

b Goldberg, Studer, *Helv.* **24** 295 E, 298 E.

a Heard, Hoffman, *J. Biol. Ch.* **138** 651, 652, 653, 656, 658 et seq.

b Heard, Hoffman, *J. Biol. Ch.* **141** 329, 332 footnote.

c Heard, Bauld, Hoffman, *J. Biol. Ch.* **141** 709.

Hughes, *J. Biol. Ch.* **140** 21.

Inhoffen, Zühlsdorff, *Ber.* **74** 1911, 1914 footnote 5.

* Société pour l'industrie chimique à Bâle; Gesellschaft für chemische Industrie in Basel.

Ito, *C. A.* **1942** 528; *Ch. Ztbl.* **1942** I 501.
King, Franks, *J. Am. Ch. Soc.* **63** 2042 et seq.
Kleiner, *J. Biol. Ch.* **138** 783.
Løvens Kemiske Fabrik ved A. Kongsted, *Dan. Pat.* 60573 (issued 1943); *C. A.* **1946**
    3858; *Ch. Ztbl.* **1943** II 1563; *Brit. Pat.* 607980 (1946, issued 1948); *C. A.* **1949** 2376;
    U.S. Pat. 2455214 (Bennekou) (1946, issued 1948); *C. A.* **1949** 1532.
Mather, *J. Biol. Ch.* **140** Proc. 84.
Schering A.-G., *Fr. Pat.* 51586 (issued 1943); *Ch. Ztbl.* **1943** II 249, 250; Schering Corp.
    (Inhoffen), *U.S. Pat.* 2361847 (1941, issued 1944); *C. A.* **1945** 2384.
Urushibara, Nitta, *Bull. Ch. Soc. Japan* **16** 179, 181.
1942 Arnaudi, *C. A.* **1946** 3152 (*Boll. ist. sieroterap. milanese* **21** 1); cf. *Ztbl. Bakt. Parasitenk.* II
    105 352 (1942/43); *C. A.* **1943** 5756.
a Bachmann, Ness, *J. Am. Ch. Soc.* **64** 536, 539, 540.
b Bachmann, Kushner, Stevenson, *J. Am. Ch. Soc.* **64** 974, 976, 979, 980.
Bennekou, Pedersen-Bjergaard, *Z. physiol. Ch.* **272** 144.
Butenandt, Friedrich, Poschmann, *Ber.* **75** 1931.
Fish, Dorfman, *J. Biol. Ch.* **143** 15.
Goldberg, Studer, *Helv.* **25** 1553, 1555.
Graubard, Pincus, *Endocrinology* **30** 265.
Huffman, *J. Am. Ch. Soc.* **64** 2235.
Konstantinova-Schlezinger, *C. A.* **1943** 6207, 6208.
Longwell, McKee, *J. Biol. Ch.* **142** 757.
Mather, *J. Biol. Ch.* **144** 617, 619, 621.
Pearlman, Pincus, *J. Biol. Ch.* **144** 569.
Pincus, Pearlman, *Endocrinology* **31** 507.
Reynolds, Ginsburg, *Endocrinology* **31** 147, 150, 152.
Westerfeld, *J. Biol. Ch.* **143** 177.
Winkler, *Klin. Wochschr.* **21** 1080.
Zondek, Sklow, *Proc. Soc. Exptl. Biol. Med.* **49** 629.
1942/43 Figge, *C. A.* **1944** 5510.
1943 Hansen, Cantarow, Rakoff, Paschkis, *Endocrinology* **33** 282, 285, 286.
Heintzberger, *Acta Brevia Neerland.* **13** 41.
Jacobsen, *J. Am. Ch. Soc.* **65** 1789, 1791.
Jayle, Crépy, Judas, *Bull. soc. chim. biol.* **25** 301.
Parkes, *J. Endocrinol.* **3** 288.
Pearlman, Pincus, *J. Biol. Ch.* **147** 379.
Ruzicka, Prelog, *Helv.* **26** 975, 979.
a Schiller, Pincus, *Arch. Biochem.* **2** 317.
b Schiller, Pincus, *Science* [N.S.] **98** 410.
Szego, Samuels, *J. Biol. Ch.* **151** 587.
Wettstein, Miescher, *Helv.* **26** 631, 641.
Zondek, Sulman, *Endocrinology* **33** 204.
1944 Butenandt, Poschmann, *Ber.* **77** 392.
Buu-Hoi, Cagniant, *Ber.* **77** 118, 120.
Cantarow, Paschkis, Rakoff, Hansen, *Endocrinology* **35** 129.
Grant, v. Seemann, *U.S. Pat.* 2392660 (issued 1946); *C. A.* **1946** 3235; *Ch. Ztbl.* **1946** I 1433;
    Brit. Pat. 590113 (to Ayerst, McKenna & Harrison Ltd.) (issued 1947); *C.A.* **1947** 7686.
Huffman, Darby, *J. Am. Ch. Soc.* **66** 150.
Koyenuma, *Naturwiss.* **32** 221.
Miescher, *Helv.* **27** 1727, 1729, 1730, 1731.
Reifenstein, Dempsey, *J. Clin. Endocrinol.* **4** 326.
Smith, *Endocrinology* **35** 146, 149 footnote 3.
Stimmel, *J. Biol. Ch.* **153** 327.
Zimmermann, *Vitamine und Hormone* **5** [1952] 1, 124 et seq., 157, 178.
1944/45 Zimmermann, May, *Ztbl. Bakt. Parasitenk.* I **151** 462.
1945 Cheymol, Carayon-Gentil, *Bull. soc. chim. biol.* **27** 180.
Courrier, Velluz, Alloiteau, Rousseau, *Compt. rend. soc. biol.* **139** 128.

Dorfman, Wise, van Wagenen, *Endocrinology* **36** 347.
Figge, *Endocrinology* **36** 178.
a Heer, Miescher, *Helv.* **28** 156, 160, 162.
b Heer, Billeter, Miescher, *Helv.* **28** 991, 995.
Jensen, Larivière, Elie, *Rev. can. biol.* **4** 535.
Mason, Kepler, *J. Biol. Ch.* **161** 235, 239, 253.
Pincus, *J. Clin. Endocrinol.* **5** 291, 292, 296.
SCHERING A.-G., *Cios Report* XXIV–16, p. 3.
a Veitch, Milone, *J. Biol. Ch.* **157** 417.
b Veitch, Milone, *J. Biol. Ch.* **158** 61.
Zondek, Finkelstein, *Endocrinology* **36** 291.
1946 Björnson, Ottesen, *Quart. J. Pharm. Pharmacol.* **19** 519.
Butenandt, Schäffler, *Z. Naturforsch.* **1** 82, 85.
Carol, Rotondaro, *J. Am. Pharm. Assoc.* **35** 176.
a Devis, *Compt. rend. soc. biol.* **140** 1083, 1085.
b Devis, Devis-van den Eeckhoudt, *Compt. rend. soc. biol.* **140** 1090.
Furchgott, Rosenkranz, Shorr, *J. Biol. Ch.* **164** 621, 623.
Heer, Miescher, *Helv.* **29** 1895, 1896, 1898, 1900, 1902, 1906.
Dr. G. HENNING, *BIOS Final Report* No. 1404, p. 146.
Jacobsen. Pincus (to G. D. SEARLE & Co.), *U.S. Pat.* 2480246 (issued 1949); *C. A.* **1950** 2564; *U.S. Pat.* 2648700 (1949, issued 1953); *C. A.* **1954** 7649.
Mayer, *Compt. rend. soc. biol.* **140** 673.
a Stimmel, *J. Biol. Ch.* **162** 99.
b Stimmel, *J. Biol. Ch.* **165** 73, 78.
1947 Hohlweg, Inhoffen, *Deut. med. Wochschr.* **72** 86.
Huffman, *J. Biol. Ch.* **167** 273, 275.
Inhoffen, *Angew. Ch.* **59** 207, 211, 212.
a Jacobsen, *J. Biol. Ch.* **171** 61.
b Jacobsen, Picha, Levy, *J. Biol. Ch.* **171** 81, 84.
Wülfert, *Acta Chem. Scand.* **1** 818, 827, 829.
1948 a Anner, Miescher, *Experientia* **4** 25.
b Anner, Miescher, *Helv.* **31** 2173.
Friedgood, Garst, Haagen-Smit, *J. Biol. Ch.* **174** 523 et seq.
Heer, Miescher, *Helv.* **31** 405, 407.
Herlant, Thomas, *C. A.* **1952** 7606.
Huffman, Lott, *J. Biol. Ch.* **172** 325, 328.
Klyne, *Nature* **161** 434.
Shoppee, *Nature* **161** 207.
a Velluz, Muller, *Compt. rend.* **226** 411.
b Velluz, Petit, *Bull. soc. chim.* **1948** 1113, 1114.
Wilds, Johnson, *J. Am. Ch. Soc.* **70** 1166, 1169, 1175.
1949 Anner, Miescher, *Helv.* **32** 1957, 1958, 1961, 1964.
Burrows, *Biological Actions of Sex Hormones*, 2nd Ed., Cambridge, p. 558.
Cohen, Bates, *Endocrinology* **45** 86.
Fieser, Fieser, *Natural Products Related to Phenanthrene*, 3rd Ed., New York, p. 480 et seq.
Grant, Glen, *J. Am. Ch. Soc.* **71** 2255.
Heer, Miescher, *Helv.* **32** 1572, 1578, 1584.
Jacques, Horeau, Courrier, *Compt. rend.* **229** 321.
Selye, *Textbook of Endocrinology*, 2nd Ed., Montreal, p. 35.
1950 Anner, Miescher, *Helv.* **33** 1379, 1380, 1383.
Hershberg, Rubin, Schwenk, *J. Org. Ch.* **15** 292, 295, 297 et seq.
Johnson, Banerjee, Schneider, Gutsche, *J. Am. Ch. Soc.* **72** 1426.
Jones, Humphries, Dobriner, *J. Am. Ch. Soc.* **72** 956, 958.
Kaufmann, Pataki, Rosenkranz, Romo, Djerassi, *J. Am. Ch. Soc.* **72** 4531, 4532, 4533.
Velluz, Muller, *Bull. soc. chim.* **1950** 166, 168.
1951 Bachmann, Controulis, *J. Am. Ch. Soc.* **73** 2636, 2638.
Hirschmann, Johnson, *J. Am. Ch. Soc.* **73** 326.

a Johnson, Chinn, *J. Am. Ch. Soc.* **73** 4987.
b Johnson, Christiansen, *J. Am. Ch. Soc.* **73** 5511.
Keller, Weiss, *J. Ch. Soc.* **1951** 1247.
1952 a Johnson, Banerjee, Schneider, Gutsche, Shelberg, Chinn, *J. Am. Ch. Soc.* **74** 2832, 2838.
b Johnson, Christiansen (to Wisconsin Alumni Research Foundation), *U.S. Pat.* 2663716
(issued 1953); *C. A.* **1955** 1827.
Mosettig, Scheer, *J. Org. Ch.* **17** 764, 767, 768.
Oneson, Cohen, *Endocrinology* **51** 173.
Picha, *J. Am. Ch. Soc.* **74** 703 footnote 8.
1953 Fried, Thoma, Klingsberg, *J. Am. Ch. Soc.* **75** 5764 footnote 7.
1954 Repke, Markwardt, *Arch. exptl. Path. Pharmakol.* **223** 271, 273.
1955 Wendler, Taub, Slates, *J. Am. Ch. Soc.* **77** 3559.
1956 Murray, Johnson, Pederson, Ott, *J. Am. Ch. Soc.* **78** 981.
1957 a Johnson, Christiansen, Ireland, *J. Am. Ch. Soc.* **79** 1995, 1999.
b Johnson, Johns, *J. Am. Ch. Soc.* **79** 2005.
1958 Johnson, David, Dehm, Highet, Warnhoff, Wood, Jones, *J. Am. Ch. Soc.* **80** 661, 665, 667,
668, 677.

---

c. Monoalkyl (except methyl)-monohydroxy-monoketo-hydro-
1.2-cyclopentenophenanthrenes

3'- Isopropyl-7-methoxy-4-keto-1.2-cyclopentano-1.4.9.10.11.12-hexahydro-
phenanthrene $C_{21}H_{26}O_2$ (not obtained pure).

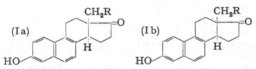

Pale yellow glass, b. 128–134°/$_{0.15}$. — **Fmn.**
From 6-methoxy-1-acetyl-3.4-dihydronaph-
thalene and the lithium derivative of 2-iso-
propylcyclopentanone on refluxing in ether
for 3 hrs., adding alcohol, and further refluxing for 1 hr.; yield, ca. 50%. —
Rns. On reduction with sodium and alcohol, followed by heating with selenium
at 320–330°, it gives 3'-isopropyl-7-methoxy-1.2-cyclopentenophenanthrene
(p. 1500 s). Reacts with methylmagnesium halides in the presence of cuprous
chloride or bromide (reaction product not described) (1944 Birch).

2-Ethyl-7-hydroxy-3'-keto-1.2-cyclopentano-1.2.3.4-tetrahydrophenanthrene,
18-Methylequilenan-3-ol-17-one
$C_{19}H_{20}O_2$. Exists in two racemic
cis- and trans-forms; the β-form
probably has the configuration
(Ia, R = CH₃) of equilenin
(p. 2527 s), the α-form the configuration (Ib, R = CH₃) of isoequilenin (page
2535 s). — **Fmn.** The two forms are obtained from the corresponding forms
of the dimethyl ester of 2-ethyl-7-methoxy-1.2.3.4-tetrahydrophenanthrene-
2-carboxylic-1-(β-propionic) acid on heating with sodium methoxide in
benzene, followed by refluxing the resulting corresponding form of 18-methyl-
3-methoxyequilenan-17-one-16-carboxylic acid methyl ester ("19-methyl-
3-methoxy-16-carbomethoxy-17-equilenone") with hydrochloric-acetic acids
for 12 hrs. (yields in the last step, 78–79%); on refluxing, in the case of the
α-form, for only 30 min., the methyl ether is obtained in 88% yield (1941a
Bachmann).

β-Form, probably dl-18-Methylequilenin, "*dl-19-Methyl-3-hydroxy-17-equi-lenone*" $C_{19}H_{20}O_2$ (Ia, R = CH$_3$). Plates (alcohol), m. 253–5⁰ (vac.); sublimes at 250⁰ and 0.01 mm. pressure (1941a Bachmann). — Has œstrogenic activity (1941a, b Bachmann). — **Fmn.** See above. — *Methyl ether* $C_{20}H_{22}O_2$. Prisms (methanol), m. 171–3⁰ (vac.). **Fmn.** From the above phenol on shaking with dimethyl sulphate in 1% aqueous NaOH (1941a Bachmann).

α-Form, probably dl-18-Methylisoequilenin, "*dl-19-Methyl-3-hydroxy-17-isoequilenone*" $C_{19}H_{20}O_2$ (Ib, R = CH$_3$). Prisms (alcohol), m. 219–229⁰ (vac.); sublimes at 240⁰ and 0.01 mm. pressure (1941a Bachmann). — Has no œstrogenic activity (1941a, b Bachmann). — **Fmn.** See above. — *Methyl ether* $C_{20}H_{22}O_2$. Needles (methanol), m. 124.5–125.5⁰ (1941a Bachmann). **Fmn.** See above under the formation of the phenol.

2-Propyl-7-hydroxy-3'-keto-1.2-cyclopentano-1.2.3.4-tetrahydrophenanthrene,
18-Ethylequilenan-3-ol-17-one $C_{20}H_{22}O_2$ (Ia and b, p. 2570 s; R = C$_2$H$_5$). For probable configuration of the two dl-forms, see the ethyl homologue (p. 2570 s). — **Fmn.** Similar to that of the two forms of the lower homologue starting from the corresponding dimethyl esters of 2-propyl-7-methoxy-1.2.3.4-tetrahydrophenanthrene-2-carboxylic-1-(β-propionic) acid; yields in the last step, 83% (1941b Bachmann).

β-Form, probably dl-18-Ethylequilenin, "*dl-19-Ethyl-3-hydroxy-17-equile-none*" $C_{20}H_{22}O_2$ (Ia, p. 2570 s; R = C$_2$H$_5$). Prisms (alcohol), m. 236–7⁰ (vac.); sublimes at 200⁰ and 0.01 mm. pressure. — Has strong œstrogenic activity (1941b Bachmann). — **Fmn.** See above. — *Methyl ether* $C_{21}H_{24}O_2$. Prisms (methanol), m. 148–149.5⁰ (vac.). **Fmn.** From the above phenol on shaking with dimethyl sulphate in 10% NaOH for 15 min.; yield, 88% (1941b Bachmann).

α-Form, probably dl-18-Ethylisoequilenin, "*dl-19-Ethyl-3-hydroxy-17-iso-equilenone*" $C_{20}H_{22}O_2$ (Ib, p. 2570 s; R = C$_2$H$_5$). Prisms (methanol), m. 153–4⁰ (vac.); sublimes at 175⁰ and 0.01 mm. pressure. — Its œstrogenic activity is $^1/_{10}$ of that of the β-form (1941b Bachmann). — **Fmn.** See above. — *Methyl ether* $C_{21}H_{24}O_2$. Prisms (methanol), m. 103.5–104.5⁰ (vac.). **Fmn.** Similar to that of the α-form of the lower homologue (above); yield, 77% (1941b Bachmann).

2-Butyl-7-hydroxy-3'-keto-1.2-cyclopentano-1.2.3.4-tetrahydrophenanthrene,
18-Propylequilenan-3-ol-17-one $C_{21}H_{24}O_2$ (Ia and b, p. 2570 s; R=CH$_2$·C$_2$H$_5$). Configuration and formation of the two forms are similar to those of the lower homologues (yield of the β-form in the last step, 86%) (1941b Bachmann).

β-Form, probably dl-18-Propylequilenin, "*dl-19-Propyl-3-hydroxy-17-equi-lenone*" $C_{21}H_{24}O_2$ (Ia, p. 2570 s; R = CH$_2$·C$_2$H$_5$). Prisms (alcohol), m. 191–2⁰ (vac.); sublimes at 200⁰ and 0.01 mm. pressure. — Shows no œstrogenic activity (1941b Bachmann). — **Fmn.** See above. — *Methyl ether* $C_{22}H_{26}O_2$.

References, p. 2573 s

Prisms (methanol), m. 141–2⁰. **Fmn.** From the above phenol on methylation; yield, 75% (1941 b Bachmann).

α-**Form**, probably **dl-18-Propylisoequilenin**, "*dl-19-Propyl-3-hydroxy-17-isoequilenone*" $C_{21}H_{24}O_2$ (Ib, p. 2570 s; R = $CH_2 \cdot C_2H_5$). Not obtained crystalline. — **Fmn.** See above. — *Methyl ether* $C_{22}H_{26}O_2$. Two modifications: prisms (methanol), m. 93–4⁰ (vac.), and prisms (methanol), m. 104–5⁰ (vac.); both are obtained from the oily phenol with dimethyl sulphate in aqueous methanol; the modification m. 104–5⁰ is the sole product when the methyl ether is prepared in the same way as the α-methyl ethers of the lower homologues (1941 b Bachmann).

2-Ethyl-9-hydroxy-3'-keto-1.2-cyclopentano-1.2.3.4-tetrahydrophenanthrene,

18-Methylequilenan-6-ol-17-one, "*19-Methyl-6-hydroxy-17-equilenone*" $C_{19}H_{20}O_2$. Exists in two racemic cis- and trans-forms (IIa and IIb), designated α and β without allotting α or β to cis- or trans-configuration. — Both forms are not œstrogenic. — **Fmn.** The two forms are obtained from the corresponding forms of the dimethyl ester of 2-ethyl-9-methoxy-1.2.3.4-tetrahydrophenanthrene-2-carboxylic-1-(β-propionic) acid on refluxing with sodium methoxide in benzene, followed by refluxing the resulting corresp. forms of 18-methyl-6-methoxyequilenan-17-one-16-carboxylic acid methyl ester ("19-methyl-6-methoxy-16-carbomethoxy-17-equilenone") with hydrochloric-acetic acids for 10 hrs.; on refluxing for only 30 min., the corresp. methyl ethers are obtained (1940 Bachmann).

α-**Form.** Prisms (acetone-petroleum ether), m. 206–8⁰ (vac.). — *Methyl ether* $C_{20}H_{22}O_2$. Needles (methanol), m. 142–142.5⁰ (1940 Bachmann).

β-**Form.** Needles (dil. alcohol), m. 121.5–123⁰; crystals with solvent of crystallization (acetone-petroleum ether), m. 109–110⁰ with evolution of gas. — *Methyl ether* $C_{20}H_{22}O_2$. Needles (methanol), m. 75–6⁰ (1940 Bachmann).

4'-Benzylidene-7-methoxy-3'-keto-1.2-cyclopentenophenanthrene $C_{25}H_{18}O_2$

(dried at 110⁰ in high vac.). Yellow needles (benzene + a little alcohol), m. 224⁰; yellow threads with 0.5 H_2O (at 100⁰), m. 223⁰. Gives a yellow solution in alcohol with a bluish green fluorescence, unchanged by NaOH. Its solution in conc. H_2SO_4 is orange-red (salmon-red when very dilute) with an intense greenish yellow fluorescence; on heating, the colour changes to bluish red and then to orange-red and the fluorescence becomes still more intense and more yellow in tone. — **Fmn.** From the methyl ether of dl-18-nor-isoequilenin ("x-norequilenin", p. 2518 s) on boiling with benzaldehyde in aqueous alcoholic NaOH for 15 hrs., followed by exposure to the air (1938, 1941 Koebner).

References, p. 2573 s

1938 Koebner, Robinson, *J. Ch. Soc.* **1938** 1994, 1997.
1940 Bachmann, Holmes, *J. Am. Ch. Soc.* **62** 2750, 2751, 2756.
1941 a Bachmann, Holmes, *J. Am. Ch. Soc.* **63** 595.
    b Bachmann, Holmes, *J. Am. Ch. Soc.* **63** 2592, 2593, 2595, 2596.
    Koebner, Robinson, *J. Ch. Soc.* **1941** 566, 574.
1944 Birch, Robinson, *J. Ch. Soc.* **1944** 503, 505, 506.

---

4. Monohydroxy-monoketo-steroids (CO and OH in the ring system) with Two Side Chains, Not Derived from Androstane or Aetiocholane

4-Methyl-$\Delta^{1.3.5(10)}$-œstratrien-1-ol-17-one, 4-Methyl-1-hydroxy-3-desoxy-œstrone $C_{19}H_{24}O_2$. Formerly described as *1-methylœstrone**;

for the true structure, see 1950, 1953 Woodward; 1950 Djerassi; 1951 Inhoffen; cf. also 1951 Herran; 1953, 1954 Dreiding; for doubtful configuration at C-9, see 1954 Dreiding. — Needles (alcohol), m. 249–251° (1948 Djerassi; cf. also 1941 Inhoffen), 251–3° cor. (1950 Hershberg); $[\alpha]_D^{25}$ +272° (chloroform) (1948 Djerassi); $[\alpha]^{25}$ in dioxane: +348°/546, +292°/D, +227.5°/643 (1950 Hershberg). Absorpt. max. in 95% alcohol at 282.5 mμ (log ε 3.37) (1948 Djerassi). Insoluble in alkali (1950 Djerassi). Is œstrogenically inactive (1950 Djerassi).

Fmn. From $\Delta^{1.4}$-androstadiene-3.17-dione (p. 2877 s) on treatment with a little conc. H_2SO_4 in acetic anhydride at room temperature, followed by hydrolysis with methanolic KOH (yield, 40%) (1941 Inhoffen; 1948 Djerassi); from the same dione on heating with p-toluenesulphonic acid in acetic anhydride on the steam-bath for $4^1/_2$ hrs. (yield, 80%) (1948 Djerassi) or, along with œstrone (p. 2543 s), on pyrolysis by heating at 300–350° in an atmosphere of CO_2 (1950 Hershberg).

Rns. Gives "1-methylœstradiol" (4-methyl-$\Delta^{1.3.5(10)}$-œstratriene-1.17β-diol, p. 1986 s) on warming with lithium aluminium hydride for a few min. (1948 Djerassi).

Benzoate $C_{26}H_{28}O_3$. Leaflets (hexane-acetone), m. 237.5–238.7° cor.; $[\alpha]^{25}$ in dioxane: +212°/546, +180°/D, +145°/643. — **Fmn.** From the preceding compound on heating with benzoyl chloride in pyridine at 90° (1950 Hershberg).

Methyl ether $C_{20}H_{26}O_2$. Crystals (hexane), m. 117–118° cor.; $[\alpha]_D^{25}$ +297° (chloroform). — **Fmn.** From the above hydroxyketone on treatment with dimethyl sulphate and NaOH in aqueous alcohol (1948 Djerassi).

16-Methylequilenan-3-ol-17-one $C_{19}H_{20}O_2$. Two racemic diastereoisomeric forms are known, dl-16-methylequilenin(Ia) and dl-16-methylisoequilenin(Ib).

HO—   (I a)   HO—   (I b)

* True 1-methylœstrone has been prepared by Djerassi (1950).

dl - 16 - Methyl - trans - equilenan - 3 - ol - 17 - one, dl - 16 - Methylequilenin
$C_{19}H_{20}O_2$ (Ia). Prisms (after sublimation at $200^0/_{0.01}$), m. $261.5-263^0$ (vac.). —
Shows only a very weak œstrogenic activity. — **Fmn.** From the β-form of
the dimethyl ester of 2-methyl-7-methoxy-1.2.3.4-tetrahydrophenanthrene-
2-carboxylic-1-(β-propionic) acid on refluxing with sodium methoxide in
benzene for 2 hrs., followed by addition of methyl iodide to the solution
containing dl-16-carbomethoxyequilenin (obtainable in 97 % yield), warming
on the water-bath for 2 hrs., and then by refluxing of the resulting methyl
ether of dl-16-methyl-16-carbomethoxy-equilenin (very small yield) with
hydrochloric-acetic acid for 11 hrs. (yield in the last step, 63 %) (1941 Bachmann).

dl - 16 - Methyl - cis - equilenan - 3 - ol - 17 - one, dl - 16 - Methylisoequilenin
$C_{19}H_{20}O_2$ (Ib). Needles (methanol), m. $183-4^0$ (vac.). — Shows no œstrogenic
activity. — **Fmn.** Similar to that of the above isomer, starting from the
α-form of the dimethyl ester; very good yields in all steps (1941 Bachmann).

2 or 4-Allylœstrone $C_{21}H_{26}O_2$ (II or III). Oil; cannot be extracted from its
ethereal solution with dil. aqueous NaOH. — Shows no œstrogenic activity. —
Fmn. From œstrone allyl ether (p. 2556 s) on refluxing in diethylaniline in
an atmosphere of nitrogen for 4 hrs. (1937 Miescher).

Benzoate $C_{28}H_{30}O_3$. Crystals (acetone-methanol), m. $155-160^0$. — **Fmn.**
From the above phenol on treatment with benzoyl chloride in pyridine (1937
Miescher).

1937 Miescher, Scholz, *Helv.* **20** 1237, 1238, 1244.
1941 Bachmann, Holmes, *J. Am. Ch. Soc.* **63** 2592, 2596.
     Inhoffen, Zühlsdorff, *Ber.* **74** 604, 614 footnote 16.
1948 Djerassi, Scholz, *J. Org. Ch.* **13** 697, 705.
1950 Djerassi, Rosenkranz, Romo, Pataki, Kaufmann, *J. Am. Ch. Soc.* **72** 4540.
     Hershberg, Rubin, Schwenk, *J. Org. Ch.* **15** 292, 297, 298.
     Woodward, Singh, *J. Am. Ch. Soc.* **72** 494.
1951 Herran, Mancera, Rosenkranz, Djerassi, *J. Org. Ch.* **16** 899.
     Inhoffen, *Angew. Ch.* **63** 297.
1953 Dreiding, Pummer, Tomasewski, *J. Am. Ch. Soc.* **75** 3159.
     Woodward, Inhoffen, Larson, Menzel, *Ber.* **86** 594.
1954 Dreiding, Voltman, *J. Am. Ch. Soc.* **76** 537.

5. Monohydroxy-monoketo-steroids (CO and OH in the ring system) with Two Side Chains, Derived from Androstane or Aetiocholane

Summary of Androstan-(Androsten-, etc.)-ol-ones

For values other than those chosen for the summary, see the compounds themselves.

Config. at C-5	Position of O	Position of OH	Double bond(s)	M. p. °C.	[α]D*	Acetate m.p.°	Acetate [α]D*	Benzoate m.p.°	Benzoate [α]D*	Page
—	3	17β	Δ1.4	170	+22° (chl.)	152	+28° (chl.)	216	—	2576 s
—	3	17β	Δ4.6	205	+77° (chl.)	144	+36° (chl.)	260	—	2577 s
norm.	3	17β	Δ1	158	+53° (alc.)	122	+47° (alc.)	200	—	2578 s
—	3	17α	Δ4	221	+72° (alc.)	116	—	139	+21° (chl.)	2579 s
—	3	17β	Δ4	156	+120° (chl.)	141	+88° (alc.)	200	+155° (chl.)	2580 s
—	3	17β	Δ5	—	—	147	−31° (alc.)	181	+23° (ben.)	2595 s
norm.	3	17α	satd.	180	—	151	—	—	—	2596 s
norm.	3	17β	satd.	182	+32° (alc.)	159	—	201	—	2597 s
iso	3	17α	satd.	161	+5° (chl.)	—	—	182	−21° (chl.)	2599 s
iso	3	17β	satd.	140	+33° (alc.)	144	+27° (alc.)	—	—	2600 s
—	4	17β	Δ5	159	−42° (alc.)	119	—	—	—	2601 s
—	6	17β	i(3–5)	183	—	110	—	—	—	2602 s
—	7	3β	Δ5	—	—	174	—	—	—	2602 s
norm.	7	3β	satd.	131	−69° (dio.)	113	—	—	—	2603 s
—	7	17β	Δ3.5	172	−375° (alc.)	222	−400° (chl.)	—	—	2603 s
—	7	17β	Δ5	143	—	216	—	—	—	2603 s
iso	12	3α	satd.	—	—	142	—	—	—	2604 s
norm.	16	3β	satd.	187	−170° (dio.)	110	—	210	—	2604 s
—	17	3	Δ2.4.6	—	—	—	—	—	—	2605 s
—	17	3	Δ3.5	—	—	129	—	180	—	2605 s
norm.	17	3	Δ2	—	—	—	—	—	—	2607 s
—	17	3α	Δ4	—	—	—	—	—	—	2608 s
—	17	3β	Δ4	130	—	—	—	—	—	141
—	17	3α	Δ5	221	0° (alc.)	174	—	—	—	2608 s
—	17	3β	Δ5	{141 / 153}	{+13° (alc.) / + 2° (chl.)}	172	{+6° (alc.) / −7° (chl.)}	254	+26° (chl.)	2609 s
norm.	17	3α	Δ9(11)	190	+140° (alc.)	192	+135° (alc.)	—	—	2626 s
norm.	17	3β	Δ9(11)	172	+126° (alc.)	102	+111° (ac.)	—	—	2626 s
norm.	17	3α	Δ11(?)	180	+122° (alc.)	180	+115° (alc.)	164	—	2627 s
iso	17	3α	Δ11	154	+90° (alc.)	84	+87° (chl.)	—	—	2628 s
norm.	17	3α	satd.	185	{+88° (dio.) / +95° (alc.)}	165	+77° (dio.)	179	—	2629 s
norm.	17	3β	satd.	176	{+81° (dio.) / +88° (me.)}	105	{+65° (dio.) / +75° (alc.)}	225	—	2633 s
iso	17	3α	satd.	152	+109° (alc.)	96	—	163	—	2637 s
iso	17	3β	satd.	{118 / 154}	+89° (alc.)	159	+83° (ac.)	—	—	2638 s
norm.†	17	3α	satd.	146	−100° (alc.)	122	—	—	—	2640 s
norm.	17	5α	satd.	—	—	—	—	—	—	2644 s
—	17	6β	i(3–5)	141	+122° (alc.)	114	+117° (alc.)	—	—	2644 s

* ac. = acetone; alc. = alcohol; ben. = benzene; chl. = chloroform; dio. = dioxane; me. = methanol.

† 13-iso.

$\Delta^{1\cdot4}$-**Androstadien-17β-ol-3-one, 1-Dehydrotestosterone** $C_{19}H_{26}O_2$. For configuration at C-17, see "Introduction", pp. 1363 s,
1364 s. — Needles (aqueous methanol), blocks (ben-
zene-benzine), m. 168–9° (1940a Inhoffen); needles
(petroleum ether-acetone), m. 168.5–170° cor. (1946
Wilds); $[\alpha]_D$ in chloroform + 22.5° (1940a Inhoffen),
+ 20° (1946 Wilds). Absorption max. in ether (from the graph) at ca. 236 mμ
(log ε ca. 4.21) (1940a Inhoffen), in abs. alcohol at 244 mμ (log ε 4.19) (1946
Wilds).

Is more active in the capon comb test than testosterone (1936, 1941 Schering A.-G.).

Fmn. From 2α.4α-dibromoandrostan-17β-ol-3-one (p. 2711 s) on refluxing
with potassium benzoate in butanol-toluene for 2 hrs., followed by heating
of the oily reaction product at 220–230° under 6.5 mm. pressure for 1$^1/_2$ hrs.
(1936 Schering A.-G.). The acetate is obtained from the acetate of the
afore-mentioned dibromo-ketone on boiling in collidine for 45 min. (yield,
69%) (1940a Inhoffen; cf. also 1941, 1945 Schering A.-G.); the hexahydrobenzoate is similarly obtained on boiling for 70 min. (max. yield, 52%) (1946
Wilds; cf. also 1941, 1945 Schering A.-G.); the same esters are obtained
from the corresp. esters of 2α-bromotestosterone (p. 2703 s) on boiling in collidine for $^1/_2$ hr. (1 hr., resp.) (1943 Inhoffen), the hexahydrobenzoate similarly
(15 min.; crude yield, 43%) from the ester of 4-bromo-Δ^1-androsten-17β-ol-
3-one (p. 2704 s) (1947 Djerassi); the esters are hydrolysed by refluxing with
5% methanolic KOH for 1.5 hrs. (1940a Inhoffen; 1946 Wilds).

Rns. Gives $\Delta^{1\cdot4}$-androstadiene-3.17-dione (p. 2877 s) on oxidation with
cyclohexanone and aluminium isopropoxide in boiling toluene (1940a In-
hoffen). Is reduced to $\Delta^{1\cdot4}$-androstadiene-3.17β-diol (p. 1992 s) on heating
with aluminium isopropoxide in isopropyl alcohol (1937 Schering A.-G.).
On treatment with acetic anhydride and conc. H_2SO_4 at room temperature
it gives the diacetate of 4-methyl-$\Delta^{1\cdot3\cdot5(10)}$-œstratriene-1.17$\beta$-diol (originally
described as 1-methylœstradiol; for revised structure, see the diol, p. 1986 s)
(1940b, 1941a Inhoffen; cf. 1946 Wilds); on heating in 9.10-dihydrophen-
anthrene in an evacuated sealed tube at 380–390° for 30 min. it gives a small
amount of this diol (7–14%) along with "α"-œstradiol (10%; p. 1966 s) and
other material (1946 Wilds); "α"-œstradiol is also formed on heating in a
sealed tube at 325° for 12 min. (ca. 5%, along with some œstrone) (1940b,
1941b Inhoffen) or, better (60%), on heating in tetralin (1947 Inhoffen; cf.
1941b Inhoffen; cf. also 1946 Wilds).

Acetate $C_{21}H_{28}O_3$. Light yellow needles (ether-petroleum ether), m. 151–2°;
$[\alpha]_D^{22}$ + 28° (chloroform) (1940a Inhoffen). Absorption max. at 235 mμ (1941
Schering A.-G.). — **Fmn.** and **Rns.** See above. — *Semicarbazone* $C_{22}H_{31}O_3N_3$,
pale yellow needles (methanol), m. 205–6°; absorption max. in chloroform
(from the graph) at ca. 244 (9950) and 295 (22800) m$\mu(\varepsilon)$ (1940a Inhoffen).

Propionate $C_{22}H_{30}O_3$. Leaflets (methanol), m. 138–9°. — **Fmn.** From the
above carbinol and propionic anhydride in pyridine at room temperature
(1940a Inhoffen; 1936, 1941 Schering A.-G.).

Butyrate $C_{23}H_{32}O_3$. Prisms (ether-benzine), m. 82–83°. — **Fmn.** From the carbinol on heating with butyric anhydride in pyridine on the water-bath for 1 hr. (1940a Inhoffen).

Valerate $C_{24}H_{34}O_3$. Cubes (ether-petroleum ether), m. 76–77°. — **Fmn.** From the carbinol on heating with valeric anhydride at 130° for 4 hrs. (1940a Inhoffen; 1941 SCHERING A.-G.).

Benzoate $C_{26}H_{30}O_3$. Needles (methanol), m. 215–6°. — **Fmn.** From the carbinol on heating with benzoic anhydride in pyridine on the water-bath for 5 hrs. (1940a Inhoffen; 1941 SCHERING A.-G.).

Hexahydrobenzoate $C_{26}H_{36}O_3$. Needles (ether-benzine), m. 126–7° (1943 Inhoffen); needles (petroleum ether), m. 127.5–128° cor.; $[\alpha]_D^{20} +45°$ (chloroform) (1946 Wilds). — **Fmn.** and **Rns.** See under the free carbinol, p. 2576 s.

1936 SCHERING A.-G. (Inhoffen), *Ger. Pat.* 722943 (issued 1942); SCHERING-KAHLBAUM A.-G., *Brit. Pat.* 500353 (1937, issued 1939); *Fr. Pat.* 835524 (1937, issued 1938); *Ch. Ztbl.* **1939** I 5010, 5011.
1937 SCHERING A.-G., *Fr. Pat.* 835526 (issued 1938); *C. A.* **1939** 4599; *Ch. Ztbl.* **1939** I 4088.
1940 a Inhoffen, Zühlsdorff, Huang-Minlon, *Ber.* **73** 451, 454, 456, 457.
    b Inhoffen, *Angew. Ch.* **53** 471, 474.
1941 a Inhoffen, Zühlsdorff, *Ber.* **74** 604, 614.
    b Inhoffen, Zühlsdorff, *Ber.* **74** 1911, 1914.
    SCHERING A.-G., *Fr. Pat.* 869676 (issued 1942); *Ch. Ztbl.* **1942** II 1266; SCHERING CORP. (Inhoffen, Zühlsdorff), *U.S. Pat.* 2422904 (issued 1947); *C. A.* **1947** 6288; *Ch. Ztbl.* **1948** II 237.
1943 Inhoffen, Zühlsdorff, *Ber.* **76** 233, 243.
1945 SCHERING A.-G., *Bios Final Report* No. 449, pp. 273, 274, 277, 278; *Fiat Final Report* No. 996, pp. 78, 79, 82.
1946 Wilds, Djerassi, *J. Am. Ch. Soc.* **68** 2125, 2129 et seq.
1947 Inhoffen, *Angew. Ch.* **59** 207, 212.
    Djerassi, Scholz, *J. Am. Ch. Soc.* **69** 2404, 2408.

Δ^{4.6}**-Androstadien-17β-ol-3-one, 6-Dehydrotestosterone** $C_{19}H_{26}O_2$ (p. 139). For

configuration at C-17, see "Introduction", pp. 1363 s, 1364 s. — Crystals (ethyl acetate), m. 209–211° cor. (1940 Wettstein; 1941 CIBA); yellowish crystals, m. 201 to 202° (1943 Inhoffen); m. 204–5°; $[\alpha]_D^{20} +77°$ (chloroform) (1950 Djerassi). Absorption max. in alcohol at 286 mμ (log ε 4.47) (1940 Wettstein; 1941 CIBA), 284 mμ (log ε 4.47) (1950 Djerassi), in ether (from the graph) at ca. 275 mμ (log ε ca. 4.42) (1943 Inhoffen). — Has a very small hormonal action in the capon comb test (1940 Wettstein).

Fmn. The benzoate is obtained from the 17-benzoate of *Δ*⁵-androstene-3β.17β-diol (p. 1999 s) on refluxing with benzoquinone and aluminium tert.-butoxide in toluene for 45 min. or, similarly, from *Δ*⁵-androsten-17β-ol-3-one benzoate (p. 2595 s) (1940 Wettstein; 1941 CIBA). The acetate is obtained from 6-bromotestosterone acetate (p. 2705 s) on refluxing with collidine for 30 min. (1943 Inhoffen; 1946 Meystre; 1950 Djerassi); the hexahydrobenzoate is similarly obtained from the corresp. ester of 6-bromotestosterone on boiling with collidine for 20 min. (1943 Inhoffen). The acetate is formed as a by-product (ca. 3%) in the preparation of Δ¹·⁴-androstadien-17β-ol-3-one (page 2576 s) from 2α.4α-dibromoandrostan-17β-ol-3-one acetate with collidine (1943 Inhoffen); the hexahydrobenzoate is similarly formed (1946 Wilds). The esters may be hydrolysed by refluxing with 2–3% methanolic KOH (1940 Wettstein; 1941 CIBA; 1943 Inhoffen).

Acetate $C_{21}H_{28}O_3$. Needles (hexane + a little acetone), m. 143–4° cor. (1940 Wettstein); crystals (ether), m. 142–4° (1943 Inhoffen); $[\alpha]_D^{22}$ +35.5°±4° (alcohol) (1946 Meystre), +36° (chloroform) (1950 Djerassi). Absorption max. in 95% alcohol at 284 mμ (log ε 4.53) (1950 Djerassi). — **Fmn.** From the above carbinol with acetic anhydride in pyridine at room temperature (1940 Wettstein; 1941 CIBA) or on the steam-bath (1943 Inhoffen). See also above.

Benzoate $C_{26}H_{30}O_3$ (p. 139). Needles (ethyl acetate), m. 257–260° cor. (1940 Wettstein; 1941 CIBA), 244–6° (1943 Inhoffen).

Hexahydrobenzoate $C_{26}H_{36}O_3$. Pale yellow plates (ether), m. 133–4°. — **Fmn.** From the above carbinol with hexahydrobenzoyl chloride in pyridine at room temperature (1943 Inhoffen). See also above.

1940 Wettstein, *Helv.* **23** 388, 393–395, 398.
1941 CIBA (Soc. pour l'ind. chim. à Bâle; Ges. f. chem. Ind. Basel), *Fr. Pat.* 877821 (issued 1943);
          Dutch Pat. 54147 (issued 1943); *Ch. Ztbl.* **1943** II 147, 148.
1943 Inhoffen, Zühlsdorff, *Ber.* **76** 233, 244, 245.
1946 Meystre, Wettstein, *Experientia* **2** 408.
          Wilds, Djerassi, *J. Am. Ch. Soc.* **68** 2125, 2130.
1950 Djerassi, Rosenkranz, Romo, Kaufmann, Pataki, *J. Am. Ch. Soc.* **72** 4534, 4538.

Δ¹-Androsten-17β-ol-3-one, "*Δ¹-Testosterone*" $C_{19}H_{28}O_2$. For a compound originally described with this structure, see Δ⁵-androsten-17β-ol-4-one, p. 2601 s. — Leaflets (acetone-hexane, and aqueous acetone), m. 150° (1940a Butenandt); needles (acetone-hexane), m. 156–158° (1943 Inhoffen); $[\alpha]_D^{18}$ +53° (alcohol) (1940a Butenandt), $[\alpha]_D^{23-24}$ +52° (chloroform) (1947b Djerassi). Absorption max. in alcohol at ca. 230 mμ (ε 10000) (1940a Butenandt).

Shows a high androgenic activity in the Fussgänger capon comb test, is less effective on male rats; shows no œstrogenic activity (1940a Butenandt).

Fmn. The acetate is obtained from the acetate of 2α-bromoandrostan-17β-ol-3-one (p. 2703 s) on boiling with collidine for 1 hr. (yield, 64%) (1940a Butenandt); the hexahydrobenzoate is similarly obtained from the corresp. ester on boiling with collidine for 30 min. (1943 Inhoffen; cf. 1947a Djerassi). Benzoate and hexahydrobenzoate are also formed when the esters of 2-bromo-Δ¹-androsten-17β-ol-3-one (p. 2702 s) are boiled with zinc dust in alcohol (1943 Inhoffen; cf. 1947a Djerassi). The esters are hydrolysed by boiling with dil. methanolic KOH (1940a Butenandt; 1943 Inhoffen). — **Rns.** On oxidation with CrO_3 in glacial acetic acid at room temperature it gives Δ¹-androstene-3.17-dione (p. 2879 s) (1940a Butenandt). Gives a poor yield of androstane-3β.17β-diol (p. 2014 s) on treatment with fermenting yeast in dil. alcohol (1940b Butenandt).

Acetate $C_{21}H_{30}O_3$. Crystals (dil. acetone), m. 122^0 (1940a Butenandt); $[\alpha]_D^{23}$ + 47^0 in alcohol (1940a Butenandt) or chloroform (1947b Djerassi). — **Fmn.** See above. — *Oxime* $C_{21}H_{31}O_3N$, crystals with 1 H_2O (dil. alcohol), m. 112^0 after sintering at 98^0 (1940a Butenandt).

Benzoate $C_{26}H_{32}O_3$. Plates (ethyl acetate-methanol), needles (methanol), m. 200^0. — **Fmn.** See above. Is also obtained from the above carbinol with benzoyl chloride and pyridine on the steam-bath (1 hr.) (1943 Inhoffen).

Hexahydrobenzoate $C_{26}H_{38}O_3$. Needles (methanol), m. 160–1^0 (1943 Inhoffen); m. 160–1^0 cor.; $[\alpha]_D^{23}$ + 45^0 (chloroform); absorption max. in alcohol at 231.5 mμ (log ε 3.83) (1947a Djerassi). — **Fmn.** See above.

1940 a Butenandt, Dannenberg, *Ber.* **73** 206; cf. SCHERING A.-G., *Fr. Pat.* 867697 (issued 1941); *Ch. Ztbl.* **1942** I 2561.
  b Butenandt, Dannenberg, Surányi, *Ber.* **73** 818.
1943 Inhoffen, Zühlsdorff, *Ber.* **76** 233, 240, 241.
1947 a Djerassi, Scholz, *J. Am. Ch. Soc.* **69** 2404, 2408.
  b Djerassi, *J. Org. Ch.* **12** 823, 826.

---

*Δ^4-*Androsten-17α-ol-3-one, 17-epi-Testosterone, *"cis-Testosterone"* $C_{19}H_{28}O_2$

(p. 140). For α-configuration at C-17, see 1949 Fieser; 1950 Shoppee. — Dipole moment in dioxane 5.13 D (1945 Kumler). Rate of hydrolysis with alcoholic KOH of the acetate at 16–19^0 and 81^0, and of the benzoate at 81^0; ratio of hydrolysed esters of epitestosterone and testosterone (below) at 16–19^0 1:2.5, at 81^0 1:1.6 (1938 Ruzicka). Epitestosterone is only weakly androgenic (1936a Ruzicka).

Fmn. The passus on p. 140 has to be replaced by: The benzoate is obtained from the 17-benzoate of Δ^5-androstene-3β.17α-diol (p. 1998 s) on bromination with 1 mol. bromine in glacial acetic acid, followed by oxidation with CrO_3 in strong acetic acid at room temperature and debromination with zinc dust in glacial acetic acid at 100^0 (yield, 78%); it is hydrolysed by refluxing with methanolic NaOH for 2 hrs. (1936a Ruzicka; 1936i, k CIBA). Epitestosterone is probably obtained, in very small amount along with testosterone, from Δ^4-androstene-3.17-dione (p. 2880 s) on boiling with aluminium tert. butoxide in sec.-butyl alcohol-benzene (1939a Miescher; 1937c CIBA).

Rns. The benzoate is converted into the benzoate of ætiocholan-17α-ol-3-one (p. 2599 s) by hydrogenation in alcohol in the presence of palladium-barium carbonate or platinum oxide, in the latter case followed by oxidation with CrO_3 in strong acetic acid. Thermal decomposition of the benzoate affords $\Delta^{4.16}$-androstadien-3-one (p. 2396 s) (1945 Prelog).

Colour reactions. With conc. H_2SO_4 + acetic anhydride in chloroform (Liebermann-Burchard reaction) it gives a dark blue-red coloration which

becomes yellow-red with yellow fluorescence on standing and, after 1 hr.,
dark yellow-red. Gives blue-red colorations with orange fluorescence on
boiling for a short time with conc. H_2SO_4 in glacial acetic acid or with $POCl_3$
in quinoline (followed by dissolution in glacial acetic acid + conc. H_2SO_4),
in both cases followed by dropwise addition of bromine (in glacial acetic acid)
or acetic anhydride (1939 Kägi). Gives an intensely blue-violet coloration
with red fluorescence with anisaldehyde + conc. H_2SO_4 in glacial acetic acid
(1946 Miescher). For the rate of reduction of phosphomolybdic acid (Folin-
Wu reagent) by epi-testosterone in glacial acetic acid at 100° to give molybdenum blue, see 1946 Heard.

Benzoate $C_{26}H_{32}O_3$ (p. 140). Needles (ether), m. 137–9° cor.; $[\alpha]_D^{17}$ +20.5°
±2° (chloroform). — **Fmn.** From epitestosterone on refluxing with benzoyl
chloride and abs. pyridine in methylene chloride for 1 hr. (1945 Prelog).
See also under formation of epitestosterone, above. — **Rns.** See above.

Δ^4-Androsten-17β-ol-3-one, Testosterone, *"trans-Testosterone"* $C_{19}H_{28}O_2$ (page

139). For β-configuration at C-17, see 1949 Fieser;
1950 Shoppee. — Crystals, m. 155–6°; $[\alpha]_D$ +116° to
120° (chloroform) (1945 Schering A.-G.); $[\alpha]_D^{23-24}$
+109° (chloroform) (1947b Djerassi). Absorption
max. in chloroform at 238 mμ (k = 3.2 mm.$^{-1}$) (1935,
1938c Butenandt; cf. also 1940 Elvidge), at 240 mμ (1939 Dannenberg); in
hexane at 230 (ε 15 100), in ether at 234 (ε 15 600), in dioxane at 236 (ε 15 200),
in methanol and ethanol at 241 (ε 15 400 and 15 800, resp.) mμ (1939 Dannen-
berg); for absorption spectra in abs. alcohol, aqueous alcohol, and conc. H_2SO_4,
see 1938 Bugyi. For infra-red absorption spectra of films of testosterone
and its propionate, see 1946 Furchgott. Dipole moment in dioxane 4.03 D
(1945 Kumler). For basic properties, see 1944, 1946 Bersin. Testosterone
is soluble in 10–20% aqueous solutions of sodium dehydrocholate (1944
Cantarow).

Rate of hydrolysis with alcoholic KOH of the acetate at 16–19° and 81°,
and of the benzoate at 81°; ratio of hydrolysed esters of testosterone and
epitestosterone (p. 2579 s) at 16–19° 2.5:1, at 81° 1.6:1 (1938 Ruzicka).

Gives a double compound with digitonin which is readily soluble in alcohol
(1939c Miescher).

Occurrence of Testosterone

For isolation of testosterone from bull testicles, see also 1937 Ciba Pharmaceutical Products; 1938 Goldberg; cf. also 1943 Ruzicka. Testosterone
occurs also in horse testes; since alkali was used during its isolation it is
not certain if it occurs in the free state or as a labile complex (1946 Tagmann;
cf. 1943 Ruzicka). Testosterone may be the testis hormone of men, since
the excretion of androsterone (the urinary androgen of men; p. 2629 s) is
increased after administration of testosterone (cf. under "Biological reactions",
p. 2585 s) (1939 Cook; 1939b Dorfman).

FORMATION OF TESTOSTERONE *

Formation from androstenol and aminoandrostenol. Testosterone acetate or propionate is obtained from the corresp. esters of Δ^4-androsten-17β-ol (page 1510 s) on oxidation with CrO_3 in strong acetic acid (1939 CIBA; 1940a Marker; 1940 PARKE, DAVIS & Co.). From 3-amino-Δ^5-androsten-17β-ol (not described) on treatment with HOCl in ether at -5^0 to 0^0 in the presence of anh. Na_2SO_4, followed by boiling with alcoholic sodium ethoxide in benzene and hydrolysis with dil. H_2SO_4 (1937a I.G. FARBENIND.).

Formation from enol-testosterone. Testosterone is obtained from its enol-benzoate ($\Delta^{3.5}$-androstadiene-3.17β-diol 3-benzoate, p. 1994 s) on refluxing with aqueous methanolic K_2CO_3 for several hrs.; its esters are similarly obtained from the corresp. 17-esters of testosterone 3-enol esters (1938 CIBA). From its enol-ethyl ether ($\Delta^{3.5}$-androstadiene-3.17β-diol 3-ethyl ether, p. 1995 s) on heating with aqueous alcoholic HCl on the water-bath for 20 min. (excellent yield) (1938 Serini; cf. 1938a SCHERING A.-G.; 1938 CIBA); the propionate and benzoate have been obtained from the corresp. 17-esters of this ether on keeping in chloroform, probably owing to traces of hydrochloric acid present in the chloroform (1938 Schwenk). For formation from its enol-ethyl ether, see also testosterone dihydrogen phosphate (p. 2588 s).

Formation from Δ^5-androstene-3β.17β-diol. Testosterone is obtained from the diol (p. 1999 s) on oxidation with CrO_3 in glacial acetic acid at 24^0 (yield, 2%) or with $KMnO_4$ (1936a N. V. ORGANON) or, after temporary protection of the double bond with bromine, on oxidation with CrO_3 in glacial acetic acid (1935 SCHERING-KAHLBAUM A.-G.); for similar formation of testosterone esters from the corresp. 17-esters of the diol by bromination, oxidation with CrO_3 in glacial acetic acid, and debromination with zinc dust in glacial acetic acid (p. 140), see also 1935h, 1936i CIBA. Testosterone or its esters may also be obtained from the diol or its corresp. 17-esters on heating with copper dust (or platinum, palladium, or nickel) in vacuo at 225^0 (1935i, 1936i CIBA) or with copper and cyclohexanone in quinoline (1935i CIBA). From the diol on heating with aluminium phenoxide in acetone-benzene (1:5) on the water-bath for 18 hrs. (yield, 40%) (1937 Kuwada); testosterone esters are obtained from the corresp. 17-esters of the diol on boiling with excess acetone or cyclohexanone and aluminium tert.-butoxide in toluene (1937 Oppenauer; 1937b N. V. ORGANON; cf. also p. 140, and 1937 SCHERING CORP.; 1937b SCHERING-KAHLBAUM A.-G.; 1945 SCHERING A.-G.; 1945 C. F. BOEHRINGER & SÖHNE). From the 3-triphenylmethyl ether (p. 2006 s) of the diol on distillation at $200-300^0$ under 15 mm. pressure (1937d CIBA). From the dibromide (p. 2002 s) of the diol on refluxing with collidine for 45 min.; yield, 25–30% (1941 Galinovsky; cf. also 1942 SCHERING A.-G.). For formation of testosterone from the diol by bacterial dehydrogenation, see under "Biological oxidation" of the diol (p. 2003 s).

Formation from a triol and a tetrol. Testosterone is obtained from Δ^5-androstene-3β.4β.17β-triol (p. 2163 s) on refluxing with HCl in alcohol for 15 min.

* For a total synthesis of dl-testosterone, carried out after the closing date for this volume, see 1955, 1956 Johnson.

(1937 f CIBA) or in methanol for 45 min. (1941 a Marker); similarly from
Δ^4-androstene-3β.6β.17β-triol (p. 2164 s) (1937 f CIBA). From androstane-
3β.5α.6β.17β-tetrol (p. 2207 s) on heating in vacuo with KHSO$_4$ (1937 f CIBA).

Formation from halogeno- and amino-ketosteroids. Testosterone acetate is
obtained from 17-chloro-Δ^4-androsten-3-one (p. 2474 s) on heating with potassium acetate in glacial acetic acid in a sealed tube at 150–180° (1935, 1936
SCHERING-KAHLBAUM A.-G.). From 17-amino-Δ^4-androsten-3-one (p. 2508 s)
on treatment with NaNO$_2$ in aqueous alcoholic acetic acid (1937 c I.G. FAR-
BENIND.). From 17-amino-ætiocholan-3-one (p. 2509 s) on treatment with
bromine and hydrobromic acid in glacial acetic acid, heating of the resulting
product with NaNO$_2$ in alcohol-acetic acid at 50–70°, and dehydrobromination with silver acetate in glacial acetic acid (1937 b I.G. FARBENIND.).

Formation from hydroxyketosteroids. Testosterone is obtained from esters
of $\Delta^{3.5}$-androstadien-3-ol-17-one (p. 2605 s) on reduction with 1 mol. hydrogen
in the presence of nickel in abs. alcohol or with fermenting yeast in aqueous
alcohol, in both cases followed by boiling with aqueous methanolic or alcoholic K$_2$CO$_3$ (1937 a CIBA). From Δ^5-androsten-17β-ol-3-one (p. 2595 s) on
gently warming with aqueous methanolic HCl; the acetate is similarly obtained from the corresp. ester (1937 a SCHERING-KAHLBAUM A.-G.). From
Δ^5-androsten-3β-ol-17-one (5-dehydroepiandrosterone, p. 2609 s) on shaking
with a mixture of oxidizing bacteria in yeast water under oxygen at 32° for
48 hrs., collection of the reaction product (Δ^4-androstene-3.17-dione, p. 2880 s),
dissolution in alcohol, removal of the bacteria, and treatment with fermenting
yeast in dil. alcohol (4 days); yield, 81 % (1938 b Mamoli); for partial dismutation of 5-dehydroepiandrosterone to testosterone by refluxing with aluminium tert.-butoxide in anh. benzene for 14 hrs., see 1937 c N. V. ORGANON;
1937 Oppenauer. From ætiocholan-17β-ol-3-one (p. 2600 s) on treatment
with bromine and hydrobromic acid in acetic acid, followed by refluxing of
the resulting crude product (m. 150–156°) with pyridine for 2 hrs. (1936
CIBA PHARMACEUTICAL PRODUCTS; 1940 b Marker). Testosterone acetate and
hexahydrobenzoate are obtained from the corresp. esters of 2α-bromotesto-
sterone (p. 2703 s) on refluxing with zinc dust in alcohol for 3 hrs. (1943 Inhoffen; 1943 SCHERING A.-G.). The propionate is obtained from that of
2α-bromoandrostan-17β-ol-3-one (p. 2703 s) on refluxing with pyridine and
thermal decomposition of the resulting pyridine addition compound under
12 mm. pressure (1937 e CIBA); the hexahydrobenzoate is obtained in 18 %
yield, along with that of Δ^1-androsten-17β-ol-3-one, from the corresp. ester of
the same bromohydroxyketone on refluxing with collidine (1947 a Djerassi).

Formation from diketosteroids. Testosterone can be obtained in ca. 5 %
yield from 4β-bromopregnane-3.20-dione (p. 2813 s) on treatment with persulphuric acid in acetic acid at 25° for 10 days, removal of HBr by boiling
with pyridine for 1c hrs., sublimation of the resulting mixture in a high
vacuum, hydrolysis of desoxycorticosterone acetate present by refluxing
with KHCO$_3$ in aqueous methanol for $^1/_2$ hr., extraction with ether, oxidation
of the neutral fraction with CrO$_3$ in strong acetic acid at room temperature
(45 min.), removal of the 3-keto-Δ^5-ætiocholenic acid formed from desoxy-

corticosterone, refluxing with 1 % methanolic KOH for 30 min., and isolation of testosterone via the hydrogen succinate (1940c Marker).

From Δ^4-androstene-3.17-dione (p. 2880) on hydrogenation in alcohol in
the presence of a nickel catalyst (1936h, i CIBA), on heating with aluminium
isopropoxide in isopropyl alcohol (1936a N. V. ORGANON), on boiling with
1.5 mols. of aluminium tert.-butoxide and 50 mols. of sec.-butyl alcohol
in benzene for 7–8 hrs. (crude yield, 70%) (1939a Miescher; 1937c CIBA),
or on reduction with sodium and 60% alcohol in ether and then with aluminium isopropoxide in boiling isopropyl alcohol, followed by oxidation of
the resulting 17.17'-dihydroxy-3.3'-pinacol (see p. 2587 s) with lead acetate
in methanol at room temperature; the acetate (propionate) may be obtained
from the diacetate (dipropionate) of this pinacol by similar oxidation or,
after protection of the double bonds with bromine, by oxidation with CrO_3
in strong acetic acid, followed by debromination with zinc in 80% acetic
acid (1937a N. V. ORGANON; 1938 ROCHE-ORGANON). From the 3-ethylene
ketal of Δ^4-androstene-3.17-dione (p. 2890 s) on reduction with sodium in
boiling abs. alcohol, followed by hydrolysis of the resulting "testosterone
ethylene ketal" (p. 2595 s) by refluxing with aqueous alcoholic HCl (1941
SQUIBB & SONS). For formation of testosterone from Δ^4-androstene-3.17-dione
by the action of fermenting yeast (p. 140), see also 1938b SCHERING A.-G.;
1938 SCHERING CORP.; cf. also 1938d Mamoli, and the formation from dehydroepiandrosterone (p. 2582 s). In small yield, along with androstane-
$3\beta.17\beta$-diol (p. 2014 s) from Δ^5-androstene-3.17-dione (p. 2890 s) by the action
of fermenting yeast in dil. alcohol (1937 Mamoli).

REACTIONS OF TESTOSTERONE

Oxidation. On treatment of testosterone acetate with 30% $H_2O_2 + OsO_4$
in ether at room temperature for 24 hrs. it gives the 17-acetate of andro-
stane-4.5.17β-triol-3-one (p. 2767 s) (1938b Butenandt). On ozonolysis in
glacial acetic acid the acetate gives 5-keto-3‖5-A-nor-androstan-17β-ol-3-oic
acid (p. 2587 s) (1938 Bolt; 1936b N. V. ORGANON). Testosterone benzoate is
not attacked by benzoquinone in boiling toluene, in the absence or presence
of aluminium tert.-butoxide (1940 Wettstein). Testosterone and its acetate
slowly give a precipitate with ammoniacal silver solution in methanol (1939b
Miescher).

Reduction. On slow distillation of testosterone with aluminium isopropoxide
in abs. isopropyl alcohol for 8 hrs. it gives Δ^4-androstene-3α.17β-diol (p. 1996 s;
yield, 11.5%) and Δ^4-androstene-3β.17β-diol (p. 1997 s; yield, 45%) (1938a
Butenandt). For reduction of testosterone propionate with sodium amalgam
(1940 Butenandt), see testosterone pinacol dipropionate, p. 2587 s. For
reduction of testosterone, see also under "Biological reactions", p. 2584 s.

Action of light. On irradiation of testosterone propionate with ultra-violet
light or intense sunlight in benzene-hexane (1:10) with exclusion of air for
72 hrs. it gives 19% yield of lumitestosterone propionate (p. 2586 s) (1939,
1940 Butenandt) and 2% yield of testosterone pinacol dipropionate (p. 2587 s);
27% of the starting material may be recuperated (1940 Butenandt).

E.O.C. XIV s. 179

Halogenation. On refluxing testosterone acetate with 1 mol. N-bromo-succinimide in carbon tetrachloride for 20–30 min. in the dark it affords 72% yield of the acetate of 6-bromotestosterone (p. 2705 s) (1950 Djerassi; cf. 1946 Meystre). For dibromination of testosterone, see 1936 SCHERING A.-G. and, after the closing date for this volume, 1950 Rosenkranz; 1950 Djerassi. On treatment of testosterone with $PCl_5 + CaCO_3$ in chloroform at 0^0 it gives 17-chloro-Δ^4-androsten-3-one (p. 2474 s; main product), testo-sterone phosphate (p. 2588 s), and 3.17-dichloro-$\Delta^{3.5}$-androstadiene (p. 1467 s) (1939 Westphal). For reaction with $POCl_3$ and pyridine, see below.

Reactions with organic compounds. Testosterone reacts with ethylene glycol and with propylene glycol in benzene in the presence of p-toluenesulphonic acid to give cyclic ketals (1939 SCHERING A.-G.) in which the double bond is shifted to the 5.6-position (see p. 2595 s). Gives the thioenol - ethyl ether (m. 112^0, $[\alpha]_D$ —200^0 in benzene) on shaking with ethyl mercaptan and benzenesulphonic acid in formic acid (1940 CHINOIN). For conversion of testosterone and its esters into diesters of the enol form ($\Delta^{3.5}$-androstadiene-3.17β-diol, p. 1992 s) by boiling with the anhydride and the anhydrous sodium salt of an aliphatic acid or with acyl chlorides, see also 1938/39 Miescher; cf. also 1936f CIBA. On heating testosterone propionate and benzoate with ethyl orthoformate, formic acid, and a little conc. H_2SO_4 for 10 min. at 50^0, they give the corresp. 17-esters of $\Delta^{3.5}$-androstadiene-3.17β-diol 3-ethyl ether (page 1995 s) (1938 Schwenk). On heating testosterone with potassium cyanide and ammonium carbonate in alcohol under CO_2 at 135^0 and 25 atm. pressure it is converted into Δ^4 - androsten - 17β-ol - 3 - spirohydantoin $C_{21}H_{30}O_3N_2$ (I) (1937g CIBA). On boiling testo-sterone with diethyl oxalate and sodium ethoxide in benzene for 1 hr. it gives ethyl testosterone-2-glyoxylate which by heating with dil. H_2SO_4 is converted into testosterone-2-aldehyde (p. 2835 s) (1937b CIBA). Testosterone reacts only to an extent of 10% with phthalic anhydride in benzene in a sealed tube at 85^0 (2 hrs.) (1942 Wettstein).

On treatment of testosterone propionate with 1 mol. $POCl_3$ and excess dry pyridine at room temperature it gives 50% yield of N-(17β-propionyloxy-$\Delta^{3.5}$-androstadien-3-yl)-pyridinium chloride $C_{27}H_{36}O_2NCl$ [lemon-yellow crys-tals, dec. 260–270^0; absorption max. in chloroform at ca. 330 mμ (log ε 4.0); soluble in alcohol and chloroform, sparingly soluble in pyridine and petroleum ether] which on heating in a high vacuum to 200^0 gives 54% yield of $\Delta^{3.5.7}$-androstatrien-17β-ol propionate (p. 1509 s) and on hydrogenation in glacial acetic acid in the presence of platinum yields 3-piperidinoandrostan-17β-yl propionate hydrochloride (1944b Müller). Testosterone does not condense with isatin in the presence of KOH (1944 Buu-Hoi).

Biological reactions. Testosterone gives small amounts of ætiocholan-17β-ol-3-one (p. 2600 s) and ætiocholane -3α.17β-diol ("epi-ætiocholanediol", p. 2017 s) on treatment with an aqueous extract of stallion testicles at 37^0 (1938 Ercoli); this reduction is due to the presence of putrefactive bacteria (1938 Schramm; 1938a Mamoli); the diol is also obtained, alone or along with androstane-3β.17β-diol ("isoandrostanediol", p. 2014 s), on using putrescent extracts of bull testicles (1938a, c Mamoli); on using Bacillus putrificus (Bienstock), isolated from a putrescent extract of bull testicles, in dil. yeast water, testosterone gives only ætiocholan-17β-ol-3-one, whereas in a more conc. aqueous yeast suspension ˙ætiocholane-3α.17β-diol (p. 2017 s) is also formed (1939 Mamoli). By the action of Proactinomyces erythropolis GRAY & THORNTON (and other Proactinomyces species) testosterone gives Δ^4-androstene-3.17-dione (p. 2880 s) (1946 Turfitt).

Testosterone propionate administered to normal men increases the excretion in the urine of androgenic substances, especially of androsterone (p. 2629 s; ca. 7 times that of normal excretion) (1939a, b Callow; cf. also 1939 Cook; 1939b, 1940b Dorfman) and of ætiocholan-3α-ol-17-one (5-iso-androsterone, p. 2637 s; ca. 5.5 times that of normal excretion) (1939a, b Callow). After administration of testosterone propionate to hypogonadal men there have been isolated androsterone (1940a Dorfman) and epiandrosterone (page 2633 s) (1941 Dorfman), the latter also after administration to an adult male guinea pig (1940c Dorfman). For increased excretion of 17-keto-steroids after administration of testosterone to dwarfed, sexually retarded boys, see 1944 Frame. Increased amounts of androsterone and ætiocholan-3α-ol-17-one have also been isolated after administration of testosterone propionate to adult male chimpanzees (1944 Fish) and to a normal woman (1945 Schiller), in both cases along with $\Delta^{2 \text{ or } 3}$-androsten-17-one (p. 2401 s) (1944 Fish; 1945 Schiller) which is considered to be formed by dehydration of androsterone during the hydrolysis and extraction of the urine (1944 Fish); from the urine of the woman a small quantity of a non-ketonic substance, m. 217–221°, probably androstane-3α.17α-diol (p. 2010 s) was also isolated (1945 Schiller). After injection of testosterone propionate into a pregnant Rhesus monkey, androsterone has been isolated from the urine (1944 Horwitt).

Strongly increased excretion of œstrogens has been observed after administration of testosterone propionate to normal men (1937 Steinach).

PHYSIOLOGICAL PROPERTIES OF TESTOSTERONE

For physiological properties of testosterone and its esters, see p. 139 and pp. 1367 s, 1368 s and the literature cited there. Cf. also "Biological reactions" above.

ANALYTICAL PROPERTIES OF TESTOSTERONE

Colour reactions and determination. When a solution of testosterone in 96% alcohol is underlayered with conc. H_2SO_4, the following interfacial colours may be observed: yellow-green with green fluorescence (after 5 min.) → green-yellow with intense grass-green fluorescence and a very pale ring (after 4 hrs.) → green-yellow with intense grass-green fluorescence, lilac (after 24 hrs.) (1939b Woker; cf. 1939 Scherrer). The Liebermann-Burchard reaction (conc. H_2SO_4 + acetic anhydride in chloroform) is negative; also on boiling with conc. H_2SO_4 in glacial acetic acid for a few sec., followed by addition of bromine or acetic anhydride, no colours appear (1939 Kägi). On boiling with $POCl_3$ in quinoline for a short time, followed by dissolution in glacial acetic acid + conc. H_2SO_4 and dropwise addition of bromine (dissolved in glacial acetic acid) or acetic anhydride, a blue-violet coloration is produced

179*

(1939 Kägi). With a highly conc. solution of $SbCl_3$ in glacial acetic acid-acetic anhydride (9:1) it gives only a faint yellowish colour (1943 Pincus). Interfacial colours are produced on underlayering alcoholic solutions of testosterone with conc. H_2SO_4 in the presence of benzaldehyde (1939 Scherrer) and furfural (1939a, b Woker); no colour reaction with anisaldehyde + conc. H_2SO_4 in glacial acetic acid (1946 Miescher). A bright green solution with max. absorption at ca. 639 mμ is obtained on heating of testosterone (dry or in 95 % alcohol) with conc. H_2SO_4 for 2 min. at 100°, addition of thiocol (potassium guaiacolsulphonate) + copper sulphate in water with cooling in ice, heating for 2 min. at 100°, cooling, and addition of 50 % H_2SO_4; testosterone propionate and oxime, and Δ^4-androstene-3.17-dione give the same colour; androsterone and dehydroepiandrosterone do not interfere appreciably; the reaction may be used for determination of testosterone (1941 Koenig). For the rate of reduction of phosphomolybdic acid (Folin-Wu reagent) by testosterone in glacial acetic acid at 100° to give molybdenum blue, see 1946 Heard.

Testosterone and its propionate slowly develop a weak reddish brown coloration in the Zimmermann reaction with m-dinitrobenzene + KOH in alcohol (1938 Callow; 1943 Wettstein); according to Zimmermann (1944) testosterone develops at once a deep violet colour which in a few seconds changes to olive-brown; for measurement of the absorption spectra of the reaction mixture, in the absence or the presence of androsterone, and for determination of the two hormones in mixtures, see 1939 Langstroth; cf. also 1940 McCullagh; 1940 Tastaldi; 1943 Hansen; 1944 Zimmermann.

Testosterone and its propionate give a slight pink coloration with 1.4-naphthoquinone and conc. HCl in glacial acetic acid at 60–70° (1939b Miescher).

For polarographic assay of testosterone propionate and for determination of testosterone in mixtures by polarographic analysis, see 1939 Eisenbrand.

REACTION PRODUCTS OF TESTOSTERONE OF INCOMPLETELY KNOWN STRUCTURE AND DEGRADATION PRODUCTS

Lumitestosterone propionate $C_{44}H_{64}O_6$. For the probable formula (II, R =CO·C_2H_5), see 1940 Butenandt; but cf. 1952 Butenandt. — Leaflets (chloroform-alcohol), begins to dec. at 300°, m. 350° to 355°; $[\alpha]_D^{22}$ +47.3° (chloroform); very sparingly soluble (1939 Butenandt). — **Fmn.** From testosterone propionate in benzene-hexane (1:10) on irradiation with ultra-violet light or intense sunlight with exclusion of air (yield, ca. 19 % after irradiation for 72 hrs.), along with 2 % yield of testosterone pinacol dipropionate (see below) (1939, 1940 Butenandt).

Testosterone pinacol 17.17'-dipropionate $C_{44}H_{66}O_6$ (III, R = $CO \cdot C_2H_5$). Needles (alcohol), m. 223–225° after previous sintering; $[\alpha]_D^{23} + 75°$ (chloroform). — **Fmn.** From testosterone propionate on treatment with sodium amalgam in 96% alcohol + acetic acid; a fraction m. 247–252°, $[\alpha]_D^{22} + 77°$ (chloroform), obtained from the mother liquor, is converted into the pinacol dipropionate on repeated crystallization from alcohol and inoculation with the crystalline dipropionate. For another formation, see under the formation of the preceding compound. — **Rns.** On repeated crystallization from alcohol, on treatment with some HCl in alcohol, or on irradiation of its solution in chloroform with sunlight, it gives a **compound** $C_{44}H_{62}O_4$ (probably IV, R = $CO \cdot C_2H_5$), m. 275–280° (begins to dec. at 230°), $[\alpha]_D^{22} - 272°$, absorption max. in chloroform at 313 mμ (ε 50000) (1940 Butenandt).

5-Keto-3∥5-A-nor-androstan-17β-ol-3-oic acid $C_{18}H_{28}O_4$. Prisms (acetone), m. 206.5–207° cor. Readily soluble in organic solvents except petroleum ether and benzine, insoluble in water. — **Fmn.** From testosterone acetate on ozonolysis in glacial acetic acid; yield 44%. — **Rns.** On treatment of its *oxime* $C_{18}H_{29}O_4N$ [needles (acetic acid or aqueous alcohol), m. 199–202° cor.] with sodium in boiling abs. alcohol it gives ca. 50% yield of **5-amino-3∥5-A-nor-androstan-17β-ol-3-oic acid lactam,** $C_{18}H_{29}O_2N$ [V, R = O; needles (acetone), m. 262–3° cor.; $[\alpha]_D^{18} + 33°$ (pyridine); insoluble in water and petroleum ether, readily soluble in other solvents] which on boiling with acetic anhydride and pyridine affords 31% yield of its *diacetyl derivative* $C_{22}H_{33}O_4N$ [needles (acetone), m. 164–7° cor.; readily soluble in organic solvents except petroleum ether and benzine]; on boiling the diacetyl derivative with sodium in amyl alcohol it gives 72% yield of **7.12-dimethyl-5'-hydroxy-7.8-cyclopentano-1-azaperhydrophenanthrene** $C_{18}H_{31}ON$ [V, R = H_2; platelets (methanol), m. 202–3° cor.; $[\alpha]_D^{18} + 0.28°$ (pyridine); soluble in organic solvents except petroleum ether, insoluble in water; *diacetyl derivative* $C_{22}H_{35}O_3N$, platelets (petroleum ether), m. 180.5–181.5° cor.] (1938 Bolt; 1936b N. V. Organon).

DERIVATIVES OF TESTOSTERONE

Keto-derivatives

Testosterone semicarbazone $C_{20}H_{31}O_2N_3$, m. 225° dec. (1941a Marker).

For the *oxime* and the *2.4-dinitrophenylhydrazone of testosterone dihydrogen phosphate dimethyl ester,* see p. 2588 s.

Testosterone acetate p-carboxyphenylhydrazone $C_{28}H_{36}O_4N_2$ (not analysed). The crude compound m. 260° dec. — **Fmn.** From testosterone acetate on refluxing with p-carboxyphenylhydrazine and some acetic acid in 95% alcohol for 2 hrs.; yield, 70%. — **Rns.** On refluxing with formaldehyde in 95% alcohol for 6 hrs., 20% is hydrolysed, further 60% on successive refluxing with pyruvic acid in 95% alcohol for 4 hrs.; 16% of the hydrazone may be recovered (1938 Butz).

For *cyclic ketals of testosterone,* see the ketals of Δ^5-androsten-17β-ol-3-one, pp. 2595 s, 2596 s.

Testosterone esters

Testosterone dihydrogen phosphate $C_{19}H_{29}O_5P$. Needles (aqueous methanol) with I H_2O which is not lost on heating in vac. over P_2O_5 at 100° for 3 days; m. 160° with frothing (1939 Westphal); dec. 155–157°; $[\alpha]_D^{20} + 71.9°$ (methanol) (1944a Müller). Very readily soluble in methanol, more sparingly in ethanol, very sparingly in acetone and in water (1939 Westphal). — Is only weakly androgenic (1939 Westphal). — **Fmn.** Along with other compounds, from testosterone on treatment with PCl_5 and $CaCO_3$ in chloroform with ice-cooling, followed by aqueous $NaHCO_3$ (1939 Westphal). From testosterone enol-ethyl ether ($\Delta^{3.5}$-androstadiene-3.17β-diol 3-ethyl ether, p. 1995 s) on keeping for 3 days with I mol. $POCl_3$ in dry pyridine, followed by treatment with 2 N HCl; crude yield, 94% (1944a Müller). Its dimethyl ester (below) is obtained from the dimethyl ester of Δ^5-androstene-3β.17β-diol 17-(dihydrogen phosphate) (p. 2003 s) on bromination in glacial acetic acid, followed by oxidation with CrO_3 at 25° (2 days) and debromination with zinc dust in acetic acid at 50° (15 min.); yield, 59% (1944a Müller).

Dimethyl ester $C_{21}H_{33}O_5P$. Needles (benzine or aqueous methanol), m. 152° to 153°; $[\alpha]_D^{20} + 78.3°$ (alcohol); shows the same absorption spectrum as testosterone. — Is weakly androgenic. — **Fmn.** See above. May also be obtained from the preceding compound with diazomethane in ether-acetone (almost quantitative yield). — **Rns.** Could not be hydrolysed with dil. alcoholic alkali. — *Oxime* $C_{21}H_{34}O_5NP$, dec. 185°. — *2.4-Dinitrophenylhydrazone* $C_{27}H_{37}O_8N_4P$, dark red needles, m. 228–230° dec. (1944a Müller).

Diethyl ester $C_{23}H_{37}O_5P$. Oil, b. 160° in a high vacuum. — **Fmn.** As above, using diazoethane (1944a Müller).

The following esters of testosterone and those described on p. 140 may be obtained from the 17-esters of Δ^5-androstene-3β.17β-diol (see p. 2581 s) or, generally, from testosterone on heating with the appropriate acid anhydride in anh. pyridine or on treatment with the appropriate acid chloride and pyridine (1935d, 1936f, h CIBA; 1937b N. V. ORGANON; 1937 Miescher). For androgenic activity of the esters, see 1937, 1938/39 Miescher.

Testosterone acetate $C_{21}H_{30}O_3$ (p. 140), $[\alpha]_D^{20} + 88°$ (alcohol) (1935 Butenandt), $+ 91°$ (chloroform) (1947b Djerassi).—*p-Carboxyphenylhydrazone,* see p. 2587 s.

Testosterone propionate $C_{22}H_{32}O_3$ (p. 140). Crystals (methanol), m. 121.3° to 122.6°; $[\alpha]_D + 97° \pm 2°$ (chloroform). — **Fmn.** From testosterone and propionic anhydride on heating in pyridine at 110–115° for 14 hrs. (1945 SCHERING A.-G.). — For its propylene ketal, see p. 2596 s.

Testosterone crotonate $C_{23}H_{32}O_3$. Leaflets or needles (hexane), m. 158–9° (1937 Miescher; 1935h CIBA).

Testosterone n-valerate $C_{24}H_{36}O_3$. Crystals, m. 109–111° cor. (1936b Ruzicka; 1935i, 1936h, i CIBA; 1937b N. V. ORGANON).

Testosterone chloroacetate $C_{21}H_{29}O_3Cl$. Needles (hexane), m. 123–4⁰. — **Fmn.** From testosterone and chloroacetic anhydride on heating at 110⁰ for $1\frac{1}{4}$ hrs. (1937 Miescher; 1935a CIBA).

Testosterone α-bromopropionate $C_{22}H_{31}O_3Br$. Crystals (ethyl acetate), m. 187⁰ to 188⁰. — **Fmn.** From testosterone and α-bromopropionyl bromide in benzene in the cold (1937 Miescher; 1935b CIBA).

Testosterone α-dimethylamino-propionate $C_{24}H_{37}O_3N$. Crystals (dil. methanol), m. 83–85⁰. — **Fmn.** From the preceding compound and dimethylamine in benzene-diisopropyl ether in a sealed tube at room temperature (12 hrs.), then at 50⁰ (3 hrs.) (1937 Miescher).

Testosterone succinate $C_{42}H_{58}O_6$. Crystals (chloroform-alcohol), m. 225⁰. — **Fmn.** From testosterone hydrogen succinate (p. 140) on treatment with thionyl chloride in benzene, followed by treatment of the resulting acid chloride with testosterone in dry pyridine (1940 SCHERING CORP.).

Testosterone glutarate $C_{43}H_{60}O_6$. Crystals (chloroform-alcohol), m. 252⁰. — **Fmn.** From testosterone and glutaric acid dichloride in dry pyridine at room temperature (1940 SCHERING CORP.).

Testosterone chlorocarbonate $C_{20}H_{27}O_3Cl$ (I, R = Cl). Prisms (diisopropyl ether), m. 139–140.5⁰. — **Fmn.** From testosterone and phosgene in chloroform with cooling in a freezing mixture (1937 Miescher). — **Rns.** See the following compounds.

Testosterone carbamate, Testosterone urethan $C_{20}H_{29}O_3N$ (I, R = NH₂). Needles (methanol) with $\frac{1}{2}$ H₂O, m. 110–120⁰ with evolution of gas; the resolidified melt remelts at 155–7⁰. — **Fmn.** From the above chlorocarbonate on treatment with gaseous ammonia in benzene (1937 Miescher).

Testosterone N-propylcarbamate $C_{23}H_{35}O_3N$ (I, R = NH·CH₂·C₂H₅). Crystals (methanol), m. 190–1⁰. — **Fmn.** From the above chlorocarbonate and propylamine (1937 Miescher).

Testosterone methyl carbonate $C_{21}H_{30}O_4$ (I, R = O·CH₃). Crystals (diisopropyl ether), m. 140.5–141.5⁰ (1937 Miescher), 141.5–142.5⁰ (1936b CIBA). — **Fmn.** From the above chlorocarbonate and methanol with pyridine in benzene at room temperature (1937 Miescher). From testosterone and methyl chloroformate with pyridine in benzene (1936b CIBA).

Testosterone ethyl carbonate $C_{22}H_{32}O_4$ (I, R = O·C₂H₅). Leaflets (alcohol), m. 141–2⁰. — **Fmn.** Similar to that of the preceding compound (1937 Miescher; 1936c CIBA).

Testosterone propyl carbonate $C_{23}H_{34}O_4$ (I, R = O·CH₂·C₂H₅). Crystals (hexane), m. 87–9⁰. — **Fmn.** From the chlorocarbonate and propanol with pyridine in benzene (1937 Miescher).

Testosterone butyl carbonate $C_{24}H_{36}O_4$ (I, R = O·CH₂·CH₂·C₂H₅). Oil. — **Fmn.** Similar to that of the preceding compound (1937 Miescher).

Testosterone phenyl carbonate $C_{26}H_{32}O_4$ (I, p. 2589 s; R = $O \cdot C_6H_5$). Leaflets (methanol), m. 144.5–145.5⁰. — **Fmn.** Similar to that of the methyl carbonate (1937 Miescher; 1936 d CIBA).

Testosterone benzyl carbonate $C_{27}H_{34}O_4$ (I, p. 2589 s; R = $O \cdot CH_2 \cdot C_6H_5$). Needles (alcohol), m. 156.5–157⁰. — **Fmn.** Similar to that of the methyl carbonate (1937 Miescher; 1936 e CIBA).

Testosterone β-diethylamino-ethyl carbonate $C_{26}H_{41}O_4N$ [I, p. 2589 s; R = $O \cdot CH_2 \cdot CH_2 \cdot N(C_2H_5)_2$]. — **Fmn.** From the chlorocarbonate (p. 2589 s) and β-diethylamino-ethanol in benzene. — *Hydrochloride* $C_{26}H_{42}O_4NCl$, needles with $\frac{1}{2}$ H₂O (diisopropyl ether-acetone), m. 178–180⁰ dec. (1937 Miescher).

Testosterone benzoate $C_{26}H_{32}O_3$ (p. 140). Crystals (hexane), m. 198–200⁰ (1935 h, 1936 h CIBA); m. 192.5–194⁰ cor., $[\alpha]_D^{20}$ + 155⁰ ± 3⁰ (chloroform) (1945 SCHERING A.-G.); $[\alpha]_D$ + 143⁰ (benzene) (1937 Ruzicka). — **Fmn.** From testosterone and benzoÿl chloride in pyridine (1935 Ruzicka; 1936 h CIBA). From the 17-benzoate of Δ^5-androstene-3β.17β-diol (p. 1999 s) on bromination in glacial acetic acid, followed by oxidation with CrO₃ and debromination by boiling with sodium iodide in abs. alcohol-benzene or by heating with zinc dust in glacial acetic acid on the water-bath (1935 Ruzicka; 1935 h, 1936 k CIBA); from the same ester on refluxing with cyclohexanone and aluminium tert.-butoxide in toluene for 3 hrs. (yield, 90%) (1937 b N. V. ORGANON).

Testosterone hexahydrobenzoate $C_{26}H_{38}O_3$. Plates (methanol), m. 127⁰ (1943 Inhoffen); $[\alpha]_D^{23}$ + 82.5⁰ (chloroform); absorption max. in alcohol at 242 mμ (log ε 4.12) (1947 a Djerassi). — **Fmn.** From testosterone and hexahydro-benzoyl chloride in pyridine at room temperature (1943 Inhoffen). From the hexahydrobenzoate of 2α-bromoandrostan-17β-ol-3-one (p. 2703 s) on boiling with collidine; yield, 18% (1947 a Djerassi).

Testosterone phenylacetate $C_{27}H_{34}O_3$. Needles (hexane), m. 129–131⁰ cor. — **Fmn.** From testosterone on treatment with phenylacetyl chloride and pyridine in benzene (1937 Miescher; 1935 f CIBA).

TESTOSTERONE GLYCOSIDES

Testosterone β-D-glucoside $C_{25}H_{38}O_7$. Crystals (methanol), m. 213–4⁰ (1939 Rabald). Solubility in water at 20⁰ 2.5⁰/₀₀, at 95⁰ 8⁰/₀₀ (1944 b Meystre). — **Fmn.** From its tetraacetate (below) on treatment with barium methoxide in methanol at 0⁰; almost quantitative yield (1939 Rabald). — **Rns.** With POCl₃ in pyridine, followed by addition of water, it gives a testosterone glucosidophosphoric acid ($[\alpha]_D^{22}$ + 24⁰) forming water-soluble salts with alkali and alkaline earth metals (1939 RARE CHEMICALS, INC.).

Tetraacetate $C_{33}H_{46}O_{11}$. Dimorphic: crystals (ether), m. 125–8⁰, resolidi-fying as needles, and remelting at 163⁰ (1944 a Meystre); needles (methanol), m. 165–6⁰; $[\alpha]_D^{18}$ + 49.5⁰ (chloroform) (1939 Rabald). — **Fmn.** From testosterone and acetobromoglucose on shaking with Ag₂O in abs. ether for 48 hrs. (yield, 19%) (1939 Rabald) or with Ag₂CO₃ in benzene (yield, 43%) (1944 a Meystre).

References, pp. 2591–2595 s

Testosterone β-maltoside $C_{31}H_{48}O_{12}$. Crystals (methanol-acetone), m. 250^0 to 255^0 cor. dec.; $[α]_D^{19}$ +73 (methanol). Unlimited solubility in water at 20^0 and 95^0. — **Fmn.** From testosterone, acetobromomaltose, and Ag_2CO_3 in boiling benzene, followed by hydrolysis of the resulting heptaacetate with barium methoxide in methanol in the cold; 39% of the testosterone has reacted giving 87% crude yield. — **Rns.** It is hydrolysed by 2 N HCl on the water-bath ($^1/_2$ hr.) to give testosterone (1944b Meystre).

Heptaacetate $C_{45}H_{62}O_{19}$. Needles (strong alcohol), m. $175-180^0$ cor.; $[α]_D^{19}$ +74°±4° (methanol). Readily soluble in ether (1944b Meystre). — **Fmn.** and **Rns.** See above.

Testosterone β-D-triacetylglucuronide, 3-Keto-$Δ^4$-androsten-17β-yl-β-D-triacetylglucuronic acid $C_{31}H_{42}O_{11}$. — *Methyl ester* $C_{32}H_{44}O_{11}$. Needles (96% alcohol) with 1 EtOH which is lost at 78^0; m. 186–9°; $[α]_D^{20}$ +28.3° (chloroform). Is inactive in the capon comb test. **Fmn.** From testosterone and methyl acetobromo-D-glucuronate on shaking with Ag_2CO_3 in benzene for 24 hrs.; yield, ca. 10% (1939 Schapiro).

1935 Butenandt, Hanisch, *Z. physiol. Ch.* **237** 89, 92.
a CIBA*, *Swiss Pat.* 199982 (issued 1938); *C. A.* **1939** 3533; *Ch. Ztbl.* **1939** I 3932.
b CIBA*, *Swiss Pat.* 199983 (issued 1938); *C. A.* **1939** 3533; *Ch. Ztbl.* **1939** I 3932.
c CIBA*, *Swiss Pat.* 203257 (issued 1939); *C. A.* **1939** 9554; *Ch. Ztbl.* **1940** I 603.
d CIBA*, *Swiss Pat.* 208889 (issued 1940); *C. A.* **1941** 3773; *Ch. Ztbl.* **1941** I 2284.
e CIBA*, *Swiss Pat.* 208890 (issued 1940); *C. A.* **1941** 3773; *Ch. Ztbl.* **1941** I 2284.
f CIBA*, *Swiss Pat.* 208892 (issued 1940); *C. A.* **1941** 3773; *Ch. Ztbl.* **1941** I 2284.
g CIBA*, *Swiss Pat.* 208894 (issued 1940); *C. A.* **1941** 3773; *Ch. Ztbl.* **1941** I 2284.
h CIBA*, *Swiss Pats.* 210762–210767 (issued 1940); *C. A.* **1941** 5650, 5651; *Ch. Ztbl.* **1941** I 3409.
i CIBA*, *Swiss Pat.* 223159 (issued 1942); *C. A.* **1949** 3572; *Ch. Ztbl.* **1943** I 2221; *Dutch Pat.* 44043 (issued 1938); *Ch. Ztbl.* **1939** I 1806; *Austrian Pat.* 160524 (issued 1941); *Ch. Ztbl.* **1942** I 778; *Fr. Pat.* 815896 (issued 1937); *C. A.* **1938** 1713; *Swiss Pat.* 229140 (issued 1944); *C. A.* **1949** 4432; *Ch. Ztbl.* **1944** II 1204.
Ruzicka, Wettstein, Kägi, *Helv.* **18** 1478, 1482.
SCHERING-KAHLBAUM A.-G., *Brit. Pat.* 464396 (issued 1937); *C. A.* **1937** 6418; *Ch. Ztbl.* **1937** II 3041.
1936 a CIBA*, *Swiss Pat.* 198133 (issued 1938); *C. A.* **1939** 3531; *Ch. Ztbl.* **1939** I 2826, 2827.
b CIBA*, *Swiss Pat.* 204748 (issued 1939); *C. A.* **1941** 3037; *Ch. Ztbl.* **1940** I 603.
c CIBA*, *Swiss Pat.* 204749 (issued 1939); *C. A.* **1941** 3037; *Ch. Ztbl.* **1940** I 603.
d CIBA*, *Swiss Pat.* 204750 (issued 1939); *C. A.* **1941** 3039; *Ch. Ztbl.* **1940** I 603.
e CIBA*, *Swiss Pat.* 204751 (issued 1939); *C. A.* **1941** 2282; *Ch. Ztbl.* **1940** I 603.
f CIBA*, *Swiss Pat.* 206119 (issued 1939); *C. A.* **1941** 3040; *Ch. Ztbl.* **1940** I 3958.
g CIBA*, *Swiss Pats.* 214329, 214332, 215548–215551, 215554–215557, 216106, 217936, 217937 (issued 1941, 1942); *C. A.* **1942** 5960; **1948** 3784, 5057, 7346; *Ch. Ztbl.* **1942** I 2296; **1943** I 1190.
h CIBA*, *Fr. Pat.* 811107 (issued 1937); *C. A.* **1938** 588; *Ch. Ztbl.* **1937** II 1619.
i CIBA*, (Ruzicka, Wettstein), *U.S. Pat.* 2143453 (issued 1939); *C. A.* **1939** 3078; *Ch. Ztbl.* **1939** I 5009, 5010; CIBA PHARMACEUTICAL PRODUCTS (Ruzicka, Wettstein), *U.S. Pat.* 2387469 (1937, issued 1945); *C. A.* **1946** 5536; *Ch. Ztbl.* **1948** I 1234.
k CIBA*, *Austrian Pat.* 160291 (issued 1941); *Ch. Ztbl.* **1941** II 2351.
CIBA PHARMACEUTICAL PRODUCTS (Miescher, Wettstein), *U.S. Pat.* 2260328 (issued 1941); *C. A.* **1942** 873; *Ch. Ztbl.* **1944** II 246.
a N. V. ORGANON, *Dutch Pat.* 41871 (issued 1937); *Ch. Ztbl.* **1938** I 2217.

* Gesellschaft für chemische Industrie in Basel; Société pour l'industrie chimique à Bâle.

b N. V. ORGANON, *Dutch Pat.* 47226 (issued 1939); *Ch. Ztbl.* **1940** I 2829.

a Ruzicka, Kägi, *Helv.* **19** 842, 844, 848.

b Ruzicka, Wettstein, *Helv.* **19** 1141, 1144.

SCHERING A.-G., *Ger. Pat.* 699248 (issued 1940); *C. A.* **1941** 6742; *Ch. Ztbl.* **1941** I 1325.

SCHERING-KAHLBAUM A.-G., *Swiss Pat.* 202972 (issued 1939); *Ch. Ztbl.* **1940** I 429, 430.

1936/37 CIBA*, *Brit. Pat.* 477400 (issued 1938); *C. A.* **1938** 3914; *Ch. Ztbl.* **1938** II 119, 120.

1937 a CIBA*, *Swiss Pat.* 202846 (issued 1939); *C. A.* **1939** 9554; *Ch. Ztbl.* **1940** I 93.

     b CIBA*, *Swiss Pat.* 207498 (issued 1940); *C. A.* **1941** 3038; *Ch. Ztbl.* **1940** II 1328; CIBA PHARMACEUTICAL PRODUCTS (Ruzicka), *U.S. Pat.* 2281622 (1938, issued 1942); *C. A.* **1942** 5958.

     c CIBA*, *Swiss Pat.* 208748 (issued 1940); *C. A.* **1941** 3773; *Fr. Pat.* 847134 (1938, issued 1939); *Ch. Ztbl.* **1940** I 2828; CIBA PHARMACEUTICAL PRODUCTS (Miescher, Fischer), *U.S. Pat.* 2253798 (1938, issued 1941); *C. A.* **1941** 8218.

     d CIBA*, *Swiss Pat.* 213178 (issued 1941); *C. A.* **1942** 4675; *Brit. Pat.* 520393 (issued 1940); *C. A.* **1942** 628; *Ch. Ztbl.* **1940** II 3226.

     e CIBA*, *Swiss Pat.* 222441 (issued 1942); *C. A.* **1949** 824; *Ch. Ztbl.* **1943** I 2113; cf. CIBA* (Ruzicka), *Swed. Pat.* 98368 (1938, issued 1940); *Ch. Ztbl.* **1940** II 2185.

     f CIBA*, *Brit. Pat.* 497394 (issued 1939); *C. A.* **1939** 3812; *Ch. Ztbl.* **1939** I 2828; CIBA PHARMACEUTICAL PRODUCTS (Miescher, Wettstein), *U.S. Pat.* 2229813 (1938, issued 1941); *C. A.* **1941** 3040; *Ch. Ztbl.* **1941** II 2467, 2468.

     g CIBA*, *Fr. Pat.* 830484 (issued 1938); *C. A.* **1939** 2542; *Ch. Ztbl.* **1939** I 2826, 2827.

CIBA PHARMACEUTICAL PRODUCTS (Laqueur, David, Dingemanse, Freud), *U.S. Pat.* 2175963 (issued 1939); *C. A.* **1940** 857; *Ch. Ztbl.* **1940** I 2984, 2985.

a I. G. FARBENIND. (Bockmühl, Ehrhart, Ruschig, Aumüller), *Ger. Pat.* 693351 (issued 1940); *C. A.* **1941** 4921; *Swiss Pat.* 216342 (1938, issued 1941); *Ch. Ztbl.* **1942** II 197; *Fr. Pat.* 842026 (1938, issued 1939); *C. A.* **1940** 4743; *Ch. Ztbl.* **1940** I 429.

b I. G. FARBENIND., *Swiss Pat.* 218515 (issued 1942); *Ch. Ztbl.* **1943** I 1389, 1390.

c I. G. FARBENIND., *Fr. Pat.* 819975 (issued 1937); *C. A.* **1938** 2956; *Ch. Ztbl.* **1938** I 2403, 2404; *Swiss Pat.* 207799 (issued 1940).

Kuwada, Joyama, *C. A.* **1938** 1709; *Ch. Ztbl.* **1938** II 1612.

Mamoli, Vercellone, *Ber.* **70** 2079, 2082.

Miescher, Kägi, Scholz, Wettstein, Tschopp, *Biochem. Z.* **294** 39, 41–45, 52–56; cf. also CIBA*, *Fr. Pat.* 830858 (1937, issued 1938); *C. A.* **1939** 1761; *Ch. Ztbl.* **1939** I 2458; *Brit. Pat.* 491638 (1937, issued 1938); *C. A.* **1939** 1106; *Ch. Ztbl.* **1939** I 2643.

a N. V. ORGANON, *Dutch Pats.* 46498 (issued 1939), 47831; *C. A.* **1939** 8923; **1940** 6414; *Ch. Ztbl.* **1939** II 4282; *Fr. Pat.* 842940 (1938, issued 1939); *C. A.* **1940** 6414; *Ch. Ztbl.* **1940** I 429; *Swiss Pats.* 219009, 223235, 223236 (1938, issued 1942), 225731, 225732 (1938, issued 1943); *Ch. Ztbl.* **1943** II 749; **1944** I 37.

b N. V. ORGANON, *Swiss Pats.* 203803 (issued 1939), 209272–209274, 209276–209278 (issued 1940); *Ch. Ztbl.* **1943** I 654.

c N. V. ORGANON, *Fr. Pat.* 823618 (issued 1938); *C. A.* **1938** 5854; *Ch. Ztbl.* **1938** II 355; *Fr. Pat.* 827623 (issued 1938); *C. A.* **1938** 8437; *Ch. Ztbl.* **1938** II 3119; *Brit. Pat.* 488966 (issued 1938); *C. A.* **1939** 324.

Oppenauer, *Acta Brevia Neerland.* **7** 176; *Rec. trav. chim.* **56** 137, 141; *U.S. Pat.* 2384335 (issued 1945); *C. A.* **1946** 179; *Ch. Ztbl.* **1948** I 1039.

Ruzicka, Goldberg, Bosshard, *Helv.* **20** 541, 546 footnote.

SCHERING CORP. (Serini, Köster, Strassberger), *U.S. Pat.* 2379832 (issued 1945); *C.A.* **1945** 5053.

a SCHERING-KAHLBAUM A.-G., *Swiss Pats.* 209119, 209120 (issued 1940); *Ch. Ztbl.* **1943** I 654; *Brit. Pat.* 492725 (issued 1938); *C. A.* **1939** 1884; *Ch. Ztbl.* **1939** I 2642.

b SCHERING-KAHLBAUM A.-G., *Fr. Pat.* 822551 (issued 1938); *C. A.* **1938** 4174; *Ch. Ztbl.* **1938** II 120.

Steinach, Kun, *Lancet* **233** 845.

1938 Bolt, *Rec. trav. chim.* **57** 905, 908–910, 1410.

     Bugyi, *Acta Lit. Sci. Univ. Hung. Sect. Chem. Mineral. Phys.* **6** 27, 33; *C. A.* **1938** 2191; *Ch. Ztbl.* **1938** I 1144.

---

* Gesellschaft für chemische Industrie in Basel; Société pour l'industrie chimique à Bâle.

a Butenandt, Heusner, *Ber.* **71** 198, 202.
b Butenandt, Wolz, *Ber.* **71** 1483, 1486.
c Butenandt, Peters, *Ber.* **71** 2688, 2690.
Butz, Hall, *J. Biol. Ch.* **126** 265, 268.
Callow, Callow, Emmens, *Biochem. J.* **32** 1312.
Ciba* (Miescher, Fischer), *Ger. Pat.* 712857 (issued 1941); *C. A.* **1943** 4534; *Ch. Ztbl.* **1942** I 1282.
Ercoli, *Ber.* **71** 650.
Goldberg, *Ergeb. Vitamin- u. Hormonforsch.* **1** 371, 388.
a Mamoli, Schramm, *Ber.* **71** 2083.
b Mamoli, *Ber.* **71** 2278.
c Mamoli, Schramm, *Ber.* **71** 2698.
d Mamoli, Vercellone, *Gazz.* **68** 311.
Roche-Organon (Oppenauer), *U.S. Pat.* 2264888 (issued 1941); *C. A.* **1942** 1740.
Ruzicka, Furter, Goldberg, *Helv.* **21** 498, 504–506.
a Schering A.-G., *Fr. Pat.* 850115 (issued 1939); *C. A.* **1942** 1740; *Ch. Ztbl.* **1940** II 931.
b Schering A.-G., *Brit. Pat.* 515411 (issued 1940); *Ch. Ztbl.* **1940** II 1180.
Schering Corp. (Mamoli), *U.S. Pat.* 2186906 (issued 1940); *C. A.* **1940** 3436; *Ch. Ztbl.* **1940** I 3426.
Schramm, Mamoli, *Ber.* **71** 1322.
Schwenk, Fleischer, Whitman, *J. Am. Ch. Soc.* **60** 1702.
Serini, Köster, *Ber.* **71** 1766, 1770.

1938/39 Miescher, Fischer, Tschopp, *Biochem. Z.* **300** 14, 15, 18 et seq.

1939 Butenandt, Wolff, *Ber.* **72** 1121, 1123.
a Callow, Callow, Emmens, *Chemistry & Industry* **1939** 147.
b Callow, *Biochem. J.* **33** 559.
Ciba*, *Swiss Pat.* 241646 (issued 1946); *C. A.* **1949** 7978; Ciba Pharmaceutical Products (Miescher, Wettstein), *U.S. Pat.* 2374369 (1940, issued 1945); *C. A.* **1945** 5412.
Cook, Hamilton, Dorfman, *Chemistry & Industry* **1939** 148.
Dannenberg, *Abhandl. Preuss. Akad. Wiss. Math.-naturwiss. Kl.* **1939** No. 21, pp. 5, 10, 53.
a Dorfman, Hamilton, *Endocrinology* **25** 28.
b Dorfman, Cook, Hamilton, *J. Biol. Ch.* **130** 285.
Eisenbrand, Picher, *Z. physiol. Ch.* **260** 83, 89 et seq.
Kägi, Miescher, *Helv.* **22** 683, 685, 687.
Langstroth, Talbot, *J. Biol. Ch.* **128** 759; cf. also Langstroth, *C. A.* **1941** 3686.
Mamoli, Koch, Teschen, *Z. physiol. Ch.* **261** 287, 295, 296.
a Miescher, Fischer, *Helv.* **22** 158, 160.
b Miescher, Wettstein, Scholz, *Helv.* **22** 894, 903, 904.
c Miescher, Klarer, *Helv.* **22** 962, 965 footnote 1.
Rabald, Dietrich, *Z. physiol. Ch.* **259** 251.
Rare Chemicals, Inc. (Johannessohn, Rabald, Hagedorn), *U.S. Pat.* 2263990 (issued 1941); *C. A.* **1942** 1741.
Schapiro, *Biochem. J.* **33** 385.
Schering A.-G. (Köster, Inhoffen), *Swed. Pat.* 109502 (issued 1944); *Ch. Ztbl.* **1944** II 247; *Dutch Pat.* 52656 (issued 1942); *Ch. Ztbl.* **1942** II 2615; *Norw. Pat.* 64579 (1940, issued 1942); *Ch. Ztbl.* **1942** II 1822; Schering Corp. (Köster, Inhoffen), *U.S. Pat.* 2302636 (issued 1942); *C. A.* **1943** 2388.
Scherrer, *Helv.* **22** 1329, 1335, 1336.
Westphal, Wang, Hellmann, *Ber.* **72** 1233, 1234, 1239, 1240.
a Woker, Antener, *Helv.* **22** 511, 517.
b Woker, Antener, *Helv.* **22** 1309, 1319.

1940 Butenandt, Poschmann, *Ber.* **73** 893, 897.
Chinoin gyógyszer és vegyészeti termékek gyára r. t. (Földi), *Hung. Pat.* 135687 (issued 1949); *C. A.* **1950** 4047, 4048.
a Dorfman, *Proc. Soc. Exptl. Biol. Med.* **45** 739.
b Dorfman, Hamilton, *J. Biol. Ch.* **133** 753.

* Gesellschaft für chemische Industrie in Basel; Société pour l'industrie chimique à Bâle.

c Dorfman, Fish, *J. Biol. Ch.* **135** 349.

Elvidge, *Quart. J. Pharm. Pharmacol.* **13** 219, 220.

McCullagh, Schneider, Emery, *Endocrinology* **27** 71.

a Marker, Wittle, Tullar, *J. Am. Ch. Soc.* **62** 223, 225.

b Marker, Rohrmann, *J. Am. Ch. Soc.* **62** 900, 902.

c Marker, *J. Am. Ch. Soc.* **62** 2543, 2547.

PARKE, DAVIS & Co. (Marker, Wittle), *U.S. Pat.* 2397424 (issued 1946); *C. A.* **1946** 3571.

SCHERING CORP. (Inhoffen), *U.S. Pat.* 2384550 (issued 1945); *C. A.* **1946** 177; *Ch. Ztbl.* **1948** I 1039.

Tastaldi, Lemmi, Lacaz de Moraes, *C. A.* **1943** 6292.

Wettstein, *Helv.* **23** 388, 397.

1941 Dorfman, *Proc. Soc. Exptl. Biol. Med.* **46** 351.

Fieser, Wolfe, *J. Am. Ch. Soc.* **63** 1485.

Galinovsky, *Ber.* **74** 1624, 1626.

Koenig, Melzer, Szego, Samuels, *J. Biol. Ch.* **141** 487.

a Marker, Crooks, Wittbecker, *J. Am. Ch. Soc.* **63** 777.

b Marker, *J. Am. Ch. Soc.* **63** 1485.

SQUIBB & SONS (Fernholz), *U.S. Pat.* 2356154 (issued 1944); *C. A.* **1945** 1515; *Ch. Ztbl.* **1945** II 1385.

1942 Jadassohn, Fierz, Pfanner, *Helv.* **25** 2, 8.

Kochakian, Clark, *J. Biol. Ch.* **143** 795.

SCHERING A.-G., *Fr. Pat.* 884085 (issued 1943); *Ch. Ztbl.* **1944** I 450.

Wettstein, Miescher, *Helv.* **25** 718, 726.

1943 Hansen, Cantarow, Rakoff, Paschkis, *Endocrinology* **33** 282.

Inhoffen, Zühlsdorff, *Ber.* **76** 233, 242, 243.

Pincus, *Endocrinology* **32** 176.

Ruzicka, Prelog, *Helv.* **26** 975, 977.

SCHERING A.-G., *Fr. Pat.* 899737 (issued 1945); *Ch. Ztbl.* **1946** I 1432, 1433.

Wettstein, Miescher, *Helv.* **26** 631, 639.

1944 Bersin, Meyer, *Angew. Ch.* **57** 117, 118.

Butenandt, Poschmann, *Ber.* **77** 394, 397.

Buu-Hoi, Cagniant, *Ber.* **77** 118, 120.

Cantarow, Paschkis, Rakoff, Hansen, *Endocrinology* **35** 129.

Fish, Dorfman, *Endocrinology* **35** 22.

Frame, Fleischmann, Wilkins, *Bull. Johns Hopkins Hosp.* **75** 95.

Horwitt, Dorfman, van Wagenen, *Endocrinology* **34** 351.

a Kochakian, *J. Biol. Ch.* **155** 579, 582, 584.

b Kochakian, Vail, *J. Biol. Ch.* **156** 779.

a Meystre, Miescher, *Helv.* **27** 231, 234.

b Meystre, Miescher, *Helv.* **27** 1153, 1156–1158.

a Müller, Langerbeck, Riedel, *Z. physiol. Ch.* **281** 29.

b Müller, Langerbeck, Neuhoff, *Ber.* **77** 141, 150.

Zimmermann, *Vitamine und Hormone* **5** [1952] 1, 124, 152, 165, 170, 237, 260, 276.

1945 C. F. BOEHRINGER & SÖHNE, *Bios Final Report* No. 766, pp. 138, 139; *Fiat Final Report* No. 71, p. 65.

Kumler, Fohlen, *J. Am. Ch. Soc.* **67** 437; **70** 4273 (1948).

Prelog, Ruzicka, Meister, Wieland, *Helv.* **28** 618, 622.

SCHERING A.-G., *Bios Final Report* No. 449, pp. 264–270, 272; *Fiat Final Report* No. 996, pp. 17, 53–59, 77.

Schiller, Dorfman, Miller, *Endocrinology* **36** 355.

1946 Bersin, *Naturwiss.* **33** 108, 110.

Furchgott, Rosenkrantz, Shorr, *J. Biol. Ch.* **163** 375, 381.

Heard, Sobel, *J. Biol. Ch.* **165** 687, 690.

Meystre, Miescher, *Experientia* **2** 408.

Miescher, *Helv.* **29** 743, 751.

Tagmann, Prelog, Ruzicka, *Helv.* **29** 440.

Turfitt, *Biochem. J.* **40** 79.

1947 a Djerassi, Scholz, *J. Am. Ch. Soc.* **69** 2404, 2408. — b Djerassi, *J. Org. Ch.* **12** 823, 826.
1949 Fieser, Fieser, *Natural Products Related to Phenanthrene*, 3rd Ed., New York, p. 325–327.
1950 Djerassi, Rosenkranz, Romo, Kaufmann, Pataki, *J. Am. Ch. Soc.* **72** 4534, 4538.
     Rosenkranz, Djerassi, Kaufmann, Pataki, Romo, *Nature* **165** 814.
   · Shoppee, Lewis, Elks, *Chemistry & Industry* **1950** 454.
1952 Butenandt, Karlson-Poschmann, Failer, Schiedt, Biekert, *Ann.* **575** 123.
1955 Johnson, Bannister, Pappo, Pike, *J. Am. Ch. Soc.* **77** 817.
1956 Johnson, Bannister, Pappo, Pike, *J. Am. Ch. Soc.* **78** 6354.

Δ^5-Androsten-17β-ol-3-one, "*Δ^5-Testosterone*" $C_{19}H_{28}O_2$ (p. 141: *Δ^5-Androsten-17-ol-3-one*). For β-configuration at C-17, see p. 1363 s.

— **Fmn.** In the preparation from the 17-acetate of Δ^5-androstene-3β.17β-diol (p. 1999 s) on bromination, followed by oxidation with CrO_3 and then by debromination with zinc dust in boiling methanol (see page 141), the acetate is obtained (1936 Butenandt); the propionate and benzoate are similarly formed (1937 Ruzicka; 1935 CIBA); the debromination (e.g., in the case of the benzoate) is better effected in boiling abs. alcohol (15 min.) (1940 Wettstein).

Rns. On refluxing the benzoate with benzoquinone and aluminium tert.-butoxide in toluene it gives the benzoate of $\Delta^{4·6}$-androstadien-17β-ol-3-one (6-dehydrotestosterone, p. 2577 s) (1940 Wettstein; 1941 CIBA). The propionate gives the 17-propionate of Δ^5-androstene-3α.17β-diol (p. 1997 s) on hydrogenation in 95 % alcohol in the presence of Raney nickel at room temperature, until 1 mol. of hydrogen has been absorbed, followed by treatment with digitonin to remove the 3β-isomer (not isolated) (1937 Ruzicka; 1936/37 CIBA). It is isomerized to testosterone (p. 2580 s) by warming gently with aqueous methanolic HCl; the acetate reacts similarly to give testosterone acetate (1937 SCHERING-KAHLBÁUM A.-G.).

Propionate $C_{22}H_{32}O_3$ (not obtained pure). Crystals (petroleum ether b. 40° to 70°), m. ca. 135°; $[\alpha]_D$ —17° (alcohol) (1937 Ruzicka). — **Fmn.** and **Rns.** See above.

Benzoate $C_{26}H_{32}O_3$ (p. 141). Prisms (acetone), m. 178–181° cor. after sintering (1940 Wettstein). — **Fmn.** and **Rns.** See above.

Ethylene ketal, "*Testosterone ethylene ketal*" $C_{21}H_{32}O_3$. For the structure, see 1941 Fernholz; 1941 SQUIBB & SONS; cf. also, e.g., 1952 Antonucci; 1953 Poos; 1953 Herzog; 1954 Dauben.— Needles (methanol), m. 183°; shows no absorption band in the ultra-violet (1939 SCHERING A.-G.; 1941 SQUIBB & SONS). — Has the same androgenic activity as testosterone, but is effective for a longer time (1939 SCHERING A.-G.). — **Fmn.** From testosterone (page 2580 s) and 1 mol. ethylene glycol in benzene in the presence of p-toluene-sulphonic acid on distillation (1939 SCHERING A.-G.). From Δ^4-androstene-3.17-dione and ethylene glycol, similarly to above, followed by reduction of the resulting 3-ethylene ketal (see p. 2890 s) with sodium in boiling abs.

alcohol (1941 SQUIBB & SONS). — **Rns.** Gives testosterone on refluxing with aqueous alcoholic HCl (1941 SQUIBB & SONS).

Propylene ketal, "*Testosterone propylene ketal*" $C_{22}H_{34}O_3$. For the structure, see the literature cited above.

Propionate $C_{25}H_{38}O_4$. Crystals (cyclohexane), m. ca. 210°; $[\alpha]_D^{20}$ —41° (dioxane). — Has the same androgenic activity as testosterone propionate, but is effective for a longer time. — **Fmn.** From testosterone propionate and 1 mol. propylene glycol in benzene in the presence of p-toluenesulphonic acid on distillation (1939 SCHERING A.-G.).

1935 CIBA*, *Swiss Pats.* 210768, 210769 (issued 1940); *C. A.* **1941** 5651; *Ch. Ztbl.* **1941** I 3409.
1936 Butenandt, Hanisch, *Ber.* **69** 2773, 2775.
1936/37 CIBA*, *Brit. Pat.* 483926 (issued 1938); *C. A.* **1938** 7218; *Ch. Ztbl.* **1938** II 3273; *Fr. Pat.* 828338 (1937, issued 1938); *C. A.* **1939** 177; *Ch. Ztbl.* **1938** II 3273; (Ruzicka), *U.S. Pat.* 2168685 (1937, issued 1939); *C. A.* **1939** 9324; *Ch. Ztbl.* **1940** I 1391.
1937 Ruzicka, Goldberg, Bosshard, *Helv.* **20** 541, 543, 544, 546.
SCHERING-KAHLBAUM A.-G., *Swiss Pats.* 209119, 209120 (issued 1940); *Ch. Ztbl.* **1943** I 654; *Brit. Pat.* 492725 (issued 1938); *Ch. Ztbl.* **1939** I 2642.
1939 SCHERING A.-G. (Köster, Inhoffen), *Swed. Pat.* 109502 (issued 1944); *Ch. Ztbl.* **1944** II 247; *Dutch Pat.* 52656 (issued 1942); *Ch. Ztbl.* **1942** II 2615; *Norw. Pat.* 64579 (1940, issued 1942); *Ch. Ztbl.* **1942** II 1822; SCHERING CORP. (Köster, Inhoffen), *U.S. Pat.* 2302636 (issued 1942); *C. A.* **1943** 2388.
1940 Wettstein, *Helv.* **23** 388, 397, 398.
1941 CIBA*, *Fr. Pat.* 877821 (issued 1943); *Dutch Pat.* 54147 (issued 1943); *Ch. Ztbl.* **1943** II 147, 148.
Fernholz, Stavely, *Abstracts of the 102nd meeting of the American Chemical Society, Atlantic City*, p. M. 39.
SQUIBB & SONS (Fernholz), *U.S. Pat.* 2356154 (issued 1944); *C. A.* **1945** 1515; *Ch. Ztbl.* **1945** II 1385.
1952 Antonucci, Bernstein, Littell, Sax, Williams, *J. Org. Ch.* **17** 1341, 1349.
1953 Herzog, Jevnik, Tully, Hershberg, *J. Am. Ch. Soc.* **75** 4425.
Poos, Arth, Beyler, Sarett, *J. Am. Ch. Soc.* **75** 422, 423 footnote 5.
1954 Dauben, Löken, Ringold, *J. Am. Ch. Soc.* **76** 1359.

Androstan-17α-ol-3-one, epi-Dihydrotestosterone, "*cis-Dihydrotestosterone*" $C_{19}H_{30}O_2$. Needles (methanol), m. 179.5–180° cor. (1937 Ruzicka). — Its androgenic activity (300 γ) is 15 times smaller than that of dihydrotestosterone; the activity of the benzoate is very small (1937 Ruzicka). Causes ovipositor growth in the female bitterling (1941, 1941/42 Duyvené de Wit).

Fmn. Its hexahydrobenzoate is obtained from the 17-hexahydrobenzoate of androstane-3β.17α-diol (p. 2013 s) on oxidation with CrO_3 in glacial acetic acid at room temperature; it is hydrolysed by boiling for some hrs. with excess 1 N methanolic NaOH (1937 Ruzicka); other esters are similarly formed (1935 b CIBA) and may also be obtained from the corresp. 17-esters of androstane-3α.17α-diol (p. 2010 s) (1935 b CIBA).

* Gesellschaft für chemische Industrie in Basel; Société pour l'industrie chimique à Bâle.

Rns. On treating the hexahydrobenzoate with perbenzoic acid in chloroform at -10^0 for 1 week it gives the **17-hexahydrobenzoate of 3∥4-androstane-4.17α-diol-3-oic acid lactone** $C_{26}H_{40}O_4$ (I) [needles (AcOEt-petrol. ether), m. 195.5–196^0 cor.; $[\alpha]_D^{16}$ $-36^0\pm3^0$ (chloroform)] which on distillation in a current of nitrogen through a tube heated at 310^0 is converted into 3∥4-$\varDelta^{16}$-androsten-4-ol-3-oic acid lactone (p. 2397 s) (1945 Prelog). Gives androstane-3α.17α-diol (p. 2010 s) and probably androstane-3β.17α-diol (p. 2013 s) on hydrogenation in glacial acetic acid at 65^0 in the presence of platinum dioxide and hydrobromic acid, followed by boiling with methanolic NaOH (1937 Ruzicka). On distillation of the hexahydrobenzoate in a stream of nitrogen through a tube heated at 300^0 it gives $\varDelta^{16}$-androsten-3-one (p. 2396 s) (1944 Prelog).

Colour reactions (cf. those of dihydrotestosterone, p. 2598 s). In the Liebermann-Burchard reaction it gives at once a red-yellow colour which becomes dirty red-brown and, after 1 hr., brown-yellow. Gives a violet coloration on brief boiling with glacial acetic acid and conc. H_2SO_4, followed by addition of bromine or acetic anhydride to the cooled solution; a violet coloration is also produced on brief boiling with $POCl_3$ and quinoline, followed by addition of glacial acetic acid, conc. H_2SO_4, and bromine (or acetic anhydride) to the cooled solution (1939 Kägi). For colour reactions with conc. H_2SO_4 in alcohol in the absence or the presence of benzaldehyde, see 1939 Scherrer.

Acetate $C_{21}H_{32}O_3$. Crystals (hexane or benzine), m. 150–1^0 cor. — **Fmn.** From the above carbinol on treatment with acetic anhydride in pyridine at room temperature for 48 hrs. (1937 Ruzicka). — **Rns.** Rate of hydrolysis with alcoholic KOH at 16–19^0 and 81^0; ratio of hydrolysed esters of dihydrotestosterone and epi-dihydrotestosterone 1:3.15 and 1:1.4, resp. (1938 Ruzicka).

Hexahydrobenzoate $C_{26}H_{40}O_3$. Needles (alcohol), m. 137.5–138^0 cor. (1937 Ruzicka). — **Fmn.** and **Rns.** See above.

Androstan-17β-ol-3-one, Dihydrotestosterone, *"trans-Dihydrotestosterone"*

$C_{19}H_{30}O_2$ (p. 141: *17-Hydroxyætioallocholan-3-one*). Crystals (ethyl acetate), m. 181.5–182.5^0 cor. (1937 Ruzicka); $[\alpha]_D^{23}$ $+30^0$ (chloroform) (1947 b Djerassi).

Fmn. For formation of its esters (hydrogen succinate, hexahydrobenzoate, and others) from the corresp. 17-esters of androstane-3β.17β-diol (p. 2014 s) with CrO_3, see also 1935 c, 1936 c CIBA; 1937, 1941 Ruzicka; 1945 SCHERING A.-G.; 1946 Wilds; the oxidation may also be carried out with benzoquinone and aluminium tert.-butoxide in boiling toluene (1940 Wettstein); the resulting esters are hydrolysed by refluxing with 2% methanolic NaOH or KOH for 2 hrs. (1937, 1941 Ruzicka; 1940 Wettstein); the esters may also be obtained from the corresp. 17-esters of androstane-3α.17β-diol (p. 2011 s) by oxidation with CrO_3 (1935 c CIBA), the acetate also from the 17-acetate of an androstane-3.17β-diol on dehydrogenation with a palladium catalyst (1936b, c CIBA). From androstane-3.17-dione 3-diethyl ketal (p. 2894 s) on reduction with sodium in propyl alcohol, followed by heating with aqueous alcoholic HCl on the water-bath for 15 min. (1938 Serini; 1945 SCHERING A.-G.); is similarly obtained from androstane-3.17-dione 3-enol ethyl ether (p. 2607 s) (1938 SCHERING A.-G.). For formation from androstane-3.17-dione by reduction with sodium in boiling methanol, see 1935 d, 1936 c CIBA.

Rns. Gives androstane-3α.17β-diol (p. 2011 s) and probably also androstane-3β.17β-diol (p. 2014 s) on hydrogenation in glacial acetic acid at 70⁰ in the presence of platinum dioxide and hydrobromic acid, followed by treatment with alcoholic NaOH (1937 Ruzicka). On treatment with 1 mol. bromine and some hydrobromic acid in glacial acetic acid at 20⁰, the acetate gives the acetate of 2α-bromoandrostan-17β-ol-3-one (p. 2703 s) (1936 CIBA PHARMA-CEUTICAL PRODUCTS; 1938 Butenandt); the hexahydrobenzoate reacts similarly on similar bromination (1943 Inhoffen) or on refluxing with 1 mol. of N-bromosuccinimide in carbon tetrachloride for 1–2 min. with exposure to strong light (1947 a, b Djerassi). When the hexahydrobenzoate is treated at 23⁰ with 2 mols. bromine in glacial acetic acid, the ester of 2.2-dibromo-androstan-17β-ol-3-one (p. 2710 s) is very rapidly formed and then immediately begins to rearrange to the ester of 2α.4α-dibromoandrostan-17β-ol-3-one (p. 2711 s); the rate of rearrangement is highly increased by raising the temperature to 48⁰ or by adding hydrogen bromide, and decreased by adding sodium acetate or 30% H_2O_2; the 2.2-dibromo-derivative can be isolated when glacial acetic acid distilled from potassium permanganate is used (it then contains 2–3% of water owing to partial oxidation); the bromination under these conditions is very slow and can be accelerated by addition of hydrogen bromide (1947 b Djerassi; 1946 Wilds; 1943 Inhoffen; 1945 SCHERING A.-G.); the acetate behaves similarly on dibromination (1947 b Djerassi; 1940, 1943 Inhoffen; 1945 SCHERING A.-G.) and also the benzoate (1943 Inhoffen).

With potassium cyanide and glacial acetic acid in alcohol at 0⁰, the acetate gives 3-cyanoandrostane-3.17β-diol 17-acetate (1943 Goldberg). The acetate gives 17β-acetoxyandrostan-3-one-2-carboxylic acid (not described) on warming with sodamide in abs. ether, followed by passing dry carbon dioxide through the suspension of the resulting sodium derivative (1938 CIBA PHARMA-CEUTICAL PRODUCTS).

Colour reactions (cf. those of epi-dihydrotestosterone, p. 2597 s). In the Liebermann-Burchard reaction it gives only a pale yellowish coloration after 1 hr. Gives no colour reaction on brief boiling with glacial acetic acid and conc. H_2SO_4, followed by addition of bromine or acetic anhydride. It gives a violet-red coloration on brief boiling with $POCl_3$ and quinoline, followed by addition of glacial acetic acid, conc. H_2SO_4, and bromine (or acetic anhydride) to the cooled solution (1939 Kägi). For colour reactions with conc. H_2SO_4 in alcohol in the absence or the presence of benzaldehyde, see 1939 Scherrer. With m-dinitrobenzene + 15% KOH in alcohol or alcoholic di-oxane it very rapidly gives a violet coloration which becomes brown (1943 Wettstein); for absorption spectrum of the reaction mixture, see 1938 Callow.

Physiological properties (see also pp. 1367–1368 s and the literature cited there). Has a strong androgenic activity (15–20 γ) (1937 Miescher; 1937 Ruzicka); for androgenic activity of its aliphatic esters, see 1937 Miescher. For the effect of high dosages on male guinea pigs, see 1938 Bottomley. Causes ovipositor growth in the female bitterling (1941, 1941/42 Duyvené de Wit). Prevents tumour formation following subcutaneous œstrogen implantation in castrated guinea pigs (1943/44 Lipschütz). Increases arginase activity in kidneys of castrated mice (1944 Kochakian). Causes mitosis damage (1942 v. Möllendorff).

Esters of Dihydrotestosterone

For their formation, see also under formation of dihydrotestosterone (page 2597 s).

Formate $C_{20}H_{30}O_3$. Crystals (hexane), m. 142–3° (1935 a CIBA), 141–142.5° (1937 Miescher). — **Fmn.** From dihydrotestosterone on heating with formic acid in pyridine at 110° for 1.5 hrs. (1935 a CIBA; 1937 Miescher).

Acetate $C_{21}H_{32}O_3$ (p. 141), m. 158.5–159.5° (1936 c CIBA); $[\alpha]_D^{23}$ +26° (chloroform) (1947 c Djerassi). For the rate of hydrolysis, see that of epi-dihydrotestosterone acetate (p. 2597 s).

Propionate $C_{22}H_{34}O_3$. Crystals (hexane), m. 121–121.7° (1937 Miescher; 1935 c CIBA). — **Fmn.** From dihydrotestosterone on heating with propionic anhydride in pyridine at 100° for 1.5 hrs. (1937 Miescher).

Butyrate $C_{23}H_{36}O_3$. Dimorphous crystals (aqueous methanol), m. 90.5–91° and 100–101° (1937 Miescher; 1935 c CIBA). — **Fmn.** From dihydrotestosterone on heating with butyric anhydride in pyridine at 100° (1937 Miescher).

n-Valerate $C_{24}H_{38}O_3$. Amorphous (aqueous alcohol), m. 102.5–103° (1937 Miescher; 1935 c CIBA). — **Fmn.** From dihydrotestosterone on heating with valeric anhydride in pyridine at 110° for 1.5 hrs. (1937 Miescher).

Hydrogen succinate $C_{23}H_{34}O_5$. Crystals (chloroform-hexane); on heating very slowly it melts at 168.5° cor.; on heating very rapidly it gives an opaque melt at 135.5° cor. becoming clear at ca. 160°; the resolidified substance m. 168.5° cor. (1941 Ruzicka).

Hexahydrobenzoate $C_{26}H_{40}O_3$. Microcrystalline powder (methanol), m. 165° to 167° cor. (1946 Wilds; cf. 1935 c, 1936 a CIBA); $[\alpha]_D^{29}$ +24° (chloroform); $[\alpha]_D^{23}$ +27° (chloroform) (1947 c Djerassi). Absorption max. in abs. alcohol at 280.5 mμ (log ε 1.49) (1946 Wilds). — **Fmn.** From dihydrotestosterone and hexahydrobenzoyl chloride in pyridine (1936 a CIBA).

5-iso-Androstan-17α-ol-3-one, Aetiocholan-17α-ol-3-one $C_{19}H_{30}O_2$. Plates

(ether-petroleum ether), m. 160–1° cor.; $[\alpha]_D^{17}$ +5.4° ±3° (chloroform). Gives an intense blue coloration on boiling with $POCl_3$ in quinoline, followed by treatment with bromine and conc. H_2SO_4 in glacial acetic acid (1945 Prelog). — **Fmn.** Its benzoate is obtained

E.O.C. XIV s. 180

References, pp. 2600 s, 2601 s

from the benzoate of epi-testosterone (p. 2579 s) on hydrogenation in alcohol in the presence of palladium on barium carbonate or in the presence of platinum oxide, in the latter case followed by oxidation with CrO_3 in strong acetic acid; it is hydrolysed by refluxing with 1N methanolic KOH for 2 hrs. (1945 Prelog).

Rns. On treatment of the benzoate with perbenzoic acid in chloroform at —10° (1 week) it gives the **17-benzoate of 3‖4-ætiocholane-4.17α-diol-3-oic acid lactone** $C_{26}H_{34}O_4$ (I) [crystals (methylene chloride-petroleum ether, or acetone), m. 208.5–209.5° cor.; $[\alpha]_D^{20}$ —20.8° ±2° (chloroform)] which on distillation through a tube heated at 300° in a stream of nitrogen is converted into 3‖4-Δ^{16}-ætiocholen-4-ol-3-oic acid lactone (p. 2398 s); similar distillation of the benzoate of ætiocholan-17α-ol-3-one gives Δ^{16}-ætiocholen-3-one (p. 2398 s) (1945 Prelog).

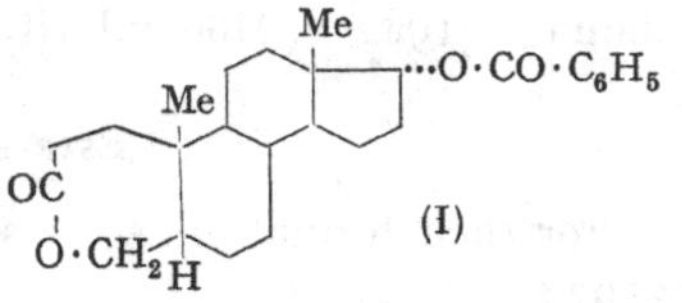

Benzoate $C_{26}H_{34}O_3$. Plates (ether), m. 181–2° cor.; $[\alpha]_D^{17}$ —21.3° ±2° (chloroform) (1945 Prelog). — **Fmn.** and **Rns.** See above.

5-iso-Androstan-17β-ol-3-one, Aetiocholan-17β-ol-3-one $C_{19}H_{30}O_2$ (p. 141:

"*17-Hydroxyætiocholan-3-one*"). — **Fmn.** Along with ætiocholan-3α-ol-17-one (p. 2637 s), from the diacetate of ætiocholane-3α.17β-diol (p. 2017 s) on partial saponification with methanolic KOH at room temperature, followed by oxidation with CrO_3 in strong acetic acid and then by hydrolysis with alcoholic KOH (1940 Marker). From testosterone (p. 2580 s) by biochemical reduction using an aqueous extract of stallion testicles at 37° (small yield) (1938 Ercoli); the reduction is due to the presence of putrefactive bacteria (1938 Schramm; cf. also 1938 a, b Mamoli) and can also be performed with Bacillus putrificus (Bienstock) in dil. yeast water or in an aqueous yeast suspension; in the latter case a small amount of ætiocholane-3α.17β-diol (p. 2017 s) is also formed (1939 Mamoli). Along with a small amount of the last-mentioned diol, from Δ^4-androstene-3.17-dione (p. 2880 s) by the action of Bacillus putrificus in yeast suspension (1939 Mamoli).

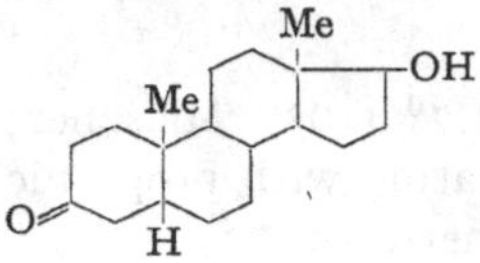

Rns. May be converted into testosterone (p. 2580 s) by treatment with bromine-hydrobromic acid in glacial acetic acid, followed by boiling with pyridine (1936 CIBA PHARMACEUTICAL PRODUCTS; 1940 Marker). The diol, obtained on reduction with sodium in boiling isopropyl alcohol (1937 Butenandt; cf. 1938 Ercoli) or with amalgamated aluminium (1937 Butenandt), is ætiocholane-3α.17β-diol (p. 2017 s).

Acetate $C_{21}H_{32}O_3$ (p. 141). Crystals (dil. methanol), m. 143° (1938 Ercoli).

1935 a CIBA*, *Swiss Pat.* 208893 (issued 1940); *Ch. Ztbl.* **1941** I 2284.
     b CIBA*, *Swiss Pat.* 210761 (issued 1940); *Ch. Ztbl.* **1941** I 3409.
     c CIBA*, *Swiss Pats.* 210770–210774 (issued 1940); *Ch. Ztbl.* **1941** I 3409.
     d CIBA*, *Swiss Pat.* 212190 (issued 1941); *Ch. Ztbl.* **1942** I 231.
1936 a CIBA*, *Swiss Pat.* 206120 (issued 1939); *C. A.* **1941** 3040.
     b CIBA*, *Dutch Pat.* 44043 (issued 1938); *Ch. Ztbl.* **1939** I 1806.

  * Gesellschaft für chemische Industrie in Basel; Société pour l'industrie chimique à Bâle.

c CIBA* (Ruzicka, Wettstein), *U.S. Pat.* 2143453 (issued 1939); *C. A.* **1939** 3078; *Ch. Ztbl.*
   1939 I 5009, 5010; CIBA PHARMACEUTICAL PRODUCTS (Ruzicka, Wettstein), *U.S. Pat.*
   2387469 (1937, issued 1945); *C. A.* **1946** 5536; *Ch. Ztbl.* **1948** I 1234.
   CIBA PHARMACEUTICAL PRODUCTS (Miescher, Wettstein), *U.S. Pat.* 2260328 (issued 1941);
   C. A. **1942** 873; *Ch. Ztbl.* **1944** II 246.
   SCHERING A.-G. (Inhoffen), *Ger. Pat.* 722943 (issued 1942); *Brit. Pat.* 500353 (1937, issued
   1939); *Fr. Pat.* 835524 (1937, issued 1938); *Ch. Ztbl.* **1939** I 5010, 5011.

1937 Butenandt, *Z. physiol. Ch.* **248** 205, 209 (footnote 10), 211.
   Miescher, Kägi, Scholz, Wettstein, Tschopp, *Biochem. Z.* **294** 39, 41, 50–52.
   Ruzicka, Kägi, *Helv.* **20** 1557, 1560, 1562–1564.

1938 Bottomley, Folley, *C. A.* **1939** 718; *Ch. Ztbl.* **1939** II 136.
   Butenandt, Dannenberg, *Ber.* **71** 1681, 1684.
   Callow, Callow, Emmens, *Biochem. J.* **32** 1312, 1322.
   CIBA PHARMACEUTICAL PRODUCTS (Ruzicka), *U.S. Pat.* 2281622 (issued 1942); *C. A.* **1942**
   5958.
   Ercoli, *Ber.* **71** 650.
   a Mamoli, Schramm, *Ber.* **71** 2083.
   b Mamoli, Schramm, *Ber.* **71** 2698.
   Ruzicka, Furter, Goldberg, *Helv.* **21** 498, 504–506.
   SCHERING A.-G., *Swiss Pat.* 215817 (issued 1941); *Ch. Ztbl.* **1942** II 75.
   Schramm, Mamoli, *Ber.* **71** 1322.
   Serini, Köster, *Ber.* **71** 1766, 1769.

1939 Kägi, Miescher, *Helv.* **22** 683, 685, 687.
   Mamoli, Koch, Teschen, *Z. physiol. Ch.* **261** 287, 295, 296.
   Scherrer, *Helv.* **22** 1329, 1334.

1940 Inhoffen, Zühlsdorff, Huang-Minlon, *Ber.* **73** 451, 455.
   Marker, Rohrmann, *J. Am. Ch. Soc.* **62** 900, 902.
   Wettstein, *Helv.* **23** 388, 398.

1941 Duyvené de Wit, *Biochem. Z.* **309** 297, 300.
   Ruzicka, Goldberg, Grob, *Helv.* **24** 1151, 1154.

1941/42 Duyvené de Wit, *Biochem. Z.* **310** 101, 108; 171, 173.

1942 v. Möllendorff, *C. A.* **1943** 5464; *Ch. Ztbl.* **1942** II 793.

1943 Goldberg, Kirchensteiner, *Helv.* **26** 288, 297.
   Inhoffen, Zühlsdorff, *Ber.* **76** 233, 234, 238–240.
   Wettstein, Miescher, *Helv.* **26** 631, 639.

1943/44 Lipschütz, *C. A.* **1944** 4990.

1944 Kochakian, *J. Biol. Ch.* **155** 579, 582.
   Prelog, Ruzicka, Wieland, *Helv.* **27** 66, 68.

1945 Prelog, Ruzicka, Meister, Wieland, *Helv.* **28** 618, 622, 623, 625, 626.
   Ruzicka, Prelog, Meister, *Helv.* **28** 1651, 1657–1659.
   SCHERING A.-G., *Bios Final Report* No. 449, pp. 272, 273, 276, 277; *Fiat Final Report* No. 996,
   pp. 77, 78, 81, 82.

1946 Wilds, Djerassi, *J. Am. Ch. Soc.* **68** 2125, 2128, 2129.

1947 a Djerassi, Scholz, *Experientia* **3** 107.
   b Djerassi, Scholz, *J. Am. Ch. Soc.* **69** 2404.
   c Djerassi, *J. Org. Ch.* **12** 823, 825.

Δ⁵-Androsten-17β-ol-4-one $C_{19}H_{28}O_2$. Originally described as Δ¹-androsten-
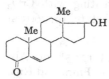
17-ol-3-one (1938 Butenandt; cf. 1936 CIBA PHARMA-
CEUTICAL PRODUCTS), then as *hetero-(h-)-Δ¹-androsten-
17-ol-3-one* (1939 Butenandt); for the true structure, see
1944 Butenandt. — Crystals (acetone-petroleum ether,
hexane, and dil. alcohol), m. 158–9°; $[\alpha]_D^{20}$ —42.3° (alcohol)

---

* Gesellschaft für chemische Industrie in Basel; Société pour l'industrie chimique à Bâle.

(1938 Butenandt). Absorption max. in chloroform at 240 mμ (ε 5600) (1938 Butenandt; 1939 Dannenberg). — Strongly œstrogenic; weakly active in the Fussgänger capon comb test (1938 Butenandt). — **Fmn.** Its acetate is obtained, along with other compounds, from 2α-bromoandrostan-17β-ol-3-one (p. 2703 s) or its acetate on heating with potassium acetate and glacial acetic acid in a sealed tube at 200° for 5 hrs.; very small yields. From Δ^5-androstene-4.17-dione (p. 2897 s) on treatment with fermenting yeast in dil. alcohol; yield, 83% (1938 Butenandt).

Acetate $C_{21}H_{30}O_3$. Crystals (dil. acetone and dil. alcohol), m. 118–9°. — **Fmn.** From the above carbinol with acetic anhydride and pyridine at room temperature. See also above. — *Oxime* $C_{21}H_{31}O_3N$, crystals (dil. alcohol), m. 213–5° dec. (1938 Butenandt).

1936 Ciba Pharmaceutical Products (Miescher, Wettstein), *U.S. Pat.* 2260328 (issued 1941);
     C. A. **1942** 873; *Ch. Ztbl.* **1944** II 246.
1938 Butenandt, Dannenberg, *Ber.* **71** 1681.
1939 Butenandt, Mamoli, Dannenberg, Masch, Paland, *Ber.* **72** 1617, 1618.
     Dannenberg, *Abhandl. Preuss. Akad. Wiss. Math.-naturwiss. Kl.* **1939** No. 21, pp. 14, 54.
1944 Butenandt, Ruhenstrotu-Bauer, *Ber.* **77** 397, 400.

---

i-Androsten-17β-ol-6-one, *i-Androstan-17β-ol-6-one* $C_{19}H_{28}O_2$. Configuration at C-17 by analogy with that of "α"-œstradiol (p. 1966 s), also obtainable (from œstrone) by fermenting yeast (Editor). — Crystals (alcohol), m. 182–3°. — **Fmn.** From i-androstene-6.17-dione (p. 2898 s) on treatment with fermenting yeast in dil. alcohol; yield, ca. 50% (1942 Butenandt).

Acetate $C_{21}H_{30}O_3$, m. 109–110°. — **Fmn.** From the above compound on treatment with acetic anhydride in pyridine (1942 Butenandt).

1942 Butenandt, Surányi, *Ber.* **75** 591, 595, 596.

Δ^5-**Androsten-3β-ol-7-one** $C_{19}H_{28}O_2$. — **Fmn.** The acetate is obtained from the acetate of Δ^5-androsten-3β-ol (p. 1504 s) on oxidation with CrO_3 in glacial acetic acid at 55° (3 hrs.); crude yield (m. 170–4°), 20%. — **Rns.** On shaking the acetate with hydrogen and a platinum oxide catalyst in glacial acetic acid, followed by treating with CrO_3 in acetic acid at 4° for 16 hrs., it gives the saturated acetate (p. 2603 s) (1941 Heard).

Acetate $C_{21}H_{30}O_3$. Crystals (ligroin and alcohol), m. 173.5–174.5° cor. Absorption max. in alcohol at 234 mμ (ε 11 200) (1941 Heard). — **Fmn.** and **Rns.** See above.

Androstan-3β-ol-7-one $C_{19}H_{30}O_2$. Leaflets (aqueous acetone), m. 130–1° cor. (after sublimation in vac.); $[\alpha]_D^{22}$ —68.7° ±6° (dioxane). Is precipitated with digitonin. — **Fmn.** For formation of the acetate from the preceding unsaturated acetate, see above (yield, 81 %); the acetate is hydrolysed by 4 % alcoholic KOH at room temperature (1941 Heard).

Acetate $C_{21}H_{32}O_3$. Crystals (ligroin), m. 110–113° (1941 Heard). — **Fmn.** and **Rns.** See above.

1941 Heard, McKay, *J. Biol. Ch.* **140** Proc. LVI; *J. Biol. Ch.* **165** 677, 684, 685 (1946).

Δ³·⁵-Androstadien-17β-ol-7-one $C_{19}H_{26}O_2$. Platelets (dil. acetone), m. 171–2°; $[\alpha]_D^{20}$ —375° (alcohol) (1938 Butenandt). Absorption max. in chloroform at 280 mμ (ε 28000) (1938 Butenandt; 1939 Dannenberg). — Is not œstrogenic (1938 Butenandt). — **Fmn.** From the diacetate of Δ⁵-androstene-3β.17β-diol-7-one (p. 2742 s) on boiling with 2 N methanolic KOH (1936 SCHERING-KAHLBAUM A.-G.; cf. 1938 Butenandt) or with 1 N methanolic HCl for 20 min. (yield, 75 %) (1938 Butenandt).

Acetate $C_{21}H_{28}O_3$. Needles (alcohol), m. 222°; $[\alpha]_D^{20}$ —400° (chloroform) (1938 Butenandt). Absorption max. in chloroform at 280 mμ (ε 27800) (1939 Dannenberg). — **Fmn.** From the above carbinol on heating with acetic anhydride to gentle boiling for 20 min. (1938 Butenandt).

Δ⁵-Androsten-17β-ol-7-one $C_{19}H_{28}O_2$. M. 143–144° (1939 Kuwada), 141.5° to 142.5° (1940 Marker). — Shows ¹/₁₀ of testosterone activity in the Korenchevsky test (1939 Kuwada). — **Fmn.** Its acetate is obtained from Δ⁵-androsten-17β-yl acetate (p. 1511 s) on oxidation with CrO_3 in acetic acid; it is hydrolysed by alcoholic KOH or alcoholic HCl (at 80°) (1939 Kuwada; 1940 Marker).

2.4-Dinitrophenylhydrazone $C_{25}H_{32}O_5N_4$, orange coloured, m. 230–2° (1940 Marker).

Acetate $C_{21}H_{30}O_3$. Needles (methanol), m. 212–3° (1939 Kuwada), 215–7° (1940 Marker). Absorption max. in hexane at 232 mμ (1939 Kuwada). — *Oxime* $C_{21}H_{31}O_3N$, m. 128–131° dec. (1939 Kuwada).

1936 SCHERING-KAHLBAUM A.-G., *Brit. Pat.* 486 596 (issued 1938); *C. A.* **1938** 8706; *Ch. Ztbl.* **1939** I 1602, 1603; SCHERING A.-G. (Butenandt, Logemann), *U.S. Pat.* 2 170 124 (issued 1939); *C. A.* **1940** 1133; *Ch. Ztbl.* **1940** I 1232.
1938 Butenandt, Hausmann, Paland, *Ber.* **71** 1316, 1319, 1320.
1939 Dannenberg, *Abhandl. Preuss. Akad. Wiss. Math.-naturwiss. Kl.* **1939** No. 21, pp. 27, 59. Kuwada, Tutihasi, *C. A.* **1939** 8209; *Ch. Ztbl.* **1939** II 2433.
1940 Marker, Wittle, Tullar, *J. Am. Ch. Soc.* **62** 223, 225.

5-iso-Androstan-3α-ol-12-one, Aetiocholan-3α-ol-12-one $C_{19}H_{30}O_2$. — Fmn.
The acetate is obtained, along with other compounds, from the 3-acetate of ætiocholane-3α.12α-diol-17-one (p. 2749 s) on hydrogenation in glacial acetic acid in the presence of platinum oxide at 75° for 20 hrs., followed by oxidation of the oily part of the reaction product with CrO_3 in glacial acetic acid at room temperature (1945 Reich).

Acetate $C_{21}H_{32}O_3$. Needles (dil. methanol), m. 141–2° cor. (1945 Reich).

1945 Reich, *Helv.* **28** 863, 871.

Androstan-3β-ol-16-one, "*Isoandrosterone-16*", $C_{19}H_{30}O_2$. For the structure, see 1951, 1954 Huffman. — Needles (benzene-ligroin), m. 187–187.5° (1939 Heard), 190.5–191° cor. (1941 Oppenauer), 186–186.5° (1951, 1954 Huffman); $[α]_D$ in dioxane —160° (1939 Heard), —157° (1941 Oppenauer), —180° (1951, 1954 Huffman). Is precipitated with digitonin (1939 Heard; 1941 Oppenauer). —
Shows no androgenic activity in the capon comb test (1941 Oppenauer).

Occ. In the urine of pregnant mares (1939 Heard; 1941 Oppenauer). —
Fmn. From epi-androsterone (p. 2633 s) on nitrosation, followed by reduction with zinc dust in acetic acid and then by Clemmensen reduction of the resulting diolone (m. 217–8°) (1951, 1954 Huffman). — **Rns.** On oxidation with CrO_3 in 90% acetic acid with cooling in ice it gives the corresponding diketone (p. 2876 s) which yields androstane (p. 1396 s) on Clemmensen reduction (1939 Heard).

Oxime $C_{19}H_{31}O_2N$, needles (aqueous alcohol), m. 194–5° dec. (1939 Heard), 199° (1951, 1954 Huffman).

Semicarbazone $C_{20}H_{33}O_2N_3$, needles (alcohol), m. 282–5° cor. dec. (1941 Oppenauer).

Acetate $C_{21}H_{32}O_3$. Needles (dil. methanol), m. 109–110° cor. — **Fmn.** From the above carbinol with acetic anhydride and pyridine at room temperature (1941 Oppenauer).

Benzoate $C_{26}H_{34}O_3$. Needles (acetone), m. 209–211° cor. (1941 Oppenauer), 208.5–209° (1951, 1954 Huffman). Sparingly soluble in ether, methanol, and ethanol (1939 Heard). — **Fmn.** From the above carbinol with benzoyl chloride and pyridine (1939 Heard; 1941 Oppenauer).

1939 Heard, McKay, *J. Biol. Ch.* **131** 371.
1941 Oppenauer, *Z. physiol. Ch.* **270** 97, 102, 103.
1951 Huffman, Lott, *J. Am. Ch. Soc.* **73** 878.
1954 Huffman, Lott, *J. Biol. Ch.* **207** 431.

Δ²·⁴·⁶ - Androstatrien - 3 - ol - 17 - one, 3 - enol of Δ⁴·⁶-Androstadiene-3.17-dione

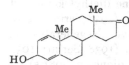

$C_{19}H_{24}O_2$. For formation of its acetate from Δ⁴·⁶-andros-tadiene-3.17-dione (p. 2878 s) by heating with acetic anhydride and fused potassium acetate, and for its pyrolysis, see 1937 SCHERING A.-G.

1937 SCHERING A.-G., *Fr. Pat.* 838704 (issued 1939), *Brit. Pat.* 508576 (issued 1939); *C. A.* **1940** 777; *Ch. Ztbl.* **1939** II 1722; *Fiat Final Report* 996, pp. 155, 156 (1947); SCHERING CORP. (Inhoffen), *U.S. Pat.* 2280828 (issued 1942); *C. A.* **1942** 5618.

Δ¹·⁴-Androstadien-3β-ol-17-one

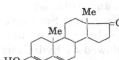

$C_{19}H_{26}O_2$. A compound possessing this struc-ture is said to be formed from Δ⁵-androsten-3β-ol-17-one (5-dehydroepiandrosterone, p. 2609 s) by mono-bromination, followed by dehydrobromination; on pas-sing over a catalyst (nickel, Raney nickel, etc.) at 500–600° in a stream of an indifferent vapour (e.g., cyclohexane), methane is split off and œstrone (p. 2543 s) is obtained (1945 SCHERING A.-G.). For these reactions cf., however, 1950 Hershberg.

1945 SCHERING A.-G., *Cios Report* XXIV–16, p. 3.
1950 Hershberg, Rubin, Schwenk, *J. Org. Ch.* **15** 292 footnote 2.

Δ³·⁵-Androstadien-3-ol-17-one, 3-enol of Δ⁴-Androstene-3.17-dione $C_{19}H_{26}O$

(cf. p. 156). — **Fmn.** For formation of its esters from Δ⁴-androstene-3.17-dione (p. 2880 s) by acylation, see also 1936, 1936/37 CIBA. See also formation of its ethers, below. — **Rns.** (See also those of the ethyl ether, below). The esters may be reduced to the cor-resp. 3-esters of Δ³·⁵-androstadiene-3.17β-diol (p. 1992 s) [which give testo-sterone (p. 2580 s) on hydrolysis with K_2CO_3 in aqueous methanol or ethanol] by refluxing with aluminium isopropoxide in isopropyl alcohol (1938a CIBA) or by treatment with fermenting yeast in aqueous alcohol at 30° (1937 CIBA); the ethers (below) are reduced to the corresp. 3-ethers of the same diol by sodium in boiling propyl alcohol (1938 Serini; 1937, 1938a CIBA; 1940a, 1941 SCHERING A.-G.; 1940 SCHERING CORP.); 3-esters of the same diol are also obtained on hydrogenation (1 mol. H_2) in abs. alcohol in the presence of a weakly active nickel catalyst at room temperature (1937 CIBA), whereas on hydrogenation (1 mol. H_2) of the butyrate or ethyl ether in ethyl acetate in the presence of Raney nickel at room temperature a mixture of the two 3-epimeric Δ⁴-androsten-3-ol-17-one butyrates (or ethyl ethers, resp.; p. 2608 s) is said to be formed (1937 SCHERING A.-G.).

On treatment of the benzoate (or other esters or ethers) with ethyl chloro-acetate and sodium ethoxide in ether it gives the corresp. derivatives of the epimeric epoxy-acid ethyl esters (partial formula I) which on hydrolysis with alkali, followed by heating, give a mixture of the epimeric Δ^4-androsten-3-one-17-aldehydes (p. 2777 s) (1938c CIBA); on treatment of the benzoate with ethyl α.α-dibromopropionate in the presence of amalgamated magnesium, followed by heating with alcoholic NaOH and decarboxylation, it gives progesterone (p. 2782 s) and neoprogesterone * (1938d CIBA); for these reactions, cf. also 1939 Miescher.

Me
O
CH·CO$_2$·C$_2$H$_5$
(I)

Ethyl ether $C_{21}H_{30}O_2$. Leaflets (alcohol-pyridine), m. 152⁰ (1938 Serini; 1937, 1938 SCHERING A.-G.; 1938a CIBA); $[\alpha]_D^{20}$ —89⁰ (dioxane) (1938 Serini), —89⁰ (chloroform) (1938 SCHERING A.-G.), $[\alpha]_D$ +34⁰ (solvent not stated) (1937 SCHERING A.-G.). — **Fmn.** From Δ^4-androstene-3.17-dione (p. 2880 s) on heating with ca. 1 mol. ethyl orthoformate or acetone diethyl ketal and alcoholic HCl in benzene at 75⁰ for 2 hrs. or 3 hrs., resp. (1938 Serini; 1938, 1939 SCHERING A.-G.). — **Rns.** For reduction, see above. Regenerates the dione on boiling with aqueous alcoholic HCl for 1 hr. (1938 Serini; 1938 SCHERING A.-G.). Gives 16-oximino-Δ^4-androstene-3.17-dione (p. 2981 s) on treatment with sodium ethoxide and butyl nitrite in alcohol, followed by hydrolysis of the resulting 16-oximino-derivative with aqueous alcoholic NaOH at room temperature and then by warming with 50% acetic acid (1942 Stodola). On treatment with methylmagnesium bromide in toluene-ether, followed by acid hydrolysis of the resulting 17α-methyl-$\Delta^{3.5}$-androstadiene-3.17β-diol 3-ethyl ether, it gives 17-methyltestosterone (p. 2645 s); reacts similarly with ethyl-, allyl-, and benzylmagnesium halides (1943 CIBA PHARMACEUTICAL PRODUCTS). Gives 17-ethynyltestosterone (p. 2648 s) on treatment in ether-benzene with a solution of potassium acetylide in liquid ammonia, followed by heating of the resulting compound with aqueous alcoholic HCl for 15 min. on the water-bath (1939 Inhoffen; 1939 SCHERING A.-G.; cf. 1939 CIBA) or on treatment in benzene or ether - benzene with acetylene in the presence of potassium tert.-amyl oxide, followed by hydrolysis with aqueous alcoholic HCl (1938b, 1939 CIBA; 1939 SCHERING A.-G.). On refluxing with benzaldehyde and sodium methoxide in methanol it gives the 16-benzylidene derivative (p. 2698 s) (1941 Stodola).

Ethyl ether diethyl ketal $C_{25}H_{40}O_3$. Crystals (alcohol), m. 91–92.5⁰; $[\alpha]_D^{20}$ —141.6⁰ (dioxane). — **Fmn.** From Δ^4-androstene-3.17-dione (p. 2880 s) on warming with ca. 3 mols. of ethyl orthoformate and alcoholic HCl in benzene at 50⁰ for 1 hr. — **Rns.** Regenerates the dione on boiling with aqueous alcoholic HCl (1938 Serini).

Benzyl ether $C_{26}H_{32}O_2$. Crystals (alcohol-pyridine), m. 184⁰ (1940a, 1941 SCHERING A.-G.), 164⁰ (1940 SCHERING CORP.); $[\alpha]_D^{20}$ in dioxane —65.3⁰ (1940a SCHERING A.-G.), —65.3⁰ (1940 SCHERING CORP.). — **Fmn.** From Δ^4-androstene-3.17-dione (p. 2880 s) on boiling with benzyl alcohol and p-toluene-sulphonic acid in benzene, with simultaneous removal of the water formed

* Neoprogesterone is a D-homo-steroid (cf. 1940 Ruzicka).

(1940a, 1941 SCHERING A.-G.; 1940 SCHERING CORP.). — Rns. For reduction, see under the reactions of the hydroxyketone, above.

Cyclohexyl ether $C_{25}H_{36}O_2$. Crystals (alcohol-pyridine), m. 164°; $[\alpha]_D^{20}$ —79° (dioxane). — Fmn. Similar to that of the preceding ether, using cyclohexanol (1940b, 1941 SCHERING A.-G.; 1940 SCHERING CORP.). — Rns. For reduction, see under the reactions of the hydroxyketone, above.

1936 CIBA*, Swiss Pat. 214327 (issued 1941); C. A. **1942** 5960.
1936/37 CIBA*, Brit. Pat. 477400 (issued 1938); Ch. Ztbl. **1938** II 119, 120.
1937 CIBA*, Swiss Pat. 202846 (issued 1939); Ch. Ztbl. **1940** I 93.
    SCHERING A.-G. (Dannenbaum), Ger. Pat. 717438 (issued 1942); C. A. **1944** 2456; Ch. Ztbl. **1942** II 75.
1938 a CIBA* (Miescher, Fischer), Ger. Pat. 712857 (issued 1941); C. A. **1943** 4534; Ch. Ztbl. **1942** I 1282.
    b CIBA*, Swiss Pat. 227455 (issued 1943); C. A. **1949** 3476; Ch. Ztbl. **1944** II 1203.
    c CIBA*, Brit. Pat. 516542 (issued 1940); Ch. Ztbl. **1940** II 1327.
    d CIBA* (Oppenauer, Miescher, Kägi), Swed. Pat. 96375 (issued 1939); Ch. Ztbl. **1940** I 1231; Ital. Pat. 362454; Ch. Ztbl. **1939** II 3151.
    SCHERING A.-G., Fr. Pat. 850115 (issued 1939); Ch. Ztbl. **1940** II 931; Swiss Pat. 215813 (issued 1941); Ch. Ztbl. **1942** II 196.
    Serini, Köster, Ber. 71 1766, 1769, 1770.
1939 CIBA*, Fr. Pat. 861536 (issued 1941); Ch. Ztbl. **1941** I 3259.
    Inhoffen, Köster, Ber. 72 595.
    Miescher, Kägi, Helv. 22 184.
    SCHERING A.-G., Swed. Pat. 99036 (issued 1940); Ch. Ztbl. **1940** II 3668; Swiss Pat. 227580 (issued 1943); Ch. Ztbl. **1944** II 1300; SCHERING CORP. (Inhoffen), U.S. Pat. 2358808 (issued 1944); C. A. **1945** 1738; Ch. Ztbl. **1945** II 1069.
1940 Ruzicka, Meldahl, Helv. 23 364.
    a SCHERING A.-G., Swiss Pat. 220206 (issued 1942); Ch. Ztbl. **1943** I 1695.
    b SCHERING A.-G., Swiss Pat. 229775 (issued 1944); Ch. Ztbl. **1944** II 1301.
    SCHERING CORP. (Köster), U.S. Pat. 2363338 (issued 1944); C. A. **1945** 3128; Ch. Ztbl. **1945** II 1638.
1941 SCHERING A.-G., Fr. Pat. 875070 (issued 1942); Ch. Ztbl. **1943** I 1190.
    Stodola, Kendall, J. Org. Ch. 6 837, 839.
1942 Stodola, Kendall, J. Org. Ch. 7 336, 339.
1943 CIBA PHARMACEUTICAL PRODUCTS (Miescher), U.S. Pat. 2386331 (issued 1945); C. A. **1946** 179; Ch. Ztbl. **1948** I 1040.

Δ²-Androsten-3-ol-17-one, 2-Dehydroandrosterone, 3-enol of Androstane-3.17-dione, $C_{19}H_{28}O_2$. For position of the double bond, cf. 1950 Inhoffen.

    Ethyl ether $C_{21}H_{32}O_2$. Crystals (alcohol-pyridine), m. 105–6°; $[\alpha]_D^{20}$ +126° (dioxane) (1938 Serini; 1938 SCHERING A.-G.). — Fmn. From the 3-diethyl ketal of androstane-3.17-dione (p. 2894 s) on boiling in xylene for 2 hrs. (1938 Serini; 1938 SCHERING A.-G.) or directly from the dione on heating with ethyl orthoformate or acetone diethyl acetal and alcoholic HCl in benzene or xylene (1938 SCHERING A.-G.). — Rns. Gives di-

---

* Gesellschaft für chemische Industrie in Basel; Société pour l'industrie chimique à Bâle.

hydrotestosterone (p. 2597 s) on heating with sodium and propanol on the water-bath, followed by heating the crystalline dihydrotestosterone enol-ethyl ether with aqueous alcoholic HCl on the water-bath for 20 min. (1938 SCHERING A.-G.). On heating with 4 N HCl in abs. alcohol on the water-bath for 10 min., androstane-3.17-dione is regenerated (1938 Serini).

1938 SCHERING A.-G., *Swiss Pat.* 215817 (issued 1941); *Ch. Ztbl.* **1942** II 75.
    Serini, Köster, *Ber.* **71** 1766, 1769.
1950 Inhoffen, Becker, Kölling, *Ann.* **568** 181.

Δ⁴-Androsten-3α-ol-17-one, 4-Dehydroandrosterone (I), and Δ⁴-Androsten-3β-ol-17-one, 4-Dehydroepiandrosterone

(p. 141: "*Δ⁴-Dehydroandrosterone*"), $C_{19}H_{28}O_2$. A mixture of the corresp. derivatives of the two 3-epimeric compounds is obtained from the butyrate and the ethyl ether of Δ³·⁵-androstadien-3-ol-17-one (p. 2605 s) on hydrogenation (1 mol. H_2) in the presence of Raney nickel at room temperature and atmospheric pressure in ethyl acetate (or abs. alcohol-ethyl acetate, resp.); from the mixture of butyrates, ca. 20 % could be isolated as needles (aqueous acetone), m. 129–135°, [α]_D +98° (solvent not stated); from the mixture of ethyl ethers, ca. 70 % formed crystals (aqueous acetone), m. 107–110°, [α]_D +77° (solvent not stated) (1937 SCHERING A.-G.). — On heating "Δ⁴-dehydroandrosterone" with copper powder in a vacuum at 180° it gives Δ⁴-androstene-3.17-dione (p. 2880 s) (1935 CIBA).

1935 CIBA, *Swiss Pat.* 201202 (issued 1939); *C. A.* **1939** 8625; *Ch. Ztbl.* **1939** II 172.
1937 SCHERING A.-G. (Dannenbaum), *Ger. Pat.* 717438 (issued 1942); *C. A.* **1944** 2456; *Ch. Ztbl.* **1942** II 75.

Δ⁵-Androsten-3α-ol-17-one, 5-Dehydroandrosterone, "*cis-Dehydroandrosterone*" $C_{19}H_{28}O_2$

(p. 142: "*epi-Δ⁵-Dehydroandrosterone*"). The name "5-dehydroandrosterone" for this compound is in accordance with that of androsterone, since both compounds have the α-configuration at C-3; this name (or merely "dehydroandrosterone"), however, was usually, but wrongly, applied to the natural product with β-configuration at C-3 (see below), which must be designated 5-dehydroepiandrosterone.

For colour reactions in 70 % alcohol with conc. H_2SO_4 alone or in the presence of benzaldehyde, see 1939 Scherer. — **Fmn.** For formation from Δ⁵-androstene-3.17-dione (p. 2890 s) on catalytic hydrogenation, see also 1936/37 CIBA. May be separated from 5-dehydroepiandrosterone via the acetates (1936b CIBA).

Δ^5-Androsten-3β-ol-17-one, 5-Dehydroepiandrosterone, "Dehydroisoandro-

sterone", "trans-Dehydroandrosterone", "Dehydroandrosterone" $C_{19}H_{28}O_2$ (p. 141). For the names, see the epimeric compound, above. — For conversion of the higher-melting form into the lower-melting, see under "Action of light", p. 2612 s. For infra-red absorption spectrum of a film, see 1946 Furchgott. $[\alpha]_D^{24} + 12^0$ to $+ 13^0$ (alcohol), $+ 2^0$ (chloroform) (1946 Barton). Dipole moment in dioxane at 25^0 2.73 D (1945 Kumler).

Occ. In normal male urine, to a considerable extent as an ester of sulphuric acid (1942 Munson); for occurrence in normal male urine, see also 1939a Callow; 1940 Hoskins; 1941 Engel. In increased amount in male castrate urine (1940 Callow; cf. 1940 Hoskins). In normal female urine in an amount comparable with that contained in male urine (1938b, 1939b Callow; but cf. 1942 Pearlman) and in approximately the same amount in the urine of ovariectomized women (1939 Hirschmann). In large amounts in the urine of women with adrenal cortical tumours (1941 Wolfe; 1936, 1944 Callow; 1939 Crooke; 1945b Mason) and in the urine of a man with an adrenal cortical tumour (1939 Crooke); in the urine of cancerous males and females (1942 Pearlman); the dehydroepiandrosterone found in these cases is probably, to some extent at any rate, an artefact produced by the action of hydrochloric acid on i-androsten-6-ol-17-one (p. 2644 s) (1948 Dingemanse; 1948 Barton; cf. also 1941 Rosenheim), but the latter compound itself may be an artefact derived from a conjugate of dehydroepiandrosterone (1949 Fieser); when the urine is extracted with an organic solvent before hydrolysis, the amount of dehydroepiandrosterone found is much smaller than when the urine is boiled immediately with acid (1948 Dingemanse). Dehydroepiandrosterone occurs also, but in small amounts, in the urine of bulls (1939a Marker; cf. also 1938 Butz), steers (1939b Marker), mares (1941 Oppenauer), and pregnant cows (1939a Marker). For the possibility that dehydroepiandrosterone secreted in urine arises from a Δ^4-3-ketosteroid [e.g. testosterone (p. 2580 s) or Δ^4-androstene-3.17-dione (p. 2880 s)], see 1938, 1941 Marker; cf., however, 1939a Callow; 1941 Wolfe; 1941 Fieser.

For its isolation from urine see, e.g., 1936 N. V. ORGANON; 1938 Dingemanse; 1941 Engel; 1941b Pincus; 1941 Wolfe; 1942 Munson; 1943 Talbot. See also under "Analytical properties", p. 2617 s.

FORMATION OF 5-DEHYDROEPIANDROSTERONE

From C_{19}- and C_{20}-steroids. The benzoate is obtained from 3β-chloro-Δ^5-androsten-17-one (p. 2474 s) on heating with potassium benzoate and benzoic acid in a sealed tube at 180^0 for 90 min. (yield, 65 %); it is hydrolysed by heating with 2 N methanolic KOH (1934 Butenandt; 1936f CIBA). The acetate is obtained from 3β-bromo-Δ^5-androsten-17-one (p. 2476 s) on refluxing with silver acetate in glacial acetic acid for 20 min. (1942a Butenandt). From the N-chloro-derivative of 17-amino-Δ^5-androsten-3β-ol (p. 1911 s) on heating with sodium ethoxide in alcohol on the water-bath, followed by acid hydrolysis

(1937 I.G. FARBENIND.; cf. 1942 Ruschig). In small yield, together with testosterone and a little epitestosterone, from Δ^4-androstene-3.17-dione (page 2880 s) on heating with aluminium tert.-butoxide and sec.-butyl alcohol in benzine (1937 d CIBA). Along with its 3-epimer (p. 2608 s), from Δ^5-androstene-3.17-dione (p. 2890 s) on hydrogenation in 95 % alcohol in the presence of Raney nickel at room temperature until 1 mol. of hydrogen has been absorbed (1936 c Ruzicka; cf. 1936/37 CIBA); separation by precipitation of 5-dehydroepiandrosterone with digitonin (1936 c Ruzicka) or via the acetates (1936 b CIBA). From its cyanohydrin on heating in vac. at 110–120° (1937 a Kuwada) or on shaking with alkali in ether (1937 a Kuwada; cf. also 1938 b Butenandt).

From C_{21}- and C_{22}-steroids. The acetate is obtained, along with the 3-acetate of 3β.17-dihydroxy-Δ^5-ætiocholenic acid, from the 3-acetate of 17β-ethynyl-Δ^5-androstene-3β.17α-diol (p. 2027 s) on bromination with 1 mol. bromine in carbon tetrachloride, followed by ozonization and then by debromination with zinc dust in glacial acetic acid-ethyl acetate (1939 a Reichstein). From Δ^5-pregnene-3β.17α.20β-triol (p. 2115 s) on oxidation with lead tetraacetate in glacial acetic acid at 20° (1939 b Butenandt); similarly from an "isomeric Δ^5-pregnene-3.17.20-triol" (m. 241°) (1939 b Butenandt), which is probably an addition compound of Δ^5-pregnene-3β.17α.20β-triol with Δ^5-pregnene-3β.16.17-triol (see p. 2116 s). The acetate is obtained from $\Delta^{5.16}$-pregnadien-3β-ol-20-one acetate oxime (p. 2232 s) on rearrangement by treatment with benzenesulphonyl or p-toluenesulphonyl chloride in pyridine, followed by hydrolysis with dil. H_2SO_4 (1941 a PARKE, DAVIS & Co.). The acetate is obtained from the 3-acetate of 17-iso-Δ^5-pregnene-3β.17-diol-20-one (p. 2337 s) on bromination with 1 mol. bromine in glacial acetic acid, followed by oxidation with CrO_3 in glacial acetic acid at 45° (4 hrs.) and then by debromination with zinc dust on the steam-bath (crude yield, 51 %) (1940 Stavely); similarly from the 3-acetate of Δ^5-pregnene-3β.17-diol-20-one (p. 2333 s) (1947 Hirschmann). The acetate is formed from the acetate of 17-isopropylidene-Δ^5-androsten-3β-ol (20-methyl-$\Delta^{5.17(20)}$-pregnadien-3β-ol, p. 1524 s) on bromination with 1 mol. bromine in chloroform at —5°, ozonization, and heating with zinc dust in acetic acid on the steam-bath (yield, 71 %) (1934 a SCHERING A.-G.; cf. 1942 Marker); it is similarly obtained from 3β-acetoxy-$\Delta^{5.17(20)}$-pregnadien-21-oic acid (yield, 26 %) (1942 Marker; 1941 b PARKE, DAVIS & Co.).

From C_{27}- and higher steroids. For formation of its acetate from the acetates of cholesterol (p. 1568 s), stigmasterol (p. 1784 s), and "β"-sitosterol (p. 1808 s) by oxidation with CrO_3 after temporary protection of the double bond with bromine (p. 142), see also 1934 b, 1938 a, 1944 SCHERING A.-G.; 1937 b Ruzicka; 1936 f, 1938 c CIBA; 1938 Mitui; 1939 Kiprianov; 1940 Hattori; 1942 CHINOIN; 1946 GLIDDEN Co. For formation of its acetate from cholesteryl acetate on anodic electrolysis (after temporary protection of the double bond with bromine), see 1938 RICHTER GEDEON. The acetate is obtained from the acetate of cholesterol dichloride (p. 1888 s) on treatment with lead tetraacetate, followed by hydrolysis, dehydration, re-acetylation, ozonization, and treatment with CrO_3; similarly from the acetates of cholesterol dibromide

(p. 1889 s) and "β"-sitosterol dichloride (p. 1898 s) (1936 LES LABORATOIRES FRANÇAIS DE CHIMIOTHÉRAPIE). From stigmasterol (p. 1784 s) on oxidation with CrO_3-sulphuric acid in acetic acid (1935 b SCHERING A.-G.).

The acetate may also be obtained from 3β-acetoxy-Δ^5-ætiobilienic acid (I, R = R′ = H, R″ = CO·CH₃) [which is one of the CrO_3-degradation products of cholesterol acetate and may be reobtained from 5-dehydro-epiandrosterone acetate (see p. 2612 s)] in the following way (1938 Kuwada):

REACTIONS OF 5-DEHYDROEPIANDROSTERONE

Oxidation. For the androstan-5-ol-3.6.17-trione obtained in small amount, along with Δ^4-androstene-3.6.17-trione (p. 2979 s), on oxidation with CrO_3 in cold acetic acid (p. 142), see also p. 2983 s. For conversion into Δ^5-androstene-3.17-dione (p. 2890 s) by oxidation of its dibromide with CrO_3 in glacial acetic acid, followed by debromination with zinc in boiling methanol, see also 1937 b SCHERING-KAHLBAUM A.-G.; for conversion into Δ^4-androstene-3.17-dione (p. 2880 s) by the same oxidation, but followed by debromination with zinc in acetic acid, see also 1936 e CIBA; the debromination, giving the Δ^4-dione, is preferably carried out with chromous chloride solution (containing $ZnCl_2$ and HCl) in acetone under CO_2 (1945 Julian; 1944 GLIDDEN COMP.). Δ^4-Androstene-3.17-dione is also formed when dehydroepiandrosterone is heated with copper powder in vac. at ca. 180⁰ (1935 CIBA), with aluminium isopropoxide and cyclohexanone at 100⁰ (1937 a SCHERING-KAHLBAUM A.-G.), or with aluminium tert.-amyl oxide and acetophenone (1937 a Oppenauer; 1937 b N. V. ORGANON); see also under "Isomerization", p. 2613 s. Gives $\Delta^{4.6}$-androstadiene-3.17-dione (p. 2878 s) on refluxing with aluminium tert.-butoxide and benzoquinone in abs. toluene (1941 CIBA).

On treatment of dehydroepiandrosterone with perbenzoic acid in chloroform at room temperature it gives mainly 5.6α-epoxyandrostan-3β-ol-17-one (p. 2769 s) (1936 Zelinskiĭ; 1937 a Ushakov; 1938 a Miescher; 1940, 1948 Ehrenstein); the acetate and benzoate react similarly with perbenzoic acid or monoperphthalic acid to give the corresp. esters of 5.6α-epoxyandrostan-3β-ol-17-one (1940 b Ruzicka; 1941, 1948 Ehrenstein) but, using the acetate and monoperphthalic acid, the acetate of 5.6β-epoxyætiocholan-3β-ol-17-one (p. 2770 s) could also be isolated (1944 Ruzicka); the acetates of both epoxy-compounds are also formed when dehydroepiandrosterone acetate is treated with $KMnO_4$ in acetic acid at 50⁰ (1940, 1948 Ehrenstein). On heating dehydroepiandrosterone acetate with 30% H_2O_2 in glacial acetic acid on the water-bath for 2 hrs. it gives the 3-acetate of androstane-3β.5.6β-triol-

17-one (p. 2768 s) (1938a Miescher; 1936d CIBA; 1939, 1948 Ehrenstein); the tetraacetylglucoside (p. 2620 s) reacts similarly (1938a Miescher). On treatment with OsO_4 in abs. ether at 30° for 3 days, dehydroepiandrosterone gives androstane-3β.5.6α-triol-17-one (p. 2767 s) (1937b, 1939 Ushakov; 1939, 1948 Ehrenstein); the acetate does not react with OsO_4 in ether at room temperature (5 days) (1939 Westphal).

On treatment with SeO_2 it gives a mixture of Δ^5-androstene-3β.4-diol-17-one and Δ^4-androstene-3β.6-diol-17-one, which on heating with sulphuric acid in alcohol yields Δ^4-androstene-3.17-dione (p. 2880 s) (1937c CIBA). On dropwise addition of n-amyl nitrite to a solution of dehydroepiandrosterone and potassium tert.-butoxide in tert.-butyl alcohol in an atmosphere of nitrogen, Δ^5-androsten-3β-ol-16.17-dione 16-oxime (p. 2950 s) is formed (1941 Stodola).

On treatment of dehydroepiandrosterone acetate with potassium hypoiodite in aqueous methanol at room temperature it gives 3-hydroxy-Δ^5-ætiobilienic acid monomethyl ester (I, p. 2611 s; R = R'' = H, R' = CH_3) and dimethyl ester (ratio 4:1) (1941 Wettstein; 1950 von Seemann); for reconversion of this acid into dehydroepiandrosterone, see p. 2611 s.

For oxidation, see also under "Biological reactions", p. 2616 s.

Reduction. Is reduced to androstan-3β-ol-17-one (p. 2633 s) on hydrogenation in alcohol in the presence of palladium black (1935 Wallis) or in glacial acetic acid in the presence of platinum black (1935c SCHERING A.-G.; 1936c SCHERING-KAHLBAUM A.-G.). On hydrogenation of its acetate in glacial acetic acid in the presence of platinum oxide, followed by oxidation with CrO_3 in glacial acetic acid, it gives the acetates of androstan-3β-ol-17-one (main product; cf. also the formation of androstane-3β.17β-diol, p. 2015 s) and ætiocholan-3β-ol-17-one (p. 2638 s) (1941 Reichstein; 1944 Wenner). For reduction, see also under "Formation" of Δ^5-androstene-3β.17α-diol (p. 1998 s) and Δ^5-androstene-3β.17β-diol (p. 2000 s).

On heating the acetate semicarbazone with sodium ethoxide in abs. alcohol for 10 hrs. in a sealed tube at 180° it gives ætiocholan-3α-ol (p. 1508 s) and, as the main product, a co-ordination complex of Δ^5-androsten-3β-ol and androstan-3β-ol (see p. 1505 s) (1946 Heard; cf. 1941 Heard; cf. also 1937 Burrows; 1938 Raoul); the reduction to Δ^5-androsten-3β-ol may also be carried out by heating dehydroepiandrosterone semicarbazone with alcoholic sodium ethoxide in a sealed tube at 145° for 6–8 hrs. (1942a Butenandt; cf. also 1946 Milas).

For reduction, see also under "Biological reactions" (p. 2616 s) and under "Isomerization" (p. 2613 s).

Action of light. The higher-melting form (148° uncor.) passes into the lower-melting (137–8° uncor.) on exposure to diffuse daylight for several days; on intense illumination with sunlight or arc light the m. p. is lowered to 123–5° (after sintering at 95°) (1935 Butenandt). By X-ray-photographic methods it has been found that on intense ultra-violet irradiation with exposure to the air an amorphous product is obtained, while on the same irradiation in vacuo a crystalline product is formed; thorium-X irradiation also causes a change (1937 Saupe).

Isomerization. Dehydroepiandrosterone is isomerized to i-androsten-6β-ol-17-one (p. 2644 s) when its p-toluenesulphonate is refluxed with potassium acetate in 50% aqueous acetone (1942a Butenandt). On refluxing with aluminium tert.-butoxide in anhydrous benzene for 14 hrs. dismutation occurs with formation of Δ⁵-androstene-3.17-diols, Δ⁴-androstene-3.17-dione (p. 2880 s) and Δ⁴-androsten-17β-ol-3-one (testosterone, p. 2580 s) (1937a N. V. Organon; see also 1937b Oppenauer).

Halogenation, Action of acids, Dehydration. With 1 mol. bromine in chloroform it gives the 5.6-dibromide, which on refluxing with collidine for 30 min. gives Δ⁴-androstene-3.17-dione (p. 2880 s) (1941 Galinovsky) and on treatment with 2 mols. silver acetate in ether-pyridine in the dark for 20 min. yields the 4-acetate of Δ⁵-androstene-3β.4β-diol-17-one (p. 2746 s) (1943 Petrow). With 1 mol. bromine in glacial acetic acid, followed by oxidation with CrO₃ in glacial acetic acid, it gives 5.6-dibromoandrostane-3.17-dione (1938a Ciba). On bromination of its acetate in chloroform in the presence of HBr it yields the acetate of 16-bromo-Δ⁵-androsten-3β-ol-17-one dibromide (p. 2705 s) (1937 Schering A.-G.). Addition of hydrogen chloride to dehydroepiandrosterone (see p. 142) affords its hydrochloride (5-chloroandrostan-3β-ol-17-one, p. 2705 s) only to an extent of ca. 60%; at the same time 3β-chloro-Δ⁵-androsten-17-one (p. 2474 s) is also formed (1950 Hershberg); the latter chloroketone is also obtained on treatment of dehydroepiandrosterone with PCl₅ in dry chloroform (1937 Wallis), with thionyl chloride in abs. ether at 14° (1935 Butenandt; cf. also 1939 Kuwada), or, but only to a very small extent, by refluxing with methanolic HCl for 2 hrs. (1935 Butenandt). Is dehydrated to give di-(17-keto-Δ⁵-androsten-3β-yl) ether (p. 2620 s) and Δ³·⁵-androstadien-17-one (p. 2400 s) by heating with anhydrous copper sulphate at 200° (1937 Burrows); the latter ketone is also obtained on boiling with P₂O₅ in benzene (1945 Ross) or on treatment with acids (1942 Dobriner; cf. 1941a Pincus); in the presence of benzene it is unchanged by boiling with aqueous HCl (1938 Dingemanse).

Action of organic compounds. For addition compounds with hexane and with petroleum ether, see 1937 Saupe. Dehydroepiandrosterone reacts with methylmagnesium iodide on boiling in ether for 7 hrs. to give 17α-methyl-Δ⁵-androstene-3β.17β-diol (p. 2022 s) and 17β-methyl-Δ⁵-androstene-3β.17α-diol (p. 2023 s) (1939c Miescher; cf. 1935 Ruzicka; 1939 Kiprianov); for the similar reaction with ethylmagnesium halides, see 17α-ethyl-Δ⁵-androstene-3β.17β-diol (p. 2036 s); the simultaneous reductive action of alkylmagnesium halides (see p. 142) is the sole reaction on using propylmagnesium iodide, giving the two C-17 epimeric Δ⁵-androstene-3β.17-diols (1936a Ciba; cf. 1936b Ruzicka); cf. also the reduction with isopropylmagnesium iodide (1934 Schering-Kahlbaum A.-G.). Does not react with vinylmagnesium bromide; the "vinylandrostenediol" (cf. p. 2032 s), the formation of which is alleged by Kuwada (1936b), is unchanged dehydroepiandrosterone (1938b, c Ruzicka; 1938c Butenandt); see, however, also 1936a Schering-Kahlbaum A.-G., where vinylandrostenediol (without indication of properties) is said to be formed in this reaction. With allylmagnesium bromide in ether it gives

17α-allyl-Δ^5-androstene-3β.17β-diol (p. 2041 s) (1936b Kuwada); the acetate reacts similarly (1938c Butenandt; cf. also 1940 v. Euw). With ethynyl-magnesium bromide in ether the acetate gives the 3-acetate of 17α-ethynyl-Δ^5-androstene-3β.17β-diol (p. 2030 s) (1936a SCHERING-KAHLBAUM A.-G.). On reaction of the acetate with bis-bromomagnesium-acetylene in ether it gives a compound m. 256°, said to be bis-(3β.17β-dihydroxy-Δ^5-androsten-17α-yl)-acetylene 3.3'-diacetate (1936a SCHERING-KAHLBAUM A.-G.) which, however, according to Sondheimer (1956) melts at 294–7°; moreover, reaction of the free dehydroepiandrosterone with bis-bromomagnesium-acetylene gave only 17α-ethynyl-Δ^5-androstene-3β.17β-diol (83 % yield) and no trace of the di-substituted acetylene (1956 Sondheimer); the latter is obtained as a by-product (2–3 % yield) in the reaction of dehydroepiandrosterone with acetylene in the presence of potassium tert.-amyl oxide in ether-benzene-tert.-amyl alcohol (1956 Sondheimer), along with the main product 17α-ethynyl-Δ^5-andro-stene-3β.17β-diol (1939 Stavely; cf. 1938 SQUIBB & SONS; 1937b CIBA) which is also obtained in the presence of sodamide in ether (1935 Claude) or with potassium acetylide (in liquid ammonia) in benzene-ether at room temperature (1937a Ruzicka; 1937b CIBA; 1937 Kathol; 1945 SCHERING A.-G.); in the last-mentioned reaction a small amount of 17β-ethynyl-Δ^5-androstene-3β.17α-diol (p. 2027 s) is also formed (1939a Reichstein).

Dehydroepiandrosterone may be converted into 17β-formyl-Δ^5-andro-sten-3β-ol (p. 2227 s) by refluxing of its acetate with methoxymethyl-mag-nesium chloride in ether, followed by treatment with sulphuric acid and heating with anhydrous formic acid at 110–115° for 3 hrs. (1936d SCHERING-KAHLBAUM A.-G.); for conversion into 17-formyl-Δ^5-androsten-3β-ol, see also the reaction with ethyl dichloroacetate (p. 2616 s). May be converted into a mixture of the C-17 epimeric 17-formyl-Δ^5-androstene-3β.17-diols and 17-formyl-Δ^4-androsten-17-ol-3-ones by reaction with styryl bromide and lithium in abs. ether, followed by ozonization of the resulting mixture of epimeric 17-styryl-Δ^5-androstene-3β.17-diols in ethyl acetate after temporary protection of the nuclear double bond with bromine (1938b SCHERING A.-G.).

Dehydroepiandrosterone and its benzoate do not react with p-tolyl iso-cyanide or o-anisyl isocyanide in the presence of benzoic acid (1945 Baker). Reacts with various acid hydrazides to give water-soluble hydrazones; for polarography of these hydrazones, see 1946 Barnett.

Gives the 16-benzylidene-derivative (p. 2699 s) on refluxing with benz-aldehyde and $KHCO_3$ in aqueous methanol for 10 hrs. (1945 Velluz) or on refluxing of its acetate with benzaldehyde and sodium methoxide in methanol for 1 hr. (1942 Stodola); reacts similarly with m-nitrobenzaldehyde (5 % meth-anolic KOH in alcohol at room temperature, 20 hrs.) and with piperonal (10 % NaOH in alcohol at room temperature for 16 hrs., sodium ethoxide in alcohol on boiling for 1.5 hrs., or K_2CO_3 in aqueous methanol on boiling for 3 hrs.) (1943 Wettstein). With acetone in ether in the presence of sodium or sodamide dehydroepiandrosterone or its acetate gives the 16-isopropylidene derivative (p. 2656 s); reacts similarly with methyl ethyl ketone (1939a Butenandt; 1939 SCHERING A.-G.). The 16-isopropylidene derivative is also formed (in 80 % yield) on refluxing the acetate with KOH in acetone-methanol

(1:1) for 1 hr.; on one occasion an unstable substance, m. 180° with evolution of gas, probably the 16-(α-hydroxyisopropyl) derivative, was obtained which on warming with methanol was converted into the isopropylidene compound (1945 Ross). For reaction with cyclohexanone diethyl acetal, see the $Δ^1$-cyclohexenyl ether (p. 2620 s).

For esterification of dehydroepiandrosterone with butyric, palmitic, stearic, and oleic acids in the presence of pancreas esterase, see 1940 Schramm. Dehydroepiandrosterone is converted into its cyanohydrin (mixture of the two C-17 epimers) with anhydrous hydrogen cyanide in a sealed tube at 50° or at room temperature (1935d Schering A.-G.; 1937a Kuwada; 1938b Miescher), with potassium cyanide and conc. HCl in ice-cold alcohol (1937a Kuwada), or with potassium cyanide and glacial acetic acid in abs. alcohol on refluxing for 15 min. (1938b Butenandt); its acetate also gives the cyanohydrin with potassium cyanide and conc. HCl (1937a Kuwada), while the cyanohydrin 3-acetate is formed with anhydrous hydrogen cyanide at 50° (1935d Schering A.-G.) or on refluxing with potassium cyanide and glacial acetic acid in alcohol for $1/_2$ hr. (1938a, b Butenandt; cf. 1945 Schering A.-G.) or at 0° (1943 Goldberg).

On heating dehydroepiandrosterone or its acetate with potassium cyanide and ammonium carbonate in 60–70% alcohol in an atmosphere of CO_2 at 120–150° and an initial pressure of 25 atm. for 15–24 hrs. it gives, depending on the conditions, $Δ^5$-ætiocholen-3β-ol-17-spirohydantoin ($Δ^5$-androsten-3β-ol-17-spirohydantoin) $C_{21}H_{30}O_3N_2$ [II, R = H; crystals (acetic acid or aqueous pyridine); dec. above 365° to above 400° depending on the rate of heating] or, on using the acetate, its 3-acetate $C_{23}H_{32}O_4N_2$ [II, R = CO·CH₃; crystals (methanol), m. 311° cor. dec.; non-androgenic] which on heating with KOH in propanol for 2 hrs. gives (II, R = H); on heating (II, R = H) with acetic anhydride and potassium acetate at 125° for 4 hrs. it gives a *diacetyl derivative* $C_{25}H_{34}O_5N_2$, crystals (acetic acid), dec. ca. 330° (1938b Miescher; cf. 1936c Ciba).

The acetate reacts with ethyl bromoacetate and zinc (activated by iodine) on refluxing in abs. benzene for 5 hrs., followed by refluxing with aqueous methanolic KOH for 30 min., to give 3β.17β-dihydroxy-17-iso-$Δ^5$-pregnen-20-oic acid (1939b Reichstein; cf. also 1935e Schering A.-G.). On refluxing of the acetate with ethyl α-bromopropionate and zinc (activated by iodine) in abs. benzene for 5 hrs. it gives a small amount of an acid* $C_{22}H_{32}O_3$ [presumably $Δ^{5.17(20)}$-pregnadien-3β-ol-20-carboxylic acid; *methyl ester* $C_{23}H_{34}O_3$, granula (ether-pentane), m. 124–6° cor., $[α]_D^{21}$ —176.5° (acetone)] and, after refluxing with aqueous methanolic KOH for 30 min., a mixture of the two C-20 epimeric 17-iso-$Δ^5$-pregnene-3β.17β-diol-20-carboxylic acids (1941 Lardon; cf. 1935e Schering A.-G.); for similar condensation in the presence of magnesium, see 1936 Schering Corp. For condensation with ethyl α-chloropropionate, see p. 2616 s.

---

* It is not clear in which phase of the reaction hydrolysis takes place.

E.O.C. XIV s. 181

Dehydroepiandrosterone reacts with ethyl dichloroacetate in the presence of amalgamated magnesium to give its dichloroacetate (p. 2619 s) (1937 Ercoli; cf. also 1939a Miescher); the same reaction carried out with dehydroepiandrosterone acetate proceeds in the same way as the following reaction, finally yielding 17-formyl-Δ^5-androsten-3β-ol (see p. 2227 s) (1939a Miescher; 1938b Ciba). On treatment of the acetate with excess ethyl $\alpha.\alpha$-dichloropropionate and amalgamated magnesium in abs. ether with cooling in ice, finally on boiling for 1 hr., followed by treatment with dil. HCl, it gives a mixture [crystalline powder (methanol), m. 161–4⁰] of 20-chloro-3β-acetoxy-17-hydroxy-Δ^5-bisnorcholenic acid ethyl esters which on careful addition of methanolic NaOH is converted into a mixture of 3β-acetoxy-17.20-epoxy-Δ^5-bisnorcholenic acid ethyl esters; on refluxing the latter mixture with aqueous methanolic NaOH for 1 hr. it gives a mixture of the four possible C-17 and C-20 epimeric 3β-hydroxy-17.20-epoxy-Δ^5-bisnorcholenic acids, which on decarboxylation (e.g., with diethylaniline or quinoline at 200⁰) yields non-ketonic compounds, Δ^5-pregnen-3β-ol-20-one (p. 2233 s), and "neopregnenolone" (1939a Miescher; 1938b Ciba); the latter ketone is 17a-methyl-D-homo-Δ^5-androsten-3β-ol-17-one (III) (1940a Ruzicka). A similar mixture of 3β-hydroxy-17.20-epoxy-Δ^5-bisnorcholenic acids (see above), Δ^5-pregnen-3β-ol-20-one,

and an "isopregnenolone" [which may be identical with the above neopregnenolone (Editor)] is obtained, along with some Δ^5-androstene-3β.17β-diol (p. 1999 s), on refluxing of dehydroepiandrosterone with a large excess of ethyl α-chloropropionate and sodium ethoxide in dry ether, followed by refluxing with alcoholic NaOH (1937, 1939 Yarnall; 1937 Merck & Co.). For condensation of dehydroepiandrosterone with β-chloro-γ-butyrolactone in the presence of activated magnesium, see 1942 Ciba.

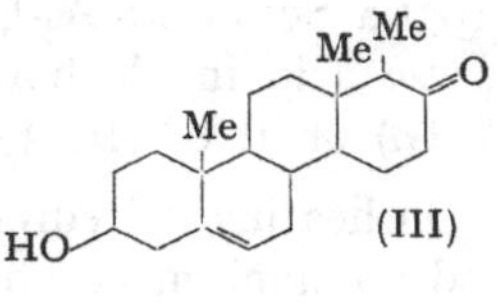

Gives the chlorocarbonate (p. 2619 s) with phosgene in chloroform (1937a Ciba). On heating with ethyl diazoacetate at 130–140⁰, followed by hydrolysis with methanolic KOH, the carboxymethyl ether (p. 2620 s) is formed (1938 Ercoli). On boiling with ¹/₂ mol. oxalic acid in ethyl acetate it gives an addition **compound** $C_{40}H_{58}O_8$ (crystals, m. 144–145.5⁰ dec., partly resolidifying, and remelting at ca. 200⁰) which is decomposed into the components by water or alcohol (1941 Miescher). On condensation of its acetate with methylmalonic-acetic anhydride in the presence of conc. H_2SO_4, followed by hydrolysis with alcoholic KOH and heating to 160⁰, it gives 17-ethylidene-Δ^5-androsten-3β-ol (p. 1518 s) (1936 Schering Corp.). Gives diethyl (3β.17β-dihydroxy-Δ^5-androsten-17-yl)-succinate on condensation with diethyl succinate in the presence of sodium ethoxide, piperidine, or sodamide (1936 Schering Corp.).

With an equal amount of epi-androsterone (p. 2633 s) it forms apparently homogeneous crystals (methanol), m. 157.5–158.5⁰ cor., $[\alpha]_D^{29}$ +51⁰ (alcohol) (1942 Pearlman).

Biological reactions. Dehydroepiandrosterone is dehydrogenated to give Δ^4-androstene-3.17-dione (p. 2880 s; almost quantitative yields) by the action of

Flavobacterium (Micrococcus) dehydrogenans (1940 Ercoli; 1942 Arnaudi), Flavobacterium carbonylicum (1944 Molina), and Proactinomyces erythropolis GRAY & THORNTON and other P. species (1946 Turfitt; cf. also 1948 Turfitt). With a mixture of aerobic bacteria in yeast water on shaking under oxygen for 46 hrs. at 32° it gives Δ^4-androstene-3.17-dione (1938b Mamoli; cf. also 1938a, c Mamoli) and some ætiocholane-3α.17β-diol (p. 2017 s) (1938 Schramm; 1938a, b Mamoli). It is reduced to Δ^5-androstene-3β.17β-diol (p. 1999 s) by fermenting yeast in dil. alcohol at 20° (1937 Mamoli; cf. also 1939 Westphal) or by Bacillus putrificus Bienstock in yeast suspension (1939 Mamoli).

After administration of dehydroepiandrosterone acetate to a man with anterior pituitary insufficiency, there could be isolated from the urine increased amounts of androsterone (p. 2629 s) and ætiocholan-3α-ol-17-one (p. 2637 s), a very small amount of Δ^5-androstene-3β.17β-diol (p. 1999 s), and other compounds (1945a Mason). For increased excretion of 17-keto-steroids after administration of dehydroepiandrosterone to men, see also 1940 Dorfman; 1944a Frame. It can be re-obtained to a considerable extent from female urine after subcutaneous administration (1942 Munson).

PHYSIOLOGICAL PROPERTIES OF 5-DEHYDROEPIANDROSTERONE

For physiological properties, see pp. 141, 1367 s, 1368 s and the literature cited there; see also under "Biological reactions", above.

ANALYTICAL PROPERTIES OF 5-DEHYDROEPIANDROSTERONE

Dehydroepiandrosterone dissolves in conc. H_2SO_4, preferably on brief boiling, to give a yellow solution which on cautious addition of water gives a blue-violet ring and, on shaking, a blue-violet solution (absorption max. at ca. 560 mμ); the colour disappears on further addition of water; sensitivity 5 γ in 1 c.c.; the reaction is also given by i-androsten-6β-ol-17-one (p. 2644 s) (1943 Dirscherl). For the colour reaction with conc. H_2SO_4 alone and in the presence of benzaldehyde or furfural, see 1939 Scherrer and 1939 Woker, resp.; dehydroepiandrosterone and its acetate give very strong Pettenkofer-Schmidt colour reactions (with furfural and sulphuric acid in acetic acid) which may be used for colorimetric estimation (1944 Kerr; 1948 Munson). Gives a faint yellowish blue colour with a highly conc. solution of $SbCl_3$ in glacial acetic acid-acetic anhydride (9:1) (1943 Pincus).

In the Zimmermann reaction with m-dinitrobenzene and 15% KOH in alcohol or alcohol-dioxane, dehydroepiandrosterone very rapidly gives a strong red colour (1943 Wettstein); for colorimetric assay and determination by the Zimmermann reaction, see 1938a Callow; 1939a, b Langstroth; 1940 McCullagh; 1940 Baumann; 1940, 1941a, b Friedgood; 1941 de Laat; 1941, 1943 Saier; 1942 Talbot; 1943 Hansen; 1944 Zimmermann; 1945 Wilson; 1946 Dingemanse.

Gives no colour reaction with 1.4-naphthoquinone + conc. HCl in glacial acetic acid (1939b Miescher).

181*

For separation from other urinary steroids by chromatography, see 1939a, 1942 Callow; 1941 de Laat. May be separated from 5-dehydroandrosterone via the acetates (1936b CIBA), from androsterone, epiandrosterone, and cholesterol by chromatography of their esters made with p-benzeneazobenzoyl chloride (1941 Coffman). May be isolated via its cyanohydrin (see under reactions, p. 2615 s) which readily regenerates the ketone on heating in vacuo at 110–120° or on shaking with alkali in ether (1937a Kuwada). May be isolated on micro-scale in 90 % yield with the aid of succinic anhydride (1941a Pincus).

For determination in the androgen fraction of urine by oxidation with aluminium tert.-butoxide + acetone, followed by polarographic determination of the resulting Δ^4-androstene-3.17-dione in the form of its Girard T derivative, see 1940 Hershberg. For polarographic curve of its Girard T derivative, and polarographic analysis of mixtures, see 1940 Wolfe.

DERIVATIVES OF 5-DEHYDROEPIANDROSTERONE
Keto-derivatives and Esters

Oxime $C_{19}H_{29}O_2N$ (p. 142). — **Rns.** Gives 17-amino-Δ^5-androsten-3β-ol (page 1910 s) on boiling with sodium in alcohol (1936a Ruzicka).

Hydrogen sulphate $C_{19}H_{28}O_5S$. — **Occ.** May be isolated from normal adult urine (1942 Munson; cf. also 1943 Talbot). — **Fmn.** From dehydroepiandrosterone and pyridine sulphur trioxide (precipitated from chlorosulphonic acid and pyridine) on heating in benzene at 72° for 90 min., followed by treatment of the resulting pyridine dehydroepiandrosterone sulphate [not analysed; precipitate (chloroform-petroleum ether), m. 194–5°] with conc. NaCl solution (1943 Talbot). — **Rns.** Its sodium salt is partly hydrolysed by water, slowly at 5°, more rapidly at 100°, and by hydrochloric acid; the hydrolysis is complete on using $BaCl_2$ at 100°, preferably at pH 6.0 (1943 Talbot). On treatment with hydrochloric acid it gives equal quantities of dehydroepiandrosterone and 3β-chloro-Δ^5-androsten-17-one (p. 2474 s) (1942 Venning). — *Sodium salt* $C_{19}H_{27}O_5SNa$. Crystals (methanol-chloroform), m. 192–3°; soluble in water, methanol, ethanol, and butanol, sparingly soluble in acetone (0.16 mg./c.c.) and ethyl acetate (0.07 mg./c.c.), insoluble in ether, benzene, chloroform, and carbon tetrachloride; distribution ratio between equal vols. of butanol and water 9.1:1.0. Is stable in the dry state for long periods of time at room temperature and for a few min. at 100° (1943 Talbot). — *Semicarbazone sodium salt* $C_{20}H_{30}O_5N_3SNa$, bead-like precipitate containing 1 EtOH (alcohol) (1942 Munson).

Acetate $C_{21}H_{30}O_3$ (p. 142), $[\alpha]_D^{24}$ +6° (alcohol), —7° (chloroform) (1946 Barton).

Acetate oxime $C_{21}H_{31}O_3N$. Microcrystals (dil. methanol, or benzene-ligroin), m. 162–3° cor. On refluxing with pyruvic acid and sodium acetate, the acetate is recovered in 75 % yield (1948 Hershberg).

Acetate semicarbazone $C_{22}H_{33}O_3N_3$ (p. 142). Crystals (abs. ethanol, or methanol), m. 279.5–280.5° cor. dec. (bath preheated to 255°) (1948 Hershberg;

cf. 1938a Ruzicka; 1942 Marker; 1944 Schering A.-G.). May be hydrolysed to give dehydroepiandrosterone acetate in practically quantitative yield by NaNO$_2$ in glacial acetic acid on gently warming, finally on boiling for $^1/_2$ min. (1946 Goldschmidt; cf. 1948 Hershberg) or by pyruvic acid and sodium acetate on refluxing in strong acetic acid for 10 min., followed by dropwise addition of water during refluxing (1948 Hershberg); on using pyruvic acid alone, the acetyl group is also hydrolysed to a small extent (1948 Hershberg).

Acetate p-nitrophenylhydrazone C$_{27}$H$_{35}$O$_4$N$_3$. Yellow needles (isopropyl alcohol), m. 291–2^0 cor. dec. (1940 Wolfe); for polarographic curve, see 1940 Wolfe.

Dichloroacetate C$_{21}$H$_{28}$O$_3$Cl$_2$. Needles (dil. acetone), m. 198–9^0. — **Fmn.** Was obtained on attempted condensation of dehydroepiandrosterone with ethyl dichloroacetate and amalgamated magnesium by the method of Darzens. — *Oxime* C$_{21}$H$_{29}$O$_3$NCl$_2$, crystals (alcohol), m. 204^0 (1937 Ercoli).

Chlorocarbonate C$_{20}$H$_{27}$O$_3$Cl, m. 126–7^0. — **Fmn.** From dehydroepiandrosterone and phosgene in chloroform with cooling in a freezing mixture (1937a Ciba).

Carbamate, 5-*Dehydroepiandrosterone urethan* C$_{20}$H$_{29}$O$_3$N, m. 207–8^0. — **Fmn.** From the above chlorocarbonate on treatment with gaseous ammonia in benzene (1937a Ciba).

Methyl carbonate C$_{21}$H$_{30}$O$_4$, m. 194–6^0. — **Fmn.** From 5-dehydroepiandro-sterone and methyl chloroformate with pyridine in benzene or from the above chlorocarbonate and methanol with pyridine in benzene (1937a Ciba).

Phenyl carbonate C$_{26}$H$_{32}$O$_4$, m. 170–2^0 (1937a Ciba).

Benzyl carbonate C$_{27}$H$_{34}$O$_4$, m. 159–160.5^0 (1937a Ciba).

Hydrogen succinate C$_{23}$H$_{32}$O$_5$. Prisms (alcohol), m. 257–9^0 cor. — **Fmn.** From dehydroepiandrosterone and succinic anhydride on refluxing in pyridine for 2 hrs. — **Rns.** With diazomethane in ether it gives the *methyl ester* C$_{24}$H$_{34}$O$_5$, prisms (dil. methanol), m. 156–7^0 cor. (1941 Wolfe).

Benzoate C$_{26}$H$_{32}$O$_3$ (p. 142). Needles (ethyl acetate) which pass into a platy form at 236–9^0; m. 253–4^0 (Kofler block) (1940 Callow), 254–5^0 cor. (1942 Munson).

For the *p-benzeneazobenzoate*, see 1941 Coffman.

p-Toluenesulphonate C$_{26}$H$_{34}$O$_4$S (p. 142). Crystals (acetone), m. 153–4^0. — **Fmn.** From dehydroepiandrosterone on treatment with 2 mols. or more p-toluenesulphonyl chloride in pyridine at room temperature; yield, 93% (1942a Butenandt).

Ethers of 5-Dehydroepiandrosterone

Methyl ether C$_{20}$H$_{30}$O$_2$ (p. 142). — **Fmn.** From cholesteryl methyl ether (p. 1663 s) on conversion into its dibromide, followed by oxidation with CrO$_3$ in 86% acetic acid at 30^0 and then by debromination with zinc dust at room temperature (1936f Ciba). — **Rns.** On treatment with potassium tert.-butoxide

and isoamyl nitrite in tert.-butyl alcohol at room temperature it gives the
16-oximino-derivative (Δ^5-androsten-3β-ol-16.17-dione methyl ether 16-oxime,
p. 2951 s) (1948 Huffman).

Carboxymethyl ether, Δ^5**-Androsten-3β-ol-17-one O-acetic acid** $C_{21}H_{30}O_4$.
Leaflets (dil. acetic acid), m. 203–5°. — Is weakly œstrogenic. — **Fmn.** From
dehydroepiandrosterone and ethyl diazoacetate on heating at 130–140° for
2 hrs., followed by hydrolysis with 8% methanolic KOH. — *Methyl ester*
$C_{22}H_{32}O_4$, crystals (alcohol), m. 137–8° (1938 Ercoli).

Δ^1**-Cyclohexenyl ether** $C_{25}H_{36}O_2$, $[\alpha]_D$ —6° (benzene). — **Fmn.** From dehydroepiandrosterone and cyclohexanone diethyl acetal on heating at 135–140°
for 30 min., then at 180–5° for 45 min., followed by cooling and addition of
methanol and some pyridine. — **Rns.** With sodium in propanol it gives
Δ^5-androstene-3β.17β-diol 3-(Δ^1-cyclohexenyl ether) (p. 2006 s) (1941 Chinoin).

Triphenylmethyl ether $C_{38}H_{42}O_2$. Crystalline powder (alcohol), m. 185–6°
(1935a Schering A.-G.; 1937 Les Laboratoires Français de Chimiothérapie), 190° (1937e Ciba). Sparingly soluble in methanol, ethanol, and
petroleum ether, readily in other organic solvents (1935a Schering A.-G.;
1937 Les Laboratoires Français de Chimiothérapie). — **Fmn.** From dehydroepiandrosterone and triphenylmethyl chloride on heating in anhydrous
pyridine for 2$^1/_2$ hrs. (1935a Schering A.-G.; 1937 Les Laboratoires Français de Chimiothérapie; 1937e Ciba). — **Rns.** Gives Δ^5-androstene-3β.17β-
diol 3-triphenylmethyl ether (p. 2006 s) with sodium in propanol (1935a
Schering A.-G.; 1937 Les Laboratoires Français de Chimiothérapie) or
on hydrogenation in alcohol in the presence of nickel (1937e Ciba). On
heating with Al_2O_3 under nitrogen at 350° it gives Δ^4-androstene-3.17-dione
(p. 2880 s) (1937e Ciba).

Di-(17-keto-Δ^5-androsten-3β-yl) ether $C_{38}H_{54}O_3$. Plates (alcohol), m. 263°.
— **Fmn.** Along with $\Delta^{3.5}$-androstadien-17-one (p. 2400 s), from dehydroepiandrosterone on heating with anhydrous copper sulphate at 200° for 15 min.
(1937 Burrows).

β**-D-Glucoside** $C_{25}H_{38}O_7$. Crystals (alcohol), m. 223–5°. Readily soluble in
hot water, sparingly in cold water, methanol, ethanol, ether, and ethyl acetate.
— **Fmn.** Its tetraacetate is obtained from dehydroepiandrosterone and
acetobromoglucose on shaking with Ag_2O in abs. ether at room temperature
for 60 hrs.; it is hydrolysed by aqueous alcoholic NaOH on the water-bath. —
Tetraacetyl glucoside $C_{33}H_{46}O_{11}$, crystals (alcohol), m. 192–193.5° (1938a
Miescher).

β - D - **Glucuronide, 17 - Keto -** Δ^5 **-androsten -3β -yl -** β **-D -glucuronic acid**
$C_{25}H_{36}O_8$. Prisms (80% or abs. alcohol); the crystals become yellow at ca.
185–190°, opaque at ca. 210°, m. 262–4° dec. (Kofler block). Somewhat
soluble in boiling water. — Is weakly androgenic. — **Fmn.** The triacetate
methyl ester (see below) is obtained from dehydroepiandrosterone and methyl
acetobromo-D-glucuronate on shaking with Ag_2CO_3 in benzene for 24 hrs.
(yield, 24%); it is hydrolysed with $Ba(OH)_2$ in hot methanol and the resulting
barium salt is decomposed with aqueous methanolic sulphuric acid on the

water-bath. — *Triacetate methyl ester* $C_{32}H_{44}O_{11}$, crystals (abs. alcohol),
m. 193–6⁰ (Kofler block); $[\alpha]_D^{20}$ —19.7⁰ (chloroform), —16.2⁰ (benzene) (1938,
1939 Schapiro).

"Dehydroandrosterone" oxide $C_{19}H_{28}O_3$ (p. 142). See 5.6α-epoxyandrostan-
3β-ol-17-one, p. 2769 s.

1934 Butenandt, Dannenbaum, *Z. physiol. Ch.* **229** 192, 203–205.
   a SCHERING A.-G. (Butenandt), *Ger. Pat.* 695774 (issued 1940); *C. A.* **1941** 5649; *Ch. Ztbl.*
     1940 II 2924; SCHERING-KAHLBAUM A.-G., *Swiss Pat.* 194942 (1935, issued 1938); *Ch. Ztbl.*
     1938 II 1995.
   b SCHERING A.-G. (Schoeller, Serini, Hildebrandt, Kathol), *Ger. Pat.* 698910 (issued 1940);
     C. A. **1941** 6603; *Ch. Ztbl.* **1941** I 1444; SCHERING-KAHLBAUM A.-G., *Brit. Pat.* 449379
     (issued 1936); *C. A.* **1936** 8533; *Ch. Ztbl.* **1937** I 3370.
   SCHERING-KAHLBAUM A.-G., *Swiss Pat.* 197784 (issued 1938); *Ch. Ztbl.* **1939** I 2251.
1935 Butenandt, Dannenbaum, Hanisch, Kudszus, *Z. physiol. Ch.* **237** 57, 59, 70, 71.
   CIBA*, *Swiss Pat.* 201201 (issued 1939); *C. A.* **1939** 8625; *Ch. Ztbl.* **1939** II 172.
   Claude, *Fr. Pat.* 809819 (issued 1937); *C. A.* **1937** 8544; *Ch. Ztbl.* **1937** II 814, 815; SCHERING-
     KAHLBAUM A.-G., *Brit. Pat.* 468123 (1936, issued 1937); *C. A.* **1937** 8836.
   Ruzicka, Goldberg, Rosenberg, *Helv.* **18** 1487, 1496.
   a SCHERING A.-G. (Serini, Strassberger, Logemann), *Ger. Pat.* 726629 (issued 1942); *C. A.*
     1943 6278; *Ch. Ztbl.* **1943** I 1297; SCHERING-KAHLBAUM A.-G., *Swiss Pat.* 200912 (1936,
     issued 1939); *Ch. Ztbl.* **1939** I 5010.
   b SCHERING A.-G., *Austrian Pat.* 160824 (issued 1942); *Ch. Ztbl.* **1942** II 2505.
   c SCHERING A.-G., *Norw. Pat.* 61267 (issued 1939); *Ch. Ztbl.* **1940** I 1710.
   d SCHERING A.-G. (Schoeller, Serini), *Ger. Pat.* 657017 (issued 1938); *C. A.* **1938** 3912; *Ch.
     Ztbl.* **1938** II 1088.
   e SCHERING A.-G. (Allardt, Strassberger), *Ger. Pat.* 695952 (issued 1940); *C. A.* **1941** 6068;
     SCHERING-KAHLBAUM A.-G., *Brit. Pat.* 483670 (1936, issued 1938); *C. A.* **1938** 7218; *Ch.
     Ztbl.* **1938** II 3273; *Swiss Pat.* 201619 (1936, issued 1939).
   Wallis, Fernholz, *J. Am. Ch. Soc.* **57** 1511.
1936 Callow, *Chemistry & Industry* **55** 1030.
   a CIBA*, *Fr. Pat.* 47466 (issued 1937); *C. A.* **1937** 7885; *Ch. Ztbl.* **1937** II 3488.
   b CIBA*, *Swiss Pat.* 190542 (issued 1937); *C. A.* **1938** 588; *Ch. Ztbl.* **1938** I 941, 942.
   c CIBA*, *Swiss Pat.* 198133 (issued 1938); *C. A.* **1939** 3531; *Fr. Pat.* 830484 (1937, issued
     1938); *C. A.* **1939** 2542; *Ch. Ztbl.* **1939** I 2826, 2827; *Brit. Pat.* 488829 (1937, issued 1938);
     C. A. **1939** 180.
   d CIBA*, *Swiss Pat.* 200055 (issued 1938); *C. A.* **1939** 3535; *Brit. Pat.* 489364 (issued 1938);
     C. A. **1939** 815; *Ch. Ztbl.* **1939** I 3932.
   e CIBA* (Ruzicka, Wettstein), *U.S. Pat.* 2194235 (issued 1940); *C. A.* **1940** 4746; *Ch. Ztbl.*
     1942 I 778; *Fr. Pat.* 805380 (issued 1936); *C. A.* **1937** 3502; *Ch. Ztbl.* **1937** I 3674.
   f CIBA*, *Fr. Pat.* 804229 (issued 1936); *C. A.* **1937** 2613; *Ch. Ztbl.* **1937** I 3674; *Indian Pat.*
     22679 (issued 1936); *Ch. Ztbl.* **1937** I 2819.
   a Kuwada, *C. A.* **1937** 2224; *Ch. Ztbl.* **1936** I 4737.
   b Kuwada, Yago, *C. A.* **1937** 2224; *Ch. Ztbl.* **1937** I 3808.
   LES LABORATOIRES FRANÇAIS DE CHIMIOTHÉRAPIE (Girard, Sandulesco), *Ger. Pat.* 703342
     (issued 1941); *C. A.* **1942** 223; *Ch. Ztbl.* **1941** I 2972, 2973; *Fr. Pat.* 817340 (1937); *C. A.*
     1938 2150; *Ch. Ztbl.* **1938** I 2217, 2218.
   N. V. ORGANON, *Dutch Pat.* 43190 (issued 1938); *Ch. Ztbl.* **1938** II 1639.
   a Ruzicka, Goldberg, *Helv.* **19** 107.
   b Ruzicka, Kägi, *Helv.* **19** 842, 845.
   c Ruzicka, Goldberg, *Helv.* **19** 1407.
   SCHERING A.-G. (Butenandt, Logemann), *U.S. Pat.* 2170124 (issued 1939); *C. A.* **1940** 1133;
     Ch. Ztbl. **1940** I 1232.

* Gesellschaft für chemische Industrie in Basel; Société pour l'industrie chimique à Bâle.

SCHERING CORP. (Schwenk, Whitman), *U.S. Pat.* 2247822 (issued 1941); *C. A.* **1941** 6396;
SCHERING-KAHLBAUM A.-G. (Schwenk, Whitman), *Brit. Pat.* 501196 (1937, issued 1939);
C. A. **1939** 5868; *Ch. Ztbl.* **1939** II 686, 687.

a SCHERING-KAHLBAUM A.-G., *Fr. Pat.* 47743 (issued 1937); *C. A.* **1938** 4607; *Ch. Ztbl.* **1937** II
3199, 3200; *Brit. Pat.* 467790 (issued 1937); *C. A.* **1937** 8835; SCHERING CORP. (Serini,
Strassberger), *U.S. Pat.* 2243887 (issued 1941); *C. A.* **1941** 5654.

b SCHERING-KAHLBAUM A.-G., *Fr. Pat.* 826330 (issued 1938); *C. A.* **1938** 7478; *Ch. Ztbl.*
1938 II 1447, 1448.

c SCHERING-KAHLBAUM A.-G., *Brit. Pat.* 471908 (issued 1937); *C. A.* **1938** 1284; *Ch. Ztbl.*
1938 I 2402.

d SCHERING-KAHLBAUM A.-G., *Brit. Pat.* 493128 (issued 1938); *Ch. Ztbl.* **1939** I 2829.

Zelinskiĭ, Ushakov, *Bull. acad. sci. U.R.S.S., Sér. chim.* **1936** 879, 890.

1936/37 CIBA*, *Brit. Pat.* 483926 (issued 1938); *C. A.* **1938** 7218; *Fr. Pat.* 828338 (1937, issued
1938); *C. A.* **1939** 177; *Ch. Ztbl.* **1938** II 3273; CIBA* (Ruzicka), *U.S. Pat.* 2168685 (1937,
issued 1939); *C. A.* **1939** 9324; *Ch. Ztbl.* **1940** I 1391.

1937 Burrows, Cook, Roe, Warren, *Biochem. J.* **31** 950, 958, 959.

Butenandt, Tscherning, Dannenberg, *Z. physiol. Ch.* **248** 205, 209 footnote 11.

a CIBA*, *Fr. Pat.* 830858 (issued 1938); *C. A.* **1939** 1761; *Ch. Ztbl.* **1939** I 2458; *Brit. Pat.*
491638 (issued 1938); *C. A.* **1939** 1106; *Ch. Ztbl.* **1939** I 2643.

b CIBA*, *Swiss Pat.* 202847 (issued 1939); *C. A.* **1939** 9554; *Ch. Ztbl.* **1940** I 249, 250.

c CIBA*, *Brit. Pat.* 497394 (issued 1939); *C. A.* **1939** 3812; *Ch. Ztbl.* **1939** I 2828; CIBA
PHARMACEUTICAL PRODUCTS (Miescher, Wettstein), *U.S. Pat.* 2229813 (1938, issued 1941);
C. A. **1941** 3040; *Ch. Ztbl.* **1941** II 2467, 2468.

d CIBA*, *Swiss Pat.* 208748 (issued 1940); *C. A.* **1941** 3773; *Fr. Pat.* 847134 (1938, issued
1939); *C. A.* **1941** 5653; *Ch. Ztbl.* **1940** I 2828.

e CIBA*, *Swiss Pat.* 205888 (issued 1939); *C. A.* **1941** 3039; *Brit. Pat.* 520393 (1938, issued
1940); *C. A.* **1942** 628; *Ch. Ztbl.* **1940** II 3226.

Ercoli, Mamoli, *Ch. Ztbl.* **1938** I 335.

I.G. FARBENIND. (Bockmühl, Ehrhart, Ruschig, Aumüller), *Ger. Pat.* 693351 (issued 1940);
C. A. **1941** 4921; *Brit. Pat.* 502666 (issued 1939); *C. A.* **1939** 7316; *Ch. Ztbl.* **1939** II 1126;
Fr. Pat. 842026 (1938, issued 1939); *C. A.* **1940** 4743; *Ch. Ztbl.* **1940** I 429.

Kathol, Logemann, Serini, *Naturwiss.* **25** 682.

a Kuwada, Miyasaka, *C. A.* **1939** 2145; *Ch. Ztbl.* **1937** II 1825.

b Kuwada, Joyama, *C. A.* **1938** 1709; *Ch. Ztbl.* **1938** II 1612.

LES LABORATOIRES FRANÇAIS DE CHIMIOTHÉRAPIE, Girard, Sandulesco, *Fr. Pat.* 817754
(issued 1937); *C. A.* **1938** 2144; *Ch. Ztbl.* **1938** I 2218.

Mamoli, Vercellone, *Z. physiol. Ch.* **245** 93.

MERCK & CO. (Wallis, Yarnall), *U.S. Pat.* 2123217 (issued 1938); *C. A.* **1938** 6808; *Ch. Ztbl.*
1939 I 2036.

a N. V. ORGANON, *Fr. Pat.* 823618 (issued 1938); *C. A.* **1938** 5854; *Ch. Ztbl.* **1938** II 355;
Brit. Pat. 488966 (issued 1938); *C. A.* **1939** 324.

b N. V. ORGANON, *Fr. Pat.* 827623 (issued 1938); *C. A.* **1938** 8437; *Ch. Ztbl.* **1938** II 3119.

a Oppenauer, *U.S. Pat.* 2384335 (issued 1945); *C. A.* **1946** 178; *Ch. Ztbl.* **1948** I 1039.

b Oppenauer, *Acta Brevia Neerland.* **7** 176; *Rec. trav. chim.* **56** 137, 141.

a Ruzicka, Hofmann, *Helv.* **20** 1280.

b Ruzicka, Fischer, *Helv.* **20** 1291, 1294, 1296.

Saupe, Klötzer, *Fortschr. Geb. Röntgenstrahlen* **56** 344.

SCHERING A.-G. (Inhoffen), *Ger. Pat.* 720015 (issued 1942); *C. A.* **1943** 2520; *Ch. Ztbl.* **1942** II
1037.

a SCHERING-KAHLBAUM A.-G., *Fr. Pat.* 822551 (issued 1938); *C. A.* **1938** 4174; *Ch. Ztbl.*
1938 II 120; SCHERING CORP. (Serini, Köster, Strassberger), *U.S. Pat.* 2379832 (issued
1945); *C. A.* **1945** 5053.

b SCHERING-KAHLBAUM A.-G., *Swiss Pat.* 199448 (issued 1938); *Ch. Ztbl.* **1939** I 3931; *Brit.
Pat.* 486992 (issued 1938); *C. A.* **1938** 8707; *Ch. Ztbl.* **1939** I 1603.

a Ushakov (Ouchakov), Lyutenberg, *J. Gen. Ch. U.S.S.R.* **7** 1821; *Bull. soc. chim.* [5] **4**
1394, 1396.

---

* Gesellschaft für chemische Industrie in Basel; Société pour l'industrie chimique à Bâle.

b Ushakov, Lyutenberg, *Nature* **140** 466.
Wallis, Fernholz, *J. Am. Ch. Soc.* **59** 764.
Yarnall, Wallis, *J. Am. Ch. Soc.* **59** 951.
1938 a Butenandt, Schmidt-Thomé, *Naturwiss.* **26** 253.
b Butenandt, Schmidt-Thomé, *Ber.* **71** 1487, 1489, 1490.
c Butenandt, Peters, *Ber.* **71** 2688, 2691.
Butz, Hall, *J. Biol. Ch.* **126** 265.
a Callow, Callow, Emmens, *Biochem. J.* **32** 1312, 1322.
b Callow, Callow, *Biochem. J.* **32** 1759.
a CIBA*, *Fr. Pat.* 833339 (issued 1938); *C. A.* **1939** 2905; *Ch. Ztbl.* **1939** I 2643.
b CIBA* (Oppenauer, Miescher, Kägi), *Swed. Pat.* 96375 (issued 1939); *Ch. Ztbl.* **1940** I 1231;
Ital. Pat. 362454; *Ch. Ztbl.* **1939** II 3151.
c CIBA* (Miescher, Fischer), *U.S. Pat.* 2172590 (issued 1939); *C. A.* **1940** 450; *Ch. Ztbl.*
1940 I 428; *Fr. Pat.* 841242 (issued 1939); *C. A.* **1940** 4240; *Ch. Ztbl.* **1940** I 916.
Dingemanse, Laqueur, *Biochem. J.* **32** 651.
Ercoli, Mamoli, *Gazz.* **68** 142, 144, 145.
Kuwada, Nakamura, *Ch. Ztbl.* **1939** I 4200.
a Mamoli, Vercellone, *Ber.* **71** 154.
b Mamoli, Vercellone, *Ber.* **71** 1686.
c Mamoli, *Ber.* **71** 2278.
Marker, *J. Am. Ch. Soc.* **60** 1725, 1728.
a Miescher, Fischer, *Helv.* **21** 336, 353, 354.
b Miescher, Wettstein, *Helv.* **21** 1317, 1321–1323.
Mitui, *C. A.* **1938** 6254; *Ch. Ztbl.* **1938** II 3813.
Raoul, Meunier, *Compt. rend.* **207** 681.
RICHTER GEDEON VEGYÉSZETI GYÁRA R. T., *Hung. Pat.* 123023 (issued 1940); *C. A.* **1941**
8217; *Ch. Ztbl.* **1941** I 1573.
a Ruzicka, Hofmann, *Helv.* **21** 88, 91.
b Ruzicka, Hofmann, Meldahl, *Helv.* **21** 371, 372.
c Ruzicka, Hofmann, Meldahl, *Helv.* **21** 597, 598 footnote 2.
Schapiro, *Nature* **142** 1036.
a SCHERING A.-G., *Fr. Pat.* 834941 (issued 1938); *C. A.* **1939** 4602; *Ch. Ztbl.* **1939** I 2827.
b SCHERING A.-G., *Fr. Pat.* 49586 (issued 1939); *C. A.* **1942** 2689; *Ch. Ztbl.* **1940** I 428, 429;
SCHERING CORP. (Logemann, Dannenbaum), *U.S. Pat.* 2344992 (issued 1944); *C. A.* **1944**
3783.
Schramm, Mamoli, *Ber.* **71** 1322.
SQUIBB & SONS (Stavely), *U.S. Pat.* 2239864 (issued 1941); *C. A.* **1941** 4921; *Ch. Ztbl.* **1942** I
1281.
1939 a Butenandt, Schmidt-Thomé, Weiss, *Ber.* **72** 417, 421, 424.
b Butenandt, Schmidt-Thomé, Paul, *Ber.* **72** 1112, 1117.
a Callow, *Biochem. J.* **33** 559, 562, 563.
b Callow, Callow, *Biochem. J.* **33** 931.
Crooke, Callow, *Quart. J. Med.* **32** 233, 242.
Ehrenstein, *J. Org. Ch.* **4** 506, 514.
Hirschmann, *J. Biol. Ch.* **130** 421; **136** 483, 490 (1940).
Kiprianov, Frenkel, *J. Gen. Ch. U.S.S.R.* **9** 1682; *C. A.* **1940** 3756; *Ch. Ztbl.* **1940** I 2802.
Kuwada, Tutihasi, *Ch. Ztbl.* **1939** II 2433.
a Langstroth, Talbot, *J. Biol. Ch.* **129** 759.
b Langstroth, Talbot, Fineman, *J. Biol. Ch.* **130** 585.
Mamoli, Koch, Teschen, *Z. physiol. Ch.* **261** 287, 296.
a Marker, *J. Am. Ch. Soc.* **61** 944.
b Marker, *J. Am. Ch. Soc.* **61** 1287, 1289.
a Miescher, Kägi, *Helv.* **22** 184, 189, 190.
b Miescher, Wettstein, Scholz, *Helv.* **22** 894, 904.
c Miescher, Klarer, *Helv.* **22** 962, 965.
a Reichstein, Meystre, *Helv.* **22** 728, 735.

* Gesellschaft für chemische Industrie in Basel; Société pour l'industrie chimique à Bâle.

b Reichstein, Müller, Meystre, Sutter, *Helv.* **22** 741, 744.

Schapiro, *Biochem. J.* **33** 385.

SCHERING A.-G. (Butenandt, Schmidt-Thomé, Schwenk), *Swed. Pat.* 99872 (issued 1940); *Ch. Ztbl.* **1940** II 2784, 2785; SCHERING CORP. (Butenandt, Schmidt-Thomé, Schwenk), *U.S. Pat.* 2353808 (issued 1944); *C. A.* **1945** 1258.

Scherrer, *Helv.* **22** 1329, 1335.

Stavely, *J. Am. Ch. Soc.* **61** 79.

Ushakov, Lyutenberg, *J. Gen. Ch. U.S.S.R.* **9** 69, 71; *C. A.* **1939** 6334.

Westphal, Wang, Hellmann, *Ber.* **72** 1233, 1237, 1239.

Woker, Antener, *Helv.* **22** 511, 515.

Yarnall, Wallis, *J. Org. Ch.* **4** 270, 279 et seq.

1940 Baumann, Metzger, *Endocrinology* **27** 664.

Callow, Callow, *Biochem. J.* **34** 276, 278.

Dorfman, Hamilton, *J. Biol. Ch.* **133** 753, 756.

Ehrenstein, Decker, *J. Org. Ch.* **5** 544, 552, 555.

Ercoli, *Boll. sci. fac. chim. ind. Bologna* **1940** 279; *C. A.* **1944** 1540; *Ch. Ztbl.* **1941** II 1028.

v. Euw, Reichstein, *Helv.* **23** 1114, 1117.

Friedgood, Whidden, *Endocrinology* **27** 242, 258.

Hattori, Nakamura, *C. A.* **1940** 7294.

Hershberg, Wolfe, Fieser, *J. Am. Ch. Soc.* **62** 3516; *J. Biol. Ch.* **140** 215 (1941).

Hoskins, Webster, *Proc. Soc. Exptl. Biol. Med.* **43** 604.

McCullagh, Schneider, Emery, *Endocrinology* **27** 71.

a Ruzicka, Meldahl, *Helv.* **23** 364.

b Ruzicka, Grob, Raschka, *Helv.* **23** 1518, 1522, 1524.

Schramm, Wolff, *Z. physiol. Ch.* **263** 73.

Stavely, *J. Am. Ch. Soc.* **62** 489.

Wolfe, Hershberg, Fieser, *J. Biol. Ch.* **136** 653, 660, 663.

1941 CHINOIN GYÓGYSZER ÉS VEGYÉSZETI TERMÉKEK GYÁRA R. T., KERESZTY, WOLF (Földi), *Hung. Pat.* 131346 (issued 1943); *C. A.* **1949** 3980; *Belg. Pat.* 446045 (1942, issued 1943); *C. A.* **1945** 709; *Ch. Ztbl.* **1943** II 1501.

CIBA*, *Fr. Pat.* 877821 (issued 1943); *Dutch Pat.* 54147 (issued 1943); *Ch. Ztbl.* **1943** II 147, 148.

Coffman, *J. Biol. Ch.* **140** Proc. XXVIII.

Ehrenstein, *J. Org. Ch.* **6** 626, 637.

Engel, Thorn, Lewis, *J. Biol. Ch.* **137** 205, 211.

Fieser, Wolfe, *J. Am. Ch. Soc.* **63** 1485.

a Friedgood, Whidden, *Endocrinology* **28** 237.

b Friedgood, Berman, *Endocrinology* **28** 248.

Galinovsky, *Ber.* **74** 1624, 1626.

Heard, McKay, *J. Biol. Ch.* **140** Proc. LVI.

de Laat, *Acta Brevia Neerland.* **11** 51.

Lardon, Reichstein, *Helv.* **24** 1127, 1130.

Marker, *J. Am. Ch. Soc.* **63** 1485.

Miescher, Kägi, *Helv.* **24** 986.

Oppenauer, *Z. physiol. Ch.* **270** 97, 103.

a PARKE, DAVIS & Co. (Tendick, Lawson), *U.S. Pat.* 2335616 (issued 1943); *C. A.* **1944** 3095; *Brit. Pat.* 563889 (issued 1944); *C. A.* **1946** 2942.

b PARKE, DAVIS & Co. (Marker, Crooks), *U.S. Pat.* 2359773 (issued 1944); *C. A.* **1945** 4198; *Ch. Ztbl.* **1945** II 1233.

a Pincus, Pearlman, *Science* [N.S.] **93** 163.

b Pincus, Pearlman, *Endocrinology* **29** 413.

Reichstein, Lardon, *Helv.* **24** 955, 958.

Rosenheim, *Nature* **147** 776.

Saier, Warga, Grauer, *J. Biol. Ch.* **137** 317.

Stodola, Kendall, McKenzie, *J. Org. Ch.* **6** 841, 842.

Wettstein, Fritzsche, Hunziker, Miescher, *Helv.* **24** 332 E, 353 E.

Wolfe, Fieser, Friedgood, *J. Am. Ch. Soc.* **63** 582, 589 et seq.

* Gesellschaft für chemische Industrie in Basel; Société pour l'industrie chimique à Bâle.

1942 Arnaudi, *C. A.* **1946** 3152; *Ztbl. Bakt. Parasitenk.* II **105** 352; *C. A.* **1943** 5756.

a Butenandt, Surányi, *Ber.* **75** 591, 594, 597.

b Butenandt, Surányi, *Ber.* **75** 597, 605.

Callow, *Biochem. J.* **36** Proc. XIX.

CHINOIN GYÓGYSZER ÉS VEGYÉSZETI TERMÉKEK GYÁRA R. T., KERESZTY, WOLF, *Fr. Pat.* 882126 (issued 1943); *Ch. Ztbl.* **1944** II 675.

CIBA*, *Fr. Pat.* 882162 (issued 1943); *Ch. Ztbl.* **1944** I 387.

Dobriner, Gordon, Rhoads, Liebermann, Fieser, *Science* [N.S.] **95** 534.

Marker, Crooks, Jones, Shabica, *J. Am. Ch. Soc.* **64** 1276.

Munson, Gallagher, Koch, *Endocrinology* **30** S 1036; *J. Biol. Ch.* **152** 67, 72, 73 (1944).

Pearlman, *Endocrinology* **30** 270.

Ruschig, *Medizin und Chemie* **4** 327, 341; *C. A.* **1944** 4954; *Ch. Ztbl.* **1943** I 2689.

Stodola, Kendall, *J. Org. Ch.* **7** 336, 338.

Talbot, Berman, MacLachlan, *J. Biol. Ch.* **143** 211.

Venning, Hoffman, Browne, *J. Biol. Ch.* **146** 369, 377.

1943 Dirscherl, Zilliken, *Naturwiss.* **31** 349.

Goldberg, Wydler, *Helv.* **26** 1142, 1148.

Hansen, Cantarow, Rakoff, Paschkis, *Endocrinology* **33** 282.

Petrow, Rosenheim, Starling, *J. Ch. Soc.* **1943** 135, 138.

Pincus, *Endocrinology* **32** 176.

Saier, Grauer, Starkey, *J. Biol. Ch.* **148** 213.

Talbot, Ryan, Wolfe, *J. Biol. Ch.* **148** 593.

Wettstein, Miescher, *Helv.* **26** 631, 639, 640.

1944 Callow, Crooke, *C. A.* **1944** 6369; *Ch. Ztbl.* **1944** II 523.

a Frame, Fleischmann, Wilkins, *Bull. Johns Hopkins Hosp.* **75** 95.

b Frame, *Endocrinology* **34** 175.

GLIDDEN Co. (Julian, Cole, Magnani, Conde), *U.S. Pat.* 2374683 (issued 1945); *C. A.* **1946** 1636; *Ch. Ztbl.* **1946** I 372.

Kerr, Hoehn, *Arch. Biochem.* **4** 155.

Molina, Ercoli, *C. A.* **1946** 3152.

Ruzicka, Muhr, *Helv.* **27** 503, 507.

SCHERING A.-G., *Bios Final Report* No. 449, pp. 213–230; *Fiat Final Report* No. 996, pp. 29–44.

Wenner, Reichstein, *Helv.* **27** 24, 30.

Zimmermann, *Vitamine und Hormone* **5** [1952] 1, 124, 152, 170, 237, 260, 276.

1945 Baker, Schlesinger, *J. Am. Ch. Soc.* **67** 1499.

Julian, Cole, Magnani, Meyer, *J. Am. Ch. Soc.* **67** 1728.

Kumler, Fohlen, *J. Am. Ch. Soc.* **67** 437; **70** 4273 (1948).

a Mason, Kepler, *J. Biol. Ch.* **160** 255.

b Mason, Kepler, *J. Biol. Ch.* **161** 235, 239, 240.

Ross, *J. Ch. Soc.* **1945** 25.

SCHERING A.-G., *Bios Final Report* No. 449, pp. 285, 305–307; *Fiat Final Report* No. 996, pp. 61, 62, 72.

Velluz, Petit, *Bull. soc. chim.* [5] **12** 949.

Wilson, Nathanson, *Endocrinology* **37** 208.

1946 Barnett, Morris, *Biochem. J.* **40** 450.

Barton, *J. Ch. Soc.* **1946** 1116, 1121.

Dingemanse, Huis in't Veld, *Acta Brevia Neerland.* **14** 34.

Furchgott, Rosenkrantz, Shorr, *J. Biol. Ch.* **163** 375, 378.

GLIDDEN Co. (Julian, Karpel, Armstrong), *U.S. Pat.* 2464236 (issued 1949); *C. A.* **1949** 4706; *Ch. Ztbl.* **1949** E 1599.

Goldschmidt, Veer, *Rec. trav. chim.* **65** 796, 798.

Heard, McKaye, *J. Biol. Ch.* **165** 677, 679, 680–682.

Milas, Milone, *J. Am. Ch. Soc.* **68** 738, 739.

Turfitt, *Biochem. J.* **40** 79.

1947 Hirschmann, Hirschmann, *J. Biol. Ch.* **167** 7, 21.

* Gesellschaft für chemische Industrie in Basel; Société pour l'industrie chimique à Bâle.

1948 Barton, Klyne, *Nature* **162** 493.
Dingemanse, Huis in't Veld, Hartogh-Katz, *Nature* **161** 848; **162** 492.
Ehrenstein, *J. Org. Ch.* **13** 214.
Hershberg, *J. Org. Ch.* **13** 542 (footnote 2), 544, 546.
Huffman, Lott, *J. Biol. Ch.* **172** 789, 791.
Munson, Jones, McCall, Gallagher, *J. Biol. Ch.* **176** 73.
Turfitt, *Biochem. J.* **42** 376, 377.
Wieland, Miescher, *Helv.* **31** 211, 212.
1949 Fieser, Fieser, *Natural Products Related to Phenanthrene*, 3rd Ed., New York, p. 494.
1950 Hershberg, Rubin, Schwenk, *J. Org. Ch.* **15** 292, 295.
von Seemann, Grant, *J. Am. Ch. Soc.* **72** 4073, 4075.
1956 Sondheimer, Mancera, Flores, Rosenkranz, *J. Am. Ch. Soc.* **78** 1742.

$Δ^{9(11)}$**-Androsten-3α-ol-17-one** $C_{19}H_{28}O_2$. For the structure, see 1943, 1946 Shoppee. — Crystals (methanol), m. 188.5–190⁰ (1946, 1950 Miller); prisms, m. 189–190⁰ cor. (Kofler block) (1946 Shoppee); m. 187–9⁰, $[α]_D^{26}$ + 140⁰ (95 % alcohol) (1945 Mason). — Its androgenic potency is one fourth that of androsterone (1945 Dorfman, 1946 Miller).

Occ. and **Fmn.** Has been isolated from the urine (hydrolysed with 10 % HCl) of a woman with (probable) adrenal cortical hyperplasia (1945 Dorfman; 1946, 1950 Miller); its acetate is obtained, along with some free hydroxyketone, from the 3-acetate of androstane-3α.11β-diol-17-one (p. 2748 s) on dehydration by refluxing with glacial acetic acid-conc. HCl (4:1) for 15 min., and is hydrolysed by aqueous methanolic NaOH at room temperature (1945 Mason; cf. also 1946, 1950 Miller; 1946 Shoppee); the aforementioned diolone was present in the non-hydrolysed urine of the same woman (1945 Dorfman; 1946, 1950 Miller) [and also occurs in the urine of other persons; see the diolone, p. 2748 s] and may also be dehydrated by POCl₃ in pyridine at room temperature (1950 Miller).

Rns. On treatment with 1 mol. CrO₃ in glacial acetic acid at 20⁰ for 16 hrs. it gives 62 % yield of $Δ^{9(11)}$-androstene-3.17-dione (p. 2891 s) and traces of 9.11-epoxyandrostane-3.17-dione (p. 2891 s) (1946, 1947 Shoppee). On hydrogenation in glacial acetic acid in the presence of reduced platinum oxide it gives androstane-3α.17β-diol (p. 2011 s) (1945 Mason).

Acetate $C_{21}H_{30}O_3$. Crystals (acetone), m.191–2⁰ (1945 Mason); prisms, m.192–3⁰ cor. (Kofler block) (1946 Shoppee); $[α]_D^{25}$ + 135⁰ (95 % alcohol) (1945 Mason).

$Δ^{9(11)}$**-Androsten-3β-ol-17-one** $C_{19}H_{28}O_2$. For the structure, see 1946, 1947 Shoppee; 1947 Reich; cf. also 1943 Alther. — Crystals (dil. methanol), m. 170–172.5⁰ cor. (Kofler block); sublimes at 130⁰/0.01; $[α]_D^{13}$ + 125.5⁰, $[α]_{5461}^{13}$ + 161⁰ (alcohol) (1946 Shoppee). — **Fmn.*** From androstane-3β.11β-diol-17-one (p. 2748 s) on refluxing with conc. HCl in strong acetic acid for 15 min. (1940 Shoppee); its acetate is obtained from the 3-acetate of the same diolone on

*For synthesis of the corresponding dl-compound, carried out after the closing date for this volume, see 1957 Johnson.

distillation with $KHSO_4$ at 135–140° (bath temperature) under 0.05 mm. pressure (1940 Shoppee) or on treatment with $POCl_3$ in pyridine at 0° and then at 18° (20 hrs.; yield, 80%) (1947 Reich), and is hydrolysed with K_2CO_3 in aqueous methanol (20°, 48 hrs.) (1946 Shoppee).

Rns. On treatment with 1 mol. of CrO_3 in glacial acetic acid at 15° for 16 hrs. it gives 84% crude yield of $\Delta^{9(11)}$-androstene-3.17-dione (p. 2891 s) (1946 Shoppee) and traces of 9.11-epoxyandrostane-3.17-dione (p. 2891 s) (1946, 1947 Shoppee). Oxidation of the acetate with 1 mol. of CrO_3 in glacial acetic acid at 27–8° (24 hrs.) affords ca. 10% yield of 9.11-epoxyandrostan-3β-ol-17-one acetate (below) and ca. 13% yield of $\Delta^{9(11)}$-androsten-3β-ol-12.17-dione acetate (p. 2950 s); 62% of the starting acetate remains unattacked (1947 Reich). Oxidation with excess perbenzoic acid in chloroform in the dark at room temperature gives 85% yield of the afore-mentioned epoxy-acetate (1947 Reich). On hydrogenation of the acetate in glacial acetic acid in the presence of platinum oxide, followed by heating with acetic anhydride and pyridine at 100°, it gives the diacetate of androstane-3β.17β-diol (p. 2014 s) (1940 Shoppee).

Acetate $C_{21}H_{30}O_3$. Prisms (pentane), m. 102° cor.; $[\alpha]_D^{14}$ +111° (acetone). Gives a strong, yellow coloration with tetranitromethane in chloroform. — **Fmn.** From the preceding compound on treatment with acetic anhydride and pyridine at 20° (1940 Shoppee). See also above. — **Rns.** See above.

9.11-Epoxyandrostan-3β-ol-17-one, $\Delta^{9(11)}$-Androsten-3β-ol-17-one oxide

$C_{19}H_{28}O_3$. Leaflets (ether-petroleum ether), m. 182° to 186° cor. (Kofler block). — **Fmn.** Its acetate is obtained from the preceding acetate on oxidation with ca. 3 mols. of perbenzoic acid in chloroform in the dark at room temperature for 20 hrs. (yield, 85%) or on oxidation with 1 mol. of CrO_3 in glacial acetic acid at 27–28° for 24 hrs. (yield, ca. 10%); it is hydrolysed by 2% methanolic KOH at room temperature (16 hrs.) to give the free epoxyandrostanolone. — **Rns.** Oxidation with 1 mol. of CrO_3 in glacial acetic acid at room temperature affords 9.11-epoxyandrostane-3.17-dione (p. 2891 s) (1947 Reich).

Acetate. $C_{21}H_{30}O_4$. Leaflets (ether-petroleum ether), m. 172–8° cor. (Kofler block); $[\alpha]_D^{16}$ +71° to +79° (acetone). The absorption max. (in alcohol) at 225 mμ (log ε 3.10) is possibly due to the presence of an impurity (1947 Reich). — **Fmn.** See above.

$\Delta^{11(?)}$-Androsten-3α-ol-17-one $C_{19}H_{28}O_2$.

Prisms (aqueous methanol), m. 181° to 183° cor. (1941 Wolfe); needles (ether - pentane), m. 178–180° cor. (Kofler block) (1946 Shoppee); crystals (acetone - ligroin), m. 176–176.5° (1948 Lieberman); $[\alpha]_D^{28}$ +122° (95% alcohol) (1941 Wolfe; 1948 Lieberman). Gives a faint, yellow coloration with tetranitromethane (1941 Wolfe; cf. 1948 Lieberman). For polarographic analysis of its Girard derivative, see 1941 Wolfe.

Occ. Has been isolated from the acid - hydrolysed urine of a ten-year-old girl with a cortico-adrenal tumour (8 mg./l.) (1941 Wolfe). Its acetate has been isolated from the urine of normal persons and that of a male eunuch with an adrenal hyperplasia, after acid hydrolysis, treatment with Girard's reagent T, hydrolysis of the condensation products in the presence of glacial acetic acid, and chromatographic adsorption on Al_2O_3 (1948 Lieberman). It may arise from a precursor which is easily dehydrated (1941 Wolfe; 1945 Dorfman; 1946 Miller; 1946 Shoppee; 1948 Lieberman).

Rns. Gives $\Delta^{11(?)}$-androstene-3.17-dione (p. 2892 s) on oxidation with CrO_3 in glacial acetic acid at 15° (1946 Shoppee). On hydrogenation of its acetate in glacial acetic acid in the presence of Adams's platinum catalyst at room temperature and atmospheric pressure it gives the 3-acetate of androstane-3α.17β-diol (p. 2011 s) (1941 Wolfe). It is not attacked by refluxing with alcoholic HCl for 2 hrs. (1941 Wolfe).

Semicarbazone $C_{20}H_{31}O_2N_3$ (no analysis given). Crystals (95 % alcohol), m. 279–280° cor. dec. (1941 Wolfe).

Acetate $C_{21}H_{30}O_3$. Needles (aqueous acetic acid, ligroin, or aqueous acetone), m. 178–180° cor.; $[\alpha]_D^{27} + 115°$ (95 % alcohol) (1941 Wolfe; 1948 Lieberman). — **Fmn.** From the above hydroxyketone on heating with acetic anhydride in pyridine at 100° for 30 min. (1941 Wolfe). See also under "Occurrence" (above). — **Rns.** Is hydrolysed by K_2CO_3 in aqueous methanol, only partially at 20° (1946 Shoppee), almost completely on refluxing for 1 hr. (1948 Lieberman).

Benzoate $C_{26}H_{32}O_3$ (no analysis given). Plates (aqueous acetone), crystals (ligroin), m. 162–4° cor. (1941 Wolfe; 1948 Lieberman). — **Fmn.** From the above hydroxyketone on heating with benzoyl chloride in pyridine at 100° for 30 min. (1941 Wolfe; 1948 Lieberman). — **Rns.** It is not hydrolysed by K_2CO_3 in aqueous methanol at 20° (48 hrs.) and only partially on refluxing (1 hr.) or by NaOH in aqueous methanol at 20° (16 hrs.) (1946 Shoppee).

5 - iso -Δ^{11}-Androsten - 3α - ol - 17 - one, Δ^{11}-Aetiocholen - 3α - ol - 17 - one $C_{19}H_{28}O_2$.

Not characterized by Reichstein (1943); has been described in detail, after the closing date for this volume, by Lardon (1947). — **Fmn.** Along with Δ^{11}-pregnen-3α-ol-20-one (p. 2239 s) and other compounds, from the dibromide of methyl 3α-acetoxy-Δ^{11}-cholenate on oxidation with CrO_3 in strong acetic acid at 70°, followed by debromination with zinc dust in 95 % acetic acid at room temperature and hydrolysis by refluxing with methanolic KOH (1943 Reichstein).

1940 Shoppee, *Helv.* **23** 740, 743–745.
1941 Wolfe, Fieser, Friedgood, *J. Am. Ch. Soc.* **63** 582, 591, 592.
1943 Alther, Reichstein, *Helv.* **26** 492, 497.
          Reichstein, *U.S. Pat.* 2387706 (issued 1945); *C. A.* **1946** 994, 995; *Brit. Pat.* 567393 (issued 1945); *C. A.* **1947** 3264.
          Shoppee, Reichstein, *Helv.* **26** 1316, 1319 footnote 1.

1945 Dorfman, Schiller, Sevringhaus, *Endocrinology* 37 262.
 Mason, Kepler, *J. Biol. Ch.* 161 235, 248, 249.
1946 Miller, Dorfman, Sevringhaus, *Endocrinology* 38 19.
 Shoppee, *J. Ch. Soc.* 1946 1134.
1947 Lardon, Lieberman, *Helv.* 30 1373.
 Reich, Lardon, *Helv.* 30 329, 700.
 Shoppee, *Helv.* 30 766.
1948 Lieberman, Dobriner, Hill, Fieser, Rhoads, *J. Biol. Ch.* 172 263, 272, 279
1950 Miller, Dorfman, *Endocrinology* 46 514, 522, 524.
1957 Johnson, Allen, *J. Am. Ch. Soc.* 79 1261.

Androstan-3α-ol-17-one, Androsterone, *"cis-Androsterone"* $C_{19}H_{30}O_2$ (p. 143).

Crystals (acetone-ether), m. 185–185.5° cor. (Kofler block); $[\alpha]_D^{15} + 88°$, $[\alpha]_{5461}^{15} + 107°$ (dioxane) (1942 v. Euw). Absorption max. in alcohol at 292 mμ (1937 Hogness); for absorption spectra in alcohol and in conc. H_2SO_4, see also 1937 Bugyi. For the infra-red absorption spectrum of a film, see 1946 Furchgott. Dipole moment in dioxane at 25° 3.70 D (1945 Kumler). Soluble in 10–20% aqueous solutions of sodium dehydrocholate (1944 Cantarow).

Occurrence and Biological Formation of Androsterone

For occurrence in normal male urine, see also 1939a, 1940 Callow; 1939 Chou; 1940 Hoskins; 1941 Engel; occurs in smaller amount also in castrate urine (1940 Callow; 1940 Hoskins). Its hydrogen sulphate has been isolated from the urine of a man with an interstitial cell tumour of the testis (1942 Venning). In urine of normal women in an amount comparable to that in normal men's urine (1938b, 1939b Callow); in almost the same amount in the urine of ovariectomized women (1939, 1940 Hirschmann) and in that of pregnant women (1938 Marker). In small amount in the urine of women with cortico-adrenal tumours (1941 Wolfe; 1945b Mason); in varying amounts in urine from patients with adrenal hyperplasia (1942 Dobriner; 1945b Mason); in the urine of post-menopausal women with cancer of the breast (1943 Hill). In small amounts in the urine of bulls (1939a Marker), steers (1939b Marker), and pregnant sows (1939c Marker).

Androsterone is excreted in increased amount after administration of testosterone propionate (p. 2588 s) to normal men (1940b Dorfman), to hypogonadal men (1939a Callow; 1939, 1940d Dorfman), normal women (1945 Schiller), male chimpanzees (1944 Fish), pregnant rhesus monkeys (1944 Horwitt), after administration of dehydroepiandrosterone (p. 2609 s) to a man with anterior pituitary insufficiency (1945a Mason), and after administration to hypogonadal men of androsterone itself (1940b Dorfman), of androstane-3.17-dione (p. 2892 s) (1940a, b Dorfman), Δ^4-androstene-3.17-dione (page 2880 s), and androstane-3α.17β-diol (p. 2011 s) (1940b Dorfman). For increased excretion of androsterone and other 17-ketosteroids after administration of various steroids, see also 1944a Frame.

Androsterone is obtained in small yield from Δ^4-androstene-3.17-dione
(p. 2880 s) by the action of bulls' testes extract at 37^0 (1939 Ercoli); the
reductive action of such extracts is due to the presence of putrefactive bacteria
(1938 Schramm; 1938a Mamoli). Along with an equal amount of epiandrosterone, from androstane-3.17-dione (p. 2892 s) with bulls' testes extract containing putrefactive bacteria (1938b Mamoli).

FORMATION OF ANDROSTERONE

The preparation of androsterone starting from cholesteryl chloride dibromide (1936 Marker; see p. 143) passes through 3β-chloro-Δ^5-androsten-
17-one (p. 2474 s) and its hydrogenation product 3β-chloroandrostan-17-one
(p. 2476 s) which is refluxed with potassium acetate and n-valeric acid for
14 hrs. and then with aqueous alcoholic NaOH for 2 hrs.; the 3-chloro-
Δ^5-androsten-17-one, similarly used by Butenandt (see p. 143) was also the
3β-isomer; for the latter formation, see also 1934a SCHERING-KAHLBAUM A.-G.
From 3β-chloroandrostan-17-one on treatment with aqueous NaOH (1935
PARKE, DAVIS & Co.).

Androsterone is obtained from the p-toluenesulphonate of epiandrosterone
(p. 2633 s) on heating with potassium acetate in glacial acetic acid (1937c
CIBA). From Δ^4-androstene-3.17-dione (p. 2880 s) on hydrogenation in glacial
acetic acid in the presence of platinum black and some 48% HBr (1936a
PFEILRING-WERKE); similarly from androstane-3.17-dione (p. 2892 s) (1936
SCHERING A.-G.; 1936a, b PFEILRING-WERKE); for formation from the two
diketones, see also under "Biological Formation", p. 2629 s.

Androsterone butyrate may be obtained from the butyrate oxime of Δ^{16}-allo-
pregnen-3α-ol-20-one (p. 2241 s) by treatment with p-toluenesulphonyl chloride
in cold pyridine, followed by hydrolysis with dil. H_2SO_4 at room temperature
(1941a PARKE, DAVIS & Co.). Androsterone acetate is obtained from allo-
pregnan-3α-ol-20-one (p. 2246 s) on treatment with phenylmagnesium bromide,
followed by dehydration and acetylation and then by ozonolysis (1934e
SCHERING-KAHLBAUM A.-G.; cf. 1936 WINTHROP CHEMICAL Co. INC.; 1936
I.G. FARBENIND.); for its formation from 3α-hydroxyætioallocholanic acid
methyl ester via 17-diphenylmethylene-androstan-3α-yl acetate (see p. 143),
see also 1935b SCHERING A.-G.; 1936 WINTHROP CHEMICAL Co. INC.; 1936
I.G. FARBENIND. From 3α-hydroxy-$\Delta^{17(20)}$-allopregnen-21-oic acid on ozonolysis in chloroform (1942c Marker; 1941b PARKE, DAVIS & Co.).

REACTIONS OF ANDROSTERONE

For reduction of androsterone and its esters to androstane-3α.17α-diol and
androstane-3α.17β-diol and their 3-esters, see under the formation of the
two diols, pp. 2010 s and 2011 s, resp. On irradiation with ultra-violet light
in dioxane, androsterone is isomerized to lumiandrosterone (p. 2640 s) (1944
Butenandt). With 1 mol. $PCl_5 + CaCO_3$ in chloroform at 0^0 it gives 3β-chloro-
androstan-17-one (p. 2476 s); on dropwise addition of an ethereal solution
at 5^0 to a solution of thionyl chloride in ether containing $CaCO_3$, a small

amount of 3α-chloroandrostan-17-one (p. 2476 s) is formed and, as main
product, a crystalline compound, m. ca. 200⁰, probably androsterone sulphite,
from which androsterone may be regenerated by hydrolysis with 4% methanolic KOH (1946 Shoppee).

Reacts with alkylmagnesium halides to give the corresp. 17α-alkylandro-
stane-3α.17β-diols (see, e.g., pp. 2023 s, 2034 s, 2038 s) (1935 a, b Ruzicka;
1934 b, c, 1936 a SCHERING-KAHLBAUM A.-G.); at the same time androstane-
3α.17β-diol (p. 2011 s) is formed (see, e.g., 1936 b Ruzicka); this reductive
action is the sole one on using isopropylmagnesium iodide (1934 d SCHERING-
KAHLBAUM A.-G.). On heating of the acetate with ethyl bromoacetate, zinc,
and a little iodine in benzene on the water-bath, followed by saponification,
it gives androstane-3α.17β-diol-17α-acetic acid (1936 SCHERING CORP.); reacts
similarly with ethyl α-bromopropionate (1937 SCHERING A.-G.). May be converted into 17β-formylandrostan-3α-ol (p. 2228 s) by reaction of its esters with
alkyl dihalogenoacetates in ether in the presence of amalgamated magnesium,
followed by treatment with alkali and decarboxylation; its benzoate reacts similarly with ethyl α.α-dibromophenylacetate in the presence of zinc (1938 CIBA).

Reacts with m-nitrobenzaldehyde and a little KOH in alcohol at room
temperature to give the 16-m-nitrobenzylidene-derivative (p. 2717 s) (1943
Wettstein). Gives the cyanohydrin on treatment with liquid hydrogen
cyanide or with potassium cyanide + conc. HCl
in cold alcohol (1938 Kuwada); the acetate
reacts similarly with potassium cyanide and
glacial acetic acid in alcohol on boiling (1939
Butenandt). On heating with potassium cyanide
and ammonium carbonate in 70% alcohol at
135⁰ under 25 atm. pressure it gives **androstan-3α-ol-17-spirohydantoin**
$C_{21}H_{32}O_3N_2$ (I) (1936 b CIBA).

Biological reactions. Androsterone administered to dwarfed, sexually retarded boys causes increased excretion of 17-ketosteroids in the urine (1944 a
Frame). After administration of androsterone or its propionate to normal
male guinea pigs, epiandrosterone (p. 2633 s) could be isolated from the urine
(1945 Dorfman). Strongly increased excretion of œstrogens has been observed
after administration of androsterone benzoate to normal men (1937 Steinach).

PHYSIOLOGICAL PROPERTIES OF ANDROSTERONE

For physiological properties, see pp. 143, 1367 s, 1368 s, and the literature
cited there; see also under "Biological reactions", above.

ANALYTICAL PROPERTIES OF ANDROSTERONE

For colour reactions in alcohol with conc. H_2SO_4 alone and in the presence
of benzaldehyde or furfural, see 1939 Scherrer and 1939 Woker, resp.

Androsterone and its hydrogen sulphate and succinate give an intense
blue colour (absorption max. at 610–620 mμ) with a highly conc. solution of
SbCl₃ in glacial acetic acid-acetic anhydride (9:1) on heating on the water-
bath for 20 min., diluting with a mixture of 19 parts of glacial acetic acid

and 1 part of water, and incubating in the dark at room temperature for 40–60 min.; the same colour in equal intensity is given by epiandrosterone, ætiocholan-3α-ol-17-one, and an androsten-17-one. With $SbCl_3$ in acetic anhydride alone a greenish colour (absorption max. at 460 and 610 mμ) is produced, with $SbCl_3$ in chloroform a colour with absorption max. at 560 mμ to 620 mμ (1943 Pincus).

For the Zimmermann test with m-dinitrobenzene and KOH (see p. 144) and its application to colorimetric determination, see also 1938a Callow; 1938 Neustadt; 1939a, b, c Langstroth; 1939, 1940a, b, c, 1941a, b Friedgood; 1940 Holtorff; 1940 McCullagh; 1940 Baumann; 1940 Tastaldi; 1941 de Laat; 1941a Pincus; 1941, 1943 Saier; 1943 Hansen; 1943 Luft; 1943 Nathanson; 1944 Zimmermann; 1945 Wilson; 1946 Dingemanse.

For chromatographic separation from other steroids, see 1937a Ciba; 1942 Callow. For separation from other urinary steroids, see also 1941b Pincus; 1944b Frame. Androsterone may be isolated from reaction mixtures as acetate oxime or acetate semicarbazone (1934b Schering A.-G.), as acetate semicarbazone, propionate semicarbazone, or benzoate semicarbazone (1935b Ciba). May be separated from epiandrosterone via the acetates or the benzoates, androsterone acetate being the less soluble in methanol and epiandrosterone benzoate the less soluble in ethyl acetate (1936a Ciba). May be separated from epiandrosterone, dehydroepiandrosterone, and cholesterol by chromatography of their esters prepared with p-benzeneazobenzoyl chloride (1941 Coffman).

For determination in the androgen fraction of urine by oxidation with aluminium tert.-butoxide + acetone, followed by polarographic determination of the resulting androstane-3.17-dione in the form of its Girard T derivative, see 1940 Hershberg.

Derivatives of Androsterone

Oxime $C_{19}H_{31}O_2N$ (p. 144). — **Rns.** Gives 17-aminoandrostan-3α-ol (page 1911 s) on boiling with sodium in alcohol or on hydrogenation in dioxane in the presence of platinum oxide or HCl (1935a Ciba).

Hydrogen sulphate $C_{19}H_{30}O_5S$. — **Occ.** In the urine of a patient with an interstitial cell tumour of the testis. — **Rns.** On refluxing its sodium salt with 8–10% HCl in carbon tetrachloride for 6 hrs. it gives androsterone and $\Delta^{2\ or\ 3}$-androsten-17-one (p. 2401 s). — *Sodium salt* $C_{19}H_{29}O_5SNa$. Needles (acetone containing a few drops of water), m. 144°; needles with 1 H_2O (alcohol), m. ca. 190° dec.; readily soluble in water, methanol, and glacial acetic acid, less soluble in ethanol, sparingly in anhydrous acetone, insoluble in ether, benzene, and chloroform. Is only weakly androgenic. — *Semicarbazone* (not analysed), crystals, m. 245°; sparingly soluble in water (1942 Venning).

Acetate $C_{21}H_{32}O_3$ (p. 144). Crystals (ether), m. 164.5–165.5° cor. (Kofler block) (1942 v. Euw), 165–6° cor. (1939 Dorfman); $[\alpha]_D^{14} +77°$, $[\alpha]_{5461}^{14} +92°$ (dioxane) (1942 v. Euw).

Butyrate $C_{23}H_{36}O_3$. Needles (methanol), m. 102–3⁰. — **Fmn.** From androsterone and butyric anhydride on heating in pyridine on the water-bath for 1 hr. (1937 Miescher). See also p. 2630 s.

Palmitate $C_{35}H_{60}O_3$ (not analysed). Butter-like mass, m. ca. 40⁰. — **Fmn.** From androsterone and palmitoyl chloride in pyridine (1937 Miescher).

Benzoate $C_{26}H_{34}O_3$ (p. 144), m. 178.5–179.5⁰; more soluble in ethyl acetate than epiandrosterone benzoate (1936a CIBA).

For the *p-benzeneazobenzoate*, see 1941 Coffman.

β-D-**Glucoside** $C_{25}H_{40}O_7$. Crystals (acetone), m. 228–9⁰ cor. — **Fmn.** Its tetraacetate (below) is obtained from androsterone and acetobromo-D-glucose on shaking with Ag_2O in abs. ether for 20 hrs.; it is hydrolysed with methanolic sodium methoxide at 20⁰ (cf. also the formation of epiandrosterone β-D-glucoside, p. 2637 s). — *Tetraacetate* $C_{33}H_{48}O_{11}$, needles (alcohol), m. 154⁰ cor., re-solidifying, and remelting at 179–181⁰ cor. (1941 b Miescher).

Androstan-3β-ol-17-one, Epiandrosterone, *"trans-Androsterone"*, *"Isoandro-*

sterone" $C_{19}H_{30}O_2$ (p. 144). Granula (ether-pentane), m. 178⁰ cor.; prisms + aq. (dil. methanol) (1938 Steiger); crystals (acetone-ether), m. 176–7⁰ cor. (Kofler block) (1942 v. Euw); prisms (ether-pentane), m. 174⁰ to 176⁰ cor. (1943b Shoppee); $[\alpha]_D^{15} +81⁰$, $[\alpha]_{5461}^{15} +101⁰$ (dioxane) (1942 v. Euw), $[\alpha]_D^{18} +88⁰\pm3⁰$ (methanol) (1943b Shoppee), $[\alpha]_D^{24} +95⁰$ (ethanol), $+90⁰$ (chloroform) (1946 Barton). For the infra-red absorption spectrum of a film, see 1946 Furchgott. Dipole moment in dioxane at 25⁰ 2.95 D (1945 Kumler).

OCCURRENCE AND BIOLOGICAL FORMATION OF EPIANDROSTERONE

Epiandrosterone occurs in the urine of normal men and women (1940, 1942 Pearlman); it was also isolated from the urine of a woman suffering from adrenal hyperplasia (1938 Butler; cf. also 1941 a Hirschmann) and from the urines of both men and women suffering from cancer (1940, 1942 Pearlman; but cf. 1943 Hill). Could be isolated from the urine of a hypogonadal man after administration of testosterone propionate (1941 Dorfman) and from the urine of normal male guinea pigs after administration of testosterone propionate (1940c Dorfman) or androsterone (or its propionate) (1945 Dorfman).

Epiandrosterone is obtained, along with an equal amount of androsterone, from androstane-3.17-dione (p. 2892 s) by the action of bulls' testes extract containing putrefactive bacteria (1938b Mamoli) and is also obtained, in 36 % yield, from the same diketone by the action of Bacillus putrificus Bienstock in aqueous yeast suspension at 36–7⁰ (6 days) (1939 Mamoli).

FORMATION OF EPIANDROSTERONE *

The acetate may be obtained from the acetate of cholestanol by oxidation with lead tetraacetate in glacial acetic acid, followed by hydrolysis with 5 %

---

* For a total synthesis of dl-epiandrosterone, carried out after the closing date for this volume, see 1953, 1956 Johnson.

182*

H_2SO_4 in alcohol, heating with H_3PO_4, acetylation, ozonization, and treatment with CrO_3 (1936 LES LABORATOIRES FRANÇAIS DE CHIMIOTHÉRAPIE).

Epiandrosterone acetate is obtained from 3α-chloroandrostan-17-one (p. 2476 s) on heating with a 20% solution of fused potassium acetate in glacial acetic acid for 2 hrs. in a sealed tube at 180°; yield, 19% (1946 Shoppee). Epiandrosterone (or its acetate) is obtained from 5-dehydroepiandrosterone (or its acetate) (p. 2609 s) on reduction with 1 mol. hydrogen in glacial acetic acid in the presence of platinum black at 20° (1935 c SCHERING A.-G.). The acetate is obtained, along with a small amount of the acetate of ætiocholan-3β-ol-17-one (p. 2638 s), from 5-dehydroepiandrosterone acetate on reduction with excess hydrogen in glacial acetic acid in the presence of platinum oxide at room temperature, followed by oxidation with CrO_3 in glacial acetic acid at room temperature (1941 Reichstein; cf. also 1938 Steiger); the two isomers are separated by hydrolysis with aqueous methanolic K_2CO_3 (refluxing for 2 hrs.), dissolution of the ketone mixture in benzene and precipitation of epiandrosterone with ether; the mother liquor is treated with acetic anhydride and pyridine at room temperature for 16 hrs., the less soluble acetate of ætiocholan-3β-ol-17-one removed by crystallization from ether-petroleum ether, and the resulting mother liquor hydrolysed as above, and reacetylated; yield of epiandrosterone, 78.5%, of ætiocholan-3β-ol-17-one acetate, 15% (1944 Wenner). From androstane-3.17-dione (p. 2892 s) on reduction with 1 mol. hydrogen in alcohol at 20° in the presence of Raney nickel (1936 b SCHERING-KAHLBAUM A.-G.) or reduced platinum oxide + some sodium ethoxide (1936 a PFEILRING-WERKE). For formation from the diketone by the action of bacteria, see under "Biological Formation", p. 2633 s.

The acetate is obtained from the 3-acetate of 17β-ethynylandrostane-3β.17α-diol (p. 2029 s) on ozonolysis in CCl_4 (1939 Reichstein). From allopregnane-3β.17α.20α- and -3β.17α.20β-triols (pp. 2118 s, 2119 s) on oxidation with periodic acid in aqueous methanolic H_2SO_4 at room temperature (1938 Steiger); similarly from 17-iso-allopregnane-3β.17.20a-triol (p. 2121 s) (1941 Stavely), 17-iso-allopregnane-3β.17.20b-triol (p. 2122 s) (1940 Prins), and allopregnane-3β.17α.20β.21-tetrol (p. 2141 s) (1938 Steiger). The acetate is obtained from the 3.21(?)-diacetate of 17-iso-allopregnane-3β.17β.20b.21-tetrol (p. 2144 s) on oxidation with CrO_3 (1942 Prins) and similarly from the 3-acetate of 20-methyl-17-iso-allopregnane-3β.17.20-triol (p. 2128 s) (1943 b Shoppee) and the 3-acetate of 17-iso-allopregnane-3β.17-diol-20-one (p. 2342 s) (1943 a Shoppee); from the 3.21-diacetate of 20-methylallopregnane-3β.20a.21-triol (p. 1956 s) on dehydration by refluxing with $POCl_3$ in pyridine for $^1/_2$ hr., followed by ozonolysis in chloroform (1941 Hegner). The benzoate may be obtained from the benzoate oxime of $Δ^{16}$-allopregnen-3β-ol-20-one (p. 2242 s) by treatment with p-toluenesulphonyl chloride in cold pyridine, followed by hydrolysis with dil. H_2SO_4 at room temperature (1941 a PARKE, DAVIS & CO.). For formation of its acetate from allopregnan-3β-ol-20-one via 20-methyl-allopregnane-3β.20-diol and the acetate of 20-methyl-$Δ^{17(20)}$-allopregnen-3β-ol (p. 1525 s) which is then ozonolysed (1935 Butenandt), see also 1934 a SCHERING A.-G.; the last-mentioned acetate may also be oxidized with $KMnO_4$ (1934 a SCHERING A.-G.). The acetate is also obtained from 3β-acetoxy-

$\Delta^{17(20)}$-allopregnen-21-oic acid on ozonolysis in chloroform (1942c Marker; 1941b PARKE, DAVIS & Co.; cf. also 1944 Koechlin).

REACTIONS OF EPIANDROSTERONE

Epiandrosterone is oxidized to androstane-3.17-dione (p. 2892 s) by aluminium phenoxide (1941 Reich) or by N-bromoacetamide in aqueous tert.-butyl alcohol (1943 Reich). The androstane-3β.17-diol obtained on reduction of epiandrosterone with sodium in propanol, with amalgamated aluminium, or with hydrogen-platinum oxide in alcoholic HCl (see p. 144) is the 17β-epimer (p. 2014 s); this epimer is also obtained on hydrogenation in alcohol in the presence of Raney nickel (1937 Steiger) or by the action of isopropylmagnesium iodide (1934d SCHERING-KAHLBAUM A.-G.); the diol, m. 178–9°, obtained by the action of ethylmagnesium iodide (1936a Ruzicka), is probably a mixture of the two 17-epimers. Epiandrosterone is isomerized to androsterone by heating of its p-toluenesulphonate with potassium acetate in glacial acetic acid (1937c CIBA).

The 3-chloroandrostan-17-one obtained on treatment of epiandrosterone with thionyl chloride (1936 Marker; see p. 144) is 3β-chloroandrostan-17-one (cf. 1946 Shoppee). With 1 mol. PCl$_5$ + CaCO$_3$ in dry chloroform at 0° epiandrosterone gives 3α-chloroandrostan-17-one (1946 Shoppee). Epiandrosterone slowly reacts with 1 mol. of bromine and a little HBr in glacial acetic acid at room temperature to give a small amount (8% yield) of 16-bromo-epiandrosterone (p. 2706 s); the acetate reacts similarly to give 27% yield of the corresponding acetate (1937/38 Dannenberg).

Reacts with methylmagnesium iodide in ether to give 17α-methylandrostane-3β.17β-diol (p. 2024 s) (1935b Ruzicka); with ethylmagnesium iodide in ether on boiling for 8 hrs. it yields ca. equal amounts of 17α-ethylandrostane-3β.17β-diol (p. 2038 s) and a mixture of the two 17-epimeric androstane-3β.17-diols (pp. 2013 s, 2014 s) (1935b, 1936a Ruzicka; cf. also 1936b Ruzicka); for the reductive action of isopropylmagnesium iodide, see above. With potassium acetylide (in liquid ammonia) in ether-benzene it gives 17α-ethynylandrostane-3β.17β-diol (p. 2034 s) (1937 Ruzicka; 1937 Kathol; 1937b CIBA) and a very small amount of 17β-ethynylandrostane-3β.17α-diol (p. 2029 s) (1939 Reichstein). On heating with propargyl alcohol and potassium tert.-amyloxide in tert.-amyl alcohol at 60–70° it yields ω-homo-17-iso-T^{20}-allopregnyne-3β.17.22-triol (p. 2125 s) (1944 Wenner).

Gives the 16-benzylidene-derivative (p. 2699 s) on refluxing with benzaldehyde and sodium methoxide in methanol (1943 Hirschmann).

Epiandrosterone is converted into its cyanohydrin by treatment with liquid hydrogen cyanide or with potassium cyanide + conc. HCl in cold alcohol (1938 Kuwada; cf. also 1935a SCHERING A.-G.); the acetate reacts similarly (1946 Ruzicka; 1946 Prins). May be converted into 17β-benzoylandrostan-3β-ol (p. 2271 s) by reaction of its benzoate with ethyl $\alpha.\alpha$-dibromophenylacetate and zinc in ether, followed by heating with alcoholic NaOH and decarboxylation in quinoline at 180–200° (1938 CIBA). Adds on $^1/_2$ mol. oxalic acid in ethyl acetate to give the **compound** $C_{40}H_{62}O_8$ (crystals, m. 139°

to 140° dec., resolidifies, and remelts at 200°) which is split into its components by water or alcohol (1941a Miescher). Gives 3β.17-dihydroxyandrostan-
17-yl-malonic acid on treatment of its acetate with malonic-acetic anhydride
in the presence of conc. H_2SO_4, followed by saponification with alcoholic
KOH (1936 SCHERING CORP.). Reacts with phthalic anhydride in abs. benzene
in a sealed tube at 85° (2 hrs.) to an extent of 27% (1942 Wettstein).

With an equal amount of 5-dehydroepiandrosterone (p. 2609 s) it forms
apparently homogeneous crystals (methanol), m. 157.5–158.5° cor., $[\alpha]_D^{29} +51°$
(alcohol) (1942 Pearlman).

ANALYTICAL PROPERTIES OF EPIANDROSTERONE

For colour reactions in alcohol with conc. H_2SO_4 alone and in the presence
of benzaldehyde or furfural, see 1939 Scherrer and 1939 Woker, resp. Gives
the same blue colour reaction with $SbCl_3$ in glacial acetic acid-acetic anhydride
as does androsterone (see p. 2631 s) (1943 Pincus). For the Zimmermann
test with m-dinitrobenzene and KOH under various conditions, see 1945
Wilson.

For chromatographic separation from other steroids, see 1941 Dorfman.
For separation from androsterone via the acetates or benzoates, see p. 2632 s.
May be separated from androsterone, dehydroepiandrosterone, and cholesterol
by chromatography of their esters prepared with p-benzeneazobenzoyl chloride
(1941 Coffman).

DERIVATIVES OF EPIANDROSTERONE

Semicarbazone $C_{20}H_{33}O_2N_3$. Crystals (chloroform-methanol), m. 282–3° cor.
(1935b Ruzicka).

Acetate $C_{21}H_{32}O_3$ (p. 144). Is polymorphous: crystals, m. 95–6°, which on
keeping for 18 months are partly converted into a form m. 114°; crystals,
m. 106.5°, not changed on recrystallization but on seeding with the isolated
material giving a precipitate which shows two points of incomplete fusion
(followed by resolidification) at 97° and 106.5° and melts completely at
115.5–116.5°; specimens obtained by resolidification after complete fusion
usually melt at 97° and occasionally at 106.5° (1941a Hirschmann). Prisms
(pentane), m. 118° cor. (1938 Steiger; cf. also 1935c SCHERING A.-G.); crystals
(ether-pentane), m. 103–4° (1941 Reichstein), 104–5° cor. (Kofler block) (1942
v. Euw), m. 101°, resolidifying and remelting at 122° (1939 Reichstein),
m. 97° cor. and 102–4° cor. (1946 Shoppee); $[\alpha]_D^{24} +75°$ (alcohol), +68 to 69°
(chloroform) (1946 Barton), $[\alpha]_D^{15} +65°$, $[\alpha]_{5461}^{15} +80°$ (dioxane) (1942 v. Euw).
Is more readily soluble in methanol than androsterone acetate (1936a CIBA).

Benzoate $C_{26}H_{34}O_3$ (p. 144). Crystals (ethyl acetate), m. 224.5–225.5°; less
soluble in ethyl acetate than androsterone benzoate (1936a CIBA).

For the *p-benzeneazobenzoate,* see 1941 Coffman.

Methyl ether $C_{20}H_{32}O_2$, m. 91°. — **Fmn.** From cholestanyl methyl ether
(p. 1701 s) on oxidation. — *Semicarbazone* $C_{21}H_{35}O_2N_3$; begins to dec. at 240°,
m. 249° (1938 SCHERING A.-G.).

For a compound, m. 42° and 73°, assumed to be the methyl ether of epiandrosterone and prepared from cholestanyl methyl ether via 3β-methoxy-Δ^{16}-androstene (? p. 1506 s) and its epoxide (?), see 1937a SCHERING-KAHL-BAUM A.-G.

β-D-**Glucoside** $C_{25}H_{40}O_7$. Crystals (alcohol) with 1 H_2O, which is lost on continued heating in a high vacuum at 130°, but is taken up again on exposure to the air (1938 Miescher); crystals (acetone-ether), m. 216–7° cor. (1938 Miescher; 1944 Meystre; 1945 CIBA). Fairly readily soluble in hot water, sparingly soluble in cold water, methanol, ethanol, ether, and ethyl acetate (1938 Miescher). — **Fmn.** From epiandrosterone with acetobromo-D-glucose and Ag_2CO_3 in boiling benzene which is allowed to distil off during the reaction (removal of the water formed); the resulting tetraacetate (yield, 51%) crystallizes from ether; the mother liquor is evaporated and the residue treated with methanolic sodium methoxide at 20° (yield, 34%) (1944 Meystre; cf. 1938, 1941b Miescher; 1945 CIBA). — *Tetraacetate* $C_{33}H_{48}O_{11}$, crystals (alcohol or ether), m. 192° cor. (1941b Miescher; 1944 Meystre; 1945 CIBA).

5-iso-Androstan-3α-ol-17-one, Aetiocholan-3α-ol-17-one, 5-iso-Androsterone

$C_{19}H_{30}O_2$ (p. 144: *3α-Hydroxyætiocholan-17-one*). Needles (dil. methanol), melting incompletely at 142° and passing into a platy crystalline form on resolidification; melts completely at 151.5–152.5° (1939a Callow; 1941b Hirschmann; cf. 1941 Wolfe); $[\alpha]_D^{28}$ + 109° (alcohol) (1939 Hirschmann), $[\alpha]_{5461}$ + 130° (alcohol) (1939a Callow). For infra-red absorption spectra of films of ætiocholan-3α-ol-17-one and its acetate, see 1946 Furchgott.

Occ. Aetiocholan-3α-ol-17-one occurs in the urine of normal men (1939a Callow; 1941 Engel; 1942 Dobriner) and, to a smaller extent, in that of castrated men (1940 Callow). In the urine of normal women in an amount comparable to that in normal men's urine (1939b Callow; 1942 Dobriner) and in approximately the same amount in the urine of ovariectomized women (1939, 1940 Hirschmann). In increased amount in the urine of women with corticoadrenal hyperplasia (1938 Butler; 1945b Mason; cf. 1942 Dobriner; 1950 Miller) or tumour (1941 Wolfe; 1945b Mason). In varying amounts in the urine of patients with cancer (1942 Dobriner).

Aetiocholan-3α-ol-17-one was excreted after administration of testosterone propionate to a normal man (1939a Callow), to men with deficient testicular secretion (1939, 1940d Dorfman), to a normal woman (1945 Schiller), to adult male chimpanzees (1944 Fish), and of dehydroepiandrosterone acetate to a man with anterior pituitary insufficiency (1945a Mason).

For chromatographic *separation* from other urinary steroids, see 1942Callow.

Fmn. The acetate is obtained from the diacetate of ætiocholane-3α.17β-diol (see p. 2017 s) on partial hydrolysis, followed by oxidation with CrO_3 in strong acetic acid at room temperature (1940 Marker), or from the acetate of 20-methyl-$\Delta^{17(20)}$-pregnen-3α-ol (p. 1524 s) on ozonolysis in chloroform (yield, 64%) (1938 Butenandt). From pregnane-3α.17α.20α-triol (p. 2117 s) on oxidation with lead tetraacetate in glacial acetic acid at room temperature in a stream of nitrogen (20 hrs.) (1937 Butler). From pregnane-3α.17α.20β-triol (p. 2118 s) on oxidation with HIO_4 in aqueous methanol at room temperature (1941 b Hirschmann). Its propionate (not described) may be obtained from the propionate oxime of Δ^{16}-pregnen-3α-ol-20-one (p. 2239 s) by treatment with p-toluenesulphonyl chloride in cold pyridine, followed by hydrolysis with dil. H_2SO_4 at room temperature (1941 a PARKE, DAVIS & Co.). The acetate is obtained from the 3-acetate of pregnane-3α.17α-diol-20-one (p. 2335 s) on oxidation with CrO_3 in glacial acetic acid at room temperature (1945 Lieberman; 1947 Mason). From the methyl ester of 3β-hydroxy-$\Delta^{17(20)}$-pregnen-21-oic acid on oxidation by refluxing with aluminium isopropoxide and acetone in toluene for 6 hrs., followed by acidification with dil. HCl and reduction of the resulting product by refluxing with aluminium isopropoxide in isopropyl alcohol for 16 hrs.; the mixture, obtained on acidification is then treated with digitonin and the non-digitonin precipitated fraction ozonolysed in chloroform (yield, 12.5 %) (1942 d Marker; 1941 b PARKE, DAVIS & Co.).

Rns. Oxidation with CrO_3 in glacial acetic acid affords ætiocholane-3.17-dione (p. 2895 s) (1941 Reichstein). On refluxing with amalgamated zinc in conc. HCl it yields ætiocholan-3α-ol (p. 1508 s) (1946 Heard). Gives ætiocholane-3α.17β-diol (see p. 2017 s) on catalytic hydrogenation (see p. 144), on reduction with sodium in alcohol (1940 Marker), or on treatment of its acetate with isopropylmagnesium iodide, followed by hydrolysis (1937 b SCHERING-KAHLBAUM A.-G.); on treatment with ethylmagnesium halide, mainly 17α-ethylætiocholane-3α.17β-diol (p. 2038 s) is obtained (1937 b SCHERING-KAHLBAUM A.-G.).

Colour reactions. Gives the same blue colour reaction with $SbCl_3$ in glacial acetic acid-acetic anhydride as does androsterone (see p. 2631 s) (1943 Pincus). For the Zimmermann test with m-dinitrobenzene and KOH under various conditions, see 1944 Zimmermann; 1945 Wilson.

Aetiocholan - 3α - ol - 17-one semicarbazone $C_{20}H_{33}O_2N_3$. Crystals (alcohol), m. 235° dec. (1942 d Marker), 264–5° (1937 Butler).

Aetiocholan-3α-ol-17-one acetate $C_{21}H_{32}O_3$. Needles (aqueous methanol), m. 94.5–96° cor. (1941 Wolfe); crystals (ligroin), m. 96–7° cor. (1945 Lieberman). — **Fmn.** From ætiocholan-3α-ol-17-one on heating with acetic anhydride in pyridine at 100° for 30 min. (1941 Wolfe; cf. 1940 d Dorfman; 1940 Hirschmann). See also under formation of the free hydroxyketone.

Aetiocholan-3α-ol-17-one benzoate $C_{26}H_{34}O_3$. Needles (aqueous or abs. methanol), m. 162–163.5° (1939 a Callow; 1940 d Dorfman; 1941 Wolfe), 165–6° (1941 b Hirschmann).

5-iso-Androstan-3β-ol-17-one, Aetiocholan-3β-ol-17-one $C_{19}H_{30}O_2$ (p. 144:

3β-*Hydroxyætiocholan-17-one*). Is dimorphous: needles (pentane or ether-pentane), m. 117–8° (1940, 1942 b Marker) and needles or prisms (ether-pentane), m. 150° to 152° (1942 a Marker), 152–4° (1941 Reichstein); is also obtained in a hydrated (?) form, sintering at 106°,

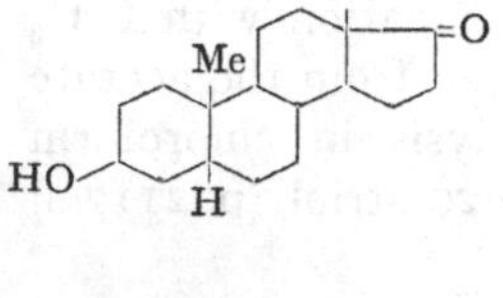

m. 133–5⁰, resolidifying, and remelting at 152⁰ (1941 Reichstein); $[\alpha]_D^{20} + 89^0$ (abs. alcohol) (1941 Reichstein).

Fmn. From the monomethyl ester of 3β-acetoxy-ætiobilianic acid (I) on conversion into the chloride with thionyl chloride in benzene, followed by treatment with diazo-methane in ether at room temperature, heating of the resulting diazoketone with Ag_2O in abs. alcohol at 70–80⁰ for 1 hr., hydrolysis with hot alcoholic KOH, refluxing with acetic anhydride for 25 min., and hydrolysis with alcoholic KOH (1940 Marker). The acetate is obtained in small yield, along with the acetate of epiandrosterone (p. 2633 s), from the acetate of 5-dehydroepiandrosterone (p. 2609 s) on treatment with excess hydrogen in glacial acetic acid in the presence of platinum oxide at room temperature, followed by oxidation with CrO_3 in glacial acetic acid at room temperature (1941 Reichstein); for separation of the two isomers (1944 Wenner), see p. 2634 s. The acetate is obtained from the acetate oxime of Δ^{16}-pregnen-3β-ol-20-one (p. 2240 s) on treatment with p-toluenesulphonyl chloride in pyridine in the cold, followed by hydrolysis with dil. H_2SO_4 at room temperature (1941 a PARKE, DAVIS & Co.). The acetate is obtained from the acetate of 20-methyl-$\Delta^{17(20)}$-pregnen-3β-ol (p. 1525 s) on ozonolysis in chloroform; yield, 67 % (1938 Butenandt). From 3β-hydroxy-$\Delta^{17(20)}$-pregnen-21-oic acid on oxidation with $KMnO_4$ in aqueous KOH (yield, 70 %) (1942 d Marker); in smaller yield from the corresp. acetate on oxidation with CrO_3 in strong acetic acid at 50–55⁰ (4 hrs.) or on ozonolysis in chloroform, followed by hydrolysis with alcoholic KOH (1942 a Marker; 1941 b PARKE, DAVIS & Co.). For formation of its acetate from coprostanyl acetate (cf. p. 144; small yield), see also 1941 Reichstein.

Rns. Gives ætiocholane-3.17-dione (p. 2895 s) on oxidation with CrO_3 in glacial acetic acid at room temperature (1941 Reichstein). On refluxing with sodium in amyl alcohol for 10 hrs. it is reduced, with inversion, to ætio-cholane-3α.17β-diol (p. 2017 s) (1940 Marker). Reduction to the 3-acetate of an ætiocholane-3.17-diol (not described) occurs when the acetate is treated with isopropylmagnesium iodide in ether, whereas mainly 17α-methylætio-cholane-3β.17β-diol (p. 2025 s) is formed when the free hydroxyketone reacts with methylmagnesium iodide in ether (1937 b SCHERING-KAHLBAUM A.-G.). On treatment with acetylene and potassium tert.-amyl oxide in anhydrous benzene-ether-tert.-amyl alcohol at room temperature it gives 17α-ethynyl-ætiocholane-3β.17β-diol (p. 2034 s) and, probably, traces of the 17β-ethynyl epimer (1944 Shoppee; cf. 1950 Shoppee). On treatment of the acetate with potassium cyanide and glacial acetic acid in alcohol, first at 0⁰ and then at 18⁰, it gives the 3-acetate of 3β.17β-dihydroxy-17-iso-ætiocholanic acid nitrile; on performing the reaction at 80⁰, a small amount of the 17-epimeric nitrile could also be isolated (1946 Meyer).

Semicarbazone $C_{20}H_{33}O_2N_3$. Crystals (aqueous methanol), m. 244⁰ dec. (1942 b Marker; cf. 1940, 1942 a Marker).

Acetate $C_{21}H_{32}O_3$. Rhombohedra (ether-pentane), m. 157–9⁰ (1941 Reich-stein), 161–3⁰ cor. (Kofler block) (1946 Meyer); $[\alpha]_D^{18}$ +82.5⁰ (acetone) (1941 Reichstein). — **Fmn.** From ætiocholan-3β-ol-17-one on treatment with acetic anhydride in abs. pyridine for 16 hrs. at room temperature and then for 30 min. at 60⁰ (1941 Reichstein). See also under formation of the free hydroxy-ketone (above). — *Oxime* $C_{21}H_{33}O_3N$, crystalline powder (dil. alcohol), m. 188–9⁰ (1938 Butenandt). — *Semicarbazone* $C_{22}H_{35}O_3N_3$ (p. 144), leaflets (alcohol), m. 236–8⁰ (1938 Butenandt), 248–250⁰ (1941 Reichstein).

13 - iso - Androstan - 3α - ol - 17 - one, 13-iso-Androsterone, Lumiandrosterone

$C_{19}H_{30}O_2$. For the configuration at C-13, see also 1948 Klyne. — Plates (hexane), m. 145–6⁰; $[\alpha]_D^{22}$ —100⁰ (alcohol). Is not androgenic (1944 Butenandt). — **Fmn.** From androsterone (p. 2629 s) in dioxane on irradiation with ultra-violet light for 3¹/₂ hrs.; yield, 34 % calculated on reacted androsterone (1944 Bute-nandt). — **Rns.** On oxidation with CrO_3 in glacial acetic acid at room tem-perature it gives lumiandrostane-3.17-dione (p. 2895 s). Does not react with Girard's reagent T and is not precipitated by digitonin (1944 Butenandt).

Oxime $C_{19}H_{31}O_2N$. Crystals (aqueous alcohol), m. 201–2⁰ (1944 Butenandt).

Acetate $C_{21}H_{32}O_3$. Crystals (aqueous alcohol and aqueous acetone), m. 121⁰ to 122⁰. — **Fmn.** From the above hydroxyketone on treatment with acetic anhydride in pyridine at room temperature for 2 days (1944 Butenandt).

1933 Ciba*, *Swiss Pats.* 188202, 188240–188243 (issued 1937); *C. A.* **1937** 5110; *Ch. Ztbl.* **1937** II 814.

1934 a Schering A.-G., *Ger. Pat.* 704818 (issued 1941); *C. A.* **1942** 2089; Schering-Kahlbaum A.-G., *Fr. Pat.* 794117 (1935, issued 1936); *C. A.* **1936** 4628; *Ch. Ztbl.* **1937** I 3184; *Swiss Pat.* 196548 (1935, issued 1938); *Ch. Ztbl.* **1939** I 471.

b Schering A.-G., *Ger. Pat.* 717224 (issued 1942); *C. A.* **1944** 2456; *Ch. Ztbl.* **1942** I 3021; Schering-Kahlbaum A.-G., *Swiss Pats.* 197780, 198068 (1935, issued 1938); *Ch. Ztbl.* **1939** I 2250.

a Schering-Kahlbaum A.-G., *Swiss Pat.* 185533 (issued 1936); *Ch. Ztbl.* **1937** I 2820.

b Schering-Kahlbaum A.-G., *Swiss Pat.* 193542 (issued 1938); *Ch. Ztbl.* **1938** II 356.

c Schering-Kahlbaum A.-G., *Swiss Pat.* 194981 (issued 1938); *Ch. Ztbl.* **1938** II 1088.

d Schering-Kahlbaum A.-G., *Swiss Pats.* 197782–197784 (issued 1938); *Ch. Ztbl.* **1939** I 2251.

e Schering-Kahlbaum A.-G., *Austrian Pat.* 160242 (issued 1941); *Ch. Ztbl.* **1941** II 2708.

1935 Butenandt, Cobler, *Z. physiol. Ch.* **234** 218, 223.

a Ciba*, *Brit. Pat.* 451352 (issued 1936); *C. A.* **1937** 219; *Ch. Ztbl.* **1937** I 1479.

b Ciba*, *Ital. Pat.* 337887; *Ch. Ztbl.* **1937** II 4069.

Parke, Davis & Co. (Marker), *U. S. Pat.* 2144726 (issued 1939); *C. A.* **1939** 2909; *Ch. Ztbl.* **1939** II 171.

a Ruzicka, Goldberg, Meyer, *Helv.* **18** 994, 996.

b Ruzicka, Goldberg, Rosenberg, *Helv.* **18** 1487, 1497, 1498.

a Schering A.-G. (Schoeller, Serini), *Ger. Pat.* 657017 (issued 1938); *C. A.* **1938** 3912; *Ch. Ztbl.* **1938** II 1088.

b Schering A.-G., *Ger. Pat.* 734563 (issued 1943); *C. A.* **1944** 3094.

c Schering A.-G., *Norwegian Pat.* 61267 (issued 1939); *Ch. Ztbl.* **1940** I 1710; Schering-Kahlbaum A.-G., *Brit. Pat.* 471908 (1936, issued 1937); *C. A.* **1938** 1284; *Ch. Ztbl.* **1938** I 2402. Tschopp, *Arch. intern. pharmacodynamie* **52** 381.

* Gesellschaft für chemische Industrie in Basel; Société pour l'industrie chimique à Bâle.

1936 a CIBA*, *Swiss Pat.* 190542 (issued 1937); *C. A.* **1938** 588; *Ch. Ztbl.* **1938** I 941, 942.

b CIBA*, *Swiss Pat.* 198133 (issued 1938); *C. A.* **1939** 3531; *Fr. Pat.* 830484; *C. A.* **1939** 2542; *Brit. Pat.* 488829 (1937, issued 1938); *C. A.* **1939** 180; *Ch. Ztbl.* **1939** I 2826, 2827.

I.G. FARBENIND., *Brit. Pat.* 478583 (issued 1938); *C. A.* **1938** 5161; *Ch. Ztbl.* **1938** II 1086, 1087.

LES LABORATOIRES FRANÇAIS DE CHIMIOTHÉRAPIE (Girard, Sandulesco), *Ger. Pat.* 703342 (issued 1941); *C. A.* **1942** 223; *Ch. Ztbl.* **1941** I 2972, 2973.

Marker, Whitmore, Kamm, Oakwood, Blatterman, *J. Am. Ch. Soc.* **58** 338.

a PFEILRING-WERKE A.-G., *Fr. Pat.* 812354 (issued 1937); *C. A.* **1938** 960; *Ch. Ztbl.* **1937** II 3347.

b PFEILRING-WERKE A.-G., *Swiss Pat.* 202156 (issued 1939); *Ch. Ztbl.* **1940** I 250.

a Ruzicka, Rosenberg, *Helv.* **19** 357, 365.

b Ruzicka, Kägi, *Helv.* **19** 842, 845.

SCHERING A.-G., *Austrian Pat.* 160484 (issued 1941); *Ch. Ztbl.* **1941** II 2973, 2974.

SCHERING CORP. (Schwenk, Whitman), *U.S. Pat.* 2247822 (issued 1941); *C. A.* **1941** 6396; SCHERING-KAHLBAUM A.-G. (Schwenk, Whitman), *Brit. Pat.* 501196 (1937, issued 1939); *C. A.* **1939** 5868; *Ch. Ztbl.* **1939** II 686, 687; *Swiss Pat.* 202242 (1937, issued 1939).

a SCHERING-KAHLBAUM A.-G., *Fr. Pat.* 47743 (issued 1937); *C. A.* **1938** 4607; *Ch. Ztbl.* **1937** II 3199, 3200; *Brit. Pat.* 467790 (issued 1937); *C. A.* **1937** 8835; SCHERING CORP. (Serini, Strassberger), *U.S. Pat.* 2243887 (issued 1941); *C. A.* **1941** 5654.

b SCHERING-KAHLBAUM A.-G., *Fr. Pat.* 811567 (issued 1937); *C. A.* **1938** 1284; *Ch. Ztbl.* **1937** II 3919; *Brit. Pat.* 464271 (issued 1937); *C. A.* **1937** 6417.

WINTHROP CHEMICAL CO. INC. (Bockmühl, Ehrhart, Ruschig), *U.S. Pat.* 2188330 (issued 1940); *C. A.* **1940** 3884; *Ch. Ztbl.* **1940** II 1053.

1937 Bugyi, *Acta Lit. Sci. Univ. Hung. Sect. Chem. Mineral. Phys.* **6** 27, 33.

Butler, Marrian, *J. Biol. Ch.* **119** 565, 569, 571.

a CIBA*, *Swiss Pat.* 199449 (issued 1938); *C. A.* **1939** 3393; *Ch. Ztbl.* **1939** I 3930, 3931.

b CIBA*, *Swiss Pat.* 202847 (issued 1939); *C. A.* **1939** 9554; *Ch. Ztbl.* **1940** I 249, 250; *Fr. Pat.* 843083 (issued 1939); *C. A.* **1940** 6413; *Ch. Ztbl.* **1940** I 1079.

c CIBA*, *Brit. Pat.* 495887 (issued 1938); *C. A.* **1939** 2905; *Ch. Ztbl.* **1939** I 4225.

Hogness, Sidwell, Zscheile, *J. Biol. Ch.* **120** 239, 243, 249.

Kathol, Logemann, Serini, *Naturwiss.* **25** 682.

Miescher, Kägi, Scholz, Wettstein, Tschopp, *Biochem. Z.* **294** 39.

Ruzicka, Hofmann, *Helv.* **20** 1280, 1282.

SCHERING A.-G., *Swiss Pat.* 211772 (issued 1941); *Ch. Ztbl.* **1942** I 643.

a SCHERING-KAHLBAUM A.-G., *Swiss Pat.* 201946 (issued 1939); *Ch. Ztbl.* **1939** II 1126.

b SCHERING-KAHLBAUM A.-G., *Brit. Pat.* 479252 (issued 1938); *C. A.* **1938** 5162; *Ch. Ztbl.* **1938** II 3426; SCHERING CORP. (Hildebrandt), *U.S. Pat.* 2238936 (issued 1941); *C. A.* **1941** 4920.

Steiger, Reichstein, *Helv.* **20** 817, 827.

Steinach, Kun, *Lancet* **233** 845.

1937/38 Dannenberg, *Über einige Umwandlungen des Androstandions und Testosterons*, Thesis, Danzig pp. 33, 34.

1938 Butenandt, Müller, *Ber.* **71** 191, 197.

Butler, Marrian, *J. Biol. Ch.* **124** 237; Marrian, Butler, *Nature* **142** 400.

a Callow, Callow, Emmens, *Biochem. J.* **32** 1312.

b Callow, Callow, *Biochem. J.* **32** 1759.

CIBA* (Oppenauer, Miescher, Kägi), *Swed. Pat.* 96375 (issued 1939); *Ch. Ztbl.* **1940** I 1231; *Ital. Pat.* 362454 (issued 1938); *Ch. Ztbl.* **1939** II 3151.

Kuwada, Miyasaka, *C. A.* **1938** 7474, 7475; *Ch. Ztbl.* **1939** I 1372.

a Mamoli, Schramm, *Ber.* **71** 2083.

b Mamoli, Schramm, *Ber.* **71** 2698.

Marker, Lawson, *J. Am. Ch. Soc.* **60** 2928, 2930.

Miescher, Fischer, *Helv.* **21** 336, 355.

Neustadt, *Endocrinology* **23** 711.

SCHERING A.-G., *Fr. Pat.* 834941 (issued 1938); *C. A.* **1939** 4602; *Ch. Ztbl.* **1939** I 2827.

* Gesellschaft für chemische Industrie in Basel; Société pour l'industrie chimique à Bâle.

Schramm, Mamoli, *Ber.* **71** 1322.
Steiger, Reichstein, *Helv.* **21** 546, 556, 558, 559.
1939 Butenandt, Mamoli, Heusner, *Ber.* **72** 1614, 1616.
  a Callow, *Biochem. J.* **33** 559.
  b Callow, Callow, *Biochem. J.* **33** 931.
  Chou, Wang, *Ch. Ztbl.* **1941** I 387.
  Dorfman, Cook, Hamilton, *J. Biol. Ch.* **130** 285.
  Ercoli, *Ber.* **72** 190.
  Friedgood, Whidden, *Endocrinology* **25** 919.
  Hirschmann, *J. Biol. Ch.* **130** 421.
  a Langstroth, Talbot, *J. Biol. Ch.* **128** 759.
  b Langstroth, Talbot, *J. Biol. Ch.* **129** 759.
  c Langstroth, Talbot, Fineman, *J. Biol. Ch.* **130** 585.
  Mamoli, Koch, Teschen, *Z. physiol. Ch.* **261** 287.
  a Marker, *J. Am. Ch. Soc.* **61** 944.
  b Marker, *J. Am. Ch. Soc.* **61** 1287, 1289.
  c Marker, Rohrmann, *J. Am. Ch. Soc.* **61** 3476.
  Mitui, *C. A.* **1940** 384; *Ch. Ztbl.* **1940** I 714.
  Reichstein, Meystre, *Helv.* **22** 728, 733, 739.
  Scherrer, *Helv.* **22** 1329, 1333.
  Woker, Antener, *Helv.* **22** 511, 515, 516.
1940 Baumann, Metzger, *Endocrinology* **27** 664.
  Callow, Callow, *Biochem. J.* **34** 276.
  a Dorfman, *J. Biol. Ch.* **132** 457.
  b Dorfman, Hamilton, *J. Biol. Ch.* **133** 753.
  c Dorfman, Fish, *J. Biol. Ch.* **135** 349.
  d Dorfman, *Proc. Soc. Exptl. Biol. Med.* **45** 739.
  a Friedgood, Whidden, *Endocrinology* **27** 242.
  b Friedgood, Whidden, *Endocrinology* **27** 249.
  c Friedgood, Whidden, *Endocrinology* **27** 258.
  Hershberg, Wolfe, Fieser, *J. Am. Ch. Soc.* **62** 3516, 3517; *J. Biol. Ch.* **140** 215 (1941).
  Hirschmann, *J. Biol. Ch.* **136** 483.
  Holtorff, Koch, *J. Biol. Ch.* **135** 377.
  Hoskins, Webster, *Proc. Soc. Exptl. Biol. Med.* **43** 604.
  McCullagh, Schneider, Emery, *Endocrinology* **27** 71.
  Marker, Rohrmann, *J. Am. Ch. Soc.* **62** 900.
  Pearlman, *J. Biol. Ch.* **136** 807.
  Prins, Reichstein, *Helv.* **23** 1490, 1497.
  Tastaldi, Lemmi, Lacaz de Moraes, *C. A.* **1943** 6292.
1941 Coffman, *J. Biol. Ch.* **140** Proc. XXVIII.
  Dorfman, *Proc. Soc. Exptl. Biol. Med.* **46** 351.
  Engel, Thorn, Lewis, *J. Biol. Ch.* **137** 205.
  a Friedgood, Whidden, *Endocrinology* **28** 237.
  b Friedgood, Berman, *Endocrinology* **28** 248.
  Hegner, Reichstein, *Helv.* **24** 828, 837.
  a Hirschmann, *Proc. Soc. Exptl. Biol. Med.* **46** 51.
  b Hirschmann, *J. Biol. Ch.* **140** 797, 805.
  de Laat, *Acta Brevia Neerland.* **11** 51.
  a Miescher, Kági, *Helv.* **24** 986.
  b Miescher, Meystre, Heer, *Helv.* **24** 988, 996, 998.
  a PARKE, DAVIS & Co. (Tendick, Lawson), *U.S. Pat.* 2335616 (issued 1943); *C. A.* **1944** 3095; *Brit. Pat.* 563889 (issued 1944); *C. A.* **1946** 2942.
  b PARKE, DAVIS & Co. (Marker, Crooks), *U.S. Pat.* 2359773 (issued 1944); *C. A.* **1945** 4198; *Ch. Ztbl.* **1945** II 1233.
  a Pincus, Pearlman, *Science* [N.S.] **93** 163.
  b Pincus, Pearlman, *Endocrinology* **29** 413.
  Reich, Reichstein, *Arch. intern. pharmacodynamie* **65** 415, 416.

Reichstein, Lardon, *Helv.* **24** 955.
Saier, Warga, Grauer, *J. Biol. Ch.* **137** 317.
Stavely, *J. Am. Ch. Soc.* **63** 3127, 3130.
Wolfe, Fieser, Friedgood, *J. Am. Ch. Soc.* **63** 582, 592.

1942 Callow, *Biochem. J.* **36** Proc. XIX.
Dobriner, Gordon, Rhoads, Lieberman, Fieser, *Science* [N.S.] **95** 534.
v. Euw, Reichstein, *Helv.* **25** 988, 1020, 1021.
a Marker, Crooks, Wagner, *J. Am. Ch. Soc.* **64** 817.
b Marker, Crooks, Jones, Shabica, *J. Am. Ch. Soc.* **64** 1276, 1278.
c Marker, Crooks, Wagner, Wittbecker, *J. Am. Ch. Soc.* **64** 2089, 2092.
d Marker, Wagner, Wittbecker, *J. Am. Ch. Soc.* **64** 2093, 2096.
Pearlman, *Endocrinology* **30** 270.
Prins, Reichstein, *Helv.* **25** 300, 309.
Venning, Hoffman, Browne, *J. Biol. Ch.* **146** 369.
Wettstein, Miescher, *Helv.* **25** 718, 726.

1943 Hansen, Cantarow, Rakoff, Paschkis, *Endocrinology* **33** 282.
Hill, Longwell, *Endocrinology* **32** 319, 321.
Hirschmann, *J. Biol. Ch.* **150** 363, 376.
Luft, *Ch. Ztbl.* **1945** II 1040.
Nathanson, Wilson, *Endocrinology* **33** 189.
Pincus, *Endocrinology* **32** 176.
Reich, Reichstein, *Helv.* **26** 562, 583.
Saier, Grauer, Starkey, *J. Biol. Ch.* **148** 213.
a Shoppee, Prins, *Helv.* **26** 185, 199.
b Shoppee, Prins, *Helv.* **26** 2089, 2094.
Wettstein, Miescher, *Helv.* **26** 631, 640.

1944 Butenandt, Poschmann, *Ber.* **77** 394.
Cantarow, Paschkis, Rakoff, Hansen, *Endocrinology* **35** 129.
Fish, Dorfman, *Endocrinology* **35** 22.
a Frame, Fleischmann, Wilkins, *Bull. Johns Hopkins Hosp.* **75** 95.
b Frame, *Endocrinology* **34** 175.
Horwitt, Dorfman, van Wagenen, *Endocrinology* **34** 351.
Koechlin, Reichstein, *Helv.* **27** 549, 555, 563.
Meystre, Miescher, *Helv.* **27** 231, 233.
Shoppee, *Helv.* **27** 8, 13.
Wenner, Reichstein, *Helv.* **27** 24, 30, 31.
Zimmermann, *Vitamine und Hormone* **5** [1952] 1, 124, 152, 170, 237, 260, 276.

1945 CIBA*, *Fr. Pat.* 910844 (issued 1946); *Ch. Ztbl.* **1946** I 2280; *Brit. Pat.* 584062 (issued 1947);
   C. A. **1947** 3120; *Ch. Ztbl.* **1948** II 518.
Dorfman, Schiller, Fish, *Endocrinology* **36** 349.
Kumler, Fohlen, *J. Am. Ch. Soc.* **67** 437.
Lieberman, Dobriner, *J. Biol. Ch.* **161** 269, 277, 278.
a Mason, Kepler, *J. Biol. Ch.* **160** 255.
b Mason, Kepler, *J. Biol. Ch.* **161** 235, 239, 247.
Schiller, Dorfman, Miller, *Endocrinology* **36** 355.
Wilson, Nathanson, *Endocrinology* **37** 208.

1946 Barton, *J. Ch. Soc.* **1946** 1116, 1121.
Dingemanse, Huis in't Veld, *Acta Brevia Neerland.* **14** 34.
Furchgott, Rosenkrantz, Shorr, *J. Biol. Ch.* **163** 375, 378, 380.
Heard, McKay, *J. Biol. Ch.* **165** 677, 681.
Meyer, *Helv.* **29** 1580, 1584.
Prins, Shoppee, *J. Ch. Soc.* **1946** 494, 498.
Ruzicka, Plattner, Heusser, Pataki, *Helv.* **29** 936, 939.
Shoppee, *J. Ch. Soc.* **1946** 1138, 1143, 1146, 1147.

1947 Mason, Strickler, *J. Biol. Ch.* **171** 543, 548.

* Gesellschaft für chemische Industrie in Basel; Société pour l'industrie chimique à Bâle.

1948 Klyne, *Nature* **161** 434.
1950 Miller, Dorfman, *Endocrinology* **46** 514, 523.
     Shoppee, *Nature* **166** 107.
1953 Johnson, Bannister, Bloom, Kemp, Pappo, Rogier, Szmuszkovicz, *J. Am. Ch. Soc.* **75** 2275.
1956 Johnson, Bannister, Pappo, *J. Am. Ch. Soc.* **78** 6331.

Androstan-5-ol-17-one $C_{19}H_{30}O_2$. — **Fmn.** Along with other compounds, from cholestan-5-ol (p. 1723 s) on oxidation with CrO_3 in strong acetic acid at room temperature. — **Rns.** Gives androstane-5.17-diol (p. 2020 s) on reduction, e.g. by hydrogenation in alcohol in the presence of a nickel catalyst until 1 mol. of hydrogen has been absorbed (1940 Ciba).

i-Androsten-6β-ol-17-one, *i-Androstan-6β-ol-17-one* $C_{19}H_{28}O_2$. For probable β-configuration at C-6, see 1952 Shoppee. — Crystals (petroleum ether), m. 136–8° (1942 Butenandt), 140.5–141° (1948 Dingemanse; cf. 1948 Barton); $[\alpha]_D^{20}$ +122° (alcohol) (1942 Butenandt; 1948 Dingemanse; 1948 Barton). Solutions of i-androsten-6β-ol-17-one and its acetate in conc. H_2SO_4 give blue-violet colorations on addition of water (1943 Dirscherl).

Occ. In normal human urine in small quantity; in considerable amount in the urine of a female patient with adenoma of the adrenal cortex (1948 Dingemanse; cf. 1948 Barton); according to Fieser (1949) the i-androstenolone isolated from urine may be an artefact derived from a conjugate of dehydroepiandrosterone.

Fmn. Its acetate is obtained from the p-toluenesulphonate of dehydroepiandrosterone (p. 2609 s) on refluxing with potassium acetate in 50% aqueous acetone for 6 hrs., followed by heating with acetic anhydride on the water-bath for 2 hrs. (yield, 54%); it is hydrolysed by boiling with 5% alcoholic KOH for 30 min. (1942 Butenandt).

Rns. Is not attacked by perbenzoic acid (1942 Butenandt). It gives i-androstene-6.17-dione (p. 2898 s) with CrO_3 in strong acetic acid at 10° (1942 Butenandt) or on Oppenauer oxidation (1948 Dingemanse; cf. 1948 Barton). Is not attacked by hydrogen in the presence of platinum oxide in alcohol, whereas its acetate on similar hydrogenation in glacial acetic acid gives androstan-17β-ol (p. 1512 s) (1942 Butenandt). On treatment with dil. aqueous alcoholic HCl at room temperature it gives 3β-chloro-Δ^5-androsten-17-one (p. 2474 s; 4 parts) and dehydroepiandrosterone (p. 2609 s; 1 part); the latter is the main product when the reaction is carried out at 100° (1948 Dingemanse; cf. 1948 Barton). On treatment of the acetate with 48% HBr in glacial acetic acid it gives 3β-bromo-Δ^5-androsten-17-one (p. 2476 s) (1942 Butenandt).

Oxime $C_{19}H_{29}O_2N$, m. 109–113° (1948 Dingemanse).

Semicarbazone $C_{20}H_{31}O_2N_3$, m. 237–240° (1942 Butenandt).

Acetate $C_{21}H_{30}O_3$. Crystals (methanol), m. 113–4°; $[\alpha]_D^{20}$ +117° (alcohol) (1942 Butenandt). — **Fmn.** and **Rns.** See above.

1940 CIBA (Miescher, Wettstein), *Swed. Pat.* 105 135 (issued 1942); *Ch. Ztbl.* **1943** I 1695, 1696;
   CIBA PHARMACEUTICAL PRODUCTS (Miescher, Wettstein), *U.S. Pat.* 2374 370 (issued 1945);
   C. A. **1945** 5412.
1942 Butenandt, Surányi, *Ber.* **75** 591, 594–596.
1943 Dirscherl, Zilliken, *Naturwiss.* **31** 349.
1948 Barton, Klyne, *Nature* **162** 493.
   Dingemanse, Huis in't Veld, Hartogh-Katz, *Nature* **161** 848; **162** 492.
1949 Fieser, Fieser, *Natural Products Related to Phenanthrene*, 3rd Ed., New York, p. 494.
1952 Shoppee, Summers, *J. Ch. Soc.* **1952** 3361, 3363.

6. Monohydroxy-monoketosteroids (CO and OH in the Ring System) C_{20}–C_{26} with Three Aliphatic Side Chains

a. HYDROXY-3-KETOSTEROIDS

9-iso-Pregnan-11β-ol-3-one $C_{21}H_{34}O_2$. This structure was ascribed by Marker (1939) to **uran-11-ol-3-one** [needles (ether-pentane), m. 169.5–171°; *semicarbazone* $C_{22}H_{37}O_2N_3$, m. 251–3° dec.; *acetate* $C_{23}H_{36}O_3$, plates (aqueous acetone), m. 195–7°] which could be isolated from mares' pregnancy urine (1938 Marker) and may be obtained from urane-3.11-diol by refluxing with aluminium isopropoxide and cyclohexanone in toluene (1939 Marker); on oxidation with CrO_3 it gives urane-3.11-dione (1938 Marker). The "urane" compounds are, according to Klyne (1950, 1952), probably D-homo-steroids (see urane, p. 1403 s).

1938 Marker, Lawson, Wittle, Crooks, *J. Am. Ch. Soc.* **60** 1559.
1939 Marker, Rohrmann, *J. Am. Ch. Soc.* **61** 2719.
1950 Klyne, *Nature* **166** 559.
1952 Klyne, Shoppee, *Chemistry & Industry* **1952** 470.

17α-Methyl-Δ¹-androsten-17β-ol-3-one $C_{20}H_{30}O_2$. No properties are given. —

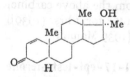

Fmn. From 17α-methylandrostan-17β-ol-3-one (page 2647 s) on treatment with 1 mol. bromine in glacial acetic acid, followed by heating of the resulting 2-bromo-derivative with dimethylaniline (1936 CIBA PHARMACEUTICAL PRODUCTS).

17α-Methyl-Δ⁴-androsten-17β-ol-3-one, 17-Methyltestosterone $C_{20}H_{30}O_2$

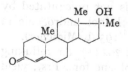

(p. 145). For configuration at C-17, cf. 1949 Miescher. Crystals (dil. methanol and dil. acetone), m. 163–4° (1939 Mamoli), 165–6° cor. (1939b Miescher); $[\alpha]_D$ +82° (alcohol) (1939b Miescher), $[\alpha]_D^{23-24}$ +80° (chloroform) (1947 Djerassi). Dipole moment in dioxane 4.17 D (1945 Kumler). Is not precipitated by digitonin (1939b Miescher).

Fmn. The diol, from which 17-methyltestosterone was obtained by oxi-
dation (see p. 145), is 17α-methyl-Δ^5-androstene-3β.17β-diol (p. 2022 s); the
oxidation may also be carried out by refluxing with aluminium isopropoxide
and acetone in benzene (25 hrs.) (1939 China'eva), with aluminium isopropoxide and cyclohexanone in toluene ($^1/_2$ hr.) (1945 C. F. BOEHRINGER & SÖHNE),
with chloromagnesium tert.-butoxide in benzene (1937 Oppenauer), or by
treatment with dehydrogenating bacteria (from yeast) in the presence of
oxygen in yeast water at 32^0 for 5 days (yield, 75%) (1939 Mamoli). The
acetate and propionate may be obtained from the corresp. 17-esters of the
aforesaid diol on oxidation with CrO_3 after temporary protection of the
double bond with bromine (1938 Kuwada; cf. also 1935b, 1936 CIBA; 1936
SCHERING A.-G.); the debromination may be effected in methanol-acetic acid
with chromous chloride solution in an atmosphere of CO_2 at room temperature
(1944 GLIDDEN Co.; see also 1945 Julian). From the 3-ethylene mercaptal
of Δ^4-androstene-3.17-dione (p. 2886 s) on boiling with methylmagnesium
bromide in ether - toluene, followed by refluxing of the resulting 17-methyl-
testosterone ethylene mercaptal with $HgCl_2$ and $CdCO_3$ in acetone (1941 CIBA).

Rns. On heating with anhydrous copper sulphate at 135–150^0 and 0.01 mm.
it gives 17-methylene-Δ^4-androsten-3-one (p. 2405 s) (1939b Miescher).

For *colour reactions* in alcohol with conc. H_2SO_4 alone and in the presence
of benzaldehyde or furfural, see 1939 Scherrer and 1939 Woker, resp. For its
absorption spectrum in the Zimmermann test with m-dinitrobenzene and
KOH, see 1940 McCullagh; 1943 Hansen. Gives a green coloration with
1.4-naphthoquinone + conc. HCl in glacial acetic acid at 60–70^0 (1939a
Miescher). For the rate of reduction of phosphomolybdic acid (Folin-Wu
reagent) by methyltestosterone in glacial acetic acid at 100^0 to give molybdenum blue, see 1946 Heard.

For polarographic *determination*, see 1944 Sartori.

For *physiological properties*, see p. 145 and pp. 1367 s – 1368 s and the
literature cited there.

Semicarbazone $C_{21}H_{33}O_2N_3$, m. 226^0 cor. dec.; 270–272^0 cor. dec. (in a closed
capillary) (1939b Miescher).

Acetate $C_{22}H_{32}O_3$. Needles (methanol), m. 177^0 (1938 Kuwada); m. 176–176.5^0
cor.; $[\alpha]_D^{20}$ + 69^0 (alcohol) (1939b Miescher). — **Fmn.** From the above carbinol
on heating with acetic anhydride in pyridine at 130–140^0 for 4 hrs. (1939b
Miescher) or on boiling with acetic anhydride for $^1/_2$ hr. (1939 Mamoli).

17β - Methyl - Δ^4 - androsten-17α-ol-3-one, 17-Methyl-17-epi-testosterone

$C_{20}H_{30}O_2$. For configuration at C-17, see 1949 Miescher.
—Crystals (benzene-hexane), m. 182–3^0 cor.; $[\alpha]_D^{21}$ + 66^0
(alcohol), + 72^0 (chloroform). Is not precipitated by
digitonin (1939b Miescher). — Is not androgenic in
the capon comb and rat tests (1939b Miescher). —
Fmn. From 17β-methyl-Δ^5-androstene-3β.17α-diol (p. 2023 s) on refluxing
with aluminium isopropoxide and cyclohexanone in toluene for 2 hrs.; yield
70–80% (1939b Miescher). — **Rns.** On heating with anhydrous copper sul-

phate at 135–150⁰ and 0.01 mm. pressure it gives 17-methylene-Δ^4-androsten-3-one (p. 2405 s). Could not be converted into a crystalline acetate by heating with acetic anhydride alone or in pyridine (1939b Miescher).

Semicarbazone $C_{21}H_{33}O_2N_3$, crystals (alcohol), m. 220–222⁰ cor. dec. (1939b Miescher).

17α-Methylandrostan-17β-ol-3-one, 17-Methyldihydrotestosterone $C_{20}H_{32}O_2$

(p. 145). Crystals (dil. alcohol), m. 192⁰ (1935a CIBA); crystals (ethyl acetate), m. 188–9⁰ cor. (1947 Ruzicka); [α]$_D$ in chloroform +11⁰ (1947 Ruzicka), +7⁰ (1947 Djerassi). — **Fmn.** From 17α-methylandrostane-3α.17β-diol (p. 2023 s) on oxidation with CrO_3 in glacial acetic acid at room temperature (1935 Ruzicka; 1935a, 1936 CIBA; cf. also 1935 SCHERING-KAHLBAUM A.-G.); similarly from 17α-methylandrostane-3β.17β-diol (p. 2024 s) (1935a CIBA; 1947 Ruzicka). For formation from 3-acetals or 3-enol esters or ethers of Δ^4-androstene-3.17-dione by the action of methylmagnesium bromide, followed by hydrolysis, see 1943 CIBA PHARMACEUTICAL PRODUCTS. — **Rns.** On distillation with anhydrous copper sulphate under 0.1 mm. pressure it gives 17-methylene-androstan-3-one (p. 2405 s) (1947 Ruzicka; for doubtful homogeneity of the reaction product, see 1957 Sondheimer). Gives 17α-methyl-Δ^1-androsten-17β-ol-3-one (p. 2645 s) on treatment with 1 mol. bromine in glacial acetic acid, followed by heating of the resulting 2-bromo-derivative with dimethylaniline (1936 CIBA PHARMACEUTICAL PRODUCTS).

Semicarbazone $C_{21}H_{35}O_2N_3$ (p. 145), crystals (abs. alcohol), m. 235⁰ dec. (1935a, 1936 CIBA).

1935 a CIBA*, *Swiss Pat.* 208080 (issued 1940); *C. A.* **1941** 4552; *Ch. Ztbl.* **1940** I 2349; *Austrian Pat.* 160293 (1936, issued 1941); *Ch. Ztbl.* **1941** II 2352.
   b CIBA*, *Swiss Pat.* 210757 (issued 1940); *Ch. Ztbl.* **1941** I 3409.
   Ruzicka, Goldberg, Rosenberg, *Helv.* **18** 1487, 1495.
   SCHERING-KAHLBAUM A.-G., *Brit. Pat.* 464396 (issued 1937); *C. A.* **1937** 6418; *Ch. Ztbl.* **1937** II 3041.
1936 CIBA* (Ruzicka, Wettstein), *U.S. Pat.* 2143453 (issued 1939); *C. A.* **1939** 3078; *Ch. Ztbl.* **1939** I 5009, 5010; CIBA PHARMACEUTICAL PRODUCTS (Ruzicka, Wettstein), *U.S. Pat.* 2387469 (1937, issued 1945); *C. A.* **1946** 5536; *Ch. Ztbl.* **1948** I 1234.
   CIBA PHARMACEUTICAL PRODUCTS (Miescher, Wettstein), *U.S. Pat.* 2260328 (issued 1941); *C. A.* **1942** 873; *Ch. Ztbl.* **1944** II 246.
   SCHERING A.-G., *Swiss Pat.* 217146 (issued 1942); *Ch. Ztbl.* **1942** II 1155.
1937 Oppenauer, *U.S. Pat.* 2384335 (issued 1945); *C. A.* **1946** 178; *Ch. Ztbl.* **1948** I 1039; N. V. ORGANON, *Fr. Pat.* 827623 (issued 1938); *C. A.* **1938** 8437; *Ch. Ztbl.* **1938** II 3119.
1938 Kuwada, Miyasaka, *C. A.* **1938** 5849; *Ch. Ztbl.* **1938** II 3095.
1939 China'eva, Ushakov, Marchevskiï, *J. Gen. Ch. U.S.S.R.* **9** 1865; *C. A.* **1940** 4073; *Ch. Ztbl.* **1940** I 2954.
   Mamoli, *Gazz.* **69** 237.
   a Miescher, Wettstein, Scholz, *Helv.* **22** 894, 904.
   b Miescher, Klarer, *Helv.* **22** 962–964, 967, 968.
   Scherrer, *Helv.* **22** 1329, 1336.
   Woker, Antener, *Helv.* **22** 1309, 1320.

* Société pour l'industrie chimique à Bâle; Gesellschaft für chemische Industrie in Basel.

1940 Dorfman, Hamilton, *J. Biol. Ch.* **133** 753, 755.
McCullagh, Schneider, Emery, *Endocrinology* **27** 71, 77.
Tanaka, Horinaga, *Ch. Ztbl.* **1941** I 66.
1941 Ciba*, *Fr. Pat.* 51887 (issued 1943); *Ch. Ztbl.* **1944** II 778.
1943 Ciba Pharmaceutical Products (Miescher), *U.S. Pat.* 2386331 (issued 1945); *C. A.* **1946**
179; *Ch. Ztbl.* **1948** I 1040.
Hansen, Cantarow, Rakoff, Paschkis, *Endocrinology* **33** 282, 285, 286.
1944 Glidden Co. (Julian, Cole, Magnani, Conde), *U.S. Pat.* 2374683 (issued 1945); *C. A.* **1946**
1636; *Ch. Ztbl.* **1946** I 372.
Sartori, Bianchi, *Gazz.* **74** 8.
1945 C. F. Boehringer & Söhne, *Bios Final Report* No. 766, p. 139; *Fiat Final Report* No. 71, p. 66.
Julian, Cole, Magnani, Meyer, *J. Am. Ch. Soc.* **67** 1728.
Kumler, *J. Am. Ch. Soc.* **67** 1901; cf. **70** 4273 (1948).
1946 Heard, Sobel, *J. Biol. Ch.* **165** 687.
1947 Djerassi, *J. Org. Ch.* **12** 823, 825, 826.
Ruzicka, Meister, Prelog, *Helv.* **30** 867, 873.
1949 Miescher, Kägi, *Helv.* **32** 761 footnote 4.
1957 Sondheimer, Mechoulam, *J. Am. Ch. Soc.* **79** 5029, 5032, 5033.

17-iso-$\Delta^4 \cdot T^{20}$-Pregnenyn-17β-ol-3-one, 17α-Ethynyl-Δ^4-androsten-17β-ol-3-one, 17-Ethynyltestosterone, Pregneninolone

$C_{21}H_{28}O_2$. For the name "pregneninolone", see 1938b Inhoffen. For configuration at C-17, see 1948 Fieser; 1950 Shoppee; cf. also pp. 1363 s, 1364 s. — Crystals (alcohol-chloroform), m. 264–6° (1938a, b Inhoffen), 265–7° (1939 Inhoffen); crystals (ethyl acetate), m. 270–2° cor. (1938a Ruzicka), 274–6° cor. (vac.) (1938d Ruzicka); $[\alpha]_D$ +22° in dioxane (1938b Inhoffen; 1938a Ruzicka) and in alcohol (1938b Ruzicka). Absorption max. in chloroform at 240 mμ (ε 16800) (1938b Inhoffen; 1939 Dannenberg). Very sparingly soluble in ether (1938b Inhoffen).

With m-dinitrobenzene and KOH in aqueous alcohol it gives a deep violet colour which in a few seconds passes to olive-brown (1944 Zimmermann); for the spectrophotometric curve of this colour, see 1940 McCullagh. Gives a weak red-violet coloration with 1.4-naphthoquinone + conc. HCl in glacial acetic acid at 60–70° (1939 b Miescher). For the rate of reduction of phosphomolybdic acid (Folin-Wu reagent) by 17-ethynyltestosterone in glacial acetic acid at 100° to give molybdenum blue, see 1946 Heard. For polarographic determination, see 1944 Sartori.

Is not androgenic (1938b Inhoffen). Has a marked progestational activity (1938a, b Inhoffen; 1939a Ruzicka). For physiological properties, see also pp. 1367 s, 1368 s and the literature cited there.

Fmn. 17-Ethynyltestosterone may be obtained from 17α-ethynyl-Δ^5-androstene-3β.17β-diol (p. 2030 s) on oxidation with CrO_3 after temporary protection of the double bond with bromine (1938a Ciba) or on Oppenauer oxidation using aluminium isopropoxide and acetone in benzene (refluxing

---

* Société pour l'industrie chimique à Bâle; Gesellschaft für chemische Industrie in Basel.

References, pp. 2652 s, 2653 s

for 15 hrs.; yield, 60%) (1938a, b Inhoffen), aluminium tert.-butoxide and acetone in benzene (refluxing for 20 hrs.; yield, 60%) (1938a Ruzicka; 1937, 1938a CIBA), aluminium isopropoxide and formaldehyde or cyclohexanone in toluene (1938b, 1945 SCHERING A.-G.); from the same diol by the action of Flavobacterium (Micrococcus) dehydrogenans in yeast water (1941 Ercoli; 1942 Arnaudi).

From the 3-enol ethyl ether of Δ^4-androstene-3.17-dione ($\Delta^{3.5}$-androstadien-3-ol-17-one ethyl ether, p. 2606 s) on treatment in ether-benzene with a solution of potassium acetylide in liquid ammonia or with acetylene and potassium tert.-amyl oxide, in both cases followed by hydrolysis with aqueous alcoholic HCl on the water-bath (yields, 80%) (1939 Inhoffen; 1939a SCHERING A.-G.; 1938c, 1939c CIBA); similarly from the 3-thioenol ethyl ether of the same dione, or on boiling the ether with ethynylmagnesium bromide in ether, followed by acid hydrolysis (1941 CIBA); similarly also from the dione 3-ethylene ketal (1939c CIBA).

Rns. 17-Ethynyltestosterone is converted into 17β-acetoxy-17-iso-progesterone (p. 2844 s) on treatment with HgO, glacial acetic acid, acetic anhydride, and BF$_3$-ether at 20^0 (1938d Ruzicka; cf. 1939c Ruzicka; 1943 Shoppee) or on shaking (of the free hydroxyketone or its acetate) with mercuric acetate in abs. alcohol or, better, ethyl acetate at room temperature for 24 hrs., followed by treatment with H$_2$S (1939c Ruzicka; 1939d CIBA; cf. 1943 Shoppee); on heating 17-ethynyltestosterone with bis-(p-toluene-sulphonamido)-mercury in 96% alcohol on the water-bath for 100 hrs., ca. 50% react to give 17β-hydroxy-17-iso-progesterone (1943 Goldberg). According to SCHERING A.-G. (1938a) 17-ethynyltestosterone gives the 20-enol acetate of 17β-hydroxy-17-iso-progesterone (p. 2720 s) on treatment with mercuric acetate and BF$_3$-ether in glacial acetic acid-acetic anhydride; cf., however, the preceding reactions and the analogous reaction of 17α-ethynyl-Δ^5-androstene-3β.17β-diol (p. 2031 s) (1938c Ruzicka). Gives $\Delta^{4.16}$-pregnadiene-3.20-dione (p. 2780 s) on refluxing with mercury acetamide in abs. alcohol, followed by decomposition with H$_2$S (1940c CIBA). On hydrogenation in pyridine in the presence of palladium-calcium carbonate at room temperature it gives 17-vinyltestosterone (p. 2650 s) (1939d Ruzicka; 1940a CIBA). On refluxing with 90% formic acid for 2 hrs. it is dehydrated to give 17-ethynyl-$\Delta^{4.16}$-androstadien-3-one (p. 2406 s) (1938b Inhoffen); for the structure of the dehydration product cf., however, 1945 Hardegger.

Oxime $C_{21}H_{29}O_2N$. Crystals (dil. alcohol), m. 234–5^0 cor. dec. (1938a Ruzicka).

Semicarbazone $C_{22}H_{31}O_2N_3$. Needles (acetone), m. 230–1^0 dec. (1938a, b Inhoffen).

Acetate $C_{23}H_{30}O_3$. Crystals (ether-pentane), m. 167–8^0 cor. (vac.). Absorption max. (from the graph) in alcohol at ca. 240 mμ (log ε 4.25). — **Fmn.** From the above hydroxyketone on heating with acetic anhydride and pyridine at 100^0 for 48 hrs. (1938d Ruzicka).

183*

Propylene ketal, *17α-Ethynyl-Δ⁵-androsten-17β-ol-3-one propylene ketal*
$C_{24}H_{34}O_3$ (partial formula I). For this structure (with shift of the double
bond to the 5.6-position), see the literature cited with
the cyclic ketals of testosterone (p. 2595 s). — Solid,
m. 153–161°. — **Fmn.** From 17-ethynyltestosterone
and propane-1.3-diol on heating with p-toluenesulphonic acid in benzene (1939 b SCHERING A.-G.).

**17-iso-Δ⁴·²⁰-Pregnadien-17β-ol-3-one, 17α-Ethenyl-Δ⁴-androsten-17β-ol-3-one,
17-Vinyltestosterone** $C_{21}H_{30}O_2$. Needles (ether-pentane), granula (ether), m. 140–1° cor. (1938b Ruzicka); crystals (acetone), m. 142° (1938b Inhoffen);
$[α]_D$ +87.6° (abs. alcohol) (1938b Ruzicka), +77.6°
(dioxane) (1938b Inhoffen). Absorption max. in
chloroform at 242 mμ (ε 16000) (1939 Dannenberg; cf. 1938b Inhoffen). Gives
no colour reaction with tetranitromethane (1938b Ruzicka). With 1.4-naph-
thoquinone and conc. HCl in glacial acetic acid at 60–70° it gives a strong
red-violet coloration (1939b Miescher).

Has a moderate progestational activity (1938a Inhoffen; 1939a Ruzicka);
for physiological properties, see also pp. 1367 s, 1368 s and the literature
cited there.

Fmn. From 17-ethynyltestosterone (p. 2648 s) on hydrogenation in pyridine
in the presence of palladium - $BaCO_3$ at room temperature for $1\frac{1}{2}$ hrs.; yield,
90% (1939d Ruzicka; 1940a CIBA). From 17α-vinyl-Δ⁵-androstene-3β.17β-
diol (p. 2032 s) on Oppenauer oxidation carried out by refluxing with aluminium isopropoxide and acetone in benzene for 15 hrs. (1938a Inhoffen;
1938b SCHERING A.-G.), with aluminium isopropoxide and cyclohexanone
in toluene for 1 hr. (1938b Inhoffen; 1938b SCHERING A.-G.), or with aluminium tert.-butoxide and acetone in benzene for 20 hrs. (1938b Ruzicka;
1938a CIBA); the acetate is obtained from the 17-acetate of the same diol
on refluxing with aluminium tert.-butoxide and acetone in benzene for 24 hrs.
(yield, 29%) (1942 Prins).

The "17-vinyltestosterone" (m. 168.5–170.5° cor.) described by Kuwada
(1936) was Δ⁴-androstene-3.17-dione, and the "17α-vinyl-Δ⁵-androstene-3β.17β-
diol", from which it was prepared by oxidation, was 5-dehydroepiandrosterone
(p. 2609 s) which had not reacted with vinylmagnesium bromide (1938a
Ruzicka; 1938b Butenandt).

Rns. Gives the 20.21-epoxy-compound (p. 2651 s) on treatment with perbenzoic acid in chloroform at room temperature in the dark (1941 Prins). On
treatment with OsO_4 in ether for 4 days, followed by refluxing with Na_2SO_3
in aqueous alcohol, it gives 17-iso-Δ⁴-pregnene-17β.20a.21-triol-3-one (page
2726 s) (1938 Serini; cf. also 1940b CIBA and the reactions of the aforesaid
triolone).

Vinyltestosterone is rearranged to Δ⁴·¹⁷⁽²⁰⁾-pregnadien-21-ol-3-one (p. 2512 s)
on heating, first with acetic anhydride at 100°, then with trichloroacetic acid
in glacial acetic acid at 60°, followed by refluxing with K_2CO_3 in 80% methanol
(1939a Miescher; 1939a CIBA). With PBr_3 and a little pyridine in abs. chloro-

form at room temperature it gives 21-bromo-$\Delta^{4 \cdot 17(20)}$-pregnadien-3-one (page 2473 s) (1939 b Ruzicka; 1939 a Ciba; 1939 c Schering A.-G.). Gives 17-iso-Δ^4-pregnene-17β.20-diol-3-one (yellow oil) by the action of dry HCl in ether followed by treatment with alkali acetate or silver acetate in alcohol and then by hydrolysis with KOH (1938 b Ciba).

Semicarbazone $C_{22}H_{33}O_2N_3$. Crystals (aqueous acetone), m. 223° dec. (1938 b Inhoffen).

Acetate $C_{23}H_{32}O_3$. Needles or leaflets (pentane, or ether-pentane), m. 120° to 122°; [α]$_D$ +83° (acetone). — **Fmn.** From the above hydroxyketone on refluxing with acetic anhydride and pyridine for 5 hrs. — **Semicarbazone** $C_{24}H_{35}O_3N_3$ (no analysis given), needles (methanol), m. 218–221° (1942 Prins).

20.21-Epoxy-17-iso-Δ^4-pregnen-17β-ol-3-one, 17-iso-$\Delta^{4 \cdot 20}$-Pregnadien-17β-ol-3-one 20.21-oxide, 17-Vinyltestosterone

epoxide $C_{21}H_{30}O_3$. Needles (ether-pentane), m. 202° to 204° cor.; sublimes at 170–5° (bath temperature) under 0.01 mm. pressure; [α]$_D^{14}$ +73° ±2.5° (acetone) (1941 Prins). — **Fmn.** From 17-vinyltestosterone on treatment with perbenzoic acid in chloroform at room temperature in the dark for 24 hrs. (1941 Prins) or on treatment with perphthalic acid (1939 b Ciba). From 17α-formyl-Δ^4-androsten-17β-ol-3-one (p. 2835 s) on treatment with diazomethane in ether-dioxane at room temperature for 60 hrs., preferably followed by oxidation of non-reacted aldehyde with periodic acid (1941 Prins). — **Rns.** For conversion into 17β-acetoxy-17-iso-progesterone (p. 2844 s) by hydrogenation in the presence of palladium - CaCO$_3$, followed by oxidation of the 17-acetate of the resulting 17-iso-Δ^4-pregnene-17β.20-diol-3-one with CrO$_3$ in acetic acid, see 1939 b Ciba.

17-iso-Δ^4-Pregnen-17β-ol-3-one, 17α-Ethyl-Δ^4-androsten-17β-ol-3-one, 17-Ethyltestosterone $C_{21}H_{32}O_2$ (p. 145). Crystals (ether-pentane), m. 143–4° cor.; [α]$_D$ +71° (abs. alcohol) (1938 b Ruzicka). Absorption max. in chloroform at 243 mμ (ε 14300) (1939 Dannenberg). Gives a green coloration with 1.4-naphthoquinone and conc. HCl in glacial acetic acid at 60–70° (1939 b Miescher). — **Fmn.** From 17α-ethyl-Δ^5-androstene-3β.17β-diol (p. 2036 s) on refluxing with aluminium tert.-but-oxide and acetone in benzene for 20 hrs. (1938 b Ruzicka); in the preparation from the same diol by bromination, CrO$_3$ oxidation, and debromination with zinc (see p. 145), the debromination must be carried out in methanol acidified with H$_2$SO$_4$ (1936 Butenandt) or the 17α-ethyl-Δ^5-androsten-17β-ol-3-one (below) formed when methanol or ethanol alone is used must be heated with aqueous alcoholic H$_2$SO$_4$ on the water-bath for some min. (1936, 1938 a Butenandt; cf. also 1936 Ruzicka; 1935 a Ciba; 1935, 1937 Schering-Kahlbaum A.-G.). From the 3-enol ethyl ether of Δ^4-androstene-3.17-dione ($\Delta^{3 \cdot 5}$-androstadien-3-ol-17-one ethyl ether, p. 2606 s) on treatment with ethylmagnesium bromide, followed by hydrolysis (1943 Ciba Pharmaceutical Products).

17-iso-Δ^5-Pregnen-17β-ol-3-one, 17α-Ethyl-Δ^5-androsten-17β-ol-3-one

Me OH $C_{21}H_{32}O_2$ (p. 145). For its formation from 17α-ethyl-
Me ⸺·C₂H₅ Δ^5-androstene-3β.17β-diol (p. 2036 s) and for its con-
O version into 17α-ethyl-Δ^4-androsten-17β-ol-3-one, see
also under the formation of the preceding compound.

17-iso-Allopregnan-17β-ol-3-one, 17α-Ethylandrostan-17β-ol-3-one $C_{21}H_{34}O_2$

(p. 145). Crystals (ether), m. 147–8⁰ cor.; $[\alpha]_D^{20}$ +52⁰
Me OH $\pm 3^0$ (chloroform) (1947 Ruzicka). — **Fmn.** From
Me ⸺·C₂H₅ 17α-ethylandrostane-3β.17β-diol (p. 2038 s) on oxi-
dation with CrO_3 in strong acetic acid with cooling
O H (1935, 1947 Ruzicka; 1935b CIBA); similarly from
17α-ethylandrostane-3α.17β-diol (p. 2038 s) (1935b
CIBA). — **Rns.** Gives $\Delta^{17(20)}$-allopregnen-3-one (p. 2411 s) on distillation with
anhydrous $CuSO_4$ at 140⁰ under 0.01 mm. pressure (1947 Ruzicka).

1935 a CIBA*, *Swiss Pat.* 210758 (issued 1940); *Ch. Ztbl.* **1941** I 3409.
  b CIBA*, *Swiss Pat.* 210759 (issued 1940); *Ch. Ztbl.* **1941** I 3409; CIBA* (Ruzicka, Wettstein),
    U.S. Pat. 2143453 (1936, issued 1939); *C. A.* **1939** 3078; *Ch. Ztbl.* **1939** I 5009, 5010; CIBA
    PHARMACEUTICAL PRODUCTS (Ruzicka, Wettstein), *U.S. Pat.* 2387469 (1937, issued 1945);
    C. A. **1946** 5536; *Ch. Ztbl.* **1948** I 1234.
  Ruzicka, Goldberg, Rosenberg, *Helv.* **18** 1487, 1496.
  SCHERING-KAHLBAUM A.-G., *Brit. Pat.* 464396 (issued 1937); *C. A.* **1937** 6418; *Ch. Ztbl.*
    1937 II 3041.
1936 Butenandt, Schmidt-Thomé, *Ber.* **69** 882, 888.
  Kuwada, Yago, *C. A.* **1937** 2224; *Ch. Ztbl.* **1937** I 3808.
  Ruzicka, Rosenberg, *Helv.* **19** 357, 366.
1937 CIBA*, *Swiss Pat.* 211653 (issued 1941); *C. A.* **1942** 3638; *Ch. Ztbl.* **1942** I 385.
  SCHERING-KAHLBAUM A.-G., *Brit. Pat.* 492725 (issued 1938); *Ch. Ztbl.* **1939** I 2642.
1938 a Butenandt, Schmidt-Thomé, Paul, *Ber.* **71** 1313, 1315.
  b Butenandt, Peters, *Ber.* **71** 2688.
  a CIBA*, *Ital. Pat.* 363163; *Ch. Ztbl.* **1939** II 4032; *Fr. Pat.* 840854 (issued 1939); *C. A.* **1940**
    1821; *Ch. Ztbl.* **1940** I 915.
  b CIBA*, *Brit. Pat.* 517942 (issued 1940); *C. A.* **1941** 7122; *Ch. Ztbl.* **1941** I 927.
  c CIBA*, *Swiss Pat.* 227455 (issued 1943); *C. A.* **1949** 3476; *Ch. Ztbl.* **1944** II 1203.
  a Inhoffen, Hohlweg, *Angew. Ch.* **51** 173.
  b Inhoffen, Logemann, Hohlweg, Serini, *Ber.* **71** 1024, 1027–1029, 1031, 1032.
  a Ruzicka, Hofmann, Meldahl, *Helv.* **21** 371.
  b Ruzicka, Hofmann, Meldahl, *Helv.* **21** 597.
  c Ruzicka, Meldahl, *Helv.* **21** 1760 footnote 5.
  d Ruzicka, Meldahl, *Helv.* **21** 1760, 1762, 1768.
  a SCHERING A.-G., *Fr. Pat.* 845473 (issued 1939); *C. A.* **1941** 1185; *Ch. Ztbl.* **1943** I 653;
    SCHERING CORP. (Inhoffen, Logemann, Dannenbaum), *U.S. Pat.* 2280236 (1939, issued
    1942); *C. A.* **1942** 5322.
  b SCHERING A.-G., *Fr. Pat.* 855730 (issued 1940); *Ch. Ztbl.* **1940** II 1327; *Dutch Pat.* 51153
    (issued 1941); *Ch. Ztbl.* **1942** I 778.
  Serini, Logemann, *Ber.* **71** 1362, 1366.
1939 a CIBA*, *Fr. Pat.* 860323 (issued 1941); *Ch. Ztbl.* **1941** I 3550.
  b CIBA*, *Fr. Pat.* 861318 (issued 1941); *Ch. Ztbl.* **1941** I 3259; *Brit. Pat.* 536621 (issued 1941);
    C. A. **1942** 1739; *Ch. Ztbl.* **1943** II 147; CIBA PHARMACEUTICAL PRODUCTS (Ruzicka), *U.S.
    Pat.* 2365292 (issued 1944); *C. A.* **1945** 4199; *Ch. Ztbl.* **1945** II 1233.

* Société pour l'industrie chimique à Bâle; Gesellschaft für chemische Industrie in Basel.

c Ciba*, *Fr. Pat.* 861536 (issued 1941); *Ch. Ztbl.* **1941** I 3259.

d Ciba*, *Brit. Pat.* 532262 (issued 1941); *C. A.* **1942** 874; *Fr. Pat.* 863162 (1940, issued 1941); *Ch. Ztbl.* **1941** II 3218; Ciba Pharmaceutical Products (Ruzicka), *U.S. Pat.* 2315817 (1940, issued 1943); *C. A.* **1943** 5558.

Dannenberg, *Abhandl. Preuss. Akad. Wiss. Math.-naturwiss. Kl.* **1939** No. 21, p. 53.

Inhoffen, Köster, *Ber.* **72** 595.

a Miescher, Scholz, *Helv.* **22** 120, 124. — b Miescher, Wettstein, Scholz, *Helv.* **22** 894, 904.

a Ruzicka, Hofmann, *Helv.* **22** 150, 152.

b Ruzicka, Müller, *Helv.* **22** 416, 419.

c Ruzicka, Goldberg, Hunziker, *Helv.* **22** 707, 710, 714, 715.

d Ruzicka, Müller, *Helv.* **22** 755.

a Schering A.-G., *Swiss Pat.* 227580 (issued 1943); *Ch. Ztbl.* **1944** II 1300; *Swed. Pat.* 99036 (issued 1940); *Ch. Ztbl.* **1940** II 3668; Schering Corp. (Inhoffen), *U.S. Pat.* 2358808 (issued 1944); *C. A.* **1945** 1738; *Ch. Ztbl.* **1945** II 1069.

b Schering A.-G. (Köster, Inhoffen), *Swed. Pat.* 109502 (issued 1944); *Ch. Ztbl.* **1944** II 247; *Dutch Pat.* 52656 (issued 1942); *Ch. Ztbl.* **1942** II 2615.

c Schering A.-G. (Logemann), *Swed. Pat.* 102639 (issued 1941); *Ch. Ztbl.* **1942** II 74; *Fr. Pat.* 868336 (1940, issued 1941); *Ch. Ztbl.* **1942** I 2038.

1940 a Ciba*, *Fr. Pat.* 864310 (issued 1941); *Ch. Ztbl.* **1942** I 899.

b Ciba*, *Fr. Pat.* 876214 (issued 1942); *Ch. Ztbl.* **1943** I 1496; Ciba Pharmaceutical Products (Miescher, Heer), *U.S. Pat.* 2372841 (issued 1945); *C. A.* **1945** 4195; *Ch. Ztbl.* **1945** II 1773.

c Ciba*, *Fr. Pat.* 876213 (issued 1942); *Ch. Ztbl.* **1943** I 2320.

McCullagh, Schneider, Emery, *Endocrinology* **27** 71, 77.

1941 Ciba*, *Fr. Pat.* 51887 (issued 1943); *Ch. Ztbl.* **1944** II 778; Ciba Pharmaceutical Products (Miescher), *U.S. Pat.* 2435013 (1942, issued 1948); *C. A.* **1948** 2996.

Ercoli, *C. A.* **1944** 2987; *Ch. Ztbl.* **1941** II 3086.

Prins, Reichstein, *Helv.* **24** 945, 953–955.

1942 Arnaudi, *C. A.* **1946** 3152; cf. *Ztbl. Bakt. Parasitenk.* II **105** 352 (1942/43); *C. A.* **1943** 5756.

Prins, Reichstein, *Helv.* **25** 300, 316 et seq.

1943 Ciba Pharmaceutical Products (Miescher), *U.S. Pat.* 2386331 (issued 1945); *C. A.* **1946** 178; *Ch. Ztbl.* **1948** I 1040.

Goldberg, Aeschbacher, Hardegger, *Helv.* **26** 680, 685.

Shoppee, Prins, *Helv.* **26** 185, 194.

1944 Sartori, Bianchi, *Gazz.* **74** 8.

Zimmermann, *Vitamine und Hormone* **5** [1952] 152, 163–165.

1945 Hardegger, Scholz, *Helv.* **28** 1355, 1356.

Schering A.-G., *Bios Final Report* No. 449, pp. 286–289; *Fiat Final Report* No. 996, pp. 73–75.

1946 Heard, Sobel, *J. Biol. Ch.* **165** 687.

1947 Ruzicka, Meister, Prelog, *Helv.* **30** 867, 876.

1948 Fieser, Fieser, *Experientia* **4** 285, 287.

1950 Shoppee, *Nature* **166** 107.

17α-Allyl-Δ^4-androsten-17β-ol-3-one, 17-Allyltestosterone ** $C_{22}H_{32}O_2$. Needles with $\frac{1}{2}$ H_2O (alcohol), m. 105–107.5°; the intensely dried substance crystallizes from ethyl acetate as needles which on keeping under the solvent change to hexagonal prisms with $\frac{1}{2}$ H_2O, m. 93°; the prisms are also obtained

---

* Société pour l'industrie chimique à Bâle; Gesellschaft für chemische Industrie in Basel.

** The compound, m. 150–153° cor., assumed to be 17-allyltestosterone and obtained from 17α-allyl-Δ^5-androstene-3β.17β-diol by bromination with 4 mols. bromine, followed by oxidation with CrO_3 and debromination with zinc in glacial acetic acid (1936 Kuwada), was according to Butenandt (1938) unchanged diol.

on crystallization from pure acetone (1938 Butenandt). A non-analysed sample formed needles (ether-petroleum ether), m. 113–4° cor. (1940 v. Euw).

Fmn. From 17α-allyl-Δ^5-androstene-3β.17β-diol (p. 2041 s) on refluxing with aluminium isopropoxide and cyclohexanone in toluene for 40 min. (yield, 80%) (1938 Butenandt) or with aluminium tert.-butoxide and acetone in benzene for 24 hrs. (yield, 70%) (1940 v. Euw). From the 3-ethylene ketal of Δ^4-androstene-3.17-dione (p. 2890 s) on treatment with allylmagnesium bromide, followed by hydrolysis with dil. H_2SO_4 (1939b CIBA).

Rns. With ammoniacal silver solution in methanol it slowly gives a precipitate (1939 Miescher). On treatment with OsO_4 in ether at 20° for 48 hrs., followed by boiling with Na_2SO_3 in aqueous alcohol, it gives a mixture of the two 20-epimeric ω-homo-17-iso-Δ^4-pregnene-17β.21.22-triol-3-ones [17-(β.γ-dihydroxypropyl)-testosterones, p. 2728 s] (1938 Butenandt; 1939 Miescher; 1939a CIBA). On refluxing with $POCl_3$ in pyridine for 20 min., followed by treatment with conc. HCl, it is dehydrated to give 17-allylidene-Δ^4-androsten-3-one (p. 2412 s) (1938 Butenandt; 1940 v. Euw).

Oxime $C_{22}H_{33}O_2N$, crystals with $^1/_2$ H_2O (aqueous alcohol), m. 144–6° (1938 Butenandt).

1936 Kuwada, Yago, *C. A.* **1937** 2224; *Ch. Ztbl.* **1937** I 3808.
1938 Butenandt, Peters, *Ber.* **71** 2688, 2692, 2693.
1939 a CIBA*, *Fr. Pat.* 857 122 (issued 1940); *Ch. Ztbl.* **1941** I 1324.
     b CIBA*, *Fr. Pat.* 861 536 (issued 1941); *Ch. Ztbl.* **1941** I 3259; CIBA PHARMACEUTICAL PRODUCTS (Miescher), *U.S. Pat.* 2 386 331 (1943, issued 1945); *C. A.* **1946** 179; *Ch. Ztbl.* **1948** I 1040.
     Miescher, Wettstein, Scholz, *Helv.* **22** 894, 901, 903.
1940 v. Euw, Reichstein, *Helv.* **23** 1114, 1118.

---

b. HYDROXY-11-KETOSTEROIDS C_{20}–C_{26} WITH THREE ALIPHATIC SIDE CHAINS

$\Delta^{16\cdot20}$-**Pregnadien-3α-ol-11-one** $C_{21}H_{30}O_2$. — **Fmn.** Its acetate is obtained as a by-product in the preparation of $\Delta^{17(20)}$-pregnene-3α.21-diol-11-one diacetate (p. 2723 s) from 21-bromo - $\Delta^{17(20)}$ - pregnen - 3α - ol - 11 - one acetate by refluxing with potassium acetate in acetone. — *Acetate* $C_{23}H_{32}O_3$, prisms (methanol and petroleum ether), m. 127–129.5° cor.; absorption max. (solvent not stated) at 238 mμ (ε 15 000) (1946a Sarett).

Δ^{16}-**Pregnen-3α-ol-11-one** (? I), $\Delta^{17(20)}$-**Pregnen-3α-ol-11-one** (II), and Δ^{20}-**Pregnen-3α-ol-11-one** (III), $C_{21}H_{32}O_2$. Pure $\Delta^{17(20)}$-pregnen-3α-ol-11-one (II) [m. 191–2° cor.; $[\alpha]_D^{25}$ +55° (acetone)] has been isolated after the closing date for this volume by Sarett (1948), and pure Δ^{20}-pregnen-3α-ol-11-one (III) [m. 153–4° cor.; $[\alpha]_D^{25}$ + 38.5° (acetone)] by Sarett (1949b). — **Fmn.** A mixture of the acetates of the three hydroxyketones is obtained from 20-aminopregnan-3α-ol-11-one 3-acetate (p. 2718 s) on heating with $NaNO_2$, aqueous

---

* Société pour l'industrie chimique à Bâle; Gesellschaft für chemische Industrie in Basel.

pyridine, and pyridine hydrochloride on the steam-bath; a small amount of pregnane-3α.20-diol-11-one 3-acetate (see p. 2722 s) formed at the same time (cf. also 1949a Sarett) may be converted into a crystalline mixture of the

acetates of (II) and (III) by treatment with p-toluenesulphonyl chloride in pyridine, followed by refluxing of the resulting 20-p-toluenesulphonate with collidine (1946a, 1948 Sarett; 1946 MERCK & Co., INC.). — **Rns.** On ozonization of the mixture of acetates in ethyl acetate-methanol, followed by oxidation with $KMnO_4$ in aqueous acetone at room temperature, it gives the acetates of ætiocholan-3α-ol-11.17-dione (p. 2949 s; derived from II), 3α-hydroxy-11-ketoætiocholanic acid (derived from III) (1946a Sarett; 1946 MERCK & Co., INC.), and after refluxing with methanolic H_2SO_4, followed by treatment with acetic anhydride and perchloric acid in glacial acetic acid, the **lactone** $C_{23}H_{32}O_5$ [IV? (derived from I); crystals (ether), m. 230–1° cor.] which on alkaline hydrolysis gives an amorphous acid (1946a Sarett). On hydrolysis of the crystalline mixture of the acetates of (II) and (III) (see above) with boiling methanolic KOH, followed by oxidation with CrO_3 in strong acetic acid, a mixture of $\Delta^{17(20)}$- and Δ^{20}-pregnene-3.11-diones (p. 2899 s) is obtained (1946b Sarett).

9-iso-Pregnan-3β-ol-11-one $C_{21}H_{34}O_2$.

This structure was ascribed by Marker (1939) to **uran-3β-ol-11-one** [crystals (acetone), m. ca. 208°; *acetate*, m. 170.5–172°] obtained from urane-3β.11-diol on oxidation with CrO_3 in acetic acid or from urane-3.11-dione on reduction by refluxing with aluminium isopropoxide in isopropyl alcohol or by shaking with hydrogen and Adams' catalyst in abs. alcohol at 25° and 3 atm. pressure; it may be re-oxidized to the dione with CrO_3 in acetic acid (1939 Marker). According to Klyne (1950, 1952) the urane compounds are probably D-homo-steroids (see urane, p. 1403 s).

1939 Marker, Rohrmann, *J. Am. Ch. Soc.* **61** 2719, 2721.
1946 MERCK & Co., INC. (Sarett), *U.S. Pat.* 2540964 (issued 1951); *C. A.* **1951** 7159; *Brit. Pat.* 630103 (issued 1949); *C. A.* **1950** 7891.
    a Sarett, *J. Biol. Ch.* **162** 601, 616–619, 623.
    b Sarett, *J. Am. Ch. Soc.* **68** 2478, 2480.

1948 Sarett, *J. Am. Ch. Soc.* **70** 1690, 1693.
1949 a Sarett, *J. Am. Ch. Soc.* **71** 1165, 1167.
    b Sarett, *J. Am. Ch. Soc.* **71** 1169, 1173.
1950 Klyne, *Nature* **166** 559.
1952 Klyne, Shoppee, *Chemistry & Industry* **1952** 470.

---

c. Hydroxy-12-ketosteroids C_{20}–C_{26} with Three Aliphatic Side Chains

Norcholan-3α-ol-12-one $C_{23}H_{38}O_2$. — **Fmn.** Its acetate is obtained from 23-bromo-norcholan-3α-ol-12-one acetate (p. 2701 s) on reduction with zinc in acetic acid; yield, 97 %. — *Acetate* $C_{25}H_{40}O_3$, crystals (methanol), m. 158.5° to 159.5°; $[\alpha]_D^{20} + 119°$ (acetone) (1946 Brink).

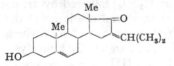

1946 Brink, Clark, Wallis, *J. Biol. Ch.* **162** 695, 703.

---

d. Hydroxy-17-ketosteroids C_{20}–C_{26} with Three Aliphatic Side Chains

16-Isopropylidene-Δ^5-androsten-3β-ol-17-one, 16-Isopropylidene-5-dehydro-epiandrosterone $C_{22}H_{32}O_2$. Needles (methanol or ethanol), m. 223° (1939 Butenandt; 1945 Ross). Absorption max. (solvent not stated) at 250 (4.17) and 338 (2.0) mμ (log ε) (1945 Ross). — **Fmn.** From the acetate of 5-dehydroepiandrosterone (p. 2609 s) and acetone in ether in the presence of sodium or sodamide (small yield) (1939 Butenandt; 1939 Schering A.-G.). From the same acetate on refluxing with KOH in acetone - methanol (1:1) for 1 hr. (yield, 80 %); on one occasion a product m. 180° (with evolution of gas), probably the 16-(α-hydroxyisopropyl)-compound, was obtained which on warming with methanol gave the isopropylidene-compound (1945 Ross).

Rns. On boiling with P_2O_5 in benzene for $^1/_2$ hr. it is dehydrated to give 16-isopropylidene-$\Delta^{3.5}$-androstadien-17-one (p. 2412 s). On heating with hydrazine hydrate and sodium ethoxide in alcohol in a sealed tube at 210° for 16 hrs. it gives a **compound** $C_{20}H_{34}O_3$, needles (ether - light petroleum b. 40–60°), m. 190–192° with evolution of gas (1945 Ross).

Acetate $C_{24}H_{34}O_3$. Needles (alcohol), m. 189°. — **Fmn.** From the above carbinol on boiling with acetic anhydride for 10 min. (1939 Butenandt; cf. also 1945 Ross).

Benzoate $C_{29}H_{36}O_3$*. Needles (acetone), m. 210–212°. — **Fmn.** From the carbinol on treatment with benzoyl chloride in pyridine (1945 Ross).

---

* In the original paper the formula $C_{27}H_{36}O_3$ is erroneously indicated; the values calculated on this formula, and also found, are C 79.4 % and H 8.9 %; for the correct formula these values are 80.5 and 8.3, resp.

16-(α-Methylpropylidene)-$\varDelta^5$-androsten-3β-ol-17-one, 16-(2-Butylidene)-$\varDelta^5$-androsten-3β-ol-17-one, 16-(α-Methylpropylidene)-5-dehydroepiandrosterone $C_{23}H_{34}O_2$.

Needles (ethyl acetate), m. 176° (1939 Butenandt). Absorption max. in ether at 245 mμ (ε 12800) (1939 Butenandt; 1939 Dannenberg).

Fmn. From the acetate of 5-dehydroepiandrosterone (p. 2609 s) and methyl ethyl ketone in the presence of sodium or sodamide in ether; crude yield 32 % (1939 Butenandt; 1939 SCHERING A.-G.).

Rns. On bromination of its acetate with 1 mol. bromine in chloroform, followed by ozonization and refluxing with zinc dust in glacial acetic acid, it gives $\varDelta^5$-androstene-3β.17β-diol-16-one 3-acetate (p. 2744 s) (1939 Butenandt; 1939 SCHERING A.-G.; for the structure, see 1947, 1949 Huffman; 1950 Gallagher) and, mainly, the **acetoxydicarboxylic acid** $C_{21}H_{30}O_6$ (I) [crystals (acetone-ethyl acetate), m. 251° dec. after sintering at 235°] which on treatment with acetic anhydride in cold pyridine gives the *anhydride* $C_{21}H_{28}O_5$ [needles (methanol), m. 186°] (1939 Butenandt). On reduction by refluxing with aluminium isopropoxide in benzene · isopropyl alcohol it gives 16-(2-butylidene)-$\varDelta^5$-androstene-3β.17-diol (p. 2043 s) (1942 Stodola).

Acetate $C_{25}H_{36}O_3$. Leaflets (alcohol), m. 148°, and needles (alcohol), m. 156°. **Fmn.** From the above carbinol on refluxing with acetic anhydride for 10 min. (1939 Butenandt).

1939 Butenandt, Schmidt-Thomé, Weiss, *Ber.* **72** 417, 421, 422, 424.
Dannenberg, *Abhandl. Preuss. Akad. Wiss. Math.-naturwiss. Kl.* **1939** No. 21, p. 15.
SCHERING A.-G. (Butenandt, Schmidt-Thomé, Schwenk), *Swed. Pat.* 98872 (issued 1940); *Ch. Ztbl.* **1940** II 2784, 2785; SCHERING CORP. (Butenandt, Schmidt-Thomé, Schwenk), *U.S. Pat.* 2353808 (issued 1944); *C. A.* **1945** 1258.
1942 Stodola, Kendall, *J. Org. Ch.* **7** 336, 338.
1945 Ross, *J. Ch. Soc.* **1945** 25.
1947 Huffman, Lott, *J. Am. Ch. Soc.* **69** 1835.
1949 Huffman, Lott, *J. Am. Ch. Soc.* **71** 719, 724.
1950 Gallagher, Kritchevsky, *J. Am. Ch. Soc.* **72** 882, 884 footnote 17.

7. Monohydroxy-monoketosteroids (CO and OH in the Ring System) C$_{27}$ with Three Aliphatic Side Chains

Summary of Hydroxy-ketones with Cholestane Skeleton, including Enolic Forms of the Corresponding Diketones

For values other than those chosen for the summary, see the compounds themselves.

Position of			M.p. °C.	$[\alpha]_D$ (°) in CHCl$_3$	Absorpt. max. (mμ) in alc.	Acetate		Page
CO	OH	double bond				m.p. °C.	$[\alpha]_D$ in CHCl$_3$	
2	3	Δ^3	145	+79	272	139	+94	2659 s
2	3β	satd.	105	+65	—	146	+73	2660 s
2	5α	Δ^3	174	+36[1])	223	—	—	2661 s
2	5α	satd.	182	+30	—	—	—	2662 s
3	2	Δ^1	169	+57	270	142	—	2662 s
3	2α	Δ^4	117	—	242[2])	142	+64	2663 s
3	2α(?)	satd.	126	+38	—	see p. 2665 s		2665 s
3	4	$\Delta^{4.6}$	163	+38	319	—	—	2666 s
3	4	Δ^1	—	—	—	—	—	2668 s
3	4	Δ^4	150	+80	278	103	+93	2669 s
3	4α	satd.	—	—	—	see p. 2670 s		2670 s
3	4(?)	satd.[3])	—	—	—	149	—	2670 s
3	5α	satd.	{ 209 230	+41	—	—	—	2671 s
3	6	$\Delta^{4.6}$	—	—	—	—	—	2671 s
3	6α	Δ^4	160	+82	240	107	+79	[4])
3	6β	Δ^4	192	+27	237	102	+36	2672 s
3	6α	satd.	—	—	—	—	—	2673 s
3	6β	satd.	190	+9	—	101	—	2673 s
4	3	$\Delta^{2.5}$	136	+31	265, 300	161	+13	2674 s
4	3β	Δ^5	—	—	240[2])	124	—77	2675 s
4(?)	3α(?)	satd.	174	+15	—	144	—8	2676 s
4	3β	satd.	—	—	—	118	—23	[5])
6	3	$\Delta^{2.4}$	—	—	317[2])	140	+27	2676 s
6	3β	Δ^4	151	—13	239, 319	110	—51	2677 s
6	3α	satd.	160	+3	—	108	—4	[6])
6	3β	satd.	140; 153	—6	280[2])	129	—15	2677 s
6	7	$\Delta^{2.4.7}$	—	—	—	—	—	2680 s
7	3	$\Delta^{3.5}$	186	—51	322	107	—185	2680 s
7	3β	$\Delta^{8.14}$	—	—··	224, 298[2])	174	—15	2681 s
7	3α	Δ^5	—	—	234[1]),[2])	119	—	2681 s
7	3β	Δ^5	171	—104	237	158	—100	2682 s
7	3β	Δ^8	—	—	253[2])	156	—32	2685 s
7	3β	$\Delta^{8(14)}$	—	—	262[2])	142	—62	2685 s
7	3β	satd.	129; 169	—35	—	149	—36	2686 s
15	3β	$\Delta^{8(14)}$	146	—	259[2])	135	+118	2687 s

[1]) In ether. [2]) For the acetate. [3]) Coprostane derivative. [4]) 1953 Sondheimer; cf. 1953a Fieser. [5]) 1950 Lieberman. [6]) 1948 Dodson.

Δ^3-**Cholesten-3-ol-2-one** $C_{27}H_{44}O_2$. *One enolic form (form A) of cholestane-2.3-dione* (p. 2902 s). For structure, see 1944a Ruzicka; cf. 1938 Stiller. — Pale yellow or colourless, prismatic needles (petrol or glacial acetic acid) (1938 Stiller), m. 144–5⁰ cor. (1944a Ruzicka; cf. 1938 Stiller); $[\alpha]_{5461}^{20}+92^0$, $[\alpha]_D^{20}+79^0$ (both in chloroform); absorption max. in 95 % alcohol at 272 mμ (log ε 3.7), in 0.005 N NaOH (in 95 % alcohol) at ca. 320 mμ (log ε 3.57) (1938 Stiller). Readily soluble in dioxane, pyridine, and ethyl acetate, sparingly soluble in cold methanol, ethanol, and glacial acetic acid (1938 Stiller).

Colour reactions. With tetranitromethane, yellow; with $FeCl_3$, purple; positive Liebermann-Burchard reaction; Salkowski reaction: colourless chloroform layer, H_2SO_4 layer yellow changing to cherry-red; with conc. H_2SO_4 alone, lemon-yellow changing to cherry-red. Titration with pyridine sulphate dibromide shows absorption of 0.89 atoms of bromine in 15 min. at room temperature (1938 Stiller).

Fmn. From cholestan-3-one (p. 2436 s) by refluxing for a short time with SeO_2 in aqueous alcohol; separation from other products via the potassium salt (see below) which is then decomposed with dil. HCl in ether; yield, 30 %. From Δ^1-cholesten-2-ol-3-one (form B of the parent 2.3-diketone, p. 2662 s) by shaking with 20 % aqueous KOH in ether at 0⁰ and decomposition of the potassium salt as above (1938 Stiller). From the (crude) 2-(p-dimethylaminophenyl)-isoxime of cholestane-2.3-dione (p. 2902 s) by treatment with 2 N aqueous HCl in ether to give a crude "form C" (p. 2663 s) (yield, 39 %) and subsequent conversion to the hydroxyketone potassium salt which is decomposed as described above (1944a Ruzicka).

Rns. Oxidation with H_2O_2 in boiling aqueous alcoholic KOH gives cholestane-2‖3-dioic acid (p. 1713 s) (1938 Stiller). Hydrogenation of the acetate in alcohol at 50⁰, using Raney nickel catalyst, affords the acetate of cholestan-3β-ol-2-one (p. 2660 s); the same product is ultimately obtained when the hydrogenation is carried out in glacial acetic acid with a platinum catalyst, with subsequent CrO_3 oxidation of the not isolated diol acetate (1944b Ruzicka). Brief heating in glacial acetic acid, containing some HCl, on the water-bath yields "form C" (p. 2663 s). Refluxing with o-phenylenediamine in alcohol gives the quinoxaline derivative of the parent dione (p. 2903 s) (1938 Stiller). See also reactions of the p-toluenesulphonate, p. 2660 s.

Sodium salt $C_{27}H_{43}O_2Na$. White crystals (alcohol), obtained from the acetate (below) by treatment with sodium ethoxide (1.2 mols.) in alcohol - ether at room temperature. Gives the "A form" on warming with glacial acetic acid (1938 Stiller).

Potassium salt (no formula given). Colourless solid, insoluble in 20 % aqueous KOH and in ether; obtained from the parent "form A" by shaking with 20 % aqueous KOH in ether at 0⁰; regenerates "form A" on treatment with aqueous HCl in ether at room temperature (1938 Stiller; cf. 1944a Ruzicka).

Acetate, 3-Acetoxy-Δ³-cholesten-2-one $C_{29}H_{46}O_3$. Crystals (85 % alcohol), plates (acetic anhydride) (1938 Stiller), crystals (glacial acetic acid), m. 138–9⁰ cor. (1944a Ruzicka); $[\alpha]_{5461}^{18} + 110^0$, $[\alpha]_D^{18} + 92^0$ (1938 Stiller), $[\alpha]_D + 96^0$ (1944a Ruzicka) (all in chloroform); absorption max. in alcohol at 238 mμ (log ε 3.87) (1938 Stiller). In contrast to the free diosphenol it gives a purple coloration with m-dinitrobenzene and alcoholic KOH, but no coloration with $FeCl_3$ (1938 Stiller). — **Fmn.** From the above hydroxyketone on refluxing with acetic anhydride for 40 min. (1938 Stiller) or on heating with acetic anhydride in abs. pyridine for 10 min. at 100⁰ (yield, 85 %) (1944a Ruzicka). — **Rns.** For saponification, see the sodium salt (p. 2659 s), for hydrogenation, see reactions of the hydroxyketone, p. 2659 s.

Benzoate, 3-Benzoyloxy-Δ³-cholesten-2-one $C_{34}H_{48}O_3$. Dimorphous: prismatic needles (ethyl acetate), which change into transparent rhombic crystals on standing; interconversion readily occurs; the two forms show identical m. p. and clearing point. Melts at 162–3⁰ to give an anisotropic liquid which changes to the isotropic melt at 193–4⁰; reverts to the anisotropic melt on cooling. — **Fmn.** From the above hydroxyketone on treatment with benzoyl chloride in pyridine at room temperature (1938 Stiller).

p - Toluenesulphonate, 3 - (p - Toluenesulphonyl - oxy) - Δ³ - cholesten - 2 - one $C_{34}H_{50}O_4S$. Crystals (ethyl acetate - methanol), m. 161–2⁰ cor.; $[\alpha]_D + 83^0$ (chloroform). — **Fmn.** From the above hydroxyketone on treatment with p-toluenesulphonyl chloride in abs. pyridine at 20⁰; yield, 75 %. — **Rns.** Hydrogenation in alcohol at 70⁰, using Raney nickel catalyst, affords cholestan-2-one (p. 2417 s). Heating with sodium iodide in dry acetone at 160⁰ (bomb), followed by reduction with zinc dust in boiling alcohol, gives $Δ^{3.5}$-cholestadien-2-one (p. 2417 s) (1944a Ruzicka).

Cholestan-3β-ol-2-one $C_{27}H_{46}O_2$. Crystals (aqueous methanol), m. 104–5⁰ cor.; $[\alpha]_D + 65^0$ (chloroform). Gives a *digitonide*, sparingly soluble in 80 % alcohol. — **Fmn.** The acetate is obtained from that of *Δ³-cholesten-3-ol-2-one* (p. 2659 s) by catalytic hydrogenation using Raney nickel in alcohol at 50⁰ (yield, 82 %), or using platinum in glacial acetic acid, followed by CrO_3 oxidation of the (not isolated) diol acetate intermediate; the acetate is saponified with NaOH (1 % excess) in benzene - methanol at 20⁰ and the oily product purified via the digitonide. — **Rns.** Oxidation with H_2O_2 in alcoholic NaOH on the water-bath gives cholestane-2‖3-dioic acid (p. 1713 s). Wolff-Kishner reduction (conditions not precisely stated) of the acetate furnishes only cholestane (p. 1429 s) (1944b Ruzicka).

Oxime $C_{27}H_{47}O_2N$. Crystals (petroleum ether), m. 207–8⁰ cor. dec.; obtained by saponification of the acetate oxime (below), using NaOH in benzene - methanol at room temperature (1944b Ruzicka).

Acetate, 3β-Acetoxycholestan-2-one $C_{29}H_{48}O_3$. Crystals (methanol), m. 145.2 to 146.5° cor., $[\alpha]_D +73°$ (chloroform). — **Fmn.** From the parent hydroxyketone by reaction with acetic anhydride and pyridine at room temperature, or as described under the formation of the hydroxyketone (above). — **Rns.** See those of the parent hydroxyketone (above). — *Oxime* $C_{29}H_{49}O_3N$, crystal clusters (petroleum ether · methanol), m. 178–179.5° cor. dec. (1944 b Ruzicka).

Δ³-Cholesten-5-ol-2-one $C_{27}H_{44}O_2$. For structure, see 1953 Conca; the structure *2β.5β-epoxy-"cholestan"-3 or 4-one* had previously been suggested by Bergmann (1939) for this product, referred to as *"Ketone B"*. — Needles (ether), crystals (alcohol), m. 173.5–174° cor., $[\alpha]_D^{25} +36°$ (ether) (1939 Bergmann). Absorption max. in abs. alcohol at 223 mμ (log ε 3.9) (1953 Conca; cf. 1939 Bergmann); infra-red absorption max. in chloroform at 2.81 and 5.98 μ (1953 Conca). — **Fmn.** From 4α.5-epoxycholestan-2-one ("Ketone A", p. 2753 s) by distillation at 210–230° and 1 mm. pressure (yield, 80%) or by refluxing with acetic anhydride (low yield). From the 4-methyl ether of cholestane-4β.5-diol-2-one ("Ketone C", p. 2754 s) by distillation at 220–230° and 1 mm. pressure (1939 Bergmann).

Rns. Gives the acid $C_{25}H_{42}O_3$ (below) on oxidation with CrO_3 in strong acetic acid at 55–60° (1953 Conca). Hydrogenation in ethyl acetate at room temperature, using platinum black catalyst, yields cholestan-5-ol-2-one (1953 Conca; cf. 1939 Bergmann). Refluxing with a 5% solution of KOH in 95% methanol gives "ketone C" (p. 2754 s) (1939 Bergmann). On heating with p-toluenesulphonic acid in anhydrous benzene on the steam-bath or on treatment with sulphuric acid in dioxane it gives approximately equal amounts of Δ³·⁵-cholestadien-2-one (p. 2417 s) and a **phenol** $C_{27}H_{42}O$(?), crystals (pentane), m. 119–120.5° cor., $[\alpha]_D^{25} +74°$ (chloroform), absorption max. in abs. alcohol at 280 mμ (1953 Conca).

Semicarbazone $C_{28}H_{47}O_2N_3$. Crystals (alcohol), m. 234° cor. dec. (1939 Bergmann).

Acid $C_{25}H_{42}O_3$ (I). Crystals (acetone and ethyl acetate), m. 166–7° cor.; $[\alpha]_D^{23} +33°$ (chloroform). Infra-red absorption max. in chloroform at 2.82 and 5.65 μ (1953 Conca). — **Fmn.** From the above Δ³-cholesten-5-ol-2-one on

oxidation with CrO_3 in strong acetic acid at 55–60°; similarly from 4α.5-epoxycholestan-2-one (p. 2753 s) (1953 Conca). — **Rns.** On refluxing with amalgamated zinc and conc. HCl in acetic acid · toluene it gives the **acid** $C_{25}H_{44}O_2$ (II), crystals (ethyl acetate and glacial acetic acid), m. 153–4° cor., $[\alpha]_D^{25} +45°$

(chloroform) (1953 Conca); for the acid (II), see also 1932 Tschesche; cf. also p. 41. — *p-Bromophenacyl ester* $C_{33}H_{47}O_4Br$, m. 126–127.5° cor.; $[\alpha]_D^{25} + 54°$ (chloroform) (1953 Conca).

Cholestan-5-ol-2-one $C_{27}H_{46}O_2$. Needles (acetone or alcohol), m. 181.5–182.5°

cor.; $[\alpha]_D^{27} + 29°$ to $+ 32°$ (chloroform). Infra-red maxima in chloroform at 2.79 and 5.86 μ. — **Fmn.** From cholestane-2α.5-diol (p. 2050 s) on oxidation with CrO_3 in strong acetic acid at room temperature; yield, 80%. From Δ^3-cholesten-5-ol-2-one (p. 2661 s) on hydrogenation in ethyl acetate in the presence of platinum black at room temperature and atmospheric pressure; yield, ca. 84%. From 4α.5-epoxycholestan-2-one (p. 2753 s) on hydrogenation in alcohol in the presence of Adams's platinum catalyst at room temperature and atmospheric pressure; yield, ca. 70%. — **Rns.** Gives cholestan-5-ol (p. 1723 s) on heating with hydrazine hydrate in methanol - diethylene glycol (1953 Conca).

Δ^1-Cholesten-2-ol-3-one $C_{27}H_{44}O_2$. *Second enolic form (form B) of cholestane-*

2.3-dione (p. 2902 s). Structure assigned by exclusion of that now accorded to "form A" (p. 2659 s) (Editor; cf. 1938 Stiller). Is considered (1938 Stiller) to be identical with the compound described by Inhoffen (1937) as cholestane-2.3-dione* (p. 157). — Needles (ethyl acetate), m. 168–9° (1938 Stiller), crystals (aqueous acetone), m. 161–2°** (1938 Inhoffen); $[\alpha]_{5461}^{20} + 67°$, $[\alpha]_D^{20} + 57°$ (both in chloroform) (1938 Stiller; cf. 1937, 1938 Inhoffen). Absorption max. in ether at 268 mμ (ε 10400) (1939 Dannenberg; cf. 1937 Inhoffen), in alcohol at 270 mμ (log ε 3.93); in 95% alcoholic 0.005 N NaOH the absorption curve is identical with that recorded under "form A" (p. 2659 s) (1938 Stiller). Sparingly soluble in cold ethyl acetate or petroleum ether (1937 Inhoffen), much less soluble than "form A" (1938 Stiller).

The *colour reactions* with tetranitromethane (yellow) and with $FeCl_3$ (purple) are the same as those given by "form A"; in contrast: the Liebermann-Burchard reaction is negative; in the Salkowski reaction the H_2SO_4 layer turns yellow, no colour in the chloroform layer; the solution in conc. H_2SO_4 has a permanent yellow colour (1938 Stiller).

* The possibility that "form B" might be the true 2.3-diketo-form, also envisaged by Stiller (1938), seems excluded by its absorption spectrum (1939 Dannenberg; cf. 1948 Fieser); in solution the presence of some diketo-form in equilibrium with the enolic form B is likely (1938 Stiller).

** Samples with m. p. ranging from 159–161° to 171–2°, but with constant $[\alpha]$ value, have been obtained and the m. p. may rise or fall on recrystallization; this behaviour may be due to the presence of keto-enol mixtures (1937 Inhoffen; 1937b SCHERING A.-G.) or to admixture with form A (1938 Stiller).

Fmn. From "form C" (below) by recrystallization from ethyl acetate (1938 Stiller). From the benzoate of Δ^4-cholesten-2-ol-3-one (p. 2664 s) by saponification with KOH in alcohol - benzene at room temperature (yield, > 8 %) or in boiling methanol (yield, ca. 11 %), when an equal amount of "cholestane-3.4-dione" (Δ^4-cholesten-4-ol-3-one, p. 2669 s) is isolated by fractional crystallization from petroleum ether (1937 Inhoffen; 1937 b SCHERING A.-G.; 1937 b SCHERING-KAHLBAUM A.-G.). Together with cholestane-2‖3-dioic acid (p. 1713 s), from the benzoate of cholestan-2-ol-3-one (p. 2665 s) or the "isomeric benzoate" (m. 145–6°; p. 2666 s) by treatment with KOH (excess) in methanol - benzene at room temperature (1938 Inhoffen).

Rns. Oxidation with H_2O_2 in boiling aqueous alcoholic KOH affords cholestane-2‖3-dioic acid (1937 Inhoffen; 1938 Stiller). By heating at 340° in a stream of carbon dioxide, elimination of methane occurs and a phenolic product can be isolated (1937 a SCHERING A.-G.; cf. 1937 b SCHERING A.-G.). Heating with o-phenylenediamine alone (1937 Inhoffen) or in boiling alcohol (1938 Stiller) gives the quinoxaline derivative of cholestane-2.3-dione (p. 2903 s). By shaking with 20 % aqueous KOH in ether at 0°, the potassium salt of "form A" (p. 2659 s) is obtained (1938 Stiller).

Equimolecular mixture (addition compound?) of Δ^1-cholesten-2-ol-3-one with Δ^3-cholesten-3-ol-2-one ("form A"; p. 2659 s), enolic *"form C"* of cholestane-2.3-dione (1938 Stiller). — Colourless prismatic needles (glacial acetic acid), m. 130–2°, clearing at 150°; $[\alpha]_{5461}^{20}$ +80°, $[\alpha]_D^{20}$ +68° (both in chloroform); absorption max. in alcohol at 271 mμ (log ε 3.80) (1938 Stiller). — **Fmn.** From "form A" (p. 2659 s) by heating for a short time with glacial acetic acid (containing small amounts of conc. HCl) on the water-bath. By recrystallizing together equal amounts of "forms A and B" (in glacial acetic acid?) (1938 Stiller). The crude "cholestane-2.3-dione" (m. 130–3°, clears at 145°) obtained by hydrolysis of its isoxime derivative (p. 2902 s) (1944 a Ruzicka) appears to be more or less pure "form C" (Editor). — **Rns.** On recrystallizing from ethyl acetate, separation into the constituents occurs; the less soluble "form B" crystallizes first (1938 Stiller).

Acetate, *2 - Acetoxy - Δ^1-cholesten - 3 - one* $C_{29}H_{46}O_3$ (p. 157: *cholestane-2.3-dione enol-acetate*). Colourless needles (methanol), m. 142° (1937 Inhoffen; cf. 1938 Stiller). Absorption max. in alcohol at 237 mμ (log ε 3.95). Gives a yellow coloration with tetranitromethane; no colour reaction with FeCl$_3$ (1938 Stiller).

Benzoate, *2-Benzoyloxy-Δ^1-cholesten-3-one* $C_{34}H_{48}O_3$. See p. 157: *cholestane-2.3-dione enol-benzoate*.

Δ^4-Cholesten-2α-ol-3-one $C_{27}H_{44}O_2$. For configuration at C-2, see 1953 d Fieser; 1953 Sondheimer. — Needles (ether-methanol), m. 116–8°; gives a purple coloration with FeCl$_3$ in alcohol (1950 Rivett; cf. 1953 d Fieser). — **Fmn.** The acetate is obtained from Δ^4-chole-

E.O.C. XIV s. 184

sten-3-one (p. 2424 s) by treatment with lead tetraacetate in glacial acetic acid containing acetic anhydride at 70°, followed by chromatographic separation from unchanged starting material (28 %); yield, 15 % of unrecovered starting material (1944 Seebeck). From 6β-chloro- or 6β-bromo-Δ^4-cholesten-3-one (p. 2484 s) by refluxing with anhydrous potassium acetate in glacial acetic acid; the acetate so formed (yield in each case, ca. 8 %) is saponified with alcoholic NaOH at room temperature (1950 Rivett; cf. 1953 d Fieser; 1953 Sondheimer).

Rns. Hydrogenation of the acetate in glacial acetic acid at room temperature using PtO_2 catalyst affords an amorphous diol acetate which on saponification with boiling methanolic KOH yields a cholestane-2.3-diol (m. 150–7°, probably a mixture of epimers), which on oxidation with CrO_3 in glacial acetic acid at room temperature gives cholestane-2‖3-dioic acid (p. 1713 s) (1944 Seebeck).

Acetate, 2α - *Acetoxy* - Δ^4 - *cholesten* - 3 - *one* $C_{29}H_{48}O_3$. Colourless needles (acetone - methanol) (1950 Rivett), colourless prisms (ether - petroleum ether), m. 141–2° cor. (1944 Seebeck); $[\alpha]_D^{15} + 65.5°$ (chloroform) (1944 Seebeck), $[\alpha]_D^{20} + 62°$ (chloroform); absorption max. in alcohol at 242 mμ (ε 14500) (1950 Rivett; cf. 1953 d Fieser). Does not reduce alkaline silver diammine solution in methanol at room temperature (1944 Seebeck). — **Fmn.** and **Rns.** See those of the hydroxy-ketone above.

Δ^4-**Cholesten-2-ol-3-one** $C_{27}H_{44}O_2$. Configuration at C-2 unknown. —*Benzoate* $C_{34}H_{48}O_3$ [p. 130: *benzoate (b), m. 137–8°*]. The alternative possibility that this substance is a mixture of the benzoates of Δ^4-cholesten-2-ol-3-one and Δ^1-cholesten-4-ol-3-one (p. 2668 s) is not excluded, but considered unlikely (1937 Inhoffen); cf., however, the saponification reaction, below (Editor). — Prisms (chloroform - alcohol), m. 137–8° to a turbid liquid (play of colours), clearing at 204°; $[\alpha]_D^{20} + 58°$ (chloroform); slightly more soluble than the benzoate of Δ^1-cholesten-4-ol-3-one (1937 Inhoffen). — **Fmn.** Along with the benzoate of Δ^1-cholesten-4-ol-3-one (p. 2668 s), from 2α.4α-dibromocholestan-3-one (p. 2495 s) by refluxing with potassium benzoate in butanol-toluene; separation from the isomeric benzoate by fractional crystallization from chloroform - alcohol; yield, 23 % (1937 Inhoffen; 1937 b Schering-Kahlbaum A.-G.; cf. 1936 Schering A.-G.). — **Rns.** By heating at 220° and 2 mm. pressure benzoic acid is eliminated; by subsequent distillation at 220° and 0.05 mm. pressure, $\Delta^{1.4}$-cholestadien-3-one (p. 2418 s) is obtained (1936 Schering A.-G.; 1937 b Schering-Kahlbaum A.-G.). Heating in a stream of carbon dioxide, first at 300–310°, then at 330°, brings about elimination of benzoic acid and methane with formation of a phenolic product (1937 b Schering A.-G.). Saponification with KOH in hot methanol, or in alcohol-benzene at room temperature, gives approximately equal amounts of cholestane-2.3-dione (p. 2902 s) and -3.4-dione (p. 2907 s) (1937 Inhoffen; cf. 1937 b Schering A.-G.).

Cholestan-2α(?)-ol-3-one $C_{27}H_{46}O_2$. The 2α-configuration is assigned to the 2-hydroxy-3-ketone constituent of the acetate addition compound described below. For Ruzicka's (1938; 1944b) free hydroxy-ketone the same structure and configuration is likely, but mixture with some isomeric 4α-hydroxy-3-ketone is probable; the 2-hydroxy-3-ketone of Inhoffen (1938) seems to be identical with that of Ruzicka (1938) (Editor). — Crystals (methanol), m. 125–7° with previous sintering (1938 Inhoffen; cf. 1938, 1944b Ruzicka); [α]$_D$ +38° (chloroform) (1944b Ruzicka). Gives no yellow coloration with tetranitromethane (1944b Ruzicka). For reducing properties of the acetate towards phosphomolybdic acid and alkaline cupric solutions, see 1946 Heard.

Fmn. The benzoate is obtained from 2α.4α-dibromocholestan-3-ol (p. 1887 s) by heating with potassium benzoate and benzoic acid at 220° (1938 Inhoffen), also from 2α-bromocholestan-3-one (p. 2480 s) by refluxing with potassium benzoate in butanol - toluene, followed by crystallization from acetone - alcohol for separation from the isomer (see below) formed simultaneously (yield, 6%) (1938 Inhoffen; cf. 1937a SCHERING A.-G.); saponification of the benzoate is effected with KOH (ca. 1 mol.) in benzene - alcohol at room temperature (1938 Inhoffen) or with KOH (excess) in boiling methanol (1937a SCHERING A.-G.). From the acetate addition compound (see below) by mild hydrolysis (1938 Ruzicka; cf. 1944b Ruzicka).

Rns. Oxidation with CrO_3 in glacial acetic acid on the water-bath affords cholestane-2∥3-dioic acid (p. 1713 s) (1938 Ruzicka). Treatment of the benzoate with excess KOH in methanol - benzene at room temperature gives a mixture of neutral and acidic products from which "cholestane-2.3-dione" (Δ¹-cholesten-2-ol-3-one, p. 2662 s) and cholestane-2∥3-dioic acid can respectively be isolated; by addition of H_2O_2 to the reaction mixture the acid product is formed exclusively (1938 Inhoffen). Heating with dehydrating agents such as H_3PO_4, P_2O_5, $KHSO_4$, or $ZnCl_2$ gives "Δ¹-cholesten-3-one" (1937a, b SCHERING A.-G.) (probably "heterocholestenone", which see p. 2450 s).

"*Acetate* $C_{29}H_{48}O_3$". The compound originally described as 2-acetoxy-cholestanone (1938 Ruzicka) has been shown to be an *addition compound (1:1) of the acetates of cholestan-2α-ol-3-one and cholestan-4α-ol-3-one* (1953d Fieser). — Crystals (glacial acetic acid or methanol) (1938 Ruzicka), m. 148.5° to 149.5° cor. (1944b Ruzicka); [α]$_D$ +27° (chloroform) (1944b Ruzicka; cf. 1953d Fieser). The complex is not resolved by fractional crystallization nor by chromatography (1953d Fieser). — **Fmn.** From 2α-bromocholestan-3-one (p. 2480 s) on refluxing with anhydrous sodium acetate in glacial acetic acid; yield, 81% (1938 Ruzicka; cf. 1953d Fieser). — **Rns.** Hydrogenation in glacial acetic acid at room temperature using platinum oxide catalyst, discontinued after absorption of 1 mol. of hydrogen, affords a complex mixture from which have been isolated: a compound thought to be the acetate of cholestan-1-ol (p. 1537 s) (1944b Ruzicka), probably a mixture of the acetates of cholestan-

184*

2α-ol (p. 1537 s) and cholestan-4α-ol (p. 1722 s) (1953 d Fieser); acetoxycholestanol-A (p. 2077 s) (1944 b Ruzicka), now considered to be a monoacetate of cholestane-3β.4α-diol (1954 Fieser); an oil, probably a monoacetate of cholestane-2β.3α-diol (p. 2049 s), together with a mixture of unidentified cholestanediol acetates (1944 b Ruzicka). Hydrogenation under otherwise similar conditions, but at 120° and 170 atm. pressure, yields small amounts of the acetoxycholestanol-A together with an isomer B (p. 2077 s) of unknown structure. Hydrogenation in alcohol - ether at room temperature using a platinum catalyst gives acetoxycholestanol-C (p. 2077 s), also of unknown structure (1944 b Ruzicka). Reduction of the hydrazone (not isolated) by heating with sodium amyl oxide in amyl alcohol at 180–200° yields cholestane (p. 1429 s), cholestan-4α-ol and "cholestan-1-ol" (1944 b Ruzicka), shown later to be a mixture of cholestan-2α- and -4α-ols (1953 d Fieser); from the less soluble fraction of the mixed reaction products two ketazines have been isolated: **azine A** $C_{54}H_{92}N_2$ (m. 235–242° cor. dec.) and **azine B** $C_{54}H_{92}N_2$ (m. 200–210° cor. dec.) (1944 b Ruzicka). By Clemmensen reduction only cholestane is obtained. Saponification with alcoholic KOH, shortly at 40° then at room temperature, affords an enolic cholestene-3.4-diol, probably Δ^3- (p. 2050 s); using K_2CO_3 in benzene - aqueous methanol at room temperature, a mixture of cholestan-3α(?)-ol-4-one (p. 2676 s) and cholestan-2α(?)-ol-3-one (above) is obtained (1944 b Ruzicka).

Cholestan - 2α(?) - ol - 3 - one benzoate $C_{34}H_{50}O_3$. Crystals (alcohol - chloroform or -acetone), m. 198–9° (1938 Inhoffen; 1937 b SCHERING A.-G.). — **Fmn.** and **Rns.** See those of the parent hydroxyketone (above).

Isomeric cholestan - 2 - ol - 3 - one(?)benzoate $C_{34}H_{50}O_3$. This compound is considered to be probably the 2-epimer of that described above, but the isomeric structure cholestan-3-ol-2-one benzoate is not excluded (1938 Inhoffen). In view of the similar mode of formation it might also be an addition compound of the benzoates of cholestan-2α-ol-3-one and cholestan-4α-ol-3-one, analogous to the "acetate" described above (Editor). — Crystals (chloroform - alcohol), m. 145–6°. — **Fmn.** From 2α-bromocholestan-3-one (p. 2480 s), together with the isomeric benzoate, as described under formation of the hydroxyketone (above); yield, 17%. — **Rns.** Treatment with excess KOH, with or without addition of H_2O_2, as for the isomer, affords the same reaction products (no information is given concerning yields). Mild saponification, as for the isomer, gives an oily product which crystallizes slowly, but no pure product has been isolated (1938 Inhoffen).

$\Delta^{4 \cdot 6}$**-Cholestadien-4-ol-3-one** $C_{27}H_{42}O_2$ (p. 157: *Diosterol*). *One enolic form (form B, 1940 Petrow) of Δ^5-cholestene-3.4-dione* (p. 2903 s). For revised structure, based on absorption spectra and chemical properties, see 1948, 1954 Fieser; cf. 1937 Krekeler. — Crystals (ether - methanol) (1937 Krekeler), needles (chloroform- or acetone - alcohol) (1937 b SCHERING-KAHLBAUM A.-G.), colourless needles (alcohol), m. 162–3⁰ cor. (1940 Petrow); $[\alpha]^{22}_{5461}$ +85⁰, $[\alpha]^{22}_D$ +57⁰ (both in benzene) (1940 Petrow), $[\alpha]^{24}_D$ +40⁰ (1950 Rivett), $[\alpha]_D$ +36⁰ (1954 Fieser) (both in chloroform); absorption max. in ether at 313 mμ (ε 19 000), in chloroform at 320 mμ (ε 21 000) (1939 Dannenberg; cf. 1936 Inhoffen; 1937 Krekeler; 1938 Wolff), in alcohol at 319 mμ (ε 23 640) (1954 Fieser; cf. 1937 Rosenheim). Readily soluble in ether, sparingly in alcohol (1937 Krekeler); is extracted from ether or benzene by strong alkali, yielding insoluble sodium or potassium salts (1937 Rosenheim). Gives an intense orange-red coloration with tetranitromethane in chloroform and a violet coloration with alcoholic FeCl₃ (1940 Petrow; cf. 1937 b SCHERING-KAHLBAUM A.-G.; 1938 Wolff). Reacts with Schiff's fuchsine reagent (1937 Rosenheim). Reaction with Zerewitinoff reagent shows the presence of one active hydrogen atom (1936 Inhoffen). Is very sensitive towards oxidants (1937 b SCHERING-KAHLBAUM A.-G.; 1938 Wolff).

Fmn. From 2α.4α-dibromocholestan-3-one (p. 2495 s) by treatment with 6% AgNO₃ in pyridine at room temperature, followed by heating of the reaction product (see p. 2495 s) with potassium acetate and glacial acetic acid at 180⁰ (sealed tube) (1938 Wolff). The yield from 2.2.5α.6β-tetrabromocholestan-3-one (p. 157: 4.4.5.6-tetrabromo-) is 29% (1937 b SCHERING-KAHLBAUM A.-G.). From $\Delta^{2.5}$-cholestadien-3-ol-4-one ("form A", p. 2674 s) by heating with glacial acetic acid containing small amounts of conc. HCl (1940 Petrow) or from its acetate by refluxing with alcoholic HCl containing some benzene (yield, 74%) (1936 Inhoffen) or with alcohol containing 10% of conc. H₂SO₄ (1937 Krekeler). From the acetate of Δ^5-cholesten-4β-ol-3-one 5.6-oxide (p. 2771 s) by refluxing with glacial acetic acid and sodium acetate (yield, 80%) or with conc. HCl in benzene - alcohol (yield, 58%) (1940 Petrow; cf. 1948 Fieser).

Rns. Hydrogenation in ether at room temperature, using a platinum catalyst, yields coprostane-3.4-diol (see p. 2668 s) (1937 Krekeler). Treatment with alcoholic KOH affords a sparingly soluble *potassium salt* (1937 Krekeler; cf. 1937 Rosenheim) from which it is regenerated by acetic acid (1940 Petrow). Treatment with acetic anhydride, with or without sodium acetate, does not yield a crystalline acetyl derivative (1940 Petrow), but reaction does occur (1936 Inhoffen; 1948 Fieser). Heating with o-phenylenediamine at 150⁰ gives the quinoxaline derivative of Δ^5-cholestene-3.4-dione (p. 2903 s) (1940 Petrow). Phenylhydrazine in ether - alcohol gives a *phenylhydrazone* [no analysis indicated; red leaflets (benzene - alcohol), m. 197⁰] (1937 Krekeler); refluxing with Brady's reagent in alcohol gives the mono-2.4-dinitrophenylhydrazone of Δ^5-cholestene-3.4-dione (p. 2903 s) (1940 Petrow). Reaction with semicarbazide hydrochloride, benzoyl chloride or its 3.5-dinitro-derivative, under the usual conditions for preparation of semicarbazones or esters, resp., gives the compounds described below.

Semicarbazone C₂₈H₄₅O₂N₃. Faintly yellow, crystalline powder, m. 252–3⁰ dec. (1948 Fieser).

Benzoate, *4-Benzoyloxy-Δ⁴·⁶-cholestadien-3-one* $C_{34}H_{46}O_3$. Colourless needles (alcohol), m. 160°; absorption max. in abs. alcohol at 232 and 287 mμ
(log ε 4.19 and 4.41, resp.) (1948 Fieser).

3.5-Dinitrobenzoate, *4-(3.5-Dinitrobenzoyl-oxy)-Δ⁴·⁶-cholestadien-3-one*
$C_{34}H_{44}O_7N_2$. Colourless needles (solvent not stated), m. 181°; regenerates the
parent hydroxyketone by saponification, followed by acidification (1937 Krekeler).

Coprostane-3.4-diol $C_{27}H_{48}O_2$. Probably a mixture of stereoisomers. —
Needles (alcohol or acetone), m. 170°
(after many recrystallizations); easily
soluble in the usual organic solvents.
— **Fmn.** From *Δ⁴·⁶*-cholestadien-4-ol-
3-one (p. 2667 s) by hydrogenation in
ether at room temperature in the
presence of a platinum catalyst. — **Rns.** Oxidation with CrO_3 in 90% acetic
acid at room temperature affords coprostane-3‖4-dioic acid (p. 1718 s). —
Monoacetate $C_{29}H_{50}O_3$, crystals (alcohol - benzene), m. 197°; obtained from
the diol by reaction with acetic anhydride in pyridine at room temperature. —
Diacetate $C_{31}H_{52}O_4$, needles (acetone), m. 135°; from the diol by refluxing
with acetic anhydride and sodium acetate. — *Monobenzoate* $C_{34}H_{52}O_3$, crystals
(benzene), m. 245° (1937 Krekeler).

Δ¹-Cholesten-4-ol-3-one $C_{27}H_{44}O_2$. Only esters have been prepared; the structure assigned to the benzoate is based
on its mode of formation (1937 Inhoffen), that of the isovalerate by analogy (Editor).

Isovalerate, *4-Isovaleryloxy-Δ¹-chole-
sten-3-one* $C_{32}H_{52}O_3$. In view of the
m. p., this product is probably impure; by its mode of formation it might
well contain the alternative 2-hydroxy-3-keto-Δ⁴-isomer and possibly also
the two corresponding benzoates (Editor). — Crystals (alcohol), m. 118–127°
with sintering at 102°, charring at 144°, colour play between 127° and 144°
(1936 Schering A.-G.; 1937b Schering-Kahlbaum A.-G.). — **Fmn.** From
2α.4α-dibromocholestan-3-one (p. 2495 s) by boiling with potassium benzoate
in isovaleric acid (1937b Schering-Kahlbaum A.-G.). — **Rns.** Heating at
220° and 2 mm. pressure, followed by distillation at 220° and 0.05 mm. pressure, affords Δ¹·⁴-cholestadien-3-one (1936 Schering A.-G.; cf. 1937b Schering-Kahlbaum A.-G.). Heating at 330–340° in a stream of carbon dioxide
causes elimination of isovaleric acid and methane with formation of a phenolic
product (1937a, b Schering A.-G.).

Benzoate, *4-Benzoyloxy-Δ¹-cholesten-3-one* $C_{34}H_{48}O_3$ [p. 130: *benzoate (a)*,
m. 177°]. Needles (chloroform - alcohol), m. 177° to a turbid melt which clears
at 216–7°; $[\alpha]_D^{20}$ +26° (chloroform); easily soluble in cold chloroform or benzene,

hot ether, acetone, or ethyl acetate, very sparingly in ethanol or methanol
(1937 Inhoffen; 1937b SCHERING-KAHLBAUM A.-G.). — **Fmn.** The yield from
2α.4α-dibromocholestan-3-one (p. 2495 s) (cf. p. 130) by reaction with potassium benzoate in toluene-butanol is 12%; separation from the benzoate of
Δ^4-cholesten-2-ol-3-one (p. 2664 s) is effected by fractional crystallization
from chloroform - alcohol (1937 Inhoffen; cf. 1936 SCHERING A.-G.; 1937b
SCHERING-KAHLBAUM A.-G.); can be obtained similarly from 2.2-dibromo-
cholestan-3-one (p. 2494 s) (yield, 16%) (1936 SCHERING A.-G.). — **Rns.** By
heating at 220° under 2 mm. pressure, benzoic acid is eliminated and after
subsequent distillation at 0.05 mm. $\Delta^{1.4}$-cholestadien-3-one (p. 2418 s) is obtained (1937b SCHERING-KAHLBAUM A.-G.). Heating at 310–320° in an atmosphere of carbon dioxide effects elimination of methane together with benzoic
acid, and after distillation at 170–200° in a high vacuum, a phenolic product
is isolated which shows a positive Allen-Doisy reaction (rats) (1937a, b
SCHERING A.-G.). Saponification with excess KOH in alcohol - benzene at
room temperature (1937 Inhoffen) or boiling methanol (1937b SCHERING-
KAHLBAUM A.-G.) gives Δ^4-cholesten-4-ol-3-one (below).

Δ^4-Cholesten-4-ol-3-one $C_{27}H_{44}O_2$ (p. 158: *Cholestane-3.4-dione*). Enolic form
of the 3.4-dione which has not yet
been obtained in the free state; for the
structure assigned, see 1940 Petrow;
1954 Fieser; cf. 1939 Dannenberg. —
Needles (alcohol-aqueous acetone)
(1937b SCHERING-KAHLBAUM A.-G.),
needles (alcohol), m. 149–150° cor. (1940 Petrow); sublimes in high vac. at
125–130° (1937b SCHERING-KAHLBAUM A.-G.; 1938 Wolff); $[\alpha]_D^{18} +80°$, $[\alpha]_{5461}^{18}$
$+94.5°$ (both in chloroform) (1940 Petrow; cf. 1938 Wolff; 1954 Fieser);
absorption max. at 280 mμ (ε 11 500) in chloroform (1939 Dannenberg; cf.
1936b Butenandt) or in dioxane (1944c Ruzicka), at 278 mμ (ε 13 000) in
alcohol (1954 Fieser). Gives a purple coloration with $FeCl_3$ (1940 Petrow).

Fmn. From 2α-bromo-Δ^4-cholesten-3-one (p. 2480 s) on refluxing with anhydrous potassium acetate in butanol (1937b SCHERING-KAHLBAUM A.-G.);
similarly from 2β.4β-dibromocoprostan-3-one (p. 2497 s) (1938 Wolff) and
from 2-bromo-5-chlorocholestan-3-one (originally thought to be 4-bromo-
5-chlorocholestanone, see p. 2497 s) by reaction in glacial acetic acid at 160°
(1937b SCHERING A.-G.; 1937b SCHERING-KAHLBAUM A.-G.). Together with
"cholestane-2.3-dione" (Δ^1-cholesten-2-ol-3-one, p. 2662 s) from the benzoate
of Δ^4-cholesten-2α-ol-3-one (p. 2663 s) by saponification with KOH in benzene -
alcohol at room temperature or in boiling methanol, followed by fractional
crystallization from petroleum ether (yields, ca. 12%); from the isomeric
benzoate of Δ^1-cholesten-4-ol-3-one (p. 2668 s) by refluxing with methanolic
KOH (1937b SCHERING A.-G.; 1937b SCHERING-KAHLBAUM A.-G.). From
the acetate of Δ^5-cholesten-3β-ol-4-one (p. 2675 s) by refluxing for a short
time with alcoholic HCl (yield, 88%) or by treatment with sodium ethoxide
in alcohol - benzene at room temperature, followed by dilution with water
(yield, 62%) (1940 Petrow).

References, pp. 2687–2690 s

Rns. Oxidation with H_2O_2 in alcoholic NaOH affords dihydro-Diels's acid (p. 1718 s) (1940 Petrow). Heating with o-phenylenediamine alone at 102–3⁰ (1940 Petrow) or in abs. alcohol - glacial acetic acid (1938 Wolff; cf. 1937b Schering-Kahlbaum A.-G.) gives the quinoxaline derivative of cholestane-3.4-dione (p. 2904 s); there is no reaction with this reagent in alcohol alone (1940 Petrow). Reaction with Brady's reagent in alcohol yields the mono-2.4-dinitrophenylhydrazone of the dione (p. 2904 s) (1940 Petrow).

Acetate, 4-Acetoxy- Δ^4-cholesten-3-one $C_{29}H_{46}O_3$ (p. 158: *enol-acetate, m. 100–1⁰*). For structure, see 1939 Dannenberg; 1954 Fieser). — Needles (85 % alcohol), m. 102–3⁰ cor.; $[\alpha]_D^{18}$ +93⁰, $[\alpha]_{5461}^{18}$ +101⁰ (both in chloroform) (1940 Petrow; cf. 1954 Fieser); absorption max. in chloroform at 248 mμ (ε 14500) (1939 Dannenberg; cf. 1936b Butenandt), in alcohol at 247 mμ (ε 15100) (1954 Fieser). — **Fmn.** and **Rns.** The acetate is prepared from the parent hydroxyketone by the usual methods (yield, 68 %) and regenerates the same compound on saponification (1940 Petrow; cf. 1938 Wolff; 1937b Schering-Kahlbaum A.-G.).

Cholestan-4α-ol-3-one $C_{27}H_{46}O_2$. — Is known with certainty only in the form of an addition compound of its *acetate* $C_{29}H_{48}O_3$ with that of cholestan-2α-ol-3-one, which see, p. 2665 s. It is probable that the *acetoxycholestanone-A*, m. 145–6⁰ cor. (p. 2077 s) is identical with the acetate of cholestan-4α-ol-3-one, as the alternative structure for this compound, 3β-acetoxycholestan-4-one, seems excluded by its m. p. (Editor; cf. 1950 Liebermann); for structure of the parent cholestanediol-A, see 1954 Fieser.

Coprostan-4-ol-3-one (?) $C_{27}H_{46}O_2$. In view of the probable structure of its reduction product "coprostane-3.4-diol" (which see p. 2056 s), the structure (I) assigned to this compound seems doubtful (1954 Georg); by analogy with

parallel reactions, isomerization is likely to occur in the course of the formation of the hydroxyketone (acetolysis of a 4-bromo-compound), so that a **2-acetoxy-cholestan-3-one** structure derived from (II) seems likely for the acetate next described (Editor).

Acetate $C_{29}H_{48}O_3$. Crystals (alcohol), m. 149⁰. — **Fmn.** From 4β-bromo-coprostan-3-one (p. 2483 s) by refluxing with fused potassium acetate in glacial acetic acid. — **Rns.** Hydrogenation in ether - alcohol at room temperature and 3 atm. pressure, using a platinum oxide catalyst, followed by saponification with boiling alcoholic KOH, affords "coprostane-3.4-diol" (p. 2056 s) (1939 b Marker).

Cholestan-5-ol-3-one $C_{27}H_{46}O_2$. For structure and configuration, see 1944 a, 1948 Plattner. — Dimorphic: crystals (acetone), m. 208–210⁰ and 229–231⁰ (1954 Fudge); crystals (aqueous acetone), m. 205–8⁰ (1944 a Plattner), crystals (ether), m. 224–6⁰ cor. (1948 Plattner; cf. 1944 a Plattner). — **Fmn.** From cholestane-3β.5-diol (p. 2057 s) by oxidation with CrO_3 in strong acetic acid at room temperature; yield, 84% (1944 a, 1948 Plattner; cf. 1943 Chuman). — **Rns.** Dehydration by boiling with acetic anhydride or with HCl gas in chloroform gives Δ⁴-cholesten-3-one (p. 2424 s) (1944 a Plattner; cf. 1943 Chuman).

Oxime $C_{27}H_{47}O_2N$. Crystals (methanol), m. 195–7⁰ cor. (1948 Plattner).

Δ⁴·⁶-Cholestadien-6-ol-3-one $C_{27}H_{42}O_2$. *Enolic form of Δ⁴-cholestene-3.6-dione* (p. 2905 s). For structure assigned see 1946 Ross (potassium salt and benzoate); 1953 b Fieser (ethyl ether).

Potassium salt $C_{27}H_{41}O_2K$. Yellow; sparingly soluble (1906 Windaus); the alcoholic solution is dark green by reflected, red by transmitted light (1946 Ross). — **Fmn.** From the parent dione by treatment with one equivalent of KOH in alcohol (1946 Ross).

Benzoate, 6-Benzoyloxy-Δ⁴·⁶-cholestadien-3-one $C_{34}H_{46}O_3$. Not obtained pure, not crystallized; $[\alpha]_D^{20}$ —16⁰ (chloroform - alcohol); absorption max. in chloroform - alcohol at 230 and 280–290 mμ (ε 34000 and 14500, resp.). — **Fmn.** From the dibenzoate of Δ²·⁴·⁶-cholestatriene-3.6-diol (p. 2058 s) by saponification with 0.1 N alcoholic KOH in chloroform (1946 Ross).

Ethyl ether, 6-Ethoxy-Δ⁴·⁶-cholestadien-3-one $C_{29}H_{46}O_2$ (p. 158: *Enol-ethylether, m. 166–7⁰*). Needles (90% alcohol) (1901 van Oordt), prisms (alcohol) (1907 Windaus; cf. 1937 Butenandt), crystals (benzene - acetone) (1937 Krekeler), prismatic needles (methanol) (1946 Ross), m. 169⁰ (1937 Krekeler; cf. 1938 Petrow), m. 163⁰ (1937 Butenandt; cf. 1946 Ross); $[\alpha]_D^{20}$ +2⁰ (chloroform) (1946 Ross). Absorption max. in ether at ca. 241 and 295 mμ (ε ca. 7800 and 16800, resp.) (1939 Dannenberg), in chloroform at 240–1 and 297–8 mμ (ε 7850 and 16600, resp.; from the curve) (1937 Butenandt). Easily soluble in most organic solvents on heating, also in cold chloroform or benzene (1901 van Oordt). Yellow coloration with $SbCl_3$ in chloroform (1946 Ross).

Fmn. From the 6-ethyl ether 3-benzoate of Δ²·⁴·⁶-cholestatriene-3.6-diol (p. 2058 s) by saponification with sodium methoxide in methanol at room temperature (1946 Ross). From the 2-epimeric 2.5.6β-tribromocholestan-3-ones (p. 2504 s) by refluxing with sodium iodide in abs. alcohol and benzene

(1937 Krekeler). From the potassium salt (p. 2671 s) by refluxing with ethyl iodide (1946 Ross). From 2-bromo-6β-ethoxy-Δ^4-cholesten-3-one (p. 2706 s) by refluxing with potassium acetate in butanol for 6 hrs. (1936a Butenandt) or with HBr - glacial acetic acid in benzene - alcohol for 2 hrs. (1937 Butenandt). From Δ^4-cholestene-3.6-dione ("oxycholestenone", p. 2905 s) by treatment with alcoholic HCl (1906 Windaus) or on refluxing with alcoholic H_2SO_4 (1938 Petrow). From cholestan-5-ol-3.6-dione ("oxycholestenediol", p. 2953 s) by treatment with HCl in abs. alcohol at room temperature or on the water-bath (1901 van Oordt), or by refluxing with alcoholic H_2SO_4 (1907 Windaus).

Rns. Conversion to the parent dione (p. 2905 s) can be effected by refluxing with 80% acetic acid (1907 Windaus) or with glacial acetic acid containing zinc acetate (1906 Windaus) or potassium acetate (1937 Krekeler). Refluxing with benzoyl chloride and pyridine affords the 6-ethyl ether 3-benzoate of $\Delta^{2.4.6}$-cholestatriene-3.6-diol (p. 2058 s) (1946 Ross). Does not give an insoluble hydrazone with phenylhydrazine (1901 van Oordt).

Page 158, line 2 from bottom: for "1936 Butenandt" *read* "1936a Butenandt".

Benzyl ether, 6-Benzyloxy-$\Delta^{4.6}$-cholestadien-3-one $C_{34}H_{48}O_2$. M. p. 132° (1906 Windaus).

Δ^4-Cholesten-6β-ol-3-one, 6β-Hydroxycoprostenone $C_{27}H_{44}O_2$ (p. 146). For

configuration at C-6, see 1939 Ellis. For identity with the "α-oxycholeste-nol" of Mauthner (1896), see 1953a Fieser; the structure previously postulated for this compound was Δ^4-cholesten-3-ol-6-one (p. 147) (1908 Windaus). Needles (95% alcohol), m. 185–6° (1901 van Oordt), (aqueous acetone), m. 192° (1939 Ellis); needles (ligroin), spars (methanol, slow crystallization) or plates (methanol, rapid crystallization), m. 194–5° (rapid), m. 176–7° (slow heating), or intermediate values (1953a Fieser); $[\alpha]_D^{23}$ + 27° (chloroform), + 31° (dioxane) (1953a Fieser). Absorption max. in alcohol at 236–7 mμ (ε 14000 average) (1953a Fieser). With the Liebermann-Burchard reagent only a yellow coloration with pale green fluorescence is observed (1896 Mauthner).

Fmn. From cholesterol (p. 1568 s) by oxidation with $KMnO_4$ in aqueous H_2SO_4 at 15° (1901 van Oordt; cf. 1908 Windaus) or by heating with HgO in glacial acetic acid at 100° (1928 Montignie). From the 6-acetate of cholestane-5.6β-diol-3-one (p. 2756 s) on treatment with thionyl chloride in pyridine, first at room temperature, then at 100°, or on refluxing with acetic anhydride, followed by hydrolysis of the resulting acetate (yield, 60%) with 1% methanolic KOH at room temperature (1939 Ellis).

Rns. Oxidation with $Na_2Cr_2O_7$ and aqueous H_2SO_4 in benzene - acetic acid at 15° yields Δ^4-cholestene-3.6-dione (p. 2905 s) (1939 Ellis). On hydrogenation of the acetate in the presence of palladium or platinum catalyst, hydrogen is rapidly absorbed, but no crystalline product could be isolated; reduction of the acetate with sodium in boiling amyl alcohol affords a yellow viscous resin from which cholestane-3β.6α-diol (p. 2062 s) can be obtained, via its dibenzoate (1943 Paige). Reaction with bromine occurs in carbon disulphide (but not in chloroform) involving substitution as well as addition (1896 Mauthner). On refluxing the hydroxyketone or its acetate with alcoholic HCl, cholestane-3.6-dione (p. 2907 s) is formed (1939 Ellis).

Page 146 line 5 from bottom: for "Δ^4-Cholesten-5-ol-3-one" read "Δ^4-Cholesten-6β-ol-3-one".

Semicarbazone $C_{28}H_{47}O_2N_3$ (p. 146). Prisms (alcohol) (1937 Dane), m. 221° (1939 Ellis).

Acetate, 6β-Acetoxy-Δ^4-cholesten-3-one $C_{29}H_{46}O_3$. Needles (dil. alcohol) (1896 Mauthner), needles (acetone - methanol), m. 101.5°; $[\alpha]_D^{19}$ +36° (chloroform) (1939 Ellis); absorption max. in alcohol at 237 mμ (ε 12600) (1953a Fieser).

Ethyl ether (?), 6-Ethoxy-Δ^4-cholesten-3-one $C_{29}H_{48}O_2$. Configuration at C-6 uncertain; in view of the isomerization of Δ^4-cholesten-6β-ol-3-one by acids to the 6α-epimer (1953 Sondheimer), a 6α-configuration seems more likely for this compound as well as for its bromo-precursor (Editor). — Needles (alcohol or dil. acetone), m. 109°. — **Fmn.** From 2-bromo-6-ethoxy-Δ^4-cholesten-3-one (p. 2706 s) on heating with zinc dust in methanol containing benzene (1937 Butenandt).

Cholestan-6α-ol-3-one $C_{27}H_{46}O_2$.

Benzoate $C_{34}H_{50}O_3$. Crystals (alcohol), m. 187° (1921 Windaus; 1941 PARKE, DAVIS & Co.). — **Fmn.** Together with the benzoate of 6α-hydroxycholestane-2‖3-dioic acid (p. 2063 s), from the 6-benzoate of cholestane-3β.6α-diol (p. 2062 s) by oxidation with CrO_3 in strong acetic acid at room temperature (1921 Windaus). From the same diol by first treating with triphenyl-methyl chloride and then with benzoyl chloride, both in dry pyridine at room temperature, followed by partial hydrolysis of the resulting crude 3β-triphenylmethyloxy-6α-benzoyloxy-cholestane with 5% H_2SO_4 in warm acetic acid and then by oxidation with CrO_3 in strong acetic acid at room temperature (1941 PARKE, DAVIS & Co.). — **Rns.** Slow distillation at 190° in high vacuo yields Δ^4-cholestenone (p. 2424 s) (1941 PARKE, DAVIS & Co.).

Cholestan-6β-ol-3-one $C_{27}H_{46}O_2$.

White leaflets (methanol), m. 190° (1940a Marker); prisms (acetone - methanol), m. 181-4° (Kofler block), $[\alpha]_D$ +9° (chloroform) (1952 Shoppee). — **Fmn.** From the diacetate of cholestane-3β.6β-diol (p. 2064 s) by partial saponification to the 6-monoacetate (not isolated), which is oxidized with CrO_3 in strong acetic acid at room temperature and then saponified by refluxing with methanolic KOH (1940a Marker; cf. 1941 PARKE, DAVIS & Co.). — **Rns.** Hydrogenation in 95% alcohol at room temperature and 3 atm. pressure, using Adams's PtO_2 catalyst, affords cholestane-3β.6β-diol (p. 2064 s) (1940b Marker; cf. 1941 PARKE, DAVIS & Co.). Heating with $KHSO_4$ at 125°, followed by distillation at 185° and 4 mm. pressure, gives Δ^4-cholesten-3-one (p. 2424 s) (1940a Marker).

Acetate, 6β-Acetoxycholestan-3-one $C_{29}H_{48}O_3$. Thick or fine needles (petroleum ether or methanol, resp.), m. 101° (1940a Marker; cf. 1941 PARKE, DAVIS & Co.). — **Fmn.** See that of the hydroxyketone (above). — *Oxime* $C_{29}H_{49}O_3N$, m. 170° (1941 PARKE, DAVIS & Co.).

$\Delta^{2.5}$**-Cholestadien-3-ol-4-one,** *Diosterol-II* $C_{27}H_{42}O_2$. One enolic form (form A, 1940 Petrow) of Δ^5-*cholestene-3.4-dione* (p. 2903 s). For the enolic structure ascribed, see 1948 Fieser; on the basis of absorption spectra the same authors suggest that in solution there is equilibrium with the triene-diol form $\Delta^{2.4.6}$**-cholestatriene-3.4-diol.** — Slightly yellow, rectangular plates (aqueous acetone) containing ½ H_2O, m. 135–6° cor.; $[\alpha]_D^{22}$ +30.5°, $[\alpha]_{5461}^{22}$ +41.3° (both in chloroform) (1940 Petrow); absorption max. in alcohol at 265 and 300 mμ (log ε 3.71 and 3.73, resp.) (1948 Fieser). Gives a yellow colour with tetranitromethane in chloroform and a purple coloration with $FeCl_3$ in alcohol (1940 Petrow).

Fmn. From 5.6-dibromocholestane-3β.4β-diol (p. 2101 s) by oxidation with CrO_3 in benzene-acetic acid at room temperature, followed by debromination with sodium iodide in alcohol at room temperature, and isolation via the potassium salt (see below), which is decomposed by acetic acid; yield, 14% (1940 Petrow). See also formation of the acetate and the benzoate (below).

Rns. Oxidation with H_2O_2 and KOH in aqueous alcohol, followed by acidification, gives Diels's acid (p. 1717 s). On warming with glacial acetic acid containing small amounts of conc. HCl the isomeric $\Delta^{4.6}$-cholestadien-4-ol-3-one (p. 2666 s) is obtained. Refluxing with o-phenylenediamine in alcohol affords the quinoxaline derivative of the parent diketone (p. 2903 s). Treatment with Brady's reagent in alcohol yields the mono-2.4-dinitrophenylhydrazone of Δ^5-cholestene-3.4-dione (1940 Petrow).

Potassium salt $C_{27}H_{41}O_2K$. Bright yellow crystals; insoluble in ether or in 20% KOH (1940 Petrow). — **Fmn.** From the hydroxyketone by shaking with 20% aqueous KOH in benzene - ether (1940 Petrow) or from the acetate (below) by reaction with a slight excess of 1% methanolic KOH (1937 Krekeler). — **Rns.** Shaking with 2 N acetic acid regenerates the hydroxyketone (1940 Petrow; cf. 1937 Krekeler).

Acetate, 3-Acetoxy-$\Delta^{2.5}$-cholestadien-4-one $C_{29}H_{44}O_3$ (p. 157: *enol-acetate, m. 158–9°*). For structure, based on absorption spectra, see 1948 Fieser; cf. 1939 Dannenberg. — Needles (benzine) (1936 Inhoffen), prisms (acetone)

(1937 Krekeler), rhombic plates (alcohol), prisms (aqueous acetone), m. 160⁰
to 161⁰ cor. to an anisotropic liquid showing a blue colour on cooling (1940
Petrow; cf. 1936 Inhoffen). Optical rotation in chloroform: $[\alpha]_D^{19}$ +14⁰,
$[\alpha]_{5461}^{19}$ +17.4⁰ (1940 Petrow), $[\alpha]_D^{24}$ +12⁰ (1950 Rivett). Absorption max. in
ether at 238 mμ (ε 9600) (1939 Dannenberg; cf. 1936 Inhoffen), in chloroform
at 248 and ca. 274 mμ (ε 7700 and ca. 5800, resp.) (1939 Dannenberg; cf.
1937 Krekeler). Readily soluble in chloroform, benzene, and hot benzine,
sparingly soluble in methanol and ethanol (1936 Inhoffen) and in ether
(1937 Krekeler). — **Fmn.** From the parent hydroxyketone by treatment
with acetic anhydride and pyridine at room temperature (1940 Petrow).
From the acetate of Δ^5-cholesten-4β-ol-3-one 5.6-oxide (or Δ^5-cholesten-3β-ol-
4-one 5.6-oxide, see p. 2771 s) by refluxing with acetic anhydride and sodium
acetate (1940 Petrow). The crude yields from 2α.6β-dibromo-Δ^4-cholesten-
3-one and from 2α.5α.6β-tribromocholestan-3-one (pp. 2499 s and 2504 s,
resp.; cf. p. 157*) are 81% and 85%, resp. (1937a, b SCHERING-KAHLBAUM
A.-G.). — **Rns.** Is converted into $\Delta^{4.6}$-cholestadien-4-ol-3-one (p. 2666 s) by
refluxing with HCl in alcohol, containing some benzene (1937b SCHERING-
KAHLBAUM A.-G.), or with alcoholic H_2SO_4 (1937 Krekeler). For alkaline
saponification see formation of the potassium salt (above). Boiling with acetic
anhydride does not bring about further acetylation (1937b SCHERING-KAHL-
BAUM A.-G.).

Benzoate, 3-Benzoyloxy-$\Delta^{2.5}$-cholestadien-4-one $C_{34}H_{46}O_3$. Colourless needles (alcohol), m. 176–7⁰; absorption max. in abs. alcohol at 234 mμ (log ε 4.30).
— **Fmn.** From 2α.6β-dibromo-Δ^4-cholesten-3-one (p. 2499 s) by refluxing with
sodium benzoate in alcohol (1948 Fieser).

Δ^5-Cholesten-3β-ol-4-one $C_{27}H_{44}O_2$. The free hydroxyketone, obtained by
acid or alkaline hydrolysis of its ace-
tate (below), is unstable and changes
smoothly into Δ^4-cholesten-4-ol-3-one
(p. 2669 s) (1940 Petrow).

*Acetate, 3β-Acetoxy-Δ^5-cholesten-
4-one* $C_{29}H_{46}O_3$. For structure, based
on absorption spectra, see 1940, 1943 Petrow; 1954 Fieser. — Needles (alcohol),
m. 123–4⁰ cor.; $[\alpha]_D^{20}$ —77⁰, $[\alpha]_{5461}^{20}$ —94⁰ (both in chloroform) (1940 Petrow);
absorption max. in alcohol at 240 mμ (ε 6350) (1954 Fieser; cf. 1943 Petrow). —
Fmn. From the 3-acetate of 5.6-dibromocholestane-3β.4β-diol (p. 2101 s) by
oxidation with CrO_3 in benzene-aqueous acetic acid at room temperature
and subsequent debromination by refluxing with alcoholic sodium iodide;
yield, 67% (1940 Petrow). — **Rns.** Refluxing for a short time with alcoholic
HCl, or treatment with sodium ethoxide in alcohol · benzene at room temperature, followed by addition of water, affords Δ^4-cholesten-4-ol-3-one (page
2669 s). Refluxing with acetic anhydride and sodium acetate gives the diacetate of $\Delta^{3.5}$-cholestadiene-3.4-diol (p. 2050 s) (1940 Petrow).

---

* Where the now discarded 4.6-dibromo- and 4.5.6-tribromo-structures for these compounds
are used.

Cholestan-3α-ol-4-one $C_{27}H_{46}O_2$ (I). The alternative 4-ol-3-one structure (II), though less likely, is not excluded; cf. the acetate (below). — Crystals (meth-

anol), m. 173–5⁰ cor. with previous sintering at 171⁰; [α]ᴅ + 14.5⁰ (chloroform); does not give any yellow coloration with tetranitromethane. — **Fmn.** From the addition compound of the acetates of cholestan-2α-ol-3-one and cholestan-4α-ol-3-one (p. 2665 s) by saponification with K_2CO_3 in benzene · aqueous methanol at room temperature; yield, 32 %. — **Rns.** Oxidation with alkaline H_2O_2 in aqueous alcohol at 100⁰ affords dihydro-Diels's acid (p. 1718 s). Treatment with HBr · glacial acetic acid (containing some chloroform) at 100⁰ (5 min.) and then at room temperature (15 hrs.) yields a brominated oily product, which after debromination with zinc dust in glacial acetic acid at 100⁰ and oxidation with CrO_3 in glacial acetic acid at 20⁰ gives cholestan-4-one (p. 2451 s) (1944b Ruzicka).

Acetate, 3α-Acetoxycholestan-4-one $C_{29}H_{48}O_3$. Assuming the 3-ol-4-one structure to be correct, the 3α-configuration may be assigned by exclusion, in view of the subsequent preparation (1950 Liebermann) of the *3β-epimer*. — Crystals (methanol), m. 143.5–144.5⁰ cor.; [α]ᴅ —7.5⁰ (chloroform). — **Fmn.** From $Δ^3$-cholestene-3.4-diol (p. 2050 s) with acetic anhydride and pyridine at 20⁰ (low yield), or similarly from the parent hydroxyketone (1944b Ruzicka).

$Δ^{2.4}$-**Cholestadien-3-ol-6-one** $C_{27}H_{42}O_2$. *Second enolic form of $Δ^4$-cholestene-3.6-dione* (p. 2905 s) *.

Acetate, 3-Acetoxy-$Δ^{2.4}$-cholestadien-6-one $C_{29}H_{44}O_3$. Pale yellow needles (acetic acid or methanol), m. 139–140⁰ (1938a Heilbron; 1938 Jackson); [α]ᴅ²⁰ + 27⁰ (chloroform); absorption max. in alcohol at 317 mμ (log ε 3.8); negative Tortelli-Jaffe reaction (1938a Heilbron). — **Fmn.** From the acetate of 4-bromo-$Δ^4$-cholesten-3β-ol-6-one (p. 2707 s) by refluxing with anhydrous pyridine (1938 Jackson); from the acetate of either of the two 7-epimeric 5.7-dibromocholestan-3β-ol-6-ones (p. 2714 s) similarly (yield, 51 %) (1938a Heilbron).—**Rns.** Saponification with sodium methoxide in methanol at 22⁰ affords the parent dione (1938a Heilbron).

Benzoate, 3-Benzoyloxy-$Δ^{2.4}$-cholestadien-6-one $C_{34}H_{46}O_3$. Considered by the authors to be the main constituent of a mixture, containing possibly also the C-8 epimer, from which no crystalline product has been isolated. — Optical

---

* The $Δ^{3.5.7}$-cholestatriene-3.6-diol structure suggested by Fieser (1949) has been withdrawn (1957 Fieser) in favour of that assigned, as originally postulated by Heilbron (1938a).

rotation in chloroform · alcohol, $[\alpha]_D^{20}$ $+97^0$. — **Fmn.** By saponification of the dibenzoate of $\Delta^{3.5.7}$-cholestatriene-3.6-diol (p. 2058 s) using 0.1 N alcoholic KOH in chloroform at room temperature (20 hrs.) (1946 Ross).

Δ^4-Cholesten-3β-ol-6-one, 6-Ketocoprosten-3β-ol $C_{27}H_{44}O_2$. This structure (cf. p. 147) was formerly erroneously assigned to Mauthner's "α-oxycholestenol", for which see p. 2672 s. — Needles (methanol), m. 150–1^0; $[\alpha]_D^2$ —13^0 (chloroform); absorption max. in alcohol at 239 and 319 mμ (log ε 3.8 and 1.93, resp.). Very soluble in methanol, insoluble in light petroleum (1937 Heilbron). — **Fmn.** From the acetate of 5α-bromocholestan-3β-ol-6-one (p. 2707 s) by refluxing in dry pyridine and saponification of the resulting acetate (yield, 64%) using 1% methanolic KOH at room temperature (1937 Heilbron). The acetate is obtained from that of 4.5-dibromo-cholestan-3β-ol-6-one (p. 2713 s) by refluxing with potassium iodide in acetone for 1 min. (quantitative yield) or with anhydrous pyridine for 2$^1/_2$ hrs. (1938 Jackson). The acetate is also formed by refluxing that of cholestane-3β.5-diol-6-one (p. 2757 s) with thionyl chloride in pyridine for 10 min. (1939 Ellis).

Rns. Hydrogenation of the acetate in ether at room temperature, using palladium black catalyst, gives cholestan-6-one (p. 2454 s) (1938 Jackson). Reduction of the acetate with aluminium isopropoxide in boiling dry isopropyl alcohol affords Δ^4-cholestene-3β.6α-diol (p. 2060 s) (1937 Heilborn). Treatment of the acetate with zinc dust in glacial acetic acid at 100^0 yields bi-(6-keto-Δ^4-cholesten-3-yl). Bromination (2 mols.) of the acetate in glacial acetic acid at 18^0 yields its 4.5-dibromide (p. 2713 s); using the acetate in ether - glacial acetic acid with dropwise addition of bromine (1 mol.), the acetate of 4-bromo-Δ^4-cholesten-3β-ol-6-one (p. 2707 s) results (1938 Jackson).

Acetate, 3β-Acetoxy-Δ^4-cholesten-6-one $C_{29}H_{46}O_3$. Six-sided plates (methanol), m. 110^0 (1937 Heilbron; 1939 Ellis); $[\alpha]_D^{22}$ —50.5^0 (chloroform) (1937 Heilbron), absorption max. in alcohol at 236 and 320 mμ (log ε 3.8 and 1.95, resp.) (1937 Heilbron; cf. 1945 Huber). Sparingly soluble in methanol, readily soluble in pyridine. Gives a yellow coloration with tetranitromethane in chloroform (1937 Heilbron). — **Fmn.** and **Rns.** See those of the parent hydroxyketone (above).

Cholestan-3β-ol-6-one, 6-Ketocholestanol $C_{27}H_{46}O_2$ (p. 147). For confirmation of configuration assigned at C-3, see 1948 Shoppee. — Crystals (methanol), m. 152.5–153^0 (1941 Chinaeva), m. 148–9^0 cor. (1944b Plattner); crystals (aqueous methanol), m. 140^0 (1948 Barton; cf. 1948 Dodson); $[\alpha]_D^{20}$ —3.1^0 (ether) (1906 Mauthner), —6^0 (chloroform) (1948 Barton; cf. 1948 Dodson). Gives a water-surface film with molecular area ca. 53 sq. Å at zero pressure, the properties of which are intermediate between those of cholestan-3β-ol (p. 1695 s) and cholestan-6-one (p. 2454 s) (1935 Adam). Shows good emulsifying properties (1940 Janistyn). Liebermann-Burchard reaction: violet (1938 Petrow).

Occ. Has been isolated from the unsaponifiable material of swine spleen; the hydroxyketone is separated in the crude state (95 mg.) from an acetone extract (15.7 kg.) of the fresh tissue (1500 kg.) and purified via the acetate (yield, 7 mg.) (1943 Prelog).

Fmn. From cholesterol (p. 1568 s) by biological oxidation using spleen tissue (1946 Churý). From 6-iodocholesterol (p. 1880 s) by heating with aqueous $NaHCO_3$ and $CuCl_2$ catalyst in a steel bomb at 225^0; the crude product is purified via the benzoate; yield, 89 % as benzoate (1940 Levin). For improved mode of preparation (cf. p. 147) from the nitrate of 6-nitro-cholesterol (p. 1905 s) (crude yield, 65 %), see 1938a Heilbron. From Δ^5-chole-stene-3β.6-diol dibenzoate (p. 2062 s) by hydrolysis with alcoholic KOH (1938 Petrow). The acetate (1944a Plattner) and the methyl succinate (1946 Reich) are respectively obtained from the 3-acetate and 3-(methyl succinate) of cholestane-3β.6β-diol (p. 2064 s) by oxidation with CrO_3 in strong acetic acid at room temperature. From cholesterol α-oxide (p. 2175 s) by treatment with a large excess of phenylmagnesium bromide in ether and then in boiling benzene; isolation from the yellow tarry reaction product is effected by con-version to the benzoate semicarbazone (yield, 21 %) which is hydrolysed with boiling alcoholic H_2SO_4 (1941 Chinaeva); the benzoate is obtained from that of cholesterol α-oxide on refluxing with P_2O_5 in xylene (yield, 8 %) or on heating with anhydrous alum at 180^0 under 1 mm. pressure (1939 Spring).

From i-cholesten-6-one (p. 2452 s) on refluxing with H_2SO_4 in acetic acid (yield, 91 %) (1939 Ladenburg). From 5α-bromocholestan-3β-ol-6-one (page 2707 s) or from 7α-bromocholestan-3β-ol-6-one (p. 2708 s) by reduction with zinc dust in boiling alcohol (1943 Sarett); the acetate is obtained from their respective acetates using aluminium amalgam in moist ether at room tempe-rature (1937 Heilbron); in the case of the 7-bromo-acetate reduction has also been carried out by refluxing with dimethylaniline (yield of acetate, 54 %) (1940 Jackson). From Δ^4-cholestene-3.6-dione (p. 2905 s) by hydro-genation using Raney nickel catalyst in alcohol at room temperature (ces-sation after absorption of 2 mols. of hydrogen); crude yield, 44 % (1941 Bretschneider).

Rns. Oxidation to cholestane-3.6-dione (cf. p. 147) can also be effected with CrO_3 in strong acetic acid at 0^0 (1941 Chinaeva). Reaction occurs with SeO_2 in glacial acetic acid at 100^0, but not in boiling alcohol (1938 Stiller)*. Hydrogenation at room temperature using a platinum oxide catalyst, in methanol at 3 atm. pressure (1940a Marker) or in alcohol at atmospheric pressure (1944b Plattner), affords cholestane-3β.6β-diol (p. 2064 s) (cf. 1941 PARKE, DAVIS & Co.; 1943 Chuman). Reduction with sodium in boiling alcohol yields cholestane-3β.6α-diol (p. 2062 s) (1917 Windaus; cf. 1944b Plattner).

---

* Analogous reaction carried out with the acetate or benzoate affords the corresponding ester of cholestane-3β.5-diol-6-one (1955 Hodinář).

Reaction with bromine (1 mol.) in ether · glacial acetic acid at room temperature, with immediate working up, gives 5α-bromocholestan-3β-ol-6-one (page
2707 s) (1943 Sarett); under similar conditions, or in ether · chloroform, but
at 0⁰ the acetate behaves similarly, but if the reaction mixture is heated
under reflux the isomeric 7α-bromocholestan-3β-ol-6-one (p. 2708 s) results
(1937 Heilbron); with bromine (2 mols.) under similar conditions at 0⁰ the
same 5-bromo-compound is isolated, but by reaction at 23⁰ in glacial acetic
acid (containing small amounts of HBr) the acetate of 5α.7α-dibromochole-
stan-3β-ol-6-one (p. 2714 s) is isolated after 1 hr., whereas the 7-epimeric
dibromo-compound (p. 2714 s) results after 18 hrs. standing, when the presence
of free bromine has been avoided (1938a Heilbron; cf. 1941 Woodward).
Treatment with PCl₅ in chloroform at room temperature affords 3α-chloro-
cholestan-6-one (see p. 2484 s) (1904 Windaus). Treatment with phenylmagnesium bromide in ether, followed by hydrolysis, affords 6-phenylchole-
stane-3β.6-diol (p. 2095 s) (1941 Chinaeva). Heating with benzoyl chloride
in pyridine yields the benzoate (see below), refluxing with benzoic anhydride
or benzoyl chloride gives Δ⁵-cholestene-3β.6-diol dibenzoate (p. 2062 s) (1938
Petrow).

p-Nitrophenylhydrazone C₃₃H₅₁O₃N₃ (p. 147). Needles (alcohol) containing
1 mol. alcohol, m. 194⁰ (1903 Windaus); yellow needles (methanol), m. 196–7⁰
(1938 Petrow).

Acetate, 3β-Acetoxycholestan-6-one C₂₉H₄₈O₃ (p. 147). Prisms (alcohol)
(1938a Heilbron), leaflets (aqueous acetone), m. 128–9⁰ cor. (1944a Plattner;
cf. 1938a Heilbron; 1943 Prelog; 1948 Barton); [α]ᴅ in chloroform —14.4⁰
(average) (1943 Prelog; 1944a Plattner), —16⁰ (1948 Barton; cf. 1948 Dodson); absorption max. (in alcohol?) at 280 mμ (log ε 1.6) (1938 Barr). The
surface film on water has a molecular area of ca. 65 sq. Å at zero pressure;
it is larger and more compressible than that of the parent hydroxyketone
(1935 Adam). — **Fmn.** and **Rns.** See those of the hydroxyketone above.

Acetate oxime C₂₉H₄₉O₃N. Colourless plates (aqueous alcohol), m. 201–2⁰
(1946b Barnett; cf. 1940 Mitui). — **Fmn.** From the acetate (above) by
refluxing with hydroxylamine hydrochloride and sodium acetate in alcohol
(1946b Barnett; cf. 1940 Mitui) or from the acetate of 6-nitrocholesterol
(p. 1905 s) by reduction with zinc dust in ether · glacial acetic acid (1:1)
(1940 Mitui). — **Rns.** With sodium in alcohol 6-aminocholestan-3β-ol (p. 1912 s)
is obtained (1946b Barnett).

Acetate p-nitrophenylhydrazone C₃₅H₅₃O₄N₃ (p. 147), yellow needles (methanol), m. 146–7⁰ (1938 Petrow).

Benzoate, 3β-Benzoyloxy-cholestan-6-one C₃₄H₅₀O₃ (p. 147). Prisms (ethyl
acetate) (1939 Spring; cf. 1941 Chinaeva), (acetone) (1938 Petrow; cf. 1903
Windaus), (methanol · acetone) (1940 Jackson); crystals (methanol · chloroform, or ethanol · ether), m. 173⁰ (1903 Windaus; 1948 Barton), m. 170–1⁰
clearing at 179⁰ (1939 Spring; cf. 1940 Jackson; 1940 Levin); the melt shows
colour play (1939 Spring); [α]²⁰ᴅ +4⁰ (1940 Levin; 1948 Barton), +6.9⁰
(average) (1939 Spring), both in chloroform. — **Fmn.** From the parent hydr-

E.O.C. XIV s. 185

oxyketone by reaction with benzoyl chloride in presence of alkali (1903 Windaus) or in pyridine at room temperature (1938 Petrow; 1940 Levin); see also under formation of the hydroxyketone (p. 2678 s). — *Benzoate semicarbazone* $C_{35}H_{53}O_3N_3$, crystals (alcohol), m. 226–7° (1941 Chinaeva).

Hydrogen succinate, 3β-(β-Carboxypropionyloxy)-cholestan-6-one $C_{31}H_{50}O_5$. — *Methyl ester* $C_{32}H_{52}O_5$. Crystals (methanol), m. 96–97° (1946 Reich); obtained as described under formation of the hydroxyketone (p. 2678 s).

$Δ^{2.4.7}$-Cholestatrien-7-ol-6-one $C_{27}H_{40}O_2$. *Enolic form of $Δ^{2.4}$-cholestadiene-6.7-dione* (p. 2911 s).

Methyl ether, 7-Methoxy-$Δ^{2.4.7}$-cholestatrien-6-one $C_{28}H_{42}O_2$. Needles (methanol), m. 119–121°; absorption max. in alcohol at 315 mμ (log ε 4.2); the pure product, in contrast to the crude, gives no colour with alcoholic $FeCl_3$. — **Fmn.** From the acetate of 5α.7β-dibromocholestan-3β-ol-6-one (p. 2714 s) by refluxing for 7 hrs. with anhydrous potassium acetate in dry butanol, followed by several recrystallizations from methanol (1938 a Heilbron).

$Δ^{3.5}$-Cholestadien-3-ol-7-one $C_{27}H_{42}O_2$. *One enolic form of $Δ^5$-cholestene-3.7-dione* (p. 2910 s).

For structure assigned, see 1952 Greenhalgh; cf. 1946a Barnett. — Rectangular leaflets (ether · light petroleum), m. 185° to 186°; unstable if exposed to light (1946a Barnett); [α]$_D^{16.5}$ —53° (1946a Barnett), —49° (1952 Greenhalgh), both in chloroform. Absorption max. in alcohol at 322 mμ, in ca. 0.005 N alcoholic NaOH at 390 mμ (log ε 4.13 and 4.53, resp., from graph) (1946a Barnett); in alcohol at 320 mμ, in alcohol containing small amounts of KOH at 392.5 mμ (ε 24300 and 62200, resp.) (1952 Greenhalgh). Soluble in methanol, ethanol, acetone, and benzene, sparingly soluble in ether, insoluble in light petroleum. Readily soluble in aqueous NaOH from which it is regenerated by acids (1946a Barnett). — **Fmn.** From 5.6-dibromocholestan-3β-ol-7-one (p.2715 s) by oxidation with CrO_3 in benzene-aqueous acetic acid at room temperature, followed by debromination of the (not isolated) diketo-dibromide with zinc in boiling glacial acetic acid (1946a Barnett). — **Rns.** Refluxing with hydroxylamine acetate in aqueous alcohol gives the dioxime of the diketo-form (p. 2910 s) (1946a Barnett).

Acetate, 3-Acetoxy-$Δ^{3.5}$-cholestadien-7-one $C_{29}H_{44}O_3$. Crystals (methanol), m. 106–8° (1946a Barnett); [α]$_D$ —185° (chloroform); absorption max. in alcohol at 282.5 mμ (ε 22500) (1952 Greenhalgh), 284 mμ (log ε 4.15 from graph) (1946a Barnett). — **Fmn.** By acetylation of the parent hydroxyketone with acetic anhydride and pyridine at 15° (1946a Barnett). — **Rns.**

Regenerates the hydroxyketone very readily on treatment with alcoholic NaOH in the cold (1946a Barnett).

Methyl ether, 3-Methoxy-$\Delta^{3.5}$-cholestadien-7-one $C_{28}H_{44}O_2$. Crystals (methanol), m. 136–7° (1946a Barnett); $[\alpha]_D$ —323° (chloroform) (1952 Greenhalgh); absorption max. in alcohol at 310 mμ (log ε 4.14) from graph (1946a Barnett), 308 mμ (log ε 4.44) (1952 Greenhalgh); obtained from the parent hydroxyketone by refluxing with conc. H_2SO_4 in methanol (1946a Barnett).

Ethyl ether, 3-Ethoxy-$\Delta^{3.5}$-cholestadien-7-one $C_{29}H_{46}O_2$, not described, except for the absorption curve which closely resembles that of the methyl ether (1946a Barnett).

$\Delta^{8.14}$-Cholestadien-3β-ol-7-one $C_{27}H_{42}O_2$. The structure ascribed to the acetate (below) in preference to the alternative $\Delta^{8(14).9(11)}$-structure also suggested (1943 b Wintersteiner) is favoured by Fieser (1953e).

Acetate, 3β-Acetoxy-$\Delta^{8.14}$-cholestadien-7-one $C_{29}H_{44}O_3$. Slightly yellowish needles (methanol), m. 172—174.5° (1953e Fieser; cf. 1943b, c Wintersteiner); $[\alpha]_D$ —14.5° (chloroform) (1953e Fieser). Absorption max. in alcohol at 224 and 298 mμ (ε 15 600 and 5060, resp.) (1953e Fieser; cf. 1943b, c Wintersteiner). — **Fmn.** From the acetate of 8α.14α-epoxy-8-iso-cholestan-3β-ol-7-one (p. 2773 s) by refluxing with alcohol containing conc. HCl, followed by reacetylation with acetic anhydride in pyridine at room temperature; yield, 33% (1943b, c Wintersteiner; cf. 1953e Fieser). — **Rns.** It readily undergoes autoxidation (1943b, c Wintersteiner). Hydrogenation in abs. alcohol using a platinum catalyst, discontinued after absorption of 1 mol. of hydrogen, affords the acetate of $\Delta^{8(14)}$-cholesten-3β-ol-7-one (p. 2685 s) (1943b Wintersteiner). Wolff-Kishner reduction gives the acetate of $\Delta^{8.14}$-cholestadien-3β-ol (p. 1557 s) (1953e Fieser).

Acetate 2.4-dinitrophenylhydrazone $C_{35}H_{48}O_6N_4$. Orange coloured needles (benzene - alcohol), m. 235–6°. — **Fmn.** Is obtained as for the acetate but starting from the 2.4-dinitrophenylhydrazone of the epoxy-compound, or from the acetate by the usual method (1943b Wintersteiner).

Δ^5-Cholesten-3α-ol-7-one, 7-Keto-epi-cholesterol $C_{27}H_{44}O_2$. **Fmn.** The acetate is obtained, together with the acetate of cholestan-5α-ol-3.6-dione (see p. 2953 s), from that of epi-cholesterol (p. 1670 s) by oxidation with CrO_3 in ca. 95% acetic acid at 50°; separation by fractional crystallization from methanol (1939 Windaus). A crude oily product has been obtained from 7-ketocholesteryl chloride (p. 2788 s) by refluxing with fused potassium acetate in valeric acid, followed by saponification with alcoholic

185*

KOH (1937 Marker); for mechanism of the Walden inversion involved, see
1946 Shoppee. — **Rns.** The semicarbazone (not described) affords epi-cholesterol by heating with sodium ethoxide in alcohol at 170–180° (1937 Marker).
By refluxing the acetate with aluminium isopropoxide in isopropyl alcohol,
followed directly by treatment with methanolic KOH, a mixture of the
7-epimeric Δ^5-cholestene-3α.7-diols (pp. 2071 s, 2072 s) is obtained. The acetate
undergoes elimination of acetic acid very readily to give $\Delta^{3.5}$-cholestadien-
7-one (p. 2454 s) (1939 Windaus).

Acetate, 3α-Acetoxy-Δ^5-cholesten-7-one $C_{29}H_{46}O_3$. Leaflets (methanol),
m. 119°, not quite pure; absorption max. in ether at 234 mμ (ε 12410) (1939
Windaus). — **Fmn.** and **Rns.** See those of the parent hydroxyketone (above).

Δ^5-**Cholesten-3β-ol-7-one, 7-Ketocholesterol** $C_{27}H_{44}O_2$ (p. 147). For identity
and structure of this compound,
originally described as *"β-oxychole-
stenol"* by Mauthner (1896), see 1920
Windaus. — Needles (methanol) (1896
Mauthner), crystals (ether - pentane),
m. 170–2° cor. (1941 Bergström; cf.
1943 DU PONT DE NEMOURS; 1952 Greenhalgh); $[\alpha]_D^{23}$ —104°(chloroform) (1941
Bergström; cf. 1948 Barton). Absorption max. in alcohol at 237 and 334.5 mμ
(ε 13400 and 50, resp.) (1952 Greenhalgh; cf. 1941 Bergström). Dipole moment in dioxane at 25°: μ = 3.79 D (1945 Kumler). Forms a sparingly soluble
digitonide in 90% alcohol. The acetate exerts only a very slight reducing
action towards phosphomolybdic acid in glacial acetic acid on the water-bath
(1946 Heard).

Occ. Has been isolated from the testicles of bull (1943 Prelog, quoting
F. Steinmann) and of swine (1947 Prelog; cf. 1943 Ruzicka), and from wool
fat (1945 Daniel; 1956a, b Milburn).

Fmn. From cholesterol (p. 1568 s) by oxygenation, during several hours,
of its colloidal aqueous solution, stabilized by addition of sodium stearate,
at 85° (yield, 40–45% by spectroscopic assay), followed by separation from
other products by means of Girard's reagent T (yield, 20%) (1941 Bergström;
1941a, b Wintersteiner); more detailed investigation of this reaction shows
that reaction also occurs at 37° but much more slowly (12 days) (yield, 65%
by spectroscopy), still more slowly at 25° and not at all at 4° (18 days);
the reaction is activated by Cu^{++}ions, inhibited by cyanide ions; variation
in the concentration of the solution (0.05–0.5%), in oxygen pressure, pH (6.5
to 9.5), or the nature of the emulsifier are without appreciable effect (1942b
Bergström); reaction proceeds also in the dark or in ultra-violet light; in
absence of all trace of Cu^{++}ions reaction does not occur (1942a Bergström);
the use of cholesteryl esters at 85° retards the oxidation very considerably
(e.g., acetate 9% after 6 hrs., palmitate 17% after 8 hrs.) (1942c Bergström);
for further details, see formation of "oxycholesterol" (p. 1594 s); for possible
mechanism, see 1942b Bergström. The acetate is obtained from that of
cholesterol by direct oxygenation of a solution in xylene at 100° in presence

of iron phthalocyanine (yield, 8.7%); neither cholesterol nor its dihydrogen phosphate are oxidized under these conditions (1938 A. Cook). The acetate can be obtained directly from cholesterol by oxidation with CrO_3 in excess acetic anhydride (1944 Boots), or from the methyl ether of i-cholesterol (p. 1723 s) by CrO_3 oxidation in 80% acetic acid at 97° (1938b Heilbron); similarly the benzoate is obtained from that of cholesterol by oxidation with CrO_3 in strong acetic acid and carbon tetrachloride at 55° (minimum yield, 24%) (1938 Du Pont de Nemours), and the methyl ether analogously, without addition of carbon tetrachloride (1943 Du Pont de Nemours).

The acetate is obtained from those of 6α- or 6β-bromocholestan-3β-ol-7-ones (p. 2709 s) by boiling with anhydrous pyridine; this elimination of HBr occurs rather more readily in the case of the 6β-epimer (originally assumed to be the 6α-epimer), from which it can also be effected, rather more quickly, in the presence of $AgNO_3$ (1938 Barr). The acetate is regenerated from its 5.6-dibromide (p. 2715 s) by boiling with potassium iodide in acetone for 1 min. (1940 Jackson).

Rns. The acetate gives that of cholestan-3β-ol-7-one (p. 2686 s) on catalytic hydrogenation, using palladium-norit in glacial acetic acid (1938 Barr), using Adams's platinum catalyst in ether at 25° and 3 atm. pressure (1939a Marker), or using platinum oxide in ethyl acetate at room temperature (1943a Wintersteiner); using platinum oxide in glacial acetic acid at room temperature, at 3 atm. pressure (1939a Marker) or at ordinary pressure (1943a Wintersteiner), the 3-acetate of cholestane-3β.7β-diol is obtained; in the latter case the 7α-epimer and the acetate of cholestan-3β-ol-7-one can also be isolated from the reaction products (1943a Wintersteiner). Boiling with sodium in dry isopropyl alcohol affords pseudocholesterol (p. 1726 s) (1938 Ogata). Reduction of the acetate with aluminium isopropoxide in boiling isopropyl alcohol, followed by saponification, gives the two 7-epimeric Δ^5-cholestene-3β.7-diols (pp. 2066 s, 2068 s) together with small amounts of a dicholestadienyl ether (p. 1545 s) and of $\Delta^{4.6}$-cholestadien-3-one (p. 2420 s) (1942 Wintersteiner); for improved procedure for this reduction, in the presence of organic nitrogenous bases, see 1942 Du Pont de Nemours; see also 1935 I.G. Farbenind.; 1936 Winthrop Chemical Co.; for the analogous reduction of the benzoate, see 1938 Du Pont de Nemours, of the methyl ether and the triphenylmethyl ether, see 1943 Du Pont de Nemours. The acetate remains unchanged on boiling with zinc dust in glacial acetic acid (1906 Windaus). Reaction with bromine in CS_2 at 20° affords the 5.6-dibromide (p. 2715 s) which by CrO_3 oxidation in benzene - aqueous acetic acid at room temperature, followed by refluxing with zinc in glacial acetic acid, yields $\Delta^{3.5}$-cholestadien-3-ol-7-one (p. 2680 s) (1946a Barnett); the acetate reacts similarly with excess bromine in glacial acetic acid at room temperature (1 hr.) to give its 5.6-dibromide (p. 2715 s), whereas bromination in glacial acetic acid in the presence of HBr at 30° for 60 hrs. yields 3.4.6-tribromo-Δ^5-cholesten-7-one (p. 2504 s) (1940 Jackson). Treatment with dry HCl in chloroform gives $\Delta^{3.5}$-cholestadien-7-one (p. 2454 s) (1938 Ogata), also formed by reaction of the acetate with HBr in glacial acetic acid at 37° (15 hrs.) or on refluxing (10 min.)

(1940 Jackson); see also under reactions of the acetate (p. 2684 s). Reaction with potassium in toluene at 100° and then with potassium iodide affords di-(7-ketocholesteryl) ether (p. 2685 s) (1943 Du Pont de Nemours).

On heating of the acetate with ethyl bromoacetate and zinc in dry benzene on the steam-bath, followed by hydrolysis with methanolic KOH, it gives Δ^5-cholestene-3β.7-diol-7-acetic acid (1937 Jones).

Physiological reactions. 7-Ketocholesterol injected subcutaneously (orally less effective) into rats subsequently exposed to ultra-violet light, produces an antirachitic effect slightly greater than that of 7-dehydrocholesterol (page 1547 s) (1943 Lassen). Treatment of the acetate with conc. H_2SO_4 and acetic anhydride in glacial acetic acid at 85–90° does not yield an antirachitically active product (1939 Eck). Exerts neither anæsthetic nor folliculoid effect (1942 Selye).

Oxime $C_{27}H_{45}O_2N$. Needles (solvent not stated), m. 235° dec. (1938 Eckhardt; cf. 1938 Ogata; 1938 Hattori); obtained from the acetate oxime (below) by saponification with methanolic KOH (1938 Eckhardt) or directly from the hydroxyketone (1938 Hattori).

Acetate, 3β-Acetoxy-Δ^5-cholesten-7-one $C_{29}H_{46}O_3$ (p. 148). Square platelets (80% alcohol) (1896 Mauthner), needles (methanol) (1938b Heilbron), plates (aqueous acetone) (1940 Jackson), m. 157–9° cor. (1941 Bergström; cf. 1938 Hattori; 1948 Barton); $[\alpha]_D^{25}$ —97° (1941 Bergström), $[\alpha]_D^{ca.\,20}$ —103° (1948 Barton), both in chloroform. Absorption max. in ether at 234 mμ (log ε 4.15) (1943 Evans; cf. 1938 Eckhardt), in alcohol at 237 mμ (log ε 4.09) (1956b Milburn). Sparingly soluble in 75% methanol; gives a positive cholestol reaction (cf. p. 1805 s) (1896 Mauthner). — **Fmn.** From the free hydroxyketone by acetylation (1938 Hattori; 1941 Bergström); see also formation of the parent hydroxyketone (above). — **Rns.** Saponification without simultaneous dehydration can be effected with sodium methoxide in boiling methanol (1896 Mauthner), with dil. methanolic KOH (1938 Hattori), or with K_2CO_3 in 80% methanol (1941 Bergström; cf. 1946a Barnett); saponification also occurs with H_2O_2 in 4 N aqueous NaOH in acetone - alcohol at room temperature (1938 Ogata). Boiling with KOH in ethanol (1896 Mauthner) or in methanol (1946a Barnett; cf. 1938 Hattori) yields $\Delta^{3.5}$-cholestadien-7-one (p. 2454 s). For other reactions, see those of the free hydroxyketone (above).

Acetate oxime $C_{29}H_{47}O_3N$ (p. 148). Needles (benzene - petroleum ether) (1906 Windaus), prisms (ethyl acetate - methanol), leaflets (alcohol) (1938 Eckhardt), m. 186° (1938 Hattori); $[\alpha]_D^{21}$ —196° (chloroform) (1938 Eckhardt); absorption max. in ether at 238 mμ (log ε 4.15) (1943 Evans; cf. 1938 Eckhardt). Readily soluble in ether, chloroform, and hot ethyl acetate, sparingly in ethanol and methanol (1938 Eckhardt). — **Rns.** Reduction by means of sodium in hot alcohol gives 7-aminocholesterol (7-epimeric mixture) (p. 1913 s) (1938 Eckhardt).

Acetate semicarbazone $C_{30}H_{49}O_3N_3$, dec. 220° (1938 Hattori).

Benzoate, *3β-Benzoyloxy-Δ⁵-cholesten-7-one* $C_{34}H_{48}O_3$. Colourless needles (alcohol), m. 159.5–161° to an anisotropic liquid, pale green at 182.5°, clearing at 183.5° (1944 J. Cook; cf. 1938 Du Pont de Nemours); crystals (acetone-methanol), m. 159°; $[\alpha]_D^{ca.\ 20}$ —54.5° (average, in chloroform) (1948 Barton). — **Fmn.** From the above hydroxyketone by treatment with benzoyl chloride in dry pyridine at room temperature (1944 J. Cook) or as described under formation of the hydroxyketone.

Methyl ether, *3β-Methoxy-Δ⁵-cholesten-7-one* $C_{28}H_{46}O_2$. Crystals (acetone), m. 121–3° (1943 Du Pont de Nemours). — **Fmn.** and **Rns.** See those of the hydroxyketone (p. 2682 s).

Triphenylmethyl ether, *3β-Triphenylmethoxy-Δ⁵-cholesten-7-one* $C_{46}H_{58}O_2$. Crystals (ethyl acetate), m. 186–188.5°. — **Fmn.** From the hydroxyketone by boiling with triphenylmethyl chloride in dry pyridine (yield, 43%) (1943 Du Pont de Nemours). — **Rns.** See those of the hydroxyketone (p. 2683 s).

Di-(7-ketocholesteryl) ether $C_{54}H_{86}O_3$. Crystals (benzene-alcohol), m. 245° to 249°. — **Fmn.** From 7-ketocholesterol by reaction with potassium in toluene at 100° and then with potassium iodide (1943 Du Pont de Nemours).

Δ⁸-Cholesten-3β-ol-7-one $C_{27}H_{44}O_2$.

Acetate, *3β-Acetoxy-Δ⁸-cholesten-7-one* $C_{29}H_{46}O_3$. Blades (methanol), m. 155–6°; $[\alpha]_D^{25}$ —32° (chloroform); absorption max. in alcohol at 253 mμ (ε 15 500); infra-red absorption max. in chloroform at 5.84, 6.04, 6.3, and 8.0 μ. — **Fmn.** From the acetate of 8α.9α-epoxy-8-iso-cholestan-3β-ol-7-one (p. 2772 s) on refluxing with zinc dust in glacial acetic acid for 3 hrs.; yield, 72% (1953c Fieser).

Δ⁸⁽¹⁴⁾-Cholesten-3β-ol-7-one $C_{27}H_{44}O_2$.

Δ⁸⁽¹⁴⁾-Structure inferred from the observed absorption (1943b Wintersteiner; cf. 1953c Fieser).

Acetate, *3β-Acetoxy-Δ⁸⁽¹⁴⁾-cholesten-7-one* $C_{29}H_{46}O_3$. Square platelets (methanol), m. 141.5–142.5°; $[\alpha]_D^{21}$ —62° (chloroform); absorption max. in alcohol at 262 mμ (ε 9500) (1943b Wintersteiner; cf. also 1953c Fieser); infra-red absorption max. in chloroform at 5.80, 5.98, 6.29, and 7.95 μ (1953c Fieser). — **Fmn.** From the acetate of Δ⁸·¹⁴-cholestadien-3β-ol-7-one (p. 2681 s) by hydrogenation in abs. alcohol using platinum catalyst, discontinued after absorption of one mol. of hydrogen; almost quantitative yield (1943b Wintersteiner). From the acetate of 8α.14α-epoxy-8-iso-cholestan-3β-ol-7-one (page 2773 s; erroneously referred to in the original paper as the 8α.9α-epoxy-isomer) on refluxing with zinc dust in glacial acetic acid; yield, 69% (1953c Fieser). — **Rns.** Hydrogenation in glacial acetic acid using palladium catalyst affords a mixture of the acetates of (chiefly) α-cholestenol (p. 1692 s) and cholestan-3β-ol-7-one (below) (1943 b Wintersteiner).

Acetate 2.4-dinitrophenylhydrazone $C_{35}H_{50}O_6N_4$. Orange coloured needles (solvent not indicated), m. 210–1° (1943 b Wintersteiner).

Cholestan-3β-ol-7-one, 7-Ketocholestanol $C_{27}H_{46}O_2$ (p. 148). For confirmation of the configuration assigned at C-3, see 1948 Shoppee. — White plates (acetone-methanol), m. 128–130°; re-solidifies and melts finally at 157° to 159° (1939 a Marker); crystals (alcohol), m. 164–5°, $[\alpha]_D^{25}$ —33° (chloroform) (1943 a Wintersteiner); crystals (acetone-methanol), m. 168.5°, $[\alpha]_D^{ca.\,20}$ —36° (chloroform) (1948 Barton). Dipole moment in dioxane at 25°: $\mu =$ 2.98 D (1945 Kumler).

Fmn. The acetate is obtained from the 3-acetate of either of the 7-epimeric cholestane-3β.7-diols (pp. 2072 s, 2073 s) by oxidation with CrO_3 in glacial acetic acid at room temperature; crude yield from the 7α-epimer 93%, from the 7β-epimer 66% (1943 a Wintersteiner). The acetate is also obtained from that of 7-ketocholesterol (p. 2682 s) by catalytic hydrogenation, which can be effected at room temperature with palladium black in glacial acetic acid - ether (1920 Windaus) or with palladium-norit in glacial acetic acid (yield, 98%) (1938 Barr), or in ether at 25° and 3 atm. pressure using Adams's platinum oxide catalyst (1939 a Marker); if the hydrogenation is carried out at room temperature with platinum oxide catalyst in ethyl acetate, 7-keto-cholestanyl acetate is the main product (yield, 66%), but using glacial acetic acid as solvent it is obtained (yield, 13%) together with the 3-monoacetates of the 7-epimeric cholestane-3β.7-diols (1943 a Wintersteiner); saponification of the acetate can be effected with boiling alcoholic KOH (1920 Windaus; 1939 a Marker) or 5% methanolic KOH (1943 b Wintersteiner). The acetate is obtained from that of $\Delta^{8(14)}$-cholesten-3β-ol-7-one (p. 2685 s) by hydrogenation in glacial acetic acid at room temperature using a palladium catalyst (yield, 11%) (1943 b Wintersteiner), or from that of 6α.8β-dibromocholestan-3β-ol-7-one (p. 2715 s) by refluxing with dimethylaniline (1940 Jackson).

Rns. Hydrogenation of the acetate in glacial acetic acid at room temperature using a platinum oxide catalyst yields a mixture of the 3-monoacetates of cholestane-3β.7α- and 3β.7β-diols (pp. 2072 s, 2073 s), the former predominating (1943 a Wintersteiner). Refluxing of the acetate with aluminium isopropoxide in isopropyl alcohol affords cholestane-3β.7β-diol (1939 a Marker). Reduction of the crude *semicarbazone* [crystals (alcohol); no constants given] by heating with sodium ethoxide in ethanol at 190° gives cholestan-3β-ol (p. 1695 s) (1938 Barr). The acetate does not react with selenium dioxide in boiling alcohol or in glacial acetic acid at 100° (1938 Stiller). In glacial acetic acid at room temperature the acetate is not attacked by bromine, but in chloroform reaction with one mol. gives the acetate of 6α-bromocholestan-3β-ol-7-one (p. 2709 s) and a smaller amount of its 6β-epimer (p. 2709 s), and with 2 mols. of bromine the acetate of 6α.8β-dibromocholestan-3β-ol-7-one (p. 2715 s) results (1938 Barr).

Oxime $C_{27}H_{47}O_2N$. White needles (aqueous alcohol), m. 232–3° (1939a Marker).

Acetate, 3β-Acetoxycholestan-7-one $C_{29}H_{48}O_3$ (p. 148). Needles (alcohol) (1920 Windaus), plates (methanol, acetone- or ether - methanol) (1940 Jackson; 1939a Marker; 1938 Barr), m. 149–149.5°, [α]ᴅ —36° (chloroform) (1943a Wintersteiner; cf. 1948 Barton). — **Fmn.** and **Rns.** See those of the above hydroxyketone. — *Semicarbazone* $C_{30}H_{51}O_3N_3$, m. 232° dec. (1948 Barton).

Benzoate, 3β-Benzoyloxy-cholestan-7-one $C_{34}H_{50}O_3$. Crystals (acetone), m. 169.5°, [α]ᴅ^{ca. 20} —17.5° (average value in chloroform) (1948 Barton).

Cholestadien-3β-ol-15-one $C_{270}H_{42}O_2$. For its acetate, see pp. 2773 s, 2774 s.

Δ^{8(14)}-Cholesten-3β-ol-15-one $C_{27}H_{44}O_2$.

For confirmation of structure, see 1954 Barton. — Needles (75 % alcohol), m. 145–6°; gives a sparingly soluble crystalline *digitonide* in 90 % alcohol (1943c Wintersteiner). — **Fmn.** The acetate is formed, together with those of Δ^{8(14)}-cholesten-3β-ol-7-one and 15-one 8.14-oxides (p. 2773 s), by oxidizing the acetate of Δ^{8(14)}-cholesten-3β-ol (α-cholestenol, p. 1692 s) with CrO_3 in strong acetic acid - benzene at room temperature; after chromatographic separation, the acetate (yield, 6 %) is saponified with boiling 5 % methanolic KOH (1943c Wintersteiner). — **Rns.** The acetate undergoes hydrogenation very slowly at room temperature in glacial acetic acid, using a palladium catalyst, to give α-cholestenyl acetate together with unchanged material (1943c Wintersteiner).

Acetate, 3β-Acetoxy-Δ^{8(14)}-cholesten-15-one $C_{29}H_{46}O_3$. Six-sided plates (methanol), m. 134–5°, [α]ᴅ +118° (chloroform), absorption max. in alcohol at 259 mμ (ε 12750) (1943c Wintersteiner; cf. 1954 Barton). —**Fmn.** and **Rns.** See those of the hydroxyketone above. — *2.4-Dinitrophenylhydrazone* $C_{35}H_{50}O_6N_4$, yellow needles (solvent not indicated), m. 208–9° (1943c Wintersteiner).

1896 Mauthner, Suida, *Monatsh.* **17** 579, 582 et seq., 594 et seq.
1901 van Oordt, *Über Cholesterin*, Thesis Univ. Freiburg i. Br., pp. 21, 29 et seq.
1903 Windaus, *Ber.* **36** 3752, 3755.
1904 Windaus, Stein, *Ber.* **37** 3699, 3702.
1906 Mauthner, *Sitzber. Akad. Wiss. Wien* IIb **115** 251, 253; *Monatsh.* **27** 421, 423.
    Windaus, *Ber.* **39** 2249, 2251 et seq., 2259, 2260.
1907 Windaus, *Ber.* **40** 257, 261.
1908 Windaus, *Arch. Pharm.* **246** 117, 140.
1917 Windaus, *Ber.* **50** 133, 136.
1920 Windaus, Kirchner, *Ber.* **53** 614, 615, 620.
1921 Windaus, Lüders, *Z. physiol. Ch.* **115** 257, 268.
1928 Montignie, *Bull. soc. chim.* [4] **43** 1403.
1932 Tschesche, *Ann.* **498** 185, 191.

1935 Adam, Askew, Danielli, *Biochem. J.* **29** 1786, 1798, 1799.

I. G. FARBENIND., *Brit. Pat.* 454260 (issued 1936); *C. A.* **1937** 1041; *Ch. Ztbl.* **1937** I 4129.

1936 a Butenandt, Schramm, *Ber.* **69** 2289, 2298.

b Butenandt, Schramm, Wolff, Kudszus, *Ber.* **69** 2779.

Inhoffen, *Ber.* **69** 1702, 1705, 1706, 1708, 1709.

SCHERING A.-G. (Inhoffen), *Ger. Pat.* 722943 (issued 1942); *C. A.* **1943** 5201; SCHERING-KAHL-
BAUM A.-G., *Brit. Pat.* 500353 (1937, issued 1939); *Fr. Pat.* 835524 (1937, issued 1938);
C. A. **1939** 5872; *Ch. Ztbl.* **1939** I 5010, 5011.

WINTHROP CHEMICAL Co. (Windaus, Schenck), *U.S. Pat.* 2098984 (issued 1937); *C. A.* **1938**
196; *Ch. Ztbl.* **1938** I 3659.

1937 Butenandt, Schramm, Kudszus, *Ann.* **531** 176, 180, 200, 203 et seq.

Dane, Wang, Schulte, *Z. physiol. Ch.* **245** 80, 88.

Heilbron, Jones, Spring, *J. Ch. Soc.* **1937** 801.

Inhoffen, *Ber.* **70** 1695, 1696, 1697 et seq.

Jones, Spring, *J. Ch. Soc.* **1937** 302.

Krekeler, *Einige Umsetzungen am Cholestenon*, Thesis Univ. Göttingen, pp. 12 et seq., 21 et seq.

Marker, Kamm, Fleming, Popkin, Wittle, *J. Am. Ch. Soc.* **59** 619.

Rosenheim, King, *Nature* **139** 1015.

a SCHERING A.-G., *Brit. Pat.* 508576 (issued 1939); *Fr. Pat.* 838704 (issued 1939); *C. A.* **1940**
777; *Ch. Ztbl.* **1939** II 1722; SCHERING CORP. (Inhoffen), *U.S. Pat.* 2280828 (issued 1942);
C. A. **1942** 5618.

b SCHERING A.-G., *Austrian Pat. Application*; *Fiat Final Report* No. 996, pp. 130, 134, 136,
147 et seq., 153, 154, 157, 160 (issued 1947).

a SCHERING-KAHLBAUM A.-G., *Ital. Pat.* 351844 (issued 1938); *Ch. Ztbl.* **1938** I 4356.

b SCHERING-KAHLBAUM A.-G., *Fr. Pat.* 835524 (issued 1938), *Brit. Pat.* 500353 (issued 1939);
C. A. **1939** 5872; *Ch. Ztbl.* **1939** I 5010; SCHERING CORP. (Inhoffen, Butenandt, Schwenk),
U.S. Pat. 2340388 (issued 1944); *C. A.* **1944** 4386.

Windaus, Kuhr, *Ann.* **532** 52, 62, 63.

1938 Barr, Heilbron, Jones, Spring, *J. Ch. Soc.* **1938** 334.

A. Cook, *J. Ch. Soc.* **1938** 1774, 1780.

DU PONT DE NEMOURS (Rosenberg, Tinker), *U.S. Pat.* 2215727 (issued 1940); *C. A.* **1941** 758;
Ch. Ztbl. **1941** I 1443; *Brit. Pat.* 537030 (issued 1940); *C. A.* **1942** 1738.

Eckhardt, *Ber.* **71** 461, 464.

Hattori, *C. A.* **1938** 7473; *Ch. Ztbl.* **1939** I 1182.

a Heilbron, Jackson, Jones, Spring, *J. Ch. Soc.* **1938** 102.

b Heilbron, Hodges, Spring, *J. Ch. Soc.* **1938** 759.

Inhoffen, Huang-Minlon, *Ber.* **71** 1720, 1726, 1728 et seq.

Jackson, Jones, *J. Ch. Soc.* **1938** 1406.

Ogata, Kawakami, *C. A.* **1939** 640.

Petrow, Rosenheim, Starling, *J. Ch. Soc.* **1938** 677.

Ruzicka, Plattner, Aeschbacher, *Helv.* **21** 866, 871.

Stiller, Rosenheim, *J. Ch. Soc.* **1938** 353.

Wolff, *Über die Bromierung von Cholestanon und Koprostanon*, Thesis Techn. Hochschule
Danzig, pp. 11, 15, 16, 18.

1939 Bergmann, Hirschmann, Skau, *J. Org. Ch.* **4** 29.

Dannenberg, *Abhandl. Preuss. Akad. Wiss., Math.-naturwiss. Kl.* **1939** No. 21, pp. 1, 14, 17,
27 et seq., 32, 36, 45, 54 et seq., 59 et seq., 65.

Eck, Thomas, *J. Biol. Ch.* **128** 257, 260.

Ellis, Petrow, *J. Ch. Soc.* **1939** 1078, 1081 et seq.

Ladenburg, Chakravorty, Wallis, *J. Am. Ch. Soc.* **61** 3483, 3486.

a Marker, Rohrmann, *J. Am. Ch. Soc.* **61** 3022.

b Marker, Wittle, Plambeck Jr., Rohrmann, Krueger, Ulshafer, *J. Am. Ch. Soc.* **61** 3317, 3319.

Spring, Swain, *J. Ch. Soc.* **1939** 1356, 1359.

Windaus, Naggatz, *Ann.* **542** 204, 206, 209, 211.

1940 Jackson, Jones, *J. Ch. Soc.* **1940** 659, 662, 663.

Janistyn, *Fette u. Seifen* **47** 351, 355.

Levin, Spielman, *J. Am. Ch. Soc.* **62** 920.

a Marker, Krueger, *J. Am. Ch. Soc.* **62** 79; PARKE, DAVIS & Co. (Marker), *U.S. Pats.* 2337563, 2337564 (issued 1943); *C. A.* **1944** 3423.
b Marker, Krueger, Adams, Jones, *J. Am. Ch. Soc.* **62** 645.
Mitui, *C. A.* **1941** 4390; *Ch. Ztbl.* **1942** I 878.
Petrow, Starling, *J. Ch. Soc.* **1940** 60.
1941 Bergström, Wintersteiner, *J. Biol. Ch.* **141** 597, 601 et seq.
Bretschneider, *Ber.* **74** 1361.
Chinaeva (Tschinajewa), Ushakov, *J. Gen. Ch. U.S.S.R.* **11** 335; *C. A.* **1941** 5903; *Ch. Ztbl.* **1942** I 757.
PARKE, DAVIS & Co. (Marker, Lawson), *U.S. Pat.* 2366204 (issued 1945); *C. A.* **1945** 1649; *Ch. Ztbl.* **1945** II 1385.
a Wintersteiner, Bergström, *J. Biol. Ch.* **137** 785.
b Wintersteiner, Bergström, *J. Biol. Ch.* **140** CXLI.
Woodward, Clifford, *J. Am. Ch. Soc.* **63** 2727, 2728.
1942 a Bergström, *Naturwiss.* **30** 684.
b Bergström, Wintersteiner, *J. Biol. Ch.* **145** 309.
c Bergström, Wintersteiner, *J. Biol. Ch.* **145** 327, 329, 330.
DU PONT DE NEMOURS (Rosenberg), *U.S. Pat.* 2376817 (issued 1945); *C. A.* **1946** 97.
Selye, *Endocrinology* **30** 437, 444.
Wintersteiner, Ruigh, *J. Am. Ch. Soc.* **64** 2453.
1943 Chuman, *C. A.* **1947** 3806.
DU PONT DE NEMOURS (Rosenberg, Turnbull), *U.S. Pat.* 2386636 (issued 1945); *C A.* **1946** 1974; *Ch. Ztbl.* **1946** I 1432.
Evans, Gillam, *J. Ch. Soc.* **1943** 565, 570.
Lassen, Geiger, *C. A.* **1943** 5114.
Paige, *J. Ch. Soc.* **1943** 437, 439, 440.
Petrow, Rosenheim, Starling, *J. Ch. Soc.* **1943** 135, 136 footnote.
Prelog, Ruzicka, Stein, *Helv.* **26** 2222, 2225, 2236.
Ruzicka, Prelog, *Helv.* **26** 975, 981.
Sarett, Chakravorty, Wallis, *J. Org. Ch.* **8** 405, 415, 416.
a Wintersteiner, Moore, *J. Am. Ch. Soc.* **65** 1503, 1505, 1506.
b Wintersteiner, Moore, *J. Am. Ch. Soc.* **65** 1507, 1509, 1512.
c Wintersteiner, Moore, *J. Am. Ch. Soc.* **65** 1513.
1944 BOOTS PURE DRUG Co. LTD. (Peak, Short), *Brit. Pat.* 558361 (issued 1944); *C. A.* **1945** 4100.
J. Cook, Paige, *J. Ch. Soc.* **1944** 336, 337.
a Plattner, Petrzilka, Lang, *Helv.* **27** 513, 519, 522.
b Plattner, Lang, *Helv.* **27** 1872, 1876, 1877.
a Ruzicka, Plattner, Furrer, *Helv.* **27** 524, 527 et seq.
b Ruzicka, Plattner, Furrer, *Helv.* **27** 727, 730, 733, 734, 736.
c Ruzicka, Jeger, Norymberski, *Helv.* **27** 1185, 1187.
Seebeck, Reichstein, *Helv.* **27** 948.
1945 Daniel, Lederer, Velluz, *Bull. soc. chim. biol.* **27** 218, 221.
Huber, Ewing, Kriger, *J. Am. Ch. Soc.* **67** 609, 612.
Kumler, Fohlen, *J. Am. Ch. Soc.* **67** 437.
1946 a Barnett, Ryman, Smith, *J. Ch. Soc.* **1946** 526.
b Barnett, Ryman, Smith, *J. Ch. Soc.* **1946** 528.
Churý, *C. A.* **1948** 7405.
Heard, Sobel, *J. Biol. Ch.* **165** 687, 689 et seq., 697.
Reich, Lardon, *Helv.* **29** 671, 678.
Ross, *J. Ch. Soc.* **1946** 737.
Shoppee, *J. Ch. Soc.* **1946** 1147, 1149.
1947 Prelog, Tagmann, Lieberman, Ruzicka, *Helv.* **30** 1080, 1086.
1948 Barton, Cox, *J. Ch. Soc.* **1948** 783, 792.
Dodson, Riegel, *J. Org. Ch.* **13** 424, 432 et seq.
Fieser, Fieser, Rajagopalan, *J. Org. Ch.* **13** 800, 802.
Plattner, Fürst, Koller, Lang, *Helv.* **31** 1455, 1460.
Shoppee, *J. Ch. Soc.* **1948** 1032.

1949 Fieser, Fieser, *Natural Products related to Phenanthrene*, 3rd. Ed., New York, pp. 269, 270.
1950 Lieberman, Fukushima, *J. Am. Ch. Soc.* **72** 5211, 5218.
     Rivett, Wallis, *J. Org. Ch.* **15** 35, 40.
1952 Greenhalgh, Henbest, Jones, *J. Ch. Soc.* **1952** 2375, 2377, 2379.
     Shoppee, Summers, *J. Ch. Soc.* **1952** 3361, 3372.
1953 Conca, Bergmann, *J. Org. Ch.* **18** 1104, 1109.
     a Fieser, *J. Am. Ch. Soc.* **75** 4377, 4381, 4383, 4384.
     b Fieser, *J. Am. Ch. Soc.* **75** 4386, 4388, 4390, 4392.
     c Fieser, *J. Am. Ch. Soc.* **75** 4395, 4402.
     d Fieser, Romero, *J. Am. Ch. Soc.* **75** 4716, 4718.
     e Fieser, Nakanishi, Huang, *J. Am. Ch. Soc.* **75** 4719.
     Sondheimer, Kaufmann, Romo, Martinez, Rosenkranz, *J. Am. Ch. Soc.* **75** 4712, 4713, 4715.
1954 Barton, Laws (with Barnes), *J. Ch. Soc.* **1954** 52, 62.
     Fieser, Stevenson, *J. Am. Ch. Soc.* **76** 1728, 1731 et seq.
     Fudge, Shoppee, Summers, *J. Ch. Soc.* **1954** 958, 963.
     Georg, *Arch. sci.* **7** 114.
1955 Hodinář, Pelc, *Chem. Listy* **49** 1733; *Collection Czechoslov.* **21** 264 (1956).
1956 a Milburn, Truter, *J. Ch. Soc.* **1956** 1736, 1739.
     b Milburn, Truter, Woodford, *J. Ch. Soc.* **1956** 1740.
1957 Fieser, Fieser, *private communication.*

---

8. Monohydroxy-monoketosteroids (CO and OH in the Ring System) C_{28} with Three Aliphatic Side Chains

$\Delta^{4.6}$-**Ergostadien-6-ol-3-one** $C_{28}H_{44}O_2$. *Enolic form of* Δ^4-*ergostene-3.6-dione* (p. 2912 s).

Ethyl ether $C_{30}H_{48}O_2$ (p. 160). For structure, based on absorption spectrum, see 1939 Dannenberg. — Lancets (alcohol), m. 161° (1934 Windaus). Absorption max. in ether (from the graph) at 247 and 295 mμ (ε 7650 and 15 950, resp.) (1934 Windaus; cf. 1939 Dannenberg). — **Fmn.** From Δ^4-ergostene-3.6-dione (p. 2912 s) on refluxing with abs. alcohol containing some conc. H_2SO_4 (1934 Windaus).

$\Delta^{8.14.22}$-**Ergostatrien-3β-ol-7-one** $C_{28}H_{42}O_2$. For revision of the $\Delta^{8(14).9(11).22}$-structure originally assigned by Stavely (1943b), see 1953 Fieser.

Acetate $C_{30}H_{44}O_3$. Crystals (methanol), m. 187–9°; $[\alpha]_D^{24}$ —47° (chloroform). Absorption max. in abs. alcohol at 300 mμ (ε 5000) and (from graph) <235 mμ (ε_{235} ca. 12000). — **Fmn.** From the acetate of either 8α.9α- or 8α.14α-epoxy-8-iso-Δ^{22}-ergosten-3β-ol-7-one (both on p. 2774 s) by refluxing with alcoholic HCl, followed by reacetylation. — **Rns.** Catalytic hydrogenation using palladium catalyst at room temperature affords the acetate of $\Delta^{8(14)}$-ergosten-3β-ol-7-one (p. 2692 s) if carried out in alcohol, or the acetates of α-ergostenol (p. 1765 s) and ergostan-3β-ol-7-one (p. 2692 s) when effected in glacial acetic acid (1943b Stavely).

Acetate 2.4-dinitrophenylhydrazone $C_{36}H_{48}O_6N_4$, orange solid, m. 222° dec. (1943b Stavely).

Δ⁸·¹⁴-Ergostadien-3β-ol-7-one $C_{28}H_{44}O_2$. For revision of the Δ⁸⁽¹⁴⁾·⁹⁽¹¹⁾-structure originally assigned by Stavely (1943a), see 1953 Fieser.

Acetate $C_{30}H_{46}O_3$. Crystals (80% alcohol), m. 176–8°; $[\alpha]_D^{24}$ —22° (chloroform). Absorption max. in alcohol at 298 mμ (ε 5100) and (from graph) <245 mμ (ε_{245} ca. 8000). — **Fmn.** From the acetate of 8α.14α-epoxy-8-iso-ergostan-3β-ol-7-one (p. 2774 s) on refluxing with alcoholic HCl, followed by reacetylation. — **Rns.** Hydrogenation in alcohol at room temperature, using palladium catalyst, affords the acetate of Δ⁸⁽¹⁴⁾-ergosten-3β-ol-7-one (p. 2692 s) (1943a Stavely).

Acetate 2.4-dinitrophenylhydrazone $C_{36}H_{50}O_6N_4$, orange solid, m. 223° dec. (1943a Stavely).

Δ⁸·²²-Ergostadien-3β-ol-7-one $C_{28}H_{44}O_2$.

Acetate $C_{30}H_{46}O_3$. Crystals (methanol), m. 206–8°; $[\alpha]_D^{24}$ —53° (chloroform). Absorption max. in abs. alcohol at 252 mμ (ε 10000). — **Fmn.** Together with the acetates of 8α.9α- and 8α.14α-epoxy-8-iso-Δ²²-ergosten-3β-ol-7-ones, from the acetate of Δ⁷·²²-ergostadien-3β-ol (p. 1758 s) by oxidation with CrO_3 in strong acetic acid · benzene at room temperature, followed by chromatographic separation; yield, 4%. — **Rns.** Hydrogenation in glacial acetic acid at room temperature, using palladium catalyst, yields the acetates of α-ergostenol (p. 1765 s) and ergostan-3β-ol-7-one (p. 2692 s). Treatment with 2.4-dinitrophenylhydrazine in acidic solution at room temperature affords an orange *2.4-dinitrophenylhydrazone* (neither constants nor analysis given); by treatment with semicarbazide in pyridine · ethyl alcohol at room temperature (3 days) no derivative has been isolated (1943b Stavely).

Δ¹⁴·²²-Ergostadien-3β-ol-7-one $C_{28}H_{44}O_2$. The alternative structure of Δ⁷·²²-ergostadien-3β-ol-15-one acetate, also envisaged by the authors for the acetate described below, is now excluded (Editor), as a compound with that structure has since been described by Barton (1954).

Acetate $C_{30}H_{46}O_3$ (p. 79: *3-acetoxyergostadienone*). Rhomb-like plates (ethyl acetate), m. 180–1°, $[\alpha]_D^{19}$ +36.5° (chloroform). Gives an intense yellow colour with sodium nitroprusside; weak red colour with fuchsine-sulphurous acid. —

Fmn. From $\Delta^{14\cdot22}$-ergostadiene-3β.7.8-triol (p. 79: ergostadienetriol, m. 227⁰) by boiling with acetic anhydride. — **Rns.** Hydrogenation in glacial acetic acid at room temperature, using platinum black catalyst, affords the acetate of ergostanol (p. 1767 s). On boiling with acetic anhydride, sodium acetate and zinc dust for a short time it yields a **compound** $C_{32}H_{50}O_4$ [leaflets (ethyl acetate - alcohol), m. 168⁰ (probably $\Delta^{14\cdot22}$-**ergostadiene-3β.7-diol diacetate**, Editor)]. By treatment with semicarbazide acetate in alcohol no derivative has been obtained (1937 Chen).

24-iso-Δ^5-Ergosten-3β-ol-7-one, 7-Ketocampesterol $C_{28}H_{46}O_2$.

Has been iso-lated only in the crude state (1942 Bergström). — **Fmn.** From cam-pesterol (p. 1772 s) by autoxida-tion in colloidal aqueous solution, stabilized by sodium stearate, at 85⁰; yield, after 3 hrs., ca. 19% (1942 Bergström). The acetate is obtained from that of campesterol by oxidation with CrO_3 in strong acetic acid at ca. 55⁰; yield, 18% (1942 Ruigh).— **Rns.** Reduction of the acetate with aluminium isopropoxide in boiling iso-propyl alcohol affords Δ^5-campestene-3β.7β-diol (p. 2080 s) (1942 Ruigh).

Acetate $C_{30}H_{48}O_3$. Needles (alcohol), m. 177–8⁰; $[\alpha]_D^{24}$ —89⁰ (chloroform) (1942 Ruigh; cf. 1942 Bergström); obtained by acetylation of the parent hydroxy-ketone (1942 Bergström) or as described under formation of the same (above).

$\Delta^{8(14)}$-Ergosten-3β-ol-7-one $C_{28}H_{46}O_2$.

Acetate $C_{30}H_{48}O_3$. Crystals (80% alcohol), m. 155⁰; $[\alpha]_D^{23}$ —64⁰ (aver-age, in chloroform). Absorption max. in abs. alcohol at 262 mμ (ε 9800). — **Fmn.** From the ace-tate of either $\Delta^{8\cdot14\cdot22}$-ergostatrien-3β-ol-7-one (p. 2690 s) or $\Delta^{8\cdot14}$-ergostadien-3β-ol-7-one (p. 2691 s) by hydro-genation in alcohol at room temperature using palladium catalyst. — **Rns.** The compound is recovered unchanged after refluxing with HCl, followed by reacetylation (1943 a, b Stavely).

Ergostan-3β-ol-7-one $C_{28}H_{48}O_2$.

Crystals (80% alcohol) containing 1 mol. H_2O, m. 154⁰. — **Fmn.** The acetate is obtained, together with that of α-ergostenol (p. 1765 s), from the acetate of $\Delta^{8\cdot14\cdot22}$-ergostatrien-3β-ol-7-one (p. 2690) or $\Delta^{8\cdot22}$-ergo-stadien-3β-ol-7-one (p. 2691 s) by hydrogenation in glacial acetic acid at room temperature, using palladium catalyst (yields, 41% and 45%, resp.); after separation the acetate is saponi-fied with boiling methanolic KOH (1943 b Stavely).

References, p. 2693 s

Acetate $C_{30}H_{50}O_3$. Crystals (80% alcohol), m. 183–4°; $[\alpha]_D^{23}$ —36° (chloroform); obtained as described under the parent hydroxyketone (above). — *Semicarbazone* $C_{31}H_{53}O_3N_3$, crystals (alcohol), m. 225–8°. — *2.4-Dinitrophenylhydrazone* $C_{36}H_{54}O_6N_4$, yellow solid, m. 216° dec. (1943b Stavely).

$\Delta^{8(14)\cdot9(11)}$-Ergostadien-3β-ol-15-one $C_{28}H_{44}O_2$.

Acetate $C_{30}H_{46}O_3$. Crystals (80% alcohol), m. 155°. Absorption max. in abs. alcohol at 307 mμ (ε 10700). — **Fmn.** From the acetate of 8α.14α-epoxy-8-iso-ergostan-3β-ol-15-one (p. 2774s) by refluxing with alcoholic HCl, followed by reacetylation (1943a Stavely).

$\Delta^{8(14)}$-Ergosten-3β-ol-15-one $C_{28}H_{46}O_2$ (p. 83: α-*ergostenolone*).

For structure see 1943a Stavely. — Colourless needles (alcohol), m. 155–6° (1932 Heilbron). — **Fmn.** The acetate is obtained from that of α-ergostenol (p. 1765 s) by oxidation with CrO_3 in strong acetic acid at 60° (1932 Heilbron) or with added benzene at room temperature, and is separated (yield, 7%) by fractional crystallization and chromatography from the acetates of 8α.14α-epoxy-8-iso-ergostan-3β-ol-7-one (p. 2774 s) and -15-one (p. 2774 s) also formed (1943a Stavely); the acetate is saponified with methanolic KOH (1932 Heilbron). — **Rns.** Hydrogenation of the acetate in glacial acetic acid with palladium catalyst at room temperature, affords the acetate of α-ergostenol; under similar conditions but in alcohol, there is no reaction (1943a Stavely).

Acetate $C_{30}H_{48}O_3$ (p. 83: *acetate, m. 170–1°*). Colourless plates (alcohol) (1932 Heilbron), crystals (80% alcohol), m. 170°, $[\alpha]_D^{23}$ +110° (chloroform). Absorption max. in abs. alcohol at 259 mμ (ε 13300) (1943a Stavely). Sparingly soluble in alcohol (1932 Heilbron). — **Fmn.** By acetylation of the parent hydroxyketone (1932 Heilbron) or as described under formation of the same (above). — **Rns.** See those of the hydroxyketone (above).

Acetate 2.4-dinitrophenylhydrazone $C_{36}H_{52}O_6N_4$, orange coloured solid, m. 220° dec. (1943a Stavely).

1932 Heilbron, Simpson, Wilkinson, *J. Ch. Soc.* **1932** 1699, 1703, 1704.
1934 Windaus, Inhoffen, v. Reichel, *Ann.* **510** 248, 257, 258.
1937 Chen, *Ber.* **70** 1432.
1939 Dannenberg, *Abhandl. Preuss. Akad. Wiss., Math.-naturwiss. Kl.* **1939** No. 21, pp. 29, 59.
1942 Bergström, Wintersteiner, *J. Biol. Ch.* **145** 327, 330 et seq.
    Ruigh, *J. Am. Ch. Soc.* **64** 1900.
1943 a Stavely, Bollenback, *J. Am. Ch. Soc.* **65** 1285.
    b Stavely, Bollenback, *J. Am. Ch. Soc.* **65** 1290.
1953 Fieser, Nakanishi, Huang, *J. Am. Ch. Soc.* **75** 4719.
1954 Barton, Laws, *J. Ch. Soc.* **1954** 52, 60.

9. Monohydroxy-monoketosteroids (CO and OH in the Ring System) C$_{29}$ with Three Aliphatic Side Chains

$\varDelta^{22}$-**Stigmasten-6α-ol-3-one** C$_{29}$H$_{48}$O$_2$. Crystals (methanol), no constants given. — **Fmn.** From $\varDelta^{22}$-stigmastene-3β.6α-diol (p. 2084 s) by acetylation, followed by partial saponification in boiling aqueous methanolic K$_2$CO$_3$ to give a crude 6-monoacetate (not described), which is refluxed with aluminium tert.-butoxide and acetone and then with added dil. alkali for saponification. — **Rns.** Slow sublimation in high vac. in the presence of fused ZnCl$_2$ affords $\varDelta^{4.22}$-stigmastadien-3-one (p. 2465 s) (1941 PARKE, DAVIS & Co.).

Stigmastan-(or Sitostan-)6β-ol-3-one C$_{29}$H$_{50}$O$_2$. Crystals (methanol), m. 190° to 192°. — **Fmn.** The diacetate of sitostane-3β.6β-diol (p. 2084 s) is partly saponified by refluxing with NaHCO$_3$ in aqueous methanol; the crude 6-monoacetate (not described) so obtained is oxidized with CrO$_3$ in strong acetic acid at room temperature and the product is saponified. — **Rns.** Heating with p-toluenesulphonic acid, followed by slow sublimation in high vac., affords $\varDelta^4$-sitosten-3-one (p. 2467 s) (1941 PARKE, DAVIS & Co.).

$\varDelta^{22}$-**Stigmasten-3β-ol-6-one, 6-Keto-$\varDelta^{22}$-stigmastenol** C$_{29}$H$_{48}$O$_2$.
Acetate C$_{31}$H$_{50}$O$_3$. Crystals (methanol-ether), m. 145–6°; [α]$_D$ —31° (chloroform) (1954 Anagnostopoulos; cf. 1939 Mitui). — **Fmn.** From the acetate of 6-nitrostigmasterol (p. 1906 s) by reduction with zinc in boiling glacial acetic acid; using zinc in glacial acetic acid-ether (1:1), the acetate oxime (below) is obtained, which can be hydrolysed to the acetate by heating with zinc in glacial acetic acid (1939, 1940 Mitui). — **Rns.** Reduction with sodium in propanol affords $\varDelta^{22}$-stigmastene-3β.6α-diol (p. 2084 s) (1939 Mitui). — *Oxime* C$_{31}$H$_{51}$O$_3$N, m. 172° (1940 Mitui).

Stigmastan-(or Sitostan-)3β-ol-6-one C$_{29}$H$_{50}$O$_2$. Crystals (alcohol), m. 138° to 140° (1940 PARKE, DAVIS & Co.); white needles (aqueous alcohol), m. 115° (1910 Gloth), from the analysis probably a hydrate (Editor). **Fmn.** From the nitrate of 6-nitrositosterol (p. 1906 s) by reduction with zinc dust in slightly aqueous acetic acid under reflux (1910 Gloth; cf. 1918 Windaus) or on the water-bath (1940 PARKE, DAVIS & Co.); the acetate is obtained from that of 6-nitrositosterol by boiling with zinc in glacial acetic acid (1939 Mitui). — **Rns.** Hydrogenation in methanol at room temperature and 3 atm. pressure, using platinum oxide catalyst, affords sitostane-$3\beta.6\beta$-diol (p. 2084 s) (1940 PARKE, DAVIS & Co.). Reduction of the acetate with sodium in propanol gives sitostane-$3\beta.6\alpha$-diol (p. 2084 s) (1939 Mitui).

Acetate $C_{31}H_{52}O_3$. M. p. 129° (1939 Mitui). — **Fmn.** and **Rns.** See those of the parent hydroxyketone (above).

$\Delta^{8.14.22}$-**Stigmastatrien-3β-ol-7-one** $C_{29}H_{44}O_2$. For revision of the $\Delta^{8(14).9(11).22}$-structure originally ascribed by Stavely (1943), see 1953 Fieser.

Acetate $C_{31}H_{46}O_3$. Crystals (aqueous acetone), m. 190–2°; $[\alpha]_D^{23}$ —24° (chloroform). Absorption max. in abs. alcohol at 299 mμ (ε 5300). — **Fmn.** From the acetate of either 8α.9α- or 8α.14α-epoxy-8-iso-Δ^{22}-stigmasten-3β-ol-7-one (bot on p. 2775 s) by refluxing with alcoholic HCl. — **Rns.** Catalytic hydrogenation at room temperature using palladium black in alcohol affords the acetate of $\Delta^{8(14)}$-stigmasten-3β-ol-7-one (p. 2697 s); on using palladium black or platinum oxide in glacial acetic acid, the acetate of α-spinasterol (p. 1823 s) results (1943 Stavely).

$\Delta^{5.22}$-**Stigmastadien-3β-ol-7-one, 7-Ketostigmasterol** $C_{29}H_{46}O_2$ (p. 148). Platelets (ether - pentane), m. 137° after sintering at 120°, probably not quite pure; absorption max. (in alcohol?) at 237 mμ (ε 13000) (1942 Bergström). — **Fmn.** From stigmasterol (p. 1784 s) by aeration of its colloidal aqueous solution, stabilized with sodium stearate, at 85°, followed by separation from non-ketonic products by means of Girard's reagent T; yield, by spectroscopy, ca. 20% (1942 Bergström). The acetate is obtained from that of stigmasterol by oxidation with CrO_3 in glacial acetic acid at 50°; yield, ca. 25% (1936 Linsert; cf. 1935 I. G. FARBENIND.; 1939 Haslewood).

Rns. Hydrogenation of the acetate in ether - glacial acetic acid, using platinum oxide catalyst, affords the acetate of 7-ketositosterol (p. 2696 s) (1939 Mitui). For reduction of the acetate with aluminium isopropoxide in boiling isopropyl alcohol to give $\Delta^{5.22}$-stigmastadiene-3β.7β-diol (p. 2085 s), see also 1935 I. G. FARBENIND.

Acetate $C_{31}H_{48}O_3$ (p. 148). Leaflets (methanol, ether - methanol, or ethanol) (1935 I. G. FARBENIND.; 1936 Linsert; 1939 Haslewood), m. 184–6° cor. (1942 Bergström; cf. 1939 Haslewood); $[\alpha]_D^{23}$ —110° (chloroform); absorption max. (in alcohol?) at 237 mμ (ε 12300) (1942 Bergström). Easily soluble in ether

E.O.C. XIV s. 186

or acetone, sparingly soluble in methanol (1936 Linsert). — **Fmn.** By acetylation of the hydroxyketone (above) (1942 Bergström) or as described under this compound. — **Rns.** See those of the parent hydroxyketone (above).

$\Delta^{8 \cdot 22}$-Stigmastadien-3β-ol-7-one $C_{29}H_{46}O_2$.

Acetate $C_{31}H_{48}O_3$. Crystals (methanol), m. 202–4°, $[\alpha]_D^{23}$ —36° (chloroform). Absorption max. in alcohol at 252 mμ (ε 8300). Unreactive towards semicarbazide acetate (1943 Stavely). — **Fmn.** Together with the acetates of 8α.9α- and 8α.14α-epoxy-8-iso-Δ^{22}-stigmasten-3β-ol-7-ones (pp. 1796 s, 1797 s; cf. p. 2775 s), by oxidation of the acetate of "α"-spinasterol (p. 1794 s) with CrO_3 in strong acetic acid at room temperature; separation by means of Girard's reagent T and then by chromatography; yield, 3% (1943 Stavely; cf. 1949 Fieser).

Δ^5-Stigmasten-3β-ol-7-one, 7-Ketositosterol $C_{29}H_{48}O_2$ (p. 148).

Acetate $C_{31}H_{50}O_3$. Crystals (alcohol) (1936 Wunderlich), leaflets (ether - methanol), m. 154° (1935 I. G. FARBENIND.), m. 170° (1939 Mitui), m. 176°, $[\alpha]_D^{17}$ —94° (chloroform) (1936 Wunderlich); readily soluble in ether or acetone, sparingly soluble in methanol (1935 I. G. FARBENIND.). — **Fmn.** From the acetate of sitosterol (p. 1801 s) by oxidation with CrO_3 in glacial acetic acid at 55–60°; yield, ca. 17% (1936 Wunderlich; cf. 1935 I. G. FARBENIND.). — **Rns.** Catalytic hydrogenation in ether - glacial acetic acid, using platinum oxide, affords the acetate of 7-ketositostanol (p. 2697 s) (1939 Mitui). Reduction with aluminium isopropoxide in boiling isopropyl alcohol gives Δ^5-stigmastene-3β.7β-diol (p. 2085 s), probably along with the 7α-epimer (1936 Wunderlich).

Acetate semicarbazone $C_{32}H_{53}O_3N_3$. Crystals (methanol), m. 250–2° dec.; $[\alpha]_D^{21}$ —265° (chloroform) (1936 Wunderlich).

24-iso-Δ^5-Stigmasten-3β-ol-7-one, 7-Ketoclionasterol $C_{29}H_{48}O_2$.

Acetate $C_{31}H_{50}O_3$. Crystals (alcohol), m. 172–3°; $[\alpha]_D^{27}$ —99° (chloroform). — **Fmn.** From the acetate of clionasterol (p. 1839 s) by oxidation with CrO_3 in strong acetic acid at 60–65°; yield, 20%. — **Rns.** Boiling with aluminium isopropoxide in isopropyl alcohol affords a mixture of the 7-epimeric Δ^5-clionastene-3β.7-diols (p. 2086 s) (1944 Bergmann).

Δ8(14)-Stigmasten-3β-ol-7-one C$_{29}$H$_{48}$O$_2$.

Me CH$_3$ C$_2$H$_5$

CH·[CH$_2$]$_2$·CH·CH(CH$_3$)$_2$

Me

HO   H O

Acetate C$_{31}$H$_{50}$O$_3$. Crystals (80%
alcohol); m. 140–1°, [α]$_D$ —53°
(chloroform). Absorption max. in
abs. alcohol at 260 mμ (ε 7800). —
Fmn. From the acetate of Δ$^{8\cdot14\cdot22}$-
stigmastatrien-3β-ol-7-one (page
2695 s) by hydrogenation in alcohol at room temperature using palladium
black catalyst (1943 Stavely).

Stigmastan-3β-ol-7-one, 7-Keto-stigmastanol (or -sitostanol) C$_{29}$H$_{50}$O$_2$.

Me CH$_3$ C$_2$H$_5$

CH·[CH$_2$]$_2$·CH·CH(CH$_3$)$_2$

Me

HO   H O

Acetate C$_{31}$H$_{52}$O$_3$. M. p. 161°. —
Fmn. From the acetate of either
7-ketostigmasterol (p. 2695 s) or
7-ketositosterol (p. 2696 s) by
hydrogenation in ether-glacial
acetic acid using platinum oxide
catalyst. — Rns. Reduction by means of sodium in propanol yields stigmastane-
3β.7-diol (p. 2086 s) together with some sitostanol (p. 1826 s) (1939 Mitui).

1910 Gloth, *Über das Cholesterin und einige Phytosterine*, Thesis Univ. Munich, p. 39.
1918 Windaus, Rahlén, *Z. physiol. Ch.* **101** 223, 225.
1935 I. G. FARBENIND., *Brit. Pat.* 454260 (issued 1936); *C. A.* **1937** 1041; *Ch. Ztbl.* **1937** I 4129;
    WINTHROP CHEMICAL Co. (Windaus, Schenck), *U.S. Pat.* 2098984 (1936, issued 1937); *C. A.*
    1938 196; *Ch. Ztbl.* **1938** I 3659.
1936 Linsert, *Z. physiol. Ch.* **241** 125.
    Wunderlich, *Z. physiol. Ch.* **241** 116, 120.
1939 Haslewood, *Biochem. J.* **33** 454.
    Mitui, *C. A.* **1940** 383^9.
1940 Mitui, *Bull. Agr. Ch. Soc. Japan* **16** 144; *C. A.* **1941** 4390; *Ch. Ztbl.* **1942** I 878.
    PARKE, DAVIS & Co. (Marker), *U.S. Pat.* 2337564 (issued 1943); *C. A.* **1944** 3423.
1941 PARKE, DAVIS & Co. (Marker, Lawson), *U.S. Pat.* 2366204 (issued 1945); *C. A.* **1945** 1649;
    Ch. Ztbl. **1945** II 1385.
1942 Bergström, Wintersteiner, *J. Biol. Ch.* **145** 327, 330 et seq.
1943 Stavely, Bollenback, *J. Am. Ch. Soc.* **65** 1600.
1944 Bergmann, Lyon, McLean, *J. Org. Ch.* **9** 290.
1949 Fieser, Fieser, Chakravarti, *J. Am. Ch. Soc.* **71** 2226, 2228.
1953 Fieser, Nakanishi, Huang, *J. Am. Ch. Soc.* **75** 4719.
1954 Anagnostopoulos, Fieser, *J. Am. Ch. Soc.* **76** 532, 535.

---

10. Monohydroxy-monoketosteroids (CO and OH in the Ring System) with Four Aliphatic Side Chains

6β-Methylcholestan-5α-ol-3-one C$_{28}$H$_{48}$O$_2$. For structure and configuration

Me CH$_3$

CH·[CH$_2$]$_3$·CH(CH$_3$)$_2$

Me

O HO CH$_3$

at C-6, see 1952 Turner. — Leaflets
(glacial acetic acid), crystals (ethyl
acetate-alcohol), m. 215.5—216° cor.,
sublimes in high vac. at 175–180° (1939
Ushakov); crystals (ethyl acetate),
m. 227–8° cor. dec., [α]$_D$ +20.5° (di-

186*

oxane) (1952 Turner). — **Fmn.** From 6β-methylcholestane-3β.5α-diol (page 2087 s) by oxidation with CrO_3 in glacial acetic acid at 35°; yield, 39%. — **Rns.** Treatment with hydrogen chloride in dry chloroform at 0° affords 6α-methyl-Δ^4-cholesten-3-one (p. 2464 s) (1939 Ushakov).

1939 Ushakov, Madaeva, *J. Gen. Ch. U.S.S.R.* **9** 436, 441; *C. A.* **1939** 9309; *Ch. Ztbl.* **1939** II 4488.
1952 Turner, *J. Am. Ch. Soc.* **74** 5362.

---

11. Monohydroxy-monoketosteroids (CO and OH in the Ring System) Containing Cyclic Radicals in Side Chain

16 - Benzylidene - Δ^4 - androsten - 17β - ol - 3 - one, 16 - Benzylidenetestosterone

$C_{26}H_{32}O_2$.

Acetate $C_{28}H_{34}O_3$. Prisms (aqueous acetone), m. 178–9°; some samples solidified above this point and remelted at 197–198°. — **Fmn.** From 16 - benzylidene - $\Delta^{3.5}$-androstadien - 3 - ol - 17-one ethyl ether (below) on refluxing with aluminium isopropoxide in benzene for several hrs. with repeated removal (by distillation) of the acetone formed, followed by hydrolysis with acetic acid and acetylation with acetic anhydride and pyridine at 60°. — **Rns.** It may be converted into Δ^4-androsten-17β-ol-3.16-dione acetate (p. 2945 s) by treatment with 1.1 mol. OsO_4 in carbon tetrachloride at room temperature in the dark for 3 days, followed by treatment with zinc dust in acetic acid at 45–50° and oxidation of the resulting 16-(α-hydroxybenzyl)-Δ^4-androstene-16.17β-diol-3-one 17-acetate with periodic acid in aqueous alcohol or, preferably, with lead tetraacetate in benzene (1941 Stodola).

17α - Benzyl - Δ^4 - androsten - 17β - ol - 3 - one, 17 - Benzyltestosterone $C_{26}H_{34}O_2$

(p. 145). Leaflets, m. 225° (1936 Fujii). — **Fmn.** From 17α-benzyl-Δ^5-androstene-3β.17β-diol [prepared from 5-dehydroepiandrosterone (p. 2609 s) and benzylmagnesium chloride] on oxidation with temporary protection of the double bond with bromine (1936 Fujii). From the 3-enol ethyl ether of Δ^4-androstene-3.17-dione ($\Delta^{3.5}$-androstadien-3-ol-17-one ethyl ether, p. 2606 s) on treatment with benzylmagnesium bromide, followed by acid hydrolysis (1943 CIBA PHARMACEUTICAL PRODUCTS).

16-Benzylidene-$\Delta^{3.5}$-androstadien-3-ol-17-one ethyl ether $C_{28}H_{34}O_2$.

Bars (acetone + a little pyridine), m. 181–186° after sintering at 177° (1941 Stodola); crystals (chloroform), m. 197–8°; [α]$_D$ —111.5° (chloroform) (1945 Velluz; 1958 Petit). — **Fmn.** From Δ^4-androstene-3.17-dione 3-enol-ethyl ether ($\Delta^{3.5}$-androstadien-3-ol-17-one ethyl ether, p. 2606 s) and benzaldehyde on refluxing with sodium methoxide in methanol (almost quantitative yield)

(1941 Stodola) or with KHCO$_3$ in aqueous methanol (1945 Velluz). — **Rns.**
Gives 16-benzylidenetestosterone (p. 2698 s) on refluxing with aluminium isopropoxide in benzene, followed by warming with acetic acid (1941 Stodola).
On rapid hydrolysis (30 min.) with methanolic H$_2$SO$_4$ it gives 16-benzylidene-
Δ^4-androstene-3.17-dione (p. 2914 s) (1945 Velluz).

16-Benzylidene-Δ^5-androsten-3β-ol-17-one, 16-Benzylidene-5-dehydroepiandrosterone

C$_{26}$H$_{32}$O$_2$. Crystals (methanol),
m. 202–5^0 (Mason in 1942 Stodola); crystals
(ethyl acetate), m. 209–210^0 (block); [α]$_D$
—26.5^0 (chloroform) (1945 Velluz). — **Fmn.**
From the acetate of Δ^5-dehydroepiandrosterone
(p. 2609 s) and benzaldehyde on refluxing with sodium methoxide in methanol
(yield, 90%) (Mason in 1942 Stodola); similarly from the unesterified hydroxy-
ketone and benzaldehyde on refluxing with KHCO$_3$ in aqueous methanol
(yield, 80%) (1945 Velluz). — **Rns.** Gives 16-benzylidene-Δ^5-androstene-
3β.17-diol (p. 2045 s) on refluxing with aluminium isopropoxide in benzene -
isopropyl alcohol (1942 Stodola). On hydrogenation in the presence of Raney
nickel it gives **16-benzyl-Δ^5-androsten-3β-ol-17-one** C$_{26}$H$_{34}$O$_2$ [no analysis;
m. 170–1^0, [α]$_D$ +75^0 (chloroform); *acetate*, m. 170–1^0] (1945 Velluz). Consumes 1.25 mols. of bromine. Does not react with semicarbazide or Girard's
T reagent (1945 Velluz).

Acetate C$_{28}$H$_{34}$O$_3$, m. 255–6^0 (block); prepared with acetic anhydride in
pyridine (1945 Velluz).

16-Benzylidene-androstan-3β-ol-17-one, 16-Benzylidene-epiandrosterone

C$_{26}$H$_{34}$O$_2$ (I). Needles (acetone), m. 181.5–182.5^0 cor. — **Fmn.** From epiandro-

sterone (p. 2633 s) and benzaldehyde on boiling with sodium methoxide in
methanol; yield, 88%. — **Rns.** On oxidation of its acetate in strong acetic
acid at 60^0, followed by hydrolysis with aqueous NaOH, it gives 3β-hydroxy-
ætioallobilianic acid (II) (1943 Hirschmann).

Acetate C$_{28}$H$_{36}$O$_3$, plates (methanol), m. 237–8^0 cor.; prepared with acetic
anhydride in pyridine at room temperature (1943 Hirschmann).

24.24-Diphenyl-Δ^{23}-cholen-3α-ol-12-one, 2.2-Diphenyl-1-(3α-hydroxy-12-keto-bisnorcholanyl)-ethylene

C$_{36}$H$_{46}$O$_2$. Plates (alcohol) containing
$^1/_2$ C$_2$H$_5$OH (retained after drying at
78^0 and 1 mm. pressure), m. 158–9^0
cor. — **Fmn.** From the 3-(hydrogen
succinate) of 24.24-diphenyl-Δ^{23}-chol-
ene-3α.12α-diol (p. 2094 s) by oxi-

dation with CrO_3 in ca. 86% acetic acid at 0–5°, followed by saponification
of the crude 3-(hydrogen succinate) with aqueous alcoholic KOH; yield, 77%.
The acetate is obtained from (3α-hydroxy-12-keto-norcholanyl)-diphenyl-
carbinol (p. 2724 s) by refluxing with a mixture of glacial acetic acid and
acetic anhydride; yield, 24% based on the methyl 3α-hydroxy-12-ketocholanate
precursor. — **Rns.** Oxidation of the acetate with CrO_3 in chloroform - acetic
acid at 35°, followed by saponification in boiling aqueous NaOH, affords
3α-hydroxy-12-keto-norcholanic acid (1943 Riegel).

Acetate $C_{38}H_{48}O_3$. Crystals (acetone), m. 181.5–182.5° cor.; obtained as
described under the hydroxyketone (above) or from that compound by reaction
with acetic anhydride and pyridine on the water-bath (yield, 74%) (1943
Riegel).

1936 Fujii, Matsukawa, *Ch. Ztbl.* **1936** II 3305.
1941 Stodola, Kendall, *J. Org. Ch.* **6** 837.
1942 Stodola, Kendall, *J. Org. Ch.* **7** 336, 338, 339.
1943 CIBA PHARMACEUTICAL PRODUCTS (Miescher), *U.S. Pat.* 2386331 (issued 1945); *C. A.* **1946**
        179; *Ch. Ztbl.* **1948** I 1040.
        Hirschmann, *J. Biol. Ch.* **150** 363, 376.
        Riegel, Moffett, *J. Am. Ch. Soc.* **65** 1971.
1945 Velluz, Petit, *Bull. soc. chim.* [5] **12** 949.
1958 Petit, *private communication.*

12. HALOGENO-MONOHYDROXY-MONOKETOSTEROIDS (CO AND OH IN THE RING SYSTEM)

a. Halogen in Side Chain

21-Bromo-$\Delta^{17(20)}$-pregnen-3α-ol-11-one $C_{21}H_{31}O_2Br$.

Acetate $C_{23}H_{33}O_3Br$. Crystals (ether-pentane), m. 116–7° cor. — **Fmn.** From the 3-acetate of 17-iso-Δ^{20}-pregnene-3α.17β-diol-11-one (p. 2752 s) on treatment with PBr_3 in dry chloroform containing a little pyridine, first at —60°, then at room temperature; crude yield, 53% (1946 Sarett; 1946a MERCK & Co., INC.). — **Rns.** Gives the diacetate of $\Delta^{17(20)}$-pregnene-3α.21-diol-11-one (p. 2723 s) on refluxing with anhydrous potassium acetate in dry acetone (1946 Sarett; 1946b MERCK & Co., INC.).

23-Bromonorcholan-3α-ol-12-one $C_{23}H_{37}O_2Br$.

Acetate $C_{25}H_{39}O_3Br$. Crystals (dil. acetone), m. 209.5–211°; $[\alpha]_D^{20} +126°$ (chloroform). — **Fmn.** From the silver salt of 3α-acetoxy-12-ketocholanic acid on treatment with 1 mol. bromine in boiling carbon tetrachloride; yield, 27%. — **Rns.** On reduction with zinc and acetic acid it gives norcholan-3α-ol-12-one acetate (p. 2656 s) (1946 Brink).

1946 Brink, Clark, Wallis, *J. Biol. Ch.* **162** 695, 702, 703.
   a MERCK & Co., INC. (Sarett), *U.S. Pat.* 2492191 (issued 1949); *C. A.* **1950** 3045; *Ch. Ztbl.* **1952** 1211.
   b MERCK & Co., INC. (Sarett), *U.S. Pat.* 2492192 (issued 1949); *C. A.* **1950** 3045; *Ch. Ztbl.* **1952** 1211.
   Sarett, *J. Biol. Ch.* **162** 601, 621.

---

b. Halogen in the Ring System

α. Monohalogeno-Compounds

dl-4-Chloro-isoequilenin methyl ether $C_{19}H_{19}O_2Cl$.

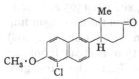

Colourless prisms (alcohol), m. 183–4°. — **Fmn.** From 16-piperonylidene-isoequilenin methyl ether (p. 2730 s) on treatment with chlorine in dioxane - carbon tetrachloride at 0°, followed by heating with sodium ethoxide in alcohol-benzene on the steam-bath for 1 hr., boiling with 15% HCl for $^1/_2$ hr., refluxing with 2% NaOH in benzene for 1 hr., treatment with dil. acid, and refluxing of the resulting 4-chloro-16-(3.4-methylenedioxybenzoyl)-isoequilenin methyl ether (see page 2864 s) with aqueous alcoholic KOH for $2^1/_2$ hrs. (1945 Birch).

2.4-Dinitrophenylhydrazone $C_{25}H_{23}O_5N_4Cl$, orange crystals, m. 290°; very sparingly soluble (1945 Birch).

For a **bromo-equilenin** $C_{18}H_{17}O_2Br$, see p. 135.

7-Chloroœstrone $C_{18}H_{21}O_2Cl$. — **Fmn.** Its benzoate is obtained from 7-hydroxyœstrone 3-benzoate (p. 2740 s) on treatment with PCl_5 in the presence of $CaCO_3$ in dry chloroform at 0°; yield, 70%. — **Rns.** On heating the benzoate with sodium iodide in dry pyridine in a sealed tube on the steam-bath for 40 hrs. it gives the benzoate of 6-dehydroœstrone ("$\varDelta^6$-isoequilin", p. 2538 s) (1940 Pearlman).

Benzoate $C_{25}H_{25}O_3Cl$. Crystals (abs. alcohol + a little benzene), m. 247–8° cor. dec. (1940 Pearlman). — **Fmn.** and **Rns.** See above.

16-Bromoœstrone $C_{18}H_{21}O_2Br$. — **Fmn.** Its acetate is obtained from the acetate of œstrone (p. 2543 s) on treatment with 1 mol. bromine and some HBr in glacial acetic acid at 10° (yield, 56%), its benzoate from the corresponding ester with 1 mol. bromine in chloroform at room temperature. — **Rns.** On refluxing of the acetate with aluminium isopropoxide in isopropyl alcohol for 6 hrs. it gives **16-bromoœstradiol** $C_{18}H_{23}O_2Br$ [crystals (ether), m. 235°] which on refluxing with potassium acetate in butanol - xylene, followed by hydrolysis with methanolic KOH, yields a 16-hydroxyœstradiol* m. 240–2° dec. On refluxing the acetate with potassium acetate in butanol - xylene it gives 16-hydroxyœstrone diacetate which on hydrolysis with methanolic KOH, followed by refluxing with aluminium isopropoxide in isopropyl alcohol, yields a 16-hydroxyœstradiol* m. 265–271° dec. On refluxing of the benzoate with potassium acetate in butanol - toluene, the 3-benzoate 16-acetate of 16-hydroxyœstrone (p. 2740 s) is obtained (1937 Schering A.-G.).

Acetate $C_{20}H_{23}O_3Br$. Needles (ether), m. 165.5° (1937 Schering A.-G.). — **Fmn.** and **Rns.** See above.

Benzoate $C_{25}H_{25}O_3Br$. Crystals (ether), dec. ca. 185° (1937 Schering A.-G.). — **Fmn.** and **Rns.** See above.

For a **bromoœstrone methyl ether** $C_{19}H_{23}O_2Br$, see p. 138.

2 - Bromo - $\varDelta^1$ - androsten - 17β - ol - 3 - one, *2 - Bromo-$\varDelta^1$-testosterone* $C_{19}H_{27}O_2Br$. For the configuration at C-17, see pp. 1363 s, 1364 s. —**Fmn.** Its esters are obtained from the corresponding esters of 2.2-dibromo-androstan-17β-ol-3-one (p. 2710s) on refluxing with collidine for 15 min. (1943 Inhoffen; 1946 Wilds; 1947b Djerassi). — **Rns.** The esters give the corresponding esters of $\varDelta^1$-androsten-17β-ol-3-one ($\varDelta^1$-testosterone, p. 2578 s) on refluxing with zinc dust in alcohol (1943

* Cf. the m. p.'s of œstriol (p. 2152 s), 17-epi- and 16-epi-œstriols (p. 2159 s).

Inhoffen; 1947b Djerassi); for reduction of the hexahydrobenzoate with zinc and acetic acid, see 1947b Djerassi. The hexahydrobenzoate is not attacked on attempted reduction with chromous chloride (1947b Djerassi). The benzoate remains unchanged on boiling with alcoholic HCl for 5 hrs., on heating with silver acetate in acetic acid on the water-bath for 2 hrs., or on boiling with potassium acetate in butanol for $2^1/_2$ hrs. (1943 Inhoffen). The hexahydrobenzoate rearranges to the corresponding ester of 4-bromo-Δ^1-androsten-17β-ol-3-one (p. 2704 s) on keeping in glacial acetic acid containing some 4 N HBr at room temperature for 23 hrs. (1947b Djerassi).

Benzoate $C_{26}H_{31}O_3Br$. Plates (glacial acetic acid), m. 225° (1943 Inhoffen). — **Fmn.** and **Rns.** See above.

Hexahydrobenzoate $C_{26}H_{37}O_3Br$. Needles (methanol), m. 155–6° cor. (1943 Inhoffen; 1946 Wilds); $[\alpha]_D^{21}$ +38° (1946 Wilds). Absorption max. in ether at 254 mμ (log ε 3.89) (1943 Inhoffen), in abs. alcohol at 255 mμ (log ε 3.88) (1946 Wilds). — **Fmn.** and **Rns.** See above.

2α-Bromo-Δ^4-androsten-17β-ol-3-one, 2α-Bromotestosterone $C_{19}H_{27}O_2Br$. For

the configuration at C-17, see pp. 1363 s, 1364 s. — **Fmn.** Its esters are obtained from the corresponding esters of 2α.4α-dibromoandrostan-17β-ol-3-one (page 2711 s) on refluxing with collidine for $1^1/_2$—5 min. (1943 Inhoffen; 1943 SCHERING A.-G.; 1947b Djerassi). — **Rns.** The esters give the corresponding esters of testosterone (p. 2580 s) on boiling with zinc dust and alcohol for 3 hrs. (1943 Inhoffen; 1943 SCHERING A.-G.). The acetate and the hexahydrobenzoate rearrange to the corresponding esters of 6-bromotestosterone (p. 2705 s) on treatment with glacial acetic acid containing some HBr at 25–8° for 24 hrs. (1943 Inhoffen). On boiling the acetate or the hexahydrobenzoate with collidine for $1/_2$ hr. or 1 hr., resp., they give the corresponding esters of $\Delta^{1.4}$-androstadien-17β-ol-3-one (p. 2576 s) (1943 Inhoffen).

Acetate $C_{21}H_{29}O_3Br$. Needles (methanol), m. 177–9° dec. (1943 Inhoffen). — **Fmn.** and **Rns.** See above.

Benzoate $C_{26}H_{31}O_3Br$. Needles (chloroform-alcohol), m. 193–5° dec. (1943 Inhoffen). — **Fmn.** and **Rns.** See above.

Hexahydrobenzoate $C_{26}H_{37}O_3Br$. Needles (ether), m. 161–2° (1943 Inhoffen), 147.5–148.5° cor. dec. (1947b Djerassi); $[\alpha]_D^{22}$ +92° (chloroform) (1947b Djerassi). Absorption max. in ether (from the graph) at 236 mμ (log ε 4.14) (1943 Inhoffen), in alcohol at 244.5 (log ε 4.08) (1947b Djerassi).

2α-Bromoandrostan-17β-ol-3-one, 2α-Bromodihydrotestosterone $C_{19}H_{29}O_2Br$

(p. 148). For α-configuration at C-2, see 1952 Jones; cf. also the literature quoted in connection with the configuration of 2α-bromocholestan-3-one (p. 2480 s). For β-configuration at C-17, see p. 1363 s. — Rotation in chloroform, $[\alpha]_D^{23}$ +33° (1947c Djerassi). —

Fmn. Its esters are obtained from the corresponding esters of androstan-17β-ol-3-one (dihydrotestosterone, p. 2597 s) on bromination with 1 mol. of bromine in the presence of HBr in glacial acetic acid at room temperature (yield, 50%) (1938 Butenandt; 1943 Inhoffen; 1947 a, b Djerassi; cf. also 1936 CIBA PHARMACEUTICAL PRODUCTS; 1937 b CIBA) or (e.g., in the case of the hexahydrobenzoate) with 1 mol. of N-bromosuccinimide by refluxing in carbon tetrachloride for 1–2 min. with exposure to strong light (yield, 52%) (1947 a, b Djerassi); the free bromo-hydroxy-ketone is obtained in 68% yield from its acetate by treatment with 3% methanolic HCl at room temperature for 24 hrs. (1938 Butenandt).

Rns. On boiling the acetate with collidine it gives the acetate of Δ^1-androsten-17β-ol-3-one (p. 2578 s) (1940 Butenandt; 1940 SCHERING A.-G.); the hexahydrobenzoate on similar treatment gives the hexahydrobenzoates of Δ^1-androsten-17β-ol-3-one (42%; p. 2578 s) (1943 Inhoffen; 1947 b Djerassi) and Δ^4-androsten-17β-ol-3-one (18%; p. 2580 s) (1947 b Djerassi); the propionate reacts on boiling with pyridine, followed by heating of the resulting crystalline addition product under 12 mm. pressure, to form the propionate of Δ^4-androsten-17β-ol-3-one (1937 b CIBA). On heating the free hydroxy-ketone or its acetate with potassium acetate in glacial acetic acid in a sealed tube at 200° for 5 hrs. there is obtained a very small amount (ca. 2%) of the acetate of an androstenolone which was originally assumed to be Δ^1-androsten-17β-ol-3-one (1938 Butenandt; 1936 CIBA PHARMACEUTICAL PRODUCTS), but which is very probably Δ^5-androsten-17β-ol-4-one (p. 2601 s) (1944 Butenandt); on using the free hydroxy-ketone there is also formed ca. 3% of a **compound** $C_{21}H_{28}O_3$ [needles (acetone - petroleum ether), m. 208°; absorption max. in chloroform at 238 and 280 mμ] (1938 Butenandt).

Acetate $C_{21}H_{31}O_3Br$. Needles (methanol), m. 177–8° (1938 Butenandt); $[\alpha]_D^{23}$ +35° (chloroform) (1947 c Djerassi). — **Fmn.** and **Rns.** See above.

Hexahydrobenzoate $C_{26}H_{39}O_3Br$. Needles (acetone), m. 181–2° dec. (1943 Inhoffen); m. 182–3° cor.; $[\alpha]_D^{23}$ +42.5° (chloroform) (1947 b Djerassi). — **Fmn.** and **Rns.** See above.

4-Bromo-Δ^1-androsten-17β-ol-3-one, *4-Bromo-Δ^1-testosterone* $C_{19}H_{27}O_2Br$. For configuration at C-17, see pp. 1363 s, 1364 s. — **Fmn.** Its hexahydrobenzoate is obtained by rearrangement of the corresponding ester of 2-bromo-Δ^1-androsten-17β-ol-3-one (p. 2702 s) on keeping in glacial acetic acid containing some 4 N HBr at room temperature for 23 hrs. — **Rns.** On refluxing the hexahydrobenzoate with collidine it gives the hexahydrobenzoate of $\Delta^{1.4}$-androstadien-17β-ol-3-one (p. 2576 s) (1947 b Djerassi).

Hexahydrobenzoate $C_{26}H_{37}O_3Br$ (not obtained pure). Amorphous solid, m. 47–65°; $[\alpha]_D$ +11° (chloroform); absorption max. in alcohol at 234 mμ (log ε 3.92) (1947 b Djerassi). — **Fmn.** and **Rns.** See above.

4-Bromo-ætiocholan-17β-ol-3-one $C_{19}H_{29}O_2Br$. See under the reactions of 17-amino-ætiocholan-3-one, p. 2509 s.

6β(?)-Bromo-Δ^4-androsten-17β-ol-3-one, 6-Bromotestosterone $C_{19}H_{27}O_2Br$ (p. 148). For probable β-configuration at C-6, see 1950 Djerassi; for β-configuration at C-17, see pp. 1363 s, 1364 s. — **Fmn.** Its acetate is obtained from the acetate of testosterone (p. 2580 s) on treatment with 1 mol. N-bromosuccinimide in carbon tetrachloride in the dark for 5 hrs. (1946 Meystre; cf. 1950 Djerassi). The acetate and the hexahydrobenzoate are obtained by rearrangement of the corresponding esters of 2α-bromotestosterone (p. 2703 s) on keeping in glacial acetic acid containing some HBr at 25–8⁰ for 24 hrs. (1943 Inhoffen). For formation of its esters (propionate, benzoate) from the corresponding esters of 5.6-dibromodihydrotestosterone (p. 2712 s) by refluxing with anhydrous sodium acetate in abs. alcohol (1937 Ruzicka; cf. p. 148), see also 1937a CIBA. — **Rns.** On refluxing with collidine the esters are converted into the corresponding esters of $\Delta^{4.6}$-androstadien-17β-ol-3-one (6-dehydrotestosterone, p. 2577 s) (1943 Inhoffen; 1946 Meystre).

Acetate $C_{21}H_{29}O_3Br$. Crystals (hexane-ether), m. 140–2⁰ (127–130⁰ cor. dec. Kofler block); $[\alpha]_D^{20}$ —16⁰ (chloroform); absorption max. in 95% alcohol at 248 mμ (log ε 4.19) (1950 Djerassi). — **Fmn.** and **Rns.** See above.

Hexahydrobenzoate $C_{26}H_{37}O_3Br$ (not analysed). Needles (ether), m. 163–5⁰ (1943 Inhoffen). — **Fmn.** and **Rns.** See above.

5-Chloroandrostan-3β-ol-17-one, 5-Dehydroepiandrosterone hydrochloride $C_{19}H_{29}O_2Cl$ (p. 148). Configuration at C-5 ascribed by analogy with 5-chlorocholestan-3β-ol (p. 1877 s), similarly obtained (Editor). — The compounds described by Ruzicka and by Fujii (see p. 148) were mixtures (1950 Hershberg). — Crystals (ether), m. 171.5–172.5⁰ cor.; $[\alpha]_D^{25}$ +62.4⁰ (dioxane) (1950 Hershberg). — **Fmn.** Along with 3β-chloro-Δ^5-androsten-17-one (p. 2474 s), from Δ^5-androsten-3β-ol-17-one (5-dehydroepiandrosterone, p. 2609 s) on treatment with dry HCl in chloroform with cooling in a CO_2-acetone bath, followed by crystallization from chloroform-hexane; yield, ca. 60% (1950 Hershberg). — **Rns.** Gives 5-chloroandrostane-3.17-dione (p. 2918 s) on oxidation with CrO_3 and strong acetic acid in ethylene chloride at room temperature (1950 Hershberg).

16-Bromo-Δ^5-androsten-3β-ol-17-one, 16-Bromo-5-dehydroepiandrosterone $C_{19}H_{27}O_2Br$. — **Fmn.** Its acetate is obtained from the acetate of dehydroepiandrosterone (p. 2609 s) on treatment with bromine and some HBr in glacial acetic acid at room temperature, followed by debromination of the resulting **5.6-dibromide** (dec. 164⁰ to 165⁰) with sodium iodide in alcohol-benzene at room temperature. —

Rns. On refluxing the acetate with aluminium isopropoxide in isopropyl alcohol for 6 hrs. it gives **16-bromo-Δ^5-androstene-3β.17-diol** $C_{19}H_{29}O_2Br$ [needles (aqueous alcohol)]. On refluxing the acetate with potassium acetate in butanol-xylene for 2 hrs. it yields the diacetate of Δ^5-androstene-3β.16-diol-17-one (1937 Schering A.-G.).

Acetate $C_{21}H_{29}O_3Br$. Needles (methanol), m. 167–9⁰ dec. (1937 Schering A.-G.). — **Fmn.** and **Rns.** See above.

16-Bromoandrostan-3β-ol-17-one, 16-Bromo-epiandrosterone $C_{19}H_{29}O_2Br$.

Needles (dil. acetone), m. 164–5⁰. — **Fmn.** From epiandrosterone (p. 2633 s) on treatment with 1 mol. of bromine and a little HBr in glacial acetic acid at room temperature (yield, 8%); the acetate is similarly formed from the corresponding acetate (yield, 27%), and is hydrolysed by 3% methanolic HCl at room temperature (yield, 84%). — **Rns.** Gives 16-bromoandrostane-3.17-dione (p. 2919 s) on oxidation with CrO_3 in glacial acetic acid at room temperature (1937/38 Dannenberg).

Acetate $C_{21}H_{31}O_3Br$. Prisms (acetone-petroleum ether, and methanol), needles (dil. acetone), m. 177–8⁰ (1937/38 Dannenberg). — **Fmn.** and **Rns.** See above.

2-Bromo-Δ^4-cholesten-4-ol-3-one $C_{27}H_{43}O_2Br$. See under the reactions of 2α.4α-dibromocholestan-3-one, p. 2495 s.

2-Bromo-Δ^4-cholesten-6-ol-3-one $C_{27}H_{43}O_2Br$.

For structure of the following ethers, their parent compounds, and their reaction products, see 1955 Fieser; cf. also 1956 Ellis.

Methyl ether, 2-Bromo-6-methoxy-Δ^4-cholesten-3-one $C_{28}H_{45}O_2Br$. Needles, dec. 101⁰ with strong, red coloration. Absorption max. (in ether?) at 235 mμ. — **Fmn.** From 2.2.6β-tribromo-Δ^4-cholesten-3-one (p. 2503 s; originally described as 4.6.6-tribromo-Δ^4-cholesten-3-one) on boiling with sodium iodide in methanol-benzene (1937 Butenandt).

Ethyl ether, 2-Bromo-6-ethoxy-Δ^4-cholesten-3-one $C_{29}H_{47}O_2Br$. Needles (dil. acetone), m. 110–111⁰ dec. (1936 Butenandt). Absorption max. in ether at 237 mμ (ε 11000) (1939 Dannenberg). — **Fmn.** As above, using ethanol (yield, 46%) (1936 Butenandt). From 2.2.5α.6β-tetrabromocholestan-3-one (p. 2505 s; originally described as 4.4.5.6-tetrabromocholestan-3-one) on boiling with sodium iodide in abs. ethanol-benzene for 2 hrs. (yield, 30%) (1936 Butenandt). — **Rns.** On boiling with zinc dust and methanol in benzene it probably gives 6-ethoxy-Δ^4-cholesten-3-one (p. 2673 s) (1937 Butenandt). It is converted into 6-ethoxy-$\Delta^{4.6}$-cholestadien-3-one (p. 2671 s) on refluxing with potassium acetate in butanol for 6 hrs. (1936 Butenandt) or with HBr - glacial acetic acid in alcohol-benzene for 2 hrs. (1937 Butenandt).

4-Bromo-Δ⁴-cholesten-3β-ol-6-one* C₂₇H₄₃O₂Br. — **Fmn.** Its acetate is obtained from the acetate of Δ⁴-cholesten-3β-ol-6-one (p. 2677 s) on treatment with 1 mol. bromine in ether-glacial acetic acid at 18⁰; crude yield, 96%. The acetate is also formed from 4.5-dibromocholestan-3β-ol-6-one acetate (p. 2713 s) on boiling in ether (yield, 39%) or on refluxing with potassium acetate in glacial acetic acid for 50 min. (yield, 29%). — **Rns.** On heating the acetate with zinc dust in glacial acetic acid at 100⁰ for 15 min. or on refluxing with zinc dust in abs. alcohol for 1½ hrs. it gives bi-(6-keto-Δ⁴-cholesten-3-yl). On refluxing the acetate with anhydrous pyridine for 7 hrs., the acetate of Δ²·⁴-cholestadien-3-ol-6-one (p. 2676 s) is formed. On treatment with sodium methoxide in methanol at 20⁰ for 18 hrs., the acetate gives 4.5-dimethoxycholestan-3β-ol-6-one (p. 2772 s). The acetate does not react with potassium acetate in glacial acetic acid at 100⁰ (1 hr.) or in alcohol on boiling for 12 hrs., or with silver nitrate in pyridine at 20⁰ (48 hrs.) (1938 Jackson).

Acetate C₂₉H₄₅O₃Br. Plates (methanol), m. 115–6⁰ (1938 Jackson). Absorption max. in alcohol at 245 (3.9) and 335 (2.3) mμ (log ε) (1938 Jackson; cf. also 1939 Dannenberg). — **Fmn.** and **Rns.** See above.

5α-Bromocholestan-3β-ol-6-one C₂₇H₄₅O₂Br. For α-configuration at C-5 (1937 Heilbron), see also 1943 Sarett; 1954 Cookson; 1954 James; 1954 Corey. — Needles (dil. acetone), m. ca. 150⁰ dec.; [α]²⁵_D −156⁰ (chloroform) (1943 Sarett). — **Fmn.** From cholestan-3β-ol-6-one (p. 2677 s) on treatment with 1 mol. bromine in ether - glacial acetic acid (1943 Sarett); the acetate is similarly obtained (at 0⁰) from the corresponding acetate (yield, 71%) or in chloroform in the presence of dil. HBr (1937 Heilbron).

Rns. Gives 5α-bromocholestane-3.6-dione (p. 2922 s) on oxidation with CrO₃ in strong acetic acid with cooling (1943 Sarett). Reduction with zinc dust and alcohol affords cholestan-3β-ol-6-one (p. 2677 s) (1943 Sarett); the acetate of this hydroxy-ketone is obtained on reduction of the acetate of the bromo-hydroxy-ketone with aluminium amalgam and water in ether (1937 Heilbron). The acetate rearranges to that of 7α-bromocholestan-3β-ol-6-one (below) on heating with HBr - glacial acetic acid on the steam-bath for 15 min. (1937 Heilbron). On bromination with 1.1–1.2 mols. of bromine in glacial acetic acid in the presence of HBr, the acetate gives the acetates of the two epimeric 5α.7-dibromocholestan-3β-ol-6-ones, the higher-melting

---

* Fieser (1958) suggests the 7-bromo-structure for this compound.

(5α.7α-) form (p. 2714 s) being obtained at 30° (30 min.), the lower-melting (5α.7β-) form (p. 2714 s) on the steam-bath (15 min.) (1938 Heilbron). On refluxing of the acetate with dry pyridine for 8 hrs. it gives that of Δ^4-cholesten-3β-ol-6-one (p. 2677 s), on refluxing with methanolic KOH for 2 hrs. it yields "cholestane-3β.5-diol-6-one m. 138°" (see p. 2758 s) (1937 Heilbron).

Acetate $C_{29}H_{47}O_3Br$. Plates (light petroleum b. 60–80°), m. 162° dec. (the m. p. varies with the rate of heating); $[\alpha]_D^{22}$ —133° (chloroform) (1937 Heilbron). Absorption max. in alcohol at 308 mμ (log ε 2.1) (1938 Barr; cf. 1954 Cookson). Readily soluble in ether and chloroform, sparingly in glacial acetic acid and light petroleum (1937 Heilbron). — **Fmn.** and **Rns.** See above.

7α-Bromocholestan-3β-ol-6-one $C_{27}H_{45}O_2Br$. For α-configuration at C-7 (1937

Heilbron), see also 1953, 1954 Corey; 1954 Cookson; 1954 James. — Crystals (dil. acetone), m. 113°; $[\alpha]_D^{20}$ +51° (chloroform) (1943 Sarett). — **Fmn.** The acetate is obtained from that of cholestan-3β-ol-6-one (p. 2677 s) by bromination with 1 mol. of bromine in ether - glacial acetic acid at 35°, followed by refluxing for 2 hrs., or from the above 5α-isomer on heating with HBr - glacial acetic acid on the steam-bath for 15 min. (1937 Heilbron). The acetate may also be obtained from the two epimeric 5α.7-dibromocholestan-3β-ol-6-one acetates: from the higher-melting (5α.7α-) form (p. 2714 s) on treatment with 2 mols. of HBr in glacial acetic acid at 22° for 24 hrs. (yield, 50%), from the lower-melting (5α.7β-) form on treatment with 1.1 mols. of HBr in glacial acetic acid on the steam-bath for 15 min. (yield, 28%) (1938 Heilbron). The acetate is hydrolysed by refluxing with conc. HCl in alcohol for 10 min. (1943 Sarett).

Rns. Gives 7α-bromocholestane-3.6-dione (p. 2923 s) on oxidation with CrO_3 in strong acetic acid with cooling (1943 Sarett). Reduction with zinc dust and alcohol affords cholestan-3β-ol-6-one (p. 2677 s) (1943 Sarett); the acetate of this hydroxy-ketone is obtained on reduction of the acetate with aluminium amalgam and water in ether (1937 Heilbron) or on refluxing of the acetate with dimethylaniline for 2 hrs. (1940 Jackson). Bromination of the acetate with 1 mol. of bromine in the presence of HBr in glacial acetic acid at 23° (40 hrs.) affords 5α.7α-dibromocholestan-3β-ol-6-one acetate (page 2714 s) (1938 Heilbron). When the acetate is refluxed with methanolic KOH for 2 hrs., cholestane-3β.7-diol-6-one (p. 2759 s) is obtained (1937 Heilbron). Treatment of the acetate with silver nitrate in anhydrous pyridine at room temperature or at 100° is without effect, but on refluxing for 5 hrs. it gives the 3-acetate of cholestan-3β-ol-6.7-dione (mono-enolic form, see p. 2954 s) (1937 Heilbron).

Acetate $C_{29}H_{47}O_3Br$. Prisms (light petroleum b. 60–80°), plates (aqueous acetic acid), m. 144–5°; does not decompose on heating to 220°; $[\alpha]_D^{22}$ +41° (chloroform) (1937 Heilbron). Absorption max. in alcohol at 310 mμ (log ε 2.2) (1938 Barr; cf. 1954 Cookson). Very soluble in ether, chloroform, and light petroleum (b. 60–80°), moderately readily soluble in glacial acetic acid and alcohol (1937 Heilbron). — **Fmn.** From the above hydroxy-ketone on heating with acetic anhydride at 100° for 2 hrs. (1943 Sarett). See also under formation of the hydroxy-ketone. — **Rns.** See those of the above hydroxy-ketone.

6α-Bromocholestan-3β-ol-7-one $C_{27}H_{45}O_2Br$. Originally assumed to be the 6β-epimer (1938 Barr); for the true configuration, see 1949 Fieser; 1953, 1954 Corey; 1954 Cookson; 1956 James.

Acetate $C_{29}H_{47}O_3Br$. Leaflets (dil. acetic acid), m. 142–3° (1938 Barr); crystals (methanol), m. 138–141° (1954 Cookson); [α]$_D$ in chloroform —9° (1938 Barr), —6° (1954 Cookson), —10.5° (1956 James). Absorption max. in alcohol at 282 mμ (log ε 1.6) (1938 Barr), in chloroform - abs. alcohol (1:4) at 279 mμ (log ε 1.86) (1954 Cookson). — **Fmn.** Along with the 6β-epimer (below; yield, 17%), from the acetate of cholestan-3β-ol-7-one (p. 2686 s) on treatment with 1 mol. of bromine in chloroform at 20° (yield, 48%); the epimers are separated by crystallization from acetone, the 6β-isomer being the less soluble. From the 6β-epimer (below) on heating with HBr - glacial acetic acid on the steam-bath for 15 min. (1938 Barr). — **Rns.** On refluxing with anhydrous pyridine for at least 8 hrs. it gives Δ^5-cholesten-3β-ol-7-one acetate (p. 2682 s) and $\Delta^{3.5}$-cholestadien-7-one (p. 2454 s). On refluxing with silver nitrate in anhydrous pyridine for 5 hrs. it gives the acetate of cholestan-3β-ol-6.7-dione (monoenolic form; see p. 2954 s) (1938 Barr). Bromination with 1 mol. bromine in the presence of HBr in ether - glacial acetic acid at room temperature affords the acetate of 6α.8β-dibromocholestan-3β-ol-7-one (p. 2715 s) (1938 Barr; cf. 1940 Jackson; 1954 Cookson).

6β-Bromocholestan-3β-ol-7-one $C_{27}H_{45}O_2Br$. Originally assumed to be the 6α-epimer (1938 Barr); for the true configuration, see 1949 Fieser; 1953, 1954 Corey; 1954 Cookson; 1956 James.

Acetate $C_{29}H_{47}O_3Br$. Plates (acetone), m. 173–5° (1938 Barr), 178° to 180° (1954 Cookson); [α]$_D$ in chloroform +35° (1938 Barr), +33.5° (1954 Cookson), +37° (1956 James). Absorption max. in alcohol at 313 mμ (log ε 2.2) (1938 Barr), in chloroform-abs. alcohol (1:4) at 309 mμ (log ε 2.26) (1954 Cookson). — **Fmn.** See under that of the 6α-epimer (above). — **Rns.** Rearranges to the 6α-epimer on heating with HBr - glacial acetic acid on the steam-bath for 15 min. Reacts on refluxing with anhydrous pyridine as does the 6α-epimer, but more easily (6 hrs.). On refluxing with silver nitrate in anhydrous pyridine for 5 hrs. it gives the acetate of Δ^5-cholesten-3β-ol-7-one (p. 2682 s). Gives cholestane-3β.6-diol-7-one (p. 2759 s) on refluxing with methanolic KOH for 2 hrs. Bromination affords the same product as that of the 6α-isomer (1938 Barr).

References, p. 2710 s

1936 Butenandt, Schramm, *Ber.* **69** 2289, 2297, 2298.
　CIBA PHARMACEUTICAL PRODUCTS (Miescher, Wettstein), *U.S. Pat.* 2260328 (issued 1941);
　C. A. **1942** 873; *Ch. Ztbl.* **1944** II 246.
1937 Butenandt, Schramm, Kudszus, *Ann.* **531** 176, 199, 200.
　a CIBA*, *Swiss Pat.* 198621 (issued 1938); *Ch. Ztbl.* **1939** I 2643.
　b CIBA*, *Swiss Pat.* 222441 (issued 1942); *Ch. Ztbl.* **1943** I 2113; cf. also CIBA* (Ruzicka),
　Swed. Pat. 98368 (1938, issued 1940); *Ch. Ztbl.* **1940** II 2185.
　Heilbron, Jones, Spring, *J. Ch. Soc.* **1937** 801.
　Ruzicka, Bosshard, *Helv.* **20** 328, 331, 332.
　SCHERING A.-G. (Inhoffen), *Ger. Pat.* 720015 (issued 1942); *C. A.* **1943** 2520; *Ch. Ztbl.* **1942** II
　1037.
1937/38 Dannenberg, *Über einige Umwandlungen des Androstandions und Testosterons*, Thesis,
　Danzig, pp. 33, 34.
1938 Barr, Heilbron, Jones, Spring, *J. Ch. Soc.* **1938** 334.
　Butenandt, Dannenberg, *Ber.* **71** 1681, 1684, 1685.
　Heilbron, Jackson, Jones, Spring, *J. Ch. Soc.* **1938** 102, 105.
　Jackson, Jones, *J. Ch. Soc.* **1938** 1406.
1939 Dannenberg, *Abhandl. Preuss. Akad. Wiss. Math.-naturwiss. Kl.* **1939** No. 21, pp. 13, 15.
1940 Butenandt, Dannenberg, *Ber.* **73** 206.
　Jackson, Jones, *J. Ch. Soc.* **1940** 659, 662.
　Pearlman, Wintersteiner, *J. Biol. Ch.* **132** 605, 609, 610.
　SCHERING A.-G., *Fr. Pat.* 867697 (issued 1941); *Ch. Ztbl.* **1942** I 2561.
1943 Inhoffen, Zühlsdorff, *Ber.* **76** 233, 235, 240–244.
　Sarett, Chakravorty, Wallis, *J. Org. Ch.* **8** 405, 407, 415, 416.
　SCHERING A.-G., *Fr. Pat.* 899737 (issued 1945); *Ch. Ztbl.* **1946** I 1432, 1433.
1944 Butenandt, Ruhenstroth-Bauer, *Ber.* **77** 397, 400.
1945 Birch, Jaeger, Robinson, *J. Ch. Soc.* **1945** 582, 584, 585.
1946 Meystre, Wettstein, *Experientia* **2** 408.
　Wilds, Djerassi, *J. Am. Ch. Soc.* **68** 2125, 2129, 2130.
1947 a Djerassi, Scholz, *Experientia* **3** 107.
　b Djerassi, Scholz, *J. Am. Ch. Soc.* **69** 2404, 2407, 2408.
　c Djerassi, *J. Org. Ch.* **12** 823, 827.
1949 Fieser, Fieser, *Natural Products Related to Phenanthrene* 3rd Ed., New York, pp. 263, 268, 270.
1950 Djerassi, Rosenkranz, Romo, Kaufmann, Pataki, *J. Am. Ch. Soc.* **72** 4534, 4536, 4538.
　Hershberg, Rubin, Schwenk, *J. Org. Ch.* **15** 292, 295.
1952 Jones, Ramsay, Herling, Dobriner, *J. Am. Ch. Soc.* **74** 2828.
1953 Corey, *Experientia* **9** 329.
1954 Cookson, *J. Ch. Soc.* **1954** 282.
　Corey, *J. Am. Ch. Soc.* **76** 175.
　James, Shoppee, *J. Ch. Soc.* **1954** 4224.
1955 Fieser, Romero, Fieser, *J. Am. Ch. Soc.* **77** 3305.
1956 Ellis, Petrow, *J. Ch. Soc.* **1956** 1179.
　James, Shoppee, *J. Ch. Soc.* **1956** 1064.
1958 M. Fieser, L. F. Fieser, *private communication.*

β. Dihalogeno-monohydroxy-monoketosteroids (CO, OH, and Hlg in the Ring System)

2.2 - Dibromoandrostan - 17β - ol - 3-one,　2.2 - Dibromodihydrotestosterone

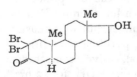

$C_{19}H_{28}O_2Br_2$. For the configuration at C-17, see pp. 1363 s, 1364 s. — **Fmn.** Its hexahydrobenzoate is obtained from the corresponding ester of androstan-17β-ol-3-one (p. 2597 s) on treatment with 2 mols. bromine in glacial acetic acid at 23°; the bromination

* Société pour l'industrie chimique à Bâle; Gesellschaft für chemische Industrie in Basel.

proceeds rapidly, but is immediately followed by rearrangement to the ester of 2α.4α-dibromoandrostan-17β-ol-3-one (below); when glacial acetic acid distilled from permanganate (and then containing water) is used, the rates of bromination and rearrangement are decreased (1947 Djerassi; cf. 1946 Wilds; 1943 Inhoffen); the acetate and benzoate are similarly formed (1947 Djerassi and 1943 Inhoffen, resp.).

Rns. Rearrangement of the esters to the corresponding esters of 2α.4α-dibromoandrostan-17β-ol-3-one (see also above) occurs on warming in glacial acetic acid for 3–5 min. (yields, up to 80%) (1947 Djerassi; cf. 1946 Wilds) or on treatment with HBr · glacial acetic acid in chloroform at 25⁰ for 20 hrs. (1943 Inhoffen); the hexahydrobenzoate is unchanged on warming quickly in glacial acetic acid and cooling immediately, or on heating in glacial acetic acid for 10 min. in the presence of sodium acetate or 30% H_2O_2, or on shaking with glacial acetic acid at room temperature for 120 hrs. in the presence or absence of sodium acetate or 30% H_2O_2 (1947 Djerassi). On refluxing the esters with collidine for 15 min. they give the corresponding esters of 2-bromo-Δ¹-androsten-17β-ol-3-one (p. 2702 s) (1943 Inhoffen; 1946 Wilds).

Acetate $C_{21}H_{30}O_3Br_2$. Needles (alcohol), m. 139–140⁰ cor.; $[\alpha]_D^{23}$ +106⁰ (chloroform) (1947 Djerassi). — **Fmn.** and **Rns.** See above.

Benzoate $C_{26}H_{32}O_3Br_2$. Needles (glacial acetic acid), m. 178–180⁰ dec. (1943 Inhoffen). — **Fmn.** and **Rns.** See above.

Hexahydrobenzoate $C_{26}H_{38}O_3Br_2$. Needles (alcohol - chloroform), m. 165.5⁰ to 166.5⁰ cor. dec. (1946 Wilds); m. p.'s (with decomposition) ranging from 152–4⁰ cor. to 185–6⁰ cor. have been observed (1947 Djerassi); $[\alpha]_D$ +110⁰ (chloroform or glacial acetic acid) (1947 Djerassi; 1946 Wilds). Slightly soluble in glacial acetic acid at room temperature (1947 Djerassi). — **Fmn.** and **Rns.** See above.

2α.4α-Dibromoandrostan-17β-ol-3-one, 2α.4α-Dibromodihydrotestosterone

$C_{19}H_{28}O_2Br_2$. For the configuration at C-17, see pp. 1363 s, 1364 s; α-configuration at C-2 and C-4 ascribed by analogy with 2α.4α-dibromocholestan-3-one (p. 2495 s) similarly obtained (Editor). — **Fmn.** For formation of its esters from the esters of androstan-17β-ol-3-one via the esters of 2.2-dibromoandrostan-17β-ol-3-one, see the latter (above); the esters may be directly obtained from the corresponding esters of the hydroxyketone on treating with 2 mols. bromine in warm glacial acetic acid, followed by adding a little 4 N HBr and keeping for 2.5 hrs. at room temperature (yield of the hexahydrobenzoate, 85–90%) (1947 Djerassi; cf. 1940, 1943 Inhoffen; 1941 b SCHERING A.-G.; 1946 Wilds). The esters are also obtained from the corresponding esters of the 2.2-dibromo-isomer (above) on warming in glacial acetic acid for 3 min. or longer, followed by cooling and keeping for several hrs. at 10⁰ or by addition of anhydrous sodium acetate or 30% H_2O_2 (yields of hexahydrobenzoate, up to 80%) (1947 Djerassi; cf. 1946 Wilds), or on treating with HBr · glacial acetic acid at 25⁰ for 20 hrs. (1943 Inhoffen).

E.O.C. XIV s. 187

References, p. 2716 s

Rns. On boiling the acetate with pyridine for 6 hrs. it gives a **compound**
$C_{26}H_{34}O_3NBr$ [crystals (acetone - petroleum ether), m. 228–9°; readily soluble
in water, methanol, ethanol, and chloroform, soluble in acetone, almost insoluble in ether and petroleum ether] (1940 Inhoffen), which according to
Inhoffen (1940) is the *hydrobromide of 2-(4-pyridyl)-Δ⁴-androsten-17β-ol-3-one
acetate or 4-(4-pyridyl)-Δ¹-androsten-17β-ol-3-one acetate*, but which according
to Fieser (1949) may also be *N-(17β-acetoxy-3-keto-Δ⁴-androsten-2-yl)-pyridi-
nium bromide or N-(17β-acetoxy-3-keto-Δ¹-androsten-4-yl)-pyridinium bromide.*
On refluxing the esters with collidine for $1^1/_2$–5 min. they give the corresponding esters of 2α-bromo-Δ⁴-androsten-17β-ol-3-one (2α-bromotestosterone,
p. 2703 s) (1943 Inhoffen; 1943 Schering A.-G.; 1947 Djerassi); on refluxing
the esters with collidine for 45 min., they are converted into the corresponding
esters of Δ¹·⁴-androstadien-17β-ol-3-one (p. 2576 s) and, to a small extent,
into those of Δ⁴·⁶-androstadien-17β-ol-3-one (p. 2577 s) (1940, 1943 Inhoffen;
1936, 1941a, 1945 Schering A.-G.; 1946 Wilds); a small amount of 2-bromo-
Δ¹-androsten-17β-ol-3-one ester (p. 2702 s), formed in this reaction, probably
owes its formation to 2.2-dibromoandrostan-17β-ol-3-one ester present as an
impurity (1946 Wilds). On refluxing of the free hydroxy-ketone (not described) with potassium benzoate in butanol-toluene for 2 hrs., followed by heating
of the resulting oily Δ⁴-androstene-2.17β-diol-3-one 2-benzoate and/or
Δ¹-androstene-4.17β-diol-3-one 4-benzoate at 220–230° under 6.5 mm. pressure for $1^1/_2$ hrs. and then by high vac. distillation, it gives Δ¹·⁴-androstadien-
17β-ol-3-one (p. 2576 s) (1936 Schering A.-G.).

Acetate $C_{21}H_{30}O_3Br_2$. Needles (alcohol), m. 194° dec. (1940 Inhoffen; 1941b
Schering A.-G.), 187–8° cor. dec. (1947 Djerassi); $[\alpha]_D^{23}$ —11° (chloroform)
(1947 Djerassi). — **Fmn.** and **Rns.** See above.

Benzoate $C_{26}H_{32}O_3Br_2$. Needles (glacial acetic acid), crystals (chloroform-
alcohol), m. 193–5° dec. (1943 Inhoffen). — **Fmn.** and **Rns.** See above.

Hexahydrobenzoate $C_{26}H_{38}O_3Br_2$. Needles (chloroform-alcohol), m. 178–180°
dec. (1943 Inhoffen); m. p.'s (with decomposition) ranging from 153–4° cor.
to 175–6° cor. have been observed; $[\alpha]_D^{23}$ —6° (chloroform or glacial acetic acid)
(1947 Djerassi; cf. 1946 Wilds).

5.6 - Dibromoandrostan - 17β - ol - 3 - one,   5.6 - Dibromodihydrotestosterone

$C_{19}H_{28}O_2Br_2$ (not described in detail). For probable
configuration at C-5 and C-6, cf. 5α.6β-dibromochole-
stan-3β-ol (p. 1889 s); for configuration at C-6, cf.
also 6-bromotestosterone (p. 2705 s) into which the
dibromo-compound may be converted (see below). —
Fmn. Its esters are obtained from the corresponding
17-esters of Δ⁵-androstene-3β.17β-diol (p. 1999 s) on bromination with 1 mol.
bromine in glacial acetic acid, followed by oxidation of the resulting dibromide
with CrO_3 in glacial acetic acid at room temperature (1935, 1936 Butenandt;
1935a, b, 1937a, b Ruzicka; 1935 Schering-Kahlbaum A.-G.; 1935a, b
Ciba). — **Rns.** The esters are converted into the corresponding esters of

Δ^4-androsten-17β-ol-3-one (testosterone, p. 2580 s) on refluxing with sodium iodide in methanol or ethanol for 3 hrs. or on heating with zinc dust in glacial acetic acid on the water-bath for 10 min. (1935 Butenandt; 1935 a, b Ruzicka; 1935 SCHERING-KAHLBAUM A.-G.; 1935 a CIBA), whereas on refluxing with zinc dust in methanol for 30–45 min. the corresponding esters of Δ^5-androsten-17β-ol-3-one (p. 2595 s) are obtained (1936 Butenandt; 1937 b Ruzicka; 1935 b CIBA); for a **by-product** $C_{22}H_{30}O_4$ [needles (dil. acetone), m. 180° after sintering at 165–170°] of the last-mentioned reaction (using the acetate), see 1936 Butenandt. On refluxing of its esters with anhydrous sodium acetate in abs. alcohol they give the corresponding esters of 6-bromo-Δ^4-androsten-17β-ol-3-one (p. 2705 s) (1937 a Ruzicka; 1937 CIBA).

Other 5.6-dibromoandrostan-17-ol-3-one compounds, which are similarly obtained by oxidation of the dibromides of Δ^5-androstene-3.17-diols (or their 17-esters) and give Δ^4-androsten-17-ol-3-ones (or their esters) on debromination with zinc dust in glacial acetic acid, are: **5.6-dibromoandrostan-17α-ol-3-one** (see 17-epi-testosterone, p. 2579 s), **17α-methyl-, 17α-ethyl-,** and **17α-benzyl-5.6-dibromoandrostan-17β-ol-3-one** [see 17-methyltestosterone (p. 2645 s), 17-ethyltestosterone (p. 2651 s), and 17-benzyltestosterone (page 2698 s), resp.].

5.6-Dibromo-androstan (or-ætiocholan)-3β-ol-17-one, 5-Dehydroepiandrosterone dibromide $C_{19}H_{28}O_2Br_2$ (not described in detail).

For its formation and its reactions, see 5-dehydroepiandrosterone under "formation" (p. 2610 s), "oxidation" (p. 2611 s), and "halogenation" (p. 2613 s).

4.5-Dibromocholestan-3β-ol-6-one, Δ^4-Cholesten-3β-ol-6-one dibromide

$C_{27}H_{44}O_2Br_2$. — **Fmn.** Its acetate is obtained from the acetate of Δ^4-cholesten-3β-ol-6-one (p. 2677 s) on treatment with 2 mols. bromine in glacial acetic acid at 18°. — **Rns.** The acetate is reconverted into the starting acetate on boiling with potassium iodide in acetone for 1 min. or on refluxing with anhydrous pyridine for 2$^1/_2$ hrs. The acetate gives the acetate of 4-bromo-Δ^4-cholesten-3β-ol-6-one (p. 2707 s) on boiling in ether or on refluxing with potassium acetate in glacial acetic acid. On treatment of the acetate with methanolic sodium methoxide at 20° for 20 hrs., 4.5-dimethoxycholestan-3β-ol-6-one (p. 2772 s) is obtained (1938 Jackson).

Acetate $C_{29}H_{46}O_3Br_2$. Plates (ether-aqueous acetic acid), m. 81–82° dec.; its absorption curve in alcohol shows an inflexion at 310 mμ (log ε 2.0). Stable in air, but readily decomposed in warm solution (1938 Jackson). — **Fmn.** and **Rns.** See above.

187*

5α.7α - Dibromocholestan - 3β - ol - 6 - one, *5.7 - Dibromocholestan - 3β - ol-6-one*

$C_{27}H_{44}O_2Br_2$. For configurations at C-5 and C-7, see 1954 Corey; 1954, 1955 Cookson.

Acetate $C_{29}H_{46}O_3Br_2$. Needles (aqueous acetone), m. 152°; $[\alpha]_D^{20}$ —140° (chloroform); dimorphic, separating on prolonged standing from acetic acid or acetone in rhombic crystals; the two forms are readily interconvertible (1938 Heilbron). Absorption max. in alcohol at 340 mμ (log ε 2.2) (1938 Barr; cf. 1955 Cookson). — **Fmn.** From the acetate of cholestan-3β-ol-6-one (p. 2677 s) on treatment with 2 mols. bromine and a little HBr in glacial acetic acid at 23°; yield 42%. From the acetates of 5α-bromocholestan-3β-ol-6-one (p. 2707 s) and 7α-bromocholestan-3β-ol-6-one (p. 2708 s) on treatment with 1.2 mols. bromine and a little HBr in glacial acetic acid at 30° or 23°, resp.; yields, 35% and 23%, resp. (1938 Heilbron). — **Rns.** Is reduced to the acetate of 7α-bromocholestan-3β-ol-6-one (p. 2708 s) on treatment with 2 mols. HBr in glacial acetic acid at 22° for 24 hrs. (1938 Heilbron). On refluxing with anhydrous pyridine for $6\frac{1}{2}$ hrs. it gives 55% yield of the acetate of $\Delta^{2.4}$-cholestadien-3-ol-6-one (p. 2676 s) and 20% yield of a compound $C_{29}H_{46}O_4$ (m. 227–9°), assumed to be the 3-acetate of Δ^4-cholestene-3β.7-diol-6-one (1938 Heilbron), but which is actually 3β-acetoxy-B-nor-Δ^5-cholestene-6-carboxylic acid (p. 1380 s) (1941 a, b Woodward). Is unchanged after heating with sodium acetate in alcohol (1938 Heilbron).

5α.7β - Dibromocholestan - 3β - ol - 6 - one, *5'.7 - Dibromocholestan - 3β-ol-6-one*

$C_{27}H_{44}O_2Br_2$. For configurations at C-5 and C-7, see 1954 Corey; 1955 Cookson. — Crystals (ethyl acetate - methanol), dec. 117–9°; $[\alpha]_D^{20}$ —50° (chloroform). — **Fmn.** From its acetate (below) on refluxing with conc. HCl in alcohol for 30 min. — **Rns.** Oxidation with CrO_3 in strong acetic acid at room temperature yields 5α.7β-dibromocholestane-3.6-dione (p. 2930 s). May be reconverted into its acetate by heating with acetic anhydride at 100° for 30 min. and then refluxing for 15 min. (1943 Sarett).

Acetate $C_{29}H_{46}O_3Br_2$. Plates (glacial acetic acid), m. 129°; $[\alpha]_D^{20}$ —51° (chloroform) (1938 Heilbron). Absorption max. in alcohol at 305 mμ (log ε 2.1) (1938 Barr; cf. 1955 Cookson). More soluble in the common organic solvents than the 7-epimer (above) (1938 Heilbron). — **Fmn.** A mixture (leaflets, m. 123–8°) with a little of the 7-epimer is obtained in 56% yield from the acetate of cholestan-3β-ol-6-one (p. 2677 s) in glacial acetic acid containing some 40% HBr on careful addition of 2 mols. bromine with shaking at room temperature, avoiding the presence of free bromine (1941 b Woodward; 1938 Heilbron). From the acetate of 5α-bromocholestan-3β-ol-6 one (p. 2707 s) on heating with 1.1 mols. bromine and HBr in glacial acetic acid on the steam-bath for 15 min.; yield, 8.5% (1938 Heilbron). For its formation from the free hydroxy-ketone, see above. — **Rns.** Is stable to HBr at room temperature, but on heating with 1.1 mols. HBr in glacial acetic acid on the steam-bath for 15 min. it gives the acetate of 7α-bromocholestan-3β-ol-6-one (p. 2708 s) (1938 Heilbron). Reacts on refluxing with anhydrous pyridine as does the 7-epimer (p. 2714 s) (1938 Heilbron; 1941 a, b Woodward). On refluxing with fused potassium acetate in dry butanol for 7 hrs., followed by repeated crystallization from methanol, it gives 7-methoxy-$\Delta^{2.4.7}$-cholestatrien-6-one (p. 2680 s) (1938 Heilbron). Gives the diacetate of 7β-bromocholestane-3β.5α-diol-6-one (p. 2763 s) on heating with fused potassium acetate in glacial acetic acid on the steam-bath for 1½ hrs.* or on refluxing with fused sodium acetate in abs. alcohol for 1 hr. (1938 Heilbron); in the latter reaction there are also obtained a small amount of a compound $C_{31}H_{50}O_4$ (m. 119–120°) assumed by Heilbron (1938) to be 3β-acetoxy-7-ethoxy-Δ^4-cholesten-6-one, but being actually the ethyl ester of 3β-acetoxy-B-nor-Δ^5-cholestene-6-carboxylic acid (p. 1380 s) (1941 a, b Woodward), and an even smaller amount of cholestan-3β-ol-6.7-dione acetate (p. 2954 s) (1938 Heilbron).

5.6-Dibromocholestan-3β-ol-7-one, 7-Ketocholesterol dibromide $C_{27}H_{44}O_2Br_2$.

Needles (ether - abs. alcohol), m. 124° to 125° dec.; $[\alpha]_D^{17}$ —5.6° (chloroform). Soluble in ether and carbon disulphide, practically insoluble in alcohol and light petroleum (1946 Barnett). — **Fmn.** From 7-ketocholesterol and bromine in carbon disulphide at 20° (1946 Barnett); the acetate is similarly formed (in glacial acetic acid) (1940 Jackson). — **Rns.** On shaking with CrO_3 and aqueous acetic acid in benzene for 5 hrs., followed by refluxing with zinc wool in glacial acetic acid for 1 hr., it gives Δ^5-cholestene-3.7-dione (mono-enolic form; see p. 2680 s) (1946 Barnett). On boiling the acetate with potassium iodide in acetone for 1 min. it gives the starting 7-ketocholesteryl acetate. Gives $\Delta^{3.5}$-cholestadien-7-one (p. 2454 s) on refluxing the acetate with dimethylaniline for 2 hrs. On refluxing of the acetate with fused potassium acetate in glacial acetic acid it gives a compound with absorption max. at 240 and 320 mμ ($E_{1cm}^{1\%}$ 110 and 1.4, resp.), probably an unsaturated bromoketone (1940 Jackson).

Acetate $C_{29}H_{46}O_3Br_2$. Needles (glacial acetic acid), m. 146–7° dec.; absorption max. in alcohol at 314 mμ (log ε 2.2) (1940 Jackson).— **Fmn.** and **Rns.** See above.

6α.8β-Dibromocholestan-3β-ol-7-one $C_{27}H_{44}O_2Br_2$.

Originally assumed to be the 6.6-dibromo-compound (1940 Jackson); for the 6α.8β-structure, see 1954 Cookson.

Acetate $C_{29}H_{46}O_3Br_2$. Plates (acetone or light petroleum b. 40–60°), m. 176–7°; $[\alpha]_D^{19}$ +38° (chloroform);

---

* According to Cookson (1955), the reaction takes place only after 7 hrs.' boiling.

absorption max. (in alcohol?) at 304 mμ (log ε 2.2) (1938 Barr). — **Fmn.**
From the acetate of cholestan-3β-ol-7-one (p. 2686 s) on treatment with
2 mols. bromine in chloroform at 20°; yield, 47%. From the acetates of
6α-bromo- and 6β-bromocholestan-3β-ol-7-ones (p. 2709 s) on treatment
with 1 mol. bromine in the presence of HBr for 48 hrs. at room temperature
in glacial acetic acid - ether or glacial acetic acid, resp.; yields 70% and 32%,
resp. (1938 Barr). — **Rns.** On refluxing with dimethylaniline for 3 hrs. it
is reconverted into the acetate of cholestan-3β-ol-7-one (1940 Jackson). Gave
only unworkable gels on treatment with pyridine, silver nitrate in pyridine,
methanolic KOH, and potassium acetate in alcohol. Does not react with
o-phenylenediamine in alcohol (1938 Barr).

1935 Butenandt, Hanisch, *Ber.* **68** 1859, 1861; *Z. physiol. Ch.* **237** 89, 92, 96.
    a CIBA, *Swiss Pats.* 210762–210767 (issued 1940); *C. A.* **1941** 5650; *Ch. Ztbl.* **1941** I 3409.
    b Ruzicka, *Swiss Pats.* 210768, 210769 (issued 1940); *C. A.* **1941** 5650; *Ch. Ztbl.* **1941** I 3409.
    a Ruzicka, Wettstein, *Helv.* **18** 1264, 1274.
    b Ruzicka, Wettstein, Kägi, *Helv.* **18** 1478, 1482.
    SCHERING-KAHLBAUM A.-G., *Brit. Pat.* 464396 (issued 1937); *C. A.* **1937** 6418; *Ch. Ztbl.*
      1937 II 3041.
1936 Butenandt, Hanisch, *Ber.* **69** 2773, 2775.
    SCHERING A.-G. (Inhoffen), *Ger. Pat.* 722943 (issued 1942); SCHERING-KAHLBAUM A.-G., *Brit.*
      Pat. 500353 (1937, issued 1939); *Fr. Pat.* 835524 (1937, issued 1938); *Ch. Ztbl.* **1939** I
      5010, 5011.
1937 CIBA, *Swiss Pat.* 198621 (issued 1938); *Ch. Ztbl.* **1939** I 2643.
    a Ruzicka, Bosshard, *Helv.* **20** 328, 331, 332.
    b Ruzicka, Goldberg, Bosshard, *Helv.* **20** 541, 543, 546.
1938 Barr, Heilbron, Jones, Spring, *J. Ch. Soc.* **1938** 334.
    Heilbron, Jackson, Jones, Spring, *J. Ch. Soc.* **1938** 102.
    Jackson, Jones, *J. Ch. Soc.* **1938** 1406.
1940 Inhoffen, Zühlsdorff, Huang-Minlon, *Ber.* **73** 451, 455, 456.
    Jackson, Jones, *J. Ch. Soc.* **1940** 659.
1941 a SCHERING A.-G., *Fr. Pat.* 869676 (issued 1942); *Ch. Ztbl.* **1942** II 1266; SCHERING CORP.
      (Inhoffen, Zühlsdorff), *U.S. Pat.* 2422904 (issued 1947); *C. A.* **1947** 6288; *Ch. Ztbl.* **1948** II
      237.
    b SCHERING A.-G., *Swiss Pat.* 228433 (issued 1943); *Ch. Ztbl.* **1945** I 196.
    a Woodward, *J. Am. Ch. Soc.* **63** 1123, 1125.
    b Woodward, Clifford, *J. Am. Ch. Soc.* **63** 2727.
1943 Inhoffen, Zühlsdorff, *Ber.* **76** 233, 238–240, 242, 243, 245.
    Sarett, Chakravorty, Wallis, *J. Org. Ch.* **8** 405, 407, 413.
    SCHERING A.-G., *Fr. Pat.* 899737 (issued 1945); *Ch. Ztbl.* **1946** I 1432, 1433.
1945 SCHERING A.-G., *Bios Final Report* No. 449, pp. 273, 277; *Fiat Final Report* No. 996, pp. 78, 82.
1946 Barnett, Ryman, Smith, *J. Ch. Soc.* **1946** 526.
    Wilds, Djerassi, *J. Am. Ch. Soc.* **68** 2125, 2128–2130.
1947 Djerassi, Scholz, *J. Am. Ch. Soc.* **69** 2404, 2406–2408.
1949 Fieser, Fieser, *Natural Products Related to Phenanthrene*, 3rd Ed., New York, pp. 268–270.
1954 Cookson, *J. Ch. Soc.* **1954** 282.
    Corey, *J. Am. Ch. Soc.* **76** 175.
1955 Cookson, Dandegaonker, *J. Ch. Soc.* **1955** 352.

5.6.16 - Tribromoandrostan - 3β - ol - 17 - one $C_{19}H_{27}O_2Br_3$.   See 16-bromo-Δ^5-androsten-3β-ol-17-one 5.6-dibromide, p. 2705 s.

13. NITRO-MONOHYDROXY-MONOKETOSTEROIDS (CO AND OH IN THE RING SYSTEM)

16-(m-Nitrobenzylidene)-œstrone methyl ether $C_{26}H_{27}O_4N$. Yellowish crystals (methanol), m. 187–8° cor. — **Fmn.** From œstrone methyl ether (p. 2555 s) by the action of excess m-nitrobenzaldehyde and some methanolic KOH in alcohol-dioxane at room temperature (40 hrs.); yield, 54% (1943 Wettstein).

16-(m-Nitrobenzylidene)-Δ⁵-androsten-3β-ol-17-one, 16-(m-Nitrobenzylidene)-5-dehydroepiandrosterone $C_{26}H_{31}O_4N$. Slightly yellowish crystals (acetone-hexane), m. 248.5–250° cor. — **Fmn.** From 5-dehydroepiandrosterone (page 2609 s) by the action of excess m-nitrobenzaldehyde and some methanolic KOH in alcohol at room temperature (20 hrs.); yield, 68% (1943 Wettstein).

16-(m-Nitrobenzylidene)-androstan-3α-ol-17-one, 16-(m-Nitrobenzylidene)-androsterone $C_{26}H_{33}O_4N$. Yellowish crystals, m. 189–190° cor. — **Fmn.** From androsterone (p. 2629 s) by the action of excess m-nitrobenzaldehyde and some methanolic KOH in alcohol (40 hrs.) (1943 Wettstein).

1943 Wettstein, Miescher, *Helv.* **26** 631, 640, 641.

14. AMINO-MONOHYDROXY-MONOKETOSTEROIDS (CO AND OH IN THE RING SYSTEM)

4′-(N-Methyl-anilinomethylene)-7-methoxy-3′-keto-1.2-cyclopentano-1.2.3.4-tetrahydrophenanthrene $C_{26}H_{25}O_2N$ (I, R = H). dl-cis-Form, **16-(N-Methylanilino-methylene)-dl-18-nor-isoequilenin methyl ether.** Yellow needles (toluene), m. 206°. — **Fmn.** From 16-hydroxy methylene-dl-18-nor-isoequilenin methyl ether (p. 2719 s) on heating with N-methylaniline in toluene on the steam-bath for 1½ hrs.; yield, 74%. — **Rns.** On refluxing with sodamide and methyl iodide in benzene-toluene it gives 16-(N-methyl-anilinomethylene)-dl-isoequilenin methyl ether (below) (1945 Birch).

2-Methyl-4'-(N-methyl-anilinomethylene)-7-methoxy-3'-keto-1.2-cyclopentano-1.2.3.4-tetrahydrophenanthrene $C_{27}H_{27}O_2N$ (I, p. 2717 s; R = CH_3).

dl-cis-Form, 16-(N-Methyl-anilinomethylene)-dl-isoequilenin methyl ether. Lemon-yellow, triclinic prisms (alcohol), m. 165°, and colourless needles, m. 152°, which are convertible into the higher-melting form. — **Fmn.** From the preceding compound on refluxing in benzene-toluene, first with sodamide, then with methyl iodide; yield, 48%. — **Rns.** Gives dl-isoequilenin methyl ether (p. 2538 s) on refluxing with aqueous alcoholic H_2SO_4, followed by refluxing with 10% aqueous alcoholic (1:1) KOH (1945 Birch).

20-Aminopregnan-3α-ol-11-one $C_{21}H_{35}O_2N$. Crystals (dil. alcohol), m. 185–6°

cor. (1946 Sarett; 1946a MERCK & Co., INC.). — **Fmn.** The O-acetate is obtained from the azide of 3α-acetoxy-11-keto-bisnorcholanic acid on heating with aqueous acetic acid on the water-bath, followed by removal of the free acid, formed at the same time, with aqueous KOH at 10° (yield, 80% corrected for recovered acid); the acetate gives the free amino-hydroxy-ketone on alkaline saponification (1946 Sarett; 1946a MERCK & Co., INC.). — **Rns.** On heating the O-acetate with sodium nitrite, aqueous pyridine, and pyridine hydrochloride on the steam-bath it gives a mixture of the acetates of Δ^{16}-, $\Delta^{17(20)}$-, and Δ^{20}-pregnen-3α-ol-11-ones (p. 2654 s) and a smaller amount of the 3-acetates of the two 20-epimeric pregnane-3α.20-diol-11-ones (p. 2722 s) (1946, 1949 Sarett; 1946b MERCK & Co., INC.). For acetylation, see below.

O-Acetate $C_{23}H_{37}O_3N$. Crystals (abs. ether), m. 163.5–165.5° cor. (1946 Sarett; 1946a MERCK & Co., INC.). — **Fmn.** and **Rns.** See above.

N-Acetyl derivative $C_{23}H_{37}O_3N$ (not analysed), m. 219.5° cor. — **Fmn.** From the O.N-diacetyl derivative (below) on alkaline saponification (1946 Sarett; 1946a MERCK & Co., INC.).

O.N-Diacetyl derivative $C_{25}H_{39}O_4N$ (not analysed), m. 235° cor. — **Fmn.** From the free amino-hydroxy-ketone or its O-acetate on acetylation with acetic anhydride in pyridine (1946 Sarett; 1946a MERCK & Co., INC.). — **Rns.** See the preceding compound.

1945 Birch, Jaeger, Robinson, *J. Ch. Soc.* **1945** 582, 585, 586.
1946 a MERCK & Co., INC. (Sarett), *U.S. Pat.* 2492188 (issued 1949); *C. A.* **1950** 3043; *Ch. Ztbl.*
      1952 1211.
      b MERCK & Co., INC. (Sarett), *U.S. Pat.* 2540964 (issued 1951); *C. A.* **1951** 7159; *Brit. Pat.*
      630103 (issued 1949); *C. A.* **1950** 7891.
      Sarett, *J. Biol. Ch.* **162** 601, 614–616.
1949 Sarett, *J. Am. Ch. Soc.* **71** 1165, 1167.

15. DIHYDROXY-MONOKETOSTEROIDS WITH CO AND ONE OH IN THE RING SYSTEM

4'-Hydroxymethylene-7-methoxy-3'-keto-1.2-cyclopentano-1.2.3.4-tetrahydro-

phenanthrene (I) and/or **4'-For-myl-7-methoxy-3'-keto-1.2-cyclo-pentano-1.2.3.4-tetrahydrophen-anthrene** (II), $C_{19}H_{18}O_3$.

dl-cis-Form, **16-Hydroxymethylene-dl-18-nor-isoequilenin methyl ether** (I) and/or **16-Formyl-dl-18-nor-isoequilenin methyl ether** (II). Colourless plates (aqueous acetone), m. 145–6° dec. Gives a purplish blue coloration with alcoholic $FeCl_3$. — **Fmn.** From dl-18-nor-isoequilenin methyl ether (p. 2518 s) on stirring with ethyl formate and sodium in benzene under nitrogen for 7 hrs. at room temperature; crude yield, 95 %. — **Rns.** Gives 16-(N-methyl-anilinomethylene)-dl-18-nor-isoequilenin methyl ether (p. 2717 s) with N-methylaniline in toluene on the steam-bath (1945 Birch).

16-Hydroxymethylene-œstrone (III) and/or **16-Formylœstrone** (IV), $C_{19}H_{22}O_3$. Solid, m. ca. 225°. Gives a water-soluble sodium salt. — **Fmn.** From œstrone (p. 2543 s) on shaking with amyl formate and sodium in abs. ether for several hrs., followed by acidification (1939 HOFFMANN-LA ROCHE).

3-Methyl ether $C_{20}H_{24}O_3$. Scales (dil. acetone), m. 170–171° (1936 Bardhan), 165° (1938 Litvan), 165–6° cor. (vac.) (1942 Goldberg). Insoluble in aqueous $NaHCO_3$, readily soluble in dil. NaOH; gives a violet coloration with $FeCl_3$ in alcohol (1936 Bardhan). — **Fmn.** From œstrone methyl ether (p. 2555 s), ethyl formate, and sodium in dry benzene (15 hrs.); almost quantitative yield (1936 Bardhan). — **Rns.** On gently warming with ca. 20 % NaOH, followed by addition of hydroxylamine hydrochloride, heating on the steam-bath for 15 min., and heating of the resulting pasty mass with excess 33 % KOH on the steam-bath, until no more ammonia is evolved, it gives the dicarboxylic acid (V) which on distillation with acetic anhydride regenerates œstrone methyl ether (1936 Bardhan; 1938 Litvan); a small amount of an alkali-insoluble product is obtained along with the dicarboxylic acid (1938 Litvan).

17 (β?)-Hydroxymethyl-Δ⁴-androsten-17 (α?)-ol-3-one $C_{20}H_{30}O_3$. For probable configuration at C-17, see 1955 Sondheimer. — Crystals (ethyl acetate), m. 238° cor.; $[\alpha]_D^{20} +51°$ (alcohol). — **Fmn.** From 17-methylene-Δ⁴-andro-sten-3-one (p. 2405 s) on treatment with OsO_4 in abs. ether at —10°, followed by boiling with Na_2SO_3 in aqueous alcohol for 2 hrs.; crude yield, 90 % (1939 Miescher; cf. 1955 Sondheimer).

17-Hydroxymethyl-androstan-17-ol-3-one $C_{20}H_{32}O_3$. One of the two possible C-17 epimers has been isolated as crystals (petroleum ether - chloroform), m. 194–9° cor., $[\alpha]_D^{20} +$ 4.0°±2° (chloroform); a mixture m. 175–187° (ether - petroleum ether), $[\alpha]_D^{20} + 22°$ (chloroform) was obtained at the same time. — **Fmn.** From 17-methylene-androstan-3-one (of doubtful homogeneity; see p. 2405 s) on treatment with OsO_4 and a little pyridine in ether at room temperature, followed by shaking with mannitol in 1 N KOH. — **Rns.** Gives androstane-3.17-dione (p. 2892 s) on oxidation with lead tetraacetate in glacial acetic acid at room temperature (1947a Ruzicka).

Pregnane-4 (?).20α-diol-3-one $C_{21}H_{34}O_3$. For the doubtful position of the 4-OH group and its possible position at C-2, see 1954 Georg.

Diacetate $C_{25}H_{38}O_5$. Crystals (alcohol), m. 247° (1937 Marker), 250° (1939b Marker). — **Fmn.** From the 4.20-diacetate of pregnane-3.4(?).20α-triol (p. 2107 s) on oxidation with CrO_3 in acetic acid (1939b Marker). From 4β-chloro- and 4β-bromopregnan-20α-ol-3-one acetate (pp. 2513 s, 2514 s) on refluxing with anhydrous potassium acetate in glacial acetic acid for 3 hrs. (1937 Marker; 1938 PARKE, DAVIS & Co.). — **Rns.** It reverts to the pregnanetriol diacetate on refluxing with aluminium isopropoxide in isopropyl alcohol with simultaneous removal (by continuous distillation) of the acetone formed in this reaction, or on shaking with hydrogen and platinum oxide in ether - alcohol under 3 atm. pressure for 4 hrs. (1939b Marker). On hydrogenation in glacial acetic acid in the presence of platinum oxide under 3 atm. pressure for 2 hrs., followed by refluxing with alcoholic KOH for 1 hr. and then by oxidation with CrO_3 in strong acetic acid at 20°, it gives an acid (needles, m. 95–7°), originally (on account of mixed m. p. determinations of the acids and their derivatives) said to be identical with the acid obtained by oxidation of "pregnanetriol-B" (1938 Marker), but which must possess another structure, since "pregnanetriol-B" was later found to be allopregnane-3β.16α.20β-triol (see p. 2112 s) and the acid obtained by oxidation was found to be 3.17.20-triketo-16‖17-allopregnan-16-oic acid (p. 2113 s) (1939a Marker).

17-iso-$\Delta^{4.20}$-Pregnadiene-17β.20-diol-3-one, 20-enol of 17β-Hydroxy-17-iso-progesterone $C_{21}H_{30}O_3$.

20-Acetate $C_{23}H_{32}O_4$. An oily compound obtained from 17-ethynyltestosterone (p. 2648 s) on treatment with mercuric acetate and BF_3-ether in glacial acetic acid - acetic anhydride, followed by distillation in a high vacuum and chromatographic adsorption, is said to possess this structure (1938 SCHERING A.-G.); cf., however, the reactions of 17-ethynyltestosterone (p. 2649 s).

Δ⁴-Pregnene-17α.20β-diol-3-one $C_{21}H_{32}O_3$. Rhombs (chloroform-ethyl acetate) m. 199⁰ (1938 Butenandt); crystals (ethyl acetate), m. 204–5⁰ cor. (1942 Ruzicka). — **Fmn.** From Δ⁴·¹⁷⁽²⁰⁾-pregnadien-3-one (p. 2406 s) on treatment with OsO_4 in abs. ether at 0⁰ for 2 days, followed by boiling with Na_2SO_3 in aqueous alcohol for 2 hrs.; yield, 70% (1938 Butenandt; 1939 Sche-RING A.-G.; 1942 Ruzicka). From the 3-enol acetate of 17α-acetoxy-17β-formyl-Δ⁴-androsten-3-one (p. 2320 s) on treatment with dimethyl zinc in ether, followed by saponification with alcoholic NaOH (1938a CIBA).

Rns. Gives Δ⁴-androstene-3.17-dione (p. 2880 s) on oxidation with lead tetraacetate in glacial acetic acid at room temperature (1938 Butenandt; 1942 Ruzicka). On refluxing its 17-esters in benzine with aluminium tert.-butoxide and acetone or with aluminium isopropoxide and cyclohexanone, they give the corresponding esters of "17-hydroxyprogesterone" (1938c, 1939 CIBA) which, under the conditions of this reaction (presence of aluminium alcoholate), is probably rearranged to 17a-methyl-D-homo-Δ⁴-androsten-17a-ol-3.17-dione (cf. 17-hydroxyprogesterone, p. 2843 s) (Editor; cf. 1941 Hegner). Gives progesterone (p. 2782 s) on heating with hydrochloric or sulphuric acid in propanol or dioxane (1938a CIBA).

For **anhydro-compounds** of **Δ⁴-pregnene-17α.20β-diol-3-one**, see 17.20-epoxy-Δ⁴-pregnen-3-ones (p. 2407 s).

17-iso-Δ⁴-Pregnene-17β.20-diol-3-one $C_{21}H_{32}O_3$. — **Fmn.** The 17-acetate 20-propionate is obtained from the correspon-ding diester of 17-iso-Δ⁵-pregnene-3β.17β.20-triol (p. 2116 s) on oxidation with aluminium isopropoxide and cyclohexanone (1938b CIBA). From 17-vinyltestosterone (17α-vinyl-Δ⁴-an-drosten-17β-ol-3-one, p. 2650 s) on treatment with dry HCl in ether, followed by treatment with silver acetate or alkali acetate in alcohol and refluxing of the resulting acetate with aqueous alcoholic KOH (1938b CIBA). The 17-acetate is obtained from 17-vinyltestosterone 20.21-epoxide (p. 2651 s) on hydrogenation in the presence of palladium - calcium carbonate, followed by acetylation and partial saponification (1939 CIBA). — **Rns.** On oxidation of the 17-acetate with CrO_3 in strong acetic acid in the cold it gives 17-acetoxy-17-iso-progesterone (p. 2844 s) (1939 CIBA).

Allopregnane-17α.20β-diol-3-one $C_{21}H_{34}O_3$. Crystals (ether-pentane); partly melts at 170⁰, resolidifies, and remelts at 181–2⁰ cor. (1941 Reich); m. 183–5⁰ cor., $[\alpha]_D^{20} + 13⁰$ (chloro-form) (1947b Ruzicka). — **Fmn.** From allopreg-nane-3β.17α.20β-triol (p. 2119 s) on refluxing with aluminium phenoxide and acetone in abs. benzene for 16 hrs.; isolated as 20-acetate (yield, 75%) which is saponified with methanolic KOH (1941 Reich). From Δ¹⁷⁽²⁰⁾-allopregnen-3-one (p. 2411 s) on treatment with OsO_4 and some pyridine in abs. ether at room temperature, followed by shaking with mannitol and 1 N KOH (1947 b Ruzicka). — **Rns.** Gives androstane-3.17-dione (p. 2892 s) on oxidation with periodic acid in slightly dil. methanol at room temperature (1941 Reich) or with lead tetraacetate in glacial acetic acid (1947 b Ruzicka).

20-Acetate $C_{23}H_{36}O_4$. Leaflets (ether-pentane), m. 189–190° cor.; $[\alpha]_D^{18}+49°$ (acetone). — **Fmn.** From the above diolone on heating with acetic anhydride in pyridine at 60° for 2 hrs. (1941 Reich).

Δ^5-Pregnene-3β.20-diol-7-one $C_{21}H_{32}O_3$. The configuration of the 20-hydroxyl group is unknown; after the closing date for this volume, Δ^5-pregnene-3β.20β-diol-7-one has been described (1952 Romo), the diacetate of which m. 192–3° (evacuated capillary) or 198–200° (Kofler block). — **Fmn.** The *diacetate* $C_{25}H_{36}O_5$ (m. 165–6°) is obtained from the diacetate of a Δ^5-pregnene-3β.20-diol (see p. 1922 s) on oxidation with CrO_3 (1936 SCHERING-KAHLBAUM A.-G.).

Pregnane-3α.20-diol-11-one $C_{21}H_{34}O_3$. The two 20-epimers have been described after the closing date for this volume by Sarett (1948, 1949). — **Fmn.** A mixture of the 3-acetates of the two epimers is obtained, along with the acetates of Δ^{16}-, $\Delta^{17(20)}$-, and Δ^{20}-pregnen-3α-ol-11-ones (page 2654 s), from the O-acetate of 20-amino-pregnan-3α-ol-11-one (p. 2718 s) on heating with sodium nitrite, pyridine hydrochloride, and aqueous pyridine on the steam-bath (1946, 1949 Sarett; 1946 c MERCK & Co., INC.). — **Rns.** On treatment of the 3-acetate with p-toluenesulphonyl chloride in pyridine it gives the 3-acetate 20-p-toluene-sulphonate which on refluxing with collidine yields the acetates of $\Delta^{17(20)}$- and Δ^{20}-pregnen-3α-ol-11-ones (1946, 1948 Sarett; 1946 c MERCK & Co., INC.).

Pregnane-3β.20β-diol-11-one $C_{21}H_{34}O_3$. The β-configuration at C-20 is very probable (1955 Reichstein). — Needles (acetone), at 150–160° changing to more slender needles m. 182–4° cor. (1944 v. Euw). — **Fmn.** The 3-acetate is obtained from the 3-acetate of pregnane-3β.11β.20β-triol (p. 2108 s) on oxidation with CrO_3 in glacial acetic acid at room temperature; the diacetate is similarly obtained from the 3.20-diacetate of the triol. Along with pregnan-3β-ol-11.20-dione (p. 2848 s), from pregnane-3.11.20-trione (p. 2967 s) on partial hydrogenation in glacial acetic acid in the presence of reduced platinum oxide; the 3-acetate is similarly obtained from the acetate of the aforesaid hydroxydiketone (1944 v. Euw).

3-Acetate $C_{23}H_{36}O_4$. Rodlets (acetone), m. 200–201° cor. (1944 v. Euw). — **Fmn.** See above.

Diacetate $C_{25}H_{38}O_5$. Plates (acetone-ether), m. 209–210° cor. — **Fmn.** From the 3-acetate on heating with acetic anhydride and pyridine at 60–70° for 2 hrs.; similarly from the free diolone (1944 v. Euw). See also above.

$\varDelta^{17(20)}$-Pregnene-3α.21-diol-11-one $C_{21}H_{32}O_3$.

Crystals (benzene or ethyl acetate), m. 200.5° cor.; $[\alpha]_D$ +45.5° (acetone) (1946 Sarett; 1946a MERCK & CO., INC.). — **Fmn.** The diacetate is obtained from 21-bromo - $\varDelta^{17(20)}$ - pregnen - 3α-ol-11-one acetate (p. 2701 s) on refluxing with anhydrous potassium acetate in dry acetone for 5 hrs. (yield, 90%); it is hydrolysed by refluxing with aqueous methanolic KOH for 15 min. (1946 Sarett; 1946a MERCK & CO., INC.). — **Rns.** On oxidation of the 21-hemisuccinate (below) with CrO_3 in strong acetic acid at 12–17°, followed by saponification with methanolic KOH, it gives $\varDelta^{17(20)}$-pregnen-21-ol-3.11-dione (p. 2934 s) (1946 Sarett; 1946b MERCK & CO., INC.; cf. 1948 Sarett).

Diacetate $C_{25}H_{36}O_5$. Crystals (acetone-petroleum ether b. 30–60°, and dil. acetone), m. 115.0–115.5° cor. (1946 Sarett; 1946a MERCK & CO., INC.). — **Fmn.** See above.

21-Hemisuccinate $C_{25}H_{36}O_6$. Crystals (ether and dil. acetone), m. 178.5° to 181.0° cor. — **Fmn.** From the diolone and succinic anhydride in pyridine at 28.5° (2 hrs.) (1946 Sarett; 1946a MERCK & CO., INC.). — **Rns.** See above.

4'-Anisylidene-4-hydroxy-3'-keto-1.2-cyclopentenophenanthrene $C_{25}H_{18}O_3$.

Yellow needles (isoamyl alcohol); softens and decomposes to a viscous tar at 305–310°. Very sparingly soluble in boiling formic and acetic acids, alcohol, and ethyl acetate, readily soluble in hot pyridine and nitrobenzene (pale yellow solution); the solution in alcoholic alkali is red, in sulphuric acid crimson. — **Fmn.** From 4-hydroxy-3'-keto-1.2-cyclopentenophenanthrene (p. 2516 s) and anisaldehyde in the presence of KOH in boiling alcohol (1938 Robinson).

$\varDelta^4$-Cholestene-16β.26-diol-3-one $C_{27}H_{44}O_3$. Crystals (ethyl acetate), m. 163–4°.

— **Fmn.** From the triacetate of "tetrahydrodiosgenin" ($\varDelta^5$-cholestene- 3β.16β.26-triol) on refluxing with SeO_2 in benzene – 97% acetic acid for 1 hr., followed by addition of potassium acetate and further refluxing for 10 min., saponification with alcoholic K_2CO_3, and refluxing of the resulting **cholestenetetrol** $C_{27}H_{46}O_4$

[crystals (acetone), m. 196°; yield, 15.6%] with conc. HCl in alcohol for 10 min. (1941 Marker).

22.22-Diphenyl-bisnorcholane-12β.22-'diol-3-one, (12β-Hydroxy-3-ketopregnan-20α-yl)-diphenyl-carbinol $C_{34}H_{44}O_3$ (I).

Anhydro-compound, 22.22-Diphenyl-12β.22-epoxy-bisnorcholan-3-one $C_{34}H_{42}O_2$ (II). Plates (benzene - petroleum ether), prisms (benzene - methanol), m. 283–4° cor.; $[\alpha]_D^{17} +218°$ (chloroform). — Fmn. From 22.22-diphenyl-12β.22-epoxy-bisnorcholan-3α-ol (p. 2135 s) on oxidation with CrO_3 in glacial acetic acid at 18°; crude yield, 94% (1945 Sorkin).

(I)   (II)

24.24-Diphenylcholane-3α.24-diol-12-one, (3α-Hydroxy-12-ketonorcholanyl)-diphenyl-carbinol $C_{36}H_{48}O_3$.

Crystals (benzene or 95% alcohol), m. 215–216.5° cor. — Fmn. From the 3-hemisuccinate of 3α.12α-dihydroxynorcholanyl-diphenyl-carbinol (p. 2137 s) on oxidation with CrO_3 in strong acetic acid at room temperature, followed by refluxing with methanolic KOH for 2 hrs.; yield, 14%. From methyl 3α-hydroxy-12-ketocholanate on refluxing with phenylmagnesium bromide in ether-benzene for 3.5 hrs.; yield, 32%. — Rns. On refluxing with acetic anhydride and glacial acetic acid for 2.5 hrs. it gives the acetate of 2.2-diphenyl-1-(3α-hydroxy-12-ketobisnorcholanyl)-ethylene (p. 2699 s) (1943 Riegel).

1936 Bardhan, J. Ch. Soc. 1936 1848, 1851.
    Schering-Kahlbaum A.-G., Brit. Pat. 486854 (issued 1938); C. A. 1938 8706; Ch. Ztbl. 1939 I 1603.
1937 Marker, Kamm, Jones, J. Am. Ch. Soc. 59 1595.
1938 Butenandt, Schmidt-Thomé, Paul, Ber. 71 1313, 1315, 1316.
    a Ciba*, Fr. Pat. 840783 (issued 1939); C. A. 1939 8627; Ch. Ztbl. 1940 I 915, 916.
    b Ciba*, Brit. Pat. 517942 (issued 1940); C. A. 1941 7122; Ch. Ztbl. 1941 I 927, 928.
    c Ciba*, Swiss Pat. 228644 (issued 1943); C. A. 1949 3476; Ch. Ztbl. 1944 II 1300.
    Litvan, Robinson, J. Ch. Soc. 1938 1997, 2000, 2001.
    Marker, Kamm, Wittle, Oakwood, Lawson, J. Am. Ch. Soc. 60 1067, 1070.
    Parke, Davis & Co. (Marker), Brit. Pat. 516846 (issued 1940); Ch. Ztbl. 1940 II 2648.
    Robinson, J. Ch. Soc. 1938 1390, 1395.
    Schering A.-G., Fr. Pat. 845473 (issued 1939); C. A. 1941 1185; Ch. Ztbl. 1943 I 653; Schering Corp. (Inhoffen, Logemann, Dannenbaum), U.S. Pat. 2280236 (1939, issued 1942); C. A. 1942 5322.
1939 Ciba*, Fr. Pat. 861318 (issued 1941); Ch. Ztbl. 1941 I 3259; Brit. Pat. 536621 (issued 1941); C. A. 1942 1739; Ch. Ztbl. 1943 II 147; Ciba Pharmaceutical Products (Ruzicka), U.S. Pat. 2365292 (issued 1944); C. A. 1945 4199; Ch. Ztbl. 1945 II 1233.

* Société pour l'industrie chimique à Bâle; Gesellschaft für chemische Industrie in Basel.

HOFFMANN-LA ROCHE & CO., *Swiss Pat.* 211292 (issued 1940); *C. A.* **1942** 3635; *Ch. Ztbl.*
1941 I 2972; *Brit. Pat.* 538478 (issued 1941); *C. A.* **1942** 3635; *Swed. Pat.* (Meyer) 100175
(1940, issued 1940); *Ch. Ztbl.* **1941** I 1572; *U.S. Pat.* (Meyer) 2257701 (1940, issued 1941);
C. A. **1942** 628.
  a Marker, Wittle, *J. Am. Ch. Soc.* **61** 855, 856.
  b Marker, Wittle, Plambeck, Rohrmann, Krueger, Ulshafer, *J. Am. Ch. Soc.* **61** 3317.
  Miescher, Klarer, *Helv.* **22** 962, 969.
  SCHERING A.-G., *Fr. Pat.* 868721 (issued 1942); *Ch. Ztbl.* **1942** II 74; *Swed. Pat.* 103355
(issued 1941); *Ch. Ztbl.* **1942** II 1603.
1941 Hegner, Reichstein, *Helv.* **24** 828, 832.
  Marker, Turner, *J. Am. Ch. Soc.* **63** 767, 769.
  Reich, Reichstein, *Arch. intern. pharmacodynamie* **65** 415, 418–420.
1942 Goldberg, Studer, *Helv.* **25** 1553, 1555.
  Ruzicka, Goldberg, Hardegger, *Helv.* **25** 1297, 1301.
1943 Riegel, Moffett, *J. Am. Ch. Soc.* **65** 1971.
1944 v. Euw, Lardon, Reichstein, *Helv.* **27** 821, 832–834.
1945 Birch, Jaeger, Robinson, *J. Ch. Soc.* **1945** 582, 585.
  Sorkin, Reichstein, *Helv.* **28** 875, 889.
1946 a MERCK & CO., INC. (Sarett), *U.S. Pat.* 2492192 (issued 1949); *C. A.* **1950** 3045; *Ch. Ztbl.*
1952 1211.
  b MERCK & CO., INC. (Sarett), *U.S. Pat.* 2492193 (issued 1949); *C. A.* **1950** 3043; *Ch. Ztbl.*
1952 1211.
  c MERCK & CO., INC. (Sarett), *U.S. Pat.* 2540964 (issued 1951); *C. A.* **1951** 7159; *Brit. Pat.*
630103 (issued 1949); *C. A.* **1950** 7891.
  Sarett, *J. Biol. Ch.* **162** 601, 621–623.
1947 a Ruzicka, Meister, Prelog, *Helv.* **30** 867, 874, 875.
  b Ruzicka, Meister, Prelog, *Helv.* **30** 867, 877, 878.
1948 Sarett, *J. Am. Ch. Soc.* **70** 1690, 1692, 1694.
1949 Sarett, *J. Am. Ch. Soc.* **71** 1165, 1167.
1952 Romo, Rosenkranz, Djerassi, *J. Org. Ch.* **17** 1413.
1954 Georg, *Arch. sci.* **7** 114.
1955 Reichstein, *private communication.*
  Sondheimer, Mancera, Urquiza, Rosenkranz, *J. Am. Ch. Soc.* **77** 4145, 4148.

---

16. TRIHYDROXY-MONOKETOSTEROIDS WITH CO AND ONE OH IN THE RING SYSTEM

Δ⁴-Pregnene-17α.20β.21-triol-3-one $C_{21}H_{32}O_4$. For configuration at C-17, see

pp. 1363 s, 1364 s. — Crystals (methanol),
m. 190° cor.; $[\alpha]_D +63°$ (dioxane) (1939 Logemann; 1939 Ruzicka). Its solution in conc.
H_2SO_4 shows a green fluorescence (1940
v. Euw). — Is inactive in the cortin test
(15 mg.) and in the corpus luteum test (1939
Logemann). — **Fmn.** From the acetate of Δ⁴·¹⁷⁽²⁰⁾-pregnadien-21-ol-3-one
(p. 2512 s) on treatment with OsO_4 in ether at room temperature for 100 hrs.,
followed by heating with Na_2SO_3 in aqueous alcohol on the steam-bath for
2 hrs. (1939 Ruzicka; cf. 1939 Logemann). The 20.21-acetonide (p. 2726 s) is
obtained from the corresponding acetonide of Δ⁵-pregnene-3β.17α.20.21-tetrol
(p. 2139 s) on refluxing with aluminium tert.-butoxide and acetone in abs.
benzene for 24 hrs. (yield, 45 %); it is hydrolysed by heating with dil. acetic
acid at 70° for 2 hrs. (1943 Koechlin).

References, pp. 2731 s, 2732 s

Rns. On oxidation with 1 mol. periodic acid in aqueous dioxane at room temperature it gives 17β-formyl-Δ^4-androsten-17α-ol-3-one (p. 2836 s) (1941 Prins; cf. 1942 Reichstein). Oxidation with H_2O_2 in aqueous dioxane in the presence of ferrous sulphate at room temperature affords Δ^4-pregnene-17α.20β-diol-3-on-21-al (p. 2863 s) (1939 CIBA PHARMACEUTICAL PRODUCTS). On oxidation of the 21-acetate with CrO_3 in strong acetic acid at 11^0 it gives the 21-acetate of Δ^4-pregnene-17α.21-diol-3.20-dione (17-hydroxydesoxy-corticosterone, p. 2859 s) and Δ^4-androstene-3.17-dione (p. 2880 s) (1946 Sarett). On refluxing the 20.21-diacetate with zinc dust in abs. toluene it gives 17-iso-desoxycorticosterone acetate (17-iso-Δ^4-pregnen-21-ol-3.20-dione acetate, p. 2829 s) (1940 Shoppee; cf. 1940 CIBA). Gives desoxycorticosterone (p. 2820 s) on refluxing with sulphuric acid in aqueous dioxane (1937 SCHE-RING A.-G.). The 21-acetate reacts with 1 mol. N-bromoacetamide, but does not give a crystalline compound (1943 Reich).

21-Acetate $C_{23}H_{34}O_5$. Crystals (dil. alcohol), m. 189–192^0 cor. — **Fmn.** From the above triolone on treatment with 1.4 mols. acetic anhydride and dry pyridine in abs. dioxane at room temperature (1946 Sarett). — **Rns.** See above.

20.21-Diacetate $C_{25}H_{36}O_6$. Prisms (acetone-ether) containing solvent of crystallization; m. ca. 100^0, resolidifies to needles m. 170–2^0 cor., and again resolidifies to crystals m. 193–4^0; the last-mentioned two forms are inter-convertible by seeding (1940 Shoppee); needles (ether), m. 196–197^0 cor. (Kofler block), and a lower-melting form (1943 Koechlin); crystals, m. 189^0 to 190^0 (1939 Logemann); $[\alpha]_D$ +125^0 (dioxane) (1939 Logemann), +136^0 to 137^0 (acetone) (1943 Koechlin). — **Fmn.** From the above triolone on treatment with acetic anhydride in pyridine at room temperature (1940 Shoppee; cf. 1943 Koechlin). — **Rns.** See above.

20.21-Acetonide $C_{24}H_{36}O_4$. Needles (ether · petroleum ether), m. 146–7^0 cor. (Kofler block); sometimes another modification, m. 200–204^0 cor. (Kofler block), was obtained; $[\alpha]_D^{22}$ +75^0 (acetone). Absorption max. in alcohol at 241 mμ (log ε 4.10) (1943 Koechlin). — **Fmn.** and **Rns.** See under formation of the above triolone.

17.20-Anhydride $C_{21}H_{30}O_3$. For its acetate, see 21-acetoxy-17.20-epoxy-Δ^4-pregnen-3-one, p. 2512 s.

17-iso-Δ^4-Pregnene-17β.20 a.21-triol-3-one $C_{21}H_{32}O_4$. Previously assigned 17α.20β-configuration; for revised configuration, see 1947 a, b Salamon; cf. also pp. 1363 s to 1365 s. — Crystals (methanol), m. 233–5^0 (1938 Serini), 234.5–235.5^0 cor. (1944 Fieser); $[\alpha]_D^{20}$ +65.6^0 (dioxane) (1938 Serini). Has essentially the same chromogenic characteristics in the Zimmermann test as an equimolar quantity of testosterone (p. 2580 s) (1944 Talbot). — **Fmn.** From 17-iso-$\Delta^{4.20}$-pregnadien-17β-ol-3-one (17-vinyltestosterone, p. 2650 s) on treatment with OsO_4 in ether at room temperature for 4 days, followed by refluxing of the resulting osmic ester with Na_2SO_3 in aqueous alcohol for 2 hrs.; yield, 60% (1938 Serini; 1940 CIBA; cf. 1944 Fieser). The 20.21-acetonide (p. 2728 s) is obtained from the corresponding acetonide of 17-iso-Δ^5-pregnene-3β.17β.20a.21-tetrol (p. 2140 s) on refluxing with aluminium tert.-butoxide and acetone in abs. benzene for 24 hrs. (yield, ca. 60%); it is hydrolysed by warming with dil. acetic acid (1941 Reich).

Rns. On oxidation with 1 mol. periodic acid in aqueous dioxane at room temperature for 4 hrs. it gives 17α-formyl-Δ^4-androsten-17β-ol-3-one (p. 2835 s) (1941 Prins); on oxidation with excess periodic acid in aqueous methanolic H_2SO_4 at room temperature it is almost quantitatively converted into Δ^4-androstene-3.17-dione (determined colorimetrically) (1944 Talbot; see also 1944 Fieser). On oxidation with H_2O_2 in aqueous dioxane in the presence of ferrous sulphate and barium carbonate it gives 17-iso-Δ^4-pregnene-17β.20a-diol-3-on-21-al (p. 2863 s) (1939 CIBA PHARMACEUTICAL PRODUCTS). Is not dehydrogenated by Corynebacterium mediolanum (1939 Mamoli).

On shaking of the 20.21-osmic ester (see above, under formation) with $KClO_3$ and aqueous H_2SO_4 in ether for $2^1/_2$ hrs., followed by treatment with acetic anhydride and abs. pyridine at 20° for 16 hrs., it gives a compound $C_{23}H_{32}O_5$ and a compound $C_{20}H_{26}O_2$ or $C_{20}H_{28}O_2$; the **compound** $C_{23}H_{32}O_5$ forms leaflets (acetone), m. 265–8°, $[\alpha]_D^{14}$ +58° (dioxane), reduces alkaline silver solution on the water-bath, is also obtainable (in very small yield) from 17-vinyltestosterone acetate on similar treatment, and on hydrolysis with $KHCO_3$ in aqueous methanol at 20° gives needles (ether - pentane), m. 108–125°, which on oxidation with periodic acid in aqueous methanol at room temperature yield mainly acid, uncrystallizable products; the **compound** $C_{20}H_{26}O_2$ or $C_{20}H_{28}O_2$ (possibly 17-formyl-$\Delta^{4.16}$-androstadien-3-one) forms needles or prisms (ether - pentane), m. 149–151°, $[\alpha]_D^{19}$ +13° (acetone), absorption max. in alcohol at 241 mμ (log ε 4.22), reduces alkaline silver solution in 2 min. at room temperature, and probably gives a polymer [leaflets (acetone - pentane), m. 270–3°] on sublimation under 0.02 mm. (1942 Prins).

The 20.21-diacetate gives desoxycorticosterone acetate (Δ^4-pregnen-21-ol-3.20-dione acetate, p. 2824 s) on distillation with zinc dust in a high vacuum at 150–200° (1939 Serini; 1939 SCHERING A.-G.) or on refluxing with zinc dust in toluene (1940 CIBA); the free triolone reacts similarly (1939 SCHERING A.-G.). Is altered by treatment with dil. acetic acid or with anhydrous $CuSO_4$; gives a resinous product on refluxing with a trace of p-toluenesulphonic acid in benzene (1944 Fieser); for reaction with m-hydroxybenzaldehyde and a trace of p-toluenesulphonic acid in benzene, see 1944 Fieser.

Semicarbazone $C_{22}H_{35}O_4N_3$. Crystals (dil. methanol), m. 216–8° dec. (1938 Serini).

20.21-Diacetate $C_{25}H_{36}O_6$. Crystals (petroleum ether - acetone), m. 178–9° (1938 Serini); needles (ether - pentane), m. 180–1° cor. (1941 Reich); $[\alpha]_D$ +43.6° (dioxane) (1938 Serini), +50° (acetone) (1941 Reich). — **Fmn.** From the above triolone on treatment with acetic anhydride in pyridine at room temperature (1938 Serini; cf. 1941 Reich).

E.O.C. XIV s. 188

References, pp. 2731 s, 2732 s

20.21-Acetonide $C_{24}H_{36}O_4$. Prisms (ether - pentane), m. 173–5° cor.; $[\alpha]_D^{17}$ + 39° (acetone) (1941 Reich). — **Fmn.** and **Rns.** See under formation of the above triolone.

17-iso-Δ^4-Pregnene-17β.20b.21-triol-3-one $C_{21}H_{32}O_4$. Previously assigned 17α.20α-configuration; for revised configuration, see 1947 a, b Salamon; cf. also pp. 1363 s to 1365 s. — Leaflets (methanol-ether), m. 233° to 235° cor. (not analysed sample) (1941 Reich). — **Fmn.** The 20.21-acetonide (below) is obtained from the corresponding acetonide of 17-iso-Δ^5-pregnene-3β.17β.20b.21-tetrol (p. 2141 s) on refluxing with aluminium tert.-butoxide and acetone in abs. benzene for 24 hrs. (yield, 55%); it is hydrolysed by heating with dil. acetic acid at 70° for 2 hrs. (1941 Reich).

20.21-Diacetate $C_{25}H_{36}O_6$. Needles (ether-pentane), m. 165–6° cor.; $[\alpha]_D^{15}$ + 21.6° ± 3° (acetone). — **Fmn.** From the above triolone on treatment with acetic anhydride in abs. pyridine, first at room temperature for 16 hrs., then at 60° for 1 hr. (1941 Reich).

20.21-Acetonide $C_{24}H_{36}O_4$. Needles (acetone-pentane), m. 220–221.5° cor.; $[\alpha]_D^{15}$ + 67° (acetone) (1941 Reich). — **Fmn.** and **Rns.** See above.

A **17-iso-Δ^4-pregnene-17β.20.21-triol-3-one** $C_{21}H_{32}O_4$ (not characterized) is obtained from the 3-enol acetate of 17.21-diacetoxy-17-iso-progesterone (p. 2357 s) on reduction, e.g. by refluxing with aluminium isobutoxide in isobutyl alcohol, followed by saponification of the resulting 3-enol acetate with 2% alcoholic alkali hydroxide (1937 CIBA).

The *21-triphenylmethyl ether* $C_{40}H_{46}O_4$ [crystals (acetone), m. 199–200°] of a **17-iso (?)-Δ^4-pregnene-17β (?).20.21-triol-3-one** is obtained from the triolone by heating with 1.1 mols. of triphenylmethyl chloride in pyridine for 1 hr. on the water-bath. It gives the corresponding ether of 17-iso(?)-Δ^4-pregnene-17β(?).21-diol-3.20-dione (p. 2859 s) on oxidation with CrO₃ in strong acetic acid at room temperature or with acetone and aluminium isopropoxide in boiling benzene (1938 SCHERING A.-G.).

For a **17-iso-Δ^4-pregnene-17β.20.21-triol-3-one 20.21-anhydride,** see 17-vinyltestosterone epoxide, p. 2651 s.

ω-**Homo-17-iso-Δ^4-pregnene-17β.21.22-triol-3-ones, 17-(β.γ-Dihydroxypropyl)-testosterones** $C_{22}H_{34}O_4$. Both C-21 epimers, (I) and (II), are known. — **Fmn.** A mixture of the two epimers is obtained from 17-allyltestosterone (p. 2653 s) on treatment with OsO₄ in abs. ether at 20° for 48 hrs., followed by boiling with Na₂SO₃ in aqueous alcohol for 2 hrs. (yield, 70.5%) (1938 Butenandt; cf. 1939 Miescher; 1940

v. Euw). — **Rns.** On oxidation of both epimers with lead tetraacetate in dry benzene in an atmosphere of nitrogen at 45° they give 17-iso-Δ^4-pregnen-17β-ol-3-on-21-al (p. 2846 s); in the presence of air or using glacial acetic acid as solvent, 17-iso-Δ^4-pregnen-17β-ol-3-on-21-oic acid is obtained (1938 Butenandt); the aforesaid aldehyde is also formed when a mixture of the two epimers is oxidized with periodic acid in aqueous methanolic H_2SO_4 in an atmosphere of nitrogen at 22° (1939 Miescher). The mixture of the two epimers, dissolved in methanol, rather slowly gives a precipitate with ammoniacal silver solution (1939 Miescher). On refluxing of the mono- triphenylmethyl ether of the epimer (I) with aluminium isopropoxide and cyclohexanone in abs. toluene for 30 min. it gives Δ^4-androstene-3.17-dione (p. 2880 s) (1938 Butenandt). With triphenylmethyl chloride in pyridine only the epimer (I) reacts to give the mono-triphenylmethyl ether (below) (1938 Butenandt).

Epimer (I) $C_{22}H_{34}O_4$. Prisms (chloroform - ethyl acetate), m. 224–5° (1938 Butenandt), 228–230° cor. (1940 v. Euw); $[\alpha]_D^{20}$ +54° (solvent not stated) (1938 Butenandt). — Both epimers have no cortin action and are also inactive in the Fussgänger test and the Allen-Doisy test (1938 Butenandt). — **Fmn.** and **Rns.** See above.

21.22-Dibenzoate $C_{36}H_{42}O_6$. Crystals (methanol), m. 169–170° cor. — **Fmn.** From the above triolone (I) on warming with benzoyl chloride in dry pyridine at 60° (1940 v. Euw).

21.22-Acetonide $C_{25}H_{38}O_4$. Crystals (ether-pentane), m. 135–6° cor.; $[\alpha]_D^{15}$ +38°, $[\alpha]_{5461}^{15}$ +45° (acetone). — **Fmn.** From the above triolone (I) on shaking with dry acetone and anhydrous copper sulphate at room temperature for 48 hrs. (1940 v. Euw).

Mono-triphenylmethyl ether $C_{41}H_{48}O_4$. Leaflets (dil. methanol and ethyl acetate), m. 197.5°. — **Fmn.** From the above epimer (I) on treatment with triphenylmethyl chloride in dry pyridine at room temperature for 48 hrs.; yield, 62% (1938 Butenandt). — **Rns.** For oxidation, see under the reactions of the mixture of epimers (above).

Epimer (II) $C_{22}H_{34}O_4$. Plates (chloroform - ethyl acetate), m. 198°, $[\alpha]_D^{20}$ +48° (solvent not stated); occurs also in an unstable modification m. 168°; is more readily soluble in chloroform - ethyl acetate than epimer (I) (1938 Butenandt). When purified through its acetonide (below), from which it is regenerated by warming with 50% acetic acid at 60° for 2 hrs., it forms crystals (ether-acetone), m. 207–207.5° cor. (1940 v. Euw). — For physiological properties, see epimer (I). — **Fmn.** and **Rns.** See p. 2728 s.

21.22-Dibenzoate $C_{36}H_{42}O_6$. Granula (acetone-ether), m. 161–2° cor. — **Fmn.** From the above epimer (II) on warming with benzoyl chloride in dry pyridine at 60° (1940 v. Euw).

21.22-Acetonide $C_{25}H_{38}O_4$. Leaflets with $^1/_2$ H_2O (aqueous methanol), m. 107–107.5° cor.; $[\alpha]_D^{16}$ +61°, $[\alpha]_{5461}^{16}$ +71° (acetone). — **Fmn.** From the above epimer (II) on shaking with dry acetone and anhydrous copper sulphate at room temperature for 48 hrs. — **Rns.** Is hydrolysed by warming with 50% acetic acid at 60° (1940 v. Euw).

188*

6-Methylallopregnane-5.20.21-triol-3-one $C_{22}H_{36}O_4$.

20.21-Diacetate $C_{26}H_{40}O_6$ (not obtained quite pure). Crystals (ether - petroleum ether), m. ca. 205–210⁰. Does not reduce alkaline silver diammine in methanol. — **Fmn.** From the 21-acetate of 6-methyl-allopregnane-3β.5.20.21-tetrol (p. 2147 s) on treatment with CrO_3 in strong acetic acid at 29⁰ for 19 hrs. — **Rns.** Gives 6-methyl-$\varDelta^4$-pregnene-20.21-diol-3-one diacetate (p. 2514 s) by the action of dry HCl in dry chloroform at a temperature not exceeding $+2^0$ (1943 Ehrenstein).

4'-Piperonylidene-7-methoxy-3'-keto-1.2-cyclopentano-1.2.3.4-tetrahydrophenanthrene, 16-Piperonylidene-dl-18-nor-isoequilenin methyl ether $C_{26}H_{22}O_4$ (III, R = R'= H).

Pale yellow, rhombic plates (benzene-light petroleum b. 40–60⁰), m. 187–8⁰. Gives an intense yellow coloration with sodium tert.-butoxide in tert.-butyl alcohol (1941 Koebner). — **Fmn.** From dl-18-nor-isoequilenin methyl ether (p. 2518 s) on heating with piperonal in aqueous alcoholic NaOH on the steam-bath (1941 Koebner; cf. 1938 Koebner). — **Rns.** For methylation, see the following compound.

(III)

16 - Piperonylidene - 3 - methoxy - 17 - keto - dl - cis - equilenan, 16-Piperonylidene-dl-isoequilenin methyl ether $C_{27}H_{24}O_4$ (III, R = CH_3, R'= H).

Pale yellow plates (benzene - light petroleum b. 40–60⁰), m. 180–1⁰. Its orange-red solution in cold conc. H_2SO_4 becomes brown on heating . Gives no coloration with alcoholic sodium ethoxide (1941 Koebner). — **Fmn.** From the preceding compound on refluxing with methyl iodide and potassium tert.-butoxide in tert.-butyl alcohol for 2 hrs. (1941 Koebner; cf. 1943 Johnson). — **Rns.** With $KMnO_4$ in ice-cold acetone it gives ld-cis ("α")-bisdehydromarrianolic acid methyl ether (IV; see also the scheme on p. 2528 s) (1945 Birch). Gives 4-chloro-16-piperonylidene-dl-isoequilenin methyl ether $C_{27}H_{23}O_4Cl$

(IV)

(III, R = CH_3, R'= Cl) [pale yellow, crystalline powder (dioxane-alcohol), m. 203–4⁰] on treatment with chlorine in ice-cold dioxane-CCl_4 and heating of the resulting yellow resin with alcoholic sodium ethoxide in benzene on the steam-bath for 1 hr.; with more chlorine in dioxane - carbon tetrachloride, followed by heating with alcoholic sodium ethoxide in benzene on the steam-bath for 1 hr., refluxing with 15 % HCl for $^1/_2$ hr. and then with 2 % NaOH in benzene for 1 hr., it gives a pale yellow sodium salt which on acidification probably yields 4-chloro-16-(3.4-methylenedioxybenzoyl)-isoequilenin methyl ether (p. 2864 s); this ether gives dl-4-chloroisoequilenin methyl ether (page 2701 s) and 2-methyl-8-chloro-7-methoxy-1-[β-(3.4-methylenedioxybenzoyl)-ethyl]-1.2.3.4-tetrahydro-2-phenanthroic acid on refluxing with KOH in aqueous alcohol (1945 Birch).

16-Piperonylidene-3-methoxy-17-keto-d-trans-equilenan, 16-Piperonylidene-d-equilenin methyl ether $C_{27}H_{24}O_4$.

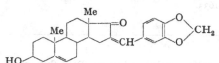

For configuration, see d-equilenin (p. 2527 s). — Colourless plates (alcohol), m. 208–9⁰. Its orange-red solution in cold conc. H_2SO_4 slowly becomes brownish green on heating. Gives no coloration with alcoholic sodium ethoxide. — **Fmn.** From d-equilenin methyl ether (p. 2533 s) and piperonal in the presence of KOH in alcohol. — **Rns.** Does not react with methyl iodide and potassium tert.-butoxide in tert.-butyl alcohol (1941 Koebner).

16-Piperonylidene-Δ^5-androsten-3β-ol-17-one, 16-Piperonylidene-5-dehydro-epiandrosterone $C_{27}H_{32}O_4$.

Crystals (hexane-acetone), m. 242–3⁰ cor. — **Fmn.** From 5-dehydroepiandrosterone (p. 2609 s) and piperonal on refluxing with sodium ethoxide in alcohol for 1½ hrs., with K_2CO_3 in dil. methanol for 3 hrs., or on treatment with NaOH in strong alcohol at room temperature for 16 hrs.; yields, 89% (1943 Wettstein).

4-Chloro-16-piperonylidene-dl-isoequilenin methyl ether $C_{27}H_{23}O_4Cl$. See p. 2730 s.

1937 CIBA*, *Swiss Pat.* 207496 (issued 1940); *C. A.* **1942** 3635; *Brit. Pat.* 517288 (1938, issued 1940); *Ch. Ztbl.* **1940** II 1328; CIBA PHARMACEUTICAL PRODUCTS (Miescher, Wettstein), *U.S. Pat.* 2239012 (1938, issued 1941); *C. A.* **1941** 4921.
  SCHERING A.-G. (Serini, Logemann), *Ger. Pat.* 736848 (issued 1943); *C. A.* **1944** 3094; *Ch.Ztbl.* **1944** I 300, 301.
1938 Butenandt, Peters, *Ber.* **71** 2688, 2693–2695.
  Koebner, Robinson, *J. Ch. Soc.* **1938** 1994, 1997.
  SCHERING A.-G., *Fr. Pat.* 857832 (issued 1940); *Ch. Ztbl.* **1941** I 1324.
  Serini, Logemann, *Ber.* **71** 1362, 1366.
1939 CIBA PHARMACEUTICAL PRODUCTS (Miescher, Fischer, Scholz, Wettstein), *U.S. Pats.* 2275790 (issued 1942), 2276543 (issued 1942); *C. A.* **1942** 4289, 4674; CIBA*, *Fr. Pat.* 857122 (issued 1940); *Ch. Ztbl.* **1941** I 1324.
  Logemann, *Naturwiss.* **27** 196.
  Mamoli, *Ber.* **72** 1863, 1865.
  Miescher, Wettstein, Scholz, *Helv.* **22** 894, 895, 901, 903.
  Ruzicka, Müller, *Helv.* **22** 755, 757.

* Société pour l'industrie chimique à Bâle; Gesellschaft für chemische Industrie in Basel.

SCHERING A.-G. (Logemann, Hildebrand), *Ger. Pat.* 737420 (issued 1943); *C. A.* **1945** 5051;
 Ch. Ztbl. **1944** I 366, 367; *Swiss Pat.* 229604 (issued 1944); *Ch. Ztbl.* **1944** II 1203; *Fr. Pat.*
 856641 (1940); *Ch. Ztbl.* **1941** I 407.
 Serini, Logemann, Hildebrand, *Ber.* **72** 391, 394.
1940 CIBA* (Miescher, Heer), *Ger. Pat.* 737570 (issued 1943); *C. A.* **1946** 179; *Fr. Pat.* 876214
 (issued 1942); *Ch. Ztbl.* **1943** I 1496; *Dutch Pat.* 55050 (issued 1943); *Ch. Ztbl.* **1944** I 367;
 CIBA PHARMACEUTICAL PRODUCTS (Miescher, Heer), *U.S. Pat.* 2372841 (issued 1945); *C. A.*
 1945 4195; *Ch. Ztbl.* **1945** II 1773.
 v. Euw, Reichstein, *Helv.* **23** 1114, 1116, 1119–1121.
 Shoppee, *Helv.* **23** 925, 928, 929; cf. *Helv.* **24** 354 (erratum).
1941 Koebner, Robinson, *J. Ch. Soc.* **1941** 566, 573.
 Prins, Reichstein, *Helv.* **24** 945, 949, 951.
 Reich, Montigel, Reichstein, *Helv.* **24** 977, 983–985.
1942 Prins, Reichstein, *Helv.* **25** 300, 318–322.
 Reichstein, *Fr. Pat.* 888228 (issued 1943); *Ch. Ztbl.* **1945** I 68.
1943 Ehrenstein, *J. Org. Ch.* **8** 83, 88, 89, 93.
 Johnson, *J. Am. Ch. Soc.* **65** 1317, 1319.
 Koechlin, Reichstein, *Helv.* **26** 1328, 1332–1334.
 Reich, Reichstein, *Helv.* **26** 562, 585.
 Wettstein, Miescher, *Helv.* **26** 631, 639, 640.
1944 Fieser, Fields, Lieberman, *J. Biol. Ch.* **156** 191, 194, 199.
 Talbot, Eitingon, *J. Biol. Ch.* **154** 605, 609–612.
1945 Birch, Jaeger, Robinson, *J. Ch. Soc.* **1945** 582, 584.
1946 Sarett, *J. Biol. Ch.* **162** 601, 626, 627.
1947 a Salamon, Reichstein, *Helv.* **30** 1616, 1623, 1635.
 b Salamon, Reichstein, *Helv.* **30** 1929.

---

17. TETRAHYDROXY-MONOKETOSTEROIDS WITH CO AND ONE OH IN THE RING SYSTEM

ω-Homo-Δ^4-pregnene-17α.20β.21.22-tetrol-3-one, "*17-(α.β.γ-Trihydroxy-propyl)-testosterone*" $C_{22}H_{34}O_5$. For configuration at C-17, see pp. 1363 s, 1364 s. — Needles (alcohol), m. 237.5° (1938 Butenandt), 239–244° cor. (1940 v. Euw). Sparingly soluble in chloroform, ethyl acetate, acetone, methanol, ethanol, and in cold water, readily soluble in hot water (1938 Butenandt). — Is inactive in the cortin test, the Fussgänger test, and the Allen-Doisy test (1938 Butenandt). — **Fmn.** From 17-allylidene-Δ^4-androsten-3-one (p. 2412 s) on treatment with OsO_4 in abs. ether at 20° for 80 hrs., followed by refluxing with Na_2SO_3 in dil. alcohol for 3 hrs.; yield, 13.5% (1938 Butenandt; cf. 1940 v. Euw).

Rns. On oxidation with periodic acid in water-dioxane (1:15) at room temperature (1 hr.) it gives Δ^4-androstene-3.17-dione (p. 2880 s) and Δ^4-pregnene-17α.20β-diol-3-on-21-al (p. 2863 s) which on heating with abs. pyridine in a sealed tube at 111° for 6 hrs. is rearranged to Δ^4-pregnene-17α.21-diol-3.20-dione (17-hydroxydesoxycorticosterone, p. 2859 s) (1941 v. Euw); on oxidation of the 20-acetate (below) with periodic acid in aqueous dioxane at room temperature, the 20-acetate of the aforesaid aldehyde is obtained

* Société pour l'industrie chimique à Bâle; Gesellschaft für chemische Industrie in Basel.

(1940 v. Euw). Gives the 21.22-acetonide (below) on shaking with acetone and anhydrous copper sulphate; reacts similarly with cyclohexanone (1940 v. Euw).

20-Acetate $C_{24}H_{36}O_6$. Rodlets (acetone), m. 210–211.5° cor.; $[\alpha]_D^{18} + 100°$ (dioxane). — **Fmn.** From its 21.22-acetonide (below) or its 21.22-cyclo-hexylidene-derivative (below) on heating for $1^1/_2$ hrs. with 55% acetic acid at 55° or with 70% acetic acid at 75°, resp. (1940 v. Euw). — **Rns.** For oxidation, see above.

21.22-Acetonide $C_{25}H_{38}O_5$. Needles (acetone), m. 235–236.5° cor.; $[\alpha]_D^{18} + 67°$ (dioxane). — **Fmn.** From the above tetrolone on shaking with acetone and anhydrous copper sulphate for 15 hrs.; yield, 84% (1940 v. Euw).

21.22 - Acetonide 20 - acetate $C_{27}H_{40}O_6$. Rhombs and hexagonal crystals (acetone-ether), m. 221–3° cor.; $[\alpha]_D^{17} + 107°$ (acetone). — **Fmn.** From the preceding compound on warming with acetic anhydride in abs. pyridine at 60° for 2 hrs. (1940 v. Euw). — **Rns.** For hydrolysis, see the 20-acetate (above).

21.22-Cyclohexylidene-derivative $C_{28}H_{42}O_5$ (not analysed). Needles (benzene-ether), m. 198.5–200° cor. — **Fmn.** From the above tetrolone on shaking with cyclohexanone and anhydrous copper sulphate for 20 hrs.; yield, 58%. — **Rns.** On treatment with acetic anhydride in abs. pyridine for 16 hrs. at room temperature and then for 3 hrs. at 60°, it gives the *20-acetyl derivative* $C_{30}H_{44}O_6$ (not analysed), leaflets (benzene-ether), m. 229–231° cor. with partial change to hexagonal plates m. 240–5° cor. (1940 v. Euw); for hydrolysis of this acetyl derivative, see the 20-acetate (above).

1938 Butenandt, Peters, *Ber.* **71** 2688, 2691, 2693.
1940 v. Euw, Reichstein, *Helv.* **23** 1114, 1122–1124; Reichstein, *Fr. Pat.* 872985 (1941, issued 1942);
    Ch. Ztbl. **1942** II 2175.
1941 v. Euw, Reichstein, *Helv.* **24** 1140; Reichstein, *Fr. Pat.* 888228 (1942, issued 1943); Ch. Ztbl.
    1945 I 68.

18. Dihydroxy-monoketosteroids with CO and OH in the Ring System

a. Compounds C_{17}

4.7 - Dihydroxy - 3' - keto - 1.2 - cyclopentenophenanthrene $C_{17}H_{12}O_3$ (I, R =

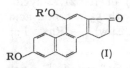

R'= H). Needles (aqueous pyridine), m. 338⁰ dec. (1939 Robinson). — **Fmn.** Its diacetate (below) is obtained by the following synthesis: 6-methoxy-2-acetylnaphthalene is condensed with furfural to give 6-methoxy-2-furfurylideneacetyl-naphthalene which is converted by hydrochloric acid into 7-(6-methoxy-2-naphthyl)-4.7-diketo-heptanoic acid; this acid is demethylated by refluxing with acetic-hydro-chloric acids and then cyclized by heating with dil. KOH on the water-bath to give 3-(6-hydroxy-2-naphthyl)-Δ^2-cyclopenten-1-one-2-acetic acid which is then refluxed with acetic anhydride for 30 min.; the diacetate is hydrolysed by refluxing with aqueous alcoholic NaOH for 2 hrs. (1938, 1939 Robinson). May also be obtained from the 7-methyl ether 4-acetate (p. 2735 s) on boiling with HI (d 1.96) for a few seconds (1938, 1941 b Robinson).

4.7-Diacetoxy-3'-keto-1.2-cyclopentenophenanthrene $C_{21}H_{16}O_5$ (I, R = R'= CO·CH₃). Needles (dil. acetic acid), m. 196-7⁰ (1939 Robinson). — **Fmn.** See above.

4-Hydroxy-7-methoxy-3'-keto-1.2-cyclopentenophenanthrene $C_{18}H_{14}O_3$ (I, R = CH₃, R'= H). Pale yellow plates (pyridine - acetic acid); darkens at 268⁰, softens at 285⁰, m. at 293–9⁰ to a brown tar. Sparingly soluble in most organic solvents; its solution in alcohol exhibits a feeble bluish violet fluor-escence (1938 Robinson). — **Fmn.** Its acetate (below) is similarly obtained to the above unmethylated compound, but without demethylation of the inter-mediate diketoheptanoic acid, and is then hydrolysed (1938, 1939 Robinson).

Rns. On reduction with sodium amalgam it gives amorphous, non-phenolic material (1941 Cardwell). On hydrogenation of the acetate in glacial acetic acid in the presence of Adams's catalyst at 70⁰ (24–26 hrs.) it probably gives 4-hydroxy-7-methoxy-3'-keto-1.2-cyclopenteno-9.10-dihydrophenanthrene (p. 2736 s) and its acetate (1939 Robinson; cf. 1941 b Robinson), the acetate of 4-hydroxy-7-methoxy-1.2-cyclopenteno-1.2.3.4-tetrahydrophenanthrene (p. 1958 s), and a liquid **mixture** $C_{29}H_{36}O$ (b. 178–182⁰/₀.₁₅, n_D^{17} 1.5764) which on heating with palladium charcoal at 280–360⁰ yields a mixture (4:1) of 1.2-cyclopentenophenanthrene (p. 1382 s) and 7-methoxy-1.2-cyclopenteno-phenanthrene (p. 1494 s) (1939 Robinson); on heating of the acetate with hydrogen and Raney nickel in glacial acetic acid at 125⁰ under 30 atm. pressure for 15 hrs. it is deacetylated and hydrogenated to a small extent giving the afore-mentioned 9.10-dihydro-derivative; using twice the relative amount of nickel catalyst at 155⁰ under 50 atm. pressure (25 hrs.) it gives much tar and a **substance** $C_{36}H_{32}O_4$*, colourless plates (dimethylaniline),

---

* The molecular formula (confirmed by analysis) does not correspond with the structural for-mula (VI), given on p. 394 of the original paper, even when the latter is corrected for the number of rings.

m. 313° after slight sintering, giving a light yellow solution in alcoholic NaOH and a mauve solution in conc. H_2SO_4 (1941 b Robinson).

The acetate gives the free dihydroxyketone (above) on boiling with HI (d 1.96) for a few seconds (1938, 1941 b Robinson). On refluxing of the acetate with HBr (d 1.50) in glacial acetic acid for 20 hrs. it gives "**apodihydroxyketocyclopentenophenanthrene**" $C_{17}H_{12}O_3$ [of unknown structure; yellow needles (pyridine), not m. at 380°; gives a red solution in aqueous NaOH; its magenta-coloured solution in cold conc. H_2SO_4 becomes dark purple on heating; gives a *2.4-dinitrophenylhydrazone* and a *dimethyl ether* $C_{19}H_{16}O_3$ (not analysed), yellow needles (nitrobenzene), m. 301°, giving a deep reddish violet solution in conc. H_2SO_4] (1941 b Robinson).

Gives a deep red p-nitrobenzeneazo-derivative which gives a violet solution in alcoholic NaOH and a bluish magenta coloration in conc. H_2SO_4 (1938 Robinson). — *Potassium salt*, orange needles (alcohol) (1938 Robinson). —

Oxime $C_{18}H_{15}O_3N$. Needles with 1 H_2O (aqueous pyridine - acetic acid), m. 268° (1939 Robinson).

4-Acetoxy-7-methoxy-3'-keto-1.2-cyclopentenophenanthrene $C_{20}H_{16}O_4$ (I, page 2734 s; R = CH_3, R' = CO·CH_3). Nearly colourless needles (glacial acetic acid), m. 254° after shrinking at 250° (1938 Robinson). — **Fmn.** and **Rns.** See under those of the preceding compound.

4.7-Dimethoxy-3'-keto-1.2-cyclopentenophenanthrene $C_{19}H_{16}O_3$ (I, p. 2734 s; R = R' = CH_3). Pale yellow needles (cyclohexanone-alcohol), m. 200–201° after softening at 195°. Its alcoholic solution exhibits a violet fluorescence (1938 Robinson). — **Fmn.** From 4-hydroxy-7-methoxy-3'-keto-1.2-cyclopentenophenanthrene (p. 2734 s) on methylation with dimethyl sulphate and aqueous alcoholic NaOH at 60° (1939 Robinson).

Rns. On shaking with hydrogen, Adams's catalyst, and FeCl$_3$ in strong acetic acid at 75° for 20 hrs. it is reduced to a small extent, giving 4.7-dimethoxy-1.2-cyclopenteno-phenanthrene (p. 1958 s). Attempted reduction with sodium and amyl alcohol, butyl alcohol, or propyl alcohol gave mainly a compound [yellow needles (nitrobenzene), m. 313°; very sparingly soluble; purple solution in conc. H_2SO_4], probably resulting from the condensation of two molecules (1941 Cardwell). On treatment with potassium tert.-butoxide and isoamyl nitrite in dry tert.-butyl alcohol it gives the 4'-oxime of 4.7-dimethoxy-3'.4'-diketo-1.2-cyclopentenophenanthrene (p. 2961 s). Gives the 4'-hydroxymethylene derivative (p. 2764 s) by the action of ethyl formate and alcoholic sodium ethoxide in pyridine; a condensation product of this derivative with 4.7-dimethoxy-3'-keto-1.2-cyclopentenophenanthrene (see p. 2764 s) is obtained when the dimethoxyketone is treated with amyl formate and potassium tert.-butoxide in dry tert.-butyl alcohol at room temperature (1939 Robinson). For attempted conversion into the cyanohydrin, see 1941 Cardwell. On refluxing with ethyl bromoacetate, zinc wool, and a little iodine in benzene - toluene (2:1) for 6–8 hrs., followed by refluxing for several hrs. with dil. H_2SO_4, it gives the ethyl ester of (4.7-dimethoxy-1.2-cyclopentenophenanthrylidene)-3'-acetic acid (1941 a Robinson).

References, *p. 2737 s*

4-Ethoxy-7-methoxy-3'-keto-1.2-cyclopentenophenanthrene $C_{20}H_{18}O_3$ (I, page 2734 s; R = CH_3, R'= C_2H_5). Needles (alcohol), m. 194° (1939 Robinson). — **Fmn.** From 4-hydroxy-7-methoxy-3'-keto-1.2-cyclopentenophenanthrene (p. 2734 s) on heating with diethyl sulphate and aqueous alcoholic NaOH on the water-bath; quantitative yield (1939 Robinson). — **Rns.** On reduction with aluminium isopropoxide and isopropyl alcohol it gives an inseparable mixture, m. 100–320°, and chiefly a self-condensation product (cf. the reactions of the preceding compound) (1941 Cardwell). Gives the 4'-hydroxymethylene derivative (p. 2765 s) with ethyl formate and alcoholic sodium ethoxide in pyridine (1939 Robinson).

Oxime $C_{20}H_{19}O_3N$. Pale yellow needles (toluene), m. 244–6°. — **Rns.** On hydrogenation in acetic anhydride in the presence of Adams's catalyst, ca. 30 % is converted into 3'-acetamino-7-methoxy-4-ethoxy-1.2-cyclopentenophenanthrene (p. 2106 s), the chief product being the *acetyl derivative of the oxime*, $C_{22}H_{21}O_4N$ [prisms (toluene), m. 209–210°] which may also be obtained from the oxime and acetic anhydride (1941 Cardwell).

4.7-Dihydroxy-3'-keto-1.2-cyclopenteno-9.10-dihydrophenanthrene $C_{17}H_{14}O_3$ (II, R = R'= H).

4-Hydroxy-7-methoxy-3'-keto-1.2-cyclopenteno-9.10-dihydrophenanthrene $C_{18}H_{16}O_3$ (II, R = CH_3, R'= H). For the (probable) structure, see 1941 b Robinson. — Needles (ethyl acetate or methanol), m. 140° (1939, 1941 b Robinson). — **Fmn.** From 4-acetoxy-7-methoxy-3'-keto-1.2-cyclopentenophenanthrene (p. 2734 s) in small yields on hydrogenation in glacial acetic acid in the presence of Adams's catalyst at 70° (1939 Robinson; cf. 1941 b Robinson) or in the presence of Raney nickel at 125° under 30 atm. pressure (1941 b Robinson). — **Rns.** Gives an orange 2.4-dinitrophenylhydrazone and a yellow piperonylidene derivative (1941 b Robinson).

4-Acetoxy-7-methoxy-3'-keto-1.2-cyclopenteno-9.10-dihydrophenanthrene $C_{20}H_{18}O_4$ (II, R = CH_3, R'= CO·CH_3). Needles (methanol), m. 145°. — **Fmn.** As above, using Adams's catalyst (1939 Robinson; cf. 1941 b Robinson).

4.7-Dimethoxy-3'-keto-1.2-cyclopenteno-9.10-dihydrophenanthrene $C_{19}H_{18}O_3$ (II, R = R'= CH_3). Laths (dil. acetone), m. 143°. Gives a bright yellow coloration with conc. H_2SO_4, a brilliant green fluorescence developing on warming. — **Fmn.** From the lead salt of 4.7-dimethoxy-9.10-dihydrophenanthrene-2-carboxylic-1-(β-propionic) acid on pyrolysis; this acid is obtained from the 4'-hydroxymethylene derivative of 4.7-dimethoxy-3'-keto-1.2-cyclopentenophenanthrene (see p. 2764 s) on heating with hydroxylamine hydrochloride in glacial acetic acid at 70°, refluxing of the resulting 4'-cyano-derivative for 10 days, hydrogenation of the dimethyl ester of the resulting 4.7-dimethoxyphenanthrene-2-carboxylic-1-(β-propionic) acid in the presence of Adams's catalyst in glacial acetic acid at 70°, and hydrolysis by refluxing with aqueous methanolic KOH (1939 Robinson).

2.4-Dinitrophenylhydrazone $C_{25}H_{22}O_6N_4$. Deep red prisms (dil. acetic acid), m. 242–3⁰ (1939 Robinson).

1938 Robinson, *J. Ch. Soc.* **1938** 1390, 1397; *Brit. Pat.* 514592 (issued 1939); *Fr. Pat.* 854270
    (1939, issued 1940); *Ch. Ztbl.* **1940** II 1651, 1652.
1939 Robinson, Rydon, *J. Ch. Soc.* **1939** 1394.
1941 Cardwell in Koebner, Robinson, *J. Ch. Soc.* **1941** 566, 574, 575.
    a Robinson, Slater, *J. Ch. Soc.* **1941** 376, 383.
    b Robinson, Willenz, *J. Ch. Soc.* **1941** 393, 396, 397.

---

b. Dihydroxy-monoketosteroids C_{18} with CO and OH in the Ring System

cis* - 1 - Methyl - 7 - methoxy - 4' - hydroxy - 5' - keto - 1.2 - cyclopentano-
1.2.3.4.9.10.11.12-octahydrophenanthrene $C_{19}H_{24}O_3$

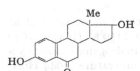

(I). Originally assumed to possess structure (I) or that of a 16-hydroxyœstrone methyl ether (cf. p. 2740 s) by Dane (1938, 1939); for the true structure (I), see 1942 E. W. J. Butz; 1949 L. W. Butz. — Colourless crystals (alcohol), m. 167⁰. Gives no colour reaction with $FeCl_3$ (1939 Dane). — **Fmn.** From 3-methyl-Δ^3-cyclopentene-1.2-dione and excess 1-vinyl-6-methoxy-3.4-dihydronaphthalene on heating in dioxane for 50 hrs. at 110–5⁰, followed by shaking of the resulting cis*-1-methyl-7-methoxy-4'.5'-diketo-1.2-cyclopentenohexahydrophenanthrene (p. 2939 s) with hydrogen and palladium-charcoal in methanol; over-all yield, 12% (1939 Dane). — **Rns.** On refluxing with 48% HBr in glacial acetic acid for $1/_2$ hr. it gives cis*-1-methyl-7-hydroxy-5'-keto-1.2-($\Delta^{3'}$-cyclopenteno)-1.2.3.4.9.10.11.12-octahydrophenanthrene (p. 2522 s) (1939 Dane).

$\Delta^{1.3.5(10)}$-Oestratriene-3.17β-diol-6-one, 6-Ketoœstradiol $C_{18}H_{22}O_3$. Plates (abs. alcohol), m. 281–3⁰ cor. with slight dec.; $[\alpha]_D^{24}$ +4⁰ (alcohol) (1940 Longwell). Absorption max. (solvent not stated) at 256 (8000) and 326 (3000) mμ (ε) (1940 Longwell; cf. 1938 Hirschmann). Its solution in conc. H_2SO_4 is greenish yellow, becoming deep purple with a bluish fluorescence on dilution with water. The solution in aqueous or methanolic KOH immediately assumes a yellow colour, disappearing on acidification. Is not precipitated by digitonin in 80% alcohol (1940 Longwell). — **Fmn.** Its diacetate is obtained, along with a diacetoxy-keto-acid (see p. 1968 s), from the diacetate of "α"-œstradiol (p. 1966 s) on oxidation with CrO_3 in strong acetic acid at 23–24⁰ (24 hrs.; yield, 20–25%); it is hydrolysed by alkali (1940 Longwell). — **Rns.** Couples with diazotized p-nitroaniline to give a dark red pigment which becomes brick-red on acidification with acetic acid (1940 Longwell).

Semicarbazone $C_{19}H_{25}O_3N_3$. Needles (abs. alcohol), dec. 280–310⁰ with evolution of gas (1940 Longwell).

* "cis" refers to the fusion of rings C and D.

Diacetate $C_{22}H_{26}O_5$. Needles (90% alcohol), m. 173–5° cor. Absorption max. (from the graph; solvent not stated) at 270 (9000) and 300 (1800) mμ (ε) (1940 Longwell). — **Fmn.** See above.

$Δ^{1.3.5(10)}$-**Oestratriene-3.17β-diol-7-one, 7-Ketoœstradiol** $C_{18}H_{22}O_3$. Colourless crystals (methanol or water), m. 127–130°. — **Fmn.** From 8-iso-$Δ^{1.3.5(10)}$-œstratriene-3.7.8.17β-tetrol (page 2206 s) on heating in a high vacuum for 5 hrs. at 200–220° (bath temperature) or on refluxing with 8% H_2SO_4. — **Rns.** On heating of its semicarbazone with alcoholic sodium ethoxide in a sealed tube at 180–190° for 12 hrs. it gives "α"-œstradiol (p. 1966 s). Gives $Δ^{1.3.5(10)}$-œstratriene-3.7.17β-triol 3.17-dibenzoate (p. 2152 s) on hydrogenation of its *dibenzoate* (prepared with benzoyl chloride in anhydrous pyridine at room temperature) in the presence of Raney nickel in methanol under atmospheric pressure (1941 SCHERING CORP.).

Semicarbazone $C_{19}H_{25}O_3N_3$, m. 187–8° (1941 SCHERING CORP.).

$Δ^{1.3.5(10)}$-**Oestratriene-3.17β-diol-16-one, 16-Ketoœstra-3.17β-diol, 16-Keto-"α"-œstradiol** $C_{18}H_{22}O_3$. Formerly supposed to be a 16-hydroxyœstrone (cf. p. 2740 s) (1942 Huffman; 1946 Butenandt); for the true structure, see 1947 b, c, 1948, 1949 Huffman. For the configuration at C-17, see 1950 Gallagher; 1950 Heusser; cf. also pp. 1363 s, 1364 s. — Plates (50% acetic acid), m. 234–7° to an orange liquid; $[α]_D^{29.5}$ —102° (alcohol) (1942, 1948 Huffman). Gives no precipitate with digitonin (1948 Huffman). — **Fmn.** From 16-oximino-œstrone (p. 2941 s) on refluxing with zinc dust in 50% acetic acid for 1 hr. (yield, 75–90%) (1942, 1944, 1948 Huffman); similarly from 16-ketoœstrone (p. 2941 s; yield, 72%) (1948 Huffman).

Rns. On reduction with sodium amalgam and alcohol in acetic acid at 40° it gives œstriol (p. 2152 s) and 16-epi-œstriol (p. 2159 s) (1947 c, 1949 Huffman); the latter is the sole reaction product on hydrogenation in 0.5 N NaOH in the presence of Adams's catalyst at room temperature (1944 Huffman). See also the reactions of the 3-methyl ether, below.

Oxime $C_{18}H_{23}O_3N$. Crystals (aqueous methanol), m. 224.5–225° dec. (1948 Huffman; cf. 1942 Huffman).

3-Benzoate $C_{25}H_{26}O_4$. Crystals (acetone), m. 241.5–243.5° to an orange liquid. — **Fmn.** From the above diolone and benzoyl chloride in cold 0.5 N NaOH (1948 Huffman; cf. 1942 Huffman).

3-Methyl ether* $C_{19}H_{24}O_3$. Needles (dil. acetone, dil. methanol, or ethyl acetate), m. 169–169.5° (1946 Butenandt), 167–8° (1947 b Huffman); the flakes (aqueous alcohol) and needles (aqueous methanol), m. 162–3° with

---

* The methyl ether, m. 174–177°, and its oxime, m. 175–177°, described by Huffman (1942) were not homogeneous substances (1948 Huffman).

slight yellowing, described by Huffman (1947 a) possibly represent another modification (1949 Huffman). — **Fmn.** From 16-oximino-œstrone methyl ether (p. 2942 s) on refluxing with zinc dust in 50 % acetic acid for 60–70 min.; yield, 57 % (1946 Butenandt), 77 % (1947 a, b Huffman). From 16-keto-œstrone methyl ether (p. 2941 s) on heating on the steam-bath with zinc dust in 50 % acetic acid for 40 min. (yield, 82 %) or with $TiCl_3$ in acetic acid for 30 min. (yield, 52 %) (1947 a Huffman).

Rns. Gives 16-ketoœstrone methyl ether (p. 2941 s) on refluxing with cupric acetate in methanol for 1 hr. (1942, 1947 a Huffman). On treatment with lead tetraacetate in acetic acid at room temperature it gives marrianolic acid 2-hemi-aldehyde methyl ether (I) (1949 Huffman). Reduction with sodium in boiling anhydrous isopropyl alcohol gives œstriol 3-methyl ether (p. 2157 s); once a very small amount of the 16-epimer (p. 2160 s) has also been obtained (1946 Butenandt); the former compound is also obtained on reduction with sodium amalgam and alcohol in acetic acid at 40° (1947 b Huffman).

3-Methyl ether oxime * $C_{19}H_{25}O_3N$. Crystals with $^1/_2 H_2O$ (alcohol), m. 211° to 213° (1946 Butenandt); crystals (aqueous methanol), m. 198.5–199.5° (1947 a Huffman; cf. 1947 b Huffman).

3-Methyl ether 17-acetate $C_{21}H_{26}O_4$. Needles (alcohol), m. 149°. — **Fmn.** From the above 3-methyl ether with acetic anhydride in pyridine on keeping for a day (1946 Butenandt).

3-Benzyl ether $C_{25}H_{28}O_3$. Needles (aqueous acetic acid), m. 196–198.5° to a yellow liquid (1948 Huffman); needles (abs. methanol), m. 198.5–200.5° (1949 Huffman). — **Fmn.** From 16-oximino-œstrone 3-benzyl ether (p. 2942 s) on refluxing with zinc dust in 66 % acetic acid for 80 min.; yield, 92 %. From the free diolone on refluxing with benzyl chloride and anhydrous K_2CO_3 in alcohol for 1.5 hrs.; yield, 79 % (1948 Huffman).

3-Benzyl ether 17-acetate $C_{27}H_{30}O_4$. Needles (aqueous acetone - alcohol), m. 141°. — **Fmn.** From the preceding compound on treatment with acetic anhydride in dry pyridine at room temperature (1948, 1949 Huffman).

7-Hydroxy-6-dehydroœstrone, 7-enol of 7-Ketoœstrone $C_{18}H_{20}O_3$.

Diacetate $C_{22}H_{24}O_5$. Needles (abs. alcohol), m. 171° to 171.5° cor. Absorption max. (solvent not stated) at 268 mμ (ε 9680). — **Fmn.** From 7-ketoœstrone (p. 2939 s) on refluxing with acetic anhydride and anhydrous sodium acetate for 1 hr. — **Rns.** Gives the diacetate of 7-hydroxyœstrone (below) on hydrogenation in glacial acetic acid in the presence of palladium black. Reverts to 7-ketoœstrone on refluxing with aqueous methanolic HCl (1939 Pearlman).

---

* See the footnote on p. 2738.

7β-Hydroxyœstrone $C_{18}H_{22}O_3$. For β-configuration at C-7, see 1958 Iriarte. — Needles (abs. alcohol), m. 265–7° cor. dec.; $[\alpha]_D^{25}$ + 134.5° (dioxane) (1939 Pearlman). — Its œstrogenic potency is $1/_{300}$ that of œstrone (1939 Pearlman). — **Fmn.** Its diacetate is obtained from the preceding diacetate on hydrogenation in glacial acetic acid in the presence of palladium black at room temperature (yield, 80%); it is hydrolysed by refluxing with aqueous methanolic alkali (1939 Pearlman). — **Rns.** The 3-benzoate gives that of 7-chloroœstrone (p. 2702 s) on treatment with PCl_5 and some dry $CaCO_3$ in dry chloroform (1940 Pearlman).

Diacetate $C_{22}H_{26}O_5$. Needles (90% alcohol), m. 131–2° cor., and prisms (90% alcohol), m. 122–3° cor.; the lower-melting form passes into the higher-melting after fusion. Absorption max. (solvent not stated) at 269 mμ (ε 660) (1939 Pearlman). — **Fmn.** See above.

3-Benzoate $C_{25}H_{28}O_4$. Plates (abs. alcohol), m. 181° cor. — **Fmn.** From the above diolone on shaking with benzoyl chloride in 1.25 N NaOH (1939 Pearlman). — **Rns.** See under those of the diolone.

8-Hydroxy-9(11)-dehydro-14-iso-œstrone, "*14-epi-Δ9(11)-8-Hydroxyequilin*"

$C_{18}H_{20}O_3$. Colourless platelets (benzene-acetone), m. 204° cor. dec. (1938 Hirschmann). Absorption max. in alcohol at 270 mμ (ε 16500) (1938 Hirschmann; 1939 Dannenberg). — **Fmn.** From the acetate of 8-dehydro-14-iso-œstrone ("isoequilin A", p. 2540 s) on treatment with OsO_4 in ether, followed by refluxing with Na_2SO_3 in dil. alcohol; yield, 25% (1938 Hirschmann). — **Rns.** Gives an orange pigment on coupling with an acid solution of p-nitrodiazobenzene (1938 Hirschmann).

Monoacetate $C_{20}H_{22}O_4$. Colourless oil. — **Fmn.** From the above diolone on treatment with acetic anhydride in dry pyridine at room temperature (1938 Hirschmann).

16-Hydroxyœstrone $C_{18}H_{22}O_3$. For a compound to which this structure was originally ascribed, see 16-keto-"α"-œstradiol (page 2738 s). — Crystals (ether), m. 231°. — **Fmn.** From 16-bromo-œstrone acetate (p. 2702 s) on refluxing with potassium acetate in butanol-xylene for $1^1/_2$ hrs. followed by refluxing of the resulting diacetate with methanolic KOH for 1 hr. (yield, 23%); similarly from 16-bromoœstrone benzoate. — **Rns.** On refluxing with aluminium isopropoxide in isopropyl alcohol for 5–6 hrs. it gives a 16-hydroxyœstradiol* m. 265–271° (1937 Schering A.-G.).

1937 Schering A.-G. (Inhoffen), *Ger. Pat.* 720015 (issued 1942); *C. A.* **1943** 2520; *Ch. Ztbl.* **1942** II 1037.
1938 Dane, Schmitt, *Ann.* **536** 196.
     Hirschmann, Wintersteiner, *J. Biol. Ch.* **126** 737, 740, 746, 747.

* Cf. the m. p.'s of œstriol (p. 2152 s) and 16-epi-œstriol (p. 2159 s).

1939 Dane, Schmitt, *Ann.* **537** 246.
   Dannenberg, *Abhandl. Preuss. Akad. Wiss. Math.-naturwiss. Kl.* **1939** No. 21, pp. 39, 63, 64.
   Pearlman, Wintersteiner, *J. Biol. Ch.* **130** 35, 39, 41, 43, 44.
1940 Longwell, Wintersteiner, *J. Biol. Ch.* **133** 219, 224–226.
   Pearlman, Wintersteiner, *J. Biol. Ch.* **132** 605, 609.
1941 SCHERING CORP. (Schwenk, Bloch, Whitman), *U.S. Pat.* 2418603 (issued 1947); *C. A.* **1947** 4518; SCHERING A.-G. (Schwenk, Bloch, Whitman), *Fr. Pat.* 892206 (1942, issued 1944);
   Ch. Ztbl. **1948** I 492.
1942 E. W. J. Butz, L. W. Butz, *J. Org. Ch.* **7** 199, 216.
   Huffman, *J. Am. Ch. Soc.* **64** 2235.
1944 Huffman, Darby, *J. Am. Ch. Soc.* **66** 150, 151.
1946 Butenandt, Schäffler, *Z. Naturforsch.* **1** 82, 84 (footnote 11), 86.
1947 a Huffman, *J. Biol. Ch.* **167** 273, 277, 279, 280.
   b Huffman, *J. Biol. Ch.* **169** 167.
   c Huffman, Lott, *J. Am. Ch. Soc.* **69** 1835.
1948 Huffman, Lott, *J. Biol. Ch.* **172** 325.
1949 L. W. Butz, Rytina, *Organic Reactions* Vol. V, New York, pp. 136, 139, 189, 190.
   Huffman, Lott, *J. Am. Ch. Soc.* **71** 719, 724, 725, 727.
1950 Gallagher, Kritchevsky, *J. Am. Ch. Soc.* **72** 882, 884 footnote 17.
   Heusser, Feurer, Eichenberger, Prelog, *Helv.* **33** 2243, 2246.
1958 Iriarte, Ringold, Djerassi, *J. Am. Ch. Soc.* **80** 6105, 6109.

c. Dihydroxy-monoketosteroids C_{19} with CO and OH in the Ring System

Summary of Dihydroxyketones with Androstane Skeleton *

Position of			Config. at C-5	M.p. °C.	$[\alpha]_D^{(0)}$	Diacetate, m.p.°C.	Page
CO	OH	double bond					
3	16β.17β	Δ^4	—	173	—	199	2742 s
7	3β.17β	Δ^5	—	201$^+$	—133$^+$ (alc.)	220	2742 s
7	3β.17β	satd.	α	199	—53 (CHCl$_3$)	194	2743 s
11	3.17β	satd.	α	248	—	163	2743 s
16	3β.17β	Δ^5	—	200	—	125	2744 s
17	2β.3α(?)	satd.	α	198	—	—	2745 s
17	3β.4β	Δ^5	—	205	—29 (CHCl$_3$)	—	2746 s
17	3β.5	satd.	α	288	+93 (methanol)	—	2746 s
17	3β.6β	Δ^4	—	270	—	164	2747 s
17	3β.6β	satd.	α	205	—	—	2747 s
17	3α.6α	satd.	β	—	—	—	2747 s
17	3α.11β	satd.	α	198	+97 (alc.)	—	2748 s
17	3β.11β	satd.	α	235	+85 (alc.)	156	2748 s
17	3α.12α	satd.	β	165	+167 (CHCl$_3$)	158	2749 s
17	3β.16	Δ^5	—	oil	—	—	2751 s

* For values other than chosen for this summary, see the compounds themselves.
$^+$ Hydrate.

Δ^4-Androstene-16β.17β-diol-3-one, 16β-Hydroxytestosterone $C_{19}H_{28}O_3$. Leaflets

(water), m. 172–3° (1939 Butenandt). Absorption max. in methanol at 241 mμ (ε 16700), in water at 250 mμ (ε 16400) (1939 Butenandt; 1939 Dannenberg). — Strongly œstrogenic; its androgenic potency (in the Fussgänger capon comb test) is $^1/_{300}$ that of testosterone (1939 Butenandt). — **Fmn.** Its acetonide (below) is obtained from that of Δ^5-androstene-3β.16β.17β-triol (p. 2167 s) by refluxing with aluminium isopropoxide and cyclohexanone in toluene for 1 hr. (yield, 63%); it is hydrolysed by refluxing with 50% acetic acid - dioxane (1:1) for 2 hrs. (yield, 68%) (1939 Butenandt). — *Diacetate* $C_{23}H_{32}O_5$. Needles (methanol), m. 199°; prepared with acetic anhydride in pyridine (1939 Butenandt).

Acetonide $C_{22}H_{32}O_3$. Crystals (methanol), m. 183–4° (1939 Butenandt). Absorption max. in ether at 234 mμ (ε 17400) (1939 Butenandt; 1939 Dannenberg). — **Fmn.** and **Rns.** See the above diolone.

1939 Butenandt, Schmidt-Thomé, Weiss, *Ber.* **72** 417, 420, 423, 424.
Dannenberg, *Abhandl. Preuss. Akad. Wiss. Math.-naturwiss. Kl.* **1939** No. 21, p. 53.

Δ^5-Androstene-3β.17β-diol-7-one $C_{19}H_{28}O_3$. Prisms with 1 H_2O (alcohol),

m. 201°; $[\alpha]_D^{20}$ —133° (alcohol) (1936, 1938 Butenandt; 1936 SCHERING A.-G.; 1936 SCHERING-KAHLBAUM A.-G.). Absorption max. in chloroform at 238 mμ (ε 7000) (1938 Butenandt; 1939 Dannenberg). — Has no androgenic activity in the capon comb test

(1936 Tscherning). — **Fmn.** The diacetate is obtained from that of Δ^5-andros-tene-3β.17β-diol (p. 1999 s) on oxidation with CrO_3 in strong acetic acid at 55° (yield, 26%); it is hydrolysed by heating with methanolic sodium meth-oxide on the water-bath for 30 min. (1936, 1938 Butenandt; 1936 SCHERING A.-G.; 1936 SCHERING-KAHLBAUM A.-G.; 1938 Ogata).

Rns. On refluxing of the diacetate with aluminium isopropoxide in abs. isopropyl alcohol it gives Δ^5-androstene-3β.7β.17β-triol* (p. 2164 s) (1936, 1938 Butenandt; 1936 SCHERING A.-G.). On shaking the diacetate in alcohol with hydrogen in the presence of Raney nickel until 1 mol. is taken up, it gives the diacetate of androstane-3β.17β-diol-7-one (below); on hydrogenation in glacial acetic acid in the presence of previously reduced platinum oxide, until 2 mols. of hydrogen are taken up, the 3.17-diacetate of androstane-3β.7.17β-triol (p. 2165 s) is formed (1936 SCHERING A.-G.; cf., after the closing date for this volume, 1952 Heusler). The diacetate gives $\Delta^{3,5}$-androstadien-17β-ol-7-one (p. 2603 s) on refluxing for 20 min. with 1 N methanolic HCl (1938 Butenandt) or with 2 N methanolic KOH (1936 SCHERING A.-G.).

Diacetate $C_{23}H_{32}O_5$. Needles (alcohol), m. 219–220° (1936, 1938 Butenandt; 1936 SCHERING A.-G.; 1936 SCHERING-KAHLBAUM A.-G.; 1938 Ogata); $[\alpha]_D^{20}$ —135° (chloroform) (1938 Butenandt). — **Fmn.** and **Rns.** See above. — *Oxime* $C_{23}H_{33}O_5N$, needles (dil. alcohol), m. 226° dec. (1938 Ogata).

Androstane-3β.17β-diol-7-one $C_{19}H_{30}O_3$. For this compound, see 1952 Heusler

(after the closing date for this volume). *Diacetate* $C_{23}H_{34}O_5$, m. 193–4° (1936 SCHERING A.-G.). — **Fmn.** From the above unsaturated di-acetate on shaking with hydrogen in alcohol in the presence of Raney nickel until 1 mol. is taken up (1936 SCHERING A.-G.; cf. 1952 Heusler).

1936 Butenandt, Riegel, *Ber.* **69** 1163, 1165 footnote.
SCHERING A.-G. (Butenandt, Logemann), *U.S. Pat.* 2170124 (issued 1939); *C. A.* **1940** 1133; *Ch. Ztbl.* **1940** I 1232; SCHERING-KAHLBAUM A.-G., *Brit. Pats.* 486596, 486854 (issued 1938); *C. A.* **1938** 8706; *Ch. Ztbl.* **1939** I 1602, 1603.
SCHERING-KAHLBAUM A.-G., *Swiss Pat.* 194756 (issued 1938); *Ch. Ztbl.* **1938** II 1087.
Tscherning, *Angew. Ch.* **49** 11, 13, 15.
1938 Butenandt, Hausmann, Paland, *Ber.* **71** 1316, 1320, 1321.
Ogata, Kawakami, *C. A.* **1938** 4172; *Ch. Ztbl.* **1938** I 4057.
1939 Dannenberg, *Abhandl. Preuss. Akad. Wiss. Math.-naturwiss. Kl.* **1939** No. 21, pp. 15, 55.
1952 Heusler, Wettstein, *Helv.* **35** 284, 290, 291.

Androstane-3.17β-diol-11-one $C_{19}H_{20}O_3$ (p. 149). For position of the keto-group, see also 1937 Reichstein. Configuration at

C-17 by analogy with similar cases; according to Heusser (1952) the compound is probably androstane-3β.17β-diol-11-one. — **Rns.** Reverts to androstane-3.11.17-trione (p. 2980 s) on oxidation with CrO_3 in

* According to Butenandt (1938) the free triol is obtained in this reaction, according to SCHERING A.-G. (1936) the triol 3.17-diacetate.

glacial acetic acid (1937 Steiger). Is reduced by the Clemmensen method, but not by the Wolff-Kishner method (1944 Shoppee).

1937 Reichstein, *Helv.* **20** 978.
Steiger, Reichstein, *Helv.* **20** 817, 825.
1944 Shoppee, *Helv.* **27** 246, 248 footnote.
1952 Heusser, Heusler, Eichenberger, Honegger, Jeger, *Helv.* **35** 295, 297, 303.

Δ⁵-Androstene-3β.17β-diol-16-one $C_{19}H_{28}O_3$. Originally assumed to be Δ^5-androstene-3β.16-diol-17-one (16-hydroxy-5-dehydro-epiandrosterone) (1939 Butenandt; cf. also 1941 Stodola); for the true structure, see 1942 Stodola; cf. also 1949 Huffman. For β-configuration at C-17, see 1950 Gallagher; cf. also pp. 1363 s, 1364 s. — Crystals (dioxane-water), m. 197⁰ (1939 Butenandt); leaflets (aqueous acetic acid), m. 197–9⁰ (1949 Huffman), 200–202⁰ (1947 Huffman). — **Fmn.** The 3-acetate is obtained from the acetate of 16-(2-butyl-idene)-Δ^5-androsten-3β-ol-17-one (p. 2657 s) on conversion into the 5.6-dibromide with bromine in chloroform, followed by ozonization in chloroform and then by refluxing with zinc dust in glacial acetic acid for 15 min. (crude yield, 12%); it is hydrolysed by warming with 4% methanolic KOH (1939 Butenandt). The diacetate is obtained from the diacetate of 16-(2-butylidene)-Δ^5-androstene-3β.17-diol (p. 2043 s) on treatment with OsO_4 in carbon tetrachloride at room temperature in the dark, followed by warming with zinc dust in ca. 65% acetic acid at 45–50⁰ and oxidation of the lower-melting isomer (m. 150–2⁰) of the resulting 16-(α-methyl-α-hydroxypropyl)-Δ^5-androstene-3β.16.17-triol 3.17-diacetates (p. 2198 s) with lead tetraacetate in dry benzene (over-all yield, 20%); similarly from the diacetate of 16-benzylidene-Δ^5-androstene-3β.17-diol (p. 2045 s) via 16-(α-hydroxybenzyl)-Δ^5-androstene-3β.16.17-triol 3.17-diacetate (p. 2200 s) (over-all yield, 9%) (1942 Stodola). From 16-oximino-Δ^5-androsten-3β-ol-17-one (p. 2950 s) on refluxing with zinc dust in 50% acetic acid for 1 hr.; crude yield, 96% (1941 Stodola).

Rns. Gives Δ^5-androstene-3β.16β.17β-triol (p. 2167 s) on hydrogenation in the presence of Raney nickel in alcohol (1941 Stodola); the 3-acetate reacts similarly in ethyl acetate (1939 Butenandt), and the diacetate in alcohol (1941 Stodola) or in alcohol - ethyl acetate at 60–5⁰ (1947, 1949 Huffman). On reduction with sodium amalgam in dil. alcohol - acetic acid at 40⁰, the two 16-epimeric Δ^5-androstene-3β.16.17β-triols (pp. 2165 s and 2167 s) are obtained (1947 Huffman). On treatment of the diacetate with ethyl mercaptan in the presence of $ZnCl_2$ and anhydrous Na_2SO_4 at 0⁰ (72 hrs.) it gives the *diethyl thioketal* $C_{27}H_{42}O_4S_2$ [flakes (aqueous acetone + a trace of pyridine), m. 136.5–138⁰] which on heating with Raney nickel in abs. alcohol-dioxane on the steam-bath for 7 hrs. gives the diacetate of Δ^5-androstene-3β.17β-diol (p. 1999 s) (1947, 1949 Huffman).

Δ⁵-Androstene-3β.17β-diol-16-one 3-acetate $C_{21}H_{30}O_4$. Needles with 1 H_2O (methanol), m. 192⁰ (1939 Butenandt). — **Fmn.** and **Rns.** See those of the free diolone, above. — *Oxime* $C_{21}H_{31}O_4N$. Leaflets (dil. alcohol), m. 244° dec. (1939 Butenandt).

Δ^5-*Androstene-3β.17β-diol-16-one diacetate* $C_{23}H_{32}O_5$. Needles (alcohol, 70% alcohol, or acetone - petroleum ether), m. 123° (1939 Butenandt), 124.5° to 125° (1941, 1942 Stodola; 1947, 1949 Huffman). — **Fmn.** From the above diolone on treatment with acetic anhydride in dry pyridine at room temperature (24 hrs.) or with acetic anhydride and some 90% phosphoric acid (very rapidly) (1947, 1949 Huffman). From the above 3-acetate on treatment with acetic anhydride in pyridine at room temperature (1939 Butenandt; 1941 Stodola). See also under formation of the free diolone. — **Rns.** See those of the free diolone.

Δ^5-*Androstene-3β.17β-diol-16-one 3-methyl ether* $C_{20}H_{30}O_3$. Crystals (aqueous alcohol), m. 188–191° (1949 Huffman). — **Fmn.** From 16-oximino-Δ^5-androsten-3β-ol-17-one methyl ether (p. 2951 s) on refluxing with zinc dust in 50% acetic acid for 1 hr.; yield, 64% (1948, 1949 Huffman). — **Rns.** Gives the 3-methyl ethers of the two 16-epimeric Δ^5-androstene-3β.16.17β-triols (pp. 2167 s and 2168 s) on reduction with sodium amalgam in alcohol - acetic acid at 40° (1949 Huffman).

Δ^5-*Androstene-3β.17β-diol-16-one 3-methyl ether 17-acetate* $C_{22}H_{32}O_4$. Needles (methanol), m. 166–9°. — **Fmn.** From the preceding compound on treatment with acetic anhydride in pyridine (1948 Huffman).

1939 Butenandt, Schmidt-Thomé, Weiss, *Ber.* **72** 417, 422, 423; SCHERING A.-G. (Butenandt, Schmidt-Thomé, Schwenk), *Swed. Pat.* 98872 (issued 1940); *Ch. Ztbl.* **1940** II 2784; SCHERING CORP. (Butenandt, Schmidt-Thomé, Schwenk), *U.S. Pat.* 2353808 (issued 1944); *C. A.* **1945** 1258.
1941 Stodola, Kendall, McKenzie, *J. Org. Ch.* **6** 841.
1942 Stodola, Kendall, *J. Org. Ch.* **7** 336.
1947 Huffman, Lott, *J. Am. Ch. Soc.* **69** 1835.
1948 Huffman, Lott, *J. Biol. Ch.* **172** 789, 792, 794.
1949 Huffman, Lott, *J. Am. Ch. Soc.* **71** 719, 723, 724, 726.
1950 Gallagher, Kritchevsky, *J. Am. Ch. Soc.* **72** 882, 884 footnote 17.

Androstane-2β(?).3α(?)-diol-17-one, 2-Hydroxyandrosterone $C_{19}H_{30}O_3$. For

probable configuration at C-2 and C-3, see the remark on the corresponding triol (p. 2163 s) obtained by reduction. — Crystals (dil. acetone) m. 195–8°; very difficult to separate from a small amount of isomers *. Not precipitated by digitonin in alcohol. — **Fmn.** From Δ^2 or Δ^3-androsten-17-one (p. 2401 s) on heating with H_2O_2 in acetic acid at 100°, followed by saponification. — **Rns.** Gives androstane-2β(?).3α(?).17β-triol (p. 2163 s) on reduction with sodium in propanol (1939 Marker).

1939 Marker, Plambeck, *J. Am. Ch. Soc.* **61** 1332.

---

* Cf. also the structure of the starting androsten-17-one (p. 2401 s).

Δ^5-Androstene-3β.4β-diol-17-one, 4β-Hydroxy-5-dehydroepiandrosterone

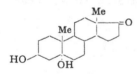

$C_{19}H_{28}O_3$. Plates (50% aqueous dioxane), m. 204–5°; $[\alpha]_D^{20}$ —28.5° (chloroform). — Is not androgenic in the capon comb test. — **Fmn.** The 4-acetate is obtained from Δ^5-androsten-3β-ol-17-one (5-dehydroepiandrosterone, p. 2609 s) on conversion into the dibromide with 1 mol. bromine in chloroform, followed by treatment with silver acetate in ether-pyridine for 20 min. in the dark (1943 Petrow).

4-Acetate $C_{21}H_{30}O_4$. Prismatic needles (50% alcohol or dioxane), m. 192–3°; $[\alpha]_D^{20}$ —60.7°, $[\alpha]_{5461}^{20}$ —69.2° (chloroform). Gives an intense blue coloration with trichloroacetic acid (1943 Petrow). — **Fmn.** See above.

A mixture of Δ^5-androstene-3β.4-diol-17-one and Δ^4-androstene-3β.6-diol-17-one is obtained by the action of selenium dioxide on Δ^5-androsten-3β-ol-17-one (5-dehydroepiandrosterone, p. 2609 s); on refluxing with 30% H_2SO_4 in alcohol it gives Δ^4-androstene-3.17-dione (p. 2880 s) (1937 CIBA).

1937 CIBA, *Brit. Pat.* 497394 (issued 1939); *C. A.* **1939** 3812; *Ch. Ztbl.* **1939** I 2828; CIBA PHARMACEUTICAL PRODUCTS (Miescher, Wettstein), *U.S. Pat.* 2229813 (1938, issued 1941); *C. A.* **1941** 3040; *Ch. Ztbl.* **1941** II 2467, 2468.
1943 Petrow, Rosenheim, Starling, *J. Ch. Soc.* **1943** 135, 138.

Androstane-3β.5-diol-17-one, 5-Hydroxy-epiandrosterone

$C_{19}H_{30}O_3$. Hexagonal plates (ethyl acetate - methanol), m. 278–288° cor. dec.; m. 281–2° cor. in an evacuated capillary, with partial sublimation; $[\alpha]_D^{15}$ +93° (methanol). Sparingly soluble in ether, acetone, ethyl acetate, and hexane, somewhat more readily in methanol. — **Fmn.** The 3-acetate is obtained from the 3-acetate of androstane-3β.5.17β-triol (p. 2163 s) on oxidation with CrO_3 in strong acetic acid at room temperature or from the acetate of 5α.6α-epoxyandrostan-3β-ol-17-one (p. 2769 s) on hydrogenation in glacial acetic acid in the presence of platinum oxide until 1 mol. hydrogen is absorbed; it is hydrolysed by refluxing with K_2CO_3 in aqueous methanol for 3 hrs. — **Rns.** Gives Δ^4-androstene-3.17-dione (p. 2880 s) on refluxing with acetone and aluminium tert.-butoxide in dioxane-benzene for 21 hrs. On hydrogenation of the 3-acetate in glacial acetic acid in the presence of platinum oxide it reverts to the 3-acetate of androstane-3β.5.17β-triol. Is not acylated by treatment in cold pyridine with acetic anhydride or benzoyl chloride (1944 Ruzicka).

3-Acetate $C_{21}H_{32}O_4$. Crystals (aqueous methanol), m. 152.5–153.5° cor., resolidifying, and remelting at 162.5–163.5° cor.; $[\alpha]_D^{18}$ +59° (chloroform) (1944 Ruzicka). — **Fmn.** and **Rns.** See above.

1944 Ruzicka, Muhr, *Helv.* **27** 503, 508–510.

Δ^4-**Androstene-3β.6β-diol-17-one** $C_{19}H_{28}O_3$. Has been described after the closing date for this volume by Davis (1949) (cf. also 1954 Amendolla). — For a mixture of a Δ^4-androstene-3β.6-diol-17-one with a Δ^5-androstene-3β.4-diol-17-one, see p. 2746 s.

Androstane-3β.6β-diol-17-one, 6β-Hydroxy-epiandrosterone $C_{19}H_{30}O_3$. Crystals

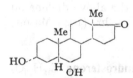

(methanol - ether - pentane), m. 205⁰ (1940 Marker; 1940a, b PARKE, DAVIS & Co.). — **Fmn.** Its diacetate is obtained from that of cholestane-3β.6β-diol (p. 2064 s) on oxidation with CrO_3 in strong acetic acid at 90⁰ for 9 hrs. (isolated as semicarbazone; yield, ca. 2%); the free diolone is obtained from the diacetate semicarbazone on refluxing with aqueous alcoholic H_2SO_4 for ¹/₂ hr. and then with 2% methanolic KOH for 1 hr. (1940 Marker; 1940a, b PARKE, DAVIS & Co.). The diacetate is also formed when the diacetate oxime of Δ^{16}-allopregnene-3β.6β-diol-20-one (p. 2324 s) is treated with p-toluenesulphonyl chloride in cold pyridine and then hydrolysed with dil. H_2SO_4 at room temperature (1941a PARKE, DAVIS & Co.). — **Rns.** Gives androstan-6β-ol-3.17-dione (p. 2947 s) on treatment of the diacetate with methanolic KOH (1 mol.) at room temperature for 2 days, followed by oxidation of the resulting 6-acetate (not described in detail) with CrO_3 in strong acetic acid at room temperature and then by refluxing with alcoholic NaOH (1941b PARKE, DAVIS & Co.).

Diacetate semicarbazone $C_{24}H_{37}O_5N_3$. Needles (methanol), m. 222⁰ (1940 Marker). — **Fmn.** See above.

5-iso-Androstane-3α.6α-diol-17-one, Aetiocholane-3α.6α-diol-17-one $C_{19}H_{30}O_3$. Crystals (dil. alcohol). — **Fmn.** From the diacetate of pregnane-3α.6α-diol-20-one (p. 2324 s) on refluxing with methylmagnesium iodide in benzene, followed by dehydration of the resulting carbinol by refluxing with acetic anhydride, ozonization of the resulting ethylene compound in chloroform, and saponification with alkali; may be similarly obtained from the methyl ester of 3α.6α-dihydroxyætio-cholanic acid (ætiohyodesoxycholic acid) (1940a PARKE, DAVIS & Co.).

1940 Marker, Krueger, Adams, Jones, *J. Am. Ch. Soc.* **62** 645.
  a PARKE, DAVIS & Co. (Marker), *U.S. Pat.* 2337563 (issued 1943); *C. A.* **1944** 3423.
  b PARKE, DAVIS & Co. (Marker), *U.S. Pat.* 2337564 (issued 1943); *C. A.* **1944** 3423.
1941 a PARKE, DAVIS & Co. (Tendick, Lawson), *U.S. Pat.* 2335616 (issued 1943); *C. A.* **1944** 3095.
  b PARKE, DAVIS & Co. (Marker, Lawson), *U.S. Pat.* 2366204 (issued 1945); *C. A.* **1945** 1649;
    Ch. Ztbl. **1945** II 1385, 1386.
1949 Davis, Petrow, *J. Ch. Soc.* **1949** 2536.
1954 Amendolla, Rosenkranz, Sondheimer, *J. Ch. Soc.* **1954** 1226.

Androstane-3α.11β-diol-17-one, 11β-Hydroxyandrosterone $C_{19}H_{30}O_3$. Crystals
(acetone), m. 197–8° (1945 Mason), 198° cor. (Kofler

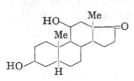

block) (1946 Shoppee), 194.5–196.5° (1946 Miller);
needles (acetone - ligroin, or benzene - ligroin), m. 199 to
200° (1948 Lieberman); hexagonal platelets (acetone-
ether), m. 195°→199° cor. (with partial change) (1953
Finkelstein); $[\alpha]_D$ +97° (alcohol) (1945 Mason; cf. also
1948 Lieberman); $[\alpha]_D^{17}$ +90° (acetone) (1953 Finkelstein). Not precipitated
by digitonin in methanol or ethanol (1945 Mason; 1946 Miller). The intensity
of colour produced in the Zimmermann test (with m-dinitrobenzene and KOH
in alcohol) is 74% of that produced by dehydroepiandrosterone (1945 Mason).
— The androgenic activity of its 3-acetate is one fourth that of androsterone
(1946 Miller).

Occ. In the urine of normal persons (1942 Dobriner*; 1946 Mason), of
patients with adrenal tumours and adrenal hyperplasia (1942 Dobriner*; 1945
Mason; 1948 Lieberman; 1946, 1950 Miller), in the urine of female pseudo-
hermaphrodites (1953 Finkelstein), and in the urine of patients with Cushing's
syndrome (1948 Lieberman); the $\Delta^{9(11)}$-androsten-3α-ol-17-one found in the
urine (hydrolysed by hydrochloric acid) of a girl with probable adrenal hyperplasia (1945 Dorfman) is very probably formed from the diolone during the
hydrolysis (see the reactions, below) (1946, 1950 Miller).

Rns. On oxidation with CrO_3 in strong acetic acid at room temperature
it gives androstane-3.11.17-trione (p. 2980 s) (1945 Mason). The 3-acetate
gives the acetate of $\Delta^{9(11)}$-androsten-3α-ol-17-one (p. 2626 s) on refluxing
with conc. HCl - glacial acetic acid (1:4) for 15–30 min. (followed by partial
reacetylation) (1945 Mason; 1946 Miller; 1946 Shoppee) or on treatment
with $POCl_3$ in pyridine at room temperature (1950 Miller). Gives a yellow
dinitrophenylhydrazone, m. 250–1° (1945 Mason).

3-Acetate $C_{21}H_{32}O_4$. Rods (acetone), crystals (abs. methanol), m. 238–240°
(1945 Mason; 1946 Miller); flat needles (acetone-ether), m. 236–240° cor.;
$[\alpha]_D^{15}$ +90° (acetone) (1953 Finkelstein). — **Fmn.** From the above diolone on
heating with acetic anhydride in pyridine at 90° for $^1/_2$–1 hr. (1945 Mason;
1946 Miller). — **Rns.** See above.

Androstane - 3β.11β - diol - 17 - one,   11β - Hydroxy - epiandrosterone $C_{19}H_{30}O_3$
(p. 149). Needles (acetone-ether), m. 234–5° cor. (Kof-
ler block) (1942 v. Euw), 235–7° cor. (1940a Shoppee);
$[\alpha]_D^{19}$ +81°, $[\alpha]_{5461}^{19}$ +105° (dioxane) (1942 v. Euw). —
Occ. For its isolation from adrenal cortex, see also
1941 v. Euw. — **Fmn.** The allopregnanepentol from
which the present diolone is obtained by oxidation
with lead tetraacetate in glacial acetic acid at 50° or with excess periodic
acid in aqueous methanol at room temperature (1936 Reichstein) is allo-
pregnane-3β.11β.17α.20β.21-pentol (see p. 2202 s). From 3β.11β.17α-trihydr-

---

* In this paper the substance was described as "compound $C_{19}H_{32}O_3$"; for the true composition
and structure, see 1945 Mason; 1948 Lieberman.

oxyætioallocholanic acid on oxidation with lead tetraacetate in glacial acetic acid at 55^0 for $1^1/_4$ hrs.; crude yield, 65 % (1942 v. Euw).

Rns. Gives androstan-11β-ol-3.17-dione (p. 2947 s) on refluxing with aluminium phenoxide and acetone in benzene for 22 hrs. (1941 Reich). The intermediate product m. 156.5–158^0 cor. obtained in the oxidation with CrO_3 to androstane-3.11.17-trione (1936 Reichstein) was impure androstan-3β-ol-11.17-dione (p. 2949 s) (1941 Reich). On refluxing with conc. HCl - glacial acetic acid (1 vol.:9 vol.) for 15 min. it gives $\Delta^{9(11)}$- androsten-3β-ol-17-one (p. 2626 s) (1940b Shoppee; cf. 1946 Shoppee for the structure of the reaction product); the acetate of this hydroxyketone is obtained when the 3-acetate of the diolone is heated with $KHSO_4$ at 135–140^0 (bath temperature) under 0.05 mm. pressure (1940b Shoppee).

3-Acetate $C_{21}H_{32}O_4$ (p. 149). Needles (acetone-ether 1:2), m. 228–9^0 cor. (Kofler block); $[\alpha]_D^{19}$ +70.5^0, $[\alpha]_{5461}^{19}$ +87^0 (dioxane) (1942 v. Euw). — **Fmn.** From the above diolone on treatment with acetic anhydride in pyridine at room temperature (1937 Steiger; 1942 v. Euw). — **Rns.** See above.

Diacetate $C_{23}H_{34}O_5$. Leaflets (ether-pentane), m. 154–6^0. — **Fmn.** From the above diolone on heating with acetic anhydride and pyridine in a sealed tube at 100^0 for 20 hrs. (1937 Steiger).

1936 Reichstein, *Helv.* **19** 402, 406–410.
1937 Steiger, Reichstein, *Helv.* **20** 817, 823, 824.
1940 a Shoppee, Reichstein, *Helv.* **23** 729, 734.
      b Shoppee, *Helv.* **23** 740, 743, 744.
1941 v. Euw, Reichstein, *Helv.* **24** 879, 880, 886.
      Reich, Reichstein, *Arch. intern. pharmacodynamie* **65** 415, 421.
1942 Dobriner, Gordon, Rhoads, Lieberman, Fieser, *Science* [N.S.] **95** 534.
      v. Euw, Reichstein, *Helv.* **25** 988, 993, 1012, 1020.
1945 Dorfman, Schiller, Sevringhaus, *Endocrinology* **37** 262.
      Mason, *J. Biol. Ch.* **158** 719; Mason, Kepler, *J. Biol. Ch.* **161** 235, 240, 248–250.
1946 Mason, *J. Biol. Ch.* **162** 745.
      Miller, Dorfman, Sevringhaus, *Endocrinology* **38** 19.
      Shoppee, *J. Ch. Soc.* **1946** 1134, 1136.
1948 Lieberman, Dobriner, Hill, Fieser, Rhoads, *J. Biol. Ch.* **172** 263, 285.
1950 Miller, Dorfman, *Endocrinology* **46** 514, 522, 523.
1953 Finkelstein, v. Euw, Reichstein, *Helv.* **36** 1266, 1268, 1274.

---

5 - iso - Androstane - 3α.12α - diol - 17 - one,   Aetiocholane - 3α.12α - diol - 17 - one

$C_{19}H_{30}O_3$. For α-configuration at C-12, see 1946 Sorkin. — Crystals (ether), m. 165.5–168^0 cor. (Kofler block) (1945 Reich; cf. 1946 Ettlinger); needles (ethyl acetate), m. 164.5–165^0 cor.; $[\alpha]_D$ +167^0 (chloroform) (1948 Marshall). — **Fmn.** From pregnane-3α.12α-diol-20-one (p. 2329 s) on treatment with ethyl nitrite and sodium ethoxide in abs. alcohol at room temperature for a week, followed by refluxing with aqueous alcoholic HCl for 3 hrs. (crude yield, 20 %) (1946 Ettlinger); from the diacetate of the same diolone on bromination with 3 mols. of bromine in glacial acetic acid not above 38^0, followed by boiling of the resulting

product (p. 2329 s footnote**) with 5 % methanolic KOH for 1 hr., methylation of the acid fraction with diazomethane, acetylation with acetic anhydride
in pyridine on the water-bath (3 hrs.), ozonization in chloroform, reduction
with zinc dust in glacial acetic acid on the water-bath, and boiling with 3 %
methanolic KOH for 1 hr. (isolated as diacetate; max. yield, 8 %) (1944
Koechlin). From desoxycholic acid methyl ester diacetate on heating with
CrO_3 in strong acetic acid at ca. 70°, refluxing of the neutral part of the
reaction product with 5 % methanolic KOH for 2 hrs., separation of the
neutral parts with Girard's reagent T, and hydrolysis of the condensation
product with hydrochloric acid at pH ca. 0; isolated as diacetate (very small
yield) which is hydrolysed with 4 % methanolic KOH at room temperature
(1943, 1945 Reich). The 3-acetate may be regenerated, in 28 % yield, from
its oxidation product (ætiocholan-3α-ol-12.17-dione acetate, p. 2950 s) by
hydrogenation in glacial acetic acid in the presence of platinum oxide at
room temperature (1945 Reich). See also the compound $C_{23}H_{34}O_5$ (p. 2751 s).

Rns. On oxidation with excess CrO_3 in glacial acetic acid at room temperature it gives ætiocholane-3.12.17-trione (p. 2980 s); on using less CrO_3,
ætiocholan-12α-ol-3.17-dione (p. 2947 s) is also obtained; the acetate of the
latter compound is obtained on similar oxidation of the 12-acetate (below),
whereas on similar oxidation of the 3-acetate, the acetate of ætiocholan-
3α-ol-12.17-dione (p. 2950 s) is formed. The 3-acetate and the diacetate are
not attacked by hydrogen in the presence of platinum oxide in glacial acetic
acid at room temperature; on hydrogenation at 75–80° (ca. 20 hrs.) the
diacetate gives a small amount of a compound m. 164.5–166.5° cor. (Kofler
block), and the 3-acetate a small amount of ætiocholane-3α.12α.17-triol
3-acetate (p. 2165 s), 25 % of unchanged material, and an oil which on oxidation with CrO_3 in glacial acetic acid at room temperature yields the acetate
of ætiocholan-3α-ol-12-one (p. 2604 s). On heating of the diacetate with
hydrazine hydrate and sodium ethoxide in abs. alcohol for 15 hrs. in an
evacuated sealed tube at 180° it gives ætiocholane-3α.12α-diol (p. 1992 s)
(1945 Reich).

3-Acetate $C_{21}H_{32}O_4$. Needles (acetone-hexane), m. 167.5–168.5° cor. (Kofler
block); sublimes at 170° under 0.01 mm. pressure. — **Fmn.** From the above
diolone on heating with 1⅓ mols. of acetic anhydride in abs. benzene on
the water-bath for 4 hrs. (1945 Reich). — **Rns.** See above.

12-Acetate $C_{21}H_{32}O_4$ (not obtained crystalline). — **Fmn.** From the diacetate
(below) on treatment with 1 % methanolic HCl at room temperature for
15 hrs. (1945 Reich). — **Rns.** For oxidation, see above.

Diacetate $C_{23}H_{34}O_5$. Prisms (dil. methanol or petroleum ether), m. 162°
to 162.5° cor. (Kofler block) (1943, 1945 Reich; cf. 1944 Koechlin); prisms
(acetone-ligroin), m. 156–7° cor. (1948 Marshall); m. 157.5–158.5° (1946 Ettlinger); $[\alpha]_D^{18}$ in acetone +176° to +179° (1943, 1945 Reich; 1946 Ettlinger),
+186° (1944 Koechlin); $[\alpha]_D$ +193° (chloroform) (1948 Marshall); $[\alpha]_{5461}^{16}$
+214° (acetone) (1943 Reich). — **Fmn.** From the above diolone on heating
with acetic anhydride in pyridine on the water-bath for 3 hrs. (1943 Reich). —

Rns. See the reactions of the free diolone and the formation of the 12-acetate (above). — *Oxime* $C_{23}H_{35}O_5N$. Microcrystalline powder (ether - petroleum ether), m. 99–102° cor. (Kofler block) (1945 Reich).

A **compound** $C_{23}H_{34}O_5$ [needles (dil. alcohol), m. 205–6°, giving a cherry-red Liebermann-Burchard reaction] said to be ætiocholane-3α.12α-diol-17-one diacetate (above) has been obtained in small yield from 24.24-diethylcholane-3α.12α.24-triol triacetate (p. 2133 s) on oxidation with CrO_3 in glacial acetic acid on the water-bath, along with nordesoxycholic acid diacetate and a compound $C_{34}H_{54}O_7$ (m. 180°) said to be the triacetate of 24.24-diethylcholane-3α.12α.24-triol-23-one (p. 2362 s) (1939 Kazuno).

1939 Kazuno, Shimizu, *J. Biochem.* **29** 421, 432.
1943 Reich, Reichstein, *Helv.* **26** 2102, 2106.
1944 Koechlin, Reichstein, *Helv.* **27** 549, 563, 564.
1945 Reich, *Helv.* **28** 863.
1946 Ettlinger, Fieser, *J. Biol. Ch.* **164** 451.
　　　Sorkin, Reichstein, *Helv.* **29** 1218.
1948 Marshall, Kritchevsky, Lieberman, Gallagher, *J. Am. Ch. Soc.* **70** 1837, 1839.

Δ⁵-Androstene-3β.16-diol-17-one, 16-Hydroxy-5-dehydroepiandrosterone

$C_{19}H_{28}O_3$. Light oil. — **Fmn.** From 16-bromo-5-dehydroepiandrosterone acetate on refluxing with potassium acetate in butanol-xylene for 2 hrs., followed by refluxing of the resulting diacetate with methanolic KOH for 2 hrs. — **Rns.** On reduction, e.g. by refluxing with aluminium isopropoxide in isopropyl alcohol, it gives a Δ⁵-androstene-3β.16.17β-triol, possibly a mixture of the two C-16 epimers (see pp. 2165 s and 2167 s).

1937 Schering A.-G. (Inhoffen), *Ger. Pat.* 720015 (issued 1942); *C. A.* **1943** 2520; *Ch. Ztbl.* **1942** II 1037.

d. Dihydroxy-monoketosteroids C_{21} with CO and OH in the Ring System

17-iso-T^{20}-Pregnyne-3α.17β-diol-11-one, 17α-Ethynylætiocholane-3α.17β-diol-11-one $C_{21}H_{30}O_3$. For reversal of the configuration at C-17, originally assigned, see 1950 Gallagher. — Crystals (ethyl acetate), m. 218.5–219° cor.; $[\alpha]_D$ —9.4° (acetone) (1946 Sarett; 1946a MERCK & Co., INC.). — **Fmn.** From ætiocholan-3α-ol-11.17-dione (p. 2949 s) on treatment in dioxane-abs. ether with a solution of potassium acetylide in liquid ammonia at room temperature; yield, 80% (1946 Sarett; 1946a MERCK & Co., INC.). — **Rns.** For catalytic hydrogenation, see the formation of the following diolone.

3-Acetate $C_{23}H_{32}O_4$. Crystals (ether · petroleum ether), m. 186–9° cor. — **Fmn.** From the above diolone on heating with acetic anhydride and pyridine for 10 min. at 100° (1946a MERCK & Co., INC.).

17-iso-Δ^{20}-Pregnene-3α.17β-diol-11-one, 17α-Vinylætiocholane-3α.17β-diol-11-one $C_{21}H_{32}O_3$. For configuration at C-17, see the preceding diolone. — Crystals with water of crystallization (dil. methanol), m. 113–5° cor. (if heated rapidly), resolidifying, and remelting at 182–4° cor.; after drying by heating at 100° in vacuo, it shows only the higher m. p. The dried compound has $[\alpha]_D$ +43° (acetone) (1946 Sarett). — **Fmn.** From the preceding diolone on shaking in abs. alcohol with hydrogen and palladium - barium carbonate catalyst until 1 mol. hydrogen has been absorbed (1946 Sarett; 1946b MERCK & Co., INC.); the 3-acetate is similarly obtained from the corresponding acetate (1946b MERCK & Co., INC.). — **Rns.** On treatment of the 3-acetate with PBr_3 in dry chloroform containing a little pyridine it gives the acetate of 21-bromo-$\Delta^{17(20)}$-pregnen-3α-ol-11-one (p. 2701 s) (1946 Sarett; 1946c MERCK & Co., INC.).

3-Acetate $C_{23}H_{34}O_4$. Crystals (ethyl acetate - pentane), m. 189.5–191.0° cor. **Fmn.** From the above diolone on treatment with acetic anhydride in pyridine at room temperature (1946 Sarett; 1946b MERCK & Co., INC.). See also under formation of the diolone. — **Rns.** See above.

1946 a MERCK & Co., INC. (Sarett), *U.S. Pat.* 2492189 (issued 1949); *C. A.* **1950** 3045; *Ch. Ztbl.* **1952** 1211; *Fr. Pat.* 942260 (1947, issued 1949); *Ch. Ztbl.* **1950** I 583, 584.
   b MERCK & Co., INC. (Sarett), *U.S. Pat.* 2492190 (issued 1949); *C. A.* **1950** 3045; *Ch. Ztbl.* **1952** 1211; *Fr. Pat.* 942260 (1947, issued 1949); *Ch. Ztbl.* **1950** I 583, 584.
   c MERCK & Co., INC. (Sarett), *U.S. Pat.* 2492191 (issued 1949); *C. A.* **1950** 3045; *Ch. Ztbl.* **1952** 1211; *Fr. Pat.* 942260 (1947, issued 1949); *Ch. Ztbl.* **1950** I 583, 584.
   Sarett, *J. Biol. Ch.* **162** 601, 619.
1950 Gallagher, Kritchevsky, *J. Am. Ch. Soc.* **72** 882, 884.

e. Dihydroxy-monoketosteroids C_{27}–C_{29} with CO and OH in the Ring System

5-Methyl-19-nor-Δ^9-coprostene (or -cholestene)-3β.6β-diol-11-one oxide, 5-Methyl-9.10-epoxy-19-nor-coprostane (or -cholestane)-3β.6β-diol-11-one

$C_{27}H_{44}O_4$. Originally described as the 8.9-epoxy-compound (1939 Petrow); for revised structure, see 1952 Ellis; cf. also structure and configuration of the parent "Westphalen's diol", p. 2046 s. — Needles (aqueous methanol), m. 219–220° cor. after sintering at 203° cor.; $[\alpha]_D^{19}$ +123° (chloroform). Readily soluble in acetone and alcohol, sparingly in light petroleum (1939 Petrow). — **Fmn.** The diacetate is obtained from the diacetate of 5-methyl-19-nor-Δ^9-coprostene-3β.6β-diol (p. 2046 s) on oxidation with CrO_3 in strong acetic acid at 55–60° (yield, 30%) or from the 3.6-diacetate of 5-methyl-19-nor-Δ^9-coprostene-3β.6β.11-triol (p. 2171 s) on oxidation with CrO_3 and 80% acetic acid in benzene at room temperature (yield, 40%) or with Kiliani's chromic acid mixture in acetic acid-benzene at 0° (yield, 60%); it is hydrolysed with 5% methanolic KOH to the diolone (yield, almost quantitative) (1939 Petrow).

Rns. Gives 5-methyl-9.10-epoxy-19-nor-coprostane-3.6.11-trione (p. 2984 s) on oxidation with CrO_3 and 70% acetic acid in benzene at room temperature. Is unaffected by lead tetraacetate. The diacetate could not be reduced with aluminium isopropoxide and gave a gum on attempted reduction with sodium and alcohol. The diacetate absorbs bromine after a short period of induction; it resinified on attempted hydrolysis of the oxide ring with alcoholic HCl, and was recovered unchanged after treatment with hydroxylamine or 2.4-dinitrophenylhydrazine (1939 Petrow).

Diacetate $C_{31}H_{48}O_6$. Needles (aqueous methanol, or aqueous acetone-methanol), m. 159.5–160.5° cor.; $[\alpha]_D^{19}$ +120° to +121° (chloroform). Shows no selective absorption in the ultra-violet. — **Fmn.** From the above diolone on acetylation (1939 Petrow); see also under formation of the diolone. — **Rns.** See above.

1939 Petrow, *J. Ch. Soc.* **1939** 998, 1002.
1952 Ellis, Petrow, *J. Ch. Soc.* **1952** 2246.

Cholestane-4α.5α-diol-2-one $C_{27}H_{46}O_3$.

4α.5α-Epoxycholestan-2-one, "*Ketone A*" $C_{27}H_{44}O_2$. For the structure, see 1953 Conca. — Crystals (abs. alcohol), m. 172° cor.; $[\alpha]_D^{25}$ +141° (chloroform) (1939 Bergmann; cf. also 1938 Skau). Infra-red absorption max. in chloroform at 5.86 μ (1953 Conca). — **Fmn.** From 2α.5α-peroxido-Δ^3-cholestene (p. 2050 s) in abs. alcohol on exposure to sunlight for 1 week (yield, 88%) (1938 Skau;

1939 Bergmann); may be similarly obtained directly from $\Delta^{2.4}$-cholestadiene (p. 1409 s) (1938 Skau).

Rns. Oxidation with CrO_3 in strong acetic acid at 55–60° affords Δ^3-cholesten-5α-ol-2-one ("Ketone B", p. 2661 s) and the acid $C_{25}H_{42}O_3$ (p. 2661 s) (1953 Conca). Gives cholestan-5α-ol-2-one (p. 2662 s) on hydrogenation in alcohol in the presence of Adams's platinum catalyst at room temperature and atmospheric pressure. Treatment with lithium aluminium hydride yields a mixture containing cholestane-2α.5α-diol (p. 2050 s) (1953 Conca). Gives Δ^3-cholesten-5α-ol-2-one (p. 2661 s) on refluxing with acetic anhydride or, in better yield, on distillation at 210–230° under 1 mm. pressure. On refluxing with a 5% solution of KOH in 95% methanol it gives 4β-methoxycholestan-5α-ol-2-one ("Ketone C", below) (1939 Bergmann). For the action of conc. H_2SO_4 in boiling abs. methanol, see 1953 Conca.

Oxime $C_{27}H_{45}O_2N$. Crystals (abs. alcohol), m. 225–229° cor. (1939 Bergmann).

Cholestane-4β.5α-diol-2-one $C_{27}H_{46}O_3$ (I, R = H).

4-Methyl ether, 4β-Methoxycholestan-5α-ol-2-one, *"Ketone C"*, $C_{28}H_{48}O_3$ (I, R= CH_3). For the structure, see 1953 Conca. — Crystals (methanol), m. 153.5–154° cor.; $[\alpha]_D^{26} + 35.5°$ (chloroform or ether) (1939 Bergmann). Infra-red absorption max. in chloroform at 2.80, 5.85, and 9.06–9.16 μ (1953 Conca). — **Fmn.** From 2α.5α-peroxido-Δ^3-cholestene (p. 2050 s) on refluxing for $1\frac{1}{2}$ hrs. with a 5% solution of KOH in 95% methanol (yield, 65%); is similarly obtained from Δ^3-cholesten-5α-ol-2-one ("Ketone B", p. 2661 s; yield, 80%) and from 4α.5α-epoxycholestan-2-one ("Ketone A", above) (1939 Bergmann). — **Rns.** Reverts to Δ^3-cholesten-5α-ol-2-one on distillation at 220–230° under 1 mm. pressure. Is unchanged by refluxing with acetic anhydride for 3 hrs. (1939 Bergmann). — *Semicarbazone* $C_{29}H_{51}O_3N_3$, crystals (abs. alcohol), m. 251–254° cor. dec. (1939 Bergmann).

1938 Skau, Bergmann, *J. Org. Ch.* **3** 166, 173, 174.
1939 Bergmann, Hirschmann, Skau, *J. Org. Ch.* **4** 29, 37, 38.
1953 Conca, Bergmann, *J. Org. Ch.* **18** 1104.

Cholestane-4.5-diol-3-one or **Coprostane-4.5-diol-3-one** $C_{27}H_{46}O_3$. For the structure, cf. the osmic ester (below). —

Leaflets (dil. alcohol), m. 206–8°; $[\alpha]_D^{22} + 44°$ (chloroform). — **Fmn.** From Δ^4-cholesten-3-one (p. 2424 s) on treatment with 30% H_2O_2 in ether in the presence of some OsO_4 at 20° for 15 hrs.; yield, 60% (1938 Butenandt).

4-Acetate $C_{29}H_{48}O_4$. Crystals (chloroform-alcohol), m. 225–7°. — **Fmn.** From the above diolone on treatment with acetic anhydride in pyridine at 20° for 24 hrs. (1938 Butenandt).

The *4.5-osmic ester* of possibly the same diolone (the hydroxyl groups of which are then in cis-position) is obtained as *complex with 2 mols. of pyridine* $C_{37}H_{54}O_5N_2Os$ (brown prisms) in 35 % yield from 1 mol. of Δ^4-cholesten-3-one, 1 mol. of OsO_4, and 4 mols. of pyridine in ether, or as *complex with 1 mol. of α.α′-dipyridyl* $C_{37}H_{52}O_5N_2Os$ (light brown prowder; insoluble in water and petroleum ether, readily soluble in chlorinated hydrocarbons and in aceto-nitrile) in 58 % yield from 1 mol. of Δ^4-cholesten-3-one, 0.5 mol. of OsO_4 and 2 mols. of α.α′-dipyridyl in ether. Such osmic ester complexes may be hydroly-sed with aqueous KOH in the presence of mannitol to give the free diols (1942 Criegee).

$\Delta^{4.6}$-Cholestadiene-4.6-diol-3-one $C_{27}H_{42}O_3$. Is the 4.6-di-enol of cholestane-3.4.6-trione (see p. 2982 s).

Cholestane-5α.6α-diol-3-one $C_{27}H_{46}O_3$.

5α.6α-Epoxycholestan-3-one, Δ^5-Chol-esten-3-one α-oxide $C_{27}H_{44}O_2$. Origi-nally referred to as *Δ^5-cholesten-3-one "β"-oxide** (1937 Ruzicka); for the true structure, see 1949 Urushibara; 1955 Mori; 1956 Ellis. — Crystals (80 % alcohol), m. 122° (1937 Ruzicka); needles, m. 120–1° (1955 Mori). Infra-red absorption max. in Nujol at 5.82 μ (1955 Mori). — **Fmn.** Along with a smaller amount of the "α-oxide"*, from Δ^5-cholesten-3-one (p. 2435 s) on treatment with perbenzoic acid in chloroform at room temperature; max. yield, 44 % (1937 Ruzicka). From cholesterol α-oxide (p. 2175 s) on oxidation with CrO_3-pyridine at 10–15° (1955 Mori); the substance (needles, m. ca. 250° cor.), said to be Δ^5-cholesten-3-one oxide and obtained from cholesterol α-oxide by oxidation with CrO_3 in acetic acid (1936 Fujii), may be cholestan-5-ol-3.6-dione (p. 2953 s) (Editor); on treatment of this substance with bromine it is converted into 2.2.7α-tribromo-Δ^4-cholestene-3.6-dione (see p. 2931 s), originally formulated as 2.4.5-tribromocholestane-3.6-dione (1936 Fujii) and then as 4.7.7-tribromo-Δ^4-cholestene-3.6-dione (1937 Butenandt). — **Rns.** Gives cholestane-3.6-dione (p. 2907 s) on refluxing with 2 N H_2SO_4 in dioxane for 24 hrs. (1937 Ruzicka).

5α.6α - Epoxycholestan - 3 - one ethylene ketal, Δ^5- Cholesten - 3 - one α - oxide ethylene ketal $C_{29}H_{48}O_3$ (no analysis and no constants given). — **Fmn.** From Δ^5-cholesten-3-one ethylene ketal (p. 2435 s) on treatment with perbenzoic acid. — **Rns.** Gives cholestan-5-ol-3.6-dione (p. 2953 s) on mild oxidation with CrO_3 (1941 Fernholz).

* For the structure of "Δ^5-*cholesten-3-one α-oxide*" (1937 Ruzicka) as 4-hydroxy-3‖4-Δ^5-cholesten-3-oic acid lactone (formula I, p. 2435 s), see 1954, 1955 Mori.

Cholestane-5α.6β-diol-3-one $C_{27}H_{46}O_3$.

6-Acetate $C_{29}H_{48}O_4$. Plates (aqueous alcohol), m. 161–2° cor.; $[\alpha]_D^{19}$ —10° (chloroform). — **Fmn.** From the 6-acetate of cholestane-3β.5α.6β-triol (page 2177 s) on oxidation with CrO_3 in strong acetic acid at 100° for 3 min. (yield, 75%); a small amount of 5α.6β-dihydroxycholestane-2‖3-dioic acid 2→5 lactone 6-acetate (p. 2178 s), formed by further oxidation of the present diolone acetate, is obtained at the same time. — **Rns.** On treatment with 1 mol. bromine in glacial acetic acid at 35° it gives the 2-bromo-derivative (p. 2762 s) and a small amount of the 2.2-dibromo-derivative (p. 2763 s); the latter is the sole reaction product when the acetate is treated with 2 mols. bromine and some HBr in glacial acetic acid at room temperature. It is converted into cholestane-3.6-dione (p. 2907 s) on refluxing with 5% methanolic sodium methoxide for 1 hr. Gives the acetate of Δ^4-cholesten-6β-ol-3-one (p. 2672 s) on refluxing with acetic anhydride for 10 hrs. or on dropwise addition of 2 mols. thionyl chloride to a solution in dry pyridine at room temperature, followed by refluxing for 10 min. (1939 Ellis).

1936 Fujii, Matsukawa, *J. Pharm. Soc. Japan* **56** 150; *C. A.* **1937** 1033; *Ch. Ztbl.* **1937** I 2616.
1937 Butenandt, Schramm, Kudszus, *Ann.* **531** 176, 188.
   Ruzicka, Bosshard, *Helv.* **20** 244, 248, 249.
1938 Butenandt, Wolz, *Ber.* **71** 1483.
1939 Ellis, Petrow, *J. Ch. Soc.* **1939** 1078.
1941 Fernholz, Stavely, *Abstracts of the 102nd meeting of the American Chemical Society, Atlantic City*, p. M 39.
1942 Criegee, Marchand, Wannowius, *Ann.* **550** 99, 102, 106, 123–125.
1949 Urushibara, Chuman, *Bull. Ch. Soc. Japan* **22** 273.
1954 Mori, Mukawa, *Bull. Ch. Soc. Japan* **27** 479.
1955 Mori, Mukawa, *Proc. Japan Acad.* **31** 532.
1956 Ellis, Petrow, *J. Ch. Soc.* **1956** 4417.

Cholestane-5.6-diol-4-one or Coprostane-5.6-diol-4-one $C_{27}H_{46}O_3$.

Originally (1938 Butenandt) described as cholestane-1.2-diol-3-one; for the true structure of this compound and the parent cholestenone, see 1944 Butenandt. — Leaflets (dil. acetone), m. 186–8° (1938 Butenandt), 184–6° (1944 Butenandt). Sparingly soluble in ether (1944 Butenandt). — **Fmn.** From Δ^5-cholesten-4-one (p. 2450 s) on treatment with 30% H_2O_2 in ether in the presence of some OsO_4 at 20° for 30 hrs. (1938 Butenandt) or, in better yield (ca. 20%), on treatment with $KMnO_4$ in boiling acetone (1944 Butenandt).

6-Acetate $C_{29}H_{48}O_4$. Crystals (aqueous methanol), m. 150–2°. — **Fmn.** From the above diolone on heating with acetic anhydride on the water-bath for 20 min. (1944 Butenandt).

1938 Butenandt, Wolz, *Ber.* **71** 1483.
1944 Butenandt, Ruhenstroth-Bauer, *Ber.* **77** 397, 401.

Δ^3-**Cholestene-3.5-diol-6-one** or Δ^3-**Coprostene-3.5-diol-6-one** $C_{27}H_{44}O_3$.

3-Acetate $C_{29}H_{46}O_4$. A compound (obtained from epi-cholesteryl acetate by CrO_3-oxidation), to which this structure was tentatively assigned (1939 Windaus), is actually the acetate of cholestan-5α-ol-3.6-dione (see p. 2953 s) (1953 Tarlton).

Cholestane-3β.5α-diol-6-one $C_{27}H_{46}O_3$ (p. 149) (cf. also "coprostane-3β.5-diol-6-one", p. 2758 s). For configuration at C-5, see 1939 Ellis; cf. also 1944 Prelog. — Crystals, m. 237° cor. (1939 Ellis)*. — **Fmn.** The 3-acetate is obtained from cholesteryl acetate on oxidation with $KMnO_4$ in 80% acetic acid at room temperature or at 50° (1940 Marker); yield (at 50°), 5–6% (1940 Ehrenstein)**. The 3-acetate is formed when the acetate of cholesterol α-oxide is oxidized with CrO_3 in acetic acid (1915 Westphalen). From cholestane-3β.5α.6β-triol (p. 2177 s) on oxidation with ³/₄ mol. of CrO_3 in strong acetic acid at room temperature (isolated as 3-acetate; yield, 65%); the acetate is saponified with sodium methoxide solution (1939 Ellis; cf. also the formation from the 3-acetate of the same triol, referred to on p. 149). The 3-acetate may be obtained from its 7α-bromo-derivative (p. 2763 s) on reduction with zinc dust in boiling alcohol (1943 Sarett)*, the diacetate from its 7β-bromo-derivative (p. 2763 s) on reduction with aluminium amalgam and water in ether at 20° (1938 Heilbron) or with zinc dust in glacial acetic acid at room temperature (1955 Cookson).

Rns. On bromination with 1 mol. bromine in glacial acetic acid at 35° it gives the 7α-bromo-derivative (p. 2762 s); the 3-acetate reacts similarly (1943 Sarett; cf. 1949 Fieser; 1955 Cookson). On dehydration of the 3-acetate by refluxing with thionyl chloride in pyridine for 10 min. it gives the acetate of Δ^4-cholesten-3β-ol-6-one (p. 2677 s); attempted dehydration by heating with $KHSO_4$ in acetic anhydride on the water-bath for 15 min. gave 35% yield of the diacetate (p. 2758 s) (1939 Ellis).

3-Acetate $C_{29}H_{48}O_4$ (p. 149). Plates (acetone or glacial acetic acid), m. 226.5° to 228.5° (1940 Ehrenstein), 231–3° (1940 Marker), 238° cor. (1939 Ellis); $[\alpha]_D^{23}$ —56° (chloroform)* (1939 Ellis). Absorption max. in alcohol at 299.5 mμ (log ε 1.77) (1955 Cookson). — **Fmn.** From the above diolone on refluxing with acetic anhydride for 45 min. (1939 Ellis). See also under the formation

---

* It is not clear if the statement "m. 234°, $[\alpha]_D^{25}$ + 19° (chloroform)" given by Sarett (1943) for the "cholestane-3β.5α-diol-6-one", obtained from the 3-acetate of its 7α-bromo-derivative on reduction with zinc dust in boiling alcohol, refers to the free diolone or its 3-acetate; cf., however, the rotation $[\alpha]_D$ —56° (chloroform) stated by Ellis (1939) for the 3-acetate.

** After the closing date for this volume, the formation of the diolone from cholesterol in 85.5% yield has been described by Fieser (1949).

References, p. 2759 s

of the diolone. — **Rns.** For bromination and dehydration, see under the reactions of the diolone.

3-Acetate oxime $C_{29}H_{49}O_4N$. Crystals (aqueous methanol), m. 204–6⁰ (1940 Marker).

Diacetate $C_{31}H_{50}O_5$ (p. 149). Needles (methanol or aqueous acetone), m. 169⁰ to 170⁰ cor. (1939 Ellis; cf. also 1938 Heilbron); $[\alpha]_D^{22}$ —11⁰ (chloroform) (1939 Ellis). Absorption max. in alcohol at 290 mμ (log ε 1.93) (1955 Cookson). **Fmn.** From the above 3-acetate on refluxing with acetic anhydride for $10^1/_2$ hrs. (1939 Ellis). See also under formation and reactions of the free diolone.

The following compound $C_{27}H_{46}O_3$ was assumed to be the 5-epimer of the above cholestane-3β.5α-diol-6-one, thus being *coprostane-3β.5-diol-6-one* (1938 Heilbron; 1939 Ellis); its oxidation product, the corresponding 5-ol-3.6-dione m. 253⁰, was thought (1939 Ellis) to be different from cholestan-5α-ol-3.6-dione m. 232⁰; for the identity of the two 5-ol-3.6-diones see, however, 1944 Prelog; cf. also 1953 Fieser. The difference in m. p.'s of the two diolones, which thus must be identical, remains unexplained. — Rectangular tables containing solvent of crystallization (methanol), m. 138⁰ after softening at 128⁰; suffers profound decomposition on attempted sublimation at 220⁰ in a high vacuum; $[\alpha]_D^{22}$ +29.3⁰ (chloroform). Sparingly soluble in methanol, very soluble in chloroform and pyridine (1937 Heilbron). — **Fmn.** From the acetate of 5α-bromocholestan-3β-ol-6-one (p. 2707 s) on refluxing with 10 % methanolic KOH for 2 hrs.; crude yield, ca. 75 % (1937 Heilbron). — **Rns.** On oxidation with CrO_3 in 95 % acetic acid at room temperature it gives cholestan-5-ol-3.6-dione (p. 2953 s) (1939 Ellis; see also the remarks above). On treatment with benzoyl chloride in pyridine at room temperature it gives the *3-benzoate* $C_{34}H_{50}O_4$, needles or plates (methanol-acetone; the two forms being interconvertible), m. 170⁰; $[\alpha]_D^{20}$ +23⁰ (chloroform); very sparingly soluble in methanol, readily in acetone (1937 Heilbron).

Δ^4-Cholestene-3β.7-diol-6-one $C_{27}H_{44}O_3$ (I). A compound (with its derivatives) to which this structure was assigned by Heilbron (1938) is actually 3β-hydroxy-B-nor-Δ^5-cholestene-6-carboxylic acid (with its derivatives; see pp. 1379 s, 1380 s) (1941 a, b Woodward).

Δ^7-Cholestene-3β.7-diol-6-one $C_{27}H_{44}O_3$ (II). For a compound which may possess this structure or may be Δ^5-cholestene-3β.6-diol-7-one (III), see cholestan-3β-ol-6.7-dione (p. 2954 s).

References, p. 2759 s

Cholestane-3β.7-diol-6-one $C_{27}H_{46}O_3$ (dried at 100° in vac. for 8 hrs.). Plates (methanol), m. 179°; sublimes in needles at 220° in a high vacuum; $[\alpha]_D^{22}$ +31.4° (chloroform). Sparingly soluble in methanol, soluble in acetone, very soluble in chloroform and pyridine. — **Fmn.** From 7α-bromo-cholestan-3β-ol-6-one acetate (p. 2708 s) on refluxing with 10% methanolic KOH for 2 hrs. (1937 Heilbron).

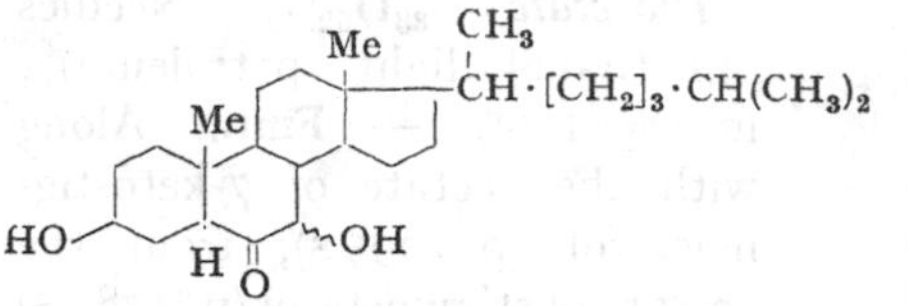

Dibenzoate $C_{41}H_{54}O_5$. Needles (methanol-acetone), m. 169–170°; $[\alpha]_D^{20}$ +62° (chloroform); sparingly soluble in methanol, very soluble in acetone. — **Fmn.** From the above diolone on treatment with benzoyl chloride in pyridine at room temperature for 20 hrs. (1937 Heilbron).

1915 Westphalen, *Ber.* **48** 1064, 1068.
1937 Heilbron, Jones, Spring, *J. Ch. Soc.* **1937** 801, 804, 805.
1938 Heilbron, Jackson, Jones, Spring, *J. Ch. Soc.* **1938** 102, 104, 106, 107.
1939 Ellis, Petrow, *J. Ch. Soc.* **1939** 1078, 1081, 1083.
    Windaus, Naggatz, *Ann.* **542** 204, 209.
1940 Ehrenstein, Decker, *J. Org. Ch.* **5** 544, 555, 556.
    Marker, Rohrmann, *J. Am. Ch. Soc.* **62** 516.
1941 a Woodward, *J. Am. Ch. Soc.* **63** 1123, 1125.
    b Woodward, Clifford, *J. Am. Ch. Soc.* **63** 2727.
1943 Sarett, Chakravorty, Wallis, *J. Org. Ch.* **8** 405, 407 (footnote), 414.
1944 Prelog, Tagmann, *Helv.* **27** 1867, 1869, 1871.
1949 Fieser, Rajagopalan, *J. Am. Ch. Soc.* **71** 3938.
1953 Fieser, *J. Am. Ch. Soc.* **75** 4386, 4391.
    Tarlton, Fieser, Fieser, *J. Am. Ch. Soc.* **75** 4423.
1955 Cookson, Dandegaonker, *J. Ch. Soc.* **1955** 352, 354.

Δ^5-Cholestene-3β.6-diol-7-one $C_{27}H_{44}O_3$ (III, p. 2758 s). For a compound which may possess this structure or may be Δ^7-cholestene-3β.7-diol-6-one (II), see cholestan-3β-ol-6.7-dione (p. 2954 s).

Cholestane-3β.6-diol-7-one $C_{27}H_{46}O_3$ (not analysed). Leaflets (methanol), m. 148–150° after softening at 140°. — **Fmn.** From the acetate of 6β-bromo-cholestan-3β-ol-7-one (see p. 2709 s) on refluxing with 10% methanolic KOH for 2 hrs. (1938 Barr).

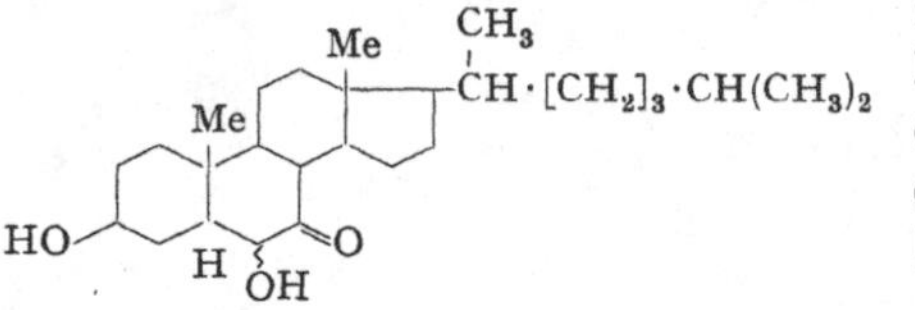

Dibenzoate $C_{41}H_{54}O_5$. Prisms (acetone-methanol), m. 184–6° after softening at 182°. — **Fmn.** From the above diolone on treatment with benzoyl chloride in pyridine at room temperature for 18 hrs. (1938 Barr).

1938 Barr, Heilbron, Jones, Spring, *J. Ch. Soc.* **1938** 334, 336.

Δ^{22}-Stigmastene-3β.5-diol-6-one, Δ^{22}-Sitostene-3β.5-diol-6-one $C_{29}H_{48}O_3$.

Diacetate $C_{33}H_{52}O_5$. Needles (methanol - light petroleum), m. 189–190°. — **Fmn.** Along with the acetate of 7-ketostigmasterol (p. 2695 s), from the acetate of stigmasterol (p. 1784 s) on oxidation with CrO_3 in strong acetic acid at 50–55° (1939 Haslewood).

Stigmastane-3β.5-diol-6-one, Sitostane-3β.5-diol-6-one $C_{29}H_{50}O_3$.

3-Acetate $C_{31}H_{52}O_4$. Needles (acetone), m. 251°. — **Fmn.** From the acetate of "β"-sitosterol (p. 1808 s) on oxidation with $KMnO_4$ in aqueous acetic acid at 50° or at room temperature (1940 Marker).

1939 Haslewood, *Biochem. J.* **33** 454.
1940 Marker, Rohrmann, *J. Am. Ch. Soc.* **62** 516.

19. HALOGENO-DIHYDROXY-MONOKETOSTEROIDS WITH CO AND OH IN THE RING SYSTEM

8-Chloro-4.7-dihydroxy-3'-keto-1.2-cyclopentenophenanthrene $C_{17}H_{11}O_3Cl$ (I, R = R'= H).

8-*Chloro-4-hydroxy-7-methoxy-3'-keto-1.2-cyclopen-tenophenanthrene* $C_{18}H_{13}O_3Cl$ (I, R = CH$_3$, R'= H). Needles (nitrobenzene), m. 335⁰ (dec. from 300⁰). Gives a yellow solution in conc. H$_2$SO$_4$. — **Fmn.** Its acetate (below) is obtained by the following synthesis: 5-chloro-6-methoxy-2-acetylnaphthalene is condensed with furfural to give 5-chloro-6-methoxy-2-furfurylideneacetyl-naphthalene which is converted by hydro-chloric·acetic acid into 7-(5-chloro-6-methoxy-2-naphthyl)-4.7-diketohepta-noic acid; this acid is cyclized by heating with dil. KOH on the steam-bath to give 3-(5-chloro-6-methoxy-2-naphthyl)-Δ^2-cyclopenten-1-one 2-acetic acid which is then refluxed with acetic anhydride for $^1/_2$ hr.; the acetate is hydrolysed by refluxing with NaOH in strong alcohol for 5 hrs. (1941 Robinson).

Rns. On hydrogenation of the acetate in glacial acetic acid in the presence of Raney nickel at 105⁰ and 37 atm. or at 158⁰ and 57 atm. pressure it is partly deacetylated and partly reduced to give the 9.10-dihydro-derivative (p. 2762 s) and more fully reduced compounds. Reduction of the acetate with sodium and isoamyl alcohol afforded (in one experiment out of seven) a phenolic, non-ketonic product [C 68.1%, H 4.7%, Cl 10.1%; pale yellow needles (nitrobenzene), m. 292⁰ after sintering from 270⁰] (1941 Robinson).

On refluxing the acetate with hydrobromic acid (d 1.5) in glacial acetic acid for 22 hrs. it gives "**apochlorodihydroxyketocyclopentenophenanthrene**" $C_{17}H_{11}O_3Cl$ (of unknown structure) [intensely yellow needles (pyridine); does not melt or decompose below 380⁰; extremely sparingly soluble in most solvents; 1 g. dissolves in more than 100 c.c. of boiling pyridine; freely soluble in aqueous NaOH to give an intensely red solution; gives a purple solution in conc. H$_2$SO$_4$] which gives no ketone derivatives and no piperonylidene derivative, but an amorphous diacetate and, with dimethyl sulphate in aqueous alcoholic NaOH, a *dimethyl ether* $C_{19}H_{15}O_3Cl$, yellow needles (nitro-benzene), m. 335⁰ dec., insoluble in aqueous alkalis, giving a blue coloration in cold conc. H$_2$SO$_4$ which is destroyed on heating (1941 Robinson).

8 - *Chloro - 4 - acetoxy - 7 - methoxy - 3' - keto - 1.2 - cyclopentenophenanthrene* $C_{20}H_{15}O_4Cl$ (I, R = CH$_3$, R'= CO·CH$_3$). Pale yellow needles (glacial acetic acid), m. 254–5⁰ dec., after sintering. Readily soluble in hot pyridine and hot dioxane, sparingly in methanol, ethanol, acetone, and benzene; 1 g. dissolves in 85 c.c. of boiling glacial acetic acid. Its solution in conc. H$_2$SO$_4$ is yellow. — **Fmn.** and **Rns.** See those of the preceding compound. — *Oxime* $C_{20}H_{16}O_4NCl$. Needles (pyridine); darkens from 280⁰ to 320⁰, does not melt at 370⁰ (1941 Robinson).

8-*Chloro-4.7-dimethoxy-3'-keto-1.2-cyclopentenophenanthrene* $C_{19}H_{15}O_3Cl$ (I, R = R'= CH$_3$). Prisms (ethyl acetate), needles (nitrobenzene), m. 247⁰. Gives a yellow solution in conc. H$_2$SO$_4$. — **Fmn.** From the monomethyl ether on treatment with dimethyl sulphate in aqueous alcoholic NaOH. — **Rns.** Gives a 2.4-dinitrophenylhydrazone (1941 Robinson).

190*

8-Chloro-4-hydroxy-7-methoxy-3'-keto-1.2-cyclopenteno-9.10-dihydrophenanthrene $C_{18}H_{15}O_3Cl$.

Colourless needles (dioxane and glacial acetic acid), m. 236–7⁰. Readily soluble in alcoholic NaOH. — **Fmn.** Along with other compounds, from 8-chloro-4-acetoxy-7-methoxy-3'-keto-1.2-cyclopentenophenanthrene (p. 2761 s) on hydrogenation in glacial acetic acid in the presence of Raney nickel at 105⁰ and 37 atm. or at 158⁰ and 57 atm. pressure. — **Rns.** Gives an orange 2.4-dinitrophenylhydrazone and a yellow piperonylidene derivative which dec. below 360⁰ and gives a wine-red solution in conc. H_2SO_4 (1941 Robinson).

2-Bromocholestane-5α.6β-diol-3-one $C_{27}H_{45}O_3Br$.

6-Acetate $C_{29}H_{47}O_4Br$. Needles (aqueous acetone), m. 186⁰ cor.; $[\alpha]_D^{21}$ +3.6⁰ (chloroform). — **Fmn.** From the 6-acetate of cholestane-5α.6β-diol-3-one (p. 2756 s) on treatment with 1 mol. bromine in glacial acetic acid at 35⁰; yield, 40%. — **Rns.** Is recovered unchanged after refluxing with pyridine for 1 hr. Gives 2.5-epoxycholestan-6β-ol-3-one (p. 2771 s) on warming with 1.5% methanolic KOH at 55–60⁰ for 1 hr. On treatment with 1 mol. bromine and a little HBr in glacial acetic acid at room temperature it gives 2.2-dibromocholestane-5α.6β-diol-3-one 6-acetate (p. 2763 s) (1939 Ellis).

7α-Bromocholestane-3β.5α-diol-6-one $C_{27}H_{45}O_3Br$.

For reversal of configuration assigned by Fieser (1949), see 1955 Cookson; cf. also 1954 Corey. — Needles (alcohol and dil. acetone), dec. 250⁰; $[\alpha]_D^{26}$ —24⁰ (chloroform) (1943 Sarett). — **Fmn.** From cholestane-3β.5α-diol-6-one (p. 2757 s) on treatment with 1 mol. of bromine in chloroform - glacial acetic acid at 35⁰ (yield, 56%) (1943 Sarett); the 3-acetate is obtained from the diolone 3-acetate in a similar way (1943 Sarett) or by warming with 1.1 mols. of bromine and BF_3-ether in glacial acetic acid for 15 min. at 60⁰ (crude yield, 71%) (1949 Fieser); the diacetate is obtained by the latter method (25 min. at 75⁰) from the diolone diacetate, probably along with a small amount of its 7-epimer (p. 2763 s) (1949 Fieser). — **Rns.** Gives 7α-bromocholestan-5α-ol-3.6-dione (p. 2957 s) on oxidation with CrO_3 in strong acetic acid at room temperature (1943 Sarett). The 3-acetate is reduced to the 3-acetate of cholestane-3β.5α-diol-6-one (cf. p. 2757 s footnote *) on boiling with zinc dust in alcohol (1943 Sarett). For dehydrobromination of the 3-acetate by triethylamine, see 1949 Fieser; the diacetate is resistant to attempted dehydrobromination with triethylamine or pyridine (1949 Fieser). On keeping the diacetate in glacial acetic acid containing 5% of hydrogen bromide it is isomerized to the 7β-epimer (p. 2763 s) (1955 Cookson).

References, pp. 2763 s, 2764 s

3-Acetate $C_{29}H_{47}O_4Br$. Needles with 1 CH_3OH (methanol), m. 172° (1943 Sarett); needles (methanol), m. 170–1° (1949 Fieser); $[\alpha]_D^{28} + 7.5°$ (dioxane) (1949 Fieser), $[\alpha]_D + 7°$ (chloroform) (1955 Cookson). Absorption max. in alcohol at 333.5 mμ (log ε 2.04) (1955 Cookson). Very readily soluble in ether, fairly soluble in petroleum ether, moderately soluble in hot methanol (ca. 30 c.c./g.) (1949 Fieser). — **Fmn.** From the above diolone on boiling with acetic anhydride (1943 Sarett). See also under formation of the diolone. — **Rns.** See p. 2762 s.

Diacetate $C_{31}H_{49}O_5Br$. Needles (methanol), m. 216.5–217.5° (1949 Fieser), 215–6° (1955 Cookson); $[\alpha]_D^{25} + 37°$ (dioxane) (1949 Fieser), $[\alpha]_D + 39°$ (chloroform) (1955 Cookson). Absorption max. in alcohol at 317 mμ (log ε 1.69) (1955 Cookson). — **Fmn.** From the above 3-acetate on treatment with hot acetic anhydride and BF_3-ether; yield, 79% (1949 Fieser). — **Rns.** See those of the diolone (p. 2762 s).

7β-Bromocholestane-3β.5α-diol-6-one $C_{27}H_{45}O_3Br$. For configuration at C-7,

see 1955 Cookson. *Diacetate* $C_{31}H_{49}O_5Br$. Prisms (glacial acetic acid), m. 198° dec. (1938 Heilbron), 198–9° (1955 Cookson). Absorption max. in alcohol at 284 mμ (log ε 1.96) (1955 Cookson). Sparingly soluble in methanol, moderately soluble in ether, acetone, and glacial acetic acid (1938 Heilbron). — **Fmn.** From the acetate of 5α.7β-dibromocholestan-3β-ol-6-one (p. 2714 s) on heating with fused potassium acetate in glacial acetic acid on the steam-bath for 1$\frac{1}{2}$ hrs. (yield, 41%)(1938 Heilbron; cf. 1955 Cookson) or on refluxing with fused sodium acetate in abs. alcohol for 1 hr. (1938 Heilbron). From its 7α-epimer (above) on keeping for 40 hrs. at room temperature in glacial acetic acid containing 5% of hydrogen bromide; yield, 45% (1955 Cookson). — **Rns.** Gives the diacetate of cholestane-3β.5α-diol-6-one (p. 2757 s) on reduction at room temperature with aluminium amalgam and water in ether (1938 Heilbron) or with zinc dust in glacial acetic acid (1955 Cookson).

2.2-Dibromocholestane-5α.6β-diol-3-one $C_{27}H_{44}O_3Br_2$.

6-Acetate $C_{29}H_{46}O_4Br_2$. Needles (aqueous acetone), m. 218° cor. dec.; $[\alpha]_D^{19} + 71°$ (chloroform). — **Fmn.** From the 6-acetate of cholestane-5α.6β-diol-3-one (p. 2756 s) on treatment with 2 mols. bromine and a little HBr in glacial acetic acid at room temperature; similarly from the 6-acetate of 2-bromocholestane-5α.6β-diol-3-one (p. 2762 s) using 1 mol. bromine. — **Rns.** Liberates bromine when warmed with NaI in benzene-alcohol (1939 Ellis).

1938 Heilbron, Jackson, Jones, Spring, *J. Ch. Soc.* **1938** 102, 106, 107.
1939 Ellis, Petrow, *J. Ch. Soc.* **1939** 1078, 1082, 1083.

1941 Robinson, Willenz, *J. Ch. Soc.* **1941** 393; cf. also Robinson, *Brit. Pat.* 514592 (1938, issued 1939); *Fr. Pat.* 854270 (1939, issued 1940); *Ch. Ztbl.* **1940** II 1651, 1652.
1943 Sarett, Chakravorty, Wallis, *J. Org. Ch.* **8** 405, 407, 414, 415.
1949 Fieser, Rajagopalan, *J. Am. Ch. Soc.* **71** 3938, 3941.
1954 Corey, *J. Am. Ch. Soc.* **76** 175.
1955 Cookson, Dandegaonker, *J. Ch. Soc.* **1955** 352.

---

20. AMINO-DIHYDROXY-MONOKETOSTEROIDS WITH CO AND OH IN THE RING SYSTEM

23-Aminonorcholane-3α.12α-diol-7-one $C_{23}H_{39}O_3N$. Needles (toluene), m. 123^0 to 127^0. — **Fmn.** From the hydrazide

hydrazone of 3α.12α-dihydroxy-7-keto-cholanic acid on treatment with $NaNO_2$ in dil. HCl at 0^0, followed by decomposition of the resulting azide with glacial acetic acid at 60–70^0 and alkalinization with NaOH; yield (isolated as hydrochloride by passing dry HCl into the cooled alcoholic solution), 23% (1946 James).

Hydrochloride $C_{23}H_{40}O_3NCl$. Prisms (aqueous acetone containing a little hydrochloric acid), m. 263^0 dec.; $[\alpha]_D^{18} \pm 0^0$ (alcohol) (1946 James).

1946 James, Smith, Stacey, Webb, *J. Ch. Soc.* **1946** 665, 668.

---

21. POLYHYDROXY-MONOKETOSTEROIDS WITH CO AND TWO OH IN THE RING SYSTEM

4'-Hydroxymethylene-4.7-dihydroxy-3'-keto-1.2-cyclopentenophenanthrene and/or **4'-Formyl-4.7-dihydroxy-3'-keto-1.2-cyclopentenophenan-threne** $C_{18}H_{12}O_4$ (I and/or II, R = R' = H).

4.7-Dimethyl ether $C_{20}H_{16}O_4$ (I and/or II, R = R' = CH_3). Pale yellow needles (ethyl acetate), dec. 195^0. Readily soluble in aqueous alcoholic NaOH. Gives a deep olive-green coloration with $FeCl_3$ in alcohol. — **Fmn.** From 4.7-dimethoxy-3'-keto-1.2-cyclo-pentenophenanthrene (p. 2735 s) on successive treatment with ethyl formate and alcoholic sodium ethoxide in pyridine at room temp.; crude yield, 97%. — **Rns.** On heating with hydroxylamine hydrochloride in glacial ace-

tic acid at 70^0 it gives 4.7-dimethoxy-3'-keto-4'-cyano-1.2-cyclopentenophen-anthrene (not obtained pure) which on refluxing with aqueous alcoholic KOH for 10 days yields 4.7-dimethoxyphenanthrene-2-carboxylic-1-(β-propionic) acid and the **condensation product with 4.7-dimethoxy-3'-keto-1.2-cyclo-pentenophenanthrene**, $C_{39}H_{30}O_6$ (III) [orange needles (nitrobenzene), m. 302^0;

gives a purple coloration with conc. H_2SO_4 and a plum-coloured solution with a bright red fluorescence in hot acetic acid containing a little HCl]; (III) is also obtained when 4.7-dimethoxy-3'-keto-1.2-cyclopentenophenanthrene is treated with isoamyl formate and potassium tert.-butoxide in dry tert.-butyl alcohol at room temperature (1939 Robinson).

4-Ethyl ether 7-methyl ether $C_{21}H_{18}O_4$ (I and/or II, p. 2764 s; R = CH_3, R' = C_2H_5). Not isolated in the pure state. — **Fmn.** and **Rns.** Similar to those of the preceding dimethyl ether (1939 Robinson).

$\varDelta^4$-Pregnene-11β.17α.20β.21-tetrol-3-one, *Reichstein's Substance E* $C_{21}H_{32}O_5$

(p. 150). For configurations, see the literature cited for those of the reduction product (p. 2202 s). — Crystals with aq. (aqueous acetone), m. 124–9° cor. (1941 Reichstein). — Has a weak cortin-like activity (1939 Waterman). — **Occ.** For isolation from adrenal glands, see also 1937, 1941 Reichstein; 1942 v. Euw.

Rns. On oxidation of the 20.21-diacetate with CrO_3 in glacial acetic acid at room temperature it gives the corresponding diacetate of $\varDelta^4$-pregnene-17α.20β.21-triol-3.11-dione (Substance U, p. 2959 s). On shaking of the 20.21-diacetate with hydrogen and platinum oxide in glacial acetic acid it consumes 3 mols. of hydrogen giving, after acetylation with acetic anhydride in abs. pyridine at room temperature, 23% yield of the 3.20.21-triacetate of allopregnane-3β.11β.17α.20β.21-pentol (p. 2202 s) and a mixture m. 170° to 190° with the molecular formula $C_{27}H_{42}O_7$ (thus containing less oxygen than the pentol triacetate) (1941 Reichstein). Gives $\varDelta^4$-pregnene-11β.21-diol-3.20-dione (corticosterone, p. 2853 s) on refluxing with 30% H_2SO_4 in propanol for 20 min. (1938 CIBA).

20.21-Diacetate $C_{25}H_{36}O_7$. Needles (acetone-ether), m. 229–230° cor.; $[\alpha]_D^{22}$ + 163° (acetone). With conc. H_2SO_4 it gives an orange solution with green fluorescence. Gives a brownish red coloration on warming with 1.4-dihydroxynaphthalene in conc. HCl - glacial acetic acid (1 vol.:5 vols.). — **Fmn.** From the above tetrolone on treatment with acetic anhydride in abs. pyridine at room temperature (16 hrs.), then at 50° (1 hr.). — **Rns.** Reacts only very slowly in methanol with alkaline silver diammine solution. Is hydrolysed by refluxing with K_2CO_3 in 76% methanol (1941 Reichstein).

Allopregnane-3β.17α.20β.21-tetrol-11-one $C_{21}H_{34}O_5$.

For configurations, see the literature cited for those of the parent compound (p. 2202 s). — Crystals (after sublimation and addition of a drop of methanol), m. 160–170°, resolidifying, and remelting at 212–6° cor.; sublimes at 230° to 240° (block temperature) and 0.002 mm. pressure. Is at once precipitated by digitonin in hot aqueous methanol (1940 Shoppee). — **Fmn.** The 3.20.21-triacetate is obtained from the corresponding triacetate of allopregnane-$3\beta.11\beta.17\alpha.20\beta.21$-pentol (p. 2202 s) on oxidation with CrO_3 in glacial acetic acid at $20°$ (16 hrs.); it is hydrolysed by refluxing with aqueous methanolic NaOH (1940 Shoppee; 1940 CIBA PHARMACEUTICAL PRODUCTS). — **Rns.** On refluxing of the 3.20.21-triacetate with zinc dust in toluene it gives the diacetate of 17-iso-allopregnane-$3\beta.21$-diol-11.20-dione (p. 2862 s) (1940 Shoppee; 1940 CIBA PHARMACEUTICAL PRODUCTS).

3.20.21-Triacetate $C_{27}H_{40}O_8$. Needles (acetone-pentane), m. 183–4° cor., resolidifying, and remelting at 211–2° cor.; the higher-melting modification is generally directly obtained on crystallization from ether; $[\alpha]_D^{14} +69° \pm 3.5°$ (acetone) (1940 Shoppee). — **Fmn. and Rns.** See above.

Coproergostane-7α.12α.24.25-tetrol-3-one, 7α.12α.24.25-Tetrahydroxy-3-keto-bufostane $C_{28}H_{48}O_5$. Scales (aqueous alcohol), m. 161°. Gives an orange-red coloration in the Liebermann-Burchard test and a red coloration in the Jaffe test. — **Fmn.** The 7.12-diacetate is obtained from the corresponding diacetate of 3α.7α.12α.24.25-pentahydroxy-bufostane (p. 2204 s) on oxidation with CrO_3 in strong acetic acid at room temperature (yield, 60%); it is hydrolysed by heating with 10% KOH on the water-bath for 2 hrs. (yield, 80%) (1940 Kazuno).

Oxime $C_{28}H_{49}O_5N$. Scales (aqueous alcohol), m. 211° (1940 Kazuno).

7.12-Diacetate $C_{32}H_{52}O_7$. Plates (aqueous acetone), m. 149–150°. Gives an orange-red coloration in the Liebermann-Burchard test and an orange-yellow coloration in the Jaffe test (1940 Kazuno). — **Fmn.** See above.

1937 Reichstein, *Helv.* **20** 953, 964.
1938 CIBA*, *Fr. Pat.* 838916 (issued 1939); *C. A.* **1939** 6875; *Ch. Ztbl.* **1939** II 169, 170.
1939 Robinson, Rydon, *J. Ch. Soc.* **1939** 1394, 1401, 1403.
  Waterman, Danby, Gaarenstroom, Spanhoff, Uyldert, *Acta Brevia Neerland.* **9** 75, 77; *C. A.* **1940** 1062; *Ch. Ztbl.* **1939** II 2346.
1940 CIBA PHARMACEUTICAL PRODUCTS (Miescher, Heer), *U.S. Pat.* 2372841 (issued 1945); *C. A.* **1945** 4195; *Ch. Ztbl.* **1945** II 1773; CIBA* (Miescher, Heer), *Ger. Pat.* 737570 (issued 1943); *C. A.* **1946** 179; *Fr. Pat.* 876214 (issued 1942); *Ch. Ztbl.* **1943** I 1496; *Dutch Pat.* 55050 (issued 1943); *Ch. Ztbl.* **1944** I 367.
  Kazuno, *Z. physiol. Ch.* **266** 11, 22, 23.
  Shoppee, Reichstein, *Helv.* **23** 729, 734, 735, 738.
1941 Reichstein, v. Euw, *Helv.* **24** 247E, 254–256E, 260–264E.
1942 v. Euw, Reichstein, *Helv.* **25** 988, 1007.

* Société pour l'industrie chimique à Bâle; Gesellschaft für chemische Industrie in Basel.

22. Tri- and Tetrahydroxy-monoketosteroids with CO and OH in the Ring System

8-iso-$\Delta^{1\cdot3\cdot5(10)}$-Oestratriene-3.7α.8α-triol-17-one, Equilin-7.8-glycol $C_{18}H_{22}O_4$.

The α-configuration at C-7 and C-8 has been proved by Fieser (1953b) for the analogous case of the glycol obtained from Δ^7-cholesten-3β-ol (see p. 2183 s); β-configuration at C-8 (and therefore, in view of a cis-diol formation, alsoat C-7) was preferred by Pearlman (1939).

Crystals (ethyl acetate), m. 245° (1938 Serini); exists in polymorphic forms (ethyl acetate or 25% alcohol), m. ca. 210–6° cor. and 251° cor.; $[\alpha]_D^{23} + 135°$ to $+139°$ (dioxane) (1939 Pearlman). Fairly readily soluble in water, sparingly in most organic solvents (1938 Serini). — Has no œstrogenic activity and only a weak androgenic activity (1938 Serini).

Fmn. From the acetate of equilin (p. 2539 s) on treatment with OsO_4 in abs. ether, followed by refluxing with Na_2SO_3 in aqueous alcohol (1938 Serini; 1939 Pearlman). — **Rns.** Gives 8-iso-$\Delta^{1\cdot3\cdot5(10)}$-œstratriene-3.7α.8α.17β-tetrol (p. 2206 s) on reduction with sodium in hot alcohol (1941 Schering Corp.). On distillation at 205–210° in a high vac. it is dehydrated to give 7-ketoœstrone (p. 2939 s) (1939 Pearlman); the *3-methyl ether* $C_{19}H_{24}O_4$ (not described in detail) reacts similarly (1943 Lettré).

3.7-Diacetate $C_{22}H_{26}O_6$. Crystals (ether-hexane), m. 214° (1938 Serini), m. 211–2° cor. (1939 Pearlman); $[\alpha]_D^{26} + 90°$ to $+92°$ (alcohol) (1939 Pearlman). — **Fmn.** From the above triolone on treatment with acetic anhydride in pyridine at room temperature (1938 Serini).

Androstane-4.5.17β-triol-3-one, Testosterone-4.5-glycol $C_{19}H_{30}O_4$.

17-Acetate $C_{21}H_{32}O_5$. Leaflets (dil. acetone), m. 185° to 188°; $[\alpha]_D^{21} + 36°$ (chloroform). — **Fmn.** From the acetate of testosterone (p. 2580 s) on treatment with 30% H_2O_2 in the presence of OsO_4 in ether at room temperature (24 hrs.); yield, 76% (1938 Butenandt). *4.17-Diacetate* $C_{23}H_{34}O_6$. Needles (dil. acetone), m. 220–222° dec.* — Has no androgenic or œstrogenic activity. — **Fmn.** From the preceding compound on treatment with acetic anhydride in pyridine at 20° (1938 Butenandt).

Androstane-3β.5α.6α-triol-17-one $C_{19}H_{30}O_4$ (p. 150).

For α-configuration at C-5 and C-6, see 1948 Ehrenstein; 1949 Davis. — Prisms (ethyl acetate), m. 243.5–244° cor. (1939 Ushakov), 243–245.5°; $[\alpha]_D^{26} + 79.5°$ (methanol) (1939 Ehrenstein). — **Fmn.** From Δ^5-androsten-3β-ol-17-one (dehydroepiandrosterone, p. 2609 s) on treatment with OsO_4 in abs. ether at ca. 30° for 3 days, followed by refluxing with Na_2SO_3 in dil. alcohol for $2\frac{1}{2}$ hrs.; yield, 34% (1939

---

* In the theoretical part of the original paper the m. p. 120–122° is given.

Ehrenstein; cf. 1939 Ushakov). — **Rns.** Gives androstan-5α-ol-3.6.17-trione (p. 2983 s) with CrO_3 in strong acetic acid at ca. 30⁰ (1939 Ehrenstein).

3.6-Diacetate $C_{23}H_{34}O_6$. Crystals (methanol), m. 248.5–249.2⁰ cor. (1939 Ushakov); crystals (95 % alcohol), m. 253–4⁰; $[\alpha]_D^{26}$ +64⁰ (acetone) (1940 Ehrenstein). — **Fmn.** From the above triolone on refluxing with acetic anhydride for $1^1/_2$ hrs. (1940 Ehrenstein; cf. 1939 Ushakov).

For its *anhydro-compound, dehydroepiandrosterone 5α.6α-oxide*, see p. 2769 s.

Androstane-3β.5α.6β-triol-17-one $C_{19}H_{30}O_4$ (p. 150: *Stereoisomeric Androstan-17-one-3.5.6-triol*). For configuration at C-5 and C-6, see 1948 Ehrenstein; 1949 Davis. — Plates (aqueous alcohol), m. 301–2⁰ cor. dec. after becoming yellow at 290⁰ (1937 Ushakov); needles (methanol or acetone), m. 298–300⁰ dec. (1938 Miescher); prisms (acetone), m. 295–8⁰ (1939 Ehrenstein). Fairly readily soluble in cold methanol, very readily in hot; readily soluble on heating in water, chloroform, acetone, and alcohol, more sparingly in benzene and petroleum ether (1938 Miescher). — **Fmn.** From the acetate of Δ^5-androsten-3β-ol-17-one (dehydroepiandrosterone, p. 2609 s) on heating with 30 % H_2O_2 in glacial acetic acid on the water-bath for 2 hrs., followed by warming with 5 % methanolic KOH; yield, 85 % (1938 Miescher; 1936 CIBA; cf. 1939 Ehrenstein). From dehydroepiandrosterone 5α.6α-oxide (p. 2769 s) on treatment with dil. H_2SO_4 in acetone at room temperature for 36 hrs. (1937 Ushakov) or on heating with water in a sealed tube at 110–5⁰ for 10 hrs. (1938 Miescher); the 3-acetate is obtained from the acetate of the same oxide on heating with water-dioxane (7:3) in a sealed tube at 145–150⁰ for 7 days (yield, 70%), and similarly the 3-benzoate (130⁰, 2 days; yield, 40%) (1940 Ruzicka); the 3-acetate is also obtained from the acetates of the 5α.6α-oxide and the 5β.6β-oxide on treatment with sulphuric acid in aqueous acetone at room temperature for 46 hrs. (1940 Ehrenstein), the 6-acetate and the 3.6-diacetate from the 5α.6α-oxide on refluxing with glacial acetic acid for 2 hrs., the 3.5-diacetate from the acetate of the 5β.6β-oxide on refluxing with glacial acetic acid for 2 hrs. (1941 Ehrenstein).

Rns. Gives androstan-5α-ol-3.6.17-trione (p. 2983 s) on oxidation with CrO_3 in glacial acetic acid at room temperature (1937 Ushakov; 1939 Ehrenstein); when only 1.1 equiv. of CrO_3 is used in aqueous acetic acid at 0⁰, androstane-3β.5α-diol-6.17-dione (p. 2962 s) is formed; the 3-esters of this dioldione are obtained when the 3-esters of the triolone are similarly oxidized (1940 Ehrenstein) or in glacial acetic acid - chloroform at room temperature (1940 Ruzicka). The 6-acetate gives the 6-acetate of androstane-5α.6β-diol-3.17-dione (page 2961 s) on oxidation with CrO_3 (1.1 equiv.) in strong acetic acid at room temperature (1941 Ehrenstein).

3-Acetate $C_{21}H_{32}O_5$. Crystals (ether), m. 234–5⁰ (1940 Ehrenstein); crystals (ethyl acetate), m. 231–2⁰ cor.; $[\alpha]_D^{19}$ +34⁰ (chloroform) (1940 Ruzicka). — **Fmn.** and **Rns.** See above.

6-Acetate $C_{21}H_{32}O_5$. Prisms (acetone), m. 276–7⁰ with decomposition to a light brown liquid; $[\alpha]_D^{24}$ +23.6⁰ (methanol) (1941 Ehrenstein). — **Fmn.** and **Rns.** See those of the free triolone (above).

3.5-Diacetate $C_{23}H_{34}O_6$. Prisms (ether - petroleum ether), m. 202.5–204⁰*; $[\alpha]_D^{26}$ +23⁰ (acetone) (1941 Ehrenstein). — **Fmn.** and **Rns.** See those of the free triolone (above).

3.6-Diacetate $C_{23}H_{34}O_6$. Needles (95 % alcohol), prisms (ether - petroleum ether), m. 216.5–217.5⁰ (1940, 1941 Ehrenstein); crystals (ether-hexane), m. 184–185⁰ (1954 Amendolla; cf. also 1936 Zelinskiĭ), which on recrystallization in the presence of a trace of the higher-melting modification melt at 216–8⁰ (1954 Amendolla). $[\alpha]_D^{26}$ ±0⁰ (acetone) (1940 Ehrenstein). — **Fmn.** From the above triolone on refluxing with acetic anhydride for $1^1/_2$ hrs.; similarly from the 3-acetate (1940 Ehrenstein) and the 6-acetate (1941 Ehrenstein). See also under formation of the free triolone.

Triacetate $C_{25}H_{36}O_7$. Crystals (aqueous methanol), m. 184–5⁰; $[\alpha]_D^{26}$ —8⁰ (acetone). — **Fmn.** From the 3.5-diacetate on refluxing with acetic anhydride for 85 min. From the 3.6-diacetate on refluxing with acetic anhydride for 75 min. with passage of dry HCl (1941 Ehrenstein).

3-Benzoate $C_{26}H_{34}O_5$. Needles (ethyl acetate), m. 262–4⁰ cor. dec. (1940 Ruzicka). — **Fmn.** and **Rns.** See under those of the free triolone.

3-β-Glucoside $C_{25}H_{40}O_9$. Needles with 2 H_2O (dil. alcohol), m. 275⁰ dec. after sintering at 180⁰; when dried it rapidly absorbs 2 H_2O from the air. Sparingly soluble in water and most organic solvents in the cold; on heating it is readily soluble in water, methanol, ethanol, and ethyl acetate, somewhat soluble in acetone and chloroform, sparingly in benzene and hexane. — **Fmn.** From the tetraacetylglucoside of Δ^5-androsten-3β-ol-17-one (dehydroepiandrosterone, p. 2620 s) on heating with 30 % H_2O_2 in glacial acetic acid on the water-bath for 2 hrs., followed by warming with 5 % methanolic KOH for 30 min.; yield, 40 % (1938 Miescher).

Anhydro-compounds of androstane-3β.5α.6α-triol-17-one and ætiocholane-3β.5β.6β-triol-17-one:

5α.6α-Epoxyandrostan-3β-ol-17-one, 5-Dehydroepiandrosterone 5α.6α-oxide $C_{19}H_{28}O_3$ (p. 142). For α-configuration at C-5 and C-6, see 1948 Ehrenstein. — Prisms (ethyl acetate), m. 227.5–228.5⁰ cor.; slowly decomposes on keeping (1937 Ushakov); crystals (ether), m. 226–228.5⁰ (1940 Ehrenstein); prisms (acetone), crystals (methanol), m. 229–230⁰ (1938 Miescher).

Fmn. From Δ^5-androsten-3β-ol-17-one (dehydroepiandrosterone, p. 2609 s) on treatment with perbenzoic acid in chloroform at room temperature (yield, 65 %) (1937 Ushakov), at +2⁰ (yield, 22 %) (1942 Butenandt), or at —10⁰ (1938 Miescher; 1940 Ehrenstein). The acetate is obtained, along with that of the 5β.6β-oxide (below), from dehydroepiandrosterone acetate on treatment

---

* After the closing date for this volume the m. p. 231⁰ cor. was observed (1949 Davis).

with perbenzoic or monoperphthalic acid in carbon tetrachloride - ether
(1940*, 1944 Ruzicka; cf. 1941 Ehrenstein) or on warming with $KMnO_4$ in
aqueous acetic acid at 50° for 1 hr. (1940 Ehrenstein). The benzoate is obtained from dehydroepiandrosterone benzoate on treatment with monoperphthalic acid in ether-chloroform at —4° (24 hrs.), then at room temperature
(20 hrs.); yield, 79% (1940 Ruzicka).

Rns. The acetate and benzoate give the corresponding 3-esters of andro-
stane-3β.5α-diol-6.17-dione (p. 2962 s) on oxidation with CrO_3 in glacial
acetic acid - chloroform at room temperature (1940 Ruzicka). On reduction
of the acetate with 1 mol. hydrogen in the presence of platinum oxide in
alcohol it gives the 3-acetate of 5α.6α-epoxyandrostane-3β.17β-diol (p. 2207 s),
whereas in glacial acetic acid the 3-acetate of androstane-3β.5α-diol-17-one
(p. 2746 s) is formed, and with excess hydrogen in glacial acetic acid the
3-acetate of androstane-3β.5α.17β-triol (p. 2163 s) (1944 Ruzicka). Is hydrolysed to androstane-3β.5α.6β-triol-17-one (p. 2768 s) by treatment with dil.
H_2SO_4 in acetone at room temperature for 36 hrs. (1937 Ushakov) or by
heating in a sealed tube with water at 110–5° for 10 hrs. (1938 Miescher);
the acetate reacts similarly to give the triolone 3-acetate on treatment with
dil. H_2SO_4 in acetone at room temperature for 2 days (1940 Ehrenstein) or
on heating with water-dioxane (7 : 3) in a sealed tube at 145–150° for 7 days
(1940 Ruzicka); the benzoate gives the triolone 3-benzoate with aqueous
dioxane on refluxing for 5 days or, better, on heating in a sealed tube at
130° for 2 days (1940 Ruzicka). Acetolysis by refluxing with glacial acetic
acid for 2 hrs. converts the oxide into androstane-3β.5α.6β-triol-17-one
6-acetate and 3.6-diacetate, and the oxide acetate into the triolone 3.6-di-
acetate (1941 Ehrenstein). Gives 6.17α-dimethylandrostane-3β.5.17β-triol
(p. 2171 s) on heating with methylmagnesium bromide in anisole on the water-
bath (1942 Butenandt).

Acetate $C_{21}H_{30}O_4$. Crystals (ether), m. 223–4° (1940, 1941 Ehrenstein);
needles (ethyl acetate), m. 222–4° cor. (1940, 1944 Ruzicka); $[\alpha]_D^{26}$ —10°
(acetone) (1940, 1941 Ehrenstein), $[\alpha]_D$ —12° (acetone or chloroform) (1940,
1944 Ruzicka). — **Fmn.** From the preceding compound on refluxing with
acetic anhydride for 1 hr. (1940 Ehrenstein). See also under formation of
the oxide (above). — **Rns.** See those of the oxide (above).

Benzoate $C_{26}H_{32}O_4$. Needles (ethyl acetate - petroleum ether b. 40–70°),
m. 218–220° cor. (1940 Ruzicka). — **Fmn.** and **Rns.** See above.

5β.6β-Epoxy-ætiocholan-3β-ol-17-one, 5-Dehydroepiandrosterone 5β.6β-

oxide $C_{19}H_{28}O_3$. For β-configuration at C-5 and C-6,
see 1948 Ehrenstein. — **Fmn.** See that of the 5α.6α-
oxide (p. 2769 s). — **Rns.** On hydrogenation of the
acetate in the presence of platinum oxide in glacial
acetic acid until 2 mols. of hydrogen are absorbed it
gives androstane-3β.6β.17β-triol 3-acetate (p. 2164 s),

---

* The oxide acetate m. 205–7° cor., $[\alpha]_D$ —28° (chloroform), once obtained from dehydroepiandrosterone acetate with perbenzoic acid (1940 Ruzicka), was probably a mixture.

androstan-17-one (p. 2403 s), and androstan-17β-ol (p. 1512 s); the last-mentioned alcohol is the sole reaction product on complete hydrogenation in glacial acetic acid or alcohol (1944 Ruzicka). On treatment of the acetate with dil. H_2SO_4 in acetone at room temperature for 46 hrs. it gives the 3-acetate of androstane-3β.5α.6β-triol-17-one (p. 2768 s) (1940 Ehrenstein); on refluxing of the oxide acetate with glacial acetic acid for 2 hrs., the 3.5-diacetate of the same triolone is obtained (1941 Ehrenstein).

Acetate $C_{21}H_{30}O_4$. Needles (benzene - petroleum ether), m. 188–190° (1940 Ehrenstein); leaflets (ethyl acetate - alcohol), m. 186–7° cor. (1944 Ruzicka); $[\alpha]_D^{26} + 58°$ (acetone) * (1940 Ehrenstein), $[\alpha]_D^{15} + 47°$ (acetone), $+40.5°$ (chloroform) (1944 Ruzicka). — **Fmn.** and **Rns.** See above.

Androstane-3β.9.11-triol-17-one $C_{19}H_{30}O_4$. For its *9.11-anhydro-derivative*, see 9.11-epoxyandrostan-3β-ol-17-one (p. 2627 s).

Cholestane-2.5.6β-triol-3-one $C_{27}H_{46}O_4$.

2.5 - Epoxycholestan - 6β - ol - 3 - one

$C_{27}H_{44}O_3$. Needles (aqueous acetone), m. 181–2° cor. Gives a crimson coloration with conc. H_2SO_4, no coloration with $FeCl_3$. — **Fmn.** From the 6-acetate of 2-bromocholestane-5α.6β-diol-3-one (p. 2762 s) on warming with 1.5% methanolic KOH at 55–60° for 1 hr.; yield, 20–30%. — **Rns.** Gives 2.5-epoxycholestane-3.6-dione (p. 2963 s) on oxidation with CrO_3 in 95% acetic acid - benzene at room temperature. Is unaffected by lead tetraacetate and by refluxing for 3 hrs. with 10% alcoholic HCl (1939 Ellis).

Acetate $C_{29}H_{46}O_4$. Plates (methanol), m. 84° cor. — **Fmn.** From the preceding compound on treatment with acetic anhydride in dry pyridine at room temperature. — **Rns.** Gives the preceding compound on warming with 2.5% methanolic KOH at 55–60° (1939 Ellis).

Cholestane-4β.5.6-triol-3-one and **Cholestane-3β.5.6-triol-4-one** $C_{27}H_{46}O_4$.

5.6-Epoxycholestan-4β-ol-3-one, Δ^5-**Cholesten-4β-ol-3-one 5.6-oxide (I)** or **5.6-Epoxycholestan-3β-ol-4-one,** Δ^5-**Cholesten-3β-ol-4-one 5.6-oxide (II),**

* The correction made by Ehrenstein (1941) of the + sign (originally given by him in 1940) to the — sign is very probably erroneous as regards the 5β.6β-oxide acetate (Editor); cf. the rotation stated by Ruzicka (1944).

$C_{27}H_{44}O_3$. Structure (II) was assigned by Petrow (1940), structure (I) suggested by Fieser (1948). — **Fmn.** The acetate and benzoate are obtained from the corresponding 3-esters of Δ^5-cholestene-3β.4β-diol (p. 2051 s) on oxidation with CrO_3 in strong glacial acetic acid at room temperature; yield of the acetate, 46% (1940 Petrow).

Rns. On refluxing of the acetate with glacial acetic acid and anhydrous sodium acetate or with conc. HCl in alcohol-benzene it gives a mono-enolic form (m. 162–3° cor.) of Δ^5-cholestene-3.4-dione (1940 Petrow) which was formulated by Petrow (1940) as $\Delta^{2.5}$-cholestadien-3-ol-4-one, but which according to Fieser (1948) is $\Delta^{4.6}$-cholestadien-4-ol-3-one (see p. 2666 s). On refluxing of the acetate with acetic anhydride and anhydrous sodium acetate it gives a cholestadienolone acetate (1940 Petrow) which was formulated by Inhoffen (1936) and Petrow (1940) as $\Delta^{4.6}$-cholestadien-4-ol-3-one acetate, but which according to Fieser (1948) (cf. also 1939 Dannenberg) is $\Delta^{2.5}$-cholestadien-3-ol-4-one acetate (see p. 2674 s). On treatment of the acetate with sodium ethoxide in alcohol-ether, after 1 min. followed by acidification with 1 N HCl, it gives a dimeric product assumed to be 6-(4-hydroxy-3-keto-$\Delta^{4.6}$-cholestadien-6-yl)-3-hydroxy-4-keto-$\Delta^{2.5}$-cholestadiene (1940 Petrow).

Acetate $C_{29}H_{46}O_4$. Needles (alcohol), m. 173–4° cor.; $[\alpha]_D^{20} + 3.8°$, $[\alpha]_{5461}^{20} + 4.5°$ (chloroform). Shows no selective absorption in the ultra-violet region (1940 Petrow). — **Fmn.** and **Rns.** See above.

Benzoate $C_{34}H_{48}O_4$. Needles changing into prisms (benzene-alcohol 1:10), m. 185–6°; optically inactive in chloroform; $[\alpha]_D^{22} + 6.4°$, $[\alpha]_{5461}^{22} + 9.3°$ (benzene) (1940 Petrow). — **Fmn.** See above.

Cholestane-3β.4.5-triol-6-one 4.5-dimethyl ether, 4.5-Dimethoxycholestan-

3β-ol-6-one $C_{29}H_{50}O_4$. Needles (aqueous methanol), m. 149–150°; the m. p. gradually falls on keeping; sublimes at 155° in a high vacuum. Exhibits no intense absorption in the ultra-violet above 220 mμ. — **Fmn.** From the acetate of 4-bromo-Δ^4-cholesten-3β-ol-6-one (p. 2707 s) on treatment with sodium methoxide in methanol at room temperature for 18 hrs. (yield, 23%); is similarly obtained from the acetate of 4.5-dibromocholestan-3β-ol-6-one (p. 2713 s) (1938 Jackson).

Benzoate $C_{36}H_{54}O_5$. Octagonal tablets (methanol), m. 129–130° after softening at 126°. — **Fmn.** From the preceding compound on treatment with benzoyl chloride in pyridine at room temperature (1938 Jackson).

8-iso-Cholestane-3β.8α.9α-triol-7-one $C_{27}H_{46}O_4$.

8α. 9α - Epoxy - 8 - iso - cholestan - 3β - ol - 7 - one, Δ^8-Cholesten-3β-ol-7-one 8.9-oxide $C_{27}H_{44}O_3$ [I, R= $CH(CH_3) \cdot [CH_2]_3 \cdot CH(CH_3)_2$]. For α-configuration at C-8 and C-9, see 1953b Fieser.

Acetate $C_{29}H_{46}O_4$. Blades (methanol), m. 177.5° to 178°; $[\alpha]_D$ —32° (chloroform). — **Fmn.** Along with the 8.14-oxide (below), from the acetate of Δ^7-cholesten-3β-ol (γ-chole-stenol, p. 1689 s) on oxidation with CrO_3 in strong acetic acid at 25°; yield, ca. 9%. — **Rns.** Gives the acetate of Δ^8-cholesten-3β-ol-7-one (p. 2685 s) on refluxing with zinc dust in glacial acetic acid for 3 hrs. (1953a Fieser).

8-iso-Cholestane-3β.8α.14α-triol-7-one $C_{27}H_{46}O_4$.

8α.14α - Epoxy - 8 - iso - cholestan - 3β - ol - 7 - one, $\Delta^{8(14)}$-Cholesten-3β-ol-7-one 8.14-oxide $C_{27}H_{44}O_3$ [II, R = $CH(CH_3) \cdot [CH_2]_3 \cdot CH(CH_3)_2$]. For α-configuration at C-8 and C-14, see 1953b Fieser.

Acetate $C_{29}H_{46}O_4$. Rhombohedral plates (80% alcohol), m. 139.5–140° (1943a Wintersteiner); needles (methanol or petroleum ether b. 30–60°), m. 142–142.5° (1953a Fieser); [α]$_D$ in chloroform —76° (1943a Wintersteiner), —70° (1953a Fieser). Shows only a weak end absorption below 240 mμ (1943a Wintersteiner). Very soluble in methanol (1953a Fieser). — **Fmn.** Along with the acetate of 8α.14α-epoxy-8-iso-cholestan-3β-ol-15-one (below) and other compounds, from the acetate of $\Delta^{8(14)}$-cholesten-3β-ol (α-cholestenol, p. 1692 s) on oxidation with CrO_3 in strong acetic acid - benzene at room temperature; yield, 17% (1943b Wintersteiner). Along with the 8.9-oxide acetate (p. 2772 s), from the acetate of Δ^7-cholesten-3β-ol (γ-cholestenol, p. 1689 s) on oxidation with CrO_3 in strong acetic acid at 25° (1953a Fieser). From the 3-acetate of 8α.14α-epoxy-8-iso-cholestane-3β.7α-diol (p. 2209 s) on oxidation with CrO_3 in strong acetic acid at room temperature; crude yield, 94% (1943a Wintersteiner).

Rns. Gives the acetate of $\Delta^{8(14)}$-cholesten-3β-ol-7-one (p. 2685 s) on refluxing with zinc dust in glacial acetic acid (1953a Fieser). On refluxing with conc. HCl in alcohol for 2 hrs., followed by reacetylation, it gives the acetate of $\Delta^{8.14}$-cholestadien-3β-ol-7-one (p. 2681 s) (1943a Wintersteiner; for the structure, see 1953c Fieser); the 2.4-dinitrophenylhydrazone of the same dienolone is obtained when the present epoxy-compound is treated at room temperature with 2.4-dinitrophenylhydrazine in abs. alcohol containing 1% HCl (1943a Wintersteiner). No crystalline semicarbazone could be obtained (1943a Wintersteiner).

8-iso-Cholestane-3β.8α.14α-triol-15-one $C_{27}H_{46}O_4$.

8α.14α - Epoxy - 8 - iso - cholestan - 3β - ol - 15 - one, $\Delta^{8(14)}$-Cholesten-3β-ol-15-one 8.14-oxide $C_{27}H_{44}O_3$ [III, R = $CH(CH_3) \cdot [CH_2]_3 \cdot CH(CH_3)_2$]. For α-configuration at C-8 and C-14, see 1953b Fieser.

Acetate $C_{29}H_{46}O_4$. Rods (methanol), m. 180–1°; [α]$_D^{22}$ +4.7° (chloroform). — **Fmn.** Along with 8α.14α-epoxy-8-iso-cholestan-3β-ol-7-one (see above) and other compounds, from the acetate of $\Delta^{8(14)}$-cholesten-3β-ol (α-cholestenol, p. 1692 s) on oxidation with CrO_3 in strong acetic acid - benzene at room temperature; yield, ca. 2%. — **Rns.** On refluxing with conc. HCl in alcohol for 2 hrs., followed by reacetylation, it gives a small amount of a compound which is probably a **cholestadien-**

References, pp. 2775 s, 2776 s

3β-ol-15-one acetate $C_{29}H_{44}O_3$ (not analysed), rosettes (methanol), m. 132° to 133.5°, absorption max. (in alcohol?) at 219 mμ (ε 6750) (1943 b Wintersteiner).

8-iso-Δ²²-Ergostene-3β.8α.9α-triol-7-one $C_{28}H_{46}O_4$.

8α.9α-Epoxy-8-iso-Δ²²-ergosten-3β-ol-7-one, Δ⁸·²²-Ergostadien-3β-ol-7-one 8.9-oxide $C_{28}H_{44}O_3$ [I, p. 2772 s; R = $CH(CH_3) \cdot CH:CH \cdot CH(CH_3) \cdot CH(CH_3)_2$]. For α-configuration at C-8 and C-9, see 1953 b Fieser; cf. also 1952 Heusser. — For its *acetate*, see p. 1758 s; the acetoxy-ketone obtained by refluxing in alcoholic HCl, followed by reacetylation (1943 b Stavely), is the acetate of Δ⁸·¹⁴·²²-ergostatrien-3β-ol-7-one (p. 2690 s) (1953 c Fieser).

8-iso-Δ²²-Ergostene-3β.8α.14α-triol-7-one $C_{28}H_{46}O_4$.

8α.14α-Epoxy-8-iso-Δ²²-ergosten-3β-ol-7-one, Δ⁸⁽¹⁴⁾·²²-Ergostadien-3β-ol-7-one 8.14-oxide $C_{28}H_{44}O_3$ [II, p. 2773 s; R = $CH(CH_3) \cdot CH:CH \cdot CH(CH_3) \cdot CH(CH_3)_2$]. For α-configuration at C-8 and C-14, see 1953 b Fieser; cf. also 1952 Heusser. — For its *acetate*, see p. 1759 s; the acetoxy-ketone obtained by refluxing in alcoholic HCl, followed by reacetylation (1943 b Stavely), is the acetate of Δ⁸·¹⁴·²²-ergostatrien-3β-ol-7-one (p. 2690 s) (1953 c Fieser).

8-iso-Ergostane-3β.8α.14α-triol-7-one $C_{28}H_{48}O_4$.

8α.14α-Epoxy-8-iso-ergostan-3β-ol-7-one, Δ⁸⁽¹⁴⁾-Ergosten-3β-ol-7-one 8.14-oxide $C_{28}H_{46}O_3$ [II, p. 2773 s; R = $CH(CH_3) \cdot CH_2 \cdot CH_2 \cdot CH(CH_3) \cdot CH(CH_3)_2$]. For α-configuration at C-8 and C-14, see 1953 b Fieser.

Acetate $C_{30}H_{48}O_4$. Crystals (80% alcohol), m. 134°; [α]$_D^{23}$ —83° (chloroform) (1943 a Stavely). — **Fmn.** Along with 8α.14α-epoxy-8-iso-ergostan-3β-ol-15-one acetate (below; yield, 2%) and other compounds, from the acetate of Δ⁸⁽¹⁴⁾-ergosten-3β-ol (α-ergostenol, p. 1765 s) on oxidation with CrO_3 in strong acetic acid - benzene at room temperature; yield, 18% (1943 a Stavely). — **Rns.** On refluxing with conc. HCl in alcohol for 2½ hrs., followed by reacetylation it gives the acetate of Δ⁸·¹⁴-ergostadien-3β-ol-7-one (p. 2691 s) (1943 a Stavely; for the structure, see 1953 c Fieser). With 2.4-dinitrophenylhydrazine in acidified alcohol it gives an orange hydrazone possibly derived from the afore-mentioned dienolone acetate (1943 a Stavely; cf. the analogous reaction of 8α.14α-epoxy-8-iso-cholestan-3β-ol-7-one, p. 2773 s). Gives no semicarbazone (1943 a Stavely).

8-iso-Ergostane-3β.8α.14α-triol-15-one $C_{28}H_{48}O_4$.

8α.14α-Epoxy-8-iso-ergostan-3β-ol-15-one, Δ⁸⁽¹⁴⁾-Ergosten-3β-ol-15-one 8.14-oxide $C_{28}H_{46}O_3$ [III, p. 2773 s; R = $CH(CH_3) \cdot CH_2 \cdot CH_2 \cdot CH(CH_3) \cdot CH(CH_3)_2$]. For α-configuration at C-8 and C-14, see 1953 b Fieser.

Acetate $C_{30}H_{48}O_4$. Crystals (80% alcohol), m. 208–210°; [α]$_D^{23}$ —6° (chloroform). Shows no absorption in the ultra-violet above 230 mμ. — **Fmn.** See that of the preceding oxide acetate. — **Rns.** On heating with conc. HCl in alcohol on the steam-bath for 2 hrs., followed by reacetylation, it gives the acetate of $\Delta^{8(14),9(11)}$-ergostadien-3β-ol-15-one (p. 2693 s). Does not react with 2.4-dinitrophenylhydrazine in acidified alcohol at room temperature (1943a Stavely).

8-iso-Δ^{22}-Stigmastene-$3\beta.8\alpha.9\alpha$-triol-7-one and **8-iso-Δ^{22}-Stigmastene-$3\beta.8\alpha.14\alpha$-triol-7-one** $C_{29}H_{48}O_4$.

8α.9α-Epoxy-8-iso-Δ^{22}-stigmasten-3β-ol-7-one, $\Delta^{8,22}$-Stigmastadien-3β-ol-7-one 8.9-oxide, and **8α.14α-Epoxy-8-iso-Δ^{22}-stigmasten-3β-ol-7-one, $\Delta^{8(14),22}$-Stigmastadien-3β-ol-7-one 8.14-oxide,** $C_{29}H_{46}O_3$ [I and II, resp., p. 2772 s, 2773 s; R = $CH(CH_3)\cdot CH:CH\cdot CH(C_2H_5)\cdot CH(CH_3)_2$]. For α-configuration at C-8, C-9, and C-14, see 1953b Fieser. — For their *acetates*, see pp. 1796 s and 1797 s, resp.; the acetoxy-ketone obtained by refluxing of the two acetates in alcoholic HCl, followed by reacetylation (1943c Stavely), is the acetate of $\Delta^{8,14,22}$-stigmastatrien-3β-ol-7-one (p. 2695 s) (1953c Fieser).

5-Methyl-19-nor-coprostane(or -cholestane)-$3\beta.6\beta.9.10$-tetrol-11-one $C_{27}H_{46}O_5$. For its *9.10-anhydro-derivative*, see 5-methyl-9.10-epoxy-19-nor-coprostane (or -cholestane)-$3\beta.6\beta$-diol-11-one, p. 2753 s.

1936 Ciba (Ges. f. chem. Ind. in Basel), *Swiss Pat.* 200055 (issued 1938); *C. A.* **1939** 3535; *Brit. Pat.* 489364 (issued 1938); *C. A.* **1939** 815; *Ch. Ztbl.* **1939** I 3932.
Inhoffen, *Ber.* **69** 1702, 1705.
Zelinskiĭ, Ushakov, *Bull. acad. sci. U.R.S.S. Sér. chim.* **1936** 879.

1937 Ushakov, Lyutenberg (Ouchakov, Lutenberg), *J. Gen. Ch. U.S.S.R.* **7** 1821; *Bull. soc. chim.* [5] **4** 1394.

1938 Butenandt, Wolz, *Ber.* **71** 1483, 1485, 1486.
Jackson, Jones, *J. Ch. Soc.* **1938** 1406, 1408.
Miescher, Fischer, *Helv.* **21** 336, 353, 354.
Serini, Logemann, *Ber.* **71** 186, 189.

1939 Dannenberg, *Abhandl. Preuss. Akad. Wiss. Math.-naturwiss. Kl.* **1939** No. 21, p. 32.
Ehrenstein, *J. Org. Ch.* **4** 506, 514, 515.
Ellis, Petrow, *J. Ch. Soc.* **1939** 1078, 1082.
Pearlman, Wintersteiner, *J. Biol. Ch.* **130** 35, 39, 40.
Ushakov, Lyutenberg, *J. Gen. Ch. U.S.S.R.* **9** 69, 71; *C. A.* **1939** 6334; *Ch. Ztbl.* **1939** II 4489.

1940 Ehrenstein, Decker, *J. Org. Ch.* **5** 544, 551–555.
Petrow, Starling, *J. Ch. Soc.* **1940** 60, 63, 65.
Ruzicka, Grob, Raschka, *Helv.* **23** 1518, 1522–1525.

1941 Ehrenstein, *J. Org. Ch.* **6** 626, 629, 637–641.
Schering Corp. (Schwenk, Bloch, Whitman), *U.S. Pat.* 2418603 (issued 1947); Schering A.-G. (Schwenk, Bloch, Whitman), *Fr. Pat.* 892206 (1942, issued 1944); *C. A.* **1947** 4518; *Ch. Ztbl.* **1948** I 492.

1942 Butenandt, Surányi, *Ber.* **75** 597, 605.

1943 Lettré, *Z. physiol. Ch.* **278** 206.
 a Stavely, Bollenback, *J. Am. Ch. Soc.* **65** 1285.
 b Stavely, Bollenback, *J. Am. Ch. Soc.* **65** 1290, 1293.
 c Stavely, Bollenback, *J. Am. Ch. Soc.* **65** 1600, 1603.

a Wintersteiner, Moore, *J. Am. Ch. Soc.* **65** 1507, 1512.
b Wintersteiner, Moore, *J. Am. Ch. Soc.* **65** 1513, 1515.
1944 Ruzicka, Muhr, *Helv.* **27** 503, 507 et seq.
1948 Ehrenstein, *J. Org. Ch.* **13** 214, 216–218.
      Fieser, Fieser, Rajagopalan, *J. Org. Ch.* **13** 800.
1949 Davis, Petrow, *J. Ch. Soc.* **1949** 2536.
1952 Heusser, Saucy, Anliker, Jeger, *Helv.* **35** 2090.
1953 a Fieser, *J. Am. Ch. Soc.* **75** 4395, 4401, 4402.
      b Fieser, Ourisson, *J. Am. Ch. Soc.* **75** 4404, 4407, 4408.
      c Fieser, Nakanishi, Huang, *J. Am. Ch. Soc.* **75** 4719.
1954 Amendolla, Rosenkranz, Sondheimer, *J. Ch. Soc.* **1954** 1226. 1231.

B. Dioxosteroids with One CO in the Ring System

I. COMPOUNDS WITHOUT OTHER FUNCTIONAL GROUPS

a. Compounds C_{18}–C_{20}

4'-Formyl-3'-keto-1.2-cyclopentano-1.2.3.4-tetrahydrophenanthrene $C_{18}H_{16}O_2$. See the enolic form, p. 2510 s.

19-Nor*-14-iso-17-iso-Δ^4-pregnene-3.20-dione, 19-Nor*-14-iso-17-iso-pro-gesterone $C_{20}H_{28}O_2$. For normal (β) configuration at C-10, see 1958 Djerassi; for the other configurations, see the diazo-compound on p. 2292 s. — Prisms (aqueous methanol and aqueous acetone), m. 90–91°; $[\alpha]_D^{21}$ +57° (chloroform); absorption max. in alcohol at 240 mμ (ε 16000) (1957 Barber). — Physiologically about as potent as progesterone (1944 Allen; 1944 Ehrenstein). — **Fmn.** From 19-nor*-14-iso-17-iso-$\Delta^{5(10?)}$-pregnen-3β-ol-20-one (p. 2226 s) on refluxing with aluminium tert.-butoxide and acetone in benzene (1944 Ehrenstein); when the acetate of the aforesaid norpregnenolone is saponified with alcoholic KOH (which may cause inversion at C-17) and the resulting hydroxyketone is oxidized as above, there are obtained needles (aqueous methanol), m. 110°, $\lambda_{max}^{alc.}$ 239 mμ (ε 17000), possibly *19-nor-14-iso-progesterone* (1957 Barber; cf. also 1952 Ehrenstein; 1954 Barber). — For "normal" *19-nor-progesterone*, see 1951 Miramontes; 1953 Djerassi; 1955 Barber.

17-Formyl-$\Delta^{4.16}$-androstadien-3-one, $\Delta^{4.16}$-Androstadien-3-one-17-aldehyde $C_{20}H_{26}O_2$. For a compound possibly possessing this structure, see the compound $C_{20}H_{26}O_2$ or $C_{20}H_{28}O_2$ under the reactions of 17-iso-Δ^4-pregnene-17β.20a.21-triol-3-one, p. 2727 s.

17β-Formyl-Δ^4-androsten-3-one, Δ^4-Androsten-3-one-17β-aldehyde, 3-Keto-Δ^4-ætiocholenaldehyde, 21-Nor-progesterone, Norpro-gesterone, "*20-Nor-progesterone*" $C_{20}H_{28}O_2$. Crystals (ether), m. 151–3° cor. vac.; $[\alpha]_D^{19}$ +159° (dioxane) (1940a Miescher). — Has a weak progestational and a weak androgenic activity (1940b Miescher). — **Fmn.** From 17β-formyl-Δ^5-androsten-3β-ol (p. 2227 s) on oxidation with aluminium tert.-butoxide and acetone (1939 CIBA). From Δ^4-pregnene-20"α".21-diol-3-one (p. 2514 s) with HIO_4 in aqueous dioxane under CO_2; yield, 89% (1940a Miescher). For formation of a mixture of stereoisomeric Δ^4-androsten-3-one-17-aldehydes from esters or ethers of $\Delta^{3.5}$-androstadien-3-ol-17-one (1938a CIBA), see pp. 2605 s, 2606 s.

Rns. On treatment with dimethylzinc it gives Δ^4-pregnen-20α-ol-3-one (p. 2510 s) (1937 CIBA). May be converted into progesterone (p. 2782 s) by the action of diazomethane in ether, or by the action of diethyl diazomalonate

---

* The original designation 10-nor was later changed to 19-nor (see 1951 Ehrenstein).

191*

or a diazoacetoacetic ester in ether, followed by refluxing with 5 % methanolic KOH and then with dil. aqueous methanolic H_2SO_4; gives 21-methylprogesterone (p. 2803 s) with diazoethane; its *3-enol esters* or *ethers* give the corresponding 3-enol derivatives of progesterone (see p. 2229 s) on treatment with diazomethane (1938b CIBA). — *Disemicarbazone* $C_{22}H_{34}O_2N_6$. Crystals, dec. 296° cor. vac. (1940a Miescher).

1937 CIBA, *Swiss Pat.* 215287 (issued 1941); *C. A.* **1948** 3538; *Ch. Ztbl.* **1942** I 2430.
1938 a CIBA, *Brit. Pat.* 516542 (issued 1940); *Ch. Ztbl.* **1940** II 1327, 1328.
    b CIBA (Miescher, Wettstein), *Ger. Pat.* 739083 (issued 1943); *C. A.* **1945** 392; *Ch. Ztbl.*
    1944 I 110, 111; *Fr. Pat.* 840513 (issued 1939); *C. A.* **1939** 8627; *Ch. Ztbl.* **1939** II 1125.
1939 CIBA, *Fr. Pat.* 851017 (issued 1940); *C. A.* **1942** 2090; *Ch. Ztbl.* **1940** II 932.
1940 a Miescher, Hunziker, Wettstein, *Helv.* **23** 400. — b id., *Helv.* **23** 1367, 1368.
1944 Allen, Ehrenstein, *Science* [N.S.] **100** 251.
    Ehrenstein, *J. Org. Ch.* **9** 435, 443, 454.
1951 Ehrenstein, Barber, Gordon, *J. Org. Ch.* **16** 349, 355, footnote 7.
    Miramontes, Rosenkranz, Djerassi, *J. Am. Ch. Soc.* **73** 3540.
1952 Ehrenstein, *Chimia (Switz.)* **6** 287.
1953 Djerassi, Miramontes, Rosenkranz, *J. Am. Ch. Soc.* **75** 4440.
1954 Barber, Ehrenstein, *J. Org. Ch.* **19** 365, 367.
1955 Barber, Ehrenstein, *J. Org. Ch.* **20** 1253, 1258.
1957 Barber, Ehrenstein, *Ann.* **603** 89, 96, 105, 106, 107, 109.
1958 Djerassi, Ehrenstein, Barber, *Ann.* **612** 93, 97.

b. Dioxosteroids C_{21} (One CO in the Ring System) with Three Side Chains
Summary of 3.20-Diketones with Pregnane Skeleton *

Configuration at			Double bond(s)	M. p. °C.	$[\alpha]_D^{(0)}$ in chloroform	Absorption max (mμ) in alcohol	Page
C-5	C-14	C-17					
—	norm.	norm.	$\Delta^{4.6}$	148	+150[1])	282[2])	2778 s
—	norm.	norm.	$\Delta^{4.9(11)}$	122	+145[3])	—	2779 s
—	norm.	norm.	$\Delta^{4.11}$	177	+181[3])	238	2779 s
—	norm.	norm.	$\Delta^{4.16}$	188	+154[1])	234[4])	2780 s
α	norm.	norm.	Δ^1	210	+126(?)	230	2781 s
—	norm.	norm.	Δ^4	130	+196	234[4])	2782 s
—	norm.	norm.	Δ^5	160	+66	—	2793 s
β	norm.	norm.	Δ^{11}	135	+85[3])	—	2793 s
β	norm.	norm.	Δ^{16}	200	+84	239	2794 s
α	norm.	norm.	Δ^{16}	211	+72	—	2795 s
β	norm.	norm.	satd.	123	+112	—	2796 s
α	norm.	norm.	satd.	205	+121	—	2797 s
—	norm.	iso	Δ^4	145	±0[1])	243	2800 s
α	norm.	iso	satd.	149	−50[1])	—	2800 s
α	iso	iso	satd.	141	+40	—	2800 s

* For values other than those chosen for this summary, see the compounds themselves.
[1]) In alcohol.   [2]) Solvent not stated.   [3]) In acetone.   [4]) In ether.

$\Delta^{4.6}$ - Pregnadiene - 3.20 - dione, 6 - Dehydroprogesterone $C_{21}H_{28}O_2$. Crystals

(hexane), m. 147–8° cor.; sublimes at 140° under 0.02 mm. pressure; $[\alpha]_D^{18}$ +150° (alcohol); absorption max. (solvent not stated) at 282 mμ (log ε 4.40), inflexion at ca. 350 mμ (1940 Wettstein; 1941 CIBA). — Its progestational activity is about $1/2$ that of progesterone (1940 Wettstein; 1941 CIBA). — **Fmn.** From Δ^5-pregnen-3β-ol-20-one (p. 2233 s) on refluxing with aluminium tert.-but-oxide and benzoquinone in abs. toluene for 1 hr., followed by removal of unchanged pregnenolone with the aid of its hemisuccinate (1940 Wettstein; 1941 CIBA). From 6β-bromo-Δ^4-pregnene-3.20-dione (p. 2813 s) on boiling with dry pyridine for $2^1/_2$ hrs.; yield, ca. 8% (1937 SCHERING CORP.; 1937 CIBA).

$\Delta^{4,9(11)}$-Pregnadiene-3.20-dione, 9(11)-Dehydroprogesterone $C_{21}H_{28}O_2$. Origi-nally described as $\Delta^{4,11}$-pregnadiene-3.20-dione; for the true structure, see 1943 Hegner; 1946 Shoppee. — Needles (ether - pentane), m. 120° to 122° cor.; $[\alpha]_D^{18} +145°$, $[\alpha]_{5461}^{18} +185°$ (both in acetone); readily soluble in cold ether, sparingly in pentane; gives a slight yellow coloration with tetranitromethane (1941 Shoppee). — Is physiologically about $1/2$ as active as progesterone (1941 Shoppee). — **Fmn.** From 11β-hydroxyprogesterone (p. 2840 s) on refluxing with glacial acetic acid - conc. HCl (4 vols.:1 vol.) for 30 min. (yield, 70% allowing for recuperated starting material) (1941 Shoppee) or, along with an isomer, on refluxing for a short time with 0.4 parts of 50% H_2SO_4 in 10 parts of acetic anhydride (1942 b Reichstein).

Rns. On shaking with hydrogen in glacial acetic acid in the presence of platinum oxide at 18°, followed by refluxing with 4% methanolic KOH for 5 min., it gives a mixture of diols, which on oxidation with CrO_3 in glacial acetic acid at 15° yields allopregnane-3.20-dione (p. 2797 s) and some preg-nane-3.20-dione (p. 2796 s) (1941 Shoppee).

$\Delta^{4,11}$-Pregnadiene-3.20-dione, 11-Dehydroprogesterone $C_{21}H_{28}O_2$. For a com-pound originally described with this structure, see above. — Prisms (ether - petroleum ether), m. 175° to 177° cor. (Kofler block) (1942 a Reichstein; 1943 Hegner; 1946 v. Euw); $[\alpha]_D^{15} +181°$ (acetone) (1943 Hegner). Absorption max. in alcohol at 238 mμ (log ε 4.34), weak inflexion at ca. 290 mμ (1948 Meystre). Gives a yellow coloration with tetranitromethane in chloroform (1948 Meystre). — Is physiologically three times as active as progesterone (1948 Meystre). — **Fmn.** Along with Δ^{11}-pregnene-3.20-dione (p. 2793 s), from the p-toluenesulphonate of pregnan-12α-ol-3.20-dione (p. 2841 s) on treatment with 1 mol. bromine in glacial acetic acid, followed by refluxing of the resulting bromo-compound with collidine for 3 hrs.; very small yield (1946 v. Euw). From the acetate or the benzoate of Δ^4-pregnen-12α-ol-3.20-dione (12α-hydroxyprogesterone, p. 2840 s) on heating at 308–312° (block temperature) under 12 mm. pressure (yield, 27% allowing for recuperated starting material) (1942 a Reichstein; 1943 Hegner); from the p-toluenesulphonate of the same hydroxydiketone on refluxing with sym. collidine in o-xylene for 24 hrs. (1948 Meystre).

References, pp. 2780 s, 2781 s

$\Delta^{4,16}$-**Pregnadiene-3.20-dione,   16-Dehydroprogesterone** $C_{21}H_{28}O_2$.   Leaflets
(aqueous acetone), crystals (ethyl acetate), m. 186°
to 188°; the crystals contain water of crystallization which may be removed by sublimation at
210–220° under ca. 15 mm. pressure (1939 Butenandt); $[\alpha]_D^{21}$ +154° (alcohol) (1944 Wettstein).
Absorption max. in ether at 234 mμ (ε 26100) (1939 Butenandt; 1939 Dannenberg).

Has a weak progesterone-like activity (1943 Selye); has no œstrogenic
activity, but shows a weak androgenic activity in the Fussgänger capon
comb test (1939 Butenandt).

Fmn. From 17-ethynyl-$\Delta^{4,16}$-androstadien-3-one (? p. 2406 s) on heating
with HgSO$_4$ and conc. H$_2$SO$_4$ in 70% acetic acid for 4 hrs., or on shaking
with HgO and BF$_3$ - ether in glacial acetic acid and some acetic anhydride
for 5 days, followed by refluxing with methanolic KOH for 2 hrs. (1938
Schering A.-G.). From 17α-ethynyl-Δ^4-androsten-17β-ol-3-one (p. 2648 s) on
refluxing with mercury acetamide in abs. alcohol, followed by decomposition
with H$_2$S (1940 Ciba). From $\Delta^{5,16}$-pregnadien-3β-ol-20-one (p. 2230 s) on
refluxing with aluminium isopropoxide and cyclohexanone in toluene for
45 min.; yield, 47% (1939 Butenandt; 1939 Schering A.-G.). From 17-hydr-
oxy-(17-iso?)-progesterone (see pp. 2843 s and 2844 s) on heating with fuller's
earth at 150–160° under 0.1 mm. pressure (1939 Ciba). From pseudo-Δ^4-tigo-
genone on oxidation with CrO$_3$ in strong acetic acid at 25–28° (1940 Marker).

Rns. Gives allopregnane-3β.20α-diol (p. 1933 s) on reduction with sodium
in abs. alcohol (1940 Marker). On partial hydrogenation in ether in the
presence of palladium-barium sulphate at room
temperature under 5 lb. pressure it gives Δ^4-preg-
nene-3.20-dione (progesterone, p. 2782 s) and
Δ^{16} - pregnene - 3.20 - dione (p. 2794 s) while on
complete hydrogenation pregnane - 3.20 - dione
(p. 2796 s) is formed (1940 Marker); for similar reduction, using Raney nickel,
see 1939 Ciba. On treatment with diazomethane in ether at room temperature
it gives the **pyrazoline compound** $C_{22}H_{30}O_2N_2$ (I) [crystals (acetone), m. 173°
cor. with strong decomposition] which on heating at 150–180° under 0.03 mm.
pressure is decomposed to give 16-methyl-$\Delta^{4,16}$-pregnadiene-3.20-dione (page
2807 s) (1944 Wettstein).

1937 Ciba*, *Brit. Pat.* 492377 (issued 1938); *Fr. Pat.* 833102 (1938, issued 1938); *Ch. Ztbl.* **1939** I
2642, 2643.
    Schering Corp. (Inhoffen, Butenandt, Schwenk), *U.S. Pat.* 2340388 (issued 1944); *C. A.*
1944 4386; Schering-Kahlbaum A.-G., *Fr. Pat.* 835524 (issued 1938); *Brit. Pat.* 500353
(issued 1939); *C. A.* **1939** 5872; *Ch. Ztbl.* **1939** I 5010, 5011.
1938 Schering A.-G., *Dutch Pat.* 52579 (issued 1942); *Ch. Ztbl.* **1942** II 2615; *Fr. Pat.* 845473
(issued 1939); *C.A.* **1941** 1185; *Ch.Ztbl.* **1943** I 653; Schering Corp. (Inhoffen, Logemann,
Dannenbaum), *U.S. Pat.* 2280236 (1939, issued 1942); *C. A.* **1942** 5322.
1939 Butenandt, Schmidt-Thomé, *Ber.* **72** 182, 184, 186.
    Ciba*, *Fr. Pat.* 860278 (issued 1941); *Ch. Ztbl.* **1941** I 3258.

* Société pour l'industrie chimique à Bâle; Gesellschaft für chemische Industrie in Basel.

Dannenberg, *Abhandl. Preuss. Akad. Wiss. Math.-naturwiss. Kl.* **1939** No. 21, pp. 46, 65.
SCHERING A.-G., *Fr. Pat.* 871942 (issued 1942); *Ch. Ztbl.* **1942** II 2504.
1940 CIBA*, *Fr. Pat.* 876213 (issued 1942); *Ch. Ztbl.* **1943** I 2320.
Marker, Tsukamoto, Turner, *J. Am. Ch. Soc.* **62** 2525, 2528, 2530; PARKE, DAVIS & CO.
(Marker), *U.S. Pat.* 2420489 (issued 1947); *C. A.* **1947** 5689; *Ch. Ztbl.* **1948** I 380.
Wettstein, *Helv.* **23** 388, 391, 393, 395.
1941 CIBA*, *Fr. Pat.* 877821 (issued 1943); *Dutch Pat.* 54147 (issued 1943); *Ch.Ztbl.***1943** II 147, 148.
Shoppee, Reichstein, *Helv.* **24** 351, 353–356.
1942 a Reichstein, *Fr. Pat.* 879972 (issued 1943); *Ch. Ztbl.* **1945** I 195; *Swed. Pat.* 110155 (issued
1944); *Ch. Ztbl.* **1945** II 263.
b Reichstein, *U.S. Pat.* 2409798 (issued 1946); *Ch. Ztbl.* **1947** 232; *Brit. Pat.* 560812 (issued
1944); *C. A.* **1946** 4486; *Ch. Ztbl.* **1945** II 1773.
1943 Hegner, Reichstein, *Helv.* **26** 715, 717, 720.
Selye, Masson, *J. Pharmacol.* **77** 301.
1944 Wettstein, *Helv.* **27** 1803, 1812.
1946 v. Euw, Reichstein, Shoppee, *Helv.* **29** 654, 669.
Shoppee, *J. Ch. Soc.* **1946** 1134 footnote.
1948 Meystre, Tschopp, Wettstein, *Helv.* **31** 1463, 1468, 1469.

Δ^1-**Allopregnene-3.20-dione** $C_{21}H_{30}O_2$ (p. 150). The compound originally des-

cribed with this structure (1935 Butenandt) and later designated hetero (h)-Δ^1-allopregnene-3.20-dione (1939 Butenandt) is actually Δ^5-pregnene-4.20-dione (p. 2801 s) (1944 Butenandt). — Crystals (acetone or ethyl acetate), m. 208–210° (1939 Marker). The sample showing $[\alpha]_D$ in chloroform $+126°$ (see p. 150) was possibly contaminated with the Δ^{16}-isomer (1947 Djerassi). Absorption max. in alcohol at ca. 230 mμ (ε 10500) (1939 Butenandt; 1939 Dannenberg). — **Fmn.** From 2α-bromoallopregnane-3.20-dione (p. 2812 s) on refluxing with collidine for $2^1/_2$ hrs. (yield, 51%) (1939 Butenandt; 1940 SCHERING A.-G.); from the pyridinium salt of the same bromodione on distillation under 14 mm. pressure (yield, ca. 8%) (1939 Butenandt; 1940 SCHERING A.-G.; see also 1939 Marker).

Rns. On hydrogenation in alcohol in the presence of Adams's platinum oxide catalyst at room temperature under 3 atm. pressure it gives allopregnane-3β.20β-diol (p. 1934 s); on using a partially inactivated platinum catalyst under the same conditions, allopregnan-3β-ol-20-one (p. 2247 s) is formed (1939 Marker). On hydrogenation in glacial acetic acid in the presence of platinum black, followed by oxidation with CrO_3 in glacial acetic acid, it gives allopregnane-3.20-dione (p. 2797 s) (1939 Butenandt). Is not attacked by fermenting yeast (1940 Butenandt).

Dioxime $C_{21}H_{32}O_2N_2$. Crystals (dil. acetone), m. 248–250° (1939 Marker).

hetero-Δ^1-Allopregnene-3.20-dione $C_{21}H_{30}O_2$ (p. 150). This compound is actually Δ^5-pregnene-4.20-dione (p. 2801 s) (1944 Butenandt).

1935 Butenandt, Mamoli, *Ber.* **68** 1850.
1939 Butenandt, Mamoli, Dannenberg, Masch, Paland, *Ber.* **72** 1617, 1620-1622.

* Société pour l'industrie chimique à Bâle; Gesellschaft für chemische Industrie in Basel.

Dannenberg, *Abhandl. Preuss. Akad. Wiss. Math.-naturwiss. Kl.* **1939** No. 21, pp. 13, 14, 54.
Marker, Wittle, Plambeck, *J. Am. Ch. Soc.* **61** 1333.
1940 Butenandt, Dannenberg, Surányi, *Ber.* **73** 818.
SCHERING A.-G., *Fr. Pat.* 867697 (issued 1941); *Ch. Ztbl.* **1942** I 2561; *Danish Pat.* 61322 (issued 1943); *Ch. Ztbl.* **1944** II 49; (Butenandt), *Swed. Pat.* 111450 (issued 1944); *Ch. Ztbl.* **1947** 1875.
1944 Butenandt, Ruhenstroth-Bauer, *Ber.* **77** 397, 400.
1947 Djerassi, *J. Org. Ch.* **12** 823, 824.

Δ^4-Pregnene-3.20-dione, Progesterone, Corpus luteum hormone $C_{21}H_{30}O_2$

(p. 151). Crystals (light petroleum or acetone), m. 121° (1938b Beall; 1938 Dirscherl); prisms (ether), m. 129.5–130.5° cor. (1937 Oppenauer); $[\alpha]_D + 193°$ (alcohol) (1938b Beall), $+ 192°$ (chloroform) (1938 Dirscherl), $+ 196°$ (chloroform) (1941b Bretschneider). Absorption max. in ether at 234 mμ (ε 17300) (1939a Butenandt; 1939 Dannenberg), in 96% H_2SO_4 at 290 mμ (log k 4.35) (1939 Bandow); for absorption in the ultra-violet, see also 1935 Tavastsherna; 1938 Bugyi. Soluble on boiling in aqueous sodium dehydrocholate solutions (1 mg. in 1 c.c. of 20% solution, in 1.4 c.c. of 10% solution, in more than 6 c.c. of 4% solution) (1944 Cantarow). For polarographic behaviour of progesterone, see 1939 Eisenbrand. For basic properties, see 1944, 1946 Bersin.

OCCURRENCE OF PROGESTERONE

For occurrence of progesterone in the placenta of women and animals, see also 1937 Ehrhardt. Progesterone could not be found in human pregnancy urine (1937 Marker). Occurs in ox adrenals (1938a, b Beall; cf. also 1941 v. Euw); for occurrence of progesterone (not isolated, but detected by progestational activity) in adrenal glands of horses, cattle, pigs, and sheep, and in bulls' and boars' testes, see 1936 Callow.

FORMATION OF PROGESTERONE

Formation from compounds with fewer C-atoms. From Δ^4-androsten-3-one-17-aldehyde (p. 2777 s) by the action of diazomethane in ether (yield, 55%) or by the action of diethyl diazomalonate or a diazoacetoacetic ester in ether, followed by refluxing with 5% methanolic KOH for 2 hrs. and then with dil. aqueous methanolic H_2SO_4 for 2 hrs. (1938f CIBA). From the 3-enol benzoate of Δ^4-androstene-3.17-dione ($\Delta^{3.5}$-androstadien-3-ol-17-one benzoate, p. 2605 s) on treatment with ethyl chloroacetate and sodium ethoxide in ether, followed by hydrolysis of the resulting 17-glycidic acid ester, decarboxylation by heat, and repetition of the process with the resulting mixture of Δ^4-androsten-3-one-17-aldehydes or their 3-enol esters (1938h CIBA); from the same enol - benzoate on treatment with ethyl $\alpha.\alpha$-dibromopropionate and amalgamated magnesium, followed by hydrolysis and decarboxylation (1938a CIBA). From the 3-enol acetate of 3-keto-Δ^4-ætiocholenic acid chloride by the action of dimethylzinc, followed by hydrolysis (1938c, d CIBA); from the aforementioned acid chloride or its 3-enol esters or ethers on treatment with diethyl sodiomalonate in alcohol - benzene, followed by hydrolysis of the resulting 20-one-21.21-dicarboxylic ester with 2% alcoholic NaOH and decarboxylation by heating in a high vacuum (1938c CIBA).

Formation from pregnenols, pregnenediols, and pregnenetriols. From Δ^4-pregnen-20-ol (p. 1487 s) on oxidation with CrO_3 in strong acetic acid at 45° (1940b PARKE, DAVIS & Co.). From Δ^4-pregnene-3.20-diols (mixture of stereoisomers; p. 1920 s) on oxidation with an aluminium alcoholate and acetone or cyclohexanone or, after temporary protection of the double bond by bromination, with CrO_3 in glacial acetic acid (1941 CIBA PHARMACEUTICAL PRODUCTS). From Δ^5-pregnene-3β.20α-diol (p. 1920 s) on heating with precipitated copper at 230° under 20 mm. pressure for 30 min., followed by sublimation in a high vacuum at 125° (1938a Marker; cf. also 1936a CIBA) or on oxidation with CrO_3 in strong acetic acid at room temperature, after protection of the double bond by bromination, followed by debromination with zinc dust in acetic acid at 95° (1938a, 1940c Marker; 1945 Hirschmann; cf. 1935 I.G. FARBENIND.). From Δ^5-pregnene-3β.20-diol with unknown configuration at C-20 (see p. 1922 s) on oxidation with an aluminium or magnesium alcoholate and acetone or acetophenone (1937a N. V. ORGANON) or, after protection of the double bond by bromination or by addition of a hydrogen halide, with CrO_3 or other oxidants, followed by debromination with zinc dust or sodium iodide in acetone or with hydrogen in the presence of platinum or nickel or by dehydrohalogenation with pyridine or alkali acetate (1936 SCHERING-KAHLBAUM A.-G.; 1938g CIBA). From a Δ^5-pregnene-3β.17.20-triol (p. 2117 s) on dehydration, followed by oxidation (1938g CIBA).

Formation from pregnenone, pregnenolones, etc. From Δ^4-pregnen-20-one (p. 2216 s) on oxidation with CrO_3 in strong acetic acid at room temperature (1940a CIBA; 1941 CIBA PHARMACEUTICAL PRODUCTS; 1940a PARKE, DAVIS & Co.). From $\Delta^{3.5}$-pregnadien-3-ol-20-one acetate (p. 2229 s) on hydrolysis with alcoholic NaOH or KOH on the water-bath (1937e, 1938g CIBA).

Progesterone is obtained from Δ^5-pregnen-3β-ol-20-one ("pregnenolone", p. 2233 s) on heating with aluminium tert.-butoxide in toluene, along with a Δ^5-pregnene-3β.20-diol (p. 1922 s) and Δ^4-pregnen-20-ol-3-one (p. 2510 s) (1937b N. V. ORGANON). From the same pregnenolone on oxidation by refluxing with aluminium tert.-butoxide and acetone in benzene for 11 hrs. (yield, 76%) (1937 Oppenauer), or with aluminium isopropoxide and cyclohexanone or acetone in toluene or xylene, resp. (1937a SCHERING-KAHLBAUM A.-G.; 1939, 1940a, c SCHERING A.-G.), on heating with copper powder (or palladium, platinum, silver) in a vacuum at 200° or with a zinc-copper couple in the presence of cyclohexanone or quinoline (1936/37 CIBA), with platinum black under CO_2 at 250–300° for 1 hr. (1940e Marker). From the same pregnenolone on oxidation with CrO_3 in glacial acetic acid after protection of the double bond with HCl or HBr, followed by dehydrohalogenation with potassium acetate or pyridine (1935b SCHERING-KAHLBAUM A.-G.), or after protection with bromine, followed by debromination of the resulting 5α.6β-dibromoallopregnane-3.20-dione (p. 2814 s) with zinc powder and sodium

iodide in acetone - benzene (1941 I.G. FARBENIND.) or with zinc dust in
glacial acetic acid on the water-bath (1935a SCHERING-KAHLBAUM A.-G.;
cf. also 1937 MERCK & Co.; 1939 Yarnall; 1945 Julian); the debromination
is best carried out by treatment with chromous chloride solution (containing
$ZnCl_2$ and hydrochloric acid) in acetone under CO_2 at 26⁰ (2 hrs.; yield, 90%)
(1945 Julian; 1944a GLIDDEN COMP.). From the same pregnenolone by the
action of dehydrogenating bacteria isolated from yeast water (Corynebacterium mediolanum, Flavobacterium dehydrogenans Arn., Micrococcus dehydrogenans) on shaking for 6 days under oxygen at 32⁰; yield, 40% (1938,
1939a, b Mamoli; cf. 1940 Ercoli; 1942 Arnaudi; 1942 Molina). From the
imide of Δ^5-pregnen-3β-ol-20-one on refluxing with aluminium tert.-butoxide
and acetone, followed by shaking of the resulting 20-imide of progesterone
with $^1/_2$ N H_2SO_4 (1938a I.G. FARBENIND.); the latter imide may also be
obtained from 3-keto-Δ^4-ternorcholenylamine (p. 2508 s) on treatment with
HOCl solution in ether, followed by treatment of the resulting N-chloro-
derivative with sodium methoxide in methanol and then with aqueous alcoholic H_2SO_4 (1935 I.G. FARBENIND.; 1939a Ehrhart; 1942 Ruschig).

From 17-iso-Δ^5-pregnen-3β-ol-20-one (p. 2249 s) on oxidation with CrO_3 in
glacial acetic acid, after temporary protection of the double bond by bromination (1936 Fleischer; 1937 SCHERING A.-G.). From 21-chloro-Δ^5-pregnen-
3β-ol-20-one (p. 2277 s) on conversion into the dibromide, followed by oxidation with CrO_3 in glacial acetic acid and then by debromination with zinc
dust, sodium iodide, or potassium iodide in glacial acetic acid (1937 Steiger;
1937d, 1938g CIBA). From Δ^5-pregnen-3β-ol-20-one dibromide (p. 2289 s) on
boiling with collidine (1942 SCHERING A.-G.).

From Δ^4-pregnen-20α-ol-3-one (p. 2510 s) on oxidation with CrO_3 in glacial
acetic acid at room temperature (1941 Marker; 1935c SCHERING-KAHLBAUM
A.-G.). From Δ^4-pregnene-17α.20β-diol-3-one (p. 2721 s) on heating with
hydrochloric or sulphuric acid in propanol or dioxane (1938i CIBA). For
formation from a pregnane-5.17.20-triol-3-one, see 1938k CIBA.

Formation from pregnadiene- and pregnenediones. From $\Delta^{4.16}$-pregnadiene-
3.20-dione (16-dehydroprogesterone, p. 2780 s) on hydrogenation in the
presence of palladium - barium sulphate in ether under 5 lb. pressure at
room temperature (20 min.; yield, 7%) (1940c Marker; 1947 PARKE, DAVIS
& Co.) or in the presence of Raney nickel (1939 CIBA). From Δ^5-pregnene-
3.20-dione (p. 2793 s) on heating with dil. aqueous alcoholic H_2SO_4 on the
water-bath; yield, 90% (1936 Westphal; 1937b SCHERING-KAHLBAUM A.-G.).
From 17-iso-progesterone (p. 2800 s) on refluxing with conc. HCl in alcohol
for 15 min.; yield, 38% (1939b Butenandt; 1940b SCHERING A.-G.). From
Δ^4-pregnen-21-ol-3.20-dione (desoxycorticosterone, p. 2820 s) on treatment
with 2 mols. of p-toluenesulphonyl chloride and 3 mols. of pyridine in chloroform, followed by boiling of the resulting mixture of desoxycorticosterone
p-toluenesulphonate and 21-chloro-Δ^4-pregnene-3.20-dione (21-chloropro-
gesterone, p. 2810 s) with sodium iodide in acetone for 5 min. and reduction
of the resulting 21-iodoprogesterone with zinc dust in glacial acetic acid
(yield, 70%) (1940b Reichstein); directly from 21-chloroprogesterone by

reduction with zinc dust in strong acetic acid at 100^0 (1937e CIBA), similarly from 21-bromoprogesterone (1937e CIBA) and 21.21-dichloroprogesterone (1941 CIBA). Along with an equal amount of Δ^1-allopregnene-3.20-dione (p. 2781 s), from the pyridinium salt of 2α-bromoallopregnane-3.20-dione (page 2812 s) on distillation under 10 mm. pressure (1939b Marker; cf. also 1938b CIBA). From 4β-bromopregnane-3.20-dione (p. 2813 s) on refluxing with pyridine for 12 hrs. (1935c SCHERING-KAHLBAUM A.-G.; cf. also 1940d Marker). From 5-chloroallopregnane-3.20-dione (p. 2813 s) on boiling with K_2CO_3 in aqueous alcohol (1947 Meystre; 1946b CIBA PHARMACEUTICAL PRODUCTS). For formation from 5.6-dibromopregnane-3.20-dione, see under formation from Δ^5-pregnen-3β-ol-20-one (p. 2783 s).

Formation from pregnanoldiones. From the benzoate of pregnan-4-ol-3.20-dione (p. 2838 s) on sublimation in a high vacuum (1937b CIBA). From pregnan-5-ol-3.20-dione [obtained from pregnane-3β.5β-diol-20-one (p. 2323 s) by oxidation with CrO_3 in glacial acetic acid at room temperature] on dehydration with a 5% solution of HCl in glacial acetic acid (1937a CIBA). From the diacetate of pregnane-3α.6α-diol-20-one (p. 2324 s) on treatment with 1 mol. KOH in methanol at 20^0 (48 hrs.), neutralization with dil. H_2SO_4, oxidation with CrO_3 in strong acetic acid at room temperature (1 hr.), refluxing of the resulting pregnan-6α-ol-3.20-dione acetate with 2% methanolic KOH for 75 min., and dehydration by heating with fused $KHSO_4$ under 4 mm. pressure at 130^0 for $1^1/_2$ hrs. and then at 180^0 for 4 hrs. (1940a Marker; 1941 PARKE, DAVIS & Co.).

Formation from compounds with more C-atoms. For its formation, along with other compounds, from Δ^4-cholestene (p. 1423 s) on oxidation with CrO_3 in glacial acetic acid, see 1940b CIBA. Progesterone is obtained from cholesterol by conversion into the dibromide, followed by oxidation and debromination with zinc dust in glacial acetic acid; the oxidation may be carried out with CrO_3 in strong acetic acid (1938a CHINOIN), with $KMnO_4$ and aqueous H_2SO_4 in benzene at room temperature (yield, 0.2%) (1939 Spielman; cf. also 1941 Salvioni), or with 30% H_2O_2, dil. NaOH, and a trace of silver oxide in benzene at $60-80^0$ (1942 Serono). From Δ^4-cholesten-3β-ol ("allocholesterol", p. 1564 s) on oxidation with CrO_3 in strong acetic acid at 50^0 (1935b C. F. BOEHRINGER & SÖHNE). From Δ^4-cholesten-3-one (p. 2424 s) on oxidation with CrO_3 in strong acetic acid at $40-50^0$ (1938 Dirscherl; cf. 1935a C. F. BOEHRINGER & SÖHNE; 1937f CIBA), in water - carbon tetrachloride at 20^0 (1939 C. F. BOEHRINGER & SÖHNE), or in sulphuric acid (more than 50%)-acetic acid not above -2^0 (1938b CHINOIN); from the same ketone on oxidation with air or oxygen at $120-130^0$, preferably in the presence of $KMnO_4$ or V_2O_5 (1939 CHINOIN), with oxygen in the presence of V_2O_5 at 170^0 (bath temperature) (1941b Bretschneider). From cholestenone dibromide (p. 2500 s) on oxidation with $KMnO_4$ and aqueous H_2SO_4 in benzene (yield, 0.7-1%) (1935 Tavastsherna; but cf. 1939 Spielman), on oxidation with CrO_3 in strong acetic acid at 20^0, followed by debromination with zinc in acetic acid at 45^0 (1938, 1945 C. F. BOEHRINGER & SÖHNE), or on oxidation with ozone in the presence of $CaCO_3$ in ethyl chloride or chloroform, or in the presence of

piperidine in chloroform - glacial acetic acid, followed by debromination with
zinc dust and glacial acetic acid in boiling ether (1943 Squibb & Sons). From
6β-bromo- or 2.2.6β-tribromo-$\varDelta^4$-cholesten-3-one (see pp. 2484 s and 2503 s,
resp.) on oxidation with CrO_3 in glacial acetic acid - carbon tetrachloride,
followed by debromination, e.g., using zinc in glacial acetic acid (1938 C. F.
Boehringer & Söhne). From 24.24 - diphenyl - $\varDelta^{4.20(22).23}$ - cholatrien - 3 - one
(p. 2472 s) on oxidation with CrO_3 and 80% acetic acid in chloroform
at 0–3° (yield, 80%) (1946, 1947 Meystre; 1943, 1946a, b, 1947 Ciba Pharmaceutical Products; cf. also 1947 Miescher); similarly from 24.24-dimethyl-
$\varDelta^{4.20(22).23}$-cholatrien-3-one (p. 2413 s) (1943 Ciba Pharmaceutical Products).

For preparation from stigmasterol (cf. p. 151), see also 1935, 1940c Schering
A.-G.; 1946 I.G. Farbenind. Progesterone may be prepared in ca. 48% yield
from (3β-hydroxy-$\varDelta^5$-ternorcholenyl)-diphenyl-carbinol (p. 1947 s) by conversion, in glacial acetic acid, into its dibromide, followed by ozonization,
oxidation with CrO_3 in aqueous sulphuric - acetic acid, and treatment with
chromous chloride; in the last step vanadous or titanous chloride may also
be used (1944a, b Glidden Co.).

Isolation from reaction mixtures. May be extracted from petroleum ether,
ligroin, or hexane solutions with conc. HCl (1938 Dirscherl; 1937 C. F. Boehringer & Söhne; 1938a, b, 1939 Chinoin; 1941b Bretschneider; 1943 Squibb
& Sons) [after separation from cholestenone by extraction with 90% methanol
(1939 Chinoin)] or from benzene solutions with 50% H_2SO_4 (1946 Meystre);
from the acid extracts, after dilution with water, it is then extracted with
ether (1938 Dirscherl; 1946 Meystre; see also 1939 Chinoin). For isolation
by chromatographic adsorption, see 1937 c Ciba; 1941a Bretschneider; 1939
Chinoin; by use of Girard's reagent T, see 1938e Ciba; by use of 2.6-lutidine-
3.5-dicarboxylic acid dihydrazide (1946 Velluz; 1946 Les Laboratoires
Français de Chimiothérapie), see under "Reactions" (p. 2788 s).

REACTIONS OF PROGESTERONE

Oxidation. Gives pregnane-4.5-diol-3.20-dione (p. 2865 s) on treatment
with 30% H_2O_2 in the presence of OsO_4 in ether at room temperature (1938
Butenandt). On treatment with ozone in chloroform at 0° (partly followed
by treatment with H_2O_2 in aqueous Na_2CO_3 at room temperature) it gives
37% yield of **2.5-dimethyl-3′-acetyl-1.2-cyclopentano-6-decalone-5-propionic
acid** $C_{20}H_{30}O_4$ (I) [crystals (strong acetic acid), m. 173–5° cor.; $[\alpha]_D^{17}$ $+108°$

(acetone)] which on heating with amalgamated zinc and conc. HCl in glacial
acetic acid in an atmosphere of hydrogen chloride on the water-bath for
9 hrs. gives 18% yield of **2.5 - dimethyl - 3′ - ethyl - 1.2 - cyclopentanodecalin-**

References, pp. 2789–2793 s

5-propionic acid $C_{20}H_{34}O_2$ (II) [needles (acetone at -80^0), m. 126^0, resolidifying to spear-like crystals m. $141-5^0$ cor.; $[\alpha]_D^{15}$
$+9^0$ (chloroform)] (1940a Reichstein). On treatment with $K_2S_2O_8$ and conc. H_2SO_4 in glacial
acetic acid at room temperature for 7 days it
gives 30% yield of **5 - hydroxy - 20 - keto - 3‖5-
A-nor-pregnan-3-oic acid lactone** $C_{20}H_{30}O_3$ (III) [crystals (ether or aqueous
acetone), m. 154^0; insoluble in Na_2CO_3 solution, soluble in aqueous KOH and
reprecipitated by dil. HCl; *semicarbazone* $C_{21}H_{33}O_3N_3$, crystals, dec. $260-4^0$]
(1942 Salamon).

(III)

On heating with lead tetraacetate in glacial acetic acid at $75-85^0$ for 7 hrs.
it gives ca. 3% yield of desoxycorticosterone acetate (p. 2824 s), along with
2α-hydroxydesoxycorticosterone diacetate (p. 2852 s) and, after refluxing with
aqueous methanolic $KHCO_3$, probably 2α-hydroxyprogesterone (p. 2837 s)
and another hydroxyprogesterone $C_{21}H_{30}O_3$ [needles (ether), m. 184^0; $[\alpha]_D^{20}$
$+40^0$ (alcohol)] of unknown structure (see p. 2845 s) (1937, 1938b, c I.G.
FARBENIND.; 1939a, b Ehrhart; 1939 Reichstein; 1953 Sondheimer; cf. also
1944 v. Euw). Progesterone is scarcely attacked by N-bromoacetamide in
tert.-butanol at 20^0 (1943 Reich).

Reduction. On shaking with hydrogen and platinum oxide in alcohol for
1 hr. under 3 atm. pressure it gives pregnane-$3\alpha.20\beta$-diol (p. 1929 s) and allo-
pregnane-$3\beta.20\beta$-diol (p. 1934 s) (1939a Marker). On hydrogenation in abs.
alcohol in the presence of palladium on zirconium oxide it gives mainly allo-
pregnane-3.20-dione (p. 2797 s) and, presumably, pregnane-3.20-dione (page
2796 s) (1946 Pearlman). Progesterone is almost quantitatively reduced to
pregnane-3.20-dione by Bacillus putrificus Bienstock in yeast water or yeast
suspension under nitrogen at $36-37^0$ (1939a Mamoli).

Action of light. On irradiation of progesterone with sunlight or light of a
quartz lamp, in benzene or benzene - hexane with exclusion of air, it gives
lumi-progesterone $C_{42}H_{60}O_4$ [($(C_{21}H_{30}O_2)_2$?; crystals, slowly dec. above 340^0;
$[\alpha]_D^{23} + 107^0$ (chloroform)] which on heating with hydroxylamine acetate in
alcohol for 4 hrs. gives a *dioxime* $C_{42}H_{62}O_4N_2$ [solid (chloroform - alcohol),
m. $390-400^0$ after decomposition beginning at 280^0] (1939c Butenandt); for
structure of analogous compounds obtained by irradiation of Δ^4-cholesten-
3-one and testosterone propionate, see 1940, 1952 Butenandt.

Halogenation and sulphonation. Gives 6-bromoprogesterone (p. 2814 s) on
refluxing with 1 mol. of N-bromosuccinimide in dry carbon tetrachloride for
1 hr. (1953 Sondheimer). Treatment of progesterone with 1 mol. of conc.
H_2SO_4 in ice-cold acetic anhydride affords a **progesteronesulphonic acid**
$C_{21}H_{30}O_5S$ [colourless crystals, m. $190-2^0$ dec., very readily soluble in water
and alcohol, sparingly in ether and glacial acetic acid; *methyl ester* $C_{22}H_{32}O_5S$,
colourless needles, m. $160-1^0$] (1937 Windaus; 1936/37 I.G. FARBENIND.); for
probable position of the SO_3H-group at C-6 (cf. Δ^4-cholesten-3-one-6-sulphonic
acid), see 1939 Kuhr.

Action of organic compounds. For the action of lead tetraacetate, see under
"Oxidation". Progesterone is converted into its 3-enol esters ($\Delta^{3.5}$-pregnadien-

3-ol-20-one esters, p. 2230 s) on heating under nitrogen with acetic anhydride -
acetyl chloride (2:3) at 100–150⁰ for 5 hrs., with propionic anhydride - propionyl chloride (1:1) at 120⁰ for 1 hr., with sodium butyrate in butyric anhydride at 195⁰ for 2 hrs. (1937 Westphal), or on refluxing with benzoyl
chloride in benzine for 40 hrs. (1936b CIBA); gives the 3-enol benzyl ether
(p. 2230 s) on boiling with benzyl alcohol and some p-toluenesulphonic acid,
with simultaneous removal (by distillation) of the water formed (1940 SCHE-
RING CORP.). Progesterone does not give a precipitate with nicotinic acid
hydrazide (1945 Velluz), but on refluxing with 2.6-lutidine-3.5-dicarboxylic
acid dihydrazide in methanol and some acetic acid for 1 hr. it gives a quantitative yield of the *monohydrazone* $C_{30}H_{41}O_3N_5$, from which progesterone may
be regenerated by treatment with benzaldehyde (1946 Velluz; 1946 LES
LABORATOIRES FRANÇAIS DE CHIMIOTHÉRAPIE).

Biological reactions. For the action of Bacillus putrificus on progesterone,
see under "Reduction" (p. 2787 s). After administration of progesterone to
women, pregnane-3α.20α-diol (p. 1923 s; as sodium glucosiduronate) is found
in the urine; for administration under various conditions, see 1938, 1940
Venning; 1940 Cope; 1940 Müller; 1943 Hoffmann; 1944 Allen. Pregnane-
3α.20α-diol could also be found in the urine after administration of progesterone
to men (1939 Buxton; 1940 Hamblen) and to rabbits (1941 Heard; 1941,
1942 Westphal; 1942 Hoffman; but cf. 1940 Knoppers), but not after administration to monkeys (1939 Westphal; 1940b Marker) or to guinea-pigs
(1942 Fish; 1942 Westphal). For metabolism of progesterone, see also 1938b
Marker; 1940 Westphal.

PHYSIOLOGICAL PROPERTIES OF PROGESTERONE

For physiological properties of progesterone, see p. 151 and pp. 1367 s,
1368 s, and the literature cited there.

ANALYTICAL PROPERTIES OF PROGESTERONE

Progesterone slowly gives a precipitate with ammoniacal silver salt solution
(1939 Miescher). It gives no colour reaction with conc. H_2SO_4 in 70% alcohol
(1939 Scherrer). With $SbCl_5$ it gives a red coloration immediately becoming
green (1946 Elvidge). For the rate of reduction of phosphomolybdic acid
(Folin-Wu reagent) by progesterone in glacial acetic acid at 100⁰ to give
molybdenum blue, see 1946 Heard. In the Zimmermann test with m-dinitrobenzene and KOH in alcohol, progesterone gives a brownish violet colour passing
to red after 20 hrs. (1935 Zimmermann); for absorption spectrum of this colour
and for use of this reaction for estimation of progesterone, see 1935, 1944 Zimmermann; 1938 Callow; 1943 Hansen. For colour reactions in alcohol with
benzaldehyde + conc. H_2SO_4, see 1939 Scherrer; with fural + conc. H_2SO_4, see
1939 Woker. Progesterone gives a red coloration with salicylaldehyde in
strongly alkaline solution (1946 Elvidge); for use of this reaction for colorimetric determination of progesterone in ethyl oleate solutions, see 1946
Elvidge. Gives a weakly pink coloration with 1.4-naphthoquinone + conc.
HCl in glacial acetic acid at 60–70⁰ (1939 Miescher). Gives a green coloration

with $ZnCl_2$ + benzoyl chloride in glacial acetic acid - chloroform at $70-100^0$ (1936 Pincus; 1946 Elvidge).

For polarographic detection of progesterone, see 1939 Eisenbrand.

For gravimetric estimation of progesterone by means of Girard's reagent T and conversion of the resulting hydrazone into its sparingly soluble salt with mercuric iodide, see 1941 Hughes.

3-Enolic derivatives (p. 151). See p. 2230 s.

1935 a C. F. Boehringer & Söhne (Dirscherl, Hanusch), *Ger. Pat.* 665 549 (issued 1938); *C. A.* **1939** 1106; *Ch. Ztbl.* **1939** I 2830.

b C. F. Boehringer & Söhne (Dirscherl, Hanusch), *Ger. Pat.* 712 591 (issued 1941); *C. A.* **1943** 4534; *Ch. Ztbl.* **1942** I 1162; Rare Chemicals Inc. (Dirscherl, Hanusch), *U.S. Pat.* 2 152 626 (1936, issued 1939); *C. A.* **1939** 5132; *Ch. Ztbl.* **1940** I 759.

I. G. Farbenind., *Austrian Pat.* 160 651 (issued 1941); *Ch. Ztbl.* **1942** I 1025; *Austrian Pat.* 160 752 (1936, issued 1942); *Ch. Ztbl.* **1942** II 1037; *Swiss Pat.* 234 347 (issued 1945); *Ch. Ztbl.* **1946** I 2107; *Fr. Pat.* 797 960 (issued 1936); *C. A.* **1936** 7123; *Ch. Ztbl.* **1937** I 663; *Brit. Pat.* 465 960 (issued 1937); *C. A.* **1937** 7444.

Schering A.-G., *Austrian Pat.* 160 824 (issued 1942); *Ch. Ztbl.* **1942** II 2505.

a Schering-Kahlbaum A.-G., *Swiss Pat.* 188 987 (issued 1937); *Ch. Ztbl.* **1937** II 3041, 3042.

b Schering-Kahlbaum A.-G., *Australian Pat.* 3962/1936 (issued 1937); *Ch. Ztbl.* **1937** I 4991, 4992; *Brit. Pat.* 464 397 (issued 1937); *C. A.* **1937** 6254; *Ch. Ztbl.* **1938** I 375.

c Schering-Kahlbaum A.-G., *Fr. Pat.* 806 467 (issued 1936); *C. A.* **1937** 4682; *Australian Pat.* 3961/1936 (issued 1936); *Ch. Ztbl.* **1937** I 4264, 4265; cf. also *Australian Pat.* 24 429/1935 (issued 1936); *Ch. Ztbl.* **1937** I 3023, 3024.

Tavastsherna, *Arch. sci. biol. U.S.S.R.* **40** 141, 143, 145; *C. A.* **1937** 6670; *Ch. Ztbl.* **1937** II 3323.

Zimmermann, *Z. physiol. Ch.* **233** 257, 263.

1936 Callow, Parkes, *J. Physiol.* **87** 28 P.

a Ciba*, *Swiss Pat.* 201 870 (issued 1939); *C. A.* **1939** 8925; *Ch. Ztbl.* **1939** II 1126.

b Ciba*, *Swiss Pat.* 214 331 (issued 1941); *C. A.* **1942** 5960; *Brit. Pat.* 477 400 (issued 1938); *C. A.* **1938** 3914; *Ch. Ztbl.* **1938** II 119, 120.

Fleischer in Butenandt, Schmidt-Thomé, Paul, *Ber.* **72** 1112 (1939).

Pincus, Wheeler, Young, Zahl, *J. Biol. Ch.* **116** 253.

Schering-Kahlbaum A.-G., *Brit. Pat.* 488 103 (issued 1938); *C. A.* **1939** 324; *Ch. Ztbl.* **1939** I 1603; *Swiss Pat.* 198 836 (issued 1938); *Ch. Ztbl.* **1939** I 2642.

Westphal, Schmidt-Thomé, *Ber.* **69** 889, 892.

1936/37 Ciba*, *Brit. Pat.* 476 749 (issued 1938); *Fr. Pat.* 823 139 (1937, issued 1938); *C. A.* **1938** 3770, 5585; *Ch. Ztbl.* **1938** II 120, 121.

I. G. Farbenind., *Brit. Pat.* 473 629 (issued 1937); *C. A.* **1938** 3096; *Ch. Ztbl.* **1938** I 2402.

1937 C. F. Boehringer & Söhne, *Fr. Pat.* 835 527 (issued 1938); *Ch. Ztbl.* **1939** I 2827.

a Ciba*, *Brit. Pat.* 482 321 (issued 1938); *C. A.* **1938** 6808; *Ch. Ztbl.* **1938** II 3272.

b Ciba*, *Brit. Pat.* 497 394 (issued 1939); *C. A.* **1939** 3812; *Ch. Ztbl.* **1939** I 2828; Ciba Pharmaceutical Products (Miescher, Wettstein), *U.S. Pat.* 2 229 813 (1938, issued 1941); *C. A.* **1941** 3040; *Ch. Ztbl.* **1941** II 2467, 2468; cf. also Ciba*, *Dutch Pat.* 51 420 (1938, issued 1941); *Ch. Ztbl.* **1942** I 2561.

c Ciba*, *Swiss Pat.* 199 449 (issued 1938); *C. A.* **1939** 3393; *Ch. Ztbl.* **1939** I 3930.

d Ciba*, *Swiss Pat.* 215 138 (issued 1941); *C. A.* **1948** 3145.

e Ciba*, *Swiss Pats.* 222 122–123 (issued 1942); *C. A.* **1949** 366; *Ch. Ztbl.* **1943** I 2113.

f Ciba*, *Fr. Pat.* 830 043 (issued 1938); *C. A.* **1939** 179; *Ch. Ztbl.* **1939** I 2035.

Ehrhardt, Hardt, *Ch. Ztbl.* **1937** II 1023.

I. G. Farbenind., *Brit. Pat.* 502 474 (issued 1939); *C. A.* **1939** 7316; *Ch. Ztbl.* **1939** II 1337.

Marker, Kamm, McGrew, *J. Am. Ch. Soc.* **59** 616.

* Société pour l'industrie chimique à Bâle; Gesellschaft für chemische Industrie in Basel.

MERCK & Co. (Wallis, Yarnall), *U.S. Pat.* 2123217 (issued 1938); *C. A.* **1938** 6808; *Ch. Ztbl.* **1939** I 2036.

a N. V. ORGANON, *Swiss Pat.* 223017 (issued 1942); *Ch. Ztbl.* **1943** I 1798.

b N. V. ORGANON, *Fr. Pat.* 823618 (issued 1938); *C. A.* **1938** 5854; *Ch. Ztbl.* **1938** II 355; *Brit. Pat.* 488987 (issued 1938); *C. A.* **1939** 324; *Ch. Ztbl.* **1939** II 474; *Swiss Pat.* 223018 (issued 1942); *Ch. Ztbl.* **1943** I 1798; *Dutch Pat.* 50942 (1938, issued 1941); *Ch. Ztbl.* **1942** I 513.

Oppenauer, *Rec. trav. chim.* **56** 137, 144.

SCHERING A.-G., *Brit. Pat.* 508305 (issued 1939); *C. A.* **1940** 59; *Ch. Ztbl.* **1940** I 1232.

a SCHERING-KAHLBAUM A.-G., *Fr. Pat.* 822551 (issued 1938); *C. A.* **1938** 4174; *Ch. Ztbl.* **1938** II 120; SCHERING CORP. (Serini, Köster, Strassberger), *U.S. Pat.* 2379832 (issued 1945); *C. A.* **1945** 5053.

b SCHERING-KAHLBAUM A.-G., *Brit. Pat.* 492725 (issued 1938); *C. A.* **1939** 1884; *Ch. Ztbl.* **1939** I 2642.

Steiger, Reichstein, *Helv.* **20** 1164, 1168.

Westphal, *Ber.* **70** 2128, 2134, 2135.

Windaus, Kuhr, *Ann.* **532** 52, 65.

1938 a Beall, Reichstein, *Nature* **142** 479.

b Beall, *Biochem. J.* **32** 1957, 1959.

C. F. BOEHRINGER & SÖHNE, *Fr. Pat.* 844850 (issued 1939); *C. A.* **1940** 7932; *Ch. Ztbl.* **1940** I 1391; cf. also *Fr. Pat.* 860492 (1939, issued 1941); *Ch. Ztbl.* **1941** I 3408.

Bugyi, *Acta Lit. Sci. Univ. Hung. Sect. Chem. Mineral. Phys.* **6** 27; *C. A.* **1938** 2191; *Ch. Ztbl.* **1938** I 1144.

Butenandt, Wolz, *Ber.* **71** 1483, 1486.

Callow, Callow, Emmens, *Biochem. J.* **32** 1312, 1323.

a CHINOIN*, *Fr. Pat.* 840964 (issued 1939); *C. A.* **1940** 1821, 5601; *Ch. Ztbl.* **1940** I 427.

b CHINOIN* (Bretschneider), *Brit. Pat.* 518266 (issued 1940); *C. A.* **1941** 7662; *Ch. Ztbl.* **1941** I 928; *Fr. Pat.* 866306 (1939, issued 1941); *Dutch Pat.* 51984 (1939, issued 1942); *Ch. Ztbl.* **1942** II 317; (Bretschneider, Salamon), *U.S. Pat.* 2246341 (issued 1941); *C. A.* **1941** 6068.

a CIBA** (Oppenauer, Miescher, Kägi), *Swed. Pat.* 96375 (issued 1939); *Ch. Ztbl.* **1940** I 1231; *Ital. Pat.* 362454; *Ch. Ztbl.* **1939** II 3151.

b CIBA** (Ruzicka); *Swed. Pat.* 98368 (issued 1940); *Ch. Ztbl.* **1940** II 2185.

c CIBA**, *Fr. Pat.* 840515 (issued 1939); *C. A.* **1939** 8627; *Ch. Ztbl.* **1939** II 1721, 1722.

d CIBA**, *Danish Pat.* 61590 (issued 1943); *Ch. Ztbl.* **1944** II 49.

e CIBA** (Miescher, Fischer), *U.S. Pat.* 2172590 (issued 1939); *C. A.* **1940** 450; *Ch. Ztbl.* **1940** I 428.

f CIBA** (Miescher, Wettstein), *Ger. Pat.* 739083 (issued 1943); *C. A.* **1945** 392; *Ch. Ztbl.* **1944** I 110, 111; *Fr. Pat.* 840513 (issued 1939); *C. A.* **1939** 8627; *Ch. Ztbl.* **1939** II 1125.

g CIBA**, *Fr. Pat.* 835814 (issued 1939); *C. A.* **1939** 5008; *Ch. Ztbl.* **1939** I 2827, 2828; *Brit. Pat.* 512954 (issued 1939); *C. A.* **1941** 1800.

h CIBA**, *Brit. Pat.* 516542 (issued 1940); *C. A.* **1941** 6395; *Ch. Ztbl.* **1940** II 1327, 1328.

i CIBA**, *Fr. Pat.* 840783 (issued 1939); *C. A.* **1939** 8627; *Ch. Ztbl.* **1940** I 915, 916.

k CIBA**, *Fr. Pat.* 838916 (issued 1939); *C. A.* **1939** 6875; *Ch. Ztbl.* **1939** II 169, 170.

Dirscherl, Hanusch, *Z. physiol. Ch.* **252** 49.

a I. G. FARBENIND., *Swiss Pat.* 225070 (issued 1943); *Ch. Ztbl.* **1944** II 675; *Brit. Pat.* 508804 (issued 1939); *C. A.* **1940** 2864; *Ch. Ztbl.* **1940** I 1391, 1392.

b I. G. FARBENIND., *Swiss Pat.* 225258 (issued 1943); *Ch. Ztbl.* **1944** II 564.

c I. G. FARBENIND., *Swiss Pats.* 234536–540 (issued 1945); *Ch. Ztbl.* **1946** I 2280.

Mamoli, *Ber.* **71** 2701.

a Marker, Rohrmann, *J. Am. Ch. Soc.* **60** 1565.

b Marker, Rohrmann, *J. Am. Ch. Soc.* **60** 1725.

Venning, Browne, *Am. J. Physiol.* **123** 209.

1939 Bandow, *Biochem. Z.* **301** 37, 44, 54.

C. F. BOEHRINGER & SÖHNE, *Fr. Pat.* 846099 (issued 1939); *C. A.* **1941** 1185; *Ch. Ztbl.* **1940** I 2830.

* CHINOIN Gyógyszer és Vegyészeti Termékek Gyára R. T. (Dr. Kereszty & Dr. Wolf).
** Société pour l'industrie chimique à Bâle; Gesellschaft für chemische Industrie in Basel.

a Butenandt, Schmidt-Thomé, *Ber.* **72** 182, 183.
b Butenandt, Schmidt-Thomé, Paul, *Ber.* **72** 1112, 1118.
c Butenandt, Wolff, *Ber.* **72** 1121.
Buxton, Westphal, *Proc. Soc. Exptl. Biol. Med.* **41** 284.
CHINOIN* (Bretschneider, Fári), *U.S. Pat.* 2283411 (issued 1942); *C. A.* **1942** 6170; *Brit. Pat.* 530559 (issued 1941); *C. A.* **1942** 98; *Ch. Ztbl.* **1942** I 1025.
CIBA**, *Fr. Pat.* 860278 (issued 1941); *Ch. Ztbl.* **1941** I 3258.
Dannenberg, *Abhandl. Preuss. Akad. Wiss. Math.-naturwiss. Kl.* **1939** No. 21, pp. 10, 53.
a Ehrhart, Ruschig, Aumüller, *Angew. Ch.* **52** 363, 366.
b Ehrhart, Ruschig, Aumüller, *Ber.* **72** 2035.
Eisenbrand, Picher, *Z. physiol. Ch.* **260** 83, 89 et seq.
Goldberg, Aeschbacher, *Helv.* **22** 1185, 1188.
Kuhr, *Ber.* **72** 929.
a Mamoli, Koch, Teschen, *Z. physiol. Ch.* **261** 287, 295, 296.
b Mamoli, *Ber.* **72** 1863.
a Marker, Lawson, *J. Am. Ch. Soc.* **61** 588.
b Marker, Wittle, Plambeck, *J. Am. Ch. Soc.* **61** 1333.
Miescher, Wettstein, Scholz, *Helv.* **22** 894, 903.
Reichstein, Montigel, *Helv.* **22** 1212, 1215, 1219.
SCHERING A.-G. (Logemann, Hildebrand), *Ger. Pat.* 737420 (issued 1943); *C. A.* **1945** 5051; *Ch. Ztbl.* **1944** I 366, 367.
Scherrer, *Helv.* **22** 1329, 1338.
Spielman, Meyer, *J. Am. Ch. Soc.* **61** 893.
Westphal, Buxton, *Proc. Soc. Exptl. Biol. Med.* **42** 749.
Woker, Antener, *Helv.* **22** 511, 518.
Yarnall, Wallis, *J. Org. Ch.* **4** 270, 282.

1940 Butenandt, Poschmann, *Ber.* **73** 893.
a CIBA**, *Swiss Pat.* 235279 (issued 1945); *C. A.* **1949** 7649; *Fr. Pat.* 886415 (issued 1943); *Ch. Ztbl.* **1944** I 450.
b CIBA** (Miescher, Wettstein), *Swed. Pat.* 103744 (issued 1942); *Ch. Ztbl.* **1942** II 2294; *Fr. Pat.* 886410 (issued 1943); *Ch. Ztbl.* **1944** I 448; *Brit. Pat.* 550478 (issued 1943); *C. A.* **1944** 1612; CIBA PHARMACEUTICAL PRODUCTS (Miescher, Wettstein), *U.S. Pat.* 2319012 (issued 1943); *C. A.* **1943** 6096.
Cope, *Clinical Science* **4** 217, 228.
Ercoli, *C. A.* **1944** 1540; *Ch. Ztbl.* **1941** II 1028.
Hamblen, Cuyler, Hirst, *Endocrinology* **27** 172.
Knoppers, *Proc. Kon. Akad. Wet. Amsterdam* **43** 1127, 1130.
a Marker, Krueger, *J. Am. Ch. Soc.* **62** 79, 81.
b Marker, Hartman, *J. Biol. Ch.* **133** 529, 535.
c Marker, Tsukamoto, Turner, *J. Am. Ch. Soc.* **62** 2525, 2530, 2531.
d Marker, *J. Am. Ch. Soc.* **62** 2543, 2547.
e Marker, Krueger, *J. Am. Ch. Soc.* **62** 3349.
Müller, *Klin. Wochschr.* **19** 318, 321.
a PARKE, DAVIS & Co. (Marker, Wittle), *U.S. Pats.* 2397424–425 (issued 1946); *C. A.* **1946** 3571.
b PARKE, DAVIS & Co. (Marker, Wittle), *U.S. Pat.* 2397426 (issued 1946); *C. A.* **1946** 3571.
a Reichstein, Fuchs, *Helv.* **23** 676, 682, 683.
b Reichstein, Fuchs, *Helv.* **23** 684, 687.
a SCHERING A.-G., *Fr. Pat.* 856641 (issued 1940); *Ch. Ztbl.* **1941** I 407.
b SCHERING A.-G., *Fr. Pat.* 868904 (issued 1942); *Ch. Ztbl.* **1942** II 317.
c SCHERING A.-G., *Bios Final Report* No. 449, pp. 294–303, 316, 317; *Fiat Final Report* No. 996, pp. 70, 71, 106–115, 125.
SCHERING CORP. (Köster), *U.S. Pat.* 2363338 (issued 1944); *C. A.* **1945** 3128; *Ch. Ztbl.* **1945** II 1638; SCHERING A.-G., *Swiss Pat.* 229776 (issued 1944); *Ch. Ztbl.* **1944** II 1301.

---

* CHINOIN Gyógyszer és Vegyészeti Termékek Gyára R. T. (Dr. Kereszty & Dr. Wolf).
** Société pour l'industrie chimique à Bâle; Gesellschaft für chemische Industrie in Basel.

Venning, Browne, *Endocrinology* **27** 707.

Westphal, *Naturwiss.* **28** 465.

1941 a Bretschneider, *Sitzber. Akad. Wiss. Wien* IIb **150** 127; *Monatsh.* **74** 53.

b Bretschneider, *Ber.* **74** 1360.

CIBA*, *Swiss Pat.* 224833 (issued 1943); *C. A.* **1949** 1916; *Ch. Ztbl.* **1944** II 563.

CIBA PHARMACEUTICAL PRODUCTS (Miescher, Wettstein), *U.S. Pats.* 2323276–277 (issued 1943); *C. A.* **1944** 222, 223; CIBA* (Miescher, Wettstein), *Swed. Pats.* 105136–137 (issued 1942); *Ch. Ztbl.* **1944** II 143.

v. Euw, Reichstein, *Helv.* **24** 879, 880 (footnote 3), 886.

Heard, Bauld, Hoffman, *J. Biol. Ch.* **141** 709.

Hughes, *J. Biol. Ch.* **140** 21.

I. G. FARBENIND., *Swiss Pat.* 227974 (issued 1943); *Ch. Ztbl.* **1945** I 196; *Fr. Pat.* 886984 (1942, issued 1943); *Ch. Ztbl.* **1944** II 1091.

Marker, Crooks, Wittbecker, *J. Am. Ch. Soc.* **63** 777, 778.

PARKE, DAVIS & Co. (Marker, Lawson), *U.S. Pat.* 2366204 (issued 1945); *C. A.* **1945** 1649; *Ch. Ztbl.* **1945** II 1385, 1386; *U.S. Pat.* 2420491 (issued 1947); *C. A.* **1947** 5659; *Ch. Ztbl.* **1948** I 379.

Salvioni, *C. A.* **1944** 1244; *Ch. Ztbl.* **1941** II 48.

Westphal, *Naturwiss.* **29** 782.

1942 Arnaudi, *C. A.* **1946** 3152; *Ztbl. Bakt. Parasitenk.* II. Abt. **105** 352.

Fish, Dorfman, Young, *J. Biol. Ch.* **143** 715.

Hoffman, *Can. Med. Assoc. J.* **47** 424, 427; *C. A.* **1944** 4295.

Molina, *C. A.* **1946** 3152.

Ruschig, *Medizin und Chemie* **4** 327, 340; *C. A.* **1944** 4954; *Ch. Ztbl.* **1943** I 2689.

Salamon, *Z. physiol. Ch.* **272** 61, 64.

SCHERING A.-G., *Fr. Pat.* 884085 (issued 1943); *Ch. Ztbl.* **1944** I 450.

Serono, Marchetti, *Gazz.* **72** 151.

Westphal, *Z. physiol. Ch.* **273** 1.

1943 CIBA PHARMACEUTICAL PRODUCTS (Miescher, Frey, Meystre, Wettstein), *U.S. Pat.* 2461563 (issued 1949); *C. A.* **1949** 3474; *Ch. Ztbl.* **1949** E 1870.

Hansen, Cantarow, Rakoff, Paschkis, *Endocrinology* **33** 282, 285, 286.

Hoffmann, v. Lám, *Zentr. Gynäkol.* **67** 1082; *C. A.* **1944** 6357; *Ch. Ztbl.* **1943** II 1021.

Reich, Reichstein, *Helv.* **26** 562, 585.

SQUIBB & SONS (Ruigh), *U.S. Pat.* 2413000 (issued 1946); *C. A.* **1947** 1396; *Ch. Ztbl.* **1947** 1102.

1944 Allen, Viergiver, Soule, *C. A.* **1944** 5562.

Bersin, Meyer, *Angew. Ch.* **57** 117, 118.

Cantarow, Paschkis, Rakoff, Hansen, *Endocrinology* **35** 129.

v. Euw, Lardon, Reichstein, *Helv.* **27** 1287, 1288.

a GLIDDEN Co. (Julian, Cole, Magnani, Conde), *U.S. Pat.* 2374683 (issued 1945); *C. A.* **1946** 1636; *Ch. Ztbl.* **1946** I 372.

b GLIDDEN Co. (Julian, Cole, Magnani, Conde), *U.S. Pat.* 2433848 (issued 1948); *C. A.* **1948** 1710; *Ch. Ztbl.* **1948** E 1126.

Zimmermann, *Vitamine und Hormone* **5** [1952] 1, 12, 13, 152, 162, 163.

1945 C. F. BOEHRINGER & SÖHNE, *Bios Final Report* No. 766, p. 140; *Fiat Final Report* No. 71, p. 66.

Hirschmann, Hirschmann, *J. Biol. Ch.* **157** 601, 608.

Julian, Cole, Magnani, Meyer, *J. Am. Ch. Soc.* **67** 1728, 1729.

Velluz, Petit, *Bull. soc. chim.* [5] **12** 951.

1946 Bersin, *Naturwiss.* **33** 108, 110.

a CIBA PHARMACEUTICAL PRODUCTS (Miescher, Frey, Meystre, Wettstein), *U.S. Pat.* 2461910 (issued 1949); *C. A.* **1949** 3474, 3475; *Ch. Ztbl.* **1949** E 1870, 1871.

b CIBA PHARMACEUTICAL PRODUCTS (Miescher, Frey, Meystre, Wettstein), *U.S. Pat.* 2461911 (issued 1949); *C. A.* **1949** 3475; *Ch. Ztbl.* **1949** E 1870, 1871.

Elvidge, *Quart. J. Pharm. Pharmacol.* **19** 260.

* Société pour l'industrie chimique à Bâle; Gesellschaft für chemische Industrie in Basel.

Heard, Sobel, *J. Biol. Ch.* **165** 687.
I. G. FARBENIND., *Fiat Final Report* No. 996, pp. 121–125.
LES LABORATOIRES FRANÇAIS DE CHIMIOTHÉRAPIE, *Swiss Pat.* 259252 (issued 1949); *Ch. Ztbl.*
    1949 E 1530.
Meystre, Frey, Neher, Wettstein, Miescher, *Helv.* **29** 627, 633.
Pearlman, *J. Biol. Ch.* **166** 473, 475.
Velluz, Rousseau, *Bull. soc. chim.* [5] **13** 288.
1947 CIBA PHARMACEUTICAL PRODUCTS (Miescher, Frey, Meystre, Wettstein), *U.S. Pat.* 2461912
    (issued 1949); *C. A.* **1949** 3475; *Ch. Ztbl.* **1949** E 1870, 1871.
Meystre, Wettstein, Miescher, *Helv.* **30** 1022.
Miescher, Schmidlin, *Helv.* **30** 1405, 1408.
PARKE, DAVIS & Co. (Marker), *U.S. Pat.* 2420489 (issued 1947); *C. A.* **1947** 5689; *Ch. Ztbl.*
    1948 I 380.
1952 Butenandt, Karlson-Poschmann, Failer, Schiedt, Biekert, *Ann.* **575** 123.
1953 Sondheimer, Kaufmann, Romo, Martinez, Rosenkranz, *J. Am. Ch. Soc.* **75** 4712.

Δ5-Pregnene-3.20-dione $C_{21}H_{30}O_2$ (p. 152). — Causes ovipositor growth in the female bitterling (1941, 1941/42 Duyvené de Wit).

— **Fmn.** For its formation from Δ5-pregnen-3β-ol-20-one (cf. p. 152), see also 1937a SCHERING-KAHLBAUM A.-G. — **Rns.** On hydrogenation in alcohol in the presence of Raney nickel, until 1 mol. hydrogen has been absorbed, it gives Δ5-pregnen-3α-ol-20-one (p. 2232 s; crude yield, 28%) and Δ5-pregnen-3β-ol-20-one (page 2233 s; yield, 53%) (1939 Butenandt). Is isomerized to progesterone (page 2782 s) by treating with aqueous alcoholic H_2SO_4 for 10 min. and probably also partially by heating at the m. p. (1936 Westphal; 1937b SCHERING-KAHLBAUM A.-G.).

1936 Westphal, Schmidt-Thomé, *Ber.* **69** 889, 892.
1937 a SCHERING-KAHLBAUM A.-G., *Swiss Pat.* 199448 (issued 1938); *Ch. Ztbl.* **1939** I 3931; *Brit.
    Pat.* 486992 (issued 1938); *Ch. Ztbl.* **1939** I 1603.
    b SCHERING-KAHLBAUM A.-G., *Brit. Pat.* 492725 (issued 1938); *Ch. Ztbl.* **1939** I 2642.
1939 Butenandt, Heusner, *Ber.* **72** 1119.
1941 Duyvené de Wit, *Biochem. Z.* **309** 297.
1941/42 Duyvené de Wit, *Biochem. Z.* **310** 83, 87.

Δ11-Pregnene-3.20-dione $C_{21}H_{30}O_2$. Crystals (acetone - ether), leaflets (ether - petroleum ether), m. 134–5° cor. (Kofler block) (1943 Hegner; 1946 v. Euw); $[\alpha]_D^{18}$ +85°, $[\alpha]_{5461}^{18}$ +104° (both in acetone) (1943 Hegner). Gives a yellow coloration with tetranitromethane in chloroform (1943 Hegner). — **Fmn.** From the benzoate of pregnan-12α-ol-3.20-dione (p. 2841 s) on heating at 310° (bath temperature) under 12 mm. pressure in a current of CO_2 (1943 Hegner); in better yield (40%) from the anthraquinone-β-carboxylate of the same hydroxydiketone on heating at 295–300° (bath temperature) under 0.05 mm. pressure (1944 v. Euw) or from the corresponding p-toluenesulphonate on refluxing with collidine for 3 hrs. (yield, 32.5%) (1946 v. Euw).

Rns. On treatment with N-bromoacetamide, sodium acetate, and acetic acid in aqueous acetone at 16° it gives mainly 12α-bromopregnan-11β-ol-3.20-dione (p. 2850 s) [which on CrO_3-oxidation gives 12α-bromopregnane-3.11.20-trione (p. 2971 s) and on subsequent debromination pregnane-3.11.20-trione (p. 2967 s)]; on oxidation of the mother liquor, followed by debromination, $\Delta^{9(11)}$-pregnene-3.12.20-trione (p. 2969 s) is obtained, along with pregnane-3.11.20-trione and regenerated Δ^{11}-pregnene-3.20-dione (1944 v. Euw; see also 1943 Hegner; 1942 Reichstein).

1942 Reichstein, *Fr. Pat.* 887641 (issued 1943); *Ch. Ztbl.* **1944** II 878; *U.S. Pat.* 2403683 (1943, issued 1946); *C. A.* **1946** 6216, 6221; *Ch. Ztbl.* **1947** 232.
1943 Hegner, Reichstein, *Helv.* **26** 721, 724, 725.
1944 v. Euw, Lardon, Reichstein, *Helv.* **27** 821, 829–831.
1946 v. Euw, Reichstein, Shoppee, *Helv.* **29** 654, 669.

Δ^{16}-**Pregnene-3.20-dione** $C_{21}H_{30}O_2$. Plates (acetone), m. 200–202° (1940a Marker); scales (dil. alcohol and acetone), m. 196°, $[\alpha]_D^{23} +84°$ (chloroform) (1938 Masch; 1939 Butenandt). Absorption max. in alcohol at 239 mμ (ε 9800) (1939 Dannenberg; cf. 1938 Masch). — Has no androgenic (1938 Masch) and no progestational activity (1943 Selye).

Fmn. From Δ^{16}-pregnen-3α-ol-20-one (p. 2239 s) on oxidation with CrO_3 in strong acetic acid at room temperature (1940j Marker); similarly from Δ^{16}-pregnen-3β-ol-20-one (p. 2240 s) (1940b, f Marker). From 17-bromopregnan-3β-ol-20-one (p. 2285 s) on oxidation with CrO_3 in glacial acetic acid at room temperature, followed by refluxing of the resulting 17-bromopregnane-3.20-dione (p. 2814 s) with dry pyridine for 6 hrs. (over-all yield, ca. 30%) (1938 Masch; cf. 1942a Marker); for formation from 17-bromopregnan-3α-ol-20-one, see 1939 Butenandt.

Δ^{16}-Pregnene-3.20-dione is obtained from the following compounds on oxidation with CrO_3 in strong acetic acid at room temperature, followed by acid or alkaline hydrolysis (see 1941, 1947 Marker): pseudosarsasapogenin (yield, 50–70%) (1939, 1940a, 1941 Marker) [also on oxidation with $KMnO_4$ in 70% acetic acid at 15–20° (1940f Marker)], epi-pseudosarsasapogenin (1940c Marker), pseudosarsasapogenone (1940c Marker), dihydropseudosarsasapogenin, along with a keto-acid which on further oxidation with CrO_3 also gives the dione (1940b Marker; cf. 1942b Marker).

Rns. Unattacked by perbenzoic acid in boiling chloroform. On oxidation with OsO_4 and a little H_2O_2 in ether, followed by refluxing with Na_2SO_3 in aqueous alcohol, it yields pregnane-16α.17α-diol-3.20-dione (p. 2866 s) (1938 Masch). Gives pregnane-3α.20α-diol (p. 1923 s) on reduction with sodium in boiling abs. alcohol (1939, 1940a Marker). On hydrogenation in abs. alcohol in the presence of Adams's catalyst under 3 atm. pressure it gives pregnane-3α.20β-diol (p. 1929 s), pregnane-3β.20α-diol (p. 1930 s), and pregnane-3β.20β-diol (p. 1931 s) (1940a Marker). Gives pregnane-3.20-dione

(p. 2796 s) on hydrogenation in abs. alcohol in the presence of palladium-barium sulphate under 1 atm. pressure (1940a Marker) or on heating with zinc dust and glacial acetic acid on the steam-bath (1942a Marker). Is unattacked by fermenting yeast (1938 Masch).

Disemicarbazone $C_{23}H_{36}O_2N_6$. Crystals (alcohol), m. 310^0 dec. (1940a Marker).

Δ^{16}-**Allopregnene-3.20-dione** $C_{21}H_{30}O_2$. Crystals (ether or acetone), m. 211–3^0 (1940d, 1941, 1942b Marker); leaflets (dil. alcohol), m. 205–8^0; $[\alpha]_D^{24}$ +72^0 (chloroform); slowly decomposes on exposure to the air (1939 Butenandt). Has no progestational activity (1939 Butenandt).

Fmn. From allopregnane-3β.16α.20β-triol (page 2112 s) on refluxing with aluminium isopropoxide and cyclohexanone in toluene for 18 hrs., followed by sublimation in a high vacuum at 130–140^0; yield, 50% (1940g Marker). From Δ^{16}-allopregnen-3α-ol-20-one (p. 2241 s) on oxidation with CrO_3 in glacial acetic acid in the cold (yield, 65%) (1939 Butenandt); similarly from Δ^{16}-allopregnen-3β-ol-20-one (p. 2242 s) (1940i Marker). From 17-bromoallopregnan-3β-ol-20-one (p. 2286 s) on oxidation with CrO_3 in strong acetic acid at room temperature, followed by refluxing of a solution of the resulting crude bromo-dione with pyridine for 6 hrs. or with anhydrous potassium acetate in glacial acetic acid for 2 hrs.; yields (from the bromo-dione), 75% (1942c Marker). From $\Delta^{4.16}$-pregnadiene-3.20-dione (p. 2780 s) on partial hydrogenation in the presence of palladium - barium sulphate (1940e Marker).

Δ^{16}-Allopregnene-3.20-dione is obtained from the following compounds on oxidation with CrO_3 in strong acetic acid at room temperature, followed by acid or alkaline hydrolysis (see 1941, 1947 Marker): pseudotigogenin (yield, 38%) (1940d, 1941 Marker), pseudo - epitigogenin (1940h Marker), pseudo-tigogenone (1940i Marker), dihydropseudotigogenin (along with a keto-acid which on further oxidation with CrO_3 also gives the dione) (1940e, 1942b Marker), and dihydropseudo-epitigogenin (1940h Marker).

Rns. Gives allopregnane-3β.20α-diol (p. 1933 s) on reduction with sodium in boiling abs. alcohol, and allopregnane-3β.20β-diol (p. 1934 s) on hydrogenation in glacial acetic acid in the presence of Adams's catalyst under 3 atm. pressure (1940d Marker). On heating with zinc dust and glacial acetic acid on the steam-bath it yields allopregnane-3.20-dione (p. 2797 s) (1942c Marker).

Dioxime $C_{21}H_{32}O_2N_2$, dec. 198–202^0 (from dil. alcohol) (1939 Butenandt).

1938 Masch, *Über Bromierungen in der Pregnanreihe*, Thesis, Danzig, pp. 14, 15, 23, 27–29.
1939 Butenandt, Mamoli, Heusner, *Ber.* **72** 1614.
   Dannenberg, *Abhandl. Preuss. Akad. Wiss. Math.-naturwiss. Kl.* **1939** No. 21, pp. 16, 56.
   Marker, Rohrmann, *J. Am. Ch. Soc.* **61** 3592.
1940 a Marker, Rohrmann, *J. Am. Ch. Soc.* **62** 518.
   b Marker, Rohrmann, *J. Am. Ch. Soc.* **62** 521, 524, 525.
   c Marker, Rohrmann, Jones, *J. Am. Ch. Soc.* **62** 648.

d Marker, Rohrmann, *J. Am. Ch. Soc.* **62** 898.
e Marker, Tsukamoto, Turner, *J. Am. Ch. Soc.* **62** 2525, 2528, 2532.
f Marker, Jones, Krueger, *J. Am. Ch. Soc.* **62** 2532, 2534, 2536.
g Marker, Turner, *J. Am. Ch. Soc.* **62** 2540.
h Marker, *J. Am. Ch. Soc.* **62** 2621.
i Marker, Turner, *J. Am. Ch. Soc.* **62** 3003.
j Marker, *J. Am. Ch. Soc.* **62** 3350, 3352.
1941 Marker, Turner, Wagner, Ulshafer, Crooks, Wittle, *J. Am. Ch. Soc.* **63** 779.
1942 a Marker, Crooks, Wagner, *J. Am. Ch. Soc.* **64** 210.
b Marker, Turner, Ulshafer, *J. Am. Ch. Soc.* **64** 1655, 1657.
c Marker, Crooks, Wagner, Wittbecker, *J. Am. Ch. Soc.* **64** 2089, 2091.
1943 Selye, Masson, *J. Pharmacol.* **77** 301.
1947 Marker, Wagner, Ulshafer, Wittbecker Goldsmith, Ruof, *J. Am. Ch. Soc.* **69** 2167, 2172.

Pregnane - 3.20 - dione $C_{21}H_{32}O_2$ (p. 152). $[\alpha]_D$ $+112^0$ (chloroform) (1955

Slomp). In the Zimmermann test with m-dinitro-benzene and KOH in alcohol it gives a brownish violet coloration changing to red after 16 hrs. (1935 Zimmermann). — Causes ovipositor growth in the female bitterling (1941/42 Duyvené de Wit). — **Occ.** In mares' pregnancy urine; small amount (1938c Marker).

Fmn. Along with allopregnane-3.20-dione (p. 2797 s), from a mixture of pregnane-3α.20α-diol (p. 1923 s) and allopregnane-3α.20α- and -3β.20α-diols (p. 1933 s), present in mares' pregnancy urine, on oxidation with CrO_3 in strong acetic acid (1937c, 1939d Marker); separation via the disemicarbazones, that of the allo-dione being the less soluble in alcohol (1937c Marker). From a pregnan-3-ol-20-one (see p. 2246 s) on heating with copper powder at 200°; similarly from pregnan-20α-ol-3-one (p. 2511 s) (1936/37 Ciba). From 17-bromopregnan-3β-ol-20-one (p. 2285 s) on oxidation with CrO_3 in strong acetic acid, followed by hydrogenation in methanol in the presence of palladium - barium sulphate and pyridine at room temperature under 3 atm. pressure; yield, 63% (1942a Marker).

From $\Delta^{4.9(11)}$-pregnadiene-3.20-dione (p. 2779 s), in small yield along with allopregnane-3.20-dione, on hydrogenation in glacial acetic acid in the presence of platinum oxide at 18°, followed by refluxing for 5 min. with 4% methanolic KOH and oxidation of the resulting mixture of diols with CrO_3 in glacial acetic acid at 15° (1941 Shoppee). From $\Delta^{4.16}$-pregnadiene-3.20-dione (page 2780 s) on hydrogenation in ether in the presence of palladium - barium sulphate at room temperature under pressure (1940b Marker). Presumably from Δ^4-pregnene-3.20-dione (progesterone, p. 2782 s), along with allopregnane-3.20-dione, on hydrogenation in abs. alcohol in the presence of palladium or zirconium oxide (1946 Pearlman); from the same compound by the action of Bacillus putrificus Bienstock in yeast water or yeast suspension at 36–37° (yields, 87%) (1939 Mamoli). From Δ^{16}-pregnene-3.20-dione (p. 2794 s) on hydrogenation in abs. alcohol in the presence of palladium - barium sulphate at room temperature under 1 atm. pressure (1940a Marker) or on heating with zinc dust in glacial acetic acid on the steam-bath for 1 hr. (1942a Marker).

References, pp. 2799 s, 2800 s

From 21-chloropregnane-3.20-dione (p. 2811 s) on reduction with zinc dust in strong acetic acid (1938 CIBA).

Rns. The **keto-dicarboxylic acid** $C_{21}H_{32}O_5$ (see p. 95), obtained on CrO_3-oxidation of pregnane-3.20-dione, melts at 281° (1939e Marker); for its structure, see 1939e Marker; but cf. pp. 1926 s, 1927 s. On shaking with hydrogen and platinum oxide in alcohol for 10 min. under 3 atm. pressure it gives pregnan-3α-ol-20-one (p. 2244 s) and a small amount of pregnan-3β-ol-20-one (p. 2245 s) (1937b Marker). Reacts with hydrogen in glacial acetic acid in the presence of previously reduced platinum oxide and some 48% HBr at room temperature under 3 atm. pressure on shaking for 20 min. to give mainly pregnan-3β-ol-20-one (p. 2245 s) (1937b Marker), on shaking for 90 min. to give pregnane-3β.20β-diol (p. 1931 s) and a very small amount of pregnane-3α.20β-diol (p. 1929 s) (1939a Marker). Gives pregnan-20-one (p. 2217 s) on refluxing with zinc and aqueous alcoholic HCl (1939b Marker). For Clemmensen reduction to pregnane (p. 1401 s), see also 1938 Steiger. On treatment with 1 mol. of bromine in chloroform at 0° it gives 4β-bromopregnane-3.20-dione (p. 2813 s) (1935 SCHERING-KAHLBAUM A.-G.) and probably 17-bromopregnane-3.20-dione (p. 2814 s) as a by-product (1942a Marker).

Disemicarbazone $C_{23}H_{38}O_2N_6$. Crystals (alcohol), m. 257° dec. (1937c Marker), 244° dec. (1940a Marker); more soluble in alcohol than the disemicarbazone of allopregnane-3.20-dione (1937c Marker).

Allopregnane-3.20-dione $C_{21}H_{32}O_2$ (p. 152). Crystals (alcohol), m. 203–5° cor. (Kofler block); $[\alpha]_D^{21}$ +108.5° (chloroform) (1946 Meystre); $[\alpha]_D$ +121° (chloroform) (Barton, see 1949 Shoppee). — Causes ovipositor growth in the female bitterling (1941/42 Duyvené de Wit). — **Occ.** In mares' pregnancy urine; small amount (1938c Marker).

Fmn. For formation from the diol-mixture contained in mares' pregnancy urine, see under formation of pregnane-3.20-dione (p. 2796 s). From allopregnane-3β.20α-diol (p. 1933 s) on oxidation with CrO_3 in glacial acetic acid at room temperature (1938d Marker; cf. also 1938 Steiger); similarly from allopregnane-3β.20β-diol (p. 1934 s; yield, 70%) (1939b Butenandt; cf. also 1938 Steiger). From allopregnane-3β.16α.20β-triol ("pregnanetriol B", see p. 2112 s) on oxidation with CrO_3 in strong acetic acid at 25°, followed by reduction of the resulting solid with sodium in alcohol and oxidation of the reduction product with CrO_3 in glacial acetic acid at 25°; yield, 11% (1938b Marker). From 17-iso-allopregnane-3β.17.20-triol (p. 2121 s) on dehydration, followed by oxidation (1938 SCHERING A.-G.). From Δ^5-pregnen-3β-ol-20-one on hydrogenation in glacial acetic acid in the presence of platinum oxide, followed by oxidation with CrO_3 in glacial acetic acid; yield, 90% (1938 Butenandt). From allopregnan-3β-ol-20-one (p. 2247 s) on oxidation with CrO_3 in glacial acetic acid at 20°; crude yield, 88% (1946 Meystre; cf. also 1939c Marker; 1939 Heard; 1943 Plattner). From 17-bromo-allopregnan-

3β-ol-20-one (p. 2286 s) on oxidation with CrO_3 in strong acetic acid at room temperature, followed by heating of the resulting crude bromodiketone (page 2814 s) with iron dust in acetic acid on the steam-bath; yield from the bromo-diketone, 67 % (1942 b Marker).

Along with small amounts of pregnane-3.20-dione and other compounds, from $\Delta^{4.9(11)}$-pregnadiene-3.20-dione (p. 2797 s) on hydrogenation in glacial acetic acid in the presence of platinum oxide at 18^0, followed by refluxing for 5 min. with 4 % methanolic KOH and oxidation of the resulting mixture of diols with CrO_3 in glacial acetic acid at 15^0 (1941 Shoppee). From Δ^1-allo-pregnene-3.20-dione (p. 2781 s) on hydrogenation in glacial acetic acid in the presence of a platinum catalyst, followed by oxidation with CrO_3 in glacial acetic acid; yield, 70 % (1939 a Butenandt). From Δ^4-pregnene-3.20-dione (progesterone, p. 2782 s) on hydrogenation in abs. alcohol in the presence of palladium or zirconium oxide; yield, 30 % (1946 Pearlman). From Δ^{16}-allo-pregnene-3.20-dione (p. 2795 s) on hydrogenation in ether in the presence of palladium - barium sulphate under 1.7 atm. pressure (1940 c Marker) or on heating with zinc dust in glacial acetic acid on the steam-bath for 1 hr. (yield, 70 %) (1942 b Marker). From 17-iso-allopregnane-3.20-dione (p. 2800 s) on refluxing with sodium methoxide in methanol for 2 hrs. (1939 c Marker).

From dihydrocholesterol (p. 1965 s) on oxidation with CrO_3 in strong acetic acid (1937 CIBA).

Rns. Gives 20-keto-allopregnane-2‖3-dioic acid (below) on oxidation with CrO_3 in glacial acetic acid at 50^0 (1937 a, 1939 e Marker). On hydrogenation in glacial acetic acid in the presence of platinum oxide at room temperature under 2.3 atm. pressure (shaking for 15 min.) it gives allopregnan-3β-ol-20-one (p. 2247 s) (1939 a Marker); this is also formed (yield, 73 %), along with allo-pregnan-3α-ol-20-one (p. 2246 s; yield, 5–10 %), on hydrogenation in abs. alcohol in the presence of Raney nickel (1938 Butenandt), whereas in glacial acetic acid in the presence of platinum black and 48 % HBr on heating, the 3α-epimer is the main product (ratio $3\alpha:3\beta = 2:1$) (1938 Fleischer); the same two olones are obtained, along with allopregnane-$3\alpha.20\beta$-diol (p. 1933 s), on shaking with hydrogen, platinum oxide, and 48 % HBr in glacial acetic acid at room temperature under 2.6 atm. pressure until 1 mol. of hydrogen has been absorbed, whereas on complete hydrogenation under these conditions allopregnane-$3\alpha.20\beta$-and-$3\beta.20\beta$-diols (7:2) are formed (1939 a Marker). Gives allopregnane-$3\beta.20\alpha$-diol (p. 1933 s) on reduction with sodium in boiling alcohol (1946 Meystre). On refluxing with zinc and aqueous alcoholic HCl, allopregnan-20-one (p. 2217 s) is obtained (1939 b Marker). Gives allopregnane (p. 1403 s) on heating with amalgamated zinc and conc. HCl in a current of hydrogen chloride on the water-bath (1938 Steiger) or on refluxing with zinc and conc. HCl in glacial acetic acid (1938 a Marker).

Is partially isomerized to 17-iso-allopregnane-3.20-dione (p. 2800 s) by refluxing with methanolic KOH or methanolic sodium methoxide (1939 c Marker).

Disemicarbazone $C_{23}H_{38}O_2N_6$, not m. below 325^0; less soluble in alcohol than the disemicarbazone of pregnane-3.20-dione (1937 c Marker).

References, pp. 2799 s, 2800 s

20-Keto-allopregnane-2‖3-dioic acid $C_{21}H_{32}O_5$. Crystals (80% acetic acid), m. 218° (1937a, 1939e Marker); crystals (alcohol), m. 219–219.5° cor.; $[\alpha]_D + 94°$ (chloroform) (1943 Plattner). — **Fmn.** Along with allopregnane-3.20-dione (p. 2797 s), from allopregnan-3β-ol-20-one (p. 2247 s) on oxidation with CrO_3 in strong acetic acid at room temperature; yield, 15% (1943 Plattner). From allopregnane-3.20-dione (p. 2797 s) on warming with CrO_3 in glacial acetic acid at 50° for 5 hrs. (1937a, 1939e Marker). — **Rns.** Gives A-nor-allopregnane-2.20-dione ("pyro-allopregnanedione", p. 1376 s) on heating with acetic anhydride, with removal of the solvent by distillation, followed by heating of the residue at 200–280° for 30 min. and then by distillation under 2 mm. pressure (1937a Marker).

1935 SCHERING-KAHLBAUM A.-G., *Australian Pat.* 24429/1935 (issued 1936); *Ch. Ztbl.* 1937 I 3023, 3024; *Fr. Pat.* 806467 (issued 1936); *Australian Pat.* 3961/1936 (issued 1936); *Ch. Ztbl.* 1937 I 4264, 4265.
　　Zimmermann, *Z. physiol. Ch.* 233 257, 264.
1936/37 CIBA*, *Brit. Pat.* 476749 (issued 1938); *Fr. Pat.* 823139 (issued 1938); *C. A.* 1938 3770, 5585; *Ch. Ztbl.* 1938 II 120, 121.
1937 CIBA*, *Fr. Pat.* 830043 (issued 1938); *C. A.* 1939 179; *Ch. Ztbl.* 1939 I 2035.
　a Marker, Kamm, Jones, *J. Am. Ch. Soc.* 59 1595.
　b Marker, Kamm, Wittle, *J. Am. Ch. Soc.* 59 1841.
　c Marker, Kamm, Crooks, Oakwood, Lawson, Wittle, *J. Am. Ch. Soc.* 59 2297.
1938 Butenandt, Heusner, *Z. physiol. Ch.* 256 236, 241.
　CIBA*, *Fr. Pat.* 835814 (issued 1939); *C. A.* 1939 5008; *Ch. Ztbl.* 1939 I 2827, 2828; *Brit. Pat.* 512954 (issued 1939); *C. A.* 1941 1800.
　Fleischer, Whitman, Schwenk, *J. Am. Ch. Soc.* 60 79.
　a Marker, Kamm, Oakwood, Wittle, Lawson, *J. Am. Ch. Soc.* 60 1061, 1066.
　b Marker, Kamm, Wittle, Oakwood, Lawson, *J. Am. Ch. Soc.* 60 1067, 1069.
　c Marker, Lawson, Wittle, Crooks, *J. Am. Ch. Soc.* 60 1559.
　d Marker, Binkley, Wittle, Lawson, *J. Am. Ch. Soc.* 60 1904.
　SCHERING A.-G., *Fr. Pat.* 846090 (issued 1939); *C. A.* 1941 1185; *Ch. Ztbl.* 1940 I 2829; SCHERING CORP. (Logemann), *U.S. Pat.* 2372440 (issued 1945); *C. A.* 1945 4199; *Ch. Ztbl.* 1945 II 1385 (in *C. A.* and *Ch. Ztbl.* erroneously referred to as *U.S. Pat.* 2372400).
　Steiger, Reichstein, *Helv.* 21 161, 170, 171.
1939 a Butenandt, Mamoli, Dannenberg, Masch, Paland, *Ber.* 72 1617, 1622.
　b Butenandt, Schmidt-Thomé, *Ber.* 72 1960, 1962.
　Heard, McKay, *J. Biol. Ch.* 131 371, 378.
　Mamoli, Koch, Teschen, *Z. physiol. Ch.* 261 287, 295.
　a Marker, Lawson, *J. Am. Ch. Soc.* 61 588.
　b Marker, Lawson, *J. Am. Ch. Soc.* 61 852.
　c Marker, Wittle, Plambeck, *J. Am. Ch. Soc.* 61 1333, 1335.
　d Marker, Rohrmann, *J. Am. Ch. Soc.* 61 2537, 2539.
　e Marker, Wittle, Plambeck, Krueger, Ulshafer, *J. Am. Ch. Soc.* 61 3317, 3319.
1940 a Marker, Rohrmann, *J. Am. Ch. Soc.* 62 518, 520.
　b Marker, Tsukamoto, Turner, *J. Am. Ch. Soc.* 62 2525, 2530, 2531.
　c Marker, Turner, *J. Am. Ch. Soc.* 62 2540.
1941 Shoppee, Reichstein, *Helv.* 24 351, 356.
1941/42 Duyvené de Wit, *Biochem. Z.* 310 83, 85, 87, 89.
1942 a Marker, Crooks, Wagner, *J. Am. Ch. Soc.* 64 210.
　b Marker, Crooks, Wagner, Wittbecker, *J. Am. Ch. Soc.* 64 2089, 2091.
1943 Plattner, Fürst, *Helv.* 26 2266, 2270.

* Société pour l'industrie chimique à Bâle; Gesellschaft für chemische Industrie in Basel.

1946 Meystre, Miescher, *Helv.* **29** 33, 43, 44.
     Pearlman, *J. Biol. Ch.* **166** 473, 475.
1949 Shoppee, *J. Ch. Soc.* **1949** 1671, 1676, 1677, 1680.
1955 Slomp, Shealy, Johnson, Donia, Johnson, Holysz, Pederson, Jensen, Ott, *J. Am. Ch. Soc.*
     77 1216, 1221.

17-iso-Δ^4-Pregnene-3.20-dione, 17-iso-Progesterone $C_{21}H_{30}O_2$. Needles (dil. alcohol), m. 145° after sintering at 142°; $[\alpha]_D^{20}$ 0° (alcohol) (1939 Butenandt). Absorption max. in alcohol at 243 mμ (ε 16300) (1939 Butenandt; 1939 Dannenberg). — Has no progestational activity (1939 Butenandt). — **Fmn.** From 17-iso-Δ^5-pregnen-3β-ol-20-one (p. 2249 s) on refluxing with aluminium isopropoxide and cyclohexanone in toluene for 1 hr. (1939 Butenandt; 1940 SCHERING A.-G.) or by the action of Corynebacterium mediolanum (isolated from yeast) (1942 Butenandt). — **Rns.** Is isomerized to progesterone (p. 2782 s) by refluxing with conc. HCl in alcohol for 15 min. (1939 Butenandt; 1940 SCHERING A.-G.).

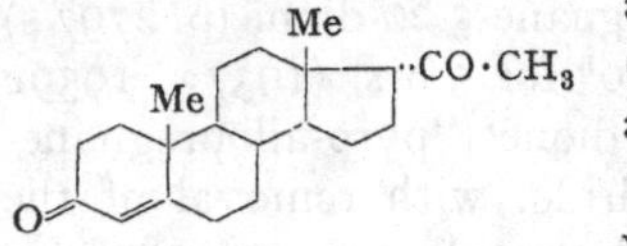

Neoprogesterone, to which the above structure was tentatively ascribed (1939 Miescher), is actually 17a-methyl-D-homo-Δ^4-androstene-3.17-dione (1940 Ruzicka; cf. also 1939 Butenandt).

17-iso-Allopregnane-3.20-dione $C_{21}H_{32}O_2$ (p. 152: *Isoallopregnanedione*). The compound described by Butenandt (1935) was impure; it was prepared from a mixture (m. 148°) of 17-iso-allopregnan-3β-ol-20-one with ca. 15 % of allopregnan-3β-ol-20-one (1948, 1949 Shoppee). Crystals, m. 148–9° (1939c Marker); prisms (acetone - pentane), m. 148–9° cor. (Kofler block); $[\alpha]_D^{21}$ —49.5° (alcohol) (1949 Shoppee). — **Fmn.** From 17-iso-allopregnan-3β-ol-20-one (p. 2250 s) on oxidation with CrO_3 in strong acetic acid at 20°; almost quantitative yield (1949 Shoppee). From allopregnane-3.20-dione (p. 2797 s) on epimerization by refluxing with methanolic sodium methoxide or with 5 % methanolic KOH; small yields (1939c Marker). — **Rns.** It largely reverts to allopregnane-3.20-dione on refluxing with methanolic sodium methoxide for 2 hrs. (1939c Marker).

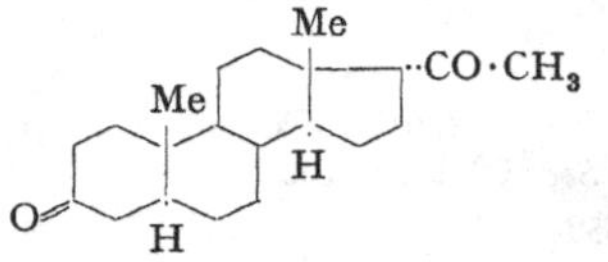

14-iso-17-iso-Allopregnane-3.20-dione*, *Diketodiginane* $C_{21}H_{32}O_2$. For the structure, see 1947 Press. — Needles (ether - pentane), m. 140–1° cor. (Kofler block) (1944 Shoppee); platelets (pentane), m. 138–141° cor. (Kofler block) (1947 Press); $[\alpha]_D^{13}$ +39.5° (acetone) (1944 Shoppee), $[\alpha]_D^{20}$ +41° (acetone), +40° (chloroform) (1947 Press). — **Fmn.** From 14-iso-17-iso-allopregnan-3β-ol-20-one** on oxidation with CrO_3 in glacial acetic acid at 18°;

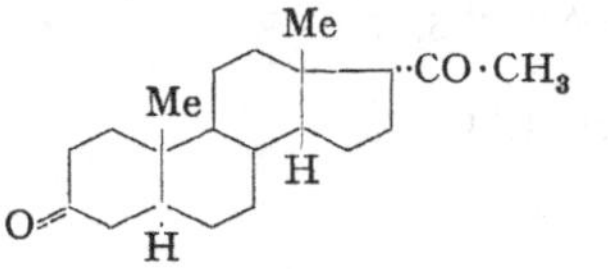

---

* For 14-iso-17-iso-pregnane-3.20-dione, see (after the closing date for this volume) 1947 Meyer.
** Described after the closing date for this volume (see 1947 Plattner; 1947 Press).

yield, 60% (1947 Press). From diginigenin $C_{21}H_{28}O_4$ (the aglycon of the cardiac glycoside diginin) on Wolff-Kishner reduction, followed by hydrogenation in glacial acetic acid in the presence of platinum oxide and oxidation of the resulting dihydroxydiginane $C_{21}H_{36}O_2$ with CrO_3 in glacial acetic acid at 20° (yield in the last step, 73%) (1944 Shoppee).

Rns. Gives 14-iso-17-iso-allopregnane (diginane, p. 1404 s) on heating with hydrazine hydrate and sodium ethoxide in abs. alcohol for 8 hrs. in a sealed tube at 180° (1944 Shoppee; 1947 Press).

Bis-(2.4-dinitrophenylhydrazone) $C_{33}H_{40}O_8N_8$. Orange-yellow, microcrystalline powder (ethyl acetate - alcohol), m. 185° cor. (Kofler block) after previous sintering (1944 Shoppee).

1935 Butenandt, Mamoli, *Ber.* **68** 1847, 1850.
1939 Butenandt, Schmidt-Thomé, Paul, *Ber.* **72** 1112, 1114, 1115, 1118.
     Dannenberg, *Abhandl. Preuss. Akad. Wiss. Math.-naturwiss. Kl.* **1939** No. 21, p. 53.
     Marker, Wittle, Plambeck, *J. Am. Ch. Soc.* **61** 1333, 1335.
     Miescher, Kägi, *Helv.* **22** 184, 189, 194.
1940 Ruzicka, Meldahl, *Helv.* **23** 364.
     SCHERING A.-G., *Fr. Pat.* 868904 (issued 1942); *Ch. Ztbl.* **1942** II 317.
1942 Butenandt, *Naturwiss.* **30** 4, 7.
1944 Shoppee, *Helv.* **27** 246, 258–260.
1947 Meyer, *Helv.* **30** 2024, 2027.
     Plattner, Ruzicka, Heusser, Angliker, *Helv.* **30** 385, 391, 393.
     Press, Reichstein, *Helv.* **30** 2127, 2131.
1948 Shoppee, *Experientia* **4** 418, 420.
1949 Shoppee, *J. Ch. Soc.* **1949** 1671, 1676, 1677, 1680.

Δ^5-Pregnene-4.20-dione $C_{21}H_{30}O_2$ (p. 150: *hetero-Δ^1-Allopregnene-3.20-dione*).

For the structure, see 1944 Butenandt. — Absorption max. in ether at 234 mμ (ε 4700) (1939 Dannenberg). — **Fmn.** For its formation from 2α-bromoallopregnane-3.20-dione (p. 2812 s) on heating with potassium acetate, see also 1935 SCHERING-KAHLBAUM A.-G.

1935 SCHERING-KAHLBAUM A.-G., *Fr. Pat.* 806467 (issued 1936); *Australian Pat.* 3961/1936 (issued 1936); *Ch. Ztbl.* **1937** I 4264, 4265; *Australian Pat.* 24429/1935 (issued 1936); *Ch. Ztbl.* **1937** I 3023, 3024.
1939 Dannenberg, *Abhandl. Preuss. Akad. Wiss. Math.-naturwiss. Kl.* **1939** No. 21, p. 54.
1944 Butenandt, Ruhenstroth-Bauer, *Ber.* **77** 397, 400.

Δ^4-Pregnene-6.20-dione $C_{21}H_{30}O_2$. Not characterized. — **Fmn.** Along with Δ^4-pregnene-3.20-dione (progesterone, p. 2782 s) and Δ^4-pregnene-3.6.20-trione (p. 2965 s), from Δ^4-pregnen-20-one (p. 2216 s) on oxidation with CrO_3 in strong acetic acid at room temperature (1940 CIBA).

1940 CIBA, *Fr. Pat.* 886415 (issued 1943); *Ch. Ztbl.* **1944** I 450; CIBA PHARMACEUTICAL PRODUCTS (Miescher, Wettstein), *U.S. Pat.* 2323277 (issued 1943); *C. A.* **1944** 222, 223.

Pregnane-11.20-dione $C_{21}H_{32}O_2$. For a compound which may possess this structure, see under the reactions of pregnane-3.11.20-trione, p. 2967 s.

9-iso-Pregnane-11.20-dione $C_{21}H_{32}O_2$. This structure was ascribed by Marker (1939) to *urane-11.20-dione* [needles (ether - pentane), m. 199–201°] obtained from uranetrione by Clemmensen reduction. According to Klyne (1950, 1952) the urane compounds are probably D-homo-steroids (see urane, p. 1403 s).

1939 Marker, Rohrmann, *J. Am. Ch. Soc.* **61** 2719, 2722.
1950 Klyne, *Nature* **166** 559.
1952 Klyne, Shoppee, *Chemistry & Industry* **1952** 470.

$\Delta^{4.17(20)}$-**Pregnadien-3-on-21-al** $C_{21}H_{28}O_2$. Needles (aqueous acetone), crystals (hexane - acetone), m. 150–2° cor.; may be sublimed in a high vacuum without decomposition; $[\alpha]_D^{18} + 141.5°$ (abs. alcohol) (1939 Miescher; 1939a CIBA). Absorption max. in alcohol at ca. 244 mμ (log ε 4.5) (1939 Miescher). Gives a precipitate with ammoniacal silver salt solution at room temperature. With hydrogen halide in alcohol it gives a red-brown coloration and sparingly soluble coloured products, with conc. H_2SO_4 in glacial acetic acid an intense, fluorescent, violet to green-brown coloration. Gives red colorations with 1.4-dihydroxynaphthalene or 1.4-naphthoquinone in the presence of conc. HCl in glacial acetic acid at 60–70° (1939 Miescher).

Fmn. From $\Delta^{4.17(20)}$-pregnadien-21-ol-3-one (p. 2512 s) on shaking in benzene with $K_2Cr_2O_7$ + aqueous H_2SO_4 with cooling in ice or on treatment with CrO_3 in glacial acetic acid. From 17-iso-Δ^4-pregnen-17β-ol-3-on-21-al (p. 2846 s) on distillation over anhydrous $CuSO_4$ at 135° in a high vacuum, or on refluxing under nitrogen with iodine in m-xylene for 1–3 hrs., with glacial acetic acid for 4 hrs. (with simultaneous removal of the water formed in this reaction by repeated addition of acetic anhydride), or with propionic acid for 1 hr. (1939 Miescher; 1939a CIBA). From its N-(p-dimethylaminophenyl)-isoxime (p. 2803 s) on shaking with 2 N NaOH in ether (1940 Reich; cf. 1940 N. V. ORGANON). Its acetal is obtained from that of $\Delta^{5.17(20)}$-pregnadien-3β-ol-21-al (p. 2251 s) by Oppenauer oxidation (1939b CIBA).

Rns. Is very easily oxidized to $\Delta^{4.17(20)}$-pregnadien-3-on-21-oic acid, e.g. on attempted crystallization from heptane + some acetone without exclusion of air, or on passing a current of air through its solution in toluene at 100° (1939 Miescher). Oxidation of its acetal (see above, under formation; also obtained from the present compound on refluxing with ethyl orthoformate and ammonium nitrate in methanol) with OsO_4 in ether at room temperature, followed by heating with alcoholic Na_2SO_3 and hydrolysis of the resulting acetal with alcoholic HCl, affords Δ^4-pregnene-17.20-diol-3-on-21-al (p. 2863 s)

(1939b CIBA). On refluxing with acetic anhydride in glacial acetic acid under nitrogen it gives $\Delta^{4.16.20}$-pregnatrien-21-ol-3-one acetate (p. 2512 s) (1939 Miescher). Gives 21-methyl-$\Delta^{4.17(20)}$-pregnadiene-3.21-dione (p. 2804 s) on treatment with diazomethane in methylene chloride (1938 CIBA).

Disemicarbazone $C_{23}H_{34}O_2N_6$. Crystals (dil. acetic acid), rapidly decomposing above 370^0; very sparingly soluble except in glacial acetic acid and pyridine (1939 Miescher).

21-N-(p-Dimethylaminophenyl)-isoxime $C_{29}H_{38}O_2N_2$. Brownish yellow crystals (aqueous alcohol), m. $152-5^0$ cor. after sintering at 148^0. — **Fmn.** From N-(3-keto-$\Delta^{4.17(20)}$-pregnadien-21-yl)-pyridinium bromide (p. 2473 s) on treatment with p-nitrosodimethylaniline in aqueous alcoholic NaOH at room temperature; yield, 46%. — **Rns.** Gives the above free keto-aldehyde on shaking with 2 N NaOH in ether (1940 Reich; cf. 1940 N. V. ORGANON).

Δ^4-**Pregnen-3-on-21-al** $C_{21}H_{30}O_2$. Crystals (dil. methanol). — **Fmn.** From 21-bromo-Δ^4-pregnen-3-one (p. 2477 s) on refluxing with hexamethylenetetramine in aqueous alcohol for 3 hrs. — **Rns.** On refluxing with acetic anhydride and sodium acetate for 6 hrs. it gives the 21-enol acetate ($\Delta^{4.20}$-pregnadien-21-ol-3-one acetate, p. 2513 s) (1939a CIBA).

1938 CIBA (Miescher, Wettstein), *Ger. Pat.* 739083 (issued 1943); *C. A.* **1945** 392; *Ch. Ztbl.* **1944** I 110, 111.

1939 a CIBA, *Fr. Pat.* 857122 (issued 1940); *Ch. Ztbl.* **1941** I 1324; CIBA PHARMACEUTICAL PRODUCTS (Miescher, Fischer, Scholz, Wettstein), *U.S. Pats.* 2275790, 2276543 (both issued 1942); *C. A.* **1942** 4289, 4674.

b CIBA (Miescher, Wettstein), *Swed. Pat.* 101201 (issued 1941); *Ch. Ztbl.* **1941** II 3101; *Norwegian Pat.* 60152 (issued 1942); *Ch. Ztbl.* **1943** I 2421; *Fr. Pat.* 863257 (issued 1941); *Ch. Ztbl.* **1942** I 81.

Miescher, Wettstein, Scholz, *Helv.* **22** 894.

1940 N. V. ORGANON, *Fr. Pat.* 864315 (issued 1941); *Ch. Ztbl.* **1942** I 642.

Reich, *Helv.* **23** 219, 222.

---

c. Dioxosteroids $C_{22}-C_{27}$ (One CO in the Ring System) with Three Aliphatic Side Chains

21-Methyl-Δ^4-pregnene-3.20-dione, ω-Homo-Δ^4-pregnene-3.20-dione, 17β-Propionyl-Δ^4-androsten-3-one, 21-Methylprogesterone $C_{22}H_{32}O_2$. Colourless crystals (benzene), m. $151-2^0$ cor. (1940 Wettstein). — Shows considerable progestational activity (1940 Wettstein). — **Fmn.** From 21-methyl-Δ^5-pregnen-3β-ol-20-one (p. 2258 s) on boiling with aluminium isopropoxide and cyclohexanone in toluene (1940 Wettstein). From 3-keto-Δ^4-ætiocholenaldehyde (p. 2777 s) by the action of diazoethane (1938a CIBA).

21 - Methyl - $\Delta^{4 \cdot 17(20)}$ - pregnadiene - 3.21 - dione, ω-Homo-$\Delta^{4\ 17(20)}$-pregnadiene-3.21 - dione, 17 - Acetonylidene - Δ^4 - androsten-3-one $C_{22}H_{30}O_2$.

Crystals (methanol), m. 176–7° cor.; $[\alpha]_D$ +87° (dioxane); absorption max. in alcohol (from the graph) at 241 (4.3), inflexion at 300 (2.05) mμ (log ε) (1941 Plattner). —
Fmn. From 21-methyl-$\Delta^{5 \cdot 17(20)}$-pregnadien-3β-ol-21-one (p. 2258 s) on boiling with aluminium tert.-butoxide and acetone in abs. benzene (1941 Plattner). From $\Delta^{4 \cdot 17(20)}$-pregnadien-3-on-21-al (p. 2802 s) by the action of diazomethane in methylene chloride (1938a CIBA).

21 - Methyl - Δ^4 - pregnene - 3.21 - dione, ω - Homo - Δ^4 - pregnene - 3.21 - dione, 17β - Acetonyl - Δ^4 - androsten - 3 - one $C_{22}H_{32}O_2$.

For the configuration at C-17, see 1944 Plattner. Crystals (methanol), m. 153–4° cor.; $[\alpha]_D$ +89° (dioxane); absorption max. in alcohol at 242.5 (4.55) and 314 (2.25) mμ (log ε). Tetranitromethane produces no yellow coloration. — Shows no progestational activity. — **Fmn.** From 21-methyl-Δ^5-pregnen-3β-ol-21-one (p. 2258 s) on boiling with aluminium tert.-butoxide and acetone in abs. benzene (1941 Plattner).

21-Ethyl-Δ^4-pregnene-3.20-dione, 17β-Butyryl-Δ^4-androsten-3-one, 21-Ethyl-progesterone $C_{23}H_{34}O_2$.

Crystals (hexane), m. 118–120° cor. (1940 Wettstein). — Shows some progestational (1942b, 1943b Selye; cf., however, 1940 Wettstein), folliculoid (1942 Clarke), and anæsthetic (1943a Selye) activities. — **Fmn.** From 21-ethyl-Δ^5-pregnen-3β-ol-20-one (p. 2259 s) on boiling with aluminium isopropoxide and cyclohexanone in toluene (1940 Wettstein).

21-Nor-Δ^4-cholestene-3.20-dione, 21-Isoamylprogesterone $C_{26}H_{40}O_2$.

No properties given. — **Fmn.** For its formation from Δ^4-cholestene (p. 1423 s), along with other diketones and with keto-carboxylic acids, by oxidation with CrO_3 in glacial acetic acid at room temperature, see 1940 CIBA.

Δ^4-Norcholene-3.22-dione, 3-Keto-Δ^4-ternorcholenyl methyl ketone, 3-Keto-Δ^4 - pregnen - 20α - yl methyl ketone $C_{23}H_{34}O_2$.

Needles (hexane - acetone), m. 213–5° cor. (1941 Wettstein); prisms (benzene), m. 206–8°; $[\alpha]_D^{29}$ +54° (chloroform) (1945 Cole). — Shows no progestational activity (1941 Wettstein; 1945 Cole). — **Fmn.** From 3β-hydroxy-Δ^5-ternor-cholenyl methyl ketone (p. 2259 s) on boiling in toluene with cyclohexanone

and aluminium isopropoxide (1941 Wettstein) or tert.-butoxide (1945 Cole).
From 2-(3-keto-Δ^4-ternorcholenyl)-propene (p. 2413 s) on ozonization in
chloroform in an ice-bath (1945 Julian). From 3-keto-Δ^4-bisnorcholenic acid
(p. 200) by interaction with thionyl chloride and one drop of pyridine in
ether at room temperature and treatment of the resulting waxy chloride with
dimethylzinc in benzene at 35° (1945 Cole).

20 - iso - Δ^4 - Norcholene - 3.22 - dione, 3-Keto-20-iso-Δ^4-ternorcholenyl methyl ketone, 3 - Keto - Δ^4 - pregnen - 20β - yl methyl ketone

$C_{23}H_{34}O_2$. Colourless plates (dry acetone), m. 165–6° (the cooled melt remelts at
180–8°); prisms (methanol), m. 189–191°; $[\alpha]_D^{25}$
+ 67° (chloroform). — Shows no progestational
activity. — **Fmn.** From 3β-hydroxy-20-iso-Δ^5-ternorcholenyl methyl ketone
(p. 2260 s) on boiling with aluminium tert.-butoxide and cyclohexanone in
toluene (1945 Cole).

Δ^4-Cholene-3.22-dione, 3-Keto-Δ^4-ternorcholenyl ethyl ketone, 3-Keto-Δ^4-pregnen-20α-yl ethyl ketone

$C_{24}H_{36}O_2$. Prisms (wet
acetone), m. 157–160°; crystalline powder (after
vac. sublimation), m. 158°, resolidifies, and
remelts at 168–170°; $[\alpha]_D^{25}$ + 59° (chloroform). —
Shows no progestational activity. — **Fmn.** From
3β - hydroxy - Δ^5 - ternorcholenyl ethyl ketone
(p. 2262 s) on refluxing with aluminium tert.-butoxide and cyclohexanone
in toluene (1945 Cole).

Δ^4-Bisnorcholestene-3.24-dione

$C_{25}H_{38}O_2$. Needles (methanol), m. 152°. —
Fmn. From Δ^5-bisnorcholesten-3β-ol-
24-one (p. 2263 s) on oxidation with
aluminium phenoxide and acetone in
benzene (1938 Kuwada).

Dioxime $C_{25}H_{40}O_2N_2$, dec. 117° (1938
Kuwada).

3-Semicarbazone $C_{26}H_{41}O_2N_3$. White needles, m. 217°. — **Fmn.** By solution
of the disemicarbazone in glacial acetic acid and dilution with water (1938
Kuwada).

Disemicarbazone $C_{27}H_{44}O_2N_6$, dec. 234° (1938 Kuwada).

27-Nor-Δ^4-cholestene-3.25-dione

$C_{26}H_{40}O_2$ (no analysis given), m. 128–9° cor.
(1937b Ruzicka). — Is inactive in the
capon, the rat, and the Corner-Clauberg
tests (1937b Ruzicka). Exerts no anæsthetic action on rats (1942a Selye) or
fish (1943a Selye). — **Fmn.** From

27-nor-Δ^5-cholesten-3β-ol-25-one (p. 2263 s) on mild oxidation with CrO_3 after temporary protection of the double bond by bromine (1937b Ruzicka), or by the action of dehydrogenating bacteria (Flavobacterium dehydrogenans Arn., Micrococcus dehydrogenans, Corynebacterium mediolanum) (1940 Ercoli; 1942 Arnaudi; 1942 Molina). For its formation, along with other compounds, from Δ^4-cholestene (p. 1423 s) and Δ^4-cholesten-3-one (page 2424 s) by oxidation with CrO_3 in glacial acetic acid, see 1938b, 1940 CIBA.

27-Norcholestane-3.25-dione $C_{26}H_{42}O_2$.

Leaflets (aqueous methanol), m. 139.5° to 140.5° cor. (after sublimation at 133° and 0.001 mm.) (1937a Ruzicka). — Is inactive in the capon comb test, the Allen-Doisy, and the Corner-Clauberg tests (1937a Ruzicka). — **Fmn.** From 27-nor-Δ^5-cholesten-3β-ol-25-one (page 2263 s) on hydrogenation using platinum oxide catalyst in glacial acetic acid, followed by oxidation with CrO_3 in acetic acid (1937b Ruzicka). From 27-norcholestan-3α-ol-25-one (p. 2264 s) on oxidation with CrO_3 in acetic acid at room temperature (1937a Ruzicka).

Disemicarbazone $C_{28}H_{48}O_2N_6$ (not quite pure), m. 240° cor. (1937a Ruzicka).

Δ^4-Cholestene-3.22-dione, 3-Keto-Δ^4-ternorcholenyl isoamyl ketone, 3-Keto-Δ^4-pregnen-20α-yl isoamyl ketone

$C_{27}H_{42}O_2$. Was not obtained crystalline; $[\alpha]_D^{25}$ +49° (chloroform). — Shows no progestational activity. **Fmn.** From 3β-hydroxy-Δ^5-ternorcholenyl isoamyl ketone (p. 2265 s) on heating with aluminium tert.-butoxide and cyclohexanone in toluene at 100° (1945 Cole).

3(?)-Semicarbazone $C_{28}H_{45}O_2N_3$ (no analysis given), m. 228–230° (1945 Cole).

Coprostane-3.24-dione $C_{27}H_{44}O_2$ (p. 152).

Page 152 line 7 from bottom: for "(1926 Reindel)" *read* "(1936 Reindel, Niederländer, Ann.* **522** 218, 231)".

1937 a Ruzicka, Oberlin, Wirz, Meyer, *Helv.* **20** 1283, 1287, 1289.
     b Ruzicka, Fischer, *Helv.* **20** 1291, 1293, 1295, 1296.
1938 a CIBA* (Miescher, Wettstein), *Ger. Pat.* 739083 (issued 1943); *C. A.* **1945** 392; *Ch. Ztbl.* **1944** I 110.
     b CIBA* (Miescher, Fischer), *U.S. Pat.* 2172590 (issued 1939); *C. A.* **1940** 450; *Ch. Ztbl.* **1940** I 428; *Fr. Pat.* 841242 (issued 1939); *Ch. Ztbl.* **1940** I 916.
     Kuwada, Yosiki, *C. A.* **1938** 8432; *Ch. Ztbl.* **1939** I 1372.
1940 CIBA* (Miescher, Wettstein), *Swed. Pat.* 103744 (issued 1942); *Ch. Ztbl.* **1942** II 2294; *Fr. Pat.* 886410 (issued 1943); *Ch. Ztbl.* **1944** I 448; *Brit. Pat.* 550478 (issued 1943); *C. A.* **1944**

* Société pour l'industrie chimique à Bâle; Gesellschaft für chemische Industrie in Basel.

1612; CIBA PHARMACEUTICAL PRODUCTS (Miescher, Wettstein), *U.S. Pat.* 2319012 (issued 1943); *C. A.* **1943** 6096.
Ercoli, *C. A.* **1944** 1540; *Ch. Ztbl.* **1941** II 1028.
Wettstein, *Helv.* **23** 1371, 1373, 1377, 1378.
1941 Plattner, Schreck, *Helv.* **24** 472, 474, 476, 477.
Wettstein, *Helv.* **24** 311, 313, 315.
1942 Arnaudi, *C. A.* **1946** 3152; cf. *Ztbl. Bakt. Parasitenk.* II. Abt. **105** 352 (1942/43); *Biological Abstracts* **21** 889 (1947).
Clarke, Selye, *Am. J. Med. Sci.* **204** 401, 405, 407.
Molina, *C. A.* **1946** 3152.
a Selye, *Endocrinology* **30** 437, 444.
b Selye, Masson, *Science* [N. S.] **96** 358.
1943 a Selye, Heard, *Anesthesiology* **4** 36, 41, 44.
b Selye, Masson, *J. Pharmacol.* **77** 301, 305, 307.
1944 Plattner, Bucher, Hardegger, *Helv.* **27** 1177, 1179.
1945 Cole, Julian, *J. Am. Ch. Soc.* **67** 1369, 1373, 1374.
Julian, Cole, Meyer, Herness, *J. Am. Ch. Soc.* **67** 1375, 1378.

---

d. Dioxosteroids C$_{22}$–C$_{28}$ (One CO in the Ring System) with Four Aliphatic Side Chains

16- Methyl-$\Delta^{4.16}$- pregnadiene -3.20- dione, 16-Methyl-16-dehydroprogesterone

C$_{22}$H$_{30}$O$_2$. Crystals (ethyl acetate or hexane), m. 176–7° cor.; $[\alpha]_D^{22}$ +86° (alcohol); absorption max. in alcohol (from the graph) at 250 (4.3) and 312.5 (2.3) mμ (log ε). Instantaneously decolorizes dil. alcoholic permanganate solution. Gives only a weak yellow coloration with tetranitromethane. — **Fmn.** From 16-methyl-$\Delta^{5.16}$-pregnadien-3β-ol-20-one (p. 2267 s) on boiling with aluminium isoprop-oxide and cyclohexanone in toluene. From the pyrazoline compound (page 2780 s), obtained from $\Delta^{4.16}$-pregnadiene-3.20-dione and diazomethane, by thermal decomposition at 170–180° (1944 Wettstein).

16α-Methyl-Δ^4-pregnene-3.20-dione, 16α-Methylprogesterone C$_{22}$H$_{32}$O$_2$.

For the configuration at C-16, see 1952 Romo. — Crystals (ether), m. 133–5° (1942b Marker). — **Fmn.** From 16α-methyl-Δ^5-pregnen-3β-ol-20-one (m. 191–2°; p. 2268 s) on refluxing with aluminium tert.-butoxide and acetone in dry toluene for 6 hrs.; yield, 61% (1942b Marker).

16β-Methyl-Δ^4-pregnene-3.20-dione, 16β-Methylprogesterone C$_{22}$H$_{32}$O$_2$.

For the configuration at C-16, see 1952 Romo. — Needles (hexane), m. 210–211° cor. (1944 Wettstein); $[\alpha]_D^{23}$ +131° (alcohol) (1944 Wettstein), $[\alpha]_D^{20}$ +133° (dioxane) (1952 Romo). Absorption max. (solvent not stated) at 236 mμ (log ε 4.56) (1944 Wettstein), 240 mμ (log ε 4.34) (1952 Romo). Gives no yellow coloration with tetranitromethane (1944 Wettstein). It instantaneously decolorizes dil. alco-

References, p. 2808 s

holic permanganate solution (1944 Wettstein). — **Fmn.** From 16β-methyl-
Δ⁵-pregnen-3β-ol-20-one (m. 205–7⁰ cor.; p. 2268 s) on boiling with aluminium isopropoxide and cyclohexanone in abs. toluene for 2 hrs. (1944 Wettstein; 1952 Romo). — **Rns.** On treatment with K_2CO_3 solution or mineral
acids, rearrangement occurs (1944 Wettstein).

Dioxime $C_{22}H_{34}O_2N_2$. Crystals (chloroform · methanol), m. 239–242⁰; $[\alpha]_D^{20}$
+159⁰ (dioxane) (1952 Romo).

16β-Methyl-17-iso-Δ⁴-pregnene-3.20-dione, 16β-Methyl-17-iso-progesterone

$C_{22}H_{32}O_2$. Crystals (methanol), m. 164–166⁰; $[\alpha]_D^{20}$
+27⁰ (dioxane). Absorption max. (solvent not
stated) at 242 mμ (log ε 4.35). — **Fmn.** From
16β-methyl-17-iso-Δ⁵-pregnen-3β-ol-20-one (page
2269 s) on Oppenauer oxidation (1952 Romo).

17α-Methylpregnane-3.20-dione $C_{22}H_{34}O_2$.

For configuration at C-17, see under
the starting compound (p. 2269 s). — White crystals (methanol), m. 131–4⁰. — **Fmn.** From 17α-
methylpregnan-3β-ol-20-one (p. 2269 s) on oxidation with CrO_3 in acetic acid at room temperature
(1942 a Marker).

16α-Isopropyl-Δ⁴-pregnene-3.20-dione, 16α-Isopropylprogesterone $C_{24}H_{36}O_2$.

For the configuration at C-16, see 1952 Romo. —
Crystals (ether · pentane), m. 106.5–108⁰ (1942 b
Marker). — **Fmn.** From 16α-isopropyl-Δ⁵-preg-
nen-3β-ol-20-one (p. 2269 s) on refluxing with
aluminium tert.-butoxide and acetone in toluene;
yield, ca. 57% (1942 b Marker).

16α-tert.-Butyl-Δ⁴-pregnene-3.20-dione, 16α-tert.-Butylprogesterone $C_{25}H_{38}O_2$.

For the configuration at C-16, see 1952 Romo. —
Crystals (ether), m. 154–5⁰ (1942 b Marker). —
Fmn. From 16α-tert.-butyl-Δ⁵-pregnen-3β-ol-
20-one (p. 2270 s) on refluxing with aluminium
tert.-butoxide and acetone in toluene; yield, 63%
(1942 b Marker).

2 or 4-Formylcholestan-3-one $C_{28}H_{46}O_2$.
See the tautomeric 2 or 4-hydroxy-
methylene-cholestan-3-one, p. 2513 s.

1942 a Marker, Wagner, *J. Am. Ch. Soc.* **64** 1273, 1275.
    b Marker, Crooks, *J. Am. Ch. Soc.* **64** 1280.
1944 Wettstein, *Helv.* **27** 1803, 1805, 1807, 1809, 1810, 1812 et seq.
1952 Romo, Lepe, Romero, *C. A.* **1954** 9399, 9400.

e. **Dioxosteroids (One CO in the Ring System) containing Cyclic Radicals in Side Chain**

17β- (2- Ketocyclohexylidene-methyl) -Δ⁴-androsten-3-one, 2-[(3-Keto-Δ⁴-androsten-17β-yl)-methylene]-cyclohexanone

$C_{26}H_{36}O_2$. Crystals (hexane), m. 190–193° cor. On heating with 1.4-dihydroxynaphthalene, glacial acetic acid, and conc. HCl it gives a strong yellow colour with an intense green fluorescence. — **Fmn.** Together with 17β-hydroxymethyl-Δ⁴-androsten-3-one (p. 2510 s), by refluxing 17β-hydroxymethyl-Δ⁵-androsten-3β-ol (p. 1917 s) with aluminium isopropoxide and cyclohexanone in toluene (1939 Miescher).

21-Benzylidene-Δ⁴-pregnene-3.20-dione, 21-Benzylideneprogesterone $C_{28}H_{34}O_2$

(I, R = CH:CH·C₆H₅). Crystals (methanol), m. 155–8°. — **Fmn.** From 21-benzylidene-Δ⁵-pregnen-3β-ol-20-one (p. 2271 s) on refluxing with aluminium tert.-butoxide and acetone in dry toluene for 5 hrs. (1942 Marker).

21-Benzyl-Δ⁴-pregnene-3.20-dione, 21-Benzylprogesterone $C_{28}H_{36}O_2$ (I, R = CH₂·CH₂·C₆H₅). Crystals (ether-ligroin), m. 86–88°. — **Fmn.** From 21-benzyl-Δ⁵-pregnen-3β-ol-20-one (p. 2272 s) as above (1942 Marker).

21-Benzylidenepregnane-3.20-dione $C_{28}H_{36}O_2$. Needles (acetone or methanol), m. 212–4°. — **Fmn.** From the two 3-epimeric 21-benzylidenepregnan-3-ol-20-ones (p. 2272 s) on oxidation with CrO_3 in acetic acid at 25° (1939 Marker).

3-Keto-Δ⁴-ternorcholenyl phenyl ketone, 3-Keto-Δ⁴-pregnen-20α-yl phenyl ketone $C_{28}H_{36}O_2$. Needles (alcohol), m. 227–8° (1937 Butenandt); needles (acetone), m. 228° to 229° (1945 Cole); $[\alpha]_D^{20}$ +87° (dioxane) (1937 Butenandt), $[\alpha]_D^{25}$ +90° (chloroform) (1945 Cole). — Shows no physiological activity (1937 Butenandt). — **Fmn.** From (3β-hydroxy-Δ⁵-ternorcholenyl)-phenyl-carbinol (p. 1947 s) by bromination in glacial acetic acid, oxidation with CrO_3 in glacial acetic acid at room temperature, and debromination using zinc in glacial acetic acid at water-bath temperature; yield, 45% (1937 Butenandt). From 3β-hydroxy-Δ⁵-ternorcholenyl phenyl ketone (p. 2273 s) on refluxing with aluminium tert.-butoxide and cyclohexanone in toluene; yield, 70% (1945 Cole). — *Dioxime* $C_{28}H_{38}O_2N_2$ (no analysis given). Crystals (dil. alcohol), m. 208–9° (1937 Butenandt).

1937 Butenandt, Fleischer, *Ber.* **70** 96, 97, 98, 100.
1939 Marker, Wittle, *J. Am. Ch. Soc.* **61** 1329.
        Miescher, Wettstein, *Helv.* **22** 1262, 1266.
1942 Marker, Wittle, Jones, Crooks, *J. Am. Ch. Soc.* **64** 1282.
1945 Cole, Julian, *J. Am. Ch. Soc.* **67** 1369, 1372.

II. HALOGENO-DIOXOSTEROIDS WITH ONE CO IN THE RING SYSTEM

a. Halogen in Side Chain Only

21 - Chloro - Δ^4 - pregnene - 3.20 - dione, 21 - Chloroprogesterone $C_{21}H_{29}O_2Cl$

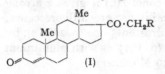

(I, R = Cl). Colourless leaflets (benzene - ether, or acetone - pentane), m. 203–5° cor. (1939 Reich), 206–8° cor. (1940d Reichstein); $[\alpha]_D^{24}$ +210°, $[\alpha]_{5461}^{24}$ +255° (both in chloroform); absorption max. in alcohol (from the graph) at 242 mμ (log ε 4.15) (1939 Reich). — Fmn. From 21-chloro-Δ^5-pregnen-3β-ol-20-one (p. 2277 s) on oxidation with aluminium tert.-butoxide and acetone in abs. benzene by refluxing for 24 hrs. (yield, 46%) (1939 Reich) or by keeping at room temperature for 20 days (yield, 73%) (1940a Reichstein). From 21-diazoprogesterone (p. 2816 s) on treatment with dry HCl in ether; yield, 68% (1940a Reichstein; 1940b N. V. ORGANON). From Δ^4-pregnen-21-ol-3.20-dione (desoxycorticosterone, p. 2820 s) on treatment with PCl$_5$ in chloroform in the presence of CaCO$_3$ at 0° (small yield) (1939 Reich) or, along with desoxycorticosterone p-toluenesulphonate, on treatment with p-toluenesulphonyl chloride (2 mols.) and pyridine (3 mols.) in dry chloroform at room temperature for 16 hrs. (1940d Reichstein); from the afore-mentioned p-toluenesulphonate by the action of ammonium chloride in acetone (1940c Reichstein).

Rns. Gives progesterone (p. 2782 s) on reduction with zinc dust in strong acetic acid at 100° (1937 CIBA) or on conversion into 21-iodo-progesterone (p. 2811 s) by refluxing with sodium iodide in acetone, followed by reduction with zinc dust in glacial acetic acid (1940d Reichstein). On refluxing with sodium acetate in glacial acetic acid it gives the acetate of Δ^4-pregnen-21-ol-3.20-dione (desoxycorticosterone, p. 2820 s) (1939 Reich); reacts similarly with sodium propionate in abs. alcohol (1938c CIBA). On heating with abs. pyridine on the water-bath for 1 hr. it gives **N-(3.20-diketo-Δ^4-pregnen-21-yl)-pyridinium chloride** $C_{26}H_{34}O_2NCl$ [I, R = NC$_5$H$_5$(Cl); colourless crystals (alcohol - benzene), m. 274–5° cor. dec.] which with p-ni-trosodimethylaniline in aqueous alcoholic NaOH at —10° gives the 21-N-(p-dimethylaminophenyl)-isoxime of Δ^4-pregnene-3.20-dion-21-al (p. 2870 s) (1939 Reich; 1940a N. V. ORGANON).

21 - Bromo - Δ^4 - pregnene - 3.20 - dione, 21 - Bromoprogesterone $C_{21}H_{29}O_2Br$ (I,

above; R=Br). Colourless leaflets (acetone - ether), m. 190–191° dec. Absorption max. in alcohol (from the graph) at 240 mμ (log ε 4.2) (1939 Reich). — **Fmn.** From 21-bromo-Δ^5-pregnen-3β-ol-20-one (p. 2278 s) on refluxing with aluminium tert.-butoxide and acetone in abs. benzene (20 hrs.) (1939 Reich). From the p-toluenesulphonate of Δ^4-pregnen-21-ol-3.20-dione (desoxycorticosterone, p. 2820 s) on refluxing with sodium bromide in acetone or methanol (1940c Reichstein). — **Rns.** Gives progesterone (p. 2782 s) on heating with zinc dust in strong acetic acid (1937 CIBA). On heating with abs. pyridine on the water-bath it gives **N-(3.20-diketo-Δ^4-pregnen-21-yl)-pyridinium bromide** $C_{26}H_{34}O_2NBr$ (not analysed) [I, p. 2810 s; R $=$ NC$_5$H$_5$(Br)], m. 265–8° cor. dec. (1939 Reich).

21-Iodo-Δ^4-pregnene-3.20-dione, 21-Iodoprogesterone $C_{21}H_{29}O_2I$ (not analysed) (I, p. 2810 s; R $=$ I). — **Fmn.** From 21-chloroprogesterone (above) or from the p-toluenesulphonate of Δ^4-pregnen-21-ol-3.20-dione (desoxycorticosterone, p. 2820 s) on treatment with sodium iodide in acetone (1940 c, d Reichstein). — **Rns.** Very readily splits off iodine, e.g. on addition of acetic acid (1940 c Reichstein). Gives progesterone (p. 2782 s) on reduction with zinc dust in glacial acetic acid (1940 d Reichstein; 1937 CIBA).

21-Chloropregnane-3.20-dione $C_{21}H_{31}O_2Cl$ (II, R $=$ Cl). Crystals (abs. alcohol), m. 185–9° cor. (1940 b Reichstein). — **Fmn.** From

21-chloropregnan-3α-ol-20-one (p. 2280 s) on oxidation with CrO$_3$ in glacial acetic acid at room temperature (16 hrs.); yield, 82 % (1940 b Reichstein). — **Rns.** Gives pregnane-3.20-dione (page 2796 s) on reduction with zinc dust in strong acetic acid (1938 a CIBA). Refluxing with potassium acetate in glacial acetic acid affords pregnan-21-ol-3.20-dione acetate (p. 2831 s) (1940 b Reichstein). On treatment with 1 mol. bromine in glacial acetic acid, followed by dehydrobromination of the resulting 4-bromo-derivative by refluxing with sodium formate in 90 % alcohol, it gives Δ^4-pregnen-21-ol-3.20-dione (desoxycorticosterone, p. 2820 s) whereas on dehydrobromination with potassium acetate, the acetate of this compound is obtained (1938 N. V. ORGANON; 1938 "DEGEWOP").

21-Bromopregnane-3.20-dione $C_{21}H_{31}O_2Br$ (II, R $=$ Br). A compound which probably possesses this structure is obtained in small amount, along with 17-bromopregnane-3.20-dione (p. 2814 s), from the mixture of bromopregnanolones [formed from pregnan-3β-ol-20-one (p. 2245 s) with 1 mol. bromine in ether] on oxidation with CrO$_3$ (1.2 at. oxygen) in glacial acetic acid in the cold; it forms colourless blocks (ethyl acetate and dil. acetone), m. 168°, [α]$_D^{20}$ +142° (chloroform) (1938 Masch).

21-Chloroallopregnane-3.20-dione $C_{21}H_{31}O_2Cl$ (III, R $=$ Cl). Crystals (acetone-ether), m. 186–194° cor. — **Fmn.** From 21-chloro-

allopregnan-3β-ol-20-one (p. 2282 s) on oxidation with CrO$_3$ in glacial acetic acid at room temperature (1940 b Reichstein).

21-Bromoallopregnane-3.20-dione $C_{21}H_{31}O_2Br$ (III, R $=$ Br). Colourless crystals (benzene), m. 177–9° cor. — **Fmn.** From 21-bromoallopregnan-3β-ol-20-one (p. 2282 s) on oxidation with CrO$_3$ in glacial acetic acid at room temperature (1940 b Reichstein).

References, p. 2812 s

21.21 - Dichloro - Δ^4 - pregnene - 3.20 - dione, 21.21 - Dichloroprogesterone

$C_{21}H_{28}O_2Cl_2$ (I, p. 2810 s; $CH_2R = CHCl_2$). Physical properties and mode of formation have not been described. — **Rns.** Gives progesterone (p. 2782 s) on reduction with zinc dust or sodium iodide in strong acetic acid on the water-bath (1941 CIBA).

21.21 - Dibromo - Δ^4 - pregnene - 3.20 - dione, 21.21 - Dibromoprogesterone

$C_{21}H_{28}O_2Br_2$ (I, p. 2810 s; $CH_2R = CHBr_2$). — **Fmn.** From 3-keto-Δ^4-ætio-cholenic acid chloride by the action of diazomethane, followed by bromination. — **Rns.** Gives Δ^4-pregnene-3.20-dion-21-al (p. 2870 s) on refluxing with $CaCO_3$ in dil. propanol for several hrs. (1938 b CIBA).

1937 CIBA, *Swiss Pat.* 222 122 (issued 1942); *C. A.* **1949** 366; *Ch. Ztbl.* **1943** I 2113.
1938 a CIBA, *Fr. Pat.* 835 814 (issued 1939); *C. A.* **1939** 5008; *Ch. Ztbl.* **1939** I 2827, 2828; *Brit. Pat.* 512 954 (issued 1939); *C. A.* **1941** 1800.
   b CIBA, *Fr. Pat.* 840 514 (issued 1939); *C. A.* **1939** 8627; *Ch. Ztbl.* **1939** II 1125, 1126.
   c CIBA, *Swiss Pat.* 222 901 (issued 1942); *C. A.* **1949** 1916; *Ch. Ztbl.* **1943** I 2114; CIBA PHARMACEUTICAL PRODUCTS (Miescher, Fischer), *U.S. Pat.* 2 265 183 (1939, issued 1941); *C. A.* **1942** 1738; *Ch. Ztbl.* **1945** I 1036.
   "DEGEWOP" GESELLSCHAFT WISSENSCHAFTLICHER ORGAN- UND HORMONPRÄPARATE (Reichstein), *Ger. Pat.* 737 539 (issued 1943); *C. A.* **1945** 5043; *Ch. Ztbl.* **1944** II 563.
   Masch, *Über Bromierungen in der Pregnanreihe*, Thesis, Danzig, pp. 14, 24.
   N. V. ORGANON, *Fr. Pat.* 835 669 (issued 1938); *Ch. Ztbl.* **1939** I 4089, 4090; *Indian Pat.* 25 074 (issued 1938); *Ch. Ztbl.* **1939** I 3034, 3035; ROCHE-ORGANON, INC. (Reichstein), *U.S. Pat.* 2 232 730 (issued 1941); *C. A.* **1941** 3773; *Ch. Ztbl.* **1941** II 2467.
1939 Reich, Reichstein, *Helv.* **22** 1124, 1127, 1134 et seq.
1940 a N. V. ORGANON, *Fr. Pat.* 864 315 (issued 1941); *Ch. Ztbl.* **1942** I 642; *Swiss Pat.* 221 311 (issued 1942); *Ch. Ztbl.* **1943** I 1297, 1298.
   b N. V. ORGANON (Reichstein), *Swed. Pat.* 102 320 (issued 1941); *Ch. Ztbl.* **1942** II 74; *Dan. Pat.* 61 323 (issued 1943); *Ch. Ztbl.* **1944** II 343.
   a Reichstein, v. Euw, *Helv.* **23** 136. — b Reichstein, Fuchs, *Helv.* **23** 658, 666, 668, 669.
   c Reichstein, Schindler, *Helv.* **23** 669, 670, 671.
   d Reichstein, Fuchs, *Helv.* **23** 684, 686, 687.
1941 CIBA, *Swiss Pat.* 224 833 (issued 1943); *C. A.* **1949** 1916; *Ch. Ztbl.* **1944** II 563.
   ROCHE-ORGANON, INC. (Reichstein), *U.S. Pat.* 2 312 483 (issued 1943); *C. A.* **1943** 4862; *Ch. Ztbl.* **1945** II 687.

---

b. Halogeno-dioxosteroids with One CO and Halogen in the Ring System

2 α-Bromoallopregnane-3.20-dione $C_{21}H_{31}O_2Br$ (p. 152). For α-configuration at C-2, assigned by Editor, see that of 2-bromo-cholestan-3-one (p. 2480 s). — Crystals, m. 199° to 201° (1939 Marker). — **Rns.** The "hetero-Δ^1-allo-pregnenedione" obtained by heating with potassium acetate (see pp. 152, 153) is actually Δ^5-pregnene-4.20-dione (p. 2801 s) (1944 Butenandt). On distillation of the pyridinium salt (below) under reduced pressure it gives approximately equal amounts of Δ^1-allopregnene-3.20-dione (p. 2781 s) and Δ^4-pregnene-3.20-dione (progesterone, p. 2782 s) (1939 Marker; cf. also 1940 SCHERING A.-G.; 1938 CIBA).

Pyridinium salt $C_{26}H_{36}O_2NBr$ (p. 153). Crystals (methanol - acetone), m. 300–302° dec.; very soluble in alcohols, sparingly soluble in acetone. — **Fmn.** From the above bromodione on refluxing with dry pyridine for 3 hrs. (1939 Marker). — **Rns.** See above.

4β-Bromopregnane-3.20-dione $C_{21}H_{31}O_2Br$ (p. 153). For β-configuration at C-4, assigned by Editor, see that of 4-bromocoprostan-3-one (p. 2483 s). — **Fmn.** For its formation from 4β-bromopregnan-20α-ol-3-one (p. 2514 s) by oxidation and from pregnane-3.20-dione (p. 2796 s) by monobromination, see also 1935 SCHERING-KAHLBAUM A.-G.; in the latter case, an isomer, probably 17-bromopregnane-3.20-dione (p. 2814 s), is obtained as a by-product (1942a Marker). — **Rns.** On oxidation with v. Baeyer's persulphate mixture in glacial acetic acid at 25° (10 days), followed by refluxing of the resulting product with dry pyridine for 10 hrs., refluxing with KHCO₃ in 94% methanol for ½ hr., and then by oxidation with CrO₃ in strong acetic acid at room temperature, it gives 3-keto-Δ⁴-ætiocholenic acid [formed from Δ⁴-pregnen-21-ol-3.20-dione (desoxycorticosterone, p. 2820 s) primarily obtained] and a neutral fraction which on refluxing with 1% methanolic KOH for ½ hr. gives Δ⁴-androsten-17β-ol-3-one (testosterone, p. 2580 s); Δ⁴-pregnene-3.20-dione (progesterone, p. 2782 s) has also been isolated from the reaction mixture (1940 Marker).

5-Chloroallopregnane-3.20-dione $C_{21}H_{31}O_2Cl$. For allo-configuration, see page 1884 s footnote **. — Needles (alcohol), m. 166° to 170° cor. with vigorous decomposition; $[\alpha]_D^{21}$ +102° (chloroform). Very sparingly soluble in ether and acetone. — **Fmn.** From 24.24-diphenyl-5-chloro-Δ²⁰⁽²²⁾·²³-allocholadien-3β-ol (p. 1884 s) on oxidation in chloroform with CrO₃ and 80% acetic acid at 20°. From 5-chloroallopregnan-3β-ol-20-one (p. 2284 s) on oxidation with CrO₃ in 90% acetic acid at 20°. — **Rns.** On refluxing with K₂CO₃ in aqueous alcohol for 1 hr. it gives Δ⁴-pregnene-3.20-dione (progesterone, page 2782 s) (1947 Meystre; 1946 CIBA PHARMACEUTICAL PRODUCTS).

6β-Bromo-Δ⁴-pregnene-3.20-dione, 6β-Bromoprogesterone $C_{21}H_{29}O_2Br$ (cf. also the following compound). Crystals (aqueous alcohol), m. 137–8° dec. — **Fmn.** From Δ⁵-pregnen-3β-ol-20-one (p. 2233 s) on conversion into the dibromide (p. 2289 s), followed by oxidation with CrO₃ in strong acetic acid and boiling of the resulting crude 5α.6β-dibromo-allopregnane-3.20-dione (p. 2814 s) with potassium acetate in alcohol - benzene for 15 min.; over-all yield, 28%. — **Rns.** Gives Δ⁴·⁶-pregnadiene-3.20-dione (p. 2778 s) on boiling with dry pyridine for 2½ hrs. (1937 SCHERING CORP.).

6-Bromo-Δ^4-pregnene-3.20-dione, 6-Bromoprogesterone, with unknown configuration at C-6 (possibly identical with the preceding compound) $C_{21}H_{29}O_2Br$. Crystals (acetone - hexane), m. 143–5⁰ dec.; $[\alpha]_D^{20}$ $+77^0$ (chloroform). Absorption max. in the ultra-violet (in 95 % alcohol) at 248 mμ (log ε 4.22), in the infra-red (in chloroform) at 1700 and 1670 cm.$^{-1}$. — **Fmn.** From progesterone (p. 2782 s) on refluxing with 1 mol. of N-bromosuccinimide in dry carbon tetrachloride for 1 hr.; yield, 41 %. — **Rns.** Gives 2α-acetoxyprogesterone (p. 2837 s) on refluxing with anhydrous potassium acetate in glacial acetic acid for 4 hrs. (1953 Sondheimer).

17-Bromopregnane-3.20-dione $C_{21}H_{31}O_2Br$. Crystals (dil. acetone and methanol), m. 145⁰; $[\alpha]_D^{21}$ —35⁰ (chloroform) (1938 Masch). — **Fmn.** From 17-bromopregnan-3β-ol-20-one (p. 2285 s) on oxidation with CrO_3 in glacial acetic acid in the cold; yield, ca. 45 % (1938 Masch; cf. 1942a Marker). Is probably formed as a byproduct, along with 4β-bromopregnane-3.20-dione (p. 2813 s), from pregnane-3.20-dione (p. 2796 s) on monobromination (1942a Marker). — **Rns.** On refluxing with anhydrous pyridine for 6 hrs. it gives Δ^{16}-pregnene-3.20-dione (p. 2794 s) (1938 Masch; 1942a Marker).

17-Bromoallopregnane-3.20-dione $C_{21}H_{31}O_2Br$. Not obtained in a pure state. — **Fmn.** From 17-bromoallopregnan-3β-ol-20-one (p. 2286 s) on oxidation with CrO_3 in strong acetic acid at room temperature. — **Rns.** On heating with iron dust and acetic acid on the steam-bath it yields allopregnane-3.20-dione (p. 2797 s). Gives Δ^{16}-allopregnene-3.20-dione (p. 2795 s) on refluxing with dry pyridine for 6 hrs. or with anhydrous potassium acetate in glacial acetic acid for 2 hrs. (1942b Marker).

21-Chloro-4-bromopregnane-3.20-dione $C_{21}H_{30}O_2ClBr$. For its formation and reactions, see under the reactions of 21-chloropregnane-3.20-dione (p. 2811 s).

5α.6β-Dibromoallopregnane-3.20-dione $C_{21}H_{30}O_2Br_2$. Colourless crystals (methanol), m. 78–80⁰ to a red liquid; after standing for 5 hrs. in a closed vial at 25⁰ the crystals become pink and give off bromine vapour (1945 Julian). — **Fmn.** From Δ^5-pregnen-3β-ol-20-one dibromide (p. 2289 s) on oxidation with CrO_3 in strong acetic acid at 20⁰ (yield from pregnenolone, 70 %) (1945 Julian; cf. 1934, 1936 Butenandt; 1937 SCHERING CORP.; 1941 I.G. FARBENIND.) or with $KMnO_4$ and dil. H_2SO_4 in benzene at room temperature (1934 Fernholz). — **Rns.** Gives Δ^4-pregnene-3.20-dione (progesterone, p. 2782 s) on heating with zinc dust and glacial acetic acid on the water-bath for 10 min. (1934, 1936 Butenandt; cf. 1934 Fernholz), on heating with zinc dust and sodium iodide in acetone (1941 I.G. FARBENIND.), on treatment with excess sodium iodide in alcohol for 10 hrs. at room temperature (yield, 65 %) or with chromous chloride solution (containing $ZnCl_2$ and hydrochloric acid) in acetone under CO_2 for 2 hrs. at 26° (yield, 90 %) (1945 Julian); on more energetic treatment with zinc dust and glacial acetic acid, $\varDelta^4$-pregnen-20-one (p. 2216 s) is also formed (1934 Butenandt). On boiling with potassium acetate in alcohol · benzene for 15 min. it gives 6β-bromo-$\varDelta^4$-pregnene-3.20-dione (p. 2813 s) (1937 Schering Corp.).

11.12-Dibromopregnane-3.20-dione $C_{21}H_{30}O_2Br_2$. Not isolated. — Fmn. Is

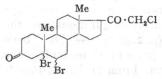

probably formed as a by-product from $\varDelta^{11}$-pregnene-3.20-dione (p. 2793 s) on treatment with N-bromoacetamide at 20°. — **Rns.** Reverts to $\varDelta^{11}$-pregnene-3.20-dione on debromination with zinc dust in glacial acetic acid (1943 Hegner).

21-Chloro-5.6-dibromopregnane-3.20-dione $C_{21}H_{29}O_2ClBr_2$. — Fmn. From

21-chloro-$\varDelta^5$-pregnen-3β-ol-20-one (p. 2277 s) on bromination with 1 mol. bromine in chloroform, followed by oxidation with CrO_3 in glacial acetic acid at room temperature (1938 N. V. Organon; 1938, 1941 Roche-Organon, Inc.; 1937 Steiger; 1937 Ciba). — **Rns.** Gives $\varDelta^4$-pregnene-3.20-dione (progesterone, p. 2782 s) on treatment with zinc dust, sodium iodide, or potassium iodide in glacial acetic acid (1937 Steiger; 1937 Ciba). On heating with fused sodium acetate in glacial acetic acid on the water-bath for several hrs., followed by debromination with zinc dust in glacial acetic acid, it gives the acetate of $\varDelta^4$-pregnen-21-ol-3.20-dione (desoxycorticosterone, p. 2820 s) (1938 N. V. Organon; 1938, 1941 Roche-Organon, Inc.).

1934 Butenandt, Westphal, *Ber.* **67** 2085.
    Fernholz, *Ber.* **67** 2027, 2029.
1935 Schering-Kahlbaum A.-G., *Fr. Pat.* 806467 (issued 1936); *C. A.* **1937** 4682; *Australian Pats.* 24429/1935 (issued 1936); 3961/1936 (issued 1936); *Ch. Ztbl.* **1937** I 3023, 3024, 4264, 4265.
1936 Butenandt, Westphal, *Ber.* **69** 443, 447.
1937 Ciba, *Swiss Pat.* 215138 (issued 1941); *C. A.* **1948** 3145.
    Schering Corp. (Inhoffen, Butenandt, Schwenk), *U.S. Pat.* 2340388 (issued 1944); *C. A.* **1944** 4386; Schering-Kahlbaum A.-G., *Fr. Pat.* 835524 (issued 1938); *Brit. Pat.* 500353 (issued 1939); *C. A.* **1939** 5872; *Ch. Ztbl.* **1939** I 5010, 5011.
    Steiger, Reichstein, *Helv.* **20** 1164, 1168.
1938 Ciba (Ruzicka), *Swed. Pat.* 98368 (issued 1940); *Ch. Ztbl.* **1940** II 2185.
    Masch, *Über Bromierungen in der Pregnanreihe*, Thesis, Danzig, pp. 22, 23.
    N. V. Organon, *Fr. Pat.* 835669 (issued 1938); *C. A.* **1939** 8626; *Ch. Ztbl.* **1939** I 4089, 4090; *Indian Pat.* 25064 (issued 1938); *Ch. Ztbl.* **1939** I 2458; *Brit. Pat.* 502289 (issued 1939); *C. A.* **1939** 6346.
    Roche-Organon, Inc. (Reichstein), *U.S. Pat.* 2312481 (issued 1943); *C. A.* **1943** 5201; *Ch. Ztbl.* **1945** II 687, 688.
1939 Marker, Wittle, Plambeck, *J. Am. Ch. Soc.* **61** 1333.
1940 Marker, *J. Am. Ch. Soc.* **62** 2543, 2547.
    Schering A.-G., *Fr. Pat.* 867697 (issued 1941); *Ch. Ztbl.* **1942** I 2561; *Dan. Pat.* 61322 (issued 1943); *Ch. Ztbl.* **1944** II 49.
1941 I. G. Farbenind., *Swiss Pat.* 227974 (issued 1943); *Ch. Ztbl.* **1945** I 196; *Fr. Pat.* 886984 (1942, issued 1943); *Ch. Ztbl.* **1944** II 1091.

ROCHE-ORGANON, INC. (Reichstein), *U.S. Pat.* 2312483 (issued 1943); *C. A.* **1943** 4862; *Ch. Ztbl.* **1945** II 687.

1942 a Marker, Crooks, Wagner, *J. Am. Ch. Soc.* **64** 210, 213.

  b Marker, Crooks, Wagner, Wittbecker, *J. Am. Ch. Soc.* **64** 2089, 2091.

1943 Hegner, Reichstein, *Helv.* **26** 721, 723, 725, 727.

  Shoppee, Prins, *Helv.* **26** 1004, 1008 footnote 2.

1944 Butenandt, Ruhenstroth-Bauer, *Ber.* **77** 397, 400.

1945 Julian, Cole, Magnani, Meyer, *J. Am. Ch. Soc.* **67** 1728, 1729; GLIDDEN Co. (Julian, Cole, Magnani, Conde), *U.S. Pat.* 2374683 (issued 1945); *C. A.* **1946** 1636; *Ch. Ztbl.* **1946** I 372.

1946 CIBA PHARMACEUTICAL PRODUCTS (Miescher, Frey, Meystre, Wettstein), *U.S. Pat.* 2461911 (issued 1949); *C. A.* **1949** 3475; *Ch. Ztbl.* **1949** E 1870, 1871.

1947 Meystre, Wettstein, Miescher, *Helv.* **30** 1022, 1026, 1027.

1953 Sondheimer, Kaufmann, Romo, Martinez, Rosenkranz, *J. Am. Ch. Soc.* **75** 4712, 4715.

III. DIAZO-DIOXOSTEROIDS WITH ONE CO IN THE RING SYSTEM

21 - Diazo - 19-nor * - 14-iso-17-iso-Δ^4-pregnene-3.20-dione, 21-Diazo-19-nor * - 14-iso-17-iso-progesterone $C_{20}H_{26}O_2N_2$. For the normal (β) configuration at C-10, see 1958 Djerassi; for the other configurations, see the parent compound (p. 2292 s). — Has been obtained crystalline (m. 143–4° dec.) after the closing date for this volume (1957 Barber). — **Fmn.** From 21-diazo-19-nor * - 14-iso-17-iso-$\Delta^{5(10?)}$-pregnen-3β-ol-20-one (p. 2292 s) on refluxing with dry acetone and aluminium tert.-butoxide in dry benzene for 10 hrs.; quantitative yield (1944 Ehrenstein). — **Rns.** On heating with glacial acetic acid on the water-bath it gives the acetate of 19-nor * - 14-iso-17-iso-11-desoxycorticosterone (p. 2817 s) (1944 Ehrenstein; 1957 Barber).

21-Diazo-Δ^4-pregnene-3.20-dione, 21-Diazoprogesterone $C_{21}H_{28}O_2N_2$. Light yellow crystals (ether), m. 182–4° cor. dec. (1940 a Reichstein), 177–178° cor. dec. (1948 Wilds). Sparingly soluble in ether, moderately soluble in acetone, more readily in benzene (1940 a Reichstein). — **Fmn.** From 21-diazo-Δ^5-pregnen-3β-ol-20-one (p. 2293 s) with aluminium tert.-butoxide and acetone in abs. benzene on keeping in a sealed tube for 20 days at room temperature or on refluxing for 14 hrs.; yield in both cases, 70% (1940 a Reichstein). From 3-keto-Δ^4-ætiocholenic acid chloride on treatment with diazomethane in ether - benzene (1937 a, 1938 CIBA; cf. 1948 Wilds).

Rns. Gives 21-chloroprogesterone (p. 2810 s) on treatment with dry HCl in abs. ether (1940 a Reichstein). On heating with 2 N H_2SO_4 in dioxane at 75° for 4 hrs. it yields Δ^4-pregnen-21-ol-3.20-dione (desoxycorticosterone, p. 2820 s) (1955 Salem; cf. 1938 CIBA; 1940 a Reichstein); boiling with glacial acetic acid for 3 min. affords desoxycorticosterone acetate (1940 a Reichstein; 1938 CIBA), and heating with p-toluenesulphonic acid in abs. benzene at 50° the corresponding p-toluenesulphonate (1940 b Reichstein).

* The original designation 10-nor was later changed to 19-nor (see 1951 Ehrenstein).

21-Diazo-Δ^{11}-pregnene-3.20-dione $C_{21}H_{28}O_2N_2$. See in the article "Δ^{11}-Pregnene-3α.21-diol-20-one 21-acetate", p. 2309 s.

3-Keto-Δ^4-ternorcholenyl diazomethyl ketone $C_{23}H_{32}O_2N_2$. Light yellow crystals (benzene - hexane). — **Fmn.** From 3-keto-Δ^4-bisnorcholenic acid chloride on treatment with diazomethane in ether - benzene (1937 b CIBA).

1937 a CIBA, *Swiss Pat.* 211650 (issued 1941); *C. A.* **1942** 3638.
   b CIBA, *Swiss Pat.* 214534 (issued 1941); *C. A.* **1942** 4978; *Ch. Ztbl.* **1942** I 2430.
1938 CIBA, *Fr. Pat.* 840417 (issued 1939); *Ch. Ztbl.* **1939** II 1125.
1940 a Reichstein, v. Euw, *Helv.* **23** 136; N. V. ORGANON (Reichstein), *Swed. Pat.* 102320 (issued 1941); *Ch. Ztbl.* **1942** II 74; *Dan. Pat.* 61323 (issued 1943); *Ch. Ztbl.* **1944** II 343.
   b Reichstein, Schindler, *Helv.* **23** 669, 673; ROCHE-ORGANON, INC. (Reichstein), *U.S. Pat.* 2357224 (issued 1944); *C. A.* **1945** 391; *Ch. Ztbl.* **1945** II 1232.
1944 Ehrenstein, *J. Org. Ch.* **9** 435, 449.
1948 Wilds, Shunk, *J. Am. Ch. Soc.* **70** 2427.
1951 Ehrenstein, Barber, Gordon, *J. Org. Ch.* **16** 349, 355 footnote 7.
1955 Salem, Zorbach, *J. Am. Ch. Soc.* **77** 1055.
1957 Barber, Ehrenstein, *Ann.* **603** 89, 97, 106.
1958 Djerassi, Ehrenstein, Barber, *Ann.* **612** 93, 97.

IV. HYDROXY-DIOXOSTEROIDS WITH ONE CO IN THE RING SYSTEM AND OH IN SIDE CHAIN

19-Nor*-14-iso-17-iso-Δ^4-pregnen-21-ol-3.20-dione, 19-Nor*-14-iso-17-iso-11-desoxycorticosterone** $C_{20}H_{28}O_3$. For configurations, see the parent compound (page 2816 s). — Platelets (ether - petroleum ether, and acetone - water), m. 135–6°; $[\alpha]_D^{23}$ +64° (chloroform); absorption max. in alcohol at 240 mμ (ε 15600) (1957 Barber). — **Fmn.** The acetate is obtained from 21-diazo-19-nor*-14-iso-17-iso-progesterone (p. 2816 s) on heating with glacial acetic acid for ¹/₂ hr. on the water-bath (1944 Ehrenstein; 1957 Barber); it is saponified with aqueous methanolic KHCO₃ at room temperature (1957 Barber).

Acetate $C_{22}H_{30}O_4$. Rodlets (aqueous methanol), m. 130–131°; $[\alpha]_D^{21.5}$ +51° (chloroform); absorption max. in alcohol at 240 mμ (ε 17000) (1957 Barber). — Has no cortin-like activity (1944 Ehrenstein). — **Fmn.** See above. — **Rns.** Reduces alkaline silver diammine in methanol after short standing at room temperature. The optical rotation remains practically unchanged after refluxing with conc. HCl in alcohol for 30 min., followed by reacetylation (1944 Ehrenstein).

---

* See the footnote on p. 2816 s.
** For "normal" 19-nor-11-desoxycorticosterone, described after the closing date for this volume, see 1953, 1955 Sandoval.

1944 Ehrenstein, *J. Org. Ch.* **9** 435, 440–443, 449–452.
1953 Sandoval, Miramontes, Rosenkranz, Djerassi, Sondheimer, *J. Am. Ch. Soc.* **75** 4117.
1955 Sandoval, Thomas, Djerassi, Rosenkranz, Sondheimer, *J. Am. Ch. Soc.* **77** 148.
1957 Barber, Ehrenstein, *Ann.* **603** 89, 107, 108.

$Δ^{4.6}$-Pregnadien-21-ol-3.20-dione, 6-Dehydrodesoxycorticosterone $C_{21}H_{28}O_3$.

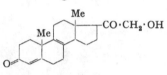

Acetate $C_{23}H_{30}O_4$. Needles (hexane + some acetone), m. 115–6° cor.; $[α]_D^{18} +151.5°$ (alcohol). Absorption max. (solvent not stated) at 283 mμ (log ε 4.53). — Shows hardly any cortin-like action. — **Fmn.** From the 21-acetate of $Δ^5$-pregnene-3β.21-diol-20-one (p. 2301 s) on refluxing with aluminium tert.-butoxide and benzoquinone in abs. toluene for 1 hr.; small yield (1940 Wettstein; 1941 CIBA).

$Δ^{4.8(?)}$-Pregnadien-21-ol-3.20-dione, 8(?)-Dehydrodesoxycorticosterone

$C_{21}H_{28}O_3$. Originally the acetate was tentatively formulated as $Δ^{4.11}$-pregnadien-21-ol-3.20-dione acetate (11-anhydrocorticosterone acetate) (1943 Shoppee; see also 1942a N. V. ORGANON; 1943 Reichstein); the latter, however, described after the closing date for this volume (1948 v. Euw; 1948 Meystre; 1949 Wettstein), shows m. p. 180–181° cor., $[α]_D +178°$ (acetone), and has a very high cortin-like potency. For formulation of the present acetate as a $Δ^{4.8}$-diene-compound, see 1942 Reichstein; cf. also 1948 Meystre; 1953 Casanova. According to Fieser (1949) it is probably $Δ^{4.14}$-pregnadien-21-ol-3.20-dione acetate; a not analysed substance to which this structure has also been tentatively ascribed [plates (acetone - ether), m. 142–4° cor., giving an orange solution in conc. H_2SO_4 with a strong green fluorescence] was obtained in small yield from the (amorphous) 21-acetate of 14-isopregnane-14.21-diol-3.20-dione with bromine in glacial acetic acid at 0°, followed by heating with pyridine at 135° (1947 Meyer).

Acetate $C_{23}H_{30}O_4$. Needles (ether), m. 142–3° cor.; $[α]_D^{15} +98°$, $[α]_{5461}^{15} +130°$ (both in acetone). Absorption max. in alcohol at 239 mμ (log ε 4.21). Gives a green fluorescence with conc. H_2SO_4 and a yellow coloration with tetranitromethane in chloroform (1943 Shoppee). — It is at least 2–3 times less potent than desoxycorticosterone acetate in the Everse - de Fremery test (1943 Shoppee). — **Fmn.** Along with the $Δ^{4.9(11)}$-isomer (below), from the 21-acetate of $Δ^4$-pregnene-11β.21-diol-3.20-dione (corticosterone, p. 2853 s) on refluxing with conc. HCl · glacial acetic acid (1:4 by vol.) for 30 min., followed by reacetylation with acetic anhydride in abs. pyridine at 20° (yield, 17%), or from the afore-mentioned isomer in the same way (yield, ca. 50%) (1943 Shoppee). From $Δ^{11}$-pregnen-21-ol-3.20-dione (p. 2830 s) on treatment with 2 mols. of bromine in glacial acetic acid, followed by refluxing with abs. pyridine for 5 hrs. or by heating with dimethylaniline for 2 hrs. at 150°, and then by heating with zinc dust in glacial acetic acid for 15 min. at 80° (1942a N. V. ORGANON; 1943 Reichstein).

$\Delta^{4 \cdot 9(11)}$ - Pregnadien - 21 - ol - 3.20 - dione, 9 (11) - Dehydrodesoxycorticosterone, 9 (11) - Anhydrocorticosterone $C_{21}H_{28}O_3$. For the structure, see also 1948 Meystre; 1953 Casanova.

Acetate $C_{23}H_{30}O_4$. Prisms (ether), m. 159° to 160° cor.; $[\alpha]_D^{18}$ +129°, $[\alpha]_{5461}^{18}$ +150° (both in acetone). Absorption max. in alcohol at 240 mμ (log ε 4.21). With conc. H_2SO_4 it gives an orange-yellow solution with a strong green fluorescence. Gives a yellow coloration with tetranitromethane in chloroform. Its solution in methanol strongly and immediately reduces alkaline silver diammine solution at room temperature (1943 Shoppee). — Has the 2–3 fold potency of desoxycorticosterone acetate in the Everse - de Fremery test (1943 Shoppee).

Fmn. From the 21-acetate of Δ^4-pregnene-11β.21-diol-3.20-dione (corticosterone, p. 2853 s) on refluxing with conc. HCl - glacial acetic acid (1:9 by vol.) for 30 min., followed by reacetylation with acetic anhydride in abs. pyridine at 20° for 16 hrs. (yield, 35–40% allowing for recovered starting material) (1943 Shoppee), along with 5–10% of a crystalline compound m. 122–5°, originally assumed to be 11-epi-corticosterone 21-acetate (1943 Shoppee), but which was probably impure corticosterone acetate (Gallagher in 1949 Fieser; cf. also 1953 Eppstein); when corticosterone 21-acetate is refluxed with conc. HCl - glacial acetic acid (1:4 by vol.) for 30 min. and the product is reacetylated, there are obtained 26% of the present acetate, 17% of the preceding isomer, and a trace of a substance crystallizing as plates m. 169° cor. (1943 Shoppee; cf. 1942 Reichstein).

From $\Delta^{9(11)}$-pregnen-21-ol-3.20-dione acetate (p. 2830 s) on treatment with 2 mols. of bromine in glacial acetic acid, followed by boiling for 5 hrs. with abs. pyridine or by heating for 2 hrs. with dimethylaniline at 150°, and then by heating for 15 min. with zinc dust in glacial acetic acid at 80° (1942b N. V. ORGANON; 1943 Reichstein).

Rns. The acetate is scarcely attacked on refluxing with conc. HCl - glacial acetic acid (1:9 by vol.) for 30 min. (followed by reacetylation); on refluxing with conc. HCl - glacial acetic acid (1:4 by vol.) for 30 min. (followed by reacetylation) it passes, to an extent of ca. 50%, into the preceding isomer (1943 Shoppee).

1940 Wettstein, *Helv.* **23** 388, 391, 394, 396.
1941 CIBA (Soc. pour l'ind. chim. à Bâle; Ges. f. chem. Ind. in Basel), *Fr. Pat.* 877821 (issued 1943); *Dutch Pat.* 54147 (issued 1943); *Ch. Ztbl.* **1943** II 147, 148.
1942 a N. V. ORGANON, *Swiss Pat.* 254992 (issued 1949); *Ch. Ztbl.* **1950** I 2388.
   b N. V. ORGANON, *Swiss Pat.* 255307 (issued 1949); *Ch. Ztbl.* **1949** E 1529.
   Reichstein, *U.S. Pat.* 2409798 (issued 1946); *C. A.* **1947** 1397; *Ch. Ztbl.* **1947** 232; *Fr. Pat.* 52035 (issued 1943); *Ch. Ztbl.* **1945** I 195, 196; *Brit. Pat.* 560812 (1944, issued 1945); *C. A.* **1946** 4486; *Ch. Ztbl.* **1945** II 1773.
1943 Reichstein, *U.S. Pat.* 2401775 (issued 1946); *C. A.* **1946** 5884; *Ch. Ztbl.* **1947** 74; *U.S. Pat.* 2404768 (1945, issued 1946); *C. A.* **1946** 6222; *Ch. Ztbl.* **1947** 233.
   Shoppee, Reichstein, *Helv.* **26** 1316.
1947 Meyer, Reichstein, *Helv.* **30** 1508, 1521.

1948 v. Euw, Reichstein, *Helv.* **31** 2076.
Meystre, Wettstein, *Helv.* **31** 1890, 1898.
1949 Fieser, Fieser, *Natural Products Related to Phenanthrene* 3rd Ed., New York, p. 409.
Wettstein, Meystre, *Helv.* **32** 880.
1953 Casanova, Shoppee, Summers, *J. Ch. Soc.* **1953** 2983.
Eppstein, Meister, Peterson, Murray, Leigh, Lyttle, Reineke, Weintraub, *J. Am. Ch. Soc.*
75 408, 409.

**Δ⁴-Pregnen-21-ol-3.20-dione, 11-Desoxycorticosterone, 21-Hydroxyprogester-
one, Cortexone,** *Reichstein's Substance Q*

$C_{21}H_{30}O_3$ (p. 153). Crystals (acetone - ether),
m. 140–3° cor.; sublimes in a high vacuum at
190° (1937h N. V. ORGANON). Absorption max.
in alcohol at 240 mμ (ε 19000) (1939 Dannenberg). Solubility in water at 18° 0.12 °/₀₀ (1942 Miescher).

Occ. For isolation of desoxycorticosterone from adrenal cortex, see 1938 a
ROCHE-ORGANON, INC.

FORMATION OF DESOXYCORTICOSTERONE*

From a mixture of 17-iso-Δ⁵-pregnene-3β.4β.17β.20.21-pentol and 17-iso-
Δ⁴-pregnene-3β.6β.17β.20.21-pentol (see p. 2201 s) on heating with conc. HCl
in alcohol for ½ hr. at 60° (1936d CIBA).

In the preparation of the acetate from the 21-acetate of Δ⁵-pregnene-
3β.21-diol-20-one (p. 2301 s) by oxidation of its dibromide with CrO_3 in
glacial acetic acid, followed by debromination (see p. 153), the latter is effected by warming with zinc dust and anhydrous sodium acetate at 60°,
followed by addition of glacial acetic acid and heating for 10 min. to boiling
(1937 Steiger; 1937h N. V. ORGANON; 1938b ROCHE-ORGANON, INC.) or, preferably, by treatment with aqueous chromous chloride solution (containing
$ZnCl_2$ and hydrochloric acid) in acetone under CO_2 at room temperature
(yield from the diolone acetate, 80%) (1945 Julian); other esters are similarly
prepared from the dibromides of the appropriate 21-esters of the diolone
by oxidation with CrO_3 in strong acetic acid at room temperature, followed
by debromination with zinc dust in methanol, first at room temperature,
then at 70° (1937a, b, c, h N. V. ORGANON); when the 21-triphenylmethyl
ether of the diolone is similarly treated, the triphenylmethyl group is for the
most part eliminated on effecting the debromination by heating with zinc
dust and sodium acetate in alcohol on the water-bath for 1 hr., and completely by subsequent warming with aqueous alcoholic HCl, to give free
desoxycorticosterone (1937h N. V. ORGANON; 1938b ROCHE-ORGANON, INC.).
Esters of desoxycorticosterone may also be obtained directly from the corresponding 21-esters of Δ⁵-pregnene-3β.21-diol-20-one by Oppenauer oxidation:
e.g., the acetate, using aluminium tert.-butoxide and acetone in benzene
(refluxing for 1½ hrs.; yield, 76%) (1937d N. V. ORGANON) or the same

---

* For a synthesis of desoxycorticosterone, described after the closing date for this volume,
see 1957 Romo.

aluminate and cyclohexanone in toluene (refluxing for 5 hrs.) (1939c SCHERING
A.-G.), the propionate and the butyrate, using aluminium isopropoxide and
cyclohexanone in toluene (refluxing for some time; yields, 70%) (1937e, f
N. V. ORGANON), the palmitate as in the preceding case (1937g N. V. ORGANON;
1939c, 1944 SCHERING A.-G.), and similarly the benzoate (1939c SCHERING
A.-G.). Free desoxycorticosterone is obtained from the 21-acetate of Δ^5-preg-
nene-3β.21-diol-20-one by the action of the Corynebacterium mediolanum
in yeast water on shaking under oxygen for 6 days at 36–37°; yield, 34%
(47%, allowing for recovered starting material) (1939 Mamoli).

From Δ^4-pregnene-17α.20β.21-triol-3-one (p. 2725 s) on refluxing with sulphuric acid in aqueous dioxane (1937 SCHERING A.-G.). From 17-iso-Δ^4-preg-
nene-17β.20a.21-triol-3-one (p. 2726 s) on sublimation with zinc dust in a
high vacuum at 150–170° (yield, 18%) or on refluxing with zinc dust in
xylene for 2 hrs. (yield, 74%) (1939a SCHERING A.-G.); the acetate is similarly
obtained from the 20.21-diacetate of the same triolone on sublimation with
zinc dust in a high vacuum at 150–200° (yield, 60%) (1939 Serini; cf. 1939a
SCHERING A.-G.) or on refluxing with zinc dust in toluene for 16 hrs. (almost
quantitative yield) (1940 CIBA).

The acetate is obtained from Δ^4-pregnene-3.20-dione (progesterone, page
2782 s) on heating with lead tetraacetate in glacial acetic acid at 75–85°
for 7 hrs.; yield, less than 3% (1939 Reichstein; 1944 v. Euw; cf. 1939a, b
Ehrhart; 1937 I.G. FARBENIND.). The acetate is also obtained from 21-chloro-
progesterone (p. 2810 s) on refluxing with fused sodium acetate in glacial
acetic acid for 1 hr. (yield, 22.5%) (1939 Reich); the propionate is similarly
obtained with sodium propionate in abs. alcohol (1938b CIBA; 1939 CIBA
PHARMACEUTICAL PRODUCTS). The acetate is formed from 21-chloro-5.6-di-
bromopregnane-3.20-dione (p. 2815 s) on heating with fused sodium acetate
in glacial acetic acid on the water-bath for several hrs., followed by addition
of powdered zinc and further heating for ca. $1/2$ hr. (1938b ROCHE-ORGANON,
INC.). From 21-chloropregnane-3.20-dione (p. 2811 s) on treatment with 1 mol.
bromine in glacial acetic acid, followed by refluxing of the resulting 4-bromo-
derivative with sodium formate in 90% alcohol for 12 hrs.; the acetate is
obtained when the bromo-compound is refluxed with potassium acetate in
glacial acetic acid (yield in both cases, 50%) (1938 "DEGEWOP"; 1938 N. V.
ORGANON).

Desoxycorticosterone and its esters are obtained from 21-diazo-Δ^4-preg-
nene-3.20-dione (21-diazoprogesterone, p. 2816 s) by the action of acids: free
desoxycorticosterone by heating with dil. H_2SO_4 in benzene (1938a CIBA)
or with 2 N H_2SO_4 in dioxane at 40° (1940 N. V. ORGANON), the acetate by
boiling with glacial acetic acid for 3 min. (1940a Reichstein; 1940 N. V.
ORGANON), the benzoate by heating with benzoic acid in toluene (1940
N. V. ORGANON), the p-toluenesulphonate by heating with p-toluenesulphonic acid in abs. benzene at 50° (yield, 50%) (1940c Reichstein; 1941
ROCHE-ORGANON, INC.), the phosphate by heating with anhydrous phosphoric
acid in anhydrous dioxane at 45–50° (1940c Reichstein; 1940 N. V. ORGANON).

The acetate is obtained from that of 17-iso-desoxycorticosterone (p. 2829 s)
on refluxing with conc. HCl in alcohol for $1/2$ hr., followed by reacetylation

with acetic anhydride in abs. pyridine at room temperature; yield, 65 % (1940 Shoppee). The acetate is also obtained from that of pregnan-21-ol-3.20-dione (p. 2831 s) on treatment with 1 mol. bromine in glacial acetic acid at room temperature, followed by refluxing of the resulting 4-bromo-derivative with abs. pyridine for 5 hrs. (yield, 10 %) (1940b Reichstein; cf. 1938 "Degewop"; 1938 N. V. Organon) or with anhydrous potassium acetate in glacial acetic acid for 4 hrs. (1938 "Degewop"). From Δ^4-pregnen-20-ol-3-on-21-al (p. 2832 s) on refluxing with abs. pyridine in an atmosphere of CO_2 for $5\frac{1}{2}$ hrs. (isolated as acetate; yield, 24 %) (1941 Schindler; 1941 Reichstein).

Free desoxycorticosterone may be obtained from its esters by hydrolysis with aqueous HCl on refluxing for 40 min. (1937 Steiger; 1937h N. V. Organon; 1938b Roche-Organon, Inc.) or with $KHCO_3$ or K_2CO_3 in aqueous methanol at room temperature (16 hrs.) or on refluxing for 1 hr. (1938a, b Reichstein; 1937h N. V. Organon).

REACTIONS OF DESOXYCORTICOSTERONE

Desoxycorticosterone is oxidized to Δ^4-pregnene-3.20-dion-21-al (p. 2870 s) on refluxing with aluminium isopropoxide and acetone or cyclohexanone in benzene or toluene, resp. (1938 Schering A.-G.) or similarly using aluminium tert.-butoxide and diethyl ketone (1939c Ciba); the tris-phenylhydrazone of the same dional is formed, when desoxycorticosterone is heated with phenyl-hydrazine in dil. acetic acid on the water-bath (1939c Ciba). When silver oxide (instead of silver carbonate) is used for the condensation of desoxy-corticosterone with acetobromocarbohydrates to give the acetylglycosides (see pp. 2825 s, 2826 s), 3-keto-Δ^4-ætiocholenic acid is formed as a by-product (1942 Miescher). For oxidation with N-bromoacetamide in aqueous tert.-butyl alcohol, see 1943 Reich. On bromination by refluxing with N-bromosuccinimide in dry carbon tetrachloride for 10 min. with illumination, followed by refluxing of the resulting crude 6-bromo-derivative with anhydrous potassium acetate in glacial acetic acid for 4 hrs., it gives a small amount of 2α-hydroxydesoxycorticosterone diacetate (p. 2852 s) (1953 Sondheimer).

On hydrogenation of the acetate in alcohol in the presence of palladium - calcium carbonate (followed by reacetylation), it gives the acetates of pregnan-21-ol-3.20-dione (main product; p. 2831 s) and allopregnan-21-ol-3.20-dione (p. 2832 s) (1940 Wettstein). The hydrogenation in the presence of platinum oxide was probably not carried out with desoxycorticosterone acetate as stated on p. 169 of the paper of Steiger and Reichstein (1938), but with the 21-acetate of Δ^5-pregnene-3β.21-diol-20-one (p. 2301 s) (see p. 163 of the same paper, and 1957 Reichstein).

For bromination, see above. On treatment of desoxycorticosterone with PCl_5 and $CaCO_3$ in chloroform at 0° it gives a small yield of 21-chloro-Δ^4-preg-nene-3.20-dione (21-chloroprogesterone, p. 2810 s) (1939 Reich); the same compound, along with desoxycorticosterone p-toluenesulphonate, is obtained on treatment of desoxycorticosterone with 2 mols. of p-toluenesulphonyl chloride and 3 mols. of pyridine in dry chloroform (1 vol. pyridine : 9 vols. chloroform) at room temperature for 16 hrs.; both the reaction products give 21-iodoprogesterone (p. 2811 s) on boiling with sodium iodide in acetone for 5 min. (1940d Reichstein).

The esters give the corresponding 3-enol esters (diesters of $\Delta^{3.5}$-pregnadiene-3.21-diol-20-one, p. 2300 s) on boiling with an acid anhydride + fused potassium acylate or on heating with an acid chloride and pyridine in an inert solvent (1936b, c CIBA).

Biological reactions. Desoxycorticosterone is converted in the organism into pregnane-3α.20α-diol (p. 1923 s) after administration of its acetate: orally to men or women with Addison's disease, to a hypogonadal man (1944 Horwitt), to an ovariectomized chimpanzee (1943 Fish; 1944 Horwitt), subcutaneously to male or female rabbits (1941, 1942 Westphal; 1943 Hoffman); the alleged conversion after intramuscular administration to a normal man (1940 Cuyler) could not be confirmed (1942 Cuyler); the conversion does not take place in normal women (1941 Hamblen).

PHYSIOLOGICAL PROPERTIES OF DESOXYCORTICOSTERONE

For physiological properties of desoxycorticosterone, see p. 153, pp. 1367 s, 1368 s, and the literature cited there.

ANALYTICAL PROPERTIES OF DESOXYCORTICOSTERONE

With conc. H_2SO_4 in alcohol the acetate gives a yellow ring in a green-fluorescent colour zone, with conc. H_2SO_4 + fural a light purple to brown-violet ring and under it a green-yellow fluorescent ring (1939 Woker). In the Zimmermann test with m-dinitrobenzene and KOH in alcohol desoxycorticosterone develops a deep violet colour which in a few seconds changes to olive-brown (1944 Zimmermann); for absorption spectrum of the colour obtained and for use of the test for photometric determination, see 1943 Hansen; 1944 Zimmermann. Desoxycorticosterone gives a weak pink coloration with 1.4-naphthoquinone + conc. HCl in glacial acetic acid at 60–70° (1939 Miescher). The acetate, dissolved in methanol, very rapidly gives a precipitate with ammoniacal silver salt solution at room temperature (1939 Miescher). Desoxycorticosterone and its acetate strongly and rapidly reduce alkaline copper solution as well as phosphomolybdic acid (Folin-Wu reagent) (1946 Heard); for the rate of the latter reaction in glacial acetic acid at 100° (giving molybdenum blue) and its use for colorimetric estimation, see 1946 Heard.

ESTERS OF DESOXYCORTICOSTERONE

Phosphate $C_{21}H_{31}O_6P$. — **Fmn.** For its formation from 21-diazoprogesterone, see p. 2821 s. — *Monosodium salt* $C_{21}H_{30}O_6PNa$. Almost colourless, amorphous or micro-crystalline powder. Readily soluble in water and alcohol, very

E.O.C. XIV s. 194

sparingly in acetone, almost insoluble in ether and benzene. Is obtained from the phosphate by treatment with sodium methoxide in methanol (1940c Reichstein).

Acetate $C_{23}H_{32}O_4$ (p. 153). Needles (acetone - ether), m. 157.5–158° cor. (1940 Shoppee); needles (by sublimation in a high vacuum), m. 157–9° (1936a Ciba; 1944 Nowacki); needles (acetone - petroleum ether), m. 158–160° cor. (1937h N. V. Organon); crystals (dil. methanol), m. 161–162.5° cor. (1941 Schindler); a 22.34, b 7.578, c 12.056 Å; n = 4; space group P$2_1 2_1 2_1$; d 1.179 (X-ray determination), d^{21} 1.180 ± 0.06 (suspension method) (1944 Nowacki); $[\alpha]^{21}$ in abs. alcohol +185°/$_D$, +224°/$_{5461}$, in acetone +169.5°/$_D$, +207°/$_{5461}$ (1940 Shoppee). Absorption max. in alcohol at 240 mμ (ε 17400)(1939 Dannenberg). Very soluble in glacial acetic acid, methanol, ethanol, acetone, benzene, ethyl acetate, and dioxane, soluble to a considerable extent in ether, very sparingly soluble in petroleum ether and water (1937 Steiger; 1938b Roche-Organon, Inc.). Soluble in aqueous sodium dehydrocholate solution on boiling: 1 mg. in 1.3 c.c. of a 20% solution, 1 mg. in > 20 c.c. of a 2% solution (1944 Cantarow). For basic properties, see 1944, 1946 Bersin.

For polarographic detection, see 1939 Eisenbrand; for other analytical properties, see under those of the free desoxycorticosterone (p. 2823 s).

Fmn. From desoxycorticosterone on treatment with acetic anhydride in pyridine at room temperature for 24 hrs. (1936a Ciba) or on passing ketene (4–5 mols.) into its solution in acetone with cooling (1936a Ciba; 1937 Roche-Organon, Inc.). See also under formation of desoxycorticosterone. — **Rns.** See under those of desoxycorticosterone (pp. 2822 s, 2823 s).

Propionate $C_{24}H_{34}O_4$. Crystals (acetone - petroleum ether), m. 163–4°; $[\alpha]_D^{25}$ +186° (chloroform) (1937a, e, h N. V. Organon; 1938b Ciba; 1939 Ciba Pharmaceutical Products). — **Fmn.** From desoxycorticosterone and propionic anhydride or propionyl chloride in pyridine at room temperature (1939 Ciba Pharmaceutical Products). See also under formation of desoxycorticosterone.

Butyrate $C_{25}H_{36}O_4$. Crystals (acetone - petroleum ether), m. 110–111°; $[\alpha]_D^{21}$ +177° (chloroform) (1937b, f, h N. V. Organon; 1939 Ciba Pharmaceutical Products). — **Fmn.** From desoxycorticosterone similarly to that of the preceding ester (1939 Ciba Pharmaceutical Products). See also under formation of desoxycorticosterone.

Valerate $C_{26}H_{38}O_4$. Crystals (acetone - petroleum ether), m. 84–85°; $[\alpha]_D^{21}$ +175° (chloroform). — **Fmn.** From desoxycorticosterone and valeric anhydride in pyridine (1939 Ciba Pharmaceutical Products).

Palmitate $C_{37}H_{60}O_4$. Crystals (petroleum ether), m. 60–61°; $[\alpha]_D^{21}$ +128° (chloroform) (1939 Ciba Pharmaceutical Products; cf. also 1939c, 1944 Schering A.-G.). — **Fmn.** From desoxycorticosterone and palmitic acid chloride in pyridine (1939 Ciba Pharmaceutical Products). See also under formation of desoxycorticosterone.

Stearate $C_{39}H_{64}O_4$. Crystals (petroleum ether), m. 63–64°; $[\alpha]_D^{21}$ + 116° (chloroform). — **Fmn.** Similar to that of the palmitate (above) (1939 CIBA PHARMACEUTICAL PRODUCTS).

Benzoate $C_{28}H_{34}O_4$. Crystals (acetone), m. 209–210°; $[\alpha]_D$ + 202° to + 204° (solvent not stated) (1937c, 1940 N. V. ORGANON; 1939 CIBA PHARMACEUTICAL PRODUCTS; cf. also 1939c SCHERING A.-G.). — **Fmn.** From desoxycorticosterone and benzoyl chloride in pyridine (1939 CIBA PHARMACEUTICAL PRODUCTS). See also under formation of desoxycorticosterone.

p-Toluenesulphonate $C_{28}H_{36}O_5S$. Leaflets (ether - pentane), m. 170–171° cor. (1940 c Reichstein; 1941 ROCHE-ORGANON, INC.). —**Fmn.** Along with 21-chloro-progesterone (p. 2810 s), from desoxycorticosterone on treatment with 2 mols. of p-toluenesulphonyl chloride and 3 mols. of pyridine in dry chloroform (1 vol. pyridine:9 vols. chloroform) at room temperature for 16 hrs. (1940d Reichstein). See also under formation of desoxycorticosterone. — **Rns.** Gives 21-iodoprogesterone (p. 2811 s) on boiling with sodium iodide in acetone for 5 min. (1940d Reichstein).

GLYCOSIDES OF DESOXYCORTICOSTERONE

β-D-Glucoside $C_{27}H_{40}O_8$. Crystals (acetone), m. 190–5° cor. (1942 Miescher), 192–5° cor. (1944a Meystre); $[\alpha]_D^{20}$ + 109° (methanol) (1942 Miescher). Soluble in chloroform and acetone; solubility in water at 18° 1.2⁰/₀₀, at 95° 25⁰/₀₀ (1942, 1943 Miescher; 1944b Meystre); for solutions in aqueous glucose, glycerol, and other hydrotropic agents, see 1943 Miescher; 1943 CIBA PHARMACEUTICAL PRODUCTS.

Fmn. The tetraacetate is obtained from desoxycorticosterone and aceto-bromo-D-glucose on boiling with Ag_2CO_3 in benzene with simultaneous removal (by azeotropic distillation) of the water formed; it is hydrolysed by barium methoxide in methanol at a low temperature to give the free glucoside (yield, 60%) (1944a Meystre; 1944 CIBA PHARMACEUTICAL PRODUCTS; 1945 CIBA; cf. also 1942 Miescher); when Ag_2O is used instead of Ag_2CO_3, the yield of tetraacetate is only 10–14% (1941 Johnson; cf. 1938c, 1939a, b CIBA) and 3-keto-Δ^4-ætiocholenic acid is formed as a by-product (1942 Miescher).

Rns. Refluxing of the tetraacetate with aluminium isopropoxide in isopropyl alcohol affords the 21-D-glucoside of a Δ^4- or Δ^5-pregnene-3.20.21-triol (p. 1952 s) (1939b SCHERING A.-G.). On exposure of the glucoside in aqueous solution to sunlight for 2 weeks it gives an isomeric **compound** $C_{27}H_{40}O_8$ [colourless, hygroscopic jelly (aqueous methanol), m. 234–240° cor., $[\alpha]_D^{18}$ + 54° (methanol); almost insoluble in water] which is scarcely attacked on heating with aqueous alcoholic HCl on the water-bath for 3 hrs. and which on treatment with acetic anhydride and pyridine at 20° gives a *tetraacetate* $C_{35}H_{48}O_{12}$ [crystals (strong alcohol), m. 145–150° cor., $[\alpha]_D^{25}$ + 54° (methanol)] (1944b Meystre).

194*

Tetraacetate $C_{35}H_{48}O_{12}$. Crystals (acetone - ether), m. 175–6° cor. (1942 Miescher; 1944b Meystre); needles (50% alcohol), m. 176–176.5° cor. (1941 Johnson); $[\alpha]_D$ +80° (acetone) (1942 Miescher), +68° (methanol) (1944b Meystre), +80° (chloroform) (1941 Johnson). Almost insoluble in ether (1942 Miescher). — **Fmn.** and **Rns.** See above.

β-D-Galactoside $C_{27}H_{40}O_8$. Crystals (abs. alcohol), m. 195–8° cor.; $[\alpha]_D^{20}$ +136° (acetone). Soluble in chloroform; solubility in water at 18° 2.2%/oo, strongly increasing on heating. — **Fmn.** From desoxycorticosterone and acetobromo-D-galactose on warming with Ag_2CO_3 in pure chloroform at 40–45°, followed by hydrolysis with aqueous methanolic K_2CO_3 at 20°; crude yield, 67% (1943 Miescher).

β-D-Maltoside $C_{33}H_{50}O_{13}$. Crystals (alcohol), m. 232–5° cor.; $[\alpha]_D^{19.5}$ +124° (methanol) (1944a Meystre). Solubility in water at 25° 6.4%/oo, at 95° >250%/oo (1944b Meystre). — **Fmn.** The heptaacetate is obtained from desoxycorticosterone and acetobromomaltose on boiling with Ag_2CO_3 in benzene with simultaneous removal (by azeotropic distillation) of the water formed; it is hydrolysed by barium methoxide in methanol at a low temperature to give the free maltoside (yield, 42.5%) (1944a Meystre; 1944 CIBA PHARMACEUTICAL PRODUCTS; 1945 CIBA; cf. 1943 Miescher).

Heptaacetate $C_{47}H_{64}O_{20}$. Needles (acetone - ether), m. 183–5° cor. (1943 Miescher). — **Fmn.** and **Rns.** See above.

β-D-Lactoside $C_{33}H_{50}O_{13}$. Needles (water) with 1 H_2O (after drying in a high vacuum for 1 hr. at 110°), m. 202–8° cor.; $[\alpha]_D^{20}$ +80° (methanol). Soluble in methanol and ethanol, insoluble in benzene, ether, chloroform, and acetone; solubility in water at 18° 3.4%/oo, strongly increasing on heating (1943 Miescher). — **Fmn.** From desoxycorticosterone and acetobromo-D-lactose in the presence of Ag_2CO_3 in abs. benzene at 20°, followed by hydrolysis of the resulting heptaacetate with aqueous methanolic K_2CO_3 at 20°; small yield (1943 Miescher; cf. 1939a, b CIBA). — **Rns.** Refluxing of the heptaacetate with aluminium isopropoxide in isopropyl alcohol affords the 21-lactoside of a Δ^4- or Δ^5-pregnene-3.20.21-triol (p. 1952 s) (1939b SCHERING A.-G.).

Heptaacetate $C_{47}H_{64}O_{20}$. Needles (abs. alcohol - ether), crystals (ether - acetone), m. 194–5° cor.; $[\alpha]_D^{20}$ +52° (acetone). — **Fmn.** From the preceding compound on treatment with acetic anhydride in pyridine at 20° (1943 Miescher). See also above. — **Rns.** See above.

6-(β-Lactosido)-D-glucoside $C_{39}H_{60}O_{18}$. Amorphous glass (alcohol) with 2 H_2O; begins to melt at ca. 160° cor., dec. at 190° cor.; very hygroscopic after drying in a high vacuum for 3 hrs. at 145°. Soluble in water, methanol, and hot ethanol, insoluble in ether, chloroform, and acetone. — **Fmn.** From desoxycorticosterone and acetobromolactosido-D-glucose on warming with dry Ag_2CO_3 in pure chloroform at 40–45°, followed by reacetylation with acetic anhydride in pyridine and hydrolysis of the resulting hendecaacetate with barium methoxide in methanol at —15° (1943 Miescher).

Hendecaacetate $C_{61}H_{82}O_{29}$. Amorphous (aqueous alcohol), m. 120–130° (1943 Miescher). — **Fmn.** and **Rns.** See above.

For a **6-gentiobiosido-D-glucoside**, see 1939b CIBA.

1936 a CIBA*, *Swiss Pat.* 204234 (issued 1939); *C. A.* **1941** 2284; *Ch. Ztbl.* **1940** I 250.
   b CIBA*, *Swiss Pats.* 215558–559 (issued 1941); *C. A.* **1948** 3784; *Ch. Ztbl.* **1942** I 2296.
   c CIBA*, *Swiss Pats.* 216104–105 (issued 1941); *C. A.* **1948** 5057; *Ch. Ztbl.* **1942** II 197, 1037**.
   d CIBA*, *Swiss Pat.* 212337 (issued 1941); *C. A.* **1942** 3637; *Ch. Ztbl.* **1942** I 643; *Brit. Pat.*
     497394 (issued 1939); *C. A.* **1939** 3812; *Ch. Ztbl.* **1939** I 2828; *Fr. Pat.* 838916 (1938,
     issued 1939); *C. A.* **1939** 6875; *Ch. Ztbl.* **1939** II 169, 170; *Dutch Pat.* 51420 (1938, issued
     1941); *Ch. Ztbl.* **1942** I 2561; CIBA PHARMACEUTICAL PRODUCTS (Miescher, Wettstein),
     U.S. Pat. 2229813 (1938, issued 1941); *C. A.* **1941** 3040; *Ch. Ztbl.* **1941** II 2467, 2468.
1937 I. G. FARBENIND., *Brit. Pat.* 502474 (issued 1939); *C. A.* **1939** 7316; *Ch. Ztbl.* **1939** II 1337;
   BIOS Final Report No. 766, p. 135.
   a N. V. ORGANON, *Swiss Pat.* 216127 (issued 1941); *Ch. Ztbl.* **1942** II 567.
   b N. V. ORGANON, *Swiss Pat.* 216128 (issued 1941); *Ch. Ztbl.* **1942** II 567.
   c N. V. ORGANON, *Swiss Pat.* 216129 (issued 1941); *Ch. Ztbl.* **1942** II 567.
   d N. V. ORGANON, *Swiss Pat.* 217990 (issued 1942); *Ch. Ztbl.* **1943** I 1389.
   e N. V. ORGANON, *Swiss Pat.* 219346 (issued 1942); *Ch. Ztbl.* **1943** I 1389.
   f N. V. ORGANON, *Swiss Pat.* 219347 (issued 1942); *Ch. Ztbl.* **1943** I 1389.
   g N. V. ORGANON, *Swiss Pat.* 225509 (issued 1943); *Ch. Ztbl.* **1944** I 36.
   h N. V. ORGANON, *Swiss Pat.* 226174 (issued 1943); *Ch. Ztbl.* **1944** I 449.
   ROCHE-ORGANON, INC. (Reichstein, Schlittler), *U.S. Pat.* 2183589 (issued 1939); *C. A.* **1940**
     2538; *Ch. Ztbl.* **1940** II 932.
   SCHERING A.-G. (Serini, Logemann), *Ger. Pat.* 736848 (issued 1943); *C. A.* **1944** 3094; *Ch. Ztbl.*
     1944 I 300, 301.
   Steiger, Reichstein, *Helv.* **20** 1164, 1177, 1178.
1938 a CIBA*, *Fr. Pat.* 840417 (issued 1939); *Ch. Ztbl.* **1939** II 1125.
   b CIBA*, *Swiss Pat.* 222901 (issued 1942); *C. A.* **1949** 1916; *Ch. Ztbl.* **1943** I 2114.
   c CIBA*, *Swiss Pat.* 226684 (issued 1943); *Ch. Ztbl.* **1944** II 675.
   "DEGEWOP" GESELLSCHAFT WISSENSCHAFTLICHER ORGAN- UND HORMONPRÄPARATE (Reich-
     stein), *Ger. Pat.* 737539 (issued 1943); *C. A.* **1945** 5043; *Ch. Ztbl.* **1944** II 563.
   N. V. ORGANON, *Fr. Pat.* 835669 (issued 1938); *Ch. Ztbl.* **1939** I 4089, 4090; *Indian Pat.* 25074
     (issued 1938); *Ch. Ztbl.* **1939** I 3034, 3035; ROCHE-ORGANON, INC. (Reichstein), *U.S. Pat.*
     2232730 (issued 1941); *C. A.* **1941** 3773; *Ch. Ztbl.* **1941** II 2467.
   a Reichstein, v. Euw, *Helv.* **21** 1181, 1183, 1184.
   b Reichstein, v. Euw, *Helv.* **21** 1197, 1206.
   a ROCHE-ORGANON, INC. (Reichstein), *U.S. Pat.* 2166877 (issued 1939); *C. A.* **1939** 8925;
     Ch. Ztbl. **1940** I 1232, 1233.
   b ROCHE-ORGANON, INC. (Reichstein), *U.S. Pats.* 2312481 (issued 1943), 2312483 (1941,
     issued 1943); *C. A.* **1943** 4862, 5201; *Ch. Ztbl.* **1945** II 687, 688; N. V. ORGANON, *Indian*
     Pat. 25064 (issued 1938); *Ch. Ztbl.* **1939** I 2458, 2459; *Fr. Pat.* 835669 (issued 1938); *C. A.*
     1939 8626; *Ch. Ztbl.* **1939** I 4089, 4090; *Brit. Pat.* 502289 (issued 1939); *C. A.* **1939** 6346.
   SCHERING A.-G., *Fr. Pat.* 857832 (issued 1940); *Ch. Ztbl.* **1941** I 1324, 1325.
   Steiger, Reichstein, *Helv.* **21** 161, 163, 169.
1939 a CIBA* (Miescher, Fischer), *Swed. Pat.* 97167 (issued 1939); *Ch. Ztbl.* **1940** I 2032, 2033.
   b CIBA*, *Fr. Pat.* 856329 (issued 1940); *Ch. Ztbl.* **1941** I 928.
   c CIBA*, *Fr. Pat.* 857122 (issued 1940); *Ch. Ztbl.* **1941** I 1324; CIBA PHARMACEUTICAL PRO-
     DUCTS (Miescher, Fischer, Scholz, Wettstein), *U.S. Pat.* 2275790 (issued 1942); *C. A.*
     1942 4289.
   CIBA PHARMACEUTICAL PRODUCTS (Miescher, Fischer), *U.S. Pat.* 2265183 (issued 1941); *C. A.*
     1942 1738; *Ch. Ztbl.* **1945** I 1036.

---

* Société pour l'industrie chimique à Bâle; Gesellschaft für chemische Industrie in Basel.
** In *Ch. Ztbl.* **1942** II 1037 (referring to *Swiss Pat.* 216104) the starting material is erroneously given as corticosterone 21-propionate instead of desoxycorticosterone 21-propionate.

Dannenberg, *Abhandl. Preuss. Akad. Wiss. Math.-naturwiss. Kl.* **1939** No. 21, pp. 10, 53.
a Ehrhart, Ruschig, Aumüller, *Angew. Ch.* **52** 363, 366.
b Ehrhart, Ruschig, Aumüller, *Ber.* **72** 2035.
Eisenbrand, Picher, *Z. physiol. Ch.* **260** 83, 89 et seq.
Mamoli, *Ber.* **72** 1863.
Miescher, Wettstein, Scholz, *Helv.* **22** 894, 903, 904.
Reich, Reichstein, *Helv.* **22** 1124, 1135.
Reichstein, Montigel, *Helv.* **22** 1212, 1219.
a SCHERING A.-G. (Logemann, Hildebrand), *Ger. Pat.* 737 420 (issued 1943); *C. A.* **1945** 5051; *Ch. Ztbl.* **1944** I 366, 367; *Swiss Pat.* 229 604 (issued 1944); *Ch. Ztbl.* **1944** II 1203; *Fr. Pat.* 856 641 (issued 1940); *Ch. Ztbl.* **1941** I 407.
b SCHERING A.-G. (Gehrke, Schoeller, Inhoffen), *Belg. Pat.* 434 173 (issued 1939); *Fr. Pat.* 856 348 (issued 1940); *Ch. Ztbl.* **1941** I 669; *Swed. Pat.* 103 866 (issued 1942); *Ch. Ztbl.* **1942** II 1720.
c SCHERING A.-G. (Serini, Eysenbach), *Swed. Pat.* 100 374 (issued 1940); *Ch. Ztbl.* **1941** I 3259, 3260; *Fr. Pat.* 856 659 (issued 1940); *Ch. Ztbl.* **1941** I 1196.
Serini, Logemann, Hildebrand, *Ber.* **72** 391, 394.
Woker, Antener, *Helv.* **22** 511, 518.
1940 CIBA* (Miescher, Heer), *Ger. Pat.* 737 570 (issued 1943); *C. A.* **1946** 179; *Fr. Pat.* 876 214 (issued 1942); *Ch. Ztbl.* **1943** I 1496; *Dutch Pat.* 55 050 (issued 1943); *Ch. Ztbl.* **1944** I 367; CIBA PHARMACEUTICAL PRODUCTS (Miescher, Heer), *U. S. Pat.* 2 372 841 (issued 1945); *C. A.* **1945** 4195; *Ch. Ztbl.* **1945** II 1773.
Cuyler, Ashley, Hamblen, *Endocrinology* **27** 177.
N. V. ORGANON (Reichstein), *Swed. Pat.* 102 320 (issued 1941); *Ch. Ztbl.* **1942** II 74; *Danish Pat.* 61 323 (issued 1943); *Ch. Ztbl.* **1944** II 343.
a Reichstein, v. Euw, *Helv.* **23** 136, 138.
b Reichstein, Fuchs, *Helv.* **23** 658, 666.
c Reichstein, Schindler, *Helv.* **23** 669, 673, 674.
d Reichstein, Fuchs, *Helv.* **23** 684, 686, 687.
SCHERING CORP. (Inhoffen), *U. S. Pat.* 2 409 043 (issued 1946); *C. A.* **1947** 1398; *Ch. Ztbl.* **1947** 620.
Shoppee, *Helv.* **23** 925, 931, 932.
Wettstein, Hunziker, *Helv.* **23** 764.
1941 Hamblen, Cuyler, Pattee, Axelson, *Endocrinology* **28** 306.
Johnson, *J. Am. Ch. Soc.* **63** 3238.
Reichstein, *Fr. Pat.* 888 825 (issued 1943); *Ch. Ztbl.* **1944** II 1203.
ROCHE-ORGANON, INC. (Reichstein), *U. S. Pat.* 2 357 224 (issued 1944); *C. A.* **1945** 391; *Ch. Ztbl.* **1945** II 1232.
Schindler, Frey, Reichstein, *Helv.* **24** 360, 364, 371.
Westphal, *Naturwiss.* **29** 782.
1942 Cuyler, Hirst, Powers, Hamblen, *J. Clin. Endocrinology* **2** 373.
Miescher, Fischer, Meystre, *Helv.* **25** 40.
Westphal, *Z. physiol. Ch.* **273** 13.
1943 CIBA PHARMACEUTICAL PRODUCTS (Miescher, Meystre), *U. S. Pat.* 2 411 631 (issued 1946); *C. A.* **1947** 2211; *Ch. Ztbl.* **1947** 893.
Fish, Horwitt, Dorfman, *Science* [N.S.] **97** 227.
Hansen, Cantarow, Rakoff, Paschkis, *Endocrinology* **33** 282, 285.
Hoffman, Kazmin, Browne, *J. Biol. Ch.* **147** 259.
Miescher, Meystre, *Helv.* **26** 224.
Reich, Reichstein, *Helv.* **26** 562, 585.
1944 Bersin, Meyer, *Angew. Ch.* **57** 117, 118.
Cantarow, Paschkis, Rakoff, Hansen, *Endocrinology* **35** 129.
CIBA PHARMACEUTICAL PRODUCTS (Miescher, Meystre), *U. S. Pat.* 2 479 761 (issued 1949); *C. A.* **1949** 1147; *Ch. Ztbl.* **1949** E 755, 756.
v. Euw, Lardon, Reichstein, *Helv.* **27** 1287, 1288.

* Société pour l'industrie chimique à Bâle; Gesellschaft für chemische Industrie in Basel.

Horwitt, Dorfman, Shipley, Fish, *J. Biol. Ch.* **155** 213.
a Meystre, Miescher, *Helv.* **27** 231, 234, 235.
b Meystre, Miescher, *Helv.* **27** 1153, 1156, 1160.
Nowacki, *Helv.* **27** 1622.
SCHERING A.-G., *FIAT Final Report* No. 996, pp. 98–101; *BIOS Final Report* No. 449, pp. 246
   to 248, 250.
Zimmermann, *Vitamine und Hormone* **5** [1952] 1, 124, 152, 163, 237, 260, 276.
1945 CIBA*, *Fr. Pat.* 910844 (issued 1946); *Ch. Ztbl.* **1946** I 2280.
   Julian, Cole, Magnani, Meyer, *J. Am. Ch. Soc.* **67** 1728; GLIDDEN COMP. (Julian, Cole, Mag-
      nani, Conde), *U.S. Pat.* 2374683 (issued 1945); *C. A.* **1946** 1636; *Ch. Ztbl.* **1946** I 372.
1946 Bersin, *Naturwiss.* **33** 108, 110.
   Heard, Sobel, *J. Biol. Ch.* **165** 687.
1953 Sondheimer, Kaufmann, Romo, Martinez, Rosenkranz, *J. Am. Ch. Soc.* **75** 4712, 4715.
1957 Reichstein, *private communication.*
   Romo, Rosenkranz, Sondheimer, *J. Am. Ch. Soc.* **79** 5034.

17-iso-Δ^4-Pregnen-21-ol-3.20-dione, 17-iso-Desoxycorticosterone $C_{21}H_{30}O_3$.

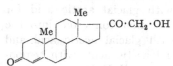

For configuration at C-17, see pp. 1363 s,
1364 s. — Prisms (acetone), m. 179–181° cor.;
$[\alpha]_D^{18}$ —6°, $[\alpha]_{5461}^{18}$ —9° (both in abs. alcohol)
(1940 Shoppee). — The acetate has practically
no cortin-like action (1940 Shoppee). — **Fmn.**
The acetate is obtained from the 20.21-diacetate of Δ^4-pregnene-17α.20β.21-triol-
3-one (p. 2725 s) on refluxing repeatedly with zinc dust in abs. toluene for
15 hrs. (yield, ca. 35 %); it is hydrolysed by $KHCO_3$ in aqueous methanol
at room temperature (40 hrs.) (1940, 1941 Shoppee; 1940 CIBA). — **Rns.**
Gives 17-iso-3-keto-Δ^4-ætiocholenic acid on oxidation with periodic acid in
aqueous methanol at 20°. Its acetate is isomerized to that of desoxycorticosterone (p. 2820 s) by refluxing with conc. HCl in alcohol for $^1/_2$ hr., followed
by reacetylation (1940 Shoppee).

Acetate $C_{23}H_{32}O_4$. Exists in two crystalline modifications: prisms (abs.
ether), m. 137–8° cor., and leaflets (acetone - ether), m. 174° cor. (1940 Shoppee); X-ray analysis of the latter form: a 19.26, b 9.888, c 10.664, space
group $P2_12_12_1$, n = 4, d 1.210 (1945 Nowacki). Optical rotation in acetone:
$[\alpha]_D^{17}$ —26°, $[\alpha]_{5461}^{17}$ —32°. Absorption max. in alcohol at 244 mμ (log ε 4.18)
(1940 Shoppee). Its solution in methanol immediately reduces alkaline silver
diammine solution at room temperature (1940 Shoppee).

1940 CIBA (Soc. pour l'ind. chim. à Bâle; Ges. f. chem. Ind. in Basel), *Fr. Pat.* 876214 (issued 1942);
      Ch. Ztbl. **1943** I 1496; *Dutch Pat.* 55050 (issued 1943); *Ch. Ztbl.* **1944** I 367; CIBA PHARMA-
      CEUTICAL PRODUCTS (Miescher, Heer), *U.S. Pat.* 2372841 (issued 1945); *C. A.* **1945** 4195;
      Ch. Ztbl. **1945** II 1773.
   Shoppee, *Helv.* **23** 925.
1941 Shoppee, Reichstein, *Helv.* **24** 354 (erratum).
1945 Nowacki, *Helv.* **28** 1373.

* Société pour l'industrie chimique à Bâle; Gesellschaft für chemische Industrie in Basel.

$\Delta^{9(11)}$-**Pregnen-21-ol-3.20-dione** $C_{21}H_{30}O_3$.

Acetate $C_{23}H_{32}O_4$. Crystals (ether - petroleum ether); an impure sample m. ca. 160°. Rapidly reduces alkaline silver diammine. Gives a yellow coloration with tetranitromethane in $CHCl_3$. — **Fmn.** From 3.11-diketo-ætiocholanic acid in the following way: the methyl ester is hydrogenated in glacial acetic acid in the presence of platinum oxide and then refluxed with aqueous methanolic KOH; the resulting mixture of the 3-epimeric 3.11-dihydroxy-ætiocholanic acids is converted into the 3-acetates by partial acetylation with acetic anhydride in glacial acetic acid and then treated with thionyl chloride at room temperature to give the 3-acetoxy-$\Delta^{9(11)}$-ætiocholenic acid chlorides which are converted into the 21-diazo-$\Delta^{9(11)}$-pregnen-3-ol-20-ones by treatment with diazomethane in dry ether - benzene, followed by hydrolysis with aqueous methanolic KOH at room temperature; the diazohydroxyketones are oxidized with aluminium phenoxide and dry acetone in dry benzene at room temperature and the resulting 21-diazo-$\Delta^{9(11)}$-pregnene-3.20-dione is heated with glacial acetic acid for 30 min. at 100°, or the diazohydroxyketones are first converted into the 3-hydroxy-21-acetoxy-20-ketones by heating with glacial acetic acid and then oxidized with chromium trioxide in glacial acetic acid, with temporary protection of the double bond by bromine (1942 N. V. ORGANON; 1943 Reichstein).

Rns. Gives $\Delta^{4.9(11)}$-pregnadien-21-ol-3.20-dione acetate (p. 2819 s) on treatment with 2 mols. of bromine in glacial acetic acid, followed by refluxing for 5 hrs. with abs. pyridine or by heating for 2 hrs. with dimethylaniline at 150°, and then by heating for 15 min. with zinc dust in glacial acetic acid at 80° (1942 N. V. ORGANON; 1943 Reichstein).

1942 N. V. ORGANON, *Swiss Pat.* 255307 (issued 1949); *Ch. Ztbl.* **1949** E 1529.
1943 Reichstein, *U.S. Pat.* 2401775 (issued 1946); *C. A.* **1946** 5884; *Ch. Ztbl.* **1947** 74; *U.S. Pat.* 2404768 (1945, issued 1946); *C. A.* **1946** 6222; *Ch. Ztbl.* **1947** 233.

Δ^{11}-**Pregnen-21-ol-3.20-dione** $C_{21}H_{30}O_3$.

For configuration at C-17, see pages 1363 s, 1364 s.

Acetate $C_{23}H_{32}O_4$. Prisms (ether - petroleum ether), m. 151–2° cor. (Kofler block); sublimes under 0.01 mm. pressure at 145–150° (bath temperature); $[\alpha]_D^{14} + 100°$ (acetone). Gives a strong yellow coloration with tetranitromethane in chloroform (1946 v. Euw). — **Fmn.** From the 21-acetate of Δ^{11}-pregnene-3α.21-diol-20-one (p. 2308 s) on oxidation with the calculated amount of CrO_3 in glacial acetic acid at room temperature or on refluxing with aluminium phenoxide and acetone in benzene (1942 Reichstein). From the 21-acetate of pregnane-12α.21-diol-3.20-dione (p. 2858 s) on treatment with p-toluenesulphonyl chloride in abs. pyridine for 4 days at 30°, followed by refluxing of the crude 12-p-toluenesulphonate with collidine for 5 hrs.; overall yield, 7% (1946 v. Euw). From 3α-acetoxy-Δ^{11}-ætiocholenic acid via 21-diazo-Δ^{11}-pregnen-3α-ol-20-one and 21-diazo-Δ^{11}-pregnene-3.20-dione (see pp. 2308 s, 2309 s) which is heated with glacial acetic acid for $1/2$ hr. at 100° (1942 N. V. ORGANON; cf. also 1943 Reichstein). — **Rns.** Gives the acetate of pregnan-21-ol-3.11.20-trione (p. 2973 s) on treatment with N-bromoacetamide in aqueous acetone at room temperature, followed by oxidation of the resulting bromohydrin with CrO_3 in strong acetic acid at 18° and heating with zinc dust and sodium acetate in glacial acetic acid on the water-bath (1942 Reichstein). Gives the acetate of $\Delta^{4.8(?)}$-pregnadien-21-ol-3.20-dione (see p. 2818 s) on treatment with 2 mols. of bromine in glacial acetic acid, followed by boiling with abs. pyridine or heating with dimethylaniline, and then by heating with zinc dust in glacial acetic acid (1942 N. V. ORGANON; cf. also 1943 Reichstein).

1942 N. V. ORGANON, *Swiss Pat.* 254992 (issued 1949); *Ch. Ztbl.* **1950** I 2388.
Reichstein, *Fr. Pat.* 887641 (issued 1943); *Ch. Ztbl.* **1944** II 878; *U.S. Pat.* 2403683 (1943, issued 1946); *C. A.* **1946** 6216, 6222; *Ch. Ztbl.* **1947** 232.
1943 Reichstein, *U.S. Pat.* 2401775 (issued 1946); *C. A.* **1946** 5884; *Ch. Ztbl.* **1947** 74; *U.S. Pat.* 2404768 (1945, issued 1946); *C. A.* **1946** 6222; *Ch. Ztbl.* **1947** 233.
1946 v. Euw, Reichstein, *Helv.* 29 654, 670.

Pregnan-21-ol-3.20-dione $C_{21}H_{32}O_3$.

Acetate $C_{23}H_{34}O_4$. Rodlets (ether), m. 150° to 151° cor. (1940 Reichstein); needles, m. 153° to 154° cor. (vac.) (1940 Wettstein); $[\alpha]_D^{17}$ + 109°, $[\alpha]_{5461}^{17}$ + 130° (both in acetone) (1940 Reichstein); $[\alpha]_D^{21}$ + 108° (chloroform) (1940 Wettstein). More readily soluble in acetone than the allo-isomer (below) (1940 Wettstein). — Shows no cortin-like activity (1940 Wettstein). — **Fmn.** From the 21-acetates of the two C-3 epimeric pregnane-3.21-diol-20-ones (pp. 2309 s, 2310 s) on oxidation with CrO_3 in glacial acetic acid at 20°. From 21-chloropregnane-3.20-dione (p. 2811 s) on refluxing with anhydrous potassium acetate in glacial acetic acid for $3^1/_2$ hrs. (1940 Reichstein). Along with a smaller amount of the allo-epimer (below), from the acetate of Δ^4-pregnen-21-ol-3.20-dione (desoxycorticosterone, page 2820 s) on hydrogenation in alcohol in the presence of palladium - calcium carbonate, followed by reacetylation (1940 Wettstein).

Rns. On treatment with 1 mol. bromine in glacial acetic acid at room temperature, followed by refluxing of the resulting 4-bromo-derivative with abs. pyridine for 5 hrs., it reverts to desoxycorticosterone acetate (1940 Reichstein; cf. 1938 "DEGEWOP"; 1938 N. V. ORGANON); the dehydrobromination may also be effected by refluxing with anhydrous potassium acetate in glacial acetic acid for 4 hrs. (1938 "DEGEWOP"; 1938 N. V. ORGANON).

Allopregnan-21-ol-3.20-dione $C_{21}H_{32}O_3$.

Acetate $C_{23}H_{34}O_4$. Needles (methanol, or acetone - ether), m. 197–9° cor. (1939 Reichstein; 1940 Wettstein); $[\alpha]_D$ +115° (chloroform); shows no selective absorption between 500 and 220 mμ; sparingly soluble in acetone (1940 Wettstein). Strongly reduces alkaline silver diammine solution at room temperature (1939 Reichstein). — Shows no cortin-like activity (1940 Wettstein). — **Fmn.** From the 21-acetate of allopregnane-3β.21-diol-20-one (p. 2311 s) on oxidation with CrO_3 in glacial acetic acid at room temperature (1939 Reichstein). Along with the preceding isomer (main product), from the acetate of Δ^4-pregnen-21-ol-3.20-dione (desoxycorticosterone, p. 2820 s) on hydrogenation in alcohol in the presence of palladium - calcium carbonate, followed by reacetylation (1940 Wettstein). — *Dioxime* $C_{23}H_{36}O_4N_2$, crystals (aqueous alcohol), dec. 212–4° cor. (1940 Wettstein).

1938 "DEGEWOP" GESELLSCHAFT WISSENSCHAFTLICHER ORGAN- UND HORMONPRÄPARATE (Reichstein), *Ger. Pat.* 737539 (issued 1943); *C. A.* **1945** 5043; *Ch. Ztbl.* **1944** II 563.
    N. V. ORGANON, *Fr. Pat.* 835669 (issued 1938); *Ch. Ztbl.* **1939** I 4089, 4090; *Indian Pat.* 25074 (issued 1938); *Ch. Ztbl.* **1939** I 3034, 3035; ROCHE-ORGANON, INC. (Reichstein), *U.S. Pat.* 2232730 (issued 1941); *C. A.* **1941** 3773; *Ch. Ztbl.* **1941** II 2467.
1939 Reichstein, v. Euw, *Helv.* **22** 1209, 1212.
1940 Reichstein, Fuchs, *Helv.* **23** 658, 665, 666, 668.
    Wettstein, Hunziker, *Helv.* **23** 764.

Δ^4-Pregnen-20β(?)-ol-3-on-21-al $C_{21}H_{30}O_3$ (Ia).

The crystalline aldehyde is probably represented by the dimolecular form (Ib) or the trimolecular form (Ic); the acetate is a derivative of the form (Ib) or, more probably, of

the form (Ic) (1941 Schindler). For probable configuration at C-20, see the remarks under Δ^5-pregnene-3β.20β(?)-diol-21-al (p. 2312 s), from which the present aldehyde may be obtained by oxidation.

Needle-like prisms (acetone - ether), m. 206–8° cor. dec.; decomposes on attempted sublimation in a high vacuum; $[\alpha]_D^{20}$ +84° (dioxane). Sparingly soluble in ether, acetone, and alcohol, more readily soluble in chloroform and dioxane. It strongly reduces alkaline silver diammine solution. Gives a red coloration with 1.4-dihydroxynaphthalene + conc. HCl in glacial acetic acid on heating (1941 Schindler). — **Fmn.** Its dimethyl acetal (below) is obtained from that of Δ^5-pregnene-3β.20β(?)-diol-21-al (p. 2312 s) by refluxing with aluminium tert.-butoxide and acetone in abs. benzene for 24 hrs. (crude yield, 60%) or with aluminium isopropoxide and cyclohexanone in abs.

toluene for $1^1/_4$ hrs. (yield, 52 %); the free aldehyde is obtained from the acetal by treatment with 2 N H_2SO_4 in glacial acetic acid at 0^0 for 16 hrs. (1941 Schindler). — **Rns.** Reverts to its acetal on treatment with methanolic HCl at room temperature (1941 Schindler). It is isomerized to Δ^4-pregnen-21-ol-3.20-dione (desoxycorticosterone, p. 2820 s) by refluxing with abs. pyridine in an atmosphere of CO_2 for $5^1/_2$ hrs. (1941 Schindler; cf. 1941 Reichstein).

Disemicarbazone $C_{23}H_{36}O_3N_6$. Solid; becomes brown from 200^0 onwards, not molten at 300^0; practically insoluble in most solvents (1941 Schindler).

Dimethyl acetal $C_{23}H_{36}O_4$. Needles (ether - pentane), m. 135–6^0 cor.; sublimes at 135^0 (bath temperature) under 0.01 mm. pressure; $[\alpha]_D^{18} + 62^0$ (acetone). Absorption max. in alcohol (from the graph) at 309 (2.0) and 225 (4.3) mμ (log ε). Gives a red coloration with 1.4-dihydroxynaphthalene + conc. HCl in glacial acetic acid on heating. — **Fmn.** See under formation of the free aldehyde, above. Is also obtained from its acetate (below) on refluxing with methanolic KOH for $^1/_2$ hr. (yield, 74 %) or with 1 % methanolic HCl for 1 hr. (yield, 44 %). From the free aldehyde on treatment with methanolic HCl at room temperature; similarly from the aldehyde acetate (below). — *Semicarbazone* $C_{24}H_{39}O_4N_3$, leaflets (aqueous alcohol), m. 220–222^0 cor. dec. (1941 Schindler).

Acetate $C_{23}H_{32}O_4$. For its structure, see under the free hydroxy-aldehyde (above). — Crystals (dioxane - ether - petroleum ether), m. 255–6^0 cor. becoming yellow; $[\alpha]_D^{20} + 56^0$ (dioxane). Somewhat more readily soluble in most solvents than the free hydroxy-aldehyde; gives the same colour reaction as the latter but more slowly. — **Fmn.** From the free hydroxy-aldehyde on treatment with acetic anhydride in dry pyridine at room temperature. — **Rns.** Gives the dimethyl acetal of the free hydroxy-aldehyde (above) with methanolic HCl in dioxane on boiling for 1 hr., whereas at room temperature the acetate of this acetal (below) is also formed to a small extent (1941 Schindler).

Acetate dimethyl acetal $C_{25}H_{38}O_5$. Needles (dil. methanol), m. 112–3^0 cor.; distils at 135^0 (bath temperature) under 0.01 mm. pressure; $[\alpha]_D^{15} + 111^0$ (methanol). Absorption max. in alcohol at 242 mμ (log ε 4.16). Gives a red coloration with 1.4-dihydroxynaphthalene + conc. HCl in glacial acetic acid on heating. — **Fmn.** From the 20-acetate dimethyl acetal of Δ^5-pregnene-$3\beta.20\beta$(?)-diol-21-al (p. 2312 s) on refluxing with aluminium tert.-butoxide and acetone in abs. benzene for 24 hrs. (yield, 67 %) or with aluminium isopropoxide and cyclohexanone in abs. toluene for $1^1/_4$ hrs. From the dimethyl acetal of the free hydroxy - aldehyde on heating with acetic anhydride in pyridine at 60^0. From the acetate (above), q.v. (1941 Schindler).

A Δ^4-**pregnen-20-ol-3-on-21-al** $C_{21}H_{30}O_3$ [needles (dil. acetone)], which may be identical with the preceding pregnenolonal, is obtained from the 21-enol acetate of Δ^4-pregnen-3-on-21-al ($\Delta^{4.20}$-pregnadien-21-ol-3-one acetate, page 2513 s) on treatment with perbenzoic acid in ether, followed by hydrolysis of the resulting oxide by heating with alcoholic HCl and then by hydrolysis of the ester group with aqueous alcoholic K_2CO_3 (1939 Ciba).

1939 Ciba (Soc. pour l'ind. chim. à Bâle; Ges. f. chem. Ind. in Basel), *Fr. Pat.* 857 122 (issued 1940); *Ch. Ztbl.* **1941** I 1324; Ciba Pharmaceutical Products (Miescher, Fischer, Scholz, Wettstein), *U.S. Pats.* 2 275 790, 2 276 543 (both issued 1942); *C. A.* **1942** 4289, 4674.
1941 Reichstein, *Fr. Pat.* 888 825 (issued 1943); *Ch. Ztbl.* **1944** II 1203.
    Schindler, Frey, Reichstein, *Helv.* **24** 360, 367–372.

Δ^4-Norcholen-23-ol-3.22-dione $C_{23}H_{34}O_3$.

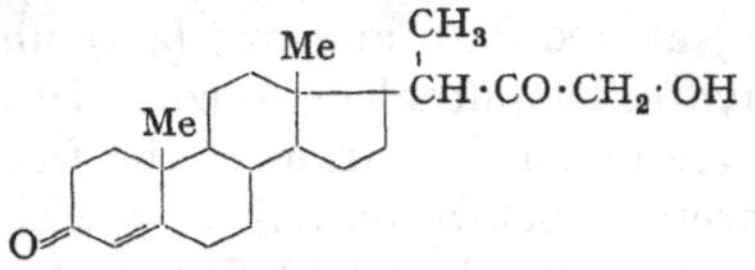

Acetate $C_{25}H_{36}O_4$. Crystals (methanol), m. 167–8° cor. Strongly reduces ammoniacal silver salt solution at room temperature. Has no cortin-like action. — **Fmn.** From the 21-acetate of Δ^5-norcholene-3β.23-diol-22-one (p. 2314 s) on refluxing with aluminium isopropoxide and cyclohexanone in abs. toluene for $^3/_4$ hr. (1941 Wettstein).

1941 Wettstein, *Helv.* **24** 311, 313, 316.

V. Hydroxy-dioxosteroids with One CO and OH in the Ring System

a. Monohydroxy-compounds

4'-Formyl-7-methoxy-3'-keto-1.2-cyclopentano-1.2.3.4-tetrahydrophenan-threne, 16-Formyl-18-nor-isoequilenin methyl ether $C_{19}H_{18}O_3$. See the tauto-meric 4'-hydroxymethylene-7-methoxy-3'-keto-1.2-cyclopentano-1.2.3.4-tetrahydrophenanthrene, 16-Hydroxymethylene-18-nor-isoequilenin methyl ether, p. 2719 s.

16-Formylœstrone $C_{19}H_{22}O_3$. See the tautomeric 16-hydroxymethylene-œstrone, p. 2719 s.

2-Formyl-Δ^4-androsten-17β-ol-3-one, 2-Formyltestosterone, Testosterone-2-aldehyde $C_{20}H_{28}O_3$. — Fmn. From testosterone (p. 2580 s) on heating with diethyl oxalate and sodium ethoxide in abs. benzene for 1 hr. on the water-bath, followed by heating of the resulting ethyl testosterone-2-glyoxylate with dil. H_2SO_4 (1937 Ciba).

17α-Formyl-Δ^4-androsten-17β-ol-3-one, 17-Formyltestosterone, Testosterone-17-aldehyde $C_{20}H_{28}O_3$. For configuration at C-17, see pp. 1363 s, 1364 s. — The crystalline aldehyde is probably a polymer (1941 b Prins). — Needle-like prisms (acetone - ether), m. 133–135° cor.; $[\alpha]_D^{13} +81°$ (acetone). Reduces alkaline silver diammine on heating slightly. Gives a strong, red colour reaction with 1.4-dihydroxynaphthalene + conc. HCl in glacial acetic acid on heating. Attempted sublimation at 130–150° (bath temperature) under 0.01 mm. pressure yielded prisms (acetone - ether), m. 200–202° cor., $[\alpha]_D^{13} +65°$ (acetone), which reduced alkaline silver diammine solution on warming, but gave no colour reaction with 1.4-dihydroxynaphthalene + conc. HCl (1941 b Prins).

Fmn. From 17-iso-Δ^4-pregnene-17β.20a.21-triol-3-one (p. 2726 s) on oxidation with 1 mol. periodic acid in aqueous dioxane at room temperature (4 hrs.); yield, 66% (1941 b Prins).

Rns. On further oxidation with periodic acid in aqueous dioxane at room temperature it gives Δ^4-androstene-3.17-dione (p. 2880 s). On treatment with diazomethane in dioxane - ether at room temperature for 60 hrs. it gives 20.21-epoxy-17-iso-Δ^4-pregnen-17β-ol-3-one (p. 2651 s) as the sole product when the unattacked aldehyde is removed with periodic acid (see above); otherwise there are also obtained prisms (ether - pentane), m. 172–4° cor., $[\alpha]_D^{13} +61°$ (acetone) [no reaction with alkaline silver diammine solution, no colour reaction with 1.4-dihydroxynaphthalene + conc. HCl, almost unattacked by CrO_3 in glacial acetic acid at room temperature] and a **compound** $C_{21}H_{30}O_3$ [platelets (acetone - ether); sublimes at 175–190° (bath temperature) under 0.01 mm. pressure and then m. 200–201° cor.; $[\alpha]_D^{13}$ +33° (acetone)] which reduces alkaline silver diammine solution on heating slightly, gives no colour reaction with 1.4-dihydroxynaphthalene + conc. hydrochloric acid, and is readily attacked by CrO_3 in glacial acetic acid at room temperature (1941 b Prins).

17β-Formyl-$\varDelta^4$-androsten-17α-ol-3-one, 17-Formyl-17-epi-testosterone, 17-epi-Testosterone-17-aldehyde $C_{20}H_{28}O_3$.

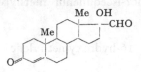

For configuration at C-17, see pp. 1363 s, 1364 s. — The crystalline aldehyde is probably a polymer (1941 b Prins). For its 3-enolic form, see 17β-formyl-$\varDelta^{3.5}$-androstadiene-3.17α-diol, p. 2320 s.

Prisms (probably anhydrous) (dioxane - ether), m. 162–4° cor., $[\alpha]_D^{13}$ +48° (acetone) (1941 b Prins); leaflets (possibly a hydrate) (acetone - ether), m. 142° to 146° cor., $[\alpha]_D^{15}$ +49° (acetone) (1941 a Prins; cf. 1941 b Prins). It strongly and rapidly reduces alkaline silver diammine solution at 40°; gives a strong, red colour reaction with 1.4-dihydroxynaphthalene + conc. HCl in glacial acetic acid on heating (1941 a, b Prins).

Fmn. From $\varDelta^4$-pregnene-17α.20β.21-triol-3-one (p. 2725 s) on oxidation with 1 mol. periodic acid in aqueous dioxane at room temperature (4 hrs.); yield of crystallized product, 52 % (1941 a, b Prins; 1942 Reichstein). — **Rns.** On further oxidation with periodic acid in aqueous dioxane at room temperature it gives $\varDelta^4$-androstene-3.17-dione (p. 2880 s). On treatment with diazomethane in dioxane - ether at room temperature it yields $\varDelta^4$-pregnen-17α-ol-3.20-dione (17-hydroxyprogesterone, p. 2843 s) (1941 b Prins).

Disemicarbazone $C_{22}H_{34}O_3N_6$ (not obtained pure). Needles (methanol); becomes brown between 280° and 300°, m. above 350° (1941 a Prins).

A mixture of the C-17 epimeric **17-formyl-$\varDelta^4$-androsten-17-ol-3-ones** and 17-formyl-$\varDelta^5$-androstene-3β.17-diols is obtained from $\varDelta^5$-androsten-3β-ol-17-one (dehydroepiandrosterone, p. 2609 s) on reaction with styryl bromide and lithium in abs. ether, followed by ozonization of the resulting mixture of epimeric 17-styryl-$\varDelta^5$-androstene-3β.17-diols in ethyl acetate after temporary protection of the nuclear double bond by bromination; the mixture is completely converted into the formyl-olones by Oppenauer oxidation (1938 Schering A.-G.).

1937 Ciba (Soc. pour l'ind. chim. à Bâle; Ges. f. chem. Ind. in Basel), *Swiss Pat.* 207 498 (issued 1940); *C. A.* **1941** 3038; *Ch. Ztbl.* **1940** II 1328; Ciba Pharmaceutical Products (Ruzicka), *U.S. Pat.* 2 281 622 (1938, issued 1942); *C. A.* **1942** 5958.

1938 Schering A.-G., *Fr. Pat.* 49 586 (issued 1939); *C. A.* **1942** 2689; *Ch. Ztbl.* **1940** I 428; Schering Corp. (Logemann, Dannenbaum), *U.S. Pat.* 2 344 992 (issued 1944); *C. A.* **1944** 3783.

1941 a Prins, Reichstein, *Helv.* **24** 396, 400.

    b Prins, Reichstein, *Helv.* **24** 945, 949–954.

1942 Reichstein, *Fr. Pat.* 888 228 (issued 1943); *Ch. Ztbl.* **1945** I 68; *U.S. Pat.* 2 389 325 (issued 1945); *C. A.* **1946** 1973.

Summary of the Monohydroxy-dioxosteroids (One CO and OH in the Ring System) with Pregnane Skeleton*

CO	OH	double bond	Config. at C-5	M. p. °C.	$[\alpha]_D(°)$ in alcohol	Absorpt.max. ($m\mu$) in alc.	Acetate m. p.°C.	Page
3.20	2α	Δ^4	—	185	+199	242	198	2837 s
3.20	4	satd.	β	158	—	—	—	2838 s
3.20	6α	Δ^4	—	193	+150[1])	240	resin	2838 s
3.20	6β	Δ^4	—	179	+107[1])	235.5	148	2838 s
3.20	6α	satd.	β	—	—	—	182	2839 s
3.20	11β	Δ^4	—	188	+223[2])	—	—	2840 s
3.20	12α	Δ^4	—	200	+205[2])	242	—	2840 s
3.20	12α	satd.	β	184	+135	—	131→136	2841 s
3.20	17α	Δ^4	—	223	+106[1])	242	—	2843 s
3.20	17β	Δ^4	—	193	+64[3])	240[4])	200	2844 s
3.20	17α	satd.	α	253	+24[1])	—	—	2845 s
3.20	x	Δ^4	—	184	+40	—	—	2845 s
3.21	17β	Δ^4	—	150	+83	243	—	2846 s
4.20	5	satd.	x	—	—	—	—	2847 s
11.20	3α	satd.	β	173	—	—	133	2847 s
11.20	3β	satd.	β	153	—	—	170	2848 s
12.20	3α	$\Delta^{9(11)}$	β	—	—	—	152→164	2849 s
12.20	3α	satd.	β	—	—	—	160	2849 s

* For values other than those chosen for this summary, see the compounds themselves.
[1]) In chloroform.   [2]) In acetone.   [3]) In dioxane.   [4]) Solvent not stated.

Δ^4-Pregnen-2α-ol-3.20-dione, 2α-Hydroxyprogesterone $C_{21}H_{30}O_3$. Prisms

(ether), m. 185° (Kofler block) (1939 Ehrhart); crystals (acetone - hexane), m. 182–3° (1953 Sondheimer); $[\alpha]_D^{20}$ +199° (alcohol); absorption max. in 95% alcohol at 242 $m\mu$ (log ε 4.22) (1953 Sondheimer; cf. also 1939 Ehrhart). — **Fmn.** The acetate is obtained, along with other compounds, from progesterone (p. 2782 s) on heating with lead tetraacetate in glacial acetic acid at 75–85° for 4 hrs. (1939 Ehrhart; cf. 1953 Sondheimer) or from 6-bromoprogesterone (p. 2814 s) on refluxing with anhydrous potassium acetate in glacial acetic acid for 4 hrs. (yield, 19%) (1953 Sondheimer); it is saponified by refluxing with $KHCO_3$ in aqueous methanol (1939 Ehrhart; 1953 Sondheimer).

Acetate $C_{23}H_{32}O_4$. Plates (ether), crystals (acetone - hexane), m. 197–8° (1939 Ehrhart; 1953 Sondheimer); $[\alpha]_D^{20}$ +164° (chloroform); absorption max. in 95% alcohol at 240 $m\mu$ (log ε 4.24); infra-red max. in chloroform at 1736, 1700, and 1684 cm.$^{-1}$ (1953 Sondheimer). — **Fmn.** From the preceding compound on treatment with acetic anhydride in pyridine (1939 Ehrhart). See also above.

1939 Ehrhart, Ruschig, Aumüller, *Ber.* **72** 2035, 2039.
1953 Sondheimer, Kaufmann, Romo, Martinez, Rosenkranz, *J. Am. Ch. Soc.* **75** 4712, 4715.

Pregnan-4-ol-3.20-dione $C_{21}H_{32}O_3$, m. 156–8⁰. — **Rns.** On heating of its benzoate (obtained with benzoyl chloride in pyridine) in a high vacuum it gives $\varDelta^4$-pregnene-3.20-dione (progesterone, p. 2782 s).

1937 CIBA (Soc. pour l'ind. chim. à Bâle; Ges f. chem. Ind. in Basel), *Brit. Pat.* 497394 (issued 1939); *C. A.* **1939** 3812; *Ch. Ztbl.* **1939** I 2828; *Fr. Pat.* 838916 (1938, issued 1939); *C. A.* **1939** 6875; *Ch. Ztbl.* **1939** II 169, 170; *Dutch Pat.* 51420 (1938, issued 1941); *Ch. Ztbl.* **1942** I 2561; CIBA PHARMACEUTICAL PRODUCTS (Miescher, Wettstein), *U.S. Pat.* 2229813 (1938, issued 1941); *C. A.* **1941** 3040; *Ch. Ztbl.* **1941** II 2467, 2468.

$\varDelta^4$-Pregnen-6α-ol-3.20-dione, 6α-Hydroxyprogesterone $C_{21}H_{30}O_3$. The acetate was originally considered an amorphous modification of the following epimer (1941 Ehrenstein); for the true structure, see 1952 Balant. — Needles (acetone - petroleum ether), m. 192–3⁰ cor.; $[\alpha]_D^{27}$ +150⁰ (chloroform). Absorption max. in alcohol at 240 mμ (ε 15570). Gives a yellow solution in conc. H_2SO_4 slowly developing a green fluorescence (1952 Balant).

The acetate has only a very weak progestational activity (1952 Balant).

Fmn. The acetate is obtained from the 6-acetate of allopregnane-5.6β-diol-3.20-dione (p. 2865 s) in dry chloroform containing 0.7 % of alcohol on passage of dry HCl at +5⁰ to —5⁰ (crude yield, 51 %) (1952 Balant; cf. 1941 Ehrenstein) or in the same way (below —5⁰) from the epimeric acetate (below) (quantitative crude yield) (1952 Balant); the acetate is saponified by methanolic KOH at room temperature (1952 Balant). — **Rns.** Gives allopregnane-3.6.20-trione (p. 2966 s) on treatment with 3 % H_2SO_4 in glacial acetic acid at room temperature for 44 hrs. (1952 Balant).

Acetate $C_{23}H_{32}O_4$. Only obtained as a resin; $[\alpha]_D^{28.5}$ +139⁰ (chloroform); absorption max. in alcohol at 235.5 mμ (ε 14910) (1952 Balant; cf. 1941 Ehrenstein). — **Fmn.** See above. May also be obtained from the above hydroxydiketone by treatment with acetic anhydride in dry pyridine (1952 Balant).

$\varDelta^4$-Pregnen-6β-ol-3.20-dione, 6β-Hydroxyprogesterone $C_{21}H_{30}O_3$. Originally (and arbitrarily) assigned 6α-configuration (1940 Ehrenstein); for the true 6β-configuration, see 1948 Ehrenstein. — Needles (acetone - petroleum ether), m. 178–9⁰ cor.; $[\alpha]_D^{28}$ +107⁰ (chloroform). Absorption max. in alcohol at 235.5 mμ (ε 12390). Gives a yellow solution in conc. H_2SO_4 slowly developing a green fluorescence (1952 Balant).

The progestational potency of the acetate is $^1/_3$ that of progesterone (1940 Ehrenstein); according to Balant (1952) it is even less active. Manifests possibly a slight cortin-like activity (1940 Ehrenstein).

References, pp. 2839 s, 2840 s

Fmn. The acetate is obtained from the 6-acetate of allopregnane-5.6β-diol-3.20-dione (p. 2865 s) in alcohol-free, dry chloroform on passage of dry HCl between —2⁰ and +2⁰, finally at room temperature (yield, 52%) or, in smaller yield, using purified carbon tetrachloride as a solvent at —5⁰ to +5⁰; it is saponified by treatment with 1.1 mols. of KOH in abs. alcohol at room temperature (1952 Balant; cf. 1940 Ehrenstein). — **Rns.** Gives allopregnane-3.6.20-trione (p. 2966 s) on heating of its acetate with 2.2 mols. of KOH in aqueous methanol on the water-bath for 1 hr. (1952 Balant; cf. 1940 Ehrenstein). The acetate gives the preceding, epimeric acetate on passage of dry HCl below —5⁰ through its solution in dry chloroform containing 0.7% of alcohol (1952 Balant).

Acetate $C_{23}H_{32}O_4$. Plates (ether - petroleum ether), m. 145–6⁰ cor. (1940 Ehrenstein); crystals (ether), m. 147–8⁰ cor. (1952 Balant); $[\alpha]_D^{17.5}$ +90⁰ (abs. alcohol) (1940 Ehrenstein), $[\alpha]_D^{32}$ + 101⁰ (chloroform) (1952 Balant). Absorption max. in alcohol at 235 mμ (ε 13970) (1952 Balant; cf. 1940 Ehrenstein). Gives a yellow solution in conc. H_2SO_4 slowly developing a green fluorescence (1952 Balant). Gives no yellow coloration with tetranitromethane in chloroform (1940 Ehrenstein). — **Fmn.** From the preceding compound on treatment with acetic anhydride in dry pyridine at room temperature (1952 Balant). See also above. — **Rns.** See above.

Pregnan-6α-ol-3.20-dione $C_{21}H_{32}O_3$. For the configuration at C-6, see 1947 Moffett. — Not isolated in a pure state. — **Fmn.** The acetate is obtained from a compound, m. 100⁰, said to be 3.6-diacetoxy-ætiocholanyl methyl ketone (pregnane-3α.6α-diol-20-one diacetate) * on hydrolysis with 0.8 mols. of KOH in methanol at 20⁰, followed by oxidation of the resulting 6-acetate with CrO_3 in strong acetic acid at room temperature; it is hydrolysed by refluxing with 2% methanolic KOH for 75 min. — **Rns.** Gives Δ⁴-pregnene-3.20-dione (progesterone, p. 2782 s) on heating with fused $KHSO_4$ at 130–180⁰ under 4 mm. pressure (1940 Marker; 1941 PARKE, DAVIS & Co.).

Acetate $C_{23}H_{34}O_4$. Prisms (ether - pentane), m. 182⁰ (1941 PARKE, DAVIS & Co.). — **Fmn.** See above.

The *benzoate* of a **pregnan-6-ol-3.20-dione** (not characterized; mode of formation not described) gives Δ⁴-pregnene-3.20-dione (progesterone, p. 2782 s) on heating in a high vacuum (1937 CIBA).

1937 CIBA (Soc. pour l'ind. chim. à Bâle; Ges. f. chem. Ind. in Basel), *Brit. Pat.* 497 394 (issued 1939); *C. A.* **1939** 3812; *Ch. Ztbl.* **1939** I 2828; *Fr. Pat.* 838 916 (1938, issued 1939); *C. A.* **1939** 6875; *Ch. Ztbl.* **1939** II 169, 170; *Dutch Pat.* 51 420 (1938, issued 1941); *Ch. Ztbl.* **1942** I 2561; CIBA PHARMACEUTICAL PRODUCTS (Miescher, Wettstein), *U.S. Pat.* 2 229 813 (1938, issued 1941); *C. A.* **1941** 3040; *Ch. Ztbl.* **1941** II 2467, 2468.

* Cf., however, the m. p. of this compound (p. 2325 s); the structure of the compound described by Marker (1940) seems doubtful (1946 Moffett).

1940 Ehrenstein, Stevens, *J. Org. Ch.* **5** 318, 323, 325–327.
Marker, Krueger, *J. Am. Ch. Soc.* **62** 79, 81.
1941 Ehrenstein, Stevens, *J. Org. Ch.* **6** 908, 917.
PARKE, DAVIS & Co. (Marker, Lawson), *U.S. Pat.* 2366204 (issued 1945); *C. A.* **1945** 1649;
Ch. Ztbl. **1945** II 1385, 1386.
1946 Moffett, Stafford, Linsk, Hoehn, *J. Am. Ch. Soc.* **68** 1857.
1947 Moffett, Hoehn, *J. Am. Ch. Soc.* **69** 1995.
1948 Ehrenstein, *J. Org. Ch.* **13** 214, 220.
1952 Balant, Ehrenstein, *J. Org. Ch.* **17** 1587, 1589 (footnote 7a), 1591–1593.

Δ⁴-Pregnen-11β-ol-3.20-dione, 11β-Hydroxyprogesterone $C_{21}H_{30}O_3$. The β-con-

figuration at C-11 is based on that of its precursor, corticosterone (p. 2853 s). — Rodlets (acetone), m. 187–8° cor.; $[\alpha]_D^{17}$ +223° (acetone) (1940 Reichstein). — Has only a very weak progestational activity, if any (1940 Reichstein). — **Fmn.** From Δ⁴-pregnene-11β.21-diol-3.20-dione (corticosterone, p. 2853 s) on treatment with 2.2 mols. of p-toluenesulphonyl chloride and 2.2 mols. of abs. pyridine in abs. chloroform at room temperature for 16 hrs., followed by refluxing of the resulting mixture with sodium iodide in acetone for 5 min. and then by reduction with zinc dust in glacial acetic acid; yield, 70% (1940 Reichstein). — **Rns.** Gives 11-ketoprogesterone (p. 2967 s) on oxidation with CrO_3 in glacial acetic acid at 20° (1940 Reichstein). On refluxing with conc. HCl·glacial acetic acid (1:4 by vol.) for 15–30 min. it gives Δ⁴·⁹⁽¹¹⁾-pregnadiene-3.20-dione (p. 2779 s) (1941 Shoppee; cf. 1943 Hegner); gives the same compound, along with an isomer*, on refluxing with 0.4 parts of 50% H_2SO_4 in 10 parts of acetic anhydride for a short time (1942 Reichstein).

1940 Reichstein, Fuchs, *Helv.* **23** 684, 687, 688.
1941 Shoppee, Reichstein, *Helv.* **24** 351, 354.
1942 Reichstein, *U.S. Pat.* 2409798 (issued 1946); *Ch. Ztbl.* **1947** 232; *Brit. Pat.* 560812 (issued 1944); *C. A.* **1946** 4486; *Ch. Ztbl.* **1945** II 1773.
1943 Hegner, Reichstein, *Helv.* **26** 715, 716.

Δ⁴-Pregnen-12α-ol-3.20-dione, 12α-Hydroxyprogesterone $C_{21}H_{30}O_3$. For the

configuration at C-12, see p. 1362 s. — Needles (ether), m. 164–7° cor., resolidifying to prisms m. 195–8° cor. (1941 Shoppee); needles (ether - petroleum ether), m. 198–200° cor. (1943a Hegner); crystals, m. 200–203° cor. (1948 Meystre); $[\alpha]_D^{15}$ +205° (acetone) (1941 Shoppee; cf. 1948 Meystre), $[\alpha]_{5461}^{15}$ +239° (acetone) (1941 Shoppee). Absorption max. in alcohol at 242 mμ (log ε 4.02) (1941 Shoppee). Gives no colour reaction with tetranitromethane in chloroform (1948 Meystre).

* Cf. also the dehydration of corticosterone 21-acetate (p. 2854 s).

Has only a very weak progestational activity, if any (1939 Ehrhart; 1948 Meystre).

Fmn. From pregnan-12α-ol-3.20-dione (below) on treatment with 0.8 mol. of bromine and a little HBr in glacial acetic acid, followed by refluxing of the resulting 4-bromo-derivative (p. 2850 s) with abs. pyridine for 5 hrs. (yield, ca. 24%) (1941 Shoppee); the acetate is similarly obtained from the corresp. acetate and is then saponified by refluxing with methanolic KOH (1939 Ehrhart; 1938 I.G. Farbenind.; 1938 Winthrop Chemical Co.; 1943a Hegner).

Rns. May be converted into $\Delta^{4.11}$-pregnadiene-3.20-dione (11-dehydroprogesterone, p. 2779 s) by thermal decomposition of its acetate or benzoate at 300–340° under 12 mm. pressure (1942 Reichstein; 1943a Hegner) or by refluxing of its p-toluenesulphonate with collidine in o-xylene for 24 hrs. (1948 Meystre).

Acetate $C_{23}H_{32}O_4$, m. 180° (1938 Winthrop Chemical Co.), 181° (1938 I.G. Farbenind.; 1949 Wagner), 186–7° cor. (1948 Meystre); $[\alpha]_D^{26} +215°$, $[\alpha]_{5461}^{26} +259°$ (both in chloroform); absorption max. in alcohol at 240 mμ (log ε 4.14) (1949 Wagner). — **Fmn.** and **Rns.** See above.

Benzoate $C_{28}H_{34}O_4$. Prisms (ether · petroleum ether), m. 164–6° cor.; $[\alpha]_D^{15} +96°$ (acetone). — **Fmn.** From the above hydroxydiketone on treatment with benzoyl chloride and abs. pyridine in abs. benzene for 20 hrs. at 20°, then for 1 hr. at 60° (1943a Hegner). — **Rns.** For thermal decomposition, see above.

p-Toluenesulphonate $C_{28}H_{36}O_5S$. Crystals (ether), m. 185–8° cor.; $[\alpha]_D^{24} +110°$ (chloroform). — **Fmn.** From the above hydroxydiketone on warming with p-toluenesulphonyl chloride in pyridine for 6 days at 40° (1948 Meystre). — **Rns.** See under those of the free hydroxydiketone.

Pregnan-12α-ol-3.20-dione $C_{21}H_{32}O_3$. For the configuration at C-12, see

p. 1362 s. — Needles (ether), crystals (acetone), m. 182–4° cor. (1941 Shoppee; 1943b Hegner); $[\alpha]_D^{17} +135°$, $[\alpha]_{5461}^{17} +164°$ (both in alcohol) (1941 Shoppee). — **Fmn.** The acetate is obtained from the 12-acetate of 24.24-diphenyl-$\Delta^{20(22).23}$-choladiene-3α.12α-diol (p. 2093 s) on oxidation with CrO_3 and strong acetic acid in chloroform at 15–20° (1944 Meystre; 1946 Ciba Pharmaceutical Products) or from the 12-acetate of pregnane-3α.12α-diol-20-one (p. 2330 s) on oxidation with CrO_3 in glacial acetic acid at 20° (yield, ca. 80%) (1939 Ehrhart; 1938 I.G. Farbenind.; 1938 Winthrop Chemical Co.; 1940 Reichstein; 1941 Shoppee; 1943a Hegner; cf. also 1944 Meystre); the acetate is hydrolysed by refluxing with methanolic KOH for 2 hrs. (1941 Shoppee; 1943b Hegner). See also the anthraquinonecarboxylate, p. 2842 s.

Rns. On treatment with 0.8 mol. of bromine and a little HBr in glacial acetic acid it gives the 4-bromo-compound (p. 2850 s), which on refluxing with abs. pyridine for 5 hrs. yields 12α-hydroxyprogesterone (p. 2840 s) (1941

195*

Shoppee); the acetate reacts similarly to give 12α-acetoxyprogesterone via 4-bromopregnan-12α-ol-3.20-dione acetate (1939 Ehrhart; 1938 I.G. FARBEN-IND.; 1938 WINTHROP CHEMICAL Co.; 1943a Hegner); when the p-toluene-sulphonate (below) is treated with 1 mol. bromine in glacial acetic acid and then refluxed with collidine for 3 hrs., it yields Δ^{11}-pregnene-3.20-dione (see the following reactions) and a smaller amount of $\Delta^{4.11}$-pregnadiene-3.20-dione (11-dehydroprogesterone, p. 2779 s) (1946 v. Euw). Pregnan-12α-ol-3.20-dione may be converted into Δ^{11}-pregnene-3.20-dione (p. 2793 s) by heating of its benzoate at 310° (bath temperature) under 12 mm. pressure in an atmosphere of CO_2 (1943b Hegner), by heating of its anthraquinone-β-carboxylate (below) at 295–300° (bath temperature) under 0.05 mm. pressure (1944 v. Euw), or by refluxing of its p-toluenesulphonate with collidine for 3 hrs. (1946 v. Euw).

Acetate $C_{23}H_{34}O_4$. Needles (ether - pentane), m. 132–4° cor. (1941 Shoppee); needles (acetone), m. 130–131° cor., resolidifying to prisms m. 136° cor. (1944 Meystre); leaflets (ether - pentane), m. 121–2°; $[\alpha]_D^{16} +141°$ (acetone) (1940 Reichstein). — **Fmn.** and **Rns.** See above.

Benzoate $C_{28}H_{36}O_4$. Plates (acetone), m. 166–7° cor.; $[\alpha]_D^{14} +92.6°$ (acetone). — **Fmn.** From the above hydroxydiketone on treatment with benzoyl chloride and abs. pyridine in abs. benzene for 20 hrs. at room temperature, then for 1 hr. at 60° (1943b Hegner). — **Rns.** For thermal decomposition, see under the reactions of the hydroxydiketone (above).

Anthraquinone-β-carboxylate $C_{36}H_{38}O_6$. Light yellow needles [benzene - methanol (1:3) + a little ether], m. 208–9° cor.; sublimes at 250–260° (bath temperature) under 0.05 mm. pressure. — **Fmn.** From the free hydroxy-diketone and anthraquinone-β-carboxylic acid chloride on brief heating with abs. pyridine in abs. benzene to boiling, followed by keeping at 20° for 16 hrs. From the corresponding ester of pregnane-3α.12α-diol-20-one (page 2330 s) on oxidation with CrO_3 in glacial acetic acid at 18° (1944 v. Euw). — **Rns.** For thermal decomposition, see above.

p-Toluenesulphonate $C_{28}H_{38}O_5S$. Prisms (ether - petroleum ether), m. 131° to 132° cor. — **Fmn.** From the free hydroxydiketone on treatment with p-toluenesulphonyl chloride and abs. pyridine in a sealed tube at 30° for 4 days (1946 v. Euw). — **Rns.** See under those of the hydroxydiketone (above).

1938 I. G. FARBENIND., *Fr. Pat.* 834072 (issued 1938); *C. A.* **1939** 3394; *Ch. Ztbl.* **1939** I 3931.
    WINTHROP CHEMICAL Co. (Bockmühl, Ehrhart, Ruschig, Aumüller), *U. S. Pat.* 2142170 (issued 1939); *C. A.* **1939** 3078; *Ch. Ztbl.* **1939** II 170.
1939 Ehrhart, Ruschig, Aumüller, *Angew. Ch.* **52** 363, 364.
1940 Reichstein, v. Arx, *Helv.* **23** 747, 750.
1941 Shoppee, Reichstein, *Helv.* **24** 351, 357–360.
1942 Reichstein, *Fr. Pat.* 879972 (issued 1943); *Ch. Ztbl.* **1945** I 195; *Swed. Pat.* 110155 (issued 1944); *Ch. Ztbl.* **1945** II 263.
1943 a Hegner, Reichstein, *Helv.* **26** 715, 719, 720.
        b Hegner, Reichstein, *Helv.* **26** 721, 723, 724.
1944 v. Euw, Lardon, Reichstein, *Helv.* **27** 821, 829, 835.
        Meystre, Frey, Wettstein, Miescher, *Helv.* **27** 1815, 1824.

1946 CIBA PHARMACEUTICAL PRODUCTS (Miescher, Frey, Meystre, Wettstein), *U.S. Pat.* 2461910
      (issued 1949); *C. A.* **1949** 3474; *Ch. Ztbl.* **1949** E 1870, 1871.
      v. Euw, Reichstein, Shoppee, *Helv.* **29** 654, 669.
1948 Meystre, Tschopp, Wettstein, *Helv.* **31** 1463.
1949 Wagner, Moore, Forker, *J. Am. Ch. Soc.* **71** 3856.

Δ⁴-Pregnen-17α-ol-3.20-dione, 17-Hydroxyprogesterone $C_{21}H_{30}O_3$. For con-

figuration at C-17, see pp. 1363 s, 1364 s. — Platelets (alcohol, acetone, or ethyl acetate), m. 212–5⁰
(Berl block) (1940, 1941 Pfiffner); leaflets (acetone - ether), m. 222–3⁰ cor. on heating not too
slowly [for partial isomerization on slow heating,
see below under reactions] (1941 v. Euw; cf. 1941 Prins; 1941 Hegner); $[\alpha]_D$
in chloroform +102⁰ (1940, 1941 Pfiffner), +106⁰ (1941 v. Euw), in acetone
+99⁰ (1941 Prins). Absorption max. in alcohol at 242 mμ (ε 18600) (1941
Pfiffner). Readily soluble in chloroform, insoluble in ether (1941 Pfiffner).
Gives a yellow solution with conc. H_2SO_4 (1941 v. Euw).

Has no progestational and only a slight if any cortin-like activity; its
androgenic potency in the rat test is of the same order as that of androsterone
(p. 2629 s) (1940, 1941 Pfiffner); it is also active in the capon comb test
(1939 Schmidt-Thomé; but cf. 1941 Pfiffner).

Occ. In adrenal glands of cattle (1940, 1941 Pfiffner); 50 mg. could be
isolated from 500 kg. of glands (1941 v. Euw).

Fmn. From 17β-formyl-Δ⁴-androsten-17α-ol-3-one (p. 2836 s) on treatment
with diazomethane in ether - dioxane at room temperature for 36 hrs.;
yield, 20–30% (1941 Prins). In very small yield, along with other compounds,
from the dibromide of Δ⁵-pregnene-3β.17α-diol-20-one (p. 2333 s) on oxidation with CrO_3 in glacial acetic acid at room temperature, followed by
debromination with zinc dust and potassium acetate on the water-bath and
then by heating with glacial acetic acid (1941 Hegner).

Rns. It is oxidized by CrO_3 in glacial acetic acid at room temperature
to give Δ⁴-androstene-3.17-dione (p. 2880 s) (1940, 1941 Pfiffner; 1941 Prins).
17-Hydroxyprogesterone is isomerized to give D-homo-compounds by the
following methods: when finely triturated crystals are slowly heated they
melt only partially at ca. 220⁰, partially changing to needles which only at
ca. 276⁰ are completely molten (cf. the following D-homo-isomer) (1941 v. Euw);
on heating in an evacuated sealed tube for some min. at 250⁰ it gives 17aβ-
methyl-D-homo-Δ⁴-androsten-17aα-ol-3.17-dione (m. 288⁰; formula V, page
2340 s) and an isomer m. 162–4⁰ cor. (1941 v. Euw) which very probably is
17β-methyl-D-homo-Δ⁴-androsten-17α-ol-3.17a-dione

(I) (1955 Fukushima); the same two isomers are
obtained under the conditions of the Oppenauer oxidation (heating with aluminium tert.-butoxide in
acetone - benzene at 100⁰ for 21 hrs.) (1941 v. Euw);
the isomer m. 288⁰ (V, p. 2340 s; yield, 18%) is ob-
tained, along with its 17a-epimer m. 182–4⁰ (VI, p. 2340 s; yield, 41%), on

refluxing 17-hydroxyprogesterone with 3 % methanolic KOH for 3 hrs. (almost no reaction on using 1.2 % methanolic KOH) (1941 v. Euw; 1943b Shoppee); when 17-hydroxyprogesterone is refluxed with conc. HCl - glacial acetic acid (1:10) for 5 min., it is for the most part unchanged and only a small amount is isomerized to the last-mentioned D-homo-isomer m. 182–4⁰ (1941 v. Euw). A not characterized "17-hydroxyprogesterone" gave $\Delta^{4.16}$-pregnadiene-3.20-dione (p. 2780 s) on heating with fuller's earth at 150–160⁰ under 0.1 mm. pressure (1939b CIBA).

17-Hydroxyprogesterone is not attacked by acetic anhydride in pyridine at room temperature (1940, 1941 Pfiffner); for acetylation performed after the closing date for this volume, see 1953 Turner.

Dioxime $C_{21}H_{32}O_3N_2$. Plates (50 % alcohol), m. 250–251⁰ dec. (Berl block) after sintering at ca. 240⁰ (1940, 1941 Pfiffner).

Disemicarbazone $C_{23}H_{36}O_3N_6$. Darkens at ca. 240⁰, sinters at 280–290⁰, and gradually blackens, but does not melt below 360⁰. Insoluble in alcohol (1940, 1941 Pfiffner).

17- iso - Δ^4- Pregnen -17β - ol -3.20- dione, 17- Hydroxy- 17- iso - progesterone

$C_{21}H_{30}O_3$. For configuration at C-17, see pp. 1363 s, 1364 s. The "*17-hydroxyprogesterone*" obtained from the acetate of the present compound by alkaline hydrolysis is actually 17aβ - methyl-D - homo - Δ^4 - androsten - 17aα - ol - 3.17 - dione (see below, under reactions).

Granula (ethyl acetate · hexane), m. 192–3⁰ cor.; [α]$_D$ +64.4⁰ (dioxane). Absorption max. (solvent not stated) at 240 mμ (log ε 4.25). — **Fmn.** From 17-ethynyltestosterone (p. 2648 s) on heating with bis-(p-toluenesulphamido)-mercury in 96 % alcohol on the water-bath for 100 hrs.; crude yield, 48 %. — **Rns.** It is isomerized to 17aα-methyl-D-homo-Δ^4-androsten-17aβ-ol-3.17-dione (m. 182–4⁰; formula VI, p. 2340 s) on chromatographic adsorption on Al$_2$O$_3$, and to the 17a-epimer (m. 288⁰; formula V, p. 2340 s) on heating with aqueous methanolic KOH on the water-bath (1943 Goldberg).

Acetate $C_{23}H_{32}O_4$. Crystals (methanol or ether), m. 198–200⁰ cor. vac.; [α]$_D$ +66⁰ (dioxane) (1938 Ruzicka; cf. 1939 Ruzicka *). —**Fmn.** From 17-ethynyltestosterone (p. 2648 s) on treatment with HgO, glacial acetic acid, acetic anhydride, and BF$_3$-ether at 20⁰ (1938 Ruzicka; 1939 a CIBA) or on shaking of the same compound or its acetate with mercuric acetate in abs. alcohol or, preferably, ethyl acetate for 24 hrs. at room temperature, followed by treatment with H$_2$S (1939 Ruzicka; 1940 CIBA; cf. 1943a Shoppee). From the 17-acetate of 17-iso-Δ^4-pregnene-17β.20-diol-3-one (p. 2721 s) on oxidation with CrO$_3$ in strong acetic acid at room temperature (1939c CIBA). From the 17-acetate of 17-iso-Δ^5-pregnene-3β.17β-diol-20-one (p. 2337 s) on refluxing

---

* The hydroxydiketones and their acetates, which were all formulated by 1938 Ruzicka as 17-hydroxypregnenediones and their acetates, resp., and then all reformulated by 1939 Ruzicka as D-homo-compounds, partly possess the first structure and partly the other (see 1943a, b Shoppee).

References, pp. 2845 s, 2846 s

with aluminium tert.-butoxide and acetone in benzene for 20 hrs. (1938 Ruzicka; 1938, 1939c CIBA) or on oxidation with CrO_3 in glacial acetic acid after temporary protection of the double bond by bromination (1938, 1939c CIBA). — **Rns.** On refluxing with methanolic KOH or with K_2CO_3 in 75% methanol it gives 17aβ-methyl-D-homo-Δ^4-androsten-17aα-ol-3.17-dione (m. 288°; see formula V on p. 2340 s) (1938, 1939 Ruzicka*; 1939c CIBA). A "17-acetoxyprogesterone" gives the diacetate of 17-iso(?)-Δ^4-pregnene-17β(?).21-diol-3.20-dione (p. 2859 s) on heating with lead tetraacetate in glacial acetic acid (1940 SCHERING A.-G.).

The *"20-anil of 17-hydroxy-17-iso-progesterone"*, obtained from 17-iso-Δ^5-pregnene-3β.17β-diol-20-one anil by Oppenauer oxidation (1939 Goldberg), is actually 17a-methyl-17a-anilino-D-homo-Δ^4-androstene-3.17-dione, since the starting anil has the corresponding D-homo-structure (see formula XI on p. 2340 s) (see 1943a, b Shoppee).

Allopregnan-17α-ol-3.20-dione $C_{21}H_{32}O_3$. A substance [leaflets (abs. alcohol), m. 270–272° cor.], which was assumed to possess this structure, is obtained along with androstane-3.17-dione from allopregnane-3β.17α-diol-20-one (substance L, p. 2335 s) on oxidation with CrO_3 in glacial acetic acid at room temperature (yield, 60%) (1938 Reichstein). After the closing date for this volume allopregnan-17α-ol-3.20-dione has been prepared from the same diolone by oxidation with N-bromoacetamide in pyridine - tert.-butyl alcohol at 20° (yield, 83%); it forms crystals (alcohol - chloroform), m. 251° to 253°, $[\alpha]_D^{20}$ +24° (chloroform), +50° (dioxane) (1950 Rosenkranz).

Δ^4-Pregnen-x-ol-3.20-dione, x-Hydroxyprogesterone $C_{21}H_{30}O_3$; x is not 2α, 2β, 6α, 6β, 17α, 17β, or 21 (1953 Sondheimer); the compound is also not identical with the 11β- and 12α-hydroxy-compounds (see p. 2840 s). — Needles (ether), m. 184°; $[\alpha]_D^{20}$ +40° (alcohol). Reduces ammoniacal silver salt solution (1939 Ehrhart). — **Fmn.** Along with other compounds, from progesterone (p. 2782 s) on heating with lead tetraacetate in glacial acetic acid at 75–85° for 7 hrs., followed by refluxing with $KHCO_3$ in aqueous methanol for 1 hr. (1939 Ehrhart).

1938 CIBA**, *Swiss Pat.* 228644 (issued 1943); *C. A.* **1949** 3476; *Ch. Ztbl.* **1944** II 1300.
    Reichstein, Gätzi, *Helv.* **21** 1497, 1503, 1504.
    Ruzicka, Meldahl, *Helv.* **21** 1760, 1767–1769.
1939 a CIBA**, *Fr. Pat.* 858127 (issued 1940); *Ch. Ztbl.* **1941** I 3258; *Brit. Pat.* 532262 (issued 1941); *C. A.* **1942** 874.
    b CIBA**, *Fr. Pat.* 860278 (issued 1941); *Ch. Ztbl.* **1941** I 3258.
    c CIBA**, *Fr. Pat.* 861318 (issued 1941); *Ch. Ztbl.* **1941** I 3259; *Brit. Pat.* 536621 (issued 1941); *C. A.* **1942** 1739; *Ch. Ztbl.* **1943** II 147; CIBA PHARMACEUTICAL PRODUCTS (Ruzicka), *U.S. Pat.* 2365292 (issued 1944); *C. A.* **1945** 4199; *Ch. Ztbl.* **1945** II 1233.
    Ehrhart, Ruschig, Aumüller, *Ber.* **72** 2035, 2039.

* See the footnote on p. 2844 s.
** Société pour l'industrie chimique à Bâle; Gesellschaft für chemische Industrie in Basel,

Goldberg, Aeschbacher, *Helv.* **22** 1188, 1190.
Ruzicka, Goldberg, Hunziker, *Helv.* **22** 707, 710, 714–716.
Schmidt-Thomé, see 1942 Butenandt.
1940 CIBA*, *Fr. Pat.* 863162 (issued 1941); *Ch. Ztbl.* **1941** II 3218; CIBA PHARMACEUTICAL PRODUCTS
      (Ruzicka), *U.S. Pat.* 2315817 (issued 1943); *C. A.* **1943** 5558.
Pfiffner, North, *J. Biol. Ch.* **132** 459.
SCHERING A.-G., *Fr. Pat.* 868396 (issued 1941); *Ch. Ztbl.* **1942** I 2560.
1941 v. Euw, Reichstein, *Helv.* **24** 879.
Hegner, Reichstein, *Helv.* **24** 828, 832, 843.
Pfiffner, North, *J. Biol. Ch.* **139** 855.
Prins, Reichstein, *Helv.* **24** 945, 950, 951.
1942 Butenandt, *Naturwiss.* **30** 4, 9.
1943 Goldberg, Aeschbacher, Hardegger, *Helv.* **26** 680, 682, 685, 686.
a Shoppee, Prins, *Helv.* **26** 185, 194.
b Shoppee, Prins, *Helv.* **26** 201, 205, 213, 216.
1950 Rosenkranz, Pataki, Kaufmann, Berlin, Djerassi, *J. Am. Ch. Soc.* **72** 4081, 4083.
1953 Sondheimer, Kaufmann, Romo, Martinez, Rosenkranz, *J. Am. Ch. Soc.* **75** 4712.
Turner, *J. Am. Ch. Soc.* **75** 3489, 3492.
1955 Fukushima, Dobriner, Heffler, Kritchevsky, Herling, Roberts, *J. Am. Ch. Soc.* **77** 6585, 6587.

17-iso-Δ^4-Pregnen-17β-ol-3-on-21-al $C_{21}H_{30}O_3$. For configuration at C-17, see

pp. 1363 s, 1364 s. — Needles (ethyl acetate), m. 142–3° cor. (1938 Butenandt); crystals (dil. acetone), m. 149–151° cor.; $[\alpha]_D^{18} +83°$ (alcohol); absorption max. in alcohol at 243 (4.2) and 315 (ca. 1.9) mμ (log ε) (1939 Miescher). Very sparingly soluble in hexane and ether, more readily in benzene, very readily in chloroform, alcohol, acetone, and ethyl acetate (1938 Butenandt). Gives an intense red coloration with 1.4-dihydroxynaphthalene + conc. HCl in glacial acetic acid at 60–70° (1939 Miescher). It rapidly and strongly reduces ammoniacal silver salt solution at room temperature (1938 Butenandt; 1939 Miescher).

Fmn. From the two C-21 epimeric ω-homo-17-iso-Δ^4-pregnene-17β.21.22-triol-3-ones (p. 2728 s) on oxidation with lead tetraacetate by shaking in dry air-free benzene under nitrogen at 45° for 15 min.; yield, 88% (1938 Butenandt). From a mixture of the aforesaid triolones on treatment with potassium periodate and sulphuric acid in aqueous methanol under nitrogen for 6 hrs. at 22° (1939 Miescher).

Rns. On attempted sublimation in a high vacuum at 145° it resinifies for the most part and only a small amount is dehydrated to give $\Delta^{4.17(20)}$-pregnadien-3-on-21-al (p. 2802 s); the dehydration proceeds more completely on distillation over anhydrous $CuSO_4$ in a high vacuum at 135° or on refluxing under nitrogen with 0.5–5% of iodine in m-xylene for 1–3 hrs., with glacial acetic acid for 4 hrs. (with simultaneous removal of the water formed in this reaction by repeated addition of acetic anhydride), or with propionic acid for 1 hr. On refluxing with acetic anhydride in glacial acetic acid under nitrogen for 2½ hrs. it gives the 21-enol acetate of the aforementioned keto-

---

* Société pour l'industrie chimique à Bâle; Gesellschaft für chemische Industrie in Basel

aldehyde ($\Delta^{4.16.20}$-pregnatrien-21-ol-3-one acetate, p. 2512 s) and a stereo-isomer(?) of this acetate [needles, (methanol), m. 262–4° cor. dec.] (1939 Miescher).

Dioxime $C_{21}H_{32}O_3N_2$. Needles with 1 H_2O (aqueous ethanol, acetone, and methanol), m. 141° with evolution of gas, resolidifies between 175° and 185°, and finally m. 208–210° dec. (1938 Butenandt); sinters at 144° cor., m. 215° cor. dec. (1939 Miescher).

1938 Butenandt, Peters, *Ber.* **71** 2688, 2695.
1939 Miescher, Wettstein, Scholz, *Helv.* **22** 894, 896, 901, 902, 905; cf. also CIBA*, *Fr. Pat.* 857 122 (issued 1940); *Ch. Ztbl.* **1941** I 1324; CIBA PHARMACEUTICAL PRODUCTS (Miescher, Fischer, Scholz, Wettstein), *U.S. Pat.* 2275790 (issued 1942); *C. A.* **1942** 4289.

Pregnan- or Allopregnan-5-ol-4.20-dione $C_{21}H_{32}O_3$. No constants given. — Fmn. From a mixture of pregnane-4.5.20-triols (see p. 2123 s) on oxidation with CrO_3 in glacial acetic acid (1941 CIBA).

1941 CIBA* (Miescher, Wettstein), *Swed. Pat.* 105 137 (issued 1942); *Ch. Ztbl.* **1944** II 143; CIBA PHARMACEUTICAL PRODUCTS (Miescher, Wettstein), *U.S. Pat.* 2323276 (issued 1943); *C. A.* **1944** 222.

Pregnan-3α-ol-11.20-dione $C_{21}H_{32}O_3$. Needles (acetone · ether, and methanol · ether), m. 172–4° cor. (Kofler block) (1944b v. Euw). — Fmn. The acetate is obtained from the 3-acetate of 1-methyl-2.2-diphenyl-1-(3α.11β-dihydroxyætiocholanyl)-ethylene (p. 2089 s) on ozonization in ethyl acetate at —80° (yield, 3%), along with the 3-acetate of pregnane-3α.11β-diol-20-one (p. 2327 s) (1944b v. Euw); from the latter acetate on oxidation with CrO_3 in glacial acetic acid at 20° (1944b v. Euw; cf. also 1942 N. V. ORGANON; 1943 Reichstein). The acetate is also obtained from 12α-bromopregnan-3α-ol-11.20-dione acetate (p. 2851 s) on heating with zinc dust and sodium acetate in glacial acetic acid for 15 min. at 70–80°; quantitative crude yield (1944b v. Euw). The acetate is saponified by treating with methanolic KOH at 20° for 20 hrs. (1944b v. Euw).

Rns. On heating with lead tetraacetate in 98% acetic acid in a sealed tube at 55° for 22 hrs. it gives a small amount of the 21-acetate of pregnane-3α.21-diol-11.20-dione (p. 2860 s) (1944c v. Euw; cf. also 1942 N. V. ORGANON; 1943 Reichstein). On hydrogenation of the acetate in glacial acetic acid in the presence of platinum oxide it yields the 3-acetate of pregnane-3α.11β.20β-triol (see p. 2108 s) (1944b v. Euw).

* Société pour l'industrie chimique à Bâle; Gesellschaft für chemische Industrie in Basel.

Acetate $C_{23}H_{34}O_4$. Needles (ether), crystals (ether · petroleum ether), m. 132⁰ to 133⁰ cor. (Kofler block); on heating very slowly it is completely molten only at 138⁰ cor.; once, hexagonal plates, m. 134–7⁰ cor. (Kofler block), were obtained; $[\alpha]_D^{13} + 122⁰$ (acetone) (1944b v. Euw). — **Fmn.** and **Rns.** See above.

Pregnan-3β-ol-11.20-dione.

$C_{21}H_{32}O_3$. Needles (ether · acetone), m. 152–3⁰ cor. (Kofler block) (1944a v. Euw; 1943 Reichstein). **Fmn.** The acetate is obtained from the 3-acetate of 1-methyl-2.2-diphenyl-1-(3β.11β-dihydroxy-ætiocholanyl)-ethylene (p. 2089 s) on ozonization in chloroform at —10⁰ (yield, 21%), along with the 3-acetate of pregnane-3β.11β-diol-20-one (p. 2328 s), or from the latter acetate on oxidation with CrO_3 in glacial acetic acid; the acetate is saponified by methanolic KOH at 20⁰ (1944a v. Euw). Along with other compounds, from pregnane-3.11.20-trione (p. 2967 s) on shaking with hydrogen and reduced platinum oxide in glacial acetic acid until 1 mol. of hydrogen has been absorbed; yield (isolated as acetate), 37% (1944a v. Euw; 1942 N. V. ORGANON; 1943 Reichstein).

Rns. On heating with lead tetraacetate in 98% acetic acid in a sealed tube at 55⁰ for 22 hrs., it gives 19% yield of the 21-acetate of pregnane-3β.21-diol-11.20-dione (p. 2861 s) (1944c v. Euw; 1942 N. V. ORGANON; 1943 Reichstein). On shaking of the acetate with hydrogen and reduced platinum oxide in glacial acetic acid until 1 mol. of hydrogen has been absorbed it gives the 3-acetate of pregnane-3β.20β-diol-11-one (see p. 2722 s); on complete hydrogenation the 3-acetate of pregnane-3β.11β.20β-triol (see p. 2108 s) is formed (1944a v. Euw).

Acetate $C_{23}H_{34}O_4$. Needles (ether·petroleum ether), m. 169–170⁰ cor. (Kofler block); $[\alpha]_D^{13} + 89⁰$ (acetone) (1944a v. Euw; cf. also 1942 N. V. ORGANON; 1943 Reichstein). — **Fmn.** and **Rns.** See above.

9-iso-Pregnan-3β-ol-11.20-dione

$C_{21}H_{32}O_3$. This structure was ascribed by Marker (1938) to **uran-3β-ol-11.20-dione** (m. 225⁰; *acetate*, m. 250⁰), obtained by partial reduction of urane-3.11.20-trione; for probable structure of urane and its derivatives as D-homo-steroids see, however, p. 1403 s.

1938 Marker, Wittle, Oakwood, *J. Am. Ch. Soc.* **60** 1567.

1942 N. V. ORGANON, *Swiss Pat.* 256509 (issued 1949); *Ch. Ztbl.* **1949** E 1869.

1943 Reichstein, *Fr. Pat.* 898140 (issued 1945); *Ch. Ztbl.* **1946** I 1747, 1748; *Brit. Pat.* 594878 (issued 1947); *C. A.* **1948** 2404; *U.S. Pat.* 2440874 (issued 1948); *C. A.* **1948** 5622; *Ch. Ztbl.* **1948** E 456.

1944 a v. Euw, Lardon, Reichstein, *Helv.* **27** 821, 826, 827, 831–833.
  b v. Euw, Lardon, Reichstein, *Helv.* **27** 821, 828, 829, 838, 839.
  c v. Euw, Lardon, Reichstein, *Helv.* **27** 1287, 1295.

$\Delta^{9(11)}$-**Pregnen-3α-ol-12.20-dione** $C_{21}H_{30}O_3$.

Acetate $C_{23}H_{32}O_4$. A compound [leaflets (ether), m. 150–152° cor., resolidifying, and remelting at 162–4° cor. (Kofler block); absorption max. in alcohol at 238 mμ (log ε 3.9)], which probably possesses this structure, is obtained in small yield, along with pregnan-3α-ol-11.20-dione acetate, from the acetate of Δ^{11}-pregnen-3α-ol-20-one (p. 2239 s) on treatment with N-bromoacetamide and sodium acetate in aqueous acetone-acetic acid at 16°, followed by oxidation of the resulting mixture* with CrO_3 in glacial acetic acid and then by heating with zinc dust and sodium acetate in glacial acetic acid at 70–80° (1944 v. Euw).

Pregnan-3α-ol-12.20-dione $C_{21}H_{32}O_3$.

Acetate $C_{23}H_{34}O_4$. Platelets (acetone - ether), m. 159–160° cor. (Kofler block); $[\alpha]_D^{20}$ +178° (acetone). — **Fmn.** From the 3-acetate of pregnane-3α.12α-diol-20-one (p. 2329 s) on oxidation with CrO_3 in glacial acetic acid at room temperature; crude yield, 95 % (1944 Katz).

1943 Hegner, Reichstein, *Helv.* **26** 721, 723.
 Reich, Reichstein, *Helv.* **26** 562, 564.
1944 v. Euw, Lardon, Reichstein, *Helv.* **27** 821, 839.
 Katz, Reichstein, *Pharm. Acta Helv.* **19** 231, 261.

3α-Hydroxy-12-ketonorcholanyl phenyl ketone $C_{30}H_{42}O_3$. Crystals (methanol), m. 176–8°; $[\alpha]_D^{25}$ +77.5° (dioxane). **Fmn.** The formate is obtained from that of 3α-hydroxy-12-ketocholanic acid chloride on treatment with diphenylcadmium in boiling benzene, followed by addition of dil. HCl; it is hydrolysed by refluxing with 5 % methanolic NaOH. — **Rns.** Gives 3.12-diketonorcholanyl phenyl ketone (p. 2970 s) on oxidation with CrO_3 in aqueous acetic acid below room temperature. On oxidation of the acetate with CrO_3 and some sulphuric acid in aqueous acetic acid at 50°, followed by refluxing with aqueous NaOH, it gives a small yield of 3α-hydroxy-12-ketonorcholanic acid (1945 Hoehn).

Dioxime $C_{30}H_{44}O_3N_2$. Crystals, m. 186–9°; $[\alpha]_D^{25}$ +125° (dioxane); moderately soluble in organic solvents (1945 Hoehn).

* Probably containing 9.11-dibromopregnane-3α.12-diol-20-one 3-acetate; for analogous reactions, see 1943 Reich; 1943 Hegner.

Formate $C_{31}H_{42}O_4$. Crystals (formic acid) (methanol), m. 183–4°; $[\alpha]_D^{25}$ + 102.5° (dioxane). — **Fmn.** See above. Is also obtained from the above hydroxydiketone on heating with formic acid (d 1.20) at 70–80° (1945 Hoehn).

Acetate $C_{32}H_{44}O_4$. Crystals, m. 196.5–197.5°; $[\alpha]_D^{25}$ + 85° (dioxane). — **Fmn.** From the above hydroxydiketone on refluxing with glacial acetic acid and acetic anhydride (1945 Hoehn). — **Rns.** See above.

1945 Hoehn, Moffett, *J. Am. Ch. Soc.* **67** 740.

12α-Bromopregnan-11β-ol-3.20-dione $C_{21}H_{31}O_3Br$. For configuration at C-11, see p. 1362 s. — Needles (chloroform), m. 245° to 246° cor. dec. (Kofler block). Sparingly soluble in acetone and ether (1944 v. Euw). — **Fmn.** From Δ^{11}-pregnene-3.20-dione (p. 2793 s) on treatment with N-bromoacetamide, sodium acetate, and acetic acid in aqueous acetone at 16°; yield, 51 % (1944 v. Euw; cf. 1943b Hegner; 1942 Reichstein). — **Rns.** Gives 12α-bromo-pregnane-3.11.20-trione (p. 2971 s) on oxidation with CrO_3 in glacial acetic acid at room temperature (1943b Hegner; 1944 v. Euw; 1942 Reichstein).

4-Bromopregnan-12α-ol-3.20-dione $C_{21}H_{31}O_3Br$. For configuration at C-12, see p. 1362 s. — Crystals (ether; not analysed), m. 156–160° cor. dec. (1941 Shoppee). — **Fmn.** From pregnan-12α-ol-3.20-dione (p. 2841 s) on treatment with less than 1 mol. of bromine and a little HBr in glacial acetic acid at room temperature (yield, 41 %) (1941 Shoppee); the acetate is similarly obtained from that of the hydroxydiketone (1938 I.G. FARBENIND.; 1938 WINTHROP CHEMICAL Co.; 1939 Ehrhart) in 85 % yield (1943a Hegner), and similarly also the p-toluenesulphonate (not isolated in a pure state) (1946 v. Euw). — **Rns.** Gives Δ^4-pregnen-12α-ol-3.20-dione (12α-hydroxy-progesterone, p. 2840 s) on refluxing with pyridine for 5 hrs. (1941 Shoppee); the acetate reacts similarly to give the corresponding acetate on refluxing with pyridine (1938 I.G. FARBENIND.; 1938 WINTHROP CHEMICAL Co.; 1939 Ehrhart; 1943a Hegner) or with collidine (1943a Hegner); refluxing of the p-toluenesulphonate with collidine for 3 hrs. affords a small amount of $\Delta^{4.11}$-pregnadiene-3.20-dione (11-dehydroprogesterone, p. 2779 s) (1946 v. Euw).

Acetate $C_{23}H_{33}O_4Br$. Leaflets (ether - petroleum ether), m. 175–7° cor. (Kofler block) (1943a Hegner; cf. also 1938 I.G. FARBENIND.; 1938 WIN-THROP CHEMICAL Co.; 1939 Ehrhart). — **Fmn.** and **Rns.** See above.

12α-Bromopregnan-3α-ol-11.20-dione $C_{21}H_{31}O_3Br$.

Acetate $C_{23}H_{33}O_4Br$. Needles (acetone - ether), m. 194–5° cor. (Kofler block). — **Fmn.** From 12α-bromopregnane-3α.11β-diol-20-one 3-acetate (p. 2351 s) on oxidation with CrO_3 in glacial acetic acid at 18°; yield, 90%. — **Rns.** Gives the acetate of pregnan-3α-ol-11.20-dione (page 2847 s) on heating with zinc dust and sodium acetate in glacial acetic acid at 70–80° for 15 min. (1944 v. Euw).

1938 I. G. FARBENIND., *Fr. Pat.* 834072 (issued 1938); *C. A.* **1939** 3394; *Ch. Ztbl.* **1939** I 3931.
WINTHROP CHEMICAL CO.(Bockmühl, Ehrhart, Ruschig, Aumüller), *U.S. Pat.* 2142170 (issued 1939); *C. A.* **1939** 3078; *Ch. Ztbl.* **1939** II 170.
1939 Ehrhart, Ruschig, Aumüller, *Angew. Ch.* **52** 363.
1941 Shoppee, Reichstein, *Helv.* **24** 351, 358.
1942 Reichstein, *Fr. Pat.* 887641 (issued 1943); *Ch. Ztbl.* **1944** II 878; *U.S. Pat.* 2403683 (1943, issued 1946); *C. A.* **1946** 6216, 6221; *Ch. Ztbl.* **1947** 232.
1943 a Hegner, Reichstein, *Helv.* **26** 715, 719.
      b Hegner, Reichstein, *Helv.* **26** 721, 725, 726.
1944 v. Euw, Lardon, Reichstein, *Helv.* **27** 821, 830, 831, 838.
1946 v. Euw, Reichstein, Shoppee, *Helv.* **29** 654, 669.

21-Diazopregnan-12α-ol-3.20-dione $C_{21}H_{30}O_3N_2$. For its *acetate*, see p. 2352 s.

21-Diazo-Δ⁵-pregnen-3β-ol-11.20-dione $C_{21}H_{28}O_3N_2$ (not analysed). Pale

yellowish leaflets (chloroform - ether), m. 161° to 163° cor. dec. (Kofler block). — **Fmn.** The acetate is obtained from 3β-acetoxy-11-keto-Δ⁵-ætiocholenic acid chloride on treatment with diazomethane in ether - benzene, first at 0°, then at 16°; it is hydrolysed by aqueous methanolic KOH at 17°. — **Rns.** On heating with glacial acetic acid at 95–102° it gives the 21-acetate of Δ⁵-pregnene-3β.21-diol-11.20-dione (p. 2860 s) (1946 v. Euw).

21-Diazopregnan-3α-ol-11.20-dione $C_{21}H_{30}O_3N_2$ (not analysed). Light yellow

prisms (acetone - ether), becoming opaque at ca. 110°, m. 169–170° cor. dec. (Kofler block).— **Fmn.** The acetate is obtained from 3α-acetoxy-11-keto-ætiocholanic acid chloride on treatment with diazomethane, first at —10°, then at 18°; it is hydrolysed by methanolic KOH. — **Rns.** On heating with 98.5% acetic acid at 95° for 40 min. it gives the 21-acetate of pregnane-3α.21-diol-11.20-dione (p. 2860 s) (1944 v. Euw).

Acetate $C_{23}H_{32}O_4N_2$ (not analysed). Light yellow platelets (ether - benzene, or benzene- acetone), m. ca. 135° cor. dec. (Kofler block) (1944 v. Euw).

21-Diazopregnan-3β-ol-11.20-dione $C_{21}H_{30}O_3N_2$. Not isolated in a pure state. —
Fmn. The acetate is obtained from 3β-acetoxy-11-keto-ætiocholanic acid chloride on treatment with diazomethane in ether - benzene, first at 0°, then at 18°; it is hydrolysed by KOH in strong methanol at 20°. — Rns. On heating with glacial acetic acid at 95–100° for ¹/₂ hr.
it gives the 21-acetate of pregnane-3β.21-diol-11.20-dione (p. 2861 s); the acetate reacts similarly on refluxing with glacial acetic acid to give the diacetate of the dioldione (1942 Reichstein; 1943 Lardon).

1942 Reichstein, *Swiss Pat.* 244 341 (issued 1947); *C. A.* 1949 5812; *U.S. Pat.* 2401 775 (1943, issued 1946); *C. A.* 1946 5884; *Ch. Ztbl.* 1947 74; *U.S. Pat.* 2404 768 (1945, issued 1946); *C. A.* 1946 6222; *Ch. Ztbl.* 1947 233.
1943 Lardon, Reichstein, *Helv.* 26 747, 752, 753.
1944 v. Euw, Lardon, Reichstein, *Helv.* 27 1287, 1296.
1946 v. Euw, Reichstein, *Helv.* 29 1913, 1919.

---

b. Polyhydroxy-dioxosteroids with One CO and One OH in the Ring System

Δ⁴-Pregnene-2α.21-diol-3.20-dione, 2α.21-Dihydroxyprogesterone, 2α-Hydroxydesoxycorticosterone $C_{21}H_{30}O_4$. Prisms (ether), m. 184° (Kofler block) (1939 Ehrhart; cf. 1944 v. Euw and 1953 Sondheimer for the structure). — Fmn. The diacetate is obtained, along with other compounds, from progesterone (p. 2782 s) on heating with lead tetraacetate in glacial acetic acid for 7 hrs. at 75–85° (1939 Ehrhart; 1953 Sondheimer); it is hydrolysed to give the above dioldione by $KHCO_3$ in aqueous methanol (1939 Ehrhart). The diacetate is also obtained, in small yield, from the acetate of desoxycorticosterone (p. 2820 s) on refluxing with N-bromosuccinimide in dry carbon tetrachloride for 10 min. with illumination, followed by refluxing of the resulting crude 6-bromo-derivative with anhydrous potassium acetate in glacial acetic acid for 4 hrs. (1953 Sondheimer).

Rns. Gives 2α-hydroxy-3-keto-Δ⁴-ætiocholenic acid on oxidation with periodic acid in aqueous methanol at room temperature (1939 Ehrhart).

Diacetate $C_{25}H_{34}O_6$. Needles (methanol), m. 198° (Kofler block) (1939 Ehrhart); crystals (acetone - hexane), m. 195–6° (1953 Sondheimer); $[\alpha]_D^{20} + 165°$ (alcohol) (1939 Ehrhart), $+154°$ to $+157°$ (chloroform) (1953 Sondheimer). Absorption max. in 95% alcohol at 240 mμ (log ε 4.22) (1953 Sondheimer; see also 1939 Ehrhart); infra-red max. in chloroform at 1736, 1700, and 1686 cm.⁻¹ (1953 Sondheimer). Reduces ammoniacal silver salt solution (1939 Ehrhart). — Fmn. See above.

1939 Ehrhart, Ruschig, Aumüller, *Ber.* 72 2035, 2038.
1944 v. Euw, Lardon, Reichstein, *Helv.* 27 1287, 1288.
1953 Sondheimer, Kaufmann, Romo, Martinez, Rosenkranz, *J. Am. Ch. Soc.* 75 4712, 4715.

Δ^4-**Pregnene-6β.21-diol-3.20-dione, 6β.21-Dihydroxyprogesterone, 6β-Hydroxydesoxycorticosterone** $C_{21}H_{30}O_4$. For configuration at C-6, see 1948 Ehrenstein.

Diacetate $C_{25}H_{34}O_6$. Crystals (acetone - petroleum ether), m. 84–88°; $[\alpha]_D^{27}$ +114° (acetone). Absorption max. in abs. alcohol at ca. 236 mμ (ε 16900) (1941 Ehrenstein). — For physiological properties, see 1941 Ehrenstein. — **Fmn.** From the 6.21-diacetate of allopregnane-5α.6β.21-triol-3.20-dione (p. 2867 s) on passage of dry hydrogen chloride through its solution in chloroform for 3 hrs. not above 2°; yield, ca. 86% (1941 Ehrenstein).

1941 Ehrenstein, *J. Org. Ch.* **6** 626, 636, 637, 644, 645.
1948 Ehrenstein, *J. Org. Ch.* **13** 214, 221.

Δ^4-**Pregnene-11α.21-diol-3.20-dione** $C_{21}H_{30}O_4$. For inversion of the 11β-configuration originally ascribed, see p. 1362 s. — Colourless crystals. — **Fmn.** The 21-acetate is obtained from that of pregnane-11α.21-diol-3.20-dione (p. 2855 s) on bromination in glacial acetic acid, followed by refluxing for 5 hrs. with abs. pyridine; it is hydrolysed by methanolic HCl or aqueous methanolic $KHCO_3$ (1942 b N. V. ORGANON).

21-Acetate $C_{23}H_{32}O_5$. Colourless needles (acetone - ether) (1942 b N. V. ORGANON). — **Fmn.** and **Rns.** See above.

Δ^4-**Pregnene-11β.21-diol-3.20-dione, Corticosterone** $C_{21}H_{30}O_4$ (p. 153). *Reichstein's Substance H, Kendall and Mason's Compound B*. For the name "corticosterone", see 1937 Reichstein. For the β-configuration at C-11, see 1947 v. Euw; cf. also p. 1362 s. — Triangular plates (acetone), m. 180–182° cor. (Kofler block) (1937 Reichstein; 1944 v. Euw). Absorption max. in alcohol at 240 mμ (ε 20000) (1937 Reichstein; 1939 Dannenberg). With conc. H_2SO_4 it gives an orange-yellow solution with an intense green fluorescence (1937 Reichstein; 1944 v. Euw).

For the Zimmermann test with m-dinitrobenzene and potassium hydroxide, see 1944 Zimmermann. For polarographic determination, see 1940 Wolfe.

Occ. For isolation of corticosterone from adrenal glands, see also 1937 Kendall; 1937 N. V. ORGANON; 1938 ROCHE-ORGANON, INC.

Fmn. Corticosterone is obtained from Δ^4-pregnene-11β.17α.20β.21-tetrol-3-one (p. 2765 s) on refluxing with 30% H_2SO_4 in propanol for 20 min. (1938 CIBA). The 21-acetate is obtained from the 21-acetate of pregnane-11β.21-diol-3.20-dione (p. 2856 s) on treatment with 1.1 mols. bromine in glacial acetic acid at 10⁰, followed by refluxing of the resulting 4-bromo-derivative with abs. pyridine for 5 hrs. (small yield); it is hydrolysed by $KHCO_3$ in aqueous methanol at room temperature (1944 v. Euw; 1943 Reichstein). See also the 11-acetate (below).

Rns. Gives 11β-hydroxy-3-keto-Δ^4-ætiocholenic acid on oxidation with periodic acid in aqueous dioxane at 22⁰ (1947 v. Euw). On hydrogenation in abs. alcohol in the presence of palladium it gives allopregnane-11β.21-diol-3.20-dione (p. 2856 s) (1937 Mason). On refluxing of the 21-acetate with conc. HCl - glacial acetic acid (1:4 by vol.) for 30 min., followed by reacetylation, it gives the acetates of $\Delta^{4\cdot9(11)}$-pregnadien-21-ol-3.20-dione [9(11)-anhydrocorticosterone, p. 2819 s; 26%] and $\Delta^{4\cdot8(?)}$-pregnadien-21-ol-3.20-dione (see p. 2818 s; 17%), and a trace of a substance crystallizing as plates m. 169⁰ cor. (1943 Shoppee; 1942 Reichstein); when conc. HCl - glacial acetic acid (1:9 by vol.) is used in the above reaction, there are formed 9(11)-anhydrocorticosterone acetate and a small amount of a compound $C_{23}H_{32}O_5$ [plates (ethyl acetate - pentane), m. 122–5⁰ cor.; $[\alpha]_D^{20} + 187^\circ$, $[\alpha]_{5461}^{20} + 222^\circ$ (both in acetone); reacts as does corticosterone acetate] assumed to be *11-epi-corticosterone acetate* (1943 Shoppee), but which was very probably impure starting material (Gallagher, see 1949 Fieser); true 11-epi-corticosterone acetate, described after the closing date for this volume, m. 163–5⁰ and shows $[\alpha]_D^{23}$ + 159⁰ in acetone (1953 Eppstein).

Corticosterone may be converted into Δ^4-pregnen-11β-ol-3.20-dione (11β-hydroxyprogesterone, p. 2840 s) by treatment with 2.2 mols. of p-toluenesulphonyl chloride and 2.2 mols. of abs. pyridine in chloroform at room temperature for 16 hrs., followed by refluxing of the resulting mixture with sodium iodide in acetone for 5 min. and then by treatment with zinc and glacial acetic acid at room temperature (1940 Reichstein).

For *corticosterone 21-dihydrogen phosphate* $C_{21}H_{31}O_7P$, made with $POCl_3$ in abs. pyridine, followed by acidification, see 1937 ROCHE-ORGANON, INC.

Corticosterone 11-acetate $C_{23}H_{32}O_5$. Colourless crystals. — **Fmn.** From the diacetate of pregnane-11β.21-diol-3.20-dione (p. 2856 s) on treatment with 1 mol. of bromine in glacial acetic acid, followed by boiling with abs. pyridine for 5 hrs. or heating with dimethylaniline for 2 hrs. at 150⁰, and hydrolysis of the resulting corticosterone diacetate with aqueous methanolic $KHCO_3$ (1942a N. V. ORGANON).

Corticosterone 21-acetate $C_{23}H_{32}O_5$ (p. 153). Granula or leaflets (acetone), m. ca. 145–146.5⁰ changing to needles, m. 152.5–153⁰ cor. (1937 Reichstein; 1937 ROCHE-ORGANON, INC.); needles (methanol), m. 148–152⁰ cor. (1940 Kuizenga); needles (acetone - ether), m. 144–5⁰ cor. (Kofler block) (1944 v. Euw); m. 147.5–148.5⁰ cor. (1943 Shoppee); $[\alpha]_D^{20} + 195^\circ$, $[\alpha]_{5461}^{20} + 236^\circ$ (1943 Shoppee). With conc. H_2SO_4 it gives an orange solution with an intense green fluorescence (1943 Shoppee). — **Fmn.** From corticosterone by the action of acetic anhydride in pyridine at room temperature (1937 Reichstein; 1937 ROCHE-ORGANON, INC.; 1940 Kuizenga). See also under formation of corticosterone itself (p. 2853 s). — **Rns.** See those of corticosterone, above.

Corticosterone 21-propionate $C_{24}H_{34}O_5$. Crystals (acetone), m. 180–182° cor. — **Fmn.** From corticosterone on heating with propionic anhydride in abs. pyridine for $1\frac{1}{2}$ hrs. at 80° (1940 Kuizenga).

Corticosterone 21-butyrate $C_{25}H_{36}O_5$ (p. 153). Crystals (ether), m. 170° (1937 ROCHE-ORGANON, INC.); crystals (acetone), m. 168–9° (1940 Kuizenga). — **Fmn.** From corticosterone on treatment with butyryl chloride in abs. pyridine with cooling (1937 ROCHE-ORGANON, INC.). From corticosterone on heating with butyric anhydride in abs. pyridine for 3 hrs. at 80° (1940 Kuizenga).

Corticosterone 21-caproate $C_{27}H_{40}O_5$. Crystals (acetone), m. 130–132°. — **Fmn.** From corticosterone and caproyl chloride in abs. pyridine at room temperature (1940 Kuizenga).

Corticosterone 21-diethylacetate $C_{27}H_{40}O_5$. Crystals (acetone and methanol), m. 179–180°. — Is four times as active as free corticosterone in adrenalectomized rats. — **Fmn.** From corticosterone and diethylacetyl chloride in abs. pyridine at room temperature (1940 Kuizenga).

Corticosterone 21-heptanoate, Corticosterone 21-œnanthate $C_{28}H_{42}O_5$. Crystals (acetone?), m. 140–142°. — **Fmn.** From corticosterone and heptanoic anhydride on heating in pyridine for 3 hrs. at 100° (1940 Kuizenga).

Corticosterone 21-palmitate $C_{37}H_{60}O_5$ (p. 153). Crystals, m. ca. 90° (1937 ROCHE-ORGANON, INC.); crystals (methanol), m. 82–4° (1940 Kuizenga). — **Fmn.** From corticosterone and palmitoyl chloride in pyridine at room temperature (1937 ROCHE-ORGANON, INC.; 1940 Kuizenga).

Corticosterone 21-(hydrogen succinate) $C_{25}H_{34}O_7$ (p. 153). Crystals (ether), m. 194° (1937 ROCHE-ORGANON, INC.); crystals (acetone), m. 195–7° (1940 Kuizenga). — **Fmn.** From corticosterone and succinic anhydride in pyridine at room temperature (1937 ROCHE-ORGANON, INC.) or at 80° (3 hrs.) (1940 Kuizenga). — Its *sodium salt* is very soluble in water (1940 Kuizenga).

Corticosterone 21-benzoate $C_{28}H_{34}O_5$ (p. 153). Crystals (methanol), m. 199° to 201°. — **Fmn.** From corticosterone and benzoyl chloride in pyridine at room temperature (1937 ROCHE-ORGANON, INC.; 1940 Kuizenga).

For *corticosterone 21-dinitrobenzyl ether* and *21-triphenylmethyl ether*, see 1937 ROCHE-ORGANON, INC.

Pregnane-11α.21-diol-3.20-dione $C_{21}H_{32}O_4$.

For inversion of the 11β-configuration originally ascribed, see p. 1362 s.

21-Acetate $C_{23}H_{34}O_5$. Colourless crystals (ether-petroleum ether). — **Fmn.** From 3β.11α-diacetoxy-ætiocholanic acid chloride on treatment with diazomethane in ether-benzene, followed by hydrolysis with KOH in strong methanol, oxidation of the resulting crude 21-diazopregnane-3β.11α-diol-20-one with acetone and aluminium phenoxide in abs. benzene, and heating with glacial acetic acid for 30 min. at 95–100°. — **Rns.** Gives the 21-acetate of Δ^4-pregnene-11α.21-diol-3.20-dione (p. 2853 s) on bromination in glacial acetic acid, followed by refluxing with abs. pyridine (1942 b N. V. ORGANON).

Pregnane-11β.21-diol-3.20-dione $C_{21}H_{32}O_4$. For the β-configuration at C-11 see, e.g., p. 1362 s.

21 - Acetate $C_{23}H_{34}O_5$. Needles (ether), m. 158–9° cor. (Kofler block); $[\alpha]_D^{18} +128°$, $[\alpha]_{5461}^{18} +146°$ (both in acetone). Gives no colour reaction with conc. H_2SO_4. It strongly reduces alkaline silver diammine solution at room temperature (1944 v. Euw). — **Fmn.** From the 21-acetate of pregnane-3β.11β.21-triol-20-one (p. 2354 s) on heating with aluminium phenoxide and acetone in abs. benzene in a slightly evacuated sealed tube on the water-bath for 26 hrs. (yield, 42%) (1944 v. Euw; 1943 Reichstein); is similarly obtained from the 21-acetate of the 3-epimeric triolone (p. 2354 s) (1944 v. Euw). — **Rns.** Gives the acetate of pregnan-21-ol-3.11.20-trione (p. 2973 s) on oxidation with CrO_3 in glacial acetic acid at room temperature (1944 v. Euw). On treatment with 1.1 mols. of bromine in glacial acetic acid at 10°, followed by refluxing of the resulting bromo-derivative with abs. pyridine for 5 hrs., it gives the 21-acetate of corticosterone (p. 2853 s) (1944 v. Euw; 1943 Reichstein).

Diacetate $C_{25}H_{36}O_6$. — **Fmn.** From 3α.11β-diacetoxy-ætiocholanic acid chloride on treatment with diazomethane in ether - benzene, followed by hydrolysis with KOH in strong methanol at room temperature, oxidation with acetone and aluminium phenoxide in abs. benzene, and heating with glacial acetic acid for 30 min. at 100° (or first heating with glacial acetic acid and then oxidation with CrO_3 in glacial acetic acid) (1942 a N. V. ORGA-NON). — **Rns.** See corticosterone 11-acetate (p. 2854 s).

Allopregnane-11β.21-diol-3.20-dione, *Dihydrocorticosterone* $C_{21}H_{32}O_4$. For the allo-configuration, cf. also 1938 Steiger; 1947 v. Euw. — Crystals (aqueous alcohol), m. 184° to 187°; $[\alpha]_{5461}^{25} +157°$ (alcohol). — Has only a very weak cortin-like activity. — **Fmn.** From corticosterone (p. 2853 s) on hydrogenation in abs. alcohol in the presence of palladium. — **Rns.** Gives 11β-hydroxy-3-keto-ætioallocholanic acid on oxidation with periodic acid in dil. H_2SO_4 at room temperature. On hydrogenation in abs. alcohol in the presence of platinum oxide it yields allopregnane-3β.11β.20.21-tetrol (p. 2138 s) (1937 Mason).

1936 CIBA*, *Swiss Pat.* 210715 (issued 1940); *C. A.* **1941** 5652; *Ch. Ztbl.* **1941** I 3259.
1937 Kendall, *Cold Spring Harbor Symp. Quant. Biol.* **5** 299.
    Mason, Hoehn, McKenzie, Kendall, *J. Biol. Ch.* **120** 719, 733.

    * Société pour l'industrie chimique à Bâle; Gesellschaft für chemische Industrie in Basel.

N. V. ORGANON (David), *U.S. Pat.* 2 115 621 (issued 1938); *C. A.* **1938** 4728; *Ch. Ztbl.* **1938** II
   892.
Reichstein, *Helv.* **20** 953, 957, 964 et seq.
ROCHE-ORGANON, INC. (Reichstein, Schlittler), *U.S. Pat.* 2 183 589 (issued 1939); *C. A.* **1940**
   2538; *Ch. Ztbl.* **1940** II 932.
1938 CIBA, *Fr. Pat.* 838 916 (issued 1939); *C. A.* **1939** 6875; *Ch. Ztbl.* **1939** II 169, 170.
ROCHE-ORGANON, INC. (Reichstein), *U.S. Pat.* 2 166 877 (issued 1939); *C. A.* **1939** 8925;
   Ch. Ztbl. **1940** I 1232, 1233.
Steiger, Reichstein, *Helv.* **21** 161.
1939 Dannenberg, *Abhandl. Preuss. Akad. Wiss. Math.-naturwiss. Kl.* **1939** No. 21, pp. 10, 53.
1940 Kuizenga, Cartland, *Endocrinology* **27** 647.
Reichstein, Fuchs, *Helv.* **23** 684, 687.
Wolfe, Hershberg, Fieser, *J. Biol. Ch.* **136** 653, 677 et seq.
1941 Reichstein, *Fr. Pat.* 888 825 (issued 1943); *Ch. Ztbl.* **1944** II 1203.
1942 a N. V. ORGANON, *Swiss Pat.* 254 994 (issued 1949); *Ch. Ztbl.* **1950** I 2388.
   b N. V. ORGANON, *Swiss Pat.* 254 995 (issued 1949); *Ch. Ztbl.* **1950** I 2388.
Reichstein, *Fr. Pat.* 52 035 (issued 1943); *Ch. Ztbl.* **1945** I 195, 196; *Brit. Pat.* 560 812 (1944,
   issued 1945); *C. A.* **1946** 4486; *Ch. Ztbl.* **1945** II 1773; *U.S. Pat.* 2 409 798 (issued 1946);
   C. A. **1947** 1397; *Ch. Ztbl.* **1947** 232.
1943 Reichstein, *Fr. Pat.* 898 140 (issued 1945); *Ch. Ztbl.* **1946** I 1747; *Brit. Pat.* 594 878 (issued
   1947); *C. A.* **1948** 2404; *U.S. Pat.* 2 440 874 (issued 1948); *C. A.* **1948** 5622; *Ch. Ztbl.* **1948** E
   456.
Shoppee, Reichstein, *Helv.* **26** 1316.
1944 v. Euw, Lardon, Reichstein, *Helv.* **27** 1287, 1293–1295.
Zimmermann, *Vitamine und Hormone* **5** [1952], pp. 1, 11, 124, 152, 163.
1947 v. Euw, Reichstein, *Helv.* **30** 205, 208, 210, 214.
1949 Fieser, Fieser, *Natural Products Related to Phenanthrene*, 3rd Ed., New York, p. 409.
1953 Eppstein, Meister, Peterson, Murray, Leigh, Lyttle, Reineke, Weintraub, *J. Am. Ch. Soc.*
   75 408.

$\varDelta^4$ - Pregnene - 12α.21 - diol - 3.20 - dione,   12α - Hydroxydesoxycorticosterone

$C_{21}H_{30}O_4$. For configuration at C-12, see page
1362 s. — Crystals (acetone - ether), probably
a hydrate, m. 98–124° with evolution of gas;
after drying in a high vacuum at 70° it shows
$[\alpha]_D^{21} + 186°$, $[\alpha]_{5461}^{21} + 221°$ (both in dioxane). —
The 21-acetate and the diacetate have a very
weak, if any, cortin-like activity in the Everse-de Fremery test; the diacetate
is inactive in the anti-insulin test of Jensen-Grattan (1943 Fuchs).

Fmn. The 21-acetate is obtained from that of pregnane-12α.21-diol-3.20-
dione (below) on treatment with 1 mol. bromine in glacial acetic acid with
cooling, followed by refluxing of the resulting 4-bromo-derivative (p. 2864 s)
with abs. pyridine for 5 hrs.; the diacetate is similarly formed starting with
the corresponding saturated diacetate; on hydrolysis of the 21-acetate with
$KHCO_3$ in aqueous methanol at room temperature it gives the free pregnene-
dioldione (1943 Fuchs; 1943 Reichstein).

12-Acetate $C_{23}H_{32}O_5$. Octahedra (acetone - ether), m. 188–192° cor. (Kofler
block); $[\alpha]_D^{19} + 185°$, $[\alpha]_{5461}^{19} + 226°$ (both in acetone). — **Fmn.** From the
diacetate (below) on hydrolysis with $KHCO_3$ in aqueous methanol at 20°
(1943 Fuchs; 1943 Reichstein).

196*

References, p. 2858 s

21-Acetate $C_{23}H_{32}O_5$. Needles (acetone - ether), m. 182–4° cor. (Kofler block);
$[\alpha]_D^{21} + 204°$, $[\alpha]_{5461}^{21} + 252°$ (both in acetone). Absorption max. in abs. alcohol
at 244 mμ (log ε 4.12) (1943 Fuchs; 1943 Reichstein). — **Fmn.** See above.

Diacetate $C_{25}H_{34}O_6$. Needles (benzene - ether), m. 158–9° cor. (Kofler block);
$[\alpha]_D^{17} + 198°$ (acetone). Absorption max. in alcohol at 244 mμ (log ε 4.15). —
Fmn. From the free dioldione (above) on heating with acetic anhydride
and abs. pyridine for 1 hr. at 95° (1943 Fuchs; 1943 Reichstein). See also
under the free dioldione. — **Rns.** For hydrolysis, see the 12-acetate (above).

21-Acetate 12-p-toluenesulphonate $C_{30}H_{38}O_7S$. Prisms (ether), m. 181–2° cor.
dec. (Kofler block); once crystals were obtained m. 158–160° cor., resolidifying, and remelting at 180° cor. — **Fmn.** From the above 21-acetate on
keeping with p-toluenesulphonyl chloride in abs. pyridine for 5 days in a
sealed tube at 30° (1946 v. Euw).

Pregnane-12α.21-diol-3.20-dione $C_{21}H_{32}O_4$.

For configuration at C-12, see
p. 1362 s. — **Fmn.** The 21-acetate is obtained
from the 21-acetate of pregnane-3α.12α.21-
triol-20-one (p. 2356 s) on refluxing with aluminium phenoxide and acetone in abs. benzene
for 20 hrs. (yield, 40%); the diacetate is obtained from the 12.21-diacetate of the same
triolone on oxidation with CrO_3 in glacial acetic acid at 20° (1943 Fuchs;
1943 Reichstein) or from 21-diazopregnan-12α-ol-3.20-dione acetate (p. 2352 s)
on heating with glacial acetic acid (1943 Reichstein).

Rns. Oxidation of the 21-acetate with CrO_3 in glacial acetic acid affords
the acetate of pregnan-21-ol-3.12.20-trione (p. 2974 s). On treatment of the
21-acetate with 1 mol. bromine in glacial acetic acid with cooling it gives
the 4-bromo-derivative (p. 2864 s) which on refluxing with abs. pyridine for
5 hrs. yields the 21-acetate of Δ^4-pregnene-12α.21-diol-3.20-dione (above);
the diacetate reacts similarly (1943 Fuchs; 1943 Reichstein). The 21-acetate
12-p-toluenesulphonate gives the acetate of Δ^{11}-pregnen-21-ol-3.20-dione (page
2830 s) on refluxing with collidine for 5 hrs. (1946 v. Euw).

21-Acetate $C_{23}H_{34}O_5$. Needles or prisms (benzene), m. 190–192° cor. (Kofler
block); $[\alpha]_D^{14} + 146°$ (acetone) (1943 Fuchs; 1943 Reichstein). — **Fmn.** and
Rns. See above.

Diacetate $C_{25}H_{36}O_6$. Crystals (ether - petroleum ether), m. 120–122° cor.
(Kofler block); $[\alpha]_D^{17} + 142°$ (chloroform). — **Fmn.** From the preceding acetate
on treatment with acetic anhydride in abs. pyridine at 20° (1943 Fuchs;
1943 Reichstein). See also above. — **Rns.** See above.

21-Acetate 12-p-toluenesulphonate $C_{30}H_{40}O_7S$. Not isolated in a pure state. —
Fmn. From the 21-acetate (above) on keeping with p-toluenesulphonyl
chloride in abs. pyridine for 4 days at 30° (1946 v. Euw). — **Rns.** See above.

1943 Fuchs, Reichstein, *Helv.* **26** 511, 514–516, 523–525, 527–530.
    Reichstein, *U.S. Pat.* 2401775 (issued 1946); *C. A.* **1946** 5884; *Ch. Ztbl.* **1947** 74; *U.S. Pat.*
    2404768 (1945, issued 1946); *C. A.* **1946** 6222; *Ch. Ztbl.* **1947** 233.
1946 v. Euw, Reichstein, *Helv.* **29** 654, 670.

Δ^4-Pregnene-17α.21-diol-3.20-dione, 17-Hydroxydesoxycorticosterone, 17.21-Dihydroxyprogesterone, 17-Hydroxycortexone,

Me OH

—·CO·CH$_2$·OH

Me

O

Reichstein's *Substance S*, $C_{21}H_{30}O_4$. For the configuration at C-17, see p. 1363 s. —Leaflets (acetone), m. 207–9° cor. with slight decomposition; needles (abs. alcohol), m. 213–7° cor. with slight decomposition (1938a, b Reichstein). Absorption max. in alcohol at 236 mμ (ε 166000) (1939 Dannenberg; cf. 1938b Reichstein). Readily soluble in acetone, methanol, and ethanol, sparingly in ether and water. Gives a carmine-red coloration with conc. H_2SO_4. Strongly reduces alkaline silver diammine solution at room temperature (1938b Reichstein).

Occ. In adrenal cortex (1938a, b Reichstein; 1939 ROCHE-ORGANON, INC.).

Fmn. The 21-acetate is obtained, along with Δ^4-androstene-3.17-dione, from the 21-acetate of Δ^4-pregnene-17α.20β.21-triol-3-one (p. 2725 s) on oxidation with CrO_3 in strong acetic acid at 11°; yield, 7% (1946 Sarett). From Δ^4-pregnene-17α.20β-diol-3-on-21-al (p. 2863 s) on refluxing with abs. pyridine for 6 hrs.; yield, ca. 50% (1940, 1941 Reichstein; cf. also 1941 v. Euw).

Rns. Gives Δ^4-androstene-3.17-dione (p. 2880 s) on oxidation with CrO_3 in glacial acetic acid at room temperature (1938b Reichstein). On oxidation with periodic acid in aqueous methanol at room temperature, 17α-hydroxy-3-keto-Δ^4-ætiocholenic acid is formed (1939 Reichstein).

21-Acetate $C_{23}H_{32}O_5$. Needles (acetone or acetone - ether), becoming opaque below 100°, m. 238–240° cor. (1938a, 1940 Reichstein; cf. also 1946 Sarett); [α]$_D$ in acetone +116° (1940 Reichstein), +118° (1946 Sarett). Absorption max. (solvent not stated) at 242 mμ (ε 14700) (1946 Sarett). Gives a carmine-red coloration with conc. H_2SO_4; strongly reduces alkaline silver diammine solution at room temperature (1938a Reichstein; 1946 Sarett). — **Fmn.** From the above dioldione with acetic anhydride in abs. pyridine at room temperature (1938a Reichstein). See also under formation of the dioldione, above. — **Rns.** Is hydrolysed by $KHCO_3$ in aqueous methanol at room temperature (1938a, b Reichstein).

17-iso-Δ^4-Pregnene-17β.21-diol-3.20-dione, 17-Hydroxy-17-iso-desoxycorticosterone, 17.21-Dihydroxy-17-iso-progesterone

Me OH

—·CO·CH$_2$·OH

Me

O

$C_{21}H_{30}O_4$. Crystals (dil. acetone). — **Fmn.** From 17-iso-Δ^5-pregnene-3β.17β.20b.21-tetrol (p. 2140 s) by the action of Acetobacter and press juice (from brewer's yeast) on a solution in aqueous dioxane at 25–30°; the *21-triphenylmethyl ether** (not characterized) is obtained from that of the same tetrol or that of a "Δ^4-pregnene-17.20.21-triol-3-one" by oxidation, e.g. with cyclohexanone in the presence of aluminium isopropoxide or with CrO_3 in acetic acid, and may be hydrolysed by hydrochloric acid in glacial acetic acid (1938 SCHERING A.-G.). The *diacetate* (not characterized) is obtained from the 17.21-diacetate of 17-iso-

---

* The identity of the ethers from the two sources is not certain.

Δ^5-pregnene-3β.17β.21-triol-20-one (p. 2358 s) on oxidation with cyclohexanone in the presence of aluminium isopropoxide (1938 CIBA PHARMACEUTICAL
PRODUCTS) or from "17-acetoxyprogesterone" [probably 17-iso-Δ^4-pregnen-
17β-ol-3.20-dione acetate, p. 2844 s (Editor)] on treatment with 1 mol. of
lead tetraacetate in glacial acetic acid (1940 SCHERING A.-G.). — **Rns.** The
diacetate is converted into its 3-enol acetate (triacetate of 17-iso-$\Delta^{3.5}$-preg-
nadiene-3.17β.21-triol-20-one, p. 2357 s) by the action of acetic anhydride
(1938 CIBA PHARMACEUTICAL PRODUCTS).

1938 CIBA PHARMACEUTICAL PRODUCTS (Miescher, Wettstein), *U.S. Pat.* 2239012 (issued 1941);
    C. A. **1941** 4921.
    a Reichstein, v. Euw, *Helv.* **21** 1197, 1209.
    b Reichstein, *Helv.* **21** 1490, 1493, 1495, 1496.
    SCHERING A.-G., *Fr. Pat.* 857832 (issued 1940); *Ch. Ztbl.* **1941** I 1324.
1939 Dannenberg, *Abhandl. Preuss. Akad. Wiss. Math.-naturwiss. Kl.* **1939** No. 21, pp. 12, 53.
    Reichstein, Meystre, v. Euw, *Helv.* **22** 1107, 1110.
    ROCHE-ORGANON, INC. (Reichstein), *U.S. Pat.* 2228706 (issued 1941); *C. A.* **1941** 3040;
    Ch. Ztbl. **1941** II 2352, 2353.
1940 Reichstein, v. Euw, *Helv.* **23** 1258.
    SCHERING A.-G., *Fr. Pat.* 868396 (issued 1941); *Ch. Ztbl.* **1942** I 2560.
1941 v. Euw, Reichstein, *Helv.* **24** 1140, 1142.
    Reichstein, *Swed. Pat.* 106103 (issued 1942); *Ch. Ztbl.* **1943** II 2077; *Fr. Pat.* 888825 (issued
    1943); *Ch. Ztbl.* **1944** II 1203.
1946 Sarett, *J. Biol. Ch.* **162** 601, 627.

Δ^5-Pregnene-3β.21-diol-11.20-dione C$_{21}$H$_{30}$O$_4$.

21-Acetate C$_{23}$H$_{32}$O$_5$ (not isolated in a pure
state). Syrup, b. 180–200^0 (bath tempera-
ture) under 0.01 mm. pressure. — **Fmn.** From
21-diazo-Δ^5-pregnen-3β-ol-11.20-dione (page
2851 s) on heating with glacial acetic acid
at 95–102^0; yield, 70%. — **Rns.** Gives the acetate of Δ^4-pregnen-21-ol-3.11.20-
trione (11-dehydrocorticosterone, p. 2972 s) on heating with aluminium tert.-
butoxide and acetone in abs. benzene in a sealed tube for 25 hrs. on the
water-bath (1946 v. Euw).

Pregnane-3α.21-diol-11.20-dione C$_{21}$H$_{32}$O$_4$.

21-Acetate C$_{23}$H$_{34}$O$_5$. Needles (acetone ·
ether, 1:5), m. 137–8^0 cor. (Kofler block);
$[\alpha]_D^{16} + 109.5^0$, $[\alpha]_{5461}^{16} + 134^0$ (both in acetone)
(1944 v. Euw). — **Fmn.** From pregnan-3α-ol-
11.20-dione (p. 2847 s) on heating with lead
tetraacetate in 98–98.5% acetic acid in a
sealed tube for 22 hrs. at 55^0; yield, 10% (1944 v. Euw; 1942 N. V. ORGANON;
1943 Reichstein). From 21-diazopregnan-3α-ol-11.20-dione (p. 2851 s) on
heating with 98.5% acetic acid for 40 min. at 95^0; yield, 15% (1944 v. Euw). —
Rns. Gives the acetate of pregnan-21-ol-3.11.20-trione (p. 2973 s) on oxidation with CrO$_3$ in glacial acetic acid at 18^0 (1944 v. Euw; 1942 N. V.
ORGANON).

Pregnane-3β.21-diol-11.20-dione $C_{21}H_{32}O_4$.

21-Acetate $C_{23}H_{34}O_5$. Leaflets (ether - petroleum ether), m. 178–181° cor. (Kofler block). It strongly and rapidly reduces alkaline silver diammine solution at room temperature (1943 Lardon). — **Fmn.** A small amount is obtained from pregnane-3β.11β-diol-20-one (p. 2328 s) on heating with lead tetraacetate in glacial acetic acid containing traces of acetic anhydride (1944 v. Euw). From pregnan-3β-ol-11.20-dione (p. 2848 s) on heating with lead tetraacetate in 98–98.5 % acetic acid in a sealed tube for 22 hrs. at 55°; yield, 19 % (1944 v. Euw; 1942 N. V. ORGANON; 1943 Reichstein). From 21-diazopregnan-3β-ol-11.20-dione (page 2852 s) on heating with glacial acetic acid for ¹/₂ hr. at 95–100°; yield, 27 % (1942 Reichstein; 1943 Lardon). — **Rns.** Gives the acetate of pregnan-21-ol-3.11.20-trione (p. 2973 s) on oxidation with CrO_3 in glacial acetic acid at 20° (1942, 1943 Reichstein; 1943 Lardon).

Diacetate $C_{25}H_{36}O_6$. Needles combined to leaflets (ether - petroleum ether), m. 169–171° cor. (Kofler block). Strongly reduces alkaline silver diammine solution at room temperature. — **Fmn.** From the above 21-acetate on treatment with acetic anhydride in abs. pyridine at 20° for 16 hrs. and then at 80° for 1 hr. From the acetate of 21-diazopregnan-3β-ol-11.20-dione (p. 2852 s) on refluxing with glacial acetic acid (1943 Lardon).

Allopregnane-3β.21-diol-11.20-dione, *Kendall and Mason's Compound H, Reichstein's Substance N*, $C_{21}H_{32}O_4$.

Crystals (aqueous alcohol), m. 172–6° (1937 Mason); plates (dry acetone, or benzene - acetone), m. 189–191° cor. (1938 Steiger); a hydrate crystallizes from aqueous acetone as leaflets or quadrangular stars, becoming opaque at ca. 120°, m. ca. 190° cor. (1938 Steiger); $[\alpha]_D^{19}$ +94° (abs. alcohol) (1938 Steiger); $[\alpha]_{5461}^{25}$ +118° (alcohol) (1937 Mason). Shows no selective absorption at 240 mμ (1938 Steiger). Gives a precipitate with digitonin in 90 % alcohol (1937 Mason). Reduces alkaline silver diammine solution at room temperature (1938 Steiger).

Occ. In adrenal cortex (1937 Kendall; 1937 Mason; 1938 Steiger); for its isolation, see also 1942 v. Euw.

Fmn. The diacetate is obtained from allopregnane-3β.11β.21-triol-20-one 3.21-diacetate (p. 2355 s) on oxidation with CrO_3 in glacial acetic acid at room temperature (1938b Reichstein). The diacetate is also obtained from its 17-epimer (below) on refluxing with hydrochloric acid (d 1.16) in alcohol for ¹/₂ hr., followed by reacetylation with acetic anhydride in abs. pyridine at room temperature; yield, 25 % (1940 Shoppee).

Rns. Gives formaldehyde and 3β-hydroxy-11-keto-ætioallocholanic acid on oxidation with periodic acid and dil. H_2SO_4 at room temperature (1937

Mason; cf. also 1947 v. Euw). The diacetate is stable towards CrO_3 in glacial acetic acid at room temperature (1938a Reichstein).

Diacetate $C_{25}H_{36}O_6$. Needles (ether - pentane, or methanol), m. 148–9° cor. (1938a, b Reichstein; 1940 Shoppee); needles (benzene - benzine), becoming opaque at 70°, m. 147–148.5° cor. (1938a Reichstein); crystals (acetone - ether), m. 144–5° cor. (Kofler block) (1942 v. Euw); $[\alpha]_D^{18} +77.5°$, $[\alpha]_{5461}^{18} +99.5°$ (both in acetone) (1940 Shoppee); $[\alpha]_D^{19} +86°$, $[\alpha]_{5461}^{18} +106°$ (both in dioxane) (1942 v. Euw). — **Fmn.** From the above dioldione on treatment with acetic anhydride in dry pyridine for 16 hrs. at 20° (1938a Reichstein). See also under formation of the dioldione (above).

17 - iso - Allopregnane - 3β.21 - diol - 11.20 - dione, *Reichstein's Substance Iso-N*,

$C_{21}H_{32}O_4$.

Diacetate $C_{25}H_{36}O_6$. Leaflets (ether - pentane), m. 131–2° cor.; $[\alpha]_D^{14} -44°$ (acetone). Strongly reduces alkaline silver diammine solution at room temperature. Gives no fluorescence with conc. H_2SO_4 (1940 Shoppee). — **Fmn.** From 17-iso-allopregnane-3β.11β.21-triol-20-one 3.21-diacetate (p. 2355 s) on oxidation with CrO_3 in glacial acetic acid at room temperature; almost quantitative crude yield (1940 Shoppee). From allopregnane-3β.17α.-20β.21-tetrol-11-one 3.20.21-triacetate (p. 2766 s) on refluxing with zinc dust in abs. toluene for 14 hrs.; yield, 22% (1940 Shoppee; 1940 CIBA PHARMACEUTICAL PRODUCTS). — **Rns.** Gives 25% yield of the 17-normal diacetate (above) on refluxing with hydrochloric acid (d 1.16) in alcohol for $^1/_2$ hr. Is unattacked by refluxing with abs. pyridine for 8 hrs. or with nitromethane for 24 hrs.; in the latter case, in the presence of piperidine, decomposition takes place (1940 Shoppee).

1937 Kendall, *Cold Spring Harbor Symp. Quant. Biol.* **5** 299.
     Mason, Hoehn, McKenzie, Kendall, *J. Biol. Ch.* **120** 719, 738, 739.
1938 a Reichstein, Gätzi, *Helv.* **21** 1185, 1189.
     b Reichstein, *Helv.* **21** 1490, 1495.
     Steiger, Reichstein, *Helv.* **21** 546, 561–564.
1940 CIBA PHARMACEUTICAL PRODUCTS (Miescher, Heer), *U.S. Pat.* 2372841 (issued 1945); *C. A.*
     1945 4195; *Ch. Ztbl.* **1945** II 1773.
     Shoppee, Reichstein, *Helv.* **23** 729, 737–739.
1942 v. Euw, Reichstein, *Helv.* **25** 988, 1006, 1022.
     N. V. ORGANON, *Swiss Pat.* 256509 (issued 1949); *Ch. Ztbl.* **1949** E 1869.
     Reichstein, *Swiss Pat.* 244341 (issued 1947); *C. A.* **1949** 5812.
1943 Lardon, Reichstein, *Helv.* **26** 747, 752–754; Reichstein, *U.S. Pat.* 2401775 (issued 1946);
     C. A. **1946** 5884; *Ch. Ztbl.* **1947** 74; *U.S. Pat.* 2404768 (1945, issued 1946); *C. A.* **1946**
     6222; *Ch. Ztbl.* **1947** 233.
     Reichstein, *Fr. Pat.* 898140 (issued 1945); *Ch. Ztbl.* **1946** I 1747, 1748; *Brit. Pat.* 594878
     (issued 1947); *C. A.* **1948** 2404; *U.S. Pat.* 2440874 (issued 1948); *C. A.* **1948** 5622; *Ch. Ztbl.*
     1948 E 456.
1944 v. Euw, Lardon, Reichstein, *Helv.* **27** 1287, 1290, 1295, 1296.
1946 v. Euw, Reichstein, *Helv.* **29** 1913, 1919.
1947 v. Euw, Reichstein, *Helv.* **30** 205, 216.

Pregnane-3α.21-diol-12.20-dione $C_{21}H_{32}O_4$.

21-Acetate $C_{23}H_{34}O_5$. Needles (benzene - ether), m. 149–151° cor. (Kofler block); $[\alpha]_D^{19}$ + 158° (acetone). — **Fmn.** Along with the 21-acetates of pregnane - 12α.21 - diol - 3.20-dione (p. 2858 s; small amount) and pregnan-21-ol-3.12.20-trione (p. 2974 s), from the 21-acetate of pregnane-3α.12α.21-triol-20-one (p. 2356 s) on oxidation with 1 equivalent of CrO_3 in glacial acetic acid at room temperature. — **Rns.** Gives the afore-mentioned pregnanoltrione acetate on further oxidation with CrO_3 in glacial acetic acid at room temperature (1943 Fuchs).

1943 Fuchs, Reichstein, *Helv.* **26** 511, 520–522.

$Δ^4$-Pregnene-17α.20β-diol-3-on-21-al $C_{21}H_{30}O_4$.

For configuration at C-17, see p. 1363 s. — Not obtained in the crystalline

state. — **Fmn.** From $Δ^4$-pregnene-17α.20β.21-triol-3-one (p. 2725 s) on oxidation with H_2O_2 in aqueous dioxane in the presence of ferrous sulphate at room temperature (1939 CIBA PHARMACEUTICAL PRODUCTS). From ω-homo-$Δ^4$-pregnene-17α.20β.21.22-tetrol-3-one (p. 2732 s) on oxidation with 1 mol. periodic acid in aqueous dioxane at room temperature (small yield) (1941 v. Euw; 1942 Reichstein); the 20-acetate is similarly obtained from that of the same tetrolone (yield, 52%) (1940 v. Euw; 1941a Reichstein), and may be hydrolysed by $KHCO_3$ in aqueous methanol at room temperature (1940, 1941a Reichstein). From the acetal of $Δ^{4.17(20)}$-pregnadien-3-on-21-al (page 2802 s) on treatment with OsO_4 in ether at room temperature, followed by heating with alcoholic Na_2SO_3 and hydrolysis with alcoholic HCl (1939 CIBA).

Rns. Gives $Δ^4$-pregnene-17α.21-diol-3.20-dione (p. 2859 s) on refluxing with abs. pyridine for 6 hrs. (1940, 1941b Reichstein; cf. 1941 v. Euw).

20-Acetate $C_{23}H_{32}O_5$. Needles (acetone - ether), m. 206–8° cor. with slight decomposition; $[\alpha]_D^{18}$ + 119° (dioxane). With conc. H_2SO_4 it gives a brown-orange solution with an intense green fluorescence. Develops an intense red colour on heating with 1.4-dihydroxy-naphthalene and conc. HCl in glacial acetic acid. It moderately rapidly reduces alkaline silver diammine solution at room temperature (1940 v. Euw).

17-iso-$Δ^4$-Pregnene-17β.20a-diol-3-on-21-al $C_{21}H_{30}O_4$.

Not characterized. —

Fmn. From 17-iso-$Δ^4$-pregnene-17β.20a.21-triol-3-one (p. 2726 s) on oxidation with H_2O_2 in aqueous dioxane in the presence of ferrous sulphate at room temperature (1939 CIBA PHARMACEUTICAL PRODUCTS).

1939 Ciba (Miescher, Wettstein); *Swed. Pat.* 101201 (issued 1941); *Ch. Ztbl.* **1941** II 1301; *Norw. Pat.* 60152 (issued 1942); *Ch. Ztbl.* **1943** I 2421; *Fr. Pat.* 863257 (issued 1941); *Ch. Ztbl.* **1942** I 81.

Ciba Pharmaceutical Products (Miescher, Fischer, Scholz, Wettstein), *U.S. Pats.* 2275790, 2276543 (both issued 1942); *C. A.* **1942** 4289, 4674; Ciba, *Fr. Pat.* 857122 (issued 1940); *Ch. Ztbl.* **1941** I 1324.

1940 v. Euw, Reichstein, *Helv.* **23** 1114, 1124.

Reichstein, v. Euw, *Helv.* **23** 1258.

1941 v. Euw, Reichstein, *Helv.* **24** 1140.

a Reichstein, *Fr. Pat.* 872985 (issued 1942); *Ch. Ztbl.* **1942** II 2175, 2176.

b Reichstein, *Fr. Pat.* 888825 (issued 1943); *Ch. Ztbl.* **1944** II 1203; *Swed. Pat.* 106103 (issued 1942); *Ch. Ztbl.* **1943** II 2077.

1942 Reichstein, *Fr. Pat.* 888228 (issued 1943); *Ch. Ztbl.* **1945** I 68.

4-Bromopregnane-12α.21-diol-3.20-dione $C_{21}H_{31}O_4Br$. — **Fmn.** The 21-acetate

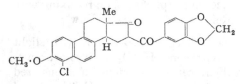

is obtained from that of pregnane-12α.21-diol-3.20-dione (p. 2858 s) on treatment with 1 mol. of bromine in glacial acetic acid (yield, 67%); the diacetate is similarly formed (yield, 79%). — **Rns.** The 21-acetate and the diacetate are converted into the corresponding derivatives of Δ^4-pregnene-12α.21-diol-3.20-dione (p. 2857 s) by refluxing with abs. pyridine (1943 Fuchs).

21-Acetate $C_{23}H_{33}O_5Br$ (not analysed). Crystals, m. 171–2° cor. dec. (Kofler block) (1943 Fuchs). — **Fmn.** and **Rns.** See above.

Diacetate $C_{25}H_{35}O_6Br$ (not analysed). Crystals, m. 165–176° cor. dec. (Kofler block) (1943 Fuchs). — **Fmn.** and **Rns.** See above.

1943 Fuchs, Reichstein, *Helv.* **26** 511, 527, 529.

---

4 - Chloro - 16 - (3.4 - methylenedioxybenzoyl) - dl - isoequilenin methyl ether $C_{27}H_{23}O_5Cl$. Prisms (ethyl acetate), m. 178°. Gives a cherry-red colour with alcoholic $FeCl_3$. — **Fmn.** From 16 - piperonylidene - dl - isoequilenin methyl ether (p. 2730 s) on treatment with 1.5 mols. of chlorine in dioxane - carbon tetrachloride, followed by heating of the resulting product with sodium ethoxide in alcohol - benzene on the steam-bath for 1 hr., refluxing with 15% HCl for $^1/_2$ hr. and then with 2% aqueous NaOH in benzene for 1 hr.; the resulting pale yellow *sodium enolate* is then hydrolysed with dil. acid. — **Rns.** On refluxing with aqueous alcoholic KOH for $2^1/_2$ hrs. it gives dl-4-chloro-isoequilenin methyl ether (p. 2701 s) and 2-methyl-8-chloro-7-methoxy-1-[β-(3.4-methylenedioxybenzoyl)-ethyl]-1.2.3.4-tetrahydro-2-phenanthroic acid (1945 Birch).

1945 Birch, Jaeger, Robinson, *J. Ch. Soc.* **1945** 582, 584, 585.

c. Polyhydroxy-dioxosteroids with One CO and Two OH in the Ring System

4.7-Dihydroxy-4′-formyl-3′-keto-1.2-cyclopentenophenanthrene $C_{18}H_{12}O_4$. For its dialkyl ethers, see those of the tautomeric 4′-hydroxymethylene-4.7-dihydroxy-3′-keto-1.2-cyclopentenophenanthrene, p. 2764 s.

Pregnane (or Allopregnane)-4.5-diol-3.20-dione $C_{21}H_{32}O_4$. Octahedra (chloroform - alcohol), m. 249–250⁰ dec.; $[\alpha]_D^{21}$ +104.5⁰ (chloroform). — **Fmn.** From Δ^4-pregnene-3.20-dione (progesterone, p. 2782 s) on treatment with 30% H_2O_2 and OsO_4 in ether at room temperature for 24 hrs.; yield, 30% (1938 Butenandt).

4-Acetate $C_{23}H_{34}O_5$. Needles (alcohol), m. 223⁰ to 225⁰. — **Fmn.** From the preceding compound on treatment with acetic anhydride in pyridine at room temperature (1938 Butenandt).

Allopregnane-5α.6β-diol-3.20-dione $C_{21}H_{32}O_4$. For configuration at C-5 and C-6, see 1948 Ehrenstein.

6-Acetate $C_{23}H_{34}O_5$. Prisms (alcohol), m. 216.5⁰ to 217.5⁰ cor.; needles, m. 215–8⁰ cor., or thick plates, m. 218–221⁰ cor., both from acetone - ether; on standing with solvent for a sufficiently long time, the needles frequently changed to the plates; $[\alpha]_D$ +20.5⁰ to +23⁰ (acetone) (1940a, 1941 Ehrenstein). — **Fmn.** From the 6-acetate of allopregnane-3β.5α.6β-triol-20-one (p. 2365 s) on oxidation with CrO_3 in strong acetic acid at room temperature; yield, 74% (1940a, 1941 Ehrenstein). — **Rns.** Gives the acetate of Δ^4-pregnen-6β-ol-3.20-dione (6β-hydroxyprogesterone, p. 2838 s) on treatment for $1\frac{1}{2}$ hrs. with dry HCl in purified carbon tetrachloride at —5⁰ to +5⁰ or in alcohol-free, dry chloroform at —2⁰ to +2⁰ and then for 20 min. at room temperature (1952 Balant; cf. 1940a Ehrenstein); the acetate of the epimeric 6α-hydroxyprogesterone (p. 2838 s) is formed when the dioldione 6-acetate is treated for $1\frac{1}{2}$ hrs. with dry HCl at —5⁰ to +5⁰ in chloroform containing 0.7% of alcohol (1952 Balant; cf. 1941 Ehrenstein).

Δ^4-Pregnene-7α.12α-diol-3.20-dione, 7α.12α-Dihydroxyprogesterone $C_{21}H_{30}O_4$. For configuration at C-7 and C-12, cf. the following compound.

Diacetate $C_{25}H_{34}O_6$. Crystals (chloroform - ether), m. 249.5–252⁰ cor. Absorption max. in abs. alcohol at ca. 240 mμ (ε ca. 26300). — **Fmn.** From the saturated diacetate (below) on treatment with 1 mol. bromine and a little 40% HBr in glacial acetic acid at room temperature, followed by refluxing of the resulting 4-bromo-derivative (p. 2867 s) with collidine for 4 hrs. (1940c Ehrenstein).

Pregnane-7α.12α-diol-3.20-dione $C_{21}H_{32}O_4$. For configuration at C-7 and C-12, see under the parent triolone (p. 2366 s). *Diacetate* $C_{25}H_{36}O_6$. Crystals (chloroform - acetone), m. 256–262⁰ cor. (1940c Ehrenstein), 266⁰ to 270⁰ cor. (Kofler block) (1946 Meystre); [α]_D in chloroform +114⁰ (1940c Ehrenstein), +117.5⁰ (1946 Meystre). Rather insoluble in acetone (1940c Ehrenstein). — **Fmn.** From the 7-acetate (formerly 12-acetate) of pregnane-3α.7α.12α-triol-20-one (p. 2367 s) on oxidation with aluminium isopropoxide and cyclohexanone by refluxing in dry toluene for 2 hrs., followed by acetylation of the resulting crude 7-acetate (yield, 47%) with acetic anhydride in pyridine on the water-bath (4 hrs.) (1940c Ehrenstein; cf. 1948 Ehrenstein) or directly from the 7.12-diacetate of the same triolone on oxidation with CrO_3 in 90% acetic acid at 20⁰ (1946 Meystre). — **Rns.** On treatment with 1 mol. bromine and a little 40% HBr in glacial acetic acid at room temperature it gives the 4-bromo-derivative (p. 2867 s) which by refluxing with collidine for 4 hrs. is dehydrobrominated to give the diacetate of $Δ^4$-pregnene-7α.12α-diol-3.20-dione (p. 2865 s) (1940c Ehrenstein).

Pregnane-16α.17α-diol-3.20-dione $C_{21}H_{32}O_4$. Configuration at C-16 and C-17 (Editor) based on Shoppee (1950). — Needles (alcohol and dil. acetone), m. 156⁰. — Has no androgenic activity in the Fussgänger capon comb test. — **Fmn.** From $Δ^{16}$-pregnene-3.20-dione (p. 2794 s) on treatment with 1 mol. of OsO_4 and a little 30% H_2O_2 in ether, followed by refluxing with Na_2SO_3 in dil. alcohol; small yield (1938 Masch).

Allopregnane-3β.5α-diol-6.20-dione $C_{21}H_{32}O_4$. For configuration at C-5, see 1948 Ehrenstein. *3-Acetate* $C_{23}H_{34}O_5$. Crystals (ether), m. 222.5⁰ to 224⁰. — **Fmn.** Along with other compounds, from the 3-acetate of $Δ^5$-pregnen-3β-ol-20-one (p. 2233 s) on oxidation with $KMnO_4$ in acetic acid at 50⁰. — *Dioxime* $C_{23}H_{36}O_5N_2$ (not analysed) crystals (aqueous alcohol), m. 262–4⁰ dec. (1940b Ehrenstein).

25-Methylcoprostane-3α.7α-diol-12.24-dione (?), Dihydroxydiketo-isobufostane $C_{28}H_{46}O_4$. For configuration at C-3 and C-7, see that of tetrahydroxy-isobufostane (p. 2199 s). — Prisms with $\frac{1}{2}$ H_2O (aqueous methanol), m. 221–4⁰. — **Fmn.** Along with tetrahydroxy-isobufostane (p. 2199 s), from tetraketo-isobufostane (p. 2985 s) on hydrogenation in glacial acetic acid in the presence of Adams's platinum oxide at room temperature (1940 Kazuno).

24-Phenylcholane-3α.12α-diol-7.24-dione, 3α.12α-Dihydroxy-7-keto-norcholanyl phenyl ketone $C_{30}H_{42}O_4$.

Configuration at C-3 and C-12 deduced from that of the ultimate precursor, cholic acid, for which see pp. 1361 s, 1362 s. — Crystals (first from moist isopropyl alcohol, then from benzene), m. 180–181.5°; $[\alpha]_D^{25}$ 0° (dioxane). — **Fmn.** From the diformate of 3α.12α-dihydroxy-7-ketocholanic acid chloride on refluxing with diphenylcadmium in dry benzene, followed by decomposition of the resulting complex with dil. HCl and hydrolysis of the crude diformate by refluxing with 5% methanolic NaOH (1945 Hoehn).

Diacetate $C_{34}H_{46}O_6$. Crystals (methanol), m. 164–6°; $[\alpha]_D^{25}$ +75° (dioxane). — **Fmn.** From the above dioldione by refluxing with glacial acetic acid - acetic anhydride (2:3 by vol.) (1945 Hoehn).

4-Bromopregnane-7α.12α-diol-3.20-dione $C_{21}H_{31}O_4Br$.

For configuration at C-7 and C-12, see under the parent dioldione, p. 2866 s.

Diacetate $C_{25}H_{35}O_6Br$ (not obtained in a pure state). Crystals (chloroform and 95% alcohol), m. 210–218° dec. — **Fmn.** From the diacetate of pregnane-7α.12α-diol-3.20-dione (p. 2866 s) on treatment with 1 mol. bromine and a little 40% HBr in glacial acetic acid at room temperature. — **Rns.** Gives Δ^4-pregnene-7α.12α-diol-3.20-dione diacetate (p. 2865 s) on refluxing with collidine for 4 hrs. (1940c Ehrenstein).

1938 Butenandt, Wolz, *Ber.* **71** 1483, 1486.
   Masch, *Über Bromierungen in der Pregnanreihe*, Thesis, Danzig, pp. 20, 29.
1940 a Ehrenstein, Stevens, *J. Org. Ch.* **5** 318, 325.
   b Ehrenstein, Decker, *J. Org. Ch.* **5** 544, 558, 559.
   c Ehrenstein, Stevens, *J. Org. Ch.* **5** 660, 667, 670, 671.
   Kazuno, *Z. physiol. Ch.* **266** 11, 26.
1941 Ehrenstein, Stevens, *J. Org. Ch.* **6** 908, 916, 917.
1945 Hoehn, Moffett, *J. Am. Ch. Soc.* **67** 740, 742.
1946 Meystre, Frey, Neher, Wettstein, Miescher, *Helv.* **29** 627, 634.
1948 Ehrenstein, *J. Org. Ch.* **13** 214, 220, 222.
1950 Shoppee, *Nature* **166** 107.
1952 Balant, Ehrenstein, *J. Org. Ch.* **17** 1587, 1591, 1592.

Allopregnane-5α.6β.21-triol-3.20-dione $C_{21}H_{32}O_5$.

For configuration at C-5 and C-6, see 1948 Ehrenstein.

6.21-Diacetate $C_{25}H_{36}O_7$. Rectangular crystals (ether?), m. 163.5–164.5° cor.; $[\alpha]_D^{26}$ +21.5° (acetone) (1941 Ehrenstein). — **Fmn.** From the 6.21-diacetate of allopregnane-3β.5α.6β.21-tetrol-20-one (p. 2372 s) on oxidation with CrO_3

in strong acetic acid at room temperature; yield, ca. 80% (1941 Ehrenstein). —
Rns. On dehydration with dry HCl in chloroform, not above 2⁰, it gives the
diacetate of Δ^4-pregnene-6β.21-diol-3.20-dione (p. 2853 s) (1941 Ehrenstein).

1941 Ehrenstein, *J. Org. Ch.* **6** 626, 644, 645.
1948 Ehrenstein, *J. Org. Ch.* **13** 214, 221.

Δ^4-Pregnene-11β.17α.21-triol-3.20-dione, 17-Hydroxycorticosterone, Hydrocortisone, Cortisol, *Reichstein's Substance M,*

Mason and Kendall's Compound F, $C_{21}H_{30}O_5$
(p. 154). For β-configuration at C-11, see also
1938 Marker. — Striated blocks (abs. ethyl
alcohol or isopropyl alcohol), m. 217–220⁰ cor.
(1938 Mason); crystals (95% alcohol or acetone), m. 216–8⁰ cor. dec.; $[\alpha]_D^{25} +164^0$
(95% alcohol) (1939, 1945 Kuizenga), $[\alpha]_{5461}^{25} +178^0$ (alcohol?) (1938 Mason).
Absorption max. in alcohol at 241 mμ (log ε 4.14) (1937a Reichstein; 1939
Dannenberg). Strongly reducing towards alkaline silver salt solution at room
temperature (1937a Reichstein).

For *physiological properties*, see 1939, 1945 Kuizenga; 1939 Waterman;
1940a, b Ingle; 1940 Grattan; 1941 Thorn; 1942 Abelin; 1943 Lardon;
cf. also pp. 1367 s, 1368 s, and the literature cited there.

Occ. For occurrence in, and isolation from, beef adrenal cortex, see also
1938 ROCHE-ORGANON, INC.; 1939 Kuizenga; 1941a Reichstein; 1942 v. Euw.
In hog adrenal cortex extract, ca. 6 times as much as in beef adrenal cortex
extract (1945 Kuizenga).

Fmn. From Δ^4-pregnene-11β.17α.20-triol-3-on-21-al on refluxing with
pyridine (1941b Reichstein).

Rns. Gives 11β.17α-dihydroxy-3-keto-Δ^4-ætiocholenic acid on oxidation
with periodic acid in aqueous dioxane at 14⁰ (1942 v. Euw; cf. also 1938
Mason).

21-Acetate $C_{23}H_{32}O_6$. Rodlets (acetone), m. 223–5⁰ cor. (1937b Reichstein;
1937 ROCHE-ORGANON, INC.), 220–222⁰ cor. (1945 Kuizenga). — **Fmn.** From
the above trioldione on treatment with acetic anhydride and pyridine at
room temperature (1937b Reichstein; 1945 Kuizenga) or on treatment with
4–5 mols. of gaseous ketene in acetone (1937 ROCHE-ORGANON, INC.).

1937 a Reichstein, *Helv.* **20** 953, 958, 968, 969.
   b Reichstein, *Helv.* **20** 978, 987, 988.
   ROCHE-ORGANON, INC. (Reichstein, Schlittler), *U.S. Pat.* 2183589 (issued 1939); *C. A.* **1940**
      2538; *Ch. Ztbl.* **1940** II 932.
1938 Marker, Kamm, Oakwood, Wittle, Lawson, *J. Am. Ch. Soc.* **60** 1061, 1062.
   Mason, Hoehn, Kendall, *J. Biol. Ch.* **124** 459, 466, 472.
   ROCHE-ORGANON, INC. (Reichstein), *U.S. Pat.* 2166877 (issued 1939); *C. A.* **1939** 8925;
      Ch. Ztbl. **1940** I 1232, 1233.
1939 Dannenberg, *Abhandl. Preuss. Akad. Wiss. Math.-naturwiss. Kl.* **1939** No. 21, p. 53.
   Kuizenga, Cartland, *Endocrinology* **24** 526, 531, 532.
   Waterman, Danby, Gaarenstroom, Spanhoff, Uyldert, *Acta Brevia Neerland.* **9** 75.

1940 Grattan, Jensen, *J. Biol. Ch.* **135** 511.
a Ingle, *Endocrinology* **26** 472.
b Ingle, *Proc. Soc. Exptl. Biol. Med.* **44** 176.
1941 a Reichstein, v. Euw. *Helv.* **24** 247 E, 256 E, 259 E.
b Reichstein, *Fr. Pat.* 888825 (issued 1943); *Ch. Ztbl.* **1944** II 1203.
Thorn, Engel, Lewis, *Science* [N.S.] **94** 348.
1942 Abelin, Althaus, *Helv.* **25** 205, 212.
v. Euw, Reichstein, *Helv.* **25** 988, 1005, 1019.
1943 Lardon, Reichstein, *Helv.* **26** 747, 748.
1945 Kuizenga, Nelson, Lyster, Ingle, *J. Biol. Ch.* **160** 15, 18–20.

Allopregnane-3α.17α.21-triol-11.20-dione $C_{21}H_{32}O_5$. For configuration at C-17, see p. 1363 s.

3.21-Diacetate $C_{25}H_{36}O_7$. Prisms (acetone-ether, 1:2), m. 222–4° cor. (Kofler block); $[\alpha]_D^{12} +94°$, $[\alpha]_{5461}^{12} +114°$ (both in dioxane). With conc. H_2SO_4 it gives an almost colourless solution changing to yellow. Rapidly and strongly reduces alkaline silver diammine solution at room temperature.— **Fmn.** From the 3.21-diacetate of allopregnane-3α.11β.17α.21-tetrol-20-one (p. 2373 s) on oxidation with CrO_3 in glacial acetic acid at 15° (1942 b v. Euw).

Allopregnane-3β.17α.21-triol-11.20-dione, *Reichstein's Substance D, Mason and Kendall's Compound G*, $C_{21}H_{32}O_5$ (p. 154).

For configuration at C-17, see p. 1363 s. — Cube-shaped crystals (acetone), m. 241–5° cor. dec. (1939 Kuizenga); crystals (abs. alcohol), m. 238–242° cor. dec. (Kofler block) (1942 a v. Euw); $[\alpha]_D^{25} +70.5°$ (abs. alcohol) (1939 Kuizenga); $[\alpha]_D^{16} +62°$, $[\alpha]_{5461}^{16} +79°$ (both in dioxane) (1942 a v. Euw); $[\alpha]_{5461}^{25} +83°$ (alcohol) (1938 Mason). With conc. H_2SO_4 it gives an almost colourless solution without fluorescence (1942 a v. Euw). — **Occ.** For isolation from adrenal cortex, see also 1939 Kuizenga; 1942 a v. Euw; cf. also p. 2374 s. — **Fmn.** The 3.21-diacetate is obtained from that of allopregnane-3β.11β.17α.21-tetrol-20-one (p. 2375 s) on oxidation with CrO_3 in glacial acetic acid; yield, 28 % (1942 a v. Euw). — **Rns.** For oxidation with periodic acid (in aqueous dioxane at 16°), giving 3β.17α-dihydroxy-11-keto-ætioallocholanic acid, see also 1942 a v. Euw. The 3.21-diacetate is scarcely attacked by CrO_3 (1 mol.) in glacial acetic acid at room temperature (1942 a v. Euw).

3.21-Diacetate $C_{25}H_{36}O_7$ (p. 154). Leaflets (chloroform - ether, 1:1), m. 223° to 224° cor. (Kofler block); $[\alpha]_D^{15} +72°$, $[\alpha]_{5461}^{15} +85°$ (both in dioxane). Gives an almost colourless solution in conc. H_2SO_4. — **Fmn.** From the above trioldione on treatment with acetic anhydride in pyridine at room temperature (1942 a v. Euw).

1938 Mason, Hoehn, Kendall, *J. Biol. Ch.* **124** 459, 473.
1939 Kuizenga, Cartland, *Endocrinology* **24** 526, 528, 529.
1942 a v. Euw, Reichstein, *Helv.* **25** 988, 1003, 1006, 1007, 1009, 1013.
b v. Euw, Reichstein, *Helv.* **25** 988, 1010.

C. Trioxosteroids with one CO in the Ring System

Δ^4-Pregnene-3.20-dion-21-al,　3-Keto-Δ^4-androsten-17β-yl-glyoxal $C_{21}H_{28}O_3$

(not obtained pure). Crystals (probably a mono-hydrate after drying in a high vacuum for 10 min. at room temperature), m. 104–6°. On heating with 1.4-dihydroxynaphthalene and conc. HCl in glacial acetic acid it gives a weakly yellow solution. Reduces alkaline silver diammine solution at room temperature only after ca. 1 min., causing in most cases a reddish brown coloration (1939 Reich; 1940b N. V. Organon).

Its cortin-like activity is weaker than that of corticosterone (1939 Reich).

Fmn. From Δ^5-pregnene-3β.21-diol-20-one (p. 2301 s) on refluxing under nitrogen with aluminium isopropoxide and acetone in benzene (1938 Schering A.-G.). Its dimethyl acetal (below) is obtained from that of Δ^5-pregnen-3β-ol-20-on-21-al (p. 2378 s) by refluxing with aluminium tert.-butoxide and acetone in abs. benzene for 24 hrs. (yield, ca. 35%); it is hydrolysed by 2 N HCl in glacial acetic acid at room temperature (1939 Reich; 1940a N. V. Organon); the diethyl mercaptal (below) is similarly obtained from that of the same olonal (refluxing for 30 hrs.; yield, 66%) (1941 Schindler). Its 21-N-(p-dimethylaminophenyl)-isoxime (below) is obtained from progesterone 21-pyridinium chloride [N-(3.20-diketo-Δ^4-pregnen-21-yl)-pyridinium chloride, p. 2810 s] by treatment with p-nitrosodimethylaniline and NaOH in aqueous alcohol at —10°; it is hydrolysed by dil. HCl in ether (1939 Reich; 1940a, b N. V. Organon). From 21.21-dibromo-Δ^4-pregnene-3.20-dione (21.21-di-bromoprogesterone, p. 2812 s) on refluxing with $CaCO_3$ in aqueous propanol for several hrs.; similarly from a 21.21-dibromoprogesterone-21-carboxylic acid ester (followed by decarboxylation) (1938 Ciba). From Δ^4-pregnen-21-ol-3.20-dione (desoxycorticosterone, p. 2820 s) on refluxing under nitrogen with aluminium isopropoxide and cyclohexanone in toluene (1938 Schering A.-G.) or with aluminium tert.-butoxide and diethyl ketone (1939 Ciba Pharmaceutical Products); from the same compound on heating with 4 parts of phenylhydrazine in dil. acetic acid on the water-bath, followed by cleavage of the resulting *tris-phenylhydrazone* with HCl (d 1.2) at 40° (1939 Ciba Pharmaceutical Products).

Trioxime $C_{21}H_{31}O_3N_3$ (no analysis given), m. 160–170° (1940b N. V. Organon).

20.21-Bis-phenylhydrazone $C_{33}H_{40}ON_4$ (no analysis given), m. 175–7° (1940b N. V. Organon).

Tris-phenylhydrazone $C_{39}H_{46}N_6$. See above.

Dimethyl acetal $C_{23}H_{34}O_4$. Needles (ether - pentane), m. 84–86°; $[\alpha]_D^2$ +170°, $[\alpha]_{5461}^{21}$ +208° (both in acetone). Absorption max. in alcohol at 244 mμ (log ε 4.22). — **Fmn.** See above. May also be obtained from the above aldehyde by treatment with methanol and dry HCl at room temperature (1939 Reich; 1940a, b N. V. Organon).

Diethyl mercaptal $C_{25}H_{38}O_2S_2$. Needles (ether - pentane), m. 94–96⁰; $[\alpha]_D^{20}$ + 258⁰ (acetone). Absorption max. in alcohol at 242 mμ (log ε ca. 4.2) (1941 Schindler). — **Fmn.** See above.

21-(p-Dimethylaminoanil) N-oxide, 21-N-(p-Dimethylaminophenyl)-isoxime $C_{29}H_{38}O_3N_2$ [side chain, $CO \cdot CH : N(O) \cdot C_6H_4 \cdot NMe_2$] (not analysed). Orange-yellow, crystalline mass, m. 112–8⁰ (1939 Reich; 1940b N. V. ORGANON). — **Fmn.** and **Rns.** See p. 2870 s.

17-iso-Δ^4-Pregnen-17β-ol-3.20-dion-21-al $C_{21}H_{28}O_4$. — Fmn. From 17-iso-

Δ^5-pregnene-3β.17β.20.21-tetrol m. 229–231⁰ [prob-ably a mixture of the two C-20 epimers, p. 2140 s (Editor)] on refluxing with aluminium isopropoxide and benzaldehyde or diethyl ketone or diisopropyl ketone in benzene for several hrs.; when acetone is used as dehydrogenating agent, a condensation prod. of the present aldehyde with acetone is formed, which may be split, e.g., with ozone after temporary protection of the endocyclic double bond, to give the aldehyde (1939 CIBA PHARMACEUTICAL PRODUCTS).

1938 CIBA (Soc. pour l'ind. chim. à Bâle; Ges. f. chem. Ind. Basel), *Fr. Pat.* 840514 (issued 1939); *C. A.* **1939** 8627; *Ch. Ztbl.* **1939** II 1125, 1126.
　SCHERING A.-G., *Fr. Pat.* 857832 (issued 1940); *Ch. Ztbl.* **1941** I 1324.
1939 CIBA PHARMACEUTICAL PRODUCTS (Miescher, Fischer, Scholz, Wettstein), *U.S. Pat.* 2275790 (issued 1942); *C. A.* **1942** 4289; CIBA, *Fr. Pat.* 857122 (issued 1940); *Ch. Ztbl.* **1941** I 1324.
　Reich, Reichstein, *Helv.* **22** 1124, 1126, 1128, 1133, 1137, 1138.
1940 a N. V. ORGANON, *Fr. Pat.* 864315 (issued 1941); *Ch. Ztbl.* **1942** I 642.
　b N. V. ORGANON, *Swiss Pat.* 221311 (issued 1942); *Ch. Ztbl.* **1943** I 1297, 1298.
1941 Schindler, Frey, Reichstein, *Helv.* **24** 360, 374.

D. Diketosteroids with Both CO Groups in the Ring System

I. Compounds without Other Functional Groups

1. Compounds C_{17}

1.2-Cyclopentenophenanthrene-9.10-quinone $C_{17}H_{12}O_2$. Red needles (chloroform - alcohol), m. 213⁰. Absorption max. in chloroform at 270 (36200) and ca. 340 (4500) mμ (ε). — **Fmn.** From 9.10(cis)-dihydroxy-1.2-cyclopenteno-9.10-dihydrophenanthrene (page 1958 s) on oxidation with CrO_3 in glacial acetic acid; yield, 70.5%. — **Rns.** On heating with methylmagnesium iodide in ether it gives 9.10-dimethyl-9.10-dihydroxy-1.2-cyclopenteno-9.10-dihydrophenanthrene (p. 1987 s) (1946 Butenandt).

4.3′-Diketo-1.2-cyclopentano-1.2.3.4-tetrahydrophenanthrene $C_{17}H_{14}O_2$. Colourless needles (alcohol), m. 115⁰ (1938 Koebner). — **Fmn.** From 3-(β-naphthyl)-cyclopentanone-2-acetic acid on addition of P_2O_5 to its solution in hot syrupy phosphoric acid (d 1.75) (yield, 63%) (1938 Koebner); only small amounts are obtained when the cyclization is effected by 80% H_2SO_4 at 98⁰, by P_2O_5 in benzene (1938 Koebner), or by anhydrous HF (1945 Birch).

Rns. On boiling with nitrobenzene and aqueous alcoholic NaOH for 3 min. it gives 4-hydroxy-3′-keto-1.2-cyclopentenophenanthrene (p. 2516 s) (1941 Koebner). Gives 3′-keto-1.2-cyclopentano-1.2.3.4-tetrahydrophenanthrene (p. 2386 s) on shaking in alcohol with hydrogen in the presence of platinum-charcoal and palladous chloride (1941 Koebner). Reduction with amalgamated zinc and dil. hydrochloric - acetic acid by refluxing in toluene for 24 hrs., followed by dehydrogenation with palladized charcoal at 330⁰, affords 1.2-cyclopentenophenanthrene (p. 1382 s) (1938 Koebner). On treatment with methyl iodide and potassium tert.-butoxide in ether - tert.-butyl alcohol, a compound [yellow needles (alcohol), m. ca. 191–2⁰] is formed, the ketonic properties of which are not well developed; it may be a **monomethyl derivative** $C_{18}H_{16}O_2$, but in its formation possibly more extensive methylation and perhaps dehydrogenation has occurred (1941 Koebner).

Monosemicarbazone $C_{18}H_{17}O_2N_3$. Colourless prisms (alcohol), m. 245⁰ (1938 Koebner).

Monohydrazone $C_{17}H_{16}ON_2$. Pale yellow prisms (methanol), m. 156⁰ (1938 Koebner).

Mono - (2.4-dinitrophenylhydrazone) $C_{23}H_{18}O_5N_4$. Deep orange-coloured, crystalline powder (ethyl acetate), m. 240⁰ (1938 Koebner).

3′.5′-Diketo-1.2-cyclopentenophenanthrene $C_{17}H_{10}O_2$ (p. 155). Pale yellow needles (benzene or acetone), m. 240⁰ cor. in a Pyrex tube; dec. above 240⁰ when heated in a soft glass capillary (1937 Cohen; cf. also 1936 Fieser). — **Fmn.** From dimethyl 1.2.3.4-tetrahydrophenanthrene-1.2-dicarboxylate on condensation with ethyl acetate by refluxing overnight in this solvent with sodium, followed by heating of the reaction product with 5 N HCl at 100° for 1 hr.; yield, 7 % (1937 Cohen).

1936 Fieser, Fieser, Hershberg, *J. Am. Ch. Soc.* **58** 2322, 2324.
1937 Cohen, Warren, *J. Ch. Soc.* **1937** 1315, 1318.
1938 Koebner, Robinson, *J. Ch. Soc.* **1938** 1994; see also Robinson, *Brit. Pats.* 514516, 514592 (issued 1939); *C. A.* **1941** 4393; *Ch. Ztbl.* **1940** II 1652; *Fr. Pat.* 854270 (1939, issued 1940); *Ch. Ztbl.* **1940** II 1652.
1941 Koebner, Robinson, *J. Ch. Soc.* **1941** 566, 569, 573, 574.
1945 Birch, Jaeger, Robinson, *J. Ch. Soc.* **1945** 582, 583.
1946 Butenandt, Dannenberg, v. Dresler, *Z. Naturforsch.* **1** 222, 224, 226.

2. Diketosteroids C_{18} (both CO groups in the ring system)

10β-Δ⁴-Oestrene-3.17-dione, 19-nor-Δ⁴-Androstene-3.17-dione $C_{18}H_{24}O_2$.

Leaflets (pentane), m. 146–8° (1940 Marker); according to Rapala (1958), the dione [m. 159–161°; absorption max. at 239 mμ (log ε 4.20)] obtained in the same way as that of Marker (1940) is identical with the latter and with a sample provided by Wilds; according to Wilds (1953), 19-nor-Δ⁴-androstene-3.17-dione, obtained from 19-nor-testosterone * by CrO_3 oxidation, forms plates (cyclohexane), m. 170–171° cor. vac. (bath preheated to 150°) after sintering at 168.5° cor., $[\alpha]_D^{27} + 137°$, $[\alpha]_{Hg}^{27} + 162°$ (both in chloroform), absorption max. in alcohol at 239 (4.23) and 298 (2.11) mμ (log ε), and may be a dimorphic form of the dione described by Marker (1940), or the latter may be an epimer or an isomer with another position of the double bond.

Fmn. From 5β.10β-œstrane-3.17-dione (œstrane-3.17-dione-B, below) on treatment with 1 mol. bromine and a little 48 % HBr in glacial acetic acid, followed by refluxing of the resulting 4β-bromo-5β.10β-œstrane-3.17-dione (p. 2916 s) with dry pyridine for 9 hrs. (1940 Marker; 1958 Rapala).

Oestrane-3.17-dione-A $C_{18}H_{26}O_2$.

Differs from œstrane-3.17-dione-B (below) at least in regard to the configuration at C-5 or C-10. — Crystals (50 % acetone), m. 124°. — **Fmn.** From œstrane-3.17-diol-A (p. 1984 s) on oxidation with CrO_3 in strong acetic acid at room temperature; yield, 72 % (1938 Marker).

5β.10β-Oestrane-3.17-dione, 19-nor-Aetiocholane-3.17-dione, Oestrane-3.17-dione-B $C_{18}H_{26}O_2$.

For configuration, see 1958 Rapala; cf. 1953 Wilds. — Exists in two interconvertible crystalline forms: crystals (dil. acetone), m. 144–6°, and needles (ether - pentane), m. 179–180° (1940 Marker); m. 179° to 181°, $[\alpha]_D^{25} + 112°$ (chloroform) (1958 Rapala). — **Fmn.** From 5β.10β-œstrane-3β.17β-diol (œstrane-3.17β-diol-B,

---

* Described after the closing date for this volume by Birch (1950) and Wilds (1953).

197*

p. 1984 s) on oxidation with CrO_3 in strong acetic acid at room temperature (1938, 1940 Marker; cf. 1958 Rapala). — **Rns.** For conversion into 10β-Δ^4-œstrene-3.17-dione via 4β-bromo-5β.10β-œstrane-3.17-dione, see the formation of the unsaturated dione (p. 2873 s).

Equilenan-11.17-dione, 3-Desoxy-11-ketoequilenin $C_{18}H_{16}O_2$.

For configuration at C-13 and C-14, see also 1950 Bachmann. — Crystals (acetone, methanol, or ethanol), m. 212–4⁰ (1939a, b Marker). — **Fmn.** From the carbinol fraction non-precipitable by digitonin, isolated from mares' pregnancy urine after hydrolysis with hydrochloric acid, on oxidation with CrO_3 in strong acetic acid at 20⁰ (1939a, b Marker); the precursor of this diketone is probably a 3-desoxydihydroequilenin (1939b Marker), but cf. also the occurrence of d-3-desoxyequilenin (p. 2391 s) in the same urine, stated by Prelog (1945).

Rns. Gives trans-equilenan-11.17-diol (p. 1985 s) on hydrogenation in ether - abs. alcohol in the presence of Adams's catalyst at room temperature under 3 atm. pressure; when some conc. HCl is also present in this reaction (in alcohol; 2 atm.), $\Delta^{5.7.9}$-œstratrien-17β-ol (p. 1498 s) is obtained (1939b Marker). On refluxing with zinc and hydrochloric acid in alcohol it gives d-3-desoxyequilenin (p. 2391 s) (1939b Marker; cf. 1950 Bachmann), whereas on using amalgamated zinc a trans-d-equilenan m. 73–75⁰ ("desoxyequilenin", p. 1386 s) is formed (1939b Marker).

Monosemicarbazone $C_{19}H_{19}O_2N_3$. Crystals (80 % alcohol), m. 255–260⁰ dec. (1939a Marker).

2-Methyl-3'.4'-diketo-1.2-($\Delta^{1(5')}$-cyclopenteno)-1.2.3.4-tetrahydrophenanthrene, 14-Dehydroequilenan-16.17-dione $C_{18}H_{14}O_2$.

Orange leaflets (benzene - alcohol), m. 203–5⁰ cor.; sublimes at 180⁰ under 0.03 mm. pressure (1944 Wilds). Absorption max. in abs. alcohol at [220 (4.26)], 236 (4.37), 270.5 (4.20), 279.5 (4.26), 291 (4.15), [332–4 (4.22)], and 345 (4.23) mμ (log ε) (1947 Wilds). Gives an olive-green solution with conc. H_2SO_4 (1944 Wilds).

Fmn. The 17-oxime (below) is obtained from 14-dehydroequilenan-16-one (p. 2388 s) on heating with potassium tert.-butoxide in dry tert.-butyl alcohol under nitrogen at 70⁰ until complete dissolution (ca. 20 min.), followed by dropwise addition of butyl nitrite with cooling and then by stirring for 5 hrs. at room temperature (yield, 98 %); it is hydrolysed by refluxing with 37 % formaldehyde solution and conc. HCl in dioxane for 1 hr. (yield, ca. 90 %) (1944 Wilds).

Rns. Gives the anhydride of syn-(2-methyl-2-carboxy-1.2.3.4-tetrahydro-1-phenanthrylidene)-acetic acid on oxidation with periodic acid in aqueous dioxane at room temperature (1944 Wilds). On hydrogenation in dioxane in the presence of palladium - charcoal at room temperature under atmospheric pressure it gives 2-methyl-3'-hydroxy-4'-keto-1.2-($\Delta^{1(5')}$-cyclopenteno)-

1.2.3.4-tetrahydrophenanthrene (p. 2525 s) (1944 Wilds; cf. 1947 Wilds for the structure). Reacts with o-phenylenediamine in boiling alcohol to give the corresponding **quinoxaline derivative** $C_{24}H_{18}N_2$ [yellow leaflets (methanol), m. 200–200.5° cor.; gives a purple solution with conc. H_2SO_4] (1944 Wilds).

17-Oxime $C_{18}H_{15}O_2N$. Yellow leaflets (methanol), sintering at 225° cor., m. 227.5–228° cor. with evolution of gas. Gives a brown solution with conc. H_2SO_4 (1944 Wilds). — **Fmn.** See above.

Dioxime $C_{18}H_{16}O_2N_2$. Light yellow leaflets (alcohol), darkening at 235° cor., m. 237–237.5° cor. with evolution of gas. Gives a reddish brown solution with conc. H_2SO_4 (1944 Wilds).

1938 Marker, Rohrmann, Lawson, Wittle, *J. Am. Ch. Soc.* **60** 1512, 1901.
1939 a Marker, Rohrmann, *J. Am. Ch. Soc.* **61** 2537.
      b Marker, Rohrmann, *J. Am. Ch. Soc.* **61** 3314, 3316.
1940 Marker, Rohrmann, *J. Am. Ch. Soc.* **62** 73.
1944 Wilds, Beck, *J. Am. Ch. Soc.* **66** 1688, 1692, 1693.
1945 Prelog, Führer, *Helv.* **28** 583, 585, 588.
1947 Wilds, Beck, Close, Djerassi, Johnson, Johnson, Shunk, *J. Am. Ch. Soc.* **69** 1985, 1987, 1988, 1993.
1950 Bachmann, Dreiding, *J. Am. Ch. Soc.* **72** 1329 footnote 4.
      Birch, *J. Ch. Soc.* **1950** 367.
1953 Wilds, Nelson, *J. Am. Ch. Soc.* **75** 5366.
1958 Rapala, Farkas, *J. Am. Ch. Soc.* **80** 1008.

3. Diketosteroids C₁₉ (both CO groups in the ring system)

Summary of Diketones with Androstane (or Aetiocholane) Skeleton

For values other than those chosen for this summary, see the compounds themselves.

Position of CO	Double bond(s)	Config. at C-5	M. p. °C.	[α]$_D$*	Absorption max. in alc.	Page
3.12	satd.	β	196	—	—	2876 s
3.16	satd.	α	158	—	—	2876 s
3.17	Δ$^{1.4.6}$	—	166	+83 (chl.), +72.5 (dio.)	222, 256, 298	2877 s
3.17	Δ$^{1.4}$	—	141	+119 (chl.), +103 (dio.)	244	2877 s
3.17	Δ$^{4.6}$	—	170	+138 (chl.)	282	2878 s
3.17	Δ$^{4.9(11)}$	—	204	+224 (chl.)	—	2879 s
3.17	Δ^1	α	140	+128 (chl.), +144 (alc.)	230	2879 s
3.17	Δ^4	—	173→144	+187 (chl.), +191 (alc.)	241	2880 s
3.17	Δ^5	—	158	—	—	2890 s
3.17	Δ$^{9(11)}$	α	154	+156 (alc.)	—	2891 s
3.17	Δ$^{11(?)}$	α	146	+142 (alc.)	—	2892 s
3.17	satd.	α	134	+100 (alc.), +105 (chl.)	—	2892 s
3.17	satd.	β	133	+110 (alc.)	—	2895 s
3.17	satd.	α[1]	166	—	—	2895 s
4.17	Δ^5	—	140	+7 (alc.)	238[2]	2897 s
6.17	Δ^2	α	191	+123 (chl.), +126 (alc.)	285–300[3]	2897 s
6.17	Δ^4	—	180	+97 (chl.)	245.5, 300	2898 s
6.17	satd.	α	135	—	—	2898 s
6.17	i(3–5)	—	183	+113 (chl.)	—	2898 s

* alc. = alcohol, chl. = chloroform, dio. = dioxane
[1]) 13-iso   [2]) in chloroform   [3]) solvent not indicated

5-iso-Androstane-3.12-dione, Aetiocholane-3.12-dione C₁₉H₂₈O₂.

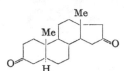

Crystals (benzene - petroleum ether 1:9), needles (acetone - hexane), m. 193–6° cor. (Kofler block). — **Fmn.** From ætiocholane-3α.12α-diol (p. 1992 s) on oxidation with CrO₃ in glacial acetic acid (1945 Reich).

1945 Reich, *Helv.* **28** 863, 872.

Androstane-3.16-dione C₁₉H₂₈O₂

(not analysed). For the structure of the parent olone, see 1951, 1954 Huffman. — Needles (ligroin), m. 157–8°; very soluble in the usual organic solvents (1939 Heard). — **Fmn.** From androstan-3β-ol-16-one (p. 2604 s) on oxidation with CrO₃ in ice-cold 90% acetic acid; yield, 75% (1939 Heard). — **Rns.** Gives androstane (p. 1396 s) on refluxing with amalgamated zinc and conc. HCl (1939 Heard).

1939 Heard, McKay, *J. Biol. Ch.* **131** 371, 377.
1951 Huffman, Lott, *J. Am. Ch. Soc.* **73** 878.
1954 Huffman, Lott, *J. Biol. Ch.* **207** 431.

$\Delta^{1\cdot4\cdot6}$-**Androstatriene-3.17-dione** $C_{19}H_{22}O_2$. Very large, hexagonal plates (ethyl acetate), m. 164.5–166° cor. (Kofler block); $[\alpha]_D^{20}$ +82.6° (chloroform), +72.5° (dioxane); absorption max. in 95% alcohol at 222 (4.19), 256 (4.13), and 298 (4.27) mμ (log ε) (1950 Kaufmann). — **Fmn.** From 6-bromo-$\Delta^{1\cdot4}$-androstadiene-3.17-dione (p. 2918 s) on refluxing with collidine for 15 min.; yield, 62% (1950 Kaufmann). From both 2α·6α(?)- and 2α·6β(?)-dibromo-Δ^4-androstene-3.17-diones (p. 2925 s) on refluxing with collidine for 30 min.; yields, 45% (1950 Djerassi; cf. 1950 Rosenkranz). — **Rns.** On passing slowly a solution in mineral oil through a glass tube filled with helices and heated to 600°, it gives 6-dehydrooestrone (p. 2538 s) (1950 Kaufmann).

1950 Djerassi, Rosenkranz, Romo, Kaufmann, Pataki, *J. Am. Ch. Soc.* **72** 4534, 4538, 4539.
Kaufmann, Pataki, Rosenkranz, Romo, Djerassi, *J. Am. Ch. Soc.* **72** 4531.
Rosenkranz, Djerassi, Kaufmann, Pataki, Romo, *Nature* **165** 814.

$\Delta^{1\cdot4}$-**Androstadiene-3.17-dione** $C_{19}H_{24}O_2$. The compound, m. 168°, described by Fujii (see p. 155) was a mixture of androstadiene-diones containing only ca. 50% of the present dione (1950 Hershberg). — Plates (dil. methanol), m. 139–140° (1940 Inhoffen); rectangular plates (hexane - acetone), m. 140–141° cor. (1948 Djerassi); square prisms (ether), m. 140.9–142.1° cor. (1950 Hershberg); $[\alpha]_D^{25}$ in chloroform +116° (1940 Inhoffen), +119° (1948 Djerassi), in dioxane +103° (1950 Hershberg). Absorption max. in isoöctane at 230–232 mμ (log ε 4.19) (1950 Hershberg), in 95% alcohol at 244 mμ (log ε 4.23) (1950 Djerassi).

Fmn. From $\Delta^{1\cdot4}$-androstadien-17β-ol-3-one (p. 2576 s) on refluxing with aluminium isopropoxide and cyclohexanone in toluene for 1 hr.; yield, 56% (1940 Inhoffen). From 2α-bromo-Δ^4-androstene-3.17-dione (p. 2916 s) on refluxing with collidine; yield, 55% (1950 Djerassi). From 2α.4α-dibromo-androstane-3.17-dione (p. 2924 s) on refluxing with pyridine, followed by vacuum distillation (1936 SCHERING A.-G.) or on refluxing with collidine for $^1/_2$ hr. (crude yield, 54%) (1948 Djerassi; cf. also 1950 Hershberg). An in-separable mixture of $\Delta^{1\cdot4}$- and $\Delta^{4\cdot6}$-androstadiene-3.17-dione is obtained from 5-chloroandrostane-3.17-dione (p. 2918 s) by treatment with 1 mol. bromine in chloroform, followed by refluxing of the resulting 2(?)-bromo-derivative with collidine for 1 hr. (1950 Hershberg).

Rns. Is reduced to $\Delta^{1\cdot4}$-androstadiene-3.17β-diol (p. 1992 s) by refluxing with aluminium isopropoxide in dry isopropyl alcohol or by the action of fermenting yeast (1937 a SCHERING A.-G.). On refluxing with 1 mol. of N-bromosuccinimide and some benzoyl peroxide in carbon tetrachloride it yields 6-bromo-$\Delta^{1\cdot4}$-androstadiene-3.17-dione (p. 2918 s) (1950 Kaufmann). Gives 4-methyl-$\Delta^{1\cdot3\cdot5(10)}$-oestratrien-1-ol-17-one ("1-methyloestrone", see page 2573 s) on treatment with a little conc. H_2SO_4 in acetic anhydride at room temperature, followed by hydrolysis of the resulting acetate with methanolic

KOH (1941 Inhoffen; 1948 Djerassi), or on heating with p-toluenesulphonic acid
in acetic anhydride on the steam-bath (1948 Djerassi); the same trienolone is
also obtained, along with œstrone (p. 2543 s), by heating of $\Delta^{1.4}$-androstadiene-
3.17-dione in an atmosphere of CO_2 at 300–350° (1950 Hershberg; cf. also
1937 b, c, 1947 SCHERING A.-G.); pyrolysis by dropwise addition of a solution
of the dione in mineral oil (b. 310–405°) to glass beads contained in a glass
tube, at 525–535° affords 21 % yield of œstrone (1950 Hershberg).

Disemicarbazone $C_{21}H_{30}O_2N_6$. Colourless needles (methanol); discolours at
ca. 320°, dec. above 350° (1941 Inhoffen).

1936 SCHERING A.-G. (Inhoffen), *Ger. Pat.* 722943 (issued 1942); *Dan. Pat.* 61375 (1937, issued
1943); *Ch. Ztbl.* **1944** II 49; SCHERING-KAHLBAUM A.-G., *Fr. Pat.* 835524 (1937, issued
1938); *Brit. Pat.* 500353 (1937, issued 1939); *C. A.* **1939** 5872; *Ch. Ztbl.* **1939** I 5010, 5011;
SCHERING CORP. (Inhoffen, Butenandt, Schwenk). *U.S. Pat.* 2340388 (1937, issued 1944);
C. A. **1944** 4386.

1937 a SCHERING A.-G., *Fr. Pat.* 835526 (issued 1938); *C. A.* **1939** 4599; *Ch. Ztbl.* **1939** I 4088;
SCHERING CORP. (Schoeller, Serini, Inhoffen), *U.S. Pat.* 2264861 (issued 1941); *C. A.* **1942**
1613.

b SCHERING A.-G., *Brit. Pat.* 508576 (issued 1939), *Fr. Pat.* 838704 (issued 1939); *C. A.* **1940**
777; *Ch. Ztbl.* **1939** II 1722; SCHERING CORP. (Inhoffen), *U.S. Pat.* 2280828 (issued 1942);
C. A. **1942** 5618.

c SCHERING A.-G., *Swiss Pat.* 230019 (issued 1944); *Ch. Ztbl.* **1944** II 1204.

1940 Inhoffen, Zühlsdorff, Huang-Minlon, *Ber.* **73** 451, 457.

1941 Inhoffen, Zühlsdorff, *Ber.* **74** 604, 614 footnote 16.

1947 SCHERING A.-G., *Fiat Final Report* No. 996, p. 150.

1948 Djerassi, Scholz, *J. Org. Ch.* **13** 697, 704, 705.

1950 Djerassi, Rosenkranz, Romo, Kaufmann, Pataki, *J. Am. Ch. Soc.* **72** 4534, 4538.
Hershberg, Rubin, Schwenk, *J. Org. Ch.* **15** 292, 296 et seq.
Kaufmann, Pataki, Rosenkranz, Romo, Djerassi, *J. Am. Ch. Soc.* **72** 4531, 4532.

$\Delta^{4.6}$-**Androstadiene-3.17-dione** $C_{19}H_{24}O_2$ (p. 155). Crystals (benzine - ethyl
acetate), m. 172° (1937, 1947 SCHERING A.-G.), 168° to
170° cor. (Kofler block); $[\alpha]_D^{20} +138°$ (chloroform) (1950
Djerassi). Absorption max. in ether at 273 mμ (ε 4.42)
(1939 Dannenberg), in 95 % alcohol at 282 mμ (ε 4.52)
(1950 Djerassi); for absorption max. (solvent not stated)
285 mμ (ε 4.7), see 1937 Ruzicka. — **Fmn.** From Δ^5-androsten-3β-ol-17-one
(5-dehydroepiandrosterone, p. 2609 s) on refluxing with aluminium tert.-
butoxide and benzoquinone in abs. toluene (1941 CIBA). From 6α(?)-bromo-
Δ^4-androstene-3.17-dione (p. 2918 s) on treatment with silver nitrate in anhydrous pyridine at room temperature (40 hrs.) (1937, 1947 SCHERING A.-G.)
or on refluxing with collidine (yield, 51 %) (1950 Djerassi). — **Rns.** Gives
its 3-enol acetate ($\Delta^{2.4.6}$-androstatrien-3-ol-17-one acetate, p. 2605 s) on boiling
with acetic anhydride and fused potassium acetate (1937, 1947 SCHERING
A.-G.).

1937 Ruzicka, Bosshard, *Helv.* **20** 328, 329.
SCHERING A.-G., *Fr. Pat.* 838704 (issued 1939), *Brit. Pat.* 508576 (issued 1939); *C. A.* **1940**
777; *Ch. Ztbl.* **1939** II 1722; SCHERING CORP. (Inhoffen), *U.S. Pat.* 2280828 (issued 1942);
C. A. **1942** 5618.

1939 Dannenberg, *Abhandl. Preuss. Akad. Wiss. Math.-naturwiss. Kl.* **1939** No. 21, pp. 27, 59.
1941 CIBA (Soc. pour l'ind. chim. à Bâle; Ges. f. chem. Ind. Basel), *Fr. Pat.* 877821 (issued 1943);
    Dutch Pat. 54147 (issued 1943); *Ch. Ztbl.* **1943** II 147, 148.
1947 SCHERING A.-G., *Fiat Final Report* No. 996, p. 155.
1950 Djerassi, Rosenkranz, Romo, Kaufmann, Pataki, *J. Am. Ch. Soc.* **72** 4534, 4538.

$\Delta^{4\cdot 9(11)}$-**Androstadiene-3.17-dione** * $C_{19}H_{24}O_2$. Crystals. — **Fmn.** Along with
a crystalline isomer **, from Δ^4-androsten-11β-ol-3.17-
dione (p. 2947 s) on refluxing with conc. HCl in glacial
acetic acid (1942 Reichstein).

> 1942 Reichstein, *Brit. Pat.* 560812 (issued 1944); *C. A.* **1946** 4486;
>   *Ch. Ztbl.* **1945** II 1773; *U.S. Pat.* 2409798 (issued 1946); *C. A.*
>   **1947** 1397, 1398.
> 1954 Bernstein, Lenhard, Williams, *J. Org. Ch.* **19** 41, 45, 46.

Δ^1-**Androstene-3.17-dione** $C_{19}H_{26}O_2$ (p. 155). Crystals (dil. acetone), m. 138°
to 139°; $[\alpha]_D^{19}$ +144° (alcohol) (1940a Butenandt); nee-
dles (hexane), m. 140–142°; $[\alpha]_D^{24}$ +128° (chloroform)
(1947 Djerassi). Absorption max. in alcohol at 230 mμ
(log ε 4.01) (1947 Djerassi; cf. also 1939 Butenandt;
1939 Dannenberg). — **Fmn.** From Δ^1-androsten-17β-ol-
3-one (p. 2578 s) on oxidation with CrO_3 in glacial acetic
acid at room temperature; yield, 27 % (1940a Butenandt; 1939 SCHERING
A.-G.). In its formation from 2α-bromoandrostane-3.17-dione (p. 2916 s) by
refluxing with collidine (see p. 155; cf. also 1939 SCHERING A.-G.) there is
obtained a mixture (3:1) of Δ^1- and Δ^4-androstene-3.17-diones (crude yield
of the Δ^1-isomer, 36%) (1947 Djerassi); the Δ^1-dione is also obtained from
the bromo-dione by refluxing with dry quinoline for 5 hrs. (1936 CIBA PHAR-
MACEUTICAL PRODUCTS). From Δ^1-allopregnene-3.20-dione (p. 2781 s) and
from Δ^1-cholesten-3-one (p. 2423 s) by oxidative degradation (1939 SCHERING
A.-G.). — **Rns.** Is reduced to androstane-3β.17β-diol (p. 2014 s) by sodium
in boiling isopropyl alcohol (1940a Butenandt) or by the action of fermenting
yeast in dil. alcohol (1940b Butenandt).

hetero-Δ^1-Androstene-3.17-dione $C_{19}H_{26}O_2$ (p. 155). Is actually Δ^5-androstene-
4.17-dione (p. 2897 s) (1944 Butenandt).

1936 CIBA PHARMACEUTICAL PRODUCTS (Miescher, Wettstein), *U.S. Pat.* 2260328 (issued 1941);
    C. A. **1942** 873; *Ch. Ztbl.* **1944** II 246.
1939 Butenandt, Mamoli, Dannenberg, Masch, Paland, *Ber.* **72** 1617, 1622.
    Dannenberg, *Abhandl. Preuss. Akad. Wiss. Math.-naturwiss. Kl.* **1939** No. 21, pp. 13, 54.

---

* This compound has been described in more detail after the closing date for this volume by
Bernstein (1954).
** Cf. the analogous dehydration of corticosterone 21-acetate (p. 2854 s).

SCHERING A.-G. (Butenandt), *Ger. Pat.* 736846 (issued 1943); *C. A.* **1944** 3094; *Swed. Pat.* 111450 (1940, issued 1944); *Ch. Ztbl.* **1947** 1875; *Dan. Pat.* 61322 (1940, issued 1943); *Ch. Ztbl.* **1944** II 49; *Fr. Pat.* 867697 (1940, issued 1941); *Ch. Ztbl.* **1942** I 2561; SCHERING CORP. (Butenandt), *U.S. Pat.* 2441560 (1940, issued 1948); *C. A.* **1948** 6862; *Ch. Ztbl.* **1949** II 681.
1940 a Butenandt, Dannenberg, *Ber.* **73** 206, 208.
  b Butenandt, Dannenberg, Surányi, *Ber.* **73** 818.
1944 Butenandt, Ruhenstroth-Bauer, *Ber.* **77** 397, 400.
1947 Djerassi, *J. Org. Ch.* **12** 823, 824, 829, 830.

Δ⁴-Androstene-3.17-dione $C_{19}H_{26}O_2$ (I) (p. 155). Crystals (dil. acetone and acetone - petroleum ether), m. 172.5–173.5° cor., resoli-

difying, and remelting at 144° cor. (1945 Hirschmann); needles (abs. ether or ether - pentane), m. 172–4° cor. (1938 Reichstein; 1941 Prins); $[\alpha]_D^{17} +197°$ (abs. chloroform) (1937 Westphal; cf. 1940, 1944 Ruzicka), $[\alpha]_D^{23}$ +187° (chloroform) (1947 Djerassi). Absorption max. in alcohol at 241 mμ (ε 15755) (1937 Hogness), in chloroform at 240 mμ (ε 16600) (1939 Dannenberg; cf. 1937 Westphal; 1938b Butenandt; 1943 Evans). For its infra-red absorption spectrum, see 1946 Furchgott. Dipole moment 3.79D (1945 Kumler). Twice as soluble in 35% alcohol as in petroleum ether (1939 Spielman). May be extracted from solutions in ligroin or benzene with conc. HCl (1938 Dirscherl; 1937 C. F. BOEHRINGER & SÖHNE). For basic properties, see 1946 Bersin.

Occ. To a very small extent in adrenal cortex extract; it is possibly formed from other compounds occurring in such extracts by oxidation in the air (1941a v. Euw).

FORMATION OF Δ⁴-ANDROSTENE-3.17-DIONE

Formation from C_{19} and C_{20} steroids. From Δ⁴-androsten-17β-ol (p. 1510 s) on oxidation with CrO_3 in strong acetic acid at 35–45° (1940a Marker; 1940 PARKE, DAVIS & Co.). From Δ⁵-androstene-3β.17β-diol (p. 1999 s) on bromination, followed by oxidation of the resulting dibromide with CrO_3 in strong acetic acid at room temperature and then by debromination with zinc dust in acetic acid on the steam-bath (1940b Marker; 1945 Hirschmann); from the same diol by the action of Flavobacterium androstenedionicum (1944 Ercoli) or Flavobacterium carbonylicum (with simultaneous formation of testosterone) (1944a Molina); for formation in small amounts from the same diol by the action of other bacteria, see under "Biological oxidation" of the diol (p. 2003 s); the formation from the diol by shaking with impoverished yeast and oxygen in water at 32° (1938 Vercellone) was probably also due to the presence of bacteria (1938c Mamoli).

From Δ⁴-androsten-17-one (p. 2402 s) on oxidation with CrO_3 in strong acetic acid at 35–45° (1940a Marker; 1940 PARKE, DAVIS & Co.). From 3-keto-Δ⁴-ætiocholenylamine (17-amino-Δ⁴-androsten-3-one, p. 2509 s) on treatment in ether with hypochlorous acid solution in the presence of Na_2SO_4

at —5° to 0°, followed by conversion of the resulting N-chloro-derivative
with sodium ethoxide in abs. alcohol on the steam-bath into the imide,
which is then hydrolysed with dil. H_2SO_4 (1937b I.G.FARBENIND.; cf. also
1939 Ehrhart).

From its 3-enol ethyl ether (ethyl ether of $\Delta^{3.5}$-androstadien-3-ol-17-one,
p. 2605 s) or the diethyl acetal of this ether on refluxing with aqueous alcoholic HCl for 1 hr. (1938 Serini; 1938b SCHERING A.-G.). From Δ^4-androsten-
17β-ol-3-one (testosterone, p. 2580 s) by the action of Proactinomyces erythropolis Gray & Thornton or other P. species; yield, 47 % (1946 Turfitt).

From Δ^4-androsten-3α and/or 3β-ol-17-one ("4-dehydroandrosterone", page
2608 s) on heating with copper powder in a vacuum at 180° (1935b CIBA);
similarly from Δ^5-androsten-3β-ol-17-one (5-dehydroepiandrosterone, p.2609 s)
(1935a CIBA). From the last-mentioned compound on heating with aluminium isopropoxide and cyclohexanone for $^1/_2$ hr. at 100° (quantitative
yield) (1937b SCHERING-KAHLBAUM A.-G.), on refluxing with aluminium
tert.-amyl oxide and acetophenone in anhydrous benzene for 14 hrs. (yield,
80–86 %) (1937 Oppenauer), or, along with Δ^5-androstene-3.17-diols and
Δ^4-androsten-17β-ol-3-one (testosterone, p. 2580 s), on refluxing with aluminium tert.-butoxide in anhydrous benzene for 14 hrs. (1937a N. V. ORGA-
NON). From 5-dehydroepiandrosterone on conversion, in chloroform, into
its dibromide, followed by oxidation with CrO_3 in strong acetic acid for
2 hrs. at 25° and then by debromination in acetone under CO_2 with 1 N
aqueous chromous chloride solution (containing $ZnCl_2$ and hydrochloric acid)
for 2 hrs. (yield, 80%) (1945 Julian; 1944 GLIDDEN Co.); when zinc dust in
acetic acid is used for debromination, the yield is only 50 % (1945 Julian;
1944 GLIDDEN Co.; cf. also 1936b CIBA); when the aforesaid dibromide is
refluxed with collidine for 30 min., 50 % yield of the dione is obtained (1941
Galinovsky). From 5-dehydroepiandrosterone on shaking in water under
oxygen with impoverished yeast containing aerobic bacteria (yield, 70 %)
(1938b, c Mamoli; cf. also 1938e Mamoli); in almost quantitative yield by
the action of dehydrogenating bacteria in the presence of air: Flavobacterium
(Micrococcus) dehydrogenans, Corynebacterium mediolanum (1940 Ercoli;
1942 Arnaudi; 1944b Molina), Flavobacterium carbonylicum (1944a Molina),
Proactinomyces erythropolis Gray & Thornton, and other P. species (1946
Turfitt; cf. also 1948 Turfitt). From 5-dehydroepiandrosterone triphenylmethyl ether (p. 2620 s) on heating with Al_2O_3 under nitrogen at 350° (1937d
CIBA). From 5-dehydroepiandrosterone on treatment with SeO_2, followed
by heating of the resulting mixture of Δ^5-androstene-3β.4-diol-17-one and
Δ^4-androstene-3β.6-diol-17-one in alcohol with sulphuric acid (1937b CIBA).
From androstane-3β.5α-diol-17-one (p. 2746 s) on refluxing with aluminium
tert.-butoxide and abs. acetone in dioxane - benzene for 21 hrs. (1944 Ruzicka).

From Δ^5-androstene-3.17-dione (p. 2890 s) on heating with 48 % HBr in
glacial acetic acid for 10 min. on the water-bath (1936 Butenandt; 1937a
SCHERING-KAHLBAUM A.-G.) or on boiling in alcohol with 2 N H_2SO_4 for
10 min. (1937a I.G. FARBENIND.). From 2α-bromo-Δ^4-androstene-3.17-dione
(p. 2916 s) on treatment in acetone with 1 N aqueous chromous chloride
solution (containing $ZnCl_2$ and hydrochloric acid) for 2 hrs. in an atmosphere

of CO_2; yield, 57% (1948 Djerassi). Is obtained in 12% crude yield, along with Δ^1-androstene-3.17-dione (crude yield, 36%), from 2α-bromoandrostane-3.17-dione (p. 2916 s) on refluxing with collidine for 1 hr. (1947 Djerassi). From 3.17-diketo-androstan-2-yl-pyridinium bromide (p. 2917 s) on heating in a sealed tube under 10 mm. pressure (1938 b Ruzicka; 1937 f CIBA). From 4β-bromo-ætiocholane-3.17-dione (p. 2917 s) on refluxing with dry pyridine for 13 hrs.; yield, 21% (1938 Ercoli; 1935 c CIBA). From androstan-6β-ol-3.17-dione (p. 2947 s) on distillation with $ZnCl_2$ in a high vacuum at 190° (1941 PARKE, DAVIS & Co.).

From 17α-formyl-Δ^4-androsten-17β-ol-3-one (p. 2835 s) on oxidation with periodic acid in aqueous dioxane at room temperature (yield, 72%); similarly from the 17-epimeric aldehyde (p. 2836 s) (1941 Prins). From 3β.17β-dihydroxy-17-iso-Δ^5-ætiocholenic acid on conversion into the dibromide, followed by oxidation with CrO_3 in glacial acetic acid at room temperature and then by debromination with zinc dust (1938 a Ruzicka).

Formation from C_{21} and C_{22} steroids. From Δ^4-pregnene-17α.20β-diol-3-one (p. 2721 s) on oxidation with lead tetraacetate in glacial acetic acid at room temperature (1938 a Butenandt; 1942 Ruzicka). Along with the 21-acetate of Δ^4-pregnene-17α.21-diol-3.20-dione (Reichstein's Substance S, p. 2859 s), from the 21-acetate of Δ^4-pregnene-17α.20β.21-triol-3-one (p. 2725 s) on oxidation with CrO_3 in strong acetic acid at 11°; yield, ca. 29% (1946 Sarett). From Δ^4-pregnen-17α-ol-3.20-dione (17-hydroxyprogesterone, p. 2843 s) on oxidation with CrO_3 in strong acetic acid at room temperature; yield, 37% (1940, 1941 Pfiffner; 1941 Prins). From Δ^4-pregnene-17α.21-diol-3.20-dione (Reichstein's Substance S, p. 2859 s) on oxidation with CrO_3 in glacial acetic acid at room temperature (1938 Reichstein).

From the mono-triphenylmethyl ether of ω-homo-17-iso-Δ^4-pregnene-17β.21.22-triol-3-one (p. 2728 s) on refluxing with aluminium isopropoxide and cyclohexanone in abs. toluene for 30 min.; yield, 42% (1938 b Butenandt). From ω-homo-Δ^4-pregnene-17α.20β.21.22-tetrol-3-one (p. 2732 s) on oxidation with periodic acid in aqueous dioxane at room temperature; small yield (1941 b v. Euw).

Formation from C_{27} steroids. For formation, along with other compounds, from Δ^4-cholestene by oxidation with CrO_3 in glacial acetic acid, see 1940 CIBA PHARMACEUTICAL PRODUCTS. Along with progesterone (p. 2782 s), from Δ^4-cholesten-3β-ol ("allocholesterol", p. 1564 s) on oxidation with CrO_3 in strong acetic acid at 50° (1935 b C. F. BOEHRINGER & SÖHNE). From Δ^5-cholesten-3β-ol (cholesterol, p. 1570 s) on oxidation with CrO_3 with temporary protection of the double bond by bromination, followed by chromatographic separation from other reaction products (1938 CHINOIN); is probably also formed as a by-product on using $KMnO_4$ in aqueous H_2SO_4 as oxidizing agent (1939 Spielman). Along with progesterone (p. 2782 s), from Δ^4-cholesten-3-one (p. 2424 s) on oxidation with CrO_3 in strong acetic acid at 40–50° (1938 Dirscherl; 1935 a C. F. BOEHRINGER & SÖHNE) or at 20° (1937 a CIBA) or in water - carbon tetrachloride at 20° (1939 a C. F. BOEHRINGER & SÖHNE); separation from unchanged cholestenone by extraction of the dione from

ligroin or benzene solutions with conc. HCl (1938 Dirscherl; 1937 C. F. Boehringer & Söhne), by chromatographic adsorption on Al_2O_3 (1937c Ciba),
or by means of Girard's reagent (1938 Ciba); separation from progesterone
based on the solubility in 35% alcohol and in petroleum ether (1939 Spielman) or by chromatographic adsorption on Al_2O_3 (1941 Bretschneider). From
6β-bromo-$Δ^4$-cholesten-3-one (p. 2484 s), 5α.6β-dibromo- and a 2α.5α.6β-tri-
bromocholestan-3-one (pp. 2500 s and 2504 s, resp.) on oxidation, followed by
debromination (1938, 1939b C. F. Boehringer & Söhne).

REACTIONS OF $Δ^4$-ANDROSTENE-3.17-DIONE

Reduction (see also under "Biological reactions", p. 2884 s). On refluxing
with aluminium tert.-butoxide and abs. sec.-butyl alcohol in benzene for
7–8 hrs. it gives $Δ^4$-androsten-17β-ol-3-one (testosterone, p. 2580 s), a little
$Δ^4$-androsten-17α-ol-3-one (epitestosterone, p. 2579 s), and a little $Δ^5$-andros-
ten-3β-ol-17-one (5-dehydroepiandrosterone, p. 2609 s) (1939a Miescher;
1937e Ciba); testosterone is also obtained on heating with aluminium isopropoxide in isopropyl alcohol (1936 N. V. Organon) or on hydrogenation
in alcohol in the presence of a nickel catalyst until 1 mol. of hydrogen has
been absorbed (1936c Ciba); when the dione is boiled with aluminium isopropoxide in abs. isopropyl alcohol, with simultaneous removal of the acetone
formed, a mixture of $Δ^5$-androstene-3.17-diols is obtained whereas on hydrogenation in decalin at 200° in the presence of nickel, a mixture of androstane-
3.17-diols is formed (1936 Pfeilringwerke A.-G.). Hydrogenation in alcohol
containing sulphuric acid in the presence of platinum oxide affords androstane-
3α.17β-diol (p. 2011 s) (1936 Pfeilringwerke A.-G.). Reduction with
sodium in alcohol yields androstane-3β.17β-diol (p. 2014 s) (1936 Pfeilring-
werke A.-G.); on treatment with
sodium amalgam and 60% alcohol
in ether, followed by boiling with
aluminium isopropoxide in isopropyl
alcohol, it gives the 17.17'-dihydr-
oxy-3.3'-pinacol (II), which on oxidation with 1.15 mols. of lead tetraacetate in methanol at room tempe-
rature yields testosterone (p. 2580 s); the 17.17'-diesters of this pinacol give
the corresponding esters of testosterone on similar oxidation or on oxidation
with CrO_3 in strong acetic acid after temporary protection of the double
bonds by bromination (1937b N. V. Organon; 1938 Roche-Organon, Inc.).

Halogenation (cf. also the reaction with acetyl chloride, p. 2884 s). Gives
6α(?)-bromo-$Δ^4$-androstene-3.17-dione (p. 2918 s) on refluxing in carbon tetrachloride with 1 mol. of N-bromosuccinimide in the absence of light for 20 min.
to 30 min. (1950 Djerassi). On treatment with 2 mols. of bromine and a
little HBr in ether - glacial acetic acid it gives 2.6β(?)-dibromo-$Δ^4$-androstene-
3.17-dione (p. 2925 s) (1950 Djerassi; cf. 1950 Rosenkranz); cf. also 1936
Schering A.-G., where the resulting product (m. 162° dec.) has been formulated
as 4.6-dibromo-$Δ^4$-androstene-3.17-dione.

References, pp. 2887–2890 s

Sulphonation. Treatment with 1 mol. of sulphuric acid in ice-cold acetic anhydride affords 41 % of a **monosulphonic acid** $C_{19}H_{26}O_5S$ [crystals (ether), dec. ca. 196°; hygroscopic; very readily soluble in water and alcohol, sparingly in ether and glacial acetic acid; its alkali salts are soluble in water; *methyl ester* $C_{20}H_{28}O_5S$, needles (aqueous acetone), m. 159–160° with darkening] (1937 Windaus; 1936/37 I.G. Farbenind.); for probable position of the SO_3H-group at C-6 (cf. Δ^4-cholesten-3-one-6-sulphonic acid), see 1939 Kuhr.

Reactions with organic compounds. For conversion into its 3-enol esters (esters of $\Delta^{3.5}$-androstadien-3-ol-17-one, p. 2605 s) by acylation (p. 156), see also 1936a, 1936/37 Ciba; 1946 Ross. On treatment with acetyl chloride at room temperature, 3-chloro-$\Delta^{3.5}$-androstadien-17-one (p. 2474 s) is obtained (1937 Kuwada). Gives the 3-enol ethyl ether (p. 2606 s) on heating with acetone diethyl ketal and abs. alcoholic HCl in benzene at 75° or, under the same conditions, with 1 mol. of ethyl orthoformate (1938 Serini; 1938b, 1939b Schering A.-G.); on using excess ethyl orthoformate, the diethyl ketal of this enol ether is formed (1938 Serini). On boiling with benzyl alcohol and p-toluenesulphonic acid in benzene, with simultaneous removal of the water formed, it gives the 3-enol benzyl ether (p. 2606 s) (1940a, 1941 Schering A.-G.; 1940 Schering Corp.); reacts similarly with cyclohexanol (1940b, 1941 Schering A.-G.; 1940 Schering Corp.), and with ethyl mercaptan to give the corresponding thio-enol ether (1941 Ciba); this ether is also obtained when androstenedione * is treated with ethyl mercaptan in ethyl formate in the presence of benzenesulphonic acid (1940b Chinoin); cf. also the reaction of the diketone with ethyl trithio-orthoformate and benzenesulphonic acid (1940a Chinoin). With 1 mol. or 2 mols. of ethylene glycol in the presence of p-toluenesulphonic acid in benzene, the dione gives the 3-ethylene ketal or the bis-(ethylene ketal) resp. (1939a Schering A.-G.; 1941 Squibb & Sons); the latter is also formed when the dione is treated with excess ethylene oxide and a little $SnCl_4$ in anhydrous carbon tetrachloride at 30° (1939a Schering A.-G.); the aforesaid ketals (p. 2890 s) are derivatives of Δ^5-androstene-3.17-dione (for the structure, see the literature cited with testosterone ethylene ketal, p. 2595 s). Gives the 3-ethylene mercaptal (p. 2886 s) on treatment with ethane-1.2-dithiol in the presence of dry hydrogen chloride (1941 Ciba) or, along with a small amount of the bis-(ethylene mercaptal) (p. 2886 s), in the presence of p-toluenesulphonic acid in acetic acid (1954 Ralls). On refluxing with nicotinic acid hydrazide and a little acetic acid in methanol for 1 hr., the dione gives 97 % yield of its bis-nicotinoylhydrazone (p. 2885 s) (1945 Velluz; 1944 Les Laboratoires Français de Chimiothérapie).

Biological reactions. For reduction of the dione to Δ^4-androsten-17β-ol-3-one (testosterone, p. 2580 s) by the action of fermenting yeast (p. 156), see also 1938a, e Mamoli; 1938a Schering A.-G.; 1938 Schering Corp. By the action of bulls' testes extract (1938d Mamoli; 1939 Ercoli) and horses' testes extract (1938 Ercoli), both containing putrefactive bacteria (1938 Schramm; 1938d

---

* Erroneously referred to as androstanedione in *C. A.* **1950** 4047, 4048.

Mamoli), there have been obtained ætiocholane-3.17-dione (p. 2895 s) (1938 d
Mamoli; 1938 Ercoli), ætiocholane-3α.17β-diol (p. 2017 s) (1938 d Mamoli;
1938 Schramm), and androstan-3α-ol-17-one (androsterone, p. 2629 s) (1939
Ercoli); ætiocholane-3.17-dione is also obtained by the action of Bacillus
putrificus Bienstock in yeast water, whereas in yeast suspension ætiocholan-
17β-ol-3-one (p. 2600 s) and ætiocholane-3α.17β-diol (p. 2017 s) are formed
(1939 Mamoli).

After oral administration of the dione to hypogonadal men, the excretion
of androsterone (p. 2629 s) is strongly increased (1940 Dorfman; cf. also 1944
Frame).

PHYSIOLOGICAL PROPERTIES OF Δ^4-ANDROSTENE-3.17-DIONE

For physiological properties of the dione, see p. 155 and pp. 1367 s, 1368 s,
and the literature cited there.

ANALYTICAL PROPERTIES OF Δ^4-ANDROSTENE-3.17-DIONE

The dione slowly gives a precipitate in methanol with ammoniacal silver
salt solution at room temperature (1939 b Miescher). For colour reactions
with conc. H_2SO_4 alone and in the presence of fural or benzaldehyde, see
1939 Woker; 1939 Scherrer. For the rate of reduction of phosphomolybdic
acid (Folin-Wu reagent) by androstenedione in glacial acetic acid at 100°
to give molybdenum blue, see 1946 Heard.   In the Zimmermann test with
m-dinitrobenzene and KOH in aqueous alcohol, the dione very rapidly and
strongly gives a dark red coloration turning brown (1943 Wettstein); for absorption spectrum of the colour formed, see 1938 Callow; 1944 Zimmermann.
Gives a weak pink coloration with 1.4-naphthoquinone + conc. HCl in glacial
acetic acid at 60–70° (1939 b Miescher). For polarographic determination,
using the compound obtained with Girard's reagent T, see 1941 Hershberg.

For separation from Δ^4-cholesten-3-one and from progesterone, see under
formation from cholestenone (p. 2882 s). The dione may be separated from
other ketosteroids by means of its bis-nicotinoylhydrazone (below) (1945
Velluz; 1944 LES LABORATOIRES FRANÇAIS DE CHIMIOTHÉRAPIE).

DERIVATIVES OF Δ^4-ANDROSTENE-3.17-DIONE

3-Semicarbazone $C_{20}H_{29}O_2N_3$ (p. 156). Absorption max. in chloroform at
270 mμ (ε 35000) (1939 Dannenberg; cf. 1937 Westphal; 1938 b Butenandt;
1943 Evans).

Bis-nicotinoylhydrazone $C_{31}H_{36}O_2N_6$. Crystals, m. above 300°; practically
insoluble in cold methanol and ethanol. — **Fmn.** From the dione and nicotinic acid hydrazide on refluxing with a little acetic acid in methanol for
1 hr.; yield, 97%. — **Rns.** Gives 90% yield of the dione on refluxing with
conc. HCl in strong alcohol for 15 min. (1945 Velluz; 1944 LES LABORA-
TOIRES FRANÇAIS DE CHIMIOTHÉRAPIE).

3-Ethylene ketal and *bis-(ethylene ketal)*, see under the derivatives of
Δ^5-androstene-3.17-dione, p. 2890 s.

References, pp. 2887–2890 s

3-Ethylene mercaptal $C_{21}H_{30}OS_2$. Crystals (ethyl acetate), m. 173–174.5°
(1954 Ralls). — **Fmn.** From Δ^4-androstene-3.17-dione and ethane-1.2-dithiol
at room temperature in the presence
of dry hydrogen chloride (1941 Ciba)
or, along with the bis-(ethylene mercaptal) (below; yield, ca. 7%), in the
presence of p-toluenesulphonic acid in
acetic acid (crude yield, 77%) (1954
Ralls). — **Rns.** On boiling with
methylmagnesium bromide in ether -
toluene it gives 17-methyltestosterone
ethylene mercaptal, which may be
converted into 17-methyl-testoster-
one (p. 2645 s) by refluxing with
$HgCl_2$ and $CdCO_3$ in acetone (1941
Ciba).

Bis-(ethylene mercaptal) $C_{23}H_{34}S_4$.
Solid (acetone), m. 174–6° (1954
Ralls). — **Fmn.** See under that of
the preceding compound.

For *3-enol esters and ethers of Δ^4-an-
drostene-3.17-dione* (cf. also p. 156), see
the corresponding esters and ethers
of $\Delta^{3.5}$-androstadien-3-ol-17-one (page
2605 s).

dl - Δ^4 - Androstene - 3.17 - dione (?)
$C_{19}H_{26}O_2$ (cf. I, p. 2880 s). Mixture
of stereoisomers; yellow, viscous oil,
b. 180–190°/$_{0.05}$. Absorption max. in
alcohol at 241.5 mμ (ε 11 500). —
Fmn. From a 2.5-dimethyl-6.3'-di-
keto - 1.2 - cyclopentano - decahydro -
naphthalene (α-seriés; III, above) on
refluxing with sodamide under nitrogen in ether for 6 hrs., followed by
addition of an alcoholic solution of
4-dimethylaminobutan-2-one methiodide with cooling in ice, keeping at
room temperature for $1\frac{1}{2}$ hrs., refluxing for $1\frac{1}{2}$ hrs., and then by decomposition with ice-cold HCl; the compound (III) is obtained from 5-methyl-
6-methoxy-1-tetralone by the steps
indicated in the scheme at the left
(1943 Martin).

1935 a C. F. BOEHRINGER & SÖHNE (Dirscherl, Hanusch), *Ger. Pat.* 665549 (issued 1938); *C. A.*
1939 1106; *Ch. Ztbl.* **1939** I 2830.

b C. F. BOEHRINGER & SÖHNE (Dirscherl, Hanusch), *Ger. Pat.* 712591 (issued 1941); *C. A.*
1943 4534; *Ch. Ztbl.* **1942** I 1162; RARE CHEMICALS INC. (Dirscherl, Hanusch), *U.S. Pat.*
2152626 (1936, issued 1939); *C. A.* **1939** 5132; *Ch. Ztbl.* **1940** I 759.

a CIBA*, *Swiss Pat.* 201201 (issued 1939); *C. A.* **1939** 8625; *Ch. Ztbl.* **1939** II 172.

b CIBA*, *Swiss Pat.* 201202 (issued 1939); *C. A.* **1939** 8625; *Ch. Ztbl.* **1939** II 172.

c CIBA*, *Swiss Pat.* 184988 (issued 1936); *Ch. Ztbl.* **1937** I 3519, 3520.

1936 Butenandt, Schmidt-Thomé, *Ber.* **69** 882, 887.

a CIBA*, *Swiss Pat.* 214327 (issued 1941); *C. A.* **1942** 5960.

b CIBA* (Ruzicka, Wettstein), *U.S. Pat.* 2194235 (issued 1940); *C. A.* **1940** 4746; *Ch. Ztbl.*
1942 I 778; *Fr. Pat.* 805380 (issued 1936); *C. A.* **1937** 3502; *Ch. Ztbl.* **1937** I 3674.

c CIBA*, (Ruzicka, Wettstein), *U.S. Pat.* 2143453 (issued 1939); *C. A.* **1939** 3078; *Ch. Ztbl.*
1939 I 5009, 5010; CIBA PHARMACEUTICAL PRODUCTS (Ruzicka, Wettstein), *U.S. Pat.*
2387469 (1937, issued 1945); *C. A.* **1946** 5536; *Ch. Ztbl.* **1948** I 1234.

N. V. ORGANON, *Dutch Pat.* 41871 (issued 1937); *Ch. Ztbl.* **1938** I 2217.

PFEILRINGWERKE A.-G., *Fr. Pat.* 812354 (issued 1937); *C. A.* **1938** 960; *Ch. Ztbl.* **1937** II
3347, 3348.

SCHERING A.-G., *Ger. Pat.* 699248 (issued 1940); *C. A.* **1941** 6742; *Ch. Ztbl.* **1941** I 1325.

1936/37 CIBA*, *Brit. Pat.* 477400 (issued 1938); *Ch. Ztbl.* **1938** II 119, 120.

I.G. FARBENIND., *Brit. Pat.* 473629 (issued 1937); *C. A.* **1938** 3096; *Ch. Ztbl.* **1938** I 2402.

1937 C. F. BOEHRINGER & SÖHNE, *Fr. Pat.* 835527 (issued 1938); *Ch. Ztbl.* **1939** I 2827.

a CIBA*, *Fr. Pat.* 830043 (issued 1938); *C. A.* **1939** 179; *Ch. Ztbl.* **1939** I 2035; *Swiss Pat.*
214603 (issued 1941); *C. A.* **1942** 4977; *Ch. Ztbl.* **1942** I 2561.

b CIBA*, *Brit. Pat.* 497394 (issued 1939); *C. A.* **1939** 3812; *Ch. Ztbl.* **1939** I 2828; *Dutch Pat.*
51420 (1938, issued 1941); *Ch. Ztbl.* **1942** I 2561; CIBA PHARMACEUTICAL PRODUCTS (Miescher,
Wettstein), *U.S. Pat.* 2229813 (1938, issued 1941); *C. A.* **1941** 3040; *Ch. Ztbl.* **1941** II
2467, 2468.

c CIBA*, *Swiss Pat.* 199449 (issued 1938); *C. A.* **1939** 3393; *Ch. Ztbl.* **1939** I 3930.

d CIBA*, *Swiss Pat.* 205888 (issued 1939); *C. A.* **1941** 3039; *Brit. Pat.* 520393 (1938, issued
1940); *C. A.* **1942** 628; *Ch. Ztbl.* **1940** II 3226.

e CIBA*, *Swiss Pat.* 208748 (issued 1940); *C. A.* **1941** 3773; *Fr. Pat.* 847134 (1938, issued
1939); *Ch. Ztbl.* **1940** I 2828; CIBA PHARMACEUTICAL PRODUCTS (Miescher, Fischer), *U.S.
Pat.* 2253798 (1938, issued 1941); *C. A.* **1941** 8218.

f CIBA*, *Swiss Pat.* 212192 (issued 1941); *Ch. Ztbl.* **1941** II 3100, 3101; (Ruzicka), *Swed. Pat.*
98368 (1938, issued 1940); *Ch. Ztbl.* **1940** II 2185; CIBA PHARMACEUTICAL PRODUCTS
(Ruzicka), *U.S. Pat.* 2232636 (1938, issued 1941); *C. A.* **1941** 3266; *Ch. Ztbl.* **1941** II 3100,
3101.

Hogness, Sidwell, Zscheile, *J. Biol. Ch.* **120** 239, 244, 250, 253.

a I.G. FARBENIND., *Brit. Pat.* 501421 (issued 1939); *C. A.* **1939** 6340, 6341; *Ch. Ztbl.* **1939** II 170.

b I.G. FARBENIND., (Bockmühl, Ehrhart, Ruschig, Aumüller), *Ger. Pat.* 693351 (issued 1940);
C. A. **1941** 4921; *Brit. Pat.* 502666 (issued 1939); *C. A.* **1939** 7316; *Ch. Ztbl.* **1939** II 1126;
Fr. Pat. 842026 (1938, issued 1939); *C. A.* **1940** 4743; *Ch. Ztbl.* **1940** I 429; *Swiss Pat.*
216340 (1938, issued 1941); *Ch. Ztbl.* **1942** II 197.

Kuwada, Miyasaka, Yosiki, *C. A.* **1938** 1275; *Ch. Ztbl.* **1938** II 1611.

a N. V. ORGANON, *Fr. Pat.* 823618 (issued 1938); *C. A.* **1938** 5854; *Ch. Ztbl.* **1938** II 355;
Brit. Pat. 488966 (issued 1938); *C. A.* **1939** 324; *Swiss Pat.* 209831 (issued 1940); *Ch. Ztbl.*
1943 I 654.

b N. V. ORGANON, *Dutch Pats.* 46498 (issued 1939), 47831; *C. A.* **1939** 8923; **1940** 6414;
Ch. Ztbl. **1939** II 4282; *Fr. Pat.* 842940 (1938, issued 1939); *C. A.* **1940** 6414; *Ch. Ztbl.*
1940 I 429; *Swiss Pats.* 219009, 223235, –236 (1938, issued 1942), 225731, –732 (1938, issued
1943); *Ch. Ztbl.* **1943** II 749; **1944** I 37.

Oppenauer, *U.S. Pat.* 2384335 (issued 1945); *C. A.* **1946** 178; *Ch. Ztbl.* **1948** I 1039; N. V.
ORGANON, *Fr. Pat.* 827623 (issued 1938); *C. A.* **1938** 8437; *Ch. Ztbl.* **1938** II 3119.

---

* Société pour l'industrie chimique à Bâle; Gesellschaft für chemische Industrie in Basel.

a SCHERING-KAHLBAUM A.-G., *Brit. Pat.* 492725 (issued 1938); *C. A.* **1939** 1884; *Ch. Ztbl.*
1939 I 2642; *Swiss Pat.* 209942 (issued 1940); *Ch. Ztbl.* **1943** I 654.

b SCHERING-KAHLBAUM A.-G., *Fr. Pat.* 822551 (issued 1938); *C. A.* **1938** 4174; *Ch. Ztbl.*
1938 II 120; SCHERING CORP. (Serini, Köster, Strassberger), *U.S. Pat.* 2379832 (issued
1945); *C. A.* **1945** 5053.

Westphal, Hellmann, *Ber.* **70** 2136.

Windaus, Kuhr, *Ann.* **532** 52, 64.

1938 C. F. BOEHRINGER & SÖHNE, *Fr. Pat.* 844850 (issued 1939); *C. A.* **1940** 7932; *Ch. Ztbl.* **1940** I
1391.

a Butenandt, Schmidt-Thomé, Paul, *Ber.* **71** 1313, 1316.

b Butenandt, Peters, *Ber.* **71** 2688, 2690, 2694.

Callow, Callow, Emmens, *Biochem. J.* **32** 1312, 1320, 1323.

CHINOIN GYÓGYSZER ÉS VEGYÉSZETI TERMÉKEK GYÁRA, Dr. KERESZTY & Dr. WOLF, *Fr. Pat.*
840964 (issued 1939); *Ch. Ztbl.* **1940** I 427; *Dutch Pat.* 55862 (issued 1944); *Ch. Ztbl.* **1944** II
675.

CIBA (Miescher, Fischer), *U.S. Pat.* 2172590 (issued 1939); *C. A.* **1940** 450; *Ch. Ztbl.* **1940** I
428; *Fr. Pat.* 841242 (issued 1939); *Ch. Ztbl.* **1940** I 916.

Dirscherl, Hanusch, *Z. physiol. Ch.* **252** 49.

Ercoli, Mamoli, *Ber.* **71** 156.

a Mamoli, Vercellone, *Gazz.* **68** 311, 316.

b Mamoli, Vercellone, *Ber.* **71** 154.

c Mamoli, Vercellone, *Ber.* **71** 1686.

d Mamoli, Schramm, *Ber.* **71** 2083.

e Mamoli, *Ber.* **71** 2278.

Reichstein, *Helv.* **21** 1490, 1496.

ROCHE-ORGANON, INC. (Oppenauer), *U.S. Pat.* 2264888 (issued 1941); *C. A.* **1942** 1740.

a Ruzicka, Hofmann, *Helv.* **21** 88, 91.

b Ruzicka, Plattner, Aeschbacher, *Helv.* **21** 866, 870.

a SCHERING A.-G., *Brit. Pat.* 515411 (issued 1940); *Ch. Ztbl.* **1940** II 1180.

b SCHERING A.-G., *Fr. Pat.* 850115 (issued 1939); *Ch. Ztbl.* **1940** II 931; *Swiss Pat.* 215813
(issued 1941); *Ch. Ztbl.* **1942** II 196.

SCHERING CORP. (Mamoli), *U.S. Pat.* 2186906 (issued 1940); *C. A.* **1940** 3436; *Ch. Ztbl.* **1940** I
3426.

Schramm, Mamoli, *Ber.* **71** 1322.

Serini, Köster, *Ber.* **71** 1766, 1769, 1770.

Vercellone, Mamoli, *Ber.* **71** 152.

1939 a C. F. BOEHRINGER & SÖHNE, *Fr. Pat.* 846099 (issued 1939); *C. A.* **1941** 1185; *Ch. Ztbl.*
1940 I 2830.

b C. F. BOEHRINGER & SÖHNE, *Fr. Pat.* 860492 (issued 1941); *Ch. Ztbl.* **1941** I 3408.

Dannenberg, *Abhandl. Preuss. Akad. Wiss. Math.-naturwiss. Kl.* **1939** No. 21, pp. 13, 53, 64.

Ehrhart, Ruschig, Aumüller, *Angew. Ch.* **52** 363, 364.

Ercoli, *Ber.* **72** 190.

Kuhr, *Ber.* **72** 929.

Mamoli, Koch, Teschen, *Z. physiol. Ch.* **261** 287, 295.

a Miescher, Fischer, *Helv.* **22** 158, 160.

b Miescher, Wettstein, Scholz, *Helv.* **22** 894, 903, 904.

a SCHERING A.-G. (Köster, Inhoffen), *Swed. Pat.* 109502 (issued 1944); *Ch. Ztbl.* **1944** II 247;
Swiss Pat. 228923 (issued 1943); *Ch. Ztbl.* **1944** II 1300; *Dutch Pat.* 52656 (issued 1942);
Ch. Ztbl. **1942** II 2615; *Norw. Pat.* 64579 (1940, issued 1942); *Ch. Ztbl.* **1942** II 1822;
SCHERING CORP. (Köster, Inhoffen), *U.S. Pat.* 2302636 (issued 1942); *C. A.* **1943** 2388.

b SCHERING A.-G., *Fr. Pat.* 856736 (issued 1940); *Ch. Ztbl.* **1943** I 652; *Swed. Pat.* 99036
(issued 1940); *Ch. Ztbl.* **1940** II 3668; *Swiss Pat.* 227580 (issued 1943); *Ch. Ztbl.* **1944** II
1300; SCHERING CORP. (Inhoffen), *U.S. Pat.* 2358808 (issued 1944); *C. A.* **1945** 1738;
Ch. Ztbl. **1945** II 1069.

Scherrer, *Helv.* **22** 1329, 1334.

Spielman, Meyer, *J. Am. Ch. Soc.* **61** 893, 895.

Woker, Antener, *Helv.* **22** 1309, 1320.

1940 a CHINOIN GYÓGYSZER ÉS VEGYÉSZETI TERMÉKEK GYARA R. T., Dr. KERESZTY & Dr. WOLF (Földi), *Hung. Pat.* 132575 (issued 1944); *C. A.* **1949** 3980, 3981.

 b CHINOIN (Földi), *Hung. Pat.* 135687 (issued 1949); *C. A.* **1950** 4047.

 CIBA PHARMACEUTICAL PRODUCTS (Miescher, Wettstein), *U.S. Pat.* 2319012 (issued 1943); *C. A.* **1943** 6096; CIBA*, *Fr. Pat.* 886410 (issued 1943); *Ch. Ztbl.* **1944** I 448; *Swed. Pat.* 103744 (issued 1942); *Ch. Ztbl.* **1942** II 2294.

 Dorfman, Hamilton, *J. Biol. Ch.* **133** 753, 755, 758.

 Ercoli, *C. A.* **1944** 1540; *Ch. Ztbl.* **1941** II 1028.

 a Marker, Wittle, Tullar, *J. Am. Ch. Soc.* **62** 223, 224, 226.

 b Marker, *J. Am. Ch. Soc.* **62** 2543, 2547.

 PARKE, DAVIS & Co. (Marker, Wittle), *U.S. Pats.* 2397424, –425 (issued 1946); *C. A.* **1946** 3571.

 Pfiffner, North, *J. Biol. Ch.* **132** 459.

 Ruzicka, Grob, Raschka, *Helv.* **23** 1518, 1521.

 a SCHERING A.-G., *Swiss Pat.* 220206 (issued 1942); *Ch. Ztbl.* **1943** I 1695.

 b SCHERING A.-G., *Swiss Pat.* 229775 (issued 1944); *Ch. Ztbl.* **1944** II 1301.

 SCHERING CORP. (Köster), *U.S. Pat.* 2363338 (issued 1944); *C. A.* **1945** 3128; *Ch. Ztbl.* **1945** II 1638.

1941 Bretschneider, *Sitzber. Akad. Wiss. Wien* IIb **150** 127; *Monatsh.* **74** 53.

 CIBA*, *Fr. Pat.* 51887 (issued 1943); *Ch. Ztbl.* **1944** II 778; CIBA PHARMACEUTICAL PRODUCTS (Miescher), *U.S. Pat.* 2435013 (1942, issued 1948); *C. A.* **1948** 2996.

 a v. Euw, Reichstein, *Helv.* **24** 879, 880, 886.

 b v. Euw, Reichstein, *Helv.* **24** 1140, 1142.

 Galinovsky, *Ber.* **74** 1624.

 Hershberg, Wolfe, Fieser, *J. Biol. Ch.* **140** 215.

 PARKE, DAVIS & Co. (Marker, Lawson), *U.S. Pat.* 2366204 (issued 1945); *C. A.* **1945** 1649; *Ch. Ztbl.* **1945** II 1385, 1386.

 Pfiffner, North, *J. Biol. Ch.* **139** 855, 860.

 Prins, Reichstein, *Helv.* **24** 945, 950, 951, 952, 954.

 SCHERING A.-G., *Fr. Pat.* 875070 (issued 1942); *Ch. Ztbl.* **1943** I 1190.

 SQUIBB & SONS (Fernholz), *U.S. Pat.* 2356154 (issued 1944); *C. A.* **1945** 1515; *Ch. Ztbl.* **1945** II 1385; *U.S. Pat.* 2378918 (issued 1945); *C. A.* **1945** 5051.

1942 Arnaudi, *C. A.* **1946** 3152; *Ztbl. Bakt. Parasitenk.* II **105** 352; *C. A.* **1943** 5756.

 Ruzicka, Goldberg, Hardegger, *Helv.* **25** 1297, 1301.

1943 Evans, Gillam, *J. Ch. Soc.* **1943** 565, 567.

 Martin, Robinson, *J. Ch. Soc.* **1943** 491, 497.

 Wettstein, Miescher, *Helv.* **26** 631, 639.

1944 Ercoli, Molina, *C. A.* **1946** 3152, 3153.

 Frame, Fleischmann, Wilkins, *Bull. Johns Hopkins Hosp.* **75** 95.

 GLIDDEN Co. (Julian, Cole, Magnani, Conde), *U.S. Pat.* 2374683 (issued 1945); *C. A.* **1946** 1636; *Ch. Ztbl.* **1946** I 372.

 LES LABORATOIRES FRANÇAIS DE CHIMIOTHÉRAPIE, *Fr. Pat.* 907312 (issued 1946); *Ch. Ztbl.* **1946** I 763.

 a Molina, Ercoli, *C. A.* **1946** 3152.

 b Molina, Ercoli, *C. A.* **1946** 3152.

 Ruzicka, Muhr, *Helv.* **27** 503, 509.

 Zimmermann, *Vitamine und Hormone* **5** [1952] 1, 10, 13.

1945 Hirschmann, Hirschmann, *J. Biol. Ch.* **157** 601, 609.

 Julian, Cole, Magnani, Meyer, *J. Am. Ch. Soc.* **67** 1728.

 Kumler, Fohlen, *J. Am. Ch. Soc.* **67** 437; **70** 4273 (1948).

 Velluz, Petit, *Bull. soc. chim.* [5] **12** 951.

1946 Bersin, *Naturwiss.* **33** 108, 110.

 Furchgott, Rosenkrantz, Shorr, *J. Biol. Ch.* **163** 375, 380.

 Heard, Sobel, *J. Biol. Ch.* **165** 687.

 Ross, *J. Ch. Soc.* **1946** 737.

---

* Société pour l'industrie chimique à Bâle; Gesellschaft für chemische Industrie in Basel.

Sarett, *J. Biol. Ch.* **162** 601, 627.
Turfitt, *Biochem. J.* **40** 79.
1947 Djerassi, *J. Org. Ch.* **12** 823, 826, 829.
1948 Djerassi, Scholz, *J. Org. Ch.* **13** 697, 704.
Turfitt, *Biochem. J.* **42** 376, 377.
1950 Djerassi, Rosenkranz, Romo, Kaufmann, Pataki, *J. Am. Ch. Soc.* **72** 4534, 4538, 4539.
Rosenkranz, Djerassi, Kaufmann, Pataki, Romo, *Nature* **165** 814.
1954 Ralls, Riegel, *J. Am. Ch. Soc.* **76** 4479.

$\varDelta^5$-**Androstene-3.17-dione** $C_{19}H_{26}O_2$ (p. 156). — **Fmn.** From $\varDelta^5$-androstene-

$3\beta.17\beta$-diol (p. 1999 s) on conversion into the dibromide in glacial acetic acid, followed by oxidation with CrO_3 in strong acetic acid at room temperature and then by heating with zinc dust in methanol for 40 min. (1937 I.G. FARBENIND.); for formation by the same method from $\varDelta^5$-androsten-3β-ol-17-one (5-dehydroepiandrosterone, p. 2609 s), see also 1937a SCHERING-KAHLBAUM A.-G. — **Rns.** For partial hydrogenation in 95 % alcohol in the presence of Raney nickel to give 5-dehydroandrosterone and 5-dehydroepiandrosterone, see also 1936/37 CIBA. By the action of fermenting yeast in dil. alcohol it gives androstane-$3\beta.17\beta$-diol (p. 2014 s) and a small amount of $\varDelta^4$-androsten-17β-ol-3-one (testosterone, p. 2580 s) (1937 Mamoli). Is isomerized to $\varDelta^4$-androstene-3.17-dione (p. 2880 s) by heating with 48 % HBr in glacial acetic acid on the water-bath for 10 min. (1936 Butenandt; 1937b SCHERING-KAHLBAUM A.-G.) or by boiling in alcohol with 2 N H_2SO_4 for 10 min. (1937 I.G. FARBENIND.).

Dioxime $C_{19}H_{28}O_2N_2$. Begins to decompose at 180°, m. ca. 205° (1936 Butenandt; 1937a SCHERING-KAHLBAUM A.-G.).

3-Ethylene ketal $C_{21}H_{30}O_3$. For the structure, cf. the corresponding ketal of testosterone (p. 2595 s). — Prisms (methanol), m. 194°, resolidifying, and remelting at 202° (1941 SQUIBB & SONS); crystals (alcohol containing pyridine), m. 199°; $[\alpha]_D^{20} + 26°$ (dioxane); shows no absorption in the ultra-violet (1939 SCHERING A.-G.). — **Fmn.** From $\varDelta^4$-androstene-3.17-dione (p. 2880 s) with 1 mol. of ethylene glycol (or preferably less than 1 mol.) and a little p-toluene-sulphonic acid in benzene on distillation with removal of the water formed (1939 SCHERING A.-G.; 1941 SQUIBB & SONS). — **Rns.** Gives "testosterone ethylene ketal" (p. 2595 s) on reduction with sodium in boiling abs. alcohol (1941 SQUIBB & SONS). On boiling with allylmagnesium bromide in ether, followed by treatment with dil. H_2SO_4 at 35°, it gives 17-allyltestosterone (1939 CIBA).

Bis-(ethylene ketal) $C_{23}H_{34}O_4$. For the structure, cf. the preceding ketal. — Crystals (alcohol containing pyridine), m. 165°; $[\alpha]_D^{20} -52°$ (dioxane); shows no absorption in the ultra-violet (1939 SCHERING A.-G.). — **Fmn.** From $\varDelta^4$-androstene-3.17-dione as above, using 2 mols. of ethylene glycol (1939 SCHERING A.-G.; cf. 1941 SQUIBB & SONS), or on treatment with excess ethylene oxide and a little $SnCl_4$ in anhydrous carbon tetrachloride at 30° (1939 SCHERING A.-G.).

1936 Butenandt, Schmidt-Thomé, *Ber.* **69** 882, 887.
1936/37 CIBA*, *Brit. Pat.* 483926 (issued 1938); *C. A.* **1938** 7218; *Ch. Ztbl.* **1938** II 3273; *Fr. Pat.*
   828338 (1937, issued 1938); *C. A.* **1939** 177; *Ch. Ztbl.* **1938** II 3273; (Ruzicka), *U.S. Pat.*
   2168685 (1937, issued 1939); *C. A.* **1939** 9324; *Ch. Ztbl.* **1940** I 1391.
1937 I.G. FARBENIND., *Brit. Pat.* 501421 (issued 1939); *C. A.* **1939** 6340, 6341; *Ch. Ztbl.* **1939** II
   170; *Fr. Pat.* 847487 (1938, issued 1939); *C. A.* **1941** 5654; *Ch. Ztbl.* **1940** I 2828.
   Mamoli, Vercellone, *Ber.* **70** 2079.
   a SCHERING-KAHLBAUM A.-G., *Swiss Pat.* 199448 (issued 1938); *Ch. Ztbl.* **1939** I 3931; *Brit.
   Pat.* 486992 (issued 1938); *C. A.* **1938** 8707; *Ch. Ztbl.* **1939** I 1603.
   b SCHERING-KAHLBAUM A.-G., *Brit. Pat.* 492725 (issued 1938); *C. A.* **1939** 1884; *Ch. Ztbl.*
   1939 I 2642; *Swiss Pat.* 209942 (issued 1940); *Ch. Ztbl.* **1943** I 654.
1939 CIBA*, *Fr. Pat.* 861536 (issued 1941); *Ch. Ztbl.* **1941** I 3259; CIBA PHARMACEUTICAL PRODUCTS
   (Miescher), *U.S. Pat.* 2386331 (1943, issued 1945); *C. A.* **1946** 179; *Ch. Ztbl.* **1948** I 1040.
   SCHERING A.-G. (Köster, Inhoffen), *Swed. Pat.* 109502 (issued 1944); *Ch. Ztbl.* **1944** II 247;
   Swiss Pat. 228923 (issued 1943); *Ch. Ztbl.* **1944** II 1300; *Dutch Pat.* 52656 (issued 1942);
   Ch. Ztbl. **1942** II 2615; *Norw. Pat.* 64579 (1940, issued 1942); *Ch. Ztbl.* **1942** II 1822;
   SCHERING CORP. (Köster, Inhoffen), *U.S. Pat.* 2302636 (issued 1942); *C. A.* **1943** 2388.
1941 SQUIBB & SONS (Fernholz), *U.S. Pat.* 2356154 (issued 1944); *C. A.* **1945** 1515; *Ch. Ztbl.*
   1945 II 1385; *U.S. Pat.* 2378918 (issued 1945); *C. A.* **1945** 5051.

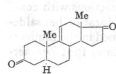

$\Delta^{9(11)}$-**Androstene-3.17-dione** $C_{19}H_{26}O_2$. Plates and prisms (ether · pentane), m. 153–4° cor. (Kofler block), resolidifying at 148–150°, and remelting at 154° cor. (Kofler block); $[\alpha]_D^{14} + 156°$, $[\alpha]_{5461}^{14} + 183°$ (both in alcohol) (1946 Shoppee). — **Fmn.** From the two 3-epimeric $\Delta^{9(11)}$-androsten-3-ol-17-ones (p. 2626 s) on oxidation with 1 mol. of CrO_3 in glacial acetic acid at room temperature (yields, ca. 60%) (1946 Shoppee), along with traces of the oxide (below) (1946, 1947 Shoppee). — **Rns.** Gives the oxide (below) on treatment with perbenzoic acid in chloroform in the dark at 20° (1947 Shoppee).

9.11-Epoxyandrostane-3.17-dione, $\Delta^{9(11)}$-**Androstene-3.17-dione oxide**

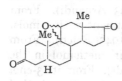

$C_{19}H_{26}O_3$. Crystals (ether · petroleum ether), at 190° to 200° changing to needles m. 207–212° cor. (Kofler block); sometimes crystals were obtained changing at 203–7°, m. 211–2° cor. (Kofler block) (1947 Reich); leaflets (ether · pentane), m. 208–212° cor. (Kofler block) (1947 Shoppee); $[\alpha]_D^{16} + 116°$ (chloroform) (1947 Reich; 1947 Shoppee). — **Fmn.** From 9.11-epoxyandrostan-3β-ol-17-one (p. 2627 s) on oxidation with 1 mol. of CrO_3 in glacial acetic acid at room temperature; yield, 50% (1947 Reich). As a by-product, in very small amounts, along with the above unsaturated dione, from the two 3-epimeric $\Delta^{9(11)}$-androsten-3-ol-17-ones (p. 2626 s) on oxidation with 1 mol. of CrO_3 in glacial acetic acid at room temperature (1946, 1947 Shoppee). From the above unsaturated dione on treatment with excess perbenzoic acid in chloroform in the dark at 20° for 22 hrs. (1947 Shoppee).

---

* Société pour l'industrie chimique à Bâle; Gesellschaft für chemische Industrie in Basel.

1946 Shoppee, *J. Ch. Soc.* **1946** 1134, 1136.
1947 Reich, Lardon, *Helv.* **30** 329, 334, 700.
     Shoppee, *Helv.* **30** 766.

$\Delta^{11(?)}$-**Androstene-3.17-dione** $C_{19}H_{26}O_2$. Prisms (ether · pentane), m. 145–6⁰ cor.
(Kofler block); $[\alpha]_D^{15}$ +141.5⁰, $[\alpha]_{5461}^{15}$ +170.5⁰ (both in alcohol). — **Fmn.** From $\Delta^{11(?)}$-androsten-3α-ol-17-one (isolated from the urine of normal and diseased persons; see p. 2627 s) on oxidation with CrO_3 in glacial acetic acid at 15⁰; yield, 46% (1946 Shoppee).

1946 Shoppee, *J. Ch. Soc.* **1946** 1134, 1137.

Androstane-3.17-dione $C_{19}H_{28}O_2$ (p. 156). Leaflets (ether · pentane), m. 132⁰ to 134⁰ cor.; $[\alpha]_D^{14}$ + 100⁰ (alcohol) (1940 Shoppee); $[\alpha]_D^{23-24}$ + 105⁰ (chloroform) (1947 Djerassi). For its infra-red absorption spectrum, see 1946 Furchgott. Dipole moment in dioxane 3.06 D (1945 Kumler). For basic properties, see 1946 Bersin. For colour reactions with conc. H_2SO_4 alone and in the presence of fural or benzaldehyde, see 1939 Woker; 1939 Scherrer. For the colour obtained in the Zimmermann test with m-dinitrobenzene and KOH, see 1938 Callow; 1944 Zimmermann.

FORMATION OF ANDROSTANE-3.17-DIONE

Formation from other C_{19}-steroids. From androstane-3β.17β-diol (p. 2014 s) on oxidation with CrO_3 in glacial acetic acid (1940b Marker; 1940 Shoppee) or with 4 mols. of N-bromoacetamide in aqueous tert.-butyl alcohol at room temperature (yield, 82%) (1943 Reich), or by the action of Flavobacterium (Micrococcus) dehydrogenans (Ercoli, quoted in 1942/43 Arnaudi). From androstan-3α-ol-17-one (androsterone, p. 2629 s) on treatment with 2 mols. of N-bromoacetamide in aqueous tert.-butyl alcohol (1943 Reich). From Δ^1-androstene-3.17-dione (p. 2879 s) on hydrogenation in methanol in the presence of palladium · calcium carbonate (1939 Butenandt). From its 3-enol ethyl ether (ethyl ether of Δ^2-androsten-3-ol-17-one, p. 2607 s) or its 3-diethyl ketal (p. 2894 s) on heating in alcohol with 4 N HCl on the water-bath (1938 Serini).

Formation from higher steroids. From a 17-hydroxymethyl-androstan-17-ol-3-one (p. 2720 s) on oxidation with lead tetraacetate in glacial acetic acid at room temperature (1947 Ruzicka). From 3β.17β-dihydroxy-17-iso-ætio-allocholanic acid on oxidation with CrO_3 in strong acetic acid at room temperature (1939 Miescher).

The dione is obtained from the following compounds by oxidation with CrO_3 in glacial acetic acid at room temperature: allopregnane-3β.17α.20α-triol

(Substance O, p. 2118 s) (1938 Steiger), allopregnane-3β.17α.20β-triol (Substance J, p. 2119 s) (1938 Steiger; cf. 1936 Reichstein), allopregnane-3β.17α. 20β.21-tetrol (Substance K, p. 2141 s) (1938 Steiger), allopregnane-3β.17α-diol-20-one (Substance L, p. 2335 s) (1938b Reichstein), allopregnane-3β.17α. 21-triol-20-one (Substance P, p. 2358 s) (1938a Reichstein), and, after hydrogenation in abs. alcohol - glacial acetic acid in the presence of platinum oxide, from ω-homo-Δ^5-pregnene-3β.17α.20β.21 b.22-pentol (p. 2149 s) (1941 Fuchs).

From the acetate of allopregnan-3β-ol-20-one (p. 2247 s) on refluxing with potassium persulphate and conc. H_2SO_4 in 90% acetic acid for 4 hrs., followed by hydrolysis with boiling alcoholic KOH and oxidation of the non-ketonic fraction with CrO_3 in strong acetic acid (1940a Marker). From allopregnan-20β-ol-3-one (p. 2511 s) on refluxing with $ZnCl_2$ in glacial acetic acid for 3 hrs., followed by ozonization of the resulting crude $\Delta^{17(20)}$-allopregnen-3-one (p. 2411 s) and then by oxidation with CrO_3 in glacial acetic acid at room temperature (1937a Marker; 1938 PARKE, DAVIS & Co.). From allopregnane-17α.20β-diol-3-one (p. 2721 s) on oxidation with periodic acid in slightly dil. methanol at room temperature (1941 Reich) or with lead tetraacetate in glacial acetic acid (1947 Ruzicka).

Along with other compounds, from cholestanol (p. 1695 s) on oxidation with CrO_3 in strong acetic acid (1935 SCHERING A.-G.; 1937 CIBA). For formation from cholestanone, sitostanone, stigmastanone, etc., see 1936 C. F. BOEHRINGER & SÖHNE; 1938 CIBA.

REACTIONS OF ANDROSTANE-3.17-DIONE

Reduction (see also under "Biological reactions", p. 2894 s, and under the 3-diethyl ketal, p. 2894 s). On boiling with 1 at. sodium in methanol it gives androstan-17β-ol-3-one (dihydrotestosterone, p. 2597 s) and other androstanolones (1935, 1936 CIBA). On hydrogenation in glacial acetic acid containing a little HBr, in the presence of a previously reduced platinum catalyst, under 3 atm. pressure (shaking for 15 min.), the dione gives androstan-3α-ol-17-one (androsterone, p. 2629 s) and a small amount of androstan-3β-ol-17-one (epiandrosterone, p. 2633 s) (1937b Marker; cf. 1936a SCHERING A.-G.; 1936a, 1936b PFEILRINGWERKE A.-G.); on complete hydrogenation in glacial acetic acid + a little HBr in the presence of platinum, androstane-3α.17β-diol (page 2011 s) is formed (1934 SCHERING A.-G.; 1935 SCHERING-KAHLBAUM A.-G.). The dione gives only epiandrosterone on hydrogenation in alcohol in the presence of sodium ethoxide and previously reduced platinum oxide (1936a PFEILRINGWERKE A.-G.) or on hydrogenation in the presence of nickel, cobalt, or nickel-cobalt alloys until 1 mol. hydrogen has been absorbed (1936a SCHERING-KAHLBAUM A.-G.); on complete hydrogenation, using nickel or cobalt catalysts, androstane-3β.17β-diol (p. 2014 s) is obtained (1936a SCHERING-KAHLBAUM A.-G.). The dione is reduced to androstane (p. 1396 s) by heating of its disemicarbazone (not described in detail) with hydrazine hydrate and sodium ethoxide in abs. alcohol for 8 hrs. at 200° (1941 Wettstein).

Halogenation. Gives 2α-bromoandrostane-3.17-dione (p. 2916 s) on treatment at 20° with 1 mol. bromine and a little HBr in glacial acetic acid (1936

Butenandt; cf. also 1936 CIBA PHARMACEUTICAL PRODUCTS) or in chloroform
or ether (1937/38 Dannenberg; cf. also 1936b SCHERING-KAHLBAUM A.-G.;
1937, 1945 SCHERING A.-G.). On treatment with 2 mols. bromine in glacial
acetic acid until the rotation has reached (after ca. 11 min.) $[\alpha]_D^{25}$ in glacial
acetic acid $+207^0$, immediately followed by pouring into water, the dione
gives 90 % crude yield of 2.2-dibromoandrostane-3.17-dione (p. 2924 s; not
obtained pure; $[\alpha]_D^{25} +158^0$ in chloroform or, similarly, in glacial acetic acid);
with 2 mols. bromine and a little HBr in glacial acetic acid at room temperature, followed after decolorization (almost instantaneous) by warming to 50⁰,
ca. 90–95 % yield of 2α.4α-dibromoandrostane-3.17-dione (p. 2924 s) is obtained (1948 Djerassi; cf. also 1937/38 Dannenberg; 1936b SCHERING A.-G.);
on treatment with 2 mols. bromine and a little HBr - glacial acetic acid in
chloroform at 20⁰, 13 % yield of 2α.16-dibromoandrostane-3.17-dione (p. 2925 s)
is obtained, which is also formed in small amount on similar bromination
in ether along with the 2.4-dibromo-dione as the main product (1937/38
Dannenberg).

Action of organic reagents. On heating with 1 mol. ethyl orthoformate and
alcoholic HCl in benzene at 75⁰ for 1 hr., the dione gives the 3-diethyl ketal
(below) (1938 Serini; 1945 SCHERING A.-G.), on heating for 2 hrs. it gives
the 3-enol ethyl ether (ethyl ether of Δ^2-androsten-3-ol-17-one, p. 2607 s);
the latter is also formed when the dione is heated with acetone diethyl ketal
and 8 % HCl in benzene at 75⁰ for 3 hrs. (1938 SCHERING A.-G.). Gives
androstane-3.17-dione dicyanohydrin on treatment with potassium cyanide
and glacial acetic acid in alcohol at 0⁰ (1943 Goldberg).

Biological reactions. By the action of fermenting yeast in dil. alcohol the
dione is reduced to androstane-3β.17β-diol (p. 2014 s) (1937 Vercellone; 1938
SCHERING CORP.). Gives equal amounts of androstan-3α-ol-17-one (androsterone, p. 2629 s) and androstan-3β-ol-17-one (epiandrosterone, p. 2633 s) by
the action of bulls' testes extract containing putrefactive bacteria (1938b
Mamoli). By the action of Bacillus putrificus Bienstock in yeast suspension
under nitrogen at 36–37⁰, equal amounts of epiandrosterone and androstane-
3β.17β-diol are formed (1939 Mamoli).

After oral administration of the dione to hypogonadal men, the excretion
of androsterone in the urine is greatly increased (1940a, b Dorfman).

For physiological properties of the dione, see p. 156; cf. also pp. 1367 s,
1368 s, and the literature cited there.

DERIVATIVES OF ANDROSTANE-3.17-DIONE

Dioxime $C_{19}H_{30}O_2N_2$, m. 260–261⁰ cor. (1947 Ruzicka).

3-Diethyl ketal $C_{23}H_{38}O_3$. Crystals (alcohol containing pyridine), m. 121–3⁰;
$[\alpha]_D^{20} +76^0$ (dioxane) (1938 Serini). — **Fmn.** From the above dione on heating
with 1 mol. ethyl orthoformate and alcoholic HCl in benzene for 1 hr. at
75⁰ (1938 Serini; 1945 SCHERING A.-G.). — **Rns.** On boiling in xylene for
2 hrs. it gives the 3-enol ethyl ether of the dione (ethyl ether of Δ^2-androsten-

3-ol-17-one, p. 2607 s) (1938 Serini; 1938 SCHERING A.-G.). Gives androstan-17β-ol-3-one (dihydrotestosterone, p. 2597 s) on heating with sodium in propyl alcohol at 100°, followed by heating with aqueous alcoholic HCl on the water-bath (1938 Serini; 1938, 1945 SCHERING A.-G.). Reverts to the dione on heating in alcohol with 4 N HCl on the water-bath for 10 min. (1938 Serini).

3-enol Ethyl ether $C_{21}H_{32}O_2$. See the ethyl ether of Δ^2-androsten-3-ol-17-one (p. 2607 s).

5 - iso - Androstane - 3.17 - dione, Aetiocholane-3.17-dione $C_{19}H_{28}O_2$ (p. 156:

3.17-Diketoætiocholane). Crystals (dil. acetone), m. 131° to 132°, $[\alpha]_D^{18}$ +113° (alcohol) (1938 Ercoli); colourless granula (ether - pentane), m. 132–4°, $[\alpha]_D^{17}$ +110° (abs. alcohol) (1941 Reichstein). — **Fmn.** From ætiocholan-3β-ol-17-one (p. 2638 s) on oxidation with CrO_3 in glacial acetic acid at room temperature; similarly from the 3α-epimer (p. 2637 s) (1941 Reichstein). From Δ^4-androstene-3.17-dione (page 2880 s) by the action of bulls' testes extract containing putrefactive bacteria (1938a Mamoli), by the action of a similar extract from horses' testes (yield, 65%) (1938 Ercoli; cf. 1938a Mamoli), or by the action of Bacillus putrificus Bienstock in yeast water under nitrogen at 36–37° (yield, 80%) (1939 Mamoli). From pregnane-3α.17α-diol-20-one (p. 2335 s) on oxidation with CrO_3 in strong acetic acid at room temperature; yield, 42% (1945 Lieberman).

Rns. Gives 4β-bromoætiocholane-3.17-dione (p. 2917 s) on treatment with 1 mol. bromine and some HBr in glacial acetic acid (1938 Ercoli).

Bis- (2.4-dinitrophenylhydrazone) $C_{31}H_{36}O_8N_8$. Yellow crystals, m. 263–4° cor. (1945 Lieberman).

13-iso-Androstane-3.17-dione, Lumiandrostane-3.17-dione $C_{19}H_{28}O_2$. Crystals

(alcohol), m. 165.5–166.5°. — **Fmn.** From 13-iso-andro-stan-3α-ol-17-one (lumiandrosterone, p. 2640 s) on oxi-dation with CrO_3 in glacial acetic acid at room tem-perature. — **Rns.** Reacts only with 1 mol. of semi-carbazide acetate in alcohol to give the *3-semicarbazone* $C_{20}H_{31}O_2N_3$, crystals (alcohol), m. 225–6° dec. (1944 Butenandt).

1934 SCHERING A.-G., *Ger. Pat.* 684498 (issued 1939); *C. A.* **1940** 3450.
1935 CIBA*, *Swiss Pat.* 212190 (issued 1941); *Ch. Ztbl.* **1942** I 231.
      SCHERING A.-G., *Dan. Pat.* 62303 (issued 1944); *Ch. Ztbl.* **1945** II 264.
      SCHERING-KAHLBAUM A.-G., *Brit. Pat.* 455019 (issued 1936); *C. A.* **1937** 1427; *Ch. Ztbl.* **1937** I 2820, 2821; *Ital. Pat.* 336417 (issued 1936); *Ch. Ztbl.* **1937** II 3348; *Swiss Pat.* 196545 (issued 1938); *Ch. Ztbl.* **1939** I 1207.
1936 C. F. BOEHRINGER & SÖHNE (Dirscherl, Hanusch), *Ger. Pat.* 695638 (issued 1940); *C. A.* **1941** 5654; *Ch. Ztbl.* **1940** II 2784.
      Butenandt, Dannenberg, *Ber.* **69** 1158, 1160.

* Société pour l'industrie chimique à Bâle; Gesellschaft für chemische Industrie in Basel.

CIBA* (Ruzicka, Wettstein), *U.S. Pat.* 2143453 (issued 1939); *C. A.* **1939** 3078; *Ch. Ztbl.* **1939** I 5009, 5010; CIBA PHARMACEUTICAL PRODUCTS (Ruzicka, Wettstein), *U.S. Pat.* 2387469 (issued 1945); *C. A.* **1946** 5536; *Ch. Ztbl.* **1948** I 1234.

CIBA PHARMACEUTICAL PRODUCTS (Miescher, Wettstein), *U.S. Pat.* 2260328 (issued 1941); *C. A.* **1942** 873; *Ch. Ztbl.* **1944** II 246.

a PFEILRINGWERKE A.-G., *Fr. Pat.* 812354 (issued 1937); *C. A.* **1938** 960; *Ch. Ztbl.* **1937** II 3347, 3348.

b PFEILRINGWERKE A.-G., *Swiss Pat.* 202156 (issued 1939); *Ch. Ztbl.* **1940** I 250.

Reichstein, *Helv.* **19** 1107, 1126.

a SCHERING A.-G., *Austrian Pat.* 160484 (issued 1941); *Ch. Ztbl.* **1941** II 2973, 2974; *Dan. Pat.* 61302 (issued 1943); *Ch. Ztbl.* **1944** II 48.

b SCHERING A.-G. (Inhoffen), *Ger. Pat.* 722943 (issued 1942); *C. A.* **1943** 5201; *Dan. Pat.* 61375 (1937, issued 1943); *Ch. Ztbl.* **1944** II 49; SCHERING-KAHLBAUM A.-G., *Fr. Pat.* 835524 (1937, issued 1938), *Brit. Pat.* 500353 (1937, issued 1939); *C. A.* **1939** 5872; *Ch. Ztbl.* **1939** I 5010, 5011; SCHERING CORP. (Inhoffen, Butenandt, Schwenk), *U.S. Pat.* 2340388 (1937, issued 1944); *C. A.* **1944** 4386.

a SCHERING-KAHLBAUM A.-G., *Brit. Pat.* 464271 (issued 1937); *C. A.* **1937** 6417; *Fr. Pat.* 811567 (issued 1937); *C. A.* **1938** 1284; *Ch. Ztbl.* **1937** II 3919.

b SCHERING-KAHLBAUM A.-G., *Brit. Pat.* 489692 (issued 1938); *C. A.* **1939** 1343; *Ch. Ztbl.* **1939** I 3590, 3591.

1937 CIBA*, *Fr. Pat.* 830043 (issued 1938); *C. A.* **1939** 179; *Ch. Ztbl.* **1939** I 2035.

a Marker, Kamm, Jones, Oakwood, *J. Am. Ch. Soc.* **59** 614.

b Marker, Kamm, Wittle, *J. Am. Ch. Soc.* **59** 1841.

SCHERING A.-G., *Brit. Pat.* 508576; *Fr. Pat.* 838704 (issued 1939); *C. A.* **1940** 777; *Ch. Ztbl.* **1939** II 1722; SCHERING CORP. (Inhoffen), *U.S. Pat.* 2280828 (issued 1942); *C. A.* **1942** 5618.

Vercellone, Mamoli, *Z. physiol. Ch.* **248** 277.

1937/38 Dannenberg, *Über einige Umwandlungen des Androstandions und Testosterons*, Thesis, Danzig, pp. 30–32.

1938 Callow, Callow, Emmens, *Biochem. J.* **32** 1312, 1323.

CIBA* (Ruzicka), *Swed. Pat.* 98368 (issued 1940); *Ch. Ztbl.* **1940** II 2185; CIBA PHARMACEUTICAL PRODUCTS (Ruzicka), *U.S. Pat.* 2232636 (issued 1941); *C. A.* **1941** 3266; *Ch. Ztbl.* **1941** II 3100, 3101.

Ercoli, Mamoli, *Ber.* **71** 156.

a Mamoli, Schramm, *Ber.* **71** 2083, 2085.

b Mamoli, Schramm, *Ber.* **71** 2698.

PARKE, DAVIS & Co. (Marker, Jones, Oakwood), *Brit. Pat.* 512940 (issued 1939); *C. A.* **1941** 1187; *Ch. Ztbl.* **1940** I 2826, 2827.

a Reichstein, Gätzi, *Helv.* **21** 1185, 1193.

b Reichstein, Gätzi, *Helv.* **21** 1497, 1503, 1504.

SCHERING A.-G., *Swiss Pat.* 215817 (issued 1941); *Ch. Ztbl.* **1942** II 75.

SCHERING CORP. (Mamoli), *U.S. Pat.* 2186906 (issued 1940); *C. A.* **1940** 3436; *Ch. Ztbl.* **1940** I 3426.

Serini, Köster, *Ber.* **71** 1766, 1768, 1769.

Steiger, Reichstein, *Helv.* **21** 546, 555, 557, 559.

1939 Butenandt, Mamoli, Dannenberg, Masch, Paland, *Ber.* **72** 1617, 1622.

Mamoli, Koch, Teschen, *Z. physiol. Ch.* **261** 287, 295, 296.

Miescher, Wettstein, *Helv.* **22** 112, 116.

Scherrer, *Helv.* **22** 1329, 1334.

Woker, Antener, *Helv.* **22** 1309, 1315.

1940 a Dorfman, *J. Biol. Ch.* **132** 457.

b Dorfman, Hamilton, *J. Biol. Ch.* **133** 753, 757.

a Marker, Rohrmann, Wittle, Crooks, Jones, *J. Am. Ch. Soc.* **62** 650.

b Marker, Turner, *J. Am. Ch. Soc.* **62** 3003, 3005.

Shoppee, *Helv.* **23** 740, 746.

* Société pour l'industrie chimique à Bâle; Gesellschaft für chemische Industrie in Basel.

1941 Fuchs, Reichstein, *Helv.* **24** 804, 811, 812.
  Reich, Reichstein, *Arch. intern. pharmacodynamie* **65** 415, 419.
  Reichstein, Lardon, *Helv.* **24** 955, 960.
  Wettstein, Fritzsche, Hunziker, Miescher, *Helv.* **24** 332 E, 357 E.
1942/43 Arnaudi, *Ztbl. Bakt. Parasitenk.* II **105** 352.
1943 Goldberg, Kirchensteiner, *Helv.* **26** 288, 301.
  Reich, Reichstein, *Helv.* **26** 562, 583, 584.
1944 Butenandt, Poschmann, *Ber.* **77** 394, 396.
  Zimmermann, *Vitamine und Hormone* **5** [1952], pp. 1,10.
1945 Kumler, *J. Am. Ch. Soc.* **67** 1901, 1902; **70** 4273 (1948).
  Lieberman, Dobriner, *J. Biol. Ch.* **161** 269, 277.
  Schering A.-G., *Fiat Final Report* No. 996, pp. 80, 81, 158.
1946 Bersin, *Naturwiss.* **33** 108, 110.
  Furchgott, Rosenkrantz, Shorr, *J. Biol. Ch.* **163** 375, 380.
1947 Djerassi, *J. Org. Ch.* **12** 823, 825.
  Ruzicka, Meister, Prelog, *Helv.* **30** 867, 875, 878.
1948 Djerassi, Scholz, *J. Org. Ch.* **13** 697, 699, 702.

Δ^5-**Androstene-4.17-dione** $C_{19}H_{26}O_2$. Originally assumed to be Δ^1-androstene-3.17-dione, then designated *hetero-Δ^1-androstene-3.17-dione* (p. 155) (1939 Butenandt); for the true structure, see 1944 Butenandt. — Absorption max. in chloroform at 238 mμ (ε 6900) (1939 Dannenberg). — **Fmn.** For its formation from 2-bromoandrostane-3.17-dione by heating with potassium acetate in glacial acetic acid at 200°, see also 1935 CIBA; 1936 SCHERING-KAHLBAUM A.-G. — **Rns.** By the action of fermenting yeast in dil. alcohol it is reduced to Δ^5-androsten-17β-ol-4-one (p. 2601 s) (1938 Butenandt).

1935 CIBA*, *Swiss Pat.* 191456 (issued 1937); *C. A.* **1938** 1280; *Ch. Ztbl.* **1938** I 1162.
1936 SCHERING-KAHLBAUM A.-G., *Brit. Pat.* 489692 (issued 1938); *C. A.* **1939** 1343; *Ch. Ztbl.* **1939** I
  3590, 3591; *Swiss Pat.* 206916 (issued 1939); *Ch. Ztbl.* **1940** I 3958.
1938 Butenandt, Dannenberg, *Ber.* **71** 1681, 1685.
1939 Butenandt, Mamoli, Dannenberg, Masch, Paland, *Ber.* **72** 1617, 1618.
  Dannenberg, *Abhandl. Preuss. Akad. Wiss. Math.-naturwiss. Kl.* **1939** No. 21, pp. 14, 54.
1944 Butenandt, Ruhenstroth-Bauer, *Ber.* **77** 397, 400.

Δ^2-**Androstene-6.17-dione** $C_{19}H_{26}O_2$. For the position of the double bond, see also 1946 Blunschy. — Crystals (alcohol, or ethyl acetate-hexane), m. 191–191.5° (1942a Butenandt; cf. 1946 Blunschy); $[\alpha]_D^{20}$ +126° (alcohol) (1942a Butenandt), +123° (chloroform) (1946 Blunschy). Absorption max. (solvent not stated) at 285–300 mμ (log ε 2.05) (1946 Blunschy). Gives a colour reaction with tetranitromethane (1946 Blunschy).

Fmn. From i-androstene-6.17-dione (p. 2898 s) on refluxing with quinoline for 1 hr. (1942a Butenandt); similarly from 3β-chloroandrostane-6.17-dione

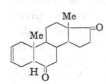

---

* Société pour l'industrie chimique à Bâle; Gesellschaft für chemische Industrie in Basel.

(p. 2919 s) (1946 Blunschy) and 3β-bromoandrostane-6.17-dione (p. 2919 s; refluxing for $^1/_2$ hr.; yield, 41%) (1942a Butenandt).

Rns. On treatment with 2 mols. of perbenzoic acid in chloroform at 2° for 3 days it gives the **epoxide** $C_{19}H_{26}O_3$ [crystals (petroleum ether - alcohol), m. 174–6°] (1942a Butenandt). The dione gives androstane-6.17-dione (see below) on shaking in alcohol with hydrogen and palladium - calcium carbonate (1942a Butenandt). Is unchanged by boiling with 5% methanolic KOH for 30 min. or with 2 N HCl in methanol for 20 min. (1946 Blunschy) or by treatment with hydrogen bromide in glacial acetic acid (1942a Butenandt). Gives 6.17α-dimethyl-$Δ^2$-androstene-6.17β-diol (p. 2044 s) on refluxing with methylmagnesium bromide in ether (1942b Butenandt).

$Δ^4$-Androstene-6.17-dione $C_{19}H_{26}O_2$. Crystals (ethyl acetate - hexane), m. 179° to 181° cor.; $[α]_D^{19}$ +97° (chloroform). Absorption max. in alcohol (from the graph) at 300 (2.25) and 245.5 (3.88) mμ (log ε). — Is androgenically 2.5 times less active than $Δ^4$-androstene-3.17-dione; not œstrogenic in the Allen-Doisy test. — **Fmn.** From the acetate of androstan-5α-ol-6.17-dione (p. 2948 s) on sublimation at 200° under 13 mm. pressure (1940 Ruzicka).

Androstane-6.17-dione $C_{19}H_{28}O_2$. Crystals (alcohol), m. 134–5°. — **Fmn.** From $Δ^2$-androstene-6.17-dione (p. 2897 s) on shaking in alcohol with hydrogen in the presence of palladium - calcium carbonate; yield, 80% (1942a Butenandt).

Dioxime $C_{19}H_{30}O_2N_2$, m. 288–290° (1942a Butenandt).

i-Androstene-6.17-dione, *i-Androstane-6.17-dione* $C_{19}H_{26}O_2$. Crystals (alcohol), m. 182–3°; $[α]_D^{20}$ +113° (chloroform) (1942a Butenandt). — Is inactive in the Fussgänger capon comb test (1942a Butenandt). — **Fmn.** From i-androsten-6β(?)-ol-17-one (p. 2644 s) on oxidation with 1.5 mols. of CrO_3 in strong acetic acid at 0–10°; yield, 68%. From 3β-bromoandrostane-6.17-dione on refluxing with collidine or with potassium acetate in glacial acetic acid (1942a Butenandt). — **Rns.** Is stable to hydrogen in alcohol in the presence of palladium - calcium carbonate. Gives i-androsten-17β-ol-6-one (p. 2602 s) by the action of fermenting yeast in dil. alcohol (1942a Butenandt). On treatment with 48% HBr in glacial acetic acid at 20° it gives 3β-bromoandrostane-6.17-dione (p. 2919 s) (1942a Butenandt; cf. 1948 Shoppee for β-configuration at C-3). Refluxing with 5 N H_2SO_4 in glacial acetic acid for 3 hrs. affords the acetate of androstan-3β-ol-6.17-dione (p. 2948 s). Gives $Δ^2$-androstene-6.17-dione (p. 2897 s) on refluxing with quinoline for 1 hr. (1942a Butenandt).

Dioxime $C_{19}H_{28}O_2N_2$, m. 269–271° (1942a Butenandt).

5-Methyl-19-nor-Δ^9-ætiocholene(or -androstene)-6.17-dione $C_{19}H_{26}O_2$.

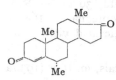

Orig-inally (and tentatively) assigned the $\Delta^{1(10)}$-structure (1940 Ruzicka). The Δ^9-structure has been given (by Editor) by analogy with that of Westphalen's diol [5-methyl-19-nor-Δ^9-coprostene(or -cholestene)-3β.6β-diol, p. 2046 s]; for the structure of this diol, see also 1952 Bladon; cf., however, 1953 Fieser.

Crystals (ethyl acetate - hexane), m. 215–6⁰ cor.; $[\alpha]_D^{19}$ +21⁰ (chloroform). Absorption max. in alcohol (from the graph) at 294 mμ (log ε 2.0). — **Fmn.** From androstan-5-ol-6.17-dione (p. 2948 s) on treatment with 1.1 mols. of p-toluenesulphonyl chloride in pyridine at room temperature, followed by refluxing of the resulting p-toluenesulphonate for 36 hrs.; yield, 27% (1940 Ruzicka).

1940 Ruzicka, Grob, Raschka, *Helv.* **23** 1518, 1521, 1522, 1528.
1942 a Butenandt, Surányi, *Ber.* **75** 591, 595 et seq.
    b Butenandt, Surányi, *Ber.* **75** 597, 604.
1946 Blunschy, Hardegger, Simon, *Helv.* **29** 199, 200, 202.
1948 Shoppee, *J. Ch. Soc.* **1948** 1043, 1044.
1952 Bladon, Henbest, Wood, *J. Ch. Soc.* **1952** 2737, 2739, 2743.
1953 Fieser, *J. Am. Ch. Soc.* **75** 4395, 4401.

---

4. Diketosteroids C_{20}–C_{24} (both CO groups in the ring system)

6α-Methyl-Δ^4-androstene-3.17-dione $C_{20}H_{28}O_2$.

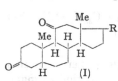

For α-configuration at C-6, see 1957 Ackroyd. — Prisms (benzine - methanol), m. 163.5–167⁰ (1940 Madaeva); needles or prisms (acetone - hexane), m. 167–8⁰; $[\alpha]_D^{22}$ +172⁰ (chloroform) (1957 Ackroyd). Insoluble in hexane (1940 Madaeva). — **Fmn.** From 6β-methylandrostan-5α-ol-3.17-dione (page 2953 s) on dehydration with dry HCl in dry chloroform, with cooling in a freezing mixture (1940 Madaeva).

$\Delta^{17(20)}$- and Δ^{20}-Pregnene-3.11-diones $C_{21}H_{30}O_2$ (I, R = :CH·CH₃ and R = ·CH:CH₂, resp.).

— **Fmn.** A mixture of the two diones is obtained from the crystalline mixture of the acetates of $\Delta^{17(20)}$- and Δ^{20}-pregnen-3α-ol-11-ones (see p. 2654 s) on oxidation with CrO_3 in strong acetic acid (1946 Sarett). — **Rns.** On treatment with OsO_4 in abs. ether and pyridine at room temperature (1 hr.), then at 0⁰ (36 hrs.), followed by refluxing with Na_2SO_3 in aqueous alcohol for 3¹/₂ hrs., the mixture gives a mixture of pregnane-17α.20β-diol-3.11-dione (p. 2958 s) and the two epimeric pregnane-20.21-diol-3.11-diones (pp. 2937 s, 2938 s) (1946 Sarett; cf. 1949 Sarett).

Pregnane-3.11-dione $C_{21}H_{32}O$. For a pregnane-dione (m. 102–103°) which may be pregnane-3.11-dione or pregnane-11.20-dione, see under the reactions of pregnane-3.11.20-trione, p. 2967 s.

9-iso-Pregnane-3.11-dione $C_{21}H_{32}O_2$. This structure was ascribed to **urane-3.11-dione** by Marker (1938 a, b, c, 1939 b); for a **urene-3.11-dione** $C_{21}H_{30}O_2$, see 1939 b Marker. According to Klyne (1950, 1952) the urane compounds are probably D-homo-steroids (see urane, p. 1403 s).

$\varDelta^{17(20)}$**-Pregnene-3.16-dione** $C_{21}H_{30}O_2$. Crystals (methanol), m. 179–182°. — **Fmn.** From pregnane-3β.16β.20α-triol (p. 2111 s) on Oppenauer oxidation by refluxing with aluminium isopropoxide and cyclohexanone in toluene (1940 Marker).

$\varDelta^{17(20)}$**-Allopregnene-3.16-dione** $C_{21}H_{30}O_2$. Crystals (acetone), m. 190–192°; distils in a molecular still at 130°. — **Fmn.** From the acetate of allopregnan-20β-ol-3.16-dione (page 2934 s) on refluxing with KOH in aqueous alcohol for $^1/_2$ hr. (yield, 48%), with $NaHCO_3$ in aqueous methanol for 2 hrs. (yield, 95%), or with conc. HCl in alcohol for 4 hrs. (yield, 36%). — **Rns.** On reduction with sodium and abs. alcohol it gives allopregnane-3.16-diol (p. 2027 s) which on oxidation with CrO_3 in strong acetic acid yields allopregnane-3.16-dione (below) (1939 a Marker).

Bis-(2.4-dinitrophenylhydrazone) $C_{33}H_{38}O_8N_8$. Red crystals, m. 190° (1939 a Marker).

Allopregnane-3.16-dione $C_{21}H_{32}O_2$. Crystals (dil. methanol), m. 128° (1939 a Marker). — **Fmn.** See under the reactions of the preceding unsaturated dione.

Bis-(2.4-dinitrophenylhydrazone) $C_{33}H_{40}O_8N_8$. Yellow crystals, m. 245° (1939 a Marker).

20-Methyl-$\varDelta^{17(20)}$-pregnene-3.16-dione $C_{22}H_{32}O_2$. Crystals (ether - pentane), m. 193–5°. — **Fmn.** From 20-methylpregnane-3β.16β.20-triol (p. 2128 s) on Oppenauer oxidation by refluxing with aluminium tert.-butoxide and dry acetone in dry benzene for 30 hrs. (1942 Marker).

22-Methyl-$\Delta^{20(22)}$-norcholene-3.12-dione $C_{24}H_{36}O_2$.

Prisms (dil. alcohol), m. 181–2°. Gives a weak, yellow-brown Liebermann colour reaction. — Has no œstrogenic or progestational activity. — **Fmn.** From 22-methyl-$\Delta^{20(22)}$-norcholene-3α.12α-diol (see p. 2043 s) on oxidation with CrO_3 in glacial acetic acid (1939 Kazuno).

1938 a Marker, Kamm, Oakwood, Wittle, Lawson, *J. Am. Ch. Soc.* **60** 1061.
   b Marker, Lawson, Wittle, Crooks, *J. Am. Ch. Soc.* **60** 1559.
   c Marker, Rohrmann, Wittle, *J. Am. Ch. Soc.* **60** 1561.
1939 Kazuno, Shimizu, *J. Biochem.* **29** 421, 423, 431.
   a Marker, Wittle, *J. Am. Ch. Soc.* **61** 855, 858.
   b Marker, Rohrmann, *J. Am. Ch. Soc.* **61** 2719.
1940 Madaeva, Ushakov, Kosheleva, *J. Gen. Ch. U.S.S.R.* **10** 213, 216; *C. A.* **1940** 7292; *Ch. Ztbl.* **1940** II 1298.
   Marker, Turner, *J. Am. Ch. Soc.* **62** 2540.
1942 Marker, Turner, *J. Am. Ch. Soc.* **64** 481.
1946 Sarett, *J. Am. Ch. Soc.* **68** 2478, 2481.
1949 Sarett, *J. Am. Ch. Soc.* **71** 1169, 1170.
1950 Klyne, *Nature* **166** 559.
1951 Fieser, Rigaudy, *J. Am. Ch. Soc.* **73** 4660.
1952 Klyne, Shoppee, *Chemistry & Industry* **1952** 470.
   Turner, *J. Am. Ch. Soc.* **74** 5362.
1957 Ackroyd, Adams, Ellis, Petrow, Stuart-Webb, *J. Ch. Soc.* **1957** 4099, 4101.

5. Diketosteroids C_{27}–C_{29} (both CO groups in the ring system) containing only aliphatic side chains

5-Methyl-19-nor-Δ^9-coprostene (or-cholestene)-3.6-dione, *Westphalen's diketone*

$C_{27}H_{42}O_2$ (p. 109: *5-Methyl-$\Delta^{8:9}$-nor-cholestene-3.6-dione*). For revised structure and probable β-configuration at C-5, see 1952 Ellis, and cf. that of Westphalen's diol (p. 2046 s). — Prismatic needles (acetone - methanol), m. 105–6°; gives a yellow coloration with tetranitromethane (1938 Petrow). — **Fmn.** From Westphalen's diol (p. 2046 s) by oxidation with CrO_3 in acetic acid at room temperature (1938 Petrow). — **Rns.** Oxidation with perhydrol in glacial acetic acid at room temperature affords 5-methyl-9.10-epoxy-19-nor-coprostane(or -cholestane)-3.6-dione (p. 2962 s) (1939 Petrow; cf. 1952 Ellis).

1938 Petrow, Rosenheim, Starling, *J. Ch. Soc.* **1938** 677, 680.
1939 Petrow, *J. Ch. Soc.* **1939** 998, 1001.
1952 Ellis, Petrow, *J. Ch. Soc.* **1952** 2246, 2249.

Cholestane-2.3-dione $C_{27}H_{44}O_2$ (p. 157). Besides the true diketo-form (I), two

enolic forms, Δ^3-cholesten-3-ol-2-one (II) and Δ^1-cholesten-2-ol-3-one (III), and a diene-diol-form (IV) are possible. The two forms actually isolated: *form A* (m. 144–5° cor.) and *form B* (m. 161–2° or 168–9°) have been identified as the enolic forms (II) and (III), resp., which see pp. 2659 s, 2662 s (1938 Stiller; 1939 Dannenberg; 1944 Ruzicka). The true diketo-form (I), not yet isolated, probably exists in solution in equilibrium with form B (1938 Stiller); the insoluble potassium salt of form A (p. 2659 s), for which no formula is given, is perhaps derived from (IV) (Editor).

Only those derivatives which are certainly or probably related to the true diketo-form, or those for which the structure is as yet unknown, are described below.

2-(*p-Dimethylaminophenyl*)-isoxime $C_{35}H_{54}O_2N_2$. Dark orange scales (chloroform - aqueous alcohol), crystals (acetone), m. 178–9° cor. — **Fmn.** From N-(3-ketocholestan-2-yl)-pyridinium bromide (p. 2482 s) in chloroform -

alcohol by treatment with p-nitrosodimethylaniline in aqueous NaOH at room temperature. — **Rns.** By shaking an ethereal suspension with aqueous 2 N HCl, crude "cholestane-2.3-dione" [probably "*form C*", equimolecular mixture of Δ^1-cholesten-2-ol-3-one and Δ^3-cholesten-3-ol-2-one, p. 2663 s (Editor)] is obtained (1944 Ruzicka).

Quinoxaline derivative $C_{33}H_{48}N_2$ (p. 157). Colourless blades (chloroform-acetone) (1938 Stiller), crystals (alcohol), m. 184° cor. (1936 Ruzicka). — **Fmn.** From 2α.4α-dibromocholestan-3-one * (p. 2495 s) by prolonged (8 hrs.) refluxing with o-phenylenediamine in alcohol, followed by elimination of the latter by distillation (1936 Ruzicka), or by heating the pure components at 140–150° (1937 Inhoffen). From either the A- or the B-form of "cholestane-2.3-dione" (see p. 2902 s) by refluxing with o-phenylenediamine in alcohol (1938 Stiller).

Δ^5-Cholestene-3.4-dione, *Diosterol* $C_{27}H_{42}O_2$ (p. 157). Besides the diketo-

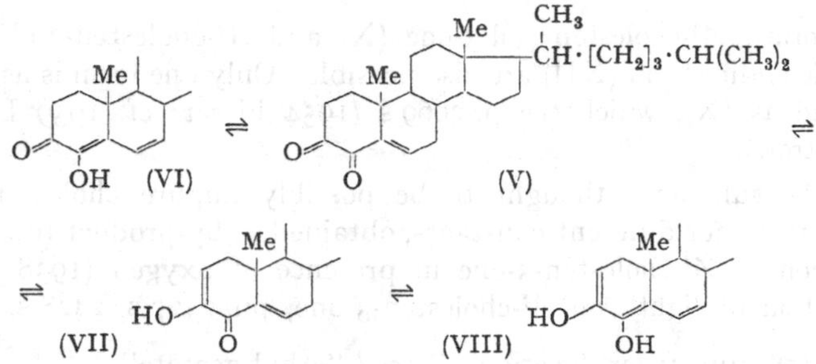

form (V), two enolic forms, $\Delta^{4.6}$-cholestadien-4-ol-3-one (VI) and $\Delta^{2.5}$-cholestadien-3-ol-4-one (VII), and a triene-diol form (VIII) are also possible. Of the two forms actually isolated, *diosterol I* or *form B* (m. 162–3°) has been identified as (VI), *diosterol II* or *form A* (m. 135–6°) as (VII), which see pp. 2666 s, 2674 s; the true diketo - form (V) is as yet unknown; the triene-diol form (VIII) is probably in equilibrium with diosterol II in solution (1948 Fieser; cf. 1940 Petrow).

Only those derivatives which are certainly or probably related to the true diketo - form, or those for which the structure is as yet unknown, are described below.

Mono-2.4-dinitrophenylhydrazone $C_{33}H_{46}O_5N_4$. Dark red microcrystals (benzene - alcohol), microprismatic needles (alcohol), m. 255° cor. — **Fmn.** From $\Delta^{2.5}$-cholestadien-3-ol-4-one ("form A") (p. 2674 s) by treating with Brady's reagent in alcohol (at room temperature?); in the same way from $\Delta^{4.6}$-cholestadien-4-ol-3-one ("form B") (p. 2666 s), but on refluxing with excess of the reagent in alcohol (1940 Petrow).

Quinoxaline derivative $C_{33}H_{46}N_2$ (p. 157). Yellowish leaflets (ethyl acetate) (1936 Inhoffen), plates (isopropanol), m. 175° cor. (1940 Petrow). — **Fmn.**

---

* In the original paper supposed to be the 2.2-dibromo-isomer.

E.O.C. XIV s. 199

From $\varDelta^{2.5}$-cholestadien-3-ol-4-one ("form A") on refluxing with o-phenylene-diamine in abs. alcohol. From $\varDelta^{4.6}$-cholestadien-4-ol-3-one ("form B") on heating with o-phenylenediamine at 150^0 (1940 Petrow; cf. 1936 Inhoffen).

Cholestane-3.4-dione $C_{27}H_{44}O_2$ (p. 158). Besides the diketo-form (IX), two

enolic forms, $\varDelta^4$-cholesten-4-ol-3-one (X) and $\varDelta^2$-cholesten-3-ol-4-one (XI), and a diene-diol form (XII) are also possible. Only one form is as yet known, identified as (X), which see p. 2669 s (1954 Fieser; cf. 1939 Dannenberg; 1940 Petrow).

For the substance thought to be possibly impure cholestane-3.4-dione despite its rather different constants, obtained as by-product (m. 157⁰) in the irradiation of $\varDelta^4$-cholesten-3-one in presence of oxygen (1938 Bergmann), see "Action of light" on $\varDelta^4$-cholesten-3-one, pp. 2427 s, 2428 s.

Page 158, line 3: for "acetic ac." read "ethyl acetate".

Page 159, 3ʳᵈ and 2ⁿᵈ lines from bottom: for "b Butenandt, Schramm, Kudszus, *Ber.* **69** 2279" *read* "b Butenandt, Schramm, Wolff, Kudszus, *Ber.* **69** 2779".

Mono-2.4-dinitrophenylhydrazone $C_{33}H_{48}O_5N_4$. Dark red microcrystals (ben-zene - alcohol), m. 252–3⁰ cor. — **Fmn.** From $\varDelta^4$-cholestene-4-ol-3-one (page 2669 s) on treatment with Brady's reagent in alcohol (1940 Petrow).

Quinoxaline derivative $C_{33}H_{48}N_2$ (p. 158). Needles (alcohol, or acetone - ether) (1938 Wolff), colourless plates (chloroform - alcohol), m. 208–9⁰ cor. (1940 Petrow), absorption max. in alcohol at 239 and 321 mμ (ε 27400 and 8870, resp.) (1954 Fieser). — **Fmn.** From $2\beta.4\beta$-dibromocoprostan-3-one* (p. 2497 s) by prolonged refluxing (8 hrs.) with o-phenylenediamine in alcohol, followed by elimination of the latter by distillation (1936 Ruzicka; cf. 1936b Butenandt), or by heating the pure components at 150⁰ (1938 Wolff). From $\varDelta^4$-cholesten-4-ol-3-one (p. 2669 s) by reaction with o-phenylenediamine, either on refluxing in abs. alcohol + glacial acetic acid (1937 Schering-Kahlbaum A.-G.; 1938 Wolff), or heating without solvent above 100⁰ (1940 Petrow; cf. 1954 Fieser).

---

* In the original paper, supposed to be the 4.4-dibromo-isomer.

Δ¹-Cholestene-3.6-dione $C_{27}H_{42}O_2$. A compound, which probably possesses this structure, is obtained from 2α-bromocholestane-3.6-dione (p. 2922 s) on refluxing with collidine for 15 min. It forms plates with $\frac{1}{2}$ H_2O (aqueous alcohol), m. 161°; absorption max. in isopropyl alcohol at 228 mμ ($E_{1\,cm}^{1\%}$. 200) (1950 Ellis).

Δ⁴-Cholestene-3.6-dione, Coprostene-3.6-dione $C_{27}H_{42}O_2$ (p. 158). For identity with Mauthner's (1896) "*Oxycholestenone*", see 1906 Windaus. — White plates (aqueous acetone) (1940a Marker), pale yellow leaflets (methanol), m. 132° cor. (1937 Ruzicka), m. 124—5° (1946 Ross; cf. 1927 Montignie; 1953 Fieser); [α]$_D^{20}$ —38° (chloroform) (1946 Ross), —40° (chloroform), —23° (dioxane) (1953 Fieser). Absorption max. in ether ca. 247 mμ (ε ca. 11 200) (from graph) (1939 Windaus), in chloroform 252 mμ (ε 11 400) (1939 Dannenberg; cf. 1934 Windaus; 1936a Butenandt), in alcohol 251.5 mμ (ε 10 600), shifting to 259 mμ (ε 10 000) on addition of alkali (1946 Heard; cf. 1953 Fieser). — Liebermann reaction negative (1927 Montignie). With conc. H_2SO_4 and acetic anhydride in chloroform the solution becomes yellow with green fluorescence, then pale red and finally (24 hrs.) indigo-blue. Heating with conc. H_2SO_4 in alcohol on the water-bath gives a pink solution with cinnabar-red fluorescence, strongly absorbing in the yellow, weakly in the green. Alcoholic ammoniacal silver solution is not reduced on short heating on the water-bath (1896 Mauthner). Is very weakly reducing towards phosphomolybdic acid in acetic acid on the water-bath (1946 Heard).

Fmn. From cholesterol (p. 1568 s) by oxidation with CrO_3 in ca. 90% acetic acid at 20—25°; yield, 29%, falling to 10—15% by reaction below 20° (1946 Ross). From cholesterol by oxidation with $Ag_2Cr_2O_7$ in boiling aqueous H_2SO_4 (1927 Montignie). Together with Δ⁴-cholesten-3-one (p. 2424 s) and Δ⁴·⁶-cholestadien-3β-ol (p. 1544 s), from cholesterol dibromide (p. 1889 s) by treatment with $AgNO_3$ in pyridine at room temperature, followed by chromatographic separation on Al_2O_3; yield, ca. 2% (1941 Spring). From the dibenzoate of Δ²·⁴·⁶-cholestatriene-3.6-diol (p. 2058 s) by refluxing with 0.1 N alcoholic KOH (1946 Ross). From Δ⁴-cholestene-3β.6β-diol (p. 2060 s) by oxidation with $Na_2Cr_2O_7$ and aqueous H_2SO_4 in benzene - acetic acid at room temperature (1938 Petrow). Together with cholestan-5-ol-3.6-dione (p. 2953 s), from cholesterol α-oxide (p. 2175 s) by oxidation with CrO_3 (1915 Westphalen). From the 6-enol ethyl ether (6-ethoxy-Δ⁴·⁶-cholestadien-3-one, p. 2671 s) by refluxing with 80% acetic acid (1907 Windaus), or with glacial acetic acid containing zinc acetate (1906 Windaus) or sodium acetate (1937 Krekeler).

199*

From the acetate of $\Delta^{2.4}$-cholestadien-3-ol-6-one (p. 2676 s) by saponification with sodium methoxide in methanol at room temperature; yield, 55% (1938 Heilbron). From Δ^4-cholesten-6β-ol-3-one (p. 2672 s) by oxidation with CrO$_3$ in acetic acid on the water-bath (yield, almost quantitative) (1896 Mauthner) or with Na$_2$Cr$_2$O$_7$ and aqueous H$_2$SO$_4$ in benzene - acetic acid at 15° (yield, 70%) (1939 Ellis). From 2α-bromocholestane-3.6-dione (p. 2922 s) on refluxing with alcoholic HCl (1950 Ellis). From 5-bromocholestane-3.6-dione (p. 2922 s) on short boiling with pyridine (1943 Sarett). From cholestan-5-ol-3.6-dione (p. 2953 s) by heating with KHSO$_4$ in a high vacuum at 150—180° (1940a Marker), by heating alone above its m. p. or, preferably, by refluxing with P$_2$O$_5$ in benzene (1946 Ross); from the acetate of the same hydroxy-diketone* by adsorption on Al$_2$O$_3$ (1939 Windaus) or by treatment with methanolic alkali (1939 Windaus; 1939 Hattori).

Rns. Gives cholestane-4.5-diol-3.6-dione (p. 2963 s) on oxidation with KMnO$_4$ in ca. 83% acetone (1937 Windaus) or with 30% H$_2$O$_2$ and small amounts of OsO$_4$ in ether (1938 Butenandt). The keto-acid C$_{27}$H$_{42}$O$_5$ obtained with CrO$_3$ (cf. p. 158) in glacial acetic acid at 70° (1937a Butenandt) has been shown to be **3-keto-Δ^4-cholestene-6‖7-dioic acid** (I) (1953 Fieser). There is no reaction with H$_2$CrO$_4$ in glacial acetic acid at room temperature (1896 Mauthner). Oxidation with SeO$_2$ occurs readily in glacial acetic acid at 100° (product not described) but not in hot alcohol (1938 Stiller).

Catalytic hydrogenation, discontinued after absorption of ca. 1.5 mols. H$_2$, at room temperature using palladium-charcoal in glacial acetic acid gives cholestane-3.6-dione (p. 2907 s) (crude yield, ca. 37%) (1941 Bretschneider); from a similar reaction carried out with palladium black in alcohol, approximately the same amount of this product has been isolated together with ca. 20% of coprostane-3.6-dione (p. 2910 s) (1947 Moffatt); using Raney nickel in alcohol, discontinued after absorption of 2 mols. H$_2$, cholestan-3β-ol-6-one (p. 2677 s) is obtained (1941 Bretschneider). Reduction with amalgamated zinc and conc. HCl in glacial acetic acid yields cholestane (p. 1429 s) (1917 Windaus).

The products of reaction with bromine (cf. p. 158) are 2α.7α-dibromo-Δ^4-cholestene-3.6-dione (p. 2929 s) by treatment with 2 mols. in glacial acetic acid - ether at room temperature, and 2.2.7α-tribromo-Δ^4-cholestene-3.6-dione (p. 2931 s) by reaction with 4.2 mols. in glacial acetic acid at 30—35° (1937b Butenandt; cf. 1957 Fieser). Sulphonation, using conc. H$_2$SO$_4$ in acetic anhydride at —10°, affords Δ^4-cholestene-3.6-dione-2-sulphonic acid (1937, 1938 Windaus; cf. 1906 Windaus). Reaction with H$_2$S gives a crystalline derivative (not described) (1906 Windaus). By treatment with hot alcoholic KOH (1 equiv.) the monopotassium salt of $\Delta^{4.6}$-cholestadien-6-ol-3-one (page 2671 s) is formed, which is converted into the ethyl ether by reaction with ethyl iodide (1946 Ross).

* Tentatively formulated by Windaus (1939) as the 3-acetate of Δ^3-cholestene-3.5-diol-6-one.

The hydrogen cyanide addition product (cf. p. 158) is stable towards alcoholic HCl (1906 Windaus) or conc. aqueous HCl, also towards 10 % H_2SO_4 or 30 % KOH in the cold, but on heating, these two reagents give dark resinous products (1901 van Oordt). With acetic anhydride no reaction has been observed; heating with acetyl chloride gives a crystalline product not identified (1896 Mauthner). On refluxing with benzoyl chloride in pyridine the 3.6-dibenzoate of $\Delta^{2.4.6}$-cholestatriene-3.6-diol (p. 2058 s) is obtained; using benzoyl chloride alone, the $\Delta^{3.5.7}$-isomer (p. 2058 s) is formed (1946 Ross). A crystalline derivative (not described) is obtained by reaction with aniline (1906 Windaus).

Page 158 line 11 from bottom: after "van Oordt" *insert* "1906 Windaus".

Δ^4-Cholestene-3.6-dione mono-o-aminoanil $C_{33}H_{48}ON_2$ (p. 158). Brick-red leaflets (ethyl acetate); sparingly soluble in alcohol and ethyl acetate, readily in chloroform, ether, and benzene; soluble in conc. H_2SO_4 or alcoholic HCl giving a purple coloration. — **Fmn.** From the dione by short heating with o-phenylenediamine at 150° (1906 Windaus).

Δ^4-Cholestene-3.6-dione monosemicarbazone $C_{28}H_{45}O_2N_3$ (p. 158). Solid (chloroform - alcohol); readily soluble in chloroform. — **Fmn.** From the dione by refluxing with semicarbazide acetate (1 mol.) in methanol (1934 Stange). — **Rns.** The diol obtained by Wolff-Kishner reduction (cf. p. 158) is cholestane-3β.6α-diol (p. 2062 s) (Editor).

A *monosemicarbazone*, crystals, m. 134°, is mentioned by Montignie (1927).

Δ^4-Cholestene-3.6-dione disemicarbazone $C_{29}H_{48}O_2N_6$ (p. 158). Crystals (benzene - alcohol); very sparingly soluble in chloroform, benzene, or alcohol (1934 Stange); absorption max. in chloroform (from the graph), ca. 317 mμ (ε ca. 23 600), <240 mμ (ε >15 000) (1939 Windaus). — **Fmn.** As for the monosemicarbazone (above) but using 2 mols. of the reagent (1934 Stange).

Δ^4-Cholestene-3.6-dione monophenylhydrazone $C_{33}H_{48}ON_2$ (p. 158). Golden yellow leaflets (chloroform - alcohol) (1896 Mauthner; cf. 1939 Windaus), m. 272° (1939 Windaus), m. 280–1° (1939 Ellis); soluble in chloroform, very sparingly soluble in alcohol; gives an intense purple coloration with conc. H_2SO_4 (1896 Mauthner).

Δ^4-Cholestene-3.6-dione 6-enol ethyl ether $C_{29}H_{46}O_2$ (p. 158). See 6-ethoxy-$\Delta^{4.6}$-cholestadien-3-one, p. 2671 s.

Cholestane-3.6-dione $C_{27}H_{44}O_2$ (p. 159). Needles (chloroform - alcohol, or methanol) (1937b Butenandt; 1938 Petrow; 1944b Prelog), tablets or needles (acetone) (1939 Ellis), crystals (light petroleum) (1948 Barton); m. 172° (1948 Barton; cf. 1939 Ellis; 1944a, b Prelog), 174–5° cor. (1936 Fujii); can be sublimed at 190° and 0.01 mm. pressure (1946 Reich); $[\alpha]_D^{ca.\ 20}$ in chloroform +4° (1948 Barton), +7.5 ± 2°, +10 ± 4° (1944a, b Prelog; cf. 1947

Moffatt). Sparingly soluble in most organic solvents, soluble with yellow coloration in hot alcoholic KOH, from which it is precipitated by dilution with water (1903 Windaus). Gives a green colour with the Liebermann-Burchard reagent (1938 Petrow), a transient violet colour with m-dinitrobenzene and KOH in alcohol (1938 Callow; cf. 1944 Zimmermann).

Page 159, line 8: for "a 7.9, b 7.62, c 19.6 Å" *read* "a 19.6, b 7.62, c 7.9 Å, β 93°; d 1.13, n = 2".

Fmn. From Δ^4-cholestene-3β.6α-diol or its 6-epimer (p. 2060 s) by oxidation with acetone and aluminium tert.-butoxide in boiling benzene (1944a Prelog). Together with 6-ketocholestane-2‖3-dioic acid (p. 1714 s), from cholestane-3β.6α-diol (p. 2062 s) by oxidation with CrO_3 in strong acetic acid (1921 Windaus); from the 6-epimeric diol (p. 2064 s) by $Na_2Cr_2O_7$ oxidation (1938 Petrow; cf. p. 159).

From 2α-bromo-Δ^4-cholesten-3-one (p. 2480 s) by treatment with potassium acetate in butanol or from 2-bromo-5-chlorocholestan-3-one (p. 2497 s) by heating with potassium acetate in glacial acetic acid at 220° (1937 SCHERING KAHLBAUM A.-G.). From either 6β-bromo-Δ^4-cholesten-3-one (p. 2484 s) or cholestenone dibromide (p. 2500 s) on refluxing with 0.1 N methanolic HCl (yield, ca. 70% in each case) (1937 Dane); for mechanism of the formation from cholestenone dibromide in boiling alcohol (cf. p. 159), see 1941 Ando. From Δ^4-cholesten-6β-ol-3-one (p. 2672 s) or its acetate by refluxing with alcoholic HCl (yield from acetate, 83%) (1939 Ellis), or from the isomeric Δ^4-cholesten-3β-ol-6-one (p. 2677 s) or its acetate on boiling with methanolic KOH (1937 Heilbron). From 5α.6α-epoxy-cholestan-3-one (p. 2755 s) on refluxing with 2 N H_2SO_4 in dioxane (1937 Ruzicka) or from the 6-acetate of cholestane-5α.6β-diol-3-one (p. 2756 s) by refluxing with sodium methoxide in methanol (1939 Ellis).

From Δ^4-cholestene-3.6-dione (p. 2905 s) by hydrogenation in glacial acetic acid at room temperature, using palladium - charcoal catalyst, discontinued after absorption of 1.5 mols. of hydrogen; yield, 37% (1941 Bretschneider; cf. 1947 Moffatt). From 2α-bromocholestane-3.6-dione (p. 2922 s) on reduction with $CrCl_2$ (1950 Ellis). From 7β-bromo-Δ^4-cholestene-3.6-dione (p. 2923 s) on refluxing with zinc dust in alcohol; from 2α-bromo-, 2α.7α-dibromo-, or 2.2.7.7-tetrabromo-Δ^4-cholestene-3.6-dione (pp. 2922 s, 2929 s, 2932 s) on boiling with iron powder in alcohol; yield from the tetrabromo - compound after 5 hrs. refluxing, 56% (1943 Sarett). From 2α.7α-dibromo-Δ^4-cholestene-3.6-dione (p. 2929 s) or 2.2.7α-tribromo-Δ^4-cholestene-3.6-dione (p. 2931 s) by reduction with zinc dust in glacial acetic acid or in alcohol (1906 Windaus; cf. 1936 Fujii); these debrominations have also been effected with zinc in benzene - alcohol under reflux (yields, 69%) or, giving smaller yields, at room temperature (1937b Butenandt). From 2.7α-dibromo-$\Delta^{1.4}$-cholestadiene-3.6-dione (p. 2927 s) by reduction with zinc dust in glacial acetic acid on the water-bath; yield, 42% (1937b Butenandt).

Rns. Oxidation with $(NH_4)_2S_2O_8$ in ca. 90% acetic acid at 70—75° affords 2-hydroxy-6-keto-2‖3-cholestan-3-oic acid (p. 1712 s) (1904 Windaus). Oxidation with SeO_2 occurs readily in glacial acetic acid at 100° (product not described) but not in hot alcohol (1938 Stiller). Hydrogenation in glacial acetic acid at room temperature, using PtO_2 catalyst, gives cholestane-$3\beta.6\beta$-diol (p. 2064 s) together with small amounts of epimeric diols, cholestan-6β-ol (p. 1725 s), and cholestan-3β-ol-6-one (p. 2677 s) (1946 Reich). Reduction with sodium in boiling abs. alcohol affords cholestane-$3\beta.6\alpha$-diol (p. 2062 s) (1940 b Marker). Treatment with amalgamated zinc and conc. HCl in glacial acetic acid yields cholestane (p. 1429 s) (1917 Windaus). By heating the disemicarbazone with sodium ethoxide in abs. alcohol at 180° (bomb), a mixture results from which cholestan-3β-ol has been isolated (1940 c Marker).

Reaction with 2 mols. of bromine* in chloroform - glacial acetic acid at room temperature, in presence of sodium acetate, gives 2.2-dibromocholestane-3.6-dione (p. 2926 s); by using 9 mols. of bromine 2.2.7.7-tetrabromo- and 2.2.7α-tribromo-Δ^4-cholestene-3.6-diones (pp. 2932 s and 2931 s, resp.) are isolated after 18 hrs. and 48 hrs. standing, resp. (1943 Sarett); the bromination leading to $2\alpha.7\alpha$-dibromo-Δ^4-cholestene-3.6-dione (cf. p. 159) is effected with 3 mols. of reagent in chloroform (containing small amounts of HBr - glacial acetic acid) at room temperature (1937 b Butenandt). Reaction with conc. H_2SO_4 in acetic anhydride at —10° yields cholestane-3.6-dione-2-sulphonic acid (1938 Windaus). Heating with alcoholic NH_3 at 120° (bomb) gives a dimolecular compound† [white needles (chloroform - alcohol), m. > 300° dec.; easily soluble in chloroform, almost insoluble in methanol or ethanol; obtained in ca. 8% yield] (1906 Windaus). Refluxing with o-phenylene-diamine (2.3 mols.) gives only a mono-o-aminoanil (see below) (1939 a Marker). For reaction with ethylene glycol, see the bis-(ethylene ketal) (below).

Cholestane-3.6-dione bis-(ethylene ketal) $C_{31}H_{52}O_4$, m. 144°; obtained from the dione by refluxing with ethylene glycol in benzene in presence of p-toluene-sulphonic acid (1941 SQUIBB & SONS).

Cholestane-3.6-dione mono-o-aminoanil $C_{33}H_{50}ON_2$. Pale yellow needles (abs. alcohol), m. 207–210° dec. to give a red oil (1939 a Marker).

Cholestane-3.6-dione dioxime $C_{27}H_{46}O_2N_2$ (p. 159). White needles (alcohol) (1937 b Butenandt), crystals (methanol), m. 208–210° cor. (1936 Ruzicka; cf. 1903 Windaus; 1946 b Barnett), m. 215° cor. dec. (1936 Fujii). — **Rns.** Reduction with sodium in abs. alcohol gives 3.6-diaminocholestane (p. 1484 s) (1946 b Barnett).

Cholestane-3.6-dione disemicarbazone $C_{29}H_{50}O_2N_6$, m. 203° dec. (1937 Dane; cf. 1940 c Marker).

Pyridazine derivative of cholestane-3.6-dione $C_{27}H_{44}N_2$ (p. 159). For evidence of the monomolecular formula, disputed by Noller (1939), see 1940 Bursian. — Six-sided leaflets (benzene - methanol or -ethanol) (1906 Windaus; 1940 Bursian); darkens at ca. 170°, melts completely >200° to a brown

---

* See also the scheme on p. 2928 s.

† The formula $C_{54}H_{85}O_3N$, given in the original, is based on a wrong formula, $C_{27}H_{42}O_2$, for the parent dione; the purity of the compound seems doubtful (Editor).

liquid (1940 Bursian); readily soluble in chloroform or benzene, very sparingly in alcohol or petroleum ether (1906 Windaus).

Coprostane-3.6-dione $C_{27}H_{44}O_2$ (p. 159). Crystals (aqueous methanol), m. 170° to 174° cor., $[\alpha]_D^{18}$ —57 ±8° (chloroform) (1944b Prelog); needles (alcohol), m. 175–9°, $[\alpha]_D^{19}$ —81° (average in chloroform) (1947 Moffatt). — **Fmn.** From coprostane-3β.6β-diol (p. 2065 s) by oxidation with CrO_3 in ca. 93%

acetic acid at room temperature (1944b Prelog). Together with cholestane-3.6-dione (p. 2907 s), from $\varDelta^4$-cholestene-3.6-dione (p. 2905 s) by hydrogenation in alcohol at room temperature, using palladium black catalyst, (cessation after uptake of 1.3 mols. of hydrogen); yield, 20% (1947 Moffatt). The yields obtained by reduction (cf. p. 159) of 2α.7α-dibromo- and 2.2.7α-tribromo-$\varDelta^4$-cholestene-3.6-diones (pp. 2929 s and 2931 s, resp.), after first crystallizing out the main product, cholestane-3.6-dione, are 2.7% and 7.4%, resp. (1937b Butenandt).

$\varDelta^5$-Cholestene-3.7-dione $C_{27}H_{42}O_2$. Besides the diketo-form (I) two enolic

forms, $\varDelta^{4.6}$-cholestadien-7-ol-3-one (II) and $\varDelta^{3.5}$-cholestadien-3-ol-7-one (III) are possible. The only form actually isolated is very probably enolic (1946a Barnett) and corresponds to (III) from which all the known enol-esters and -ethers are certainly derived (1952 Greenhalgh); for their description see $\varDelta^{3.5}$-cholestadien-3-ol-7-one, p. 2680 s. Only the dioxime, for which the structure has not been elucidated, is described here.

Dioxime $C_{27}H_{44}O_2N_2$. Crystals (aqueous alcohol), needles (alcohol or dioxane), m. 229–230° dec.; sparingly soluble in hot dioxane, less so in hot alcohol, insoluble in ether or benzene. — **Fmn.** From $\varDelta^{3.5}$-cholestadien-3-ol-7-one (p. 2680 s) on refluxing with hydroxylamine acetate in aqueous alcohol (1946a Barnett). — **Rns.** Reduction with sodium in abs. alcohol affords 3.7-diamino-$\varDelta^{5(?)}$-cholestene (p. 1485 s) (1946b Barnett).

Cholestane-3.7-dione $C_{27}H_{44}O_2$ (p. 159). White needles (acetone) (1939b Marker); crystals (light petroleum-benzene), m. 190°, $[\alpha]_D^{ca. 20}$ —19° (chloroform) (1948 Barton). — **Fmn.** From cholestane-3β.7β-diol (p. 2073 s) by oxidation with CrO_3 in 80% acetic acid at room temperature (1939b Marker).

Δ^5-**Cholestene-4.7-dione (?)** $C_{27}H_{42}O_2$. Structure only tentatively assigned. —

Pale yellow plates (ethanol, or ether - methanol), m. 160–1°, $[\alpha]_D^{21}$ —52° (carbon tetrachloride). — **Fmn.** From $\Delta^{4.6}$-cholestadiene (p. 1417 s) by oxidation with CrO_3 and aqueous H_2SO_4 in benzene - glacial acetic acid at room temperature; crude yield, 20%. — **Rns.** By heating the disemicarbazone with sodium ethoxide and abs. alcohol at 200° (sealed tube), with added hydrazine hydrate, a mixture results from which coprostene (p. 1423 s) has been isolated (1941 Eck).

Disemicarbazone $C_{29}H_{48}O_2N_6$. M. p. 322° dec. (1941 Eck).

Cholestane-4.7-dione $C_{27}H_{44}O_2$. Crystals (alcohol), m. 146–7°. — **Fmn.** From cholestan-7-one-4-sulphonic acid by oxidation with CrO_3 in 70% acetic acid at 65°. — **Rns.** Further oxidation under the same conditions as used in formation affords 7-ketocholestane-3‖4-dioic acid (page 1720 s). Treatment with hydrazine hydrate in alcohol gives a **pyridazine derivative** $C_{27}H_{44}N_2$ [leaflets (alcohol), m. 167–9°] (1938 Windaus).

$\Delta^{2.4}$-**Cholestadiene-6.7-dione** $C_{27}H_{40}O_2$. Known only in the form of the 7-enol methyl ether, $\Delta^{2.4.7}$-cholestatrien-7-ol-6-one methyl ether (p. 2680 s).

1896 Mauthner, Suida, *Monatsh.* **17** 579, 584 et seq.
1901 van Oordt, *Über Cholesterin*, Thesis Univ. Freiburg i. Br., pp. 18, 19.
1903 Windaus, *Ber.* **36** 3752, 3755, 3756.
1904 Windaus, *Ber.* **37** 2027, 2029.
1906 Windaus, *Ber.* **39** 2249, 2250, et seq.
1907 Windaus, *Ber.* **40** 257, 261.
1915 Westphalen, *Ber.* **48** 1064, 1067.
1917 Windaus, *Ber.* **50** 133, 137.
1921 Windaus, Lüders, *Z. physiol. Ch.* **115** 257, 269.
1927 Montignie, *Bull. soc. chim.* [4] **41** 947, 949.
1934 Stange, *Z. physiol. Ch.* **223** 245, 247.
        Windaus, Inhoffen, v. Reichel, *Ann.* **510** 248, 257.
1936 a Butenandt, Riegel, *Ber.* **69** 1163, 1164.
        b Butenandt, Schramm, Wolff, Kudszus, *Ber.* **69** 2779, 2781.
        Fujii, Matsukawa, *J. Pharm. Soc. Japan* **56** 150 (German abstract); *C. A.* **1937** 1033; *Ch. Ztbl.*
             1937 I 2616.
        Inhoffen, *Ber.* **69** 1702, 1709.
        Ruzicka, Bosshard, Fischer, Wirz, *Helv.* **19** 1147, 1151, 1152.
1937 a Butenandt, Hausmann, *Ber.* **70** 1154.
        b Butenandt, Schramm, Kudszus, *Ann.* **531** 176, 200 et seq.
        Dane, Wang, Schulte, *Z. physiol. Ch.* **245** 80, 86.
        Heilbron, Jones, Spring, *J. Ch. Soc.* **1937** 801, 804.
        Inhoffen, *Ber.* **70** 1695.
        Krekeler, *Einige Umsetzungen am Cholesterin*, Thesis Univ. Göttingen, pp. 12, 21.
        Ruzicka, Bosshard, *Helv.* **20** 244, 246, 249.

SCHERING-KAHLBAUM A.-G., *Fr. Pat.* 835524 (issued 1938), *Brit. Pat.* 500353 (issued 1939); *C. A.* **1939** 5872; *Ch. Ztbl.* **1939** I 5010; SCHERING CORP. (Inhoffen, Butenandt, Schwenk), *U.S. Pat.* 2340388 (issued 1944); *C. A.* **1944** 4386.
Windaus, Kuhr, *Ann.* **532** 52, 62, 63.
1938 Bergmann, Hirshberg, *Nature* **142** 1037.
Butenandt, Wolz, *Ber.* **71** 1483, 1486.
N. H. Callow, R. K. Callow, Emmens, *Biochem. J.* **32** 1312, 1323.
Heilbron, Jackson, Jones, Spring, *J. Ch. Soc.* **1938** 102, 106.
Petrow, Rosenheim, Starling, *J. Ch. Soc.* **1938** 677, 680, 681.
Stiller, Rosenheim, *J. Ch. Soc.* **1938** 353.
Windaus, Mielke, *Ann.* **536** 116, 119, 120, 123, 124.
Wolff, *Über die Bromierung von Cholestanon und Koprostanon*, Thesis Techn. Hochschule Danzig, p. 16.
1939 Dannenberg, *Abhandl. Preuss. Akad.Wiss., Math.-naturwiss. Kl.* **1939** No. 21, pp. 17, 19,30,60.
Ellis, Petrow, *J. Ch. Soc.* **1939** 1078, 1081, 1082.
Hattori, *C. A.* **1939** 8622; *Ch. Ztbl.* **1940** I 380.
a Marker, Rohrmann, *J. Am. Ch. Soc.* **61** 946, 948.
b Marker, Rohrmann, *J. Am. Ch. Soc.* **61** 3022.
Noller, *J. Am. Ch. Soc.* **61** 2976.
Windaus, Naggatz, *Ann.* **542** 204, 209, 210.
1940 Bursian, *Ber.* **73** 922.
a Marker, Rohrmann, *J. Am. Ch. Soc.* **62** 516.
b Marker, Jones, Turner, *J. Am. Ch. Soc.* **62** 2537, 2539.
c Marker, Turner, Ulshafer, *J. Am. Ch. Soc.* **62** 3009.
Petrow, Starling, *J. Ch. Soc.* **1940** 60.
1941 Ando, *C. A.* **1950** 4016.
Bretschneider, *Ber.* **74** 1361.
Eck, Hollingsworth, *J. Am. Ch. Soc.* **63** 107, 110.
Spring, Swain, *J. Ch. Soc.* **1941** 320.
SQUIBB & SONS (Fernholz), *U.S. Pat.* 2378918 (issued 1945); *C. A.* **1945** 5051.
1943 Sarett, Chakravorty, Wallis, *J. Org. Ch.* **8** 405, 411 et seq.
1944 a Prelog, Tagmann, *Helv.* **27** 1867, 1871.
b Prelog, Tagmann, *Helv.* **27** 1880.
Ruzicka, Plattner, Furrer, *Helv.* **27** 524, 527.
Zimmermann, *Vitamine und Hormone* **5** [1952] 1, 12.
1946 a Barnett, Ryman, Smith, *J. Ch. Soc.* **1946** 526.
b Barnett, Ryman, Smith, *J. Ch. Soc.* **1946** 528.
Heard, Sobel, *J. Biol. Ch.* **165** 687, 689 et seq.
Reich, Lardon, *Helv.* **29** 671, 675.
Ross, *J. Ch. Soc.* **1946** 737.
1947 Moffatt, *J. Ch. Soc.* **1947** 812.
1948 Barton, Cox, *J. Ch. Soc.* **1948** 783, 792.
L. F. Fieser, M. Fieser, Rajagopalan, *J. Org. Ch.* **13** 800, 802.
1950 Ellis, Petrow, *J. Ch. Soc.* **1950** 2194, 2198.
1952 Greenhalgh, Henbest, Jones, *J. Ch. Soc.* **1952** 2375, 2377.
1953 Fieser, *J. Am. Ch. Soc.* **75** 4386, 4388, 4390.
1954 Fieser, Stevenson, *J. Am. Ch. Soc.* **76** 1728, 1733.
1957 M. Fieser, L. F. Fieser, *private communication.*

$\varDelta$⁴-**Ergostene-3.6-dione** $C_{28}H_{44}O_2$ (p. 160). Absorption max. in ether at 254 mμ (ε 9650) (1934 Windaus); for solvent effect on position of absorption max., see 1939 Dannenberg. Very soluble in chloroform (1934 Windaus). — **Rns.** Heating with H_2SO_4 in abs. alcohol on the water-bath gives the 6-enol ethyl ether, $\Delta^{4.6}$-ergostadien-6-ol-3-one ethyl ether (p. 2690 s) (1934 Windaus).

$\Delta^{8(14)}$-Ergostene-3.15-dione, α-*Ergostenedione* $C_{28}H_{44}O_2$ (p. 160). The 3.15-dione structure may be deduced from that of its immediate precursor, $\Delta^{8(14)}$-ergosten-3β-ol-15-one, which see p. 2693 s; cf. p. 83: α-ergostenol-one (Editor).

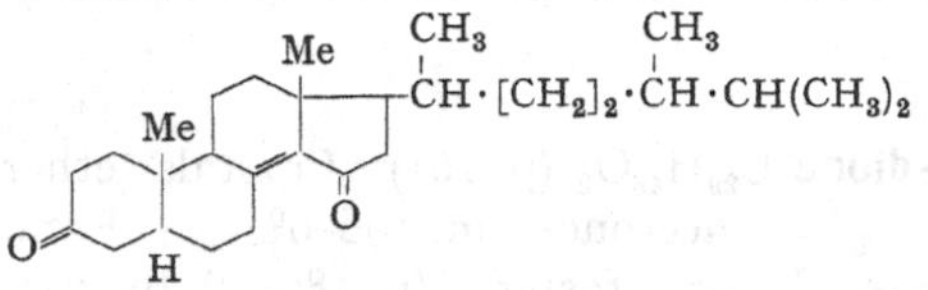

1934 Windaus, Inhoffen, v. Reichel, *Ann.* **510** 248, 256, 257.
1939 Dannenberg, *Abhandl. Preuss. Akad. Wiss., Math.-naturwiss. Kl.* **1939** No. 21, pp. 31, 32, 60.

$\Delta^{4.22}$-Stigmastadiene-3.6-dione $C_{29}H_{44}O_2$ (p. 160). Absorption max. in ether at 254 mμ (ε 8750) (1934 Windaus); for effect of solvent on position of absorption max., see 1939 Dannenberg*. — **Rns.** By refluxing with zinc dust in strong acetic acid Δ^{22}-stigmastene-3.6-dione (below) is obtained (1934 Fernholz).

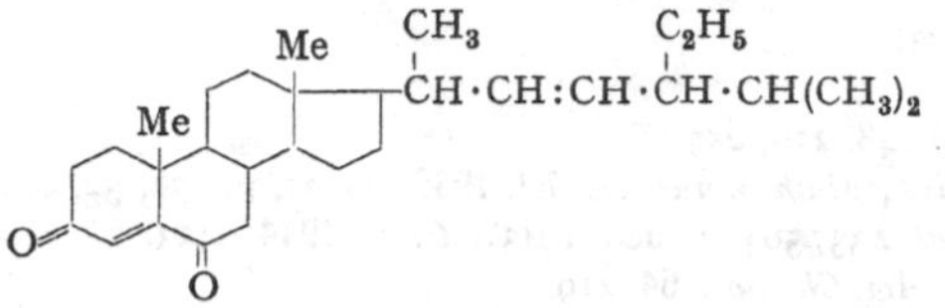

Δ^4-Stigmastene-3.6-dione, Δ^4-Sitostene-3.6-dione, α-*Fucostenedione* $C_{29}H_{46}O_2$ (p. 160). Yellowish leaflets (ethanol or methanol), m. 129° (1934 Fernholz) (allotropic form? — Editor). Soluble in most of the usual solvents. Gives no coloration with ethereal $FeCl_3$ (1908 Pickard).

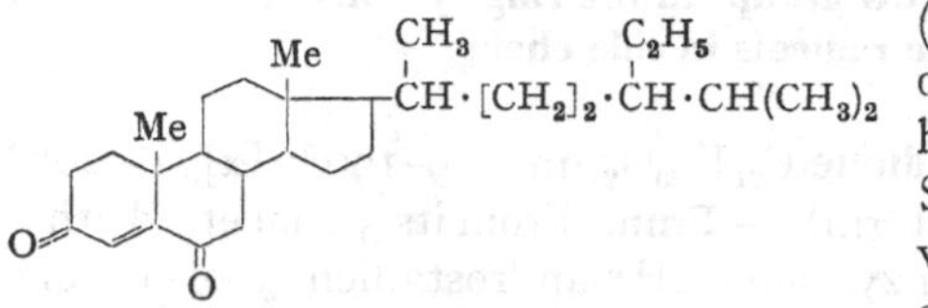

Δ^{22}-Stigmastene-3.6-dione $C_{29}H_{46}O_2$. Crystals (acetone or ether) (1942 Marker), needles (alcohol), m. 197°; very soluble in chloroform, readily soluble in ethyl acetate, less readily in alcohol (1934 Fernholz). — **Fmn.** From stigmasterol (page 1784 s) by oxidation with CrO_3 in strong acetic acid at 20–22°, followed by refluxing the acetic acid solution with zinc dust; yield, 53 % (1942 Marker). From $\Delta^{4.22}$-stigmastadiene-3.6-dione (above) by refluxing with zinc dust in strong acetic acid; yield, 66 % (1934 Fernholz). —**Rns.** On reduction with sodium in alcohol it gives Δ^{22}-stigmastene-3.6-diol [not characterized; mixture of epimers? (Editor)], from the dibenzoate of which that of 3β.6β-dihydroxy-bisnor-allocholanic acid has been obtained

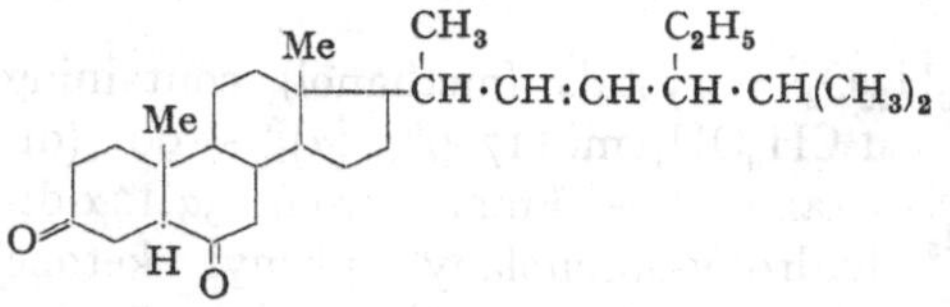

---

* This author mistakenly refers to the compound as Δ^4-stigmastenedione (Editor).

on heating with nitric acid on the steam-bath (1940 PARKE, DAVIS & Co.). Hydrogenation in acetic acid under 3 atm. pressure, using platinum oxide catalyst, gives sitostane-3β.6β-diol (p. 2084 s) (1942 Marker). Heating with hydrazine hydrate in aqueous alcohol on the water-bath gives a **pyridazine** **derivative** $C_{29}H_{46}N_2$ [needles (benzene-methanol, or chloroform-ethanol), m. ca. 260° dec.] (1934 Fernholz).

Stigmastane-3.6-dione, Sitostane-3.6-dione $C_{29}H_{48}O_2$ (p. 161). Crystals (ether or acetone), m. 196–9°. — **Fmn.** From sitosterol (p. 1801 s) by oxidation with CrO_3 in acetic acid at 15–20°, followed by refluxing the acetic acid solution with zinc dust; yield, 50%. — **Rns.** Hydrogenation in glacial acetic acid under 3 atm. pressure, using platinum oxide catalyst, affords sitostane-3β.6β-diol (p. 2084 s) (1942 Marker).

1908 Pickard, Yates, *J. Ch. Soc.* **93** 1928, 1931.
1934 Fernholz, *Ann.* **508** 215, 218, 221 et seq.
Windaus, Inhoffen, v. Reichel, *Ann.* **510** 248, 256, 257.
1939 Dannenberg, *Abhandl. Preuss. Akad. Wiss., Math.-naturwiss. Kl.* **1939** No. 21, pp. 31, 32, 60.
1940 PARKE, DAVIS & Co. (Marker), *U.S. Pat.* 2337564 (issued 1943); *C. A.* **1944** 3423.
1942 Marker, Crooks, Jones, Wittbecker, *J. Am. Ch. Soc.* **64** 219.

6. Diketosteroids (both CO groups in the ring system) containing cyclic radicals in side chain

16-Benzylidene-Δ^4-androstene-3.17-dione $C_{26}H_{30}O_2$, m. 189–190°; $[\alpha]_D$ —20°* (chloroform). — **Fmn.** From its 3-enol ethyl ether (16-benzylidene-$\Delta^{3.5}$-androstadien-3-ol-17-one ethyl ether, p. 2698 s) on rapid hydrolysis (30 min.) with methanolic H_2SO_4; quantitative yield (1945 Velluz).

24-Phenylcholane-3.12-dione $C_{30}H_{42}O_2$. Crystals (methanol) containing 1 CH_3OH, m. 117–9°; $[\alpha]_D^{25}$ +90° (dioxane). — **Fmn.** From 3α.12α-dihydroxy-norcholanyl phenyl ketone (p. 2350 s) by reduction with hydrazine hydrate and sodium methoxide in methanol at 180–190° (bomb), followed by oxidation of the resulting crude 24-phenylcholane-3α.12α-diol (p. 2088 s) with CrO_3 in strong acetic acid at room temperature; yield from the crude diol, 81% (1945 Hoehn).

* This value does not correspond to that (ca. +170°) anticipated for a compound of this structure (1958 Petit).

22.22-Diphenyl-$\Delta^{20(22)}$-bisnorcholene-3.12-dione, 1-Methyl-2.2-diphenyl-1-(3.12-diketo-ætiocholanyl)-ethylene $C_{34}H_{40}O_2$. Needles (ether - petroleum ether), rhombs (methanol), m. 243–245° cor., $[\alpha]_D^{16}$ +254° (acetone). **Fmn.** From (3.12-diketopregnan-20α-yl)-diphenyl-carbinol (p. 2935 s) by refluxing with glacial acetic acid (yield, 82%); similarly from (3.12-diketopregnan-20β-yl)-diphenyl-carbinol (p. 2936 s; yield, 74%) (1945 Sorkin).

1945 Hoehn, Moffatt, *J. Am. Ch. Soc.* **67** 740.
    Sorkin, Reichstein, *Helv.* **28** 875, 888.
    Velluz, Petit, *Bull. soc. chim.* [5] **12** 949, 950.
1958 Petit, *private communication.*

II. HALOGENO-DIKETOSTEROIDS WITH BOTH CO GROUPS IN THE RING SYSTEM

1. Monohalogeno-compounds C_{18} and C_{19}

4β - Bromo - 5β.10β - œstrane - 3.17 - dione, 4β - Bromo - 19 - nor - ætiocholane -

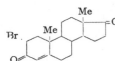

3.17-dione $C_{18}H_{25}O_2Br$. Colourless crystals (ether), m. 170–172° (1940 Marker); m. 186–8° (1958 Rapala). — **Fmn.** From 5β.10β-œstrane-3.17-dione ("œstrane-3.17-dione-B", p. 2873 s) on treatment with 1 mol. of bromine and a little 48% HBr in glacial acetic acid at room temperature (1940 Marker; cf. 1958 Rapala). — **Rns.** On refluxing with dry pyridine for 9 hrs. it gives 10β-Δ⁴-œstrene-3.17-dione (19-nor-Δ⁴-androstene-3.17-dione, p. 2873 s) (1940 Marker; 1958 Rapala).

1940 Marker, Rohrmann, *J. Am. Ch. Soc.* **62** 73, 75.
1958 Rapala, Farkas, *J. Am. Ch. Soc.* **80** 1008.

2-Bromo-Δ¹-androstene-3.17-dione $C_{19}H_{25}O_2Br$. Prismatic needles (hexane - acetone), m. 175–7° cor.; $[\alpha]_D^{25} +85°$ (chloroform). Absorption max. in 95% alcohol at 256 mμ (log ε 3.84). — **Fmn.** From 2.2-dibromoandrostane-3.17-dione (p. 2924 s) on refluxing with collidine for 10 min.; crude yield, 59% (1948 Djerassi).

2α-Bromo-Δ⁴-androstene-3.17-dione $C_{19}H_{25}O_2Br$. For 2α-configuration, see the parent compound. — Needles (hexane - acetone), m. 167–8° cor. dec.; $[\alpha]_D^{25} +170°$ (chloroform). Absorption max. in 95% alcohol at 243 mμ (log ε 4.07) (1948 Djerassi). — **Fmn.** From 2α.4α-dibromoandrostane-3.17-dione (p. 2924 s) on refluxing with collidine for 30 sec.; yield, 39% (1948 Djerassi). — **Rns.** Monobromination in ether - glacial acetic acid affords 2α.6α(?)-dibromo-Δ⁴-androstene-3.17-dione (page 2925 s) (1950 Djerassi). Gives Δ⁴-androstene-3.17-dione (p. 2880 s) on treatment of its solution in acetone with 1 N aqueous chromous chloride solution (containing $ZnCl_2$ and hydrochloric acid) in an atmosphere of CO_2 (1948 Djerassi). Refluxing with collidine yields Δ¹·⁴-androstadiene-3.17-dione (p. 2877 s) (1950 Djerassi).

2α-Bromoandrostane-3.17-dione $C_{19}H_{27}O_2Br$ (I, R = Br) (p. 161). For α-con-

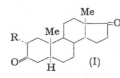

figuration at C-2, see the literature quoted for the configuration of 2α-bromocholestan-3-one (p. 2480 s). Needles (alcohol or dil. acetone), m. 216° (1936 CIBA PHARMACEUTICAL PRODUCTS); $[\alpha]_D^{23} +90°$ (chloroform) (1947 Djerassi). — **Fmn.** From androstane-3.17-dione (page 2892 s) on treatment at room temperature with 1 mol. of bromine and a little HBr in glacial acetic acid (1936 Butenandt) or in chloroform or ether (1937/38 Dannenberg; cf. also 1936 Schering-Kahlbaum A.-G.; 1937, 1947 Schering A.-G.).

Rns. On refluxing with pyridine for 1 hr. it gives 3.17-diketoandrostan-2-yl-pyridinium bromide (below) (1938 Ruzicka; 1937 Ciba). On refluxing with collidine for 1 hr. it gives 36 % crude yield of Δ^1-androstene-3.17-dione (p. 2879 s) and 12 % crude yield of Δ^4-androstene-3.17-dione (p. 2880 s) (1947 Djerassi; for obtainment of the Δ^1-isomer, see also p. 161, and 1936 Ciba Pharmaceutical Products; 1940 Schering A.-G.). The "hetero-Δ^1-androstene-3.17-dione" obtained by heating with potassium acetate in glacial acetic acid at 200° (see p. 161; cf. also 1936 Schering-Kahlbaum A.-G.) is actually Δ^5-androstene-4.17-dione (p. 2897 s) (1944 Butenandt); when this reaction was performed at lower temperature, two compounds $C_{21}H_{30}O_4$ were obtained: prisms (dil. acetone), m. 241°, $[\alpha]_D^{20}$ +80° (chloroform) and needles (dil. acetone), m. 187°, $[\alpha]_D^{20}$ +143° (alcohol), assumed to be the two epimeric 2-acetoxyandrostane-3.17-diones [1937/38 Dannenberg; see also 1937, 1947 Schering A.-G.; cf., however, the reaction of 2α-bromocholestan-3-one with sodium acetate (see p. 2481 s) giving a mixture of 2α- and 4α-acetoxycholestan-3-ones]. On refluxing with potassium benzoate in butanol-toluene it gives the benzoate of androstan-2-ol-3.17-dione (p. 2945 s) (1937, 1947 Schering A.-G.).

3.17-Diketoandrostan-2-yl-pyridinium hydroxide $C_{24}H_{33}O_3N$ (not analysed) [I, R = $N(C_5H_5)\cdot OH$]. Golden yellow needles becoming brick-red at ca.90°, m. ca. 120° cor. dec. — **Fmn.** The bromide is obtained from the above bromodione by refluxing with pyridine for 1 hr.; it is hydrolysed by treatment with 1 mol. alkali hydroxide. — **Rns.** On heating the bromide in a sealed tube under 10 mm. pressure it gives Δ^4-androstene-3.17-dione (p. 2880 s). — *Bromide* $C_{24}H_{32}O_2NBr$ [I, R = $N(C_5H_5)Br$]. Needles (water), m. ca. 315 cor. dec., which contain 1 H_2O after drying for 11 hrs. in a high vacuum at 105° (1938 Ruzicka; 1937 Ciba).

4-Bromo-Δ^4-androstene-3.17-dione $C_{19}H_{25}O_2Br$. A compound to which this structure is assigned, but the formation of which is not indicated, shows an absorption max. in ether at 248 mμ (ε 14400) (1939 Dannenberg). In view of the absorption max. quoted, the structure assigned seems very unlikely; the compound might be structurally related to Barkow's "4-bromocholestenone" (p. 2483 s) (Editor).

4β-Bromo-5-iso-androstane-3.17-dione, 4β-Bromoætiocholane-3.17-dione $C_{19}H_{27}O_2Br$. The β-configuration at C-4 is assumed by analogy with that of 4β-bromocoprostan-3-one (p. 2483 s) (Editor). — Crystals (alcohol), m. 195° dec. (1938 Ercoli). — **Fmn.** From ætiocholane-3.17-dione (p. 2895 s) on treatment with 1 mol. of bromine and a little HBr in glacial acetic acid; yield, 47 % (1938 Ercoli; cf. 1935

CIBA). — **Rns.** On refluxing with dry pyridine for 13 hrs. it gives Δ^4-andro-stene-3.17-dione (p. 2880 s) (1938 Ercoli; cf. 1935 CIBA).

5-Chloroandrostane-3.17-dione $C_{19}H_{27}O_2Cl$ (p. 161). For α-configuration at C-5, see under the parent 5-chloroandrostan-3β-ol-17-one (p. 2705 s). The compound described by 1936 Fujii (see p. 161) was made from impure 5-chloroandro-stan-3β-ol-17-one (see p. 2705 s) (1950 Hershberg). — Needles (chloroform - pentane), m. 99–102° cor. with evolution of gas (bath preheated to 90°), resolidifying, and remelting at 160–167° cor.; the initial decomposition point is dependent upon the temperature at immersion, e.g., m. 112° cor. dec. when the bath has been preheated to 110°. Decomposes readily in warm solutions with evolution of HCl, even at room temperature in chloroform solution upon treatment with charcoal (1950 Hershberg).

Fmn. From 5-chloroandrostan-3β-ol-17-one (p. 2705 s) on oxidation with CrO_3 and strong acetic acid in ethylene chloride at room temperature; yield, 92% (1950 Hershberg). — **Rns.** With 1 mol. of bromine in chloroform it gives the unstable 2(?)-bromo-5-chloroandrostane-3.17-dione (see p. 2924 s) (1950 Hershberg).

6-Bromo-$\Delta^{1.4}$-androstadiene-3.17-dione $C_{19}H_{23}O_2Br$. Crystals (ether - hexane), m. 189–192° cor. dec. (Kofler block); $[\alpha]_D^{20}$ +118° (chloro-form), +116° (dioxane); absorption max. in 95% alcohol at 250 mμ (log ε 4.34). — **Fmn.** From $\Delta^{1.4}$-androstadiene-3.17-dione (p. 2877 s) on refluxing in carbon tetrachloride with 1 mol. of N-bromosuccinimide in the presence of benzoyl peroxide for 75 min.; yield, 89%. — **Rns.** Gives $\Delta^{1.4.6}$-androstatriene-3.17-dione (p. 2877 s) on refluxing with collidine for 15 min. (1950 Kaufmann).

6α(?)-Bromo-Δ^4-androstene-3.17-dione $C_{19}H_{25}O_2Br$ (p. 161). Crystals, m. 175° to 177° cor. dec.; $[\alpha]_D^{20}$ +108° (chloroform); absorption max. in 95% alcohol at 240 mμ (log ε 4.23) (1950 Djerassi). — **Fmn.** From 6-bromo-Δ^5-androstene-3.17-diol (not described) on oxidation (1936 CIBA). For formation from 5-dehydroepiandrosterone dibromide on oxidation, followed by dehydrobromination (p. 161), see also 1936, 1938 CIBA. From Δ^4-androstene-3.17-dione (p. 2880 s) on refluxing for 20–30 min. in carbon tetrachloride with 1 mol. of N-bromo-succinimide in the absence of light; yield, 67% (1950 Djerassi).

Rns. On monobromination in ether - glacial acetic acid it gives 2α.6α(?)-dibromo-Δ^4-androstene-3.17-dione (p. 2925 s) (1950 Djerassi). Gives $\Delta^{4.6}$-androstadiene-3.17-dione (p. 2878 s) on treatment with an 8% solution of $AgNO_3$ in anhydrous pyridine at room temperature for 40 hrs., followed by acidification (1937 SCHERING A.-G.) or on refluxing with collidine (1950 Djerassi).

References, pp. 2919 s, 2920 s

16-Bromoandrostane-3.17-dione $C_{19}H_{27}O_2Br$. Needles (dil. acetone), m. 194°

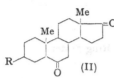

to 195°. — **Fmn.** From 16-bromoandrostan-3β-ol-17-one (16-bromo-epiandrosterone, p. 2706 s) on oxidation with CrO_3 in glacial acetic acid at room temperature; yield, 74%. — **Rns.** Gives 2α.16-dibromo-androstane-3.17-dione (p. 2925 s) on treatment with 1 mol. of bromine and a little HBr in glacial acetic acid at room temperature (1937/38 Dannenberg).

3β-Chloroandrostane-6.17-dione $C_{19}H_{27}O_2Cl$ (II, R = Cl). For β-configuration at C-3, see the parent 3β-chloro-Δ^5-androsten-17-one (p. 2474 s). — Crystals (methanol), m. 210–211°. — **Fmn.** From 3β-chloro-6-nitro-Δ^5-androsten-17-one (p. 2507 s) on heating with zinc dust in glacial acetic acid for 20 hrs. at 100°; yield, 68%. — **Rns.** Gives Δ^2-androstene-6.17-dione (p. 2897 s) on refluxing with quinoline under nitrogen for 1 hr. (1946 Blunschy).

Dioxime $C_{19}H_{29}O_2N_2Cl$. Crystals (methanol), m. 214–7° dec. (1946 Blunschy).

3β-Bromoandrostane-6.17-dione $C_{19}H_{27}O_2Br$ (II, R = Br). For β-configuration at C-3, see 1948 Shoppee. — Crystals (alcohol), m. 184° (1942 Butenandt). — **Fmn.** From i-androstene-6.17-dione (p. 2898 s) on treatment in glacial acetic acid with 48% HBr at 20°; quantitative yield (1942 Butenandt). — **Rns.** Gives the parent dione on refluxing with collidine or with potassium acetate - glacial acetic acid. On refluxing with quinoline under nitrogen for $^1/_2$ hr., Δ^2-androstene-6.17-dione (p. 2897 s) is obtained (1942 Butenandt).

1935 CIBA*, *Swiss Pat.* 184988 (issued 1936); *Ch. Ztbl.* **1937** I 3519, 3520.
1936 Butenandt, Dannenberg, *Ber.* **69** 1158, 1160.
    CIBA* (Ruzicka), *U.S. Pat.* 2085474 (issued 1937); *C. A.* **1937** 5950; *Ch. Ztbl.* **1938** I 374.
    CIBA PHARMACEUTICAL PRODUCTS (Miescher, Wettstein), *U.S. Pat.* 2260328 (issued 1941); *C. A.* **1942** 873; *Ch. Ztbl.* **1944** II 246.
    Fujii, Matsukawa, *C. A.* **1936** 8237; *Ch. Ztbl.* **1936** II 3305.
    SCHERING-KAHLBAUM A.-G., *Brit. Pat.* 489692 (issued 1938); *C. A.* **1939** 1343; *Ch. Ztbl.* **1939** I 3590, 3591; *Swiss Pat.* 206916 (issued 1939); *Ch. Ztbl.* **1940** I 3958.
1937 CIBA*, *Swiss Pat.* 212192 (issued 1941); *Ch. Ztbl.* **1941** II 3100, 3101; (Ruzicka), *Swed. Pat.* 98368 (1938, issued 1940); *Ch. Ztbl.* **1940** II 2185; CIBA PHARMACEUTICAL PRODUCTS (Ruzicka), *U.S. Pat.* 2232636 (1938, issued 1941); *C. A.* **1941** 3266; *Ch. Ztbl.* **1941** II 3100, 3101.
    SCHERING A.-G., *Fr. Pat.* 838704 (issued 1939); *Brit. Pat.* 508576; *C. A.* **1940** 777; *Ch. Ztbl.* **1939** II 1722; SCHERING CORP. (Inhoffen), *U.S. Pat.* 2280828 (issued 1942); *C. A.* **1942** 5618.
1937/38 Dannenberg, *Über einige Umwandlungen des Androstandions und Testosterons*, Thesis, Danzig, pp. 11, 29, 30, 34, 35.
1938 CIBA*, *Fr. Pat.* 833339 (issued 1938); *C. A.* **1939** 2905; *Ch. Ztbl.* **1939** I 2643.
    Ercoli, Mamoli, *Ber.* **71** 156, 158.
    Ruzicka, Plattner, Aeschbacher, *Helv.* **21** 866, 870.

* Société pour l'industrie chimique à Bâle; Gesellschaft für chemische Industrie in Basel.

1939 Dannenberg, *Abhandl. Preuss. Akad. Wiss. Math.-naturwiss. Kl.* **1939** No. 21, pp. 12, 53.
1940 SCHERING A.-G. (Butenandt), *Swed. Pat.* 111450 (issued 1944); *Ch. Ztbl.* **1947** 1875; *Dan. Pat.*
        61322 (issued 1943); *Ch. Ztbl.* **1944** II 49; *Fr. Pat.* 867697 (issued 1941); *Ch. Ztbl.* **1942** 12561.
1942 Butenandt, Surányi, *Ber.* **75** 591, 596.
1944 Butenandt, Ruhenstroth-Bauer, *Ber.* **77** 397, 400.
1946 Blunsch y, Hardegger, Simon, *Helv.* **29** 199, 202.
1947 Djerassi, *J. Org. Ch.* **12** 823, 825, 829.
        SCHERING A.-G., *Fiat Final Report* No. 996, pp. 158, 161.
1948 Djerassi, Scholz, *J. Org. Ch.* **13** 697, 704.
        Shoppee, *J. Ch. Soc.* **1948** 1043, 1044.
1950 Djerassi, Rosenkranz, Romo, Kaufmann, Pataki, *J. Am. Ch. Soc.* **72** 4534, 4536, 4538.
        Hershberg, Rubin, Schwenk, *J. Org. Ch.* **15** 292, 295, 296.
        Kaufmann, Pataki, Rosenkranz, Romo, Djerassi, *J. Am. Ch. Soc.* **72** 4531.

---

2. Monohalogeno-diketosteroids (Both CO Groups in the Ring System) with Cholestane Skeleton

2-Bromo-$\Delta^{1.4}$-cholestadiene-3.6-dione $C_{27}H_{39}O_2Br$ (p. 162: *7-bromo-$\Delta^{4.7}$-chole-stadiene-3.6-dione*). The structure orig-inally ascribed (1937 Butenandt) has been revised (Editor; 1958 Fieser) in view of the obvious identity of this Butenandt compound with that de-scribed by Ellis and Petrow (1950), for which these authors afford evidence for the 2-bromo-$\Delta^{1.4}$-diene structure by their mode of synthesis.

White needles (chloroform - methanol) (1937 Butenandt); m. 182°, $[\alpha]_D^{20}$ —141° (chloroform) (1943 Sarett); leaflike plates (aqueous acetone), m. 182°, $[\alpha]_D^{23}$ —139° (chloroform) (1950 Ellis). Absorption max. in chloroform at 259 mμ (ε 11700) (1939 Dannenberg; cf. 1937 Butenandt), in isopropyl alcohol at 255 mμ ($E_{1\,cm.}^{1\,\%} = 234$) (1950 Ellis).

Fmn. From 2.7α-dibromo-$\Delta^{1.4}$-cholestadiene-3.6-dione (p. 2927 s) by re-duction with iron powder in boiling alcohol (cf. p. 162); yield, 64% (1937 Butenandt). From 2.7.7-tribromo-$\Delta^{1.4}$-cholestadiene-3.6-dione (p. 2932 s) by refluxing with iron powder in benzene - alcohol; yield, 67% (1943 Sarett). From 2.2.7α-tribromo-Δ^4-cholestene-3.6-dione (p. 2931 s) by boiling with abs. pyridine; yield, 19% (1937 Butenandt).

Rns. Reduction with $CrCl_2$ in alcohol - acetone at room temperature or by refluxing with Raney nickel in alcohol affords 2-bromo-Δ^1-cholestene-3.6-dione (p. 2921 s). The compound is not attacked by refluxing (10 hrs.) with zinc dust in alcohol (1950 Ellis). Bromination in chloroform, containing small amounts of glacial acetic acid - HBr, regenerates 2.7α-dibromo-$\Delta^{1.4}$-cholestadiene-3.6-dione. Is recovered unchanged after refluxing with acetyl chloride in acetic anhydride (1937 Butenandt).

---

* See also the scheme on p. 2928 s.

2-Bromo-Δ^1-cholestene-3.6-dione $C_{27}H_{41}O_2Br$ (p. 162: *7-bromo-Δ^7-cholestene-3.6-dione*). The structure originally ascribed (1937 Butenandt) has been revised (Editor; 1958 Fieser) in view of the genetic relationship of this compound to 2.7α-dibromo-$\Delta^{1.4}$-cholestadiene-3.6-dione, which see, p. 2927 s. For its identity with the compound described by Sarett (1943) as *2-bromo-Δ^4-cholestene-3.6-dione*, see 1950 Ellis.

White needles (chloroform - alcohol) (1937 Butenandt); crystals (chloroform - methanol, or ether), m. 204–7^0 (1943 Sarett); dimorphous: *form A*, silvery platelets (aqueous acetone), m. 221^0, $[\alpha]_D^{19}$ —61^0 (chloroform); *form B*, fine needles (from the mother liquors of A), m. 160–2^0, $[\alpha]_D^{21}$ —63^0 (chloroform) (1950 Ellis). Absorption max. in chloroform at 256 mμ (ε 8000) (1939 Dannenberg; cf. 1937 Butenandt), in isopropyl alcohol at 257 mμ ($E_{1\,cm.}^{1\,\%} = 156$), identical for both forms (1950 Ellis).

Fmn.* From 2-bromo-$\Delta^{1.4}$-cholestadiene-3.6-dione (p. 2920 s) by reduction with $CrCl_2$ in alcohol - acetone at room temperature in an atmosphere of CO_2 (yield of form A, 30%, and of form B, 50%), or by refluxing with Raney nickel in alcohol (1950 Ellis). The yield from 2.7α-dibromo-$\Delta^{1.4}$-cholestadiene-3.6-dione (p. 2927 s) by reduction with zinc in boiling alcohol (cf. p. 162) is 23% (1937 Butenandt). From 2.2-dibromocholestane-3.6-dione (p. 2926 s) by refluxing with dry pyridine (1943 Sarett) or better with collidine (crude yield, 21%) (1950 Ellis).

Rns.* Regeneration of 2.7α-dibromo-$\Delta^{1.4}$-cholestadiene-3.6-dione (cf. p. 162) occurs on treatment with bromine (2 mols.) in chloroform - glacial acetic acid (containing small amounts of HBr) at room temperature (1937 Butenandt). The compound is not attacked by zinc dust in boiling alcohol, nor by HBr in glacial acetic acid at room temperature (24 hrs.), nor by prolonged heating in collidine (1950 Ellis).

2α - Bromo - Δ^4 - cholestene - 3.6 - dione $C_{27}H_{41}O_2Br$. The structure originally ascribed to this compound, *4-bromo-Δ^4-cholestene-3.6-dione* (1943 Sarett), has been revised in view of its formation from 2α.7α-dibromo-Δ^4-cholestene-3.6-dione, which see, p. 2929 s; a 7α-bromo structure, though not excluded (1958 Fieser), is unlikely in view of the optical rotation which seems too close to that of 7β-bromo-Δ^4-cholestene-3.6-dione (p. 2923 s) (cf. the rotations of the two C-7 epimeric 5α.7-dibromocholestan-3β-ol-6-one acetates, p. 2714 s) (Editor). — The compound originally described by Sarett (1943) as *2-bromo-Δ^4-cholestene-3.6-dione* has been identified as 2-bromo-Δ^1-cholestene-3.6-dione (above) (1950 Ellis).

* See also the scheme on p. 2928 s.

200*

White needles (petroleum ether), m. 169.5°, $[\alpha]_D^{20}$ —38° (chloroform). Absorption max. in 95 % alcohol at 254 mμ (ε 9070). — **Fmn.*** From 2α.7α-dibromo-Δ^4-cholestene-3.6-dione (p. 2929 s) or from 2.2.7α-tribromo-Δ^4-cholestene-3.6-dione (p. 2931 s) by refluxing with iron powder in benzene - alcohol (ca. 35:3 vol.) (yields, 45 % and 40 %, resp.); from 2.2.7.7-tetrabromo-Δ^4-cholestene-3.6-dione (p. 2932 s) similarly (solvent mixture ca. 35:1 vol.; yield, 37 %). — **Rns.*** Refluxing with iron powder in alcohol alone affords cholestane-3.6-dione (p. 2907 s). Refluxing with o-phenylenediamine in alcohol gives the quinoxaline derivative of cholestane-3.4.6-trione (p. 2982 s); direct hydrolysis to the trione has not been successful (1943 Sarett).

2α-Bromocholestane-3.6-dione $C_{27}H_{43}O_2Br$. 2α-Configuration is ascribed by Editor in view of the formation (presence of HBr, implying production of the more stable, equatorial 2-bromo-epimer) and of the molecular rotation difference between 2.2-dibromocholestane-3.6-dione (p. 2927 s) and the present ketone, similar to that between 2.2-dibromo- and 2α-bromocholestan-3-ones (pp. 2494 s and 2480 s, resp.).

Crystals (aqueous acetone), dec. between 140° and 180°, according to sample and rate of heating; $[\alpha]_D^{20}$ —10.4° (chloroform); a weak absorption max. at 252 mμ is ascribed to the presence of an impurity. — **Fmn.*** From 2.2-dibromocholestane-3.6-dione (p. 2927 s) on shaking in chloroform - glacial acetic acid containing some HBr; yield, 20 %. — **Rns.*** Reduction with $CrCl_2$ gives cholestane-3.6-dione (p. 2907 s) in good yield. Bromination, using 1.1 mols. of bromine in chloroform - glacial acetic acid containing sodium acetate, regenerates the dibromodione. Refluxing with alcoholic HCl affords Δ^4-cholestene-3.6-dione (p. 2905 s). Refluxing with collidine for 15 min. yields collidine hydrobromide, a **compound** $C_{35}H_{52}O_2NBr$** [plates (acetic acid), m. > 300°], and an unsaturated diketone, probably Δ^1-cholestene-3.6-dione (p. 2905 s) (1950 Ellis).

5α-Bromocholestane-3.6-dione $C_{27}H_{43}O_2Br$. Needles (aqueous acetic acid), crystals (aqueous acetone), dec. at 80–85°; $[\alpha]_D^{26}$ —140° (chloroform); undergoes slow decomposition, with loss of HBr, at room temperature. — **Fmn.** From 5α-bromo-cholestan-3β-ol-6-one (p. 2707 s) by CrO_3 oxidation in strong acetic acid, first at 0°, then at room temperature; yield, 84 %. — **Rns.** Short boiling with pyridine gives Δ^4-cholestene-3.6-dione (p. 2905 s) (1943 Sarett).

---

* See also the scheme on p. 2928 s.
** This compound might be N-(3.6-diketocholestan-2x-yl)-collidinium bromide which, however, has the molecular formula $C_{35}H_{51}O_2NBr$ (Editor).

7-Bromo-$\Delta^{4.7}$-cholestadiene-3.6-dione $C_{27}H_{39}O_2Br$ (p. 162). See 2-bromo-$\Delta^{1.4}$-cholestadiene-3.6-dione (p. 2920 s).

7β-Bromo-Δ^4-cholestene-3.6-dione $C_{27}H_{41}O_2Br$. Configuration at C-7 based

on that of its precursor, 5α.7β-dibromo-cholestane - 3.6 - dione (p. 2930 s); in view of the conditions of formation, properties, and reactions of the compound, a change of position or configuration of the 7β-bromine atom during the dehydrobromination of the precursor seems unlikely [Editor, in concurrence with Fieser (1958)].

Yellow plates (dil. acetic acid containing sodium acetate), crystals (ethyl acetate · methanol at —10°), m. 130–1°; $[\alpha]_D^{20}$ —41° (chloroform). Absorption max. in 95% alcohol at 259 mμ (ε 7250). Very sensitive towards mineral acids, and towards boiling organic solvents such as benzene, chloroform, acetone, etc. — **Fmn.** From 5α.7β-dibromocholestane-3.6-dione (p. 2930 s) on heating with potassium acetate in aqueous acetic acid up to 95° (15 min.); yield, 44%. — **Rns.** Reduction with zinc dust in boiling alcohol affords cholestane-3.6-dione (p. 2907 s). Bromination in chloroform - glacial acetic acid, in presence of sodium acetate, gives a dibromo-compound, either 2.7β- or 7.7-dibromo-Δ^4-cholestene-3.6-dione (p. 2930 s) (1943 Sarett).

7-Bromo-Δ^7-cholestene-3.6-dione $C_{27}H_{41}O_2Br$ (p. 162). See 2-bromo-Δ^1-cholestene-3.6-dione (p. 2921 s).

7α-Bromocholestane-3.6-dione $C_{27}H_{43}O_2Br$. The 7α-configuration is ascribed

on the basis of that of its precursor, 7α- bromocholestan - 3β - ol - 6 - one (page 2708 s); the possibility of inversion at C-7 during the oxidation is precluded on the basis of molecular rotation values, as the displacement which accompanies the olone → dione oxidation is comparable with that observed in the analogous oxidations of 5α-bromocholestan-3β-ol-6-one (p. 2707 s) and 5α.7β-dibromocholestan-3β-ol-6-one (p. 2714 s) (Editor).

White crystals (aqueous acetone, or benzene - petroleum ether), m. 135°; $[\alpha]_D^{20}$ +76° (chloroform). — **Fmn.** From 7α-bromocholestan-3β-ol-6-one (page 2708 s) on oxidation with CrO_3 in strong acetic acid, first at 0°, then at room temperature; yield, 97% (1943 Sarett).

1937 Butenandt, Schramm, Kudszus, *Ann.* **531** 176, 190 et seq., 206 et seq.
1939 Dannenberg, *Abhandl. Preuss. Akad. Wiss. Math.-naturwiss. Kl.* **1939** No. 21, pp. 15, 37, 55, 62.
1943 Sarett, Chakravorty, Wallis, *J. Org. Ch.* **8** 405, 408, 409, 412 et seq.
1950 Ellis, Petrow, *J. Ch. Soc.* **1950** 2194.
1958 M. Fieser, L. F. Fieser, *private communication.*

3. Dihalogeno-diketosteroids C₁₉ (Both CO Groups in the Ring System)

2.2-Dibromoandrostane-3.17-dione $C_{19}H_{26}O_2Br_2$. Not obtained in a pure state; a crystalline sample (alcohol), m. 148–150⁰ cor. dec.,

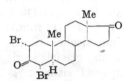

$[\alpha]_D^{25} + 158^0$ (chloroform), contained a tribromo- and probably also 2.4-dibromo-androstane-3.17-dione. — **Fmn.** From androstane-3.17-dione (p. 2892 s) on treatment with 2 mols. of bromine in glacial acetic acid on pouring immediately into water when the rotation has reached its highest value ($+207^0$, after ca. 11 min.; crude yield, 90%); easily isomerizes on bromination for a longer time to give 2α.4α-dibromo-androstane-3.17-dione (below). — **Rns.** Gives 2-bromo-Δ^1-androstene-3.17-dione (p. 2916 s) on refluxing with collidine for 10 min. (1948 Djerassi).

2α.4α-Dibromoandrostane-3.17-dione $C_{19}H_{26}O_2Br_2$. The α-configuration at C-2

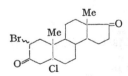

and C-4 is assumed by analogy with 2α.4α-dibromo-cholestan-3-one, p. 2495 s (Editor; see also under "Formation"). — Needles (alcohol, ethyl acetate, and acetone), m. 223–5⁰ dec. (1937/38 Dannenberg); crystals (chloroform · alcohol), m. 209–210⁰ cor. dec. (bath preheated to 200⁰) (1948 Djerassi); $[\alpha]_D$ in chloroform $+41^0$ (1937/38 Dannenberg), $+40^0$ (1948 Djerassi). — **Fmn.** From androstane-3.17-dione (p. 2892 s) on bromination with 2 mols. of bromine and a little HBr in glacial acetic acid at room temperature, followed after decolorization (almost instantaneous) by warming at ca. 50⁰; almost quantitative crude yield (1948 Djerassi; cf. also 1937/38 Dannenberg; 1936b SCHERING A.-G.). From 2α.4α-dibromocholestan-3-one (p. 2495 s) on oxidation with CrO_3 and aqueous H_2SO_4 in acetic acid - ethylene chloride at 18⁰; yield, 16% (1942 SCHERING A.-G.). — **Rns.** Gives $\Delta^{1.4}$-androstadiene-3.17-dione (page 2877 s) on refluxing with dry pyridine for 4 hrs., followed by distillation in vacuo (1936b SCHERING A.-G.), or on refluxing with collidine for 30 min. (1948 Djerassi), whereas on refluxing with collidine for 30 sec. 2α-bromo-Δ^4-androstene-3.17-dione (p. 2916 s) is obtained (1948 Djerassi). For reaction with potassium benzoate in boiling toluene - butanol to give an oily *unsaturated benzoate* $C_{26}H_{30}O_4$ which may be converted into an unsaturated diketone by thermal decomposition, see 1936b SCHERING A.-G.

2(?)-Bromo-5-chloroandrostane-3.17-dione $C_{19}H_{26}O_2ClBr$. The compound described on p. 161 was made from impure 5-chloro-androstane-3.17-dione (1950 Hershberg). — Not isolated in a pure state; unstable. — **Fmn.** From 5-chloro-androstane-3.17-dione (p. 2918 s) on treatment with 1 mol. of bromine in chloroform. — **Rns.** On refluxing with collidine for 1 hr. it gives a mixture of approximately equal amounts of $\Delta^{1.4}$-androstadiene-3.17-dione (p. 2877 s) and $\Delta^{4.6}$-androstadiene-3.17-dione (p. 2878 s) (1950 Hershberg).

References, p. 2926 s

2α.6α(?)-Dibromo-Δ⁴-androstene-3.17-dione $C_{19}H_{24}O_2Br_2$. Crystals, m. 161°

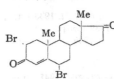

to 163° cor. dec. (Kofler block); $[\alpha]_D^{20}$ +107° (chloro-form). Absorption max. in 95% alcohol at 250 mμ (log ε 4.20). — **Fmn.** From 2α-bromo- and 6α(?)-bromo-Δ⁴-androstene-3.17-diones (pp. 2916 s and 2918 s, resp.) on monobromination in ether-glacial acetic acid (2 g. of ketone in 70 c.c. of ether and 25 c.c. of glacial acetic acid; yields, ca. 60%). — **Rns.** Gives Δ¹·⁴·⁶-androstatriene-3.17-dione (p. 2877 s) on refluxing with dry collidine for 30 min. (1950 Djerassi).

2α.6β(?)-Dibromo-Δ⁴-androstene-3.17-dione $C_{19}H_{24}O_2Br_2$. Colourless needles

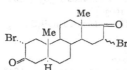

(chloroform-alcohol), m. 172–5° dec., 157–160° cor. (Kofler block); $[\alpha]_D^{20}$ +116° (chloroform). Absorption max. in 95% alcohol at 240 mμ (log ε 4.19) (1950 Djerassi). — **Fmn.** From Δ⁴-androstene-3.17-dione (p. 2880 s) on treatment with 2 mols. of bromine and a little HBr in ether-glacial acetic acid; yield, ca. 80% (1950 Djerassi; cf. 1950 Rosenkranz; see also 1936a SCHERING A.-G. where the compound was formulated as *4.6-dibromo-Δ⁴-androstene-3.17-dione*). — **Rns.** Gives Δ¹·⁴·⁶-androstatriene-3.17-dione (p. 2877 s) on refluxing with dry collidine for 30 min. (1950 Djerassi; cf. 1950 Rosenkranz).

2α.16-Dibromoandrostane-3.17-dione $C_{19}H_{26}O_2Br_2$. Configuration at C-2 assum-

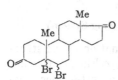

ed by Editor, by analogy with 2α-bromoandrostane-3.17-dione (p. 2916 s). — Crystals (dil. acetone), m. 219–220° dec.; $[\alpha]_D^{20}$ +42° (chloroform). Readily soluble in acetone, ethyl acetate, and chloroform. — **Fmn.** From androstane-3.17-dione (p. 2892 s) on treatment with 2 mols. of bromine and a little HBr-glacial acetic acid in chloroform at 20°; yield, 13%. From 16-bromoandrostane-3.17-dione (p. 2919 s) on treatment with 1 mol. of bromine and a little HBr in glacial acetic acid at 20°; yield, 34%. — **Rns.** Is not attacked by $AgNO_3$ in pyridine at room temperature (1937/38 Dannenberg).

5.6-Dibromo-androstane (or-ætiocholane)-3.17-dione $C_{19}H_{26}O_2Br_2$. Not isolated in a pure state. — **Fmn.** From the dibromide of Δ⁵-androstene-3β.17β-diol (p. 1999 s) on oxidation with CrO_3 in strong acetic acid at room temperature (1937 I. G. FARBENIND.; 1940 Marker; 1945 Hirschmann); similarly from the dibromide of Δ⁵-androsten-3β-ol-17-one (5-dehydroepiandrosterone, p. 2609 s) (1935 Bute-nandt; 1935, 1936, 1937 Ruzicka; 1936a, b CIBA; 1937 SCHERING-KAHLBAUM A.-G.; 1945 Julian; 1944 GLIDDEN Co.). — **Rns.** On boiling with zinc dust in methanol it gives Δ⁵-androstene-3.17-dione (p. 2890 s) (1936 Butenandt; 1937 SCHERING-KAHLBAUM A.-G.; 1937 I. G. FARBENIND.), whereas Δ⁴-an-drostene-3.17-dione (p. 2880 s) is obtained on heating with zinc dust and glacial acetic acid on the water-bath (1935 Ruzicka; 1935 Butenandt; 1935 Wallis; 1936 b CIBA; 1945 Julian; 1944 GLIDDEN Co.; 1945 Hirschmann) or, in better yield, on treatment in acetone with 1 N chromous chloride solution (containing $ZnCl_2$ and hydrochloric acid) under CO_2 (1945 Julian; 1944 GLIDDEN Co.). Gives 6α(?)-bromo-Δ^4-androstene-3.17-dione (p. 2918 s) on heating with potassium acetate and water in benzene at 160° (bath temperature) (1936 Ruzicka) or on refluxing with anhydrous sodium acetate in abs. alcohol for 1 hr. (1937 Ruzicka; 1936a CIBA).

1935 Butenandt, Kudszus, *Z. physiol. Ch.* **237** 75, 85.
     Ruzicka, Wettstein, *Helv.* **18** 986, 993.
     Wallis, Fernholz, *J. Am. Ch. Soc.* **57** 1511, 1512.
1936 Butenandt, Schmidt-Thomé, *Ber.* **69** 882, 886.
     a CIBA* (Ruzicka), *U.S. Pat.* 2085474 (issued 1937); *C. A.* **1937** 5950; *Ch. Ztbl.* **1938** I 374; *Fr. Pat.* 833339 (1938, issued 1938); *C. A.* **1939** 2905; *Ch. Ztbl.* **1939** I 2643.
     b CIBA* (Ruzicka, Wettstein), *U.S. Pat.* 2194235 (issued 1940); *C. A.* **1940** 4746; *Ch. Ztbl.* **1942** I 778.
     Ruzicka, Bosshard, Fischer, Wirz, *Helv.* **19** 1147, 1151.
     a SCHERING A.-G., *Ger. Pat.* 699248 (issued 1940); *C. A.* **1941** 6742; *Ch. Ztbl.* **1941** I 1325.
     b SCHERING A.-G. (Inhoffen), *Ger. Pat.* 722943 (issued 1942); *C. A.* **1943** 5201; *Dan. Pat.* 61375 (1937, issued 1943); *Ch. Ztbl.* **1944** II 49; SCHERING-KAHLBAUM A.-G., *Fr. Pat.* 835524 (1937, issued 1938); *Brit. Pat.* 500353 (1937, issued 1939); *C. A.* **1939** 5872; *Ch.Ztbl.* **1939** I 5010, 5011; SCHERING CORP. (Inhoffen, Butenandt, Schwenk), *U.S. Pat.* 2340388 (1937, issued 1944); *C. A.* **1944** 4386.
1937 I. G. FARBENIND., *Brit. Pat.* 501421 (issued 1939); *C. A.* **1939** 6340, 6341; *Ch. Ztbl.* **1939** II 170; *Fr. Pat.* 847487 (1938, issued 1939); *C. A.* **1941** 5654; *Ch. Ztbl.* **1940** I 2828.
     Ruzicka, Bosshard, *Helv.* **20** 328, 330.
     SCHERING-KAHLBAUM A.-G., *Swiss Pat.* 199448 (issued 1938); *Ch. Ztbl.* **1939** I 3931; *Brit. Pat.* 486992 (issued 1938); *C. A.* **1938** 8707; *Ch. Ztbl.* **1939** I 1603.
1937/38 Dannenberg, *Über einige Umwandlungen des Androstandions und Testosterons*, Thesis, Danzig, pp. 30 et seq.
1940 Marker, *J. Am. Ch. Soc.* **62** 2547.
1942 SCHERING A.-G., *Fr. Pat.* 878993 (issued 1943); *Ch. Ztbl.* **1943** II 250.
1944 GLIDDEN Co. (Julian, Cole, Magnani, Conde), *U.S. Pat.* 2374683 (issued 1945); *C. A.* **1946** 1636; *Ch. Ztbl.* **1946** I 372.
1945 Hirschmann, Hirschmann, *J. Biol. Ch.* **157** 601, 609.
     Julian, Cole, Magnani, Meyer, *J. Am. Ch. Soc.* **67** 1728.
1948 Djerassi, Scholz, *J. Org. Ch.* **13** 697, 703, 704.
1950 Djerassi, Rosenkranz, Romo, Kaufmann, Pataki, *J. Am. Ch. Soc.* **72** 4534, 4538, 4539.
     Hershberg, Rubin, Schwenk, *J. Org. Ch.* **15** 292, 293, 296.
     Rosenkranz, Djerassi, Kaufmann, Pataki, Romo, *Nature* **165** 814.

---

4. Dihalogeno-diketosteroids C_{27} (Both CO Groups in the Ring System)

2.2-Dibromocholestane-3.6-dione $C_{27}H_{42}O_2Br_2$. For confirmation of the structure assigned, see 1950 Ellis. — Needles (chloroform - ethyl acetate), dec. between 175° and 195° (1943 Sarett; 1950 Ellis); $[\alpha]_D^{24}$ +65° (1943 Sarett), $[\alpha]_D^{20}$ +68° (1950 Ellis), both in chloroform. Slowly decomposes at room tempera-

* Société pour l'industrie chimique à Bâle; Gesellschaft für chemische Industrie in Basel.

ture. Insoluble in all the usual organic solvents except chloroform and pyridine (1943 Sarett). — **Fmn.*** From cholestane-3.6-dione (p. 2907 s) by reaction with bromine (2 mols.) in chloroform · glacial acetic acid containing sodium acetate, at room temperature (yield, 59%) (1943 Sarett); under similar conditions, but using 1.1 mols. of bromine, from 2α-bromocholestane-3.6-dione (p. 2922 s; yield, 27%) (1950 Ellis). — **Rns.*** It undergoes partial debromination, giving 2α-bromocholestane-3.6-dione on shaking a suspension in chloroform · glacial acetic acid containing some HBr (1950 Ellis). Dehydrobromination to give 2-bromo-Δ^1-cholestene-3.6-dione (p. 2921 s) occurs on refluxing with pyridine (1943 Sarett) or better with collidine (1950 Ellis). Treatment with silver nitrate in pyridine gives an oily product. No quinoxaline derivative is obtained on treatment with o-phenylenediamine (1943 Sarett).

2.7α-Dibromo-$\Delta^{1.4}$-cholestadiene-3.6-dione $C_{27}H_{38}O_2Br_2$ (p. 162: *4.7-Dibromo-$\Delta^{4.7}$-cholestadiene-3.6-dione*). The revi-

sed structure assigned is based on that ascribed to its partial debromination product, 2-bromo-$\Delta^{1.4}$-cholestadiene-3.6-dione, which see p. 2920 s, and on the revised structure assigned to its precursor, 2.2.7α-tribromo-Δ^4-cholestene-3.6-dione, which see p. 2931 s (Editor; 1958 Fieser); the alternative structure, 2α.7-dibromo-$\Delta^{4.7}$-cholestadiene-3.6-dione, would require an absorption max. of at least 275 mμ (1958 Fieser).

Needles (chloroform · methanol), m. 182°; $[\alpha]_D^{20}$ —18° (chloroform) (1943 Sarett). Absorption max. in chloroform at 259 mμ (ε 11000) (1939 Dannenberg; cf. 1937 Butenandt). — **Fmn.*** The bromination (cf. p. 162) of 2-bromo-$\Delta^{1.4}$-cholestadiene-3.6-dione (p. 2920 s) is carried out with 1.1 mols. of reagent in chloroform, containing traces of HBr · glacial acetic acid, at room temperature (yield, 94%); that of 2-bromo-Δ^1-cholestene-3.6-dione (p. 2921 s) similarly, but in chloroform · glacial acetic acid and using 2 mols. of bromine (yield, 31%) (1937 Butenandt). From 2.7.7-tribromo-$\Delta^{1.4}$-cholestadiene-3.6-dione (p. 2932 s) by treatment with HBr · glacial acetic acid in chloroform at room temperature; quantitative yield (1943 Sarett). The debromination (cf. p. 162) of 2.2.7α-tribromo-Δ^4-cholestene-3.6-dione (p. 2931 s), effected with AgNO$_3$ · pyridine, is carried out at room temperature; yield, 38% (1937 Butenandt).

Rns.* The products of reduction (cf. p. 162) by use of zinc or iron in boiling alcohol are now recognized to be 2-bromo-Δ^1-cholestene-3.6-dione (p. 2921 s) and 2-bromo-$\Delta^{1.4}$-cholestadiene-3.6-dione (p. 2920 s), resp. (Editor; 1958 Fieser). The compound is recovered unchanged after refluxing with acetyl chloride in acetic anhydride (1937 Butenandt).

---

* See also the scheme on p. 2928 s.

Bromo-derivatives of Δ⁴-cholestene-3.6-dione and cholestane-3.6-dione obtained directly, or via intermediate bromo-compounds, by bromination of these diones

Δ⁴-Cholestene-3.6-dione, p. 2905 s

2α.7α-Dibromo-Δ⁴-cholestene-3.6-dione, p. 2929 s

2.2.7.7-Tetrabromo-Δ⁴-cholestene-3.6-dione, p. 2932 s

2.2.7α-Tribromo-Δ⁴-cholestene-3.6-dione, p. 2931 s

2α-Bromo-Δ⁴-cholestene-3.6-dione, p. 2921 s

2.7.7-Tribromo-Δ¹·⁶-cholestadiene-3.6-dione, p. 2932 s

2α-Bromocholestane-3.6-dione, p. 2922 s

Cholestane-3.6-dione, p. 2907 s

2.7α-Dibromo-Δ¹·⁴-cholestadiene-3.6-dione, p. 2927 s

2.2-Dibromocholestane-3.6-dione, p. 2926 s

2-Bromo-Δ¹-cholestene-3.6-dione, p. 2921 s

2-Bromo-Δ¹·⁴-cholestadiene-3.6-dione, p. 2920 s

2 Br₂   Fe in C₆H₆-alc.   AgNO₃ in pyr.   HCl in alc.   4 Br₂ + HBr   1 Br₂ + HBr   HBr · AcOH   Fe in C₆H₆ - alc.   9 Br₂ + HBr   Zn in alc.   Fe or Zn in alc.   3 Br₂ + HBr   Fe in alc.   9 Br₂   HBr in AcOH   CrCl₂   Zn in AcOH   Fe in C₆H₆ - alc.   HBr in AcOH   1 Br₂ + NaOAc   2 Br₂ + NaOAc   AgNO₃ in pyr.   Zn in alc.   2 Br₂ + HBr   pyridine   1 Br₂ + HBr   Fe in alc.   pyridine or collidine   CrCl₂ or Ni

2α.7α-Dibromo-Δ^4-cholestene-3.6-dione $C_{27}H_{40}O_2Br_2$ (p. 162: *4.7-Dibromo-Δ^4-cholestene-3.6-dione*). Originally described as *oxycholestenone dibromide* (1896 Mauthner), this compound was first assigned a *4.5-dibromocholestane-3.6-dione* structure (1906 Windaus), subsequently revised to the 4.7-dibromo-Δ^4-structure (cf. p. 162) by Butenandt (1937) on the basis of analogy with bromination products of Δ^4-cholestenone, thought to be substituted at C-4. The validity of this basis has recently been shown wanting (see bromination of Δ^4-cholestenone, p. 2428 s), so that for the compound now described, and for those genetically derived from it (see reaction scheme, p. 2928 s), revision is also necessary; the 2-bromo-structure assigned is in agreement with the observed absorption spectrum, with which a 4-bromo-structure is incompatible (Editor; 1957 Fieser). In view of the conditions of formation (presence of HBr), the thermodynamically most stable configuration should be ascribed to this compound; for position 2, this is certainly the equatorial 2α-configuration (cf. 1953 Corey); for position 7, there is no a priori evidence, but as the compound which is very probably 7β-bromo-Δ^4-cholestene-3.6-dione (p. 2923 s) is unstable, the greater stability should be ascribed to the 7α-configuration [Editor, in concurrence with Fieser (1958)].

White prisms (80% alcohol) (1904 Windaus), white needles (chloroform - methanol), m. 175°, $[\alpha]_D^{20} + 82°$ (chloroform) (1943 Sarett); monoclinic, a 15.35, b 7.58, c 11.55 Å, β 93.7°, n = 2, d 1.36 (1934 Schulze). Absorption max. in chloroform at 254 mμ (ε 9100) (1939 Dannenberg; cf. 1937 Butenandt). The alcoholic solution gives on addition of one drop of KOH or NaOH a deep red coloration, dissipated by acidification (1904 Windaus).

Fmn.* The yield obtained by bromination (cf. p. 162) of Δ^4-cholestene-3.6-dione (p. 2905 s), in ether - glacial acetic acid, is 29%, those from cholestane-3.6-dione (p. 2907 s) and coprostane-3.6-dione (p. 2910 s) in chloroform - glacial acetic acid (containing some HBr) are 75% and 66%, resp., all these reactions carried out at room temperature (1937 Butenandt). Together with 2α-bromo-Δ^4-cholestene-3.6-dione (p. 2921 s), from 2.2.7.7-tetrabromo-Δ^4-cholestene-3.6-dione (p. 2932 s) on refluxing (2 hrs.) with iron powder in benzene - alcohol (40:1.2 vol.); yield, 23% (1943 Sarett).

Rns.* Reduction by boiling (6 hrs.) with iron powder in benzene - alcohol (70:5.2 vol.) affords 2α-bromo-Δ^4-cholestene-3.6-dione (p. 2921 s). Reaction occurs with bromine (1 mol.) in acetic acid in presence of sodium acetate, but no crystalline product has been isolated (1943 Sarett). Bromination (cf. p. 162) to give 2.2.7α-tribromo-Δ^4-cholestene-3.6-dione is carried out in chloroform containing small amounts of HBr - glacial acetic acid, at room temperature. Treatment with $AgNO_3$ and pyridine at room temperature gives oily products containing bromine, together with unchanged material (1937 Butenandt). Attempts to isolate a sparingly soluble phenylhydrazone were unsuccessful (1896 Mauthner).

---

* See also the scheme on p. 2928 s.

2.7β (or 7.7)-Dibromo-Δ^4-cholestene-3.6-dione $C_{27}H_{40}O_2Br_2$. Of the two formu-

læ postulated, the 2.7-dibromo-structure (I) is favoured in view of the ready formation of a diquinoxaline derivative, which does not however exclude the 7.7-dibromo-structure (II) (1943 Sarett); but only the latter can account for the strong dextro shift in optical rotation which accompanies its formation from the 7β-monobromo-precursor (see below) (Editor).

Yellow needles (methanol or petroleum ether), m. 119°; $[\alpha]_D^{16} + 118°$ (chloroform). Absorption max. in 95 % alcohol at 260 mμ (ε 10180). Some decomposition occurs on warming in alcohol or other polar solvents. — **Fmn.** From 7β-bromo-Δ^4-cholestene-3.6-dione (p. 2923 s) by reaction with bromine in chloroform - glacial acetic acid, in presence of sodium acetate, at room temperature; yield, 62 %. — **Rns.** Attempted further bromination by prolonged treatment under the conditions of formation gives, besides 50 % of unchanged material, products with increased bromine content from which no crystalline substance has been isolated. Refluxing with o-phenylenediamine in abs. alcohol affords a **diquinoxaline compound** $C_{39}H_{46}N_4$ [formulated as derivative of Δ^4-cholestene-2.3.6.7-tetrone; red crystals (ethyl acetate - alcohol), m. 194°; gives a m. p. depression with that obtained from 2α.7α-dibromo-Δ^4-cholestene-3.6-dione] (1943 Sarett).

4.7-Dibromo-$\Delta^{4.7}$-cholestadiene-3.6-dione $C_{27}H_{38}O_2Br_2$ (p. 162). — See 2.7α-dibromo-$\Delta^{1.4}$-cholestadiene-3.6-dione, p. 2927 s.

4.7-Dibromo-Δ^4-cholestene-3.6-dione $C_{27}H_{40}O_2Br_2$ (p. 162). — See 2α.7α-dibromo-Δ^4-cholestene-3.6-dione, p. 2929 s.

5α.7β-Dibromocholestane-3.6-dione $C_{27}H_{42}O_2Br_2$. The 5β-configuration origi-

nally assigned by Sarett (1943) has been revised and the 7β-configuration introduced on the basis of those now ascribed to the precursor, 5α.7β-dibromocholestan-3β-ol-6-one (p. 2714 s) (Editor). — White crystals (aqueous acetone), dec. at 100°, $[\alpha]_D^{20} -41°$ (chloroform). — **Fmn.** From 5α.7β-dibromocholestan-3β-ol-6-one by CrO$_3$ oxidation in strong acetic acid at room temperature; yield, 61 %. — **Rns.** Heating with potassium acetate in aqueous acetic acid up to 95° (15 min.) affords 7β-bromo-Δ^4-cholestene-3.6-dione (p. 2923 s). Boiling in benzene effects rapid decomposition with evolution of HBr and formation of a yellow oil (1943 Sarett).

1896 Mauthner, Suida, *Monatsh.* **17** 579, 588, 589.
1904 Windaus, *Ber.* **37** 2027, 2031, 2032.
1906 Windaus, *Ber.* **39** 2249, 2254, 2255.
1934 Schulze, *Z. physik. Ch.* A **171** 436, 443, 444.
1937 Butenandt, Schramm, Kudszus, *Ann.* **531** 176, 179, 182, 185 et seq., 200 et seq.
1939 Dannenberg, *Abhandl. Preuss. Akad. Wiss., Math.-naturwiss. Kl.* **1939** No. 21, pp. 32, 37, 60, 62.
1943 Sarett, Chakravorty, Wallis, *J. Org. Ch.* **8** 405, 408, 409, 412 et seq.
1950 Ellis, Petrow, *J. Ch. Soc.* **1950** 2194, 2196, 2198.
1953 Corey, *Experientia* **9** 329.
1957 M. Fieser, L. F. Fieser, *private communication.*
1958 M. Fieser, L. F. Fieser, *private communication.*

5. Tri- and Tetrahalogeno-diketosteroids (Both CO Groups in the Ring System)

2.2.7α - Tribromo - Δ^4 - cholestene - 3.6 - dione $C_{27}H_{39}O_2Br_3$ (p. 162: *4.7.7 - Tri-*

bromo - Δ^4 - cholestene - 3.6 - dione). Origi- nally formulated as *2.4.5-tribromochol- estane-3.6-dione* (1936 Fujii), then as 4.7.7-tribromo-Δ^4-cholestene-3.6-dione (1937 Butenandt); a further revision of structure was necessitated for this compound in view of that now assigned to its precursor, 2α.7α-dibromo-Δ^4-cholestene-3.6-dione (p. 2929 s). Of the two revised structures recently postulated, 2.2.7- or 2.7.7-tribromo-Δ^4-cholestene-3.6-dione (1957 Fieser), the former is now accepted in view of the 2-bromo-$\Delta^{1.4}$-cholestadiene-3.6-dione (p. 2920 s) structure assigned to the product which results by dehydro- bromination, followed by partial debromination (Editor; 1958 Fieser).

Needles (acetone) (1906 Windaus), pale yellow crystals (benzene - methanol), m. 195° (1943 Sarett), m. 197° cor. (1936 Fujii); $[\alpha]_D^{20}$ + 16° (chloroform) (1943 Sarett; cf. 1950 Ellis). Absorption max. in chloroform at 257 mμ (ε 10 200) (1939 Dannenberg; cf. 1937 Butenandt). Less soluble in alcohol than is 2α.7α-dibromo-Δ^4-cholestene-3.6-dione (1906 Windaus).

Fmn.* The yield obtained from Δ^4-cholestene-3.6-dione (p. 2905 s) by bromination in glacial acetic acid (cf. p. 162), containing some HBr and effected at 30–35°, is 65% (1937 Butenandt). The bromination of chole- stane-3.6-dione (cf. p. 162) is effected with excess of bromine in chloroform- glacial acetic acid with long standing at room temperature (1943 Sarett). From 2α.7α-dibromo-Δ^4-cholestene-3.6-dione (p. 2929 s) by treatment with bromine (1 mol.) in chloroform containing small amounts of HBr - glacial acetic acid; yield, 86% (1937 Butenandt). From 2.2.7.7-tetrabromo-Δ^4-chol- estene-3.6-dione (p. 2932 s) on prolonged standing at room temperature in chloroform containing some HBr - glacial acetic acid, even in the presence of free bromine or iodine (1943 Sarett). The starting material (m. ca. 250°) utilized by Fujii (1936) (cf. p. 162) was not Δ^5-cholestenone oxide (which see,

---

* See also the scheme on p. 2928 s.

p. 2755 s), but appears to have been cholestan-5-ol-3.6-dione (p. 2953 s) (Editor), which in fact does afford 2.2.7α-tribromo-Δ^4-cholestene-3.6-dione by reaction with bromine in chloroform - glacial acetic acid at 28° (1950 Ellis).

Rns.* Reduction to cholestane-3.6-dione (p. 2907 s) has been effected with zinc in glacial acetic acid (1906 Windaus; 1936 Fujii); the similar reduction by which coprostane-3.6-dione (p. 2910 s) has also been isolated (cf. p. 162) is effected in benzene - alcohol under reflux (1 hr.) or at room temperature (2 days) (1937 Butenandt). Using iron powder in benzene - alcohol under reflux (6 hrs.) reduction to 2α-bromo-Δ^4-cholestene-3.6-dione (p. 2921 s) occurs (1943 Sarett). Dehydrobromination with AgNO₃ and pyridine (cf. p. 162) to give 2.7α-dibromo-$\Delta^{1.4}$-cholestadiene-3.6-dione (p. 2927 s) occurs on long standing at room temperature; on boiling with abs. pyridine alone, 2-bromo-$\Delta^{1.4}$-cholestadiene-3.6-dione (p. 2920 s) results. The *diquinoxaline compound* (cf. p. 162 and p. 2986 s) is obtained by reaction with o-phenylene-diamine at 150° (1937 Butenandt).

2.7.7-Tribromo-$\Delta^{1.4}$-cholestadiene-3.6-dione C₂₇H₃₇O₂Br₃.

The structure origi-nally assigned, *2.4.7 - tribromo - $\Delta^{4.7}$ - cholestadiene-3.6-dione* (1943 Sarett), has been revised to accord with that now ascribed to its reduction product, 2 - bromo - $\Delta^{1.4}$ - cholestadiene - 3.6 - dione (p. 2920 s; see also reaction scheme, p. 2928 s) (Editor; 1958 Fieser).

Yellowish plates (chloroform - methanol), m. 164°; $[\alpha]_D^{20}$ —38° (chloroform). Absorption max. in 95 % alcohol at 265 mμ (ε 14230). — **Fmn.*** From 2.2.7.7-tetrabromo-Δ^4-cholestene-3.6-dione (below) on standing with AgNO₃ in pyri-dine at room temperature; yield, 45 %. — **Rns.*** Refluxing with iron powder in benzene - alcohol affords 2-bromo-$\Delta^{1.4}$-cholestadiene-3.6-dione (p. 2920 s). Debromination to give 2.7α-dibromo-$\Delta^{1.4}$-cholestadiene-3.6-dione (p. 2927 s) occurs on standing at room temperature in chloroform containing HBr - glacial acetic acid (1943 Sarett).

2.2.7.7-Tetrabromo-Δ^4-cholestene-3.6-dione C₂₇H₃₈O₂Br₄.

The structure orig-inally assigned, *2.4.7.7 - tetrabromo - Δ^4-cholestene-3.6-dione* (1943 Sarett), has been revised to accord with those now ascribed to its partial debromi-nation products, 2.2.7α-tribromo- and 2α.7α- dibromo-Δ^4- cholestene-3.6- dione (pp. 2931 s and 2929 s, resp.) (Editor; 1957 Fieser).

White needles (chloroform- or benzene - methanol), m. 190°; $[\alpha]_D^{20}$ +22° (chloroform). Absorption max. in 95 % alcohol at 254 mμ (ε 8850). — **Fmn.*** From cholestane - 3.6-dione (p. 2907 s) by bromination (9 mols. of bromine) in chloroform - glacial acetic acid at room temperature (18 hrs.); yield, 59 %.

* See also the scheme on p. 2928 s.

This tetrabromo-compound could not be obtained by bromination of either Δ^4-cholestene-3.6-dione or its $2\alpha.7\alpha$-dibromo-derivative. — **Rns.*** Partial debromination to the $2.2.7\alpha$-tribromo-3.6-dione (p. 2931 s) occurs on prolonged standing in chloroform, in presence of some HBr - glacial acetic acid, even if a large excess of bromine and some iodine are added to the solution. Refluxing for ca. 2 hrs. with iron powder in benzene - alcohol (35 : 1 vol.) yields $2\alpha.7\alpha$-dibromo-Δ^4-cholestene-3.6-dione (p. 2929 s) and 2α-bromo-Δ^4-cholestene-3.6-dione (p. 2921 s). Refluxing for 5 hrs. with iron powder in alcohol alone gives cholestane-3.6-dione (p. 2907 s). By treatment with $AgNO_3$ in pyridine at room temperature $2.7.7$-tribromo-$\Delta^{1.4}$-cholestadiene-3.6-dione (p. 2932 s) is obtained (1943 Sarett).

1906 Windaus, *Ber.* **39** 2249, 2255.
1936 Fujii, Matsukawa, *J. Pharm. Soc. Japan* **56** 150 (German abstract); *C. A.* **1937** 1033; *Ch. Ztbl.* **1937** I 2616.
1937 Butenandt, Schramm, Kudszus, *Ann.* **531** 176, 182, 187, 189, 201 et seq.
1939 Dannenberg. *Abhandl. Preuss. Akad. Wiss., Math.-naturwiss. Kl.* **1939** No. 21, pp. 32, 60.
1943 Sarett, Chakravorty, Wallis, *J. Org. Ch.* **8** 405, 411, 412.
1950 Ellis, Petrow, *J. Ch. Soc.* **1950** 2194, 2197.
1957 M. Fieser, L. F. Fieser, *private communication.*
1958 M. Fieser, L. F. Fieser, *private communication.*

---

* See also the scheme on p. 2928 s.

III. Hydroxy-diketosteroids with Both CO Groups in the Ring System

1. Compounds Containing OH in Side Chain Only

$\Delta^{17(20)}$-**Pregnen-21-ol-3.11-dione** $C_{21}H_{30}O_3$. Crystals (ether), m. 128–128.5° cor. (1948 Sarett); a sample (not analytically pure) forming crystals (ether - pentane), m. ca. 150° cor., $[\alpha]_D +56°$ (acetone), was once obtained, probably a crystalline modification (1946, 1948 Sarett; 1946a MERCK & Co., INC.). — **Fmn.** From the 21-hemisuccinate of $\Delta^{17(20)}$-pregnene-3α.21-diol-11-one (p. 2723 s) on oxidation with CrO_3 in strong acetic acid at 12–17°, followed by heating with aqueous NaOH · K_2CO_3 and then with methanolic KOH; yield, 58% (1946, 1948 Sarett; 1946a MERCK & Co., INC.). For another mode of formation, described after the closing date for this volume, see 1948 Sarett. — **Rns.** On treatment with acetic anhydride and pyridine at room temperature, followed by treatment of the resulting amorphous acetate with OsO_4 and dry pyridine in abs. ether and then by refluxing with sodium disulphite in aqueous alcohol, it gives pregnane-17α.20β.21-triol-3.11-dione (p. 2960 s) (1946, 1948, 1949 Sarett; 1946b MERCK & Co., INC.).

1946 a MERCK & Co., INC. (Sarett), *U.S. Pat.* 2492 193 (issued 1949); *C. A.* **1950** 3043; *Ch. Ztbl.* **1952** 1211.

b MERCK & Co., INC. (Sarett), *U.S. Pat.* 2492 194 (issued 1949); *C. A.* **1950** 3044; *Ch. Ztbl.* **1952** 1211.

Sarett, *J. Biol. Ch.* **162** 601, 623, 624.

1948 Sarett, *J. Am. Ch. Soc.* **70** 1690, 1692, 1694.

1949 Sarett, *J. Am. Ch. Soc.* **71** 1169, 1171.

Allopregnan-20β-ol-3.16-dione $C_{21}H_{32}O_3$. For the structure, see 1939 Marker; 1949 Hirschmann.

Acetate $C_{23}H_{34}O_4$. Needles (aqueous methanol), m. 191–2° (1938 Odell; 1939 Marker), 191–193.5° cor. (1949 Hirschmann); $[\alpha]_D^{27} -79°$ (95% alcohol) (1949 Hirschmann). In the Zimmermann test with m-dinitrobenzene and KOH it gives an intense purple colour which changes to brown within ca. 8 min. (1949 Hirschmann). — **Fmn.** From the 20-acetate of allopregnane-3β.16α.20β-triol (page 2112 s) on oxidation with CrO_3 in strong acetic acid at room temperature; yield, 80% (1939 Marker; 1938 Odell; cf. also 1938 Marker). — **Rns.** Gives $\Delta^{17(20)}$-allopregnene-3.16-dione (p. 2900 s) on attempted hydrolysis by refluxing with KOH in aqueous alcohol for $^1/_2$ hr. (a small amount of a crystalline substance, m. 205–7° is formed at the same time), with $NaHCO_3$ in aqueous methanol for 2 hrs., or with conc. HCl in alcohol for 4 hrs. (1939 Marker).

On refluxing with amalgamated zinc and conc. HCl in glacial acetic acid
it gives allopregnane (p. 1403 s) and, as the major product, an oily unsaturated
hydrocarbon (cf. the reaction with conc. HCl alone, above) which is convertible into allopregnane by catalytic hydrogenation (1939 Marker; cf. also
1938 Marker). On heating of the disemicarbazone with sodium ethoxide in
alcohol for 12 hrs. in a sealed tube at 168–170⁰ it gives a **compound** $C_{21}H_{36}O$
[blade-shaped crystals (benzene · methanol), m. 82–83⁰*; *benzoate* $C_{28}H_{40}O_2$,
crystals, m. 141⁰] (1938 Odell). On heating with hydrazine hydrate in 95 %
alcohol for 4 hrs. at 80⁰, an amorphous, nitrogen-containing mixture is obtained (1939 Marker; cf. 1938 Odell).

Acetate disemicarbazone $C_{25}H_{40}O_4N_6$. Amorphous solid; undergoes very
slight decomposition at 220–223⁰, does not melt below 305⁰ (1938 Odell;
cf. also 1938 Marker).

1938 Marker, Kamm, Wittle, Oakwood, Lawson, *J. Am. Ch. Soc.* **60** 1067, 1069.
    Odell, Marrian, *J. Biol. Ch.* **125** 333, 338, 339.
1939 Marker, Wittle, *J. Am. Ch. Soc.* **61** 855, 858, 859.
1949 Hirschmann, Hirschmann, Daus, *J. Biol. Ch.* **178** 751, 753, 764.

17β-(α-Hydroxybenzhydryl)-ætiocholane-3.12-dione, (3.12-Diketo-ætiochol-

anyl)-diphenyl-carbinol $C_{32}H_{38}O_3$. A compound [needles (ether · pentane), m. 200–201⁰,
sublimes in a high vacuum at 215⁰ (bath temperature)] which probably possesses this structure has been obtained in very small yield on
attempted preparation of ætiocholane-3.12.17-
trione from ætiodesoxycholic acid methyl ester by refluxing with phenylmagnesium bromide in ether · benzene, followed by refluxing with aqueous
methanolic KOH, acetylation with acetic anhydride and pyridine at 100⁰,
refluxing with glacial acetic acid for 1 hr., ozonization in chloroform, refluxing
with aqueous methanolic KOH, and oxidation with CrO_3 in glacial acetic acid
(1940 Reichstein).

22.22-Diphenyl-bisnorcholan-22-ol-3.12-dione, (3.12-Diketopregnan-20α-yl)-

diphenyl-carbinol $C_{34}H_{42}O_3$. Plates (acetone-
methanol), cubes (ether · petroleum ether),
m. 240–242⁰ cor. (Kofler block); $[\alpha]_D^{16} +8⁰$
(chloroform). Readily soluble in chloro-
form, sparingly in alcohol, acetone, and
ether. — **Fmn.** From (3α.12α-dihydroxy-
pregnan-20α-yl)-diphenyl-carbinol (p. 2134 s) on oxidation with CrO_3 in glacial
acetic acid at 20⁰ (yield, 85 %); similarly from (3α.12β-dihydroxypregnan-
20α-yl)-diphenyl-carbinol (p. 2135 s; yield, ca. 32 %). — **Rns.** Is dehydrated
by refluxing with glacial acetic acid for 1 hr. to give 1-methyl-2.2-diphenyl-
1-(3.12-diketo-ætiocholanyl)-ethylene (p. 2915 s) (1945 Sorkin).

---

* Allopregnan-20β-ol melts at 140⁰ (see p. 1488 s).

22.22 - Diphenyl -20- iso - bisnorcholan -22- ol-3.12-dione, (3.12-Diketopregnan-20β-yl)-diphenyl-carbinol $C_{34}H_{42}O_3$.

Rectangular plates (chloroform - ether), m. 260° to 263° cor. (Kofler block); $[\alpha]_D^{16}$ +51° (chloroform). Readily soluble in chloroform, sparingly in alcohol, acetone, and ether. — **Fmn.** From (3α.12α-dihydroxypregnan-20β-yl)-diphenyl-carbinol (p. 2135 s) on oxidation with CrO_3 in glacial acetic acid at 20° (yield, 70%); similarly from (3α.12β-dihydroxypregnan-20β-yl)-diphenyl-carbinol (p. 2136 s; yield, 70%). — **Rns.** Gives the same ethylene compound on dehydration with glacial acetic acid as the above 20-epimer (1945 Sorkin).

1940 Reichstein, v. Arx, *Helv.* **23** 747, 752.
1945 Sorkin, Reichstein, *Helv.* **28** 875, 885, 888.

Δ^4- Pregnene - 20α.21 - diol - 3.11 - dione $C_{21}H_{30}O_4$.

Crystals (acetone - ether), m. 194–5° cor.; $[\alpha]_D^{20}$ +176.5° (acetone). — **Fmn.** The diacetate is obtained from that of pregnane-20α.21-diol-3.11-dione (p. 2937 s) on treatment with 1 mol. of bromine in glacial acetic acid at room temperature, followed by refluxing of the resulting 4-bromo-derivative with pyridine for 10 hrs.; it is hydrolysed by K_2CO_3 + $KHCO_3$ in aqueous methanol at room temperature (over-all yield, ca. 11%). — **Rns.** On oxidation of the 21-acetate (below) with CrO_3 in strong acetic acid at 18° it gives the acetate of Δ^4-pregnen-21-ol-3.11.20-trione (dehydrocorticosterone, p. 2972 s) (1946 Sarett).

21-Acetate $C_{23}H_{32}O_5$. Crystals (methanol), m. 221–7° cor. — **Fmn.** From the above dioldione on treatment with 1.3 mols. of acetic anhydride and dry pyridine in abs. dioxane at room temperature for 60 hrs.; yield, 21% (1946 Sarett). — **Rns.** See above.

Diacetate $C_{25}H_{34}O_6$. Crystals (dil. alcohol), m. 153.5–154.5° cor.; $[\alpha]_D^{20}$ +133° (acetone). Absorption max. in alcohol at 237.5 mμ (E % 403) (1946 Sarett). — **Fmn.** From the above dioldione on warming with acetic anhydride and pyridine for 15 min. on the steam-bath (1946 Sarett). See also under formation of the free dioldione.

Δ^4 - Pregnene - 20β.21-diol - 3.11 - dione, *Reichstein's Substance T*, $C_{21}H_{30}O_4$.

Crystals (acetone), m. 223.5–224.5° cor.; $[\alpha]_D^{20}$ +176° (acetone); slightly hygroscopic (1946 Sarett). — **Occ.** In adrenal cortex (1939 Reichstein). — **Fmn.** The diacetate is obtained from that of pregnane-20β.21-diol-

3.11-dione (p. 2938 s) on treatment with 1 mol. of bromine in glacial acetic acid at room temperature, followed by refluxing of the resulting 4-bromo-derivative with pyridine for 10 hrs. (crude over-all yield, 63%); it is hydrolysed by K_2CO_3 + $KHCO_3$ in aqueous methanol at room temperature (1946 Sarett). — **Rns.** Gives 3.11-diketo-Δ^4-ætiocholenic acid on oxidation with CrO_3 in glacial acetic acid at room temperature (1939 Reichstein). On oxidation of the 21-acetate (below) with CrO_3 in strong acetic acid at 18°, the acetate of Δ^4-pregnen-21-ol-3.11.20-trione (dehydrocorticosterone, page 2972 s) is obtained (1946 Sarett).

21-Acetate $C_{23}H_{32}O_5$. Crystals (methanol), m. 161.5–162.5° cor. — **Fmn.** From the above dioldione on treatment with 1.2 mols. of acetic anhydride and dry pyridine in abs. dioxane at room temperature for 60 hrs.; yield, 36% (1946 Sarett). — **Rns.** See above.

Diacetate $C_{25}H_{34}O_6$. Crystals (methanol), m. 212–3° cor. (1939 Reichstein); crystals (acetone), m. 207–8° cor.; $[\alpha]_D^{20}$ +170° (acetone); absorption max. in alcohol at 237.5 mμ (E% 357) (1946 Sarett). — **Fmn.** See under formation of the free dioldione (above).

Pregnane-20α.21-diol-3.11-dione $C_{21}H_{32}O_4$. Crystals (acetone - ether), m. 182° to 183° cor.; $[\alpha]_D^{20}$ +68.5° (acetone). — **Fmn.** Along with the 20β-epimer (below) and pregnane-17α.20β-diol-3.11-dione (p. 2958 s) from a mixture of $\Delta^{17(20)}$- and Δ^{20}-pregnene-3.11-diones (p. 2899 s) on treatment with OsO_4 and pyridine in abs. ether at room temperature (1 hr.), then at 0° (36 hrs.), followed by refluxing with Na_2SO_3 in aqueous alcohol for $3^1/_2$ hrs.; the resulting mixture of the three dioldiones is treated with succinic anhydride in pyridine on the steam-bath (15 min.), the reaction product dissolved in chloroform and extracted with 10% aqueous K_2CO_3; the extract containing the 21-hemisuccinate of the two epimeric 20.21-diol-3.11-diones is then saponified by K_2CO_3 + NaOH in aqueous methanol at room temperature, acetylated with acetic anhydride in pyridine on the steam-bath (30 min.), and the two epimeric diacetates separated chromatographically (over-all yields, 11% and 9%, resp.); the diacetates are hydrolysed by K_2CO_3 + $KHCO_3$ in water on warming for a few min. (1946 Sarett).

Rns. On bromination of the diacetate with 1 mol. of bromine in glacial acetic acid at room temperature it gives the *4-bromo-derivative* (amorphous) which on refluxing with pyridine for 10 hrs. yields the diacetate of Δ^4-pregnene-20α.21-diol-3.11-dione (p. 2936 s) (1946 Sarett).

Diacetate $C_{25}H_{36}O_6$. Crystals (ethyl acetate - petroleum ether), m. 181° cor.; $[\alpha]_D$ +45° (acetone) (1946 Sarett). — **Fmn.** and **Rns.** See above.

201*

Pregnane-20β.21-diol-3.11-dione $C_{21}H_{32}O_4$. Crystals, m. 167.5–168.0° cor.; $[\alpha]_D^{20}$ +61.5° (acetone). — **Fmn.** See under formation of the 20α-epimer (above). — **Rns.** On bromination of the diacetate with 1 mol. of bromine in glacial acetic acid at room temperature it gives the *4-bromo-derivative* (amorphous) which on refluxing with pyridine for 10 hrs. yields the diacetate of Δ^4-pregnene-20β.21-diol-3.11-dione (p. 2936 s) (1946 Sarett).

Diacetate $C_{25}H_{36}O_6$. Crystals (dil. acetone), m. 174.5–175.0° cor.; $[\alpha]_D^{20}$ +74° (acetone) (1946 Sarett). — **Fmn.** and **Rns.** See above.

1939 Reichstein, v. Euw, *Helv.* **22** 1222.
1946 Sarett, *J. Am. Ch. Soc.* **68** 2478, 2481, 2482.

2. Monohydroxy-diketosteroids with OH and Both CO in the Ring System

a. Compounds C_{17} and C_{18}

7 - Methoxy - 4.3' - diketo - 1.2 - cyclopentano - 1.2.3.4 - tetrahydrophenanthrene

$C_{18}H_{16}O_3$. Colourless prisms (methanol), m. 133^0 (1945 Birch). — **Fmn.** From 3-(6-methoxy-2-naphthyl)-cyclopentanone-2-acetic acid on heating with P_2O_5 and H_3PO_4 (d 1.75) at 120–125^0 for 2–3 min. with vigorous stirring, cooling to 60^0, and decomposing with water; yield, 25 % (1938 Koebner; 1945 Birch). — **Rns.** Hydrogenation in the presence of a platinum - palladium - charcoal catalyst in alcohol at room temperature gives the methyl ether of 7-hydroxy-3'-keto-1.2-cyclopentano-1.2.3.4-tetrahydrophenanthrene (dl-18-nor-isoequilenin, p. 2518 s) (1938 Koebner).

Mono - (2.4 - dinitrophenylhydrazone) $C_{24}H_{20}O_6N_4$. Dark red crystalline powder (ethyl acetate), m. 243^0 dec. (1938 Koebner).

cis *- 1 - Methyl-7-methoxy-4'.5'-diketo-1.2-cyclopentano-1.2.3.9.10.11-hexahydrophenanthrene $C_{19}H_{20}O_3$ (I). Originally assumed to possess structure (I) or that of *2-methyl-7-methoxy-3'.4'-diketo-1.2 - cyclopentano - 1.2.3.9.10.11 - hexahydrophenanthrene* (II) by Dane (1938, 1939); for the true structure (I), see 1942 E. W. J. Butz; 1949 L. W. Butz. — Pale flesh-coloured needles (methanol), m. 170^0 (red melt). Soluble in alkali; gives a red-brown enol reaction (1938 Dane).

Shows androgenic activity (1 mg.) in the rat test (1938 Dane).

Fmn. From 3-methyl-Δ^3-cyclopentene-1.2-dione and excess 1-vinyl-6-methoxy-3.4-dihydronaphthalene on heating in dioxane for 50 hrs. at 110–115^0; crude yield, 30 % (1938, 1939 Dane). — **Rns.** Hydrogenation in methanol in the presence of palladium-charcoal gives cis *-1-methyl-7-methoxy - 4'- hydroxy - 5'- keto - 1.2 - cyclopentano-1.2.3.4.9.10.11.12-octahydrophenanthrene (p. 2737 s) (1939 Dane). Hydrogenation in alcohol in the presence of a Rupe catalyst, until 3 mols. of hydrogen have been taken up, yields cis *-1-methyl-7-methoxy-4'.5'-dihydroxy-1.2-cyclopentano-1.2.3.4.9.10.11.12-octahydrophenanthrene (not described) which by boiling with HBr - glacial acetic acid may be converted into an isomer of œstrone (no properties given) (1938a, b SCHERING CORP.).

$\Delta^{1.3.5(10)}$-Oestratrien-3-ol-7.17-dione, 7-Ketoœstrone $C_{18}H_{20}O_3$. For proof that the configuration at the asymmetric centre at C-8 is the same as for œstrone, see also 1940 Pearlman. — Crystals (alcohol), m. 212–212.5^0 cor. dec.; $[\alpha]_D^{22} + 167^0$ (dioxane); absorption max. (solvent not indicated) 283 mμ (ε 2130) (1939 Pearlman).

* "cis" refers to the fusion of rings C and D.

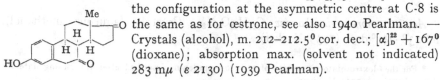

The œstrogenic potency is only $1/300$ that of œstrone (1939 Pearlman).

Fmn. From the diacetate of its 7-enolic form (7-hydroxy-6-dehydroœstrone, p. 2739 s) on refluxing with aqueous methanolic HCl in an atmosphere of nitrogen. From equilin-7.8-glycol (p. 2767 s) on distillation at 205–210° and 0.003 mm. pressure (1939 Pearlman).

Rns. Undergoes decomposition in alkaline solution in the presence of air. Refluxing with acetic anhydride and fused sodium acetate gives the diacetate of its 7-enolic form (7-hydroxy-6-dehydroœstrone, p. 2739 s). The disemicarbazone (below), heated with sodium ethoxide in abs. alcohol at 185°, yields $\Delta^{1.3.5(10)}$-œstratrien-3-ol (p. 1496 s) (1939 Pearlman).

Dioxime $C_{18}H_{22}O_3N_2$. Plates (alcohol), m. 252–3° cor. dec. (1939 Pearlman). **Rns.** Reduction with sodium and alcohol yields 7.17-diamino-$\Delta^{1.3.5(10)}$-œstratrien-3-ol (p. 1915 s) (1943 Lettré).

Disemicarbazone $C_{20}H_{26}O_3N_6$. Amorphous granula (alcohol), m. $> 295°$ (1939 Pearlman). — **Rns.** See above.

Methyl ether $C_{19}H_{22}O_3$ (not analysed and not described in detail). — **Fmn.** From the 3-methyl ether of equilin-7.8-glycol (p. 2767 s) on distillation in a high vacuum. — **Rns.** Its *dioxime* $C_{19}H_{24}O_3N_2$ (not analysed and not described in detail) gives the methyl ether of 7.17-diamino-$\Delta^{1.3.5(10)}$-œstratrien-3-ol (p. 1915 s) on reduction with sodium and alcohol (1943 Lettré).

trans-Equilenan-3-ol-11.17-dione, 11-Keto-d-equilenin * $C_{18}H_{18}O_3$. Crystals methanol), m. 271–3 °cor. (red melt); $[\alpha]_D^{24} 0°$ (dioxane); infra-red spectrum bands at 1658, 1735, and 3460 cm.$^{-1}$. —**Fmn.** From the lævorotatory acetate (below) on refluxing for 1 hr. with 7.7% methanolic KOH (1954 McNiven).

Acetate $C_{20}H_{18}O_4$. Crystals (ether), m. 195–7° (1939 Marker); acicular crystals (methanol), m. 198.5–199.5° cor.; $[\alpha]_D^{24} -31°$ (chloroform); ultra-violet absorption max. in methanol at 246 (4.40) and 314 (3.86) mμ (log ε); infra-red spectrum bands at 1511, 1598, 1666, and 1742 cm.$^{-1}$ (1954 McNiven). — **Fmn.** From the acetate of d-equilenin (p. 2527 s) on oxidation with CrO_3 in strong acetic acid at 20°; small yield ** (1939 Marker; 1954 McNiven). — **Rns.** For hydrolysis, see above.

Acetate monosemicarbazone $C_{21}H_{21}O_4N_3$. Crystals (aqueous methanol), m. 238–241° dec. (1939 Marker).

---

* For the (dextrorotatory) acetate of 11-keto-l-equilenin, described after the closing date for this volume, see 1954 McNiven.

** The main product of this oxidation is the 3-acetate of a 14-hydroxyequilenin (1954 McNiven).

Δ$^{1.3.5(10)}$**-Oestratrien-3-ol-16.17-dione, 16-Ketoœstrone** $C_{18}H_{20}O_3$ (I, R = H).

Pale yellow micro-crystals, dec. 234–8⁰. — **Fmn.** From its 16-oxime (below) on heating on the steam-bath with anhydrous Na_2SO_3 in glacial acetic acid for 15 min., then in 80% acetic acid for 45 min., addition of aqueous $NaHSO_3$, extraction with ether, heating of the aqueous phase with hydrochloric acid, and salting out (on the steam-bath) with sodium chloride; yield, ca. 60% (1948 Huffman). — **Rns.** On heating with zinc dust in 50% acetic acid on the steam-bath it is reduced to 16-keto-"α"-œstradiol (16-ketoœstra-3.17β-diol, p. 2738 s) (1948 Huffman; for 17β-configuration, see 1950 Gallagher; 1950 Heusser).

16-Ketoœstrone 16-oxime, 16-Oximinoœstrone, 16-Isonitrosoœstrone $C_{18}H_{21}O_3N$. Colourless needles (aqueous methanol), m. 214–5⁰ dec. Its solution in ether is colourless, its alkaline solution is yellow (1944 Huffman). — **Fmn.** From the benzoate of œstrone (p. 2543 s) on treatment at room temperature with isoamyl nitrite and potassium tert.-butoxide in tert.-butyl alcohol under nitrogen, followed by saponification with aqueous KOH at room temperature; yield, 80%, once 96% (1944 Huffman; see also 1942 Huffman). — **Rns.** On reduction by refluxing with zinc dust in 50% acetic acid it gives 16-keto-"α"-œstradiol (16-ketoœstra-3.17β-diol, p. 2738 s) (1942, 1944, 1947 c, 1948, 1949 Huffman; for 17β-configuration, see 1950 Gallagher; 1950 Heusser).

16-Ketoœstrone dioxime $C_{18}H_{22}O_3N_2$. Crystals (aqueous propanol), m. 218⁰ to 219⁰ dec. (1948 Huffman); occasionally a crystalline modification, m. 230⁰ to 231⁰ dec., was obtained (1942, 1948 Huffman). On addition of a very weak aqueous solution of cupric acetate to the solution in a small amount of alcohol, a yellowish green complex is formed which may be extracted with chloroform; no coloured complexes are obtained with nickelous or co-baltous ions (1942, 1948 Huffman). — **Fmn.** From the above diketone or its 16-oxime on refluxing with hydroxylamine hydrochloride and sodium acetate in aqueous alcohol (1948 Huffman).

16-Ketoœstrone di-methoxime $C_{20}H_{26}O_3N_2$. Faintly yellow needles (abs. methanol), m. 204–5⁰ dec. — **Fmn.** From the above diketone on treatment with O-methylhydroxylamine hydrochloride and sodium acetate in aqueous alcohol, first at room temperature (1 hr.) and then on refluxing (1 hr.) (1948 Huffman).

16-Ketoœstrone methyl ether $C_{19}H_{22}O_3$ (I, above; R = CH_3). Yellow crystals, m. 177–8⁰ with previous shrinking. Gives an intense violet coloration with conc. H_2SO_4; no colour is produced with alcoholic $FeCl_3$ (1942, 1947 a Huffman). — **Fmn.** From its 16-oxime (below) on refluxing with zinc dust in 50% acetic acid, followed by refluxing of the resulting 16-ketoœstra-3.17β-diol 3-methyl ether (p. 2738 s) with cupric acetate in methanol for 1 hr. (over-all yield, 12%) (1942, 1947 a Huffman); in better yield (ca. 55%) from the same oxime on heating on the steam-bath with anhydrous Na_2SO_3 in glacial acetic acid for 15 min., then in 80% acetic acid for 45 min., addition of aqueous $NaHSO_3$, extraction with ether, and heating of the aqueous phase

with hydrochloric acid on the steam-bath for 25–30 min. (1947a Huffman). —
Rns. Gives 16-ketoœstra-3.17β-diol 3-methyl ether (see p. 2738 s) on heating
on the steam-bath with zinc dust and 50% acetic acid for 40 min. or with
TiCl₃ in glacial acetic acid for 30 min. (1947a Huffman).

16-Ketoœstrone methyl ether 16-oxime, 16-Oximinoœstrone methyl ether, 16-Iso-
nitrosoœstrone methyl ether C₁₉H₂₃O₃N. Needles and plates (light petroleum
b. 60–80⁰), m. 161–2⁰ dec. (1938 Litvan; cf. 1947a Huffman); almost colourless
needles (aqueous methanol), m. 180–183⁰ dec. (1947a Huffman); faintly yellow leaflets with 1 H₂O (dil. alcohol or ethyl acetate), m. 198⁰ (1946 Butenandt). Readily soluble in dil. aqueous NaOH to give a faintly yellow
solution (1938 Litvan). — **Fmn.** From œstrone methyl ether (p. 2555 s) on
treatment with isoamyl nitrite and potassium tert.-butoxide in tert.-butyl
alcohol under nitrogen; 89% yield (1938 Litvan), 75% yield (1947a Huffman). — **Rns.** On refluxing with zinc dust in 50% acetic acid it gives the
3-methyl ether of 16-ketoœstra-3.17β-diol (p. 2738 s) (1946 Butenandt; 1947b,
c, 1949 Huffman; cf. also 1942, 1947a Huffman; for 17β-configuration, see
1950 Gallagher; 1950 Heusser). On treatment with 1.2 mols. of PCl₅ in
acetyl chloride at room temperature (2 hrs.), followed by refluxing of the
resulting product with 30% alcoholic KOH for ca. 14 days, it gives the methyl
ether of "œstric acid" [d-trans-marrianolic acid; see formula (I) on p. 2546 s]
(1938 Litvan).

16-Ketoœstrone methyl ether dioxime C₁₉H₂₄O₃N₂. White crystals (butanol),
m. 230⁰ dec. (1942, 1947a Huffman). Almost insoluble in alcohol and ethyl
acetate. With alcoholic cupric acetate it gives a yellowish green complex
which may be extracted with chloroform (1947a Huffman). — **Fmn.** From
16-ketoœstrone methyl ether or its 16-oxime (above) on oximation (1942,
1947a Huffman).

16-Ketoœstrone methyl ether di-methoxime C₂₁H₂₈O₃N₂. Faintly yellow needles
(chloroform - alcohol), dec. 231–3⁰ after previous softening. — **Fmn.** From
the diketone methyl ether (above) on treatment with O-methylhydroxylamine
hydrochloride and sodium acetate in aqueous alcohol, first at room temperature (2 hrs.) and then on refluxing (1 hr.) (1947a Huffman).

16-Ketoœstrone methyl ether p-nitrophenylhydrazone C₂₅H₂₇O₄N₃. Canary-
yellow crystals (chloroform - alcohol) softening at 250⁰, dec. 257–9⁰ (1947a
Huffman).

16-Ketoœstrone benzyl ether C₂₅H₂₆O₃ (I, p. 2941 s; R = CH₂·C₆H₅). —
16-Oxime, 16-Oximinoœstrone benzyl ether, 16-Isonitrosoœstrone benzyl ether
C₂₅H₂₇O₃N. Very pale yellow needles (alcohol), dec. 193.5–195.5⁰. **Fmn.**
Similar to that of the corresponding methyl ether (above). **Rns.** On refluxing
with zinc dust in 50% acetic acid it gives the 3-benzyl ether of 16-keto-
œstra-3.17β-diol (see p. 2739 s) (1948 Huffman).

1938 Dane, Schmitt, *Ann.* **536** 196, 197, 199, 200.
Koebner, Robinson, *J. Ch. Soc.* **1938** 1994, 1996.
Litvan, Robinson, *J. Ch. Soc.* **1938** 1997, 2000.

a SCHERING CORP. (Dane), *U.S. Pat.* 2230233 (issued 1941); *C. A.* **1941** 3037; *Ch. Ztbl.*
  1941 II 2229.
b SCHERING CORP. (Dane), *U.S. Pat.* 2249748 (issued 1941); *C. A.* **1941** 7122; *Ch. Ztbl.*
  1942 I 2037.
1939 Dane, Schmitt, *Ann.* **537** 246, 248.
  Marker, Rohrmann, *J. Am. Ch. Soc.* **61** 3314, 3316.
  Pearlman, Wintersteiner, *J. Biol. Ch.* **130** 35–42.
1940 Pearlman, Wintersteiner, *J. Biol. Ch.* **132** 605, 607.
1942 E. W. J. Butz, L. W. Butz, *J. Org. Ch.* **7** 199, 216.
  Huffman, *J. Am. Ch. Soc.* **64** 2235, 2236.
1943 Lettré, *Z. physiol. Ch.* **278** 206.
1944 Huffman, Darby, *J. Am. Ch. Soc.* **66** 150, 151.
1945 Birch, Jaeger, Robinson, *J. Ch. Soc.* **1945** 582, 585.
1946 Butenandt, Schäffler, *Z. Naturforsch.* **1** 82, 85.
1947 a Huffman, *J. Biol. Ch.* **167** 273 et seq.
  b Huffman, *J. Biol. Ch.* **169** 167, 169.
  c Huffman, Lott, *J. Am. Ch. Soc.* **69** 1835.
1948 Huffman, Lott, *J. Biol. Ch.* **172** 325.
1949 L. W. Butz, Rytina, *Organic Reactions* V 136, 139.
  Huffman, Lott, *J. Am. Ch. Soc.* **71** 719, 724, 725.
1950 Gallagher, Kritchevsky, *J. Am. Ch. Soc.* **72** 882, 884 footnote 17.
  Heusser, Feurer, Eichenberger, Prelog, *Helv.* **33** 2243, 2247.
  Stork, Singh, *Nature* **165** 816.
1954 McNiven, *J. Am. Ch. Soc.* **76** 1725, 1727.

b. Monohydroxy-diketosteroids C_{19} with OH and Both CO Groups in the Ring System

Summary of Compounds with Androstane (or Aetiocholane) Skeleton

For values other than those chosen for this summary, see the compounds themselves.

Position of			Config. at C-5	M.p. °C.	$[\alpha]_D$ * (°)	Acetate		Page
CO	OH	double bond(s)				m.p. °C.	$[\alpha]_D$ * (°)	
3.6	17β	Δ^4	—	213	−53 (ac.)	201	−47 (ac.)	2944 s
3.16	17β	Δ^4	—	$\begin{cases}189 \\ 152\text{–}8\end{cases}$	−52 (chl.)	195	−56 (alc.), −29 (chl.)	2945 s
3.17	2	satd.	α	—	—	$\begin{cases}241 \\ 187\end{cases}$	+80 (chl.) +143 (alc.)	2945 s 2945 s
3.17	6α	Δ^4	—	230	+182 (chl.)	177	+164 (chl.), +154 (ac.)	2945 s
3.17	6β	Δ^4	—	194	+109 (chl.)	201	+111 (chl.)	2946 s
3.17	6β	satd.	α	—	—	—	—	2947 s
3.17	11β	Δ^4	—	199	+220 (chl.), +203 (alc.)	—	—	2947 s
3.17	11β	satd.	α	225	+100 (chl.)	—	—	2947 s
3.17	12α	satd.	β	180	+158 (chl.)	198	+179 (chl.)	2947 s
6.17	3β	satd.	α	178	—	204	+39 (chl.)	2948 s
6.17	5	Δ^3	α	240	—	—	—	2948 s
6.17	5	satd.	α	228	—	187	—	2948 s
11.17	3β	satd.	α	168	—	164	+96 (dio.)	2949 s
11.17	3α	satd.	β	188	+138 (alc. or ac.)	163	+145 (alc.), +157 (ac.)	2949 s
12.17	3β	$\Delta^{9(11)}$	α	—	—	190	+201 (ac.)	2950 s
12.17	3α	satd.	β	—	—	172	—	2950 s
16.17	3	$\Delta^{3.5}$	—	For derivatives, see p. 2950 s				
16.17	3β	Δ^5	—	For derivatives, see p. 2950 s				

* ac = acetone; alc. = alcohol; chl. = chloroform; dio. = dioxane.

Δ^4-**Androsten-17β-ol-3.6-dione, 6-Ketotestosterone** $C_{19}H_{26}O_3$ (p. 162: Δ^4-*Androstene - 3.6 - dion - 17 - ol*). Crystals (acetone - ether), m. 212–3°; $[\alpha]_D^{20}$ −53° (acetone); absorption max. in 95 % alcohol at 250 mμ (ε 11 200); infra-red absorption max. in chloroform at 1684 cm.$^{-1}$ and a free-hydroxyl band (1954 Amendolla). — **Fmn.** As propionate (no properties given), by oxidation of the propionate of Δ^4-androsten-17β-ol (p. 1510 s) with CrO_3 in glacial acetic acid (1939 Ciba). For formation of its acetate from the 17-acetate of Δ^5-androstene-3β.17β-diol (p. 1999 s) by shaking for 10 hrs. with 4 mols. of CrO_3 in glacial acetic acid and a trace of water (yield, 20 %) and for hydrolysis of the acetate by heating with anhydrous 0.1 % methanolic HCl for 2 hrs. on the water-bath (yield, 40 %) (p. 162), see also 1936 Schering A.-G.

Acetate $C_{21}H_{28}O_4$ (p. 162). Absorption max. in chloroform at 252 mμ (ε 11400) (1939 Dannenberg), in alcohol - NaOH at 262 mμ (1951 Devis).

Δ⁴-Androsten-17β-ol-3.16-dione, 16-Ketotestosterone $C_{19}H_{26}O_3$. For β-con-figuration at C-17, see 1947, 1949 Huffman; 1950 Heus-ser. — Needles (benzene-light petroleum), m. 189⁰ (1951 Antaki); crystals (acetone - ether), m.152–8⁰cor.; $[\alpha]_D^{22}$ —52⁰ (chloroform); absorption max. in 95 % alcohol at 240.5 mμ (ε 16600); infra-red spectrum (1954 Meyer). — **Fmn.** From 16-oximino-Δ⁴-androstene-3.17-dione (p. 2981 s) on refluxing with zinc dust in 79 % acetic acid (1942 Stodola; cf. also 1951 Antaki); yield, ca. 50 % (1954 Meyer). The acetate is obtained from that of 16-benzylidenetestosterone (p. 2698 s) on treatment with 1.1 mols. of OsO_4 in carbon tetrachloride at room temperature in the dark for 3 days, followed by warming with zinc dust in strong acetic acid at 45–50⁰ and oxidation of the resulting 16-(α-hydroxybenzyl)-Δ⁴-androstene-16.17β-diol-3-one 17-ace-tate (not isolated in a pure state) with periodic acid in aqueous alcohol (small yield) or, in better yield, with lead tetraacetate in benzene (1941a Stodola).

Acetate $C_{21}H_{28}O_4$. Crystals (acetone - petroleum ether), m. 194–5⁰ (1941a Stodola); crystals (ether - pentane), m. 195–9⁰ cor. (1954 Meyer); $[\alpha]_{4561}^{25}$ —56⁰ (95 % alcohol) (1941a Stodola), $[\alpha]_D^{25}$ —29⁰ (chloroform); infra-red spectrum (1954 Meyer). Produces a deep purple colour with conc. H_2SO_4 (1941a Stodola). — **Fmn.** From the above hydroxydiketone with acetic anhydride in pyridine at room temperature (1941a Stodola; 1954 Meyer). See also under formation of the preceding compound.

Androstan-2-ol-3.17-dione $C_{19}H_{28}O_3$.

2-Acetoxyandrostane-3.17-dione $C_{21}H_{30}O_4$. The two C-2 epimeric acetoxyandrostane-3.17-diones are said to be formed from 2α-bromoandrostane-3.17-dione (p. 2916 s) by heating with potassium acetate in glacial acetic acid for 5 hrs. in a sealed tube at 200⁰ (bath temperature) (1937/38 Dannenberg; cf. also 1937, 1947 SCHERING A.-G.) [cf., however, the similar reaction of 2α-bromocholestan-3-one (see p. 2481 s) giving a mixture of 2α- and 4α-acetoxycholestan-3-ones]; the one (yield, ca. 17 %) forms crystals (dil. acetone), m. 241⁰, $[\alpha]_D^{20}$ +80⁰ (chloroform); the other forms needles (dil. acetone), m. 187⁰, $[\alpha]_D^{20}$ +143⁰ (alcohol) (1937/38 Dannenberg).

2-Benzoyloxy-androstane-3.17-dione $C_{26}H_{32}O_4$. Needles (ether); begins to melt at 188–9⁰ after sintering, the melt becoming clear at 220⁰. — **Fmn.** From 2α-bromoandrostane-3.17-dione (p. 2916 s) on refluxing with potassium benzoate in butanol-toluene for 1 hr.; yield, 36 % (1937, 1947 SCHERING A.-G.).

Δ⁴-Androsten-6α-ol-3.17-dione $C_{19}H_{26}O_3$. Colourless needles (acetone - petro-leum ether), m. 229–230⁰; $[\alpha]_D^{27}$ +182⁰ (chloroform); absorption max. in alcohol at 239.5 mμ (ε 16170) (1952 Balant). — The hydroxydiketone and its acetate show no androgenic effect in the comb-growing test (1952 Balant); according to Ehrenstein (1941) the acetate is one-fifth as active as androsterone in the capon comb test. Shows some œstrogenic activity (1942 Clarke; 1942 Selye), but no progestational activity (1943 Selye).

Fmn. The acetate is obtained from the 6-acetate of androstane-5α.6β-diol-3.17-dione (p. 2961 s) by passage of dry HCl at 0⁰ through a solution in dry chloroform containing 3.3% (by vol.) of alcohol; yield, ca. 55% (1952 Balant; cf. also 1941 Ehrenstein). The acetate may also be obtained from the 6β-epimer (below) by passage of dry HCl at —10⁰ through a solution in dry HCl containing 0.7% (by vol.) of alcohol; yield, ca. 67% (1952 Balant). The acetate is hydrolysed by treatment with 1.1 mols. of KOH in methanol at room temperature (1952 Balant).

Rns. Gives androstane-3.6.17-trione (p. 2979 s) on treatment of the acetate with glacial acetic acid in the presence of a small amount of dil. H_2SO_4 (1952 Balant).

Acetate $C_{21}H_{28}O_4$. Crystals (ether), m. 174–6⁰ (1941 Ehrenstein); colourless needles (very little acetone + much petroleum ether, or aqueous methanol), m. 176.5–177.5⁰ (1952 Balant); $[\alpha]_D^{27}$ +153.5⁰ (acetone) (1941 Ehrenstein); $[\alpha]_D^{32}$ +164⁰ (chloroform) (1952 Balant). Absorption max. in alcohol at 235 mμ (ε 16020 to 15520) (1941 Ehrenstein; 1952 Balant). — For physiological effects, see above under the free hydroxydiketone. — **Fmn.** From the above hydroxydiketone on treatment with acetic anhydride in dry pyridine at room temperature (1952 Balant). See also above. — **Rns.** See above.

Δ⁴-Androsten-6β-ol-3.17-dione * $C_{19}H_{26}O_3$.

Colourless plates (acetone · petroleum ether), m. 193.5–194.5⁰; $[\alpha]_D^{28}$ +109⁰ (chloroform); absorption max. in alcohol at 235.5 mμ (ε 13550). Gives a yellow solution in conc. H_2SO_4 slowly developing a green fluorescence. — The hydroxydiketone and its acetate show no androgenic effect in the comb-growing test (1952 Balant).

Fmn. The acetate is obtained from the 6-acetate of androstane-5α.6β-diol-3.17-dione (p. 2961 s) by passage of dry HCl through a solution in alcohol-free chloroform at —10⁰ or in purified carbon tetrachloride at —15⁰ (yields, ca. 50%); it is hydrolysed by 1.1 mols. of KOH in abs. alcohol at room temperature. — **Rns.** On heating the acetate with 2.2 mols. of KOH in dil. methanol on the water-bath for 1 hr. it gives androstane-3.6.17-trione (page 2979 s). The acetate gives the epimeric acetate (above) on passage of dry HCl at —10⁰ through its solution in chloroform containing 0.7% (by vol.) of alcohol (1952 Balant).

Acetate $C_{21}H_{28}O_4$. Colourless plates (acetone · petroleum ether), m. 201⁰ to 201.5⁰; $[\alpha]_D^{27}$ +111⁰ (chloroform); absorption max. in alcohol at 235 mμ (ε 12480). The crystals turn deep yellow upon exposure to direct sunlight; the colour change is not reversible in the dark, but on recrystallization

---

* Δ⁴-Androsten-6β-ol-3.17-dione, then believed to be Δ⁴-androsten-3β-ol-6.17-dione, has also been described (after the closing date for this volume) by Davis (1949a); for the true structure, see 1954 Amendolla.

the colourless material is reobtained. Gives a yellow solution in conc. H_2SO_4 slowly developing a green fluorescence. — **Fmn.** From the above hydroxydiketone on treatment with acetic anhydride in pyridine at room temperature (1952 Balant). See also above. — **Rns.** See above.

Androstan-6β-ol-3.17-dione $C_{19}H_{28}O_3$. White crystals (acetone); soluble in alcohol and acetone, less soluble in benzene. — **Fmn.**

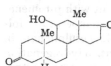

From the diacetate of androstane-3β.6β-diol-17-one (p. 2747 s) on treatment with 1 mol. of KOH in methanol at room temperature for 2 days, followed by oxidation of the resulting 6-acetate with CrO_3 in strong acetic acid at room temperature and then by refluxing with alcoholic NaOH. — **Rns.** Gives Δ^4-androstene-3.17-dione (p. 2880 s) on slow distillation with $ZnCl_2$ in a high vacuum at 190° (1941 PARKE, DAVIS & Co.).

Δ^4-Androsten-11β-ol-3.17-dione $C_{19}H_{26}O_3$ (p. 162). For configuration at C-11, see 1947 v. Euw. — Needles (ether - pentane), m.* 197° to 199.5° (1953 Jeanloz); crystals (acetone - petroleum ether), m. 199–200° (1953 Bernstein; see also 1953 Hayano); $[\alpha]_D^{23}$ +220° (chloroform) (1953 Herr); $[\alpha]_D^{30}$ +208° (chloroform) (1953 Jeanloz); $[\alpha]_D^{28}$ +203° (abs. alcohol) (1953 Bernstein); absorption max. in alcohol at 242 mμ (log ε 4.19) (1953 Herr; see also 1953 Bernstein; 1953 Salamon).

Rns. Refluxing with a mixture of glacial acetic acid and conc. HCl (4:1) gives $\Delta^{4.9(11)}$-androstadiene-3.17-dione (p. 2879 s) and a crystalline isomer (1942 Reichstein; cf. also the analogous dehydration of corticosterone 21-acetate, p. 2854 s).

Androstan-11β-ol-3.17-dione $C_{19}H_{28}O_3$. For configuration at C-11, see 1947 v. Euw. — Crystals (benzene - pentane), sinters at 212°, m. (unsharp) 225°; $[\alpha]_D^{18}$ +100° (chloroform). — **Fmn.** From androstane - 3β.11β - diol - 17 - one (p. 2748 s) on boiling with aluminium phenoxide and acetone in benzene for 22 hrs. (1941 Reich).

Dioxime $C_{19}H_{30}O_3N_2$. Needles (alcohol-water), m. 254° to 256° cor. dec. (1941 Reich).

5-iso-Androstan-12α-ol-3.17-dione, Aetiocholan-12α-ol-3.17-dione $C_{19}H_{28}O_3$. For configuration at C-12, see 1946 Sorkin. — Prisms (ether), m. 175.5–177.5° cor. (1945 Reich); needles (acetone - diisopropyl ether), m. 179–180° cor.; $[\alpha]_D^{23}$ +158° (chloroform) (1949 Meystre). — **Fmn.** In poor yield, together with ætiocholane-3.12.17-trione (p. 2980 s), on oxidation of ætiocholane-3α.12α-diol-17-one (p. 2749 s)

---

* The m. p. 225° given by Reichstein (1942) is probably an error; it is not likely that the epimeric Δ^4-androsten-11α-ol-3.17-dione, m. 226–7°, was used, since this was first prepared in 1952 (see 1952 UPJOHN Co.; see also 1954 Eppstein), and is not dehydrated on refluxing with glacial acetic acid - HCl (see 1954 Bernstein).

to the trione with insufficient amounts of CrO_3 in glacial acetic acid. As acetate, by partial saponification of ætiocholane-3α.12α-diol-17-one diacetate with 1 % methanolic HCl at room temperature, followed by oxidation of the resulting amorphous 12-acetate with CrO_3 in glacial acetic acid; the acetate is saponified with methanolic KOH at room temperature (1945 Reich).

Acetate $C_{21}H_{30}O_4$. Crystals (acetone - hexane), m. 197–198.5° cor.; sublimes at 150° and 0.02 mm. pressure (1945 Reich); $[\alpha]_D^{24} + 179°$ (chloroform) (1949 Meystre). — **Fmn.** See formation of ætiocholan-12α-ol-3.17-dione (above).

Anthraquinone-β-carboxylate $C_{34}H_{34}O_6$. Yellow leaflets (acetone - petroleum ether), m. 250.5–252.5° cor.; sublimes at 240° and 0.005 mm. pressure. — **Fmn.** By heating ætiocholan-12α-ol-3.17-dione with anthraquinone-β-carboxylic acid chloride and pyridine in toluene on the water-bath. — **Rns.** Thermal decomposition gives only unidentifiable products (1945 Reich).

Androstan-3β-ol-6.17-dione $C_{19}H_{28}O_3$*.

Acetate $C_{21}H_{30}O_4$. Needles (petroleum ether - alcohol), m. 197–8° (1942 Butenandt); needles (hexane - ethyl acetate), m. 203–5° cor.; $[\alpha]_D^{15} + 39°$ (chloroform) (1944 Ruzicka). — **Fmn.** From androstane-3β.6β.17β-triol 3-acetate (p. 2164 s) on oxidation with CrO_3 in acetic acid at room temperature (1944 Ruzicka). From i-androstene-6.17-dione (p. 2898 s) on refluxing with glacial acetic acid and 5N H_2SO_4; yield, 52 % (1942 Butenandt).

Δ^3-Androsten-5-ol-6.17-dione $C_{19}H_{26}O_3$.

Crystals (ethyl acetate - hexane), m. 238–240° cor. — **Fmn.** From androstane-3β.5-diol-6.17-dione (p. 2962 s) on sublimation with fuller's earth at 150° in a high vac. (yield, 46 %) or on treatment with p-toluene-sulphonyl chloride and pyridine at room temperature, followed by boiling of the resulting 3-p-toluenesulphonate in pyridine for 36 hrs. (yield, 70 %). — **Rns.** Hydrogenation using Pd - $CaCO_3$ catalyst in alcohol at room temperature, gives androstan-5-ol-6.17-dione (below) (1940 Ruzicka).

Androstan-5-ol-6.17-dione $C_{19}H_{28}O_3$.

Crystals (ethyl acetate), m. 225–8° cor. — **Fmn.** From Δ^3-androsten-5-ol-6.17-dione (above) on hydrogenation, using Pd - $CaCO_3$ catalyst in alcohol at room temperature. — **Rns.** Sublimation of the acetate at 200° and 13 mm. pressure gives Δ^4-androstene-6.17-dione (page 2898 s), whereas esterification with p-toluenesulphonyl chloride in pyridine at room temperature, followed by refluxing, yields an isomeric diketone, possibly 5-methyl-19-nor-Δ^9-ætio-cholene(or -androstene)-6.17-dione (see p. 2899 s) (1940 Ruzicka).

* After 1946, the closing date for this volume, androstan-3β-ol-6.17-dione (m. 176–8°) was prepared by MacPhillamy (1952).

Acetate $C_{21}H_{30}O_4$. Crystals (ethyl acetate), m. 187⁰ cor. — **Fmn.** By boiling androstan-5-ol-6.17-dione with acetic anhydride (1940 Ruzicka). — **Rns.** See reactions of androstan-5-ol-6.17-dione, above.

Androstan-3β-ol-11.17-dione $C_{19}H_{28}O_3$ (p. 163: *Androstane-11.17-dion-3-ol*).

For configuration at C-3, see 1940 Shoppee. — **Fmn.** Was obtained in poor yield in impure form (m. 156.5⁰ to 158⁰ cor.) containing some androstane-3.11.17-trione (p. 2980 s) (see 1941 Reich) by oxidation of andro-stane-3β.11β-diol-17-one (p. 2748 s) with CrO_3 in glacial acetic acid at room temperature (1936 Reichstein; cf. 1937 Steiger).

Page 163 line 4: For "From the acetate" *read* "as acetate from the 3.21-di-acetate".

Acetate $C_{21}H_{30}O_4$ (p. 163). Rotation in dioxane: $[\alpha]_D^{20} +96^0$, $[\alpha]_{5461}^{20} +118^0$ (1942 v. Euw).

5-iso-Androstan-3α-ol-11.17-dione, Aetiocholan-3α-ol-11.17-dione $C_{19}H_{28}O_3$.

Crystals (acetone - pentane), m. 187–8⁰ cor. (1946a Sarett; 1946b MERCK & Co., INC.); needles (ether-ligroin), m. 188–9⁰ cor. (1946, 1948 Lieberman); need-les (ether), m. 183–5⁰ cor.→188–9⁰ cor. (1953 Finkel-stein); $[\alpha]_D^{26} +138^0$ (acetone) (1953 Herzog), $[\alpha]_D^{25} +138^0$ (alcohol) (1953 Finkelstein).

Occ. In the urine of normal and diseased men and women (1946, 1948 Lieberman).

Fmn. The acetate is obtained from that of $\Delta^{17(20)}$-pregnen-3α-ol-11-one (mixture with the acetates of Δ^{16}-pregnen-3α-ol-11-one and Δ^{20}-pregnen-3α-ol-11-one; see p. 2654 s) on ozonolysis in ethyl acetate - methanol, followed by oxidation with $KMnO_4$ in aqueous acetone at room temperature, and reacet-ylation by heating with acetic anhydride in pyridine on the steam-bath (yield from the above-mentioned mixture, 22%); it is hydrolysed by refluxing with dil. methanolic KOH for 15 min. (1946a Sarett; 1946b MERCK & Co., INC.). The acetate is also obtained from 12α-bromo-3α-acetoxy-11-ketocholanic acid methyl ester on oxidation with CrO_3 in acetic acid at 85–90⁰, followed by debromination with zinc dust in acetic acid; small yield (1946 Clark).

Rns. Oxidation with CrO_3 in acetic acid at room temperature affords ætiocholane-3.11.17-trione (p. 2980 s) (1946b Sarett; 1946, 1948 Lieberman). On treatment with potassium acetylide in dry dioxane - abs. ether at room temperature it gives 17-iso-T^{20}-pregnyne-3α.17β-diol-11-one (p. 2752 s) (1946a Sarett; 1946a MERCK & Co., INC.; cf. 1950 Gallagher for configuration at C-17).

Acetate $C_{21}H_{30}O_4$. Needles (ethyl acetate - pentane), m. 162.5–163⁰ cor. (1946a Sarett); crystals (ether - ligroin), m. 163–4⁰ cor. (1946, 1948 Lieber-man); $[\alpha]_D^{23} +157^0$ (acetone) (1948 Lieberman), $[\alpha]_D^{18} +145^0$ (alcohol) (1946, 1948 Lieberman). — **Fmn.** From the above hydroxydiketone with acetic anhydride in pyridine (1946a Sarett; 1948 Lieberman). See also under for-mation of the above hydroxydiketone.

References, pp. 2951 s, 2952 s

$\Delta^{9(11)}$-Androsten-3β-ol-12.17-dione $C_{19}H_{26}O_3$.

Acetate $C_{21}H_{28}O_4$. Needles (ether - petroleum ether), m. 188–191° cor. (Kofler block); $[\alpha]_D^{18}$ +201° (acetone). Absorption max. in alcohol at 238 mμ (log ε ca. 4.2). — **Fmn.** From the acetate of $\Delta^{9(11)}$-androsten-3β-ol-17-one (p. 2626 s) on oxidation with CrO$_3$ in glacial acetic acid at 27–28°; yield, 13% or, allowing for recuperated starting material, 35%. — **Rns.** On attempted hydrolysis with K$_2$CO$_3$ in aqueous methanol it gave an **acid** $C_{19}H_{28}O_4$ (I) [not analysed; crystals (ether), m. 210–214° cor. (Kofler block); absorption max. in alcohol below 220 mμ (log ε ca. 3.8)] (1947 Reich).

5-iso-Androstan-3α-ol-12.17-dione, Aetiocholan-3α-ol-12.17-dione $C_{19}H_{28}O_3$.

Acetate $C_{21}H_{30}O_4$. Needles (ether - petroleum ether or acetone - hexane), m. 172–172.5° cor. (Kofler block); sublimes at 180° and 0.01 mm. pressure. — **Fmn.** From the 3-acetate of ætiocholane-3α.12α-diol-17-one (page 2749 s) on oxidation with CrO$_3$ in glacial acetic acid at room temperature. — **Rns.** On hydrogenation in glacial acetic acid at room temperature in the presence of platinum oxide it reverts to an extent of 28% to the starting acetate and gives ca. 50% yield of an ætiocholane-3α.12.17-triol 3-acetate, m. 148–9° cor. (p. 2165 s); a few crystals m. ca. 118° were also obtained (1945 Reich).

$\Delta^{3.5}$-Androstadien-3-ol-16.17-dione ethyl ether $C_{21}H_{28}O_3$ (II, R = O).

16-Oxime, 16-Oximino-$\Delta^{3.5}$-androstadien-3-ol-17-one ethyl ether, 16-Isonitroso-$\Delta^{3.5}$-androstadien-3-ol-17-one ethyl ether $C_{21}H_{29}O_3N$ (II, R = N·OH). Not obtained in a pure state. — **Fmn.** From Δ^4-androstene-3.17-dione 3-enol ethyl ether ($\Delta^{3.5}$-androstadien-3-ol-17-one ethyl ether, p. 2606 s) on treatment with sodium ethoxide and butyl nitrite in alcohol at room temperature in the dark. — **Rns.** Gives 16-oximino-Δ^4-androstene-3.17-dione (p. 2981 s) on treating with aqueous alcoholic NaOH at room temperature, followed by warming with 50% acetic acid (1942 Stodola).

Δ^5-Androsten-3β-ol-16.17-dione, 16-Keto-5-dehydroepiandrosterone $C_{19}H_{26}O_3$ (III, R = H, R' = O).

16-Oxime, 16-Oximino-Δ^5-androsten-3β-ol-17-one, 16-Oximino-5-dehydroepiandrosterone, 16-Isonitroso-5-dehydroepiandrosterone $C_{19}H_{27}O_3N$ (III, R = H, R' = N·OH). Almost colourless needles (isopropyl alcohol), m. 248–9° dec., after sintering at 240°. Difficultly soluble in alcohol and chloroform, readily in a mixture of chloroform with a little alcohol. — **Fmn.**

From 5-dehydroepiandrosterone (p. 2609 s) on treatment with potassium tert.-butoxide and n-amyl nitrite in tert.-butyl alcohol under nitrogen. — **Rns.** On refluxing with zinc dust in 50% acetic acid it gives Δ^5-androstene-3β.17β-diol-16-one (p. 2744 s; see there also for structure and configuration) (1941 b Stodola).

Acetate 16-oxime $C_{21}H_{29}O_4N$ (III, R $=$ CO·CH$_3$, R$'=$ N·OH). Needles (aqueous acetone), m. 183–4⁰. — **Fmn.** From the preceding compound on treatment with acetic anhydride and pyridine at room temperature (1941 b Stodola).

Methyl ether 16-oxime $C_{20}H_{29}O_3N$ (III, R $=$ CH$_3$, R$'=$ N·OH). Almost colourless crystals (aqueous methanol), dec. 180–183.5⁰ (1948 Huffman). — **Fmn.** From the methyl ether of 5-dehydroepiandrosterone (p. 2619 s) on treatment with potassium tert.-butoxide and isoamyl nitrite in tert.-butyl alcohol at room temperature; crude yield, 68% (1948 Huffman). — **Rns.** Gives the 3-methyl ether of Δ^5-androstene-3β.17β-diol-16-one (see p. 2744 s) on refluxing with zinc dust in 50% acetic acid (1948, 1949 Huffman).

1936 Reichstein, *Helv.* **19** 402, 404, 409.
    SCHERING A.-G. (Butenandt, Logemann), *U.S. Pat.* 2170124 (issued 1939); *C. A.* **1940** 1133; *Ch. Ztbl.* **1940** I 1232.
1937 SCHERING A.-G., *Fr. Pat.* 838704 (issued 1939); *Brit. Pat.* 508576 (issued 1939); *C. A.* **1940** 777; *Ch. Ztbl.* **1939** II 1722, 1723; SCHERING CORP. (Inhoffen), *U.S. Pat.* 2280828 (issued 1942); *C. A.* **1942** 5618.
    Steiger, Reichstein, *Helv.* **20** 817, 820.
1937/38 Dannenberg, *Über einige Umwandlungen des Androstandions und Testosterons*, Thesis, Danzig, p. 29.
1939 CIBA (Ges. f. chem. Ind. Basel), *Swiss Pat.* 241646 (issued 1946); *C. A.* **1949** 7978; (Miescher, Wettstein), *Swed. Pat.* 105134 (1940, issued 1942); *Ch. Ztbl.* **1943** I 1695; *Brit. Pat.* 550684 (1940, issued 1943); *C. A.* **1944** 1611; CIBA PHARMACEUTICAL PRODUCTS (Miescher, Wettstein), *U.S. Pat.* 2374369 (1940, issued 1945); *C. A.* **1945** 5412.
    Dannenberg, *Abhandl. Preuss. Akad. Wiss. Math.-naturwiss. Kl.* **1939** No. 21, pp. 30, 60.
1940 Ruzicka, Grob, Raschka, *Helv.* **23** 1518, 1520, 1522, 1526–1528.
    Shoppee, *Helv.* **23** 740, 742.
1941 Ehrenstein, *J. Org. Ch.* **6** 626, 630, 632, 641, 642.
    PARKE, DAVIS & Co. (Marker, Lawson), *U.S. Pat.* 2366204 (issued 1945); *C. A.* **1945** 1649; *Ch. Ztbl.* **1945** II 1385, 1386.
    Reich, Reichstein, *Arch. intern. pharmacodynamie* **65** 415, 417, 421, 422.
    a Stodola, Kendall, *J. Org. Ch.* **6** 837.
    b Stodola, Kendall, McKenzie, *J. Org. Ch.* **6** 841.
1942 Butenandt, Surányi, *Ber.* **75** 591, 595.
    Clarke, Selye, *Am. J. Med. Sci.* **204** 401, 406.
    v. Euw, Reichstein, *Helv.* **25** 988, 998, 1022.
    Reichstein, *U.S. Pat.* 2409798 (issued 1946); *C. A.* **1947** 1397, 1398; *Brit. Pat.* 560812 (issued 1944); *C. A.* **1946** 4486; *Ch. Ztbl.* **1945** II 1773.
    Selye, *Endocrinology* **30** 437, 441.
    Stodola, Kendall, *J. Org. Ch.* **7** 336, 339, 340.
1943 Selye, Masson, *J. Pharmacol.* **77** 301, 303.
1944 Ruzicka, Muhr, *Helv.* **27** 503, 512.
1945 Reich, *Helv.* **28** 863, 866, 868–871.
1946 Clark, Brink, Wallis, *J. Biol. Ch.* **162** 663.
    Lieberman, Dobriner, *J. Biol. Ch.* **166** 773.
    a MERCK & Co., INC. (Sarett), *U.S. Pat.* 2492189 (issued 1949); *C. A.* **1950** 3045; *Ch. Ztbl.* **1952** 1211; *Fr. Pat.* 942260 (1947, issued 1949); *Ch. Ztbl.* **1950** I 583, 584.

b Merck & Co., Inc. (Sarett), *U.S. Pat.* 2540964 (issued 1951); *C. A.* **1951** 7159; *Brit. Pat.*
  630103 (issued 1949); *C. A.* **1950** 7891.
a Sarett, *J. Biol. Ch.* **162** 601, 610, 616–619.
b Sarett, *J. Am. Ch. Soc.* **68** 2478, 2482.
Sorkin, Reichstein, *Helv.* **29** 1218.
1947 v. Euw, Reichstein, *Helv.* **30** 205, 206.
  Huffman, Lott, *J. Am. Ch. Soc.* **69** 1835.
  Reich, Lardon, *Helv.* **30** 329, 333, 700.
  Schering A.-G., *Fiat Final Report* No. 996, pp. 158, 161.
1948 Ehrenstein, *J. Org. Ch.* **13** 214, 218.
  Huffman, Lott, *J. Biol. Ch.* **172** 789, 791, 792.
  Lieberman, Dobriner, Hill, Fieser, Rhoads, *J. Biol. Ch.* **172** 263, 286.
1949 a Davis, Petrow, *J. Ch. Soc.* **1949** 2536, 2539.
  b Davis, Petrow, *J. Ch. Soc.* **1949** 2973.
  Huffman, Lott, *J. Am. Ch. Soc.* **71** 719, 724, 726.
  Meystre, Wettstein, *Helv.* **32** 1978, 1986, 1987.
1950 Gallagher, Kritchevsky, *J. Am. Ch. Soc.* **72** 882, 884.
  Heusser, Feurer, Eichenberger, Prelog, *Helv.* **33** 2243, 2247.
1951 Antaki, Petrow, *J. Ch. Soc.* **1951** 901, 904.
  Devis, *C. A.* **1952** 11375, 11376.
1952 Balant, Ehrenstein, *J. Org. Ch.* **17** 1587, 1590, 1591, 1594, 1595.
  MacPhillamy, Scholz, *J. Am. Ch. Soc.* **74** 5512, 5513.
  Upjohn Co. (Murray, Peterson), *U.S. Pat.* 2602769 (issued 1952); *C. A.* **1952** 8331, 8333;
    Ch. Ztbl. **1953** 9243; *U.S. Pat.* 2656370 (1952, issued 1953); *C. A.* **1954** 11504; *Ch. Ztbl.*
    1955 3690.
1953 Bernstein, Lenhard, Williams, *J. Org. Ch.* **18** 1166, 1172.
  Finkelstein, v. Euw, Reichstein, *Helv.* **36** 1266, 1269, 1276.
  Hayano, Dorfman, *J. Biol. Ch.* **201** 175, 185.
  Herr, Heyl, *J. Am. Ch. Soc.* **75** 5927, 5929.
  Herzog, Jevnik, Perlman, Nobile, Hershberg, *J. Am. Ch. Soc.* **75** 266, 267.
  Jeanloz, Levy, Jacobsen, Hechter, Schenker, Pincus, *J. Biol. Ch.* **203** 453, 459.
  Salamon, Dobriner, *J. Biol. Ch.* **204** 487, 488.
1954 Amendolla, Rosenkranz, Sondheimer, *J. Ch. Soc.* **1954** 1226, 1228, 1231, 1232.
  Bernstein, Lenhard, Williams, *J. Org. Ch.* **19** 41, 45.
  Eppstein, Meister, Leigh, Peterson, Murray, Reineke, Weintraub, *J. Am. Ch. Soc.* **76** 3174,
    3177.
  Meyer, Lindberg, *J. Am. Ch. Soc.* **76** 3033.

c. Monohydroxy-diketosteroids C_{20} with OH and Both CO Groups in the Ring System

6β-Methylandrostan-5-ol-3.17-dione $C_{20}H_{30}O_3$. By analogy with similar compounds in the cholestane series (see 1951 Fieser; 1952 Turner), 6β-configuration is probable for the present compound and the parent triol, and 6α-configuration for the dehydrated product (Editor; see also 1957 Ackroyd). — Needles (ethyl acetate), rodlets (methanol), m. 187–8° (1940 Madaeva). — **Fmn.** From 6β-methylandrostane-3β.5.17β-triol (p. 2170 s) on oxidation with CrO_3 in glacial acetic acid in the cold (1940 Madaeva). — **Rns.** Treatment with dry HCl in chloroform, in a freezing mixture, gives 6α-methyl-Δ^4-androstene-3.17-dione (page 2899 s) (1940 Madaeva).

1940 Madaeva, Ushakov, Kosheleva, *J. Gen. Ch. U.S.S.R.* **10** 213, 215, 216; *C. A.* **1940** 7292; *Ch. Ztbl.* **1940** II 1298.
1951 Fieser, Rigaudy, *J. Am. Ch. Soc.* **73** 4660.
1952 Turner, *J. Am. Ch. Soc.* **74** 5362.
1957 Ackroyd, Adams, Ellis, Petrow, Stuart-Webb, *J. Ch. Soc.* **1957** 4099.

d. Monohydroxy-diketosteroids C_{27}–C_{29} with OH and Both CO Groups in the Ring System

Δ^4-Cholesten-4-ol-3.6-dione, 4-Enol of cholestane-3.4.6-trione $C_{27}H_{42}O_3$. See under the trione, p. 2982 s.

Cholestan-5-ol-3.6-dione $C_{27}H_{44}O_3$ (p. 163: *Cholestane-3.6-dion-5-ol*). For identity of the compounds believed earlier (see p. 163; cf. also 1939 Ellis) to be stereoisomers, see 1944 Prelog. — Crystals (chloroform - methanol), m. 232° cor. (1944 Prelog), 241–3° cor. (1954 Plattner); crystals (ethyl acetate), m. 241–243.5° (1939 Ushakov); (acetone), m. 248–251° (1940 Marker); (glacial acetic acid), m. 253° cor. (1939 Ellis); that the m. p. (dec.) may vary between 232° and 253°, has been confirmed by Fieser (1953). $[\alpha]_D^{21}$ —13° (dioxane) (1944 Prelog); $[\alpha]_D^{18}$ —22° (chloroform) (1954 Plattner; see also 1953 Fieser). X-ray determination of crystal structure: a 8.32, b 7.7, c 20.1 Å, β 91°, space group $P2_1$, n = 2 (1940 Bernal). For measurements of surface films, see 1935 Adam.

Fmn. From cholesterol (p. 1570 s) on oxidation with $KMnO_4$ in acetic acid at room temperature (1940 Marker). From cholestane-3β.5α.6α-triol (p. 2175 s) (1939 Ushakov; 1944 Prelog) and from cholestane-3β.5α.6β-triol (p. 2177 s) (1944 Prelog) on oxidation with CrO_3 in acetic acid at room temperature. From cholesterol α-oxide (p. 2175 s) on oxidation with CrO_3

in acetic acid at room temperature (1915 Westphalen; 1937 Ruzicka); similarly from Δ^5-cholesten-3-one oxide ethylene ketal (p. 2755 s) (1941 Fernholz). From a cholestane-3β.5-diol-6-one, m. 138° (see p. 2758 s) on oxidation with CrO_3 in acetic acid at room temperature (1939 Ellis). From 7α-bromo-cholestan-5α-ol-3.6-dione (p. 2957 s) on reduction with zinc dust in boiling alcohol (1943 Sarett). See also formations of the acetate, below.

Rns. Gives Δ^4-cholestene-3.6-dione (p. 2905 s) on heating with $KHSO_4$ in high vac. at 150–180° (1940 Marker), on heating above its m. p., or on refluxing with P_2O_5 in benzene (1946 Ross).

Acetate $C_{29}H_{46}O_4$. Prisms (methanol), m. 165.5–167° (1939 Hattori), 163° (1939 Windaus); [α]$_D$ +3.7° (chloroform) (1953 Tarlton). — **Fmn.** Together with the acetate of 7-keto-epi-cholesterol (p. 2681 s) (1939 Windaus), from epi-cholesteryl acetate (p. 1670 s) on oxidation with CrO_3 at 50° in acetic acid (1939 Windaus; cf. 1953 Tarlton) or in propionic acid (1953 Tarlton). From the 5-acetate of cholestane-3β.5α.6β-triol (p. 2177 s) on oxidation with CrO_3 (1939 Hattori). — **Rns.** Gives Δ^4-cholestene-3.6-dione (p. 2905 s) on adsorption on Al_2O_3 (1939 Windaus) or on treatment with methanolic alkali (1939 Hattori; 1939 Windaus).

Cholestan-3β-ol-6.7-dione $C_{27}H_{44}O_3$. Exists in a mono-enolic form, (I) or (II);

the Zerewitinoff determination, applied to the acetate, indicates the presence of one active hydrogen atom (1937 Heilbron); with alcoholic $FeCl_3$ the free hydroxydiketone (or its enol) gives a deep green coloration (1946 Barnett), and the acetate a green-violet one (1937 Heilbron).

Needles (aqueous methanol), m. 152–3°. — **Fmn.** From the acetate (below) on refluxing with 2% methanolic NaOH for 15 min. (1946 Barnett).

Acetate $C_{29}H_{46}O_4$. Needles (methanol), m. 156–7°; [α]$_D^{20}$ −108° (chloroform) (1937 Heilbron). Absorption max. in alcohol at 274.5 mμ (log ε 4.03) (1937 Heilbron; cf. also 1939 Dannenberg). Dissolves in aqueous alcoholic NaOH to give a yellow solution (1937 Heilbron). — **Fmn.** From the acetate of 7α-bromocholestan-3β-ol-6-one (see p. 2708 s) on refluxing with $AgNO_3$ and pyridine for 5 hrs. (crude yield, 18%) (1937 Heilbron); similarly from the acetate of 6α-bromocholestan-3β-ol-7-one (see p. 2709 s; yield, 11%) (1938 Barr). In very small yield (ca. 4%), along with other compounds, from the acetate of 5α.7β-dibromocholestan-3β-ol-6-one (see p. 2714 s) on refluxing with fused sodium acetate in abs. alcohol for 1 hr. (1938 Heilbron). — **Rns.** On fusion with o-phenylenediamine at 140–150° it gives the corresponding **quinoxaline derivative** $C_{35}H_{50}O_2N_2$ [colourless needles (acetone), m. 186–7°] (1937 Heilbron). See also under the dioxime, below.

Acetate dioxime $C_{29}H_{48}O_4N_2$. Needles (aqueous methanol), m. 245–7°. — **Fmn.** From the above acetate on refluxing with four times the theoretical amount of hydroxylamine in alcohol for 48 hrs. — **Rns.** Gives 6.7-diamino-cholestan-3β-ol (p. 1915 s) on reduction with sodium and alcohol (1946 Barnett).

1915 Westphalen, *Ber.* **48** 1064, 1067.
1935 Adam, Askew, Danielli, *Biochem. J.* **29** 1786, 1798, 1799.
1937 Heilbron, Jones, Spring, *J. Ch. Soc.* **1937** 801, 802, 804, 805.
     Ruzicka, Bosshard, *Helv.* **20** 244, 246.
1938 Barr, Heilbron, Jones, Spring, *J. Ch. Soc.* **1938** 334, 336.
     Heilbron, Jackson, Jones, Spring, *J. Ch. Soc.* **1938** 102, 107.
1939 Dannenberg, *Abhandl. Preuss. Akad. Wiss. Math.-naturwiss. Kl.* **1939** No. 21, pp. 17, 56.
     Ellis, Petrow, *J. Ch. Soc.* **1939** 1078, 1081, 1083.
     Hattori, *C. A.* **1939** 8622; *Ch. Ztbl.* **1940** I 380.
     Ushakov, Lyutenberg, *J. Gen. Ch. U.S.S.R.* **9** 69, 71; *C. A.* **1939** 6334; *Ch. Ztbl.* **1939** II 4489.
     Windaus, Naggatz, *Ann.* **542** 204, 209.
1940 Bernal, Crowfoot, Fankuchen, *Philos. Trans. Roy. Soc. London* A **239** [1946] 135, 141, 153, 168.
     Marker, Rohrmann, *J. Am. Ch. Soc.* **62** 516.
1941 Fernholz, Stavely, *Abstracts of the 102nd meeting of the American Chemical Society, Atlantic City*, p. M39.
1943 Sarett, Chakravorty, Wallis, *J. Org. Ch.* **8** 405, 415.
1944 Prelog, Tagmann, *Helv.* **27** 1867, 1869, 1871.
1946 Barnett, Ryman, Smith, *J. Ch. Soc.* **1946** 528, 530.
     Ross, *J. Ch. Soc.* **1946** 737, 739.
1953 Fieser, *J. Am. Ch. Soc.* **75** 4386, 4391.
     Tarlton, Fieser, Fieser, *J. Am. Ch. Soc.* **75** 4423.
1954 Plattner, Fürst, Koller, Kuhn, *Helv.* **37** 258, 265, 266.

$\Delta^{7.22}$-**Ergostadien-5-ol-3.6-dione** $C_{28}H_{42}O_3$ (p. 163). Plates (chloroform - methanol), m. 249° dec. (1933 Heilbron). Absorption max. in chloroform at 252 mμ (ε 13450) and 332.5 mμ (ε 160) (1937 Burawoy). Sparingly soluble in cold chloroform (1933 Heilbron).

Page 163 lines 9–20 from bottom: Delete all data with the source "(1930 Windaus)".

Δ^7-**Ergosten-5-ol-3.6-dione** $C_{28}H_{44}O_3$. Dannenberg (1939) indicates absorption max. in chloroform at 252 mμ (ε 13500) for "Δ^7-ergosten-5-ol-3.6-dione" quoting Burawoy (1937); in the latter paper, however, this ergostenoldione does not appear, but $\Delta^{7.22}$-ergostadien-5-ol-3.6-dione with the same absorption max. (see the preceding compound); therefore it is almost certain that the value given by Dannenberg refers to the ergostadienoldione (Editor).

Δ⁸⁽¹⁴⁾-Ergosten-5-ol-3.6-dione $C_{28}H_{44}O_3$. Crystals (methanol, or chloroform-petroleum ether), m. 251–2⁰ dec.; $[\alpha]_D^{20}$ —13⁰ (chloroform). Rather sparingly soluble in chloroform, very sparingly in methanol and petroleum ether. — **Fmn.** From Δ⁸⁽¹⁴⁾-ergostene-3β.5α.6α-triol (page 2190 s) on oxidation with CrO_3 in strong acetic acid at 35⁰; small yield (1930 Windaus).

Dioxime $C_{28}H_{46}O_3N_2$. Needles (methanol), dec. above 205⁰ (1930 Windaus).

Δ⁸⁽¹⁴⁾-Ergosten-3β-ol-7.15-dione $C_{28}H_{44}O_3$.

Acetate $C_{30}H_{46}O_4$. Crystals (alcohol), m. 149–151⁰; $[\alpha]_D^{24}$ —24⁰ (chloroform); absorption max. in abs. alcohol at 255 mμ (ε 5000). — **Fmn.** Along with other compounds from the acetate of Δ⁸⁽¹⁴⁾-ergosten-3β-ol (p. 1765 s) on oxidation with CrO_3 in strong acetic acid and some benzene at room temperature; yield, 2%. — **Rns.** Refluxing with hydrazine hydrate in alcohol gives the corresponding **pyridazine derivative** $C_{30}H_{46}O_2N_2$ [crystals (alcohol), m. 220–5⁰ dec.; absorption max. in abs. alcohol at 262 mμ (ε 1800)] (1943 Stavely).

1930 Windaus, Lüttringhaus, *Ann.* **481** 119, 131.
1933 Heilbron, Morrison, Simpson, *J. Ch. Soc.* **1933** 302, 305.
1937 Burawoy, *J. Ch. Soc.* **1937** 409.
1939 Dannenberg, *Abhandl. Preuss. Akad. Wiss. Math.-naturwiss. Kl.* **1939** No. 21, pp. 15, 55.
1943 Stavely, Bollenbach, *J. Am. Ch. Soc.* **65** 1285, 1288, 1289.

Stigmastan-5-ol-3.6-dione, Sitostan-5-ol-3.6-dione $C_{29}H_{48}O_3$ (p. 164: *Stigmastane-3.6-dion-5-ol*). White crystals (acetone), m. 240⁰. — **Fmn.** From sitosterol (p. 1808 s) on oxidation with $KMnO_4$ in acetic acid at room temperature (1940 Marker).

Clionastan-5-ol-3.6-dione, 24-iso-Stigmastan-5-ol-3.6-dione $C_{29}H_{48}O_3$. Needles (acetone), m. 189–191⁰. — **Fmn.** From clionastane-3β.5α.6β-triol (p. 2193 s) on refluxing with aluminium isopropoxide and dry acetone in dry benzene for 10 hrs. (1942 Kind).

1940 Marker, Rohrmann, *J. Am. Ch. Soc.* **62** 516.
1942 Kind, Bergmann, *J. Org. Ch.* **7** 341, 344.

HALOGENO-MONOHYDROXY-DIKETOSTEROIDS WITH OH AND BOTH CO IN THE RING SYSTEM

7α-Bromocholestan-5α-ol-3.6-dione $C_{27}H_{43}O_3Br$ (not analysed). For the con-

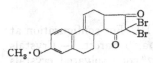

figuration at C-7, see that of the parent bromodiolone (p. 2762 s). — The partially solvated product was obtained with difficulty from warm dil. acetone as fluffy white needles; has a strong tendency to give gels instead of crystals; after drying at 100° it melts between 165° and 171°. — **Fmn.** From 7α-bromocholestane-3β.5α-diol-6-one (p. 2762 s) on oxidation with CrO_3 in 98% acetic acid, first at 15°, then at room temperature. — **Rns.** On reduction with zinc dust in boiling alcohol it gives cholestan-5α-ol-3.6-dione (p. 2953 s). Attempted dehydration with 95% formic acid (at room temperature or at 100°), with dry HCl and dry HBr in chloroform, and with hot acetic anhydride, led to uncrystallizable oils (1943 Sarett).

4′.4′-Dibromo-7-methoxy-3′.5′-diketo-1.2-cyclopentano-1.2.3.9.10.11-hexahydrophenanthrene $C_{18}H_{16}O_3Br_2$. Pale yellow leaflets (acetone), m. 166°. — **Fmn.** From 4.4-dibromo-$\varDelta^1$-cyclopentene-3.5-dione and an excess of 1-vinyl-6-methoxy-3.4-dihydronaphthalene on heating in dioxane for 24 hrs. at 110–115°; yield, ca. 12% (1939 Dane; 1938 SCHERING A.-G.).

1938 SCHERING A.-G. (Dane), *Swed. Pat.* 98608 (issued 1940); *Ch. Ztbl.* **1940** II 2959.
1939 Dane, Eder, *Ann.* **539** 207, 212.
1943 Sarett, Chakravorty, Wallis, *J. Org. Ch.* **8** 405, 415.

3. Polyhydroxy-diketosteroids with One OH and Both CO in the Ring System

Δ^4-Pregnene-17α.20β-diol-3.11-dione $C_{21}H_{30}O_4$. For configuration at C-17 and C-20, see 1949a, b Sarett. — Crystals (water) containing water of crystallization, m. 107–110° cor. with loss of water (1946b Sarett). — **Fmn.** The 20-acetate is obtained from that of pregnane-17α.20β-diol-3.11-dione (below) on treatment with 1 mol. of bromine in chloroform - glacial acetic acid at room temperature, followed by refluxing of the resulting 4-bromo-derivative (p. 2959 s) with pyridine for 8 hrs. (crude over-all yield, 22%); it is hydrolysed by $KHCO_3$ + K_2CO_3 in aqueous methanol at room temperature (1946b Sarett; 1947 Merck & Co., Inc.). — **Rns.** On oxidation with periodic acid in 80% methanol at room temperature it gives Δ^4-androstene-3.11.17-trione (adrenosterone, p. 2979 s) (1946b Sarett; 1947 Merck & Co., Inc.).

20-Acetate $C_{23}H_{32}O_5$. Crystals (chloroform - ether), m. 219–220° cor.; solvated crystals (benzene), m. 104° cor. with loss of solvent, remelting at 215–8° cor. (1946b Sarett). — **Fmn.** See above.

Pregnane-17α.20β-diol-3.11-dione $C_{21}H_{32}O_4$ (I, R = H). For configuration at C-17 and C-20, see 1949a, b Sarett. — Crystals (ethyl acetate), m. 186–7° cor.; solvated crystals (acetone - ether), m. 150° cor. dec.; $[\alpha]_D^{20}$ +47° (acetone) (1946b Sarett). — **Fmn.** Along with the two pregnane-20.21-diol-3.11-diones (pp. 2937 s, 2938 s), from a crude mixture of $\Delta^{17(20)}$- and Δ^{20}-pregnen-3α-ol-11-ones (see p. 2654 s) on oxidation with CrO_3 in strong acetic acid at 16°, followed by treatment of the resulting oily mixture of $\Delta^{17(20)}$- and Δ^{20}-pregnene-3.11-diones with OsO_4 and pyridine in abs. ether at room temperature, refluxing with Na_2SO_3 in aqueous alcohol for $3\frac{1}{2}$ hrs., and separation from the two 20.21-diol-3.11-diones by their conversion into the 21-hemisuccinates which are extracted from a solution in chloroform with aqueous K_2CO_3; isolated as 20-acetate (below; crude over-all yield, 22%) from which the free pregnane-17α.20β-diol-3.11-dione is obtained by refluxing with K_2CO_3 + $KHCO_3$ for 45 min. in aqueous methanol (1946b Sarett).

Rns. On oxidation with CrO_3 in strong acetic acid at room temperature it gives ætiocholane-3.11.17-trione (p. 2980 s) and a smaller amount of pregnan-17α-ol-3.11.20-trione (p. 2976 s) (1946b Sarett). On bromination with 1 mol. of bromine in chloroform - glacial acetic acid at room temperature, the 20-acetate gives the 4-bromo-derivative (p. 2959 s) (1946b Sarett; 1947 Merck & Co., Inc.).

20-Acetate $C_{23}H_{34}O_5$. Hygroscopic crystals (abs. toluene), m. 228–230° cor. (1946b Sarett); crystals (methanol), m. 222.0–224.5° cor.; the m. p. rises after long heating in a vacuum (1949a Sarett); $[\alpha]_D$ +75° (acetone) (1946b Sarett). — **Fmn.** From the above dioldione on heating with acetic anhydride and pyridine for $\frac{1}{2}$ hr. on the steam-bath (1946b Sarett). — **Rns.** See above.

4-Bromopregnane-17α.20β-diol-3.11-dione $C_{21}H_{31}O_4Br$ (Formula I, p. 2958 s; R = Br).

20-Acetate $C_{23}H_{33}O_5Br$ (not analysed). Crystals with EtOH (alcohol), m. 150° to 155° cor. with loss of solvent. **Fmn.** From the preceding acetate on treatment with 1 mol. of bromine in chloroform - glacial acetic acid at room temperature; yield, 68%. **Rns.** Refluxing with pyridine for 8 hrs. affords the 20-acetate of $Δ^4$-pregnene-17α.20β-diol-3.11-dione (p. 2958 s) (1946b Sarett; 1947 MERCK & Co., INC.).

$Δ^4$-Pregnene-17α.20β.21-triol-3.11-dione, *Reichstein's Substance U*, $C_{21}H_{30}O_5$.

For configuration at C-17 and C-20, see 1949a, b Sarett. — Needles (acetone - ether), m. 208° cor. (1941 Reichstein), 208.5–209.5° cor. (1946a Sarett; 1946c MERCK & Co., INC.); $[α]_D$ +140° (acetone) (1946a Sarett; 1946c MERCK & Co., INC.). — **Occ.** In adrenal cortex (1941 Reichstein). — **Fmn.** The 20.21-diacetate is obtained from that of $Δ^4$-pregnene-11β.17α.20β.21-tetrol-3-one (Substance E, p. 2765 s) on oxidation with CrO_3 in glacial acetic acid at room temperature (almost quantitative yield) (1941 Reichstein). The 20.21-diacetate is also obtained from that of pregnane-17α.20β.21-triol-3.11-dione (p. 2960 s) on treatment with 1 mol. of bromine in glacial acetic acid, followed by refluxing of the resulting 4-bromo-derivative (p. 2960 s; 93% yield) with pyridine for 10 hrs. (yield, 46%) (1946a Sarett; 1946c MERCK & Co., INC.). The free trioldione is obtained from its diacetate by refluxing with K_2CO_3 in aqueous methanol (1941 Reichstein) or by treating with $K_2CO_3 + KHCO_3$ in aqueous methanol at 30° (1946a Sarett; 1946c MERCK & Co., INC.).

Rns. On oxidation of the 21-acetate with CrO_3 in strong acetic acid at 11–24° it gives $Δ^4$-androstene-3.11.17-trione (adrenosterone, p. 2979 s) and the 21-acetate of $Δ^4$-pregnene-17α.21-diol-3.11.20-trione (p. 2977 s) (1946a Sarett; 1946a MERCK & Co., INC.).

21-Acetate $C_{23}H_{32}O_6$. Crystals (acetone - ether), m. 172–4° cor. — **Fmn.** From the above trioldione on treatment with ca. 1.3 mols. of acetic anhydride and pyridine in dioxane at room temperature (1946a Sarett; 1946a MERCK & Co., INC.). — **Rns.** For its oxidation, see above.

20.21-Diacetate $C_{25}H_{34}O_7$. Needles (acetone - ether or chloroform - ether), crystals (methanol), m. 252–3° cor.; $[α]_D^{21}$ +178.5° (acetone) (1941 Reichstein; 1946a Sarett; 1946c MERCK & Co., INC.). Absorption max. in alcohol at 239 mμ (log ε 4.1) (1941 Reichstein; cf. 1946a Sarett). Gives a weak, green fluorescence with conc. H_2SO_4. On warming with 1.4-dihydroxynaphthalene and conc. HCl in glacial acetic acid it gives a red coloration. Reduces alkaline silver diammine solution extremely slowly at room temperature (1941 Reichstein). — **Fmn.** See under that of the free trioldione (above).

Pregnane-17α.20β.21-triol-3.11-dione $C_{21}H_{32}O_5$ (II, R = H). For configuration at C-17 and C-20, see 1949a, b Sarett.

20.21-Diacetate $C_{25}H_{36}O_7$ (poor agreement of the analysis with theory). Needles (ethyl acetate), m. 212–3° cor.; $[\alpha]_D$ +93° (acetone). — **Fmn.** From the acetate of $\Delta^{17(20)}$-pregnen-21-ol-3.11-dione (p. 2934 s) on treatment with OsO_4 and dry pyridine in abs. ether at room temperature, followed by refluxing with Na_2SO_3 in dil. alcohol and then by acetylation with acetic anhydride in pyridine at room temperature; yield, 86% (1946a Sarett; 1946b MERCK & Co., INC.). — **Rns.** Gives the 4-bromo-derivative (below) with 1 mol. of bromine in glacial acetic acid (1946a Sarett; 1946c MERCK & Co., INC.).

4-Bromopregnane-17α.20β.21-triol-3.11-dione $C_{21}H_{31}O_5Br$ (II, R = Br). — **20.21-Diacetate** $C_{25}H_{35}O_7Br$ (not analysed). Crystals (abs. ether), m. 188–9° cor. dec. **Fmn.** From the preceding diacetate on treatment with 1 mol. of bromine in glacial acetic acid; yield, 93%. **Rns.** On refluxing with pyridine for 10 hrs. it gives the 20.21-diacetate of Δ^4-pregnene-17α.20β.21-triol-3.11-dione (p. 2959 s) (1946a Sarett; 1946c MERCK & Co., INC.).

1941 Reichstein, v. Euw, *Helv.* **24** 247 E, 248 E, 256 E, 260 E, 262 E.
1946 a MERCK & Co., INC. (Sarett), *U.S. Pat.* 2462133 (issued 1949); *C. A.* **1949** 4432; *Fr. Pat.* 942260 (1947, issued 1949); *Ch. Ztbl.* **1950** I 583, 584; *Brit. Pat.* 631238 (issued 1949); *C. A.* **1950** 4048, 4050.
  b MERCK & Co., INC. (Sarett), *U.S. Pat.* 2492194 (issued 1949); *C. A.* **1950** 3044; *Ch. Ztbl.* **1952** 1211.
  c MERCK & Co., INC. (Sarett), *U.S. Pat.* 2492195 (issued 1949); *C. A.* **1950** 3046; *Ch. Ztbl.* **1952** 1211.
  a Sarett, *J. Biol. Ch.* **162** 601, 624–628.
  b Sarett, *J. Am. Ch. Soc.* **68** 2478.
1947 MERCK & Co., INC. (Sarett), *Fr. Pat.* 956655 (issued 1950); *Ch. Ztbl.* **1950** II 560; *Brit. Pat.* 651719 (issued 1951); *C. A.* **1952** 150.
1949 a Sarett, *J. Am. Ch. Soc.* **71** 1169, 1170, 1172.
  b Sarett, *J. Am. Ch. Soc.* **71** 1175.

4. POLYHYDROXY-DIKETOSTEROIDS WITH OH AND CO IN THE RING SYSTEM

4.7-Dimethoxy-3'.4'-diketo-1.2-cyclopentenophenanthrene $C_{19}H_{14}O_4$ (I, R = O).

4'-Oxime, 4.7-Dimethoxy-3'-keto-4'-oximino (iso-nitroso)-1.2-cyclopentenophenanthrene $C_{19}H_{15}O_4N$ (I, R = N·OH). Orange prisms (nitrobenzene), m. 248° to 249° dec. gives a brilliant indigo-blue coloration in conc. H_2SO_4. — **Fmn.** From 4.7-dimethoxy-3'-keto-1.2-cyclopentenophenanthrene (p. 2735 s) on refluxing under nitrogen with 1 mol. potassium tert.-butoxide in dry tert.-butyl alcohol for $1\frac{1}{2}$ hrs., followed by addition of 1 mol. of isoamyl nitrite and stirring at room temperature overnight; yield, 93 % (1939 Robinson).

Androstane-5α.17β-diol-3.6-dione $C_{19}H_{28}O_4$. For the structure, and for the configuration at C-5, cf. the analogous formation of androstan-5α-ol-3.6.17-trione (p. 2983 s) (see 1939, 1948 Ehrenstein).

17-Acetate $C_{21}H_{30}O_5$. Colourless needles (alcohol); begins to melt at 229–231°, completely liquid at 239° to 241°; $[α]_D^{22}$ —30° (acetone). — **Fmn.** Along with the acetate of $Δ^4$-androsten-17β-ol-3.6-dione (6-ketotestosterone, p. 2944 s), from the 17-acetate of $Δ^5$-androstene-3β.17β-diol (p. 1999 s) on shaking with CrO_3 (6 at. of oxygen) in strong acetic acid at room temperature; small yield (1936 Butenandt; 1936 SCHERING A.-G.).

Androstane-5α.6β-diol-3.17-dione $C_{19}H_{28}O_4$. For configuration at C-5 and C-6, see 1948 Ehrenstein.

6-Acetate $C_{21}H_{30}O_5$. Needle-shaped crystals (ether), m. 219–220.5° cor.; $[α]_D^{26}$ +45° (acetone) (1941 Ehrenstein). — **Fmn.** From the 6-acetate of androstane-3β. 5α.6β-triol-17-one (see p. 2768 s) on oxidation with CrO_3 (1.15 at. of O) in strong acetic acid at room temperature; yield, 72 % (1941 Ehrenstein). — **Rns.** Gives the acetate of $Δ^4$-androsten-6β-ol-3.17-dione (p. 2946 s) on treatment with dry HCl in alcohol-free chloroform at —10° or in purified carbon tetrachloride at —15°, and that of $Δ^4$-androsten-6α-ol-3.17-dione (p. 2945 s) on treatment with dry HCl in chloroform containing 3.3 % (by vol.) of alcohol at ca. 0° (1952 Balant; cf. 1941 Ehrenstein).

Androstane-9.11-diol-3.17-dione $C_{19}H_{28}O_4$. For its *anhydro-derivative* $C_{19}H_{26}O_3$, see 9.11-epoxyandrostane-3.17-dione (p. 2891 s).

References, p. 2964 s

Androstane-3β.5α-diol-6.17-dione $C_{19}H_{28}O_4$. For configuration at C-5, see 1948 Ehrenstein. — Crystals (acetone), m. 282–4° (1940 Ehrenstein); crystals (methanol - ethyl acetate, or dil. methanol), m. 297–8° cor. dec. (high vac.) (1940 Ruzicka). Very difficultly soluble in the common solvents (1940 Ehrenstein). —**Fmn.** From androstane-3β.5α.6β-triol-17-one (see p. 2768 s) on oxidation with CrO_3 (1.1 at. of O) in strong acetic acid at 0° (crude yield, 75%) (1940 Ehrenstein); its 3-esters (acetate and benzoate) are similarly obtained from the corresponding esters of the same triolone (1940 Ehrenstein) or in glacial acetic acid at room temperature (yields, 68%) (1940 Ruzicka) or, directly, from the corresponding esters of 5-dehydroepiandrosterone 5α.6α-oxide (p. 2769 s) on oxidation with CrO_3 in glacial acetic acid at room temperature (yields, 67% and 58%, resp.) (1940 Ruzicka); the free dioldione is obtained from its 3-acetate by refluxing with a 5% solution of K_2CO_3 in aqueous methanol for 1 hr. or from its 3-benzoate by refluxing with 1 N methanolic KOH for 2 hrs. (1940 Ruzicka).

Rns. Gives $Δ^3$-androsten-5α-ol-6.17-dione (p. 2948 s) on sublimation with fuller's earth in a high vacuum at 150° or on boiling of its 3-p-toluenesulphonate (below) with abs. pyridine for 36 hrs. (1940 Ruzicka).

Dioxime $C_{19}H_{30}O_4N_2$. Needles (abs. alcohol), m. 245–7° cor. dec. (1940 Ruzicka).

3-Acetate $C_{21}H_{30}O_5$. Needles (ether), m. 197.5–199° (1940 Ehrenstein); crystals (ethyl acetate), m. 212–3° cor. (1940 Ruzicka); $[α]_D^{28} +17°$ (acetone) (1940 Ehrenstein). — **Fmn.** From the above dioldione on treatment with acetic anhydride in pyridine at room temperature (1940 Ehrenstein). See also above.

3-Benzoate $C_{26}H_{32}O_5$. Needles (ethyl acetate - petroleum ether b. 40–70°), m. 256–7° cor. (1940 Ruzicka).

3-p-Toluenesulphonate $C_{26}H_{34}O_6S$ (not analysed). Solid (ethyl acetate), m. 133° cor. dec. — **Fmn.** From the above dioldione on treatment with 1.1 mols. of p-toluenesulphonyl chloride in abs. pyridine at room temperature (1940 Ruzicka). — **Rns.** See those of the dioldione, above.

5-Methyl-19-nor-coprostane (or -cholestane)-9.10-diol-3.6-dione $C_{27}H_{44}O_4$.

Anhydro-derivative, 5-Methyl-9.10-epoxy-19-nor-coprostane (or -cholestane)-3.6-dione, 5-Methyl-19-nor-$Δ^9$-coprostene (or -cholestene)-3.6-dione oxide $C_{27}H_{42}O_3$. Originally described by Petrow (1939) as 5-methyl-19-nor-$Δ^8$-cholestene-3.6-dione oxide; for the true structure, see 1952 Ellis; cf. also that of the parent "Westphalen's diol", p. 2046 s. — Needles (aqueous acetone), m. 132–3° cor. after softening at 120°; $[α]_D^{19} -35°$ (chloroform).

Readily soluble in most organic solvents (1939 Petrow). —**Fmn.** From 5-methyl-19-nor-Δ^9-coprostene(or-cholestene)-3.6-dione (Westphalen's diketone, page 2902 s) on treatment with perhydrol in acetic acid at room temperature; yield, 55 %. From 5-methyl-9.10-epoxy-19-nor-coprostane-(or-cholestane)-3β.6β-diol (p. 2047 s) on oxidation with CrO_3 and 90 % acetic acid in benzene at room temperature; yield, 65 % (1939 Petrow). — *Mono-o-tolylsemicarbazone* $C_{35}H_{51}O_3N_3$, plates (chloroform - alcohol), m. 224–5° cor. dec. (1939 Petrow).

Cholestane-2.5-diol-3.6-dione $C_{27}H_{44}O_4$.

2.5 - Epoxycholestane - 3.6 - dione $C_{27}H_{42}O_3$. Pearly plates (aqueous acetone), m. 115–6° cor. — **Fmn.** From 2.5-epoxycholestan-6β-ol-3-one (page 2771 s) on shaking with CrO_3 in 95 % acetic acid at room temperature for 6 hrs.; yield, 60 %. — **Rns.** Is stable both to lead tetraacetate and to alcoholic HCl (1939 Ellis).

Bis - (2.4 - dinitrophenylhydrazone) $C_{39}H_{50}O_9N_8$. Plates (acetone - chloroform), m. 171° cor. (1939 Ellis).

Cholestane (or Coprostane)-4.5-diol-3.6-dione $C_{27}H_{44}O_4$.

Needles (ether), prisms (glacial acetic acid, or chloroform - methanol), m. 220–225° dec., after sintering at 210° (1937 Windaus); needles (chloroform - alcohol), slowly decomposing above 200°, m. 243–5°; $[\alpha]_D^{21}$ —16° (chloroform) (1938 Butenandt).

Fmn. From Δ^4-cholestene-3.6-dione (p. 2905 s) on oxidation with potassium permanganate in ca. 83 % acetone (1937 Windaus) or with 30 % H_2O_2 in the presence of osmium tetroxide in ether (1938 Butenandt). Along with the tricarboxylic acid $C_{26}H_{44}O_6$ (p. 2964 s), from Δ^4-cholesten-3-one-6-sulphonic acid on oxidation with aqueous potassium permanganate at room temperature; yield, 12 % (1937 Windaus). From the lithium salt of cholesterylene-6-sulphonic acid in aqueous solution on dropwise addition of aqueous potassium permanganate (1938 Windaus).

Rns. On brief treatment with alcoholic KOH it gives the *potassium salt* (yellow needles) of the *enolic form*, from which the dioldione can be regenerated by immediate addition of acetic acid; on treatment with alcoholic KOH for a longer time, decomposition occurs. On refluxing with 5 % alcoholic HCl for 3 hrs., the dioldione is almost quantitatively converted into cholestane-3.4.6-trione (p. 2982 s) (1937 Windaus).

4-Acetate $C_{29}H_{46}O_5$. Needles (dil. alcohol or chloroform - alcohol), m. 224° to 226°. —**Fmn.** From the above dioldione on treatment with acetic anhydride in pyridine at room temperature (1938 Butenandt).

References, p. 2964 s

A-Nor-cholestane-3‖5‖6-trioic acid $C_{26}H_{44}O_6$. Crystals with 1 $CH_3 \cdot CO_2H$ (glacial acetic acid), m. 97–99°; loses the acetic acid on heating in a vacuum for 3 hrs. at 115°, and then m. 136–141° (1937 Windaus). — **Fmn.** Along with the above dioldione, from Δ^4-cholesten-3-one-6-sulphonic acid on oxidation with aqueous $KMnO_4$ at room temperature (1937 Windaus). From 5-keto-3‖5-A-nor-cholestan-3-oic acid (p. 1710 s) on oxidation with potassium hypobromite (1906, 1917, 1937 Windaus).

Δ^8**-Cholestene-3β.14-diol-7.15-dione** $C_{27}H_{42}O_4$ and **Cholestane-3β.8.14-triol-7.15-dione** $C_{27}H_{44}O_5$. For the 3-acetates of these compounds (1943 Wintersteiner), see under the reactions of $\Delta^{8(14)}$-cholesten-3β-ol (p. 1692 s).

1906 Windaus, *Ber.* **39** 2008, 2013.
1917 Windaus, *Ber.* **50** 133.
1936 Butenandt, Riegel, *Ber.* **69** 1163, 1166.
      SCHERING A.-G. (Butenandt, Logemann), *U.S. Pat.* 2170124 (issued 1939); *C. A.* **1940** 1133;
      Ch. Ztbl. **1940** I 1232.
1937 Windaus, Kuhr, *Ann.* **532** 52, 60–62.
1938 Butenandt, Wolz, *Ber.* **71** 1483, 1486.
      Windaus, Mielke, *Ann.* **536** 116, 127.
1939 Ehrenstein, *J. Org. Ch.* **4** 506, 513.
      Ellis, Petrow, *J. Ch. Soc.* **1939** 1078, 1080, 1082, 1083.
      Petrow, *J. Ch. Soc.* **1939** 998, 1001.
      Robinson, Rydon, *J. Ch. Soc.* **1939** 1394, 1401.
1940 Ehrenstein, Decker, *J. Org. Ch.* **5** 544, 551, 552, 555.
      Ruzicka, Grob, Raschka, *Helv.* **23** 1518, 1523, 1525–1527.
1941 Ehrenstein, *J. Org. Ch.* **6** 626, 641.
1943 Wintersteiner, Moore, *J. Am. Ch. Soc.* **65** 1513, 1515, 1516.
1948 Ehrenstein, *J. Org. Ch.* **13** 214, 217, 218.
1952 Balant, Ehrenstein, *J. Org. Ch.* **17** 1587, 1594, 1595.
      Ellis, Petrow, *J. Ch. Soc.* **1952** 2246.

E. Triketosteroids with Two CO Groups in the Ring System

I. COMPOUNDS WITHOUT OTHER FUNCTIONAL GROUPS

Summary of Triketo-compounds with Pregnane Skeleton

For values other than those chosen for this summary, see the compounds themselves

Position of		Configuration at		M. p. °C.	$[\alpha]_D$ * (°)	Page
CO	double bond	C-5	C-17			
3.6.20	Δ^4	—	β	185–8	—	2965 s
3.6.20	Δ^{16}	α	—	223–8	—	2965 s
3.6.20	satd.	β	β	190	—46.5 (dio.)	2966 s
3.6.20	satd.	α	β	233	+47 (alc.), +53 (dio.), +61 (chl.)	2966 s
3.11.20	Δ^4	—	β	175	+244 (ac.)	2967 s
3.11.20	satd.	β	β	162	+120 (ac.)	2967 s
3.11.20	satd.	α	β	216	+133 (alc.)	2968 s
3.12.20	$\Delta^{9(11)}$	β	β	186	—	2969 s
3.12.20	satd.	β	β	205	+169 (ac.)	2969 s
3.12.20	satd.	β	α	153	+58 (ac.)	2969 s

* ac. = acetone; alc. = alcohol; chl. = chloroform; dio. = dioxane

Δ^4**-Pregnene-3.6.20-trione, 6-Ketoprogesterone** $C_{21}H_{28}O_3$. Lemon-yellow need-

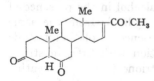

les (95 % alcohol - ether), crystals (acetone - ether), m. 185–8° cor. (1939 Ehrenstein; 1946 Moffett). Absorption max. in abs. alcohol (from the graph) at 241 mμ (K 34.6) (1939 Ehrenstein). — Shows a moderate œstrogenic activity, but no progestational activity (1939 Ehrenstein).

Fmn. For its formation from Δ^4-pregnen-20-one (p. 2216 s) by oxidation with CrO_3 in strong acetic acid, see 1940 CIBA. From 4-bromopregnane-3.6.20-trione (p. 2971 s) on refluxing with dry pyridine for 2 hrs.; yield, ca. 4 % (1946 Moffett). From allopregnan-5α-ol-3.6.20-trione (p. 2976 s) suspended in pure chloroform on passage of dry HCl at 4–5°; yield, 52 % (1939 Ehrenstein).

Δ^{16}**-Allopregnene-3.6.20-trione** $C_{21}H_{28}O_3$. Crystals (aqueous acetone), melting between 223° and 228° (1940 Marker).—**Fmn.** From Δ^{16}-allopregnene-3β.6β-diol-20-one (p. 2324 s) on oxidation with CrO_3 in strong acetic acid at room temperature (1942b Marker); is similarly obtained from pseudochlorogenin, dihydropseudochloroge-nin, pseudochlorogenone, pseudodiosgenin (1940 Marker), and pseudo-β-chlorogenin (followed by refluxing with K_2CO_3 in alcohol) (1942b Marker).

Rns. On hydrogenation in the presence of platinum oxide under 3 atm. pressure at room temperature in glacial acetic acid it gives allopregnane-3β.6β.20β-triol (p. 2108 s) (1942b Marker); similar hydrogenation in abs.

alcohol, followed by oxidation with CrO_3 in strong acetic acid, affords allo-pregnane-3.6.20-trione (below) which is also formed on hydrogenation of the present trione in the presence of palladium - barium sulphate under 1 atm. pressure in abs. alcohol (1940 Marker).

Pregnane-3.6.20-trione $C_{21}H_{30}O_3$. Crystals (benzene - ether), m. 189–190.5°; $[\alpha]_D^{27}$ —46.5° (dioxane). — **Fmn.** From pregnane-3α.6α-diol-20-one (see p. 2324 s) on oxidation with CrO_3 in strong acetic acid at 30°; yield, 83%. — **Rns.** Is rearranged to allopregnane-3.6.20-trione (below) by the action of alkali or by heating with a little conc. HCl in glacial acetic acid on the steam-bath for 1 hr. Gives the 4-bromo-derivative (p. 2971 s) with 1 mol. of bromine in glacial acetic acid at 15° (1946 Moffett).

Allopregnane-3.6.20-trione $C_{21}H_{30}O_3$. For configuration at C-5, see also 1948 Ehrenstein. — Crystals (acetone - ether), needles (acetone - petroleum ether), m. 232.5–233° cor. (1952 Balant; cf. 1940 Ehrenstein; 1940 Marker; 1946 Moffett); crystals (ethyl acetate - ligroin), m. 235–7° (1948 Lieberman); $[\alpha]_D^{23}$ +53°.(dioxane) (1946 Moffett), $[\alpha]_D^{25}$ +47° (alcohol) (1948 Lieberman), +61° (chloroform) (1952 Balant). Shows no absorption of ultra-violet light between 222 and 300 mμ (1952 Balant).

Fmn. From Δ^5-pregnen-3β-ol-20-one (p. 2233 s) on oxidation with CrO_3 in strong acetic acid at 20°, followed by refluxing with zinc dust in acetic acid (1940 Marker). From allopregnane-3β.6α-diol-20-one (p. 2326 s) on oxidation with CrO_3 in strong acetic acid at room temperature (1940 Marker; cf. 1954 Salamon; 1948 Lieberman); similarly from allopregnane-3α.6α-diol-20-one (see p. 2325 s) (1948 Lieberman). From Δ^4-pregnen-6α-ol-3.20-dione (6α-hydroxyprogesterone, p. 2838 s) on treatment with a little 3% H_2SO_4 in glacial acetic acid at room temperature for 44 hrs.; yield, 63% (1952 Balant). From the acetate of 6β-hydroxyprogesterone (p. 2838 s) on heating with 2.2 mols. of KOH in aqueous methanol on the water-bath for 1 hr.; yield, ca. 76% (1952 Balant; see also 1940 Ehrenstein). From Δ^{16}-allopregnene-3.6.20-trione (p. 2965 s) on hydrogenation in abs. alcohol in the presence of palladium-barium sulphate under 1 atm. pressure (3 hrs.); from the same trione on hydrogenation in abs. alcohol in the presence of platinum oxide under 3 atm. pressure (2 hrs.), followed by oxidation with CrO_3 in strong acetic acid (1940 Marker). From pregnane-3.6.20-trione (above) on heating with a little conc. HCl in glacial acetic acid on the steam-bath for 1 hr.; yield, 36% (1946 Moffett).

Rns. Gives allopregnane (p. 1403 s) on boiling with amalgamated zinc and conc. HCl in alcohol (1940 Marker). On hydrogenation in glacial acetic acid in the presence of platinum oxide under 3 atm. pressure it yields allopregnane-3β.6β.20β-triol (p. 2108 s) (1942a Marker).

Tris- (2.4-dinitrophenylhydrazone) $C_{39}H_{42}O_{12}N_{12}$ (not analysed). Crystals (chloroform and methanol), m. 270–275° (1948 Lieberman).

1939 Ehrenstein, *J. Org. Ch.* **4** 506, 512, 513, 517.
1940 CIBA (Soc. pour l'ind. chim. à Bâle; Ges. f. chem. Ind. Basel), *Fr. Pat.* 886 415 (issued 1943); *Ch. Ztbl.* **1944** I 450; (Miescher, Wettstein), *Swed. Pats.* 105 136–137 (1941, issued 1942); *Ch. Ztbl.* **1944** II 143; CIBA PHARMACEUTICAL PRODUCTS (Miescher, Wettstein), *U.S. Pat.* 2 323 277 (issued 1943); *C. A.* **1944** 222, 223.
     Ehrenstein, Stevens, *J. Org. Ch.* **5** 318, 323, 327.
     Marker, Jones, Turner, Rohrmann, *J. Am. Ch. Soc.* **62** 3006.
1942 a Marker, Crooks, Jones, Wittbecker, *J. Am. Ch. Soc.* **64** 219, 220.
     b Marker, Turner, Wittbecker, *J. Am. Ch. Soc.* **64** 809, 812.
1946 Moffett, Stafford, Linsk, Hoehn, *J. Am. Ch. Soc.* **68** 1857, 1859, 1860.
1948 Ehrenstein, *J. Org. Ch.* **13** 214, 220.
     Lieberman, Dobriner, Hill, Fieser, Rhoads, *J. Biol. Ch.* **172** 263, 288, 292.
1952 Balant, Ehrenstein, *J. Org. Ch.* **17** 1587, 1593.
1954 Salamon, Dobriner, *J. Biol. Ch.* **207** 323, 327.

$\varDelta^4$-Pregnene-3.11.20-trione, 11-Ketoprogesterone $C_{21}H_{28}O_3$. Rodlets (ether), m. 173–5° cor.; $[\alpha]_D^{18} +244°$, $[\alpha]_{5461}^{18} +283°$ (both in acetone) (1940, 1942 Reichstein; 1943 Hegner). — **Fmn.** From 11β-hydroxyprogesterone (p. 2840 s) on oxidation with CrO_3 in glacial acetic acid at 20° (1940 Reichstein). From 4-bromopregnane-3.11.20-trione (p. 2971 s) on refluxing with abs. pyridine for 6 hrs., followed by distillation in a high vacuum and chromatographic adsorption on Al_2O_3; yield, 5 % (1942 Reichstein; 1943 Hegner).

Pregnane-3.11.20-trione $C_{21}H_{30}O_3$. Hexagonal leaflets (acetone - ether), m. 161–2° cor. (Kofler block) (1944 v. Euw); need- les (ether, or ethyl acetate - petroleum ether), m. 154–6° (1943 Hegner), 157.5–159° cor. (1946 Long); $[\alpha]_D +120°$ to $+121°$ (acetone) (1943 Heg- ner; 1946 Long). — **Fmn.** From pregnane-3α.11α- diol-20-one (p. 2327 s) on oxidation with CrO_3 in 85 % acetic acid at 4° (1946 Long). From 12α-bromopregnane-3.11.20-trione (p. 2971 s) on heating with zinc dust and sodium acetate in glacial acetic acid for 15 min. at 70°; yield, 91 % (1944 v. Euw; cf. 1942 Reichstein; 1943 Hegner).

Rns. On hydrogenation in glacial acetic acid in the presence of previously reduced platinum oxide, until 1 mol. of hydrogen has been absorbed, it gives pregnan-3β-ol-11.20-dione (p. 2848 s) and small amounts of pregnane-3β.20β- diol-11-one (see p. 2722 s), of **pregnane-3.11-or-11.20-dione** $C_{21}H_{32}O_2$ [leaflets (petroleum ether), m. 102–3° cor. (Kofler block)], and probably also of pregnan- 3α-ol-11.20-dione (p. 2847 s) (1944 v. Euw; see also 1942 N. V. ORGANON; 1943 Reichstein). Gives the 4-bromo-derivative (p. 2971 s) on treatment with 1 mol. of bromine and a little HBr in glacial acetic acid at room temperature (1942 Reichstein; 1943 Hegner).

Allopregnane-3.11.20-trione $C_{21}H_{30}O_3$. Granula (ethyl acetate), m. 212–6⁰ cor.; sublimes in a high vacuum at 190⁰ (bath temperature); $[\alpha]_D^{20} + 133^0$ (abs. alcohol). — **Fmn.** From allopregnane-3β.11β.20-triol (see p. 2109 s) on oxidation with CrO_3 in glacial acetic acid at 16⁰; yield, 47 %. — **Rns.** On heating with amalgamated zinc and conc. HCl on the water-bath, followed by hydrogenation in glacial acetic acid in the presence of platinum oxide, it gives allopregnane (p. 1403 s) (1938 Steiger).

An **allopregnanetrione** $C_{21}H_{30}O_3$ [needles (methanol), m. 127–9⁰], assumed to be allopregnane-3.11.20-trione, was obtained by oxidation of the digitonin-precipitable carbinol fraction of mares' pregnancy urine with CrO_3 in strong acetic acid after temporary protection of double bonds with bromine; on refluxing with amalgamated zinc and conc. HCl in glacial acetic acid it gives allopregnane (p. 1403 s). Gives a *disemicarbazone* $C_{23}H_{36}O_3N_6 + \frac{1}{2} H_2O$, m. above 300⁰ (1938c, 1939a Marker).

9-iso-Pregnane-3.11.20-trione $C_{21}H_{30}O_3$. This structure was ascribed to **urane-3.11.20-trione** by Marker (1938a, b, d; 1939a, b) (see also 1939 PARKE, DAVIS & Co.); for a **urene-3.11.20-trione** $C_{21}H_{28}O_3$, see 1938a Marker; 1939 PARKE, DAVIS & Co. According to Klyne (1950, 1952) the urane compounds are probably D-homo-steroids (see urane, p. 1403 s).

1938 a Marker, Kamm, Oakwood, Wittle, Lawson, *J. Am. Ch. Soc.* **60** 1061.
   b Marker, Lawson, Rohrmann, Wittle, *J. Am. Ch. Soc.* **60** 1555, 1558.
   c Marker, Rohrmann, Wittle, *J. Am. Ch. Soc.* **60** 1561, 1563.
   d Marker, Wittle, Oakwood, *J. Am. Ch. Soc.* **60** 1567.
   Steiger, Reichstein, *Helv.* **21** 161, 167, 168.
1939 a Marker, Rohrmann, *J. Am. Ch. Soc.* **61** 2537, 2539.
   b Marker, Rohrmann, *J. Am. Ch. Soc.* **61** 2719.
   PARKE, DAVIS & Co. (Marker), *Brit. Pat.* 522066 (issued 1940); *Ch. Ztbl.* **1941** I 1443.
1940 Reichstein, Fuchs, *Helv.* **23** 684, 688.
1942 N. V. ORGANON, *Swiss Pat.* 256509 (issued 1949); *Ch. Ztbl.* **1949** E 1869.
   Reichstein, *Fr. Pat.* 887641 (issued 1943); *Ch. Ztbl.* **1944** II 878; *U.S. Pat.* 2403683 (1943, issued 1946); *C. A.* **1946** 6216; *Ch. Ztbl.* **1947** 232.
1943 Hegner, Reichstein, *Helv.* **26** 721, 727–729.
   Reichstein, *Fr. Pat.* 898140 (issued 1945); *Ch. Ztbl.* **1946** I 1747, 1748; *Brit. Pat.* 594878 (issued 1947); *C. A.* **1948** 2404; *U.S. Pat.* 2440874 (issued 1948); *C. A.* **1948** 5622; *Ch. Ztbl.* **1948** E 456.
1944 v. Euw, Lardon, Reichstein, *Helv.* **27** 821, 831, 832.
1946 Long, Marshall, Gallagher, *J. Biol. Ch.* **165** 197, 203.
1950 Klyne, *Nature* **166** 559.
1952 Klyne, Shoppee, *Chemistry & Industry* **1952** 470.

Δ⁹⁽¹¹⁾-Pregnene-3.12.20-trione $C_{21}H_{28}O_3$. Granula (acetone - petroleum ether), m. 184–6⁰ cor. (Kofler block) (1943 Hegner); prisms, m. 182–3⁰ cor. (Kofler block) (1944 v. Euw). Absorption max. in alcohol at 239 mμ (log ε 4.0) (1943 Hegner). — **Fmn.** Is obtained in small yield, along with other compounds, from Δ¹¹-pregnene-3.20-dione (p. 2793 s) on treatment with N-bromoacetamide and sodium acetate in aqueous acetone - acetic acid at 16⁰, collection of the main product (12α-bromopregnan-11β-ol-3.20-dione, p. 2850 s), oxidation of the mother liquor with CrO₃ in chloroform - glacial acetic acid, and warming with zinc dust and sodium acetate in glacial acetic acid at 70⁰ (1942 Reichstein; 1943 Hegner; 1944 v. Euw).

Pregnane-3.12.20-trione $C_{21}H_{30}O_3$. Plates (acetone), m. 204–6⁰ cor. (Kofler block) (1945 Sorkin; cf. 1940 Reichstein); after drying in a high vacuum for ¹/₂ hr. it shows $[\alpha]_D^{13}$ + 169⁰ (acetone) (1944 Katz; 1945 Sorkin; cf. 1940 Reichstein). — **Fmn.** From pregnane-3α.12α-diol-20-one (see p. 2329 s) on oxidation with CrO₃ in glacial acetic acid at 20⁰ (yield, 80%) (1940 Reichstein); is similarly obtained from pregnane-3α.12β-diol-20-one (see p. 2331 s; yield, 61%) (1945 Sorkin). — **Rns.** After refluxing with 5% methanolic KOH for 10 min., it is partially (12%) isomerized to 17-iso-pregnane-3.12.20-trione (below) (1945 Sorkin).

17-iso-Pregnane-3.12.20-trione $C_{21}H_{30}O_3$. Needles (ether - petroleum ether), m. 152–3⁰ cor. (Kofler block); $[\alpha]_D^{20}$ + 58⁰ (acetone). — **Fmn.** From 17-iso-pregnane-3α.12β-diol-20-one (see p. 2331 s) on oxidation with CrO₃ in glacial acetic acid at 20⁰; yield, 35%. From the above normal pregnanetrione on refluxing with 5% methanolic KOH for 10 min.; yield, 12% (1945 Sorkin).

1940 Reichstein, v. Arx, *Helv.* **23** 747, 750.
1942 Reichstein, *Fr. Pat.* 887641 (issued 1943); *Ch. Ztbl.* **1944** II 878; *U.S. Pat.* 2403683 (1943, issued 1946); *C. A.* **1946** 6216; *Ch. Ztbl.* **1947** 232.
1943 Hegner, Reichstein, *Helv.* **26** 721, 728.
1944 v. Euw, Lardon, Reichstein, *Helv.* **27** 821, 830, 831.
     Katz, Reichstein, *Pharm. Acta Helv.* **19** 231, 240, 261.
1945 Sorkin, Reichstein, *Helv.* **28** 875, 891.

Δ¹⁴-Cholene-3.7.22-trione (?), "*Triketocholene*" $C_{24}H_{34}O_3$. The structure was only provisionally assigned (1939/40 Kurauti); the position of the double bond and the configuration at C-5 have been confirmed on the basis of optical rotation of the parent triol (1946 Barton). — Needles (abs. alcohol or acetone), plates (benzene), m. 240–242⁰ be-

203*

coming brown. Gives a weak yellow Liebermann reaction, but no Jaffe reaction with picric acid. Instantly decolorizes $KMnO_4$ solution; adds on bromine (1939/40 Kurauti). — **Fmn.** From Δ^{14}-cholene-3.7.22-triol (see page 2129 s) on oxidation with CrO_3 in glacial acetic acid at 17° (1939/40 Kurauti). **Rns.** On refluxing with amalgamated zinc and conc. HCl in glacial acetic acid it gives an unsaturated oily compound which on hydrogenation in the presence of Adams's platinum oxide yields an oil (1939/40 Kurauti).

Trioxime $C_{24}H_{37}O_3N_3$. Scales (abs. alcohol), dec. 247° (1939/40 Kurauti).

Cholane-3.7.22-trione(?), *"Triketocholane"* $C_{24}H_{36}O_3$. For the structure, cf. that of the preceding compound. — Crystals (alcohol), dec. 245°. Gives no Jaffe reaction with picric acid. — **Fmn.** From cholane-3.7.22-triol (see p. 2130 s) on oxidation with CrO_3 in glacial acetic acid at room temperature (1939/40 Kurauti).

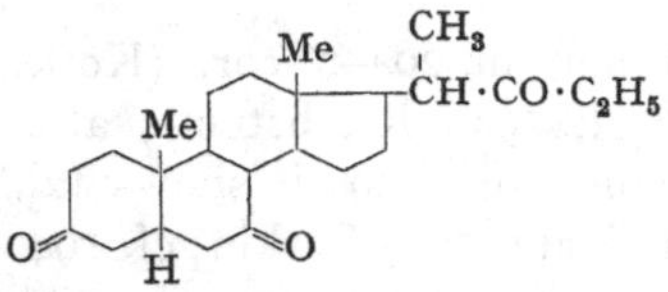

1939/40 Kurauti, Kazuno, *Z. physiol. Ch.* **262** 53, 58–60.
1946 Barton, *J. Ch. Soc.* **1946** 1116, 1120.

25-Methylcoprostane-7.12.24-trione (?), Triketo-isobufostane $C_{28}H_{44}O_3$. Needles (aqueous acetone), m. 163°; insoluble in ether. — **Fmn.** Along with isobufostane (p. 1439 s), from tetraketo-isobufostane (p. 2985 s) on boiling with amalgamated zinc and conc. HCl in glacial acetic acid for ca. 3 hrs.; yield, 10% (1940 Kazuno).

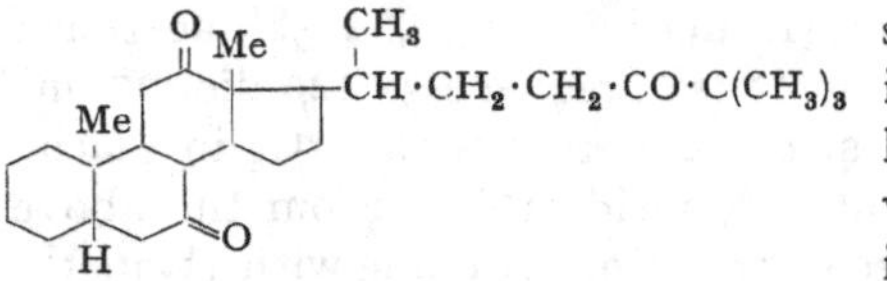

1940 Kazuno, *Z. physiol. Ch.* **266** 11, 29.

3.12-Diketonorcholanyl phenyl ketone $C_{30}H_{40}O_3$. Crystals (isopropyl alcohol), m. 170–171.5°; $[\alpha]_D^{25}$ +87.5° (dioxane). — **Fmn.** From 3α.12α - dihydroxynorcholanyl phenyl ketone (p. 2350 s) on oxidation with CrO_3 in strong acetic acid below room temperature (quantitative yield); may be similarly obtained from 3α-hydroxy-12-ketonorcholanyl phenyl ketone (p. 2849 s) (1945 Hoehn).

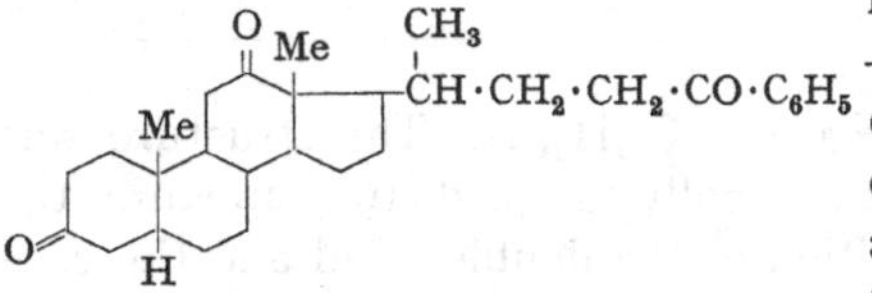

1945 Hoehn, Moffett, *J. Am. Ch. Soc.* **67** 740, 742.

II. HALOGENO-TRIKETOSTEROIDS WITH TWO CO GROUPS IN THE RING SYSTEM

21-Chloro-Δ^4-pregnene-3.6.20-trione $C_{21}H_{27}O_3Cl$. Needles or granula (acetone), m. 215–220° cor.; sublimes at 190° (bath temperature) under 0.01 mm. pressure. Absorption max. in alcohol (from the graph) at 312 and 248 mμ (log ε 3.2 and 4.1, resp.). Reduces ammoniacal silver salt solution at room temperature. — **Fmn.** From 21-chloro-Δ^5-pregnen-3β-ol-20-one (p. 2277 s) on oxidation with CrO_3 in glacial acetic acid at room temperature; yield, 18% (1937 Steiger).

4-Bromopregnane-3.6.20-trione $C_{21}H_{29}O_3Br$ (I). Crystals (ether), m. 132–132.5° dec.; $[\alpha]_D^{28}$ +26° (dioxane). — **Fmn.** From pregnane-3.6.20-trione (p. 2966 s) on treatment with 1 mol. of bromine in glacial acetic acid at 15°; yield, 70%. — **Rns.** Gives Δ^4-pregnene-3.6.20-trione (6-ketoprogesterone, p. 2965 s) on refluxing with dry pyridine for 2 hrs. (1946 Moffett).

4-Bromopregnane-3.11.20-trione $C_{21}H_{29}O_3Br$ (II) (not analysed). Needles (abs. ether), m. 158–160° (Kofler block). — **Fmn.** From pregnane-3.11.20-trione (p. 2967 s) on treatment with 1 mol. of bromine and a little HBr in glacial acetic acid at room temperature; yield, 62%. — **Rns.** Gives a small yield of Δ^4-pregnene-3.11.20-trione (11-ketoprogesterone, p. 2967 s) on refluxing with abs. pyridine for 6 hrs., followed by distillation in a high vacuum (1942 Reichstein; 1943 Hegner).

12α-Bromopregnane-3.11.20-trione $C_{21}H_{29}O_3Br$ (III). Crystals (acetone - ether), m. 192–3° cor. (Kofler block) (1944 v. Euw). — **Fmn.** From 12α-bromopregnan-11β-ol-3.20-dione (see p. 2850 s) on oxidation with CrO_3 in alcohol-free chloroform · glacial acetic acid at room temperature; yield, 67% (1944 v.Euw; cf. 1942 Reichstein; 1943 Hegner). — **Rns.** Gives pregnane-3.11.20-trione (p. 2967 s) on heating with zinc dust and sodium acetate in glacial acetic acid for 15 min. at 70° (1944 v. Euw; cf. 1942 Reichstein; 1943 Hegner).

For **bromo-uranetrione** $C_{21}H_{29}O_3Br$ (1938 Marker; 1939 PARKE, DAVIS & Co.), see the remarks under uranetrione (p. 2968 s).

1937 Steiger, Reichstein, *Helv.* **20** 1164, 1175.
1938 Marker, Kamm, Oakwood, Wittle, Lawson, *J. Am. Ch. Soc.* **60** 1061, 1065.
1939 PARKE, DAVIS & Co. (Marker), *Brit. Pat.* 522066 (issued 1940); *Ch. Ztbl.* **1941** I 1443.
1942 Reichstein, *Fr. Pat.* 887641 (issued 1943); *Ch. Ztbl.* **1944** II 878; *U.S. Pat.* 2403683 (1943, issued 1946); *C. A.* **1946** 6216; *Ch. Ztbl.* **1947** 232.
1943 Hegner, Reichstein, *Helv.* **26** 721, 726, 728.
1944 v. Euw, Lardon, Reichstein, *Helv.* **27** 821, 831.
1946 Moffett, Stafford, Linsk, Hoehn, *J. Am. Ch. Soc.* **68** 1857, 1860.

III. HYDROXY-TRIKETOSTEROIDS WITH TWO CO GROUPS IN THE RING SYSTEM,
AND ONE CO GROUP AND OH IN SIDE CHAIN

Δ^4-Pregnen-21-ol-3.11.20-trione, 11-Dehydrocorticosterone, Dehydrocorticosterone, *Kendall and Mason's Compound A,*

$C_{21}H_{28}O_4$ (p. 164). For the name "dehydrocorticosterone", see 1937 Reichstein. For the structure, see also 1937 Mason. — Prisms (isopropyl alcohol), crystals (abs. alcohol), m. 177° to 180° (1937 Mason); crystals (acetone - ether), m. 174–180° cor. (1938a Reichstein); prisms (acetone), m. 178–180° cor. (1939 Kuizenga); $[\alpha]_D^{25} + 258°$ (95 % alcohol) (1939 Kuizenga), $[\alpha]_{5461}^{25}$ in 95 % alcohol $+ 297°$ (1939 Kuizenga), $+ 299°$ (1937 Mason).

For *physiological properties*, see pp. 1367 s, 1368 s, and the literature cited there.

Occ. For isolation from beef adrenal cortex extracts, see also 1937 Kendall; 1937 Mason; 1939 Kuizenga.

Fmn. The acetate is obtained from the 21-acetate of Δ^5-pregnene-3β.21-diol-11.20-dione (p. 2860 s) on heating with aluminium tert.-butoxide and acetone in abs. benzene i nan evacuated sealed tube on the water-bath for 25 hrs.; yield, 15 % (1946 v. Euw). The acetate is also obtained from the 21-acetate of the two 20-epimeric Δ^4-pregnene-20.21-diol-3.11-diones (p. 2936 s) by oxidation with CrO_3 in strong acetic acid at 18°; yields, 80 % (1946 Sarett). The "11-epi-corticosterone 21-acetate", from which dehydrocorticosterone acetate may be obtained by oxidation with CrO_3 in glacial acetic acid (1943 Shoppee), was probably impure corticosterone acetate (Gallagher, see 1949 Fieser). Dehydrocorticosterone acetate is obtained from the acetate of 4-bromopregnan-21-ol-3.11.20-trione (p. 2975 s) by refluxing with abs. pyridine for 5 hrs., followed by chromatographic adsorption on Al_2O_3; yield, 10 % (1943 Lardon; 1942 N. V. ORGANON; 1942a, b, 1943 Reichstein).

Free dehydrocorticosterone is obtained from its acetate by refluxing with aqueous alcoholic HCl (1937 Reichstein), with methanolic HCl (1942 N. V. ORGANON; 1942a, 1943 Reichstein), or in better yield (87 %) by treatment with $KHCO_3$ in aqueous methanol for 40 hrs. at room temperature (1938a, 1940, 1942a, 1943 Reichstein; 1942 N. V. ORGANON).

Rns. Gives 3.11-diketo-Δ^4-ætiocholenic acid on oxidation with periodic acid at room temperature in aqueous alcoholic H_2SO_4 (1937 Mason) or in aqueous methanol (1940 Reichstein). On shaking with hydrogen and palladium black in abs. alcohol it gives allopregnan-21-ol-3.11.20-trione (p. 2973 s) (1937 Mason).

Analytical properties. Dehydrocorticosterone reduces phosphomolybdic acid (Folin-Wu reagent) in glacial acetic acid at 100° to give molybdenum blue; for the rate of this reaction and for its use for colorimetric estimation of dehydrocorticosterone, see 1946 Heard.

References, pp. 2973 s, 2974 s

Acetate $C_{23}H_{30}O_5$ (p. 165). Needles (dil. methanol or acetone - ether), m. 181°
to 181.5° cor. (1940 Reichstein; cf. 1942, 1946 v. Euw; 1942 N. V. ORGANON;
1942a, 1943 Reichstein; 1943 Lardon; 1943 Shoppee), 182.5–183.5° cor. (1946
Sarett; cf. also 1943 Lardon; 1946 v. Euw); $[\alpha]_D$ in acetone $+211°$ (1942a
Reichstein; cf. also 1946 v. Euw), in dioxane $+234°$ (1942 v. Euw); $[\alpha]_{5461}$
in acetone $+263°$ (1943 Shoppee), $+257°$ (1946 v. Euw), in dioxane $+285°$
(1942 v. Euw). Absorption max. in alcohol at 237.5 mμ (E% 386) (1946
Sarett). — **Fmn.** See that of the free dehydrocorticosterone (p. 2972 s).

Pregnan-21-ol-3.11.20-trione $C_{21}H_{30}O_4$.

Acetate $C_{23}H_{32}O_5$. Needles (ether, ether -
petroleum ether, or acetone - ether 1:5), m. 157°
to 158° cor. (Kofler block) (1944 v. Euw), 153°
to 155° cor. (Kofler block); $[\alpha]_D^{22} +107°$ (acetone); it strongly and rapidly reduces alkaline
silver diammine solution at room temperature
(1943 Lardon; 1942b, 1943 Reichstein; 1942 N. V. ORGANON). — **Fmn.** Is
obtained by oxidation with CrO_3 in glacial acetic acid at room temperature
from the 21-acetates of pregnane-3β.11β.21-triol-20-one (see p. 2354 s) (1943
Reichstein), pregnane-11β.21-diol-3.20-dione (see p. 2856 s) (1944 v. Euw),
pregnane-3α.21-diol-11.20-dione (p. 2860 s) (1944 v. Euw; 1942 N. V. ORGA-
NON; 1943 Reichstein), and pregnane-3β.21-diol-11.20-dione (p. 2861 s) (1943
Lardon; 1942 N. V. ORGANON; 1942a, 1943 Reichstein). From the acetate of Δ^{11}-
pregnen-21-ol-3.20-dione (p. 2830 s) on treatment with N-bromoacetamide in
aqueous acetone at room temperature (16 hrs.), followed by oxidation of
the resulting bromohydrin with CrO_3 in strong acetic acid at 18° and heating
with zinc dust and anhydrous sodium acetate in glacial acetic acid on the
water-bath; may be similarly obtained from the 21-acetate of Δ^{11}-pregnene-
3α.21-diol-20-one (see p. 2308s) (1942b Reichstein). — **Rns.** Gives the 4-bromo-
derivative (p. 2975 s) on treatment with 1 mol. of bromine in glacial acetic
acid (1943 Lardon; 1942a, b, 1943 Reichstein; 1942 N. V. ORGANON); for
doubtful homogeneity of the reaction product, see 1946 v. Euw.

Allopregnan-21-ol-3.11.20-trione $C_{21}H_{30}O_4$. For the allo-configuration at C-5,

cf. also 1938 Steiger; 1938b Reichstein. —
Crystals (aqueous acetone), m. 174–6°; $[\alpha]_{5461}^{25}$
$+163°$ (alcohol) (1937 Mason). — Has a very
weak cortin-like activity (1937 Mason). — **Fmn.**
From Δ^4-pregnen-21-ol-3.11.20-trione (dehydrocorticosterone, p. 2972 s) on hydrogenation in
abs. alcohol in the presence of palladium black at room temperature; yield,
81% (1937 Mason). — **Rns.** Gives 3.11-diketo-ætioallocholanic acid on oxidation with periodic acid in dil. H_2SO_4 (1937 Mason; cf. 1938 Steiger; 1938b
Reichstein).

1937 Kendall, *Cold Spring Harbor Symp. Quant. Biol.* **5** 299.
    Mason, Hoehn, McKenzie, Kendall, *J. Biol. Ch.* **120** 719, 728, 732.
    Reichstein, *Helv.* **20** 953, 960, 967.

1938 a Reichstein, v. Euw, *Helv.* **21** 1181, 1183.
    b Reichstein, *Helv.* **21** 1490, 1491.
    Steiger, Reichstein, *Helv.* **21** 161, 165.
1939 Kuizenga, Cartland, *Endocrinology* **24** 526, 532.
1940 Reichstein, Fuchs, *Helv.* **23** 676, 678.
1942 v. Euw, Reichstein, *Helv.* **25** 988, 1022.
    N. V. Organon, *Swiss Pats.* 254993, 256509 (both issued 1949); *Ch. Ztbl.* **1950** I 2388; **1949** E
    1869.
    a Reichstein, *Swiss Pat.* 244341 (issued 1947); *C. A.* **1949** 5812.
    b Reichstein, *Fr. Pat.* 887641 (issued 1943); *Ch. Ztbl.* **1944** II 878; *U.S. Pat.* 2403683 (1943,
    issued 1946); *C. A.* **1946** 6216; *Ch. Ztbl.* **1947** 232.
1943 Lardon, Reichstein, *Helv.* **26** 747, 754, 755; Reichstein, *U.S. Pat.* 2401775 (issued 1946);
    C. A. **1946** 5884; *Ch. Ztbl.* **1947** 74; *U.S. Pat.* 2404768 (1945, issued 1946); *C. A.* **1946**
    6222; *Ch. Ztbl.* **1947** 233.
    Reichstein, *Fr. Pat.* 898140 (issued 1945); *Ch. Ztbl.* **1946** I 1747, 1748; *Brit. Pat.* 594878
    (issued 1947); *C. A.* **1948** 2404; *U.S. Pat.* 2440874 (issued 1948); *C. A.* **1948** 5622; *Ch. Ztbl.*
    1948 E 456.
    Shoppee, Reichstein, *Helv.* **26** 1316, 1327, 1328.
1944 v. Euw, Lardon, Reichstein, *Helv.* **27** 1287, 1293, 1295.
1946 v. Euw, Reichstein, *Helv.* **29** 1913, 1914, 1919.
    Heard, Sobel, *J. Biol. Ch.* **165** 687.
    Sarett, *J. Am. Ch. Soc.* **68** 2478, 2482.
1949 Fieser, Fieser, *Natural Products Related to Phenanthrene*, 3rd Ed., New York, p. 409.

Δ⁴-Pregnen-21-ol-3.12.20-trione $C_{21}H_{28}O_4$. Needles (benzene - ether), m. 180°
to 183° cor. (Kofler block); $[\alpha]_D^{22}$ +239° (dioxane), +215° (acetone); $[\alpha]_{5461}^{22}$ +298° (dioxane), +266° (acetone) (1943 Fuchs). — **Fmn.**
The acetate is obtained from that of 4-bromopregnan-21-ol-3.12.20-trione (p. 2975 s) on refluxing with abs. pyridine for 5 hrs.; it is hydrolysed by $KHCO_3$ in aqueous methanol at room temperature (1943 Fuchs;
1943 Reichstein).

Acetate $C_{23}H_{30}O_5$. Prisms (acetone - ether), m. 182–4° cor. (Kofler block);
$[\alpha]_D^{14}$ +229° (acetone) (1943 Fuchs; 1943 Reichstein). Absorption max. in
abs. alcohol at 240 mμ (log ε 4.13) (1943 Fuchs). — Shows a very weak, if any,
activity in the Everse-de Fremery test (1943 Fuchs). — **Fmn.** See above. —
Rns. Reduces phosphomolybdic acid (Folin-Wu reagent) in glacial acetic acid
at 100° to give molybdenum blue; for the rate of this reaction and for its
use for colorimetric determination, see 1946 Heard.

Pregnan-21-ol-3.12.20-trione $C_{21}H_{30}O_4$.

Acetate $C_{23}H_{32}O_5$. Prisms (benzene - ether-petroleum ether), m. 189–191° cor. (Kofler block); $[\alpha]_D^{17}$ +153° (acetone) (1943 Fuchs). —
Fmn. From the 21-acetate of pregnane-3α.
12α.21-triol-20-one (see p. 2356 s) on oxidation
with CrO_3 (2 mols.) in glacial acetic acid at 18°
(yield, 73%) (1943 Fuchs; 1943 Reichstein); similarly from the 21-acetates

of pregnane-12α.21-diol-3.20-dione (p. 2858 s; yield, 75 %) and pregnane-3α.21-diol-12.20-dione (p. 2863 s; yield, 67 %) (1943 Fuchs). — **Rns.** Gives the 4-bromo-derivative (below) on treatment with 1 mol. of bromine in glacial acetic acid (1943 Fuchs; 1943 Reichstein).

1943 Fuchs, Reichstein, *Helv.* **26** 511, 515, 520, 522, 524–526.
  Reichstein, *U.S. Pats.* 2401775, 2404768 (both issued 1946); *C. A.* **1946** 5884, 6222; *Ch. Ztbl.*
  1947 74, 233.

IV. Halogeno-hydroxy-triketosteroids with Two CO Groups in the Ring System, and One CO Group and OH in Side Chain

4-Bromopregnan-21-ol-3.11.20-trione $C_{21}H_{29}O_4Br$.

Acetate $C_{23}H_{31}O_5Br$ (not analysed). For doubtful homogeneity, see 1946 v. Euw. — **Crystals**, m. 180–185° cor. (Kofler block). — **Fmn.** From pregnan-21-ol-3.11.20-trione (p. 2973 s) on treatment with 1 mol. of bromine in glacial acetic acid. — **Rns.** Gives the acetate of Δ⁴-pregnen-21-ol-3.11.20-trione (dehydrocorticosterone, p. 2972 s) on refluxing with abs. pyridine for 5 hrs. (1943 Lardon; 1942, 1943a, b Reichstein; 1942 N. V. Organon).

4-Bromopregnan-21-ol-3.12.20-trione $C_{21}H_{29}O_4Br$.

Acetate $C_{23}H_{31}O_5Br$ (not analysed). Crystals, m. 182–187.5° cor. (Kofler block). — **Fmn.** From pregnan-21-ol-3.12.20-trione (p. 2974 s) on treatment with 1 mol. of bromine in glacial acetic acid with cooling; yield, 83 %. — **Rns.** Gives the acetate of Δ⁴-pregnen-21-ol-3.12.20-trione (p. 2974 s) on refluxing with abs. pyridine for 5 hrs. (1943 Fuchs; 1943 b Reichstein).

1942 N. V. Organon, *Swiss Pats.* 254993, 256509 (both issued 1949); *Ch. Ztbl.* **1950** I 2388; **1949** E
  1869.
  Reichstein, *Swiss Pat.* 244341 (issued 1947); *C. A.* **1949** 5812; *Fr. Pat.* 887641 (issued 1943);
  Ch. Ztbl. **1944** II 878; *U.S. Pat.* 2403683 (1943, issued 1946); *C. A.* **1946** 6216; *Ch. Ztbl.*
  1947 232.
1943 Fuchs, Reichstein, *Helv.* **26** 511, 525, 526.
  Lardon, Reichstein, *Helv.* **26** 747, 754.
  a Reichstein, *Fr. Pat.* 898140 (issued 1945); *Ch. Ztbl.* **1946** I 1747, 1748; *Brit. Pat.* 594878
    (issued 1947); *C. A.* **1948** 2404; *U.S. Pat.* 2440874 (issued 1948); *C. A.* **1948** 5622; *Ch. Ztbl.*
    1948 E 456.
  b Reichstein, *U.S. Pats.* 2401775, 2404768 (both issued 1946); *C. A.* **1946** 5884, 6222;
    Ch. Ztbl. **1947** 74, 233.
1946 v. Euw, Reichstein, *Helv.* **29** 1913, 1914.
  Heard, Sobel, *J. Biol. Ch.* **165** 687.

V. HYDROXY-TRIKETOSTEROIDS WITH TWO CO GROUPS AND ONE OH IN THE RING SYSTEM

Allopregnan-5α-ol-3.6.20-trione $C_{21}H_{30}O_4$. For the configuration at C-5, see 1948 Ehrenstein. — Prismatic crystals or platelets (95 % alcohol), m. 271° to a dark brown liquid (1939 Ehrenstein). — **Fmn.** From Δ^5-pregnen-3β-ol-20-one (p. 2233 s) on oxidation with CrO_3 in strong acetic acid at 27° (yield, 9%) (1939 Ehrenstein); similarly from its α-oxide (5α.6α-epoxyallopregnan-3β-ol-20-one, p. 2364 s; yield, ca. 16%) or on oxidation with $KMnO_4$ in aqueous acetic acid at 50° (small yield) (1941 Ehrenstein). From allopregnane-3β.5α.6β-triol-20-one (p. 2365 s) on oxidation with CrO_3 in strong acetic acid at 27° (yield, ca. 60%); similarly from the corresponding 6α-epimer (p. 2364 s; yield, ca. 30%) (1939 Ehrenstein). — **Rns.** Gives Δ^4-pregnene-3.6.20-trione (6-ketoprogesterone, p. 2965 s) on passage of dry HCl through its suspension in pure chloroform at 4–5° (1939 Ehrenstein).

1939 Ehrenstein, *J. Org. Ch.* **4** 506, 515–517.
1941 Ehrenstein, Stevens, *J. Org. Ch.* **6** 908, 918.
1948 Ehrenstein, *J. Org. Ch.* **13** 214, 219.

Pregnan-17α-ol-3.11.20-trione $C_{21}H_{30}O_4$. For the configuration at C-17, see 1949 Sarett. — Crystals (dil. acetone), m. 204.5° to 205.5° cor.; $[\alpha]_D^{20} +75°$ (acetone) (1946 Sarett). — **Fmn.** From pregnane-17α.20β-diol-3.11-dione (p. 2958 s) on oxidation with CrO_3 in strong acetic acid at room temperature; crude yield, 26% (1946 Sarett).

1946 Sarett, *J. Am. Ch. Soc.* **68** 2478, 2482.
1949 Sarett, *J. Am. Ch. Soc.* **71** 1169.

Pregnan-7α-ol-3.12.20-trione $C_{21}H_{30}O_4$. Originally assumed to be *pregnan-12α-ol-3.7.20-trione* (1940 Ehrenstein); for the true structure, see 1948 Ehrenstein; cf. also 1947 Lardon; 1951 Ruff. — **Fmn.** Its acetate is obtained from the 7-acetate of pregnane-3α.7α.12α-triol-20-one (see p. 2366 s) on oxidation with CrO_3 (2.2 at. of oxygen) in strong acetic acid at room temperature; yield, 69% (1940 Ehrenstein).

Acetate $C_{23}H_{32}O_5$. Needles (ether), m. 160.5–163.5°; $[\alpha]_D^{26} +126°$ (acetone) (1940 Ehrenstein). — **Fmn.** See above.

1940 Ehrenstein, Stevens, *J. Org. Ch.* **5** 660, 670.
1947 Lardon, *Helv.* **30** 597, 598 footnote 5.
1948 Ehrenstein, *J. Org. Ch.* **13** 214, 222.
1951 Ruff, Reichstein, *Helv.* **34** 70, 78.

Δ⁴-Pregnene-17α.21-diol-3.11.20-trione, 17-Hydroxy-11-dehydrocortico-sterone, Cortisone, *Reichstein's Substance Fa,*

$Δ⁴$-**Pregnene-17α.21-diol-3.11.20-trione, 17-Hydroxy-11-dehydrocortico-**
sterone, Cortisone, *Reichstein's Substance Fa,*
Wintersteiner's Compound F, Kendall-Mason's
Compound E, $C_{21}H_{28}O_5$ (p. 165). For the structure, see also 1938a Mason.

Rhombohedral plates and flat needles (abs. alcohol), m. 213–7° cor. (1938 Reichstein); rhombohedra (95 % alcohol or acetone), m. 215–8° cor. dec. (1939 Kuizenga); a sample m. 201–8° dec. showed $[α]_{5461}^{25}$ +269° (benzene) (1936 Mason). Absorption max. in alcohol at 240 mμ (ε 14000) (1936 Wintersteiner). Fairly soluble in cold methanol, ethanol, and acetone, much less so in ether, benzene, and chloroform, slightly soluble in water (1936 Wintersteiner).

Occ. For occurrence in, and isolation from, beef adrenal cortex, see also 1937 Kendall; 1939 Kuizenga. Occurs in higher amount in hog adrenal cortex (1945 Kuizenga).

Fmn. The 21-acetate is obtained from that of $Δ⁴$-pregnene-11β.17α.21-triol-3.20-dione (Reichstein's Substance M, p. 2868 s) on oxidation with CrO_3 in glacial acetic acid at room temperature (yield, 50%) (1937 Reichstein) or, along with $Δ⁴$-androstene-3.11.17-trione (adrenosterone, p. 2979 s), from that of $Δ⁴$-pregnene-17α.20β.21-triol-3.11-dione (p. 2959 s) on oxidation with CrO_3 in strong acetic acid at 11–24° (1946 Sarett; 1946 MERCK & Co., INC.); the acetate is hydrolysed by $KHCO_3$ in aqueous methanol at room temperature (1938 Reichstein).

For *syntheses* of cortisone, carried out after the closing date for this volume, see, e.g., 1953 Rosenkranz.

Rns. On standing with calcium hydroxide in aqueous alcohol under nitrogen for 48 hrs. it gives $Δ⁴$-androstene-3.11.17-trione (adrenosterone, p. 2979 s) and an acid fraction which on oxidation with dichromate-sulphuric acid in aqueous acetone yielded a further amount of adrenosterone and 3.11-diketo-$Δ⁴$-ætioecholenic acid (1938b Mason). On heating with excess p-nitrobenzoyl chloride in pyridine at 100° for 2 hrs. it gave prisms (alcohol), m. 220–221° cor. dec., probably a mixture of a *mono-p-nitrobenzoate* $C_{28}H_{31}O_8N$ and the *di-p-nitro-benzoate* $C_{35}H_{34}O_{11}N_2$ (1936 Wintersteiner).

Physiological properties. For the effects of cortisone on rheumatoid arthritis and for other physiological effects, see, e.g., 1949, 1950 Hench; 1950 Sprague.

Analytical properties. Reduces phosphomolybdic acid (Folin-Wu reagent) in glacial acetic acid at 100° to give molybdenum blue; for the rate of this reaction and for its use for colorimetric determination of cortisone, see 1946 Heard.

Disemicarbazone $C_{23}H_{34}O_5N_6$ (p. 165). Absorption max. in alcohol at 270 mμ (ε 29000) and ca. 240 mμ (ε 19000) (1936 Wintersteiner).

21-Acetate $C_{23}H_{30}O_6$ (p. 165). Crystals (acetone), m. 235–8° cor. after slight sintering at 230°; $[α]_D$ +164° (acetone) (1946 Sarett; 1946 MERCK & Co., INC.). Absorption max. (solvent not stated) at 238 mμ (ε 15800–15900) (1946 Sarett).

1936 Mason, Myers, Kendall, *J. Biol. Ch.* **114** 613, 626.
Wintersteiner, Pfiffner, *J. Biol. Ch.* **116** 291.

1937 Kendall, *Cold Spring Harbor Symp. Quant. Biol.* **5** 299.
Reichstein, *Helv.* **20** 978, 989.

1938 a Mason, Hoehn, Kendall, *J. Biol. Ch.* **124** 459.
     b Mason, *J. Biol. Ch.* **124** 475; *Proc. Staff Meetings Mayo Clinic* **13** 235.
Reichstein, v. Euw, *Helv.* **21** 1181, 1182.

1939 Kuizenga, Cartland, *Endocrinology* **24** 526, 531.

1945 Kuizenga, Nelson, Lyster, Ingle, *J. Biol. Ch.* **160** 15, 18.

1946 Heard, Sobel, *J. Biol. Ch.* **165** 687.
MERCK & Co., INC. (Sarett), *U.S. Pat.* 2462133 (issued 1949); *C. A.* **1949** 4432; *Fr. Pat.* 942260 (1947, issued 1949); *Ch. Ztbl.* **1950** I 583, 584; *Brit. Pat.* 631238 (issued 1949); *C. A.* **1950** 4048, 4050.
Sarett, *J. Biol. Ch.* **162** 601, 628–630.

1949 Hench, Kendall, Slocumb, Polley, *Trans. Assoc. Am. Physicians* **62** 64.

1950 Hench, Kendall, Slocumb, Polley, *Arch. Internal Med.* **85** 545–666.
Sprague, Power, Mason, Albert, Mathieson, Hench, Kendall, Slocumb, Polley, *Arch. Internal Med.* **85** 199–258.

1953 Rosenkranz, Sondheimer, *Syntheses of Cortisone*, in *Progress in the Chemistry of Organic Natural Products* Vol. **10**, Vienna, pp. 274–389.

F. Triketosteroids with All CO Groups in the Ring System

I. COMPOUNDS WITHOUT OTHER FUNCTIONAL GROUPS

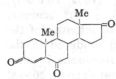

Δ^4-**Androstene-3.6.17-trione** $C_{19}H_{24}O_3$ (p. 165). Needles (96% alcohol), m. 221⁰ to 222⁰ cor. (1937 Ushakov). Absorption max. in chloroform at 252 mμ (ε 10800) (1939 Dannenberg); in alcohol- NaOH at 260 mμ (1951 Devis). — **Fmn.** For its formation from Δ^5-androsten-3β-ol-17-one (5-dehydroepiandrosterone, p. 2609 s) by oxidation, see also 1936 SCHERING A.-G. From androstan-5α-ol-3.6.17-trione (p. 2983 s) on treatment with dry HCl in chloroform at 0⁰; yield, 76% (1937 Ushakov). — **Rns.** Boiling with zinc dust in strong acetic acid yields androstane-3.6.17-trione (below) (1939 Ushakov).

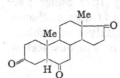

Androstane-3.6.17-trione $C_{19}H_{26}O_3$. Needles (methanol), m. 191–2⁰ (1939 Ushakov); $[\alpha]_D^{28}$ +71⁰ (chloroform); does not absorb ultra-violet light between 222 and 300 mμ (1952 Balant). —**Fmn.** From Δ^4-androstene-3.6.17-trione (above) on refluxing with zinc dust in strong acetic acid for 4 hrs. (1939 Ushakov). From the acetate of Δ^4-androsten-6α-ol-3.17-dione (p. 2945 s) on treatment with glacial acetic acid and some dil. H_2SO_4 at room temperature (yield, 66%); from the 6-epimeric acetate (p. 2946 s) on heating with 2.2 mols. of KOH in dil. methanol on the water-bath for 1 hr. (yield, 60%) (1952 Balant).

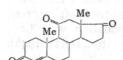

Δ^4-**Androstene-3.11.17-trione, Adrenosterone** $C_{19}H_{24}O_3$ (p. 165). $[\alpha]_D^{25}$ +281⁰ (acetone) (1946a Sarett; 1946a MERCK & Co.); $[\alpha]_D^{24}$ +284⁰ (chloroform) (1952 Fried); absorption max. in alcohol at 235 mμ (ε 12500) (1939 Dannenberg).

Occ. For isolation from adrenal cortex, see also 1941 v. Euw.

Fmn. From Δ^4-pregnene-17α.20β-diol-3.11-dione (p. 2958 s) on oxidation with periodic acid in 80% methanol at room temperature (1946b Sarett; 1947 MERCK & Co.). Together with the 21-acetate of Δ^4-pregnene-17α.21-diol-3.11.20-trione (Kendall's Compound E, p. 2977 s), from the 21-acetate of Δ^4-pregnene-17α.20β.21-triol-3.11-dione (Reichstein's Substance U, p. 2959 s) on oxidation with CrO_3 in strong acetic acid at room temperature (1946a Sarett; 1946a MERCK & Co.). From Δ^4-pregnene-17α.21-diol-3.11.20-trione dissolved in alcohol on treatment with a saturated solution of $Ca(OH)_2$ under nitrogen, or from the acidic fraction, obtained at the same time in this reaction, on oxidation with CrO_3 (1938b Mason). From 11β.17-dihydroxy-3-keto-Δ^4-ætiocholenic acid on oxidation with CrO_3 (1938a Mason). From 17-hydroxy-3.11-diketo-Δ^4-ætiocholenic acid on oxidation with $K_2Cr_2O_7$ and H_2SO_4 in aqueous acetone at room temperature (1936b Mason).

Disemicarbazone $C_{21}H_{30}O_3N_6$ (p. 165). Absorption max. (solvent not stated) at 233 and 271 mμ (ε 48000 and 55000, resp.) (1939 Dannenberg).

Androstane-3.11.17-trione $C_{19}H_{26}O_3$ (p. 165). Colourless leaflets (acetone-ether), m. 182–3° cor.; $[\alpha]_D^{12}$ +152° (acetone) (1942 v. Euw); $[\alpha]_D^{23}$ +160° (chloroform) (1952 Heusser); $[\alpha]_{5461}^{25}$ +191° (alcohol) (1938a Mason), +229° (benzene) (1936a Mason; cf. 1938a Mason). The intensity of the colour (absorption max. at 520 mμ) produced in the Zimmermann reaction with m-dinitrobenzene in alcoholic KOH is 117 % of that developed by 5-dehydroepiandrosterone (p. 2609 s) (1945a, b Mason).

Fmn. From the following compounds on oxidation with CrO_3 in acetic acid at room temperature: allopregnane-3β.11β.17.21-tetrol-20-one (Reichstein's Substance V, p. 2375 s) (1942 v. Euw), androstane-3.17β-diol-11-one (p. 2743 s) (1937 Steiger; cf. 1952 Heusser), androstane-3α.11β-diol-17-one (p. 2748 s) (1945a, b Mason), androstan-3β-ol-11.17-dione (p. 2949 s) (1937 Steiger; see also 1936 Reichstein; 1941 Reich). From 17-hydroxy-3.11-diketo-Δ^4-ætio-cholenic acid on hydrogenation, using platinum oxide catalyst in alcoholic NaOH, followed by oxidation with $K_2Cr_2O_7$ and H_2SO_4 in aqueous acetone (1938a Mason).

5-iso-Androstane-3.11.17-trione, Aetiocholane-3.11.17-trione $C_{19}H_{26}O_3$. Crystals (ether), m. 134–5° cor.; $[\alpha]_D^{20}$ +155° (acetone) (1946b Sarett; 1946b MERCK & Co.); $[\alpha]_D^{18}$ +148.5° (alcohol) (1946, 1948 Lieberman); $[\alpha]_D^{23}$ +148° (chloroform) (1953 Peterson). — **Fmn.** From ætiocholan-3α-ol-11.17-dione (p. 2949 s) on oxidation with CrO_3 in acetic acid at room temperature (1946b Sarett; 1946c MERCK & Co.; 1946, 1948 Lieberman). Together with pregnan-17α-ol-3.11.20-trione (p. 2976 s), from pregnane-17α.20β-diol-3.11-dione (p. 2958 s) on oxidation with CrO_3 in acetic acid at room temperature; yield, 61 % (1946b Sarett; 1946b MERCK & Co.).

5-iso-Androstane-3.12.17-trione, Aetiocholane-3.12.17-trione $C_{19}H_{26}O_3$. Crystals (benzene - petroleum ether), m. 276–280° cor. with some decomposition; sublimes at 170° and 0.01 mm. pressure; $[\alpha]_D^{19}$ +232° (chloroform). — **Fmn.** From ætiocholane-3α.12α-diol-17-one (p. 2749 s) on oxidation with excess CrO_3 in glacial acetic acid. — **Rns.** Treatment with 1 mol. bromine in glacial acetic acid gives crystals m. 184–6° cor. dec. Boiling with 5 % methanolic KOH yields equal parts of acids and neutral compounds (1945 Reich).

Trioxime $C_{19}H_{29}O_3N_3 + \frac{1}{2} H_2O$. White crystals (acetone - petroleum ether), m. ca. 200° cor. (1945 Reich).

Δ^4-Androstene-3.16.17-trione $C_{19}H_{24}O_3$.

16 - Oxime, 16 - Oximino - Δ^4 - androstene - 3.17 - dione, *16-Isonitroso-Δ^4-androstene-3.17-dione* $C_{19}H_{25}O_3N$. Pale yellow blocks (ethanol - methanol); sinters at 230°, m. 237–8° dec. (1942 Stodola); yellow-brown solid (benzene-light petroleum b. 60–80°) darkening above 220°, m. 244° dec.; $[\alpha]_D^{26} + 241°$ (chloroform) (1951 Antaki). — **Fmn.** From Δ^4-androstene-3.17-dione 3-enol ethyl ether ($\Delta^{3.5}$-androstadien-3-ol-17-one ethyl ether, p. 2606 s) on treatment with butyl nitrite and sodium ethoxide in benzene - alcohol, followed by treatment with aqueous alcoholic NaOH and then by warming with 50% acetic acid (1942 Stodola). From Δ^4-androstene-3.17-dione (p. 2880 s) on heating with amyl nitrite and sodium ethoxide in abs. alcohol at 60–70° (1951 Antaki). — **Rns.** Refluxing with zinc dust and acetic acid, followed by acetylation of the reduction product with acetic anhydride and pyridine at room temperature, gives 16-ketotestosterone acetate (p. 2945 s) (1942 Stodola; see also 1951 Antaki; 1954 Meyer).

1936 a Mason, Myers, Kendall, *J. Biol. Ch.* **114** 613, 625.
   b Mason, Myers, Kendall, *J. Biol. Ch.* **116** 267, 274.
   Reichstein, *Helv.* **19** 402, 410.
   SCHERING A.-G. (Butenandt, Logemann), *U.S. Pat.* 2170124 (issued 1939); *C. A.* **1940** 1133; *Ch. Ztbl.* **1940** I 1232.
1937 Steiger, Reichstein, *Helv.* **20** 817, 824, 825.
   Ushakov, Lyutenberg, *J. Gen. Ch. U.S.S.R.* **7** 1821, 1823, 1824; *Bull. soc. chim.* [5] **4** 1394, 1397; *C. A.* **1938** 577; *Ch. Ztbl.* **1938** I 4659, 4660.
1938 a Mason, Hoehn, Kendall, *J. Biol. Ch.* **124** 459, 465, 469, 472.
   b Mason, *J. Biol. Ch.* **124** 475, 477; *Proc. Staff Meetings Mayo Clinic* **13** 235.
1939 Dannenberg, *Abhandl. Preuss. Akad. Wiss. Math.-naturwiss. Kl.* **1939** No. 21, pp. 12, 30, 49, 53, 60, 65.
   Ushakov, Lyutenberg, *J. Gen. Ch. U.S.S.R.* **9** 69, 72; *C. A.* **1939** 6334; *Ch. Ztbl.* **1939** II 4489.
1941 v. Euw, Reichstein, *Helv.* **24** 879, 886.
   Reich, Reichstein, *Arch. intern. pharmacodynamie* **65** 415, 417 footnote.
1942 v. Euw, Reichstein, *Helv.* **25** 988, 992, 998, 1010.
   Stodola, Kendall, *J. Org. Ch.* **7** 336, 339, 340.
1945 a Mason, *J. Biol. Ch.* **158** 719.
   b Mason, Kepler, *J. Biol. Ch.* **161** 235, 248, 250.
   Reich, *Helv.* **28** 863, 867, 868.
1946 Lieberman, Dobriner, *J. Biol. Ch.* **166** 773.
   a MERCK & Co. (Sarett), *U.S. Pat.* 2462133 (issued 1949); *C. A.* **1949** 4432; *Fr. Pat.* 942260 (1947, issued 1949); *Ch. Ztbl.* **1950** I 583, 584; *Brit. Pat.* 631238 (issued 1949); *C. A.* **1950** 4048, 4050.
   b MERCK & Co. (Sarett), *U.S. Pat.* 2516259 (issued 1950); *C. A.* **1951** 2032.
   c MERCK & Co. (Sarett), *U.S. Pat.* 2540964 (issued 1951); *C. A.* **1951** 7159, 7160; *Fr. Pat.* 942260 (1947, issued 1949); *Ch. Ztbl.* **1950** I 583, 584; *Brit. Pat.* 630103 (issued 1949); *C. A.* **1950** 7891.
   a Sarett, *J. Biol. Ch.* **162** 601, 628, 629.
   b Sarett, *J. Am. Ch. Soc.* **68** 2478, 2482, 2483.
1947 MERCK & Co. (Sarett), *Fr. Pat.* 956655 (issued 1950); *Ch. Ztbl.* **1950** II 560; *Brit. Pat.* 651719 (issued 1951); *C. A.* **1952** 150.
1948 Lieberman, Dobriner, Hill, Fieser, Rhoads, *J. Biol. Ch.* **172** 263, 268.
1951 Antaki, Petrow, *J. Ch. Soc.* **1951** 901, 904.
   Devis, *C. A.* **1952** 11375, 11376.

1952 Balant, Ehrenstein, *J. Org. Ch.* **17** 1587, 1595, 1596.
Fried, Thoma, Gerke, Herz, Donin, Perlman, *J. Am. Ch. Soc.* **74** 3962.
Heusser, Heusler, Eichenberger, Honegger, Jeger, *Helv.* **35** 295, 297, 303, 304.
1953 Peterson, Eppstein, Meister, Magerlein, Murray, Leigh, Weintraub, Reineke, *J. Am. Ch. Soc.* **75** 412, 414.
1954 Meyer, Lindberg, *J. Am. Ch. Soc.* **76** 3033.

Cholestane-3.4.6-trione $C_{27}H_{42}O_3$ (I). According to Windaus (1937) it exists as its *4-enol*, *Δ⁴-Cholesten-4-ol-3.6-dione* (II); according to Fieser (1949) it is possibly a mixture of the enol (II) and the *dienol*, *Δ⁴·⁶-cholestadiene-4.6-diol-3-one* (III). — Yellow prisms (alcohol), m. 148–9⁰ (1937 Windaus). Absorption max. in chloroform at 275 and 335 mμ (ε 5000 and 5800) (1937 Windaus; 1939 Dannenberg*). — **Fmn.** From cholestane-4.5-diol-3.6-dione (p. 2963 s) on refluxing with 5 % alcoholic HCl for 3 hrs.; almost quantitative yield (1937 Windaus). — **Rns.** Fusion with o-phenylenediamine gives the quinoxaline derivative (below) (1943 Sarett).

Quinoxaline derivative $C_{33}H_{46}ON_2$. Orange needles (chloroform · alcohol), m. 143⁰ to an opaque liquid, clearing at 157⁰. — **Fmn.** From 2α-bromo-Δ⁴-cholestene-3.6-dione (originally formulated as the 4-bromo-compound; see p. 2921 s) on refluxing with o-phenylenediamine in abs. alcohol. From cholestane-3.4.6-trione (above) on fusion with o-phenylenediamine on the steam-bath (1943 Sarett).

1937 Windaus, Kuhr, *Ann.* **532** 52, 54, 59, 61.
1939 Dannenberg, *Abhandl. Preuss. Akad. Wiss. Math.-naturwiss. Kl.* **1939** No. 21, pp. 33, 61.
1943 Sarett, Chakravorty, Wallis, *J. Org. Ch.* **8** 405, 413.
1949 Fieser, Fieser, *Natural Products Related to Phenanthrene*, 3rd Ed., New York, p. 266.

II. Hydroxy-triketosteroids with All CO Groups in the Ring System and OH in Side Chain

Cholane-22.23-diol-3.7.12-trione $C_{24}H_{36}O_5$.

22.23 - Epoxycholane - 3.7.12 - trione (?), *"Dehydrotetrahydroxycholane"* $C_{24}H_{34}O_4$. Crystals (abs. alcohol), m. 242⁰. — **Fmn.** From "tetrahydroxycholane" (22.23-epoxycholane-3α.7α.12α-triol?, p. 2204 s) on oxidation with CrO_3 in 66% acetic acid at room temperature (1940 Kazuno).

* Dannenberg, quoting Windaus (1937), probably erroneously gives ether as the solvent.

Coproergostane-24.25-diol-3.7.12-trione, 24.25-Dihydroxy-3.7.12-triketobufo-

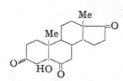

stane $C_{28}H_{44}O_5$. Needles (aqueous methanol), m. 198.5 to 199°. Readily soluble in organic solvents except ether and petroleum ether. Gives a weak, yellow Liebermann-Burchard reaction and a positive Jaffe reaction. — **Fmn.** Along with tetra-keto-isobufostane (25-methylcoprostane-3.7.12.24-tetrone, p. 2985 s), from pentahydroxybufostane (p. 2204 s) on oxidation with CrO_3 in glacial acetic acid at 20°. — **Rns.** Catalytic hydrogenation in glacial acetic acid in the presence of Adams's platinum oxide regenerates pentahydroxybufostane. Attempted Clemmensen reduction yielded an oily mass. Is rearranged, to an extent of ca. 30%, by heating with glacial acetic acid in the presence of CrO_3 at 40–50°; is decomposed on attempted rearrangement by heating in a sealed tube for 2 hrs. with 25% H_2SO_4 at 150° or with acetyl chloride at 120° (1940 Kazuno).

Trioxime $C_{28}H_{47}O_5N_3$. Crystals (aqueous alcohol), dec. 234° (1940 Kazuno).

1940 Kazuno, *Z. physiol. Ch.* **266** 11, 23–25, 29, 30.

III. Hydroxy-triketosteroids with All CO and OH Groups in the Ring System

Androstan-5α-ol-3.6.17-trione $C_{19}H_{26}O_4$.

For configuration at C-5, see 1948 Ehrenstein. — Colourless needles (alcohol), m. 248–9° (1936 Butenandt; 1936 Schering A.-G.; see also 1939 Ehrenstein), 246° dec. (1937 Ushakov); sublimes in a high vacuum at 165° without decomposition (1936 Bute-nandt); $[\alpha]_D^2 + 54.6°$ (acetone) (1936 Butenandt; see also 1939 Ehrenstein). — **Fmn.** In small amount, together with $\varDelta^4$-androstene-3.6.17-trione (p. 2979 s), from 5-dehydroepiandrosterone (p. 2609 s) on oxidation with CrO_3 in glacial acetic acid at room temperature (1936 Butenandt; 1936 Schering A.-G.; see also 1939 Ehrenstein). From both androstane-3β.5α.6α-triol-17-one (p. 2767 s) (1939 Ehrenstein; see also 1939 Ushakov) and androstane-3β.5α.6β-triol-17-one (p. 2768 s) (1937 Usha-kov; 1939 Ehrenstein) on oxidation with CrO_3 in strong acetic acid at room temperature. — **Rns.** Is stable when boiled with acetic anhydride (1936 Butenandt; 1936 Schering A.-G.). Reduction with aluminium isopropoxide in isopropanol gives androstane-3β.5α.6β.17β-tetrol (p. 2207 s) (1936 Ciba). Treatment with dry HCl in chloroform at 0° yields $\varDelta^4$-androstene-3.6.17-trione (p. 2979 s) (1937 Ushakov).

1936 Butenandt, Riegel, *Ber.* **69** 1163, 1167.
    Ciba (Soc. pour l'ind. chim. à Bâle; Ges. f. chem. Ind. Basel), *Brit. Pat.* 489364 (issued 1938); *C. A.* **1939** 815; *Ch. Ztbl.* **1939** I 3932.
    Schering A.-G. (Butenandt, Logemann), *U.S. Pat.* 2170124 (issued 1939); *C. A.* **1940** 1133; *Ch. Ztbl.* **1940** I 1232.

1937 Ushakov, Lyutenberg, *J. Gen. Ch. U.S.S.R.* **7** 1821, 1823; *Bull. soc. chim.* [5] **4** 1394, 1396, 1397; *C. A.* **1938** 577; *Ch. Ztbl.* **1938** I 4659, 4660.
1939 Ehrenstein, *J. Org. Ch.* **4** 506, 508, 513–515.
Ushakov, Lyutenberg, *J. Gen. Ch. U.S.S.R.* **9** 69; *C. A.* **1939** 6334; *Ch. Ztbl.* **1939** II 4489.
1948 Ehrenstein, *J. Org. Ch.* **13** 214, 217.

5-Methyl-19-nor-coprostane (or -cholestane)-9.10-diol-3.6.11-trione $C_{27}H_{42}O_5$.

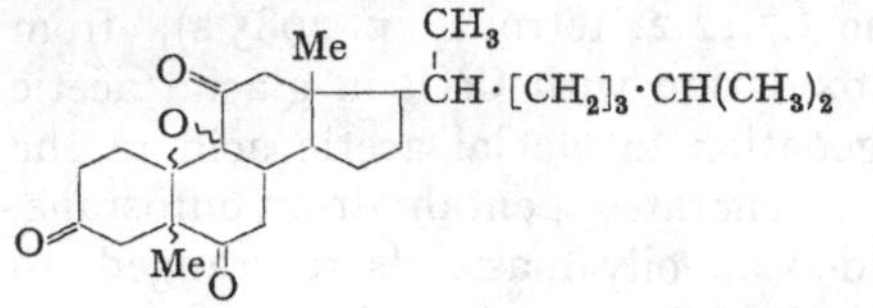

5-Methyl-9.10-epoxy-19-nor-coprostane (or -cholestane)-3.6.11-trione, 5-Methyl-19-nor-Δ^9-coprostene (or -cholestene)-3.6.11-trione oxide $C_{27}H_{40}O_4$. For structure, see 1952 Ellis; cf. also that of the parent "Westphalen's diol", p. 2046 s. — Needles (aqueous acetone - methanol), m. 165.5° to 166.5° cor.; $[\alpha]_D^{19}$ + 134° (chloroform) (1939 Petrow). Shows no significant absorption in the region 230–400 mμ in both neutral and alkaline isopropyl alcohol (1952 Ellis). Readily soluble in the usual solvents (1939 Petrow). — **Fmn.** From 5-methyl-9.10-epoxy-19-nor-coprostane(or-cholestane)-3β.6β-diol-11-one (p. 2753 s) on oxidation with CrO_3 and 70% acetic acid in benzene at room temperature (1939 Petrow). — **Rns.** Does not give a coloration with $FeCl_3$. Fails to give a quinoxaline derivative on heating with o-phenylenediamine at 140–150° (1939 Petrow).

1939 Petrow, *J. Ch. Soc.* **1939** 998, 1002.
1952 Ellis, Petrow, *J. Ch. Soc.* **1952** 2246, 2248, 2252.

G. Tetraketosteroids with Three CO Groups in the Ring System

Pregnane-3.7.12.20-tetrone $C_{21}H_{28}O_4$. Plates (ether), m. 238–242⁰; $[\alpha]_D^{26} +76^0$ (acetone). — **Fmn.** From pregnane-3α.7α.12α-

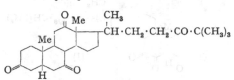

triol-20-one (p. 2366 s) on oxidation with CrO_3 in acetic acid at room temperature (1940 Ehrenstein).

For a **pregnane-3.7.12.20-tetrone** $C_{21}H_{28}O_4$ [needles (aqueous methanol), m. 185–7⁰] obtained by oxidation of a pregnane-3α.7α.12α-triol-20-one melting at 210–211⁰ (1942 Kanemitu), see this triolone (p. 2368 s).

25 - Methylcoprostane - 3.7.12.24 - tetrone, Tetraketo - isobufostane $C_{28}H_{42}O_4$.

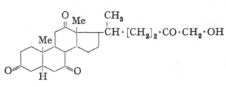

Plates (methanol), m. 248–250⁰. Sparingly soluble in all solvents. — **Fmn.** From tetrahydroxy-isobufostane (25-methylcoprostane-3α.7α.12α.24-tetrol, p. 2199 s) on oxidation with CrO_3. Along with 24.25-dihydroxy-3.7.12-triketobufostane (p. 2983 s), from pentahydroxybufostane (p. 2204 s) on oxidation with CrO_3 in glacial acetic acid at 20⁰. From 3α.7α.12α-trihydroxy-24-keto-isobufostane (p. 2368 s) on oxidation with CrO_3 in glacial acetic acid. From 24.25-dihydroxy-3.7.12-triketobufostane (p. 2983 s) on heating with glacial acetic acid in the presence of CrO_3 at 40–50⁰; yield, ca. 30%. — **Rns.** On hydrogenation in glacial acetic acid in the presence of Adams's platinum oxide it gives tetrahydroxy-isobufostane (p. 2199 s) and 3α.7α-dihydroxy-12.24-diketo-isobufostane (p. 2866 s). On refluxing with amalgamated zinc and conc. HCl in glacial acetic acid for 3 hrs. it yields 7.12.24-triketo-isobufostane (p. 2970 s) and isobufostane (p. 1439 s) (1940 Kazuno).

Tetroxime $C_{28}H_{46}O_4N_4$. Hexagonal plates with 1 H_2O (aqueous alcohol), m. 244⁰ (1940 Kazuno).

1940 Ehrenstein, Stevens, *J. Org. Ch.* **5** 660, 669.
    Kazuno, *Z. physiol. Ch.* **266** 11, 24–26, 28, 29.
1942 Kanemitu (Kanemitsu), *J. Biochem.* **35** 173, 180.

3.7.12-Triketonorcholanyl hydroxymethyl ketone $C_{25}H_{36}O_5$.

Acetate $C_{27}H_{38}O_6$. Needles (aqueous acetic acid); may also be recrystallized from acetone or methanol; m. 222⁰. — **Fmn.** From the acid chloride of 3.7.12 - triketo-cholanic acid (dehydrocholic acid) on treatment with diazomethane in ether - acetone, followed by refluxing of the resulting diazo-ketone for ¹/₂ hr. with glacial acetic acid; yield (from

dehydrocholic acid), 31 % (1938 I.G. FARBENIND.; 1938 WINTHROP CHEMI-
CAL Co.).

1938 I.G. FARBENIND., *Fr. Pat.* 847 129 (issued 1939); *C. A.* **1941** 5653; *Ch. Ztbl.* **1940** I 2827.
    WINTHROP CHEMICAL Co. (Bockmühl, Ehrhart, Ruschig, Aumüller), *U.S.Pat.* 2 202 619 (issued
    1940); *C. A.* **1940** 6772.

H. Tetraketosteroids with All CO Groups in the Ring System

Δ^4-**Cholestene-2.3.6.7-tetrone** $C_{27}H_{38}O_4$ (I). For its *diquinoxaline derivative* obtained from 2.7β (or 7.7)-dibromo-Δ^4-cholestene-3.6-dione (1943 Sarett), see
p. 2930 s.

Cholestane - 3.4.6.7 - tetrone $C_{27}H_{40}O_4$ (II). The *diquinoxaline compound*
$C_{39}H_{48}N_4$ (slightly pink leaflets, m. 209°) (p. 162), obtained by the action
of o-phenylenediamine on 2α.7α-dibromo- and 2.2.7α-tribromo-Δ^4-cholestene-
3.6-dione (originally formulated as 4.7-dibromo- and 4.7.7-tribromo-Δ^4-chole-
stene-3.6-dione, pp. 2929 s and 2931 s, resp.) by refluxing in glacial acetic
acid - abs. alcohol or by heating at 150°, resp. (1937 Butenandt), is possibly
derived from this tetrone.

1937 Butenandt, Schramm, Kudszus, *Ann.* **531** 176, 201, 202.
1943 Sarett, Chakravorty, Wallis, *J. Org. Ch.* **8** 405, 414.

Additions and Corrections

Volume 14

LIST V**

Page	Line	
215	4, 6–7	*For* "1932 c Wiel." *read* "1932 d Wiel."
215	8	*For* "dehydroxylithobilianic" *read* "hydroxylithobilianic"
216	6	*For* "With $CrO_3 \rightarrow$" *read* "With CrO_3 the trimethyl ester (m. 117–8⁰)→"

For "With CrO_3→" *read* "With CrO_3 the trimethyl ester (m. 117–8^0)→"

217   14–15   *Delete* "Clemmensen redn. . . . . (1926 a Wind.)."

218   11   *For* "1932 c Wieland" *read* "1932 d Wieland"

275   12   *For* "ether" *read* "glacial acetic acid"

1354 s   13   *For* "13 and 12" *read* "11 and 12"

1374 s   11   *After* "(pp. 1372, 1373 s)" *insert* "(1937 a Marker); a mixture of these two alcohols is also obtained by hydrogenation at room temperature in alcohol-ether containing some NaOH, using a slightly platinized Raney nickel catalyst (1937 Délépine, Horeau, *Bull. soc. chim.* [5] **4** 31, 43)."

1409 s   Footnote   *For* "cf. the footnote on p. 1414 s" *read* "cf. footnote** on p. 1413 s"

1487 s   8   *For* "5-hydroxypregnan-20-one" *read* "5-hydroxyallopregnan-20-one"

1511 s   17*   *For* "androsten-3β-ol" *read* "androsten-17β-ol"

1512 s   13*   *Delete* "followed by saponification"

1524 s   1*⎫
1525 s   9 ⎭   *Before* "ætiocholan" *insert* "the acetate of"

1548 s   10   *For* "hydrogenation" *read* "dehydrogenation"

1585 s   7   *For* "(m. 272–3⁰)" *read* "(m. 248–250⁰)"

1591 s   16*   *Insert* "When conc. H_2SO_4 is carefully added to a dilute solution of cholesterol in AcOH containing a little mercuric acetate, coloured rings develop at the interface, the lower being violet, the upper ring blue; limit of sensitivity, 0.02 mg. in 1 c.c. (1942 Nath, *C. A.* **1943** 3701)."

1592 s   17   *After* "1937 v. Novák" *add* "1942 Nath, Chakraborty, *C. A.* **1943** 3701."

1670 s   *Under* **Rns.** *of epi-cholesterol insert* "On oxidation of the acetate with CrO_3 in strong acetic acid at 50⁰ it gives 7-keto-epi-cholesteryl acetate (p. 2682 s) and a compound $C_{29}H_{46}O_4$ to which the structure Δ^3-cholestene-3.5α-diol-6-one 3-acetate was tentatively assigned (1939 Windaus, Naggatz, *Ann.* **542** 204, 209, 211), but which is actually cholestan-5α-ol-3.6-dione acetate (p. 2954 s) (1953 Tarlton, Fieser, Fieser, *J. Am. Ch. Soc.* **75** 4423)."

---

* From bottom of the text.
** For list I, see *vol.* **14** 709; lists II and III were added as loose sheets to *vols.* **13** and **12** A, resp.; for list IV, see p. 2212 s.

Page   Line

1699 s   *Between lines 13* and 12* insert* "*Dicholestanyl sulphite* $C_{54}H_{94}O_3S$. Iridescent plates (acetone), m. 194° cor. — **Fmn.** From cholestanol on reaction with thionyl chloride; quantitative yield (1946 Shoppee)."

1701 s   Footnote   *For* "p. 1505 s" *read* "p. 1507 s"

1727 s   10*   *For* "unchanged starting material" *read* "cholestan-7-one simultaneously formed"

1778 s   2   *After* "(1937 Weinhouse)." *insert* "From Δ^5-cholestene-3.7-diol-7-acetic acid on refluxing with acetic anhydride for 4 hrs., followed by refluxing with methanolic KOH for 30 min. (1937 Jones, Spring, *J. Ch. Soc.* **1937** 302)."

1844 s   10   *Delete* "with acetic anhydride"

1879 s   24   *For* "HCl" *read* "HBr"

1885 s   11   *For* "1937 Meystre" *read* "1947 Meystre"

1909 s   14*   *For* "cholestan-4-one" *read* "cholestan-3 or 4-one"

1910 s   10*   *For* "50 %" *read* "strong"

8*   *Before* "benzene" *insert* "acetic acid-"

8*   *For* "Δ^5" *read* "Δ^4"

1937 s   3–7   *Replace the article* "**Pregnane-5.20-diol**" *by* "**Allopregnane-5α.20-diol**" $C_{21}H_{36}O_2$. — **Fmn.** A mixture [crystals (diisopropyl ether)] of the two C-20 epimers is obtained from allopregnan-5α-ol-20-one (p. 2250 s) on reduction, e.g., by hydrogenation in alcohol in the presence of a nickel catalyst until 1 mol. of hydrogen has been absorbed. — **Rns.** Gives a mixture of the two epimeric Δ^4-pregnen-20-ols (see p. 1487 s) on heating with 5 % methanolic HCl; a mixture of the propionates of these alcohols and other compounds are obtained when the present mixture is boiled with propionic anhydride (1941 Ciba)."

1940 s   11–15   *Delete* "No analysis given" *and add* "$[\alpha]_D$ + 31° (abs. alcohol); gives a *diacetate* $C_{27}H_{44}O_4$, m. 135–7°, $[\alpha]_D$ + 34.5° (abs. alcohol)."

1942 s   2*   *For* "3β-hydroxy" *read* "3β-acetoxy"

1946 s   8–9   *Delete* "*3(?)-Acetate* $C_{26}H_{42}O_3$ .... Marker)."

1947 s   13*   *After* "ether" *add* "followed by refluxing with methanolic KOH"

1956 s   6*   *For* "200–220°" *read* "185–220°"

1968 s   10   *Add* "From 7-ketoœstradiol (p. 2738 s) on heating of its semicarbazone with sodium ethoxide in alcohol at 180–190° (1941 Schering Corp.)."

1985 s   14   *For* "**11.17-diol**" *read* "**11.17β-diol**"; *in the structural formula replace* "⌇OH" *by* "—OH"

Page   Line

1994 s   5*   *For* "KOH" *read* "K_2CO_3"
1995 s   12   *For* "diethyl" *read* "triethyl"
2011 s   12*   *For* "androstan" *read* "androsten"
2034 s   13   *For* "it yields" *read* "the 3-acetate yields"
2056 s   15   *After* "Georg)." *insert* "For an authentic coprostane-3.4-diol,
                see p. 2668 s."
2060 s   18   *For* "6α-acetoxy-Δ^4-cholesten-3-one" *read* "3β-acetoxy-Δ^4-
                cholesten-6-one (p. 2677 s)"
2063 s   10   *After* "(1946 Reich)" *insert* "the present diol is the sole
                reaction product when the same dione is reduced with
                sodium in boiling abs. alcohol (1940 c Marker)."
2064 s   11*   *For* "diacetate" *read* "3-acetate"
2088 s   *Replace the article* **"24-Phenylcholane-3.12-diol"** *by*

24-Phenylcholane-3α.12α-diol, 24-Phenyl-3α.12α-dihydroxycholane $C_{30}H_{46}O_2$ (no analysis given). Amorphous solid. — **Fmn.** From 3α.12α-dihydroxy-norcholanyl phenyl ketone (p. 2350 s) on heating with hydrazine hydrate and methanolic sodium methoxide for 3 hrs. at 180–190°. — **Rns.** Gives 24-phenylcholane-3.12-dione (p. 2914 s) on oxidation with CrO_3 in strong acetic acid at room temperature (1945 Hoehn).

2093 s   21⎫   *After* "(1945 Meystre)" *insert* "; the bromination with
2094 s   25⎭   N-bromosuccinimide may also be carried out in boiling
                benzene, and the elimination of HBr by heating with
                collidine at 130° (1946 Ettlinger, Fieser, *J. Biol. Ch.* **164**
                451)."

2093 s   14*   *After* "**Rns.**" *insert* "Gives pregnane-3α.12α-diol-20-one
                (p. 2329 s) on ozonization of the diacetate in methanol-
                ethyl acetate, reduction of the resulting ozonide with zinc
                dust and acetic acid, reoxidation with CrO_3, and saponi-
                fication with methanolic KOH (1946 Ettlinger, Fieser,
                J. Biol. Ch. **164** 451)."

2097 s   11   *For* "(not described)" *read* "[crystals (aqueous methanol or
                methanol-ether), m. 220–227° dec.]"

2103 s   11*   *Add* "The acetyl ester acetate is also obtained from the acetate
                of androsterone cyanohydrin by shaking with hydrogen
                and platinum in glacial acetic acid at 20° (1946 Prins,
                Shoppee, *J. Ch. Soc.* **1946** 494, 498)."

2104 s   3   *Add* "Prisms (methanol-ether), becoming opaque at 160° and
                being transformed at 225° to iridescent plates, m. 243–6°
                cor. (Kofler block) with immediate recrystallization of the
                melt (1946 Prins, Shoppee, *J. Ch. Soc.* **1946** 494, 498)."

Page	Line	
2113 s	8	*Before* "Treatment" *insert* "Gives Δ^{16}-allopregnene-3.20-dione (p. 2795 s) on refluxing with aluminium isopropoxide and cyclohexanone in toluene, followed by sublimation of the resulting product in a high vacuum at 130–140° (1940 e Marker)."
2115 s	5*–4*	*For* "Butenandt), with .... esterification)", *read* "Butenandt). The "17-hydroxyprogesterone" obtained by Oppenauer oxidation of the triol (1938 a, 1939 CIBA) is a D-homo-compound (see, e. g., the reaction of true 17-hydroxy-progesterone with aluminium tert.-butoxide, p. 2843 s)."
2117 s	2	*For* "Δ^4-pregnene" *read* "17-iso-Δ^4-pregnene"
2123 s		*Reference list* **1938** b SCHERING A.-G.⎫
2146 s		*Reference list* **1938** a SCHERING A.-G.⎭ *for* "U.S. Pat. 2372400" *read*
		"*U.S. Pat.* 2372440" (error in both, *C. A.* and *Ch. Ztbl.*)
2134 s	Footnote***	*For* "p. 2030 s" *read* "p. 2130 s"
2135 s	1*	The oxidation refers to the free alcohol and not to its acetate.
2141 s	6–11	*For* "Oxidation to .... similarly" *read* "Gives 17-iso-Δ^4-preg-nene-17β.21-diol-3.20-dione (p. 2859 s) by the action of Acetobacter and press juice (from brewer's yeast) in aqueous dioxane at 25–30°, or by oxidation of the 21-tri-phenylmethyl ether (not described), followed by acid hydrolysis"
	12	*For* "Δ^4-pregnene" *read* "17-iso-Δ^4-pregnene"
2146 s	19*	*For* "**1944** I 1203" *read* "**1944** II 1203"
2148 s	2	*For* "21-acetate" *read* "20.21-diacetate"
	3	*For* "6-methylpregnane" *read* "6-methylallopregnane"
2152 s	4–16	The substance originally described as 4.3'-dihydroxy-7-meth-oxy-1.2-cyclopentenophenanthrene (1939 Robinson) is probably 4-hydroxy-7-methoxy-3'-keto-1.2-cyclopenteno-9.10-dihydro-phenanthrene (p. 2736 s) (1941 Robinson, Willenz, *J. Ch. Soc.* **1941** 393, 394).
2157 s		*Between lines 6* and 7* insert* "Oestriol 3-benzoate C$_{25}$H$_{28}$O$_4$. Platelets (benzene), m. 225°. — **Fmn.** From œstriol by the Schotten-Baumann procedure. — **Rns.** Oxidation with lead tetra-acetate in acetic acid at 20–30° gives a product of aldehydic nature (1940 Huffman, Mac Corquodale, Thayer, Doisy, Smith, Smith, *J. Biol. Ch.* **134** 591, 596, 597)."
2168 s	14	*For* "(no constants given)" *read* "[needles (abs. methanol + a trace of pyridine), m. 146.5–148°]"
2194 s	14*	*For* "glacial acetic acid" *read* "80 % acetic acid"
2199 s	24–25	*For* "ætiocholane-3α.7α.12α-triol-17-one" *read* "pregnane-3α.7α.12α-triol-20-one"
2199 s		*Add a methyl group at C–13 in the last structural formula.*
2200 s	4	*Add* "**Rns.** Reverts to tetraketo-isobufostane on oxidation with CrO$_3$ (1940 Kazuno)."
2204 s	12*	*For* "21.22-epoxycholane" *read* "22.23-epoxycholane"

SUBJECT INDEX

(This cumulative index comprises material from the Main Volume and the Supplement. Pages with an "S" refer to the Supplement, the others to the Main Volume)

The following order of substituting radicals is used in the names:

Alkyls and **alkylidenes:** methyl, ethyl, vinyl, ethynyl, propyl..., methylene, ethylidene..., phenyl, tolyl, benzyl..., benzylidene..., naphthyl..., anthryl..., phenanthryl, etc.;

Halogen: fluoro, chloro, bromo, iodo;

Nitrogen-containing functions: e.g., nitro, amino (acetamino, anilino), hydroxylamino, azo, hydrazino, etc.;

Hydroxy (acetoxy..., methoxy...);

Acyl and other **radicals with keto groups:** e.g., acetyl, propionyl..., acetonyl..., benzoyl, toluyl..., phenacyl..., naphthoyl, etc.;

Keto;

Carboxy (cyano);

Thio (Mercapto).

As a rule, esters, ethers, acyl derivatives, oximes, hydrazones, etc., are not included in the subject index but only in the formula index.

The prefixes epi- and iso-, when referring to configuration, are disregarded in the alphabetical order of indexing.

Pages with an "S" refer to the Supplement, the others to the Main Volume

Pages with an "S" refer to the Supplement, the others to the Main Volume

Pages with an "S" refer to the Supplement, the others to the Main Volume

Pages with an "S" refer to the Supplement, the others to the Main Volume

205*

Homocholan-24-one ("Norcholyl methyl ketone"), 114; 2218 s
Homocholene-3.25-diol-24-one, 2315 s
ω-Homo-pregnadiene-3.21-dione, 2804 s
ω-Homo-pregnatrien-3-one, 2412 s
ω-Homo-pregnene-3.22-diol-21-one, 2314 s
ω-Homo-pregnene-3.20-dione, 2803 s
ω-Homo-pregnene-3.21-dione, 2804 s
ω-Homo-pregnene-17.20.21.22-tetrol-3-one, 2732 s
ω-Homo-17-iso-pregnene-17.21.22-triol-3-ones, 2728 s
Hormones, see Corticoids, Androgenic, Œstrogenic, and Progestational compounds
Hydrocortisone (Pregnenetrioldione), 154; 2868 s
3-Hydroxy-ætioallocholanaldehyde, 2228 s
(3-Hydroxy-ætiocholan-17-yl)-glyoxal, 2380 s
3-Hydroxy-ætiocholenaldehyde, 2227 s
(3-Hydroxy-ætiocholen-17-yl)-glyoxal, 2378 s
3-Hydroxy-allocholanaldehyde, 2262 s
3-Hydroxy-allopregnan-20-yl methyl ketone, 2261 s
N-(17-Hydroxy-androstadien-3-yl)-pyridinium chloride, 2584 s
(3-Hydroxy-androstan-17-yl)-glyoxal, 2380 s
3-Hydroxy-androstan-17-yl phenyl ketones, 2271 s
(3-Hydroxy-androsten-17-yl)-glyoxal, 2378 s
2-Hydroxyandrosterone, 2745 s
11-Hydroxyandrosterone, 2748 s
17-(α-Hydroxybenzhydryl)-ætiocholane-3.12-dione, 2935 s
Hydroxybenzyl-cholestanone, 2440 s
3-Hydroxy-bisnorcholanaldehyde, 2259 s
3-Hydroxybisnorcholanyl diazomethyl ketone, 2297 s
3- — hydroxymethyl ketone, 2315 s
3- — phenyl ketone, 2275 s
3-Hydroxy-bisnorcholesten-24-one, 2263 s
3-Hydroxy-cholenyl diazomethyl ketone, 2298 s
3-Hydroxycholestane-6‖7-dioic acids, 42; 2485 s, 2488 s
3- — 6‖7-dioic acid lactone, 42; 2486 s
4-Hydroxy-3‖4-Δ^5-cholesten-3-oic acid lactone (Δ^5-Cholesten-3-one "α"-oxide), 121; 2435 s
N-(3-Hydroxy-cholesten-2-yl)-pyridinium bromide, 2496 s
23-Hydroxycholophenone, 2377 s
6-Hydroxycoprostenone, 146; 2672 s
17-Hydroxycortexone, 2859 s
17-Hydroxycorticosterone (Pregnenetrioldione), 154; 2868 s
17-Hydroxy-11-dehydrocorticosterone (Reichstein's Substance Fa), 165; 2977 s

4-Hydroxy-5-dehydroepiandrosterone, 2746 s
16- — 5-dehydroepiandrosterone, 2751 s
7- — 6-dehydroœstrone, 2739 s
8- — 9(11)-dehydro-14-iso-œstrone, 2740 s
2- — desoxycorticosterone, 2852 s
6- — desoxycorticosterone, 2853 s
12- — desoxycorticosterone, 2857 s
17- — desoxycorticosterone, 2859 s
17- — 17-iso-desoxycorticosterone, 2859 s
7- — 4.3′-diketo-1.2-cyclopentano-tetrahydrophenanthrene, 2939 s
5- — epiandrosterone, 2746 s
6- — epiandrosterone, 2747 s
11- — epiandrosterone, 149; 2748 s
cis-3-Hydroxy-17-equilenone, 2535 s, 2536 s
trans-3-Hydroxy-17-equilenone (Equilenin), 134; 2527 s, 2533 s, 2534 s
6-Hydroxy-17-equilenone, 2563 s
"14-epi-$\Delta^{9(11)}$-8-Hydroxyequilin", 2740 s
N-(3-Hydroxy-20-keto-allopregnan-21-yl)-pyridinium chloride, 2282 s
7-Hydroxy-3-keto-1.2-cyclopentano-hexahydrophenanthrene, 133; 2516 s
7- — 3′ or 5′-keto-1.2-cyclopentano-hexahydrophenanthrene, 2520 s
7- — 3-keto-1.2-cyclopentano-octahydrophenanthrene, 133; 2516 s
7- — 3′-keto-1.2-cyclopentano-octahydrophenanthrene, 2520 s
7- — 3′-keto-1.2-cyclopentano-tetrahydrophenanthrene, 2518 s
7- — 3′-keto-1.2-cyclopenteno-dihydrophenanthrene, 2517 s
4- — -3′-keto-1.2-cyclopenteno-phenanthrene, 2516 s
(3-Hydroxy-12-ketonorcholanyl)-diphenylcarbinol, 2724 s
— phenyl ketone, 2849 s
5-Hydroxy-20-keto-3‖5-A-nor-pregnan-3-oic acid lactone, 2787 s
N-(3-Hydroxy-20-ketopregnan-21-yl)-pyridinium halides, 2280 s, 2281 s
N-(3-Hydroxy-20-ketopregnen-21-yl)-pyridinium salts, 2279 s, 2280 s
21-Hydroxymethyl-allopregnan-3-ol-21-one, 2314 s
17- — androstan-17-ol-3-one, 2720 s
17- — androsten-17-ol-3-one, 2719 s
17- — androsten-3-one, 2510 s
Hydroxymethylene-cholestan-3-one, 2513 s
4′- — 4.7-dihydroxy-3′-keto-1.2-cyclopentenophenanthrene, 2764 s
4′- — 7-hydroxy-3′-keto-1.2-cyclopentanotetrahydrophenanthrene, 2719 s
4′- — 3′-keto-1.2-cyclopentano-tetrahydrophenanthrene, 2510 s
16- — cis-norequilenan-17-one, 2510 s

Pages with an "S" refer to the Supplement, the others to the Main Volume

Naturally occurring compounds:

in plants

in animals (or their excreta)

Retropregnadien-20-ol-3-one, 2409 s
— 20-ol-3-one oxide, 2409 s
Retropregnane-3.20-dione, 2409 s
Retropregnene-13.14.20-triol-3-one, 2410 s

Sex hormones, see Androgenic, Œstrogenic, and
    Progestational compounds
Sitostadien-7-one, 2470 s
Sitostane-3.5-diol-6-one, 2760 s
Sitostane-3.6-dione, 161; 2914 s
Sitostan-5-ol-3.6-dione, 164; 2956 s
Sitostan-6-ol-3-one, 2694 s
Sitostan-3-ol-6-one, 2694 s
β-Sitostanone, 127; 2469 s
γ-Sitostanone, 127; 2470 s
Sitostan-6-one, 127; 2470 s
Sitostene-3.5-diol-6-one, 2760 s
Sitostene-3.6-dione, 160; 2913 s
Sitosten-3-one, 2467 s
β-Sitostenone, 126; 2467 s
β-Sitostenone dibromide, 2502 s
γ-Sitostenone, 2468 s
i-Sitosten-6-one, 2470 s
α-Spinastadienone, 87; 2467 s
Spinastanone, 2469 s
Stigmastadiene-3.6-diones, 160; 2913 s
Stigmastadien-3-ol-7-ones, 148; 2695 s, 2696 s
Stigmastadien-3-ol-7-ones, oxide, 2775 s
Stigmastadien-3-ones, 126; 2465 s, 2466 s,
    2467 s
Stigmastadien-3-one 22.23-dibromide, 2473 s
24-iso-Stigmastadien-3-one, 2466 s
Stigmastane-3.5-diol-6-one, 2760 s
Stigmastane-3.6-dione, 161; 2914 s
Stigmastan-5-ol-3.6-dione, 164; 2956 s
24-iso-Stigmastan-5-ol-3.6-dione, 2956 s
Stigmastan-6-ol-3-one, 2694 s
Stigmastan-3-ol-6-one, 2694 s
Stigmastan-3-ol-7-one, 2697 s
Stigmastan-3-one, 127; 2469 s
24-iso-Stigmastan-3-one, 2470 s
Stigmastan-6-one, 127; cf. 2470 s
Stigmastatrien-3-ol-7-one, 2695 s
Stigmastene-3.5-diol-6-one, 2760 s
Stigmastene-3.6-diones, 160; 2913 s
8-iso-Stigmastene-3.8.9-triol-7-one, 2775 s
8-iso-Stigmastene-3.8.14-triol-7-one, 2775 s
Stigmasten-5-ol-3.6-dione, 164
Stigmasten-6-ol-3-one, 2694 s
Stigmasten-3-ol-6-one, 2694 s
Stigmasten-3-ol-7-ones, 2696 s, 2697 s
24-iso-Stigmasten-3-ol-7-one, 2696 s
"Stigmastenone", 126; 2465 s
α-Stigmastenone, 2469 s
Stigmasten-3-one, 2469 s
24-iso-Stigmasten-3-one, 2468 s

3-Styrylcholestan-3-ol, 2438 s
Substance . . . ,
    see under Reichstein's Substance . . .

Testalolone, 117; 2381 s
Testosterone, 139; 2580 s
Testosterone, enolic form, 139; 1992 s
Testosterone, esters, 140; 2588–2590 s
Testosterone, ethylene ketal, 2595 s
Testosterone, glycosides, 2590 s
Testosterone, propylene ketal, 2596 s
"Δ¹-Testosterone", 2578 s
"Δ⁵-Testosterone", 2595 s
"cis-Testosterone", 140; 2579 s
"trans-Testosterone" (Testosterone), 139; 2580 s
17-epi-Testosterone ("cis-Testosterone"), 140;
    2579 s
Testosterone-aldehydes, 2835 s
17-epi-Testosterone-aldehyde, 2836 s
Testosterone-4.5-glycol, 2767 s
5.6.17.21-Tetrabromo-allopregnan-3-ol-20-one,
    2290 s
Tetrabromocholestan-3-one, 132; 2505 s
Tetrabromocholestene-3.6-dione, 2932 s
Tetrahydro-3-desoxyequilenin, 2392 s, 2393 s
— 3-desoxyisoequilenin, 2392 s, 2393 s
— equilenan-17-ones, 2392 s, 2393 s
— 17-equilenones, 2392 s, 2393 s
3.11.17.20-Tetrahydroxy-allopregnan-21-oic
    acid, 2374 s
3.7.12.23- — norcholanyl phenyl ketone, 2377 s
7.12.24.25- — 3-ketobufostane, 2766 s
Tetraketo-isobufostane, 2985 s
Theelin, 135; 2543 s
"Thelykinin", see Œstrone
Tigone diacetate, 2333 s
"Tokokin", see Œstrone
5.6.16-Tribromoandrostan-3-ol-17-one, 2716 s
Tribromo-cholestadiene-3.6-dione, 2932 s
— cholestadien-3-one, 120; 2503 s
— cholestan-3-ones, 131; 2504 s
— cholestene-3.6-dione, 162; 2931 s
— cholesten-3-ones, 132; 2502 s, 2503 s
— cholesten-7-one, 2504 s
"Tribromo-coprostanone", 130; 2502 s
17.21.21-Tribromopregnane-3.12-diol-20-one,
    2351 s
3.7.12-Trihydroxyætiocholan-17-yl methyl
    ketone (Pregnan-20-one-3.7.12-triol), 117;
    2366 s
3.7.12-Trihydroxy-24-keto-isobufostane, 2368 s
(3.7.12-Trihydroxynorcholanyl) chloromethyl
    ketone, 2370 s
— diazomethyl ketone, 2371 s
— hydroxymethyl ketone, 2376 s
— phenyl ketone, 2369 s

FORMULA INDEX

(This cumulative index comprises material from the Main Volume and the Supplement. Pages with an „S" refer to the Supplement, the others to the Main Volume)

Pages with an "S" refer to the Supplement, the others to the Main Volume

$C_{19}H_{23}NO_3$ 16-Oximinoœstrone methyl ether,
2942 s
$C_{19}H_{23}N_3O$ 6-Methyl-3-keto-1.2-cyclopentano-
hexahydrophenanthrene semicarbazone,
2387 s
Semicarbazone of 2-methyl-3′-keto-
1.2-cyclopentano-hexahydrophen-
anthrene or 2-methyl-3′-keto-1.2-cyclo-
penteno-octahydrophenanthrene, 2393 s
Œstratetraen-17-one semicarbazone, 2392 s
$C_{19}H_{23}N_3O_2$ Equilin semicarbazone, 135
8-Dehydro-14-iso-œstrone semicarbazone,
2540 s
$C_{19}H_{24}Br_2O_2$ 2.6-Dibromoandrostene-
3.17-diones, 2925 s
$C_{19}H_{24}N_2O_3$ 16-Ketoœstrone methyl ether
dioxime, 2942 s
$C_{19}H_{24}O_2$ 2-Methyl-7-methoxy-3-keto-1.2-cyclo-
pentano-octahydrophenanthrene, 134;
2525 s
1-Methyl-7-methoxy-5′-keto-1.2-cyclo-
pentano-octahydrophenanthrene, 2522 s
Œstrone methyl ether, 138; 2555 s
13-iso-Œstrone methyl ether, 2560 s
Œstrone-a methyl ether, 2561 s
6-Methoxytetrahydro-17-equilenones,
2564 s
4-Methylœstratrien-1-ol-17-one, 2573 s
Androstatrien-3-ol-17-one, 2605 s
$\Delta^{1.4}$-Androstadiene-3.17-dione, 2877 s
$\Delta^{4.6}$-Androstadiene-3.17-dione, 155; 2878 s
$\Delta^{4.9(11)}$-Androstadiene-3.17-dione, 2879 s
$C_{19}H_{24}O_3$ 1-Methyl-7-methoxy-4′-hydroxy-
5′-keto-1.2-cyclopentano-octahydro-
phenanthrene, 2737 s
16-Ketoœstradiol 3-methyl ether, 2738 s
Androstene-3.6.17-trione, 165; 2979 s
Androstene-3.11.17-trione, 165; 2979 s
Androstene-3.16.17-trione, 2981 s
Œstrololactone methyl ether, 2553 s
$C_{19}H_{24}O_4$ Equilin-7.8-glycol 3-methyl ether,
2767 s
$C_{19}H_{25}BrO_2$ 2-Bromo-Δ^1-androstene-3.17-dione,
2916 s
2-Bromo-Δ^4-androstene-3.17-dione, 2916 s
4-Bromo-Δ^4-androstene-3.17-dione, 2917 s
6-Bromo-Δ^4-androstene-3.17-dione, 161;
2918 s
$C_{19}H_{25}ClO$ 3-Chloroandrostadien-17-one, 2474 s
$C_{19}H_{25}NO_3$ 16-Ketoœstradiol 3-methyl ether
oxime, 2738 s
16-Oximino-androstene-3.17-dione, 2981 s
$C_{19}H_{25}N_3O$ Tetrahydroequilenan-17-ones, semicarbazone, 2393 s
$C_{19}H_{25}N_3O_2$ Œstrone semicarbazone, 137
8-iso-Œstrone semicarbazone, 2559 s
13-iso-Œstrone semicarbazone, 2559 s

Folliculosterone semicarbazone, 2562 s
$C_{19}H_{25}N_3O_3$ 6-Ketoœstradiol semicarbazone,
2737 s
7-Ketoœstradiol semicarbazone, 2738 s
$C_{19}H_{25}N_3O_5S$ Œstrone hydrogen sulphate semicarbazone, 2554 s
$C_{19}H_{26}BrClO_2$ 2-Bromo-5-chloroandrostane-
3.17-dione, 2924 s
$C_{19}H_{26}Br_2O_2$ 2.2-Dibromoandrostane-3.17-dione,
2924 s
2.4-Dibromoandrostane-3.17-dione, 2924 s
2.16-Dibromoandrostane-3.17-dione, 2925 s
5.6-Dibromoandrostane-3.17-dione, 2925 s
$C_{19}H_{26}ClNO_2$ 3-Chloro-6-nitro-androsten-17-one,
2507 s
$C_{19}H_{26}O$ Androstadien-3-one, 2396 s
Androstadien-11-one, 2400 s
Androstadien-17-one, 2400 s
$C_{19}H_{26}O_2$ $\Delta^{1.4}$-Androstadien-17-ol-3-one, 2576 s
$\Delta^{4.6}$-Androstadien-17-ol-3-one, 139; 2577 s
$\Delta^{3.5}$-Androstadien-17-ol-7-one, 2603 s
Androstadien-3-ol-17-ones, 2605 s
Δ^1-Androstene-3.17-dione, 155; 2879 s
Δ^4-Androstene-3.17-dione, 155; 2880 s,
2886 s
Δ^5-Androstene-3.17-dione, 156; 2890 s
$\Delta^{9(11)}$-Androstene-3.17-dione, 2891 s
Δ^{11}-Androstene-3.17-dione, 2892 s
Δ^5-Androstene-4.17-dione, 2897 s
Δ^2-Androstene-6.17-dione, 2897 s
Δ^4-Androstene-6.17-dione, 2898 s
i-Androstene-6.17-dione, 2898 s
5-Methyl-19-nor-ætiocholene-6.17-dione,
2899 s
$C_{19}H_{26}O_3$ Δ^5-Androstene-3.17-dione oxide, 156
$\Delta^{9(11)}$-Androstene-3.17-dione oxide, 2891 s
Δ^2-Androstene-6.17-dione oxide, 2898 s
Androsten-17-ol-3.6-dione, 162; 2944 s
Androsten-17-ol-3.16-dione, 2945 s
Androsten-6-ol-3.17-diones, 2945 s, 2946 s
Androsten-11-ol-3.17-dione, 162; 2947 s
Androsten-5-ol-6.17-dione, 2948 s
Androsten-3-ol-12.17-dione, 2950 s
Androsten-3-ol-16.17-dione, 2950 s
Androstane-3.6.17-trione, 2979 s
Androstane-3.11.17-trione, 165; 2980 s
Ætiocholane-3.11.17-trione, 2980 s
Ætiocholane-3.12.17-trione, 2980 s
$C_{19}H_{26}O_4$ Androstan-5-ol-3.6.17-trione, 2983 s
Œstrolic acid methyl ester, 2551 s
Œstrolic acid methyl ether, 2552 s
$C_{19}H_{27}BrO$ 3-Bromoandrosten-17-one, 2476 s
$C_{19}H_{27}BrO_2$ 2-Bromo-Δ^1-androsten-17-ol-3-one,
2702 s
2-Bromo-Δ^4-androsten-17-ol-3-one, 2703 s
4-Bromo-Δ^1-androsten-17-ol-3-one, 2704 s
6-Bromo-Δ^4-androsten-17-ol-3-one, 2705 s

Pages with an "S" refer to the Supplement, the others to the Main Volume

16-Bromo-Δ^5-androsten-3-ol-17-one, 2705 s
2-Bromoandrostane-3.17-dione, 161; 2916 s
4-Bromoætiocholane-3.17-dione, 2917 s
16-Bromoandrostane-3.17-dione, 2919 s
3-Bromoandrostane-6.17-dione, 2919 s
C$_{19}$H$_{27}$Br$_3$O$_2$ 5.6.16-Tribromoandrostan-3-ol-17-one, 2716 s
C$_{19}$H$_{27}$ClO 17-Chloroandrosten-3-one, 2474 s
  3-Chloroandrosten-17-one, 128; 2474 s
C$_{19}$H$_{27}$ClO$_2$ 5-Chloroandrostane-3.17-dione, 161; 2918 s
  3-Chloroandrostane-6.17-dione, 2919 s
C$_{19}$H$_{27}$NO Androstadien-17-one oxime, 2401 s
C$_{19}$H$_{27}$NO$_3$ Androsten-3-ol-16.17-dione 16-oxime, 2950 s
C$_{19}$H$_{27}$N$_3$O$_3$ Δ^4-Androstene-3.6.17-trione trioxime, 165
C$_{19}$H$_{28}$Br$_2$O Androsten-17-one dibromide, 2493 s
C$_{19}$H$_{28}$Br$_2$O$_2$ 2.2-Dibromoandrostan-17-ol-3-one, 2710 s
  2.4-Dibromoandrostan-17-ol-3-one, 2711 s
  5.6-Dibromoandrostan-17-ol-3-one, 2712 s
  5.6-Dibromoandrostan-3-ol-17-one, 2713 s
C$_{19}$H$_{28}$ClNO 3-Chloro-androsten-17-one oxime, 128
C$_{19}$H$_{28}$N$_2$O$_2$ Δ^1-Androstene-3.17-dione dioxime, 155
  Δ^4-Androstene-3.17-dione dioxime, 156
  Δ^5-Androstene-3.17-dione dioxime, 2890 s
  Δ^5-Androstene-4.17-dione ("hetero-Δ^1-Androstene-3.17-dione") dioxime, 155
  i-Androstene-6.17-dione dioxime, 2898 s
C$_{19}$H$_{28}$N$_2$O$_3$ Androstane-3.11.17-trione dioxime, 166
C$_{19}$H$_{28}$O 2.13-Dimethyl-7-keto-1.2-cyclopentanododecahydrophenanthrene, 2395 s
  2.13-Dimethyl-9-keto-1.2-cyclopentanododecahydrophenanthrene, 2395 s
  Androsten-3-one, 2396 s
  Ætiocholen-3-one, 2398 s
  i-Androsten-6-one, 2400 s
  Androsten-17-ones, 2401 s, 2402 s, 2403 s
C$_{19}$H$_{28}$O$_2$ Δ^1-Androsten-17β-ol-3-one, 2578 s
  Δ^4-Androsten-17α-ol-3-one, 140; 2579 s
  Δ^4-Androsten-17β-ol-3-one, 139; 2580 s
  Δ^5-Androsten-17β-ol-3-one, 141; 2595 s
  Δ^5-Androsten-17-ol-4-one, 2601 s
  i-Androsten-17-ol-6-one, 2602 s
  Δ^5-Androsten-3-ol-7-one, 2602 s
  Δ^5-Androsten-17-ol-7-one, 2603 s
  Δ^2-Androsten-3-ol-17-one, 2607 s
  Δ^4-Androsten-3-ol-17-one, 141; 2608 s
  Δ^5-Androsten-3-ol-17-ones 141, 142; 2608 s, 2609 s
  $\Delta^{9(11)}$-Androsten-3-ol-17-ones, 2626 s

206*

Δ^{11}-Androsten-3-ol-17-one, 2627 s
Δ^{11}-Ætiocholen-3-ol-17-one, 2628 s
i-Androsten-6-ol-17-one, 2644 s
Ætiocholane-3.12-dione, 2876 s
Androstane-3.16-dione, 2876 s
Androstane-3.17-dione, 156; 2892 s
Ætiocholane-3.17-dione, 156; 2895 s
Lumiandrostane-3.17-dione, 2895 s
Androstane-6.17-dione, 2898 s
3‖4-Androstene-4-ol-3-oic acid lactone, 2397 s
3‖4-Androsten-4-ol-3-oic acid lactone, 2398 s
C$_{19}$H$_{28}$O$_3$ Δ^4-Androstene-16.17-diol-3-one, 2742 s
  Δ^5-Androstene-3.17-diol-7-one, 2742 s
  Δ^5-Androstene-3.17-diol-16-one, 2744 s
  Δ^5-Androstene-3.4-diol-17-one, 2746 s
  Δ^4-Androstene-3.6-diol-17-one, 2747 s
  Δ^5-Androstene-3.16-diol-17-one, 2751 s
  Dehydroepiandrosterone oxides, 142; 2769 s, 2770 s
  $\Delta^{9(11)}$-Androsten-3-ol-17-one oxide, 2627 s
  Androstan-2-ol-3.17-dione, 2945 s
  Androstan-6-ol-3.17-dione, 2947 s
  Androstan-11-ol-3.17-dione, 2947 s
  Ætiocholan-12-ol-3.17-dione, 2947 s
  Androstan-3-ol-6.17-dione, 2948 s
  Androstan-5-ol-6.17-dione, 2948 s
  Androstan-3-ol-11.17-dione, 2949 s
  Ætiocholan-3-ol-11.17-dione, 2949 s
  Ætiocholan-3-ol-12.17-dione, 2950 s
C$_{19}$H$_{28}$O$_4$ Androstane-5.17-diol-3.6-dione, 2961 s
  Androstane-5.6-diol-3.17-dione, 2961 s
  Androstane-9.11-diol-3.17-dione, 2961 s
  Androstane-3.5-diol-6.17-dione, 2962 s
C$_{19}$H$_{28}$O$_5$S Androsten-3-ol-17-one hydrogen sulphate, 2618 s
C$_{19}$H$_{29}$BrO 3-Bromoandrostan-17-one, 128; 2477 s
C$_{19}$H$_{29}$BrO$_2$ 16-Bromo-Δ^5-androstene-3.17-diol, 2706 s
  2-Bromoandrostan-17-ol-3-one, 148; 2703 s
  16-Bromoandrostan-3-ol-17-one, 2706 s
C$_{19}$H$_{29}$ClN$_2$O$_2$ 3-Chloroandrostane-6.17-dione dioxime, 2919 s
C$_{19}$H$_{29}$ClO 3-Chloroandrostan-17-ones, 128; 2476 s
  5-Chloroandrostan-17-one, 2477 s
C$_{19}$H$_{29}$ClO$_2$ 5-Chloroandrostan-3-ol-17-one, 148; 2705 s
C$_{19}$H$_{29}$NO Androsten-3-one oxime, 2397 s
  Androsten-17-one oxime, 2402 s
  17-Aminoandrosten-3-one, 2508 s
C$_{19}$H$_{29}$NO$_2$ Δ^4-Androsten-17-ol-3-ones, oxime, 140

Δ^5-Androsten-3-ol-17-ones, oxime, 142, 143;
  2618 s
i-Androsten-6-ol-17-one oxime, 2644 s
$C_{19}H_{29}N_3O_3$ Ætiocholane-3.12.17-trione
  trioxime, 2980 s
$C_{19}H_{29}O_5P$ Testosterone dihydrogen phosphate,
  2588 s
$C_{19}H_{30}N_2O_2$ Androstane-3.17-dione dioxime,
  2894 s
  Androstane-6.17-dione dioxime, 2898 s
$C_{19}H_{30}N_2O_4$ Androstane-3.5-diol-6.17-dione
  dioxime, 2962 s
$C_{19}H_{30}O$ Androstan-3-one, 2397 s
  Ætiocholan-3-one, 2398 s
  Androstan-11-one, 119
  Androstan-17-one, 119; 2403 s
  Ætiocholan-17-one, 119
$C_{19}H_{30}O_2$ Androstan-17-ol-3-ones, 141; 2596 s,
  2597 s
  Ætiocholan-17-ol-3-ones, 141; 2599 s,
  2600 s
  Androstan-3-ol-7-one, 2603 s
  Ætiocholan-3-ol-12-one, 2604 s
  Androstan-3-ol-16-one, 2604 s
  Androstan-3-ol-17-ones, 143, 144; 2629 s,
  2633 s
  Ætiocholan-3-ol-17-ones, 144; 2637 s,
  2638 s
  Androstan-5-ol-17-one, 2644 s
  Lumiandrosterone, 2640 s
  3‖4-Androstan-4-ol-3-oic acid lactone,
  2397 s
  3‖4-Ætiocholan-4-ol-3-oic acid lactone,
  2398 s
$C_{19}H_{30}O_3$ Androstane-3.17-diol-7-one, 2743 s
  Androstane-3.17-diol-11-one, 149; 2743 s
  Androstane-2.3-diol-17-one, 2745 s
  Androstane-3.5-diol-17-one, 2746 s
  Androstane-3.6-diol-17-one, 2747 s
  Ætiocholane-3.6-diol-17-one, 2747 s
  Androstane-3.11-diol-17-ones, 149; 2748 s
  Ætiocholane-3.12-diol-17-one, 2749 s
$C_{19}H_{30}O_4$ Androstane-4.5.17-triol-3-one, 2767 s
  Androstane-3.5.6-triol-17-ones, 150; 2767 s,
  2768 s
  Androstane-3.9.11-triol-17-one, 2771 s
  Androstane-3‖4-dioic acid, 2398 s
  Ætiocholane-3‖4-dioic acid, 2400 s
$C_{19}H_{30}O_5S$ Androsterone hydrogen sulphate,
  2632 s
$C_{19}H_{31}NO$ 17-Amino-ætiocholan-3-one, 2509 s
$C_{19}H_{31}NO_2$ Androstan-17-ol-3-one oxime, 141
  Ætiocholan-17-ol-3-one oxime, 141
  Androstan-3-ol-16-one oxime, 2604 s
  Androstan-3-ol-17-ones, oxime, 144;
  2632 s
  Lumiandrosterone oxime, 2640 s

$C_{19}H_{31}N_3O$ 17-Keto-œstrane semicarbazone,
  118
$C_{19}H_{31}N_3O_2$ Œstran-3-ol-17-one semicarbazone,
  2563 s

$C_{20}H_{15}ClO_4$ 8-Chloro-4-acetoxy-7-methoxy-
  3′-keto-1.2-cyclopentenophenanthrene,
  2761 s
$C_{20}H_{16}ClNO_4$ 8-Chloro-4-acetoxy-7-methoxy-
  3′-keto-1.2-cyclopentenophenanthrene
  oxime, 2761 s
$C_{20}H_{16}O_3$ 7-Methyl-4-acetoxy-3′-keto-1.2-cyclo-
  pentenophenanthrene, 2521 s
$C_{20}H_{16}O_4$ 4-Acetoxy-7-methoxy-3′-keto-
  1.2-cyclopentenophenanthrene, 2735 s
  4′-Hydroxymethylene-4.7-dimethoxy-
  3′-keto-1.2-cyclopentenophenanthrene,
  2764 s
  5.9-Dihydroxy-3′.5′-diketo-1.2-cyclo-
  pentenophenanthrene methyl ether
  ethyl ether, 164
$C_{20}H_{18}O$ 3′-Isopropyl-5′-keto-1.2-cyclopenteno-
  phenanthrene, 2394 s
$C_{20}H_{18}O_3$ 4-Ethoxy-7-methoxy-3′-keto-
  1.2-cyclopentenophenanthrene, 2736 s
$C_{20}H_{18}O_4$ 4-Acetoxy-7-methoxy-3′-keto-
  1.2-cyclopenteno-dihydrophenanthrene,
  2736 s
  11-Ketoequilenin acetate, 2940 s
$C_{20}H_{19}NO$ 3′-Isopropyl-5′-keto-1.2-cyclo-
  pentenophenanthrene oxime, 2394 s
$C_{20}H_{19}NO_3$ 4-Ethoxy-7-methoxy-3′-keto-
  1.2-cyclopentenophenanthrene oxime,
  2736 s
$C_{20}H_{20}O_3$ Equilenin acetate, 135; 2533 s, 2534 s
  Isoequilenin acetate, 2535 s, 2538 s
$C_{20}H_{20}O_4$ Equilenin-O-acetic acid, 2533 s
$C_{20}H_{21}NO_2$ 3-Acetamino-desoxyequilenin, 2508 s
$C_{20}H_{22}O_2$ 18-Methylequilenan-3-ol-17-ones,
  methyl ether, 2571 s
  18-Methylequilenan-6-ol-17-ones, methyl
  ether, 2572 s
  18-Ethylequilenan-3-ol-17-ones, 2571 s
$C_{20}H_{22}O_3$ 6-Dehydroœstrone acetate, 2538 s
  8-Dehydro-14-iso-œstrone acetate, 2540 s
  Equilin acetate, 2540 s
$C_{20}H_{22}O_4$ 8-Hydroxy-9(11)-dehydro-14-iso-
  œstrone monoacetate, 2740 s
$C_{20}H_{23}BrO_3$ 16-Bromoœstrone acetate, 2702 s
$C_{20}H_{23}N_3O_2$ Equilenin methyl ether semicarbazone, 2533 s
$C_{20}H_{24}O$ 3′-Isopropyl-4-keto-1.2-cyclopentano-
  hexahydrophenanthrene, 2394 s
$C_{20}H_{24}O_3$ Œstrone acetate, 137; 2554 s
  13-iso-Œstrone acetate, 2560 s
  Œstratrien-3-ol-17-one acetate, 2561 s
  Folliculosterone acetate, 2562 s

Pages with an "S" refer to the Supplement, the others to the Main Volume

$C_{20}H_{33}NO$ 17-Aminomethyl-androstan-3-one, 2507 s
$C_{20}H_{33}N_3O$ Androstan-3-one semicarbazone, 2397 s
Androstan-17-one semicarbazone, 119; 2404 s
Ætiocholan-17-one semicarbazone, 119
$C_{20}H_{33}N_3O_2$ Androstan-17-ol-3-one semicarbazone, 141
Androstan-3-ol-16-one semicarbazone, 2604 s
Androstan-3-ol-17-ones, semicarbazone, 144; 2636 s
Ætiocholan-3-ol-17-ones, semicarbazone, 2638 s, 2639 s
$C_{20}H_{33}N_3O_3$ Androstane-3.11-diol-17-one semicarbazone, 149
$C_{20}H_{34}O_2$ 2.5-Dimethyl-3'-ethyl-1.2-cyclo-pentanodecalin-5-propionic acid, 2786 s, 2787 s

$C_{21}H_{16}O_5$ 4.7-Diacetoxy-3'-keto-1.2-cyclo-pentenophenanthrene, 2734 s
$C_{21}H_{17}N$ Compound from indolocholestene, 2441 s
$C_{21}H_{18}O_4$ 4'-Hydroxymethylene-4-ethoxy-7-methoxy-3'-keto-1.2-cyclopenteno-phenanthrene, 2765 s
$C_{21}H_{21}N_3O_4$ 11-Ketoequilenin acetate mono-semicarbazone, 2940 s
$C_{21}H_{22}O_4$ Equilenin-O-acetic acid methyl ester, 2533 s
$C_{21}H_{24}O_2$ 18-Ethylequilenan-3-ol-17-ones, methyl ether, 2571 s
18-Propylequilenan-3-ol-17-ones, 2571 s, 2572 s
$C_{21}H_{26}O$ 17-Ethynylandrostadien-3-one, 2406 s
$C_{21}H_{26}O_2$ Œstrone allyl ether, 138; 2556 s
3'-Isopropyl-7-methoxy-4-keto-1.2-cyclo-pentanohexahydrophenanthrene, 2570 s
2 or 4-Allylœstrone, 2574 s
$C_{21}H_{26}O_3$ Œstrone propionate, 137
$C_{21}H_{26}O_4$ Œstrone ethyl carbonate, 137
Œstrone-O-acetic acid methyl ester, 2556 s
Œstrone-O-α-propionic acid, 2556 s
16-Ketoœstradiol 3-methyl ether 17-acetate, 2739 s
Œstrololactone propionate, 2552 s
$C_{21}H_{27}ClO_3$ 21-Chloropregnene-3.6.20-trione, 2971 s
$C_{21}H_{27}N_3O_3$ Œstrone acetate semicarbazone, 137
$C_{21}H_{28}Br_2O_2$ 21.21-Dibromopregnene-3.20-dione, 2812 s
$C_{21}H_{28}Cl_2O_2$ 21.21-Dichloropregnene-3.20-dione, 2812 s
$C_{21}H_{28}Cl_2O_3$ Androsten-3-ol-17-one dichloro-acetate, 2619 s

$C_{21}H_{28}N_2O_2$ 21-Diazopregnadien-3-ol-20-one, 2293 s
21-Diazo-Δ^4-pregnene-3.20-dione, 2816 s
21-Diazo-Δ^{11}-pregnene-3.20-dione, 2309 s
$C_{21}H_{28}N_2O_3$ 21-Diazopregnen-3-ol-11.20-dione, 2851 s
16-Ketoœstrone methyl ether di-methoxime, 2942 s
$C_{21}H_{28}O$ Pregnatrien-20-one, 2215 s
17-Vinylandrostadien-3-one, 2406 s
$C_{21}H_{28}O_2$ 17-Ethynyltestosterone, 2648 s
Pregnatrien-21-ol-3-one, 2512 s
Pregnadiene-3.20-diones, 2778 s, 2779 s, 2780 s
Pregnadien-3-on-21-al, 2802 s
$C_{21}H_{28}O_3$ $\Delta^{1.4}$-Androstadien-17-ol-3-one acetate, 2576 s
$\Delta^{4.6}$-Androstadien-17-ol-3-one acetate, 2578 s
$\Delta^{3.5}$-Androstadien-17-ol-7-one acetate, 2603 s
$\Delta^{3.5}$-Androstadien-3-ol-17-one acetate, 156
Pregnadien-21-ol-3.20-diones, 2818 s, 2819 s
Androstadien-3-ol-16.17-dione ethyl ether, 2950 s
Pregnene-3.20-dion-21-al, 2870 s
Δ^4-Pregnene-3.11.20-trione, 2967 s
$\Delta^{9(11)}$-Pregnene-3.12.20-trione, 2969 s
Urene-3.11.20-trione, 2968 s
$C_{21}H_{28}O_4$ Androsten-17-ol-3.6-dione acetate, 162; 2944 s
Androsten-17-ol-3.16-dione acetate, 2945 s
Androsten-6-ol-3.17-diones, acetate, 2946 s
Androsten-3-ol-12.17-dione acetate, 2950 s
17-iso-Pregnen-17-ol-3.20-dion-21-al, 2871 s
Pregnen-21-ol-3.11.20-trione, 164; 2972 s
Pregnen-21-ol-3.12.20-trione, 2974 s
Pregnane-3.7.12.20-tetrone, 2985 s
$C_{21}H_{28}O_5$ Pregnene-17.21-diol-3.11.20-trione, 165; 2977 s
Acetoxydicarboxylic anhydride from 16-(α-methylpropylidene)-androsten-3-ol-17-one, 2657 s
$C_{21}H_{29}BrO$ 21-Bromopregnadien-3-one, 2473 s
$C_{21}H_{29}BrO_2$ 21-Bromopregnadien-3-ol-20-one, 2277 s
6-Bromopregnene-3.20-dione, 2813 s, 2814 s
21-Bromopregnene-3.20-dione, 2810 s
$C_{21}H_{29}BrO_3$ 2-Bromo-Δ^4-androsten-17-ol-3-one acetate, 2703 s
6-Bromo-Δ^4-androsten-17-ol-3-one acetate, 2705 s
16-Bromo-Δ^5-androsten-3-ol-17-one acetate, 2706 s
4-Bromopregnane-3.6.20-trione, 2971 s
4-Bromopregnane-3.11.20-trione, 2971 s
12-Bromopregnane-3.11.20-trione, 2971 s

$C_{21}H_{29}BrO_4$ 4-Bromopregnan-21-ol-3.11.20-
trione, 2975 s
4-Bromopregnan-21-ol-3.12.20-trione, 2975 s
$C_{21}H_{29}Br_2ClO_2$ 21-Chloro-5.6-dibromopregnane-
3.20-dione, 2815 s
$C_{21}H_{29}ClO$ 3-Chloropregnadien-20-one, 2221 s
$C_{21}H_{29}ClO_2$ 21-Chloropregnadien-3-ol-20-one,
2277 s
21-Chloropregnene-3.20-dione, 2810 s
$C_{21}H_{29}ClO_3$ Testosterone chloroacetate, 2589 s
$C_{21}H_{29}IO_2$ 21-Iodopregnadien-3-ol-20-one,
2277 s
21-Iodopregnene-3.20-dione, 2811 s
$C_{21}H_{29}NO_2$ 17-Ethynyltestosterone oxime,
2649 s
$C_{21}H_{29}NO_3$ Androstadien-3-ol-16.17-dione ethyl
ether 16-oxime, 2950 s
$C_{21}H_{29}NO_4$ Androsten-3-ol-16.17-dione acetate
16-oxime, 2951 s
$C_{21}H_{30}BrIO_2$ 17-Bromo-21-iodopregnen-3-ol-
20-one, 2287 s
$C_{21}H_{30}Br_2O_2$ 17.21-Dibromopregnen-3-ol-20-one,
2287 s
21.21-Dibromopregnen-3-ol-20-one, 2283 s
5.6-Dibromoallopregnane-3.20-dione, 2814 s
11.12-Dibromopregnane-3.20-dione, 2815 s
$C_{21}H_{30}Br_2O_3$ 2.2-Dibromoandrostan-17-ol-3-one
acetate, 2711 s
2.4-Dibromoandrostan-17-ol-3-one acetate,
2712 s
$C_{21}H_{30}Br_4O_2$ 5.6.17.21-Tetrabromo-allopregnan-
3-ol-20-one, 2290 s
$C_{21}H_{30}N_2O$ 21-Diazo-Δ^2-allopregnen-20-one,
2224 s
$C_{21}H_{30}N_2O_2$ 21-Diazo-Δ^5-pregnen-3-ol-20-one,
117; 2293 s
21-Diazo-Δ^{11}-pregnen-3-ol-20-one, 2308 s
$C_{21}H_{30}N_2O_3$ 21-Diazopregnan-12-ol-3.20-dione,
2851 s
21-Diazopregnan-3-ol-11.20-diones, 2851 s,
2852 s
Androsten-3-ol-17-spirohydantoin, 2615 s
Androsten-17-ol-3-spirohydantoin, 2584 s
$C_{21}H_{30}N_6O_2$ Androstadiene-3.17-dione disemicarbazone, 2878 s
$C_{21}H_{30}N_6O_3$ Androstene-3.11.17-trione disemicarbazone, 165; 2980 s
$C_{21}H_{30}O$ Pregnadien-3-ones, 145; 2406 s, 2410 s
Pregnadien-20-ones, 2215 s
$C_{21}H_{30}OS_2$ Androstene-3.17-dione 3-ethylene
mercaptal, 2886 s
$C_{21}H_{30}O_2$ 17.20-Epoxy-Δ^4-pregnen-3-one, 2407 s
Androstadien-3-ol-17-one ethyl ether,
2606 s
17-Vinyltestosterone, 2650 s
Retropregnadien-20-ol-3-one(?), 2409 s
Pregnadien-21-ol-3-ones, 2512 s, 2513 s

Pregnadien-3-ol-11-one, 2654 s
Pregnadien-3-ol-20-ones, 151; 2229 s, 2230 s
Pregnadien-3-ol-21-al, 2251 s
Pregnene-3.11-diones, 2899 s
Pregnene-3.16-dione, 2900 s
Allopregnene-3.16-dione, 2900 s
Δ^1-Allopregnene-3.20-dione, 150; 2781 s
Δ^4-Pregnene-3.20-dione, 151; 2782 s
Δ^5-Pregnene-3.20-dione, 152; 2793 s
Δ^{11}-Pregnene-3.20-dione, 2793 s
Δ^{16}-Pregnene-3.20-dione, 2794 s
Δ^{16}-Allopregnene-3.20-dione, 2795 s
17-iso-Δ^4-Pregnene-3.20-dione, 2800 s
Δ^5-Pregnene-4.20-dione, 150; 2801 s
Δ^4-Pregnene-6.20-dione, 2801 s
Pregnen-3-on-21-al, 2803 s
Urene-3.11-dione, 2900 s
"C-Isomer" from pregnadien-3-one, 2410 s
$C_{21}H_{30}O_3$ Δ^1-Androsten-17-ol-3-one acetate,
2579 s
Δ^4-Androsten-17-ol-3-ones, acetate, 140;
2588 s
Δ^5-Androsten-17-ol-3-one acetate, 141
Δ^5-Androsten-17-ol-4-one acetate, 2602 s
i-Androsten-17-ol-6-one acetate, 2602 s
Δ^5-Androsten-3-ol-7-one acetate, 2602 s
Δ^5-Androsten-17-ol-7-one acetate, 2603 s
Δ^5-Androsten-3-ol-17-ones, acetate, 142,
143; 2618 s
$\Delta^{9(11)}$-Androsten-3-ol-17-ones, acetate,
2626 s, 2627 s
Δ^{11}-Androsten-3-ol-17-one acetate, 2628 s
i-Androsten-6-ol-17-one acetate, 2645 s
17-Vinyltestosterone 20.21-oxide, 2651 s
Epoxyretropregnen-20-ol-3-one(?), 2409 s
17-iso-Pregnadiene-17.20-diol-3-one, 2720 s
17-Ethynylætiocholane-3.17-diol-11-one,
2752 s
Pregnadiene-3.21-diol-20-ones, 2300 s
Pregnene-17.20.21-triol-3-one 17.20-an-
hydride, 2726 s
Androstene-3.17-dione 3-ethylene ketal,
2890 s
Pregnen-21-ol-3.11-dione, 2934 s
Δ^4-Pregnen-2-ol-3.20-dione, 2837 s
Δ^4-Pregnen-6-ol-3.20-diones, 2838 s
Δ^4-Pregnen-11-ol-3.20-dione, 2840 s
Δ^4-Pregnen-12-ol-3.20-dione, 2840 s
Δ^4-Pregnen-17-ol-3.20-dione, 2843 s
17-iso-Δ^4-Pregnen-17-ol-3.20-dione, 2844 s
Δ^4-Pregnen-21-ol-3.20-dione, 153; 2820 s
17-iso-Δ^4-Pregnen-21-ol-3.20-dione, 2829 s
$\Delta^{9(11)}$-Pregnen-21-ol-3.20-dione, 2830 s
Δ^{11}-Pregnen-21-ol-3.20-dione, 2830 s
Δ^4-Pregnen-x-ol-3.20-dione, 2845 s
$\Delta^{9(11)}$-Pregnen-3-ol-12.20-dione, 2849 s
17-iso-Pregnen-17-ol-3-on-21-al, 2846 s

$C_{21}H_{31}O_6P$ Desoxycorticosterone phosphate, 2823 s
$C_{21}H_{31}O_7P$ Corticosterone 21-dihydrogen phosphate, 2854 s
$C_{21}H_{32}Br_2O$ 17.21-Dibromo-allopregnan-20-one, 2223 s
$C_{21}H_{32}Br_2O_2$ 17.21-Dibromopregnan-3-ol-20-one, 2288 s
　17.21-Dibromo-allopregnan-3-ol-20-one, 2288 s
　5.6-Dibromo-allopregnan-3-ol-20-one, 2289 s
　16.17-Dibromopregnan-3-ol-20-one, 2289 s
$C_{21}H_{32}Br_2O_3$ 17.21-Dibromopregnane-3.12-diol, 2351 s
$C_{21}H_{32}ClNO$ 3-Chloropregnen-20-one oxime, 114
$C_{21}H_{32}N_2O$ 21-Diazo-pregnan-20-one, 2224 s
　21-Diazo-allopregnan-20-one, 2224 s
$C_{21}H_{32}N_2O_2$ 21-Diazopregnan-3-ol-20-ones, 2294 s, 2295 s
　21-Diazoallopregnan-3-ol-20-ones, 2295 s
　Δ^4-Pregnene-3.20-dione dioxime, 151
　Δ^5-Pregnene-3.20-dione dioxime, 152
　Δ^1-Allopregnene-3.20-dione dioxime, 2781 s
　Δ^{16}-Allopregnene-3.20-dione dioxime, 2795 s
$C_{21}H_{32}N_2O_3$ 21-Diazopregnane-3.12-diol-20-one, 2352 s
　Pregnen-17-ol-3.20-dione dioxime, 2844 s
　17-iso-Pregnen-17-ol-3-on-21-al dioxime, 2847 s
　Pregnen-3-ol-20-on-21-al dioxime, 2379 s
　Androstan-3-ol-17-spirohydantoin, 2631 s
$C_{21}H_{32}O$ Pregnen-3-one, 2410 s
　Allopregnen-3-one, 2411 s
　Pregnen-20-ones, 113; 2216 s
　Allopregnen-20-ones, 2216 s, 2217 s
$C_{21}H_{32}O_2$ Δ^2-Androsten-3-ol-17-one ethyl ether, 2607 s
　17-Ethylandrosten-17-ol-3-ones, 145; 2651 s, 2652 s
　Pregnen-20-ol-3-one, 132; 2510 s
　Pregnen-3-ol-11-ones, 2654 s
　Δ^5-Pregnen-3α-ol-20-one, 2232 s
　Δ^5-Pregnen-3β-ol-20-one, 114; 2233 s
　17-iso-Δ^5-Pregnen-3β-ol-20-one, 115; 2249 s
　Δ^{11}-Pregnen-3α-ol-20-one, 2239 s
　Δ^{16}-Pregnen-3α-ol-20-one, 2239 s
　Δ^{16}-Pregnen-3β-ol-20-one, 2240 s
　Δ^{16}-Allopregnen-3-ol-20-ones, 2241 s, 2242 s
　i-Pregnen-6-ol-20-one, 2251 s
　Pregnen-12-ol-20-one, 2251 s
　Δ^2-Allopregnen-21-ol-20-one, 2225 s
　Pregnen-3-ol-21-al, 2252 s
　Pregnane-3.11-dione, 2967 s
　Allopregnane-3.16-dione, 2900 s
　Pregnane-3.20-dione, 152; 2796 s
　Allopregnane-3.20-dione, 152; 2797 s

　17-iso-Allopregnane-3.20-dione, 152; 2800 s
　14-iso-17-iso-Allopregnane-3.20-dione, 2800 s
　Pregnane-11.20-dione, 2967 s
　Retropregnane-3.20-dione(?), 2409 s
　Urane-3.11-dione, 2900 s
　Urane-11.20-dione, 2802 s
$C_{21}H_{32}O_3$ Androstan-17-ol-3-ones, acetate, 141; 2597 s, 2599 s
　Ætiocholan-17-ol-3-one acetate, 141; 2600 s
　Androstan-3-ol-7-one acetate, 2603 s
　Ætiocholan-3-ol-12-one acetate, 2604 s
　Androstan-3-ol-16-one acetate, 2604 s
　Androstan-3α-ol-17-one acetate, 144; 2632 s
　Androstan-3β-ol-17-one acetate, 144; 2636 s
　Ætiocholan-3α-ol-17-one acetate, 2638 s
　Ætiocholan-3β-ol-17-one acetate, 2640 s
　Lumiandrosterone acetate, 2640 s
　Testosterone ethylene ketal, 2595 s
　5.6-Epoxy-allopregnan-3-ol-20-one, 2364 s
　16.17-Epoxypregnan-3-ol-20-one, 2241 s
　16.17-Epoxy-allopregnan-3-ol-20-one, 2243 s
　Pregnene-17.20-diol-3-ones, 2721 s
　Pregnene-20.21-diol-3-ones, 2514 s
　Pregnene-3.20-diol-7-one, 2722 s
　17-iso-Pregnene-3.17-diol-11-one, 2752 s
　Pregnene-3.21-diol-11-one, 2723 s
　Allopregnene-3.6-diol-20-ones, 2323 s, 2324 s
　Pregnene-3.16-diol-20-one, 2332 s
　Pregnene-3.17-diol-20-one, 2333 s
　17-iso-Pregnene-3.17-diol-20-one, 2337 s
　Δ^5-Pregnene-3.21-diol-20-one, 116; 2301 s
　Δ^{11}-Pregnene-3.21-diol-20-one, 2308 s
　Allopregnene-3.21-diol-20-ones, 2308 s, 2309 s
　17-iso-Pregnene-3.17-diol-21-al, 2345 s
　Pregnene-3.20-diol-21-al, 2312 s
　Testalolone, 117; 2381 s
　Allopregnan-20-ol-3.16-dione, 2934 s
　Allopregnan-2-ol-3.20-dione, 153
　Pregnan-4-ol-3.20-dione, 2838 s
　Pregnan-6-ol-3.20-dione, 2839 s
　Pregnan-12-ol-3.20-dione, 2841 s
　Allopregnan-17-ol-3.20-dione, 2845 s
　Pregnan-21-ol-3.20-dione, 2831 s
　Allopregnan-21-ol-3.20-dione, 2832 s
　Pregnan-5-ol-4.20-dione, 2847 s
　Pregnan-3-ol-11.20-diones, 2847 s, 2848 s
　Pregnan-3-ol-12.20-dione, 2849 s
　Pregnan-3-ol-20-on-21-al, 2380 s
　Allopregnan-3-ol-20-on-21-al, 2380 s, 2381 s
　Uran-3-ol-11.20-dione, 2848 s
$C_{21}H_{32}O_4$ Androstane-3.5-diol-17-one 3-acetate, 2746 s
　Androstane-3.11-diol-17-ones, 3-acetate, 149; 2748 s, 2749 s
　5-iso-Androstane-3.12-diol-17-one, 3- and 12-acetate, 2750 s

Pregnane-3.5-diol-20-one, 2323 s
Pregnane-3.6-diol-20-one, 2324 s
Pregnane-3.7-diol-20-one, 2327 s
Pregnane-3.11-diol-20-ones, 2327 s, 2328 s
Pregnane-3.12-diol-20-ones, 2329 s, 2331 s
Pregnane-3.17-diol-20-ones, 2335 s
Pregnane-3.21-diol-20-ones, 2309 s, 2310 s
Allopregnane-3.4-diol-20-one, 2322 s
Allopregnane-3.6-diol-20-ones, 2325 s, 2326 s
Allopregnane-3.16-diol-20-one, 2333 s
Allopregnane-3.17-diol-20-one, 2335 s
Allopregnane-3.21-diol-20-ones, 2310 s
17-iso-Pregnane-3.12-diol-20-one, 2331 s
17-iso-Pregnane-3.17-diol-20-one, 2342 s
17-iso-Allopregnane-3.14-diol-20-one, 2332 s
17-iso-Allopregnane-3.17-diol-20-one, 2342 s
Allopregnane-3.20-diol-21-al, 2313 s
Dihydroxyketone from adrenal cortex, 2336 s
$C_{21}H_{34}O_4$ Pregnane-3.7.12-triol-20-ones, 117; 2366 s, 2368 s
Pregnane-3.11.21-triol-20-ones, 2354 s
Pregnane-3.12.21-triol-20-one, 2356 s
Pregnane-3.16.17-triol-20-one, 2368 s
Allopregnane-3.5.6-triol-20-ones, 2364 s, 2365 s
Allopregnane-3.5.21-triol-20-one, 2353 s
Allopregnane-3.11.21-triol-20-one, 2354 s
Allopregnane-3.16.17-triol-20-one, 2368 s
Allopregnane-3.17.21-triol-20-one, 2358 s
14-iso-Pregnane-3.5.14-triol-20-one, 2366 s
17-iso-Allopregnane-3.11.21-triol-20-one, 2355 s
17-iso-Allopregnane-3.17.21-triol-20-one, 2360 s
Allopregnane-3.17.20-triol-21-al, 2361 s, 2362 s
$C_{21}H_{34}O_5$ Allopregnane-3.17.20.21-tetrol-11-one, 2765 s
Pregnane-3.5.6.21-tetrol-20-one, 2373 s
Allopregnane-3.5.6.21-tetrol-20-ones, 2372 s
Allopregnane-3.11.17.21-tetrol-20-ones, 117; 2373 s, 2375 s
Compounds from adrenal glands, 2374 s, 2375 s
$C_{21}H_{34}O_6$ Acid from allopregnane-3.11.17.21-tetrol-20-one, 2374 s
$C_{21}H_{35}NO$ 3-Hydroxyternorcholenylamine, 2508 s
$C_{21}H_{35}NO_2$ Pregnan-3-ol-20-ones, oxime, 2245 s, 2246 s
Allopregnan-3-ol-20-one oxime, 116
17-iso-Pregnan-3-ol-20-one oxime, 2250 s
20-Aminopregnan-3-ol-11-one, 2718 s
$C_{21}H_{35}NO_3$ Pregnane-3.17-diol-20-one oxime, 2335 s

Allopregnane-3.6-diol-20-ones, oxime, 2324 s, 2326 s
$C_{21}H_{35}NO_4$ Pregnane-3.7.12-triol-20-one oxime, 2368 s
$C_{21}H_{35}N_3O_2$ Epiandrosterone methyl ether semicarbazone, 2636 s
17-Methylandrostan-17-ol-3-one semicarbazone, 2647 s
$C_{21}H_{36}N_2$ Allopregnan-3-one hydrazone, 2412 s
$C_{21}H_{36}O$ Compound from allopregnan-20-ol-3.16-dione, 2935 s

$C_{22}H_{21}NO_4$ 4-Ethoxy-7-methoxy-3'-keto-1.2-cyclopentenophenanthrene oxime, acetyl derivative, 2736 s
$C_{22}H_{24}O_4$ Equilenin-O-acetic acid ethyl ester, 2533 s
$C_{22}H_{24}O_5$ 7-Hydroxy-6-dehydroœstrone diacetate, 2739 s
$C_{22}H_{26}O_2$ 18-Propylequilenan-3-ol-17-ones, methyl ether, 2571 s, 2572 s
$C_{22}H_{26}O_5$ 6-Ketoœstradiol diacetate, 2738 s
7-Hydroxyœstrone diacetate, 2740 s
$C_{22}H_{26}O_6$ Equilin-7.8-glycol 3.7-diacetate, 2767 s
$C_{22}H_{28}O_3$ Œstrone butyrate, 137
Œstrone isobutyrate, 137
$C_{22}H_{28}O_4$ Œstrone-O-α-propionic acid methyl ester, 2556 s
$C_{22}H_{30}N_2O_2$ Pyrazoline compound from $\Delta^{4.16}$-pregnadiene-3.20-dione, 2780 s
$C_{22}H_{30}N_2O_3$ 21-Diazo-19-nor-14-iso-17-iso-pregnen-3-ol-20-one acetate, 2293 s
$C_{22}H_{30}O$ 17-Allylidene-androsten-3-one, 2412 s
16-Isopropylidene-androstadien-17-one, 2412 s
$C_{22}H_{30}O_2$ Methyl 1-(3-hydroxy-œstratrienyl)-ethyl ketone, 2227 s
16-Methylpregnadiene-3.20-dione, 2807 s
21-Methylpregnadiene-3.21-dione, 2804 s
$C_{22}H_{30}O_3$ 1-Dehydrotestosterone propionate, 2576 s
6-Dehydrotestosterone propionate, 139
$C_{22}H_{30}O_4$ 19-Nor-14-iso-17-iso-pregnen-21-ol-3.20-dione acetate, 2817 s
$C_{22}H_{31}BrO_3$ Testosterone α-bromopropionate, 2589 s
$C_{22}H_{31}N_3O_2$ 17-Ethynyltestosterone semicarbazone, 2649 s
$C_{22}H_{31}N_3O_3$ 1-Dehydrotestosterone acetate semicarbazone, 2576 s
$C_{22}H_{32}N_2O_2$ 21-Diazomethyl-pregnen-3-ol-21-one, 2296 s
Pyrazoline compound from pregnadien-3-ol-20-one, 2231 s
$C_{22}H_{32}O_2$ 17-Allyltestosterone, 2653 s
16-Isopropylidene-androsten-3-ol-17-one, 2656 s

Pages with an "S" refer to the Supplement, the others to the Main Volume

$C_{23}H_{34}O_5$ Dihydrotestosterone hydrogen
  succinate, 2599 s
  Androsterone hydrogen succinate, 144
  Androstane-3.17-diol-7-one diacetate, 2743 s
  Androstane-3.17-diol-11-one diacetate, 149
  Androstane-3.11-diol-17-one diacetate,
    2749 s
  5-iso-Androstane-3.12-diol-17-one diacetate,
    2750 s
  5.6-Epoxy-allopregnane-3.21-diol-20-one
    21-acetate, 2372 s
  Pregnene-17.20.21-triol-3-one 21-acetate,
    2726 s
  Pregnene-3.11.21-triol-20-one 21-acetate,
    2353 s
  Retropregnene-13.14.20-triol-3-one(?)
    monoacetate, 2410 s
  Pregnane-17.20-diol-3.11-dione 20-acetate,
    2958 s
  Pregnane-4.5-diol-3.20-dione 4-acetate,
    2865 s
  Allopregnane-5.6-diol-3.20-dione 6-acetate,
    2865 s
  Pregnane-11.21-diol-3.20-diones, 21-acetate,
    2855 s, 2856 s
  Pregnane-12.21-diol-3.20-dione 21-acetate,
    2858 s
  Allopregnane-3.5-diol-6.20-dione 3-acetate,
    2866 s
  Pregnane-3.21-diol-11.20-diones, 21-acetate,
    2860 s, 2861 s
  Pregnane-3.21-diol-12.20-dione 21-acetate,
    2863 s
  Compound from 24.24-diethylcholane-
    3.12.24-triol triacetate, 2751 s
$C_{23}H_{34}O_6$ Androstane-4.5.17-triol-3-one
    4.17-diacetate, 2767 s
  Androstane-3.5.6-triol-17-ones, diacetates,
    2768 s, 2769 s
$C_{23}H_{34}S_4$ Androstene-3.17-dione bis-(ethylene
    mercaptal), 2886 s
$C_{23}H_{35}BrO_3$ 4-Bromopregnan-20-ol-3-one
    acetate, 133
  17-Bromopregnan-3-ol-20-one acetate, 2286 s
  17-Bromo-allopregnan-3-ol-20-one acetate,
    2287 s
  21-Bromopregnan-3-ol-20-one acetate, 2281 s
$C_{23}H_{35}BrO_4$ Pregnane-3.21-diol-20-one
    21-bromoacetate, 2310 s
  Allopregnane-3.21-diol-20-one 21-bromo-
    acetate, 2312 s
  12-Bromopregnane-3.11-diol-20-one
    3-acetate, 2351 s
$C_{23}H_{35}ClO_3$ 21-Chloropregnan-3-ol-20-one
    acetate, 2280 s
  5-Chloro-allopregnan-3-ol-20-one acetate,
    2285 s

$C_{23}H_{35}NO_3$ Testosterone N-propylcarbamate,
    2589 s
  Δ^{16}-Pregnen-3-ol-20-one acetate oxime,
    2241 s
$C_{23}H_{35}NO_4$ 17-iso-Pregnene-3.17-diol-20-one
    3-acetate oxime, 2339 s
  5.6-Epoxyallopregnan-3-ol-20-one acetate
    oxime, 2365 s
$C_{23}H_{35}N_3O_3$ 17-Acetoxymethyl-androsten-3-one
    semicarbazone, 2510 s
$C_{23}H_{36}N_2O_5$ Allopregnane-3.5-diol-6.20-dione
    3-acetate dioxime, 2866 s
$C_{23}H_{36}N_6O_2$ Δ^{16}-Pregnene-3.20-dione disemi-
    carbazone, 2795 s
$C_{23}H_{36}N_6O_3$ Pregnen-17-ol-3.20-dione disemi-
    carbazone, 2844 s
  Pregnen-20-ol-3-on-21-al disemicarbazone,
    2833 s
  Allopregnanetrione disemicarbazone, 2968 s
$C_{23}H_{36}O_2$ 21-Ethylpregnen-3-ol-20-one, 2259 s
  Norcholen-3-ol-22-one, 2259 s
  20-iso-Norcholen-3-ol-22-one, 2260 s
$C_{23}H_{36}O_3$ Dihydrotestosterone butyrate, 2599 s
  Androsterone butyrate, 2633 s
  Pregnan-20-ol-3-one acetate, 132
  Allopregnan-20-ol-3-one acetate, 133
  Pregnan-3-ol-20-ones, acetate, 115; 2245 s,
    2246 s
  Allopregnan-3-ol-20-ones, acetate, 115, 116;
    2247 s, 2249 s
  17-iso-Pregnan-3-ol-20-one acetate, 2250 s
  17-iso-Allopregnan-3-ol-20-one acetate, 116;
    2250 s
  Allopregnan-21-ol-20-one acetate, 2225 s
  Allopregnan-3-ol-21-al acetate, 2252 s
  Retropregnan-20-ol-3-one(?) acetate,
    2409 s
  Uran-11-ol-3-one acetate, 2645 s
  Norcholene-3.23-diol-22-one, 2314 s
$C_{23}H_{36}O_4$ Allopregnane-17.20-diol-3-one
    20-acetate, 2722 s
  Pregnane-3.20-diol-11-one 3-acetate, 2723 s
  Pregnane-3.11-diol-20-ones, 3-acetate, 2328 s
  Pregnane-3.12-diol-20-one 3-acetate, 2330 s
  Pregnane-3.12-diol-20-one 12-acetate, 2330 s
  Pregnane-3.17-diol-20-ones, 3-acetate,
    2335 s, 2336 s
  Pregnane-3.21-diol-20-ones, 21-acetate,
    2309 s, 2310 s
  Allopregnane-3.21-diol-20-one, 21-acetate,
    2312 s
  17-iso-Pregnane-3.17-diol-20-one 3-acetate,
    2342 s
  17-iso-Allopregnane-3.14-diol-20-one
    3-acetate, 2332 s
  17-iso-Allopregnane-3.17-diol-20-one
    3-acetate, 2344 s

Pages with an "S" refer to the Supplement, the others to the Main Volume

Pages with an "S" refer to the Supplement, the others to the Main Volume

E.O.C. XIV s. 207

Folliculosterone benzoate, 2562 s
16-Ketoœstrone benzyl ether, 2942 s
C₂₅H₂₆O₄ 16-Ketoœstradiol 3-benzoate, 2738 s
7-Hydroxyœstrone 3-benzoate, 2740 s
Œstrololactone benzoate, 2552 s
C₂₅H₂₇NO₃ 16-Oximinoœstrone benzyl ether,
2942 s
C₂₅H₂₇NO₄ Œstrone p-nitrobenzyl ether,
2557 s
C₂₅H₂₇N₃O₄ 16-Ketoœstrone methyl ether
p-nitrophenylhydrazone, 2942 s
C₂₅H₂₈N₄O₅ 2-Methyl-7-methoxy-3-keto-
1.2-cyclopentano-octahydrophen-
anthrene dinitrophenylhydrazone, 134;
2525 s
C₂₅H₂₈N₄O₆ Δ⁴-Androstene-3.11.17-trione
mono-2.4-dinitrophenylhydrazone, 165
C₂₅H₂₈O₂ Œstrone benzyl ether, 2557 s
C₂₅H₂₈O₃ 16-Ketoœstradiol 3-benzyl ether,
2739 s
C₂₅H₃₀N₄O₂ Œstrone p-aminophenyl ether
semicarbazone, 2557 s
C₂₅H₃₂N₄O₅ Androsten-17-ol-7-one 2.4-dinitro-
phenylhydrazone, 2603 s
C₂₅H₃₄N₂O₅ Androsten-3-ol-17-spirohydantoin,
diacetyl derivative, 2615 s
C₂₅H₃₄O₃ Œstrone œnanthate, 2555 s
C₂₅H₃₄O₅ Pregnadiene-3.21-diol-20-one
diacetate, 2300 s
C₂₅H₃₄O₆ Pregnene-20.21-diol-3.11-diones,
diacetate, 2936 s, 2937 s
Pregnene-2.21-diol-3.20-dione diacetate,
2852 s
Pregnene-6.21-diol-3.20-dione diacetate,
2853 s
Pregnene-7.12-diol-3.20-dione diacetate,
2865 s
Pregnene-12.21-diol-3.20-dione diacetate,
2858 s
C₂₅H₃₄O₇ Corticosterone 21-hydrogen succinate,
153; 2855 s
Pregnene-17.20.21-triol-3.11-dione
20.21-diacetate, 2959 s
C₂₅H₃₅BrO₅ Pregnene-3.21-diol-20-one
3-acetate 21-bromoacetate, 2306 s
C₂₅H₃₅BrO₆ 4-Bromopregnane-7.12-diol-
3.20-dione diacetate, 2867 s
4-Bromopregnane-12.21-diol-3.20-dione
diacetate, 2864 s
C₂₅H₃₅BrO₇ 4-Bromopregnane-17.20.21-triol-
3.11-dione 20.21-diacetate, 2960 s
C₂₅H₃₅Br₃O₅ 17.21.21-Tribromopregnane-
3.12-diol-20-one diacetate, 2351 s
C₂₅H₃₆Br₂O₅ 17.21-Dibromopregnane-3.12-diol-
20-one diacetate, 2351 s
C₂₅H₃₆N₂O₃ 23-Diazo-norcholen-3-ol-22-one
acetate, 2297 s

C₂₅H₃₆N₂O₅ 21-Diazopregnane-3.12-diol-20-one
diacetate, 2352 s
C₂₅H₃₆O₂ Androstadien-3-ol-17-one cyclohexyl
ether, 2607 s
Androsten-3-ol-17-one cyclohexenyl ether,
2620 s
C₂₅H₃₆O₃ Progesterone 3-enol-butyrate, 151
16-(α-Methylpropylidene)-androsten-3-ol-
17-one acetate, 2657 s
C₂₅H₃₆O₄ Δ⁴-Pregnen-21-ol-3.20-dione butyrate,
2824 s
Norcholen-23-ol-3.22-dione acetate, 2834 s
C₂₅H₃₆O₅ Pregnene-3.20-diol-7-one diacetate,
2722 s
Pregnene-3.21-diol-11-one diacetate, 2723 s
Allopregnene-3.6-diol-20-ones, diacetate,
2323 s, 2324 s
17-iso-Pregnene-3.17-diol-20-one diacetate,
2339 s
Pregnene-3.21-diol-20-one diacetate, 2306 s
Allopregnene-3.21-diol-20-ones, diacetate,
2308 s, 2309 s
Corticosterone 21-butyrate, 153; 2855 s
3.7.12-Triketonorcholanyl hydroxymethyl
ketone, 2985 s
C₂₅H₃₆O₆ Pregnene-3.21-diol-11-one 21-hemi-
succinate, 2723 s
Pregnene-3.21-diol-20-one oxides, diacetate,
2372 s, 2373 s
Pregnene-17.20.21-triol-3-ones,
20.21-diacetate, 2726 s, 2728 s
Pregnene-3.17.21-triol-20-one 3.21-diacetate,
2357 s
Pregnene-3.17.20-triol-21-al 3.20-diacetate,
2361 s
Pregnane-20.21-diol-3.11-diones, diacetate,
2937 s, 2938 s
Pregnane-7.12-diol-3.20-dione diacetate,
2866 s
Pregnane-11.21-diol-3.20-dione diacetate,
2856 s
Pregnane-12.21-diol-3.20-dione diacetate,
2858 s
Pregnane-3.21-diol-11.20-dione diacetate,
2861 s
Allopregnane-3.21-diol-11.20-dione
diacetate, 2862 s
17-iso-Allopregnane-3.21-diol-11.20-dione
diacetate, 2862 s
C₂₅H₃₆O₇ Androstane-3.5.6-triol-17-one
triacetate, 2769 s
Pregnene-11.17.20.21-tetrol-3-one
20.21-diacetate, 2765 s
Pregnane-17.20.21-triol-3.11-dione
20.21-diacetate, 2960 s
Allopregnane-5.6.21-triol-3.20-dione
6.21-diacetate, 2867 s

$C_{25}H_{41}ClO$ Norcholanyl chloromethyl ketone, 2221 s

$C_{25}H_{41}ClO_4$ 3.7.12-Trihydroxynorcholanyl chloromethyl ketone, 2370 s

$C_{25}H_{41}N_3O_2$ 3-Methoxy-20-iso-norcholen-22-one, semicarbazone, 2261 s

$C_{25}H_{42}O$ Norcholanyl methyl ketone, 114; 2218 s

Norallocholanyl (Norallocholyl) methyl ketone, 114

23-Methylcholan-3-one, 2413 s

$C_{25}H_{42}O_2$ Norcholanyl hydroxymethyl ketone, 2226 s

3-Hydroxynorcholanyl (Norlithocholyl) methyl ketone, 116

Bisnor-coprostan-23-ol-3-one, 133

$C_{25}H_{42}O_3$ Homocholane-3.6-diol-24-one, 2346 s

Homocholane-3.7-diol-24-one, 2347 s

Homocholane-3.12-diol-24-one, 2347 s

Acid from Δ^3-cholesten-5-ol-2-one, 2661 s

$C_{25}H_{42}O_5$ 3.7.12-Trihydroxynorcholanyl hydroxymethyl ketone, 2376 s

$C_{25}H_{43}NO$ Norcholanyl (Norcholyl) methyl ketone oxime, 114

$C_{25}H_{43}N_3O$ Bisnorcholanyl (Bisnorcholyl) methyl ketone semicarbazone, 114

$C_{25}H_{44}O_2$ Acid from acid $C_{25}H_{42}O_3$ (from Δ^3-cholesten-5-ol-2-one), 2661 s

$C_{26}H_{19}ClO_3$ "Dimethoxyphenanthracyclopentadienochromylium chloride", 2517 s

$C_{26}H_{22}O_3$ 16-Piperonylidene-cis-equilenan-17-one, 2515 s

$C_{26}H_{22}O_4$ 16-Piperonylidene-18-nor-isoequilenin methyl ether, 2730 s

$C_{26}H_{25}NO$ 16-(N-Methyl-anilinomethylene)-3-desoxyisoequilenin, 2507 s

$C_{26}H_{25}NO_2$ 16-(N-Methyl-anilinomethylene)-18-nor-isoequilenin methyl ether, 2717 s

$C_{26}H_{25}NO_3$ Acid from œstrone and isatin, 2549 s

$C_{26}H_{27}NO_4$ 16-(m-Nitrobenzylidene)-œstrone methyl ether, 2717 s

$C_{26}H_{28}N_4O_4$ 3'-Isopropyl-4-keto-1.2-cyclopentano-hexahydrophenanthrene 2.4-dinitrophenylhydrazone, 2394 s

$C_{26}H_{28}O_3$ 4-Methylœstratrien-1-ol-17-one benzoate, 2573 s

$C_{26}H_{29}NO_3$ Œstrone p-acetaminophenyl ether, 2557 s

$C_{26}H_{29}NO_4$ Œstrone p-carbomethoxyaminophenyl ether, 2557 s

$C_{26}H_{29}N_3O_3$ Œstrone benzoate semicarbazone, 137

$C_{26}H_{30}N_4O_4$ Œstrone p-nitrobenzyl ether semicarbazone, 2557 s

$C_{26}H_{30}O_2$ 16-Benzylidene-androstene-3.17-dione, 2914 s

$C_{26}H_{30}O_3$ $\Delta^{1.4}$-Androstadien-17-ol-3-one benzoate, 2577 s

$\Delta^{4.6}$-Androstadien-17-ol-3-one benzoate, 139; 2578 s

$\Delta^{3.5}$-Androstadien-3-ol-17-one benzoate (Δ^5-Androstene-3.17-dione enol-benzoate), 156

$C_{26}H_{30}O_4$ Unsaturated benzoate from 2.4-dibromoandrostane-3.17-dione, 2924 s

$C_{26}H_{31}BrO_3$ 2-Bromo-Δ^1-androsten-17-ol-3-one benzoate, 2703 s

2-Bromo-Δ^4-androsten-17-ol-3-one benzoate, 2703 s

6-Bromo-Δ^4-androsten-17-ol-3-one benzoate, 148

$C_{26}H_{31}NO_4$ 16-(m-Nitrobenzylidene)-androsten-3-ol-17-one, 2717 s

$C_{26}H_{32}Br_2O_3$ 2.2-Dibromoandrostan-17-ol-3-one benzoate, 2711 s

2.4-Dibromoandrostan-17-ol-3-one benzoate, 2712 s

$C_{26}H_{32}N_2O_8$ Androstane-3.11-diol-17-one 3-(3.5-dinitrobenzoate), 149

$C_{26}H_{32}O_2$ Androstadien-3-ol-17-one benzyl e her, 2606 s

16-Benzylidene-androsten-17-ol-3-one, 2698 s

16-Benzylidene-androsten-3-ol-17-one, 2699 s

$C_{26}H_{32}O_3$ Δ^1-Androsten-17-ol-3-one benzoate, 2579 s

Δ^4-Androsten-17α-ol-3-one benzoate, 140; 2580 s

Δ^4-Androsten-17β-ol-3-one benzoate, 140; 2590 s

Δ^5-Androsten-17-ol-3-one benzoate, 141; 2595 s

Δ^5-Androsten-3-ol-17-one benzoate, 142; 2619 s

Δ^{11}-Androsten-3-ol-17-one benzoate, 2628 s

$C_{26}H_{32}O_4$ Androsten-17-ol-3-one phenyl carbonate, 2590 s

Androsten-3-ol-17-one phenyl carbonate, 2619 s

Dehydroepiandrosterone oxide benzoate, 2770 s

Androstan-2-ol-3.17-dione benzoate, 2945 s

$C_{26}H_{32}O_5$ Androstane-3.5-diol-6.17-dione 3-benzoate, 2962 s

$C_{26}H_{33}NO_4$ 16-(m-Nitrobenzylidene)-androstan-3-ol-17-one, 2717 s

$C_{26}H_{34}BrNO$ N-(3-Ketopregnadien-21-yl)-pyridinium bromide, 2473 s

$C_{26}H_{34}BrNO_2$ N-(3.20-Diketopregnen-21-yl)-pyridinium bromide, 2811 s

$C_{26}H_{41}N_3O_5$ Pregnane-3.7-diol-20-one diacetate semicarbazone, 2327 s

Diacetate semicarbazone of dihydroxy-ketone $C_{21}H_{34}O_3$ from adrenal cortex, 2336 s

$C_{26}H_{42}Br_2O_2$ 5.6-Dibromonorcholestan-3-ol-25-one, 2290 s

$C_{26}H_{42}ClN_3O_2$ Pregnadien-3-ol-20-one Girard T hydrazone, 2232 s

$C_{26}H_{42}O$ 21-Nor-Δ^4-cholesten-20-one, 2219 s
27-Nor-Δ^4-cholesten-25-one, 2219 s

$C_{26}H_{42}O_2$ 21-Nor-cholesten-3-ol-20-one, 2265 s
27-Nor-cholesten-3-ol-24-one, 2263 s
27-Nor-cholesten-3-ol-25-one, 2263 s
27-Nor-cholestane-3.25-dione, 2806 s

$C_{26}H_{42}O_3$ 3-Acetoxyallocholanaldehyde, 2262 s
Norcholestene-3.25-diol-24-one, 2316 s
5-Keto-3‖5-A-nor-Δ^1-cholesten-3-oic acid, 2419 s

$C_{26}H_{43}ClO$ 5-Chloro-21-nor-cholestan-20-one, 2222 s
5-Chloro-27-nor-cholestan-25-one, 2222 s

$C_{26}H_{43}N_3O_2$ Bisnorcholesten-3-ol-24-one semi-carbazone, 2263 s

$C_{26}H_{44}O$ Norcholanyl ethyl ketone, 114; 2219 s
24.24-Dimethylcholan-3-one, 2414 s

$C_{26}H_{44}O_2$ Bisnorcholanyl α-hydroxyisopropyl ketone, 2226 s
Norcoprostan-24-ol-3-one, 133
Norcholestan-3-ol-25-one, 2264 s

$C_{26}H_{44}O_6$ A-Nor-cholestane-3‖5‖6-trioic acid, 2964 s

$C_{26}H_{45}N_3O$ Norcholanyl (Norcholyl) methyl ketone semicarbazone, 114; 2219 s
Norallocholanyl (Norallocholyl) methyl ketone semicarbazone, 114
23-Methylcholan-3-one semicarbazone, 2413 s

$C_{26}H_{45}N_3O_2$ 3-Hydroxynorcholanyl (Norlitho-cholyl) methyl ketone semicarbazone, 116
Bisnor-coprostan-23-ol-3-one semicarbazone, 133

$C_{26}H_{46}$ "Bisnor-sterocholane", 2219 s

$C_{27}H_{23}ClO_4$ 4-Chloro-16-piperonylidene-iso-equilenin methyl ether, 2730 s

$C_{27}H_{23}ClO_5$ 4-Chloro-16-(3.4-methylenedioxy-benzoyl)-isoequilenin methyl ether, 2864 s

$C_{27}H_{24}O_4$ 16-Piperonylidene-equilenin methyl ether, 2731 s
16-Piperonylidene-isoequilenin methyl ether, 2730 s

$C_{27}H_{27}NO_2$ 16-(N-Methyl-anilinomethylene)-isoequilenin methyl ether, 2718 s

$C_{27}H_{29}NO_2$ Œstrone + quinoline, 137; 2553 s

$C_{27}H_{30}O_2$ Œstrone cinnamyl ether, 138

$C_{27}H_{30}O_4$ 16-Ketoœstradiol 3-benzyl ether 17-acetate, 2739 s

$C_{27}H_{31}NO_4$ Œstrone p-carbethoxyaminophenyl ether, 2558 s

$C_{27}H_{32}N_4O_8$ Δ^4-Pregnene-17.21-diol-3.11.20-trione mono-2.4-dinitrophenylhydrazone, 165

$C_{27}H_{32}O_4$ 16-Piperonylidene-androsten-3-ol-17-one, 2731 s

$C_{27}H_{34}N_2O$ Quinoxaline compound from preg-nen-3-ol-20-on-21-al, 2379 s

$C_{27}H_{34}O_3$ Testosterone phenylacetate, 2590 s

$C_{27}H_{34}O_4$ Androsten-17-ol-3-one benzyl carbonate, 2590 s
Androsten-3-ol-17-one benzyl carbonate, 2619 s

$C_{27}H_{35}N_3O_4$ Androsten-3-ol-17-one acetate p-nitrophenylhydrazone, 2619 s

$C_{27}H_{36}ClNO_2$ N-(17-Propionyloxy-androstadien-17-yl)-pyridinium chloride, 2584 s

$C_{27}H_{36}N_4O_4$ Pregnen-3-one 2.4-dinitrophenyl-hydrazone, 2410 s

$C_{27}H_{36}O_7$ 17-iso-Pregnadiene-3.17.21-triol-20-one triacetate, 2357 s

$C_{27}H_{37}Br_3O_2$ 2.7.7-Tribromocholestadiene-3.6-dione, 2932 s

$C_{27}H_{37}NO_2$ 17-iso-Pregnene-3.17-diol-20-one anil, 2339 s

$C_{27}H_{37}N_3O_3$ Epiandrosterone benzoate semi-carbazone, 144

$C_{27}H_{37}N_4O_8P$ Testosterone dimethyl phosphate 2.4-dinitrophenylhydrazone, 2588 s

$C_{27}H_{38}Br_2O$ Dibromocholestatrien-3-one, 120; 2497 s

$C_{27}H_{38}Br_2O_2$ 2.7-Dibromocholestadiene-3.6-dione, 2927 s

$C_{27}H_{38}Br_4O_2$ 2.2.7.7-Tetrabromocholestene-3.6-dione, 2932 s

$C_{27}H_{38}N_4O_4$ Pregnan-3-one 2.4-dinitrophenyl-hydrazone, 2411 s
Pregnan-20-one 2.4-dinitrophenyl-hydrazone, 2217 s
Allopregnan-20-one 2.4-dinitrophenyl-hydrazone, 2218 s

$C_{27}H_{38}N_4O_5$ Pregnan-3-ol-20-one 2.4-dinitro-phenylhydrazone, 2245 s

$C_{27}H_{38}N_4O_6$ Pregnane-3.6-diol-20-one 2.4-dinitro phenylhydrazone, 2325 s
Allopregnane-3.6-diol-20-one 2.4-dinitro-phenylhydrazone, 2325 s

$C_{27}H_{38}O$ 19-Nor-ergostapentaen-3-ol, 2416 s

$C_{27}H_{38}O_4$ Cholestene-2.3.6.7-tetrone, 2986 s

$C_{27}H_{38}O_5$ Pregnadiene-3.21-diol-20-one dipro-pionate, 2300 s

$C_{27}H_{38}O_6$ 3.7.12-Triketonorcholanyl acetoxy-methyl ketone, 2985 s

$C_{27}H_{38}O_7$ 17-iso-Pregnadiene-3.17.20.21-tetrol
3.17.21-triacetate, 2357 s
Pregnene-3.17.21-triol-20-one triacetate,
2358 s
$C_{27}H_{39}BrO$ 2-Bromocholestatrien-3-one, 2479 s
$C_{27}H_{39}BrO_2$ 2-Bromocholestadiene-3.6-dione,
2920 s
$C_{27}H_{39}BrO_5$ Norcholene-3.23-diol-22-one
3-acetate 23-bromoacetate, 2315 s
$C_{27}H_{39}Br_2NO$ 2.6-Dibromocholestatrien-3-one
oxime, 2498 s
$C_{27}H_{39}Br_3O$ Tribromocholestadien-3-one, 120;
2503 s
$C_{27}H_{39}Br_3O_2$ 2.2.7-Tribromocholestene-
3.6-dione, 2931 s
$C_{27}H_{40}Br_2O$ Dibromocholestadien-3-one, 2498 s
Dibromocholestadien-7-one, 2501 s
$C_{27}H_{40}Br_2O_2$ 2.7-Dibromocholestene-3.6-diones,
2929 s, 2930 s
$C_{27}H_{40}D_4O$ Deuteriocoprostenone, 2431 s
$C_{27}H_{40}N_2O_3$ 25-Diazo-bisnorcholesten-3-ol-
24-one acetate, 2297 s
$C_{27}H_{40}O$ Cholestatrien-3-one, 2418 s
19-Nor-ergostatrien-3-one, 2416 s
$C_{27}H_{40}O_2$ Cholestatrien-7-ol-6-one, 2680 s
Cholestadiene-6.7-dione, 2911 s
$C_{27}H_{40}O_4$ 5-Methyl-9.10-epoxy-19-nor-copro-
stane-3.6.11-trione, 2984 s
Cholestane-3.4.6.7-tetrone, 2986 s
$C_{27}H_{40}O_5$ Norcholene-3.23-diol-22-one diacetate,
2315 s
Corticosterone 21-caproate, 2855 s
Corticosterone 21-diethylacetate, 2855 s
$C_{27}H_{40}O_6$ ω-Homo-pregnene-17.20.21.22-tetrol-
3-one 21.22-acetonide 20-acetate,
2733 s
$C_{27}H_{40}O_7$ Pregnane-3.7.12-triol-20-one
triacetate, 2368 s
Pregnane-3.12.21-triol-20-one triacetate,
2357 s
Allopregnane-3.17.21-triol-20-one triacetate,
2359 s
17-iso-Allopregnane-3.17.21-triol-20-one
triacetate, 2360 s
$C_{27}H_{40}O_8$ Allopregnane-3.17.20.21-tetrol-11-one
3.20.21-triacetate, 2766 s
Allopregnane-3.5.6.21-tetrol-20-one 3.6.21-
triacetate, 2373 s
$Δ^4$-Pregnen-21-ol-3.20-dione glucoside,
2825 s
$Δ^4$-Pregnen-21-ol-3.20-dione galactoside,
2826 s
$C_{27}H_{41}BrO$ 6-Bromocholestadien-7-one, 2489 s
$C_{27}H_{41}BrO_2$ 2-Bromo-$Δ^1$-cholestene-3.6-dione,
2921 s
2-Bromo-$Δ^4$-cholestene-3.6-dione, 2921 s
7-Bromo-$Δ^4$-cholestene-3.6-dione, 2923 s

$C_{27}H_{41}Br_3O$ Tribromocholesten-3-ones, 130, 132;
2502 s, 2503 s
Tribromocholesten-7-one, 2504 s
$C_{27}H_{41}ClO_3$ 25-Chloro-bisnorcholesten-3-ol-
24-one acetate, 2283 s
$C_{27}H_{42}Br_2O$ Dibromocholesten-3-ones, 131;
2499 s
Cholestadien-7-one dibromide, 123
$C_{27}H_{42}Br_2O_2$ 2.2-Dibromocholestane-3.6-dione,
2926 s
5.7-Dibromocholestane-3.6-dione, 2930 s
$C_{27}H_{42}Br_4O$ Tetrabromocholestan-3-one, 132;
2505 s
$C_{27}H_{42}O$ Cholestadien-2-one, 2417 s
Cholestadien-3-ones, 119; 2418 s, 2420 s,
2421 s, 2422 s
Cholestadien-6-one, 2452 s
Cholestadien-7-one, 123; 2454 s
Phenol from $Δ^3$-cholesten-5-ol-2-one, 2661 s
$C_{27}H_{42}O_2$ Cholestadien-4-ol-3-one, 157; 2666 s
Cholestadien-6-ol-3-one, 2671 s
Cholestadien-3-ol-4-one, 2674 s
Cholestadien-3-ol-6-one, 2676 s
Cholestadien-3-ol-7-ones, 2680 s, 2681 s
Cholestadien-3-ol-15-one, 2687 s
$Δ^4$-Cholestene-2.3-dione, 2427 s
$Δ^5$-Cholestene-3.4-dione, 157; 2903 s
$Δ^1$-Cholestene-3.6-dione, 2905 s
$Δ^4$-Cholestene-3.6-dione, 158; 2905 s
$Δ^5$-Cholestene-3.7-dione, 2910 s
$Δ^4$-Cholestene-3.22-dione, 2806 s
$Δ^5$-Cholestene-4.7-dione, 2911 s
5-Methyl-19-nor-cholestene-3.6-dione, 109;
2902 s
$C_{27}H_{42}O_3$ 16-tert.-Butyl-pregnen-3-ol-20-one
acetate, 2270 s
Bisnorcholesten-3-ol-24-one acetate, 2263 s
5-Methyl-9.10-epoxy-19-nor-coprostane-
3.6-dione, 2962 s
2.5-Epoxycholestane-3.6-dione, 2963 s
Cholestane-3.4.6-trione, 2982 s
$C_{27}H_{42}O_3S_2$ Pregnen-3-ol-20-on-21-al acetate
diethyl mercaptal, 2379 s
$C_{27}H_{42}O_4$ Cholestene-3.14-diol-7.15-dione,
2964 s
$C_{27}H_{42}O_5$ 5-Methyl-19-nor-coprostane-9.10-diol-
3.6.11-trione, 2984 s
3-Ketocholestene-6‖7-dioic acid, 2906 s
$C_{27}H_{42}O_6$ Pregnene-3.20-diol-21-al diacetate
dimethyl acetal, 2313 s
$C_{27}H_{42}O_8$ Pregnan-3-ol-20-one glucuronide,
2245 s
$C_{27}H_{42}O_9$ Pregnane-3.17-diol-20-one glucuronide,
2335 s
$C_{27}H_{43}BrO$ 2-Bromocholesten-3-one, 2480 s
4-Bromocholesten-3-ones, 2482 s, 2483 s
6-Bromocholesten-3-one, 129; 2484 s

Pages with an "S" refer to the Supplement, the others to the Main Volume

$C_{27}H_{43}BrO_2$ 2-Bromocholesten-4-ol-3-one, 2495 s
   2-Bromocholesten-6-ol-3-one, 2706 s
   4-Bromocholesten-3-ol-6-one, 2707 s
   2-Bromocholestane-3.6-dione, 2922 s
   5-Bromocholestane-3.6-dione, 2922 s
   7-Bromocholestane-3.6-dione, 2923 s

$C_{27}H_{43}BrO_3$ 7-Bromocholestan-5-ol-3.6-dione,
   2957 s

$C_{27}H_{43}Br_3O$ Tribromocholestan-3-ones, 131;
   2504 s

$C_{27}H_{43}ClO$ 6-Chlorocholesten-3-one, 2484 s
   3-Chlorocholesten-7-one, 130; 2488 s

$C_{27}H_{43}ClO_3$ 3-Chlorocholestane-6∥7-dioic acid
   anhydride, 2485 s
   Compound from 3-chlorocholestane-6∥7-dioic
   acid, 2487 s

$C_{27}H_{43}NO$ Cholestadien-3-ones, oxime, 2421 s,
   2422 s
   Cholestadien-7-one oxime, 2455 s

$C_{27}H_{43}NO_3$ 3-Acetoxy-bisnorcholesten-24-one
   oxime, 2263 s

$C_{27}H_{43}N_3O_3$ 3-Acetoxy-ternorcholenyl ethyl
   ketone semicarbazone, 2262 s

$C_{27}H_{44}BrClO$ Bromochlorocholestan-3-one, 129;
   2497 s
   Bromochlorocholestan-6-one, 2485 s

$C_{27}H_{44}Br_2O$ 1.2-Dibromocholestan-3-one, 2494 s
   2.2-Dibromocholestan-3-one, 2494 s
   2.4-Dibromocholestan-3-one, 130; 2495 s
   5.6-Dibromocholestan-3-one, 131; 2500 s
   2.4-Dibromocoprostan-3-one, 130; 2497 s

$C_{27}H_{44}Br_2O_2$ 4.5-Dibromocholestan-3-ol-6-one,
   2713 s
   5.7-Dibromocholestan-3-ol-6-ones, 2714 s
   5.6-Dibromocholestan-3-ol-7-one, 2715 s
   6.8-Dibromocholestan-3-ol-7-one, 2715 s

$C_{27}H_{44}Br_2O_3$ 2.2-Dibromocholestane-5.6-diol-
   3-one, 2763 s

$C_{27}H_{44}Cl_2O$ 5.6-Dichlorocholestan-3-one, 2500 s

$C_{27}H_{44}N_2$ Pyridazine derivative of cholestane-
   3.6-dione, 159; 2909 s
   Pyridazine derivative of cholestane-
   4.7-dione, 2911 s

$C_{27}H_{44}N_2O$ Δ^4-Cholestene-3.6-dione monohydrazone, 158

$C_{27}H_{44}N_2O_2$ Cholestene-3.4-dione dioxime, 157
   Cholestene-3.7-dione dioxime, 2910 s

$C_{27}H_{44}N_6O_2$ Bisnorcholestene-3.24-dione
   disemicarbazone, 2805 s

$C_{27}H_{44}O$ Δ^1-Cholesten-3-one, 119; 2423 s
   Δ^4-Cholesten-3-one, 120; 2424 s
   Δ^5-Cholesten-3-one, 121; 2435 s
   Δ^8-Cholesten-3-one, 2436 s
   Δ^1-Coprosten-3-one, 2423 s
   Δ^5-Cholesten-4-one ("hetero-Δ^1-Cholesten-
   3-one"), 119; 2450 s
   Δ^2-Cholesten-6-one, 2452 s

i-Cholesten-6-one, 129; 2452 s
   Δ^5-Cholesten-7-one, 123; 2455 s
   Δ^8-Cholesten-7-one, 2456 s

$C_{27}H_{44}O_2$ Δ^3-Cholesten-3-ol-2-one, 2659 s
   Δ^3-Cholesten-5-ol-2-one, 2661 s
   Δ^1Cholesten-2-ol-3-one, 2662 s
   Δ^4-Cholesten-2-ol-3-one, 2663 s, 2664 s
   Δ^1-Cholesten-4-ol-3-one, 2668 s
   Δ^4-Cholesten-4-ol-3-one, 2669 s
   Δ^4-Cholesten-6-ol-3-one, 146; 2672 s
   Δ^5-Cholesten-3-ol-4-one, 2675 s
   Δ^4-Cholesten-3-ol-6-one, 2677 s
   Δ^5-Cholesten-3-ol-7-ones, 147; 2681 s, 2682 s
   Δ^8-Cholesten-3-ol-7-one, 2685 s
   $\Delta^{8(14)}$-Cholesten-3-ol-7-one, 2685 s
   $\Delta^{8(14)}$-Cholesten-3-ol-15-one, 2687 s
   Δ^5-Cholesten-3-ol-22-one, 2265 s
   Δ^5-Cholesten-3-ol-24-one, 2266 s
   i-Cholesten-6-ol-24-one, 2266 s
   4.5-Epoxycholestan-2-one, 2753 s
   5.6-Epoxycholestan-3-one, 2755 s
   Cholestane-2.3-dione, 157; 2902 s
   Cholestane-3.4-dione, 158; 2904 s
   Cholestane-3.6-dione, 159; 2907 s
   Coprostane-3.6-dione, 159; 2910 s
   Cholestane-3.7-dione, 159; 2910 s
   Cholestane-4.7-dione, 2911 s
   Coprostane-3.24-dione, 152; 2806 s
   4-Hydroxy-3∥4-Δ^5-cholesten-3-oic acid
   lactone ("Δ^5-Cholesten-3-one α-oxide")
   121; 2435 s

$C_{27}H_{44}O_3$ Norcholanyl acetoxymethyl ketone,
   2226 s
   3-Hydroxynorcholanyl (Norlithocholyl)
   methyl ketone acetate, 116
   Δ^4-Cholestene-16.26-diol-3-one, 2723 s
   2.5-Epoxycholestan-6-ol-3-one, 2771 s
   5.6-Epoxycholestan-4-ol-3-one, 2771 s
   5.6-Epoxycholestan-3-ol-4-one, 2771 s
   8.9-Epoxy-8-iso-cholestan-3-ol-7-one, 2772 s
   8.14-Epoxy-8-iso-cholestan-3-ol-7-one,
   2773 s
   8.14-Epoxy-8-iso-cholestan-3-ol-15-one,
   2773 s
   Cholestan-5-ol-3.6-dione, 163; 2953 s
   Cholestan-3-ol-6.7-dione, 2954 s

$C_{27}H_{44}O_4$ 5-Methyl-9.10-epoxy-19-nor-
   coprostane-3.6-diol-11-one, 2753 s
   Cholestane-2.5-diol-3.6-dione, 2963 s
   Cholestane-4.5-diol-3.6-dione, 2963 s
   5-Methyl-19-nor-coprostane-9.10-diol-
   3.6-dione, 2962 s
   Δ^4-Cholestene-6∥7-dioic acid, 2487 s
   i-Cholestene-6∥7-dioic acid, 2453 s; cf. also
   2486 s, 2487 s
   3-Hydroxycholestane-6∥7-dioic acid lactone,
   2486 s

Lactone acid from coprostenone, 120
C₂₇H₄₄O₅ 3-Ketocholestane-6‖7-dioic acid, 42; 2486 s
C₂₇H₄₄O₈ Cholestane-2‖3.6‖7-tetraoic acid, 42; 2486 s
C₂₇H₄₅BrO 2-Bromocholestan-3-one, 128; 2480 s
  5-Bromocholestan-3-one, 2483 s
  3-Bromocholestan-6-one, 2488 s
  4-Bromocoprostan-3-one, 129; 2483 s
C₂₇H₄₅BrO₂ 5-Bromocholestan-3-ol-6-one, 2707 s
  7-Bromocholestan-3-ol-6-one, 2708 s
  6-Bromocholestan-3-ol-7-ones, 2709 s
C₂₇H₄₅BrO₃ 2-Bromocholestane-5.6-diol-3-one, 2762 s
  7-Bromocholestane-3.5-diol-6-ones, 2762 s, 2763 s
C₂₇H₄₅ClO 2-Chlorocholestan-3-one, 2482 s
  5-Chlorocholestan-3-one, 129; 2483 s
  3-Chlorocholestan-6-ones, 129; 2484 s, 2486 s
  3-Chlorocholestan-7-one, 2489 s
C₂₇H₄₅ClO₄ 3-Chlorocholestane-6‖7-dioic acids, 129; 2485 s, 2487 s
C₂₇H₄₅NO Δ⁴-Cholesten-3-one oxime, 121; 2433 s
  Δ⁵-Cholesten-3-one oxime, 121
  Δ⁵-Cholesten-4-one ("hetero-Δ¹-cholesten-3-one") oxime, 119
  Δ²-Cholesten-6-one oxime, 2452 s
  i-Cholesten-6-one oxime, 2453 s
  Δ⁵-Cholesten-7-one oxime, 2456 s
C₂₇H₄₅NO₂ Δ⁵-Cholesten-3-ol-7-one oxime, 2684 s
  4.5-Epoxycholestan-2-one oxime, 2754 s
C₂₇H₄₅NO₅ 3-Ketocholestane-6‖7-dioic acid oxime, 42; 2486 s
C₂₇H₄₅N₃O₃ 3-Acetoxyallocholanaldehyde semicarbazone, 2262 s
C₂₇H₄₆ClNO 3-Chlorocholestan-6-ones, oxime, 129, 130; 2487 s
  3-Chlorocholestan-7-one oxime, 2489 s
C₂₇H₄₆N₂O₂ Cholestane-3.6-dione dioxime, 159; 2909 s
  Coprostane-3.6-dione dioxime, 159
  Cholestane-3.7-dione dioxime, 159
C₂₇H₄₆O Cholestan-1-one, 2417 s
  Cholestan-2-one, 2417 s
  Cholestan-3-one, 122; 2436 s
  Cholestan-4-one, 122; 2451 s
  Cholestan-6-one, 123; 2454 s
  Cholestan-7-one, 123; 2456 s
  Coprostan-3-one, 121; 2443 s
  Coprostan-24-one, 114
C₂₇H₄₆O₂ Cholestan-3-ol-2-one, 2660 s
  Cholestan-5-ol-2-one, 2662 s
  Cholestan-2-ol-3-one, 2665 s
  Cholestan-4-ol-3-one, 2670 s
  Coprostan-4-ol-3-one, 2670 s
  Cholestan-5-ol-3-one, 2671 s
  Cholestan-6-ol-3-ones, 2673 s

Cholestan-3-ol-4-one, 2676 s
Cholestan-3-ol-6-one, 147; 2677 s
Cholestan-3-ol-7-one, 148; 2686 s
Coprostan-3-ol-24-one, 116
C₂₇H₄₆O₃ Cholestane-4.5-diol-2-ones, 2753 s, 2754 s
  Cholestane-4.5-diol-3-one, 2754 s
  Cholestane-5.6-diol-3-ones, 2755 s, 2756 s
  Cholestane-5.6-diol-4-one, 2756 s
  Cholestane-3.5-diol-6-one, 149; 2757 s, 2758 s
  Cholestane-3.7-diol-6-one, 2759 s
  Cholestane-3.6-diol-7-one, 2759 s
C₂₇H₄₆O₄ Cholestenetetrol, 2723 s
  Cholestane-2.5.6-triol-3-one, 2771 s
  Cholestane-4.5.6-triol-3-one, 2771 s
  Cholestane-3.5.6-triol-4-one, 2771 s
  Cholestane-3.4.5-triol-6-one, 2772 s
  8-iso-Cholestane-3.8.9-triol-7-one, 2772 s
  8-iso-Cholestane-3.8.14-triol-7-one, 2773 s
  8-iso-Cholestane-3.8.14-triol-15-one, 2773 s
  Cholestane-6‖7-dioic acid, 64; 2486 s
C₂₇H₄₆O₅ 5-Methyl-19-nor-coprostane-3.6.9.10-tetrol-11-one, 2775 s
  3-Hydroxycholestane-6‖7-dioic acids, 42; 2485 s, 2488 s
C₂₇H₄₇NO Cholestan-3-one oxime, 2442 s
  Cholestan-4-one oxime, 122; 2452 s
  Cholestan-6-one oxime, 123; 2454 s
  Cholestan-7-one oxime, 2456 s
C₂₇H₄₇NO₂ Cholestenone + hydroxylamine, 2433 s
  Cholestan-3-ol-2-one oxime, 2660 s
  Cholestan-5-ol-3-one oxime, 2671 s
  Cholestan-3-ol-7-one oxime, 2687 s
C₂₇H₄₇N₃O Norcholanyl ethyl ketone semicarbazone, 114; 2219 s
  24.24-Dimethylcholan-3-one semicarbazone, 2414 s
C₂₇H₄₇N₃O₂ Norcoprostan-24-ol-3-one semicarbazone, 133
  Norcholestan-3-ol-25-one semicarbazone, 2265 s
C₂₇H₄₈N₂ Cholestanone hydrazone, 2443 s
C₂₇H₄₈N₂O Coprostan-3-ol-24-one hydrazone, 116
C₂₇H₄₈O₂ Coprostane-3.4-diol, 2668 s; cf. 2056 s, 2670 s
C₂₇H₄₈O₃ Hydroxy-acid from 3-chlorocholestane-6‖7-dioic acid, 2485 s

C₂₈H₂₇N Compound from indolocholestadiene, 2432 s
C₂₈H₂₈O₃ 21-Benzylidenepregnane-3.11-diol-20-one, 2348 s
  21-Benzylidenepregnane-3.12-diol-20-one, 2348 s

$C_{28}H_{30}O_3$ 2 or 4-Allylœstrone benzoate, 2574 s

$C_{28}H_{31}NO_8$ Pregnene-17.21-diol-3.11.20-trione mono-p-nitrobenzoate, 2977 s

$C_{28}H_{34}O_2$ 16-Benzylidene-androstadien-3-ol-17-one ethyl ether, 2698 s
21-Benzylidene-pregnene-3.20-dione, 2809 s

$C_{28}H_{34}O_3$ Pregnadien-3-ol-20-one benzoate, 2230 s
16-Benzylidene-androsten-17-ol-3-one acetate, 2698 s
16-Benzylidene-androsten-3-ol-17-one acetate, 2699 s

$C_{28}H_{34}O_4$ Pregnen-12-ol-3.20-dione benzoate, 2841 s
Pregnen-21-ol-3.20-dione benzoate, 2825 s

$C_{28}H_{34}O_5$ Corticosterone 21-benzoate, 153; 2855 s

$C_{28}H_{36}N_2O_4$ Testosterone acetate p-carboxy-phenylhydrazone, 2587 s

$C_{28}H_{36}O_2$ Pregnadien-3-ol-20-one benzyl ether, 2230 s
21-Benzylidenepregnen-3-ol-20-one, 2271 s
21-Benzylidenepregnane-3.20-dione, 2809 s
21-Benzylpregnene-3.20-dione, 2809 s
3-Ketoternorcholenyl phenyl ketone, 2809 s

$C_{28}H_{36}O_3$ Pregnen-3-ol-20-one benzoate, 2238 s
16-Benzylidene-androstan-3-ol-17-one acetate, 2699 s

$C_{28}H_{38}O_4$ Pregnene-3.21-diol-20-one 21-benzoate, 2307 s
Allopregnan-2-ol-3.20-dione benzoate, 153
Pregnan-12-ol-3.20-dione benzoate, 2842 s
Testalolone benzoate, 2381 s

$C_{28}H_{36}O_5S$ Pregnen-12-ol-3.20-dione p-toluene-sulphonate, 2841 s
Pregnen-21-ol-3.20-dione p-toluenesul-phonate, 2825 s

$C_{28}H_{37}NO_5$ Allopregnan-3-ol-20-one p-nitro-benzoate, 116

$C_{28}H_{38}Br_2O_2$ 5.6-Dibromo-3-hydroxy-ternor-allocholanyl phenyl ketone, 2290 s

$C_{28}H_{38}ClN_3O_2$ Pregnadien-3-ol-20-one Girard P hydrazone, 2232 s

$C_{28}H_{38}N_2O_2$ 3-Ketoternorcholenyl phenyl ketone dioxime, 2809 s

$C_{28}H_{38}O_2$ 21-Benzylpregnen-3-ol-20-one, 2272 s
21-Benzylidenepregnan-3-ol-20-ones, 2272 s
21-Benzylideneallopregnan-3-ol-20-ones, 2273 s
3-Hydroxypregnen-20-yl phenyl ketones, 2273 s, 2274 s

$C_{28}H_{38}O_3$ Allopregnan-3-ol-20-one benzoate, 2249 s
Pregnan-21-ol-20-one benzoate, 2225 s

$C_{28}H_{38}O_4$ Pregnane-3.12-diol-20-one 12-benzoate, 2330 s

$C_{28}H_{38}O_4S$ Pregnen-3-ol-20-one p-toluene-sulphonate, 2238 s

$C_{28}H_{38}O_5S$ Pregnene-3.21-diol-20-one 21-p-toluenesulphonate, 2307 s
Pregnan-12-ol-3.20-dione p-toluene-sulphonate, 2842 s

$C_{28}H_{39}NO_3$ Allopregnan-3-ol-20-one phenyl-urethan, 116

$C_{28}H_{40}N_2O_7$ 3.7.12-Trihydroxynorcholanyl diazo-methyl ketone triformate, 2371 s

$C_{28}H_{40}O$ Ergostatetraenone, 2458 s

$C_{28}H_{40}O_2$ 3-Hydroxypregnan-20-yl phenyl ketone, 2274 s
Benzoate of compound $C_{21}H_{36}O$ from allo-pregnan-20-ol-3.16-dione, 2936 s
Compound from ergostadien-5-ol-3.6-dione, 163

$C_{28}H_{40}O_3$ Œstrone caprinate, 137
3.12-Dihydroxyternorcholanyl phenyl ketone, 2349 s

$C_{28}H_{40}O_4$ Compound from ergostadien-5-ol-3.6-dione, 163

$C_{28}H_{41}ClO_7$ 3.7.12-Trihydroxynorcholanyl chloromethyl ketone triformate, 2370 s

$C_{28}H_{41}NO_4$ Oxime of compound $C_{28}H_{40}O_4$ from ergostadien-5-ol-3.6-dione, 163

$C_{28}H_{41}NO_5S$ Δ^{16}-Allopregnen-3-ol-20-one hydrogen sulphate p-toluidine salt, 2243 s

$C_{28}H_{42}N_2O_3$ 25-Diazo-norcholesten-3-ol-24-one acetate, 2297 s
26-Diazo-norcholesten-3-ol-25-one acetate, 2298 s

$C_{28}H_{42}O$ Ergostatrien-3-ones, 124; 2459 s, 2460 s, 2461 s
Coproergostatrien-3-ones, 124; 2461 s

$C_{28}H_{42}O_2$ 7-Methoxycholestatrien-6-one, 2680 s
Ergostatrien-3-ol-7-one, 2690 s

$C_{28}H_{42}O_3$ Ergostadien-5-ol-3.6-dione, 163; 2955 s
Lumistadien-5-ol-3.6-dione, 163

$C_{28}H_{42}O_4$ Tetraketo-isobufostane, 2985 s

$C_{28}H_{42}O_5$ ω-Homo-pregnene-17.20.21.22-tetrol-3-one 21.22-cyclohexylidene-derivative, 2733 s
Corticosterone 21-heptanoate, 2855 s

$C_{28}H_{43}BrO$ Bromodehydroergostenone, 125

$C_{28}H_{43}Br_2N_3O$ Dibromocholestadien-3-one semicarbazone, 2499 s

$C_{28}H_{43}ClO_3$ 26-Chloro-27-nor-cholesten-3-ol-25-one acetate, 2283 s

$C_{28}H_{43}NO$ Ergostatrienone-B$_1$ oxime, 124
Ergostatrienone-D oxime, 124
Coproergostatrien-3-one (u-Ergosta-trienone-B) oxime, 124

$C_{28}H_{43}NO_3$ Ergostadien-5-ol-3.6-dione monoxime, 163

$C_{28}H_{43}N_3O$ Compound from cholestenone semi-carbazone, 2427 s

Pages with an "S" refer to the Supplement, the others to the Main Volume

Ergostadien-3-ol-15-one acetate, 2693 s

$C_{30}H_{46}O_4$ Ergosten-3-ol-7.15-dione acetate, 2956 s

$C_{30}H_{48}O_2$ Ergostadien-6-ol-3-one ethyl ether, 160; 2690 s

$C_{30}H_{48}O_3$ Ergosten-3-ol-7-one acetate, 2692 s
24-iso-Ergosten-3-ol-7-one acetate, 2692 s
Ergosten-3-ol-15-one acetate, 83; 2693 s

$C_{30}H_{48}O_4$ 8.14-Epoxy-8-iso-ergostan-3-ol-7-one acetate, 2774 s
8.14-Epoxy-8-iso-ergostan-3-ol-15-one acetate, 2774 s

$C_{30}H_{49}NO_2$ 2-Formyl-Δ^2-cholestene oxime acetate, 2220 s

$C_{30}H_{49}N_3O$ Stigmastadien-3-ones, semicarbazone, 2466 s, 2467 s

$C_{30}H_{49}N_3O_3$ 3-Acetoxy-Δ^5-cholesten-7-one semicarbazone, 2684 s

$C_{30}H_{50}O_2$ Cholestenone propylene ketal, 2435 s
Cholestenone trimethylene ketal, 2435 s

$C_{30}H_{50}O_3$ Ergostan-3-ol-7-one acetate, 2693 s

$C_{30}H_{51}NO_2S$ "Cholestanone thiazolidine", 2440 s

$C_{30}H_{51}N_3O$ Sitostenone semicarbazone, 126; 2468 s
Stigmastenone semicarbazone, 2469 s

$C_{30}H_{51}N_3O_3$ 3-Acetoxycholestan-7-one semicarbazone, 2687 s

$C_{30}H_{52}ClN_3O$ 3-Chlorositostan-6-one semicarbazone, 2490 s

$C_{30}H_{53}N_3O$ Stigmastan-3-one semicarbazone, 2469 s

$C_{30}H_{53}N_3O_2$ 3-Acetylcholestan-3-ol semicarbazone, 2270 s

$C_{31}H_{30}N_2O_2$ Œstrone p-benzeneazobenzoate, 2555 s

$C_{31}H_{36}N_6O_2$ Androstene-3.17-dione bis-nicotinoylhydrazone, 2885 s

$C_{31}H_{36}N_8O_8$ Ætiocholane-3.17-dione bis-(2.4-dinitrophenylhydrazone), 2895 s

$C_{31}H_{38}O_{11}$ Methyl (œstrone triacetylglucosid)-uronate, 2558 s

$C_{31}H_{41}NO_4$ 17-iso-Pregnene-3.17-diol-20-one diacetate anil, 2339 s

$C_{31}H_{42}N_2O_3$ 3-Acetoxypregnadien-21-al N-(p-dimethylaminophenyl)-isoxime, 2252 s

$C_{31}H_{42}O_3$ 3-Formoxy-norcholenyl phenyl ketone, 2275 s

$C_{31}H_{42}O_4$ 3-Formoxy-12-ketonorcholanyl phenyl ketone, 2850 s

$C_{31}H_{42}O_{11}$ Testosterone triacetylglucuronide, 2591 s

$C_{31}H_{44}O_2$ 3-Hydroxyternorcholenyl mesityl ketone, 2275 s

$C_{31}H_{44}O_3$ 3-Acetoxybisnorcholanyl phenyl ketone, 2275 s

$C_{31}H_{46}O_3$ Stigmastatrien-3-ol-7-one acetate, 2695 s

$C_{31}H_{46}O_7$ Pregnene-3.16-diol-20-one 3-acetate 16-(δ-acetoxyisocaproate), 2332 s

$C_{31}H_{47}ClO_7$ 3.7.12-Triacetoxynorcholanyl chloromethyl ketone, 2370 s

$C_{31}H_{48}O_3$ Stigmastadien-3-ol-7-ones, acetate, 148; 2695 s, 2696 s

$C_{31}H_{48}O_6$ 5-Methyl-9.10-epoxy-19-nor-coprostane-3.6-diol-11-one diacetate, 2753 s

$C_{31}H_{48}O_7$ Allopregnane-3.16-diol-20-one 3-acetate 16-(δ-acetoxyisocaproate), 2333 s

$C_{31}H_{48}O_{12}$ Testosterone maltoside, 2591 s

$C_{31}H_{49}BrO_5$ 7-Bromocholestane-3.5-diol-6-ones, diacetate, 2763 s

$C_{31}H_{50}O_2$ Stigmastadien-3-one ethylene ketal, 2466 s

$C_{31}H_{50}O_3$ Stigmasten-3-ol-6-one acetate, 2694 s
Stigmasten-3-ol-7-ones, acetate, 2696 s, 2697 s

$C_{31}H_{50}O_5$ Cholestan-3-ol-6-one hydrogen succinate, 2680 s
Cholestane-3.5-diol-6-one diacetate, 2758 s

$C_{31}H_{51}NO_3$ Stigmasten-3-ol-6-one acetate oxime, 2694 s

$C_{31}H_{51}N_3O_3$ α-Ergostenolone acetate semicarbazone, 83; cf. 2693 s

$C_{31}H_{52}O_3$ Stigmastan-3-ol-6-one acetate, 2695 s
Stigmastan-3-ol-7-one acetate, 2697 s

$C_{31}H_{52}O_4$ Coprostane-3.4-diol diacetate, 2668 s
Stigmastane-3.5-diol-6-one 3-acetate, 2760 s
Cholestane-3.6-dione bis-(ethylene ketal), 2909 s

$C_{31}H_{52}O_5$ Dicarboxylic acid dimethyl ester from stigmastane-3.6-dione, 161

$C_{31}H_{53}NO_5$ Dicarboxylic acid dimethyl ester oxime from stigmastane-3.6-dione, 161

$C_{31}H_{53}N_3O_3$ Ergostan-3-ol-7-one acetate semicarbazone, 2693 s

$C_{31}H_{56}O_2$ Cholestanone diethyl ketal, 2442 s

$C_{31}H_{56}S_2$ Cholestanone di-ethylthio ketal, 2442 s

$C_{32}H_{38}O_3$ 17-(α-Hydroxybenzhydryl)-ætiocholane-3.12-dione, 2935 s

$C_{32}H_{40}O_{11}$ Œstrone tetraacetylglucoside, 2558 s

$C_{32}H_{42}O_5$ 21-Benzylidenepregnane-3.11-diol-20-one diacetate, 2348 s
21-Benzylidenepregnane-3.12-diol-20-one diacetate, 2348 s

$C_{32}H_{44}O_3$ 3-Acetoxy-norcholenyl phenyl ketone, 2275 s

$C_{32}H_{44}O_4$ 3-Acetoxy-12-ketonorcholanyl phenyl ketone, 2850 s
Ergostatrien-3-one, maleic anhydride adduct, 124; 2460 s

Pages with an "S" refer to the Supplement, the others to the Main Volume

Pages with an "S" refer to the Supplement, the others to the Main Volume

Printed in the United States
By Bookmasters